The HUTCHINSON
DICTIONARY OF
SCIENCE
2nd Edition

Edited by
Peter Lafferty and
Julian Rowe

Helicon

Copyright © Helicon Publishing Ltd 1993, 1994, 1998

First published 1993
Revised and updated 1994
Second edition 1998
Reprinted 1999, 2001

Helicon Publishing Ltd
42 Hythe Bridge Street
Oxford OX1 2EP

e-mail: admin@helicon.co.uk
Web site: http://www.helicon.co.uk

Printed and bound in Great Britain by
The Bath Press Ltd, Bath, Somerset

ISBN 1-85986-243-8

British Cataloguing in Publication Data

A catalogue record of this book is available
from the British Library

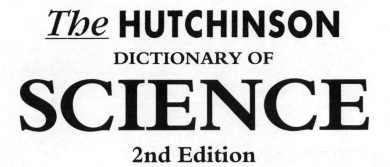

The HUTCHINSON
DICTIONARY OF
SCIENCE
2nd Edition

Introduction

Most dictionaries of science are aimed at students. Worthy as this aim may be, it provides an excuse for the publishers if the material is dull. As long as it is correct, the book has served its purpose; no matter that nobody would chose to open it for pleasure. *The Hutchinson Dictionary of Science* is different. It has been designed to be useful and appropriate for academic purposes, but at the same time to be interesting and accessible to the general reader. To this end we have built upon the success of the first edition and have extended our coverage to an even wider range of scientific topics to include medicine, photography, and archaeology, as well as the ever popular quotations, feature essays, and fact boxes that distinguish this dictionary from the rest. Our intentions remain the same and that is to show science not as an unchanging repository of facts, but rather as a number of evolving and exciting areas of research.

Arrangement of entries

Entries are ordered alphabetically, as if there were no spaces between words. Thus, entries for words beginning "sulphur" follow the order:

sulphur
sulphur dioxide
sulphuric acid
sulphur trioxide

Common or technical names?

Terms are usually placed under the better known name rather than the technical name (thus chloroform is placed under C and not under its technically correct name trichloromethane), but the technical term is also given. To aid comprehension for the non-specialist, terms are frequently explained when used within the text of an entry, even though they may have their own entry elsewhere.

Cross-references

These are shown by a ◊ symbol immediately preceding the reference. Cross-referencing is selective: a cross-reference is shown when another entry contains material directly relevant to the subject matter of an entry, in cases where the reader may not otherwise think of looking.

Units

SI and metric units are used throughout. Commonly used measurements of distances, temperatures, sizes, and so on, include an approximate imperial equivalent.

CONTRIBUTORS & CONSULTANTS
FOR THE FIRST AND SECOND EDITIONS

R.D. Bagnall
David Bradley
John Broad
Thomas Day
Dougal Dixon
Nigel Dudley
Barry Fox
James Le Fanu
Wendy Grossman
Tony Jones
Peter Lafferty
Carol Lister
Graham Littler
Paul Meehan
Pamela Morley

Chris Pellant
Paulette Pratt
Ian Ridpath
Peter Rodgers
Julian Rowe
P.M. Rowntree
Jack Schofield
Steve Smyth
Ian Stewart
Sue Solton
Chris Stringer
Michael Thain
Catherine Thompson
Colin Tudge
Martin Walters

EDITORS

Managing Editor
Sharon Brimblecombe

Project and Text Editors
Diana Gallannaugh
Barbara Newson
Catherine Thompson

Production
Tony Ballsdon

Art and Design Manager
Terence Caven

A in physics, symbol for ◊ampere, a unit of electrical current.

abacus ancient calculating device made up of a frame of parallel wires on which beads are strung. The method of calculating with a handful of stones on a 'flat surface' (Latin *abacus*) was familiar to the Greeks and Romans, and used by earlier peoples, possibly even in ancient Babylon; it survives in the more sophisticated bead-frame form of the Russian *schoty* and the Japanese *soroban*. The abacus has been superseded by the electronic calculator.

The wires of a bead-frame abacus define place value (for example, in the decimal number system each successive wire, counting from right to left, would stand for ones, tens, hundreds, thousands, and so on) and beads are slid to the top of each wire in order to represent the digits of a particular number. On a simple decimal abacus, for example, the number 8,493 would be entered by sliding three beads on the first wire (three ones), nine beads on the second wire (nine tens), four beads on the third wire (four hundreds), and eight beads on the fourth wire (eight thousands).

abdomen in vertebrates, the part of the body below the ◊thorax, containing the digestive organs; in insects and other arthropods, it is the hind part of the body. In mammals, the abdomen is separated from the thorax by the ◊diaphragm, a sheet of muscular tissue; in arthropods, commonly by a narrow constriction. In mammals, the female reproductive organs are in the abdomen. In insects and spiders, it is characterized by the absence of limbs.

aberration of starlight apparent displacement of a star from its true position, due to the combined effects of the speed of light and the speed of the Earth in orbit around the Sun (about 30 km per sec/18.5 mi per sec).

Aberration, discovered in 1728 by English astronomer James Bradley (1693–1762), was the first observational proof that the Earth orbits the Sun.

aberration, optical any of a number of defects that impair the image in an optical instrument. Aberration occurs because of minute variations in lenses and mirrors, and because different parts of the light ◊spectrum are reflected or refracted by varying amounts.

In **chromatic aberration** the image is surrounded by coloured fringes, because light of different colours is brought to different focal points by a lens. In **spherical aberration** the image is blurred because different parts of a spherical lens or mirror have different focal lengths. In **astigmatism** the image appears elliptical or cross-

shaped because of an irregularity in the curvature of the lens. In **coma** the images appear progressively elongated towards the edge of the field of view. Elaborate computer programs are now used to design lenses in which the aberrations are minimized. *See illustration overleaf.*

abiotic factor a nonorganic variable within the ecosystem, affecting the life of organisms. Examples include temperature, light, and soil structure. Abiotic factors can be harmful to the environment, as when sulphur dioxide emissions from power stations produce acid rain.

abortion ending of a pregnancy before the fetus is developed sufficiently to survive outside the uterus. Loss of a fetus at a later gestational age is termed premature stillbirth. Abortion may be accidental (◊miscarriage) or deliberate (termination of pregnancy).

In the first nine weeks of pregnancy, medical termination may be carried out using the 'abortion pill' (◊mifepristone) in conjunction with a ◊prostaglandin. There are also various procedures for surgical termination, such as ◊dilatation and curettage, depending on the length of the pregnancy.

Worldwide, an estimated 150,000 unwanted pregnancies are terminated each day by induced abortion. One-third of these abortions are performed illegally and unsafely, and cause one in eight of all maternal deaths.

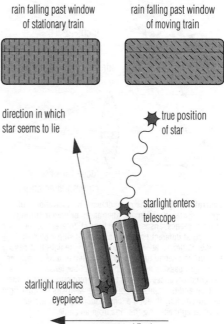

aberration of starlight The aberration of starlight is an optical illusion caused by the motion of the Earth. Rain falling appears vertical when seen from the window of a stationary train; when seen from the window of a moving train, the rain appears to follow a sloping path. In the same way, light from a star 'falling' down a telescope seems to follow a sloping path because the Earth is moving. This causes an apparent displacement, or aberration, in the position of the star.

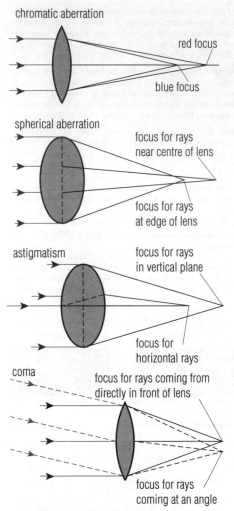

chromatic aberration

red focus

blue focus

spherical aberration

focus for rays
near centre of lens

focus for rays
at edge of lens

astigmatism

focus for rays
in vertical plane

focus for
horizontal rays

coma

focus for rays coming from
directly in front of lens

focus for rays
coming at an angle

aberration, optical The main defects, or aberrations, of optical systems. Chromatic aberration, or coloured fringes around images, arises because light of different colours is focused at different points by a lens, causing a blurred image. Spherical aberration arises because light that passes through the centre of the lens is focused at a different point from light passing through the edge of the lens. Astigmatism arises if a lens has different curvatures in the vertical and horizontal directions. Coma arises because light passing directly through a lens is focused at a different point to light entering the lens from an angle.

Abortion as a means of birth control has long been controversial. The argument centres largely upon whether a woman should legally be permitted to have an abortion and, if so, under what circumstances. Another aspect is whether, and to what extent, the law should protect the fetus. Those who oppose abortion generally believe that human life begins at the moment of conception, when a sperm fertilizes an egg. This is the view held, for example, by the Roman Catholic Church. Those who support unrestricted legal abortion may believe in a woman's right to choose whether she wants a child, and may take into account the large numbers of deaths and injuries from unprofessional backstreet abortions. Others approve abortion for specific reasons. For example, if a woman's life or health is jeopardized, abortion may be recommended; and if there is a strong likelihood that the child will be born with severe mental or physical disability. Other grounds for abortion include pregnancy resulting from sexual assault such as rape or incest.

abrasion in earth science, the effect of ◊corrasion, a type of erosion in which rock fragments scrape and grind away a surface. The rock fragments may be carried by rivers, wind, ice, or the sea. Striations, or grooves, on rock surfaces are common abrasions, caused by the scratching of rock debris embedded in glacier ice.

abrasive substance used for cutting and polishing or for removing small amounts of the surface of hard materials. There are two types: natural and artificial abrasives, and their hardness is measured using the ◊Mohs' scale. Natural abrasives include quartz, sandstone, pumice, diamond, emery, and corundum; artificial abrasives include rouge, whiting, and carborundum.

abscess collection of ◊pus in solid tissue forming in response to infection. Its presence is signalled by pain and inflammation.

abscissa in ◊coordinate geometry, the x-coordinate of a point – that is, the horizontal distance of that point from the vertical or y-axis. For example, a point with the co-ordinates (4, 3) has an abscissa of 4. The y-coordinate of a point is known as the ◊ordinate.

abscissin or *abscissic acid* plant hormone found in all higher plants. It is involved in the process of ◊abscission and also inhibits stem elongation, germination of seeds, and the sprouting of buds.

abscission in botany, the controlled separation of part of a plant from the main plant body – most commonly, the falling of leaves or the dropping of fruit controlled by ◊abscissin. In ◊deciduous plants the leaves are shed before the winter or dry season, whereas ◊evergreen plants drop their leaves continually throughout the year. Fruitdrop, the abscission of fruit while still immature, is a naturally occurring process.

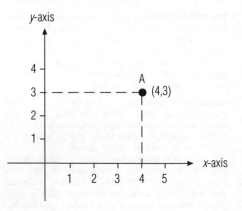

abscissa The abscissa of A is 4.

Abscission occurs after the formation of an abscission zone at the point of separation. Within this, a thin layer of cells, the abscission layer, becomes weakened and breaks down through the conversion of pectic acid to pectin. Consequently the leaf, fruit, or other part can easily be dislodged by wind or rain. The process is thought to be controlled by the amount of ◊auxin present. Fruitdrop is particularly common in fruit trees such as apples, and orchards are often sprayed with artificial auxin as a preventive measure.

absolute (of a value) in computing, real and unchanging. For example, an *absolute address* is a location in memory and an *absolute cell reference* is a single fixed cell in a spreadsheet display. The opposite of absolute is ◊relative.

absolute value or *modulus* in mathematics, the value, or magnitude, of a number irrespective of its sign. The absolute value of a number n is written $|n|$ (or sometimes as mod n), and is defined as the positive square root of n^2. For example, the numbers -5 and 5 have the same absolute value:

$$|5| = |-5| = 5$$

For a ◊complex number, the absolute value is its distance to the origin when it is plotted on an ◊Argand diagram, and can be calculated (without plotting) by applying ◊Pythagoras' theorem. By definition, the absolute value of any complex number $a + bi$ (where a and b are real numbers and i is $\sqrt{-1}$) is given by the expression:

$$|a + bi| = \sqrt{(a^2 + b^2)}$$

absolute zero lowest temperature theoretically possible, zero kelvin (0K), equivalent to $-273.15°C/-459.67°F$, at which molecules are in their lowest energy state. Although the third law of ◊thermodynamics indicates the impossibility of reaching absolute zero exactly, a temperature of 2.8×10^{-10}K (0.28 billionths of a degree above absolute zero) was produced in 1993 at Lancaster University, England. Near absolute zero, the physical properties of some materials change substantially; for example, some metals lose their electrical resistance and become superconducting.

absorption the taking up of one substance by another, such as a liquid by a solid (ink by blotting paper) or a gas by a liquid (ammonia by water). In physics, absorption is the phenomenon by which a substance retains radiation of particular wavelengths; for example, a piece of blue glass absorbs all visible light except the wavelengths in the blue part of the spectrum; it also refers to the partial loss of energy resulting from light and other electromagnetic waves passing through a medium. In nuclear physics, absorption is the capture by elements, such as boron, of neutrons produced by fission in a reactor.

absorption lines in astronomy, dark line in the spectrum of a hot object due to the presence of absorbing material along the line of sight. Absorption lines are caused by atoms absorbing light from the source at sharply defined wavelengths. Numerous absorption lines in the spectrum of the Sun (Fraunhofer lines) allow astronomers to study the composition of the Sun's outer layers. Absorption lines in the spectra of stars give clues to the composition of interstellar gas.

absorption spectroscopy or *absorptiometry* in analytical chemistry, a technique for determining the identity or amount present of a chemical substance by measuring the amount of electromagnetic radiation the substance absorbs at specific wavelengths; see ◊spectroscopy.

abyssal plain broad expanse of sea floor lying 3–6 km/2–4 mi below sea level. Abyssal plains are found in all the major oceans, and they extend from bordering continental rises to mid-oceanic ridges.

abyssal zone dark ocean region 2,000–6,000 m/6,500–19,500 ft deep; temperature 4°C/39°F. Three-quarters of the area of the deep-ocean floor lies in the abyssal zone, which is too far from the surface for photosynthesis to take place. Some fish and crustaceans living there are blind or have their own light sources. The region above is the bathyal zone; the region below, the hadal zone.

abzyme in biotechnology, an artificially created antibody that can be used like an enzyme to accelerate reactions.

AC in physics, abbreviation for ◊alternating current.

accelerated freeze drying common method of food preservation. See ◊food technology.

acceleration rate of change of the velocity of a moving body. It is usually measured in metres per second per second (m s^{-2}) or feet per second per second (ft s^{-2}). Because velocity is a ◊vector quantity (possessing both magnitude and direction) a body travelling at constant speed may be said to be accelerating if its direction of motion changes. According to Newton's second law of motion, a body will accelerate only if it is acted upon by an unbalanced, or resultant, ◊force.

Acceleration due to gravity is the acceleration of a body falling freely under the influence of the Earth's gravitational field; it varies slightly at different latitudes and altitudes. The value adopted internationally for gravitational acceleration is 9.806 m s^{-2}/32.174 ft s^{-2}.

Acceleration

As a flea jumps, its rate of acceleration is 20 times that of the space shuttle during launching. It reaches a speed of 100 m per sec within the first 500th of a second.

---

acceleration, secular in astronomy, the continuous and nonperiodic change in orbital velocity of one body around another, or the axial rotation period of a body.

An example is the axial rotation of the Earth. This is gradually slowing down owing to the gravitational effects of the Moon and the resulting production of tides, which have a frictional effect on the Earth. However, the angular ◊momentum of the Earth–Moon system is maintained, because the momentum lost by the Earth is passed to the Moon. This results in an increase in the Moon's orbital period and a consequential moving away from the Earth. The overall effect is that the Earth's axial rotation period is increasing by about 15 millionths of a second a year, and the Moon is receding from the Earth at about 4 cm/1.5 in a year.

accelerator in physics, a device to bring charged particles (such as protons and electrons) up to high speeds and energies, at which they can be of use in industry, medicine, and pure physics. At low energies, accelerated particles can be used to produce the image on a television screen and generate X-rays (by means of a ◊cathode-ray

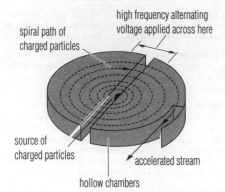

spiral path of charged particles

high frequency alternating voltage applied across here

source of charged particles

accelerated stream

hollow chambers

accelerator The cyclotron, an early accelerator, consisted of two D-shaped hollow chambers enclosed in a vacuum. An alternating voltage was applied across the gap between the hollows. Charged particles spiralled outward from the centre, picking up energy and accelerating each time they passed through the gap.

tube), destroy tumour cells, or kill bacteria. When high-energy particles collide with other particles, the fragments formed reveal the nature of the fundamental forces.

The first accelerators used high voltages (produced by ◊van de Graaff generators) to generate a strong, unvarying electric field. Charged particles were accelerated as they passed through the electric field. However, because the voltage produced by a generator is limited, these accelerators were replaced by machines where the particles passed through regions of alternating electric fields, receiving a succession of small pushes to accelerate them.

The first of these accelerators was the **linear accelerator** or **linac**. The linac consists of a line of metal tubes, called drift tubes, through which the particles travel. The particles are accelerated by electric fields in the gaps between the drift tubes.

Another way of making repeated use of an electric field is to bend the path of a particle into a circle so that it passes repeatedly through the same electric field. The first accelerator to use this idea was the **cyclotron** pioneered in the early 1930s by US physicist Ernest Lawrence. A cyclotron consists of an electromagnet with two hollow metal semicircular structures, called dees, supported between the poles of an electromagnet. Particles such as protons are introduced at the centre of the machine and travel outwards in a spiral path, being accelerated by an oscillating electric field each time they pass through the gap between the dees. Cyclotrons can accelerate particles up to energies of 25 MeV (25 million electron volts); to produce higher energies, new techniques are needed.

In the ◊synchrotron, particles travel in a circular path of constant radius, guided by electromagnets. The strengths of the electromagnets are varied to keep the particles on an accurate path. Electric fields at points around the path accelerate the particles.

Early accelerators directed the particle beam onto a stationary target; large modern accelerators usually collide beams of particles that are travelling in opposite directions. This arrangement doubles the effective energy of the collision.

The world's most powerful accelerator is the 2 km/1.25 mi diameter machine at ◊Fermilab, Illinois, USA.

This machine, the Tevatron, accelerates protons and antiprotons and then collides them at energies up to a thousand billion electron volts (or 1 TeV, hence the name of the machine). The largest accelerator is the ◊Large Electron Positron Collider at ◊CERN near Geneva, which has a circumference of 27 km/16.8 mi around which electrons and positrons are accelerated before being allowed to collide. The world's longest linac is also a colliding beam machine: the Stanford Linear Collider, in California, in which electrons and positrons are accelerated along a straight track, 3.2 km/2 mi long, and then steered to a head-on collision with other particles, such as protons and neutrons. Such experiments have been instrumental in revealing that protons and neutrons are made up of smaller elementary particles called ◊quarks.

accelerometer apparatus, either mechanical or electromechanical, for measuring ◊acceleration or deceleration – that is, the rate of increase or decrease in the ◊velocity of a moving object.

The mechanical types have a spring-supported mass with a damper system, with indication of acceleration on a scale on which a light beam is reflected from a mirror on the mass. The electromechanical types use (1) a slide wire, (2) a strain gauge, (3) variable inductance, or (4) a piezoelectric or similar device that produces electrically measurable effects of acceleration.

Accelerometers are used to measure the efficiency of the braking systems on road and rail vehicles; those used in aircraft and spacecraft can determine accelerations in several directions simultaneously. There are also accelerometers for detecting vibrations in machinery.

access time or **reaction time** in computing, the time taken by a computer, after an instruction has been given, to read from or write to ◊memory.

acclimation or **acclimatization** the physiological changes induced in an organism by exposure to new environmental conditions. When humans move to higher altitudes, for example, the number of red blood cells rises to increase the oxygen-carrying capacity of the blood in order to compensate for the lower levels of oxygen in the air.

In evolutionary terms, the ability to acclimate is an important adaptation as it allows the organism to cope with the environmental changes occurring during its lifetime.

accommodation in biology, the ability of the ◊eye to focus on near or far objects by changing the shape of the lens.

For an object to be viewed clearly its image must be precisely focused on the retina, the light-sensitive layer of cells at the rear of the eye. Close objects can be seen when the lens takes up a more spherical shape, far objects when the lens is flattened. These changes in shape are caused by the movement of ligaments attached to a ring of ciliary muscles lying beneath the iris.

From about the age of 40, the lens in the human eye becomes less flexible, causing the defect of vision known as **presbyopia** or lack of accommodation. People with this defect need different spectacles for reading and distance vision.

accretion in astrophysics, a process by which an object gathers up surrounding material by gravitational attraction, so simultaneously increasing in mass and releasing gravitational energy. Accretion on to compact objects

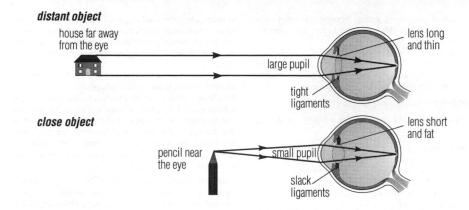

distant object

house far away from the eye

large pupil

tight ligaments

lens long and thin

close object

pencil near the eye

small pupil

slack ligaments

lens short and fat

accommodation The mechanism by which the shape of the lens in the eye is changed so that clear images of objects, whether distant or near, can be focused on the retina.

such as ◊white dwarfs, ◊neutron stars and ◊black holes can release large amounts of gravitational energy, and is believed to be the power source for active galaxies. Accreted material falling towards a star may form a swirling disc of material known as an ◊accretion disc that can be a source of X-rays.

accretion disc in astronomy, a flattened ring of gas and dust orbiting an object in space, such as a star or ◊black hole. The orbiting material is accreted (gathered in) from a neighbouring object such as another star. Giant accretion discs are thought to exist at the centres of some galaxies and ◊quasars.

If the central object of the accretion disc has a strong gravitational field, as with a neutron star or a black hole, gas falling onto the accretion disc releases energy, which heats the gas to extreme temperatures and emits short-wavelength radiation, such as X- rays.

accumulator in computing, a special register, or memory location, in the ◊arithmetic and logic unit of the computer processor. It is used to hold the result of a calculation temporarily or to store data that is being transferred.

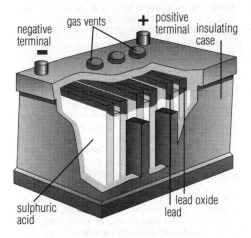

negative terminal

gas vents

+ positive terminal

insulating case

sulphuric acid

lead oxide

lead

accumulator

accumulator in electricity, a storage ◊battery – that is, a group of rechargeable secondary cells. A familiar example is the lead–acid car battery.

An ordinary 12-volt car battery consists of six lead–acid cells which are continually recharged when the motor is running, by the car's alternator or dynamo. It has electrodes of lead and lead oxide in an electrolyte of sulphuric acid. Another common type of accumulator is the 'nife' or Ni Fe cell, which has electrodes of nickel and iron in a potassium hydroxide electrolyte.

accuracy in mathematics, a measure of the precision of a number. The degree of accuracy depends on how many figures or decimal places are used in rounding off the number. For example, the result of a calculation or measurement (such as 13.429314) might be rounded off to three decimal places (13.429), to two decimal places (13.43), to one decimal place (13.4), or to the nearest whole number (13 s). The first answer is more accurate than the second, the second more accurate than the third, and so on. Accuracy also refers to a range of errors. For example, an accuracy of ± 5% means that a value may lie between 95% and 105% of a given answer.

acesulfame-K non-carbohydrate sweetener that is up to 300 times as sweet as sugar. It is used in soft drinks and desserts.

acetaldehyde common name for ◊ethanal.

acetate common name for ◊ethanoate.

acetic acid common name for ◊ethanoic acid.

acetone common name for ◊propanone.

acetylcholine (ACh) chemical that serves as a ◊neurotransmitter, communicating nerve impulses between the cells of the nervous system. It is largely associated with the transmission of impulses across the ◊synapse (junction) between nerve and muscle cells, causing the muscles to contract.

ACh is produced in the synaptic knob (a swelling at the end of a nerve cell) and stored in vesicles until a nerve impulse triggers its discharge across the synapse. When the ACh reaches the membrane of the receiving cell it binds with a specific site and brings about depolarization – a reversal of the electric charge on either side of the membrane – causing a fresh impulse (in nerve cells) or a contraction (in muscle cells). Its action is short-lived because it is quickly destroyed by the enzyme cholinesterase.

Anticholinergic drugs have a number of uses in medicine to block the action of ACh, thereby disrupting the passage of nerve impulses and relaxing certain muscles, for example in premedication before surgery.

acetylene common name for ◊ethyne.

acetylsalicylic acid chemical name for the painkilling drug ◊aspirin.

achene dry, one-seeded ◊fruit that develops from a single ◊ovary and does not split open to disperse the seed. Achenes commonly occur in groups – for example, the fruiting heads of buttercup *Ranunculus* and clematis. The outer surface may be smooth, spiny, ribbed, or tuberculate, depending on the species.

Achernar or **Alpha Eridani** brightest star in the constellation Eridanus, and the ninth brightest star in the sky. It is a hot, luminous, blue star with a true luminosity 250 times that of the Sun. It is 125 light years away.

Achilles tendon tendon at the back of the ankle attaching the calf muscles to the heel bone. It is one of the largest tendons in the human body, and can resist great tensional strain, but is sometimes ruptured by contraction of the muscles in sudden extension of the foot.

Ancient surgeons regarded wounds in this tendon as fatal, probably because of the Greek legend of Achilles, which relates how the mother of the hero Achilles dipped him when an infant into the river Styx, so that he became invulnerable except for the heel by which she held him.

achromatic lens combination of lenses made from materials of different refractive indexes, constructed in such a way as to minimize chromatic aberration (which in a single lens causes coloured fringes around images because the lens diffracts the different wavelengths in white light to slightly different extents).

acid compound that, in solution in an ionizing solvent (usually water), gives rise to hydrogen ions (H^+ or protons). In modern chemistry, acids are defined as substances that are proton donors and accept electrons to form ◊ionic bonds. Acids react with ◊bases to form salts, and they act as solvents. Strong acids are corrosive; dilute acids have a sour or sharp taste, although in some organic acids this may be partially masked by other flavour characteristics.

Acids can be detected by using coloured indicators such as ◊litmus and methyl orange. The strength of an acid is measured by its hydrogen-ion concentration, indicated by the ◊pH value. Acids are classified as monobasic, dibasic, tribasic, and so forth, according to the number of hydrogen atoms, replaceable by bases, in a molecule. The first known acid was vinegar (ethanoic or acetic acid). Inorganic acids include boric, carbonic, hydrochloric, hydrofluoric, nitric, phosphoric, and sulphuric. Organic acids include acetic, benzoic, citric, formic, lactic, oxalic, and salicylic, as well as complex substances such as ◊nucleic acids and ◊amino acids.

acidosis in medicine, a condition characterized by the body's production of abnormal acids or by a reduction in the alkali reserve of the blood. Acidosis is usually caused by faulty fat metabolism and is associated with untreated or inadequately treated ◊diabetes, starvation, persistent vomiting, and the final stages of kidney failure. A mild form of acidosis may occur in children with severe fevers. Symptoms include vomiting, thirst, restlessness, and lassitude. Acidosis may be detected by the presence of ◊ketones in the urine.

The main aim of treatment is to treat the underlying disease. Insulin should be given to patients with diabetes, otherwise a state of coma may occur that can be fatal. Alkalis should be given to treat patients with acidosis due to other causes. They can be given by mouth or, if the patient is unconscious or experiencing persistent vomiting, they can be given by intravenous infusion. Adequate fluid intake is necessary to ensure acidosis is corrected.

acid rain acidic precipitation thought to be caused principally by the release into the atmosphere of sulphur dioxide (SO_2) and oxides of nitrogen. Sulphur dioxide is formed by the burning of fossil fuels, such as coal, that contain high quantities of sulphur; nitrogen oxides are contributed from various industrial activities and from car exhaust fumes.

Acid deposition occurs not only as wet precipitation (mist, snow, or rain), but also comes out of the atmosphere as dry particles or is absorbed directly by lakes, plants, and masonry as gases. Acidic gases can travel over 500 km/310 mi a day so acid rain can be considered an example of transboundary pollution.

Acid rain is linked with damage to and the death of forests and lake organisms in Scandinavia, Europe, and eastern North America. It also results in damage to buildings and statues. US and European power stations that burn fossil fuels release about 8 g/0.3 oz of sulphur dioxide and 3 g/0.1 oz of nitrogen oxides per kilowatt-hour. According to the UK Department of the Environment figures, emissions of sulphur dioxide from power stations would have to be decreased by 81% in order to arrest damage.

acid salt chemical compound formed by the partial neutralization of a dibasic or tribasic ◊acid (one that contains two or three hydrogen atoms). Although a salt, it contains replaceable hydrogen, so it may undergo the typical reactions of an acid. Examples are sodium hydrogen sulphate ($NaHSO_4$) and acid phosphates.

aclinic line the magnetic equator, an imaginary line near the Equator, where a compass needle balances horizontally, the attraction of the north and south magnetic poles being equal.

acne skin eruption, mainly occurring among adolescents and young adults, caused by inflammation of the sebaceous glands which secrete an oily substance (sebum), the natural lubricant of the skin. Sometimes the openings of the glands become blocked, causing the formation of pus-filled swellings. Teenage acne is seen mainly on the face, back, and chest.

There are other, less common types of acne, sometimes caused by contact with irritant chemicals (chloracne).

acoustic coupler device that enables computer data to be transmitted and received through a normal telephone handset; the handset rests on the coupler to make the connection. A small speaker within the device is used to convert the computer's digital output data into sound signals, which are then picked up by the handset and transmitted through the telephone system. At the receiving telephone, a second acoustic coupler or modem

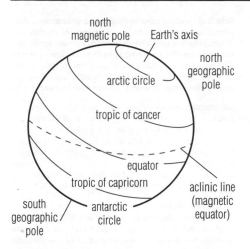

aclinic line The magnetic equator, or the line at which the attraction of both magnetic poles is equal. Along the aclinic line, a compass needle swinging vertically will settle in a horizontal position.

converts the sound signals back into digital data for input into a computer.

Unlike a ◊modem, an acoustic coupler does not require direct connection to the telephone system. However, interference from background noise means that the quality of transmission is poorer than with a modem, and more errors are likely to arise.

acoustics in general, the experimental and theoretical science of sound and its transmission; in particular, that branch of the science that has to do with the phenomena of sound in a particular space such as a room or theatre. In architecture, the sound-reflecting character of an internal space.

Acoustical engineering is concerned with the technical control of sound, and involves architecture and construction, studying control of vibration, soundproofing, and the elimination of noise. It also includes all forms of sound recording and reinforcement, the hearing and perception of sounds, and hearing aids.

acquired character feature of the body that develops during the lifetime of an individual, usually as a result of repeated use or disuse, such as the enlarged muscles of a weightlifter.

French naturalist Jean Baptiste Lamarck's theory of evolution assumed that acquired characters were passed from parent to offspring. Modern evolutionary theory does not recognize the inheritance of acquired characters because there is no reliable scientific evidence that it occurs, and because no mechanism is known whereby bodily changes can influence the genetic material. The belief that this does not occur is known as ◊central dogma.

acquired immune deficiency syndrome full name for the disease ◊AIDS.

acre traditional English land measure equal to 4,840 square yards (4,047 sq m/0.405 ha). Originally meaning a field, it was the size that a yoke of oxen could plough in a day.

As early as Edward I's reign the acre was standardized by statute for official use, although local variations in Ireland, Scotland, and some English counties continued. It may be subdivided into 160 square rods (one square rod equalling 25.29 sq m/30.25 sq yd).

acridine $C_{13}H_9N$ heterocyclic organic compound that occurs in coal tar. It is crystalline, melting at 108°C/226.4°F. Acridine is extracted by dilute acids but can also be obtained synthetically. It is used to make dyes and drugs.

acrophobia ◊phobia involving fear of heights.

acrylic acid common name for ◊propenoic acid.

actinide any of a series of 15 radioactive metallic chemical elements with atomic numbers 89 (actinium) to 103 (lawrencium). Elements 89 to 95 occur in nature; the rest of the series are synthesized elements only. Actinides are grouped together because of their chemical similarities (for example, they are all bivalent), the properties differing only slightly with atomic number. The series is set out in a band in the ◊periodic table of the elements, as are the ◊lanthanides.

actinium white, radioactive, metallic element, the first of the actinide series, symbol Ac, atomic number 89, relative atomic mass 227; it is a weak emitter of high-energy alpha particles.

Actinium occurs with uranium and radium in ◊pitchblende and other ores, and can be synthesized by bombarding radium with neutrons. The longest-lived isotope, Ac-227, has a half-life of 21.8 years (all the other isotopes have very short half-lives). Chemically, it is exclusively trivalent, resembling in its reactions the lanthanides and

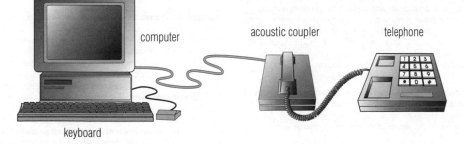

acoustic coupler The acoustic coupler converts digital output from a computer to sound signals that can be sent via a telephone line.

the other actinides. Actinium was discovered in 1899 by the French chemist André Debierne (1874–1949).

action potential in biology, a change in the ◊potential difference (voltage) across the membrane of a nerve cell when an impulse passes along it. A change in potential (from about −60 to +45 millivolts) accompanies the passage of sodium and potassium ions across the membrane.

activation energy in chemistry, the energy required in order to start a chemical reaction. Some elements and compounds will react together merely by bringing them into contact (spontaneous reaction). For others it is necessary to supply energy in order to start the reaction, even if there is ultimately a net output of energy. This initial energy is the activation energy.

active galaxy in astronomy, a type of galaxy that emits vast quantities of energy from a small region at its centre called the active galactic nucleus (AGN). Active galaxies are subdivided into ◊radio galaxies, ◊Seyfert galaxies, BL Lacertae objects, and ◊quasars.

Active galaxies are thought to contain black holes with a mass some 108 times that of the Sun, drawing stars and interstellar gas towards it in a process of accretion. The gravitational energy released by the in-falling material is the power source for the AGN. Some of the energy may appear as a pair of opposed jets emerging from the nucleus. The orientation of the jets to the line of sight and their interaction with surrounding material determines the type of active galaxy that is seen by observers. See also ◊starburst galaxy.

active transport in cells, the use of energy to move substances, usually molecules or ions, across a membrane.

Energy is needed because movement occurs against a concentration gradient, with substances being passed into a region where they are already present in significant quantities. Active transport thus differs from ◊diffusion, the process by which substances move towards a region where they are in lower concentration, as when oxygen passes into the blood vessels of the lungs. Diffusion requires no input of energy.

acupuncture in alternative medicine, a system of inserting long, thin metal needles into the body at predetermined points to relieve pain, as an anaesthetic in surgery, and to assist healing. The needles are rotated manually or electrically. The method, developed in ancient China and increasingly popular in the West, is thought to work by stimulating the brain's own painkillers, the ◊endorphins.

Acupuncture is based on a theory of physiology that posits a network of life-energy pathways, or 'meridians', in the human body and some 800 'acupuncture points' where metal needles may be inserted to affect the energy flow for purposes of preventive or remedial therapy or to produce a local anaesthetic effect. Numerous studies and surveys have attested the efficacy of the method, which is widely conceded by orthodox practitioners despite the lack of an acceptable scientific explanation.

acute in medicine, term used to describe a disease of sudden and severe onset which resolves quickly; for example, pneumonia and meningitis. In contrast, a *chronic* condition develops and remains over a long period.

acute angle an angle between 0° and 90°; that is, an amount of turn that is less than a quarter of a circle.

Ada high-level computer-programming language, developed and owned by the US Department of Defense, designed for use in situations in which a computer directly controls a process or machine, such as a military aircraft. The language took more than five years to specify, and became commercially available only in the late 1980s. It is named after English mathematician Ada Augusta Byron.

adaptation in biology, any change in the structure or function of an organism that allows it to survive and reproduce more effectively in its environment. In ◊evolution, adaptation is thought to occur as a result of random variation in the genetic make-up of organisms coupled with ◊natural selection. Species become extinct when they are no longer adapted to their environment – for instance, if the climate suddenly becomes colder.

adaptive radiation in evolution, the formation of several species, with ◊adaptations to different ways of life, from a single ancestral type. Adaptive radiation is likely to occur whenever members of a species migrate to a new habitat with unoccupied ecological niches. It is thought that the lack of competition in such niches allows sections of the migrant population to develop new adaptations, and eventually to become new species.

The colonization of newly formed volcanic islands has led to the development of many unique species. The 13 species of Darwin's finch on the Galápagos Islands, for example, are probably descended from a single species from the South American mainland. The parent stock evolved into different species that now occupy a range of diverse niches.

addiction state of dependence caused by habitual use of drugs, alcohol, or other substances. It is characterized by uncontrolled craving, tolerance, and symptoms of withdrawal when access is denied. Habitual use produces changes in body chemistry and treatment must be geared to a gradual reduction in dosage.

Initially, only opium and its derivatives (morphine, heroin, codeine) were recognized as addictive, but many other drugs, whether therapeutic (for example, tranquillizers) or recreational (such as cocaine and alcohol), are now known to be addictive.

Research points to a genetic predisposition to addiction; environment and psychological make-up are other factors. Although physical addiction always has a psychological element, not all psychological dependence is accompanied by physical dependence. A carefully controlled withdrawal programme can reverse the chemical changes of habituation. Cure is difficult because of the many other factors contributing to addiction.

Addison's disease rare condition caused by destruction of the outer part of the ◊adrenal glands, leading to reduced secretion of corticosteroid hormones; it is treated by replacement of these hormones. The condition, formerly fatal, is mostly caused by autoimmune disease or tuberculosis. Symptoms include weight loss, anaemia, weakness, low blood pressure, digestive upset, and brownish pigmentation of the skin. Addison's disease is rare in children and in those over 60 years of age. It is commonest in the second and third decades. There is no cure, but patients can live normal lives when given regular treatment with adrenal hormones. It is named after Thomas Addison, the London physician who first described it in 1849 and more fully in 1855.

addition in arithmetic, the operation of combining two numbers to form a sum; thus, $7 + 4 = 11$. It is one of the four basic operations of arithmetic (the others are subtraction, multiplication, and division).

addition reaction chemical reaction in which the atoms of an element or compound react with a double bond or triple bond in an organic compound by opening up one of the bonds and becoming attached to it, for example

$$CH_2=CH_2 + HCl \rightarrow CH_3CH_2Cl$$

An example is the addition of hydrogen atoms to ◊unsaturated compounds in vegetable oils to produce margarine. Addition reactions are used to make useful polymers from ◊alkenes.

address in a computer memory, a number indicating a specific location.

At each address, a single piece of data can be stored. For microcomputers, this normally amounts to one ◊byte (enough to represent a single character, such as a letter or digit).

The maximum capacity of a computer memory depends on how many memory addresses it can have. This is normally measured in units of 1,024 bytes (known as kilobytes, or K).

adenoids masses of lymphoid tissue, similar to ◊tonsils, located in the upper part of the throat, behind the nose. They are part of a child's natural defences against the entry of germs but usually shrink and disappear by the age of ten.

Adenoids may swell and grow, particularly if infected, and block the breathing passages. If they become repeatedly infected, they may be removed surgically (*adenoidectomy*), usually along with the tonsils.

ADH abbreviation for *antidiuretic hormone* in biology, part of the system maintaining a correct salt/water balance in vertebrates.

Its release is stimulated by the ◊hypothalamus in the brain, which constantly receives information about salt concentration from receptors situated in the neck. In conditions of water shortage, increased ADH secretion from the brain will cause more efficient conservation of water in the kidney, so that fluid is retained by the body. When an animal is able to take in plenty of water, decreased ADH secretion will cause the urine to become dilute and plentiful. The system allows the body to compensate for a varying fluid intake and maintain a correct balance.

adhesion in medicine, the abnormal binding of two tissues as a result of inflammation or damage. The moving surfaces of joints or internal organs may merge together if they have been inflamed and tissue fluid has been present between the surfaces.

Adhesions sometimes occur after abdominal operations and may lead to colicky pains and intestinal obstruction.

adhesive substance that sticks two surfaces together. Natural adhesives (glues) include gelatin in its crude industrial form (made from bones, hide fragments, and fish offal) and vegetable gums. Synthetic adhesives include thermoplastic and thermosetting resins, which are often stronger than the substances they join; mixtures of ◊epoxy resin and hardener that set by chemical reaction; and elastomeric (stretching) adhesives for flexible joints. Superglues are fast-setting adhesives used in very small quantities.

Adhesive

The Florida leaf beetle *Hemisphaerota cyanen* has 60,000 adhesive pads on its feet. These enable it to resist the pull of a 2 g/0.07 oz weight (the equivalent of a human hanging on to 200 grand pianos) so that its predators are unable to move it.

---

adiabatic in physics, a process that occurs without loss or gain of heat, especially the expansion or contraction of a gas in which a change takes place in the pressure or volume, although no heat is allowed to enter or leave.

adipose tissue type of ◊connective tissue of vertebrates that serves as an energy reserve, and also pads some organs. It is commonly called fat tissue, and consists of large spherical cells filled with fat. In mammals, major layers are in the inner layer of skin and around the kidneys and heart.

Fatty acids are transported to and from it via the blood system. An excessive amount of adipose tissue is developed in the course of some diseases, especially obesity.

adolescence in the human life cycle, the period between the beginning of puberty and adulthood.

ADP abbreviation for *adenosine diphosphate*, the chemical product formed in cells when ◊ATP breaks down to release energy.

adrenal gland or *suprarenal gland* triangular gland situated on top of the ◊kidney. The adrenals are soft and yellow, and consist of two parts: the cortex and medulla.

The *cortex* (outer part) secretes various steroid hormones and other hormones that control salt and water metabolism and regulate the use of carbohydrates, proteins, and fats.

The *medulla* (inner part) secretes the hormones adrenaline and noradrenaline which, during times of stress, cause the heart to beat faster and harder, increase blood flow to the heart and muscle cells, and dilate airways in the lungs, thereby delivering more oxygen to cells throughout the body and in general preparing the body for 'fight or flight'.

adrenaline or *epinephrine* hormone secreted by the medulla of the ◊adrenal glands. Adrenaline is synthesized from a closely related substance, noradrenaline, and the two hormones are released into the bloodstream in situations of fear or stress.

Adrenaline's action on the liver raises blood-sugar levels by stimulating glucose production and its action on adipose tissue raises blood fatty-acid levels; it also increases the heart rate, increases blood flow to muscles, reduces blood flow to the skin with the production of sweat, widens the smaller breathing tubes (bronchioles) in the lungs, and dilates the pupils of the eyes.

adsorption taking up of a gas or liquid at the surface of another substance, most commonly a solid (for example, activated charcoal adsorbs gases). It involves molecular attraction at the surface, and should be distinguished from ◊absorption (in which a uniform solution results from a gas or liquid being incorporated into the bulk structure of a liquid or solid).

aerenchyma plant tissue with numerous air-filled spaces between the cells. It occurs in the stems and roots of many aquatic plants where it aids buoyancy and facilitates transport of oxygen around the plant.

aerial or **antenna** in radio and television broadcasting, a conducting device that radiates or receives electromagnetic waves. The design of an aerial depends principally on the wavelength of the signal. Long waves (hundreds of metres in wavelength) may employ long wire aerials; short waves (several centimetres in wavelength) may employ rods and dipoles; microwaves may also use dipoles – often with reflectors arranged like a toast rack – or highly directional parabolic dish aerials. Because microwaves travel in straight lines, giving line-of-sight communication, microwave aerials are usually located at the tops of tall masts or towers.

aerial oxidation in chemistry, a reaction in which air is used to oxidize another substance, as in the contact process for the manufacture of sulphuric acid, and in the ◊souring of wine.

$$2SO_2 + O_2 \rightleftharpoons 2SO_3$$

aerobic in biology, term used to describe those organisms that require oxygen (usually dissolved in water) for the efficient release of energy contained in food molecules, such as glucose. They include almost all organisms (plants as well as animals) with the exception of certain bacteria, yeasts, and internal parasites.

Aerobic reactions occur inside every cell and lead to the formation of energy-rich ◊ATP, subsequently used by the cell for driving its metabolic processes. Oxygen is used to convert glucose to carbon dioxide and water, thereby releasing energy.

Most aerobic organisms die in the absence of oxygen, but certain organisms and cells, such as those found in muscle tissue, can function for short periods anaerobically (without oxygen).

aerodynamics branch of fluid physics that studies the forces exerted by air or other gases in motion. Examples include the airflow around bodies moving at speed through the atmosphere (such as land vehicles, bullets, rockets, and aircraft), the behaviour of gas in engines and furnaces, air conditioning of buildings, the deposition of snow, the operation of air-cushion vehicles (hovercraft), wind loads on buildings and bridges, bird and insect flight, musical wind instruments, and meteorology. For maximum efficiency, the aim is usually to design the shape of an object to produce a streamlined flow, with a minimum of turbulence in the moving air. The behaviour of aerosols or the pollution of the atmosphere by foreign particles are other aspects of aerodynamics.

aerogel light, transparent, highly porous material composed of more than 90% air. Such materials are formed from silica, metal oxides, and organic chemicals, and are produced by drying gels – networks of linked molecules suspended in a liquid – so that air fills the spaces previously occupied by the liquid. They are excellent heat insulators and have unusual optical, electrical, and acoustic properties.

Aerogels were first produced by US scientist Samuel Kristler in the early 1930s by drying silica gels at high temperatures and pressures.

aeronautics science of travel through the Earth's atmosphere, including aerodynamics, aircraft structures, jet and rocket propulsion, and aerial navigation.

In **subsonic aeronautics** (below the speed of sound), aerodynamic forces increase at the rate of the square of the speed.

Transsonic aeronautics covers the speed range from just below to just above the speed of sound and is crucial to aircraft design. Ordinary sound waves move at about 1,225 kph/760 mph at sea level, and air in front of an aircraft moving slower than this is 'warned' by the waves so that it can move aside. However, as the flying speed approaches that of the sound waves, the warning is too late for the air to escape, and the aircraft pushes the air aside, creating shock waves, which absorb much power and create design problems. On the ground the shock waves give rise to a ◊sonic boom. It was once thought that the speed of sound was a speed limit to aircraft, and the term ◊sound barrier came into use.

Supersonic aeronautics concerns speeds above that of sound and in one sense may be considered a much older study than aeronautics itself, since the study of the flight of bullets, known as ◊ballistics, was undertaken soon after the introduction of firearms. **Hypersonics** is the study of airflows and forces at speeds above five times that of sound (Mach 5); for example, for guided missiles, space rockets, and advanced concepts such as HOTOL (horizontal takeoff and landing). For all flight speeds streamlining is necessary to reduce the effects of air resistance.

Aeronautics is distinguished from astronautics, which is the science of travel through space. Astronavigation (navigation by reference to the stars) is used in aircraft as well as in ships and is a part of aeronautics.

aeroplane (US **airplane**) powered heavier-than-air craft supported in flight by fixed wings. Aeroplanes are propelled by the thrust of a jet engine or airscrew (propeller). They must be designed aerodynamically, since streamlining ensures maximum flight efficiency. The Wright brothers flew the first powered plane (a biplane) in Kitty Hawk, North Carolina, USA, in 1903. For the history of aircraft and aviation, see ◊flight.

Efficient streamlining prevents the formation of shock waves over the body surface and wings, which would cause instability and power loss. The wing of an aeroplane has the cross-sectional shape of an aerofoil, being broad and curved at the front, flat underneath, curved on top, and tapered to a sharp point at the rear. It is so shaped that air passing above it is speeded up, reducing pressure below atmospheric pressure. This follows from ◊Bernoulli's principle and results in a force acting vertically upwards, called lift, which counters the plane's weight. In level flight lift equals weight. The wings develop sufficient lift to support the plane when they move quickly through the air. The thrust that causes propulsion comes from the reaction to the air stream accelerated backwards by the propeller or the gases shooting backwards from the jet exhaust. In flight the engine thrust must overcome the air resistance, or ◊drag. Drag depends on frontal area (for example, large, airliner; small, fighter plane) and shape (drag coefficient); in level flight, drag equals thrust. The drag is reduced by streamlining the plane, resulting in higher speed and reduced fuel consumption for a given power. Less fuel need be carried for a given distance of travel, so a larger payload (cargo or passengers) can be carried.

Aerogels

There is a new material that can be used to catch micrometeorites in space, that can replace CFC-filled insulating foams, or can be part of a detector for subatomic particles. It is also transparent, and the list of its uses is growing.

The story started 60 years ago, when the American chemist Samuel Kristler attempted to dry wet silica gels without shrinking them in the process. Silica gels are useful drying agents because they readily absorb moisture from the air. The problem that Kristler was trying to overcome can be seen when mud in a puddle dries out in the sun. Cracks develop and the broken surface warps. His aim was to replace the liquid within wet silica gels with air, without causing their silica skeletons to collapse. He was successful, and called the new, highly porous materials 'aerogels'. Kristler found that these new materials had many unusual properties.

What are aerogels?

Ordinary glass, which consists mainly of silica, is a heavy material. Silica aerogels are full of holes encircled by strings of silica and so they are extraordinarily light: they are composed of 90% air and can even float on whisked egg white. Aerogels can be made from many different substances, as Kristler discovered during the ten years he was carrying out his pioneering work at Stanford University. He substituted alcohol for water in making his gels, which were then dried at high temperatures and pressures. This procedure was slow and potentially very dangerous, but Kristler succeeded brilliantly in preparing aerogels from alumina, iron, tungsten, and tin oxide as well as from silica, the oxide of the element silicon. Now there are much faster and safer methods of making aerogels in a few hours, rather than in weeks. These were developed by a group working at the Claude Bernard University at Lyons in France, and since the mid 1980s the number of researchers working with aerogels around the world has increased dramatically.

Following an accident in 1984, when the pressure vessel used to dry the aerogels leaked enough alcohol to cause an explosion that wrecked the entire facility, scientists at BASF in Germany decided to use liquid carbon dioxide instead. This method enabled pellets of aerogels to be made. While not perfectly transparent, the pellets make excellent insulators because, like other forms of aerogel, they do not burn and can withstand temperatures up to 600°C.

Aerogels as insulators and conductors

Research into better methods of insulation received a boost during the energy crisis in the 1970s, and aerogels, because of their unique microstructure and chemical properties, are likely to be the insulators of choice in the future: they are five times more effective than expanded polystyrene. In aerogels, the air-containing pores vary between 1 and 100 nanometres in diameter. This is about the distance that an air molecule travels before it collides with another air molecule. So, on average, air molecules collide with each other as often as they do with the walls of pores in an aerogel in which they are trapped. This is why aerogels insulate so well.

Aerogels can replace CFC-filled insulating foams, which have been effectively banned since Jan 1996 by international agreement. Aerogels have impressive insulating properties and have already found use as insulating materials. Because they are transparent, pelleted aerogels can be sandwiched between two layers of glass, minimizing heat loss. Since the pellets also scatter light, they can provide an excellent alternative to frosted glass. A layer of aerogel also improves the performance of solar panels.

Fast micrometeorites, which are tiny particles of interstellar debris, burn up as they enter the Earth's atmosphere. But low-density aerogels can trap them in outer space. The micrometeorites are slowed to a stop as they penetrate the porous material, which because of its transparency enables scientists to inspect them where they come to rest. Transparent tiles made of aerogels are used at the DESY accelerator, Geneva, in place of the compressed gases or low-density liquids in Cherenkov detectors which detect pions, protons, and muons moving at speeds near to the speed of light. Silica aerogels, combined with phosphors, the substances that provide the colour in colour televisions, and tiny amounts of radioactivity can provide a source of light without electricity. Aerogels prepared from organic chemicals are less brittle than those made from inorganic chemicals, and some of these are good conductors of electricity. Sound travels more slowly through aerogels than it does in air. Research in this field has already lead to novel applications of aerogels in ultrasonics.

Julian Rowe

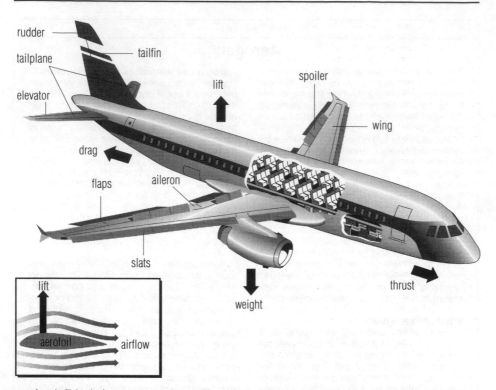

aeroplane In flight, the forces on an aeroplane are lift, weight, drag, and thrust. The lift is generated by the air flow over the wings, which have the shape of an aerofoil. The engine provides the thrust. The drag results from the resistance of the air to the aeroplane's passage through it. Various moveable flaps on the wings and tail allow the aeroplane to be controlled. The rudder is moved to turn the aeroplane. The elevators allow the craft to climb or dive. The ailerons are used to bank the aeroplane while turning. The flaps, slats, and spoilers are used to reduce lift and speed during landing.

The shape of a plane is dictated principally by the speed at which it will operate (see ◊aeronautics). A plane operating at well below the speed of sound (about 965 kph/600 mph) need not be particularly well streamlined, and it can have broad wings projecting at right angles from the fuselage. An aircraft operating close to the speed of sound must be well streamlined and have swept-back wings. This prevents the formation of shock waves over the body surface and wings, which would result in instability and high power loss. Supersonic planes (faster than sound) need to be severely streamlined, and require a needle nose, extremely swept-back wings, and what is often termed a 'Coke-bottle' (narrow-waisted) fuselage, in order to pass through the sound barrier without suffering undue disturbance.

To give great flexibility of operation at low as well as high speeds, some supersonic planes are designed with variable geometry, or ◊swing wings. For low-speed flight the wings are outstretched; for high-speed flight they are swung close to the fuselage to form an efficient ◊delta-wing configuration.

Aircraft designers experiment with different designs in ◊wind tunnel tests, which indicate how their designs will behave in practice. Fighter jets in the 1990s are being deliberately designed to be aerodynamically unstable, to ensure greater agility; an example is the European Fighter Aircraft under development by the UK, Germany, Italy, and Spain. This is achieved by a main wing of continuously modifiable shape, the airflow over which is controlled by a smaller tilting foreplane. New aircraft are being made lighter and faster (to Mach 3) by the use of heat-resistant materials, some of which are also radar-absorbing, making the aircraft 'invisible' to enemy defences.

Planes are constructed from light but strong aluminium alloys such as duralumin (with copper, magnesium, and so on). For supersonic planes special stainless steel and titanium may be used in areas subjected to high heat loads. The structure of the plane, or the airframe (wings, fuselage, and so on), consists of a surface skin of alloy sheets supported at intervals by struts known as ribs and stringers. The structure is bonded together by riveting or by powerful adhesives such as ◊epoxy resins. In certain critical areas, which have to withstand very high stresses (such as the wing roots), body panels are machined from solid metal for extra strength.

On the ground a plane rests on wheels, usually in a tricycle arrangement, with a nose wheel and two wheels behind, one under each wing. For all except some light planes the landing gear, or undercarriage, is retracted in flight to reduce drag.

Seaplanes, which take off and land on water, are fitted with nonretractable hydrofoils.

Wings by themselves are unstable in flight, and a plane

requires a tail to provide stability. The tail comprises a horizontal tailplane and vertical tailfin, called the horizontal and vertical stabilizer respectively. The tailplane has hinged flaps at the rear called elevators to control pitch (attitude). Raising the elevators depresses the tail and inclines the wings upwards (increases the angle of attack). This speeds the airflow above the wings until lift exceeds weight and the plane climbs. However, the steeper attitude increases drag, so more power is needed to maintain speed and the engine throttle must be opened up. Moving the elevators in the opposite direction produces the reverse effect. The angle of attack is reduced, and the plane descends. Speed builds up rapidly if the engine is not throttled back. Turning (changing direction) is effected by moving the rudder hinged to the rear of the tailfin, and by banking (rolling) the plane. It is banked by moving the ailerons, interconnected flaps at the rear of the wings which move in opposite directions, one up, the other down. In planes with a delta wing, such as ◊Concorde, the ailerons and elevators are combined. Other movable control surfaces, called flaps, are fitted at the rear of the wings closer to the fuselage. They are extended to increase the width and camber (curve) of the wings during takeoff and landing, thereby creating extra lift, while movable sections at the front, or leading edges, of the wing, called slats, are also extended at these times to improve the airflow. To land, the nose of the plane is brought up so that the angle of attack of the wings exceeds a critical point and the airflow around them breaks down; lift is lost (a condition known as stalling), and the plane drops to the runway. A few planes (for example, the Harrier) have a novel method of takeoff and landing, rising and dropping vertically by swivelling nozzles to direct the exhaust of their jet engines downwards. The ◊helicopter and ◊convertiplane use rotating propellers (rotors) to obtain lift to take off vertically.

The control surfaces of a plane are operated by the pilot on the flight deck, by means of a control stick, or wheel, and by foot pedals (for the rudder). The controls are brought into action by hydraulic power systems. An automatic pilot enables a plane to cruise on a given course at a fixed speed. Advanced experimental high-speed craft known as control-configured vehicles use a sophisticated computer-controlled system. The pilot instructs the computer which manoeuvre the plane must perform, and the computer, informed by a series of sensors around the craft about the altitude, speed, and turning rate of the plane, sends signals to the control surface and throttle to enable the manoeuvre to be executed.

aerosol particles of liquid or solid suspended in a gas. Fog is a common natural example. Aerosol cans contain a substance such as scent or cleaner packed under pressure with a device for releasing it as a fine spray.

Most aerosols used chlorofluorocarbons (CFCs) as propellants until these were found to cause destruction of the ◊ozone layer in the stratosphere.

The international community has agreed to phase out the use of CFCs, but most so-called 'ozone-friendly' aerosols also use ozone-depleting chemicals, although they are not as destructive as CFCs. Some of the products sprayed, such as pesticides, can be directly toxic to humans.

aestivation in zoology, a state of inactivity and reduced metabolic activity, similar to ◊hibernation, that occurs during the dry season in species such as lungfish and

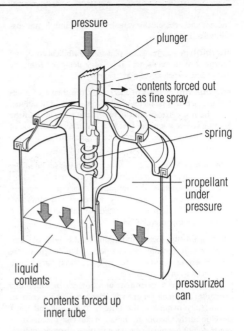

pressure
plunger
contents forced out as fine spray
spring
propellant under pressure
liquid contents
pressurized can
contents forced up inner tube

aerosol The aerosol can produces a fine spray of liquid particles, called an aerosol. When the top button is pressed, a valve is opened, allowing the pressurized propellant in the can to force out a spray of the liquid contents. As the liquid sprays from the can, the small amount of propellant dissolved in the liquid vaporizes, producing a fine spray of small droplets.

snails. In botany, the term is used to describe the way in which flower petals and sepals are folded in the buds. It is an important feature in ◊plant classification.

aetiology in medicine, the systematic investigation into the causes of disease.

affinity in chemistry, the force of attraction (see ◊bond) between atoms that helps to keep them in combination in a molecule. The term is also applied to attraction between molecules, such as those of biochemical significance (for example, between ◊enzymes and substrate molecules). This is the basis for affinity ◊chromatography, by which biologically important compounds are separated.

The atoms of a given element may have a greater affinity for the atoms of one element than for another (for example, hydrogen has a great affinity for chlorine, with which it easily and rapidly combines to form hydrochloric acid, but has little or no affinity for argon).

afforestation planting of trees in areas that have not previously held forests. (**Reafforestation** is the planting of trees in deforested areas.) Trees may be planted (1) to provide timber and wood pulp; (2) to provide firewood in countries where this is an energy source; (3) to bind soil together and prevent soil erosion; and (4) to act as windbreaks.

Afforestation is a controversial issue because while many ancient woodlands of mixed trees are being lost, the new plantations consist almost exclusively of conifers. It is claimed that such plantations acidify the soil and conflict with the interests of ◊biodiversity (they replace more ancient and ecologically valuable species and do not sustain wildlife).

afterbirth in mammals, the placenta, umbilical cord, and ruptured membranes, which become detached from the uterus and expelled soon after birth.

afterburning method of increasing the thrust of a gas turbine (jet) aeroplane engine by spraying additional fuel into the hot exhaust duct between the turbojet and the tailpipe where it ignites. Used for short-term increase of power during takeoff, or during combat in military aircraft.

afterimage persistence of an image on the retina of the eye after the object producing it has been removed. This leads to persistence of vision, a necessary phenomenon for the illusion of continuous movement in films and television. The term is also used for the persistence of sensations other than vision.

after-ripening process undergone by the seeds of some plants before germination can occur. The length of the after-ripening period in different species may vary from a few weeks to many months.

It helps seeds to germinate at a time when conditions are most favourable for growth. In some cases the embryo is not fully mature at the time of dispersal and must develop further before germination can take place. Other seeds do not germinate even when the embryo is mature, probably owing to growth inhibitors within the seed that must be leached out or broken down before germination can begin.

agar jellylike carbohydrate, obtained from seaweeds. It is used mainly in microbiological experiments as a culture medium for growing bacteria and other microorganisms. The agar is resistant to breakdown by microorganisms, remaining a solid jelly throughout the course of the experiment.

agate banded or cloudy type of ◊chalcedony, a silica, SiO_2, that forms in rock cavities. Agates are cryptocrystalline (with crystals too small to be seen with an optical microscope) and are used as ornamental stones and for art objects.

Agate stones, being hard, are also used to burnish and polish gold applied to glass and ceramics.

ageing in common usage, the period of deterioration of the physical condition of a living organism that leads to death; in biological terms, the entire life process.

Three current theories attempt to account for ageing. The first suggests that the process is genetically determined, to remove individuals that can no longer reproduce. The second suggests that it is due to the accumulation of mistakes during the replication of ◊DNA at cell division. The third suggests that it is actively induced by fragments of DNA that move between cells, or by cancer-causing viruses; these may become abundant in old cells and induce them to produce unwanted ◊proteins or interfere with the control functions of their DNA.

Agent Orange selective ◊weedkiller, notorious for its use in the 1960s during the Vietnam War by US forces to eliminate ground cover which could protect enemy forces. It was subsequently discovered to contain highly poisonous ◊dioxin. Thousands of US troops who had handled it, along with many Vietnamese people who

came into contact with it, later developed cancer or produced deformed babies.

Agent Orange, named after the distinctive orange stripe on its packaging, combines equal parts of 2,4-D (2,4-dichlorophenoxyacetic acid) and 2,4,5-T (2,4,5-trichlorophenoxyacetic acid), both now banned in the USA. Companies that had manufactured the chemicals faced an increasing number of lawsuits in the 1970s. All the suits were settled out of court in a single class action, resulting in the largest ever payment of its kind ($180 million) to claimants.

agglutination in biology, the clumping together of ◊antigens, such as red blood cells or bacteria, to form larger, visible clumps, under the influence of ◊antibodies. As each antigen clumps only in response to its particular antibody, agglutination provides a way of determining blood groups and the identity of unknown bacteria.

aggression in biology, behaviour used to intimidate or injure another organism (of the same or of a different species), usually for the purposes of gaining territory, a mate, or food. Aggression often involves an escalating series of threats aimed at intimidating an opponent without having to engage in potentially dangerous physical contact. Aggressive signals include roaring by red deer, snarling by dogs, the fluffing-up of feathers by birds, and the raising of fins by some species of fish.

agonist in biology, a ◊muscle that contracts and causes a movement. Contraction of an agonist is complemented by relaxation of its ◊antagonist. For example, the biceps (in the front of the upper arm) bends the elbow whilst the triceps (lying behind the biceps) straightens the arm.

agonist in medicine, a drug or other substance that has a similar effect to normal chemical messengers in the body through its actions at receptor sites of cells. Examples are sympathomimetic drugs that mimic the actions of adrenaline and are used in the treatment of certain heart disorders.

agoraphobia ◊phobia involving fear of open spaces and public places. The anxiety produced can be so severe that some sufferers are unable to leave their homes for many years.

Agoraphobia affects 1 person in 20 at some stage in their lives. The most common time of onset is between the ages of 18 and 28.

agribusiness ◊commercial farming on an industrial scale, often financed by companies whose main interests lie outside agriculture; for example, multinational corporations. Agribusiness farms are mechanized, large in size, highly structured, and reliant on chemicals.

agricultural revolution sweeping changes that took place in British agriculture over the period 1750–1850 in response to the increased demand for food from a rapidly expanding population. Recent research has shown these changes to be only part of a much larger, ongoing process of development.

Changes of the latter half of the 18th century included the enclosure of open fields, the introduction of four-course rotation together with new fodder crops such as turnips, and the development of improved breeds of livestock.

agriculture the practice of farming, including the cultivation of the soil (for raising crops) and the raising of

Agriculture: Chronology

10000– 8000 BC	Holocene (post-glacial) period of hunters and gatherers. Harvesting and storage of wild grains in southwest Asia. Herding of reindeer in northern Eurasia. Domestic sheep in northern Iraq.
8000	Neolithic revolution with cultivation of domesticated wheats and barleys, sheep, and goats in southwest Asia. Domestication of pigs in New Guinea.
7000–6000	Domestic goats, sheep, and cattle in Anatolia, Greece, Persia, and the Caspian basin. Planting and harvesting techniques transferred from Asia Minor to Europe.
5000	Beginning of Nile valley civilization. Millet cultivated in China.
3400	Flax used for textiles in Egypt. Widespread corn production in the Americas.
3200	Records of ploughing, raking, and manuring by Egyptians.
3000	First record of asses used as beasts of burden in Egypt. Sumerian civilization used barley as main crop with wheat, dates, flax, apples, plums, and grapes.
2900	Domestication of pigs in eastern Asia.
2500	Domestic elephants in the Indus valley. Potatoes a staple crop in Peru.
2350	Wine-making in Egypt.
2250	First known irrigation dam.
1600	Important advances in the cultivation of vines and olives in Crete.
1500	*Shadoof* (mechanism for raising water) used for irrigation in Egypt.
1300	Aqueducts and reservoirs used for irrigation in Egypt.
1200	Domestic camels in Arabia.
1000–500	Evidence of crop rotation, manuring, and irrigation in India.
600	First windmills used for corn grinding in Persia.
350	Rice cultivation well established in parts of western Africa.
c. 200	Use of gears to create ox-driven water wheel for irrigation. Archimedes screw used for irrigation.
100	Cattle-drawn iron ploughs in use in China.
AD 65	*De Re Rustica/On Rural Things*, Latin treatise on agriculture and irrigation.
500	'Three fields in two years' rotation used in China.
630	Cotton introduced into Arabia.
800	Origins of the 'open field' system in northern Europe.
900	Wheeled ploughs in use in western Europe.
1000	Frisians (NW Netherlanders) began to build dykes and reclaim land. Chinese began to introduce Champa rice which cropped much more quickly than other varieties.
11th century	Three-field system replaced the two-field system in western Europe. Concentration on crop growing.
1126	First artesian wells, at Artois, France.
12th–14th centuries	Expansion of European population brought more land into cultivation. Crop rotations, manuring, and new crops such as beans and peas helped increase productivity. Feudal system at its height.
16th century	Decline of the feudal system in western Europe. More specialist forms of production were now possible with urban markets. Manorial estates and serfdom remained in eastern Europe. Chinese began cultivation of non-indigenous crops such as corn, sweet potatoes, potatoes, and peanuts.
17th century	Potato introduced into Europe. Norfolk crop rotation became widespread in England, involving wheat, turnips, barley and then ryegrass/clover.
1700–1845	Agricultural revolution began in England.
c. 1701	Jethro Tull developed the seed drill and the horse-drawn hoe.
1747	First sugar extracted from sugar beet in Prussia.
1762	Veterinary school founded in Lyons, France.
1783	First plough factory in England.
1785	Cast-iron ploughshare patented.
1793	Invention of the cotton gin.
1800	Early threshing machines developed in England.
1820s	First nitrates for fertilizer imported from South America.
1830	Reaping machines developed in Scotland and the US. Steel plough made by John Deere in Illinois, USA.
1840s	Extensive potato blight in Europe.
1850s	Use of clay pipes for drainage well established throughout Europe.
1862	First steam plough used in the Netherlands.
1850– 1890s	Major developments in transport and refrigeration technology altered the nature of agricultural markets with crops, dairy products, and wheat being shipped internationally.
1892	First petrol-driven tractor in the USA.
1921	First attempt at crop dusting with pesticides from an aeroplane near Dayton, Ohio, USA.
1938	First self-propelled grain combine harvester used in the USA.
1942–62	Huge increase in the use of pesticides, later curbed by disquiet about their effects and increasing resistance of pests to standard controls such as DDT. Increasing use of scientific techniques, crop specialization and larger scale of farm enterprises.
1985	First cases of bovine spongiform encephalopathy (BSE) recorded by UK vets.
1995	Increase in the use of genetic engineering with nearly 3,000 transgenic crops being field-tested.
1996	Organic farming was on the increase in EU countries. The rise was 11% per year in Britain, 50% in Germany, and 40% in Italy.

domesticated animals. The units for managing agricultural production vary from smallholdings and individually owned farms to corporate-run farms and collective farms run by entire communities.

Crops are for human or animal food, or commodities such as cotton and sisal. For successful production, the land must be prepared (ploughed, cultivated, harrowed, and rolled). Seed must be planted and the growing plants nurtured. This may involve ◊fertilizers, ◊irrigation, pest control by chemicals, and monitoring of acidity or nutrients. When the crop has grown, it must be harvested and, depending on the crop, processed in a variety of ways before it is stored or sold. Greenhouses allow cultivation of plants that would otherwise find the climate too harsh. ◊Hydroponics allows commercial cultivation of crops using nutrient-enriched solutions instead of soil. Special methods, such as terracing, may be adopted to allow cultivation in hostile terrain and to retain topsoil in mountainous areas with heavy rainfall.

Animals are raised for wool, milk, leather, dung (as fuel), or meat. They may be semi-domesticated, such as reindeer, or fully domesticated but nomadic (where naturally growing or cultivated food supplies are sparse), or kept in one location. Animal farming involves accommodation (buildings, fencing, or pasture), feeding, breeding, gathering the produce (eggs, milk, or wool), slaughtering, and further processing such as tanning.

The Farmer will never be happy again; / He carries his heart in his boots; / For either the rain is destroying his grain / Or the drought is destroying his roots.

On **agriculture** Alan Patrick Herbert
'The Farmer'

---

agronomy study of crops and soils, a branch of agricultural science. Agronomy includes such topics as selective breeding (of plants and animals), irrigation, pest control, and soil analysis and modification.

AI abbreviation for ◊artificial intelligence.

AI(D) abbreviation for ◊artificial insemination (by donor). AIH is *artificial insemination by husband*.

AIDS acronym for *acquired immune deficiency syndrome*, the gravest of the sexually transmitted diseases, or ◊STDs. It is caused by the human immunodeficiency virus (HIV), now known to be a ◊retrovirus. HIV is transmitted in body fluids, mainly blood and genital secretions.

Worldwide, heterosexual activity accounts for three-quarters of all HIV infections. In addition to heterosexual men and women, high-risk groups are homosexual and bisexual men, prostitutes, intravenous drug-users sharing needles, and haemophiliacs and other patients treated with contaminated blood products. The virus has a short life outside the body, which makes transmission of the infection by methods other than sexual contact, blood transfusion, and shared syringes extremely unlikely.

US researchers in 1995 developed an explanation of why HIV is transmitted mainly by heterosexual sex in Africa and Asia, and by homosexual sex and intravenous drug use in Europe and the USA. They found that the

HIV variant subtype B – responsible for 90% of European and US cases – did not grow well in reproductive tract cells, whereas subtype E – common in developing countries – did grow well. If subtype E becomes more prevalent in Europe and the USA, infection patterns will probably change. The first case of subtype E in Britain was documented in May 1996.

The effect of the virus in those who become ill is the devastation of the immune system, leaving the victim susceptible to diseases that would not otherwise develop. Diagnosis of AIDS is based on the appearance of rare tumours or opportunistic infections in unexpected candidates. *Pneumocystis carinii* pneumonia, for instance, normally seen only in the malnourished or those whose immune systems have been deliberately suppressed, is common among AIDS victims and, for them, a leading cause of death.

Many people who have HIV in their blood are not ill; in fact, it was initially thought that during the delay between infection with HIV and the development of AIDS the virus lay dormant. However, HIV reproduces at an estimated rate of a billion viruses a day, even in individuals with no symptoms, but is held at bay by the immune system producing enough white blood cells (CD4 cells) to destroy them. Gradually, the virus mutates so much that the immune system is unable to continue to counteract; people with advanced AIDS have virtually no CD4 cells remaining. These results indicate the importance of treating HIV-positive individuals before symptoms develop, rather than delaying treatment until the onset of AIDS.

In the West the time-lag between infection with HIV and the development of AIDS seems to be about ten years, but progression is far more rapid in developing countries. Some AIDS victims die within a few months of the outbreak of symptoms, some survive for several years; roughly 50% are dead within three years. There is no cure for the disease and the antiviral drugs currently in use against AIDS have not lived up to expectations. Trials began in 1994 using a new AIDS drug called 3TC in conjunction with zidovudine (formerly ◊AZT). Although individually the drugs produce little effect, when the drugs were used together, the levels of virus in the blood were ten times lower than at the beginning of the trial. Treatment of opportunistic infections extended the average length of survival with AIDS (in Western countries) from about 11 months in 1985 to 23 months in 1994.

The United Nations AIDS programme (UNAIDS) released the figures from its annual survey in November 1997, concluding that there were 5.8 million new cases of infection in the year, bringing the world total to just under 30 million (1% of the world's adult population). In sub-Saharan Africa there are an estimated 20 million HIV sufferers (7% of the adult population and two thirds of the world total as a whole); 2.3 million will die of AIDS in the region. The figures for HIV incidence in other areas include: south and southeast Asia 6 million; South America 1.3 million; North America 860,000; and western Europe 150,000.

air the mixture of gases making up the Earth's ◊atmosphere.

airbrush small fine spray-gun used by artists, graphic designers, and photographic retouchers. Driven by air pressure from a compressor or pressurized can, it can apply a thin, very even layer of ink or paint, allowing for subtle gradations of tone.

air conditioning system that controls the state of the air inside a building or vehicle. A complete air-conditioning unit controls the temperature and humidity of the air, removes dust and odours from it, and circulates it by means of a fan. US inventor Willis Haviland Carrier (1876–1950) developed the first effective air-conditioning unit in 1902 for a New York printing plant.

The air in an air conditioner is cooled by a type of ◊refrigeration unit comprising a compressor and a condenser. The air is cleaned by means of filters and activated charcoal. Moisture is extracted by condensation on cool metal plates. The air can also be heated by electrical wires or, in large systems, pipes carrying hot water or steam; and cool, dry air may be humidified by circulating it over pans of water or through a water spray.

A specialized air-conditioning system is installed in spacecraft as part of the life-support system. This includes the provision of oxygen to breathe and the removal of exhaled carbon dioxide.

Outdoor spaces may also be cooled using overhead cool air jets.

aircraft any aeronautical vehicle capable of flying through the air. It may be lighter than air (supported by buoyancy) or heavier than air (supported by the dynamic action of air on its surfaces). ◊Balloons and ◊airships are lighter-than-air craft. Heavier-than-air craft include the ◊aeroplane, glider, autogiro, and helicopter.

air-cushion vehicle (ACV) craft that is supported by a layer, or cushion, of high-pressure air. The ◊hovercraft is one form of ACV.

airglow faint and variable light in the Earth's atmosphere produced by chemical reactions (the recombination of ionized particles) in the ionosphere.

airlock airtight chamber that allows people to pass between areas of different pressure; also an air bubble in a pipe that impedes fluid flow. An airlock may connect an environment at ordinary pressure and an environment that has high air pressure (such as a submerged caisson used for tunnelling or building dams or bridge foundations).

An airlock may also permit someone wearing breathing apparatus to pass into an airless environment (into water from a submerged submarine or into the vacuum of space from a spacecraft).

air mass large body of air with particular characteristics of temperature and humidity. An air mass forms when air rests over an area long enough to pick up the conditions of that area. When an air mass moves to another area it affects the ◊weather of that area, but its own characteristics become modified in the process. For example, an air mass formed over the Sahara will be hot and dry, becoming cooler as it moves northwards.

air passages in biology, the nose, pharynx, larynx, trachea, and bronchi. When a breath is taken, air passes through high narrow passages on each side of the nose where it is warmed and moistened and particles of dust are removed. Food and air passages meet and cross in the pharynx. The larynx lies in front of the lower part of the pharynx and it is the organ where the voice is produced using the vocal cords. The air passes the glottis (the opening between the vocal cords) and enters the trachea. The trachea leads into the chest and divides above the heart into two bronchi. The bronchi carry the air to the lungs and they subdivide to form a succession of fine tubes and, eventually, a network of capillaries that allow the exchange of gases between the inspired air and the blood.

air pollution contamination of the atmosphere caused by the discharge, accidental or deliberate, of a wide range of toxic airborne substances. Often the amount of the released substance is relatively high in a certain locality, so the harmful effects become more noticeable. The cost of preventing any discharge of pollutants into the air is prohibitive, so attempts are more usually made to reduce the amount of discharge gradually and to disperse it as quickly as possible by using a very tall chimney, or by intermittent release.

air sac in birds, a thin-walled extension of the lungs. There are nine of these and they extend into the abdomen and bones, effectively increasing lung capacity. In mammals, it is another name for the alveoli in the lungs, and in some insects, for widenings of the trachea.

The sacs subdivide into further air spaces which partially replace the marrow in many of the bird's bones. The air space in these bones assists flight by making them lighter.

airship or ***dirigible*** any aircraft that is lighter than air and power-driven, consisting of an elliptical balloon that forms the streamlined envelope or hull and has below it the propulsion system (propellers), steering mechanism, and space for crew, passengers, and/or cargo. The balloon section is filled with lighter-than-air gas, either the nonflammable helium or, before helium was industrially available in large enough quantities, the easily ignited and

Air Pollution: Major Pollutants

Pollutant	Sources	Effects
sulphur dioxide SO_2	oil, coal combustion in power stations	acid rain formed, which damages plants, trees, buildings, and lakes
oxides of nitrogen NO, NO_2	high-temperature combustion in cars, and to some extent power stations	acid rain formed
lead compounds	from leaded petrol used by cars	nerve poison
carbon dioxide CO_2	oil, coal, petrol, diesel combustion	greenhouse effect
carbon monoxide CO	limited combustion of oil, coal, petrol, diesel fuels	poisonous, leads to photochemical smog in some areas
nuclear waste	nuclear power plants, nuclear weapon testing, war	radioactivity, contamination of locality, cancers, mutations, death

flammable hydrogen. The envelope's form is maintained by internal pressure in the nonrigid (blimp) and semirigid (in which the nose and tail sections have a metal framework connected by a rigid keel) types. The rigid type (zeppelin) maintains its form using an internal metal framework. Airships have been used for luxury travel, polar exploration, warfare, and advertising.

Rigid airships predominated from about 1900 until 1940. As the technology developed, the size of the envelope was increased from about 45 m/150 ft to more than 245 m/800 ft for the last two zeppelins built. In 1852 the first successful airship was designed and flown by Henri Giffard of France. In 1900 the first successful rigid type was designed by Count (*Graf*) Ferdinand von Zeppelin of Germany. Airships were used by both sides during World War I, but they were not seriously used for military purposes after that as they were largely replaced by aeroplanes. The British mainly used small machines for naval reconnaissance and patrolling the North Sea; Germany used Schutte-Lanz and Zeppelin machines for similar patrol work and also for long-range bombing attacks against English and French cities, mainly Paris and London.

In 1919 the first nonstop transatlantic round trip flight was completed by a rigid airship, the British R34. In the early 1920s a large source of helium was discovered in the USA and was substituted for hydrogen, reducing the danger of fire. The US military attempted to use zeppelins but abandoned the effort early on. In the 1920s and early 1930s luxury zeppelin services took passengers across the Atlantic faster and in greater comfort than the great ocean liners. The successful German airship *Graf Zeppelin*, completed in 1927, was used for transatlantic, cruise, and round-the-world trips. In 1929 it travelled 32,000 km/20,000 mi around the world. It was retired and dismantled after years of trouble-free service, and was replaced by the *Hindenburg* in 1936.

Several airship accidents were caused by structural break-up during storms and by fire. The last and best known was the *Hindenburg*, which had been forced to return to the use of flammable hydrogen by a US embargo on helium; it exploded and burned at the mooring mast at Lakehurst, New Jersey, USA, in 1937. The last and largest rigid airship was the German *Graf Zeppelin II*, completed just before World War II; it never saw commercial service but was used as a reconnaissance station off the English coast early in the war (it was the only zeppelin used in the war) and was soon retired and dismantled. Rigid airships, predominant from World War II, are no longer in use but blimps continued in use for coastal and antisubmarine patrol until the 1960s, and advertising blimps can be seen to this day. Recent interest in all types of airship has surfaced (including some with experimental and nontraditional shapes for the envelopes), since they are fuel-efficient, quiet, and capable of lifting enormous loads over great distances.

air transport means of conveying goods or passengers by air from one place to another. See ◊flight.

alabaster naturally occurring fine-grained white or light-coloured translucent form of ◊gypsum, often streaked or mottled. A soft material, it is easily carved, but seldom used for outdoor sculpture.

albedo the fraction of the incoming light reflected by a body such as a planet. A body with a high albedo, near 1, is very bright, while a body with a low albedo, near 0, is dark. The Moon has an average albedo of 0.12, Venus 0.76, Earth 0.37.

albinism rare hereditary condition in which the body has no tyrosinase, one of the enzymes that form the pigment ◊melanin, normally found in the skin, hair, and eyes. As a result, the hair is white and the skin and eyes are pink. The skin and eyes are abnormally sensitive to light, and vision is often impaired. The condition occurs among all human and animal groups.

albumin or *albumen* any of a group of sulphur-containing ◊proteins. The best known is in the form of egg white; others occur in milk, and as a major component of serum. They are soluble in water and dilute salt solutions, and are coagulated by heat.

The presence of serum albumin in the urine, termed albuminuria or proteinuria, may be indicative of kidney or heart disease.

alchemy supposed technique of transmuting base metals, such as lead and mercury, into silver and gold by the philosopher's stone, a hypothetical substance, to which was also attributed the power to give eternal life.

This aspect of alchemy constituted much of the chemistry of the Middle Ages. More broadly, however, alchemy was a system of philosophy that dealt both with the mystery of life and the formation of inanimate substances. Alchemy was a complex and indefinite conglomeration of chemistry, astrology, occultism, and magic, blended with obscure and abstruse ideas derived from various religious systems and other sources. It was practised in Europe from ancient times to the Middle Ages but later fell into disrepute when ◊chemistry and ◊physics developed.

What is accomplished with fire is alchemy, whether in the furnace or the kitchen stove.

On **alchemy** Paracelsus, quoted in J Bronowski *The Ascent of Man* 1975

alcohol any member of a group of organic chemical compounds characterized by the presence of one or more aliphatic OH (hydroxyl) groups in the molecule, and which form ◊esters with acids. The main uses of alcohols are as solvents for gums, resins, lacquers, and varnishes; in the making of dyes; for essential oils in perfumery; and for medical substances in pharmacy. The alcohol produced naturally in the ◊fermentation process and consumed as part of alcoholic beverages is called ◊ethanol.

Alcohols may be liquids or solids, according to the size and complexity of the molecule. The five simplest alcohols form a series in which the number of carbon and hydrogen atoms increases progressively, each one having an extra CH_2 (methylene) group in the molecule: methanol or wood spirit (methyl alcohol, CH_3OH); ethanol (ethyl alcohol, C_2H_5OH); propanol (propyl alcohol, C_3H_7OH); butanol (butyl alcohol, C_4H_9OH); and pentanol (amyl alcohol, $C_5H_{11}OH$). The lower alcohols are liquids that mix with water; the higher alcohols, such as pentanol, are oily liquids immiscible with water; and the highest are waxy solids – for example, hexadecanol (cetyl alcohol, $C_{16}H_{33}OH$) and melissyl alcohol ($C_{30}H_{61}OH$), which occur in sperm-whale oil and beeswax respectively.

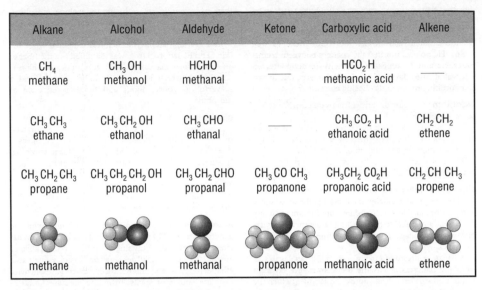

Alkane	Alcohol	Aldehyde	Ketone	Carboxylic acid	Alkene
CH_4 methane	CH_3OH methanol	HCHO methanal	—	HCO_2H methanoic acid	—
CH_3CH_3 ethane	CH_3CH_2OH ethanol	CH_3CHO ethanal	—	CH_3CO_2H ethanoic acid	CH_2CH_2 ethene
$CH_3CH_2CH_3$ propane	$CH_3CH_2CH_2OH$ propanol	CH_3CH_2CHO propanal	CH_3COCH_3 propanone	$CH_3CH_2CO_2H$ propanoic acid	CH_2CHCH_3 propene
methane	methanol	methanal	propanone	methanoic acid	ethene

alcohol The systematic naming of simple straight-chain organic molecules.

Alcohols containing the CH_2OH group are primary; those containing CHOH are secondary; while those containing COH are tertiary.

Alcohol

In 1632, the weekly food ration for each child in a children's hospital in Norwich, England, included two gallons of beer. In the days before reliable water-purification systems, the potential bad effects of alcohol were far outweighed by its sterilizing properties.

---

alcoholism dependence on alcohol. It is characterized as an illness when consumption of alcohol interferes with normal physical or emotional health. Excessive alcohol consumption, whether through sustained ingestion or irregular drinking bouts or binges, may produce physical and psychological addiction and lead to nutritional and emotional disorders. Long-term heavy consumption of alcohol leads to diseases of the heart, liver, and peripheral nerves. Support groups such as Alcoholics Anonymous are helpful.

Aldebaran or *Alpha Tauri* brightest star in the constellation Taurus and the 14th brightest star in the sky; it marks the eye of the 'bull'. Aldebaran is a red giant 60 light years away, shining with a true luminosity of about 100 times that of the Sun.

aldehyde any of a group of organic chemical compounds prepared by oxidation of primary alcohols, so that the OH (hydroxyl) group loses its hydrogen to give an oxygen joined by a double bond to a carbon atom (the aldehyde group, with the formula CHO).

Alexander technique in alternative medicine, a method of correcting bad habits of posture, breathing, and muscular tension, which Australian therapist F M Alexander maintained cause many ailments. The technique is also used to promote general health and relaxation and enhance vitality.

Back troubles, migraine, asthma, hypertension, and some gastric and gynaecological disorders are among the conditions said to be alleviated by the technique, which is also said to be effective in the prevention of disorders, particularly those of later life.

algebra system of mathematical calculations applying to any set of non-numerical symbols (usually letters), and the axioms and rules by which they are combined or operated upon; sometimes known as *generalized arithmetic*.

'Algebra' was originally the name given to the study of equations. In the 9th century, the Arab mathematician Muhammad ibn-Mūsā al-Khwārizmī used the term *al-jabr* for the process of adding equal quantities to both sides of an equation. When his treatise was later translated into Latin, *al-jabr* became 'algebra' and the word was adopted as the name for the whole subject.

In ordinary algebra the same operations are carried on as in arithmetic, but, as the symbols are capable of a more generalized and extended meaning than the figures used in arithmetic, it facilitates calculation where the numerical values are not known, or are inconveniently large or small, or where it is desirable to keep them in an analysed form.

alginate salt of alginic acid $(C_6H_8O_6)_n$ obtained from brown seaweeds and used in textiles, paper, food products, and pharmaceuticals.

ALGOL (acronym for *algo*rithmic *l*anguage) in computing, an early high-level programming language, developed in the 1950s and 1960s for scientific applications. A general-purpose language, ALGOL is best suited to mathematical work and has an algebraic style. Although no longer in common use, it has greatly influenced more recent languages, such as Ada and PASCAL.

Algol or *Beta Persei* ☿ eclipsing binary, a pair of orbiting stars in the constellation Perseus, one of which eclipses the other every 69 hours, causing its brightness to drop by two-thirds.

The brightness changes were first explained in 1782 by English amateur astronomer John Goodricke (1764–1786). He pointed out that the changes between magnitudes 2.2 and 3.5 repeated themselves exactly after an interval of 2.867 days and supposed this to be due to two stars orbiting round and eclipsing each other.

algorithm procedure or series of steps that can be used to solve a problem.

In computer science, it describes the logical sequence of operations to be performed by a program. A ◊flow chart is a visual representation of an algorithm.

The word derives from the name of 9th-century Arab mathematician Muhammad ibn-Mūsā al-Khwārizmī.

alimentary canal in animals, the tube through which food passes; it extends from the mouth to the anus. It is a complex organ, adapted for ◊digestion. In human adults, it is about 9 m/30 ft long, consisting of the mouth cavity, pharynx, oesophagus, stomach, and the small and large intestines.

A constant stream of enzymes from the canal wall and from the pancreas assists the breakdown of food molecules into smaller, soluble nutrient molecules, which are absorbed through the canal wall into the bloodstream and carried to individual cells. The muscles of the alimentary canal keep the incoming food moving, mix it with the enzymes and other juices, and slowly push it in the direction of the anus, a process known as ◊peristalsis. The wall of the canal receives an excellent supply of blood and is folded so as to increase its surface area. These two adaptations ensure efficient absorption of nutrient molecules.

aliphatic compound any organic chemical compound in which the carbon atoms are joined in straight chains, as in hexane (C_6H_{14}), or in branched chains, as in 2-methyl-pentane ($CH_3CH(CH_3)CH_2CH_2CH_3$).

Aliphatic compounds have bonding electrons localized within the vicinity of the bonded atoms. ◊Cyclic compounds that do not have delocalized electrons are also aliphatic, as in the alicyclic compound cyclohexane (C_6H_{12}) or the heterocyclic piperidine ($C_5H_{11}N$). Compare ◊aromatic compound.

alkali metal any of a group of six metallic elements with similar chemical properties: lithium, sodium, potassium, rubidium, caesium, and francium. They form a linked group (Group One) in the ◊periodic table of the elements. They are univalent (have a valency of one) and of very low density (lithium, sodium, and potassium float on water); in general they are reactive, soft, low-melting-point metals. Because of their reactivity they are only found as compounds in nature.

alkaline-earth metal any of a group of six metallic elements with similar bonding properties: beryllium, magnesium, calcium, strontium, barium, and radium. They form a linked group in the ◊periodic table of the elements. They are strongly basic, bivalent (have a valency of two), and occur in nature only in compounds.

alkaloid any of a number of physiologically active and frequently poisonous substances contained in some plants. They are usually organic bases and contain nitrogen. They form salts with acids and, when soluble, give alkaline solutions.

Substances in this group are included by custom rather than by scientific rules. Examples include morphine, cocaine, quinine, caffeine, strychnine, nicotine, and atropine.

In 1992, epibatidine, a chemical extracted from the skin of an Ecuadorian frog, was identified as a member of an entirely new class of alkaloid. It is an organochlorine compound, which is rarely found in animals, and a powerful painkiller, about 200 times as effective as morphine.

alkalosis in medicine, condition characterized by an increase in the alkalinity of the blood or, more accurately, by a decrease in hydrogen ions in the blood. Alkalosis is due to an excessive loss of hydrogen ions due to vomiting. It can also be due to excessive retention of bicarbonate ions by the kidney or to the ingestion of large amounts of alkalis, such as antacids. The treatment of alkalosis is dependent upon removing the circumstances that lead to the disturbance in the acid-base balance of the body.

alkane member of a group of ◊hydrocarbons having the general formula C_nH_{2n+2}, commonly known as **paraffins**. As they contain only single ◊covalent bonds, alkanes are said to be saturated. Lighter alkanes, such as methane, ethane, propane, and butane, are colourless gases; heavier ones are liquids or solids. In nature they are found in natural gas and petroleum. *See illustration opposite.*

alkene member of the group of ◊hydrocarbons having the general formula C_nH_{2n}, formerly known as *olefins*. Alkenes are unsaturated compounds, characterized by one or more double bonds between adjacent carbon atoms. Lighter alkenes, such as ethene and propene, are gases, obtained from the ◊cracking of oil fractions. Alkenes react by addition, and many useful compounds, such as poly(ethene) and bromoethane, are made from them.

alkyne member of the group of ◊hydrocarbons with the general formula C_nH_{2n-2}, formerly known as the *acetylenes*. They are unsaturated compounds, characterized by one or more triple bonds between adjacent carbon atoms. Lighter alkynes, such as ethyne, are gases; heavier ones are liquids or solids.

allele one of two or more alternative forms of a ◊gene at a given position (locus) on a chromosome, caused by a difference in the ◊DNA. Blue and brown eyes in humans are determined by different alleles of the gene for eye colour.

Organisms with two sets of chromosomes (diploids) will have two copies of each gene. If the two alleles are identical the individual is said to be ◊homozygous at that locus; if different, the individual is ◊heterozygous at that locus. Some alleles show ◊dominance over others.

allergy special sensitivity of the body that makes it react with an exaggerated response of the natural immune defence mechanism to the introduction of an otherwise harmless foreign substance (*allergen*).

allometry in biology, a regular relationship between a given feature (for example, the size of an organ) and the size of the body as a whole, when this relationship is not a simple proportion of body size. Thus, an organ may increase in size proportionately faster, or slower, than body size does. For example, a human baby's head is much larger in relation to its body than is an adult's.

allopathy in ◊homoeopathy a term used for orthodox

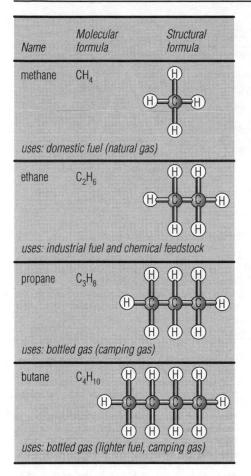

Name	Molecular formula	Structural formula
methane	CH_4	
		uses: domestic fuel (natural gas)
ethane	C_2H_6	
		uses: industrial fuel and chemical feedstock
propane	C_3H_8	
		uses: bottled gas (camping gas)
butane	C_4H_{10}	
		uses: bottled gas (lighter fuel, camping gas)

alkane

medicine, using therapies designed to counteract the manifestations of the disease. In strict usage, allopathy is the opposite of homoeopathy.

allotropy property whereby an element can exist in two or more forms (allotropes), each possessing different physical properties but the same state of matter (gas, liquid, or solid). The allotropes of carbon are diamond and graphite. Sulphur has several allotropes (flowers of sulphur, plastic, rhombic, and monoclinic). These solids have different crystal structures, as do the white and grey forms of tin and the black, red, and white forms of phosphorus.

Oxygen exists as two gaseous allotropes: one used by organisms for respiration (O_2), and the other a poisonous pollutant, ozone (O_3).

alloy metal blended with some other metallic or non-metallic substance to give it special qualities, such as resistance to corrosion, greater hardness, or tensile strength. Useful alloys include bronze, brass, cupronickel, duralumin, German silver, gunmetal, pewter, solder, steel, and stainless steel.

Among the oldest alloys is bronze, the widespread use of which ushered in the Bronze Age. Complex alloys are now common; for example, in dentistry, where a cheaper alternative to gold is made of chromium, cobalt, molybdenum, and titanium. Among the most recent alloys are superplastics: alloys that can stretch to double their length at specific temperatures, permitting, for example, their injection into moulds as easily as plastic.

alluvial deposit layer of broken rocky matter, or sediment, formed from material that has been carried in suspension by a river or stream and dropped as the velocity of the current decreases. River plains and deltas are made entirely of alluvial deposits, but smaller pockets can be found in the beds of upland torrents.

Alluvial deposits can consist of a whole range of particle sizes, from boulders down through cobbles, pebbles, gravel, sand, silt, and clay. The raw materials are the rocks and soils of upland areas that are loosened by erosion and washed away by mountain streams. Much of the world's richest farmland lies on alluvial deposits. These deposits can also provide an economic source of minerals. River currents produce a sorting action, with particles of heavy material deposited first while lighter materials are washed downstream.

Hence heavy minerals such as gold and tin, present in the original rocks in small amounts, can be concentrated and deposited on stream beds in commercial quantities. Such deposits are called 'placer ores'.

alluvial fan roughly triangular sedimentary formation found at the base of slopes. An alluvial fan results when a sediment-laden stream or river rapidly deposits its load of gravel and silt as its speed is reduced on entering a plain.

The surface of such a fan slopes outward in a wide arc from an apex at the mouth of the steep valley. A small stream carrying a load of coarse particles builds a shorter, steeper fan than a large stream carrying a load of fine particles. Over time, the fan tends to become destroyed piecemeal by the continuing headward and downward erosion levelling the slope.

Almagest book compiled by the Greek astronomer Ptolemy during the 2nd century AD, which included the idea of an Earth-centred universe; it was translated into Arabic in the 9th century. Some medieval books on astronomy, astrology, and alchemy were given the same title.

Each of the 13 sections of the book deals with a different branch of astronomy. The introduction describes the universe as spherical and contains arguments for the Earth being stationary at the centre. From this mistaken assumption, it goes on to describe the motions of the Sun, Moon, and planets; eclipses; and the positions, brightness, and precession of the 'fixed stars'. The book drew on the work of earlier astronomers such as Hipparchus.

Alpha Centauri or **Rigil Kent** brightest star in the constellation Centaurus and the third-brightest star in the sky. It is actually a triple star (see ◊binary star); the two brighter stars orbit each other every 80 years, and the third, Proxima Centauri, is the closest star to the Sun, 4.2 light years away, 0.1 light years closer than the other two.

alpha decay disintegration of the nucleus of an atom to produce an ◊alpha particle. See also ◊radioactivity.

alpha particle positively charged, high-energy particle emitted from the nucleus of a radioactive atom. It is one of the products of the spontaneous disintegration of

Alloy: Common Alloys

Name	Approximate composition	Uses
brass	35–10% zinc, 65–90% copper	decorative metalwork, plumbing fittings, industrial tubing
bronze – common	2% zinc, 6% tin, 92% copper	machinery, decorative work
bronze – aluminium	10% aluminium, 90% copper	machinery castings
bronze – coinage	1% zinc, 4% tin, 95% copper	coins
cast iron	2–4% carbon, 96–98% iron	decorative metalwork, engine blocks, industrial machinery
dentist's amalgam	30% copper, 70% mercury	dental fillings
duralumin	0.5 % magnesium, 0.5% manganese, 5% copper, 95% aluminium	framework of aircraft
gold – coinage	10% copper, 90% gold	coins
gold – dental	14–28% silver, 14–28% copper, 58% gold	dental fillings
lead battery plate	6% antimony, 94% lead	car batteries
manganin	1.5% nickel, 16% manganese, 82.5% copper	resistance wire
nichrome	20% chromium, 80% nickel	heating elements
pewter	20% lead, 80% tin	utensils
silver – coinage	10% copper, 90% silver	coins
solder	50% tin, 50% lead	joining iron surfaces
steel – stainless	8–20% nickel, 10–20% chromium, 60–80% iron	kitchen utensils
steel – armour	1–4% nickel, 0.5–2% chromium, 95–98% iron	armour plating
steel – tool	2–4% chromium, 6–7% molybdenum, 90–95% iron	tools

radioactive elements (see ◊radioactivity) such as radium and thorium, and is identical with the nucleus of a helium atom – that is, it consists of two protons and two neutrons. The process of emission, **alpha decay**, transforms one element into another, decreasing the atomic (or proton) number by two and the atomic mass (or nucleon number) by four.

Because of their large mass alpha particles have a short range of only a few centimetres in air, and can be stopped by a sheet of paper. They have a strongly ionizing effect (see ◊ionizing radiation) on the molecules that they strike, and are therefore capable of damaging living cells. Alpha particles travelling in a vacuum are deflected slightly by magnetic and electric fields.

Alps, Lunar conspicuous mountain range on the Moon, NE of the Sea of Showers (Mare Imbrium), cut by a valley 150 km/93 mi long. The highest peak is Mont Blanc, about 3,660 m/12,000 ft.

Altair or **Alpha Aquilae** brightest star in the constellation Aquila and the 12th brightest star in the sky. It is a white star 16 light years away and forms the so-called Summer Triangle with the stars Deneb (in the constellation Cygnus) and Vega (in Lyra).

alternate angles pair of angles that lie on opposite sides and at opposite ends of a transversal (a line that cuts two or more lines in the same plane). The alternate angles formed by a transversal of two parallel lines are equal. *See illustration opposite.*

alternating current (AC) electric current that flows for an interval of time in one direction and then in the opposite direction, that is, a current that flows in alternately reversed directions through or around a circuit. Electric energy is usually generated as alternating current in a power station, and alternating currents may be used for both power and lighting.

The advantage of alternating current over direct current (DC), as from a battery, is that its voltage can be raised or lowered economically by a transformer: high voltage for generation and transmission, and low voltage for safe utilization. Railways, factories, and domestic appliances, for example, use alternating current.

alternation of generations typical life cycle of terrestrial plants and some seaweeds, in which there are two distinct forms occurring alternately: **diploid** (having two sets of chromosomes) and **haploid** (one set of chromosomes). The diploid generation produces haploid spores by ◊meiosis, and is called the sporophyte, while the haploid generation produces gametes (sex cells), and is called the gametophyte. The gametes fuse to form a diploid ◊zygote which develops into a new sporophyte; thus the sporophyte and gametophyte alternate.

alternative medicine see ◊medicine, alternative.

alternator electricity ◊generator that produces an alternating current.

altimeter instrument used in aircraft that measures altitude, or height above sea level. The common type is a form of aneroid ◊barometer, which works by sensing the differences in air pressure at different altitudes. This must continually be recalibrated because of the change in air pressure with changing weather conditions. The ◊radar altimeter measures the height of the aircraft above the ground, measuring the time it takes for radio pulses emitted by the aircraft to be reflected. Radar altimeters are essential features of automatic and blind-landing systems.

altitude or **elevation** in astronomy, the angular distance of an object above the horizon, ranging from 0° on the horizon to 90° at the zenith. Together with ◊azimuth, it

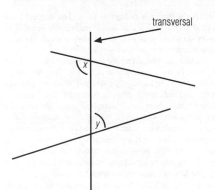

x and y are alternate angles

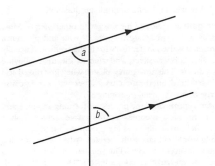

where a transversal cuts through a pair of parallel lines the alternate angles *a* and *b* are equal

alternate angles

forms the system of horizontal coordinates for specifying the positions of celestial bodies.

altitude in geometry, the perpendicular distance from a ◊vertex (corner) of a figure, such as a triangle, to the base (the side opposite the vertex).

altitude measurement of height, usually given in metres above sea level.

altruism in biology, helping another individual of the same species to reproduce more effectively, as a direct result of which the altruist may leave fewer offspring itself. Female honey bees (workers) behave altruistically by rearing sisters in order to help their mother, the queen bee, reproduce, and forgo any possibility of reproducing themselves.

ALU abbreviation for ◊arithmetic and logic unit.

alum any double sulphate of a monovalent metal or radical (such as sodium, potassium, or ammonium) and a trivalent metal (such as aluminium, chromium, or iron). The commonest alum is the double sulphate of potassium and aluminium, $K_2Al_2(SO_4)_4.24H_2O$, a white crystalline powder that is readily soluble in water. It is used in curing animal skins. Other alums are used in papermaking and to fix dye in the textile industry.

alumina or ***corundum*** Al_2O_3 oxide of aluminium, widely distributed in clays, slates, and shales. It is formed by the decomposition of the feldspars in granite and used as an abrasive. Typically it is a white powder, soluble in most strong acids or caustic alkalis but not in water. Impure alumina is called 'emery'. Rubies, sapphires, and topaz are corundum gemstones.

aluminium lightweight, silver-white, ductile and malleable, metallic element, symbol Al, atomic number 13, relative atomic mass 26.9815, melting point 658°C. It is the third most abundant element (and the most abundant metal) in the Earth's crust, of which it makes up about 8.1% by mass. It is non-magnetic, an excellent conductor of electricity, and oxidizes easily, the layer of oxide on its surface making it highly resistant to tarnish. In the USA the original name suggested by the scientist Humphry Davy, 'aluminum', is retained.

Aluminium is a reactive element with stable compounds, so a great deal of energy is needed in order to separate aluminium from its ores, and the pure metal was not readily obtainable until the middle of the 19th century. Commercially, it is prepared by the electrolysis of alumina (aluminium oxide), which is obtained from the ore ◊bauxite. In its pure state aluminium is a weak metal, but when combined with elements such as copper, silicon, or magnesium it forms alloys of great strength.

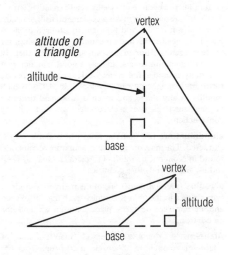

altitude of a triangle

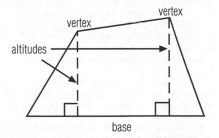

two altitudes of a quadrilateral

altitude The altitude of a figure is the perpendicular distance from a vertex (corner) to the base (the side opposite the vertex).

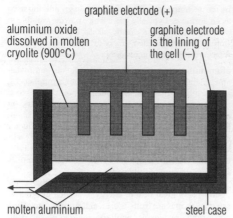

graphite electrode (+)

aluminium oxide
dissolved in molten
cryolite (900°C)

graphite electrode
is the lining of
the cell (–)

molten aluminium   steel case

aluminium Aluminium is the most common of all metals, but a great deal of energy is needed to separate it from its ores. Commercially, aluminium is extracted by electrolysis from purified bauxite (aluminium oxide, Al_2O_3) dissolved in molten cryolite. When an electric current is passed through the mixture, pure aluminium collects on the cathode.

Aluminium is widely used in the shipbuilding and air-craft industries because of its light weight (relative density 2.70). It is also used in making cooking utensils, cans for beer and soft drinks, and foil. It is much used in steel-cored overhead cables and for canning uranium slugs for nuclear reactors. Aluminium is an essential constituent in some magnetic materials; and, as a good conductor of electricity, is used as foil in electrical capacitors. A plastic form of aluminium, was developed in 1976, which moulds to any shape and extends to several times its original length, has uses in electronics, cars, and building construction.

aluminium ore raw material from which aluminium is extracted. The main ore is bauxite, a mixture of minerals, found in economic quantities in Australia, Guinea, West Indies, and several other countries.

alveolus (plural **alveoli**) one of the many thousands of tiny air sacs in the ◊lungs in which exchange of oxygen and carbon dioxide takes place between air and the bloodstream.

Alzheimer's disease common manifestation of ◊dementia, thought to afflict one in 20 people over 65. After heart disease, cancer, and strokes it is the most common cause of death in the Western world. Attacking the brain's 'grey matter', it is a disease of mental processes rather than physical function, characterized by memory loss and progressive intellectual impairment. It was first described by Alois Alzheimer in 1906. It affects up to 4 million people in the USA and around 600,000 in Britain.

Various factors have been implicated in causing Alzheimer's disease including high levels of aluminium in drinking water and the presence in the brain of an abnormal protein, known as beta-amyloid.

In 1993 the gene coding for apolipoprotein (APoE) was implicated. US researchers established that people who carry a particular version of this gene are at greatly increased risk of developing the disease. It is estimated that one person in thirty carries this protein mutation; in

the USA the figure is as high as 15%. The suspect gene can be detected with a test, so it is technically possible to identify those most at risk. As no cure is available such testing is unlikely to be widespread.

US researchers began trialling a simple eye test in 1994 that could be used to diagnose sufferers. The drug tropicamide causes marked pupil dilation in those with the disease, and only slight dilation in healthy individuals.

Some researchers are convinced that, whatever its cause, Alzheimer's disease is essentially an inflammatory condition, similar to rheumatoid arthritis. Although there is no cure, trials of anti-inflammatory drugs have shown promising results. Also under development are drugs which block the toxic effects of beta-amyloid. A 1996 study by US neuroscientists found that oestrogen skin patches were also beneficial in the treatment of female Alzheimer's patients, improving concentration and memory.

AM abbreviation for ◊amplitude modulation.

amalgam any alloy of mercury with other metals. Most metals will form amalgams, except iron and platinum. Amalgam is used in dentistry for filling teeth, and usually contains copper, silver, and zinc as the main alloying ingredients. This amalgam is pliable when first mixed and then sets hard, but the mercury leaches out and may cause a type of heavy-metal poisoning.

Amalgamation, the process of forming an amalgam, is a technique sometimes used to extract gold and silver from their ores. The ores are ground to a find sand and brought into contact with mercury, which dissolves the gold and silver particles. The amalgam is then heated to distil the mercury, leaving a residue of silver and gold. The mercury is recovered and reused.

amblyopia dimness of vision without apparent eye disorder.

americium radioactive metallic element of the ◊actinide series, symbol Am, atomic number 95, relative atomic mass 243.13; it was first synthesized in 1944. It occurs in nature in minute quantities in ◊pitchblende and other uranium ores, where it is produced from the decay of neutron-bombarded plutonium, and is the element with the highest atomic number that occurs in nature. It is synthesized in quantity only in nuclear reactors by the bombardment of plutonium with neutrons. Its longest-lived isotope is Am-243, with a half-life of 7,650 years.

The element was named by Glenn Seaborg, one of the team who first synthesized it in 1944, after the United States of America. Ten isotopes are known.

Ames Research Center US space-research (NASA) installation at Mountain View, California, USA, for the study of aeronautics and life sciences. It has managed the Pioneer series of planetary probes and is involved in the search for extraterrestrial life.

amethyst variety of ◊quartz, SiO_2, coloured violet by the presence of small quantities of impurities such as manganese or iron; used as a semiprecious stone. Amethysts are found chiefly in the Ural Mountains, India, the USA, Uruguay, and Brazil.

amide any organic chemical derived from a fatty acid by the replacement of the hydroxyl group (–OH) by an amino group ($-NH_2$).

One of the simplest amides is acetamide (CH_3CONH_2), which has a strong mousy odour.

amine any of a class of organic chemical compounds in which one or more of the hydrogen atoms of ammonia (NH_3) have been replaced by other groups of atoms.

Methyl amines have unpleasant ammonia odours and occur in decomposing fish. They are all gases at ordinary temperature.

Aromatic amine compounds include aniline, which is used in dyeing.

amino acid water-soluble organic ◊molecule, mainly composed of carbon, oxygen, hydrogen, and nitrogen, containing a basic amino group (NH_2) and an acidic carboxyl (COOH) group. They are small molecules able to pass through membranes. When two or more amino acids are joined together, they are known as ◊peptides; ◊proteins are made up of peptide chains folded or twisted in characteristic shapes.

Many different proteins are found in the cells of living organisms, but they are all made up of the same 20 amino acids, joined together in varying combinations (although other types of amino acid do occur infrequently in nature). Eight of these, the **essential amino acids**, cannot be synthesized by humans and must be obtained from the diet. Children need a further two amino acids that are not essential for adults. Other animals also need some preformed amino acids in their diet, but green plants can manufacture all the amino acids they need from simpler molecules, relying on energy from the Sun and minerals (including nitrates) from the soil.

ammeter instrument that measures electric current, usually in ◊amperes.

ammonia NH_3 colourless pungent-smelling gas, lighter than air and very soluble in water. It is made on an industrial scale by the ◊Haber (or Haber-Bosch) process, and used mainly to produce nitrogenous fertilizers, nitric acid, and some explosives.

In aquatic organisms and some insects, nitrogenous waste (from the breakdown of amino acids and so on) is excreted in the form of ammonia, rather than as urea in mammals.

amnesia loss or impairment of memory. As a clinical condition it may be caused by disease or injury to the brain, by some drugs, or by shock; in some cases it may be a symptom of an emotional disorder.

amniocentesis sampling the amniotic fluid surrounding a fetus in the womb for diagnostic purposes. It is used to detect Down's syndrome and other genetic abnormalities. The procedure carries a 1 in 200 risk of miscarriage.

amniotic fluid in mammals, fluid consisting mainly of water that is produced by the amnion, a fibrous membrane that lines the cavity of the uterus during pregnancy, in which the fetus floats and is protected from external pressure. It is swallowed by the fetus and excreted by the kidneys back into the amniotic sac. In humans, there is about 0.5–1l/0.8–1.75 pt of amniotic fluid. The amniotic sac normally ruptures in early labour to release the fluid ('waters').

amoeba (plural **amoebae**) one of the simplest living animals, consisting of a single cell and belonging to the ◊protozoa group. The body consists of colourless protoplasm. Its activities are controlled by the nucleus, and it feeds by flowing round and engulfing organic debris. It reproduces by ◊binary fission. Some species of amoeba are harmful parasites.

ampere SI unit (abbreviation amp, symbol A) of electrical current. Electrical current is measured in a similar way to water current, in terms of an amount per unit time; one

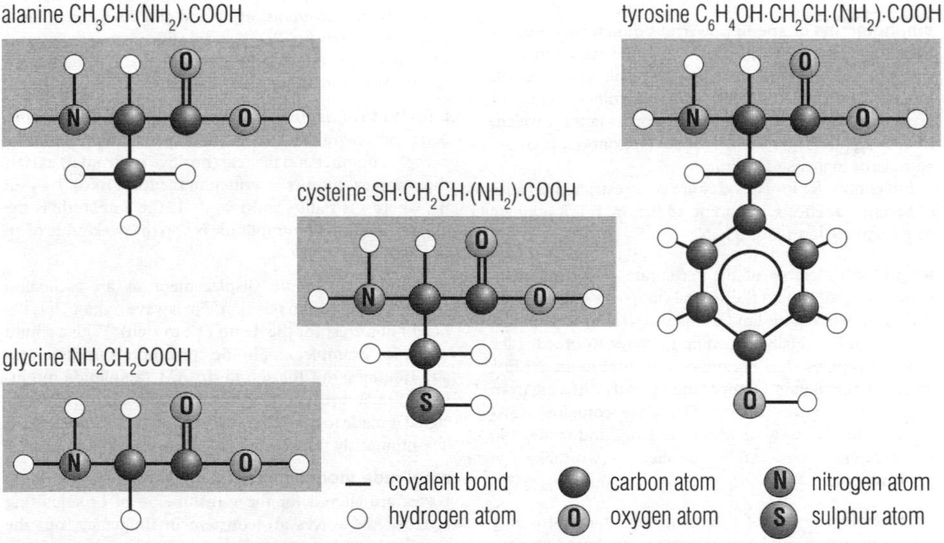

alanine $CH_3CH\cdot(NH_2)\cdot COOH$

tyrosine $C_6H_4OH\cdot CH_2CH\cdot(NH_2)\cdot COOH$

cysteine $SH\cdot CH_2CH\cdot(NH_2)\cdot COOH$

glycine NH_2CH_2COOH

— covalent bond
○ hydrogen atom
● carbon atom
Ⓞ oxygen atom
Ⓝ nitrogen atom
Ⓢ sulphur atom

amino acid Amino acids are natural organic compounds that make up proteins and can thus be considered the basic molecules of life. There are 20 different common amino acids. They consist mainly of carbon, oxygen, hydrogen, and nitrogen. Each amino acid has a common core structure (consisting of two carbon atoms, two oxygen atoms, a nitrogen atom, and four hydrogen atoms) to which is attached a variable group, known as the R group. In glycine, the R group is a single hydrogen atom; in alanine, the R group consists of a carbon and three hydrogen atoms.

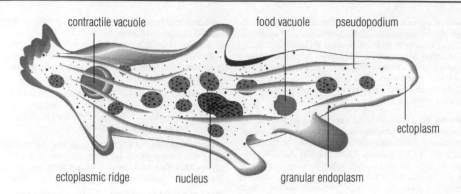

contractile vacuole   food vacuole   pseudopodium

ectoplasm

ectoplasmic ridge   nucleus   granular endoplasm

amoeba The amoebae are among the simplest living organisms, consisting of a single cell. Within the cell, there is a nucleus, which controls cell activity, and many other microscopic bodies and vacuoles (fluid-filled spaces surrounded by a membrane) with specialized functions. Amoebae eat by flowing around food particles, engulfing the particle, a process called phagocytosis.

ampere represents a flow of about 6.28×10^{18} ◊electrons per second, or a rate of flow of charge of one coulomb per second.

The ampere is defined as the current that produces a specific magnetic force between two long, straight, parallel conductors placed 1m/3.3 ft apart in a vacuum. It is named after the French scientist André Ampère.

Ampère's rule rule developed by French physicist André Ampère connecting the direction of an electric current and its associated magnetic currents. It states that if a person were travelling along a current-carrying wire in the direction of conventional current flow (from the positive to the negative terminal), and carrying a magnetic compass, then the north pole of the compass needle would be deflected to the left-hand side.

amphetamine or **speed** powerful synthetic ◊stimulant. Benzedrine was the earliest amphetamine marketed, used as a 'pep pill' in World War II to help soldiers overcome fatigue, and until the 1970s amphetamines were prescribed by doctors as an appetite suppressant for weight loss; as an antidepressant, to induce euphoria; and as a stimulant, to increase alertness.

Indications for its use today are very restricted because of severe side effects, including addiction. It is a sulphate or phosphate form of $C_9H_{13}N$.

amphibian member of the vertebrate class Amphibia, which generally spend their larval (tadpole) stage in fresh water, transferring to land at maturity (after ◊metamorphosis) and generally returning to water to breed. Like fish and reptiles, they continue to grow throughout life, and cannot maintain a temperature greatly differing from that of their environment. The class contains 4,553 known species, 4,000 of which are frogs and toads, 390 salamanders, and 163 caecilians (wormlike in appearance).

amphibole any one of a large group of rock-forming silicate minerals with an internal structure based on double chains of silicon and oxygen, and with a general formula $X_2Y_5Si_8O_{22}(OH)_2$; closely related to ◊pyroxene. Amphiboles form orthorhombic, monoclinic, and triclinic ◊crystals.

Amphiboles occur in a wide range of igneous and meta-morphic rocks. Common examples are ◊hornblende ($X =$ Ca, $Y =$ Mg, Fe, Al) and tremolite ($X =$ Ca, $Y =$ Mg).

amphoteric term used to describe the ability of some chemical compounds to behave either as an ◊acid or as a ◊base depending on their environment. For example, the metals aluminium and zinc, and their oxides and hydroxides, act as bases in acidic solutions and as acids in alkaline solutions.

Amino acids and proteins are also amphoteric, as they contain both a basic (amino, $-NH_2$) and an acidic (carboxyl, $-COOH$) group.

amplifier electronic device that magnifies the strength of a signal, such as a radio signal. The ratio of output signal strength to input signal strength is called the **gain** of the amplifier. As well as achieving high gain, an amplifier should be free from distortion and able to operate over a range of frequencies. Practical amplifiers are usually complex circuits, although simple amplifiers can be built from single transistors or valves.

amplitude or **argument** in mathematics, the angle in an ◊Argand diagram between the line that represents the complex number and the real (positive horizontal) axis. If the complex number is written in the form $r(\cos \theta + i \sin \theta)$, where r is radius and $i = \sqrt{-1}$, the amplitude is the angle θ (theta). The amplitude is also the peak value of an oscillation.

amplitude maximum displacement of an oscillation from the equilibrium position. For a wave motion, it is the height of a crest (or the depth of a trough). With a sound wave, for example, amplitude corresponds to the intensity (loudness) of the sound. In AM (amplitude modulation) radio broadcasting, the required audio-frequency signal is made to modulate (vary slightly) the amplitude of a continuously transmitted radio carrier wave.

amplitude modulation (AM) method by which radio waves are altered for the transmission of broadcasting signals. AM waves are constant in frequency, but the amplitude of the transmitting wave varies in accordance with the signal being broadcast.

amyl alcohol former name for ◊pentanol.

amylase one of a group of ◊enzymes that break down starches into their component molecules (sugars) for use in the body. It occurs widely in both plants and animals. In humans, it is found in saliva and in pancreatic juices.

Human amylase has an optimum pH of 7.2–7.4. Like most enzymes amylase is denatured by temperatures above 60°C.

anabolic steroid any ◊hormone of the ◊steroid group that stimulates tissue growth. Its use in medicine is limited to the treatment of some anaemias and breast cancers; it may help to break up blood clots. Side effects include aggressive behaviour, masculinization in women, and, in children, reduced height.

anabolism process of building up body tissue, promoted by the influence of certain hormones. It is the constructive side of ◊metabolism, as opposed to ◊catabolism.

anaemia condition caused by a shortage of haemoglobin, the oxygen-carrying component of red blood cells. The main symptoms are fatigue, pallor, breathlessness, palpitations, and poor resistance to infection. Treatment depends on the cause.

Anaemia arises either from abnormal loss or defective production of haemoglobin. Excessive loss occurs, for instance, with chronic slow bleeding or with accelerated destruction (◊haemolysis) of red blood cells. Defective production may be due to iron deficiency, vitamin B_{12} deficiency (pernicious anaemia), certain blood diseases (sickle-cell disease and thalassaemia), chronic infection, kidney disease, or certain kinds of poisoning. Untreated anaemia taxes the heart and may prove fatal.

anaerobic (of living organisms) not requiring oxygen for the release of energy from food molecules such as glucose. Anaerobic organisms include many bacteria, yeasts, and internal parasites.

Obligate anaerobes, such as certain primitive bacteria, cannot function in the presence of oxygen; but *facultative anaerobes*, like the fermenting yeasts and most bacteria, can function with or without oxygen. Anaerobic organisms release much less of the available energy from their food than do ◊aerobic organisms.

anaesthetic drug that produces loss of sensation or consciousness; the resulting state is **anaesthesia**, in which the patient is insensitive to stimuli. Anaesthesia may also happen as a result of nerve disorder.

Ever since the first successful operation in 1846 on a patient rendered unconscious by ether, advances have been aimed at increasing safety and control. Sedatives may be given before the anaesthetic to make the process easier. The level and duration of unconsciousness are managed precisely. Where general anaesthesia may be inappropriate (for example, in childbirth, for a small procedure, or in the elderly), many other techniques are available. A topical substance may be applied to the skin or tissue surface; a local agent may be injected into the tissues under the skin in the area to be treated; or a regional block of sensation may be achieved by injection into a nerve. Spinal anaesthetic, such as epidural, is injected into the tissues surrounding the spinal cord, producing loss of feeling in the lower part of the body.

analgesic agent for relieving ◊pain. ◊Opiates alter the perception or appreciation of pain and are effective in controlling 'deep' visceral (internal) pain. Non-opiates, such as ◊aspirin, ◊paracetamol, and ◊NSAIDs (nonsteroidal anti-inflammatory drugs), relieve musculoskeletal pain and reduce inflammation in soft tissues.

Pain is felt when electrical stimuli travel along a nerve pathway, from peripheral nerve fibres to the brain via the spinal cord.

An anaesthetic agent acts either by preventing stimuli from being sent (local), or by removing awareness of them (general). Analgesic drugs act on both.

Temporary or permanent analgesia may be achieved by injection of an anaesthetic agent into, or the severing of, a nerve. Implanted devices enable patients to deliver controlled electrical stimulation to block pain impulses. Production of the body's natural opiates, ◊endorphins, can be manipulated by techniques such as relaxation and biofeedback. However, for the severe pain of, for example, terminal cancer, opiate analgesics are required.

US researchers found in 1996 that some painkillers were more effective and provided longer-lasting relief for women than men.

analogous in biology, term describing a structure that has a similar function to a structure in another organism, but not a similar evolutionary path. For example, the wings of bees and of birds have the same purpose – to give powered flight – but have different origins.

Compare ◊homologous.

analogue (of a quantity or device) changing continuously; by contrast a ◊digital quantity or device varies in series of distinct steps. For example, an analogue clock measures time by means of a continuous movement of hands around a dial, whereas a digital clock measures time with a numerical display that changes in a series of discrete steps.

Most computers are digital devices. Therefore, any signals and data from an analogue device must be passed through a suitable ◊analogue-to-digital converter before they can be received and processed by computer. Similarly, output signals from digital computers must be passed through a digital-to-analogue converter before they can be received by an analogue device.

analogue computer computing device that performs calculations through the interaction of continuously varying physical quantities, such as voltages (as distinct from the more common ◊digital computer, which works with discrete quantities). An analogue computer is said to operate in real time (corresponding to time in the real world), and can therefore be used to monitor and control other events as they happen.

Although common in engineering since the 1920s, analogue computers are not general-purpose computers, but specialize in solving ◊differential calculus and similar mathematical problems. The earliest analogue computing device is thought to be the flat, or planispheric, astrolabe, which originated in about the 8th century.

analogue signal in electronics, current or voltage that conveys or stores information, and varies continuously in the same way as the information it represents (compare ◊digital signal). Analogue signals are prone to interference and distortion.

The bumps in the grooves of a vinyl record form a mechanical analogue of the sound information stored, which is then is converted into an electrical analogue signal by the record player's pick-up device.

analogue-to-digital converter (ADC) electronic circuit that converts an analogue signal into a digital one. Such a circuit is needed to convert the signal from an

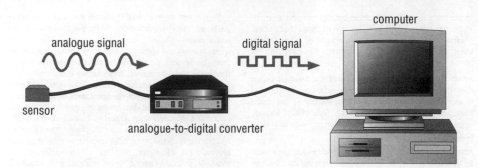

analogue-to-digital converter An ADC converts a continuous analogue signal produced by a sensor to a digital ('off and on') signal for computer processing.

analogue device into a digital signal for input into a computer. For example, many ◊sensors designed to measure physical quantities, such as temperature and pressure, produce an analogue signal in the form of voltage and this must be passed through an ADC before computer input and processing. A ◊digital-to-analogue converter performs the opposite process.

analysis branch of mathematics concerned with limiting processes on axiomatic number systems; ◊calculus of variations and infinitesimal calculus is now called analysis.

> *It takes a very unusual mind to*
> *undertake the analysis of the obvious.*
> On **analysis** Alfred North Whitehead
> *Science and the Modern World*

analytical chemistry branch of chemistry that deals with the determination of the chemical composition of substances. *Qualitative analysis* determines the identities of the substances in a given sample; *quantitative analysis* determines how much of a particular substance is present.

Simple qualitative techniques exploit the specific, easily observable properties of elements or compounds – for example, the flame test makes use of the different flame-colours produced by metal cations when their compounds are held in a hot flame. More sophisticated methods, such as those of ◊spectroscopy, are required where substances are present in very low concentrations or where several substances have similar properties.

Most quantitative analyses involve initial stages in which the substance to be measured is extracted from the test sample, and purified. The final analytical stages (or 'finishes') may involve measurement of the substance's mass (gravimetry) or volume (volumetry, titrimetry), or a number of techniques initially developed for qualitative analysis, such as fluorescence and absorption spectroscopy, chromatography, electrophoresis, and polarography. Many modern methods enable quantification by means of a detecting device that is integrated into the extraction procedure (as in gas–liquid chromatography).

analytical geometry another name for ◊coordinate geometry.

anaphylaxis in medicine, a severe allergic response. Typically, the air passages become constricted, the blood

pressure falls rapidly, and the victim collapses. A rare condition, anaphylaxis most often occurs following wasp or bee stings, or treatment with some drugs.

anatomy study of the structure of the body and its component parts, especially the ◊human body, as distinguished from physiology, which is the study of bodily functions. Herophilus of Chalcedon (c. 330–c. 260 BC) is regarded as the founder of anatomy. In the 2nd century AD, the Graeco-Roman physician Galen produced an account of anatomy that was the only source of anatomical knowledge until *On the Working of the Human Body* was published in 1543 by Belgian physician Andreas Vesalius. In 1628, English physician William Harvey published his demonstration of the circulation of the blood. With the invention of the microscope, Italian physiologist Marcello Malpighi and Dutch microscopist Anton van Leeuwenhoek were able to found the study of ◊histology. In 1747, Albinus (1697–1770), with the help of the artist Wandelaar (1691–1759), produced the most exact account of the bones and muscles, and between 1757–65 Swiss biologist Albrecht von Haller gave the most complete and exact description of the organs that had yet appeared. Among the anatomical writers of the early 19th century are the surgeon Charles Bell (1774–1842), Jonas Quain (1796–1865), and Henry Gray (1825–1861). Radiographic anatomy (using X-rays; see ◊radiography) has been one of the triumphs of the 20th century, which has also been marked by immense activity in embryological investigation.

andalusite aluminium silicate, Al_2SiO_5, a white to pinkish mineral crystallizing as square- or rhombus-based prisms. It is common in metamorphic rocks formed from clay sediments under low pressure conditions. Andalusite, kyanite, and sillimanite are all polymorphs of Al_2SiO_5.

andesite volcanic igneous rock, intermediate in silica content between rhyolite and basalt. It is characterized by a large quantity of feldspar ◊minerals, giving it a light colour. Andesite erupts from volcanoes at destructive plate margins (where one plate of the Earth's surface moves beneath another; see ◊plate tectonics), including the Andes, from which it gets its name.

AND gate in electronics, a type of ◊logic gate.

androecium male part of a flower, comprising a number of ◊stamens.

androgen general name for any male sex hormone, of which ◊testosterone is the most important.

They are all ◊steroids and are principally involved in the production of male ◊secondary sexual characteristics (such as beard growth).

Andromeda major constellation of the northern hemisphere, visible in autumn. Its main feature is the Andromeda galaxy. The star Alpha Andromedae forms one corner of the Square of Pegasus. It is named after the princess of Greek mythology.

Andromeda galaxy galaxy 2.2 million light years away from Earth in the constellation Andromeda, and the most distant object visible to the naked eye. It is the largest member of the ◊Local Group of galaxies.

Like the Milky Way, it is a spiral orbited by several companion galaxies but contains about twice as many stars. It is about 200,000 light years across.

Andromeda galaxy

If our Sun were 2.5 cm/1 in across, the star nearest to Earth would be 716 km/445 mi away. On this scale, the most distant star in our galaxy would be over 13 million km/8 million mi away. The nearest large neighbouring galaxy, the Andromeda galaxy, would be 360 million km/227 million mi away.

AND rule rule used for finding the combined probability of two or more independent events both occurring. If two events E_1 and E_2 are independent (have no effect on each other) and the probabilities of their taking place are p_1 and p_2, respectively, then the combined probability p that both E_1 and E_2 will happen is given by:

$$p = p_1 \times p_2$$

For example, if a blue die and a red die are thrown together, the probability of a blue six is 1/6, and the probability of a red six is 1/6. Therefore, the probability of both a red six and a blue six being thrown is $1/6 \times 1/6 = 1/36$. By contrast, the **OR rule** is used for finding the probability of either one event or another taking place.

anechoic chamber room designed to be of high sound absorbency. All surfaces inside the chamber are covered by sound-absorbent materials such as rubber. The walls are often covered with inward-facing pyramids of rubber, to minimize reflections. It is used for experiments in ◊acoustics and for testing audio equipment.

anemometer device for measuring wind speed and liquid flow. The most basic form, the **cup-type anemometer**, consists of cups at the ends of arms, which rotate when the wind blows. The speed of rotation indicates the wind speed. **Vane-type anemometers** have vanes, like a small windmill or propeller, that rotate when the wind blows. **Pressure-tube anemometers** use the pressure generated by the wind to indicate speed. The wind blowing into or across a tube develops a pressure, proportional to the wind speed, that is measured by a manometer or pressure gauge. **Hot-wire anemometers** work on the principle that the rate at which heat is transferred from a hot wire to the surrounding air is a measure of the air speed.

Wind speed is determined by measuring either the electric current required to maintain a hot wire at a constant temperature, or the variation of resistance while a constant current is maintained.

anemophily type of ◊pollination in which the pollen is carried on the wind. Anemophilous flowers are usually unscented, have either very reduced petals and sepals or lack them altogether, and do not produce nectar. In some species they are borne in ◊catkins. Male and female reproductive structures are commonly found in separate flowers. The male flowers have numerous exposed stamens, often on long filaments; the female flowers have long, often branched, feathery stigmas.

aneroid barometer kind of ◊barometer.

aneurysm weakening in the wall of an artery, causing it to balloon outwards with the risk of rupture and serious, often fatal, blood loss. If detected in time, some accessible aneurysms can be repaired by bypass surgery, but such major surgery carries a high risk for patients in poor health.

angel dust popular name for the anaesthetic *phencyclidine*, a depressant drug.

angina or *angina pectoris* severe pain in the chest due to impaired blood supply to the heart muscle because a coronary artery is narrowed. Faintness and difficulty in breathing accompany the pain. Treatment is by drugs or bypass surgery.

angiosperm flowering plant in which the seeds are enclosed within an ovary, which ripens into a fruit. Angiosperms are divided into ◊monocotyledons (single seed leaf in the embryo) and ◊dicotyledons (two seed leaves in the embryo). They include the majority of flowers, herbs, grasses, and trees except conifers.

There are over 250,000 different species of angiosperm, found in a wide range of habitats. Like ◊gymnosperms, they are seed plants, but differ in that ovules and seeds are protected within the carpel. Fertilization occurs by male gametes passing into the ovary from a pollen tube. After fertilization the ovule develops into the seed while the ovary wall develops into the fruit.

angle in mathematics, the amount of turn or rotation; it may be defined by a pair of rays (half-lines) that share a common endpoint but do not lie on the same line. Angles are measured in ◊degrees (°) or ◊radians (rads) – a complete turn or circle being 360° or 2π rads.

Angles are classified generally by their degree measures: *acute angles* are less than 90°; *right angles* are exactly 90° (a quarter turn); *obtuse angles* are greater than 90° but less than 180°; *reflex angles* are greater than 180° but less than 360°. *See illustration overleaf.*

angle of declination angle at a particular point on the Earth's surface between the direction of the true or geographic North Pole and the magnetic north pole. The angle of declination has varied over time because of the slow drift in the position of the magnetic north pole.

angle of dip or *angle of inclination* angle at a particular point on the Earth's surface between the direction of the Earth's magnetic field and the horizontal; see ◊magnetic dip.

angst emotional state of anxiety without a specific cause. In existentialism, the term refers to general human anxiety at having free will, that is, of being responsible for one's actions.

angstrom unit (symbol Å) of length equal to 10^{-10} metres or one-ten-millionth of a millimetre, used for atomic measurements and the wavelengths of electromagnetic radiation.

It is named after the Swedish scientist A J Ångström.

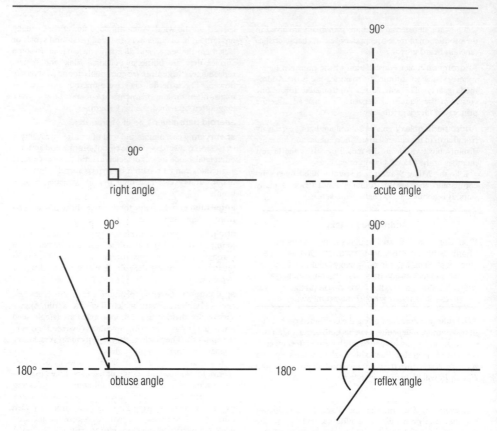

angle The four types of angle, as classified by their degree measures. No angle is classified as having a measure of 180°, as by definition such an 'angle' is actually a straight line.

anhydride chemical compound obtained by the removal of water from another compound; usually a dehydrated acid. For example, sulphur(VI) oxide (sulphur trioxide, SO_3) is the anhydride of sulphuric acid (H_2SO_4).

anhydrous of a chemical compound, containing no water. If the water of crystallization is removed from blue crystals of copper(II) sulphate, a white powder (anhydrous copper sulphate) results. Liquids from which all traces of water have been removed are also described as being anhydrous.

aniline $C_6H_5NH_2$ or **phenylamine** one of the simplest aromatic chemicals (a substance related to benzene, with its carbon atoms joined in a ring). When pure, it is a colourless oily liquid; it has a characteristic odour, and turns brown on contact with air. It occurs in coal tar, and is used in the rubber industry and to make drugs and dyes. It is highly poisonous.

Aniline was discovered in 1826, and was originally prepared by the dry distillation of indigo, hence its name from the Portuguese *anil* for 'indigo'.

animal or **metazoan** member of the ◊kingdom Animalia, one of the major categories of living things, the science of which is **zoology**. Animals are all ◊heterotrophs (they obtain their energy from organic substances produced by other organisms); they have eukaryotic cells (the genetic material is contained within a distinct nucleus) bounded by a thin cell membrane rather than the thick cell wall of plants. Most animals are capable of moving around for at least part of their life cycle.

In the past, it was common to include the single-celled ◊protozoa with the animals, but these are now classified as protists, together with single-celled plants. Thus all animals are multicellular. The oldest land animals known date back 440 million years. Their remains were found in 1990 in a sandstone deposit in Shropshire, UK, and included fragments of two centipedes a few centimetres long and a primitive spider measuring about 1 mm/0.04 in. *See illustration opposite.*

animism in anthropology, the belief that everything, whether animate or inanimate, possesses a soul or spirit. It is a fundamental system of belief in certain religions, particularly those of some pre-industrial societies. Linked with this is the worship of natural objects such as stones and trees, thought to harbour spirits (naturism); fetishism; and ancestor worship.

In psychology and physiology, animism is the view of human personality that attributes human life and behaviour to a force distinct from matter. In developmental psychology, an animistic stage in the early thought and

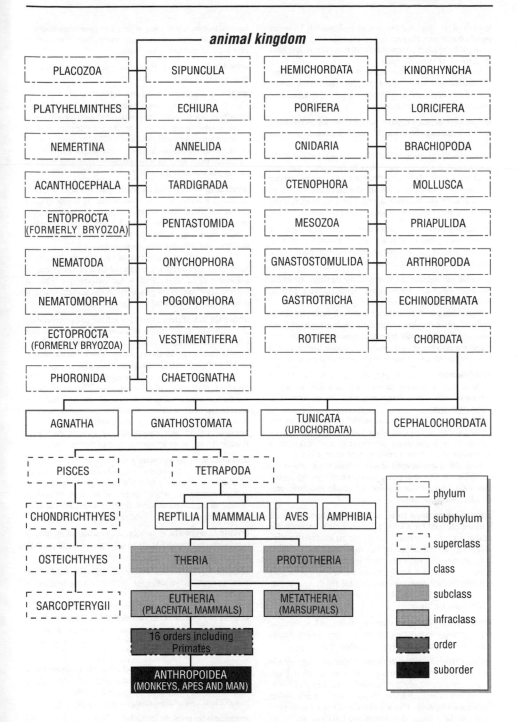

animal kingdom

PLACOZOA	SIPUNCULA	HEMICHORDATA	KINORHYNCHA
PLATYHELMINTHES	ECHIURA	PORIFERA	LORICIFERA
NEMERTINA	ANNELIDA	CNIDARIA	BRACHIOPODA
ACANTHOCEPHALA	TARDIGRADA	CTENOPHORA	MOLLUSCA
ENTOPROCTA (FORMERLY BRYOZOA)	PENTASTOMIDA	MESOZOA	PRIAPULIDA
NEMATODA	ONYCHOPHORA	GNASTOSTOMULIDA	ARTHROPODA
NEMATOMORPHA	POGONOPHORA	GASTROTRICHA	ECHINODERMATA
ECTOPROCTA (FORMERLY BRYOZOA)	VESTIMENTIFERA	ROTIFER	CHORDATA
PHORONIDA	CHAETOGNATHA		

AGNATHA — GNATHOSTOMATA — TUNICATA (UROCHORDATA) — CEPHALOCHORDATA

PISCES — TETRAPODA

CHONDRICHTHYES

REPTILIA | MAMMALIA | AVES | AMPHIBIA

OSTEICHTHYES

THERIA — PROTOTHERIA

SARCOPTERYGII

EUTHERIA (PLACENTAL MAMMALS) — METATHERIA (MARSUPIALS)

16 orders including Primates

ANTHROPOIDEA (MONKEYS, APES AND MAN)

Legend:
- phylum
- subphylum
- superclass
- class
- subclass
- infraclass
- order
- suborder

animal The animal kingdom is divided into 34 major groups or phyla. The large phylum Chordata (animals that have, at some time in their life, a notochord, or stiff rod of cells running along the length of their body) is subdivided into four subphyla, two of which are Vertebrates (animals with backbones), which is divided into superclasses, classes and subclasses, infraclasses and orders.

speech of the child has been described, notably by Swiss psychologist Jean Piaget.

In philosophy, the view that in all things consciousness or something mindlike exists.

In religious theory, the conception of a spiritual reality behind the material one: for example, beliefs in the soul as a shadowy duplicate of the body capable of independent activity, both in life and death.

anion ion carrying a negative charge. During electrolysis, anions in the electrolyte move towards the anode (positive electrode).

An electrolyte, such as the salt zinc chloride ($ZnCl_2$), is dissociated in aqueous solution or in the molten state into doubly charged Zn^{2+} zinc ◊cations and singly-charged Cl^- anions. During electrolysis, the zinc cations flow to the cathode (to become discharged and liberate zinc metal) and the chloride anions flow to the anode.

annealing process of heating a material (usually glass or metal) for a given time at a given temperature, followed by slow cooling, to increase ductility and strength. It is a common form of ◊heat treatment.

annelid any segmented worm of the phylum Annelida. Annelids include earthworms, leeches, and marine worms such as lugworms.

They have a distinct head and soft body, which is divided into a number of similar segments shut off from one another internally by membranous partitions, but there are no jointed appendages. Annelids are noted for their ability to regenerate missing parts of their bodies.

annihilation in nuclear physics, a process in which a particle and its 'mirror image' particle called an ***antiparticle*** collide and disappear, with the creation of a burst of energy.

The energy created is equivalent to the mass of the colliding particles in accordance with the ◊mass–energy equation. For example, an electron and a positron annihilate to produce a burst of high-energy X-rays.

Not all particle–antiparticle interactions result in annihilation; the exception concerns the group called ◊mesons, which are composed of ◊quarks and their antiquarks. See ◊antimatter.

annual percentage rate (APR) the true annual rate of interest charged for a loan. Lenders usually increase the return on their money by compounding the interest payable on a loan to that loan on a monthly or even daily basis. This means that each time that interest is payable on a loan it is charged not only on the initial sum (principal) but also on the interest previously added to that principal. As a result, APR is usually approximately double the flat rate of interest, or simple interest.

annual plant plant that completes its life cycle within one year, during which time it germinates, grows to maturity, bears flowers, produces seed, and then dies.

annual rings or ***growth rings*** concentric rings visible on the wood of a cut tree trunk or other woody stem. Each ring represents a period of growth when new ◊xylem is laid down to replace tissue being converted into wood (secondary xylem). The wood formed from xylem produced in the spring and early summer has larger and more numerous vessels than the wood formed from xylem produced in autumn when growth is slowing down. The result is a clear boundary between the pale spring wood and the denser, darker autumn wood. Annual rings may

be used to estimate the age of the plant (see ◊dendrochronology), although occasionally more than one growth ring is produced in a given year.

annulus in geometry, the plane area between two concentric circles, making a flat ring.

anode in chemistry, the positive electrode of an electrolytic ◊cell, towards which negative particles (anions), usually in solution, are attracted. See ◊electrolysis.

An anode is given its positive charge by the application of an external electrical potential, unlike the positive electrode of an electrical (battery) cell, which acquires its charge in the course of a spontaneous chemical reaction taking place within the cell.

anodizing process that increases the resistance to ◊corrosion of a metal, such as aluminium, by building up a protective oxide layer on the surface. The natural corrosion resistance of aluminium is provided by a thin film of aluminium oxide; anodizing increases the thickness of this film and thus the corrosion protection.

It is so called because the metal becomes the ◊anode in an electrolytic bath containing a solution of, for example, sulphuric or chromic acid as the ◊electrolyte. During ◊electrolysis oxygen is produced at the anode, where it combines with the metal to form an oxide film.

anorexia lack of desire to eat, or refusal to eat, especially the pathological condition of ***anorexia nervosa***, most often found in adolescent girls and young women. Compulsive eating, or ◊bulimia, distortions of body image, and depression often accompany anorexia.

Anorexia nervosa is characterized by severe self-imposed restriction of food intake. The consequent weight loss may lead, in women, to absence of menstruation. Anorexic patients sometimes commit suicide. Anorexia nervosa is often associated with increased physical activity and symptoms of mental disorders. Psychotherapy is an important part of the treatment.

anoxaemia or ***hypoxaemia*** shortage of oxygen in the blood; insufficient supply of oxygen to the tissues. It may be due to breathing air deficient in oxygen (for instance, at high altitude or where there are noxious fumes), a disease of the lungs, or some disorder where the oxygen-carrying capacity of the blood is impaired.

anoxia or ***hypoxia*** in biology, deprivation of oxygen, a condition that rapidly leads to collapse or death, unless immediately reversed.

antacid any substance that neutralizes stomach acid, such as sodium bicarbonate or magnesium hydroxide ('milk of magnesia').

Antacids are weak ◊bases, swallowed as solids or emulsions. They may be taken between meals to relieve symptoms of hyperacidity, such as pain, bloating, nausea, and 'heartburn'. Excessive or prolonged need for antacids should be investigated medically.

antagonist in biology, a ◊muscle that relaxes in response to the contraction of its agonist muscle. The biceps, in the front of the upper arm, bends the elbow whilst the triceps, lying behind the biceps, straightens the arm.

antagonistic muscle in the body, one of a pair of muscles allowing coordinated movement of the skeletal joints. The extension of the arm, for example, requires one set of muscles to relax, while another set contracts. The individual components of antagonistic pairs can be classified into

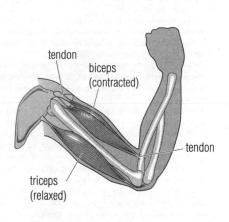

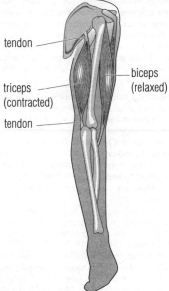

antagonistic muscle Even simple movements like bending and straightening your arm requires muscles to contract and relax simultaneously.

extensors (muscles that straighten a limb) and flexors (muscles that bend a limb).

Antarctic Circle imaginary line that encircles the South Pole at latitude 66° 32′ S. The line encompasses the continent of Antarctica and the Antarctic Ocean.

The region south of this line experiences at least one night in the southern summer during which the Sun never sets, and at least one day in the southern winter during which the Sun never rises.

Antares or *Alpha Scorpii* brightest star in the constellation Scorpius and the 15th brightest star in the sky. It is a red supergiant several hundred times larger than the Sun and perhaps 10,000 times as luminous, lies about 300 light years away, and fluctuates slightly in brightness.

antenatal in medicine, before birth. Antenatal care refers to health services provided to ensure the health of pregnant women and their unborn babies.

antenna in radio and television, another name for ◊aerial.

antenna in zoology, an appendage ('feeler') on the head. Insects, centipedes, and millipedes each have one pair of antennae but there are two pairs in crustaceans, such as shrimps. In insects, the antennae are involved with the senses of smell and touch; they are frequently complex structures with large surface areas that increase the ability to detect scents.

anther in a flower, the terminal part of a stamen in which the ◊pollen grains are produced. It is usually borne on a slender stalk or filament, and has two lobes, each containing two chambers, or pollen sacs, within which the pollen is formed.

antheridium organ producing the male gametes, ◊antherozoids, in algae, bryophytes (mosses and liverworts),

and pteridophytes (ferns, club mosses, and horsetails). It may be either single-celled, as in most algae, or multicellular, as in bryophytes and pteridophytes.

antherozoid motile (or independently moving) male gamete produced by algae, bryophytes (mosses and liverworts), pteridophytes (ferns, club mosses, and horsetails), and some gymnosperms (notably the cycads). Antherozoids are formed in an antheridium and, after being released, swim by means of one or more ◊flagella, to the female gametes. Higher plants have nonmotile male gametes contained within ◊pollen grains.

anthracene white, glistening, crystalline, tricyclic, aromatic hydrocarbon with a faint blue fluorescence when pure. Its melting point is about 216°C/421°F and its boiling point 351°C/664°F. It occurs in the high-boiling-point fractions of coal tar, where it was discovered in 1832 by the French chemists Auguste Laurent (1808–1853) and Jean Dumas (1800–1884).

anthracite hard, dense, shiny variety of ◊coal, containing over 90% carbon and a low percentage of ash and impurities, which causes it to burn without flame, smoke, or smell. Because of its purity, anthracite gives off relatively little sulphur dioxide when burnt.

Anthracite gives intense heat, but is slow-burning and slow to light; it is therefore unsuitable for use in open fires. Its characteristic composition is thought to be due to the action of bacteria in disintegrating the coal-forming material when it was laid down during the ◊Carboniferous period.

Among the chief sources of anthracite coal are Pennsylvania in the USA; S Wales, UK; the Donbas, Ukraine and Russia; and Shanxi province, China.

anthrax disease of livestock, occasionally transmitted to humans, usually via infected hides and fleeces. It may develop as black skin pustules or severe pneumonia. Treatment is with antibiotics. Vaccination is effective.

Anthrax is caused by a bacillus (*Bacillus anthracis*). In the 17th century, some 60,000 cattle died in a European pandemic known as the Black Bane, thought to have been anthrax. The disease is described by the Roman poet Virgil and may have been the cause of the biblical fifth plague of Egypt.

anthropology study of humankind. It investigates the cultural, social, and physical diversity of the human species, both past and present. It is divided into two broad categories: biological or physical anthropology, which attempts to explain human biological variation from an evolutionary perspective; and the larger field of social or cultural anthropology, which attempts to explain the variety of human cultures. This differs from sociology in that anthropologists are concerned with cultures and societies other than their own.

Biological anthropology is concerned with human ◊palaeontology, primatology, human adaptation, demography, population genetics, and human growth and development.

Social or cultural anthropology is divided into three subfields: social or cultural anthropology proper, ◊prehistory or prehistoric archaelogy, and anthropological linguistics. The term 'anthropology' is frequently used to refer solely to social anthropology. With a wide range of theoretical perspectives and topical interests, it overlaps with many other disciplines. It is a uniquely Western social science.

Anthropology's primary method involves the researcher living for a year or more in another culture, speaking the local language and participating in all aspects of everyday life; and writing about it afterwards. By comparing these accounts, anthropologists hope to understand who we are.

The awe and dread with which the untutored savage contemplates his mother-in-law are amongst the most familiar facts of anthropology.

On **anthropology** James Frazer *Golden Bough*

---

antibiotic drug that kills or inhibits the growth of bacteria and fungi. It is derived from living organisms such as fungi or bacteria, which distinguishes it from synthetic antimicrobials.

The earliest antibiotics, the ◊penicillins, came into use from 1941 and were quickly joined by ◊chloramphenicol, the ◊cephalosporins, erythromycins, tetracyclines, and aminoglycosides. A range of broad-spectrum antibiotics, the 4-quinolones, was developed in 1989, of which ciprofloxacin was the first. Each class and individual antibiotic acts in a different way and may be effective against either a broad spectrum or a specific type of disease-causing agent. Use of antibiotics has become more selective as side effects, such as toxicity, allergy, and resistance, have become better understood. Bacteria have the ability to develop resistance following repeated or subclinical (insufficient) doses, so more advanced and synthetic antibiotics are continually required to overcome them.

antibody protein molecule produced in the blood by ◊lymphocytes in response to the presence of foreign or invading substances (◊antigens); such substances include the proteins carried on the surface of infecting micro-organisms. Antibody production is only one aspect of ◊immunity in vertebrates. Each antibody acts against only one kind of antigen, and combines with it to form a 'complex'. This action may render antigens harmless, or it may destroy micro-organisms by setting off chemical changes that cause them to self-destruct.

In other cases, the formation of a complex will cause antigens to form clumps that can then be detected and engulfed by white blood cells, such as ◊macrophages and ◊phagocytes.

Each bacterial or viral infection will bring about the manufacture of a specific antibody, which will then fight the disease. Many diseases can only be contracted once because antibodies remain in the blood after the infection has passed, preventing any further invasion. Vaccination boosts a person's resistance by causing the production of antibodies specific to particular infections.

Antibodies were discovered in 1890 by German physician Emil von Behring and Japanese bacteriologist Shibasaburo Kitasato. Large quantities of specific antibodies can now be obtained by the monoclonal technique (see ◊monoclonal antibody).

anticline in geology, a fold in the rocks of the Earth's crust in which the layers or beds bulge upwards to form an arch (seldom preserved intact).

The fold of an anticline may be undulating or steeply curved. A steplike bend in otherwise gently dipping or horizontal beds is a **monocline**. The opposite of an anticline is a **syncline.**

anticoagulant substance that inhibits the formation of blood clots. Common anticoagulants are heparin, produced by the liver and some white blood cells, and derivatives of coumarin. Anticoagulants are used medically in the prevention and treatment of thrombosis and heart attacks. Anticoagulant substances are also produced by blood-feeding animals, such as mosquitoes, leeches, and vampire bats, to keep the victim's blood flowing.

Most anticoagulants prevent the production of thrombin, an enzyme that induces the formation from blood plasma of fibrinogen, to which blood platelets adhere and form clots.

anticyclone area of high atmospheric pressure caused by descending air, which becomes warm and dry. Winds radiate from a calm centre, taking a clockwise direction in the northern hemisphere and an anticlockwise direction in the southern hemisphere. Anticyclones are characterized by clear weather and the absence of rain and violent winds. In summer they bring hot, sunny days and in winter they bring fine, frosty spells, although fog and low cloud are not uncommon in the UK. *Blocking anticyclones*, which prevent the normal air circulation of an area, can cause summer droughts and severe winters.

antidepressant any drug used to relieve symptoms in depressive illness. The two main groups are the tricyclic antidepressants (TCADs) and the monoamine oxidase inhibitors (MAOIs), which act by altering chemicals available to the central nervous system. Both may produce serious side effects and are restricted.

antidiarrhoeal any substance that controls diarrhoea.

Choice of treatment depends on the underlying cause. One group, including opiates, codeine, and atropine, produces constipation by slowing down motility (muscle activity of the intestine wall).

Bulking agents, such as vegetable fibres (for example, methylcellulose), absorb fluid. Antibiotics may be appropriate for certain systemic bacterial infections, such as ◊typhoid fever, salmonella, and infective enteritis. Current therapy of acute diarrhoea is based on fluid and ◊electrolyte replacement. Chronic diarrhoea, a feature of some bowel disorders (for example, Crohn's disease, colitis, coeliac disease) responds to drugs such as ◊antispasmodics and corticosteroids, and special diet.

antiemetic any substance that counteracts nausea or vomiting.

antifreeze substance added to a water-cooling system (for example, that of a car) to prevent it freezing in cold weather.

antigen any substance that causes the production of ◊antibodies by the body's immune system. Common antigens include the proteins carried on the surface of bacteria, viruses, and pollen grains. The proteins of incompatible blood groups or tissues also act as antigens, which has to be taken into account in medical procedures such as blood transfusions and organ transplants.

antihistamine any substance that counteracts the effects of ◊histamine. Antihistamines may occur naturally or they may be synthesized.

H$_1$ antihistamines are used to relieve allergies, alleviating symptoms such as runny nose, itching, swelling, or asthma. H$_2$ antihistamines suppress acid production by the stomach, and are used in the treatment of peptic ulcers, often making surgery unnecessary.

anti-inflammatory any substance that reduces swelling in soft tissues. Antihistamines relieve allergic reactions; aspirin and ◊NSAIDs are effective in joint and musculoskeletal conditions; and rubefacients (counterirritant liniments) ease painful joints, tendons, and muscles.

◊Steroids, because of their severe side effects, are only prescribed if other therapy is ineffective, or if a condition is life-threatening. A ◊corticosteroid injection into the affected joint usually gives prolonged relief from inflammation.

antiknock substance added to ◊petrol to reduce knocking in car engines. It is a mixture of dibromoethane and tetraethyl lead. Its use in leaded petrol has resulted in atmospheric pollution by lead compounds.

antilogarithm or *antilog* the inverse of ◊logarithm, or the number whose logarithm to a given base is a given number. If $y = \log_a x$, then $x = \text{antilog}_a y$.

antimatter in physics, a form of matter in which most of the attributes (such as electrical charge, magnetic moment, and spin) of ◊elementary particles are reversed. Such particles (◊antiparticles) can be created in particle accelerators, such as those at ◊CERN in Geneva, Switzerland, and at ◊Fermilab in the USA. In 1996 physicists at CERN created the first atoms of antimatter: nine atoms of antihydrogen survived for 40 nanoseconds.

antimony silver-white, brittle, semimetallic element (a metalloid), symbol Sb (from Latin *stibium*), atomic number 51, relative atomic mass 121.75. It occurs chiefly as the ore stibnite, and is used to make alloys harder; it is also used in photosensitive substances in colour photography, optical electronics, fireproofing, pigment, and medicine. It was employed by the ancient Egyptians in a mixture to protect the eyes from flies.

antinode in physics, the position in a ◊standing wave pattern at which the amplitude of vibration is greatest (compare ◊node). The standing wave of a stretched string vibrating in the fundamental mode has one antinode at its midpoint. A vibrating air column in a pipe has an antinode at the pipe's open end and at the place where the vibration is produced.

antioxidant any substance that prevents deterioration of fats, oils, paints, plastics, and rubbers by oxidation. When used as food additives, antioxidants prevent fats and oils from becoming rancid when exposed to air, and thus extend their shelf life.

Vegetable oils contain natural antioxidants, such as vitamin E, which prevent spoilage, but antioxidants are nevertheless added to most oils. They are not always listed on food labels because if a food manufacturer buys an oil to make a food product, and the oil has antioxidant already added, it does not have to be listed on the label of the product.

antiparticle in nuclear physics, a particle corresponding in mass and properties to a given ◊elementary particle but with the opposite electrical charge, magnetic properties, or coupling to other fundamental forces. For example, an electron carries a negative charge whereas its antiparticle, the positron, carries a positive one.

When a particle and its antiparticle collide, they destroy each other, in the process called 'annihilation', their total energy being converted to lighter particles and/or photons. A substance consisting entirely of antiparticles is known as ◊antimatter.

antipodes places at opposite points on the globe.

antipruritic any skin preparation or drug administered to relieve itching.

antipyretic any drug, such as aspirin, used to reduce fever.

antiseptic any substance that kills or inhibits the growth of microorganisms. The use of antiseptics was pioneered by Joseph Lister. He used carbolic acid (◊phenol), which is a weak antiseptic; antiseptics such as TCP are derived from this.

antispasmodic any drug that reduces motility, the spontaneous action of the intestine wall. Anticholinergics are a type of antispasmodic that act indirectly by way of the autonomic nervous system, which controls involuntary movement.

Other drugs act directly on the smooth muscle to relieve spasm (contraction).

antitussive any substance administered to suppress a cough. Coughing, however, is an important reflex in clearing secretions from the airways; its suppression is usually unnecessary and possibly harmful, unless damage is being done to tissue during excessive coughing spasms.

antiviral any drug that acts against viruses, usually preventing them from multiplying. Most viral infections are not susceptible to antibiotics. Antivirals have been difficult drugs to develop, and do not necessarily cure viral diseases.

anus or **anal canal** the opening at the end of the alimentary canal that allows undigested food and other waste materials to pass out of the body, in the form of faeces. In humans, the term is also used to describe the last 4 cm/1.5 in of the alimentary canal. The anus is found in all types of multicellular animal except the coelenterates (sponges) and the platyhelminths (flatworms), which have a mouth only.

It is normally kept closed by rings of muscle called sphincters. The commonest medical condition associated with the anus is haemorrhoids (piles).

anxiety unpleasant, distressing emotion usually to be distinguished from fear. Fear is aroused by the perception of actual or threatened danger; anxiety arises when the danger is imagined or cannot be identified or clearly perceived. It is a normal response in stressful situations, but is frequently experienced in many mental disorders.

Anxiety is experienced as a feeling of suspense, helplessness, or alternating hope and despair together with excessive alertness and characteristic bodily changes such as tightness in the throat, disturbances in breathing and heartbeat, sweating, and diarrhoea.

In psychiatry, an anxiety state is a type of neurosis in which the anxiety either seems to arise for no reason or else is out of proportion to what may have caused it. 'Phobic anxiety' refers to the irrational fear that characterizes ◊phobia.

anxiolytic any drug that relieves an anxiety state.

aorta the body's main ◊artery, arising from the left ventricle of the heart in birds and mammals. Carrying freshly oxygenated blood, it arches over the top of the heart and descends through the trunk, finally splitting in the lower abdomen to form the two iliac arteries. Loss of elasticity in the aorta provides evidence of ◊atherosclerosis, which may lead to heart disease.

In fish a ventral aorta carries deoxygenated blood from the heart to the ◊gills, and the dorsal aorta carries oxygenated blood from the gills to other parts of the body.

a.p. in physics, abbreviation for **atmospheric pressure**.

Apache Point Observatory US observatory in the Sacramento Mountains of New Mexico containing a 3.5m/138 in reflector, opened in 1994, and operated by the Astrophysical Research Consortium (the universities of Washington, Chicago, Princeton, New Mexico, and Washington State).

apastron the point at which an object travelling in an elliptical orbit around a star is at its furthest from the star. The term is usually applied to the position of the minor component of a ◊binary star in relation to the primary. Its opposite is periastron.

apatite common calcium phosphate mineral, $Ca_5(PO_4)_3$ (F,OH,Cl). Apatite has a hexagonal structure and occurs widely in igneous rocks, such as pegmatite, and in contact metamorphic rocks, such as marbles. It is used in the manufacture of fertilizer and as a source of phosphorus. Carbonate hydroxyapatite, $Ca_5(PO_4CO_3)_3(OH)_2$, is the chief constituent of tooth enamel and the chief inorganic constituent of bone marrow. Apatite ranks 5 on the ◊Mohs' scale of hardness.

aperture in photography, an opening in the camera that allows light to pass through the lens to strike the film. Controlled by the iris diaphragm, it can be set mechanically or electronically at various diameters.

The **aperture ratio** or **relative aperture**, more commonly known as the ◊f-number, is a number defined as the focal length of the lens divided by the effective diameter of the aperture. A smaller f-number implies a larger diameter lens and therefore more light available for high-speed photography, or for work in poorly illuminated areas. However, small f-numbers involve small depths of focus.

aperture synthesis in astronomy, a technique used in ◊radio astronomy in which several small radio dishes are linked together to simulate the performance of one very large radio telescope, which can be many kilometres in diameter. See ◊radio telescope.

apex the highest point of a triangle, cone, or pyramid – that is, the vertex (corner) opposite a given base.

aphasia general term for the many types of disturbance in language that are due to brain damage, especially in the speech areas of the dominant hemisphere.

aphelion the point at which an object, travelling in an elliptical orbit around the Sun, is at its furthest from the Sun. The Earth is at its aphelion on 5 July.

aphrodisiac from Aphrodite, the Greek goddess of love Any substance that arouses or increases sexual desire.

apogee the point at which an object, travelling in an elliptical orbit around the Earth, is at its furthest from the Earth.

Apollo asteroid member of a group of ◊asteroids whose orbits cross that of the Earth. They are named after the first of their kind, Apollo, discovered in 1932 and then lost until 1973. Apollo asteroids are so small and faint that they are difficult to see except when close to Earth (Apollo is about 2 km/1.2 mi across).

Apollo project US space project to land a person on the Moon, was achieved on 20 July 1969, when Neil Armstrong was the first to set foot there. He was accompanied on the Moon's surface by 'Buzz' Aldrin; Michael Collins remained in the orbiting command module.

The programme was announced in 1961 by President Kennedy. The world's most powerful rocket, *Saturn V*, was built to launch the Apollo spacecraft, which carried three astronauts. When the spacecraft was in orbit around the Moon, two astronauts would descend to the surface in a lunar module to take samples of rock and set up experiments that would send data back to Earth. After three other preparatory flights, *Apollo 11* made the first lunar landing. Five more crewed landings followed, the last was in 1972. The total cost of the programme was over $24 billion.

aposematic coloration in biology, the technical name for ◊warning coloration markings that make a dangerous, poisonous, or foul-tasting animal particularly conspicuous and recognizable to a predator. Examples include the yellow and black stripes of bees and wasps, and the bright red or yellow colours of many poisonous frogs. See also ◊mimicry.

apothecaries' weights obsolete units of mass, formerly used in pharmacy: 20 grains equal one scruple; three scruples equal one dram; eight drams equal an apothecary's ounce (oz apoth.), and 12 such ounces equal an apothecary's pound (lb apoth.). There are 7,000 grains in one pound avoirdupois (0.454 kg).

apothecary person who prepares and dispenses medicines; a pharmacist.

The word 'apothecary' retains its original meaning in the USA and other countries, but in England it came to mean a licensed medical practitioner.

apparent depth depth that a transparent material such as water or glass appears to have when viewed from above. This is less than its real depth because of the ◊refraction that takes place when light passes into a less dense medium. The ratio of the real depth to the apparent depth of a transparent material is equal to its ◊refractive index.

appendicitis inflammation of the appendix, a small, blind extension of the bowel in the lower right abdomen. In an acute attack, the pus-filled appendix may burst, causing a potentially lethal spread of infection. Treatment is by removal (appendicectomy).

appendix a short, blind-ended tube attached to the ◊caecum. It has no known function in humans, but in herbivores it may be large, containing millions of bacteria that secrete enzymes to digest grass (as no vertebrate can secrete enzymes that will digest cellulose, the main constituent of plant cell walls).

Appleton layer band containing ionized gases in the Earth's upper atmosphere, above the ◊E layer (formerly the Kennelly–Heaviside layer). It can act as a reflector of radio signals, although its ionic composition varies with the sunspot cycle. It is named after the English physicist Edward Appleton.

application in computing, a program or job designed for the benefit of the end user, such as a payroll system or a ◊word processor. The term is used to distinguish such programs from those that control the computer (◊systems programs) or assist the programmer, such as a ◊compiler.

APR abbreviation for ◊annual percentage rate.

aquaculture the cultivation of fish and shellfish for human consumption; see ◊fish farming.

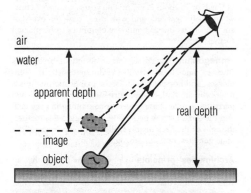

apparent depth When viewed from above, a transparent material such as water or glass appears to be less than its real depth because of refraction – the bending of light when it passes into a less dense medium. Refraction occurs because the light waves travel at different speeds in different media.

aqualung or **scuba** underwater breathing apparatus worn by divers, developed in the early 1940s by French diver Jacques Cousteau. Compressed-air cylinders strapped to the diver's back are regulated by a valve system and by a mouth tube to provide air to the diver at the same pressure as that of the surrounding water (which increases with the depth).

aquamarine blue variety of the mineral ◊beryl. A semi-precious gemstone, it is used in jewellery.

aquaplaning phenomenon in which the tyres of a road vehicle cease to make direct contact with the road surface, owing to the presence of a thin film of water. As a result, the vehicle can go out of control (particularly if the steered wheels are involved).

aqua regia mixture of three parts concentrated hydrochloric acid and one part concentrated nitric acid, which dissolves all metals except silver.

aquarium tank or similar container used for the study and display of living aquatic plants and animals. The same name is used for institutions that exhibit aquatic life. These have been common since Roman times. The first modern public aquarium was opened in Regent's Park, London, in 1853. A recent development is the oceanarium or seaquarium, a large display of marine life forms.

Aquarius zodiacal constellation a little south of the celestial equator near Pegasus. Aquarius is represented as a person pouring water from a jar. The Sun passes through Aquarius from late Feb to early March. In astrology, the dates for Aquarius, the 11th sign of the zodiac, are between about 20 Jan and 18 Feb (see ◊precession).

aquatic living in water. All life on Earth originated in the early oceans, because the aquatic environment has several advantages for organisms. Dehydration is almost impossible, temperatures usually remain stable, and the density of water provides physical support.

Life forms that cannot exist out of water, amphibians that take to the water on occasions, animals that are also perfectly at home on land, and insects that spend a stage of their life cycle in water can all be described as aquatic. Aquatic plants are known as ◊hydrophytes.

aqueduct any artificial channel or conduit for water, often an elevated structure of stone, wood, or iron built for conducting water across a valley. The Greeks built a tunnel 1,280 m/4,200 ft long near Athens some 2,500 years ago. Many Roman aqueducts are still standing, for example the one carried by the Pont du Gard at Nîmes in S France, built about 8 BC (48 m/160 ft high).

The largest Roman aqueduct, at Carthage in Tunisia, is 141 km/87 mi long and was built during the reign of Publius Aelius Hadrianus between AD 117 and 138. A recent aqueduct is the California State Water Project taking water from Lake Oroville in the north, through two power plants and across the Tehachapi Mountains, more than 177 km/110 mi to S California.

aqueous humour watery fluid found in the chamber between the cornea and lens of the vertebrate eye. Similar to blood serum in composition, it is constantly renewed.

aqueous solution solution in which the solvent is water.

aquifer any rock formation containing water. The rock of an aquifer must be porous and permeable (full of interconnected holes) so that it can absorb water. Aquifers are an important source of fresh water, for example, for drinking and irrigation, in many arid areas of the world and are exploited by the use of ◊artesian wells.

An aquifer may be underlain, overlain, or sandwiched between impermeable layers, called *aquicludes*, which impede water movement. Sandstones and porous limestones make the best aquifers.

Aquila constellation on the celestial equator (see ◊celestial sphere). Its brightest star is first-magnitude ◊Altair, flanked by the stars Beta and Gamma Aquilae. It is represented by an eagle.

Nova Aquilae, which appeared in June 1918, shone for a few days nearly as brightly as Sirius.

arable farming cultivation of crops, as opposed to the keeping of animals. Crops may be ◊cereals, vegetables, or plants for producing oils or cloth. Arable farming generally requires less attention than livestock farming. In a ◊mixed farming system, crops may therefore be found farther from the farm centre than animals.

arboretum collection of trees. An arboretum may contain a wide variety of species or just closely related species or varieties – for example, different types of pine tree.

arc in geometry, a section of a curved line or circle. A circle has three types of arc: a *semicircle*, which is exactly half of the circle; *minor arcs*, which are less than the semicircle; and *major arcs*, which are greater than the semicircle.

An arc of a circle is measured in degrees, according to the angle formed by joining its two ends to the centre of that circle. A semicircle is therefore 180°, whereas a minor arc will always be less than 180° (acute or obtuse) and a major arc will always be greater than 180° but less than 360° (reflex).

arch in geomorphology, any natural bridgelike land feature formed by erosion. A sea arch is formed from the wave erosion of a headland where the backs of two caves have met and broken through. The roof of the arch eventually collapses to leave part of the headland isolated in the sea as a ◊stack. A natural bridge is formed by wind or water erosion and spans a valley or ravine.

Archaea group of microorganisms that are without a nucleus and have a single chromosome. All are strict anaerobes, that is, they are killed by oxygen. This is thought to be a primitive condition and to indicate that Archaea are related to the earliest life forms, which appeared about 4 billion years ago, when there was little oxygen in the Earth's atmosphere. They are found in undersea vents, hot springs, the Dead Sea, and salt pans, and have even adapted to refuse tips.

Archaea was originally classified as bacteria, but in 1996 when the genome of *Methanococcus jannaschii*, an archaeaon that lives in undersea vents at temperatures around 100°C/212°F, was sequenced, US geneticists found that 56% of its genes were unlike those of any other organism, making Archaea unique.

Archaeans are found in the Antarctic (where they make up 30% of the single-celled marine biomass), Arctic, Mediterranean, and Baltic Sea.

Archaean or *Archaeozoic* the earliest eon of geological time; the first part of the Precambrian, from the formation of Earth up to about 2,500 million years ago. It was a time when no life existed, and with every new discovery of ancient life its upper boundary continues to be pushed further back.

archaeology study of prehistory and history, based on the examination of physical remains. Principal activities include preliminary field (or site) surveys, ◊excavation (where necessary), and the classification, ◊dating, and interpretation of finds.

A museum found at the ancient Sumerian city of Ur indicates that interest in the physical remains of the past stretches back into prehistory. In the Renaissance this interest gained momentum among dealers in and collectors of ancient art and was further stimulated by discoveries made in Africa, the Americas, and Asia by Europeans during the period of imperialist colonization in the 16th–19th centuries, such as the antiquities discovered during Napoleon's Egyptian campaign in the 1790s. Romanticism in Europe stimulated an enthusiasm for the mouldering skull, the ancient potsherds, ruins, and dolmens; relating archaeology to a wider context of art and literature.

Towards the end of the 19th century archaeology became an academic study, making increasing use of scientific techniques and systematic methodologies such as aerial photography. Since World War II new developments within the discipline include medieval, post-medieval, landscape, and industrial archaeology; underwater reconnaissance enabling the excavation of underwater sites; and rescue archaeology (excavation of sites risking destruction).

Useful in archaeological studies are ◊dendrochronology (tree-ring dating), ◊geochronology (science of m easuring geological time), ◊stratigraphy (study of geological strata), palaeobotany (study of ancient pollens, seeds, and grains), archaeozoology (analysis of animal remains), epigraphy (study of inscriptions), and numismatics (study of coins).

An archaeologist is the best husband any woman can have: the older she gets, the more interested he is in her.

On **archaeology** Agatha Christie

Archean or *Archeozoic* the earliest eon of geological time; the first part of the Precambrian, from the formation of Earth up to about 2,500 million years ago. It was a time when no life existed, and with every new discovery of ancient life its upper boundary continues to be pushed further back.

archegonium female sex organ found in bryophytes (mosses and liverworts), pteridophytes (ferns, club mosses, and horsetails), and some gymnosperms. It is a multicellular, flask-shaped structure consisting of two parts: the swollen base or venter containing the egg cell, and the long, narrow neck. When the egg cell is mature, the cells of the neck dissolve, allowing the passage of the male gametes, or ◊antherozoids.

Archimedes' principle in physics, law stating that an object totally or partly submerged in a fluid displaces a volume of fluid that weighs the same as the apparent loss in weight of the object (which, in turn, equals the upwards force, or upthrust, experienced by that object). It was discovered by the Greek mathematician Archimedes.

If the weight of the object is less than the upthrust exerted by the fluid, it will float partly or completely above the surface; if its weight is equal to the upthrust, the object

Archaeology: Chronology

14th–16th centuries	Interest revived in Classical Greek and Roman art and architecture, including ruins and buried art and artefacts.
1748	The Roman city of Pompeii was discovered buried under volcanic ash from Vesuvius.
1784	Thomas Jefferson excavated an Indian burial mound on the Rivanna River in Virginia, USA.
1790	John Frere identified Old Stone Age (Palaeolithic) tools together with large extinct animals.
1822	Jean François Champollion deciphered Egyptian hieroglyphics.
1836	Christian Thomsen devised the Stone, Bronze, and Iron Age classification.
1840s	Austen Layard excavated the Assyrian capital of Nineveh.
1871	Heinrich Schliemann began excavations at Troy.
1879	Ice Age paintings were first discovered at Altamira, Spain.
1880s	Augustus Pitt-Rivers developed the concept of stratigraphy (identification of successive layers of soil within a site with successive archaeological stages, the most recent being at the top).
1891	Flinders Petrie began excavating Akhetaton in Egypt.
1899–1935	Arthur Evans excavated Minoan Knossos in Crete.
1900–44	Max Uhle began the systematic study of the civilizations of Peru.
1911	The Inca city of Machu Picchu was discovered by Hiram Bingham in the Andes.
1911–12	The Piltdown skull was 'discovered'; it was proved to be a fake 1949.
1914–18	Osbert Crawford developed the technique of aerial survey of sites.
1917–27	John Eric Thompson (1898–1975) investigated the great Maya sites in Yucatán, Mexico.
1922	Tutankhamen's tomb in Egypt was opened by Howard Carter.
1935	Dendrochronology (dating events by counting tree rings) was developed by A E Douglass.
1939	An Anglo-Saxon ship-burial treasure was found at Sutton Hoo, England.
1947	The first of the Dead Sea Scrolls was discovered.
1948	The 'Proconsul' prehistoric ape was discovered by Mary Leakey in Kenya.
1950s–1970s	Several early hominid fossils were found by Louis and Mary Leakey in Olduvai Gorge.
1953	Michael Ventris deciphered Minoan Linear B.
1960s	Radiocarbon and thermoluminescence measurement techniques were developed.
1963	Walter Emery pioneered rescue archaeology at Abu Simbel before the site was flooded by the Aswan Dam.
1969	Human remains found at Lake Mungo, Australia, were dated at 26,000 years; earliest evidence of ritual cremation.
1974	The tomb of Shi Huangdi, with its terracotta army, was discovered in China; the partial skeleton of a 3.18-million year old hominid nicknamed 'Lucy' was found in Ethiopia, and hominid footprints, 3.8 million years old, at Laetoli in Tanzania.
1978	The tomb of Philip II of Macedon (Alexander the Great's father) was discovered in Greece.
1979	The Aztec capital Tenochtitlán was excavated beneath a zone of Mexico City.
1982	English king Henry VIII's warship *Mary Rose* of 1545 was raised and studied with new techniques in underwater archaeology.
1985	The tomb of Maya, Tutankhamen's treasurer, was discovered at Sakkara, Egypt.
1988	The Turin Shroud was established as being of medieval origin by radiocarbon dating.
1989	The remains of the Globe and Rose Theatres, where many of Shakespeare's plays were originally performed, were discovered in London.
1991	Clothed body of a man from 5,300 years ago, with bow, arrows, copper axe, and other implements, was found preserved in ice in the Italian Alps.
1992	The world's oldest surviving wooden structure, a well 15 m/49 ft deep made of huge oak timbers at Kückhoven, Germany, was dated by tree-rings to 5090 BC. The world's oldest sea-going vessel, dating from about 1400 BC, was discovered at Dover, southern England.
1993	Drawings done in charcoal on the walls of the Cosquer Cave (near Marseille, France, discovered 1991) were dated by radiocarbon to be 27,110 and *c.* 19,000 years old; a fragment of cloth found on a tool handle unearthed in Çayönü, southeast Turkey in 1988 was carbon-dated at 9,000 years old, making it the oldest cloth ever found.
1994	The earliest known ancestral human *Ardipithecus ramidus*, from Ethiopia, was announced (4.4 million years old); and a major new Ice Age cave, the Grotte Chauvet, was found in southeast France, a network of hundreds of Palaeolithic cave drawings, dating from 30,000 years ago.
1995	A vast underground tomb believed to be the burial site of 50 of the sons of Ramses II was discovered. It was the largest yet found in the Valley of the Kings. Researchers in Spain found human remains dating back more than 780,000 years, in a cave site at Gran Dolina at Atapuerca in northern central Spain. Previously it was thought humans reached Europe around 300,000 years later than this. Scientific tests revealed that certain of the paintings within the Chauvet complex, discovered 1994, were between 30,340 and 32,410 years old, making them the world's oldest known paintings.
1997	Spanish palaeontologists, studying 8,000-year-old fossilized bones in northern Spain, claim to have discovered a new human species *Homo antecessor*.

will come to equilibrium below the surface; if its weight is greater than the upthrust, it will sink.

Archimedes screw one of the earliest kinds of pump, thought to have been invented by Archimedes. It consists of an enormous spiral screw revolving inside a close-fitting cylinder. It is used, for example, to raise water for irrigation.

The lowest portion of the screw just dips into the water, and as the cylinder is turned a small quantity of water is scooped up. The inclination of the cylinder is such that at the next revolution the water is raised above the next thread, whilst the lowest thread scoops up another quantity. The successive revolutions, therefore, raise the water thread by thread until it emerges at the top of the cylinder.

archipelago group of islands, or an area of sea containing a group of islands. The islands of an archipelago are usually volcanic in origin, and they sometimes represent the tops of peaks in areas around continental margins flooded by the sea.

Volcanic islands are formed either when a hot spot within the Earth's mantle produces a chain of volcanoes on the surface, such as the Hawaiian Archipelago or at a destructive plate margin (see ◊plate tectonics) where the subduction of one plate beneath another produces an arc-shaped island group called an 'island arc', such as the Aleutian Archipelago. Novaya Zemlya in the Arctic Ocean, the northern extension of the Ural Mountains, resulted from continental flooding.

arc lamp or **arc light** electric light that uses the illumination of an electric arc maintained between two electrodes. The British scientist Humphry Davy developed an arc lamp in 1808, and its main use in recent years has been in cinema projectors. The lamp consists of two carbon electrodes, between which a very high voltage is maintained. Electric current arcs (jumps) between the two, creating a brilliant light.

arc minute, arc second units for measuring small angles, used in geometry, surveying, mapmaking, and

Archimedes screw The Archimedes screw, a spiral screw turned inside a cylinder, was once commonly used to lift water from canals. The screw is still used to lift water in the Nile delta in Egypt, and is often used to shift grain in mills and powders in factories.

astronomy. An arc minute (symbol ') is one-sixtieth of a degree, and an arc second (symbol ") is one-sixtieth of an arc minute. Small distances in the sky, as between two close stars or the apparent width of a planet's disc, are expressed in minutes and seconds of arc.

Arctic Circle imaginary line that encircles the North Pole at latitude 66° 32' N. Within this line there is at least one day in the summer during which the Sun never sets, and at least one day in the winter during which the Sun never rises.

Arcturus or **Alpha Boötis** brightest star in the constellation Boötes and the fourth-brightest star in the sky. Arcturus is a red giant about 28 times larger than the Sun and 70 times more luminous, 36 light years away from Earth.

are metric unit of area, equal to 100 square metres (119.6 sq yd); 100 ares make one ◊hectare.

area the size of a surface. It is measured in square units, usually square centimetres (cm^2), square metres (m^2), or square kilometres (km^2). Surface area is the area of the outer surface of a solid.

The areas of geometrical plane shapes with straight edges are determined using the area of a rectangle. Integration may be used to determine the area of shapes enclosed by curves.

Arecibo site in Puerto Rico of the world's largest single-dish ◊radio telescope, 305 m/1,000 ft in diameter. It is built in a natural hollow and uses the rotation of the Earth to scan the sky. It has been used both for radar work on the planets and for conventional radio astronomy, and is operated by Cornell University, USA.

In 1996 it received a $25 million upgrade, increasing the sensitivity of the disc tenfold. Two new mirrors were also added and the observation frequency increased from 3,000 megahertz to up to 10,000 megahertz.

arête (German **grat**; North American **combe-ridge**) sharp narrow ridge separating two ◊glacial troughs (valleys), or ◊corries. The typical U-shaped cross sections of glacial troughs give arêtes very steep sides. Arêtes are common in glaciated mountain regions such as the Rockies, the Himalayas, and the Alps.

Argand diagram in mathematics, a method for representing complex numbers by Cartesian coordinates (x, y). Along the x-axis (horizontal axis) are plotted the real numbers, and along the y-axis (vertical axis) the nonreal, or ◊imaginary, numbers.

argon colourless, odourless, nonmetallic, gaseous element, symbol Ar, atomic number 18, relative atomic mass 39.948. It is grouped with the ◊inert gases, since it was long believed not to react with other substances, but observations now indicate that it can be made to combine with boron fluoride to form compounds. It constitutes almost 1% of the Earth's atmosphere, and was discovered in 1894 by British chemists John Rayleigh and William Ramsay after all oxygen and nitrogen had been removed chemically from a sample of air. It is used in electric discharge tubes and argon lasers.

argument in computing, the value on which a ◊function operates. For example, if the argument 16 is operated on by the function 'square root', the answer 4 is produced.

argument in mathematics, a specific value of the inde-

areas of common plane shapes

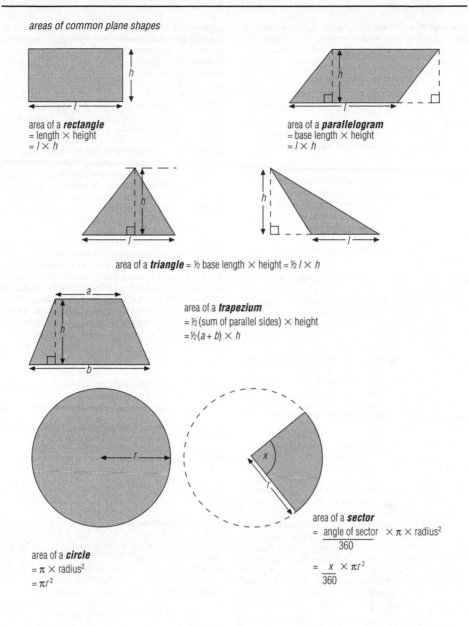

area of a **rectangle**
= length × height
= $l \times h$

area of a **parallelogram**
= base length × height
= $l \times h$

area of a **triangle** = ½ base length × height = ½ $l \times h$

area of a **trapezium**
= ½ (sum of parallel sides) × height
= ½ $(a + b) \times h$

area of a **circle**
= π × radius²
= πr^2

area of a **sector**
= $\dfrac{\text{angle of sector}}{360}$ × π × radius²

= $\dfrac{x}{360}$ × πr^2

area

pendent variable of a ◊function of x. It is also another name for ◊amplitude.

Ariane launch vehicle built in a series by the European Space Agency (first flight 1979). The launch site is at Kourou in French Guiana. Ariane is a three-stage rocket using liquid fuels. Small solid-fuel and liquid-fuel boosters can be attached to its first stage to increase carrying power.

Since 1984 it has been operated commercially by Arianespace, a private company financed by European banks and aerospace industries. A more powerful version, *Ariane 5*, was launched on 4 June 1996, and was intended to carry astronauts aboard the Hermes spaceplane. However, it went off course immediately after takeoff , turned on its side, broke into two and disintegrated. A fault in the software controlling the takeoff trajectory was to blame.

arid region in earth science, a region that is very dry and has little vegetation. Aridity depends on temperature, rainfall, and evaporation, and so is difficult to quantify, but an arid area is usually defined as one that receives less than 250 mm/10 in of rainfall each year. (By comparison, New York City receives 1,120 mm/44 in per year). There are arid regions in North Africa, Pakistan, Australia, the USA, and elsewhere. Very arid regions are ◊deserts.

Aries zodiacal constellation in the northern hemisphere between Pisces and Taurus, near Auriga, represented as the legendary ram whose golden fleece was sought by Jason and the Argonauts.

Its most distinctive feature is a curve of three stars of decreasing brightness. The brightest of these is Hamal or Alpha Arietis, 65 light years from Earth.

The Sun passes through Aries from late April to mid-May. In astrology, the dates for Aries, the first sign of the zodiac, are between about 21 March and 19 April. The spring ◊equinox once lay in Aries, but has now moved into Pisces through the effect of the Earth's ◊precession (wobble).

aril accessory seed cover other than a ◊fruit; it may be fleshy and sometimes brightly coloured, woody, or hairy. In flowering plants (◊angiosperms) it is often derived from the stalk that originally attached the ovule to the ovary wall. Examples of arils include the bright-red, fleshy layer surrounding the yew seed (yews are ◊gymnosperms so they lack true fruits), and the network of hard filaments that partially covers the nutmeg seed and yields the spice known as mace.

Another aril, the horny outgrowth found towards one end of the seed of the castor-oil plant *Ricinus communis*, is called a caruncle. It is formed from the integuments (protective layers enclosing the ovule) and develops after fertilization.

arithmetic branch of mathematics concerned with the study of numbers and their properties. The fundamental operations of arithmetic are addition, subtraction, multiplication, and division. Raising to powers (for example, squaring or cubing a number), the extraction of roots (for example, square roots), percentages, fractions, and ratios are developed from these operations.

Forms of simple arithmetic existed in prehistoric times. In China, Egypt, Babylon, and early civilizations generally, arithmetic was used for commercial purposes, records of taxation, and astronomy. During the Dark Ages in Europe, knowledge of arithmetic was preserved in India and later among the Arabs. European mathematics revived with the development of trade and overseas exploration. Hindu-Arabic numerals replaced Roman numerals, allowing calculations to be made on paper, instead of by abacus.

The essential feature of this number system was the introduction of zero, which allows us to have a *place-value* system. The decimal numeral system employs ten numerals (0,1,2,3,4,5,6,7,8,9) and is said to operate in 'base ten'. In a base-ten number, each position has a value ten times that of the position to its immediate right; for example, in the number 23 the numeral 3 represents three units (ones), and the number 2 represents two tens. The Babylonians, however, used a complex base-sixty system, residues of which are found today in the number of minutes in each hour and in angular measurement (6 x 60 degrees). The Mayas used a base-twenty system.

There have been many inventions and developments to make the manipulation of the arithmetic processes easier, such as the invention of ◊logarithms by Scottish mathematician John Napier in 1614 and of the slide rule in the period 1620–30. Since then, many forms of ready reckoners, mechanical and electronic calculators, and computers have been invented.

Modern computers fundamentally operate in base two, using only two numerals (0,1), known as a binary system. In binary, each position has a value twice as great as the position to its immediate right, so that for example binary 111 (or 111_2) is equal to 7 in the decimal system, and binary 1111 (or 1111_2) is equal to 15. Because the main operations of subtraction, multiplication, and division can be reduced mathematically to addition, digital computers carry out calculations by adding, usually in binary numbers in which the numerals 0 and 1 can be represented by off and on pulses of electric current.

Modular or modulo arithmetic, sometimes known as residue arithmetic or clock arithmetic, can take only a specific number of digits, whatever the value. For example, in modulo 4 (mod 4) the only values any number can take are 0, 1, 2, or 3. In this system, 7 is written as 3 mod 4, and 35 is also 3 mod 4. Notice 3 is the residue, or remainder, when 7 or 35 is divided by 4. This form of arithmetic is often illustrated on a circle. It deals with events recurring in regular cycles, and is used in describing the functioning of petrol engines, electrical generators, and so on. For example, in the mod 12, the answer to a question as to what time it will be in five hours if it is now ten o'clock can be expressed $10 + 5 = 3$.

arithmetic and logic unit (ALU) in a computer, the part of the ◊central processing unit (CPU) that performs the basic arithmetic and logic operations on data.

arithmetic mean the average of a set of n numbers, obtained by adding the numbers and dividing by n. For example, the arithmetic mean of the set of 5 numbers 1, 3, 6, 8, and 12 is $(1 + 3 + 6 + 8 + 12)/5 = 30/5 = 6$.

The term 'average' is often used to refer only to the arithmetic mean, even though the mean is in fact only one form of average (the others include ◊median and ◊mode).

arithmetic progression or *arithmetic sequence* sequence of numbers or terms that have a common difference between any one term and the next in the sequence. For example, 2, 7, 12, 17, 22, 27, ... is an arithmetic sequence with a common difference of 5.

The nth term in any arithmetic progression can be found using the formula:

$$n\text{th term} = a + (n - 1)d$$

where a is the first term and d is the common difference.

An *arithmetic series* is the sum of the terms in an arithmetic sequence. The sum S of n terms is given by:

$$S = \frac{n}{2}[2a + (n - 1)d]$$

armature in a motor or generator, the wire-wound coil that carries the current and rotates in a magnetic field. (In alternating-current machines, the armature is sometimes stationary.) The pole piece of a permanent magnet or electromagnet and the moving, iron part of a ◊solenoid, especially if the latter acts as a switch, may also be referred to as armatures.

armillary sphere earliest known astronomical device, in use from 3rd century BC. It showed the Earth at the centre of the universe, surrounded by a number of movable metal rings representing the Sun, Moon, and planets. The armillary sphere was originally used to observe the heavens and later for teaching navigators about the arrangements and movements of the heavenly bodies.

armoured personnel carrier (APC) wheeled or tracked military vehicle designed to transport up to ten people. Armoured to withstand small-arms fire and shell splinters, it is used on battlefields.

aromatherapy in alternative medicine, use of oils and essences derived from plants, flowers, and wood resins. Bactericidal properties and beneficial effects upon physiological functions are attributed to the oils, which are generally massaged into the skin.

Aromatherapy was first used in ancient Greece and Egypt, but became a forgotten art until the 1930s, when a French chemist accidentally spilt lavender over a cut and found that the wound healed without a scar. However, it was not until the 1970s that it began to achieve widespread popularity.

aromatic compound organic chemical compound in which some of the bonding electrons are delocalized (shared among several atoms within the molecule and not localized in the vicinity of the atoms involved in bonding). The commonest aromatic compounds have ring structures, the atoms comprising the ring being either all carbon or containing one or more different atoms (usually nitrogen, sulphur, or oxygen). Typical examples are benzene (C_6H_6) and pyridine (C_5H_5N).

arrhythmia disturbance of the normal rhythm of the heart. There are various kinds of arrhythmia, some benign, some indicative of heart disease. In extreme cases, the heart may beat so fast as to be potentially lethal and surgery may be used to correct the condition.

Extra beats between the normal ones are called **extrasystoles**; abnormal slowing is known as **bradycardia** and speeding up is known as **tachycardia**.

arsenic brittle, greyish-white, semimetallic element (a metalloid), symbol As, atomic number 33, relative atomic mass 74.92. It occurs in many ores and occasionally in its elemental state, and is widely distributed, being present in minute quantities in the soil, the sea, and the human body. In larger quantities, it is poisonous. The chief source of arsenic compounds is as a by-product from metallurgical processes. It is used in making semiconductors, alloys, and solders.

As it is a cumulative poison, its presence in food and drugs is very dangerous. The symptoms of arsenic poisoning are vomiting, diarrhoea, tingling and possibly numbness in the limbs, and collapse. It featured in some drugs, including Salvarsan, the first specific treatment for syphilis.

arteriosclerosis hardening of the arteries, with thickening and loss of elasticity. It is associated with smoking, ageing, and a diet high in saturated fats. The term is used loosely as a synonym for ◊atherosclerosis.

artery vessel that carries blood from the heart to the rest of the body. It is built to withstand considerable pressure, having thick walls which contain smooth muscle fibres. During contraction of the heart muscle, arteries expand in diameter to allow for the sudden increase in pressure that occurs; the resulting ◊pulse or pressure wave can be felt at the wrist. Not all arteries carry oxygenated (oxygen-rich) blood; the pulmonary arteries convey deoxygenated (oxygen-poor) blood from the heart to the lungs.

Arteries are flexible, elastic tubes, consisting of three

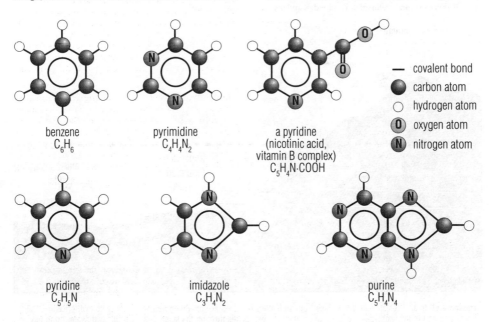

benzene
C_6H_6

pyrimidine
$C_4H_4N_2$

a pyridine
(nicotinic acid,
vitamin B complex)
$C_5H_4N \cdot COOH$

— covalent bond
● carbon atom
○ hydrogen atom
Ⓞ oxygen atom
Ⓝ nitrogen atom

pyridine
C_5H_5N

imidazole
$C_3H_4N_2$

purine
$C_5H_4N_4$

aromatic compound Compounds whose molecules contain the benzene ring, or variations of it, are called aromatic. The term was originally used to distinguish sweet-smelling compounds from others.

layers, the middle of which is muscular; its rhythmic contraction aids the pumping of blood around the body. In middle and old age, the walls degenerate and are vulnerable to damage by the build-up of fatty deposits. These reduce elasticity, hardening the arteries and decreasing the internal bore. This condition, known as ◊atherosclerosis, can lead to high blood pressure, loss of circulation, heart disease, and death.

Research indicates that a typical Western diet, high in saturated fat, increases the chances of arterial disease developing.

artesian well well that is supplied with water rising naturally from an underground water-saturated rock layer (◊aquifer). The water rises from the aquifer under its own pressure. Such a well may be drilled into an aquifer that is confined by impermeable rocks both above and below. If the water table (the top of the region of water saturation) in that aquifer is above the level of the well head, hydrostatic pressure will force the water to the surface.

Artesian wells are often overexploited because their water is fresh and easily available, and they eventually become unreliable. There is also some concern that pollutants such as pesticides or nitrates can seep into the aquifers.

arthritis inflammation of the joints, with pain, swelling, and restricted motion. Many conditions may cause arthritis, including gout, infection, and trauma to the joint. There are three main forms of arthritis: ◊rheumatoid arthritis, osteoarthritis, and septic arthritis.

arthropod member of the phylum Arthropoda; an invertebrate animal with jointed legs and a segmented body with a horny or chitinous casing (exoskeleton), which is shed periodically and replaced as the animal grows. Included are arachnids such as spiders and mites, as well as crustaceans, millipedes, centipedes, and insects.

artificial insemination (AI) introduction by instrument of semen from a sperm bank or donor into the female reproductive tract to bring about fertilization. Originally used by animal breeders to improve stock with sperm from high-quality males, in the 20th century it has been developed for use in humans, to help the infertile. See ◊in vitro fertilization.

artificial intelligence (AI) branch of science concerned with creating computer programs that can perform actions comparable with those of an intelligent human. Current AI research covers such areas as planning (for robot behaviour), language understanding, pattern recognition, and knowledge representation.

The possibility of artificial intelligence was first proposed by the English mathematician Alan Turing in 1950. Early AI programs, developed in the 1960s, attempted simulations of human intelligence or were aimed at general problem-solving techniques. By the mid-1990s, scientists were concluding that AI was more difficult to create than they had imagined. It is now thought that intelligent behaviour depends as much on the knowledge a system possesses as on its reasoning power. Present emphasis is on ◊knowledge-based systems, such as ◊expert systems, while research projects focus on ◊neural networks, which attempt to mimic the structure of the human brain.

One notably successful AI project is IBM's Deep Blue, which in 1996 was the first chess-playing computer to defeat a human grand master, the Russian Gary Kasparov.

artificial radioactivity natural and spontaneous radioactivity arising from radioactive isotopes or elements that are formed when elements are bombarded with subatomic particles – protons, neutrons, or electrons – or small nuclei.

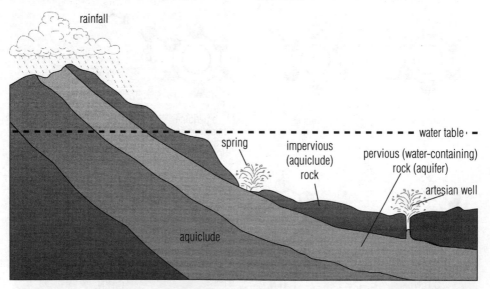

artesian well In an artesian well, water rises from an underground water-containing rock layer under its own pressure. Rain falls at one end of the water-bearing layer, or aquifer, and percolates through the layer. The layer fills with water up to the level of the water table. Water will flow from a well under its own pressure if the well head is below the level of the water table.

artificial respiration emergency procedure to restart breathing once it has stopped; in cases of electric shock or apparent drowning, for example, the first choice is the expired-air method, the *kiss of life* by mouth-to-mouth breathing until natural breathing is restored.

artificial selection in biology, selective breeding of individuals that exhibit the particular characteristics that a plant or animal breeder wishes to develop. In plants, desirable features might include resistance to disease, high yield (in crop plants), or attractive appearance. In animal breeding, selection has led to the development of particular breeds of cattle for improved meat production (such as the Aberdeen Angus) or milk production (such as Jerseys).

Artificial selection was practised by the Sumerians at least 5,500 years ago and carried on through the succeeding ages, with the result that all common vegetables, fruit, and livestock are long modified by selective breeding. Artificial selection, particularly of pigeons, was studied by the English evolutionist Charles Darwin who saw a similarity between this phenomenon and the processes of natural selection.

asbestos any of several related minerals of fibrous structure that offer great heat resistance because of their non-flammability and poor conductivity. Commercial asbestos is generally either made from serpentine ('white' asbestos) or from sodium iron silicate ('blue' asbestos). The fibres are woven together or bound by an inert material. Over time the fibres can work loose and, because they are small enough to float freely in the air or be inhaled, asbestos usage is now strictly controlled; exposure to its dust can cause cancer.

ASCII (acronym for *American standard code for information interchange*) in computing, a coding system in which numbers are assigned to letters, digits, and punctuation symbols. Although computers work in

ASCII Codes

Character	Binary code
A	1000001
B	1000010
C	1000011
D	1000100
E	1000101
F	1000110
G	1000111
H	1001000
I	1001001
J	1001010
K	1001011
L	1001100
M	1001101
N	1001110
O	1001111
P	1010000
Q	1010001
R	1010010
S	1010011
T	1010100
U	1010101
V	1010110
W	1010111
X	1011000
Y	1011001
Z	1011010

code based on the ◊binary number system, ASCII numbers are usually quoted as decimal or ◊hexadecimal numbers. For example, the decimal number 45 (binary 0101101) represents a hyphen, and 65 (binary 1000001) a capital A. The first 32 codes are used for control functions, such as carriage return and backspace.

Strictly speaking, ASCII is a 7-bit binary code, allowing 128 different characters to be represented, but an eighth bit is often used to provide ◊parity or to allow for extra characters. The system is widely used for the storage of text and for the transmission of data between computers.

ascorbic acid $C_6H_8O_6$ or *vitamin C* relatively simple organic acid found in citrus fruits and vegetables. It is soluble in water and destroyed by prolonged boiling, so soaking or overcooking of vegetables reduces their vitamin C content. Lack of ascorbic acid results in scurvy.

In the human body, ascorbic acid is necessary for the correct synthesis of ◊collagen. Lack of vitamin C causes skin sores or ulcers, tooth and gum problems, and burst capillaries (scurvy symptoms) owing to an abnormal type of collagen replacing the normal type in these tissues.

Ascorbic acid

The Australian billygoat plum, *Terminalia ferdiandiana*, is the richest natural source of vitamin C, containing 100 times the concentration found in oranges.

---

asepsis practice of ensuring that bacteria are excluded from open sites during surgery, wound dressing, blood sampling, and other medical procedures. Aseptic technique is a first line of defence against infection.

asexual reproduction in biology, reproduction that does not involve the manufacture and fusion of sex cells, nor the necessity for two parents. The process carries a clear advantage in that there is no need to search for a mate nor to develop complex pollinating mechanisms; every asexual organism can reproduce on its own. Asexual reproduction can therefore lead to a rapid population build-up.

In evolutionary terms, the disadvantage of asexual reproduction arises from the fact that only identical individuals, or clones, are produced – there is no variation.

In the field of horticulture, where standardized production is needed, this is useful, but in the wild, an asexual population that cannot adapt to a changing environment or evolve defences against a new disease is at risk of extinction. Many asexually reproducing organisms are therefore capable of reproducing sexually as well.

Asexual processes include ◊binary fission, in which the parent organism splits into two or more 'daughter' organisms, and ◊budding, in which a new organism is formed initially as an outgrowth of the parent organism. The asexual reproduction of spores, as in ferns and mosses, is also common and many plants reproduce asexually by means of runners, rhizomes, bulbs, and corms; see also ◊vegetative reproduction.

ashen light in astronomy, a faint glow occasionally reported in the dark hemisphere of ◊Venus when the planet is in a crescent phase. Its origin is unknown, but it may be related to the terrestrial airglow caused by interaction of high-energy solar radiation with the upper atmosphere of the Earth.

invention of **glassphalt**, asphalt that is 15% crushed glass. It is used to pave roads in New York.

Considerable natural deposits of asphalt occur around the Dead Sea and in the Philippines, Cuba, Venezuela, and Trinidad. Bituminous limestone occurs at Neufchâtel, France.

asphyxia suffocation; a lack of oxygen that produces a potentially lethal build-up of carbon dioxide waste in the tissues.

Asphyxia may arise from any one of a number of causes, including inhalation of smoke or poisonous gases, obstruction of the windpipe (by water, food, vomit, or a foreign object), strangulation, or smothering. If it is not quickly relieved, brain damage or death ensues.

aspirin acetylsalicylic acid, a popular pain-relieving drug (◊analgesic) developed in the late 19th century as a household remedy for aches and pains. It relieves pain and reduces inflammation and fever. It is derived from the white willow tree *Salix alba*, and is the world's most widely used drug.

Aspirin was first refined from salicylic acid by German chemist Felix Hoffman, and marketed in 1899. Although salicylic acid occurs naturally in willow bark (and has been used for pain relief since 1763) the acetyl derivative is less bitter and less likely to cause vomiting.

assay in chemistry, the determination of the quantity of a given substance present in a sample. Usually it refers to determining the purity of precious metals.

The assay may be carried out by 'wet' methods, when the sample is wholly or partially dissolved in some reagent (often an acid), or by 'dry' or 'fire' methods, in which the compounds present in the sample are combined with other substances.

assembler in computing, a program that translates a program written in an assembly language into a complete ◊machine code program that can be executed by a computer. Each instruction in the assembly language is translated into only one machine-code instruction.

assembly language low-level computer-programming language closely related to a computer's internal codes. It consists chiefly of a set of short sequences of letters (mnemonics), which are translated, by a program called an assembler, into ◊machine code for the computer's ◊central processing unit (CPU) to follow directly. In assembly language, for example, 'JMP' means 'jump' and 'LDA' means 'load accumulator'. Assembly code is used by programmers who need to write very fast or efficient programs.

Because they are much easier to use, high-level languages are normally used in preference to assembly languages. An assembly language may still be used in some cases, however, particularly when no suitable high-level language exists or where a very efficient machine-code program is required.

assembly line method of mass production in which a product is built up step-by-step by successive workers adding one part at a time. It is commonly used in industries such as the car industry.

US inventor Eli Whitney pioneered the concept of industrial assembly in the 1790s, when he employed unskilled labour to assemble muskets from sets of identical precision-made parts. In 1901 Ransome Olds in the USA began mass-producing motor cars on an assembly-

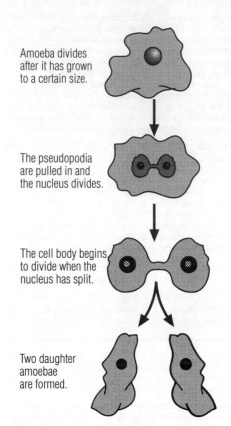

Amoeba divides after it has grown to a certain size.

The pseudopodia are pulled in and the nucleus divides.

The cell body begins to divide when the nucleus has split.

Two daughter amoebae are formed.

asexual reproduction Asexual reproduction is the simplest form of reproduction, occurring in many simple plants and animals. Binary fission, shown here occurring in an amoeba, is one of a number of asexual reproduction processes.

aspartame noncarbohydrate sweetener used in foods under the tradename Nutrasweet. It is about 200 times as sweet as sugar and, unlike saccharine, has no aftertaste.

The aspartame molecule consists of two amino acids (aspartic acid and phenylalanine) linked by a methylene (–CH_2–) group. It breaks down slowly at room temperature and rapidly at higher temperatures. It is not suitable for people who suffer from ◊phenylketonuria.

aspect in Earth sciences the direction in which a slope faces. In the northern hemisphere a slope with a southerly aspect receives more sunshine than other slopes and is therefore better suited for growing crops that require many hours of sunshine in order to ripen successfully. Vineyards in northern Europe are usually situated on south-facing slopes.

asphalt mineral mixture containing semisolid brown or black ◊bitumen, used in the construction industry. Asphalt is mixed with rock chips to form paving material, and the purer varieties are used for insulating material and for waterproofing masonry. It can be produced artificially by the distillation of ◊petroleum.

The availability of recycled coloured glass led to the

line principle, a method further refined by the introduction of the moving conveyor belt by Henry Ford in 1913 and the time-and-motion studies of Fredriech Winslow Taylor. On the assembly line human workers now stand side by side with ◊robots.

assimilation in animals, the process by which absorbed food molecules, circulating in the blood, pass into the cells and are used for growth, tissue repair, and other metabolic activities. The actual destiny of each food molecule depends not only on its type, but also on the body requirements at that time.

assisted conception in medicine, the use of medical techniques to aid the fusion of the ovum and the sperm to form the embryo, as in ◊artificial insemination and ◊in vitro fertilization.

associative operation in mathematics, an operation in which the outcome is independent of the grouping of the numbers or symbols concerned. For example, multiplication is associative, as $4 \times (3 \times 2) = (4 \times 3) \times 2 = 24$; however, division is not, as $12 \div (4 \div 2) = 6$, but $(12 \div 4) \div 2 = 1.5$. Compare ◊commutative operation and ◊distributive operation.

assortative mating in population genetics, selective mating in a population between individuals that are genetically related or have similar characteristics. If sufficiently consistent, assortative mating can theoretically result in the evolution of new species without geographical isolation (see ◊speciation).

astatine nonmetallic, radioactive element, symbol At, atomic number 85, relative atomic mass 210. It is a member of the ◊halogen group, and is very rare in nature. Astatine is highly unstable, with at least 19 isotopes; the longest lived has a half-life of about eight hours.

asteroid or *minor planet* any of many thousands of small bodies, composed of rock and iron, that orbit the Sun. Most lie in a belt between the orbits of Mars and Jupiter, and are thought to be fragments left over from the formation of the ◊Solar System. About 100,000 may exist, but their total mass is only a few hundredths the mass of the Moon.

They include ◊Ceres (the largest asteroid, 940 km/ 584 mi in diameter), Vesta (which has a light-coloured surface, and is the brightest as seen from Earth), ◊Eros, and ◊Icarus. Some asteroids are in orbits that bring them close to Earth, and some, such as the ◊Apollo asteroids, even cross Earth's orbit; at least some of these may be remnants of former comets. One group, the Trojans, moves along the same orbit as Jupiter, 60° ahead and behind the planet. One unusual asteroid, ◊Chiron, orbits beyond Saturn.

NASA's Near Earth Asteroid Rendezvous (NEAR) was launched in Feb 1996 to study Eros to ascertain what asteroids are made of and whether they are similar in structure to meteorites. In 1997 it flew past asteroid Mathilde, revealing a 25 km/15.5 mi crater covering the 53 km asteroid. The Near Earth Asteroid Tracking (NEAT) system detected about 200 new asteroids during March 1996, its first full month in operation.

The first asteroid was discovered by the Italian astronomer Giuseppe Piazzi at the Palermo Observatory, Sicily, on 1 Jan 1801. The first asteroid moon was observed by the space probe ◊*Galileo* in 1993 orbiting asteroid Ida.

Bifurcated asteroids, first discovered in 1990, are in fact two chunks of rock that touch each other. It may be that at least 10% of asteroids approaching the Earth are bifurcated.

asthenosphere division of the Earth's structure lying beneath the ◊lithosphere, at a depth of approximately 70 km/45 mi to 260 km/160 mi.

It is thought to be the soft, partially molten layer of the ◊mantle on which the rigid plates of the Earth's surface move to produce the motions of ◊plate tectonics.

asthma chronic condition characterized by difficulty in breathing due to spasm of the bronchi (air passages) in the lungs. Attacks may be provoked by allergy, infection, and stress. The incidence of asthma may be increasing as a result of air pollution and occupational hazard. Treatment is with ◊bronchodilators to relax the bronchial muscles and thereby ease the breathing, and in severe cases by inhaled ◊steroids that reduce inflammation of the bronchi.

Extrinsic asthma, which is triggered by exposure to irritants such as pollen and dust, is more common in children and young adults. In Feb 1997 Brazilian researchers reported two species of dust mite actually living on children's scalps. It explains why vacuuming of bedding sometimes fail to prevent asthma attacks. The use of antidandruff shampoo should keep numbers of mites down by reducing their food supply.

Less common, intrinsic asthma tends to start in the middle years. Approximately 5–10% of children suffer from asthma, but about a third of these will show no symptoms after adolescence, while another 5–10% of people develop the condition as adults. Growing evidence that the immune system is involved in both forms of asthma has raised the possibility of a new approach to treatment.

Although the symptoms are similar to those of bronchial asthma, **cardiac asthma** is an unrelated condition and is a symptom of heart deterioration.

The Largest Asteroids

Name	Diameter		Average distance from Sun (Earth = 1)	Orbital period (yrs)
	km	mi		
Ceres	940	584	2.77	4.6
Pallas	588	365	2.77	4.6
Vesta	576	358	2.36	3.6
Hygeia	430	267	3.13	5.5
Interamnia	338	210	3.06	5.4
Davida	324	201	3.18	5.7

astigmatism aberration occurring in the lens of the eye. It results when the curvature of the lens differs in two perpendicular planes, so that rays in one plane may be in focus while rays in the other are not. With astigmatic eyesight, the vertical and horizontal cannot be in focus at the same time; correction is by the use of a cylindrical lens that reduces the overall focal length of one plane so that both planes are seen in sharp focus.

astrolabe ancient navigational instrument, forerunner of the sextant. Astrolabes usually consisted of a flat disc with a sighting rod that could be pivoted to point at the Sun or bright stars. From the altitude of the Sun or star above the horizon, the local time could be estimated.

astrometry measurement of the precise positions of stars, planets, and other bodies in space. Such information is needed for practical purposes including accurate timekeeping, surveying and navigation, and calculating orbits and measuring distances in space. Astrometry is not concerned with the surface features or the physical nature of the body under study.

Before telescopes, astronomical observations were simple astrometry. Precise astrometry has shown that stars are not fixed in position, but have a ◊proper motion caused as they and the Sun orbit the Milky Way galaxy. The nearest stars also show ◊parallax (apparent change in position), from which their distances can be calculated. Above the distorting effects of the atmosphere, satellites such as ◊*Hipparcos* can make even more precise measurements than ground telescopes, so refining the distance scale of space.

astronaut person making flights into space; the term **cosmonaut** is used in the West for any astronaut from the former Soviet Union.

Astronaut

Astronauts in space cannot belch. It is gravity that causes bubbles to rise to the top of a liquid, so space shuttle crews were forced to request less gas in their fizzy drinks to avoid discomfort.

---

astronautics science of space travel. See ◊rocket; ◊satellite; ◊space probe.

Astronomer Royal honorary post in British astronomy. Originally it was held by the director of the Royal Greenwich Observatory; since 1972 the title of Astronomer Royal has been awarded separately as an honorary title to an outstanding British astronomer. The Astronomer Royal from 1995 is Martin Rees. There is a separate post of Astronomer Royal for Scotland.

Astronomical Almanac in astronomy, an international work of reference published jointly every year by the ◊Royal Greenwich Observatory and the ◊US Naval Observatory containing detailed tables of planetary motions, ◊eclipses, and other astronomical phenomena.

astronomical unit unit (symbol AU) equal to the mean distance of the Earth from the Sun: 149,597,870 km/ 92,955,800 mi. It is used to describe planetary distances. Light travels this distance in approximately 8.3 minutes.

astronomy science of the celestial bodies: the Sun, Moon, and the planets; the stars and galaxies; and all other objects in the universe. It is concerned with their positions, motions, distances, and physical conditions and with their origins and evolution. Astronomy thus divides into fields such as astrophysics, celestial mechanics, and ◊cosmology. See also ◊gamma-ray astronomy, ◊infrared astronomy, ◊radio astronomy, ◊ultraviolet astronomy, and ◊X-ray astronomy.

I ask you to look both ways. For the road to a knowledge of the stars leads through the atom; and important knowledge of the atom has been reached through the stars.

On **astronomy** Arthur Eddington

Astronomy is perhaps the oldest recorded science; there are observational records from ancient Babylonia, China, Egypt, and Mexico. The first true astronomers, however, were the Greeks, who deduced the Earth to be a sphere and attempted to measure its size. Ancient Greek astronomers included Thales and Pythagoras. Eratosthenes of Cyrene measured the size of the Earth with considerable accuracy. Star catalogues were drawn up, the most celebrated being that of Hipparchus. The *Almagest*, by Ptolemy of Alexandria, summarized Greek astronomy and survived in its Arabic translation. The Greeks still regarded the Earth as the centre of the universe, although this was doubted by some philosophers, notably Aristarchus of Samos, who maintained that the Earth moves around the Sun.

Ptolemy, the last famous astronomer of the Greek school, died about AD 180, and little progress was made for some centuries.

The Arabs revived the science, developing the astrolabe and producing good star catalogues. Unfortunately, a general belief in the pseudoscience of astrology continued until the end of the Middle Ages (and has been revived from time to time).

The dawn of a new era came in 1543, when a Polish canon, Copernicus, published a work entitled *De revolutionibus orbium coelestium/On the Revolutions of the Heavenly Spheres*, in which he demonstrated that the Sun, not the Earth, is the centre of our planetary system. (Copernicus was wrong in many respects – for instance, he still believed that all celestial orbits must be perfectly circular.) Tycho Brahe, a Dane, increased the accuracy of observations by means of improved instruments allied to his own personal skill, and his observations were used by German mathematician Johannes Kepler to prove the validity of the Copernican system. Considerable opposition existed, however, for removing the Earth from its central position in the universe; the Catholic church was openly hostile to the idea, and, ironically, Brahe never accepted the idea that the Earth could move around the Sun. Yet before the end of the 17th century, the theoretical work of Isaac Newton had established celestial mechanics.

The refracting telescope was invented in about 1608, by Hans Lippershey in Holland, and was first applied to astronomy by Italian scientist Galileo in the winter of 1609–10. Immediately, Galileo made a series of spectacular discoveries. He found the four largest satellites of Jupiter, which gave strong support to the Copernican theory; he saw the craters of the Moon, the phases of Venus, and the myriad faint stars of our ◊Galaxy, the Milky Way.

Galileo's most powerful telescope magnified only 30 times, but it was not long before larger telescopes were built and official observatories were established.

Astronomy: Chronology

2300 BC	Chinese astronomers made their earliest observations.
2000	Babylonian priests made their first observational records.
1900	Stonehenge was constructed: first phase.
434	Anaxagoras claims the Sun is made up of hot rock.
365	The Chinese observed the satellites of Jupiter with the naked eye.
3rd century	Aristarchus argued that the Sun is the centre of the Solar System.
2nd century AD	Ptolemy's complicated Earth-centred system was promulgated, which dominated the astronomy of the Middle Ages.
1543	Copernicus revived the ideas of Aristarchus in *De Revolutionibus*.
1608	Hans Lippershey invented the telescope, which was first used by Galileo in 1609.
1609	Johannes Kepler's first two laws of planetary motion were published (the third appeared in 1619).
1632	The world's first official observatory was established in Leiden in the Netherlands.
1633	Galileo's theories were condemned by the Inquisition.
1675	The Royal Greenwich Observatory was founded in England.
1687	Isaac Newton's *Principia* was published, including his 'law of universal gravitation'.
1705	Edmond Halley correctly predicted that the comet that had passed the Earth in 1682 would return in 1758; the comet was later to be known by his name.
1781	William Herschel discovered Uranus and recognized stellar systems beyond our Galaxy.
1796	Pierre Laplace elaborated his theory of the origin of the Solar System.
1801	Giuseppe Piazzi discovered the first asteroid, Ceres.
1814	Joseph von Fraunhofer first studied absorption lines in the solar spectrum.
1846	Neptune was identified by Johann Galle, following predictions by John Adams and Urbain Leverrier.
1859	Gustav Kirchhoff explained dark lines in the Sun's spectrum.
1887	The earliest photographic star charts were produced.
1889	Edward Barnard took the first photographs of the Milky Way.
1908	Fragment of comet fell at Tunguska, Siberia.
1920	Arthur Eddington began the study of interstellar matter.
1923	Edwin Hubble proved that the galaxies are systems independent of the Milky Way, and by 1930 had confirmed the concept of an expanding universe.
1930	The planet Pluto was discovered by Clyde Tombaugh at the Lowell Observatory, Arizona, USA.
1931	Karl Jansky founded radio astronomy.
1945	Radar contact with the Moon was established by Z Bay of Hungary and the US Army Signal Corps Laboratory.
1948	The 5 m/200 in Hale reflector telescope was installed at Mount Palomar, California, USA.
1957	The Jodrell Bank telescope dish in England was completed.
1957	The first Sputnik satellite (USSR) opened the age of space observation.
1962	The first X-ray source was discovered in Scorpius.
1963	The first quasar was discovered.
1967	The first pulsar was discovered by Jocelyn Bell and Antony Hewish.
1969	The first crewed Moon landing was made by US astronauts.
1976	A 6 m/240 in reflector telescope was installed at Mount Semirodniki, USSR.
1977	Uranus was discovered to have rings.
1977	The spacecraft *Voyager 1* and *2* were launched, passing Jupiter and Saturn 1979–1981.
1978	The spacecraft *Pioneer Venus 1* and *2* reached Venus.
1978	A satellite of Pluto, Charon, was discovered by James Christy of the US Naval Observatory.
1986	Halley's comet returned. *Voyager 2* flew past Uranus and discovered six new moons.
1987	Supernova SN1987A flared up, becoming the first supernova to be visible to the naked eye since 1604. The 4.2 m/165 in William Herschel Telescope on La Palma, Canary Islands, and the James Clerk Maxwell Telescope on Mauna Kea, Hawaii, began operation.
1988	The most distant individual star was recorded – a supernova, 5 billion light years away, in the AC118 cluster of galaxies.
1989	*Voyager 2* flew by Neptune and discovered eight moons and three rings.
1990	Hubble Space Telescope was launched into orbit by the US space shuttle.
1991	The space probe *Galileo* flew past the asteroid Gaspra, approaching it to within 26,000 km/16,200 mi.
1992	COBE satellite detected ripples from the Big Bang that mark the first stage in the formation of galaxies.
1994	Fragments of comet Shoemaker–Levy struck Jupiter.
1996	US astronomers discovered the most distant galaxy so far detected. It is in the constellation Virgo and is 14 billion light years from Earth.
1997	Data from the satellite *Hippacos* improved estimates of the age of the universe, and the distances to many nearby stars.

Galileo's telescope was a refractor; that is to say, it collected its light by means of a glass lens or object glass. Difficulties with his design led Newton, in 1671, to construct a reflector, in which the light is collected by means of a curved mirror.

Theoretical researches continued, and astronomy made rapid progress in many directions. Uranus was discovered in 1781 by William Herschel, and this was soon followed by the discovery of the first four asteroids, Ceres (1801), Pallas (1802), Juno (1804), and Vesta (1807). In 1846 Neptune was located by Johann Galle, following calculations by British astronomer John Couch Adams and French astronomer Urbain Jean Joseph Leverrier. Also significant was the first measurement of the distance of a star, when in 1838 the German astronomer Friedrich Bessel measured the ◊parallax of the star 61 Cygni, and calculated that it lies at a distance of about 6 light years (about half the correct value).

Astronomical spectroscopy was developed, first by Fraunhofer in Germany and then by people such as Pietro Angelo Secchi and William Huggins, while Gustav Kirchhoff successfully interpreted the spectra of the Sun and stars. By the 1860s good photographs of the Moon had been obtained, and by the end of the century photographic methods had started to play a leading role in research.

William Herschel, probably the greatest observer in the history of astronomy, investigated the shape of our Galaxy during the latter part of the 18th century and concluded that its stars are arranged roughly in the form of a double-convex lens. Basically Herschel was correct, although he placed our Sun near the centre of the system; in fact, it is well out towards the edge, and lies 25,000 light years from the galactic nucleus. Herschel also studied the luminous 'clouds' or nebulae, and made the tentative suggestion that these nebulae capable of resolution into stars might be separate galaxies, far outside our own Galaxy.

It was not until 1923 that US astronomer Edwin Hubble, using the 2.5 m/100 in reflector at the Mount Wilson Observatory, was able to verify this suggestion. It is now known that the 'spiral nebulae' are galaxies in their own right, and that they lie at immense distances. The most distant galaxy visible to the naked eye, the Great Spiral in ◊Andromeda, is 2.2 million light years away; the most remote galaxy so far measured lies over 14 billion light years away. It was also found that galaxies tended to form groups, and that the groups were apparently receding from each other at speeds proportional to their distances.

This concept of an expanding and evolving universe at first rested largely on Hubble's law, relating the distance of objects to the amount their spectra shift towards red – the ◊red shift. Subsequent evidence derived from objects studied in other parts of the ◊electromagnetic spectrum, at radio and X-ray wavelengths, has provided confirmation. ◊Radio astronomy established its place in probing the structure of the universe by demonstrating in 1954 that an optically visible distant galaxy was identical with a powerful radio source known as Cygnus A. Later analysis of the comparative number, strength, and distance of radio sources suggested that in the distant past these, including the ◊quasars discovered in 1963, had been much more powerful and numerous than today. This fact suggested that the universe has been evolving from an origin, and is not of infinite age as expected under a ◊steady-state theory.

The discovery in 1965 of microwave background radiation suggested that a residue survived the tremendous thermal power of the giant explosion, or Big Bang, that brought the universe into existence.

Although the practical limit in size and efficiency of optical telescopes has apparently been reached, the siting of these and other types of telescope at new observatories in the previously neglected southern hemisphere has opened fresh areas of the sky to search. Australia has been in the forefront of these developments.

The most remarkable recent extension of the powers of astronomy to explore the universe is in the use of rockets, satellites, space stations, and space probes. Even the range and accuracy of the conventional telescope may be greatly improved free from the Earth's atmosphere. When the USA launched the Hubble Space Telescope into permanent orbit in 1990, it was the most powerful optical telescope yet constructed, with a 2.4 m/94.5 in mirror. It detects celestial phenomena seven times more distant (up to 14 billion light years) than any Earth-based telescope.

See also ◊black hole and ◊infrared radiation.

astrophotography use of photography in astronomical research. The first successful photograph of a celestial object was the daguerreotype plate of the Moon taken by John W Draper (1811–1882) of the USA in March 1840. The first photograph of a star, Vega, was taken by US astronomer William C Bond (1789–1859) in 1850. Modern-day astrophotography uses techniques such as ◊charge-coupled devices (CCDs).

astrophysics study of the physical nature of stars, galaxies, and the universe. It began with the development of spectroscopy in the 19th century, which allowed astronomers to analyse the composition of stars from their light. Astrophysicists view the universe as a vast natural laboratory in which they can study matter under conditions of temperature, pressure, and density that are unattainable on Earth.

asymptote in ◊coordinate geometry, a straight line that a curve approaches progressively more closely but never reaches. The x and y axes are asymptotes to the graph of xy = constant (a rectangular ◊hyperbola).

If a point on a curve approaches a straight line such that its distance from the straight line is d, then the line is an asymptote to the curve if limit d tends to zero as the point moves towards infinity. Among ◊conic sections (curves obtained by the intersection of a plane and a double cone), a hyperbola has two asymptotes, which in the case of a rectangular hyperbola are at right angles to each other.

atavism in genetics, the reappearance of a characteristic not apparent in the immediately preceding generations; in psychology, the manifestation of primitive forms of behaviour.

ataxia loss of muscular coordination due to neurological damage or disease.

atheroma furring-up of the interior of an artery by deposits, mainly of cholesterol, within its walls.

Associated with atherosclerosis, atheroma has the effect of narrowing the lumen (channel) of the artery, thus restricting blood flow. This predisposes to a number of conditions, including thrombosis, angina, and stroke.

atherosclerosis thickening and hardening of the walls of the arteries, associated with ◊atheroma.

atlas book of maps. The atlas was introduced in the 16th century by Mercator, who began work on it in 1585; it was completed by his son in 1594.

Atlas rocket US rocket, originally designed and built as an intercontinental missile, but subsequently adapted for space use. Atlas rockets launched astronauts in the Mercury series into orbit, as well as numerous other satellites and space probes.

atmosphere mixture of gases surrounding a planet. The Earth's atmosphere is prevented from escaping by the pull of the Earth's gravity.

Atmospheric pressure decreases with height in the atmosphere. In its lowest layer, the atmosphere consists of nitrogen (78%) and oxygen (21%), both in molecular form (two atoms bonded together). The other 1% is largely argon, with very small quantities of other gases, including water vapour and carbon dioxide. The atmosphere plays a major part in the various cycles of nature (the ◊water cycle, ◊carbon cycle, and ◊nitrogen cycle). It is the principal industrial source of nitrogen, oxygen, and argon, which are obtained by fractional distillation of liquid air.

The lowest level of the atmosphere, the ◊troposphere, is heated by the Earth, which is warmed by infrared and visible radiation from the Sun. Warm air cools as it rises in the troposphere, causing rain and most other weather phenomena. However, infrared and visible radiations form only a part of the Sun's output of electromagnetic radiation. Almost all the shorter-wavelength ultraviolet radiation is filtered out by the upper layers of the atmosphere. The filtering process is an active one: at heights above about 50 km/31 mi ultraviolet photons collide with atoms, knocking out electrons to create a ◊plasma of electrons and positively charged ions. The resulting *ionosphere* acts as a reflector of radio waves, enabling radio transmissions to 'hop' between widely separated points on the Earth's surface.

Waves of different wavelengths are reflected best at different heights. The collisions between ultraviolet photons and atoms lead to a heating of the upper atmosphere, although the temperature drops from top to bottom within the zone called the *thermosphere* as high-energy photons are progressively absorbed in collisions. Between the thermosphere and the *tropopause* (at which the warming effect of the Earth starts to be felt) there is a 'warm bulge' in the graph of temperature against height, at a level called the *stratopause*. This is due to longer-wavelength ultraviolet photons that have survived their journey through the upper layers; now they encounter molecules and split them apart into atoms. These atoms eventually bond together again, but often in different combinations. In particular, many ◊ozone molecules (oxygen atom triplets, O_3) are formed. Ozone is a better absorber of ultraviolet than ordinary (two-atom) oxygen, and it is the *ozone layer* that prevents lethal amounts of ultraviolet from reaching the Earth's surface.

Atmosphere

If the Earth were the size of an apple, the atmosphere would be no thicker than the apple skin. Three-quarters of the atmosphere's mass lies below a height of 10 km/35,000 ft. The air at the top of Mount Everest is only one-third as thick as at sea level.

Far above the atmosphere, as so far described, lie the *Van Allen radiation belts*. These are regions in which high-energy charged particles travelling outwards from

Atmosphere: Composition

Gas	Symbol	Volume (%)	Role
nitrogen	N_2	78.08	cycled through human activities and through the action of microorganisms on animal and plant waste
oxygen	O_2	20.94	cycled mainly through the respiration of animals and plants and through the action of photosynthesis
carbon dioxide	CO_2	0.03	cycled through respiration and photosynthesis in exchange reactions with oxygen. It is also a product of burning fossil fuels
argon	Ar	0.093	chemically inert and with only a few industrial uses
neon	Ne	0.0018	as argon
helium	He	0.0005	as argon
krypton	Kr	trace	as argon
xenon	Xe	trace	as argon
ozone	O_3	0.00006	a product of oxygen molecules split into single atoms by the Sun's radiation and unaltered oxygen molecules
hydrogen	H_2	0.00005	unimportant

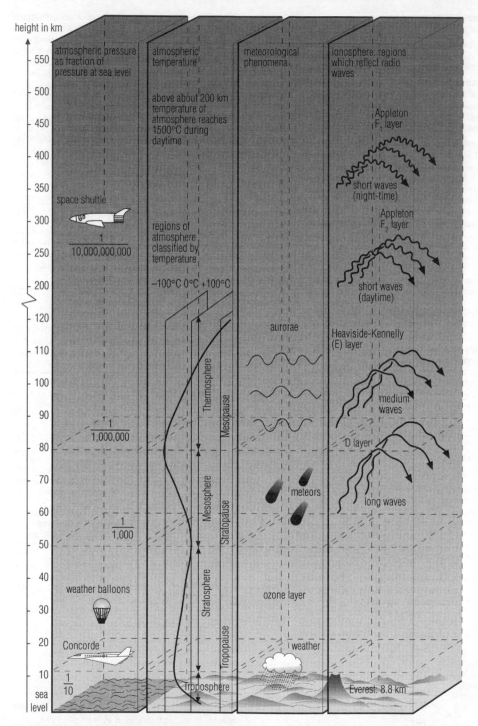

height in km

atmospheric pressure
as fraction of
pressure at sea level

atmospheric
temperature

meteorological
phenomena

ionosphere: regions
which reflect radio
waves

above about 200 km
temperature of
atmosphere reaches
1500°C during
daytime

space shuttle

$\dfrac{1}{10,000,000,000}$

regions of
atmosphere
classified by
temperature

−100°C 0°C +100°C

Appleton
F₁ layer

short waves
(night-time)

Appleton
F₂ layer

short waves
(daytime)

aurorae

Heaviside-Kennelly
(E) layer

Thermosphere

Mesopause

medium
waves

$\dfrac{1}{1,000,000}$

D layer

Mesosphere

Stratopause

meteors

long waves

$\dfrac{1}{1,000}$

Stratosphere

weather balloons

ozone layer

Tropopause

Concorde

weather

$\dfrac{1}{10}$

Troposphere

Everest: 8.8 km

sea
level

atmosphere All but 1% of the Earth's atmosphere lies in a layer 30 km/19 mi above the ground. At a height of
5,500 m/18,000 ft, air pressure is half that at sea level. The temperature of the atmosphere varies greatly with height; this
produces a series of layers, called the troposphere, stratosphere, mesosphere, and thermosphere.

the Sun (as the so-called solar wind) have been captured by the Earth's magnetic field. The outer belt (at about 1,600 km/1,000 mi) contains mainly protons, the inner belt (at about 2,000 km/1,250 mi) contains mainly electrons. Sometimes electrons spiral down towards the Earth, noticeably at polar latitudes, where the magnetic field is strongest. When such particles collide with atoms and ions in the thermosphere, light is emitted. This is the origin of the glows visible in the sky as the *aurora borealis* (northern lights) and the *aurora australis* (southern lights).

A fainter, more widespread, *airglow* is caused by a similar mechanism.

During periods of intense solar activity, the atmosphere swells outwards; there is a 10–20% variation in atmosphere density. One result is to increase drag on satellites. This effect makes it impossible to predict exactly the time of re-entry of satellites.

atmosphere or *standard atmosphere* in physics, a unit (symbol atm) of pressure equal to 760 torr, 1013.25 millibars, or 1.01325×10^5 newtons per square metre. The actual pressure exerted by the atmosphere fluctuates around this value, which is assumed to be standard at sea level and 0°C/32°F, and is used when dealing with very high pressures.

atmospheric pollution contamination of the atmosphere with the harmful by-products of human activity; see ◊air pollution.

atmospheric pressure the pressure at any point on the Earth's surface that is due to the weight of the column of air above it; it therefore decreases as altitude increases. At sea level the average pressure is 101 kilopascals (1,013 millibars, 760 mmHg, or 14.7 lb per sq in).

Changes in atmospheric pressure, measured with a barometer, are used in weather forecasting. Areas of relatively high pressure are called ◊anticyclones; areas of low pressure are called ◊depressions.

atoll continuous or broken circle of ◊coral reef and low coral islands surrounding a lagoon.

atom smallest unit of matter that can take part in a chemical reaction, and which cannot be broken down chemically into anything simpler. An atom is made up of protons and neutrons in a central nucleus surrounded by electrons (see ◊atomic structure). The atoms of the various elements differ in atomic number, relative atomic mass, and chemical behaviour. There are 112 different types of atom, corresponding with the 112 known elements as listed in the ◊periodic table of the elements.

Atoms are much too small to be seen even by the most powerful optical microscope (the largest, caesium, has a diameter of 0.0000005 mm/0.00000002 in), and they are in constant motion. However, modern electron microscopes, such as the ◊scanning tunnelling microscope (STM) and the ◊atomic force microscope (AFM), can produce images of individual atoms and molecules.

atomic clock timekeeping device regulated by various periodic processes occurring in atoms and molecules, such as atomic vibration or the frequency of absorbed or emitted radiation.

The first atomic clock was the *ammonia clock*, invented at the US National Bureau of Standards 1948. It was regulated by measuring the speed at which the nitrogen atom in an ammonia molecule vibrated back and forth. The rate of molecular vibration is not affected by temperature, pressure, or other external influences, and can be used to regulate an electronic clock.

A more accurate atomic clock is the *caesium clock*. Because of its internal structure, a caesium atom produces or absorbs radiation of a very precise frequency (9,192,631,770 Hz) that varies by less than one part in 10 billion. This frequency has been used to define the second, and is the basis of atomic clocks used in international timekeeping.

Hydrogen maser clocks, based on the radiation from hydrogen atoms, are the most accurate. The hydrogen maser clock at the US Naval Research Laboratory, Washington, DC, is estimated to lose one second in 1,700,000 years. Cooled hydrogen maser clocks could theoretically be accurate to within one second in 300 million years.

Atomic clocks are so accurate that minute adjustments must be made periodically to the length of the year to keep the calendar exactly synchronized with the Earth's rotation, which has a tendency to speed up or slow down. There have been 17 adjustments made since 1972. In 1992 the northern hemisphere's summer was longer than usual – by one second. An extra second was added to the world's time at precisely 23 hours, 59 minutes, and 60 seconds on 30 June 1992. The adjustment was called for by the International Earth Rotation Service in Paris, which monitors the difference between Earth time and atomic time.

atomic energy another name for ◊nuclear energy.

atomic force microscope (AFM) microscope developed in the late 1980s that produces a magnified image using a diamond probe, with a tip so fine that it may consist of a single atom, dragged over the surface of a specimen to 'feel' the contours of the surface. In effect, the tip acts like the stylus of a record player, reading the surface. The tiny up-and-down movements of the probe are converted to an image of the surface by computer and displayed on a screen. The AFM is useful for examination of biological specimens since, unlike the ◊scanning tunnelling microscope, the specimen does not have to be electrically conducting.

atomicity number of atoms of an ◊element that combine together to form a molecule. A molecule of oxygen (O_2) has atomicity 2; sulphur (S_8) has atomicity 8.

atomic mass see ◊relative atomic mass.

atomic mass unit or *dalton unit* (symbol amu or u) unit of mass that is used to measure the relative mass of atoms and molecules. It is equal to one-twelfth of the mass of a carbon-12 atom, which is equivalent to the mass of a

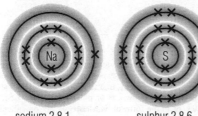

sodium 2.8.1   sulphur 2.8.6

atom

Smaller than the atom?

What is an atom? Until about 100 years ago, scientists were pretty confident in regarding atoms as the permanent bricks of which the whole universe was built. All the changes of the universe amounted to nothing more drastic than simple rearrangements of permanent, indestructible atoms. Atoms seemed like the bricks in a child's box of toys, which could be used to build many different buildings in turn.

Chipped bricks ...

This comfortable picture was changed by the investigations of British physicist Joseph John Thomson (1856–1940) into cathode rays. Thomson showed conclusively that not only could atomic 'bricks' be chipped, but that the fragments produced were identical, no matter what atom they came from. They were of equal weight or mass, and carried the same negative electrical charge. The fragments were called electrons.

Thomson's experiments enabled him to calculate the ratio of the mass of the electron to its charge: it was about 1,000 times smaller than the value that had already been calculated for a hydrogen ion in the electrolysis of liquids. This was the proof, announced in April 1897, that electrons were fundamental particles of matter, far smaller than any atom. But atoms could not consist of negatively charged electrons alone. Particles with like electrical charge repel one another and, anyway, atoms are electrically neutral.

Thomson, who built up the Cavendish Laboratory at Cambridge into a great research school, was succeeded as professor by New Zealand physicist Ernest Rutherford (1871–1937), his pupil and one of the greatest pioneers of subatomic physics. Rutherford, while professor of physics at Manchester University, had shown that α-particles were doubly ionized helium atoms, using a Geiger–Müller counter. His assistant, German physicist Hans Wilhelm Geiger (1882–1945), with a colleague Walther Müller, invented this instrument, which is still used to measure ionizing radiation.

... and indestructible tissue paper?

Rutherford's idea for attacking the problem of the nature of an atom was to study the scattering of α-particles passing through thin metal foils. This painstaking work was carried out by Geiger in 1910 with his colleague Marsden, with astonishing results. They found that most of the particles were only slightly deflected as they passed through the foil. But a very small proportion, about 1 in 8,000, were widely deflected. Rutherford described this result as ' ... quite the most incredible event that has ever happened to me in my life ... It was almost as if you fired a 15 inch shell at a piece of tissue paper and it came back and hit you.'

The nuclear atom

Rutherford concluded when he published the results in 1911 that almost all the mass of the atom was concentrated in a very small region and that most of the atom was 'empty space'. This crucial experiment established the idea of the nuclear atom. The obstacle that deflected the α-particles could only be the missing positive charges of the atom. These were carried by the minute central nucleus. The electrons, Rutherford supposed, must be in motion around the nucleus, otherwise they would be drawn to it.

The number of positive charges in the nucleus equalled the number of electrons, and so accounted for the electrical neutrality of the atom. The proton, later discovered as part of the nucleus, carried an equal but opposite charge to the electron. But on theoretical grounds, Rutherfords's planetary model of the atom was unstable. Since electrons are charged particles, an atom ought to radiate energy by virtue of their motion. Danish theoretical physicist Niels Bohr (1885–1962), who was Rutherford's pupil, provided the theory that accounted for the stability of the atom. He showed that electrons must reside in 'stationary' orbits, in which they did not radiate energy.

Julian Rowe

---

proton or 1.66×10^{-27} kg. The ◊ relative atomic mass of an atom has no units; thus oxygen-16 has an atomic mass of 16 daltons but a relative atomic mass of 16.

atomic number or ***proton number*** the number (symbol Z) of protons in the nucleus of an atom. It is equal to the positive charge on the nucleus.

In a neutral atom, it is also equal to the number of electrons surrounding the nucleus. The 112 elements are arranged in the ◊ periodic table of the elements

according to their atomic number. See also ◊ nuclear notation.

atomic radiation energy given out by disintegrating atoms during ◊ radioactive decay, whether natural or synthesized. The energy may be in the form of fast-moving particles, known as ◊ alpha particles and ◊ beta particles, or in the form of high-energy electromagnetic waves known as ◊ gamma radiation. Overlong exposure to atomic radiation can lead to ◊ radiation sickness.

Radiation biology studies the effect of radiation on living organisms. Exposure to atomic radiation is linked to chromosomal damage, cancer, and, in laboratory animals at least, hereditary disease.

atomic size or **atomic radius** size of an atom expressed as the radius in ◊angstroms or other units of length.

The sodium atom has an atomic radius of 1.57 angstroms (1.57×10^{-8} cm). For metals, the size of the atom is always greater than the size of its ion. For non-metals the reverse is true.

atomic structure internal structure of an ◊atom.

The core of the atom is the **nucleus**, a dense body only one ten-thousandth the diameter of the atom itself. The simplest nucleus, that of hydrogen, comprises a single stable positively charged particle, the **proton**. Nuclei of other elements contain more protons and additional particles, called **neutrons**, of about the same mass as the proton but with no electrical charge. Each element has its own characteristic nucleus with a unique number of protons, the atomic number. The number of neutrons may vary. Where atoms of a single element have different numbers of neutrons, they are called ◊isotopes. Although some isotopes tend to be unstable and exhibit ◊radioactivity, they all have identical chemical properties.

The nucleus is surrounded by a number of moving **electrons**, each of which has a negative charge equal to the positive charge on a proton, but which weighs only 1/1,839 times as much. In a neutral atom, the nucleus is surrounded by the same number of electrons as it contains protons. According to ◊quantum theory, the position of an electron is uncertain; it may be found at any point. However, it is more likely to be found in some places than others. The region of space in which an electron is most likely to be found is called an orbital (see ◊orbital, atomic). The chemical properties of an element are determined by the ease with which its atoms can gain or lose electrons from its outer orbitals.

Atoms are held together by the electrical forces of attraction between each negative electron and the positive protons within the nucleus. The latter repel one another with enormous forces; a nucleus holds together only because an even stronger force, called the **strong nuclear force**, attracts the protons and neutrons to one another. The strong force acts over a very short range – the protons and neutrons must be in virtual contact with one another (see ◊forces, fundamental). If, therefore, a fragment of a complex nucleus, containing some protons, becomes only slightly loosened from the main group of neutrons and protons, the natural repulsion between the protons will cause this fragment to fly apart from the rest of the nucleus at high speed. It is by such fragmentation of atomic nuclei (nuclear ◊fission) that nuclear energy is released.

atomic time time as given by ◊atomic clocks, which are regulated by natural resonance frequencies of particular atoms, and display a continuous count of seconds.

In 1967 a new definition of the second was adopted in the SI system of units: the duration of 9,192,631,770 periods of the radiation corresponding to the transition between two hyperfine levels of the ground state of the caesium-133 atom. The International Atomic Time Scale is based on clock data from a number of countries; it is a continuous scale in days, hours, minutes, and seconds from the origin on 1 Jan 1958, when the Atomic Time

Scale was made 0 h 0 min 0 sec when Greenwich Mean Time was at 0 h 0 min 0 sec.

atomic weight another name for ◊relative atomic mass.

atomizer device that produces a spray of fine droplets of liquid. A vertical tube connected with a horizontal tube dips into a bottle of liquid, and at one end of the horizontal tube is a nozzle, at the other a rubber bulb. When the bulb is squeezed, air rushes over the top of the vertical tube and out through the nozzle. Following ◊Bernoulli's principle, the pressure at the top of the vertical tube is reduced, allowing the liquid to rise. The air stream picks up the liquid, breaks it up into tiny drops, and carries it out of the nozzle as a spray. Scent spray, paint spray guns, and carburettors all use the principle of the atomizer.

ATP abbreviation for **adenosine triphosphate**, a nucleotide molecule found in all cells. It can yield large amounts of energy, and is used to drive the thousands of biological processes needed to sustain life, growth, movement, and reproduction. Green plants use light energy to manufacture ATP as part of the process of ◊photosynthesis. In animals, ATP is formed by the breakdown of glucose molecules, usually obtained from the carbohydrate component of a diet, in a series of reactions termed ◊respiration. It is the driving force behind muscle contraction and the synthesis of complex molecules needed by individual cells.

atrium either of the two upper chambers of the heart. The left atrium receives freshly oxygenated blood from the lungs via the pulmonary vein; the right atrium receives deoxygenated blood from the ◊vena cava. Atrium walls are thin and stretch easily to allow blood into the heart. On contraction, the atria force blood into the thick-walled ventricles, which then give a second, more powerful beat.

atrophy in medicine, a diminution in size and function, or output, of a body tissue or organ. It is usually due to nutritional impairment, disease, or disuse (muscle).

atropine alkaloid derived from ◊belladonna, a plant with toxic properties. It acts as an anticholinergic, inhibiting the passage of certain nerve impulses. It is used in premedication, to reduce bronchial and gastric secretions. It is also administered as a mild antispasmodic drug, and to dilate the pupil of the eye.

attar of roses perfume derived from the essential oil of roses (usually damask roses), obtained by crushing and distilling the petals of the flowers.

attention-deficit hyperactivity disorder (ADHD) psychiatric condition occurring in young children characterized by impaired attention and hyperactivity. The disorder, associated with disruptive behaviour, learning difficulties, and under-achievement, is more common in boys. It is treated with methylphenidate (Ritalin). There was a 50% increase in the use of the drug in the USA 1994–1996, with an estimated 5% of school-age boys diagnosed as suffering from ADHD.

In 1996, US researchers found that 50% of children diagnosed as ADHD sufferers carry a gene that affects brain cell response to the neurotransmitter dopamine. The same gene has also been linked to impulsiveness in adults. Diagnosis requires the presence, for at least six months, of eight behavioural problems, first developing before the age of seven. In addition to their hyperactivity, such children are found to be reckless, impulsive, and accident prone; they are often aggressive and tend to be unpopular with other children. The outlook for ADHD sufferers varies, with up to a quarter being diagnosed with antisocial personality disorder as adults.

attrition in earth science, the process by which particles of rock being transported by river, wind, or sea are rounded and gradually reduced in size by being struck against one another.

The rounding of particles is a good indication of how far they have been transported. This is particularly true for particles carried by rivers, which become more rounded as the distance downstream increases.

audiometer electrical apparatus used to test hearing.

auditory canal tube leading from the outer ◊ear opening to the eardrum. It is found only in animals whose eardrums are located inside the skull, principally mammals and birds.

aura diagnosis in alternative medicine, ascertaining a person's state of health from the colour and luminosity of the aura, the 'energy envelope' of the physical body commonly claimed to be seen by psychics. A study carried out by the Charing Cross Hospital Medical School (London) confirmed that the aura can be viewed by high frequency electrophotography techniques and is broadly indicative of states of health, but concluded that aura diagnosis cannot identify specific abnormalities.

Auriga constellation of the northern hemisphere, represented as a charioteer. Its brightest star is the first-magnitude ◊Capella, about 45 light years from Earth; Epsilon Aurigae is an ◊eclipsing binary star with a period of 27 years, the longest of its kind (last eclipse was in 1983).

aurora coloured light in the night sky near the Earth's magnetic poles, called **aurora borealis** ('northern lights') in the northern hemisphere and **aurora australis** in the southern hemisphere. Although aurorae are usually restricted to the polar skies, fluctuations in the ◊solar wind occasionally cause them to be visible at lower latitudes. An aurora is usually in the form of a luminous arch with its apex towards the magnetic pole followed by arcs, bands, rays, curtains, and coronas, usually green but often showing shades of blue and red, and sometimes yellow or white. Aurorae are caused at heights of over 100 km/60 mi by a fast stream of charged particles from solar flares and low-density 'holes' in the Sun's corona.

These are guided by the Earth's magnetic field towards the north and south magnetic poles, where they enter the upper atmosphere and bombard the gases in the atmosphere, causing them to emit visible light.

auscultation evaluation of internal organs by listening, usually with the aid of a stethoscope.

Australia Telescope giant radio telescope in New South Wales, Australia, operated by the Commonwealth Scientific and Industrial Research Organization (CSIRO). It consists of six 22 m/72 ft antennae at Culgoora, a similar antenna at Siding Spring Mountain, and the 64 m/210 ft ◊Parkes radio telescope – the whole simulating a dish 300 m/186 mi across.

autism, infantile rare disorder, generally present from birth, characterized by a withdrawn state and a failure to develop normally in language or social behaviour. Although the autistic child may, rarely, show signs of high intelligence (in music or with numbers, for example), many have impaired intellect. The cause is unknown, but is thought to involve a number of factors, possibly including an inherent abnormality of the child's brain. Special education may bring about some improvement.

Autism was initially defined by four common traits – preference for aloneness; insistence on sameness; need for elaborate routines; and the possession of some abilities that seem exceptional compared with other deficits – but current clinical diagnosis involves wider criteria.

autochrome in photography, a single-plate additive colour process devised by the Lumière brothers in 1903. It was the first commercially available process, in use 1907–35.

autoclave pressurized vessel that uses superheated steam to sterilize materials and equipment such as surgical instruments. It is similar in principle to a pressure cooker.

autogenics in alternative medicine, system developed in the 1900s by German physician Johannes Schultz, designed to facilitate mental control of biological and physiological functions generally considered to be involuntary. Effective in inducing relaxation, assisting healing processes and relieving psychosomatic disorders, autogenics is regarded as a precursor of biofeedback.

autogenous in medicine, self-generated by the body. For example, autogenous blood transfusion involves the collection of blood from a patient, prior to undergoing surgery, for use during and after the operation. It is a valuable procedure for operations that may require large transfusions or where a person has a rare blood group.

autogiro or **autogyro** heavier-than-air craft that supports itself in the air with a rotary wing, or rotor. The Spanish aviator Juan de la Cierva designed the first successful autogiro in 1923. The autogiro's rotor provides only lift and not propulsion; it has been superseded by the helicopter, in which the rotor provides both. The autogiro is propelled by an orthodox propeller.

The three- or four-bladed rotor on an autogiro spins in a horizontal plane on top of the craft, and is not driven by the engine. The blades have an aerofoil cross section, as a plane's wings. When the autogiro moves forward, the rotor starts to rotate by itself, a state known as autorotation. When travelling fast enough, the rotor develops enough lift from its aerofoil blades to support the craft.

autoimmunity in medicine, condition where the body's immune responses are mobilized not against 'foreign' matter, such as invading germs, but against the body itself. Diseases considered to be of autoimmune origin include ◊myasthenia gravis, ◊rheumatoid arthritis, and ◊lupus erythematous.

In autoimmune diseases T-lymphocytes reproduce to excess to home in on a target (properly a foreign disease-causing molecule); however, molecules of the body's own tissue that resemble the target may also be attacked, for example insulin-producing cells, resulting in insulin-dependent diabetes; if certain joint membrane cells are attacked, then rheumatoid arthritis may result; and if myelin, the basic protein of the nervous system, then multiple sclerosis results. In 1990 in Israel a T-cell vaccine was produced that arrests the excessive reproduction of T-lymphocytes attacking healthy target tissues.

autolysis in biology, the destruction of a ◊cell after its death by the action of its own ◊enzymes, which break down its structural molecules.

automatic pilot or **autopilot** control device that keeps an aeroplane flying automatically on a given course at a given height and speed.

Devised by US business executive Lawrence Sperry in 1912, the automatic pilot contains a set of ◊gyroscopes that provide references for the plane's course. Sensors detect when the plane deviates from this course and send signals to the control surfaces – the ailerons, elevators, and rudder – to take the appropriate action. Autopilot is also used in missiles. Most airliners cruise on automatic pilot for much of the time.

automation widespread use of self-regulating machines in industry. Automation involves the addition of control devices, using electronic sensing and computing techniques, which often follow the pattern of human nervous and brain functions, to already mechanized physical processes of production and distribution; for example, steel processing, mining, chemical production, and road, rail, and air control.

automatism performance of actions without awareness or conscious intent. It is seen in sleepwalking and in some (relatively rare) psychotic states.

automaton mechanical figure imitating human or animal performance. Automatons are usually designed for aesthetic appeal as opposed to purely functional robots. The earliest recorded automaton is an Egyptian wooden pigeon of 400 BC.

autonomic nervous system in mammals, the part of the nervous system that controls those functions not controlled voluntarily, including the heart rate, activity of the intestines, and the production of sweat.

There are two divisions of the autonomic nervous system. The **sympathetic** system responds to stress, when it speeds the heart rate, increases blood pressure, and generally prepares the body for action. The **parasympathetic** system is more important when the body is at rest, since it slows the heart rate, decreases blood pressure, and stimulates the digestive system.

At all times, both types of autonomic nerves carry signals that bring about adjustments in visceral organs. The actual rate of heartbeat is the net outcome of opposing signals.

Today, it is known that the word 'autonomic' is misleading – the reflexes managed by this system are actually integrated by commands from the brain and spinal cord (the central nervous system).

autopsy or **postmortem** examination of the internal organs and tissues of a dead body, performed to try to establish the cause of death.

autoradiography in biology, a technique for following the movement of molecules within an organism, especially a plant, by labelling with a radioactive isotope that can be traced on photographs. It is used to study ◊photosynthesis, where the pathway of radioactive carbon dioxide can be traced as it moves through the various chemical stages.

autosome any ◊chromosome in the cell other than a sex chromosome. Autosomes are of the same number and kind in both males and females of a given species.

autosuggestion conscious or unconscious acceptance of an idea as true, without demanding rational proof, but with potential subsequent effect for good or ill. Pioneered by French psychotherapist Emile Coué (1857–1926) in healing, it is sometimes used in modern psychotherapy to conquer nervous habits and dependence on addictive substances such as tobacco and alcohol.

autotroph any living organism that synthesizes organic substances from inorganic molecules by using light or chemical energy. Autotrophs are the **primary producers** in all food chains since the materials they synthesize and store are the energy sources of all other organisms.

All green plants and many planktonic organisms are autotrophs, using sunlight to convert carbon dioxide and water into sugars by ◊photosynthesis.

The total ◊biomass of autotrophs is far greater than that of animals, reflecting the dependence of animals on plants, and the ultimate dependence of all life on energy from the Sun – green plants convert light energy into a form of chemical energy (food) that animals can exploit. Some bacteria use the chemical energy of sulphur compounds to synthesize organic substances. It is estimated that 10% of the energy in autotrophs can pass into the next stage of the ◊food chain, the rest being lost as heat or indigestible matter. See also ◊heterotroph.

autumnal equinox see ◊equinox.

auxin plant ◊hormone that promotes stem and root growth in plants. Auxins influence many aspects of plant growth and development, including cell enlargement, inhibition of development of axillary buds, ◊tropisms, and the initiation of roots.

Synthetic auxins are used in rooting powders for cuttings, and in some weedkillers, where high auxin concentrations cause such rapid growth that the plants die. They are also used to prevent premature fruitdrop in orchards.

The most common naturally occurring auxin is known as indoleacetic acid, or IAA. It is produced in the shoot apex and transported to other parts of the plant.

avalanche fall or flow of a mass of snow and ice down a steep slope under the force of gravity. Avalanches occur because of the unstable nature of snow masses in mountain areas.

Changes of temperature, sudden sound, or earth-borne vibrations may trigger an avalanche, particularly on slopes of more than 35°. The snow compacts into ice as it moves, and rocks may be carried along, adding to the damage caused.

Avebury Europe's largest stone circle (diameter 412 m/ 1,350 ft), in Wiltshire, England. This megalithic henge monument, probably a ritual complex, contains 650 massive blocks of stone, arranged in circles and avenues. It was probably constructed around 3,500 years ago, and is linked with nearby Silbury Hill.

The henge, an earthen bank and interior ditch with opposed entrances, originally rose 15 m/49 ft above the bottom of the ditch. This earthwork and an outer ring of stones surround the inner circles. The stones vary in size from 1.5 m/5 ft to 5.5 m/18 ft high and 1 m/3 ft to 3.65 m/ 12 ft broad. They were erected by a late Neolithic or early Bronze Age culture. Visible remains seen today may cover an earlier existing site – a theory applicable to a number of prehistoric sites.

average in statistics, a term used inexactly to indicate the typical member of a set of data. It usually refers to the ◊arithmetic mean. The term is also used to refer to the middle member of the set when it is sorted in ascending or descending order (the ◊median), and the most commonly occurring item of data (the ◊mode), as in 'the average family'.

aviation term used to describe both the science of powered ◊flight and also aerial navigation by means of an aeroplane.

Avogadro's hypothesis in chemistry, the law stating that equal volumes of all gases, when at the same temperature and pressure, have the same numbers of molecules. It was first propounded by Amedeo Avogadro.

Avogadro's number or **Avogadro's constant** the number of carbon atoms in 12 g of the carbon-12 isotope (6.022045×10^{23}). The relative atomic mass of any element, expressed in grams, contains this number of atoms. It is named after Amedeo Avogadro.

avoirdupois system of units of mass based on the pound (0.45 kg), which consists of 16 ounces (each of 16 drams) or 7,000 grains (each equal to 65 mg).

axil upper angle between a leaf (or bract) and the stem from which it grows. Organs developing in the axil, such as shoots and buds, are termed axillary, or lateral.

axiom in mathematics, a statement that is assumed to be true and upon which theorems are proved by using logical deduction; for example, two straight lines cannot enclose a space. The Greek mathematician Euclid used a series of axioms that he considered could not be demonstrated in terms of simpler concepts to prove his geometrical theorems.

axis (plural **axes**) in geometry, one of the reference lines by which a point on a graph may be located. The horizontal axis is usually referred to as the x-axis, and the vertical axis as the y-axis. The term is also used to refer to the imaginary line about which an object may be said to be symmetrical (**axis of symmetry**) – for example, the diagonal of a square – or the line about which an object may revolve (**axis of rotation**).

axon long threadlike extension of a ◊nerve cell that conducts electrochemical impulses away from the cell body towards other nerve cells, or towards an effector organ such as a muscle. Axons terminate in ◊synapses, junctions with other nerve cells, muscles, or glands.

Ayurveda basically naturopathic system of medicine widely practised in India and based on principles derived from the ancient Hindu scriptures, the Vedas. Hospital treatments and remedial prescriptions tend to be nonspecific and to coordinate holistic therapies for body, mind, and spirit.

Azilian archaeological period following the close of the Old Stone (Palaeolithic) Age and regarded as the earliest culture of the Mesolithic Age in W Europe. It was first recognized at Le Mas d'Azil, a cave in Ariège, France.

azimuth in astronomy, the angular distance of an object eastwards along the horizon, measured from due north, between the astronomical ◊meridian (the vertical circle passing through the centre of the sky and the north and south points on the horizon) and the vertical circle containing the celestial body whose position is to be measured.

azo dye synthetic dye containing the azo group of two nitrogen atoms (N=N) connecting aromatic ring compounds. Azo dyes are usually red, brown, or yellow, and make up about half the dyes produced. They are manufactured from aromatic ◊amines.

AZT drug used in the treatment of AIDS; see ◊zidovudine.

Babbit metal soft, white metal, an ◊alloy of tin, lead, copper, and antimony, used to reduce friction in bearings, developed by the US inventor Isaac Babbit in 1839.

Bach flower healing a homoeopathic system of medical therapy developed in the 1920s by English physician Edward Bach. Based on the healing properties of wild flowers, it seeks to alleviate mental and emotional causes of disease rather than their physical symptoms.

bacille Calmette-Guérin tuberculosis vaccine ◊BCG.

bacillus member of a group of rodlike ◊bacteria that occur everywhere in the soil and air. Some are responsible for diseases such as ◊anthrax or for causing food spoilage.

backcross breeding technique used to determine the genetic makeup of an individual organism.

background radiation radiation that is always present in the environment. By far the greater proportion (87%) of it is emitted from natural sources. Alpha and beta particles, and gamma radiation are radiated by the traces of radioactive minerals that occur naturally in the environment and even in the human body, and by radioactive gases such as radon and thoron, which are found in soil and may seep upwards into buildings. Radiation from space (◊cosmic radiation) also contributes to the background level.

backing storage in computing, memory outside the ◊central processing unit used to store programs and data that are not in current use. Backing storage must be nonvolatile – that is, its contents must not be lost when the power supply to the computer system is disconnected.

back pain aches in the region of the spine. Low back pain can be caused by a very wide range of medical conditions. About half of all episodes of back pain will resolve within a week, but severe back pain can be chronic and disabling. The causes include muscle sprain, a prolapsed intervertebral disc, and vertebral collapse due to ◊osteoporosis or cancer. Treatment methods include rest, analgesics, physiotherapy, osteopathy, and exercises.

backup system in computing, a duplicate computer system that can take over the operation of a main computer system in the event of equipment failure. A large interactive system, such as an airline's ticket-booking system, cannot be out of action for even a few hours without causing considerable disruption. In such cases a complete duplicate computer system may be provided to take over and run the system should the main computer develop a fault or need maintenance.

Backup systems include *incremental backup* and *full backup*.

bacteria (singular *bacterium*) microscopic single-celled organisms lacking a nucleus. Bacteria are widespread, present in soil, air, and water, and as parasites on and in other living things. Some parasitic bacteria cause disease by producing toxins, but others are harmless and may even benefit their hosts. Bacteria usually reproduce by ◊binary fission (dividing into two equal parts), and this may occur approximately every 20 minutes. It is thought that 1–10% of the world's bacteria have been identified.

Bacteria are now classified biochemically, but their varying shapes provide a rough classification; for example, *cocci* are round or oval, *bacilli* are rodlike, *spirilla* are spiral, and *vibrios* are shaped like commas. Exceptionally, one bacterium has been found, *Gemmata obscuriglobus*, that does have a nucleus. Unlike ◊viruses, bacteria do not necessarily need contact with a live cell to become active.

Bacteria can be classified into two broad classes (called Gram positive and negative) according to their reactions to certain stains, or dyes, used in microscopy. The staining technique, called the Gram test after Danish

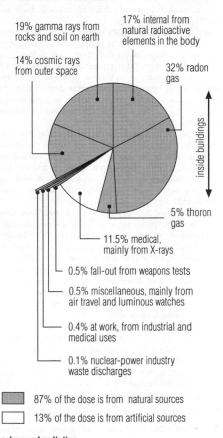

19% gamma rays from rocks and soil on earth

17% internal from natural radioactive elements in the body

14% cosmic rays from outer space

32% radon gas

inside buildings

5% thoron gas

11.5% medical, mainly from X-rays

0.5% fall-out from weapons tests

0.5% miscellaneous, mainly from air travel and luminous watches

0.4% at work, from industrial and medical uses

0.1% nuclear-power industry waste discharges

87% of the dose is from natural sources

13% of the dose is from artificial sources

background radiation

bacteriologist Hans Gram, allows doctors to identify many bacteria quickly.

Bacteria have a large loop of ◊DNA, sometimes called a bacterial chromosome. In addition there are often small, circular pieces of DNA known as ◊plasmids that carry spare genetic information. These plasmids can readily move from one bacterium to another, even though the bacteria may be of different species. In a sense, they are parasites within the bacterial cell, but they survive by coding characteristics that promote the survival of their hosts. For example, some plasmids confer antibiotic resistance on the bacteria they inhabit. The rapid and problematic spread of antibiotic resistance among bacteria is due to plasmids, but they are also useful to humans in ◊genetic engineering. There are ten times more bacterial cells than human cells in the human body.

Certain types of bacteria are vital in many food and industrial processes, while others play an essential role in the ◊nitrogen cycle, which maintains soil fertility. For example, bacteria are used to break down waste products, such as sewage; make butter, cheese, and yoghurt; cure tobacco; tan leather; and (by virtue of the ability of certain bacteria to attack metal) clean ships' hulls and derust their tanks, and even extract minerals from mines. In 1995 a US veterinary toxicologist identified several species of bacteria in the stomach of a bowhead whale capable of digesting pollutants (naphthalene and anthracene, two carcinogenic fractions of oil difficult to break down, and PCBs, also carcinogenic).

Bacteria cannot normally survive temperatures above 100°C/212°F, such as those produced in pasteurization, but those in deep-sea hot vents in the eastern Pacific are believed to withstand temperatures of 350°C/662°F. *Thermus aquaticus*, or taq, grows freely in the boiling waters of hot springs, and an enzyme derived from it is used in genetic engineering to speed up the production of millions of copies of any DNA sequence, a reaction requiring very high temperatures.

Certain bacteria can influence the growth of others; for example, lactic acid bacteria will make conditions unfavourable for salmonella bacteria. Other strains produce nisin, which inhibits growth of listeria and botulism organisms. Plans in the food industry are underway to produce super strains of lactic acid bacteria to avoid food poisoning.

An estimated 99% of bacteria live in *biofilms* rather than in single-species colonies. These are complex colonies made up of a number of different species of bacteria structured on a layer of slime produced by the bacteria. Fungi, algae, and protozoa may also inhabit the biofilms.

In 1990 a British team of food scientists announced a new, rapid (five-minute) test for contamination of food by listeria or salmonella bacteria. Fluorescent dyes, added to a liquidized sample of food, reveal the presence of bacteria under laser light.

Bacterial spores 40 million years old were extracted from a fossilized bee and successfully germinated by US scientists in 1995. It is hoped that prehistoric bacteria can be tapped as a source of new chemicals for use in the drugs industry. Any bacteria resembling extant harmful pathogens will be destroyed, and all efforts are being to made to ensure no bacteria escape the laboratory.

bacteriology the study of ◊bacteria.

bacteriophage virus that attacks ◊bacteria. Such viruses are now of use in genetic engineering.

badlands barren landscape cut by erosion into a maze of ravines, pinnacles, gullies and sharp-edged ridges. Areas in South Dakota and Nebraska, USA, are examples.

Baikonur launch site for spacecraft, located at Tyuratam, Kazakhstan, near the Aral Sea: the first satellites and all Soviet space probes and crewed Soyuz missions were launched from here. It covers an area of 12,200 sq km/4,675 sq mi, much larger than its US equivalent, the ◊Kennedy Space Center in Florida.

Baily's beads bright spots of sunlight seen around the edge of the Moon for a few seconds immediately before and after a total ◊eclipse of the Sun, caused by sunlight shining between mountains at the Moon's edge. Sometimes one bead is much brighter than the others, producing the so-called *diamond ring* effect. The effect was described 1836 by the English astronomer Francis Baily (1774–1844), a wealthy stockbroker who retired in 1825 to devote himself to astronomy.

Bakelite first synthetic ◊plastic, created by Leo Baekeland in 1909. Bakelite is hard, tough, and heatproof, and is used as an electrical insulator. It is made by the reaction of phenol with formaldehyde, producing a powdery resin that sets solid when heated. Objects are made by subjecting the resin to compression moulding (simultaneous heat and pressure in a mould).

It is one of the thermosetting plastics, which do not remelt when heated, and is often used for electrical fittings.

baking powder mixture of ◊bicarbonate of soda, an acidic compound, and a nonreactive filler (usually starch or calcium sulphate), used in baking as a raising agent. It gives a light open texture to cakes and scones, and is used as a substitute for yeast in making soda bread.

Several different acidic compounds (for example, tartaric acid, cream of tartar, sodium or calcium acid phosphates, and glucono–delta–lactone) may be used, any of which will react with the sodium hydrogencarbonate, in the presence of water and heat, to release the carbon dioxide that causes the cake mix or dough to rise.

balance apparatus for weighing or measuring mass. The various types include the *beam balance*, consisting of a centrally pivoted lever with pans hanging from each end, and the *spring balance*, in which the object to be weighed stretches (or compresses) a vertical coil spring fitted with a pointer that indicates the weight on a scale. Kitchen and bathroom scales are balances.

balance of nature in ecology, the idea that there is an inherent equilibrium in most ◊ecosystems, with plants and animals interacting so as to produce a stable, continuing system of life on Earth. The activities of human beings can, and frequently do, disrupt the balance of nature.

Organisms in the ecosystem are adapted to each other – for example, waste products produced by one species are used by another and resources used by some are replenished by others; the oxygen needed by animals is produced by plants while the waste product of animal respiration, carbon dioxide, is used by plants as a raw material in photosynthesis. The nitrogen cycle, the water cycle, and the control of animal populations by natural predators are other examples.

baldness loss of hair from the scalp, common in older men. Its onset and extent are influenced by genetic make-up and the level of male sex ◊hormones. There is no cure, and expedients such as hair implants may have no lasting effect. Hair loss in both sexes may also occur as a result of ill health or radiation treatment, such as for cancer. *Alopecia*, a condition in which the hair falls out, is different from the 'male-pattern baldness' described above.

Baldness

The best way for a man to know whether he will go bald is to look at his mother's father. Baldness is hereditary, but the gene controlling it is on the sex-linked chromosome and skips one generation.

ball-and-socket joint joint allowing considerable movement in three dimensions, for instance the joint between the pelvis and the femur. To facilitate movement, such joints are rimmed with cartilage and lubricated by synovial fluid. The bones are kept in place by ligaments and moved by muscles.

ballistics study of the motion and impact of projectiles such as bullets, bombs, and missiles. For projectiles from a gun, relevant exterior factors include temperature, barometric pressure, and wind strength; and for nuclear missiles these extend to such factors as the speed at which the Earth turns.

balloon lighter-than-air craft that consists of a gasbag filled with gas lighter than the surrounding air and an attached basket, or gondola, for carrying passengers and/or instruments. In 1783, the first successful human ascent was in Paris, in a hot-air balloon designed by the Montgolfier brothers Joseph Michel and Jacques Etienne. In 1785, a hydrogen-filled balloon designed by French physicist Jacques Charles travelled across the English Channel.

ball valve valve that works by the action of external pressure raising a ball and thereby opening a hole.

bar unit of pressure equal to 10^5 pascals or 10^6 dynes/cm^2, approximately 750 mmHg or 0.987 atm. Its diminutive, the *millibar* (one-thousandth of a bar), is commonly used by meteorologists.

barbiturate hypnosedative drug, commonly known as a 'sleeping pill', consisting of any salt or ester of barbituric acid $C_4H_4O_3N_2$. It works by depressing brain activity. Most barbiturates, being highly addictive, are no longer prescribed and are listed as controlled substances.

Tolerance develops quickly in the user so that increasingly large doses are required to induce sleep. A barbiturate's action persists for hours or days, causing confused, aggressive behaviour or disorientation. Overdosage causes death by inhibiting the breathing centre in the brain. Short-acting barbiturates are used as ◊anaesthetics to induce general anaesthesia; slow-acting ones may be prescribed for epilepsy.

bar code pattern of bars and spaces that can be read by a computer. Bar codes are widely used in retailing, industrial distribution, and public libraries. The code is read by a scanning device; the computer determines the code from the widths of the bars and spaces.

The technique was patented in 1949 but became popular only in 1973, when the food industry in North America adopted the Universal Product Code system.

barium soft, silver-white, metallic element, symbol Ba,

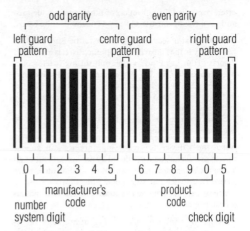

bar code The bars of varying thicknesses and spacings represent two series of numbers, identifying the manufacturer and the product. Two longer, thinner bars mark the beginning and end of the manufacturer and product codes. The bar code is used on most articles for sale in shops.

atomic number 56, relative atomic mass 137.33. It is one of the alkaline-earth metals, found in nature as barium carbonate and barium sulphate. As the sulphate it is used in medicine: taken as a suspension (a 'barium meal'), its movement along the gut is followed using X-rays. The barium sulphate, which is opaque to X-rays, shows the shape of the gut, revealing any abnormalities of the alimentary canal. Barium is also used in alloys, pigments, and safety matches and, with strontium, forms the emissive surface in cathode-ray tubes. It was first discovered in barytes or heavy spar.

bark protective outer layer on the stems and roots of woody plants, composed mainly of dead cells. To allow for expansion of the stem, the bark is continually added to from within, and the outer surface often becomes cracked or is shed as scales. Trees deposit a variety of chemicals in their bark, including poisons. Many of these chemical substances have economic value because they can be used in the manufacture of drugs. Quinine, derived from the bark of the *Cinchona* tree, is used to fight malarial infections; curare, an anaesthetic used in medicine, comes from the *Strychnus toxifera* tree in the Amazonian rainforest.

Bark technically includes all the tissues external to the vascular ◊cambium (the ◊phloem, cortex, and periderm), and its thickness may vary from 2.5 mm/0.1 in to 30 cm/12 in or more, as in the giant redwood *Sequoia* where it forms a thick, spongy layer.

barn farm building traditionally used for the storage and processing of cereal crops and hay. On older farmsteads, the barn is usually the largest building. It is often characterized by ventilation openings rather than windows and has at least one set of big double doors for access. Before mechanization, wheat was threshed by hand on a specially prepared floor inside these doors.

Tithe barns were used in feudal England to store the produce paid as a tax to the parish priest by the local occupants of the land. In the Middle Ages, monasteries often controlled the collection of tithes over a wide area and, as a result, constructed some enormous tithe barns.

Barnard's star second-closest star to the Sun, six light years away in the constellation Ophiuchus. It is a faint red dwarf of 10th magnitude, visible only through a telescope. It is named after the US astronomer Edward E Barnard (1857–1923), who discovered in 1916 that it has the fastest proper motion of any star, crossing 1 degree of sky every 350 years.

Some observations suggest that Barnard's star may be accompanied by planets.

barograph device for recording variations in atmospheric pressure. A pen, governed by the movements of an aneroid ◊barometer, makes a continuous line on a paper strip on a cylinder that rotates over a day or week to create a *barogram*, or permanent record of variations in atmospheric pressure.

barometer instrument that measures atmospheric pressure as an indication of weather. Most often used are the *mercury barometer* and the *aneroid barometer*.

In a mercury barometer a column of mercury in a glass tube, roughly 0.75 m/2.5 ft high (closed at one end, curved upwards at the other), is balanced by the pressure of the atmosphere on the open end; any change in the height of the column reflects a change in pressure. In an aneroid barometer, a shallow cylindrical metal box containing a partial vacuum expands or contracts in response to changes in pressure.

baroreceptor in biology, a specialized nerve ending that

is sensitive to pressure. There are baroreceptors in various regions of the heart and circulatory system (carotid sinus, aortic arch, atria, pulmonary veins, and left ventricle). Increased pressure in these structures stimulates the baroreceptors, which relay information to the medulla providing an important mechanism in the control of blood pressure.

barrel unit of liquid capacity, the value of which depends on the liquid being measured. It is used for petroleum, a barrel of which contains 159 litres/35 imperial gallons; a barrel of alcohol contains 189 litres/41.5 imperial gallons.

barrier island long island of sand, lying offshore and parallel to the coast.

Some are over 100 km/60 mi in length. Most barrier islands are derived from marine sands piled up by shallow longshore currents that sweep sand parallel to the seashore. Others are derived from former spits, connected to land and built up by drifted sand, that were later severed from the mainland.

Often several islands lie in a continuous row offshore. Coney Island and Jones Beach near New York City are well-known examples, as is Padre Island, Texas. The Frisian Islands are barrier islands along the coast of the Netherlands.

barrier reef ◊coral reef that lies offshore, separated from the mainland by a shallow lagoon.

Barringer Crater or *Arizona Meteor Crater* or *Coon Butte* impact crater near Winslow in Arizona caused by the impact of a 50 m/165 ft iron ◊meteorite some 25,000

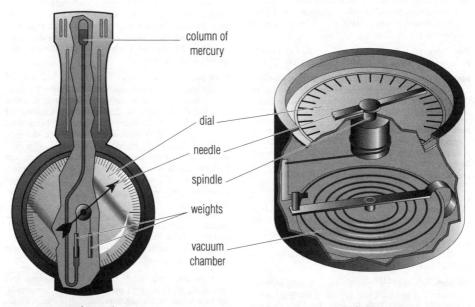

mercury barometer   aneroid barometer

barometer The mercury barometer and the aneroid barometer. In the mercury barometer, the weight of the column of mercury is balanced by the pressure of the atmosphere on the lower end. A change in height of the column indicates a change in atmospheric pressure. In the aneroid barometer, any change of atmospheric pressure causes the metal box which contains the vacuum to be squeezed or to expand slightly. The movements of the box sides are transferred to a pointer and scale via a chain of levers.

years ago. It is 1.2 km/0.7 mi in diameter, 200 m/660 ft deep and the walls are raised 50–60 m/165–198 ft above the surrounding desert.

It is named after the US mining engineer Daniel Barringer who proposed in 1902 that it was an impact crater rather than a volcanic feature, an idea confirmed in the 1960s by US geologist and astronomer Eugene Shoemaker.

barrow burial mound, usually composed of earth but sometimes of stones. Examples are found in many parts of the world. The two main types are *long*, dating from the Neolithic (New Stone Age), and *round*, dating from the early Bronze Age. Barrows made entirely of stones are known as cairns.

Long barrows may be mere mounds, typically higher and wider at one end. They usually contain a chamber of wood or stone slabs, or a turf-lined cavity, in which the body or bodies of the deceased were placed. Secondary chambers may be added in the sides of the mound.

Round barrows belong mainly to the Bronze Age, although in historic times there are examples from the Roman period, and some of the Saxon and most of the Danish invaders were barrow-builders. In northern Europe, round barrows were sometimes built above a tree-trunk coffin in which waterlogged conditions have preserved nonskeletal material, such as those found in Denmark dating from around 1000 BC.

In eastern European and Asiatic areas where mobility was afforded by the horse and wagon, a new culture developed of pit graves marked by a *kurgan*, or round mound, in which a single body lay, often accompanied by grave goods which might include a wagon. These date from around 3000 BC.

The placing of a great person's body in a ship is seen in Viking burials, such as the Oseberg ship in Norway, which was buried and sealed around AD 800. Barrows were erected over boat burials during the Saxon period, and the Sutton Hoo boat burial excavated in Suffolk, UK during 1938–39 was that of an East Anglian king of Saxon times.

baryon in nuclear physics, a heavy subatomic particle made up of three indivisible elementary particles called quarks. The baryons form a subclass of the ◊hadrons and comprise the nucleons (protons and neutrons) and hyperons.

baryte barium sulphate, $BaSO_4$, the most common mineral of barium. It is white or light-coloured, and has a comparatively high density (specific gravity 4.6); the latter property makes it useful in the production of high-density drilling muds (muds used to cool and lubricate drilling equipment). Baryte occurs mainly in ore veins, where it is often found with calcite and with lead and zinc minerals. It crystallizes in the orthorhombic system and can form tabular crystals or radiating fibrous masses.

basal metabolic rate (BMR) minimum amount of energy needed by the body to maintain life. It is measured when the subject is awake but resting, and includes the energy required to keep the heart beating, sustain breathing, repair tissues, and keep the brain and nerves functioning. Measuring the subject's consumption of oxygen gives an accurate value for BMR, because oxygen is needed to release energy from food.

A cruder measure of BMR estimates the amount of heat given off, some heat being released when food is used

up. BMR varies from one species to another, and from males to females. In humans, it is highest in children and declines with age. Disease, including mental illness, can make it rise or fall. Hormones from the ◊thyroid gland control the BMR.

basalt commonest volcanic ◊igneous rock, and the principal rock type on the ocean floor; it is basic, that is, it contains relatively little silica: about 50%. It is usually dark grey but can also be green, brown, or black.

The groundmass may be glassy or finely crystalline, sometimes with large ◊crystals embedded. Basaltic lava tends to be runny and flows for great distances before solidifying. Successive eruptions of basalt have formed the great plateaus of Colorado and the Indian Deccan. In some places, such as Fingal's Cave in the Inner Hebrides of Scotland and the Giant's Causeway in Antrim, Northern Ireland, shrinkage during the solidification of the molten lava caused the formation of hexagonal columns.

bascule bridge type of drawbridge in which one or two counterweighted deck members pivot upwards to allow shipping to pass underneath. One example is the double bascule Tower Bridge, London.

base in chemistry, a substance that accepts protons, such as the hydroxide ion (OH^-) and ammonia (NH_3). Bases react with acids to give a salt. Those that dissolve in water are called alkalis.

Inorganic bases are usually oxides or hydroxides of metals, which react with dilute acids to form a salt and water. A number of carbonates also react with dilute acids, additionally giving off carbon dioxide. Many organic compounds that contain nitrogen are bases.

base in mathematics, the number of different single-digit symbols used in a particular number system. In our usual (decimal) counting system of numbers (with symbols 0, 1, 2, 3, 4, 5, 6, 7, 8, 9) the base is 10. In the ◊binary number system, which has only the symbols 1 and 0, the base is two. A base is also a number that, when raised to a particular power (that is, when multiplied by itself a par-

Binary (base 2)	Octal (base 8)	Decimal (base 10)	Hexadecimal (base 16)
0	0	0	0
1	1	1	1
10	2	2	2
11	3	3	3
100	4	4	4
101	5	5	5
110	6	6	6
111	7	7	7
1000	10	8	8
1001	11	9	9
1010	12	10	A
1011	13	11	B
1100	14	12	C
1101	15	13	D
1110	16	14	E
1111	17	15	F
10000	20	16	10
11111111	377	255	FF
11111010001	3721	2001	7D1

base

ticular number of times as in $10^2 = 10 \times 10 = 100$), has a ◊logarithm equal to the power. For example, the logarithm of 100 to the base ten is 2.

In geometry, the term is used to denote the line or area on which a polygon or solid stands.

base pair in biochemistry, the linkage of two base (purine or pyrimidine) molecules in ◊DNA. They are found in nucleotides, and form the basis of the genetic code.

One base lies on one strand of the DNA double helix, and one on the other, so that the base pairs link the two strands like the rungs of a ladder. In DNA, there are four bases: adenine and guanine (purines) and cytosine and thymine (pyrimidines). Adenine always pairs with thymine, and cytosine with guanine.

BASIC (acronym for *beginner's all-purpose symbolic instruction code*) high-level computer-programming language, developed in 1964, originally designed to take advantage of ◊multiuser systems (which can be used by many people at the same time). The language is relatively easy to learn and is popular among microcomputer users.

Most versions make use of an ◊interpreter, which translates BASIC into ◊machine code and allows programs to be entered and run with no intermediate translation. Some more recent versions of BASIC allow a ◊compiler to be used for this process.

basicity number of replaceable hydrogen atoms in an acid. Nitric acid (HNO_3) is monobasic, sulphuric acid (H_2SO_4) is dibasic, and phosphoric acid (H_3PO_4) is tribasic.

basic–oxygen process most widely used method of steelmaking, involving the blasting of oxygen at supersonic speed into molten pig iron.

Pig iron from a blast furnace, together with steel scrap, is poured into a converter, and a jet of oxygen is then

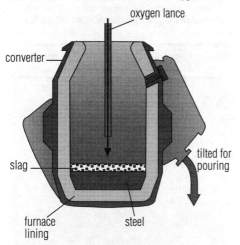

basic–oxygen process The basic–oxygen process is the primary method used to produce steel. Oxygen is blown at high pressure through molten pig iron and scrap steel in a converter lined with basic refractory materials. The impurities, principally carbon, quickly burn out, producing steel.

projected into the mixture. The excess carbon in the mix and other impurities quickly burn out or form a slag, and the converter is emptied by tilting. It takes only about 45 minutes to refine 350 tonnes/400 tons of steel. The basic–oxygen process was developed in 1948 at a steelworks near the Austrian towns of Linz and Donawitz. It is a version of the ◊Bessemer process.

basidiocarp spore-bearing body, or 'fruiting body', of all basidiomycete fungi (see ◊fungus), except the rusts and smuts. A well known example is the edible mushroom *Agaricus brunnescens*. Other types include globular basidiocarps (puffballs) or flat ones that project from tree trunks (brackets). They are made up of a mass of tightly packed, intermeshed ◊hyphae.

The tips of these hyphae develop into the reproductive cells, or *basidia*, that form a fertile layer known as the hymenium, or *gills*, of the basidiocarp. Four spores are budded off from the surface of each basidium.

batch processing in computing, a system for processing data with little or no operator intervention. Batches of data are prepared in advance to be processed during regular 'runs' (for example, each night). This allows efficient use of the computer and is well suited to applications of a repetitive nature, such as a company payroll.

In ◊*interactive computing*, by contrast, data and instructions are entered while the processing program is running.

Bates eyesight training method developed by US ophthalmologist William Bates (1860–1931) to enable people to correct problems of vision without wearing glasses. The method is of proven effectiveness in relieving all refractive conditions, correcting squints, lazy eyes, and similar problems, but does not claim to treat eye disease.

batholith large, irregular, deep-seated mass of intrusive ◊igneous rock, usually granite, with an exposed surface of more than 100 sq km/40 sq mi. The mass forms by the intrusion or upswelling of magma (molten rock) through the surrounding rock. Batholiths form the core of all major mountain ranges.

According to plate tectonic theory, magma rises in subduction zones along continental margins where one plate sinks beneath another. The solidified magma becomes the central axis of a rising mountain range, resulting in the deformation (folding and overthrusting) of rocks on either side. Gravity measurements indicate that the downward extent or thickness of many batholiths is some 6–9 mi/10–15 km.

bathyal zone upper part of the ocean, which lies on the continental shelf at a depth of between 200 m/650 ft and 2,000 m/6,500 ft.

Bathyal zones (both temperate and tropical) have greater biodiversity than coral reefs, according to a 1995 study by the Natural History Museum in London. Maximum biodiversity occurs between 1,000 m/3,280 ft and 3,000 m/9,800 ft.

bathyscaph or *bathyscaphe* or *bathyscape* deep-sea diving apparatus used for exploration at great depths in the ocean. In 1960, Jacques Piccard and Don Walsh took the bathyscaph *Trieste* to a depth of 10,917 m/35,820 ft in the Challenger Deep in the ◊Mariana Trench off the island of Guam in the Pacific Ocean.

battery any energy-storage device allowing release of electricity on demand. It is made up of one or more

In the figure: oxygen lance / converter / slag / tilted for pouring / furnace lining / steel

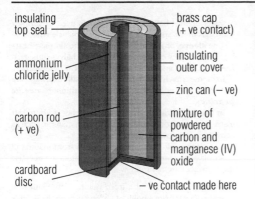

insulating top seal

brass cap (+ ve contact)

ammonium chloride jelly

insulating outer cover

zinc can (– ve)

carbon rod (+ ve)

mixture of powdered carbon and manganese (IV) oxide

cardboard disc

– ve contact made here

battery The common dry cell relies on chemical changes occurring between the electrodes – the central carbon rod and the outer zinc casing – and the ammonium chloride electrolyte to produce electricity. The mixture of carbon and manganese is used to increase the life of the cell.

electrical ◊cells. Primary-cell batteries are disposable; secondary-cell batteries, or ◊accumulators, are rechargeable. Primary-cell batteries are an extremely uneconomical form of energy, since they produce only 2% of the power used in their manufacture.

The common **dry cell** is a primary-cell battery based on the Leclanché cell and consists of a central carbon electrode immersed in a paste of manganese dioxide and ammonium chloride as the electrolyte. The zinc casing forms the other electrode. It is dangerous to try to recharge a primary-cell battery.

The lead–acid **car battery** is a secondary-cell battery. The car's generator continually recharges the battery. It consists of sets of lead (positive) and lead peroxide (negative) plates in an electrolyte of sulphuric acid (◊battery acid). Hydrogen cells and sodium–sulphur batteries were developed in 1996 to allow cars to run entirely on battery power for up to 60 km/100 mi.

The introduction of rechargeable nickel–cadmium batteries has revolutionized portable electronic news gathering (sound recording, video) and information processing (computing). These batteries offer a stable, short-term source of power free of noise and other electrical hazards.

battery acid ◊sulphuric acid of approximately 70% concentration used in lead–acid cells (as found in car batteries).

The chemical reaction within the battery that is responsible for generating electricity also causes a change in the acid's composition. This can be detected as a change in its specific gravity: in a fully charged battery the acid's specific gravity is 1.270–1.290; in a half-charged battery it is 1.190–1.210; in a flat battery it is 1.110–1.130.

baud in computing, a unit that measures the speed of transmission of data over an asynchronous connection.

A baud is actually a measure of the maximum number of times the electrical state of a communications circuit can change per second. Bauds were used as a measure to identify the speed of ◊modems until the early 1990s because at the lower modem speeds available then the baud rate generally equalled the rate of transmission measured in bps (bits per second). At higher speeds, this is not the case, and modem speeds now are generally quoted in bps.

baud in engineering, a unit of electrical signalling speed equal to one pulse per second, measuring the rate at which signals are sent between electronic devices such as telegraphs and computers.

bauxite principal ore of ◊aluminium, consisting of a mixture of hydrated aluminium oxides and hydroxides, generally contaminated with compounds of iron, which give it a red colour. It is formed by the ◊chemical weathering of rocks in tropical climates. Chief producers of bauxite are Australia, Guinea, Jamaica, Russia, Kazakhstan, Surinam, and Brazil.

To extract aluminium from bauxite, high temperatures (about 800°C/1,470°F) are needed to make the ore molten. Strong electric currents are then passed through the molten ore. The process is only economical if cheap electricity is readily available, usually from a hydroelectric plant.

Bayesian statistics form of statistics that uses the knowledge of prior probability together with the probability of actual data to determine posterior probabilities, using Bayes' theorem.

Bayes' theorem in statistics, a theorem relating the ◊probability of particular events taking place to the probability that events conditional upon them have occurred.

For example, the probability of picking an ace at random out of a pack of cards is 4/52. If two cards are picked out, the probability of the second card being an ace is conditional on the first card: if the first card is an ace the probability of drawing a second ace will be 3/51; if not it will be 4/51. Bayes' theorem gives the probability that given that the second card is an ace, the first card is also.

BCE abbreviation for *before the Common Era*, used with dates instead of BC.

B cell or *B lymphocyte* immune cell that produces antibodies. Each B cell produces just one type of ◊antibody, specific to a single ◊antigen. Lymphocytes are related to ◊T cells.

BCG abbreviation for *bacille Calmette-Guérin*, bacillus injected as a vaccine to confer active immunity to ◊tuberculosis (TB).

BCG was developed by Albert Calmette and Camille Guérin in France in 1921 from live bovine TB bacilli. These bacteria were bred in the laboratory over many generations until they became attenuated (weakened). Each inoculation contains just enough live, attenuated bacilli to provoke an immune response: the formation of specific antibodies. The vaccine provides protection for 50–80% of infants vaccinated.

beach strip of land bordering the sea, normally consisting of boulders and pebbles on exposed coasts or sand on sheltered coasts. It is usually defined by the high- and low-water marks. A berm, a ridge of sand and pebbles, may be found at the farthest point that the water reaches.

The material of the beach consists of a rocky debris eroded from exposed rocks and headlands, or carried in by rivers. The material is transported to the beach, and along the beach, by waves that hit the coastline at an angle, resulting in a net movement of the material in one particular direction. This movement is known as *longshore drift*. Attempts are often made to halt longshore drift by erecting barriers (◊groynes, at right angles to the movement. Pebbles are worn into round shapes by being battered against one another by wave action and the result is called **shingle**. The finer material, the **sand**, may be subsequently moved about by the wind to form sand dunes. Apart from the natural process of longshore drift, a beach may be threatened by the commercial use of sand and aggregate, by the mineral industry – since particles of metal ore are often concentrated into workable deposits by the wave action – and by pollution (for example, by oil spilled or dumped at sea).

Beaker people prehistoric people thought to have been of Iberian origin, who spread out over Europe from the 3rd millennium BC. They were skilled in metalworking, and are identified by their use of distinctive earthenware drinking vessels with various designs.

A type of beaker with an inverted bell-shaped profile was widely distributed throughout Europe. These bell beakers are associated with the spread of alcohol consumption, probably mead. The Beaker people favoured individual inhumation (burial of the intact body), often in round ◊barrows, with an associated set of small stone and metal artefacts, or secondary burials in some form of chamber tomb. A beaker typically accompanied male burials, possibly to hold a drink for the deceased on their final journey. The inclusion of flint, later metal, daggers in grave goods may signify a warrior, and suggests that the incursion of Bell Beaker culture may have come as an intrusion into traditional pre-existing cultures.

Bear, Great and Little common names (and translations of the Latin) for the constellations ◊Ursa Major and ◊Ursa Minor respectively.

bearing device used in a machine to allow free movement between two parts, typically the rotation of a shaft in a housing. **Ball bearings** consist of two rings, one fixed to a housing, one to the rotating shaft. Between them is a set,

or race, of steel balls. They are widely used to support shafts, as in the spindle in the hub of a bicycle wheel.

The **sleeve**, or **journal bearing**, is the simplest bearing. It is a hollow cylinder, split into two halves. It is used for the big-end and main bearings on a car ◊crankshaft.

In some machinery the balls of ball bearings are replaced by cylindrical rollers or thinner **needle bearings**.

In precision equipment such as watches and aircraft instruments, bearings may be made from material such as ruby and are known as **jewel bearings**.

For some applications bearings made from nylon and other plastics are used. They need no lubrication because their surfaces are naturally waxy.

bearing the direction of a fixed point, or the path of a moving object, from a point of observation on the Earth's surface, expressed as an angle from the north. Bearings are taken by ◊compass and are measured in degrees (°), given as three-digit numbers increasing clockwise. For instance, north is 000°, northeast is 045°, south is 180°, and southwest is 225°.

True north differs slightly from magnetic north (the direction in which a compass needle points), hence NE may be denoted as 045M or 045T, depending on whether the reference line is magnetic (M) or true (T) north. True north also differs slightly from grid north since it is impossible to show a spherical Earth on a flat map.

beat frequency in musical acoustics, fluctuation produced when two notes of nearly equal pitch or ◊frequency are heard together. Beats result from the ◊interference between the sound waves of the notes. The frequency of the beats equals the difference in frequency of the notes.

Beaufort scale system of recording wind velocity (speed), devised by Francis Beaufort in 1806. It is a numerical scale ranging from 0 to 17, calm being indicated by 0 and a hurricane by 12; 13–17 indicate degrees of hurricane force.

In 1874 the scale received international recognition; it

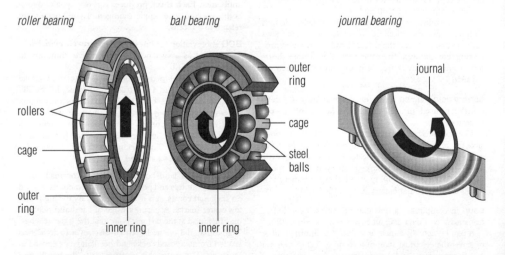

roller bearing   *ball bearing*   *journal bearing*

rollers

cage

outer ring

inner ring

outer ring

cage

steel balls

inner ring

journal

bearing Three types of bearing. The roller and the ball bearing are similar, differing only in the shape of the parts that roll when the middle shaft turns. The simpler journal bearing consists of a sleeve, or journal, lining the surface of the rotating shaft. The bearing is lubricated to reduce friction and wear.

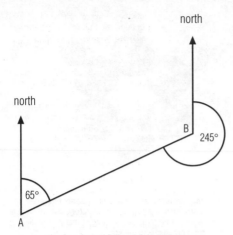

the bearing of B from A is 065°
the backbearing, or bearing of A from B, is 245°

bearing A bearing is the direction of a fixed point, or the path of a moving object, from a point of observation on the Earth's surface, expressed as an angle from the north. In the diagram, the bearing of a point A from an observer at B is the angle between the line BA and the north line through B, measured in a clockwise direction from the north line.

was modified in 1926. Measurements are made at 10 m/33 ft above ground level.

becquerel SI unit (symbol Bq) of ◊radioactivity, equal to one radioactive disintegration (change in the nucleus of an atom when a particle or ray is given off) per second.

The becquerel is much smaller than the previous standard unit, the curie (3.7×10^{10} Bq). It is named after French physicist Henri Becquerel.

bed in geology, a single ◊sedimentary rock unit with a distinct set of physical characteristics or contained fossils, readily distinguishable from those of beds above and below. Well-defined partings called **bedding planes** separate successive beds or strata.

The depth of a bed can vary from a fraction of a centimetre to several metres or yards, and can extend over

any area. The term is also used to indicate the floor beneath a body of water (lake bed) and a layer formed by a fall of particles (ash bed).

bedford level peat portion of the Fens.

bedwetting or **nocturnal enuresis** in medicine, the involuntary passing of urine during sleep. It occurs most commonly in children and it is usually due to inadequate bladder training or a psychological disorder. Most children gain full control of bladder function by the age of three years but this can occur much later in a few children. Treatment should consist of ensuring the condition is not due to an organic cause, such as a urinary tract infection. The child should be given reassurance and further bladder training. An alarm system which wakes the child if urine is being passed may also be useful. Drugs can be used if these measures fail but their long-term use is not recommended due to the occurrence of adverse effects.

behaviourism school of psychology originating in the USA, of which the leading exponent was John B Watson.

Behaviourists maintain that all human activity can ultimately be explained in terms of conditioned reactions or reflexes and habits formed in consequence. Leading behaviourists include Ivan Pavlov and B F Skinner.

The idea of man as a dominant animal of the earth whose whole behaviour tends to be dominated by his own desire for dominance gripped me. It seemed to explain almost everything.

Macfarlane Burnet,
Dominant Manual, 1970

behaviour therapy in psychology, the application of behavioural principles, derived from learning theories, to the treatment of clinical conditions such as ◊phobias, ◊obsessions, and sexual and interpersonal problems.

The symptoms of these disorders are regarded as learned patterns of behaviour that therapy can enable the patient to unlearn. For example, in treating a phobia, the patient is taken gradually into the feared situation in about 20 sessions until the fear noticeably reduces.

Beaufort Scale

Number and description		Features	Air speed kph	mph
0	calm	smoke rises vertically; water smooth	0–2	0–1
1	light air	smoke shows wind direction; water ruffled	2–5	1–3
2	light breeze	leaves rustle; wind felt on face	6–11	4–7
3	gentle breeze	loose paper blows around	12–19	8–12
4	moderate breeze	branches sway	20–29	13–18
5	fresh breeze	small trees sway, leaves blown off	30–39	19–24
6	strong breeze	whistling in telephone wires; sea spray from waves	40–50	25–31
7	near gale	large trees sway	51–61	32–38
8	gale	twigs break from trees	62–74	39–46
9	strong gale	branches break from trees	75–87	47–54
10	storm	trees uprooted; weak buildings collapse	88–101	55–63
11	violent storm	widespread damage	102–117	64–73
12	hurricane	widespread structural damage	above 118	above 74

The Beaufort Scale is a system of recording wind velocity (speed) devised in 1806 by Francis Beaufort (1774–1857). It is a numerical scale ranging from 0 for calm to 12 for a hurricane.

bel unit of sound measurement equal to ten ◊decibels. It is named after Scottish scientist Alexander Graham Bell.

belladonna or *deadly nightshade* (*Atropa belladonna*, family Solanaceae). Poisonous plant belonging to the nightshade family, found in Europe and Asia. It grows to 1.5 m/5 ft in height, with dull green leaves growing in unequal pairs, up to 20 cm/8 in long, and single purplish flowers that produce deadly black berries. Drugs are made from the leaves.

The dried powdered leaves are used to produce the drugs atropine and hyoscine. Belladonna extract acts medicinally as an anticholinergic (blocking the passage of certain nerve impulses), and is highly poisonous in large doses.

bells nautical term applied to half-hours of watch. A day is divided into seven watches, five of four hours each and two, called dogwatches, of two hours. Each half-hour of each watch is indicated by the striking of a bell, eight bells signalling the end of the watch.

benchmark in computing, a measure of the performance of a piece of equipment or software, usually consisting of a standard program or suite of programs. Benchmarks can indicate whether a computer is powerful enough to perform a particular task, and so enable machines to be compared. However, they provide only a very rough guide to practical performance, and may lead manufacturers to design systems that get high scores with the artificial benchmark programs but do not necessarily perform well with day-to-day programs or data.

bends or *compressed-air sickness* or *caisson disease* popular name for a syndrome seen in deep-sea divers, arising from too rapid a release of nitrogen from solution in their blood. If a diver surfaces too quickly, nitrogen that had dissolved in the blood under increasing water pressure is suddenly released, forming bubbles in the bloodstream and causing pain (the 'bends') and paralysis. Immediate treatment is gradual decompression in a decompression chamber, whilst breathing pure oxygen.

benzaldehyde C_6H_5CHO colourless liquid with the characteristic odour of almonds. It is used as a solvent and in the making of perfumes and dyes. It occurs in certain leaves, such as the cherry, laurel, and peach, and in a combined form in certain nuts and kernels. It can be extracted from such natural sources, but is usually made from ◊toluene.

Benzedrine trade name for ◊amphetamine, a stimulant drug.

benzene C_6H_6 clear liquid hydrocarbon of characteristic odour, occurring in coal tar. It is used as a solvent and in the synthesis of many chemicals.

The benzene molecule consists of a ring of six carbon atoms, all of which are in a single plane, and it is one of the simplest of ◊cyclic compounds. Benzene is the simplest of a class of compounds collectively known as *aromatic compounds*. Some are considered carcinogenic (cancer-inducing).

benzodiazepine any of a group of mood-altering drugs (tranquillizers), for example Librium and Valium. They are addictive and interfere with the process by which information is transmitted between brain cells, and various side effects arise from continued use. They were

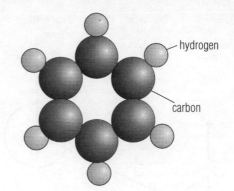

benzene The molecule of benzene consists of six carbon atoms arranged in a ring, with six hydrogen atoms attached. The benzene ring structure is found in many naturally occurring organic compounds.

originally developed as muscle relaxants, and then excessively prescribed in the West as anxiety-relieving drugs.

Today the benzodiazepines are recommended only for short-term use in alleviating severe anxiety or insomnia.

benzoic acid C_6H_5COOH white crystalline solid, sparingly soluble in water, that is used as a preservative for certain foods and as an antiseptic. It is obtained chemically by the direct oxidation of benzaldehyde and occurs in certain natural resins, some essential oils, and as hippuric acid.

benzpyrene one of a number of organic compounds associated with a particular polycyclic ring structure. Benzpyrenes are present in coal tar at low levels and are considered carcinogenic (cancer-inducing). Traces of benzpyrenes are present in wood smoke, and this has given rise to some concern about the safety of naturally smoked foods.

beriberi nutritional disorder occurring mostly in the tropics and resulting from a deficiency of vitamin B_1 (◊thiamine). The disease takes two forms: in one ◊oedema (waterlogging of the tissues) occurs; in the other there is severe emaciation. There is nerve degeneration in both forms and many victims succumb to heart failure.

Beringia or *Bering Land Bridge* former land bridge 1,600 km/1,000 mi wide between Asia and North America; it existed during the ice ages that occurred before 35000 BC and during the period 24000–9000 BC. It is now covered by the Bering Strait and Chukchi Sea.

berkelium synthesized, radioactive, metallic element of the actinide series, symbol Bk, atomic number 97, relative atomic mass 247.

It was first produced in 1949 by Glenn Seaborg and his team, at the University of California at Berkeley, USA, after which it is named.

Bernoulli's principle law stating that the speed of a fluid varies inversely with pressure, an increase in speed producing a decrease in pressure (such as a drop in hydraulic pressure as the fluid speeds up flowing through a constriction in a pipe) and vice versa. The principle also explains the pressure differences on each surface of an aerofoil, which gives lift to the wing of an aircraft. The principle was named after Swiss mathematician and physicist Daniel Bernoulli.

berry fleshy, many-seeded ◊fruit that does not split open to release the seeds. The outer layer of tissue, the exocarp, forms an outer skin that is often brightly coloured to attract birds to eat the fruit and thus disperse the seeds. Examples of berries are the tomato and the grape.

A *pepo* is a type of berry that has developed a hard exterior, such as the cucumber fruit. Another is the *hesperidium*, which has a thick, leathery outer layer, such as that found in citrus fruits, and fluid-containing vesicles within, which form the segments.

beryl mineral, beryllium aluminium silicate, $Be_3Al_2Si_6O_{18}$, which forms crystals chiefly in granite. It is the chief ore of beryllium. Two of its gem forms are aquamarine (light-blue crystals) and emerald (dark-green crystals).

beryllium hard, light-weight, silver-white, metallic element, symbol Be, atomic number 4, relative atomic mass 9.012. It is one of the ◊alkaline-earth metals, with chemical properties similar to those of magnesium. In nature it is found only in combination with other elements and occurs mainly as beryl ($3BeO.Al_2O_3.6SiO_2$). It is used to make sturdy, light alloys and to control the speed of neutrons in nuclear reactors. Beryllium oxide was discovered in 1798 by French chemist Louis-Nicolas Vauquelin (1763–1829), but the element was not isolated until 1828, by Friedrich Wöhler and Antoine-Alexandre-Brutus Bussy independently.

In 1992 large amounts of beryllium were unexpectedly discovered in six old stars in the Milky Way.

Bessemer process the first cheap method of making ◊steel, invented by Henry Bessemer in England in 1856. It has since been superseded by more efficient steel-making processes, such as the ◊basic–oxygen process. In the Bessemer process compressed air is blown into the bottom of a converter, a furnace shaped like a cement mixer, containing molten pig iron. The excess carbon in the iron burns out, other impurities form a slag, and the furnace is emptied by tilting.

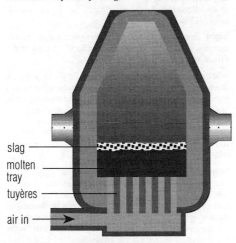

slag
molten
tray
tuyères

air in

Bessemer process In a Bessemer converter, a blast of high-pressure air oxidizes impurities in molten iron and converts it to steel.

beta-blocker any of a class of drugs that block impulses that stimulate certain nerve endings (beta receptors) serving the heart muscle. This reduces the heart rate and the force of contraction, which in turn reduces the amount of oxygen (and therefore the blood supply) required by the heart. Beta-blockers may be useful in the treatment of angina, arrhythmia (abnormal heart rhythms), and raised blood pressure, and following heart attacks. They must be withdrawn from use gradually.

beta decay the disintegration of the nucleus of an atom to produce a beta particle, or high-speed electron, and an electron-antineutrino. During beta decay, a neutron in the nucleus changes into a proton, thereby increasing the atomic number by one while the mass number stays the same. The mass lost in the change is converted into kinetic (movement) energy of the beta particle.

Beta decay is caused by the weak nuclear force, one of the fundamental ◊forces of nature operating inside the nucleus.

beta index mathematical measurement of the connectivity of a transport network. If the network is represented as a simplified topological map, made up of nodes (junctions or places) and edges (links), the beta index may be calculated by dividing the number of nodes by the number of edges. If the number of nodes is n and the number of edges is e, then the beta index β is given by the formula:

$$\beta = n/e$$

The higher the index number, the better connected the network is. If β is greater than 1, then a complete circuit exists.

beta particle electron ejected with great velocity from a radioactive atom that is undergoing spontaneous disintegration. Beta particles do not exist in the nucleus but are created on disintegration, beta decay, when a neutron converts to a proton to emit an electron.

Beta particles are more penetrating than ◊alpha particles, but less so than ◊gamma radiation; they can travel several metres in air, but are stopped by 2–3 mm of aluminium. They are less strongly ionizing than alpha particles and, like cathode rays, are easily deflected by magnetic and electric fields.

Betelgeuse or *Alpha Orionis* red supergiant star in the constellation of ◊Orion. It is the tenth brightest star in the night sky, although its brightness varies. It is 1,100 million km/700 million mi across, about 800 times larger than the Sun, roughly the same size as the orbit of Mars. It is over 10,000 times as luminous as the Sun, and lies 650 light years from Earth. Light takes 60 minutes to travel across the giant star.

Its magnitude varies irregularly between 0.4 and 1.3 in a period of 5.8 years. It was the first star whose angular diameter was measured with the Mount Wilson ◊interferometer in 1920. The name is a corruption of the Arabic, describing its position in the shoulder of Orion.

Bézier curve curved line invented by Pierre Bézier that connects a series of points (or 'nodes') in the smoothest possible way. The shape of the curve is governed by a series of complex mathematical formulae. They are used in ◊computer graphics and ◊CAD.

bhp abbreviation for *brake horsepower*.

bicarbonate common name for ◊hydrogencarbonate

bicarbonate of soda or *baking soda* (technical name *sodium hydrogencarbonate*) $NaHCO_3$ white crystalline solid that neutralizes acids and is used in medicine to treat acid indigestion. It is also used in baking powders and effervescent drinks.

bicuspid in biology, a structure with two cusps. Examples include the premolars (bicuspid teeth) and the mitral (bicuspid) valve of the heart.

bicuspid valve or *mitral valve* in the left side of the ◊heart, a flap of tissue that prevents blood flowing back into the atrium when the ventricle contracts.

bicycle Pedal-driven two-wheeled vehicle used in cycling. It consists of a tubular metal frame mounted on two large wire-spoked wheels, with handlebars in front and a seat above the back wheel. The bicycle is an energy-efficient, nonpolluting form of transport, and it is estimated that 800 million bicycles are in use throughout the world – outnumbering cars three to one. To manufacture a bicycle requires only 1% of the energy and materials used to build a car. China, India, Denmark, and the Netherlands are countries with a high use of bicycles. More than 10% of road spending in the Netherlands is on cycleways and bicycle parking.

The first bicycles were simple frames and wheels, propelled by the feet and known as hobby horses, widely in use in the eighteenth century. The first major improvement was made by the German Friehers Karl Drair von Sauerbronn (1758–1851) around 1818. It was fitted with a moveable front wheel, controlled by handlebars. The first treadle-propelled cycle was designed by the Scottish blacksmith Kirkpatrick Macmillan in 1839. By the end of the 19th century wire wheels, metal frames (replacing wood), and pneumatic tyres (invented by the Scottish veterinary surgeon John B Dunlop in 1888) had been added. Among the bicycles of that time was the front-wheel-driven penny farthing with a large front wheel. The first successful modern bicycle was the Rover Safety Bicycle of 1885 with wheels of equal size and a geared chain drive from the pedals to the rear wheel. In recent years the light but strong carbon-fibre frame and multiple gearing has created the popular mountain bike, that can be used on rough terrains as well as conventional roads and cycleways.

Recent technological developments have been related to reducing wind resistance caused by the frontal area and the turbulent drag of the bicycle. Most of an Olympic cyclist's energy is taken up in fighting wind resistance in a sprint. The first major innovation was the solid wheel, first used in competitive cycling in 1984, but originally patented as long ago as 1878. Further developments include handlebars that allow the cyclist to crouch and use the shape of the hands and forearms to divert air away from the chest. Modern racing bicycles now have a monocoque structure produced by laying carbon fibre around an internal mould and then baking them in an oven. Using all these developments, Chris Boardman set a speed record of 54.4 km/h (34 mph) on his way to winning a gold medal at the 1992 Barcelona Olympics.

biennial plant plant that completes its life cycle in two years. During the first year it grows vegetatively and the surplus food produced is stored in its ◊perennating organ, usually the root. In the following year these food reserves are used for the production of leaves, flowers, and seeds, after which the plant dies. Many root vegetables are biennials, including the carrot *Daucus carota* and parsnip *Pas-*

tinaca sativa. Some garden plants that are grown as biennials are actually perennials, for example, the wallflower *Cheiranthus cheiri*.

bifocal lens in medicine, lenses consisting of an upper section that improves distant vision and a lower section that facilitates reading. They are used most commonly to correct defects in vision that occur as part of the natural ageing process. Similar improvements in vision may be obtained using varifocal lenses.

Big Bang in astronomy, the hypothetical 'explosive' event that marked the origin of the universe as we know it. At the time of the Big Bang, the entire universe was squeezed into a hot, superdense state. The Big Bang explosion threw this compact material outwards, producing the expanding universe (see ◊red shift). The cause of the Big Bang is unknown; observations of the current rate of expansion of the universe suggest that it took place about 10–20 billion years ago. The Big Bang theory began modern ◊cosmology.

According to a modified version of the Big Bang, called the *inflationary theory*, the universe underwent a rapid period of expansion shortly after the Big Bang, which accounts for its current large size and uniform nature. The inflationary theory is supported by the most recent observations of the ◊cosmic background radiation.

Big Bang

Scientists have calculated that one million-million-million-million-million-millionth of a second after the Big Bang, the Universe was the size of a pea, and the temperature was 10,000 billion million million°C (18,000 billion million million million°F). One second after the Big Bang, the temperature was about 10,000 billion°C (18,000 billion°F).

Big Crunch in cosmology, a possible fate of the universe in which it ultimately collapses to a point following the halting and reversal of the present expansion. See also ◊Big Bang and ◊critical density.

Big Dipper north American name for the Plough, the seven brightest and most prominent stars in the constellation ◊Ursa Major.

bight coastal indentation, crescent-shaped or gently curving, such as the Bight of Biafra in W Africa and the Great Australian Bight.

bile brownish alkaline fluid produced by the liver. Bile is stored in the gall bladder and is intermittently released into the duodenum (small intestine) to aid digestion. Bile consists of bile salts, bile pigments, cholesterol, and lecithin. *Bile salts* assist in the breakdown and absorption of fats; *bile pigments* are the breakdown products of old red blood cells that are passed into the gut to be eliminated with the faeces.

bilharzia or *schistosomiasis* disease that causes anaemia, inflammation, formation of scar tissue, dysentery, enlargement of the spleen and liver, cancer of the bladder, and cirrhosis of the liver. It is contracted by bathing in water contaminated with human sewage. Some 200 million people are thought to suffer from this disease in the tropics, and 750,000 people a year die.

Freshwater snails act as host to the first larval stage of blood flukes of the genus *Schistosoma*; when these larvae leave the snail in their second stage of development, they are able to pass through human skin, become sexually mature, and produce quantities of eggs, which pass to the intestine or bladder. Numerous eggs are excreted from the body in urine or faeces to continue the cycle. Treatment is by means of drugs, usually containing antimony, to kill the parasites.

billion the cardinal number represented by a 1 followed by nine zeros (1,000,000,000 or 10^9), equivalent to a thousand million.

binary fission in biology, a form of ◊asexual reproduction, whereby a single-celled organism, such as the amoeba, divides into two smaller 'daughter' cells. It can also occur in a few simple multicellular organisms, such as sea anemones, producing two smaller sea anemones of equal size.

binary number system system of numbers to ◊base two, using combinations of the digits 1 and 0. Codes based on binary numbers are used to represent instructions and data in all modern digital computers, the values of the binary digits (contracted to 'bits') being stored or transmitted as, for example, open/closed switches, magnetized/unmagnetized discs and tapes, and high/low voltages in circuits.

The value of any position in a binary number increases by powers of 2 (doubles) with each move from right to left (1, 2, 4, 8, 16, and so on). For example, 1011 in the binary number system means $(1 \times 1) + (0 \times 2) + (1 \times 4) + (1 \times 8)$, which adds up to 13 in the decimal system.

binary search in computing, a rapid technique used to find any particular record in a list of records held in sequential order. The computer is programmed to compare the record sought with the record in the middle of the ordered list. This being done, the computer discards the half of the list in which the record does not appear, thereby reducing the number of records left to search by half. This process of selecting the middle record and discarding the unwanted half of the list is repeated until the required record is found.

binary star pair of stars moving in orbit around their common centre of mass. Observations show that most stars are binary, or even multiple – for example, the nearest star system to the Sun, ◊Alpha Centauri. One of the stars in the binary system Epsilon Aurigae may be the largest star known. Its diameter is 2,800 times that of the Sun. If it were in the position of the Sun, it would engulf Mercury, Venus, Earth, Mars, Jupiter, and Saturn. A spectroscopic binary is a binary in which two stars are so close together that they cannot be seen separately, but their separate light spectra can be distinguished by a spectroscope. Another type is the ◊eclipsing binary.

binding energy in physics, the amount of energy needed to break the nucleus of an atom into the neutrons and protons of which it is made.

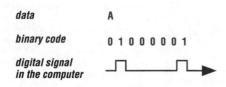

binary number system The capital letter A represented in binary form.

binoculars optical instrument for viewing an object in magnification with both eyes; for example, field glasses and opera glasses. Binoculars consist of two telescopes containing lenses and prisms, which produce a stereoscopic effect as well as magnifying the image.

Use of prisms has the effect of 'folding' the light path, allowing for a compact design.

The first binocular telescope was constructed by the Dutch inventor Hans Lippershey in 1608. Later development was largely due to the German Ernst Abbe of Jena, who at the end of the 19th century designed prism binoculars that foreshadowed the instruments of today, in which not only magnification but also stereoscopic effect is obtained. *See illustration overleaf.*

binomial in mathematics, an expression consisting of two terms, such as $a + b$ or $a - b$.

binomial system of nomenclature in biology, the system in which all organisms are identified by a two-part Latinized name. Devised by the biologist Linnaeus, it is also known as the Linnaean system. The first name is capitalized and identifies the ◊genus; the second identifies the ◊species within that genus.

binomial theorem formula whereby any power of a binomial quantity may be found without performing the progressive multiplications.

It was discovered by Isaac Newton and first published in 1676.

biochemistry science concerned with the chemistry of living organisms: the structure and reactions of proteins (such as enzymes), nucleic acids, carbohydrates, and lipids.

Its study has led to an increased understanding of life processes, such as those by which organisms synthesize essential chemicals from food materials, store and generate energy, and pass on their characteristics through their genetic material. A great deal of medical research is concerned with the ways in which these processes are disrupted. Biochemistry also has applications in agriculture and in the food industry (for instance, in the use of enzymes). *See chronology on page 73.*

biodegradable capable of being broken down by living organisms, principally bacteria and fungi. In biodegradable substances, such as food and sewage, the natural processes of decay lead to compaction and liquefaction, and to the release of nutrients that are then recycled by the ecosystem.

This process can have some disadvantageous side effects, such as the release of methane, an explosive greenhouse gas. However, the technology now exists for waste tips to collect methane in underground pipes, drawing it off and using it as a cheap source of energy. Non-biodegradable substances, such as glass, heavy metals, and most types of plastic, present serious problems of disposal.

biodiversity (contraction of *biological diversity*) measure of the variety of the Earth's animal, plant, and microbial species; of genetic differences within species; and of the ecosystems that support those species. Its maintenance is important for ecological stability and as a resource for research into, for example, new drugs and crops. In the 20th century, the destruction of habitats is believed to have resulted in the most severe and rapid loss of biodiversity in the history of the planet.

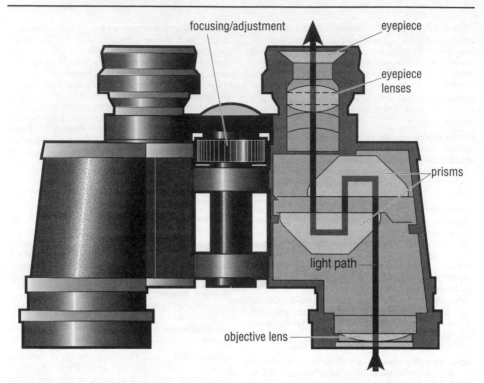

binoculars An optical instrument that allows the user to focus both eyes on the magnified image at the same time. The essential components of binoculars are objective lenses, eye pieces, and a system of prisms to invert and reverse the image. A focusing system provides a sharp image by adjusting the relative positions of these components.

Estimates of the number of species vary widely because many species-rich ecosystems, such as tropical forests, contain unexplored and unstudied habitats. Especially among small organisms, many are unknown; for instance, it is thought that only 1–10% of the world's bacterial species have been identified.

The most significant threat to biodiversity comes from the destruction of rainforests and other habitats in the southern hemisphere. It is estimated that 7% of the Earth's surface hosts 50–75% of the world's biological diversity. Costa Rica, for example, has an area less than 10% of the size of France but possesses three times as many vertebrate species.

biodynamic farming agricultural practice based on the

principle of ◊homoeopathy: tiny quantities of a substance are applied to transmit vital qualities to the soil. It is a form of ◊organic farming, and was developed by the Austrian holistic mystic Rudolf Steiner and Ehrenfried Pfiffer.

bioenergetics in alternative medicine, extension of ◊Reichian therapy principles developed in the 1960s by US physician Alexander Lowen, and designed to promote, by breathing, physical exercise, and the elimination of muscular blockages, the free flow of energy in the body and thus restore optimum health and vitality.

bioengineering the application of engineering to biology and medicine. Common applications include the

Biodiversity: Number of Species Worldwide

	Number identified	% of estimated total number of species
microorganisms	5,800	3–27%
invertebrates	1,021,000	3–27%
plants	322,500	67–100%
fish	19,100	83–100%
birds	9,100	94–100%
reptiles and amphibians	12,000	90–95%
mammals	4,000	90–95%
total	1,393,500	
	number of species	% identified
low estimate of all species	4.4 million	31
high estimate of all species	80 million	2

Biochemistry: Chronology

c. 1830	Johannes Müller discovered proteins.
1833	Anselme Payen and J F Persoz first isolated an enzyme.
1862	Haemoglobin was first crystallized.
1869	The genetic material DNA (deoxyribonucleic acid) was discovered by Friedrich Mieschler.
1899	Emil Fischer postulated the 'lock-and-key' hypothesis to explain the specificity of enzyme action.
1913	Leonor Michaelis and M L Menten developed a mathematical equation describing the rate of enzyme-catalysed reactions.
1915	The hormone thyroxine was first isolated from thyroid gland tissue.
1920	The chromosome theory of heredity was postulated by Thomas H Morgan; growth hormone was discovered by Herbert McLean Evans and J A Long.
1921	Insulin was first isolated from the pancreas by Frederick Banting and Charles Best.
1926	Insulin was obtained in pure crystalline form.
1927	Thyroxine was first synthesized.
1928	Alexander Fleming discovered penicillin.
1931	Paul Karrer deduced the structure of retinol (vitamin A); vitamin D compounds were obtained in crystalline form by Adolf Windaus and Askew, independently of each other.
1932	Charles Glen King isolated ascorbic acid (vitamin C).
1933	Tadeus Reichstein synthesized ascorbic acid.
1935	Richard Kuhn and Karrer established the structure of riboflavin (vitamin B_2).
1936	Robert Williams established the structure of thiamine (vitamin B_1); biotin was isolated by Kogl and Tonnis.
1937	Niacin was isolated and identified by Conrad Arnold Elvehjem.
1938	Pyridoxine (vitamin B_6) was isolated in pure crystalline form.
1939	The structure of pyridoxine was determined by Kuhn.
1940	Hans Krebs proposed the Krebs (citric acid) cycle; Hickman isolated retinol in pure crystalline form; Williams established the structure of pantothenic acid; biotin was identified by Albert Szent-Györgyi, Vincent Du Vigneaud, and co-workers.
1941	Penicillin was isolated and characterized by Howard Florey and Ernst Chain.
1943	The role of DNA in genetic inheritance was first demonstrated by Oswald Avery, Colin MacLeod, and Maclyn McCarty.
1950	The basic components of DNA were established by Erwin Chargaff; the alpha-helical structure of proteins was established by Linus Pauling and R B Corey.
1953	James Watson and Francis Crick determined the molecular structure of DNA.
1956	Mahlon Hoagland and Paul Zamecnik discovered transfer RNA (ribonucleic acid); mechanisms for the biosynthesis of RNA and DNA were discovered by Arthur Kornberg and Severo Ochoa.
1957	Interferon was discovered by Alick Isaacs and Jean Lindemann.
1958	The structure of RNA was determined.
1960	Messenger RNA was discovered by Sidney Brenner and François Jacob.
1961	Marshall Nirenberg and Ochoa determined the chemical nature of the genetic code.
1965	Insulin was first synthesized.
1966	The immobilization of enzymes was achieved by Chibata.
1968	Brain hormones were discovered by Roger Guillemin and Andrew Schally.
1975	J Hughes and Hans Kosterlitz discovered encephalins.
1976	Guillemin discovered endorphins.
1977	J Baxter determined the genetic code for human growth hormone.
1978	Human insulin was first produced by genetic engineering.
1979	The biosynthetic production of human growth hormone was announced by Howard Goodman and J Baxter of the University of California, and by D V Goeddel and Seeburg of Genentech.
1982	Louis Chedid and Michael Sela developed the first synthesized vaccine.
1983	The first commercially available product of genetic engineering (Humulin) was launched.
1985	Alec Jeffreys devised genetic fingerprinting.
1993	UK researchers introduced a healthy version of the gene for cystic fibrosis into the lungs of mice with induced cystic fibrosis, restoring normal function.
1996	Japanese chemists successfully synthesized cellulose.
1997	US geneticists constructed the first artificial human chromosome.

design and use of artificial limbs, joints, and organs, including hip joints and heart valves.

biofeedback in biology, modification or control of a biological system by its results or effects. For example, a change in the position or ◊trophic level of one species affects all levels above it.

Many biological systems are controlled by negative feedback. When enough of the hormone thyroxine has been released into the blood, the hormone adjusts its own level by 'switching off' the gland that produces it. In ecology, as the numbers in a species rise, the food supply available to each individual is reduced. This acts to reduce the population to a sustainable level.

biofeedback in medicine, the use of electrophysiological monitoring devices to 'feed back' information about internal processes and thus facilitate conscious control. Developed in the USA in the 1960s, independently by neurophysiologist Barbara Brown and neuropsychiatrist Joseph Kamiya, the technique is effective in alleviating hypertension and preventing associated organic and physiological dysfunctions.

biofuel any solid, liquid, or gaseous fuel produced from organic (once living) matter, either directly from plants or indirectly from industrial, commercial, domestic, or agricultural wastes. There are three main methods for the development of biofuels: the burning of dry organic wastes (such as household refuse, industrial and agricultural wastes, straw, wood, and peat); the fermentation of wet wastes (such as animal dung) in the absence of oxygen to produce biogas (containing up to 60% methane), or the fermentation of sugar cane or corn to produce alcohol and esters; and energy forestry (producing fast-growing wood for fuel).

Fermentation produces two main types of biofuels: alcohols and esters. These could theoretically be used in place of fossil fuels but, because major alterations to engines would be required, biofuels are usually mixed with fossil fuels. The EU allows 5% ethanol, derived from wheat, beet, potatoes, or maize, to be added to fossil fuels. A quarter of Brazil's transportation fuel 1994 was ethanol.

Biofuel

The flatulence of a single sheep could power a small lorry for 40 km/25 mi a day. The digestive process produces methane gas, which can be burnt as fuel. According to one New Zealand scientist, the methane from 72 million sheep could supply the entire fuel needs of his country.

---

biogenesis biological term coined in 1870 by English scientist Thomas Henry Huxley to express the hypothesis that living matter always arises out of other similar forms of living matter. It superseded the opposite idea of ◊spontaneous generation or abiogenesis (that is, that living things may arise out of nonliving matter).

biogeography study of how and why plants and animals are distributed around the world, in the past as well as in the present; more specifically, a theory describing the geographical distribution of ◊species developed by Robert MacArthur and US zoologist Edward O Wilson. The theory argues that for many species, ecological specializations mean that suitable habitats are patchy in their occurrence. Thus for a dragonfly, ponds in which to breed are separated by large tracts of land, and for edelweiss adapted to alpine peaks the deep valleys between cannot be colonized.

biological clock regular internal rhythm of activity, produced by unknown mechanisms, and not dependent on external time signals. Such clocks are known to exist in almost all animals, and also in many plants, fungi, and unicellular organisms; the first biological clock gene in plants was isolated in 1995 by a US team of researchers. In higher organisms, there appears to be a series of clocks of graded importance. For example, although body temperature and activity cycles in human beings are normally 'set' to 24 hours, the two cycles may vary independently, showing that two clock mechanisms are involved.

biological control control of pests such as insects and fungi through biological means, rather than the use of chemicals. This can include breeding resistant crop strains; inducing sterility in the pest; infecting the pest species with disease organisms; or introducing the pest's natural predator. Biological control tends to be naturally self-regulating, but as ecosystems are so complex, it is difficult to predict all the consequences of introducing a biological controlling agent.

The introduction of the cane toad to Australia 50 years ago to eradicate a beetle that was destroying sugar beet provides an example of the unpredictability of biological control. Since the cane toad is poisonous it has few Australian predators and it is now a pest, spreading throughout eastern and northern Australia at a rate of 35 km/22 mi a year.

biological oxygen demand (BOD) the amount of dissolved oxygen taken up by microorganisms in a sample of water. Since these microorganisms live by decomposing organic matter, and the amount of oxygen used is proportional to their number and metabolic rate, BOD can be used as a measure of the extent to which the water is polluted with organic compounds.

biological shield shield around a nuclear reactor that is intended to protect personnel from the effects of ◊radiation. It usually consists of a thick wall of steel and concrete.

biological weathering form of ◊weathering caused by the activities of living organisms – for example, the growth of roots or the burrowing of animals. Tree roots are probably the most significant agents of biological weathering as they are capable of prising apart rocks by growing into cracks and joints.

biology science of life. Biology includes all the life sciences – for example, anatomy and physiology (the study of the structure of living things), cytology (the study of cells), zoology (the study of animals) and botany (the study of plants), ecology (the study of habitats and the interaction of living species), animal behaviour, embryology, and taxonomy, and plant breeding. Increasingly in the 20th century biologists have concentrated on molecular structures: biochemistry, biophysics, and genetics (the study of inheritance and variation).

Biological research has come a long way towards understanding the nature of life, and during the 1990s our knowledge will be further extended as the international ◊Human Genome Project attempts to map the entire genetic code contained in the 23 pairs of human chromosomes.

bioluminescence production of light by living organisms. It is a feature of many deep-sea fishes, crustaceans, and other marine animals. On land, bioluminescence is seen in some nocturnal insects such as glow-worms and fireflies, and in certain bacteria and fungi. Light is usually produced by the oxidation of luciferin, a reaction catalysed by the ◊enzyme luciferase. This reaction is unique, being the only known biological oxidation that does not produce heat. Animal luminescence is involved in communication, camouflage, or the luring of prey, but its function in other organisms is unclear.

biomass the total mass of living organisms present in a given area. It may be specified for a particular species (such as earthworm biomass) or for a general category (such as herbivore biomass). Estimates also exist for the entire global plant biomass. Measurements of biomass can be used to study interactions between organisms, the stability of those interactions, and variations in population numbers. Where dry biomass is measured, the material is dried to remove all water before weighing.

Some two-thirds of the world's population cooks and

Biology: Chronology

c. 500 BC	First studies of the structure and behaviour of animals, by the Greek Alcmaeon of Croton.
c. 450	Hippocrates of Kos undertook the first detailed studies of human anatomy.
c. 350	Aristotle laid down the basic philosophy of the biological sciences and outlined a theory of evolution.
c. 300	Theophrastus carried out the first detailed studies of plants.
c. AD 175	Galen established the basic principles of anatomy and physiology.
c. 1500	Leonardo da Vinci studied human anatomy to improve his drawing ability and produced detailed anatomical drawings.
1628	William Harvey described the circulation of the blood and the function of the heart as a pump.
1665	Robert Hooke used a microscope to describe the cellular structure of plants.
1672	Marcello Malphigi undertook the first studies in embryology by describing the development of a chicken egg.
1677	Anton van Leeuwenhoek greatly improved the microscope and used it to describe spermatozoa as well as many microorganisms.
1736	Carolus (Carl) Linnaeus published his systematic classification of plants, so establishing taxonomy.
1768–79	James Cook's voyages of discovery in the Pacific revealed an undreamed-of diversity of living species, prompting the development of theories to explain their origin.
1796	Edward Jenner established the practice of vaccination against smallpox, laying the foundations for theories of antibodies and immune reactions.
1809	Jean-Baptiste Lamarck advocated a theory of evolution through inheritance of acquired characteristics.
1839	Theodor Schwann proposed that all living matter is made up of cells.
1857	Louis Pasteur established that microorganisms are responsible for fermentation, creating the discipline of microbiology.
1859	Charles Darwin published *On the Origin of Species*, expounding his theory of the evolution of species by natural selection.
1865	Gregor Mendel pioneered the study of inheritance with his experiments on peas, but achieved little recognition.
1883	August Weismann proposed his theory of the continuity of the germ plasm.
1900	Mendel's work was rediscovered and the science of genetics founded.
1935	Konrad Lorenz published the first of many major studies of animal behaviour, which founded the discipline of ethology.
1953	James Watson and Francis Crick described the molecular structure of the genetic material, DNA.
1964	William Hamilton recognized the importance of inclusive fitness, so paving the way for the development of sociobiology.
1975	Discovery of endogenous opiates (the brain's own painkillers) opened up a new phase in the study of brain chemistry.
1976	Har Gobind Khorana and his colleagues constructed the first artificial gene to function naturally when inserted into a bacterial cell, a major step in genetic engineering.
1982	Gene databases were established at Heidelberg, Germany, for the European Molecular Biology Laboratory, and at Los Alamos, USA, for the US National Laboratories.
1985	The first human cancer gene, retinoblastoma, was isolated by researchers at the Massachusetts Eye and Ear Infirmary and the Whitehead Institute, Massachusetts.
1988	The Human Genome Organization (HUGO) was established in Washington, DC, with the aim of mapping the complete sequence of DNA.
1991	BioSphere 2, an experiment that attempted to reproduce the world's biosphere in miniature within a sealed glass dome, was launched in Arizona, USA.
1992	Researchers at the University of California, USA, stimulated the multiplication of isolated brain cells of mice, overturning the axiom that mammalian brains cannot produce replacement cells once birth has taken place. The world's largest organism, a honey fungus with underground hyphae (filaments) spreading across 600 hectares/1,480 acres, was discovered in Washington State, USA.
1994	Scientists from Pakistan and the USA unearthed a 50-million-year-old fossil whale with hind legs that would have enabled it to walk on land.
1995	New phylum identified and named Cycliophora. It contains a single known species, *Symbion pandora*, a parasite of the lobster.
1996	The sequencing of the genome of brewer's yeast *Saccharomyces cerevisiae* is completed, the first time this has been achieved for an organism more complex than a bacterium. The 12 million base pairs took 300 scientists six years to map. A new muscle was discovered by two US dentists. It is 3 cm/1 in long, and runs from the jaw to behind the eye socket.
1997	The first mammal to be cloned from a non-reproductive cell was born. The lamb had been cloned from an udder cell from a six-year-old ewe.

heats water by burning biomass, usually wood. Plant biomass can be a renewable source of energy as replacement supplies can be grown relatively quickly. Fossil fuels however, originally formed from biomass, accumulate so slowly that they cannot be considered renewable. The burning of biomass (defined either as natural areas of the ecosystem or as forest, grasslands, and fuelwoods) produces 3.5 million tonnes of carbon in the form of carbon dioxide each year, accounting for up to 40% of the world's annual carbon dioxide production.

biome broad natural assemblage of plants and animals

shaped by common patterns of vegetation and climate. Examples include the tundra biome and the desert biome.

biomechanics application of mechanical engineering principles and techniques in the field of medicine and surgery, studying natural structures to improve those produced by humans. For example, mother-of-pearl is structurally superior to glass fibre, and deer antlers have outstanding durability because they are composed of microscopic fibres. Such natural structures may form the basis of high-tech composites. Biomechanics has been responsible for many recent advances in ◊orthopaedics, anaesthesia, and intensive care. Biomechanical assessment of the requirements for replacement of joints, including evaluation of the stresses and strains between parts, and their reliability, has allowed development of implants with very low friction and long life.

biometry literally, the measurement of living things, but generally used to mean the application of mathematics to biology. The term is now largely obsolete, since mathematical or statistical work is an integral part of most biological disciplines.

bionics (contraction of *biological electronics*) design and development of electronic or mechanical artificial systems that imitate those of living things. The bionic arm, for example, is an artificial limb (◊prosthesis) that uses electronics to amplify minute electrical signals generated in body muscles to work electric motors, which operate the joints of the fingers and wrist.

The first person to receive two bionic ears was Peter Stewart, an Australian journalist in 1989. His left ear was fitted with an array of 22 electrodes, replacing the hairs that naturally convert sounds into electrical impulses. Five years previously he had been fitted with a similar device in his right ear.

biophysics application of physical laws to the properties of living organisms. Examples include using the principles of ◊mechanics to calculate the strength of bones and muscles, and ◊thermodynamics to study plant and animal energetics.

biopsy removal of a living tissue sample from the body for diagnostic examination.

biorhythm rhythmic change, mediated by ◊hormones, in the physical state and activity patterns of certain plants and animals that have seasonal activities. Examples include winter hibernation, spring flowering or breeding, and periodic migration. The hormonal changes themselves are often a response to changes in day length (◊photoperiodism); they signal the time of year to the animal or plant. Other biorhythms are innate and continue even if external stimuli such as day length are removed. These include a 24-hour or ◊circadian rhythm, a 28-day or circalunar rhythm (corresponding to the phases of the Moon), and even a year-long rhythm in some organisms.

Such innate biorhythms are linked to an internal or ◊biological clock, whose mechanism is still poorly understood.

Often both types of rhythm operate; thus many birds have a circalunar rhythm that prepares them for the breeding season, and a photoperiodic response. There is also a nonscientific and unproven theory that human activity is governed by three biorhythms: the *intellectual* (33 days), the *emotional* (28 days), and the *physical* (23 days). Certain days in each cycle are regarded as 'critical', even more so if one such day coincides with that of another cycle.

biosensor device based on microelectronic circuits that can directly measure medically significant variables for the purpose of diagnosis or monitoring treatment. One such device measures the blood sugar level of diabetics using a single drop of blood, and shows the result on a liquid crystal display within a few minutes.

biosphere the narrow zone that supports life on our planet. It is limited to the waters of the Earth, a fraction of its crust, and the lower regions of the atmosphere.

BioSphere 2 (BS2) ecological test project, a 'planet in a bottle', in Arizona, USA. Under a sealed glass dome, several different habitats are recreated, with representatives of nearly 4,000 species, including humans, to see how well air, water, and waste can be recycled in an enclosed environment and whether a stable ecosystem can be created.

Experiments with biospheres that contain relatively simple life forms have been carried out for decades, and a 21-day trial period in 1989 that included humans preceded the construction of BS2. However, BS2 is not in fact the second in a series: the Earth is considered to be BioSphere 1.

The sealed area covers a total of 3.5 acres and contains tropical rainforest, salt marsh, desert, coral reef, and savanna habitats, as well as a section for intensive agriculture. The main problem has been in maintaining satisfactory oxygen levels, which in late 1993 became so erratic that additional oxygen had to be pumped in. The rainforest was thriving in 1994 but the plankton levels in the ocean were too low to sustain the corals; the climate was too moist for the desert plants and the desert was being overrun by grasses and shrubs.

The people within were entirely self-sufficient, except for electricity, which is supplied by a 3.7 megawatt power station on the outside (solar panels were considered too expensive). The original team of eight in residence between 1991–1993 was replaced in March 1994 with a new team of six people for ten months; in 1995 it was decided that further research would not involve sealing people within the biosphere.

It is run by a private company, Space Biospheres Ventures, whose investors expect to find commercial applications for the techniques that are developed in the course of the project.

The cost of setting up and maintaining the project has been estimated at $100 million, some of which will be covered by paying visitors, who can view the inhabitants through the geodesic glass dome.

biosynthesis synthesis of organic chemicals from simple inorganic ones by living cells – for example, the conversion of carbon dioxide and water to glucose by plants during ◊photosynthesis.

Other biosynthetic reactions produce cell constituents including proteins and fats.

biotechnology industrial use of living organisms to manufacture food, drugs, or other products. The brewing and baking industries have long relied on the yeast microorganism for ◊fermentation purposes, while the dairy industry employs a range of bacteria and fungi to convert milk into cheeses and yoghurts. ◊Enzymes, whether extracted from cells or produced artificially, are central to most biotechnological applications.

Recent advances include ◊genetic engineering, in which single-celled organisms with modified ◊DNA are used to produce insulin and other drugs.

In 1993 two-thirds of biotechnology companies were concentrating on human health developments, whilst only 1 in 10 were concerned with applications for food and agriculture.

biotin or **vitamin H** vitamin of the B complex, found in many different kinds of food; egg yolk, liver, legumes, and yeast contain large amounts. Biotin is essential to the metabolism of fats. Its absence from the diet may lead to dermatitis.

biotite dark mica, $K(Mg, Fe)_3Al Si_3O_{10}(OH, F)_2$, a common silicate mineral. It is brown to black with shiny surfaces, and like all micas, it splits into very thin flakes along its one perfect cleavage. Biotite is a mineral found in igneous rocks, such as granites, and metamorphic rocks such as schists and gneisses.

bird backboned animal of the class Aves, the biggest group of land vertebrates, characterized by warm blood, feathers, wings, breathing through lungs, and egg-laying by the female. Birds are bipedal; feet are usually adapted for perching and never have more than four toes. Hearing and eyesight are well developed, but the sense of smell is usually poor. No existing species of bird possesses teeth.

Most birds fly, but some groups (such as ostriches) are flightless, and others include flightless members. Many communicate by sounds (nearly half of all known species are songbirds) or by visual displays, in connection with which many species are brightly coloured, usually the males. Birds have highly developed patterns of instinctive behaviour. There are nearly 8,500 species of birds.

Bird

Rifleman, short-tailed pygmy tyrant, frilled coquette, bobwhite, tawny frogmouth, trembler, wattle-eye, fuscous honeyeater, dickcissel, common grackle, and forktailed drongo are all common names for species of bird.

According to the Red List of endangered species published by the World Conservation Union (IUCN) for 1996, 11% of bird species are threatened with extinction.

The wing consists of the typical bones of a forelimb, the humerus, radius and ulna, carpus, metacarpus, and digits. The first digit is the pollex, or thumb, to which some feathers, known as ala spuria, or bastard wing, are attached; the second digit is the index, which bears the large feathers known as the primaries or manuals, usually ten in number. The primary feathers, with the secondaries or cubitals, which are attached to the ulna, form the large wing-quills, called remiges, which are used in flight.

The sternum, or breastbone, of birds is affected by their powers of flight: those birds which are able to fly have a keel projecting from the sternum and serving as the basis of attachment of the great pectoral muscles which move the wings.

In birds that do not fly the keel is absent or greatly reduced. The vertebral column is completed in the tail region by a flat plate known as the pygostyle, which forms a support for the rectrices, or steering tailfeathers. The legs are composed of the femur, tibia and fibula, and the

bones of the foot; the feet usually have four toes, but in many cases there are only three. In swimming birds the legs are placed well back. The uropygial gland on the pygostyle (bone in the tail) is an oil gland used by birds in preening their feathers, as their skin contains no sebaceous glands. The eyes have an upper and a lower eyelid and a semi-transparent nictitating membrane with which the bird can cover its eyes at will.

The **vascular system** contains warm blood, which is kept usually at a higher temperature (about 41°C/106°F) than that of mammals; death from cold is rare unless the bird is starving or ill. The aortic arch (main blood vessel leaving the heart) is on the right side of a bird, whereas it is on the left in a mammal. The heart of a bird consists of a right and a left half with four chambers.

The **lungs** are small and prolonged into air-sacs connected to a number of air-spaces in the bones. These air-spaces are largest in powerful fliers, but they are not so highly developed in young, small, aquatic, and terrestrial birds. These air-spaces increase the efficiency of the respiratory system and reduce the weight of the bones. The lungs themselves are more efficient than those of mammals; the air is circulated through a system of fine capillary tubes, allowing continuous gas exchange to take place, whereas in mammals the air comes to rest in blind air sacs. The organ of voice is not the larynx, but usually the syrinx, a peculiarity of this class formed at the bifurcation of the trachea (windpipe) and the modulations are effected by movements of the adjoining muscles.

digestion Digestion takes place in the oesophagus, stomach, and intestines in a manner basically similar to mammals. The tongue aids in feeding, and there is frequently a **crop**, a dilation of the oesophagus, where food is stored and softened. The stomach is small with little storage capacity and usually consists of the proventriculus, which secretes digestive juices, and the gizzard, which is tough and muscular and grinds the food, sometimes with the aid of grit and stones retained within it. Digestion is completed, and absorption occurs, in the intestine and the digestive caeca. The intestine ends in a cloaca through which both urine and faeces are excreted.

Typically eggs are brooded in a nest and, on hatching, the young receive a period of parental care. The collection of nest material, nest building, and incubation may be carried out by the male, female, or both. The cuckoo neither builds a nest nor rears its own young, but places the eggs in the nest of another bird and leaves the foster parents to care for them. The study of birds is called ◊ornithology. *See illustration overleaf.*

birth act of producing live young from within the body of female animals. Both viviparous and ovoviviparous animals give birth to young. In viviparous animals, embryos obtain nourishment from the mother via a ◊placenta or other means. In ovoviviparous animals, fertilized eggs develop and hatch in the oviduct of the mother and gain little or no nourishment from maternal tissues. See also ◊pregnancy.

birth control another name for ◊family planning; see also ◊contraceptive.

bismuth hard, brittle, pinkish-white, metallic element, symbol Bi, atomic number 83, relative atomic mass 208.98. It has the highest atomic number of all the stable elements (the elements from atomic number 84 up are radioactive). Bismuth occurs in ores and occasionally as a

bird This diagram shows a representative species from each of the 29 orders. There are nearly 8,500 species of birds, of which the largest is the N African ostrich, reaching a height of 2.74 m/9 ft and weighing 156 kg/345 lb. The smallest bird, the bee hummingbird of Cuba and the Isle of Pines, measures 57 mm/2.24 in in length and weighs a mere 1.6 g/0.056 oz.

free metal ($\lozenge$native metal). It is a poor conductor of heat and electricity, and is used in alloys of low melting point and in medical compounds to soothe gastric ulcers. The name comes from the Latin *besemutum*, from the earlier German *Wismut*.

bit (contraction of ***binary digit***) in computing, a single binary digit, either 0 or 1. A bit is the smallest unit of data stored in a computer; all other data must be coded into a pattern of individual bits. A $\lozenge$byte represents sufficient computer memory to store a single character of data, and usually contains eight bits. For example, in the $\lozenge$ASCII code system used by most microcomputers the capital letter A would be stored in a single byte of memory as the bit pattern 01000001.

The maximum number of bits that a computer can normally process at once is called a ***word***. Microcomputers are often described according to how many bits of information they can handle at once. For instance, the first microprocessor, the Intel 4004 (launched in 1971), was a 4-bit device. In the 1970s several different 8-bit computers, many based on the Zilog Z80 or Rockwell 6502 processors, came into common use. During the early 1980s, the IBM personal computer (PC) was introduced, using the Intel 8088 processor, which combined a 16-bit processor with an 8-bit $\lozenge$data bus. Business micros of the later 1980s began to use 32-bit processors such as the Intel 80386 and Motorola 68030. Machines based on the first 64-bit microprocessor appeared in 1993.

The higher the number of bits a computer can process simultaneously, the more powerful the computer is said to be. However, other factors influence the overall speed of a computer system, such as the $\lozenge$clock rate of the processor and the amount of $\lozenge$RAM available. Tasks that require a high processing speed include sorting a database or doing long, complex calculations in spreadsheets. A system running slowly with a $\lozenge$graphical user interface may benefit more from the addition of extra RAM than from a faster processor.

In the PC industry software development lags behind hardware development, so that Windows 3.1, a 16-bit operating system, runs best on the 32-bit microprocessors and upwards, while Windows 95, OS/2, and Windows NT are all 32-bit operating systems and run best on the 64-bit microprocessors and upwards.

bit map in computing, a pattern of bits used to describe the organization of data. Bit maps are used to store typefaces or graphic images, with 1 representing black (or a colour) and 0 white.

A separate set of bit maps is required for each typesize in a bit-mapped typeface. A vector font, by contrast, can be held as one set of data and scaled as required. Bit-mapped graphics are not recommended for images that require scaling or those stored in the form of geometric formulas.

bit pad computer input device; see $\lozenge$graphics tablet.

bitumen impure mixture of hydrocarbons, including such deposits as petroleum, asphalt, and natural gas, although sometimes the term is restricted to a soft kind of pitch resembling asphalt.

Solid bitumen may have arisen as a residue from the evaporation of petroleum. If evaporation took place from a pool or lake of petroleum, the residue might form a pitch or asphalt lake, such as Pitch Lake in Trinidad. Bitumen was used in ancient times as a mortar, and by the Egyptians for embalming.

bivalent in biology, a name given to the pair of homologous chromosomes during reduction division ($\lozenge$meiosis). In chemistry, the term is sometimes used to describe an element or group with a $\lozenge$valency of two, although the term 'divalent' is more common.

black box popular name for the unit containing an aeroplane's flight and voice recorders. These monitor the plane's behaviour and the crew's conversation, thus providing valuable clues to the cause of a disaster. The box is nearly indestructible and usually painted orange for easy recovery. The name also refers to any compact electronic device that can be quickly connected or disconnected as a unit.

The maritime equivalent is the ***voyage recorder***, installed in ships from 1989. It has 350 sensors to record the performance of engines, pumps, navigation lights, alarms, radar, and hull stress.

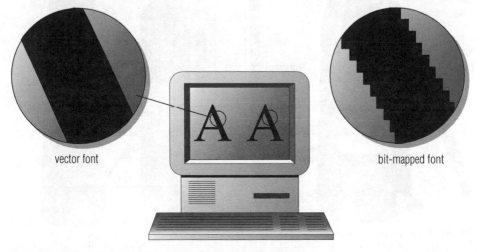

vector font   bit-mapped font

bit map The difference in close-up between a bit-mapped and a vector font.

Black Death great epidemic of bubonic ◊plague that ravaged Europe in the mid-14th century, killing between one-third and half of the population (about 75 million people). The cause of the plague was the bacterium *Yersinia pestis*, transmitted by fleas borne by migrating Asian black rats. The name Black Death was first used in England in the early 19th century.

black earth exceedingly fertile soil that covers a belt of land in NE North America, Europe, and Asia.

In Europe and Asia it extends from Bohemia through Hungary, Romania, S Russia, and Siberia, as far as Manchuria, having been deposited when the great inland ice sheets melted at the close of the last ◊ice age. In North America, it extends from the Great Lakes E through New York State, having been deposited when the last glaciers melted and retreated from the terminal moraine.

black hole object in space whose gravity is so great that nothing can escape from it, not even light. Thought to form when massive stars shrink at the end of their lives, a black hole sucks in more matter, including other stars, from the space around it. Matter that falls into a black hole is squeezed to infinite density at the centre of the hole. Black holes can be detected because gas falling towards them becomes so hot that it emits X-rays.

Black holes containing the mass of millions of stars are thought to lie at the centres of ◊quasars. Satellites have detected X-rays from a number of objects that may be black holes, but only four likely black holes in our Galaxy have been identified.

bladder hollow elastic-walled organ which stores the urine produced in the kidneys. It is present in the ◊urinary systems of some fish, most amphibians, some reptiles, and all mammals. Urine enters the bladder through two ureters, one leading from each kidney, and leaves it through the urethra.

Bladder

An English surgeon received £3,000 and a knighthood for removing a bladder stone from Leopold I, uncle to Queen Victoria. The surgeon, Henry Thompson, removed the stone using a lithotrite – an instrument inserted into the bladder through the penis, which crushes the stone so that it can be expelled naturally.

---

blast freezing industrial method of freezing substances such as foods by blowing very cold air over them. See ◊deep freezing.

blast furnace smelting furnace used to extract metals from their ores, chiefly pig iron from iron ore. The temperature is raised by the injection of an air blast.

In the extraction of iron the ingredients of the furnace are iron ore, coke (carbon), and limestone. The coke is the fuel and provides the carbon monoxide for the reduction of the iron ore; the limestone acts as a flux, removing impurities.

blastocyst in mammals, the hollow ball of cells which is an early stage in the development of the ◊embryo, roughly equivalent to the ◊blastula of other animal groups.

blastomere in biology, a cell formed in the first stages of embryonic development, after the splitting of the fertil-

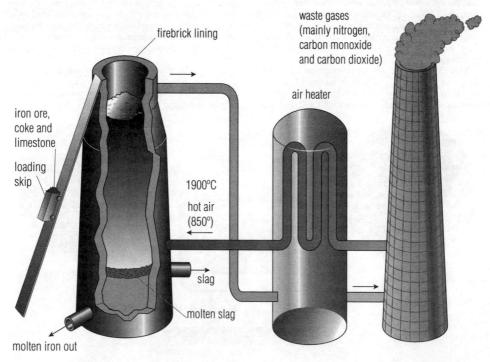

firebrick lining

waste gases (mainly nitrogen, carbon monoxide and carbon dioxide)

air heater

iron ore, coke and limestone

loading skip

1900°C

hot air (850°)

slag

molten slag

molten iron out

blast furnace

ized ovum, but before the formation of the ◊blastula or blastocyst.

blastula early stage in the development of a fertilized egg, when the egg changes from a solid mass of cells (the morula) to a hollow ball of cells (the blastula), containing a fluid-filled cavity (the blastocoel). See also ◊embryology.

bleaching decolorization of coloured materials. The two main types of bleaching agent are the *oxidizing bleaches*, which bring about the ◊oxidation of pigments and include the ultraviolet rays in sunshine, hydrogen peroxide, and chlorine in household bleaches, and the *reducing bleaches*, which bring about ◊reduction and include sulphur dioxide.

Bleaching processes have been known from antiquity, mainly those acting through sunlight. Both natural and synthetic pigments usually possess highly complex molecules, the colour property often being due only to a part of the molecule. Bleaches usually attack only that small part, yielding another substance similar in chemical structure but colourless.

bleeding loss of blood from the circulation; see ◊haemorrhage.

blight any of a number of plant diseases caused mainly by parasitic species of ◊fungus, which produce a whitish appearance on leaf and stem surfaces; for example, *potato blight* *Phytophthora infestans*. General damage caused by aphids or pollution is sometimes known as blight.

blindness complete absence or impairment of sight. It may be caused by heredity, accident, disease, or deterioration with age.

Age-related macular degeneration (AMD), the commonest form of blindness, occurs as the retina gradually deteriorates with age. It affects 1% of people over the age of 70, with many more experiencing marked reduction in sight.

Retinitis pigmentosa, a common cause of blindness, is a hereditary disease affecting 1.2 million people worldwide.

Education of the blind was begun by Valentin Haüy, who published a book with raised lettering in 1784, and founded a school. Aids to the blind include the use of the Braille and Moon alphabets in reading and writing. Guide dogs for the blind were first trained in Germany for soldiers blinded in World War I.

blind spot area where the optic nerve and blood vessels pass through the retina of the ◊eye. No visual image can be formed as there are no light-sensitive cells in this part of the retina.

Thus the organism is blind to objects that fall in this part of the visual field.

blood fluid circulating in the arteries, veins, and capillaries of vertebrate animals; the term also refers to the corresponding fluid in those invertebrates that possess a closed ◊circulatory system. Blood carries nutrients and oxygen to each body cell and removes waste products, such as carbon dioxide. It is also important in the immune response and, in many animals, in the distribution of heat throughout the body.

In humans blood makes up 5% of the body weight, occupying a volume of 5.5 l/10 pt in the average adult. It is composed of a colourless, transparent liquid called

plasma, in which are suspended microscopic cells of three main varieties:

Red cells (erythrocytes) form nearly half the volume of the blood, with about 6 million red cells in every millilitre of an adult's blood. Their red colour is caused by ◊haemoglobin.

White cells (leucocytes) are of various kinds. Some (phagocytes) ingest invading bacteria and so protect the body from disease; these also help to repair injured tissues. Others (lymphocytes) produce antibodies, which help provide immunity.

Blood *platelets* (thrombocytes) assist in the clotting of blood.

Blood cells constantly wear out and die and are replaced from the bone marrow. Red blood cells die at the rate of 200 billion per day but the body produces new cells at an average rate of 9,000 million per hour.

Blood

There are about 6 million red cells in every millilitre of an adult's blood – a millilitre is about the size of an o. The body produces red blood cells at an average rate of 9,000 million per hour. Every day 200,000 million red blood cells die and break up.

---

blood bank in medicine, institution in which blood products are prepared, tested, and stored prior to transfusion into patients. Blood is screened for infections such as hepatitis and HIV and sorted by blood group. Whole blood may be stored for three weeks when refrigerated. It can also be used to obtain blood products, such as platelets and blood cells.

blood–brain barrier theoretical term for the defence mechanism that prevents many substances circulating in the bloodstream (including some germs) from invading the brain.

The blood–brain barrier is not a single entity, but a defensive complex comprising various physical features and chemical reactions to do with the permeability of cells. It ensures that 'foreign' proteins, carried in the blood vessels supplying the brain, do not breach the vessel walls and enter the brain tissue. Many drugs are unable to cross the blood–brain barrier.

blood clotting complex series of events (known as the blood clotting cascade) that prevents excessive bleeding after injury. It is triggered by ◊vitamin K. The result is the formation of a meshwork of protein fibres (fibrin) and trapped blood cells over the cut blood vessels.

When platelets (cell fragments) in the bloodstream come into contact with a damaged blood vessel, they and the vessel wall itself release the enzyme *thrombokinase*, which brings about the conversion of the inactive enzyme *prothrombin* into the active *thrombin*. Thrombin in turn catalyses the conversion of the soluble protein *fibrinogen*, present in blood plasma, to the insoluble *fibrin*. This fibrous protein forms a net over the wound that traps red blood cells and seals the wound; the resulting jellylike clot hardens on exposure to air to form a scab. Calcium, vitamin K, and a variety of enzymes called factors are also necessary for efficient blood clotting. ◊Haemophilia is one of several diseases in which the clotting mechanism is impaired.

blood group any of the types into which blood is classified according to the presence or otherwise of certain

---

From 'sluice gates' to valves

After completing a preliminary medical course at the University of Cambridge, where would an ambitious young man in the 17th century go to get a really good medical training? To the University of Padua in Italy, where the great Italian anatomist Hieronymous Fabricius (1537–1619) taught. So this is where English physician William Harvey (1578–1657) naturally went.

William Harvey had a consuming interest in the movement of the blood in the body. In 1579, Fabricius had publicly demonstrated the valves, which he termed 'sluice gates', in the veins: his principal anatomical work was an accurate and detailed description of them.

Galen's theory

Galen, a Greek physician (c.130– c.200), had 1,500 years previously written a monumental treatise covering every aspect of medicine. In this work, he asserted that food turned to blood in the liver, ebbed and flowed in vessels and, on reaching the heart, flowed through pores in the septum (the dividing wall) from the right to left side, and was sent on its way by heart spasms. The blood did not circulate. This doctrine was still accepted and taught well into the 16th century.

Harvey was unconvinced. He had done a simple calculation. He worked out that for each human heart beat, about 60 cm^3 of blood left the heart, which meant that the heart pumped out 259 litres every hour. This is more than three times the weight of the average man.

Harvey examined the heart and blood vessels of 128 mammals and found that the valve which separated the left side of the heart from the right ventricle is a one-way structure, as were the valves in the veins discovered by his tutor Fabricius. For this reason he decided that the blood in the veins must flow only towards the heart.

Harvey's experiment

Harvey was now in a position to do his famous experiment. He tied a tourniquet round the upper part of his arm. It was just tight enough to prevent the blood from flowing through the veins back into his heart – but not so tight that arterial blood could not enter the arms. Below the tourniquet, the veins swelled up; above it, they remained empty. This showed that the blood could be entering the arm only through the arteries. Further, by carefully stroking the blood out of a short length of vein, Harvey showed that it could fill up only when blood was allowed to enter it from the end that was furthest away from the heart. He had proved that blood in the veins must flow only towards the heart.

Galen's pores in the septum of the heart had never been found. Belgian physician Andreas Versalius (1514–1564) was another alumnus of Padua University. Although brought up in the Galen tradition, he had carried out secret dissections to discover the pores, and had failed. He did, however, show that men and women had the same number of ribs!

Harvey clinched his researches into the movement of the blood when he demonstrated that no blood seeps through the septum of the heart. He reasoned that blood must pass from the right side of the heart to the left through the lungs. He had discovered the circulation of the blood, and thus, some 20 years after he left Padua, became the father of modern physiology. In 1628 Harvey published his proof of the circulation of the blood in his classic book *De motu cardis/On the Motion of the Heart and Blood in Animals*. A new age in medicine and biology had begun.

Julian Rowe

◊antigens on the surface of its red cells. Red blood cells of one individual may carry molecules on their surface that act as antigens in another individual whose red blood cells lack these molecules. The two main antigens are designated A and B. These give rise to four blood groups: having A only (A), having B only (B), having both (AB), and having neither (O). Each of these groups may or may not contain the ◊rhesus factor. Correct typing of blood groups is vital in transfusion, since incompatible types of donor and recipient blood will result in coagulation, with possible death of the recipient.

The ABO system was first described by Austrian scientist Karl Landsteiner in 1902. Subsequent research revealed at least 14 main types of blood group systems, 11 of which are involved with induced ◊antibody production.

Blood typing is also of importance in forensic medicine, cases of disputed paternity, and in anthropological studies.

blood poisoning presence in the bloodstream of quantities of bacteria or bacterial toxins sufficient to cause serious illness.

blood pressure pressure, or tension, of the blood against the inner walls of blood vessels, especially the arteries, due to the muscular pumping activity of the heart. Abnormally high blood pressure (◊hypertension) may be associated with various conditions or arise with no obvious cause; abnormally low blood pressure (hypotension) occurs in ◊shock and after excessive fluid or blood loss from any cause.

In mammals, the left ventricle of the ◊heart pumps blood into the arterial system. This pumping is assisted by waves of muscular contraction by the arteries themselves, but resisted by the elasticity of the inner and outer walls of the same arteries. Pressure is greatest when the heart ventricle contracts (*systole*) and lowest when the ventricle relaxes (*diastole*), and pressure is solely maintained by the elasticity of the arteries. Blood pressure is measured in millimetres of mercury (the height of a column on the measuring instrument, a sphygmomanometer). Normal human blood pressure varies with age, but in a young healthy adult it is around 120/80 mm Hg; the first number represents the systolic pressure and the second the diastolic. Large deviations from this reading usually indicate ill health.

blood test laboratory evaluation of a blood sample. There are numerous blood tests, from simple typing to establish the ◊blood group to sophisticated biochemical assays of substances, such as hormones, present in the blood only in minute quantities.

The majority of tests fall into one of three categories: *haematology* (testing the state of the blood itself), *microbiology* (identifying infection), and *blood chemistry* (reflecting chemical events elsewhere in the body). Before operations, a common test is haemoglobin estimation to determine how well a patient might tolerate blood loss during surgery.

blood transfusion see ◊transfusion.

blood vessel tube that conducts blood either away from or towards the heart in multicellular animals. Freshly oxygenated blood is carried in the arteries – major vessels which give way to the arterioles (small arteries) and finally capillaries; deoxygenated blood is returned to the heart by way of capillaries, then venules (small veins) and veins.

bloom whitish powdery or waxlike coating over the surface of certain fruits that easily rubs off when handled. It often contains ◊yeasts that live on the sugars in the fruit. The term bloom is also used to describe a rapid increase in number of certain species of algae found in lakes, ponds, and oceans.

Such blooms may be natural but are often the result of nitrate pollution, in which artificial fertilizers, applied to surrounding fields, leach out into the waterways. This type of bloom can lead to the death of almost every other organism in the water; because light cannot penetrate the algal growth, the plants beneath can no longer photosynthesize and therefore do not release oxygen into the water. Only those organisms that are adapted to very low levels of oxygen survive.

blubber thick layer of ◊fat under the skin of marine mammals, which provides an energy store and an effective insulating layer, preventing the loss of body heat to the surrounding water. Blubber has been used (when boiled down) in engineering, food processing, cosmetics, and printing, but all of these products can now be produced synthetically.

blue-green algae or *cyanobacteria* single-celled, primitive organisms that resemble bacteria in their internal cell organization, sometimes joined together in colonies or filaments. Blue-green algae are among the oldest known living organisms and, with bacteria, belong to the kingdom Monera; remains have been found in rocks up to 3.5 billion years old. They are widely distributed in aquatic habitats, on the damp surfaces of rocks and trees, and in the soil.

Blue-green algae and bacteria are prokaryotic organisms. Some can fix nitrogen and thus are necessary to the nitrogen cycle, while others follow a symbiotic existence – for example, living in association with fungi to form lichens. Fresh water can become polluted by nitrates and phosphates from fertilizers and detergents. This eutrophication, or overenrichment, of the water causes multiplication of the algae in the form of algae blooms. The algae multiply and cover the water's surface, remaining harmless until they give off toxins as they decay. These toxins kill fish and other wildlife and can be harmful to domestic animals, cattle, and people.

blueprint photographic process used for copying engineering drawings and architectural plans, so called because it produces a white copy of the original against a blue background.

The plan to be copied is made on transparent tracing paper, which is placed in contact with paper sensitized with a mixture of iron ammonium citrate and potassium hexacyanoferrate. The paper is exposed to ◊ultraviolet radiation and then washed in water. Where the light reaches the paper, it turns blue (Prussian blue). The paper underneath the lines of the drawing is unaffected, so remains white.

blue shift in astronomy, a manifestation of the ◊Doppler effect in which an object appears bluer when it is moving towards the observer or the observer is moving towards it (blue light is of a higher frequency than other colours in the spectrum). The blue shift is the opposite of the ◊red shift.

BMA abbreviation for *British Medical Association*.

BMR abbreviation for ◊basal metabolic rate.

Bode's law numerical sequence that gives the approximate distances, in astronomical units (distance between Earth and Sun = one astronomical unit), of the planets from the Sun by adding 4 to each term of the series 0, 3, 6, 12, 24, ... and then dividing by 10. Bode's law predicted the existence of a planet between ◊Mars and ◊Jupiter, which led to the discovery of the asteroids.

The 'law' breaks down for ◊Neptune and ◊Pluto. The relationship was first noted in 1772 by the German mathematician Johann Titius (1729–1796) 1772 (it is also known as the Titius–Bode law).

body mass index (BMI) in medicine, calculation of weight to height ratio that is a useful method of assessing whether an individual is underweight or obese. It is calculated using the formula:

$$BMI = weight\ (kg)/height\ (m^2)$$

Normal values for the BMI in adults are between 20 and 25. Values below 20 indicate that an individual is underweight and values above 30 indicate that an individual is obese. These values can be compared with extensive data obtained from life insurance companies to determine the risk of morbidity or mortality of a particular individual.

Boeing US military and commercial aircraft manufacturer. Among the models Boeing has produced are the B-17 Flying Fortress, 1935; the B-52 Stratofortress, 1952; the Chinook helicopter, 1961; the jetliner, the Boeing 707, 1957; the ◊jumbo jet or Boeing 747, 1969; and the ◊jetfoil, 1975.

The company was founded in 1916 near Seattle, Washington, by William E Boeing (1881–1956) as the Pacific Aero Products Company. Renamed the following year, the company built its first seaplane and in 1919 set up an airmail service between Seattle and Victoria, British Columbia. The company announced in Dec 1996 that they would merge with US aircraft manufacturers McDonnell Douglas to create the world's largest aerospace company, with sales of $48 billion/£29 billion, some 200,000 employees, and an order book of civil and military aircraft worth $100 billion/£60 billion. The new US group would manufacture about three-quarters of the world's commercial airliners. It would operate under the Boeing name and with its principal headquarters in Seattle, WA. The single aerospace giant would transform the whole industry and threaten all of its rivals, including the European consortium Airbus Industrie, which had a 20% share of the commercial airline market. The merger of Boeing and McDonnell Douglas was approved by the European Union in 1997.

bog type of wetland where decomposition is slowed down and dead plant matter accumulates as ◊peat. Bogs develop under conditions of low temperature, high acidity, low nutrient supply, stagnant water, and oxygen deficiency. Typical bog plants are sphagnum moss, rushes, and cotton grass; insectivorous plants such as sundews and bladderworts are common in bogs (insect prey make up for the lack of nutrients).

bohrium synthesized, radioactive element of the ◊transactinide series, symbol Bh, atomic number 107, relative atomic mass 262. It was first synthesized by the Joint Institute for Nuclear Research in Dubna, Russia, in 1976; in 1981 the Laboratory for Heavy Ion Research in Darmstadt, Germany, confirmed its existence. It was named in 1997 after Danish physicist Niels Bohr. Its temporary name was unnilseptium.

Bohr model model of the atom conceived by Danish physicist Neils Bohr in 1913. It assumes that the following rules govern the behaviour of electrons: (1) electrons revolve in orbits of specific radius around the nucleus without emitting radiation; (2) within each orbit, each electron has a fixed amount of energy; electrons in orbits farther away from the nucleus have greater energies; (3) an electron may 'jump' from one orbit of high energy to another of lower energy causing the energy difference to be emitted as a ◊photon of electromagnetic radiation such as light. The Bohr model has been superseded by wave mechanics (see ◊quantum theory).

boiler any vessel that converts water into steam. Boilers are used in conventional power stations to generate steam to feed steam ◊turbines, which drive the electricity generators. They are also used in steamships, which are propelled by steam turbines, and in steam locomotives. Every boiler has a furnace in which fuel (coal, oil, or gas) is burned to produce hot gases, and a system of tubes in which heat is transferred from the gases to the water.

The common kind of boiler used in ships and power stations is the **water-tube** type, in which the water circulates in tubes surrounded by the hot furnace gases. The water-tube boilers at power stations produce steam at a pressure of up to 300 atmospheres and at a temperature of up to 600°C/1,100°F to feed to the steam turbines. It is more efficient than the **fire-tube** type that is used in

steam locomotives. In this boiler the hot furnace gases are drawn through tubes surrounded by water.

boiling point for any given liquid, the temperature at which the application of heat raises the temperature of the liquid no further, but converts it into vapour.

The boiling point of water under normal pressure is 100°C/212°F. The lower the pressure, the lower the boiling point and vice versa.

bolometer sensitive ◊thermometer that measures the energy of radiation by registering the change in electrical resistance of a fine wire when it is exposed to heat or light.

The US astronomer Samuel Langley devised it in 1880 for measuring radiation from stars.

bolometric magnitude in astronomy, a measure of the brightness of a star over all wavelengths. Bolometric magnitude is related to the total radiation output of the star. See ◊magnitude.

Boltzmann constant in physics, the constant (symbol k) that relates the kinetic energy (energy of motion) of a gas atom or molecule to temperature. Its value is 1.38066×10^{-23} joules per Kelvin. It is equal to the gas constant R, divided by ◊Avogadro's number.

bond in chemistry, the result of the forces of attraction that hold together atoms of an element or elements to form a molecule. The principal types of bonding are ◊ionic, ◊covalent, ◊metallic, and ◊intermolecular (such as hydrogen bonding).

bone hard connective tissue comprising the ◊skeleton of most vertebrate animals. Bone is composed of a network of collagen fibres impregnated with mineral salts (largely calcium phosphate and calcium carbonate), a combination that gives it great density and strength, comparable in some cases with that of reinforced concrete. Enclosed within this solid matrix are bone cells, blood vessels, and nerves. The interior of the long bones of the limbs consists of a spongy matrix filled with a soft marrow that produces blood cells.

There are two types of bone: those that develop by replacing ◊cartilage and those that form directly from connective tissue. The latter, which includes the bones of the cranium, are usually platelike in shape and form in the skin of the developing embryo. Humans have about 206 distinct bones in the skeleton, of which the smallest are the three ossicles in the middle ear.

bone marrow substance found inside the cavity of bones. In early life it produces red blood cells but later on lipids (fat) accumulate and its colour changes from red to yellow.

Bone marrow may be transplanted in the treatment of some diseases, such as leukaemia, using immunosuppressive drugs in the recipient to prevent rejection. Transplants to adult monkeys from early aborted monkey fetuses have successfully bypassed rejection.

booster first-stage rocket of a space-launching vehicle, or an additional rocket strapped to the main rocket to assist takeoff.

The US Delta rocket, for example, has a cluster of nine strap-on boosters that fire on liftoff. Europe's Ariane 3 rocket uses twin strap-on boosters, as does the US space shuttle.

boot or **bootstrap** in computing, the process of starting up a computer. Most computers have a small, built-in

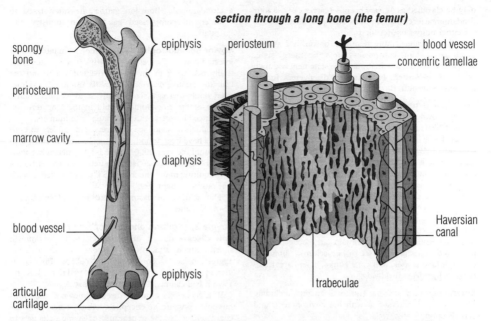

section through a long bone (the femur)

spongy bone

epiphysis

periosteum

blood vessel

concentric lamellae

periosteum

marrow cavity

diaphysis

blood vessel

Haversian canal

epiphysis

trabeculae

articular cartilage

bone Bone is a network of fibrous material impregnated with mineral salts and as strong as reinforced concrete. The upper end of the thighbone or femur is made up of spongy bone, which has a fine lacework structure designed to transmit the weight of the body. The shaft of the femur consists of hard compact bone designed to resist bending. Fine channels carrying blood vessels, nerves, and lymphatics interweave even the densest bone.

boot program that starts automatically when the computer is switched on – its only task is to load a slightly larger program, usually from a hard disc, which in turn loads the main ◊operating system.

In microcomputers the operating system is often held in the permanent ◊ROM memory and the boot program simply triggers its operation.

Some boot programs can be customized so that, for example, the computer, when switched on, always loads and runs a program from a particular backing store or always adopts a particular mode of screen display.

Boötes constellation of the northern hemisphere represented by a herdsman driving a bear (◊Ursa Major) around the pole. Its brightest star is ◊Arcturus (or Alpha Boötis), which is about 37 light years from Earth. The herdsman is assisted by the neighbouring ◊Canes Venatici, 'the Hunting Dogs'.

borax hydrous sodium borate, $Na_2B_4O_7.10H_2O$, found as soft, whitish crystals or encrustations on the shores of hot springs and in the dry beds of salt lakes in arid regions, where it occurs with other borates, halite, and ◊gypsum. It is used in bleaches and washing powders.

A large industrial source is Borax Lake, California. Borax is also used in glazing pottery, in soldering, as a mild antiseptic, and as a metallurgical flux.

bore surge of tidal water up an estuary or a river, caused by the funnelling of the rising tide by a narrowing river mouth. A very high tide, possibly fanned by wind, may build up when it is held back by a river current in the river mouth. The result is a broken wave, a metre or a few feet high, that rushes upstream.

Famous bores are found in the rivers Severn (England), Seine (France), Hooghly (India), and Chang Jiang (China), where bores of over 4 m/13 ft have been reported.

boric acid or **boracic acid** $B(OH)_3$ acid formed by the combination of hydrogen and oxygen with nonmetallic boron. It is a weak antiseptic and is used in the manufacture of glass and enamels. It is also an efficient insecticide against ants and cockroaches.

boron nonmetallic element, symbol B, atomic number 5, relative atomic mass 10.811. In nature it is found only in compounds, as with sodium and oxygen in borax. It exists in two allotropic forms (see ◊allotropy): brown amorphous powder and very hard, brilliant crystals. Its compounds are used in the preparation of boric acid, water softeners, soaps, enamels, glass, and pottery glazes. In alloys it is used to harden steel. Because it absorbs slow neutrons, it is used to make boron carbide control rods for nuclear reactors. It is a necessary trace element in the human diet. The element was named by Humphry Davy, who isolated it in 1808, from borax.

Bose–Einstein condensate hypothesis put forward in 1925 by Albert Einstein and Indian physicist Satyendra Bose, suggesting that when a dense gas is cooled to a little over absolute zero it will condense and its atoms will lose their individuality and act as an organized whole. The first Bose–Einstein condensate was produced in June 1995 by US physicists cooling rubidinum atoms to 10 billionths of a degree above zero. The condensate existed for about a minute before becoming rubidinum ice.

boson in physics, an elementary particle whose spin can only take values that are whole numbers or zero. Bosons may be classified as ◊gauge bosons (carriers of the four fundamental forces) or ◊mesons. All elementary particles are either bosons or ◊fermions.

Unlike fermions, more than one boson in a system (such as an atom) can possess the same energy state. When developed mathematically, this statement is known as the Bose–Einstein law, after its discoverers Indian physicist Satyendra Bose and Albert Einstein.

botanical garden place where a wide range of plants is grown, providing the opportunity to see a botanical diversity not likely to be encountered naturally. Among the earliest forms of botanical garden was the *physic garden*, devoted to the study and growth of medicinal plants; an example is the Chelsea Physic Garden in London, established in 1673 and still in existence. Following increased botanical exploration, botanical gardens were used to test the commercial potential of new plants being sent back from all parts of the world.

Today a botanical garden serves many purposes: education, science, and conservation. Many are associated with universities and also maintain large collections of preserved specimens (see ◊herbarium), libraries, research laboratories, and gene banks. There are 1,600 botanical gardens worldwide.

botany the study of living and fossil ◊plants, including form, function, interaction with the environment, and classification.

Botany is subdivided into a number of specialized studies, such as the identification and classification of plants (taxonomy), their external formation (plant morphology), their internal arrangement (plant anatomy), their microscopic examination (plant histology), their functioning and life history (plant physiology), and their distribution over the Earth's surface in relation to their surroundings (plant ecology). Palaeobotany concerns the study of fossil plants, while economic botany deals with the utility of plants. ◊Horticulture, ◊agriculture, and ◊forestry are branches of botany.

botulism rare, often fatal type of ◊food poisoning. Symptoms include vomiting, diarrhoea, muscular paralysis, breathing difficulties and disturbed vision.

It is caused by a toxin produced by the bacterium *Clostridium botulinum*, found in soil and sometimes in improperly canned foods.

Thorough cooking destroys the toxin, which otherwise suppresses the cardiac and respiratory centres of the brain. In neurology, botulinum toxin is sometimes used to treat rare movement disorders.

Bouguer anomaly in geophysics, an increase in the Earth's gravity observed near a mountain or dense rock mass. This is due to the gravitational force exerted by the rock mass. It is named after its discoverer, the French mathematician Pierre Bouguer (1698–1758), who first observed it in 1735.

boulder clay another name for ◊till, a type of glacial deposit.

Bourdon gauge instrument for measuring pressure, patented by French watchmaker Eugène Bourdon in 1849. The gauge contains a C-shaped tube, closed at one end. When the pressure inside the tube increases, the tube uncurls slightly causing a small movement at its closed end. A system of levers and gears magnifies this movement and turns a pointer, which indicates the pressure on a circular scale. Bourdon gauges are often fitted to cylinders of compressed gas used in industry and hospitals.

bovine somatotropin (BST) hormone that increases an injected cow's milk yield by 10–40%. It is a protein naturally occurring in milk and breaks down within the human digestive tract into harmless amino acids. However, doubts have arisen recently as to whether such a degree of protein addition could in the long term be guaranteed harmless either to cattle or to humans.

Although no evidence of adverse side effects to consumers have been found, BST was banned in Europe in 1993 until the year 2000. In the USA genetically engineered BST has been in use since Feb 1994; in Vermont a law requiring milk containing BST to be labelled as such was passed in Sept 1995.

The incidence of mastitis in herds injected with BST is 15–45% higher.

bovine spongiform encephalopathy (BSE) or *mad cow disease* disease of cattle, related to ◊scrapie in sheep, which attacks the nervous system, causing aggression, lack of coordination, and collapse. First identified in 1985, it is almost entirely confined to the UK. By 1996 it had claimed 158,000 British cattle.

BSE is one of a group of diseases known as the transmissible spongiform encephalopathies, since they are characterized by the appearance of spongy changes in brain tissue. Some scientists believe that all these conditions, including Creutzfeldt–Jakob disease (CJD) in humans, are in effect the same disease, and in 1996 a link was established between the deaths of 10 young people from CJD and the consumption of beef products.

The cause of these universally fatal diseases is not fully understood, but they may be the result of a rogue protein called a ◊prion. A prion may be inborn or it may be transmitted in contaminated tissue.

According to an official European Commission Report released in March 1997, consumers throughout Europe were being exposed to BSE-infected meat. The report also highlighted lax health controls, and supported the view that the extent of BSE throughout the EU was much wider then governments were prepared to admit.

It was also revealed in March 1997 that the British

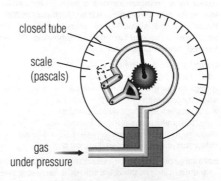

Bourdon gauge The most common form of Bourdon gauge is the C-shaped tube. However, in high-pressure gauges spiral tubes are used; the spiral rotates as pressure increases and the tip screws forwards.

Bovine Spongiform Encephalopathy: Chronology

1970s	Feeds containing animal products, including sheep remains, were developed.
1985	First cases of bovine spongiform encephalopathy (BSE) recorded by vets. UK government epidemiologists announced that the source was probably scrapie in sheep.
1988	The practice of feeding cattle on the brains of other cattle and sheep abandoned in July.
1989	Diseased cattle brains and spinal cords continue to be used in human food products till Nov.
1990	The CJD Surveillance Unit is set up in Edinburgh to monitor any changes in CJD occurrence that may be linked to BSE.
1992	Peak in the number of cases of BSE in cattle, with 700 cases per week.
1995	Around 147,000 cases of BSE had been recorded in UK cows – 18,000 of these had been born since the feed ban – with about 220 new cases per week.
1996	The government announcement in March that 10 young people had died from a variant of Creutzfeldt–Jakob disease (CJD) probably caused by eating infected meat. A worldwide ban on UK beef products was put in place by the European Union. Government scientists admitted that cows can pass BSE to their calves.
1997	Research published by British pathologists proves that the new variant of CJD (vCJD), is caused by the same agent that causes BSE, indicating that the disease has jumped species, from cattle to humans.

government had allowed more than 6,000 carcasses suspected of having BSE to be buried in landfill sites across Britain – in direct contravention of its own regulations. Because of fears that BSE could get into drinking water, or the food chain, both the government and the EU have insisted the carcasses should be incinerated.

Boyle's law law stating that the volume of a given mass of gas at a constant temperature is inversely proportional to its pressure. For example, if the pressure of a gas doubles, its volume will be reduced by a half, and vice versa. The law was discovered in 1662 by Irish physicist and chemist Robert Boyle. See also ◊gas laws.

brachiopod or **lamp shell** any member of the phylum Brachiopoda, marine invertebrates with two shells, resembling but totally unrelated to bivalves.

There are about 300 living species; they were much more numerous in past geological ages. They are suspension feeders, ingesting minute food particles from water. A single internal organ, the lophophore, handles feeding, aspiration, and excretion.

bract leaflike structure in whose ◊axil a flower or inflorescence develops. Bracts are generally green and smaller than the true leaves. However, in some plants they may be brightly coloured and conspicuous, taking over the role of attracting pollinating insects to the flowers, whose own petals are small; examples include poinsettia *Euphorbia pulcherrima* and bougainvillea.

A whorl of bracts surrounding an ◊inflorescence is termed an **involucre**. A **bracteole** is a leaflike organ that arises on an individual flower stalk, between the true bract and the ◊calyx.

brain in higher animals, a mass of interconnected ◊nerve cells forming the anterior part of the ◊central nervous system, whose activities it coordinates and controls. In ◊vertebrates, the brain is contained by the skull. At the base of the ◊brainstem, the **medulla oblongata** contains centres for the control of respiration, heartbeat rate and strength, and blood pressure. Overlying this is the **cerebellum**, which is concerned with coordinating complex muscular processes such as maintaining posture and moving limbs.

The cerebral hemispheres (**cerebrum**) are paired outgrowths of the front end of the forebrain, in early vertebrates mainly concerned with the senses, but in higher

vertebrates greatly developed and involved in the integration of all sensory input and motor output, and in thought, emotions, memory, and behaviour.

In vertebrates, many of the nerve fibres from the two sides of the body cross over as they enter the brain, so that the left cerebral hemisphere is associated with the right side of the body and vice versa. In humans, a certain asymmetry develops in the two halves of the cerebrum. In right-handed people, the left hemisphere seems to play a greater role in controlling verbal and some mathematical skills, whereas the right hemisphere is more involved in spatial perception. In general, however, skills and abilities are not closely localized. In the brain, nerve impulses are passed across ◊synapses by neurotransmitters, in the same way as in other parts of the nervous system.

If the cells and fibre of the human brain were stretched out end to end, they would certainly reach to the Moon and back. Yet the fact that they are not arranged end to end enabled man to go there himself. The astonishing tangle within our heads makes us what we are.

On the **brain** Colin Blakemore
BBC Reith Lecture 1976

In mammals the cerebrum is the largest part of the brain, carrying the **cerebral cortex**. This consists of a thick surface layer of cell bodies (grey matter), below which fibre tracts (white matter) connect various parts of the cortex to each other and to other points in the central nervous system. As cerebral complexity grows, the surface of the brain becomes convoluted into deep folds. In higher mammals, there are large unassigned areas of the brain that seem to be connected with intelligence, personality, and higher mental faculties. Language is controlled in two special regions usually in the left side of the brain: **Broca's area** governs the ability to talk, and **Wernicke's area** is responsible for the comprehension of spoken and written words. In 1990, scientists at Johns Hopkins University, Baltimore, succeeded in culturing human brain cells. *See illustration on page 89.*

brain damage impairment which can be caused by trauma (for example, accidents) or disease (such as encephalitis), or which may be present at birth. Depend-

BSE: the debate goes on

'Mad cow disease' sounds like a joke, but is in fact a deadly serious disease that threatens livestock, people, and a whole tradition of beef farming.

BSE (bovine spongiform encephalopathy) is the cattle version of a group of diseases known as TSEs (transmissible spongiform encephalopathies), which degenerate the nerve cells in mammals. The human version is known as Creutzfeldt-Jakob disease or CJD, and causes brain degeneration and death.

The background

BSE was first identified in 1985. In 1988, UK government epidemiologists announced that the source was probably scrapie, the form of TSE in sheep. They linked the emergence of the cattle disease to changes in industrial plants where protein was recovered from animal carcasses to produce cattle feed. Feeds containing animal products were developed in the 1970s, when scientists found that they could bypass the ruminant process in cows and get protein – from fishmeal or animal remains – into the animals' first 'true' stomach, thus maximizing production. As the price of soya and fishmeal rose, feed-makers turned to sheep remains.

Government action to deal with the initial crisis involved a ban on animal remains (except fishmeal) being included in sheep and cattle feed, the slaughter of animals suspected of having the disease, and the removal from the human food chain of offal believed to carry the disease such as the spinal cord and brain. However, beef itself is not the only product affected. Food which contains beef, beef bone stock, mechanically recovered meat, suet, gelatin, or animal fats includes such products as chicken gravy granules, Christmas pudding, frankfurters, baby food, cakes, and jellies.

The government was initially confident that these steps would contain the problem. By April 1995, however, there had been approximately 147,000 cases of BSE recorded in UK cows – 18,000 of which had been born since the feed ban – with about 220 new cases being reported each week. It was obvious that things were not as simple as had been hoped. One problem was the UK government's failure to carry out research to assess the scrapie/BSE link. The government's reported view was that a single, highly resilient strain of scrapie must have escaped inactivation in the rendering plants, and was thus passed onto cattle.

However, a variety of other hypotheses are now emerging. Robert Rohwer, who studies spongiform encephalopathies in the US, suggests that BSE is a new disease that first arose in a few British cows and then built up over several years as the carcasses were repeatedly recycled in animal feed. Another theory is that the routine use of organophosphate insecticides to eradicate warble fly has affected the nervous system of treated cattle and made them more susceptible to disease.

The human link

A new crisis was triggered by the government announcement in March 1996 that 10 young people had died from a variant of CJD probably caused by eating infected meat. Stephen Dorrell, the Health Secretary at the time, stated that although there was no actual 'scientific proof' that BSE can be transmitted to humans, the scientists concluded that 'the most likely explanation' for the 10 cases was exposure to BSE before the offal ban in 1989. Although the government tried to persuade processors, retailers, and consumers that eating beef had become progressively safer since the controls introduced in 1989, the statements that the risk from eating beef was only 'extremely small' did little to reduce people's fears. Nor did it ally fears in the European Union, which imposed a bitterly contested ban on beef from the UK.

By November 1997 a further 12 cases of new variant CJD had come to light and the Spongiform Encephalopathy Advisory Committee in the UK recommended that white blood cells be removed from donor blood, by a process known as leucodepletion. This advice follows research published by Professor Adriano Aguzzi of Zurich University, Switzerland – who had been working on the problem of how the disease could transmit itself from the gut to the brain – which suggests that white blood cells, lymphocytes, could play an important role in transmitting the infection. Other scientists, notably Dr Bruno Oesch of the company Prionics in Zurich, Switzerland, have been busy developing tests that could screen for the disease agent – abnormal prion proteins – in milk and blood, but a lot of work still needs to be done.

The situation remains unclear. There have been over 400 times as many reported cases of BSE in the UK compared to the rest of the world put together, but there is no evidence that CJD-related deaths are more prevalent in the UK. This, however, could be because of the long lead time for development of CJD.

Researchers are still not precisely sure how BSE developed in the British cattle herd. The number of cases are rising elsewhere and some observers believe that countries have been under-reporting the level of incidence to prevent a public panic.

Undoubtedly, earlier British complacency about the risks to humans in government assurances that the situation was under control, in conjunction with poor implementation of controls, has not helped further our knowledge of the disease. The Spongiform Encephalopathy Research Campaign has argued that by the year 2000 the British could have eaten as many as 1.8 million infected beasts. The disease has already cost the government an estimated £3,300 million.

The reaction

Fears over safety of beef has brought the £500 million per annum UK beef industry to its knees, and had repercussions throughout the world. Schools have taken beef off dinner menus and burger chains are sourcing beef products from outside the UK. A worldwide ban on UK beef products was put in place by the EU. Much discussion has also ensued about wide-scale slaughter policies and who will provide the farmers with financial compensation. Perhaps more importantly, in the long term, the issue has focused attention on the the the safety aspects of modern, intensive agricultural practice.

Sue Stolton and Nigel Dudley

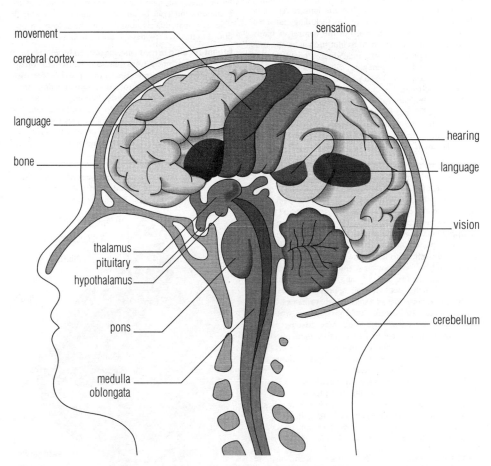

brain The structure of the human brain. At the back of the skull lies the cerebellum, which coordinates reflex actions that control muscular activity. The medulla controls respiration, heartbeat, and blood pressure. The hypothalamus is concerned with instinctive drives and emotions. The thalamus relays signals to and from various parts of the brain. The pituitary gland controls the body's hormones. Distinct areas of the large convoluted cerebral hemispheres that fill most of the skull are linked to sensations, such as hearing and sight, and voluntary activities, such as movement.

ing on the area of the brain that is affected, language, movement, sensation, judgement, or other abilities may be impaired.

brainstem region where the top of the spinal cord merges with the undersurface of the brain, consisting largely of the medulla oblongata and midbrain.

The oldest part of the brain in evolutionary terms, the brainstem is the body's life-support centre, containing regulatory mechanisms for vital functions such as breathing, heart rate, and blood pressure. It is also involved in controlling the level of consciousness by acting as a relay station for nerve connections to and from the higher centres of the brain.

In many countries, death of the brainstem is now formally recognized as death of the person as a whole. Such cases are the principal donors of organs for transplantation. So-called 'beating-heart donors' can be maintained for a limited period by life-support equipment.

brainstem death in medicine, criterion for determining that 'death' has occurred when brain damage results in the irreversible loss of brain function, including that of the brainstem (which is important in the control of many processes which are essential to life, such as respiration and heart rate), so that the individual is unable to live without the aid of a ventilator. Many countries allow artificial ventilation to be stopped once brainstem death has been determined.

Artificial ventilation and heart beat are maintained after brainstem death has been established if organs are to be removed for use in transplantation procedures. This ensures that the organs are in the best condition for transplantation.

brake device used to slow down or stop the movement of a moving body or vehicle. The mechanically applied calliper brake used on bicycles uses a scissor action to press hard rubber blocks against the wheel rim. The main braking system of a car works hydraulically: when the driver depresses the brake pedal, liquid pressure forces pistons to apply brakes on each wheel.

Two types of car brakes are used. **Disc brakes** are used on the front wheels of some cars and on all wheels of sports and performance cars, since they are the more efficient and less prone to fading (losing their braking power) when they get hot. Braking pressure forces brake pads against both sides of a steel disc that rotates with the wheel. **Drum brakes** are fitted on the rear wheels of some cars and on all wheels of some passenger cars. Braking pressure forces brake shoes to expand outwards into contact with a drum rotating with the wheels. The brake pads and shoes have a tough ◊friction lining that grips well and withstands wear.

Many trucks and trains have **air brakes**, which work by compressed air. On landing, jet planes reverse the thrust of their engines to reduce their speed quickly. Space vehicles use retrorockets for braking in space and use the air resistance, or drag of the atmosphere, to slow down when they return to Earth. *See illustration opposite.*

brass metal ◊alloy of copper and zinc, with not more than 5% or 6% of other metals. The zinc content ranges from 20% to 45%, and the colour of brass varies accordingly from coppery to whitish yellow. Brasses are characterized by the ease with which they may be shaped and machined; they are strong and ductile, resist many forms

of corrosion, and are used for electrical fittings, ammunition cases, screws, household fittings, and ornaments.

Brasses are usually classed into those that can be worked cold (up to 25% zinc) and those that are better worked hot (about 40% zinc).

brazing method of joining two metals by melting an ◊alloy into the joint. It is similar to soldering (see ◊solder) but takes place at a much higher temperature. Copper and silver alloys are widely used for brazing, at temperatures up to about 900°C/1,650°F.

breast one of a pair of organs on the chest of the human female, also known as a ◊mammary gland. Each of the two breasts contains milk-producing cells and a network of tubes or ducts that lead to openings in the nipple.

Milk-producing cells in the breast do not become active until a woman has given birth to a baby. Breast milk is made from substances extracted from the mother's blood as it passes through the breasts, and contains all the nourishment a baby needs. Breast-fed newborns develop fewer infections than bottle-fed babies because of the antibodies and white blood cells contained in breast milk. These are particularly abundant in the colostrum produced in the first few days of breast-feeding.

breast-feeding in medicine, the process by which a baby gains its nourishment from milk secreted from the breast, by sucking on the nipple. Breast milk provides complete nutrition and additional protection from infection for the infant, and helps to establish 'bonding' between the mother and the baby. It also helps the mother to return to her previous body weight and encourages the uterus to shrink to normal size following childbirth. The early establishment of breast feeding maintains and improves the milk supply.

Nursing mothers need to increase their intake of food and drinks to ensure that the milk supply is maintained. The extra food intake should be in the form of bread, cereals, pulses, and rice, which also provide generous

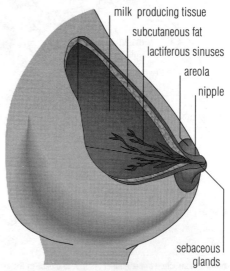

milk producing tissue
subcutaneous fat
lactiferous sinuses
areola
nipple
sebaceous glands

breast The human breast or mammary gland. Milk produced in the tissue of the breast after a woman has given birth feeds the baby along ducts which lead to openings in the nipple.

disc brake

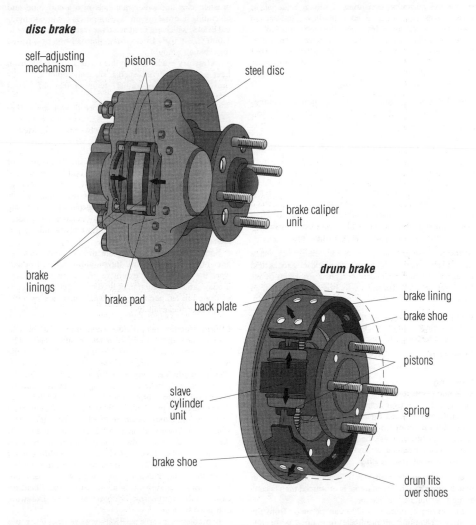

self–adjusting mechanism

pistons

steel disc

brake caliper unit

brake linings

brake pad

back plate

drum brake

brake lining

brake shoe

pistons

slave cylinder unit

spring

brake shoe

drum fits over shoes

brake Two common braking systems: the disc brake (top) and the drum brake (bottom). In the disc brake, increased hydraulic pressure of the brake fluid in the pistons forces the brake pads against the steel disc attached to the wheel. A self-adjusting mechanism balances the force on each pad. In the drum brake, increased pressure of the brake fluid within the slave cylinder forces the brake pad against the brake drum attached to the wheel.

amounts of the vitamins and minerals that are essential for infant health. Infants are usually weaned between four and six months but milk remains a crucial component of their diet even after the introduction of solid foods. Breast-feeding can be continued as long as the mother wants to do so.

breast screening in medicine, examination of the breast to detect the presence of breast cancer at an early stage. Screening methods include self-screening by monthly examination of the breasts and formal programmes of screening by palpation (physical examination) and mammography in special clinics. Screening may be offered to older women on a routine basis and it is important in women with a family history of breast cancer.

Breathalyzer trademark for an instrument for on-the-spot checking by police of the amount of alcohol consumed by a suspect driver. The driver breathes into a plastic bag connected to a tube containing a chemical (such as a diluted solution of potassium dichromate in 50% sulphuric acid) that changes colour in the presence of alcohol. Another method is to use a gas chromatograph, again from a breath sample.

breathing in terrestrial animals, the muscular movements whereby air is taken into the lungs and then expelled, a form of ◊gas exchange. Breathing is sometimes referred to as external respiration, for true respiration is a cellular (internal) process.

Lungs are specialized for gas exchange but are not

themselves muscular, consisting of spongy material. In order for oxygen to be passed to the blood and carbon dioxide removed, air is drawn into the lungs (inhaled) by the contraction of the diaphragm and intercostal muscles; relaxation of these muscles enables air to be breathed out (exhaled). The rate of breathing is controlled by the brain. High levels of activity lead to a greater demand for oxygen and an increased rate of breathing.

breathing rate the number of times a minute the lungs inhale and exhale. The rate increases during exercise because the muscles require an increased supply of oxygen and nutrients. At the same time very active muscles produce a greater volume of carbon dioxide, a waste gas that must be removed by the lungs via the blood.

The regulation of the breathing rate is under both voluntary and involuntary control, although a person can only forcibly stop breathing for a limited time. The regulatory system includes the use of chemoreceptors, which can detect levels of carbon dioxide in the blood. High concentrations of carbon dioxide, occurring for example during exercise, stimulate a fast breathing rate.

breccia coarse-grained clastic ◊sedimentary rock, made up of broken fragments (clasts) of pre-existing rocks held together in a fine-grained matrix. It is similar to ◊conglomerate but the fragments in breccia are jagged in shape.

breed recognizable group of domestic animals, within a species, with distinctive characteristics that have been produced by ◊artificial selection.

breeder reactor or **fast breeder** alternative names for ◊fast reactor, a type of nuclear reactor.

breeding in biology, the crossing and selection of animals and plants to change the characteristics of an existing ◊breed or ◊cultivar (variety), or to produce a new one.

Cattle may be bred for increased meat or milk yield, sheep for thicker or finer wool, and horses for speed or stamina. Plants, such as wheat or maize, may be bred for disease resistance, heavier and more rapid cropping, and hardiness to adverse weather.

breeding in nuclear physics, a process in a reactor in which more fissionable material is produced than is consumed in running the reactor.

For example, plutonium-239 can be made from the relatively plentiful (but nonfissile) uranium-238, or uranium-233 can be produced from thorium. The Pu-239 or U-233 can then be used to fuel other reactors. The French breeder reactor Superphénix, one of the most successful, generates 250 megawatts of electrical power.

brewing making of beer, ale, or other alcoholic beverage, from ◊malt and barley by steeping (mashing), boiling, and fermenting.

Mashing the barley releases its sugars. Yeast is then added, which contains the enzymes needed to convert the sugars into ethanol (alcohol) and carbon dioxide. Hops are added to give a bitter taste.

brewster unit (symbol B) for measuring the reaction of optical materials to stress, defined in terms of the slowing down of light passing through the material when it is stretched or compressed.

brick common block-shaped building material, with all opposite sides parallel. It is made of clay that has been fired in a kiln. Bricks are made by kneading a mixture of crushed clay and other materials into a stiff mud and extruding it into a ribbon. The ribbon is cut into individual bricks, which are fired at a temperature of up to about 1,000°C/1,800°F. Bricks may alternatively be pressed into shape in moulds.

Refractory bricks used to line furnaces are made from heat-resistant materials such as silica and dolomite. They must withstand operating temperatures of about 1,500°C/2,700°F or more.

Sun-dried bricks of mud reinforced with straw were first used in Mesopotamia some 8,000 years ago. Similar mud bricks, called adobe, are still used today in Mexico and other areas where the climate is warm and dry.

brickwork method of construction using bricks made of fired clay or sun-dried earth. In wall building, bricks are either laid out as stretchers (long side facing out) or as headers (short side facing out). The two principle patterns of brickwork are **English bond** in which alternate courses, or layers, are made up of stretchers or headers only, and **Flemish bond** in which stretchers and headers alternate within courses.

Some evidence exists of the use of fired bricks in ancient Mesopotamia and Egypt, although the Romans were the first to make extensive use of this technology. Today's mass-production of fired bricks tends to be concentrated in temperate regions where there are plentiful supplies of fuel available.

bridge structure that provides a continuous path or road over water, valleys, ravines, or above other roads. The basic designs and composites of these are based on the way they bear the weight of the structure and its load. **Beam**, or **girder**, bridges are supported at each end by the ground with the weight thrusting downwards. **Cantilever** bridges are a complex form of girder. **Arch** bridges thrust outwards but downwards at their ends; they are in compression. **Suspension** bridges use cables under tension to pull inwards against anchorages on either side of the span, so that the roadway hangs from the main cables by the network of vertical cables. The **cable-stayed** bridge relies on diagonal cables connected directly between the bridge deck and supporting towers at each end. Some bridges are too low to allow traffic to pass beneath easily, so they are designed with movable parts, like swing and draw bridges.

In prehistory, people used logs or wove vines into ropes that were thrown across the obstacle. By 4000 BC arched structures of stone and/or brick were used in the Middle East, and the Romans built long arched spans, many of which are still standing. Wooden bridges proved vulnerable to fire and rot and many were replaced with cast and wrought iron, but these were disadvantaged by low tensile strength. The ◊Bessemer process produced steel that made it possible to build long-lived framed structures that support great weight over long spans.

brine common name for a solution of sodium chloride (NaCl) in water.

Brines are used extensively in the food-manufacturing industry for canning vegetables, pickling vegetables (sauerkraut manufacture), and curing meat. Industrially, brine is the source from which chlorine, caustic soda (sodium hydroxide), and sodium carbonate are made.

British Standards Institute (BSI) UK national standards body. Although government funded, the institute is independent. The BSI interprets international technical standards for the UK, and also sets its own.

For consumer goods, it sets standards which products should reach (the BS standard), as well as testing products to see that they conform to that standard (as a result of which the product may be given the BSI 'kite' mark).

British thermal unit imperial unit (symbol Btu) of heat, now replaced in the SI system by the ◊joule (one British thermal unit is approximately 1,055 joules). Burning one cubic foot of natural gas releases about 1,000 Btu of heat.

One British thermal unit is defined as the amount of heat required to raise the temperature of 0.45 kg/1 lb of water by 1°F. The exact value depends on the original temperature of the water.

broadcasting the transmission of sound and vision programmes by ◊radio and ◊television. Broadcasting may be organized under private enterprise, as in the USA, or may operate under a compromise system, as in Britain, where a television and radio service controlled by the state-regulated British Broadcasting Corporation (BBC) operates alongside the commercial Independent Television Commission (known as the Independent Broadcasting Authority before 1991).

In the USA, broadcasting is limited only by the issue of licences from the Federal Communications Commission to competing commercial companies; in Britain, the BBC is a centralized body appointed by the state and responsible to Parliament, but with policy and programme content not controlled by the state; in Japan, which ranks next to the USA in the number of television sets owned, there is a semigovernmental radio and television broadcasting corporation (NHK) and numerous private television companies.

Television broadcasting entered a new era with the introduction of high-powered communications satellites in the 1980s. The signals broadcast by these satellites are sufficiently strong to be picked up by a small dish aerial located, for example, on the roof of a house. Direct broadcast by satellite thus became a feasible alternative to land-based television services. See also ◊cable television.

broad-leaved tree another name for a tree belonging to the ◊angiosperms, such as ash, beech, oak, maple, or birch. The leaves are generally broad and flat, in contrast to the needlelike leaves of most conifers. See also ◊deciduous tree.

bromide salt of the halide series containing the Br⁻ ion, which is formed when a bromine atom gains an electron.

The term 'bromide' is sometimes used to describe an organic compound containing a bromine atom, even though it is not ionic. Modern naming uses the term 'bromo-' in such cases. For example, the compound C_2H_5Br is now called bromoethane; its traditional name, still used sometimes, is ethyl bromide.

bromine dark, reddish-brown, nonmetallic element, a volatile liquid at room temperature, symbol Br, atomic number 35, relative atomic mass 79.904. It is a member of the ◊halogen group, has an unpleasant odour, and is very irritating to mucous membranes. Its salts are known as bromides.

Bromine was formerly extracted from salt beds but is now mostly obtained from sea water, where it occurs in small quantities. Its compounds are used in photography and in the chemical and pharmaceutical industries.

bromocriptine drug that mimics the actions of the naturally occurring biochemical substance dopamine, a neurotransmitter. Bromocriptine acts on the pituitary gland to inhibit the release of prolactin, the hormone that regulates lactation, and thus reduces or suppresses milk production. It is also used in the treatment of ◊Parkinson's disease.

Bromocriptine may also be given to control excessive prolactin secretion and to treat prolactinoma (a hormone-producing tumour). Recent research has established its effectiveness in reversing some cases of infertility.

bronchiole small-bore air tube found in the vertebrate lung responsible for delivering air to the main respiratory surfaces. Bronchioles lead off from the larger bronchus and branch extensively before terminating in the many thousand alveoli that form the bulk of lung tissue.

bronchitis inflammation of the bronchi (air passages) of the lungs, usually caused initially by a viral infection, such as a cold or flu. It is aggravated by environmental pollutants, especially smoking, and results in a persistent cough, irritated mucus-secreting glands, and large amounts of sputum.

bronchodilator drug that relieves obstruction of the airways by causing the bronchi and bronchioles to relax and widen. It is most useful in the treatment of ◊asthma.

bronchus one of a pair of large tubes (bronchii) branching off from the windpipe and passing into the vertebrate lung. Apart from their size, bronchii differ from the bronchioles in possessing cartilaginous rings, which give rigidity and prevent collapse during breathing movements.

Numerous glands in the wall of the bronchus secrete a slimy mucus, which traps dust and other particles; the mucus is constantly being propelled upwards to the mouth by thousands of tiny hairs or cilia. The bronchus is adversely effected by several respiratory diseases and by smoking, which damages the cilia and therefore the lung-cleansing mechanism.

bronze alloy of copper and tin, yellow or brown in colour. It is harder than pure copper, more suitable for ◊casting, and also resists ◊corrosion. Bronze may contain as much as 25% tin, together with small amounts of other metals, mainly lead.

Bronze is one of the first metallic alloys known and used widely by early peoples during the period of history known as the ◊Bronze Age.

Bell metal, the bronze used for casting bells, contains 15% or more tin. *Phosphor bronze* is hardened by the addition of a small percentage of phosphorus. *Silicon bronze* (for telegraph wires) and *aluminium bronze* are similar alloys of copper with silicon or aluminium and small amounts of iron, nickel, or manganese, but usually no tin.

Bronze Age stage of prehistory and early history when copper and bronze (an alloy of tin and copper) became the first metals worked extensively and used for tools and weapons. One of the classifications of the Danish archaeologist Christian Thomsen's ◊Three Age System, it developed out of the Stone Age and generally preceded the Iron Age. It first began in the Far East and may be dated 5000–1200 BC in the Middle East and about 2000–500 BC in Europe.

Mining and metalworking were the first specialized

industries, and the invention of the wheel during this time revolutionized transport.

Agricultural productivity (which began during the New Stone Age, or Neolithic period, about 6000 BC) was transformed by the ox-drawn plough, increasing the size of the population that could be supported by farming.

In some areas, including most of Africa, there was no Bronze Age, and ironworking was introduced directly into the Stone Age economy.

brown dwarf in astronomy, an object less massive than a star, but heavier than a planet. Brown dwarfs do not have enough mass to ignite nuclear reactions at their centres, but shine by heat released during their contraction from a gas cloud. Some astronomers believe that vast numbers of brown dwarfs exist throughout the Galaxy. Because of the difficulty of detection, none were spotted until 1995, when US astronomers discovered a brown dwarf, GI229B, in the constellation Lepus. It is about 20–40 times as massive as Jupiter but emits only 1% of the radiation of the smallest known star. In 1996 UK astronomers discovered four possible brown dwarfs within 150 light years of the Sun.

browser in computing, any program that allows the user to search for and view data. Browsers are usually limited to a particular type of data, so, for example, a graphics browser will display graphics files stored in many different file formats. Browsers do not permit the user to edit data, but are sometimes able to convert data from one file format to another.

Web browsers allow access to the World Wide Web. Netscape and Microsoft's Internet Explorer were the leading Web browsers in 1996. They act as a graphical interface to information available on the Internet – they read ◊HTML (hypertext markup language) documents and display them as graphical documents which may include images, video, sound, and ◊hypertext links to other documents.

The first widespread browser for personal computers (PCs) was the text-based program Lynx, which is still used via gateways from text-based on-line systems such as Delphi and CIX. Browsers using ◊graphical user interfaces became widely available from 1993 with the release of Mosaic, written by Marc Andreessen. For some specialist applications such as viewing the virtual reality sites beginning to appear on the Web, a special virtual reality modelling language (VRML) browser is needed.

brucellosis disease of cattle, goats, and pigs, also known when transmitted to humans as **undulant fever** since it remains in the body and recurs. It was named after Australian doctor David Bruce (1855–1931), and is caused by bacteria (genus *Brucella*). It is transmitted by contact with an infected animal or by drinking contaminated milk.

It has largely been eradicated in the West through vaccination of livestock and pasteurization of milk. Brucellosis is especially prevalent in the Mediterranean and in Central and South America. It can be treated with antibacterial drugs.

Brundtland Report the findings of the World Commission on Environment and Development, published in 1987 as *Our Common Future*. It stressed the necessity of environmental protection and popularized the phrase 'sustainable development'. The commission was chaired by the Norwegian prime minister Gro Harlem Brundtland.

bryophyte member of the Bryophyta, a division of the plant kingdom containing three classes: the Hepaticae (◊liverwort), Musci (◊moss), and Anthocerotae (◊hornwort). Bryophytes are generally small, low-growing, terrestrial plants with no vascular (water-conducting)

Netscape Navigator

Microsoft Internet Explorer

browser Two popular World Wide Web browsers, Netscape Navigator and Microsoft Internet Explorer, which provide the user with a straightforward method of accessing information available online.

system as in higher plants. Their life cycle shows a marked ◊alternation of generations. Bryophytes chiefly occur in damp habitats and require water for the dispersal of the male gametes (◊antherozoids).

In bryophytes, the ◊sporophyte, consisting only of a spore-bearing capsule on a slender stalk, is wholly or partially dependent on the ◊gametophyte for water and nutrients. In some liverworts the plant body is a simple ◊thallus, but in the majority of bryophytes it is differentiated into stem, leaves, and ◊rhizoids.

BSE abbreviation for ◊bovine spongiform encephalopathy.

BSI abbreviation for ◊British Standards Institute.

BST abbreviation for *British Summer Time*; also for ◊bovine somatotropin.

Btu symbol for ◊British thermal unit.

bubble chamber in physics, a device for observing the nature and movement of atomic particles, and their interaction with radiation. It is a vessel filled with a superheated liquid through which ionizing particles move and collide. The paths of these particles are shown by strings of bubbles, which can be photographed and studied. By using a pressurized liquid medium instead of a gas, it overcomes drawbacks inherent in the earlier ◊cloud chamber. It was invented by US physicist Donald Glaser in 1952. See ◊particle detector.

bubble memory in computing, a memory device based on the creation of small 'bubbles' on a magnetic surface. Bubble memories typically store up to 4 megabits (4 million ◊bits) of information. They are not sensitive to shock and vibration, unlike other memory devices such as disc drives, yet, like magnetic discs, they are nonvolatile and do not lose their information when the computer is switched off.

bubble sort in computing, a technique for ◊sorting data. Adjacent items are continually exchanged until the data are in sequence.

bubonic plague epidemic disease of the Middle Ages; see ◊plague and ◊Black Death.

buckminsterfullerene form of carbon, made up of molecules (buckyballs) consisting of 60 carbon atoms arranged in 12 pentagons and 20 hexagons to form a perfect sphere. It was named after the US architect and engineer Richard Buckminster Fuller because of its structural similarity to the geodesic dome that he designed. See ◊fullerene.

buckyballs popular name for molecules of ◊buckminsterfullerene.

bud undeveloped shoot usually enclosed by protective scales; inside is a very short stem and numerous undeveloped leaves, or flower parts, or both. Terminal buds are found at the tips of shoots, while axillary buds develop in the ◊axils of the leaves, often remaining dormant unless the terminal bud is removed or damaged. Adventitious buds may be produced anywhere on the plant, their formation sometimes stimulated by an injury, such as that caused by pruning.

budding type of ◊asexual reproduction in which an outgrowth develops from a cell to form a new individual. Most yeasts reproduce in this way.

In a suitable environment, yeasts grow rapidly, forming long chains of cells as the buds themselves produce further buds before being separated from the parent. Simple invertebrates, such as hydra, can also reproduce by budding.

In horticulture, the term is used for a technique of plant propagation whereby a bud (or scion) and a sliver of bark from one plant are transferred to an incision made in the bark of another plant (the stock). This method of ◊grafting is often used for roses.

buffer in chemistry, mixture of compounds chosen to maintain a steady ◊pH. The commonest buffers consist of a mixture of a weak organic acid and one of its salts or a mixture of acid salts of phosphoric acid. The addition of either an acid or a base causes a shift in the ◊chemical equilibrium, thus keeping the pH constant.

buffer in computing, a part of the ◊memory used to store data temporarily while it is waiting to be used. For example, a program might store data in a printer buffer until the printer is ready to print it.

bug in computing, an ◊error in a program. It can be an error in the logical structure of a program or a syntax error, such as a spelling mistake. Some bugs cause a program to fail immediately; others remain dormant, causing problems only when a particular combination of events occurs. The process of finding and removing errors from a program is called *debugging*.

bulb underground bud with fleshy leaves containing a reserve food supply and with roots growing from its base. Bulbs function in vegetative reproduction and are characteristic of many monocotyledonous plants such as the daffodil, snowdrop, and onion. Bulbs are grown on a commercial scale in temperate countries, such as England and the Netherlands.

bulbil small bud that develops above ground from a bud. Bulbils may be formed on the stem from axillary buds, as in members of the saxifrage family, or in the place of flowers, as seen in many species of onion *Allium*. They drop off the parent plant and develop into new individuals, providing a means of ◊vegetative reproduction and dispersal.

bulimia eating disorder in which large amounts of food are consumed in a short time ('binge'), usually followed by depression and self-criticism. The term is often used for *bulimia nervosa*, an emotional disorder in which eating is followed by deliberate vomiting and purging. This may be a chronic stage in ◊anorexia nervosa.

bulldozer earth-moving machine widely used in construction work for clearing rocks and tree stumps and levelling a site. The bulldozer is a kind of tractor with a powerful engine and a curved, shovel-like blade at the front, which can be lifted and forced down by hydraulic rams. It usually has ◊caterpillar tracks so that it can move easily over rough ground.

Bunsen burner gas burner used in laboratories, consisting of a vertical metal tube through which a fine jet of fuel gas is directed. Air is drawn in through airholes near the base of the tube and the mixture is ignited and burns at the tube's upper opening.

The invention of the burner is attributed to German chemist Robert von Bunsen in 1855 but English chemist and physicist Michael Faraday is known to have produced a similar device at an earlier date. A later refine-

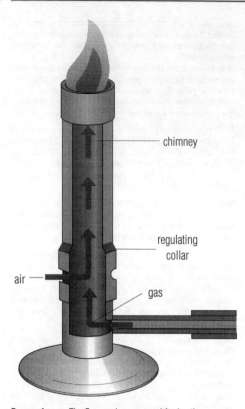

Bunsen burner The Bunsen burner, used for heating laboratory equipment and chemicals. The flame can reach temperatures of 1,500°C/2,732°F and is at its hottest when the collar is open.

ment was the metal collar that can be turned to close or partially close the airholes, thereby regulating the amount of air sucked in and hence the heat of the burner's flame.

buoy floating object used to mark channels for shipping or warn of hazards to navigation. Buoys come in different shapes, such as a pole (spar buoy), cylinder (car buoy), and cone (nun buoy). Light buoys carry a small tower surmounted by a flashing lantern, and bell buoys house a bell, which rings as the buoy moves up and down with the waves. Mooring buoys are heavy and have a ring on top to which a ship can be tied.

buoyancy lifting effect of a fluid on a body wholly or partly immersed in it. This was studied by Archimedes in the 3rd century BC.

bur or **burr** in botany, a type of 'false fruit' or ◊pseudo-carp, surrounded by numerous hooks; for instance, that of burdock *Arctium*, where the hooks are formed from bracts surrounding the flowerhead. Burs catch in the feathers or fur of passing animals, and thus may be dispersed over considerable distances.

burn in medicine, destruction of body tissue by extremes of temperature, corrosive chemicals, electricity, or radiation. *First-degree burns* may cause reddening; *second-degree burns* cause blistering and irritation but usually heal spontaneously; *third-degree burns* are disfiguring and may be life-threatening.

Burns cause plasma, the fluid component of the blood, to leak from the blood vessels, and it is this loss of circulating fluid that engenders ◊shock. Emergency treatment is needed for third-degree burns in order to replace the fluid volume, prevent infection (a serious threat to the severely burned), and reduce the pain. Plastic, or reconstructive, surgery, including skin grafting, may be required to compensate for damaged tissue and minimize disfigurement. If a skin graft is necessary, dead tissue must be removed from a burn (a process known as debridement) so that the patient's blood supply can nourish the graft.

burning common name for ◊combustion.

bus in computing, the electrical pathway through which a computer processor communicates with some of its parts and/or peripherals. Physically, a bus is a set of parallel tracks that can carry digital signals; it may take the form of copper tracks laid down on the computer's ◊printed circuit boards (PCBs), or of an external cable or connection.

A computer typically has three internal buses laid down on its main circuit board: a *data bus*, which carries data between the components of the computer; an *address bus*, which selects the route to be followed by any particular data item travelling along the data bus; and a *control bus*, which is used to decide whether data is written to or read from the data bus. An external *expansion bus* is used for linking the computer processor to peripheral devices, such as modems and printers.

bushel dry or liquid measure equal to eight gallons or four pecks (2,219.36 cu in/36.37 litres) in the UK; some US states have different standards according to the goods measured.

butane C_4H_{10} one of two gaseous alkanes (paraffin hydrocarbons) having the same formula but differing in

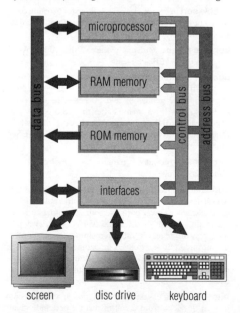

bus The communication path used between the component parts of a computer.

structure. Normal butane is derived from natural gas; isobutane is a by-product of petroleum manufacture. Liquefied under pressure, it is used as a fuel for industrial and domestic purposes (for example, in portable cookers).

butte steep-sided, flat-topped hill, formed in horizontally layered sedimentary rocks, largely in arid areas. A large butte with a pronounced tablelike profile is a ◊mesa.

Buttes and mesas are characteristic of semi-arid areas where remnants of resistant rock layers protect softer rock underneath, as in the plateau regions of Colorado, Utah, and Arizona, USA.

by-product substance formed incidentally during the manufacture of some other substance; for example, slag is a by-product of the production of iron in a ◊blast furnace. For industrial processes to be economical, by-products must be recycled or used in other ways as far as possible; in this example, slag is used for making roads.

Often, a poisonous by-product is removed by trans-forming it into another substance, which although less harmful is often still inconvenient. For example, the sulphur dioxide produced as a by-product of electricity generation can be removed from the smoke stack using ◊flue-gas desulphurization. This process produces large amounts of gypsum, some of which can be used in the building industry.

byte sufficient computer memory to store a single character of data. The character is stored in the byte of memory as a pattern of ◊bits (binary digits), using a code such as ◊ASCII. A byte usually contains eight bits – for example, the capital letter F can be stored as the bit pattern 01000110.

A single byte can specify 256 values, such as the decimal numbers from 0 to 255; in the case of a single-byte ◊pixel (picture element), it can specify 256 different colours. Three bytes (24 bits) can specify 16,777,216 values. Computer memory size is measured in *kilobytes* (1,024 bytes) or *megabytes* (1,024 kilobytes).

°C symbol for degrees ◊Celsius, sometimes called centigrade.

C in computing, a high-level, general-purpose programming language popular on minicomputers and microcomputers. Developed in the early 1970s from an earlier language called BCPL, C was first used as the language of the operating system ◊UNIX, though it has since become widespread beyond UNIX. It is useful for writing fast and efficient systems programs, such as operating systems (which control the operations of the computer).

cable unit of length, used on ships, originally the length of a ship's anchor cable or 120 fathoms (219 m/720 ft), but now taken as one-tenth of a ◊nautical mile (185.3 m/608 ft).

cable car method of transporting passengers up steep slopes by cable. In the **cable railway**, passenger cars are hauled along rails by a cable wound by a powerful winch. A pair of cars usually operates together on the funicular principle, one going up as the other goes down. The other main type is the **aerial cable car**, where the passenger car is suspended from a trolley that runs along an aerial cableway.

A unique form of cable-car system has operated in San Francisco since 1873. The streetcars travel along rails and are hauled by moving cables under the ground.

cable television distribution of broadcast signals through cable relay systems.

Narrow-band systems were originally used to deliver services to areas with poor regular reception; systems with wider bands, using coaxial and fibreoptic cable, are increasingly used for distribution and development of home-based interactive services, typically telephones.

In 1997, the USA had 65 million cable television subscribers using more than 11,000 cable systems. The systems were to spend more than $3.5 billion on basic programming in 1997, up 14% from 1996.

cache memory in computing, a reserved area of the ◊immediate access memory used to increase the running speed of a computer program.

The cache memory may be constructed from ◊SRAM, which is faster but more expensive than the normal ◊DRAM. Most programs access the same instructions or data repeatedly. If these frequently used instructions and data are stored in a fast-access SRAM memory cache, the program will run more quickly. In other cases, the memory cache is normal DRAM, but is used to store frequently used instructions and data that would normally be accessed from ◊backing storage. Access to DRAM is faster than access to backing storage so, again, the program runs more quickly. This type of cache memory is often called a **disc cache**.

CAD (acronym for **computer-aided design**) the use of computers in creating and editing design drawings. CAD also allows such things as automatic testing of designs and multiple or animated three-dimensional views of designs. CAD systems are widely used in architecture, electronics, and engineering, for example in the motor-vehicle industry, where cars designed with the assistance of computers are now commonplace.

A related development is ◊CAM (computer-assisted manufacturing).

cadmium soft, silver-white, ductile, and malleable metallic element, symbol Cd, atomic number 48, relative atomic mass 112.40. Cadmium occurs in nature as a sulphide or carbonate in zinc ores. It is a toxic metal that, because of industrial dumping, has become an environmental pollutant. It is used in batteries, electroplating, and as a constituent of alloys used for bearings with low coefficients of friction; it is also a constituent of an alloy with a very low melting point.

Cadmium is also used in the control rods of nuclear reactors, because of its high absorption of neutrons. It was named in 1817 by the German chemist Friedrich Strohmeyer (1776–1835) after the Greek mythological character Cadmus.

caecum in the ◊digestive system of animals, a blind-ending tube branching off from the first part of the large intestine, terminating in the appendix. It has no function in humans but is used for the digestion of cellulose by some grass-eating mammals.

The rabbit caecum and appendix contains millions of bacteria that produce cellulase, the enzyme necessary for the breakdown of cellulose to glucose. In order to be able to absorb nutrients released by the breakdown of cellulose, rabbits pass food twice down the intestine. They egest soft pellets which are then re-eaten. This is known as coprophagy.

Caesarean section surgical operation to deliver a baby by way of an incision in the mother's abdominal and uterine walls. It may be recommended for almost any obstetric complication implying a threat to mother or baby.

Caesarean section was named after the Roman emperor Julius Caesar, who was born this way. In medieval Europe, it was performed mostly in attempts to save the life of a child whose mother had died in labour. The Christian Church forbade cutting open the mother before she was dead.

caesium soft, silvery-white, ductile metallic element, symbol Cs, atomic number 55, relative atomic mass 132.905. It is one of the ◊alkali metals, and is the most electropositive of all the elements. In air it ignites spontaneously, and it reacts vigorously with water. It is used in the manufacture of photocells. The name comes from the blueness of its spectral line.

The rate of vibration of caesium atoms is used as the standard of measuring time. Its radioactive isotope Cs-137 (half-life 30.17 years) is a product of fission in nuclear explosions and in nuclear reactors; it is one of the most dangerous waste products of the nuclear industry, being a highly radioactive biological analogue for potassium.

caffeine ◊ alkaloid organic substance found in tea, coffee, and kola nuts; it stimulates the heart and central nervous system. When isolated, it is a bitter crystalline compound, $C_8H_{10}N_4O_2$. Too much caffeine (more than six average cups of tea or coffee a day) can be detrimental to health.

caisson hollow cylindrical or boxlike structure, usually of reinforced ◊ concrete, sunk into a riverbed to form the foundations of a bridge.

An **open caisson** is open at the top and at the bottom, where there is a wedge-shaped cutting edge. Material is excavated from inside, allowing the caisson to sink. A **pneumatic caisson** has a pressurized chamber at the bottom, in which workers carry out the excavation. The air pressure prevents the surrounding water entering; the workers enter and leave the chamber through an airlock, allowing for a suitable decompression period to prevent ◊ decompression sickness (the so-called bends).

cal symbol for ◊ **calorie**.

CAL (acronym for **computer-assisted learning**) the use of computers in education and training: the computer displays instructional material to a student and asks questions about the information given; the student's answers determine the sequence of the lessons.

calamine $ZnCO_3$ zinc carbonate, an ore of zinc. The term also refers to a pink powder made of a mixture of zinc oxide and iron(II) oxide used in lotions and ointments as an astringent for treating, for example, sunburn, eczema, measles rash, and insect bites and stings.

In the USA the term refers to zinc silicate $Zn_4Si_2O_7(OH)_2.H_2O$.

calcification in medicine, the deposition of calcium salts in bone cells as part of the growth process. Bone is comprised of fibrous tissue and a matrix containing calcium phosphate and calcium carbonate. Bones grow in thickness from the fibrous tissue and the calcium salts deposited in their cells by calcification.

calcination ◊ oxidation of metals by burning in air.

calcite colourless, white, or light-coloured common rock-forming mineral, calcium carbonate, $CaCO_3$. It is the main constituent of ◊ limestone and marble and forms many types of invertebrate shell.

Calcite often forms ◊ stalactites and stalagmites in caves and is also found deposited in veins through many rocks because of the ease with which it is dissolved and transported by groundwater; ◊ oolite is a rock consisting of spheroidal calcite grains. It rates 3 on the ◊ Mohs' scale of hardness. Large crystals up to 1 m/3 ft have been found in Oklahoma and Missouri, USA. ◊ Iceland spar is a transparent form of calcite used in the optical industry; as limestone it is used in the building industry.

calcium soft, silvery-white metallic element, symbol Ca, atomic number 20, relative atomic mass 40.08. It is one of the ◊ alkaline-earth metals. It is the fifth most abundant element (the third most abundant metal) in the Earth's crust. It is found mainly as its carbonate $CaCO_3$, which occurs in a fairly pure condition as chalk and limestone (see ◊ calcite). Calcium is an essential component of bones, teeth, shells, milk, and leaves, and it forms 1.5% of the human body by mass.

Calcium ions in animal cells are involved in regulating muscle contraction, blood clotting, hormone secretion, digestion, and glycogen metabolism in the liver. It is acquired mainly from milk and cheese, and its uptake is facilitated by vitamin D. Calcium deficiency leads to chronic muscle spasms (tetany); an excess of calcium may lead to the formation of stones (see ◊ calculus) in the kidney or gall bladder.

The element was discovered and named by the English chemist Humphry Davy in 1808. Its compounds include slaked lime (calcium hydroxide, $Ca(OH)_2$); plaster of Paris (calcium sulphate, $CaSO_4.2H_2O$); calcium phosphate ($Ca_3(PO_4)_2$), the main constituent of animal bones; calcium hypochlorite ($CaOCl_2$), a bleaching agent; calcium nitrate ($Ca(NO_3)_2.4H_2O$), a nitrogenous fertilizer; calcium carbide (CaC_2), which reacts with water to give ethyne (acetylene); calcium cyanamide ($CaCN_2$), the basis of many pharmaceuticals, fertilizers, and plastics, including melamine; calcium cyanide ($Ca(CN)_2$), used in the extraction of gold and silver and in electroplating; and others used in baking powders and fillers for paints.

calcium carbonate $CaCO_3$ white solid, found in nature as limestone, marble, and chalk. It is a valuable resource, used in the making of iron, steel, cement, glass, slaked lime, bleaching powder, sodium carbonate and bicarbonate, and many other industrially useful substances.

calculus branch of mathematics which uses the concept of a derivative (see ◊ differentiation) to analyse the way in which the values of a ◊ function vary. Calculus is probably the most widely used part of mathematics. Many real-life problems are analysed by expressing one quantity as a function of another – position of a moving object as a function of time, temperature of an object as a function of distance from a heat source, force on an object as a function of distance from the source of the force, and so on – and calculus is concerned with such functions.

There are several branches of calculus. Differential and integral calculus, both dealing with small quantities which during manipulation are made smaller and smaller, compose the **infinitesimal calculus**. **Differential equations** relate to the derivatives of a set of variables and may include the variables. Many give the mathematical models for physical phenomena such as ◊ simple harmonic motion. Differential equations are solved generally by ◊ integration, depending on their degree. If no analytical processes are available, integration can be performed numerically. Other branches of calculus include calculus of variations and calculus of errors.

> *The mathematician's patterns, like the painter's or the poets, must be beautiful; the ideas, like the colours or the words, must fit together in a harmonious way. Beauty is the first test: there is no permanent place in the world for ugly mathematics.*
>
> On **calculus** Godfrey Hardy
> *A Mathematician's Apology* 1940

caldera in geology, a very large basin-shaped ◊ crater. Calderas are found at the tops of volcanoes, where the original peak has collapsed into an empty chamber beneath. The basin, many times larger than the original volcanic vent, may be flooded, producing a crater lake, or the flat floor may contain a number of small volcanic cones, produced by volcanic activity after the collapse.

Typical calderas are Kilauea, Hawaii; Crater Lake, Oregon, USA; and the summit of Olympus Mons, on Mars. Some calderas are wrongly referred to as craters, such as Ngorongoro, Tanzania.

calendar division of the ◊year into months, weeks, and days and the method of ordering the years. From year one, an assumed date of the birth of Jesus, dates are calculated backwards (BC 'before Christ' or BCE 'before common era') and forwards (AD, Latin *anno Domini* 'in the year of the Lord', or CE 'common era'). The **lunar month** (period between one new moon and the next) naturally averages 29.5 days, but the Western calendar uses for convenience a **calendar month** with a complete number of days, 30 or 31 (Feb has 28). For adjustments, since there are slightly fewer than six extra hours a year left over, they are added to Feb as a 29th day every fourth year (**leap year**), century years being excepted unless they are divisible by 400. For example, 1896 was a leap year; 1900 was not.

The **month names** in most European languages were probably derived as follows: January from Janus, Roman god; February from *Februar*, Roman festival of purification; March from Mars, Roman god; April from Latin *aperire*, 'to open'; May from Maia, Roman goddess; June from Juno, Roman goddess; July from Julius Caesar, Roman general; August from Augustus, Roman emperor; September, October, November, December (originally the seventh to tenth months) from the Latin words meaning seventh, eighth, ninth, and tenth, respectively.

The **days of the week** are Monday named after the Moon; Tuesday from Tiu or Tyr, Anglo-Saxon and Norse god; Wednesday from Woden or Odin, Norse god; Thursday from Thor, Norse god; Friday from Freya, Norse goddess; Saturday from Saturn, Roman god; and Sunday named after the Sun.

All early calendars except the ancient Egyptian were lunar. The word calendar comes from the Latin *Kalendae* or *calendae*, the first day of each month on which, in ancient Rome, solemn proclamation was made of the appearance of the new moon.

The **Western** or **Gregorian calendar** derives from the **Julian calendar** instituted by Julius Caesar 46 BC. It was adjusted by Pope Gregory XIII in 1582, who eliminated the accumulated error caused by a faulty calculation of the length of a year and avoided its recurrence by restricting century leap years to those divisible by 400. Other states only gradually changed from Old Style to New Style; Britain and its colonies adopted the Gregorian calendar in 1752, when the error amounted to 11 days, and 3 Sept 1752 became 14 Sept (at the same time the beginning of the year was put back from 25 March to 1 Jan). Russia did not adopt it until the October Revolution of 1917, so that the event (then 25 Oct) is currently celebrated on 7 Nov.

The **Jewish calendar** is a complex combination of lunar and solar cycles, varied by considerations of religious observance. A year may have 12 or 13 months, each of which normally alternates between 29 and 30 days; the New Year (Rosh Hashanah) falls between 5 Sept and 5 Oct. The calendar dates from the hypothetical creation of the world (taken as 7 Oct 3761 BC).

The **Chinese calendar** is lunar, with a cycle of 60 years. Both the traditional and, from 1911, the Western calendar are in use in China.

The **Muslim calendar**, also lunar, has 12 months of alternately 30 and 29 days, and a year of 354 days. This results in the calendar rotating around the seasons in a 30-year cycle. The era is counted as beginning on the day Muhammad fled from Mecca AD 622.

calibration the preparation of a usable scale on a measuring instrument. A mercury ◊thermometer, for example, can be calibrated with a Celsius scale by noting the heights of the mercury column at two standard temperatures – the freezing point (0°C) and boiling point (100°C) of water – and dividing the distance between them into 100 equal parts and continuing these divisions above and below.

California current cold ocean ◊current in the E Pacific Ocean flowing southwards down the West coast of North America. It is part of the North Pacific ◊gyre (a vast, circular movement of ocean water).

californium synthesized, radioactive, metallic element of the actinide series, symbol Cf, atomic number 98, relative atomic mass 251. It is produced in very small quantities and used in nuclear reactors as a neutron source. The longest-lived isotope, Cf-251, has a half-life of 800 years.

It is named after the state of California, where it was first synthesized in 1950 by US nuclear chemist Glenn Seaborg and his team at the University of California at Berkeley.

callipers measuring instrument used, for example, to measure the internal and external diameters of pipes. Some callipers are made like a pair of compasses, having two legs, often curved, pivoting about a screw at one end. The ends of the legs are placed in contact with the object to be measured, and the gap between the ends is then measured against a rule. The slide calliper looks like an adjustable spanner, and carries a scale for direct measuring, usually with a ◊vernier scale for accuracy.

Callisto second-largest moon of Jupiter, 4,800 km/3,000 mi in diameter, orbiting every 16.7 days at a distance of 1.9 million km/1.2 million mi from the planet. Its surface is covered with large craters.

The space probe *Galileo* detected molecules containing both carbon and nitrogen atoms on the surface of Callisto, US astronomers announced in March 1997. Their presence may indicate that Callisto harboured life at some time.

callus in botany, a tissue that forms at a damaged plant surface. Composed of large, thin-walled ◊parenchyma cells, it grows over and around the wound, eventually covering the exposed area.

In animals, a callus is a thickened pad of skin, formed where there is repeated rubbing against a hard surface. In humans, calluses often develop on the hands and feet of those involved in heavy manual work.

calomel Hg_2Cl_2 (technical name **mercury(I) chloride**) white, heavy powder formerly used as a laxative, now used as a pesticide and fungicide.

calorie c.g.s. unit of heat, now replaced by the ◊joule (one calorie is approximately 4.2 joules). It is the heat required to raise the temperature of one gram of water by 1°C. In dietetics, the Calorie or kilocalorie is equal to 1,000 calories.

The kilocalorie measures the energy value of food in terms of its heat output: 28 g/1 oz of protein yields 120 kilocalories, of carbohydrate 110, of fat 270, and of alcohol 200.

calorific value the amount of heat generated by a given mass of fuel when it is completely burned. It is measured in joules per kilogram. Calorific values are measured experimentally with a bomb calorimeter.

calorimeter instrument used in physics to measure heat. A simple calorimeter consists of a heavy copper vessel that is polished (to reduce heat losses by radiation) and covered with insulating material (to reduce losses by convection and conduction).

In a typical experiment, such as to measure the heat capacity of a piece of metal, the calorimeter is filled with water, whose temperature rise is measured using a thermometer when a known mass of the heated metal is immersed in it. Chemists use a bomb calorimeter to measure the heat produced by burning a fuel completely in oxygen.

calotype paper-based photograph using a wax paper negative, the first example of the ◊negative/positive process invented by the English photographer Fox Talbot in around 1834.

calyptra in mosses and liverworts, a layer of cells that encloses and protects the young ◊sporophyte (spore capsule), forming a sheathlike hood around the capsule. The term is also used to describe the root cap, a layer of ◊parenchyma cells covering the end of a root that gives protection to the root tip as it grows through the soil. This is constantly being worn away and replaced by new cells from a special ◊meristem, the calyptrogen.

calyx collective term for the ◊sepals of a flower, forming the outermost whorl of the ◊perianth. It surrounds the other flower parts and protects them while in bud. In some flowers, for example, the campions *Silene*, the sepals are fused along their sides, forming a tubular calyx.

CAM (acronym for *computer-aided manufacturing*) the use of computers to control production processes; in particular, the control of machine tools and ◊robots in factories. In some factories, the whole design and production system has been automated by linking ◊CAD (computer-aided design) to CAM.

Linking flexible CAD/CAM manufacturing to computer-based sales and distribution methods makes it possible to produce semicustomized goods cheaply and in large numbers.

cam part of a machine that converts circular motion to linear motion or vice versa. The *edge cam* in a car engine is in the form of a rounded projection on a shaft, the camshaft. When the camshaft turns, the cams press against linkages (plungers or followers) that open the valves in the cylinders.

A *face cam* is a disc with a groove in its face, in which the follower travels. A *cylindrical cam* carries angled parallel grooves, which impart a to-and-fro motion to the follower when it rotates.

cambium in botany, a layer of actively dividing cells (lateral ◊meristem), found within stems and roots, that gives rise to ◊secondary growth in perennial plants, causing an increase in girth. There are two main types of cambium: *vascular cambium*, which gives rise to

secondary ◊xylem and ◊phloem tissues, and *cork cambium* (or phellogen), which gives rise to secondary cortex and cork tissues (see ◊bark).

Cambrian period of geological time 570–510 million years ago; the first period of the Palaeozoic era. All invertebrate animal life appeared, and marine algae were widespread. The *Cambrian Explosion* 530–520 million years ago saw the first appearance in the fossil record of all modern animal phyla; the earliest fossils with hard shells, such as trilobites, date from this period.

The name comes from Cambria, the medieval Latin name for Wales, where Cambrian rocks are typically exposed and were first described.

camcorder another name for a ◊video camera.

camera apparatus used in ◊photography, consisting of a lens system set in a light-proof box inside of which a sensitized film or plate can be placed. The lens collects rays of light reflected from the subject and brings them together as a sharp image on the film; it has marked numbers known as ◊apertures, or f-stops, that reduce or increase the amount of light that can enter. Apertures also control depth of field. A shutter controls the amount of time light has to affect the film. There are small-, medium-, and large-format cameras; the format refers to the size of recorded image and the dimensions of the print obtained.

A simple camera has a fixed shutter speed and aperture, chosen so that on a sunny day the correct amount of light is admitted. More complex cameras allow the shutter speed and aperture to be adjusted; most have a built-in exposure meter to help choose the correct combination of shutter speed and aperture for the ambient conditions and subject matter. The most versatile camera is the single lens reflex (◊SLR) which allows the lens to be removed and special lenses attached. A pin-hole camera has a small (pin-sized) hole instead of a lens. It must be left on a firm support during exposures, which are up to ten seconds with slow film, two seconds with fast film and five minutes for paper negatives in daylight. The pin-hole camera gives sharp images from close-up to infinity. *See illustration overleaf.*

camera obscura darkened box with a tiny hole for projecting the inverted image of the scene outside on to a screen inside. For its development as a device for producing photographs, see ◊photography.

camouflage colours or structures that allow an animal to blend with its surroundings to avoid detection by other animals. Camouflage can take the form of matching the background colour, of countershading (darker on top, lighter below, to counteract natural shadows), or of irregular patterns that break up the outline of the animal's body. More elaborate camouflage involves closely resembling a feature of the natural environment, as with the stick insect; this is closely akin to ◊mimicry. Camouflage is also important as a military technique, disguising either equipment, troops, or a position in order to conceal them from an enemy.

camphor $C_{10}H_{16}O$ volatile, aromatic ◊ketone substance obtained from the camphor tree *Cinnamomum camphora*. It is distilled from chips of the wood, and is used in insect repellents and medicinal inhalants and liniments, and in the manufacture of celluloid.

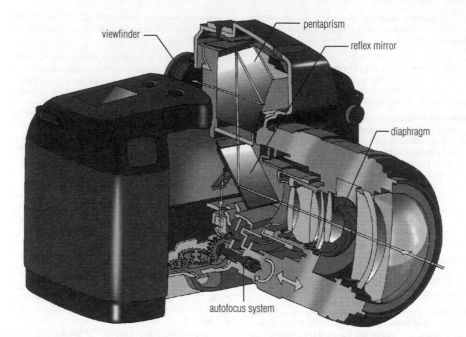

viewfinder — pentaprism

— reflex mirror

— diaphragm

autofocus system

camera The single-lens reflex (SLR) camera in which an image can be seen through the lens before a picture is taken. The reflex mirror directs light entering the lens to the viewfinder. The SLR allows different lenses, such as close-up or zoom, to be used because the photographer can see exactly what is being focused on.

The camphor tree, a member of the family Lauraceae, is native to China, Taiwan, and Japan.

Campylobacter genus of bacteria that cause serious outbreaks of gastroenteritis. They grow best at 43°C/109°F, and so are well suited to the digestive tract of birds. Poultry is therefore the most likely source of a *Campylobacter* outbreak, although the bacteria can also be transmitted via beef or milk. *Campylobacter* can survive in water for up to 15 days, so may be present in drinking water if supplies are contaminated by sewage or reservoirs are polluted by seagulls.

canal artificial waterway constructed for drainage, irrigation, or navigation. *Irrigation canals* carry water for irrigation from rivers, reservoirs, or wells, and are designed to maintain an even flow of water over the whole length. *Navigation and ship canals* are constructed at one level between ◊locks, and frequently link with rivers or sea inlets to form a waterway system. The Suez Canal 1869 and the Panama Canal 1914 eliminated long trips around continents and dramatically shortened shipping routes.

The river Nile has fed canals to maintain life in Egypt since the earliest times. The division of the waters of the Upper Indus and its tributaries, which form an extensive system in Pakistan and Punjab, India, was, for more than ten years, a major cause of dispute between India and Pakistan, settled by a treaty in 1960. The Murray basin, Victoria, Australia, and the Imperial and Central Valley projects in California, USA, are examples of 19th- and 20th-century irrigation-canal development. Excessive extraction of water for irrigation from rivers and lakes can cause environmental damage.

Probably the oldest ship canal to be still in use, as well as the longest, is the Grand Canal in China, which links Tianjin and Hangzhou and connects the Huang He (Yellow River) and Chang Jiang (Yangtze). It was originally built in three stages 485 BC–AD 283, reaching a total length of 1,780 km/1,110 mi. Large sections silted up in later years, but the entire system was dredged, widened, and rebuilt (1958–72) in conjunction with work on flood protection, irrigation, and hydroelectric schemes. It carries millions of tonnes of freight every year.

Where speed is not a prime factor, the cost-effectiveness of transporting goods by canal has encouraged a revival; Belgium, France, Germany, and the countries of the former USSR are among countries that have extended and streamlined their canals. The Baltic–Volga waterway links the Lithuanian port of Klaipeda with Kahovka, at the mouth of the Dnieper on the Black Sea, a distance of 2,430 km/1,510 mi. A further canal cuts across the north Crimea, thus shortening the voyage of ships from the Dnieper through the Black Sea to the Sea of Azov. In Central America, the Panama Canal (1904–14) links the Atlantic and Pacific oceans (64 km/40 mi). In North America, the Erie Canal (1825) links the Great Lakes with the Hudson River and opened up the northeast and Midwest to commerce; the St Lawrence Seaway (1954–59) extends from Montréal to Lake Ontario (290 km/180 mi) and, with the deepening of the Welland Ship Canal and some of the river channels, provides a waterway that enables ocean going vessels to travel (during the ice-free months) between the Atlantic and Duluth, Minnesota, USA, at the western end of Lake Superior, some 3,770 km/2,342 mi.

Canaries current cold ocean current in the North

Atlantic Ocean flowing SW from Spain along the NW coast of Africa. It meets the northern equatorial current at a latitude of 20° N.

cancel in mathematics, to simplify a fraction or ratio by dividing both numerator and denominator by the same number (which must be a common factor of both of them). For example, $5x/25$ cancels to $x/5$ when divided top and bottom by 5.

cancer group of diseases characterized by abnormal proliferation of cells. Cancer (malignant) cells are usually degenerate, capable only of reproducing themselves (tumour formation). Malignant cells tend to spread from their site of origin by travelling through the bloodstream or lymphatic system. Cancer kills about 6 million people a year worldwide.

There are more than 100 types of cancer. Some, like lung or bowel cancer, are common; others are rare. The likely causes remain unexplained. Triggering agents (◊carcinogens) include chemicals such as those found in cigarette smoke, other forms of smoke, asbestos dust, exhaust fumes, and many industrial chemicals. Some viruses can also trigger the cancerous growth of cells (see ◊oncogenes), as can X-rays and radioactivity. Dietary factors are important in some cancers; for example, lack of fibre in the diet may predispose people to bowel cancer and a diet high in animal fats and low in fresh vegetables and fruit increases the risk of breast cancer. Psychological ◊stress may increase the risk of cancer, more so if the person concerned is not able to control the source of the stress.

In some families there is a genetic tendency towards a particular type of cancer. In 1993 researchers isolated the first gene that predisposes individuals to cancer. About 1 in 200 people in the West carry the gene. If the gene mutates, those with the altered gene have a 70% chance of developing colon cancer, and female carriers have a 50% chance of developing cancer of the uterus. This accounts for an estimated 10% of all colon cancer.

In Sept 1994 a gene that triggers breast cancer was identified. *BRCA1* was found to be responsible for almost half the cases of inherited breast cancer, and most cases of ovarian cancer. In Nov 1995 a link between BRCA1 and non-inherited breast cancer was discovered. Women with the gene have an 85% chance of developing breast or ovarian cancer during their lifetime. A second breast cancer gene *BRCA2* was identified in Dec 1995.

Cancer faintest of the zodiacal constellations (its brightest stars are fourth magnitude). It lies in the northern hemisphere between ◊Leo and ◊Gemini, and is represented as a crab. The Sun passes through the constellation during late July and early Aug. In astrology, the dates for Cancer are between about 22 June and 22 July (see ◊precession).

Cancer's most distinctive feature is the open star cluster Praesepe, popularly known as the Beehive, visible to the naked eye as a nebulous patch.

candela SI unit (symbol cd) of luminous intensity, which replaced the old units of candle and standard candle. It measures the brightness of a light itself rather than the amount of light falling on an object, which is called *illuminance* and measured in ◊lux.

One candela is defined as the luminous intensity in a given direction of a source that emits monochromatic radiation of frequency 540×10^{-12} Hz and whose radiant energy in that direction is 1/683 watt per steradian.

Candida albicans yeastlike fungus present in the human digestive tract and in the vagina, which causes no harm in most healthy people. However, it can cause problems if it multiplies excessively, as in vaginal candidiasis or ◊thrush, the main symptom of which is intense itching.

The most common form of thrush is oral, which often occurs in those taking steroids or prolonged courses of antibiotics.

Newborn babies may pick up the yeast during birth and suffer an infection of the mouth and throat. There is also some evidence that overgrowth of *Candida* may occur in the intestines, causing diarrhoea, bloating, and other symptoms such as headache and fatigue, but this is not yet proven. Occasionally, *Candida* can infect immunocompromised patients, such as those with AIDS. Treatment for candidiasis is based on antifungal drugs.

candidiasis in medicine, infections due to the fungus ◊*Candida albicans*.

Canes Venatici constellation of the northern hemisphere near ◊Ursa Major, identified with the hunting dogs of ◊Boötes, the herder. Its stars are faint, and it contains the Whirlpool galaxy (M51), the first spiral galaxy to be recognized.

It contains many objects of telescopic interest, including the relatively bright ◊globular cluster M3. The brightest star, a third magnitude double, is called Cor Caroli or Alpha Canum Venaticorum.

canine in mammalian carnivores, any of the long, often pointed teeth found at the front of the mouth between the incisors and premolars. Canine teeth are used for catching prey, for killing, and for tearing flesh. They are absent in herbivores such as rabbits and sheep, and are much reduced in humans.

Canis Major brilliant constellation of the southern hemisphere, represented (with Canis Minor) as one of the two dogs following at the heel of ◊Orion. Its main star, ◊Sirius, is the brightest star in the night sky.

Epsilon Canis Majoris is also of the first magnitude, and there are three second magnitude stars.

Canis Minor small constellation along the celestial equator (see ◊celestial sphere), represented as the smaller of the two dogs of ◊Orion (the other dog being ◊Canis Major). Its brightest star is the first magnitude ◊Procyon.

Procyon and Beta Canis Minoris form what the Arabs called 'the Short Cubit', in contrast to 'the Long Cubit' formed by ◊Castor and ◊Pollux (Alpha and Beta Geminorum).

canker in medicine, small ◊ulcers that form around the mouth or lips. They often occur when a person is debilitated generally.

cannabis dried leaves and female flowers (marijuana) and ◊resin (hashish) of certain varieties of ◊hemp, which are smoked or swallowed to produce a range of effects, including feelings of great happiness and altered perception. (*Cannabis sativa*, family Cannabaceae.)

Cannabis is a soft drug in that any dependence is psychological rather than physical. It is illegal in many countries and has not been much used in medicine since the 1930s. However, recent research has led to the discovery of cannabis receptors (sensory nerve endings) in the brain, and the discovery of a naturally occurring brain chemical which produces the same effects as smoking cannabis. Researchers believe this work could lead to the use of cannabis-like compounds to treat physical illness without affecting the mind. Cannabis is claimed to have beneficial effects in treating chronic diseases such as AIDS and ◊multiple sclerosis. The main psychoactive ingredient in cannabis is delta-9-tetrahydrocannabinol (THC) which is available legally as a prescribed drug in capsule form.

canning food preservation in hermetically sealed containers by the application of heat. Originated by Nicolas Appert in France in 1809 with glass containers, it was developed by Peter Durand in England in 1810 with cans made of sheet steel thinly coated with tin to delay corrosion. Cans for beer and soft drinks are now generally made of aluminium.

Canneries were established in the USA before 1820, but the US canning industry expanded considerably in the 1870s when the manufacture of cans was mechanized and factory methods of processing were used. The quality and taste of early canned food was frequently inferior but by the end of the 19th century, scientific research made greater understanding possible of the food-preserving process, and standards improved. More than half the aluminium cans used in the USA are now recycled.

Canopus or *Alpha Carinae* second brightest star in the night sky (after Sirius), lying in the southern constellation ◊Carina. It is a first-magnitude yellow-white supergiant about 120 light years from Earth, and thousands of times more luminous than the Sun.

cantilever beam or structure that is fixed at one end only, though it may be supported at some point along its length; for example, a diving board. The cantilever principle, widely used in construction engineering, eliminates the need for a second main support at the free end of the beam, allowing for more elegant structures and reducing the amount of materials required. Many large-span bridges have been built on the cantilever principle.

A typical cantilever bridge consists of two beams cantilevered out from either bank, each supported part way along, with their free ends meeting in the middle. The multiple-cantilever Forth Rail Bridge (completed in 1890) across the Firth of Forth in Scotland has twin main spans of 521 m/1,710 ft.

canyon deep, narrow valley or gorge running through mountains. Canyons are formed by stream down-cutting, usually in arid areas, where the rate of down-cutting is greater than the rate of weathering, and where the stream or river receives water from outside the area.

There are many canyons in the western USA and in Mexico, for example the Grand Canyon of the Colorado River in Arizona, the canyon in Yellowstone National Park, and the Black Canyon in Colorado.

cap another name for a ◊diaphragm contraceptive.

capacitance, electrical property of a capacitor that determines how much charge can be stored in it for a given potential difference between its terminals. It is equal to the ratio of the electrical charge stored to the potential difference. It is measured in ◊farads.

capacitor or *condenser* device for storing electric charge, used in electronic circuits; it consists of two or more metal plates separated by an insulating layer called a dielectric.

Its *capacitance* is the ratio of the charge stored on either plate to the potential difference between the plates. The SI unit of capacitance is the farad, but most capacitors have much smaller capacitances, and the microfarad (a millionth of a farad) is the commonly used practical unit.

Cape Canaveral promontory on the Atlantic coast of Florida, USA, 367 km/228 mi N of Miami, used as a rocket launch site by ◊NASA.

Capella or *Alpha Aurigae* brightest star in the constellation ◊Auriga and the sixth brightest star in the night sky. It is a visual and spectroscopic binary that consists of a pair of yellow-giant stars 45 light years from Earth, orbiting each other every 104 days.

It is a first-magnitude star, whose Latin name means the 'the Little Nanny Goat': its kids are the three adjacent stars Epsilon, Eta, and Zeta Aurigae.

capillarity spontaneous movement of liquids up or down narrow tubes, or capillaries. The movement is due to unbalanced molecular attraction at the boundary between the liquid and the tube. If liquid molecules near the boundary are more strongly attracted to molecules in the material of the tube than to other nearby liquid molecules, the liquid will rise in the tube. If liquid molecules are less attracted to the material of the tube than to other liquid molecules, the liquid will fall.

capillary narrowest blood vessel in vertebrates, 0.008–0.02 mm in diameter, barely wider than a red blood cell. Capillaries are distributed as *beds*, complex networks connecting arteries and veins. Capillary walls are extremely thin, consisting of a single layer of cells, and so nutrients, dissolved gases, and waste products can easily pass through them. This makes the capillaries the main area of exchange between the fluid (◊lymph) bathing body tissues and the blood.

capillary in physics, a very narrow, thick-walled tube, usually made of glass, such as in a thermometer. Properties of fluids, such as surface tension and viscosity, can be studied using capillary tubes.

capitulum in botany, a flattened or rounded head (inflorescence) of numerous, small, stalkless flowers. The capitulum is surrounded by a circlet of petal-like bracts and has the appearance of a large, single flower.

Capricornus zodiacal constellation in the southern hemisphere next to ◊Sagittarius. It is represented as a fish-tailed goat, and its brightest stars are third magnitude. The Sun passes through it late Jan to mid-Feb. In astrology, the dates for Capricornus (popularly known as Capricorn) are between about 22 Dec and 19 Jan (see ◊precession).

capsule in botany, a dry, usually many-seeded fruit formed from an ovary composed of two or more fused ◊carpels, which splits open to release the seeds. The same term is used for the spore-containing structure of mosses and liverworts; this is borne at the top of a long stalk or seta.

Capsules burst open (dehisce) in various ways, including lengthwise, by a transverse lid – for example, scarlet pimpernel *Anagallis arvensis* – or by a number of pores, either towards the top of the capsule, as in the poppy

Papaver, or near the base, as in certain species of bell-flower *Campanula*.

captured rotation or *synchronous rotation* in astronomy, the circumstance in which one body in orbit around another, such as the moon of a planet, rotates on its axis in the same time as it takes to complete one orbit. As a result, the orbiting body keeps one face permanently turned towards the body about which it is orbiting. An example is the rotation of our own ⌕Moon, which arises because of the tidal effects of the Earth over a long period of time.

car small, driver-guided, passenger-carrying motor vehicle; originally the automated version of the horse-drawn carriage, meant to convey people and their goods over streets and roads.

Over 50 million motor cars are produced each year worldwide. The number of cars in the world in 1997 exceeded 500 million. Most are four-wheeled and have water-cooled, piston-type internal-combustion engines fuelled by petrol or diesel. Variations have existed for decades that use ingenious and often nonpolluting power plants, but the motor industry long ago settled on this general formula for the consumer market. Experimental and sports models are streamlined, energy-efficient, and hand-built.

Although it is recorded that in 1479 Gilles de Dom was paid 25 livres (the equivalent of 25 pounds of silver) by the treasurer of Antwerp in the Low Countries for supplying a self-propelled vehicle, the ancestor of the automobile is generally agreed to be the cumbersome steam carriage made by Nicolas-Joseph Cugnot in 1769, still preserved in Paris. Steam was an attractive form of power to the English pioneers, and in 1808 Richard Trevithick built a working steam carriage. Later in the 19th century, practical steam coaches were used for public transport until stifled out of existence by punitive road tolls and legislation.

Although a Frenchman, Jean Etienne Lenoir, patented the first internal-combustion engine (gas-driven) in 1860, and an Austrian, Siegfried Marcus, built a vehicle which was shown at the Vienna Exhibition (1873), two Germans, Gottlieb Daimler and Karl Benz are generally regarded as the creators of the motorcar.

In 1885 Daimler and Benz built and ran the first petrol-driven motorcar. The pattern for the modern motorcar was set by Panhard in 1890 (front radiator, engine under bonnet, sliding-pinion gearbox, wooden ladder-chassis) and Mercedes in 1901 (honeycomb radiator, in-line four-cylinder engine, gate-change gearbox, pressed-steel chassis) set the pattern for the modern car. Emerging with Haynes and Duryea in the early 1890s, US demand was so fervent that 300 makers existed by 1895; only 109 were left by 1900.

In England, cars were still considered to be light loco-motives in the eyes of the law and, since the Red Flag Act in 1865, had theoretically required someone to walk in front with a red flag (by night, a lantern). Despite these obstacles, which put UK development another ten years behind all others, in 1896 Frederick Lanchester produced an advanced and reliable vehicle, later much copied.

The period 1905–06 inaugurated a world motorcar boom continuing to the present day. Among the legendary cars of the early 20th century are: De Dion Bouton, with the first practical high-speed engines; Mors, notable first for racing and later as a silent tourer; Napier, the 24-hour record-holder at Brooklands in 1907, unbeaten

for 17 years; the incomparable Silver Ghost Rolls-Royce; the enduring Model T Ford; and the many types of Bugatti and Delage, from record-breakers to luxury tourers. After World War I popular motoring began with the era of cheap, light (baby) cars made by Citroën, Peugeot, and Renault (France); Austin, Morris, Clyno, and Swift (England); Fiat (Italy); Volkswagen (Germany); and the cheap though bigger Ford, Chevrolet, and Dodge in the USA. During the interwar years a great deal of racing took place, and the experience gained benefited the everyday motorist in improved efficiency, reliability, and safety. There was a divergence between the lighter, economical European car, with its good handling, and the heavier US car, cheap, rugged, and well adapted to long distances on straight roads at speed. By this time motoring had become a universal pursuit.

After World War II small European cars tended to fall into three categories, in about equal numbers: front engine and rear drive, the classic arrangement; front engine and front-wheel drive; rear engine and rear-wheel drive. Racing cars have the engine situated in the middle for balance. From the 1950s a creative resurgence produced in practical form automatic transmission for small cars, rubber suspension, transverse engine mounting, self-levelling ride, disc brakes, and safer wet-weather tyres.

By the mid-1980s, Japan was building 8 million cars a year, on par with the US. The largest Japanese manufacturer, Toyota, was producing 2.5 million cars per year.

A typical present-day medium-sized saloon car has a semi-monocoque construction in which the body panels, suitably reinforced, support the road loads through independent front and rear sprung suspension, with seats located within the wheelbase for comfort. It is usually powered by a ⌕petrol engine using a carburettor to mix petrol and air for feeding to the engine cylinders (typically four or six), and the engine is usually water cooled. In the 1980s high-performance diesel engines were being developed for use in private cars, and it is anticipated that this trend will continue for reasons of economy. From the engine, power is transmitted through a clutch to a four- or five-speed gearbox and from there, in a front-engine rear-drive car, through a drive (propeller) shaft to a ⌕differential gear, which drives the rear wheels. In a front-engine, front-wheel drive car, clutch, gearbox, and final drive are incorporated with the engine unit. An increasing number of high-performance cars are being offered with four-wheel drive, giving superior roadholding in wet and icy conditions and allowing off-road driving.

Cars are responsible for almost a quarter of the world's carbon dioxide emissions. The drive against pollution from the 1960s and the fuel crisis from the 1970s led to experiments with steam cars (cumbersome), diesel engines (slow and heavy, though economical), solar-powered cars, and hybrid cars using both electricity (in town centres) and petrol (on the open road). The industry brought on the market the stratified-charge petrol engine, using a fuel injector to achieve 20% improvement in petrol consumption (the average US car in 1991 did only 27 mi/gal); weight reduction in the body by the use of aluminium and plastics; and 'slippery' body designs with low air resistance, or drag. In 1996 Daimler-Benz unveiled the world's first car to be powered by fuel-cell, which may become the industry's most practical pollution-free alternative. It can cover 250 km/155 mi and reach speeds of over 160 kph/100 mph.

Car: Chronology

1769	Nicholas-Joseph Cugnot in France built a steam tractor.
1801	Richard Trevithick built a steam coach.
1860	Jean Etienne Lenoir built a gas-fuelled internal-combustion engine.
1865	The British government passed the Red Flag Act, requiring a person to precede a 'horseless carriage' with a red flag.
1876	Nikolaus August Otto improved the gas engine, making it a practical power source.
1885	Gottlieb Daimler developed a successful lightweight petrol engine and fitted it to a bicycle to create the prototype of the present-day motorcycle; Karl Benz fitted his lightweight petrol engine to a three-wheeled carriage to pioneer the motorcar.
1886	Gottlieb Daimler fitted his engine to a four-wheeled carriage to produce a four-wheeled motorcar.
1891	René Panhard and Emile Levassor established the present design of cars by putting the engine in front.
1896	Frederick Lanchester introduced epicyclic gearing, which foreshadowed automatic transmission.
1899	C Jenatzy broke the 100 kph barrier in an electric car *La Jamais Contente* at Achères, France, reaching 105.85 kph/65.60 mph.
1901	Ransome Olds in the USA introduced mass production on an assembly line.
1904	Louis Rigolly broke the 100 mph barrier, reaching 166.61 kph/103.55 mph in a Gobron-Brillé at Nice, France.
1908	Henry Ford also used assembly-line production to manufacture his celebrated Model T.
1911	Cadillac introduced the electric starter and dynamo lighting.
1913	Ford introduced the moving conveyor belt to the assembly line, further accelerating production.
1920	Duesenberg began fitting four-wheel hydraulic brakes.
1922	The Lancia Lambda featured unitary (all-in-one) construction and independent front suspension.
1927	Henry Segrave broke the 200 mph barrier in a Sunbeam, reaching 327.89 kph/203.79 mph.
1928	Cadillac introduced the synchromesh gearbox, greatly facilitating gear changing.
1934	Citroën pioneered front-wheel drive in their 7CV model.
1938	Germany produced its 'people's car', the Volkswagen Beetle.
1948	Michelin introduced the radial-ply tyre; Goodrich produced the tubeless tyre.
1950	Dunlop announced the disc brake.
1951	Buick and Chrysler introduced power steering.
1952	Rover's gas-turbine car set a speed record of 243 kph/152 mph.
1954	Carl Bosch introduced fuel injection for cars.
1955	Citroën produced the advanced DS-19 'shark-front' car with hydropneumatic suspension.
1957	Felix Wankel built his first rotary petrol engine.
1959	BMC (now Rover) introduced the Issigonis-designed Mini.
1965	US car manufacturers were forced to add safety features after the publication of Ralph Nader's *Unsafe at Any Speed*.
1966	California introduced legislation regarding air pollution by cars.
1970	American Gary Gabelich drove a rocket-powered car, *Blue Flame*, to a new record speed of 1,001.473 kph/622.287 mph.
1972	Dunlop introduced safety tyres, which seal themselves after a puncture.
1979	American Sam Barrett exceeded the speed of sound in the rocket-engined *Budweiser Rocket*, reaching 1,190.377 kph/ 739.666 mph, a speed not officially recognized as a record because of timing difficulties.
1980	The first mass-produced car with four-wheel drive, the Audi Quattro, was introduced; Japanese car production overtook that of the USA.
1981	BMW introduced the on-board computer, which monitored engine performance.
1983	British driver Richard Noble set an official speed record in the jet-engined *Thrust 2* of 1,019.4 kph/633.5 mph.
1987	The solar-powered *Sunraycer* travelled 3,000 km/1,864 mi from Darwin to Adelaide, Australia, in six days. Toyota Corona production topped 6 million in 29 years.
1988	California introduced stringent controls on car emissions.
1989	The first mass-produced car with four-wheel steering, the Mitsubishi Galant, was launched.
1990	Fiat of Italy and Peugeot of France launched electric passenger cars on the market.
1991	Satellite-based car navigation systems were launched in Japan. European Parliament voted to adopt stringent control of car emissions.
1992	Mazda and NEC of Japan developed an image-processing system for cars, which views the road ahead through a video camera.
1993	A Japanese electric car, the *IZA*, built by the Tokyo Electric Power Company, reached a speed of 176 kph/109 mph (10 kph/6 mph faster than the previous record for an electric car).
1995	Greenpeace designed its own environmentally-friendly car to show the industry how 'it could be done'. It produced a modified Renault Twingo with 30% less wind resistance, capable of doing 67–78 mi to the gallon (100 km per 3–3.5 litres).
1996	Daimler–Benz unveiled the first fuel-cell-powered car. It is virtually pollution-free.
1997	RAF pilot Andy Green broke the sound barrier in Richard Noble's *Thrust SCC*, setting a speed record of 1,149 kph/714 mph.

Microprocessors were also developed to measure temperature, engine speed, pressure, and oxygen/CO_2 content of exhaust gases, and readjust the engine accordingly. In 1992, General Motors and Ford joined forces to develop a battery to propel pollution-free vehicles. $130 million was allocated for research on the project, with an equal amount to be provided by the US Department of Energy. Many developments in the fight against pollution were introduced first by the large and vigorous Japanese motor industry.

In Feb 1997 Chrysler announced that it was developing an electric car that would convert petrol to hydrogen to run the vehicle. The car will seat six, need no maintenance, and emit almost no pollution. It could be available to consumers by 2015.

People can have the Model T in any colour – so long as it's black.

Henry Ford, quoted in A Nevins *Ford*

---

carat unit for measuring the mass of precious stones; it is equal to 0.2 g/0.00705 oz, and is part of the troy system of weights. It is also the unit of purity in gold (US karat). Pure gold is 24-carat; 22-carat (the purest used in jewellery) is 22 parts gold and two parts alloy (to give greater strength).

Originally, one carat was the weight of a carob seed.

carbide compound of carbon and one other chemical element, usually a metal, silicon, or boron.

Calcium carbide (CaC_2) can be used as the starting material for many basic organic chemical syntheses, by the addition of water and generation of ethyne (acetylene). Some metallic carbides are used in engineering because of their extreme hardness and strength. Tungsten carbide is an essential ingredient of carbide tools and high-speed tools. The 'carbide process' was used during World War II to make organic chemicals from coal rather than from oil.

carbohydrate chemical compound composed of carbon, hydrogen, and oxygen, with the basic formula $C_m(H_2O)_n$, and related compounds with the same basic structure but modified ◊functional groups. As sugar and starch, carbohydrates form a major energy-providing part of the human diet.

The simplest carbohydrates are sugars (***monosaccharides***), such as glucose and fructose, and ***disaccharides***, such as sucrose), which are soluble compounds, some with a sweet taste. When these basic sugar units are joined together in long chains or branching structures they form ***polysaccharides***, such as starch and glycogen, which often serve as food stores in living organisms. Even more complex carbohydrates are known, including ◊chitin, which is found in the cell walls of fungi and the hard outer skeletons of insects, and ◊cellulose, which makes up the cell walls of plants. Carbohydrates form the chief foodstuffs of herbivorous animals.

carbolic acid common name for the aromatic compound ◊phenol.

carbon nonmetallic element, symbol C, atomic number 6, relative atomic mass 12.011. It occurs on its own as diamond, graphite, and as fullerenes (the allotropes), as compounds in carbonaceous rocks such as chalk and limestone, as carbon dioxide in the atmosphere, as hydrocarbons in petroleum, coal, and natural gas, and as a constituent of all organic substances.

In its amorphous form, it is familiar as coal, charcoal, and soot. The atoms of carbon can link with one another in rings or chains, giving rise to innumerable complex compounds. Of the inorganic carbon compounds, the chief ones are ***carbon dioxide***, a colourless gas formed when carbon is burned in an adequate supply of air; and ***carbon monoxide*** (CO), formed when carbon is oxidized in a limited supply of air. ***Carbon disulphide*** (CS_2) is a dense liquid with a sweetish odour. Another group of compounds is the ***carbon halides***, including carbon tetrachloride (tetrachloromethane, CCl_4).

When added to steel, carbon forms a wide range of alloys with useful properties. In pure form, it is used as a moderator in nuclear reactors; as colloidal graphite it is a good lubricant and, when deposited on a surface in a vacuum, obviates photoelectric and secondary emission of electrons. Carbon is used as a fuel in the form of coal or coke. The radioactive isotope carbon-14 (half-life 5,730 years) is used as a tracer in biological research. Analysis of interstellar dust has led to the discovery of discrete carbon molecules, each containing 60 carbon atoms. The C_{60} molecules have been named ◊buckminsterfullerenes because of their structural similarity to the geodesic

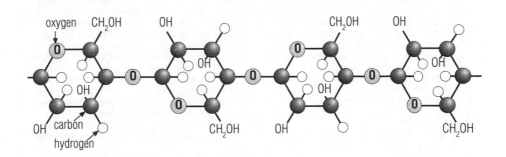

carbohydrate A molecule of the polysaccharide glycogen (animal starch) is formed from linked glucose ($C_6H_{12}O_6$) molecules. A typical glycogen molecule has 100–1,000 glucose units.

From pencil 'lead' to buckyballs

Chemists regard a diamond as just another form of carbon, an element familiar to everyone as soot, which is practically pure carbon. Equally everyday is the graphite in a 'lead' pencil, another form of carbon. The chemical difference between graphite and diamond is that the carbon atoms in each substance are arranged differently. The carbon atoms of graphite are arranged in flat, hexagonal patterns, rather like the cells of a honeycomb. Because graphite molecules are flat, they slide over one another easily. In contrast, the carbon atoms in a diamond are interlinked three-dimensionally, giving the substance its extraordinary hardness. And there until recently the matter rested: carbon was an element that came in two forms – diamond and graphite. Now chemists are excited about a third form of carbon, in which the atoms are linked together in a molecule that looks very like a soccer ball. The new form of carbon is a cagelike molecule consisting of 60 carbon atoms that make a perfect sphere. It has been named 'buckminsterfullerene' in honour of the US architect Buckminster Fuller (1895–1983), whose work included spherical domes.

Soot, space, and lasers

The story of the discovery of these exotic new molecules involves soot, outer space, and lasers. It starts when two scientists, Donald Huffman from the University of Arizona, Tucson, and Wolfgang Kratschmer, were working at the Max Planck Institute for Nuclear Physics in Heidelberg, Germany. They were heating graphite rods under special conditions and examining the soot made in the process: they speculated that a similar process might take place in outer space, contributing to clouds of interstellar dust.

Meanwhile the team of Harold Kroto and David Walton at the University of Sussex had been on the trail of interstellar molecules made up of long chains of carbon atoms that might have originated in the atmosphere that surrounds red giant stars. Enlisting the help of researchers at Rice University in Houston, Texas, who were using a giant laser to blast atoms from the surface of different target substances, Kroto and his team soon found the long-chained carbon molecules. But they were struck by a surprising discovery of a very stable molecule that contained exactly 60 carbon atoms.

Now Richard Smalley of the Rice team set out to make a model of the new molecule, using scissors, sticky tape, and paper. He soon found that a hexagonal arrangement of carbon atoms was impossible, but that a perfect sphere could be formed from 20 hexagons and 12 pentagons. Such a sphere has 60 vertices. Chemists attribute the stability of the new form of carbon to this closed cage structure. However, chemists like to prove the structures of the molecules they make: the evidence so far was merely speculative.

The buckyball's fingerprint

Ordinary soot absorbs ultraviolet light in a characteristic way, and the new molecule also showed a characteristic ultraviolet fingerprint. The problem was to make enough of the new carbon so that exact measurements could be made. If it could be crystallized, then an X-ray analysis would enable the precise distances between the carbon atoms to be determined. The Heidelberg team forged ahead, and produced milligrams of red-brown crystals by evaporating a solution of their product in benzene. The X-ray results confirmed that the molecules were indeed spherical, and that the paper model of 20 hexagons and 12 pentagons was correct.

This result was clinched when Kroto and his colleagues, using nuclear magnetic resonance spectroscopy, not only confirmed the new 60-atom structure but also provided evidence for a family of fullerenes, as the new forms of carbon are now called. Structures containing 28, 32, 50, 60, and 70 carbon atoms are known. Chemists affectionately term such molecules 'buckyballs'.

Promising future

In many ways, these new discoveries in carbon chemistry are as important as the key discovery more than a century ago of the structure of benzene. When in 1865 German chemist Friederich Kekulé (1829–1896) proposed a ring structure for this important organic molecule, the whole field of aromatic chemistry opened up, leading to dyestuffs in the first instance, and millions of new substances since. The fullerene family holds similar promise.

Chemists at Exxon's laboratories in New Jersey have already played a part in the fullerene story, and are interested in the lubricating properties of the new materials. Sumio Iijima, a Japanese scientist, has synthesized tubelike structures based on the fullerene idea, which are naturally called 'buckytubes'. Other teams have now done work that suggests that such molecules may have interesting electrical properties: they may have semiconducting abilities. Cage-like molecules can contain other atoms, such as metals: a group of researchers from the University of California at Los Angeles have produced a 'doped' fullerene that behaves as a superconductor. No evidence has been found that buckminsterfullerene exists in space, but some is almost certainly produced every time you light a candle.

Julian Rowe

domes designed by US architect and engineer Buckminster Fuller.

carbonate CO_3^{2-} ion formed when carbon dioxide dissolves in water; any salt formed by this ion and another chemical element, usually a metal.

Carbon dioxide (CO_2) dissolves sparingly in water (for example, when rain falls through the air) to form carbonic acid (H_2CO_3), which unites with various basic substances to form carbonates. Calcium carbonate ($CaCO_3$) (chalk, limestone, and marble) is one of the most abundant carbonates known, being a constituent of mollusc shells and the hard outer skeletons of crustaceans.

carbon cycle sequence by which ◊carbon circulates and is recycled through the natural world. The carbon element from carbon dioxide, released into the atmosphere by living things as a result of ◊respiration, is taken up by plants during ◊photosynthesis and converted into carbohydrates; the oxygen component is released back into the atmosphere. Some of this carbon becomes locked up in coal and petroleum and other sediments. The simplest link in the carbon cycle occurs when an animal eats a plant and carbon is transferred from, say, a leaf cell to the animal body. The oceans absorb 25–40% of all carbon dioxide released into the atmosphere.

Today, the carbon cycle is in danger of being disrupted by the increased consumption and burning of fossil fuels, and the burning of large tracts of tropical forests, as a result of which levels of carbon dioxide are building up in the atmosphere and probably contributing to the ◊greenhouse effect.

You will die but the carbon will not; its career does not end with you ... it will return to the soil, and there a plant may take it up again in time, sending it once more on a cycle of plant and animal life.

On the **carbon cycle** Jacob Bronowski
'Biography of an Atom – and the Universe'
New York Times 13 Oct 1968

carbon dating alternative name for ◊radiocarbon dating.

carbon dioxide CO_2 colourless, odourless gas, slightly soluble in water and denser than air. It is formed by the complete oxidation of carbon.

It is produced by living things during the processes of respiration and the decay of organic matter, and plays a vital role in the carbon cycle. It is used as a coolant in its solid form (known as 'dry ice'), and in the chemical

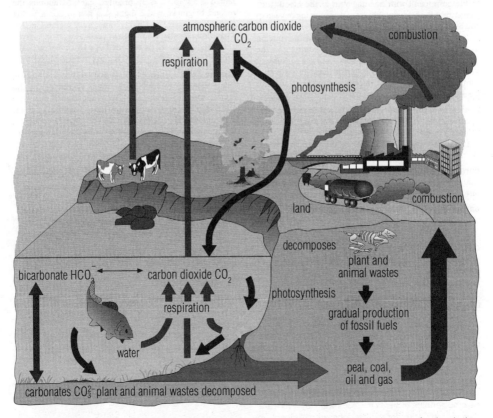

carbon cycle The carbon cycle is necessary for the continuation of life. Since there is only a limited amount of carbon in the Earth and its atmosphere, carbon must be continuously recycled if life is to continue. Other chemicals necessary for life – nitrogen, sulphur, and phosphorus, for example – also circulate in natural cycles.

industry. Its increasing density contributes to the ◊greenhouse effect and ◊global warming. Britain has 1% of the world's population, yet it produces 3% of CO_2 emissions; the USA has 5% of the world's population and produces 25% of CO_2 emissions. Annual releases of carbon dioxide reached 23 billion tones in 1997.

carbon fibre fine, black, silky filament of pure carbon produced by heat treatment from a special grade of Courtelle acrylic fibre and, used for reinforcing plastics. The resulting composite is very stiff and, weight for weight, has four times the strength of high-tensile steel. It is used in the aerospace industry, cars, and electrical and sports equipment.

Carboniferous period of geological time 362.5–290 million years ago, the fifth period of the Palaeozoic era. In the USA it is divided into two periods: the Mississippian (lower) and the Pennsylvanian (upper).

Typical of the lower-Carboniferous rocks are shallow-water ◊limestones, while upper-Carboniferous rocks have ◊delta deposits with ◊coal (hence the name). Amphibians were abundant, and reptiles evolved during this period.

carbon monoxide CO colourless, odourless gas formed when carbon is oxidized in a limited supply of air. It is a poisonous constituent of car exhaust fumes, forming a stable compound with haemoglobin in the blood, thus preventing the haemoglobin from transporting oxygen to the body tissues.

In industry, carbon monoxide is used as a reducing agent in metallurgical processes – for example, in the extraction of iron in ◊blast furnaces – and is a constituent of cheap fuels such as water gas. It burns in air with a luminous blue flame to form carbon dioxide.

Carborundum trademark for a very hard, black abrasive, consisting of silicon carbide (SiC), an artificial compound of carbon and silicon. It is harder than ◊corundum but not as hard as ◊diamond.

It was first produced in 1891 by US chemist Edward Acheson (1856–1931).

carboxyl group –COOH in organic chemistry, the acidic functional group that determines the properties of fatty acids (carboxylic acids) and amino acids.

carbuncle in medicine, a collection of boils forming an abscess, caused by bacterial infection. It is usually treated with antibiotics.

carburation mixing of a gas, such as air, with a volatile hydrocarbon fuel, such as petrol, kerosene, or fuel oil, in order to form an explosive mixture. The process, which ensures that the maximum amount of heat energy is released during combustion, is used in internal-combustion engines. In most petrol engines the liquid fuel is atomized and mixed with air by means of a device called a *carburettor*.

carcinogen any agent that increases the chance of a cell becoming cancerous (see ◊cancer), including various chemical compounds, some viruses, X-rays, and other forms of ionizing radiation. The term is often used more narrowly to mean chemical carcinogens only.

carcinogenesis in medicine, the means by which the changes responsible for the development of ◊cancer are brought about.

carcinoma malignant ◊tumour arising from the skin, the

glandular tissues, or the mucous membranes that line the gut and lungs.

cardiac pacemaker in medicine, another name for ◊pacemaker.

cardinal number in mathematics, one of the series of numbers 0, 1, 2, 3, 4, ... Cardinal numbers relate to quantity, whereas ordinal numbers (first, second, third, fourth, ...) relate to order.

cardioid heart-shaped curve traced out by a point on the circumference of a circle, resulting from the circle rolling around the edge of another circle of the same diameter.

The polar equation of the cardioid is of the form:

$$r = a(1 + \cos \theta)$$

cardiopulmonary bypass in medicine, another term for bypass surgery.

caries decay and disintegration, usually of the substance of teeth (cavity) or bone. It is caused by acids produced when the bacteria that live in the mouth break down sugars in the food. Fluoride, a low sugar intake, and regular brushing are all protective. Caries form mainly in the 45 minutes following consumption of sugary food.

Carina constellation of the southern hemisphere, represented as a ship's keel. Its brightest star is ◊Canopus, the second brightest in the night sky; it also contains Eta Carinae, a massive and highly luminous star embedded in a gas cloud, perhaps 8,000 light years away.

Carina was formerly regarded as part of Argo, and is situated in one of the brightest parts of the ◊Milky Way.

Carnac site of prehistoric ◊megaliths in Brittany, France, where remains of tombs and stone alignments of the period 2000–1500 BC (Neolithic and early Bronze Age) are found. Stones removed for local building have left some gaps in the alignments.

There are various groups of menhirs (standing stones)

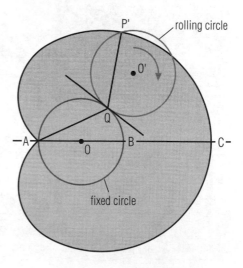

cardioid The cardioid is the curve formed when one circle rolls around the edge of another circle of the same size. It is named after its heart shape.

round the village of Carnac in the *département* of Morbihan, situated at Kermario (place of the dead), Kerlescan (place of burning), Erdeven, and St-Barbe. The largest of the stone alignments has 1,000 blocks of grey granite up to 4 m/13 ft high, extending over 2 km/1.2 mi. These menhirs are arranged in 11 parallel rows, with a circle at the western end.

Stone circles and alignments are thought to be associated with astronomical and religious ritual, and those at Carnac may possibly have been used for calculating the phases of the moon.

carnassial tooth one of a powerful scissorlike pair of molars, found in all mammalian carnivores except seals. Carnassials are formed from an upper premolar and lower molar, and are shaped to produce a sharp cutting surface. Carnivores such as dogs transfer meat to the back of the mouth, where the carnassials slice up the food ready for swallowing.

carnivore in zoology, mammal of the order Carnivora. Although its name describes the flesh-eating ancestry of the order, it includes pandas, which are herbivorous, and civet cats, which eat fruit.

Carnivores have the greatest range of body size of any mammalian order, from the 100 g/3.5 oz weasel to the 800 kg/1,764 lb polar bear.

The characteristics of the Carnivora are sharp teeth, small incisors, a well-developed brain, a simple stomach, a reduced or absent caecum, and incomplete or absent clavicles (collarbones); there are never less than four toes on each foot; the scaphoid and lunar bones are fused in the hand; and the claws are generally sharp and powerful.

Carnot cycle series of changes in the physical condition of a gas in a reversible heat engine, necessarily in the following order: (1) isothermal expansion (without change of temperature), (2) adiabatic expansion (without change of heat content), (3) isothermal compression, and (4) adiabatic compression.

The principles derived from a study of this cycle are important in the fundamentals of heat and ◊thermodynamics.

carotene naturally occurring pigment of the ◊carotenoid group. Carotenes produce the orange, yellow, and red colours of carrots, tomatoes, oranges, and crustaceans.

carotenoid any of a group of yellow, orange, red, or brown pigments found in many living organisms, particularly in the ◊chloroplasts of plants. There are two main types, the **carotenes** and the **xanthophylls**. Both types are long-chain lipids (◊fats).

Some carotenoids act as accessory pigments in ◊photosynthesis, and in certain algae they are the principal light-absorbing pigments functioning more efficiently than ◊chlorophyll in low-intensity light. Carotenoids can also occur in organs such as petals, roots, and fruits, giving them their characteristic colour, as in the yellow and orange petals of wallflowers *Cheiranthus*. They are also responsible for the autumn colours of leaves, persisting longer than the green chlorophyll, which masks them during the summer.

carotid artery one of a pair of major blood vessels, one on each side of the neck, supplying blood to the head.

carpel female reproductive unit in flowering plants (◊angiosperms). It usually comprises an ◊ovary containing one or more ovules, the stalk or style, and a ◊stigma at

its top which receives the pollen. A flower may have one or more carpels, and they may be separate or fused together. Collectively the carpels of a flower are known as the ◊gynoecium.

carrier in medicine, anyone who harbours an infectious organism without ill effects but can pass the infection to others. The term is also applied to those who carry a recessive gene for a disease or defect without manifesting the condition.

carrying capacity in ecology, the maximum number of animals of a given species that a particular area can support. When the carrying capacity is exceeded, there is insufficient food (or other resources) for the members of the population. The population may then be reduced by emigration, reproductive failure, or death through starvation.

Cartesian coordinates in ◊coordinate geometry, components used to define the position of a point by its perpendicular distance from a set of two or more axes, or reference lines. For a two-dimensional area defined by two axes at right angles (a horizontal *x*-axis and a vertical *y*-axis), the coordinates of a point are given by its perpendicular distances from the *y*-axis and *x*-axis, written in the form (x,y). For example, a point P that lies three units from the *y*-axis and four units from the *x*-axis has Cartesian coordinates (3,4) (see ◊abscissa and ◊ordinate). In three-dimensional coordinate geometry, points are located with reference to a third, *z*-axis, mutually at right angles to the *x* and *y* axes.

The Cartesian coordinate system can be extended to any finite number of dimensions (axes), and is used thus in theoretical mathematics. It is named after the French mathematician, René Descartes. The system is useful in creating technical drawings of machines or buildings, and in computer-aided design (◊CAD).

cartilage flexible bluish-white ◊connective tissue made up of the protein collagen. In cartilaginous fish it forms the skeleton; in other vertebrates it forms the greater part of the embryonic skeleton, and is replaced by ◊bone in the course of development, except in areas of wear such as bone endings, and the discs between the backbones. It also forms structural tissue in the larynx, nose, and external ear of mammals.

Cartilage does not heal itself, so where injury is severe the joint may need to be replaced surgically. In a 1994 trial, Swedish doctors repaired damaged knee joints by implanting cells cultured from the patient's own cartilage.

cartography art and practice of drawing ◊maps.

caryopsis dry, one-seeded ◊fruit in which the wall of the seed becomes fused to the carpel wall during its development. It is a type of ◊achene, and therefore develops from one ovary and does not split open to release the seed. Caryopses are typical of members of the grass family (Gramineae), including the cereals.

casein main protein of milk, from which it can be separated by the action of acid, the enzyme rennin, or bacteria (souring); it is also the main protein in cheese. Casein is used as a protein supplement in the treatment of malnutrition. It is used commercially in cosmetics, glues, and as a sizing for coating paper.

cash crop crop grown solely for sale rather than for the farmer's own use, for example, coffee, cotton, or sugar beet. Many Third World countries grow cash crops to meet their debt repayments rather than grow food for their own people. The price for these crops depends on financial interests, such as those of the multinational companies and the International Monetary Fund.

Countries dependent on cash crops for income include Uganda, Nicaragua, and Somalia.

Cassegrain telescope or *Cassegrain reflector* type of reflecting ◊telescope in which light collected by a concave primary mirror is reflected on to a convex secondary mirror, which in turn directs it back through a hole in the primary mirror to a focus behind it. As a result, the telescope tube can be kept short, allowing equipment for analyzing and recording starlight to be mounted behind the main mirror. All modern large astronomical telescopes are of the Cassegrain type.

It is named after the 17th century French astronomer, Cassegrain who first devised it as an improvement to the simpler ◊Newtonian telescope.

Cassini joint space probe of the US agency NASA and the European Space Agency to the planet Saturn. *Cassini* was launched in October 1997 and is due to go into orbit around Saturn 2004, dropping off a sub-probe, *Huygens*, to land on Saturn's largest moon, Titan.

Cassiopeia prominent constellation of the northern hemisphere, named after the mother of Andromeda. It has a distinctive W-shape, and contains one of the most powerful radio sources in the sky, Cassiopeia A. This is the remains of a ◊supernova (star explosion) that occurred c. 1667, too far away to be seen from Earth.

It was in Cassiopeia that Tycho Brahe observed a new star in 1572, probably a supernova, since it was visible in daylight and outshone ◊Venus for ten days.

cassiterite or *tinstone* mineral consisting of reddish-brown to black stannic oxide (SnO_2), usually found in granite rocks. It is the chief ore of tin. When fresh it has a bright ('adamantine') lustre. It was formerly extensively mined in Cornwall, England; today Malaysia is the world's main supplier. Other sources of cassiterite are Africa, Indonesia, and South America.

casting process of producing solid objects by pouring molten material into a shaped mould and allowing it to cool. Casting is used to shape such materials as glass and plastics, as well as metals and alloys.

The casting of metals has been practised for more than 6,000 years, using first copper and bronze, then iron, and now alloys of zinc and other metals. The traditional method of casting metal is *sand casting*. Using a model of the object to be produced, a hollow mould is made in a damp sand and clay mix. Molten metal is then poured into the mould, taking its shape when it cools and solidifies. The sand mould is broken up to release the casting. Permanent metal moulds called *dies* are also used for casting, in particular, small items in mass-production processes where molten metal is injected under pressure into cooled dies. *Continuous casting* is a method of shaping bars and slabs that involves pouring molten metal into a hollow, water-cooled mould of the desired cross section.

cast iron cheap but invaluable constructional material, most commonly used for car engine blocks. Cast iron is partly refined pig (crude) ◊iron, which is very fluid when molten and highly suitable for shaping by casting; it contains too many impurities (for example, carbon) to be readily shaped in any other way. Solid cast iron is heavy and can absorb great shock but is very brittle.

Castor or *Alpha Geminorum* second brightest star in the constellation ◊Gemini and the 23rd brightest star in the night sky. Along with the brighter ◊Pollux, it forms a prominent pair at the eastern end of Gemini, representing the head of the twins.

Second-magnitude Castor is 45 light years from Earth, and is one of the finest ◊binary stars in the sky for small telescopes. The two main components orbit each other over a period of 467 years. A third, much fainter, star orbits the main pair over a period probably exceeding 10,000 years. Each of the three visible components is a spectroscopic binary, making Castor a sextuple star system.

castration removal of the sex glands (either ovaries or testes). Male domestic animals may be castrated to prevent reproduction, to make them larger or more docile, or to eradicate disease.

Castration of humans was used in ancient and medieval times and occasionally later to preserve the treble voice of boy singers or, by Muslims, to provide eunuchs, trustworthy harem guards. If done in childhood, it inhibits sexual development: for instance, the voice remains high, and growth of hair on the face and body is reduced, owing to the absence of the hormones normally secreted by the testes.

catabolism in biology, the destructive part of ◊metabolism where living tissue is changed into energy and waste products.

It is the opposite of ◊anabolism. It occurs continuously in the body, but is accelerated during many disease processes, such as fever, and in starvation.

catalepsy in medicine, an abnormal state in which the patient is apparently or actually unconscious and the muscles become rigid.

There is no response to stimuli, and the rate of heartbeat and breathing is slow. A similar condition can be drug-induced or produced by hypnosis, but catalepsy as ordinarily understood occurs spontaneously in epilepsy, schizophrenia, and other nervous disorders. It is associated with catatonia.

catalyst substance that alters the speed of, or makes possible, a chemical or biochemical reaction but remains unchanged at the end of the reaction. ◊Enzymes are natural biochemical catalysts. In practice most catalysts are used to speed up reactions.

catalytic converter device fitted to the exhaust system of a motor vehicle in order to reduce toxic emissions from the engine. It converts harmful exhaust products to relatively harmless ones by passing the exhaust gases over a mixture of catalysts coated on a metal or ceramic honeycomb (a structure that increases the surface area and therefore the amount of active catalyst with which the exhaust gases will come into contact). *Oxidation catalysts* (small amounts of precious palladium and platinum metals) convert hydrocarbons (unburnt fuel) and carbon monoxide into carbon dioxide and water, but do not affect nitrogen oxide emissions. *Three-way catalysts* (platinum and rhodium metals) convert nitrogen oxide gases into nitrogen and oxygen.

Over the lifetime of a vehicle, a catalytic converter can reduce hydrocarbon emissions by 87%, carbon monoxide emissions by 85%, and nitrogen oxide emissions by 62%, but will cause a slight increase in the amount of carbon dioxide emitted.

Catalytic converters are destroyed by emissions from leaded petrol and work best at a temperature of 300°C. The benefits of catalytic converters are offset by any increase in the number of cars in use.

catamaran twin-hulled sailing vessel, based on the aboriginal craft of South America and the Indies, made of logs lashed together, with an outrigger. A similar vessel with three hulls is known as a trimaran. Car ferries with a wave-piercing catamaran design are also in use in parts of Europe and North America. They have a pointed main hull and two outriggers and travel at a speed of 35 knots (84.5 kph/52.5 mph).

cataract eye disease in which the crystalline lens or its capsule becomes cloudy, causing blindness. Fluid accumulates between the fibres of the lens and gives place to deposits of ◊albumin. These coalesce into rounded bodies, the lens fibres break down, and areas of the lens or the lens capsule become filled with opaque products of degeneration. The condition is estimated to have blinded more than 25 million people worldwide, and 150,000 in the UK.

The condition nearly always affects both eyes, usually one more than the other. In most cases, the treatment is replacement of the opaque lens with an artificial implant.

catarrh inflammation of any mucous membrane, especially of the nose and throat, with increased production of mucus.

catastrophe theory mathematical theory developed by René Thom in 1972, in which he showed that the growth of an organism proceeds by a series of gradual changes that are triggered by, and in turn trigger, large-scale changes or 'catastrophic' jumps. It also has applications in engineering – for example, the gradual strain on the structure of a bridge that can eventually result in a sudden collapse – and has been extended to economic and psychological events.

catch crop crop such as turnip that is inserted between two principal crops in a rotation in order to provide some quick livestock feed or soil improvement at a time when the land would otherwise be lying idle.

In the gap between harvesting a crop of winter-sown wheat and sowing a spring variety of barley, for example, an additional catch crop of turnips or ryegrass can be produced for animal feed in the late winter period when other green fodder is scarce. When the catch crop is ploughed under, the succeeding spring crop benefits from the improvement to the soil.

catchment area in earth sciences, the area from which water is collected by a river and its tributaries. In the social sciences the term may be used to denote the area from which people travel to obtain a particular service or product, such as the area from which a school draws its pupils.

catecholamine chemical that functions as a ◊neurotransmitter or a ◊hormone. Dopamine, adrenaline (epinephrine), and noradrenaline (norepinephrine) are catecholamines.

catenary curve taken up by a flexible cable suspended between two points, under gravity; for example, the curve

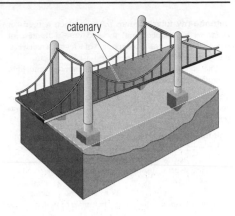

catenary

of overhead suspension cables that hold the conductor wire of an electric railway or tramway.

caterpillar track trade name for an endless flexible belt of metal plates on which certain vehicles such as tanks and bulldozers run, which takes the place of ordinary tyred wheels and improves performance on wet or uneven surfaces.

A track-laying vehicle has a track on each side, and its engine drives small cogwheels that run along the top of the track in contact with the ground. The advantage of such tracks over wheels is that they distribute the vehicle's weight over a wider area and are thus ideal for use on soft and waterlogged as well as rough and rocky ground.

catheter fine tube inserted into the body to introduce or remove fluids. The urinary catheter, passed by way of the urethra (the duct that leads urine away from the bladder) was the first to be used. In today's practice, catheters can be inserted into blood vessels, either in the limbs or trunk, to provide blood samples and local pressure measurements, and to deliver drugs and/or nutrients directly into the bloodstream.

cathode in chemistry, the negative electrode of an electrolytic ◊cell, towards which positive particles (cations), usually in solution, are attracted. See ◊electrolysis.

A cathode is given its negative charge by connecting it to the negative side of an external electrical supply. This is in contrast to the negative electrode of an electrical (battery) cell, which acquires its charge in the course of a spontaneous chemical reaction taking place within the cell.

cathode in electronics, the part of an electronic device in which electrons are generated. In a thermionic valve, electrons are produced by the heating effect of an applied current; in a photocell, they are produced by the interaction of light and a semiconducting material. The cathode is kept at a negative potential relative to the device's other electrodes (anodes) in order to ensure that the liberated electrons stream away from the cathode and towards the anodes.

cathode-ray oscilloscope (CRO) instrument used to measure electrical potentials or voltages that vary over time and to display the waveforms of electrical oscillations or signals. Readings are displayed graphically on the screen of a ◊cathode-ray tube.

cathode-ray tube vacuum tube in which a beam of electrons is produced and focused onto a fluorescent screen. It is an essential component of television receivers, computer visual display units, and oscilloscopes.

The electrons' kinetic energy is converted into light energy as they collide with the screen.

cation ◊ion carrying a positive charge. During electrolysis, cations in the electrolyte move to the cathode (negative electrode).

catkin in flowering plants (◊angiosperms), a pendulous inflorescence, bearing numerous small, usually unisexual flowers. The tiny flowers are stalkless and the petals and sepals are usually absent or much reduced in size. Many types of trees bear catkins, including willows, poplars, and birches. Most plants with catkins are wind-pollinated, so the male catkins produce large quantities of pollen. Some ◊gymnosperms also have catkin-like structures that produce pollen, for example, the swamp cypress *Taxodium*.

CAT scan or *CT scan* (acronym for *computerized axial tomography scan*) sophisticated method of X-ray imaging. Quick and noninvasive, CAT scanning is used in medicine as an aid to diagnosis, helping to pinpoint problem areas without the need for exploratory surgery. It is also used in archaeology to investigate mummies.

The CAT scanner passes a narrow fan of X-rays through successive slices of the suspect body part. These slices are picked up by crystal detectors in a scintillator and converted electronically into cross-sectional images displayed on a viewing screen. Gradually, using views taken from various angles, a three-dimensional picture of the organ or tissue can be built up and irregularities analysed.

caustic soda former name for ◊sodium hydroxide (NaOH).

cauterization in medicine, the use of special instruments to burn or fuse small areas of body tissue to destroy dead cells, prevent the spread of infection, or seal tiny blood vessels to minimize blood loss during surgery.

Cauterization

King Charles II of England died five days after medical treatment that was intended to cure him. He was bled, his scalp was cauterized (burned with a hot iron), and he was given an emetic, a rectal purge and numerous draughts and concoctions to drink.

---

cave roofed-over cavity in the Earth's crust usually produced by the action of underground water or by waves on a seacoast. Caves of the former type commonly occur in areas underlain by limestone, such as Kentucky and many Balkan regions, where the rocks are soluble in water. A *pothole* is a vertical hole in rock caused by water descending a crack; it is thus open to the sky.

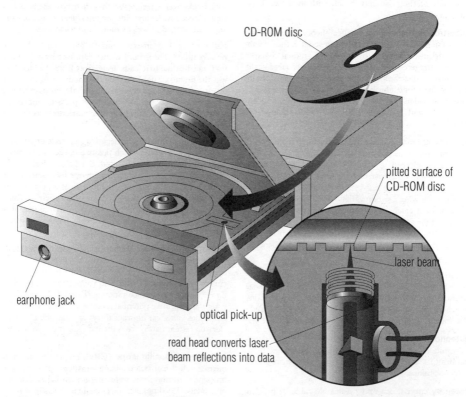

CD-ROM disc

pitted surface of CD-ROM disc

laser beam

earphone jack

optical pick-up

read head converts laser beam reflections into data

CD-ROM drive Data is obtained by the CD-ROM drive by converting the reflections from a disc's surface into digital form.

The future of the CD-ROM

The growing size of popular computer programs has made the CD-ROM (compact disc, read-only memory) the distribution medium of choice for software. A single CD can hold more than 600 megabytes of data, the equivalent of more than 400 standard 3.5 in floppy discs. Thanks to high volume production of audio CDs, CD-ROMs are also very cheap to produce, and it is not unusual to find them given away free with computer magazines. The CD-ROM has thus become the standard format for operating systems (Microsoft Windows 95 and Windows NT), for suites of programs (Microsoft Office, Corel Office), for large books and encyclopedias (Oxford English Dictionary, Microsoft's *Encarta*, The Hutchinson Multimedia Encyclopedia), and for computer games. Even the games console business is moving from cartridges to CD-ROMs. Cartridges were used by all the early machines from the Atari VCS through the Sega Master and MegaDrive to the Super Nintendo Entertainment System (SNES). But most third-generation consoles – including the Sega Saturn and Sony PlayStation – use CDs instead.

Hybrids

But the CD-ROM's advantage – that it stores a large, fixed mass of data – can also be a disadvantage. For example, a CD-ROM encyclopedia may be up to date on publication, but become out of date. At best it will be incomplete. The solution is to produce *hybrid CD-ROMs*, where the bulk of the data is delivered on disc then updated via an online communications system such as the Internet.

Even an operating system such as Windows 95 is really a hybrid: most of the code usually comes on a CD, but updates and new versions of software drivers must be downloaded from bulletin boards or World Wide Web sites. Many encyclopedias are hybrids: users can download monthly updates, and follow hypertext links from the CD-ROM to various Web sites. Film and music encyclopedias also benefit from similar updates. Hybrids are now becoming popular in the games world. CD-based titles are bought and played in the usual way on a single computer or games console, but many can also be played in multi-user mode by connecting to other users via an on-line system such as BT's WirePlay. Often, the program code for three-dimensional virtual worlds like 3DO's *Meridian 59* will be delivered on CD-ROM to avoid the costs and time-delays of downloading many megabytes of data from the Internet, but the game is played over the Internet.

Here for now, at least

In an ideal world, every computer would be permanently connected to a network that could deliver tens of megabytes of data per second; hard drives and CD-ROMs would then be unnecessary. However, outside of large corporations, most people have very slow dial-up connections via modems and ordinary phone lines, and they have to pay for every second they spend online. Under these circumstances, hybrid CD-ROMs have a useful part to play, and seem unlikely to disappear in the near future.

Jack Schofield

Cave animals often show loss of pigmentation or sight, and under isolation, specialized species may develop. The scientific study of caves is called **speleology**. During the ◊ice age, humans began living in caves leaving many layers of debris that archaeologists have unearthed and dated in the Old World and the New. They also left cave art, paintings of extinct animals often with hunters on their trail. Celebrated caves include the Mammoth Cave in Kentucky, USA, 6.4 km/4 mi long and 38 m/125 ft high; the Caverns of Adelsberg (Postumia) near Trieste, Italy, which extend for many miles; Carlsbad Cave, New Mexico, the largest in the USA; the Cheddar Caves, England; Fingal's Cave, Scotland, which has a range of basalt columns; and Peak Cavern, England.

cavitation ◊erosion of rocks caused by the forcing of air into cracks. Cavitation results from the pounding of waves on the coast and the swirling of turbulent river currents, and exerts great pressure, eventually causing rocks to break apart.

The process is particularly common at waterfalls, where the turbulent falling water contains many air bubbles, which burst to send shock waves into the rocks of the river bed and banks.

cc symbol for **cubic centimetre**; abbreviation for **carbon copy/copies**.

CD abbreviation for ◊ **compact disc**; **Corps Diplomatique** (French 'Diplomatic Corps'); **certificate of deposit**.

CD-ROM (abbreviation for **compact disc read-only memory**) computer storage device developed from the technology of the audio ◊compact disc. It consists of a plastic-coated metal disc, on which binary digital information is etched in the form of microscopic pits. This can then be read optically by passing a light beam over the disc. CD-ROMs typically hold about 650 ◊megabytes of data, and are used in distributing large amounts of text, graphics, audio, and video, such as encyclopedias, catalogues, technical manuals, and games.

Standard CD-ROMs cannot have information written onto them by computer, but must be manufactured from a master. Although recordable CDs, called CD-R discs, have been developed for use as computer discs, they are rather expensive for widespread use. A compact disc that can be overwritten repeatedly has also been developed; see ◊optical disc. The compact disc, with its enormous storage capability, may eventually replace the magnetic disc as the most common form of backing store for computers.

The technology is being developed rapidly: a standard CD-ROM disc spins at between 240–1,170 rpm, but faster discs have been introduced which speed up data retrieval to many times the standard speed. Research is being conducted into high density CDs capable of storing many ◊gigabytes of data, made possible by using multiple layers on the surface of the disc, and by using double-sided discs. Such improved storage capacity make products such as interactive movies a possibility.

PhotoCD, developed by Kodak and released in 1992, transfers ordinary still photographs onto CD-ROM discs.

Ceefax 'see facts' one of Britain's two ◊teletext systems (the other is Teletext), or 'magazines of the air', developed by the BBC and first broadcast in 1973.

In 1995 the BBC began testing a scheme to allow Ceefax (repackaged in HTML, hypertext markup language, to enable it to behave like Web pages) to be viewed on a PC by connecting a DAB (digital audio broadcasting) radio to the PC like a modem.

celestial mechanics the branch of astronomy that deals with the calculation of the orbits of celestial bodies, their gravitational attractions (such as those that produce the Earth's tides), and also the orbits of artificial satellites and space probes. It is based on the laws of motion and gravity laid down by Isaac Newton.

celestial sphere imaginary sphere surrounding the Earth, on which the celestial bodies seem to lie. The positions of bodies such as stars, planets, and galaxies are specified by their coordinates on the celestial sphere. The equivalents of latitude and longitude on the celestial sphere are called ◊declination and ◊right ascension (which is measured in hours from 0 to 24). The *celestial poles* lie directly above the Earth's poles, and the *celestial equator* lies over the Earth's Equator. The cel-

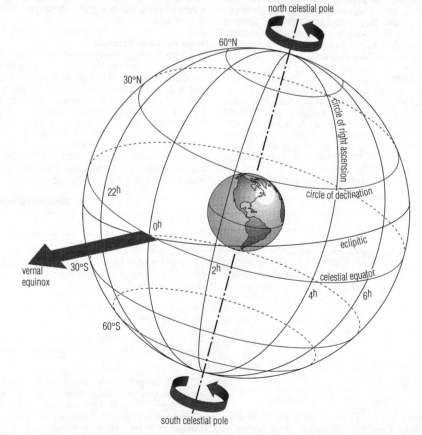

celestial sphere The main features of the celestial sphere. The equivalents of latitude and longitude on the celestial sphere are declination and right ascension. Declination runs from 0° at the celestial equator to 90° at the celestial poles. Right ascension is measured in hours eastwards from the vernal equinox, one hour corresponding to 15° of longitude.

estial sphere appears to rotate once around the Earth each day, actually a result of the rotation of the Earth on its axis.

celestine or **celestite** mineral consisting of strontium sulphate, $SrSO_4$, occurring as white or light blue crystals. Celestine occurs in cavity linings associated with calcite, dolomite, or fluorite. It is the principal source of strontium.

Celestine is found in small quantities in Germany, Italy, and the USA.

cell in biology, a discrete, membrane-bound portion of living matter, the smallest unit capable of an independent existence. All living organisms consist of one or more cells, with the exception of ◊viruses. Bacteria, protozoa, and many other microorganisms consist of single cells, whereas a human is made up of billions of cells. Essential features of a cell are the membrane, which encloses it and restricts the flow of substances in and out; the jellylike material within, the ◊cytoplasm; the ◊ribosomes, which carry out protein synthesis; and the ◊DNA, which forms the hereditary material.

cell differentiation in developing embryos, the process by which cells acquire their specialization, such as heart cells, muscle cells, skin cells, and brain cells. The seven-day-old human pre-embryo consists of thousands of individual cells, each of which is destined to assist in the formation of individual organs in the body.

Research has shown that the eventual function of a cell, in for example, a chicken embryo, is determined by the cell's position. The embryo can be mapped into areas corresponding with the spinal cord, the wings, the legs, and many other tissues. If the embryo is relatively young, a cell transplanted from one area to another will develop according to its new position. As the embryo develops the cells lose their flexibility and become unable to change their destiny.

cell division the process by which a cell divides, either ◊meiosis, associated with sexual reproduction, or ◊mitosis, associated with growth, cell replacement, or repair. Both forms involve the duplication of DNA and the splitting of the nucleus.

cell, electrical or **voltaic cell** or **galvanic cell** device in which chemical energy is converted into electrical energy; the popular name is ◊'battery', but this actually refers to a collection of cells in one unit. The reactive chemicals of a **primary cell** cannot be replenished, whereas **secondary cells** – such as storage batteries – are rechargeable: their chemical reactions can be reversed and the original condition restored by applying an electric current. It is dangerous to attempt to recharge a primary cell. *See illustration overleaf.*

cell, electrolytic device to which electrical energy is applied in order to bring about a chemical reaction; see ◊electrolysis.

cell membrane or **plasma membrane** thin layer of protein and fat surrounding cells that controls substances passing between the cytoplasm and the intercellular space. The cell membrane is semipermeable, allowing some substances to pass through and some not.

Generally, small molecules such as water, glucose, and amino acids can penetrate the membrane, while large molecules such as starch cannot. Membranes also play a part in ◊active transport, hormonal response, and cell metabolism.

cellophane transparent wrapping film made from wood ◊cellulose, widely used for packaging, first produced by Swiss chemist Jacques Edwin Brandenberger in 1908.

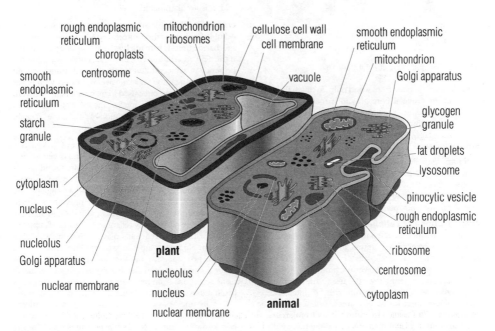

rough endoplasmic reticulum
mitochondrion
ribosomes
cellulose cell wall
cell membrane
smooth endoplasmic reticulum
mitochondrion
choroplasts
centrosome
vacuole
Golgi apparatus
smooth endoplasmic reticulum
starch granule
glycogen granule
fat droplets
lysosome
cytoplasm
nucleus
pinocytic vesicle
rough endoplasmic reticulum
nucleolus
Golgi apparatus
plant
ribosome
centrosome
nuclear membrane
nucleolus
nucleus
cytoplasm
nucleus
nuclear membrane
animal

cell

basic principle

lamp lights   lamp does
                               not light

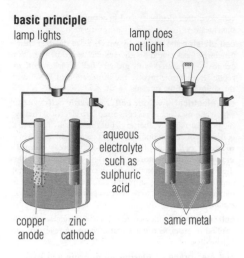

aqueous
electrolyte
such as
sulphuric
acid

copper   zinc   same metal
anode   cathode

a simple cell

electron flow

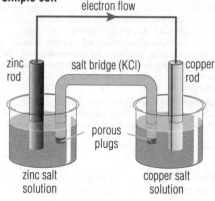

zinc   salt bridge (KCl)   copper
rod   rod

porous
plugs

zinc salt   copper salt
solution   solution

cell The basic principles of the electric cell.

Cellophane is made from wood pulp, in much the same way that the artificial fibre ◊rayon is made: the pulp is dissolved in chemicals to form a viscose solution, which is then pumped through a long narrow slit into an acid bath where the emergent viscose stream turns into a film of pure cellulose.

cell sap dilute fluid found in the large central vacuole of many plant cells. It is made up of water, amino acids, glucose, and salts. The sap has many functions, including storage of useful materials, and provides mechanical support for non-woody plants.

cellular phone or *cellphone* mobile radio telephone, one of a network connected to the telephone system by a computer-controlled communication system. Service areas are divided into small 'cells', about 5 km/3 mi across, each with a separate low-power transmitter.

The cellular system allows the use of the same set of frequencies with the minimum risk of interference. Nevertheless, in crowded city areas, cells can become overloaded. This has led to a move away from analogue transmissions to digital methods that allow more calls to be made within a limited frequency range.

cellulite fatty compound alleged by some dietitians to be produced in the body by liver disorder and to cause lumpy deposits on the hips and thighs. Medical opinion generally denies its existence, attributing the lumpy appearance to a type of subcutaneous fat deposit.

cellulitis inflammation of ◊connective tissue. It is usually due to bacterial infection.

celluloid transparent or translucent, highly flammable, plastic material (a ◊thermoplastic) made from cellulose nitrate and camphor. It was once used for toilet articles, novelties, and photographic film, but has now been replaced by the nonflammable substance ◊cellulose acetate.

cellulose complex ◊carbohydrate composed of long chains of glucose units, joined by chemical bonds called glycosidic links. It is the principal constituent of the cell wall of higher plants, and a vital ingredient in the diet of many ◊herbivores. Molecules of cellulose are organized into long, unbranched microfibrils that give support to the cell wall. No mammal produces the enzyme cellulase, necessary for digesting cellulose; mammals such as rabbits and cows are only able to digest grass because the bacteria present in their gut can manufacture it.

Cellulose is the most abundant substance found in the plant kingdom. It has numerous uses in industry: in rope-making; as a source of textiles (linen, cotton, viscose, and acetate) and plastics (cellophane and celluloid); in the manufacture of nondrip paint; and in such foods as whipped dessert toppings. Japanese chemists produced the first synthetic cellulose in 1996.

cellulose acetate or *cellulose ethanoate* chemical (an ◊ester) made by the action of acetic acid (ethanoic acid) on cellulose. It is used in making transparent film, especially photographic film; unlike its predecessor, celluloid, it is not flammable.

cellulose nitrate or *nitrocellulose* series of esters of cellulose with up to three nitrate (NO_3) groups per monosaccharide unit. It is made by the action of concentrated nitric acid on cellulose (for example, cotton waste) in the presence of concentrated sulphuric acid. Fully nitrated cellulose (gun cotton) is explosive, but esters with fewer nitrate groups were once used in making lacquers, rayon, and plastics, such as coloured and photographic film, until replaced by the nonflammable cellulose acetate. ◊Celluloid is a form of cellulose nitrate.

cell wall in plants, the tough outer surface of the cell. It is constructed from a mesh of ◊cellulose and is very strong and relatively inelastic. Most living cells are turgid (swollen with water; see ◊turgor) and develop an internal hydrostatic pressure (wall pressure) that acts against the cellulose wall. The result of this turgor pressure is to give the cell, and therefore the plant, rigidity. Plants that are not woody are particularly reliant on this form of support.

The cellulose in cell walls plays a vital role in global nutrition. No vertebrate is able to produce cellulase, the enzyme necessary for the breakdown of cellulose into sugar. Yet most mammalian herbivores rely on cellulose, using secretions from microorganisms living in the gut to break it down. Humans cannot digest the cellulose of the cell walls; they possess neither the correct gut microorganisms nor the necessary grinding teeth. However, cellulose still forms a necessary part of the human diet as ◊fibre (roughage).

Celsius scale of temperature, previously called centigrade, in which the range from freezing to boiling of water is divided into 100 degrees, freezing point being 0 degrees and boiling point 100 degrees.

The degree centigrade (°) was officially renamed Celsius in 1948 to avoid confusion with the angular measure known as the centigrade (one hundredth of a grade). The Celsius scale is named after the Swedish astronomer Anders Celsius (1701–1744), who devised it in 1742 but in reverse (freezing point was 100°; boiling point 0°).

cement any bonding agent used to unite particles in a single mass or to cause one surface to adhere to another. *Portland cement* is a powder obtained from burning together a mixture of lime (or chalk) and clay, and when mixed with water and sand or gravel, turns into mortar or concrete.

In geology, cement refers to a chemically precipitated material such as carbonate that occupies the interstices of clastic rocks.

The term 'cement' covers a variety of materials, such as fluxes and pastes, and also bituminous products obtained from tar. In 1824 English bricklayer Joseph Aspdin (1779–1855) created and patented the first Portland cement, so named because its colour in the hardened state resembled that of Portland stone, a limestone used in building.

Cenozoic or *Caenozoic* era of geological time that began 65 million years ago and continues to the present day. It is divided into the Tertiary and Quaternary periods. The Cenozoic marks the emergence of mammals as a dominant group, including humans, and the formation of the mountain chains of the Himalayas and the Alps.

Centaurus large, bright constellation of the southern hemisphere, represented as a centaur. Its brightest star, ◊Alpha Centauri, is a triple star, and contains the closest star to the Sun, Proxima Centauri, which is only 4.3 light years away. Omega Centauri, which is just visible to the naked eye as a hazy patch, is the largest and brightest ◊globular cluster of stars in the sky, 16,000 light years away.

Alpha and Beta Centauri are both of the first magnitude and, like Alpha and Beta Ursae Majoris, are known as 'the Pointers', as a line joining them leads to ◊Crux. Centaurus A, a galaxy 15 million light years away, is a strong source of radio waves and X-rays.

centigrade former name for the ◊Celsius temperature scale.

central dogma in genetics and evolution, the fundamental belief that ◊genes can affect the nature of the physical body, but that changes in the body (◊acquired character, for example, through use or accident) cannot be translated into changes in the genes.

central heating system of heating from a central source, typically of a house, larger building, or group of buildings, as opposed to heating each room individually. Steam heat and hot-water heat are the most common systems in use. Water is heated in a furnace burning oil, gas, or solid fuel, and, as steam or hot water, is then pumped through radiators in each room. The level of temperature can be selected by adjusting a ◊thermostat on the burner or in a room.

Central heating has its origins in the hypocaust heating system introduced by the Romans nearly 2,000 years ago. From the 18th century, steam central heating, usually by pipe, was available in the West and installed in individual houses on an ad hoc basis. The Scottish engineer James Watt heated his study with a steam pipe connected to a boiler, and Matthew Boulton installed steam heating in a friend's Birmingham house. Not until the latter half of the 20th century was central heating in general use. Central heating systems are usually switched on and off by a time switch. Another kind of central heating system uses hot air, which is pumped through ducts (called risers) to grills in the rooms. Underfloor heating (called radiant heat) is used in some houses, the heat coming from electric elements buried in the floor. New energy-efficient houses use heat from the Sun and good insulation to replace some central heating.

central nervous system (CNS) the brain and spinal cord, as distinct from other components of the ◊nervous system. The CNS integrates all nervous function.

In invertebrates it consists of a paired ventral nerve cord with concentrations of nerve-cell bodies, known as ◊*ganglia* in each segment, and a small brain in the head. Some simple invertebrates, such as sponges and jellyfishes, have no CNS but a simple network of nerve cells called a *nerve net*. *See illustration overleaf.*

central processing unit (CPU) main component of a computer, the part that executes individual program instructions and controls the operation of other parts. It is sometimes called the central processor or, when contained on a single integrated circuit, a microprocessor.

The CPU has three main components: the *arithmetic and logic unit* (ALU), where all calculations and logical operations are carried out; a *control unit*, which decodes, synchronizes, and executes program instructions; and the *immediate access memory*, which stores the data and programs on which the computer is currently working. All these components contain ◊registers, which are memory locations reserved for specific purposes. *See illustration on page 121.*

centre of mass point in or near an object about which the object would turn if allowed to rotate freely. A symmetrical homogeneous object such as a sphere or cube has its centre of mass at its geometrical centre; a hollow object (such as a cup) may have its centre of mass in space inside the hollow.

For an object to be in stable equilibrium, a vertical line down through its centre of mass must run within the boundaries of its base; if tilted until this line falls outside the base, the object becomes unstable and topples over.

centrifugal force useful concept in physics, based on an apparent (but not real) force. It may be regarded as a force that acts radially outward from a spinning or orbiting object, thus balancing the ◊centripetal force (which is real). For an object of mass m moving with a velocity v in a circle of radius r, the centrifugal force F equals mv^2/r (outward).

centrifuge apparatus that rotates containers at high speeds, creating centrifugal forces. One use is for separating mixtures of substances of different densities.

The mixtures are usually spun horizontally in balanced containers ('buckets'), and the rotation sets up centrifugal forces, causing their components to separate according to their densities. A common example is the separation of the lighter plasma from the heavier blood corpuscles in certain blood tests. The **ultracentrifuge** is a very high-speed centrifuge, used in biochemistry for separating ⬦colloids and organic substances; it may operate at several million revolutions per minute. The centrifuges used in the industrial separation of cream from milk, and yeast from fermented wort (infused malt), operate by having mixtures pumped through a continually rotating chamber, the components being tapped off at different points. Large centrifuges are used for physiological research – for example, in astronaut training where bodily responses to gravitational forces many times the normal level are tested.

centriole structure found in the ⬦cells of animals that

plays a role in the processes of ⬦meiosis and ⬦mitosis (cell division).

centripetal force force that acts radially inward on an object moving in a curved path. For example, with a weight whirled in a circle at the end of a length of string, the centripetal force is the tension in the string. For an object of mass m moving with a velocity v in a circle of radius r, the centripetal force F equals mv^2/r (inward). The reaction to this force is the ⬦centrifugal force.

cephalopod any predatory marine mollusc of the class Cephalopoda, with the mouth and head surrounded by tentacles. Cephalopods are the most intelligent, the fastest-moving, and the largest of all animals without backbones, and there are remarkable luminescent forms which

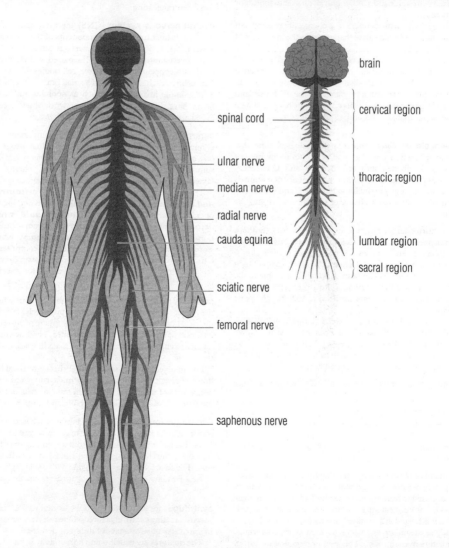

brain

cervical region

spinal cord

thoracic region

ulnar nerve

median nerve

radial nerve

cauda equina

lumbar region

sacral region

sciatic nerve

femoral nerve

saphenous nerve

central nervous system The central nervous system (CNS) with its associated nerves. The CNS controls and integrates body functions. In humans and other vertebrates it consists of a brain and a spinal cord, which are linked to the body's muscles and organs by means of the peripheral nervous system.

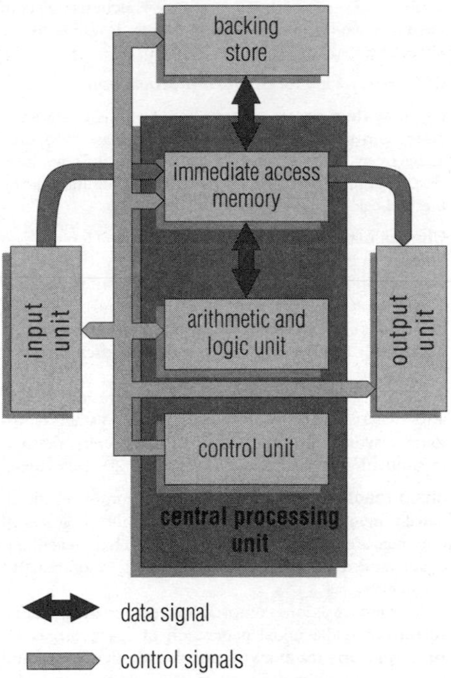
backing store
immediate access memory
input unit
output unit
arithmetic and logic unit
control unit
central processing unit
data signal
control signals

central processing unit The relationship between the three main areas of a computer's central processing unit. The arithmetic and logic unit (ALU) does the arithmetic, using the registers to store intermediate results, supervised by the control unit. Input and output circuits connect the ALU to external memory, input, and output devices.

swim or drift at great depths. They have the most highly developed nervous and sensory systems of all invertebrates, the eye in some closely paralleling that found in vertebrates. Examples include squid, octopus, and cuttlefish. Shells are rudimentary or absent in most cephalopods.

Typically, they move by swimming with the mantle (fold of outer skin) aided by the arms, but can squirt water out of the siphon (funnel) to propel themselves backwards by jet propulsion. Squid, for example, can escape predators at speeds of 11 kph/7 mph. Cephalopods grow very rapidly and may be mature in a year. The female common octopus lays 150,000 eggs after copulation, and stays to brood them for as long as six weeks. After they hatch the female dies, and, although reproductive habits of many cephalopods are not known, it is thought that dying after spawning may be typical.

cephalosporin any of a class of broad-spectrum antibiotics derived from a fungus (genus *Cephalosporium*). They are similar to penicillins and are used on penicillin-resistant infections.

The first cephalosporin was extracted from sewage-contaminated water, and other naturally occurring ones have been isolated from moulds taken from soil samples. Side effects include allergic reactions and digestive upsets. Synthetic cephalosporins can be designed to be effective against a particular ◊pathogen.

Cepheid variable yellow supergiant star that varies regularly in brightness every few days or weeks as a result of pulsations. The time that a Cepheid variable takes to pulsate is directly related to its average brightness; the longer the pulsation period, the brighter the star.

This relationship, the ***period luminosity law*** (discovered by US astronomer Henrietta Leavitt), allows astronomers to use Cepheid variables as 'standard candles' to measure distances in our Galaxy and to nearby galaxies. They are named after their prototype, Delta Cephei, whose light variations were observed in 1784 by English astronomer John Goodricke (1764–1786).

Cepheus constellation of the north polar region, named after King Cepheus of Greek mythology, husband of Cassiopeia and father of Andromeda. It contains the Garnet Star (Mu Cephei), a red supergiant of variable brightness that is one of the reddest-coloured stars known, and Delta Cephei, prototype of the ◊Cepheid variables, which are important both as distance indicators and for the information they give about stellar evolution.

cereal grass grown for its edible, nutrient-rich, starchy seeds. The term refers primarily to wheat, oats, rye, and barley, but may also refer to maize (corn), millet, and rice. Cereals contain about 75% complex carbohydrates and 10% protein, plus fats and fibre (roughage). They store well. If all the world's cereal crop were consumed as whole-grain products directly by humans, everyone could obtain adequate protein and carbohydrate; however, a large proportion of cereal production in affluent nations is used as animal feed to boost the production of meat, dairy products, and eggs.

The term also refers to breakfast foods prepared from the seeds of cereal crops. Some cereals require cooking (porridge oats), but most are ready to eat. Mass-marketed cereals include refined and sweetened varieties as well as whole cereals such as muesli. Whole cereals are more nutritious and provide more fibre than the refined cereals, which often have vitamins and flavourings added to replace those lost in the refining process.

cerebellum part of the brain of ◊vertebrate animals which controls muscle tone, movement, balance, and coordination. It is relatively small in lower animals such as newts and lizards, but large in birds since flight demands precise coordination. The human cerebellum is also well developed, because of the need for balance when walking or running, and for finely coordinated hand movements.

cerebral haemorrhage or ***apoplectic fit*** in medicine, a form of ◊stroke in which there is bleeding from a cerebral blood vessel into the surrounding brain tissue. It is generally caused by degenerative disease of the arteries and high blood pressure. Depending on the site and extent of bleeding, the symptoms vary from transient weakness and numbness to deep coma and death. Damage to the brain is permanent, though some recovery can be made. Strokes are likely to recur.

cerebral palsy any nonprogressive abnormality of the brain occurring during or shortly after birth. It is caused by oxygen deprivation, injury during birth, haemorrhage, meningitis, viral infection, or faulty development. Premature babies are at greater risk of being born with cerebral palsy, and in 1996 US researchers linked this to low levels of the thyroid hormone thyroxine. The condition is characterized by muscle spasm, weakness, lack of coordination, and impaired movement; or there may be spastic paralysis, with fixed deformities of the limbs. Intelligence is not always affected.

cerebrum part of the vertebrate ◊brain, formed from the two paired cerebral hemispheres, separated by a central fissure. In birds and mammals it is the largest and most developed part of the brain. It is covered with an infolded layer of grey matter, the cerebral cortex, which integrates brain functions. The cerebrum coordinates all voluntary activity.

Ceres the largest asteroid, 940 km/584 mi in diameter, and the first to be discovered (by Italian astronomer Giuseppe Piazzi in 1801). Ceres orbits the Sun every 4.6 years at an average distance of 414 million km/257 million mi. Its mass is about one-seventieth of that of the Moon.

cerium malleable and ductile, grey, metallic element, symbol Ce, atomic number 58, relative atomic mass 140.12. It is the most abundant member of the lanthanide series, and is used in alloys, electronic components, nuclear fuels, and lighter flints. It was discovered in 1804 by the Swedish chemists Jöns Berzelius and Wilhelm Hisinger (1766–1852), and, independently, by Martin Klaproth. The element was named after the then recently discovered asteroid Ceres.

cermet (contraction of *ceramics and metal*) bonded material containing ceramics and metal, widely used in jet engines and nuclear reactors. Cermets behave much like metals but have the great heat resistance of ceramics. Tungsten carbide, molybdenum boride, and aluminium oxide are among the ceramics used; iron, cobalt, nickel, and chromium are among the metals.

CERN particle physics research organization founded in 1954 as a cooperative enterprise among European governments. It has laboratories at Meyrin, near Geneva, Switzerland. It was originally known as the *Conseil Européen pour la Recherche Nucléaire* but subsequently renamed *Organisation Européenne pour la Recherche Nucléaire*, although still familiarly known as CERN. It houses the world's largest particle ◊accelerator, the ◊Large Electron Positron Collider (LEP), with which notable advances have been made in ◊particle physics.

In 1965 the original laboratory was doubled in size by extension across the border from Switzerland into France. In Dec 1994 the 19 member nations of CERN approved the construction of the Large Hadron Collider. It is expected to cost £1.25 million and to be fully functional by 2005.

cervical cancer ◊cancer of the cervix (neck of the womb).

cervical smear removal of a small sample of tissue from the cervix (neck of the womb) to screen for changes implying a likelihood of cancer. The procedure is also known as the *Pap test* after its originator, George Papanicolau.

cervix abbreviation for *cervix uteri*, the neck of the womb; see ◊uterus.

Cetus large constellation on the celestial equator (see ◊celestial sphere), represented as a sea monster or a whale. Cetus contains the long-period variable star ◊Mira, and Tau Ceti, one of the nearest stars visible with the naked eye.

It is named after the sea monster sent to devour Andro-

meda. Mira is sometimes the most conspicuous object in the constellation, but it is more usually invisible to the naked eye.

CFC abbreviation for ◊*chlorofluorocarbon*.

c.g.s. system system of units based on the centimetre, gram, and second, as units of length, mass, and time respectively. It has been replaced for scientific work by the ◊SI units to avoid inconsistencies in definition of the thermal calorie and electrical quantities.

Chagas's disease disease common in Central and South America, infecting approximately 18 million people worldwide. It is caused by a trypanosome parasite, *Trypanosoma cruzi*, transmitted by several species of blood-sucking insect; it results in incurable damage to the heart, intestines, and brain. It is named after Brazilian doctor Carlos Chagas (1879–1934).

It is the world's third most prevalent parasitic disease, after malaria and schistosomiasis. In its first stage symptoms resemble flu but 20–30% of sufferers develop inflammation of the heart muscles up to 20 years later.

chain reaction in chemistry, a succession of reactions, usually involving ◊free radicals, where the products of one stage are the reactants of the next. A chain reaction is characterized by the continual generation of reactive substances.

A chain reaction comprises three separate stages: *initiation* – the initial generation of reactive species; *propagation* – reactions that involve reactive species and generate similar or different reactive species; and *termination* – reactions that involve the reactive species but produce only stable, nonreactive substances. Chain reactions may occur slowly (for example, the oxidation of edible oils) or accelerate as the number of reactive species increases, ultimately resulting in explosion.

chain reaction in nuclear physics, a fission reaction that is maintained because neutrons released by the splitting of some atomic nuclei themselves go on to split others, releasing even more neutrons. Such a reaction can be controlled (as in a nuclear reactor) by using moderators to absorb excess neutrons. Uncontrolled, a chain reaction produces a nuclear explosion (as in an atom bomb).

chalaza glutinous mass of transparent albumen supporting the yolk inside birds' eggs. The chalaza is formed as the egg slowly passes down the oviduct, when it also acquires its coiled structure.

chalcedony form of the mineral quartz, SiO_2, in which the crystals are so fine-grained that they are impossible to distinguish with a microscope (cryptocrystalline). Agate, onyx, and carnelian are ◊gem varieties of chalcedony.

chalcopyrite copper iron sulphide mineral, $CuFeS_2$, the most common ore of copper. It is brassy yellow in colour and may have an iridescent surface tarnish. It occurs in many different types of mineral vein, in rocks ranging from basalt to limestone.

chalk soft, fine-grained, whitish sedimentary rock composed of calcium carbonate, $CaCO_3$, extensively quarried for use in cement, lime, and mortar, and in the manufacture of cosmetics and toothpaste. *Blackboard chalk* in fact consists of ◊gypsum (calcium sulphate, $CaSO_4.2H_2O$).

Chalk was once thought to derive from the remains of microscopic animals or foraminifera. In 1953, however, it was seen under the electron microscope to be composed chiefly of ◊coccolithophores, unicellular lime-secreting algae, and hence primarily of plant origin. It is formed from deposits of deep-sea sediments called oozes.

chance likelihood, or ◊probability, of an event taking place, expressed as a fraction or percentage. For example, the chance that a tossed coin will land heads up is 50%.

As a science, it originated when the Chevalier de Méré consulted Blaise Pascal about how to reduce his gambling losses. In 1664, in correspondence with another mathematician, Pierre de Fermat, Pascal worked out the foundations of the theory of chance. This underlies the science of statistics.

I have been trying to point out that in our lives chance may have an astonishing influence and, if I may offer advice to the young laboratory worker, it would be this – never to neglect an extraordinary appearance or happening. It may be – usually is, in fact – a false alarm that leads to nothing, but it may on the other hand be the clue provided by fate to lead you to some important advance.

On **chance** Alexander Fleming, lecture at Harvard

---

chancroid or *soft sore* acute localized, sexually transmitted ulcer on or about the genitals caused by the bacterium *Haemophilus ducreyi*.

It causes painful enlargement and suppuration of lymph nodes in the groin area.

Chandrasekhar limit or *Chandrasekhar mass* in astrophysics, the maximum possible mass of a ◊white dwarf star. The limit depends slightly on the composition of the star but is equivalent to 1.4 times the mass of the Sun. A white dwarf heavier than the Chandrasekhar limit would collapse under its own weight to form a ◊neutron star or a ◊black hole. The limit is named after the Indian-US astrophysicist Subrahmanyan Chandrasekhar who developed the theory of white dwarfs in the 1930s.

change of state a change in the physical state (solid, liquid, or gas) of a material. For instance, melting, boiling, evaporation, and their opposites, solidification and condensation, are changes of state. The former set of changes are brought about by heating or decreased pressure; the latter by cooling or increased pressure.

These changes involve the absorption or release of heat energy, called ◊latent heat, even though the temperature of the material does not change during the transition between states.

See also ◊states of matter. In the unusual change of state called **sublimation**, a solid changes directly to a gas without passing through the liquid state. For example, solid carbon dioxide (dry ice) sublimes to carbon dioxide gas.

Channel Tunnel tunnel built beneath the English Channel, linking Britain with mainland Europe. It comprises twin rail tunnels, 50 km/31 mi long and 7.3 m/24 ft in diameter, located 40 m/130 ft beneath the seabed. Construction began in 1987, and the French and English sections were linked in Dec 1990. It was officially opened on 6 May 1994. The shuttle train service, Le Shuttle, opened to lorries in May 1994 and to cars in Dec 1994. The tunnel's high-speed train service, Eurostar, linking London to Paris and Brussels, opened in Nov 1994.

chaos theory or *chaology* or *complexity theory* branch of mathematics that attempts to describe irregular, unpredictable systems – that is, systems whose behaviour is difficult to predict because there are so many variable or unknown factors. Weather is an example of a chaotic system.

Chaos theory, which attempts to predict the *probable* behaviour of such systems, based on a rapid calculation of the impact of as wide a range of elements as possible, emerged in the 1970s with the development of sophisticated computers. First developed for use in meteorology, it has also been used in such fields as economics.

characteristic in mathematics, the integral (whole-number) part of a ◊logarithm. The fractional part is the ◊mantissa.

For example, in base ten, $10^0 = 1$, $10^1 = 10$, $10^2 = 100$, and so on; the powers to which 10 is raised are the characteristics. To determine the power to which 10 must be raised to obtain a number between 10 and 100, say 20 (2×10, or log 2 + log 10), the logarithm for 2 is found (0.3010), and the characteristic 1 added to make 1.3010.

character set in computing, the complete set of symbols that can be used in a program or recognized by a computer. It may include letters, digits, spaces, punctuation marks, and special symbols.

charcoal black, porous form of ◊carbon, produced by heating wood or other organic materials in the absence of air. It is used as a fuel in the smelting of metals such as copper and zinc, and by artists for making black line drawings. *Activated charcoal* has been powdered and dried so that it presents a much increased surface area for adsorption; it is used for filtering and purifying liquids and gases – for example, in drinking-water filters and gas masks.

Charcoal was traditionally produced by burning dried wood in a kiln, a process lasting several days. The kiln was either a simple hole in the ground, or an earth-covered mound. Today kilns are of brick or iron, both of which allow the waste gases to be collected and used. Charcoal had many uses in earlier centuries. Because of the high temperature at which it burns (2,012°F/1,100°C), it was used in furnaces and blast furnaces before the development of ◊coke. It was also used in an industrial process for obtaining ethanoic acid (acetic acid), in producing wood tar and ◊wood pitch, and (when produced from alder or willow trees) as a component of gunpowder.

charge see ◊electric charge.

charge-coupled device (CCD) device for forming images electronically, using a layer of silicon that releases electrons when struck by incoming light. The electrons are stored in ◊pixels and read off into a computer at the end of the exposure. CCDs have now almost entirely replaced photographic film for applications such as astrophotography where extreme sensitivity to light is paramount.

The mathematics of chaos

Why are tides predictable years ahead, whereas weather forecasts often go wrong within a few days?

Both tides and weather are governed by natural laws. Tides are caused by the gravitational attraction of the Sun and Moon; the weather by the motion of the atmosphere under the influence of heat from the Sun. The law of gravitation is not noticeably simpler than the laws of fluid dynamics; yet for weather the resulting behaviour seems to be far more complicated.

The reason for this is 'chaos', which lies at the heart of one of the most exciting and most rapidly expanding areas of mathematical research, the theory of nonlinear dynamic systems.

Random behaviour in dynamic systems

It has been known for a long time that dynamic systems – systems that change with time according to fixed laws – can exhibit regular patterns, such as repetitive cycles. Thanks to new mathematical techniques, emphasizing shape rather than number, and to fast and sophisticated computer graphics, we now know that dynamic systems can also behave randomly. The difference lies not in the complexity of the formulae that define their mathematics, but in the geometrical features of the dynamics. This is a remarkable discovery: random behaviour in a system whose mathematical description contains no hint whatsoever of randomness.

Simple geometric structure produces simple dynamics. For example, if the geometry shrinks everything towards a fixed point, then the motion tends towards a steady state. But if the dynamics keep stretching things apart and then folding them together again, the motion tends to be chaotic – like food being mixed in a bowl. The motion of the Sun and Moon, on the kind of timescale that matters when we want to predict the tides, is a series of regular cycles, so prediction is easy. The changing patterns of the weather involve a great deal of stretching and folding, so here chaos reigns.

Fractals

The geometry of chaos can be explored using theoretical mathematical techniques such as topology – 'rubber-sheet geometry' – but the most vivid pictures are obtained using computer graphics. The geometric structures of chaos are 'fractals': they have detailed form on all scales of magnification. Order and chaos, traditionally seen as opposites, are now viewed as two aspects of the same basic process, the evolution of a system in time. Indeed, there are now examples where both order and chaos occur naturally within a single geometrical form.

Predicting the unpredictable

Does chaos make randomness predictable? Sometimes. If what looks like random behaviour is actually governed by a dynamic system, then short-term prediction becomes possible. Long-term prediction is not as easy, however. In chaotic systems any initial error of measurement, however small, will grow rapidly and eventually ruin the prediction. This is known as the butterfly effect: if a butterfly flaps its wings, a month later the air disturbance created may cause a hurricane.

Chaos can be applied to many areas of science, such as chemistry, engineering, computer science, biology, electronics, and astronomy. For example, although the short-term motions of the Sun and Moon are not chaotic, the long-term motion of the Solar System *is* chaotic. It is impossible to predict on which side of the Sun Pluto will lie in 200 million years' time. Saturn's satellite Hyperion tumbles chaotically. Chaos caused by Jupiter's gravitational field can fling asteroids out of orbit, towards the Earth. Disease epidemics, locust plagues, and irregular heartbeats are more down-to-earth examples of chaos, on a more human timescale.

Making sense of chaos

Chaos places limits on science: it implies that even when we know the equations that govern a system's behaviour, we may not in practice be able to make effective predictions. On the other hand, it opens up new avenues for discovery, because it implies that apparently random phenomena may have simple, non-random explanations. So chaos is changing the way scientists think about what they do: the relation between determinism and chance, the role of experiment, the computability of the world, the prospects for prediction, and the interaction between mathematics, science, and nature. Chaos cuts right across traditional subject boundaries, and distinctions between pure and applied mathematicians, between mathematicians and physicists, between physicists and biologists, become meaningless when compared to the unity revealed by their joint efforts.

Ian Stewart

Charles's law law stating that the volume of a given mass of gas at constant pressure is directly proportional to its absolute temperature (temperature in kelvin). It was discovered by French physicist Jacques Charles in 1787, and independently by French chemist Joseph Gay-Lussac in 1802. The gas increases by 1/273 of its volume at 0°C for each °C rise of temperature. This means that the coefficient of expansion of all gases is the same. The law is only approximately true and the coefficient of expansion is generally taken as 0.003663 per °C.

We have no right to assume that any physical laws exist, or if they have existed up to now, that they will continue to exist in a similar manner in the future.

On **physical laws** Max Planck
The Universe in the Light of Modern Physics

charm in physics, a property possessed by one type of ◊quark (very small particles found inside protons and neutrons), called the charm quark. The effects of charm are only seen in experiments with particle ◊accelerators. See ◊elementary particles.

Chavín de Huantar archaeological site in the Peruvian Andes, 3,135 m/10,000 ft above sea level, thought to be 'the womb of Andean civilization'. Its influence peaked between 1000 and 300 BC.

chelate chemical compound whose molecules consist of one or more metal atoms or charged ions joined to chains of organic residues by coordinate (or dative covalent) chemical ◊bonds. The parent organic compound is known as a **chelating agent** – for example, EDTA (ethylene-diaminetetraacetic acid), used in chemical analysis. Chelates are used in analytical chemistry, in agriculture and horticulture as carriers of essential trace metals, in water softening, and in the treatment of thalassaemia by removing excess iron, which may build up to toxic levels in the body. Metalloproteins (natural chelates) may influence the performance of enzymes or provide a mechanism for the storage of iron in the spleen and plasma of the human body.

chemical change change that occurs when two or more substances (reactants) interact with each other, resulting in the production of different substances (products) with different chemical compositions. A simple example of chemical change is the burning of carbon in oxygen to produce carbon dioxide.

chemical element alternative name for ◊element.

chemical equation method of indicating the reactants and products of a chemical reaction by using chemical symbols and formulae. A chemical equation gives two basic pieces of information: (1) the reactants (on the left-hand side) and products (right-hand side); and (2) the reacting proportions (stoichiometry) – that is, how many units of each reactant and product are involved. The equation must balance; that is, the total number of atoms of a particular element on the left-hand side must be the same as the number of atoms of that element on the right-hand side.

$$Na_2CO_3 + 2HCl \rightarrow 2NaCl + CO_2 + H_2O$$

reactants → products

This equation states that one molecule of sodium carbonate combines with two molecules of hydrochloric acid to form two molecules of sodium chloride, one of carbon dioxide, and one of water. Double arrows indicate that the reaction is reversible – in the formation of ammonia from hydrogen and nitrogen, the direction depends on the temperature and pressure of the reactants.

$$3H_2 + N_2 \rightleftharpoons 2NH_3$$

chemical equilibrium condition in which the products of a reversible chemical reaction are formed at the same rate at which they decompose back into the reactants, so that the concentration of each reactant and product remains constant.

The amounts of reactant and product present at equilibrium are defined by the *equilibrium constant* for that reaction and specific temperature.

chemical family collection of elements that have very similar chemical and physical properties. In the ◊periodic table of the elements such collections are to be found in the vertical columns (groups). The groups that contain the most markedly similar elements are group I, the ◊alkali metals; group II, the ◊alkaline-earth metals; group VII, the ◊halogens; and group 0, the noble or ◊inert gases.

chemical oxygen demand (COD) measure of water and effluent quality, expressed as the amount of oxygen (in parts per million) required to oxidize the reducing substances present.

Under controlled conditions of time and temperature, a chemical oxidizing agent (potassium permanganate or dichromate) is added to the sample of water or effluent under consideration, and the amount needed to oxidize the reducing materials present is measured. From this the chemically equivalent amount of oxygen can be calculated. Since the reducing substances typically include remains of living organisms, COD may be regarded as reflecting the extent to which the sample is polluted. Compare ◊biological oxygen demand.

chemical weathering form of ◊weathering brought about by a chemical change in the rocks affected. Chemical weathering involves the 'rotting', or breakdown, of the minerals within a rock, and usually produces a claylike residue (such as china clay and bauxite). Some chemicals are dissolved and carried away from the weathering source.

A number of processes bring about chemical weathering, such as carbonation (breakdown by weakly acidic rainwater), ◊hydrolysis (breakdown by water), ◊hydration (breakdown by the absorption of water), and ◊oxidation (breakdown by the oxygen in water).

chemiluminescence the emission of light from a substance as a result of a chemical reaction (rather than raising its temperature). See ◊luminescence.

chemisorption the attachment, by chemical means, of a single layer of molecules, atoms, or ions of gas to the surface of a solid or, less frequently, a liquid. It is the basis of catalysis (see ◊catalyst) and is of great industrial importance.

chemistry branch of science concerned with the study of the structure and composition of the different kinds of

Chemistry: Chronology

c. 3000 BC	Egyptians were producing bronze – an alloy of copper and tin.
c. 450 BC	Greek philosopher Empedocles proposed that all substances are made up of a combination of four elements – earth, air, fire, and water – an idea that was developed by Plato and Aristotle and persisted for over 2,000 years.
c. 400 BC	Greek philosopher Democritus theorized that matter consists ultimately of tiny, indivisible particles, *atomos*.
AD 1	Gold, silver, copper, lead, iron, tin, and mercury were known.
200	The techniques of solution, filtration, and distillation were known.
7th–17th centuries	Chemistry was dominated by alchemy, the attempt to transform nonprecious metals such as lead and copper into gold. Though misguided, it led to the discovery of many new chemicals and techniques, such as sublimation and distillation.
12th century	Alcohol was first distilled in Europe.
1242	Gunpowder introduced to Europe from the Far East.
1620	Scientific method of reasoning expounded by Francis Bacon in his *Novum Organum*.
1650	Leyden University in the Netherlands set up the first chemistry laboratory.
1661	Robert Boyle defined an element as any substance that cannot be broken down into still simpler substances and asserted that matter is composed of 'corpuscles' (atoms) of various sorts and sizes, capable of arranging themselves into groups, each of which constitutes a chemical substance.
1662	Boyle described the inverse relationship between the volume and pressure of a fixed mass of gas (Boyle's law).
1697	Georg Stahl proposed the erroneous theory that substances burn because they are rich in a certain substance, called phlogiston.
1755	Joseph Black discovered carbon dioxide.
1774	Joseph Priestley discovered oxygen, which he called 'dephlogisticated air'. Antoine Lavoisier demonstrated his law of conservation of mass.
1777	Lavoisier showed air to be made up of a mixture of gases, and showed that one of these – oxygen – is the substance necessary for combustion (burning) and rusting to take place.
1781	Henry Cavendish showed water to be a compound.
1792	Alessandro Volta demonstrated the electrochemical series.
1807	Humphry Davy passed electric current through molten compounds (the process of electrolysis) in order to isolate elements, such as potassium, that had never been separated by chemical means. Jöns Berzelius proposed that chemicals produced by living creatures should be termed 'organic'.
1808	John Dalton published his atomic theory, which states that every element consists of similar indivisible particles – called atoms – which differ from the atoms of other elements in their mass; he also drew up a list of relative atomic masses. Joseph Gay-Lussac announced that the volumes of gases that combine chemically with one another are in simple ratios.
1811	Publication of Amedeo Avogadro's hypothesis on the relation between the volume and number of molecules of a gas, and its temperature and pressure.
1813–14	Berzelius devised the chemical symbols and formulae still used to represent elements and compounds.
1828	Franz Wöhler converted ammonium cyanate into urea – the first synthesis of an organic compound from an inorganic substance.
1832–33	Michael Faraday expounded the laws of electrolysis, and adopted the term 'ion' for the particles believed to be responsible for carrying current.
1846	Thomas Graham expounded his law of diffusion.
1853	Robert Bunsen invented the Bunsen burner.
1858	Stanislao Cannizzaro differentiated between atomic and molecular weights (masses).
1861	Organic chemistry was defined by German chemist Friedrich Kekulé as the chemistry of carbon compounds.
1864	John Newlands devised the first periodic table of the elements.
1869	Dmitri Mendeleyev expounded his periodic table of the elements (based on atomic mass), leaving gaps for elements that had not yet been discovered.
1874	Jacobus van't Hoff suggested that the four bonds of carbon are arranged tetrahedrally, and that carbon compounds can therefore be three-dimensional and asymmetric.
1884	Swedish chemist Svante Arrhenius suggested that electrolytes (solutions or molten compounds that conduct electricity) dissociate into ions, atoms or groups of atoms that carry a positive or negative charge.
1894	William Ramsey and Lord Rayleigh discovered the first inert gases, argon.
1897	The electron was discovered by J J Thomson.
1901	Mikhail Tsvet invented paper chromatography as a means of separating pigments.
1909	Sören Sörensen devised the pH scale of acidity.
1912	Max von Laue showed crystals to be composed of regular, repeating arrays of atoms by studying the patterns in which they diffract X-rays.
1913–14	Henry Moseley equated the atomic number of an element with the positive charge on its nuclei, and drew up the periodic table, based on atomic number, that is used today.
1916	Gilbert Newton Lewis explained covalent bonding between atoms as a sharing of electrons.

Chemistry: Chronology – *continued*

1927	Nevil Sidgwick published his theory of valency, based on the numbers of electrons in the outer shells of the reacting atoms.
1930	Electrophoresis, which separates particles in suspension in an electric field, was invented by Arne Tiselius.
1932	Deuterium (heavy hydrogen), an isotope of hydrogen, was discovered by Harold Urey.
1940	Edwin McMillan and Philip Abelson showed that new elements with a higher atomic number than uranium can be formed by bombarding uranium with neutrons, and synthesized the first transuranic element, neptunium.
1942	Plutonium was first synthesized by Glenn T Seaborg and Edwin McMillan.
1950	Derek Barton deduced that some properties of organic compounds are affected by the orientation of their functional groups (the study of which became known as conformational analysis).
1954	Einsteinium and fermium were synthesized.
1955	Ilya Prigogine described the thermodynamics of irreversible processes (the transformations of energy that take place in, for example, many reactions within living cells).
1962	Neil Bartlett prepared the first compound of an inert gas, xenon hexafluoroplatinate; it was previously believed that inert gases could not take part in a chemical reaction.
1965	Robert B Woodward synthesized complex organic compounds.
1981	Quantum mechanics applied to predict course of chemical reactions by US chemist Roald Hoffmann and Kenichi Fukui of Japan.
1982	Element 109, unnilennium, synthesized.
1985	Fullerenes, a new class of carbon solids made up of closed cages of carbon atoms, were discovered by Harold Kroto and David Walton at the University of Sussex, England.
1987	US chemists Donald Cram and Charles Pederson, and Jean-Marie Lehn of France created artificial molecules that mimic the vital chemical reactions of life processes.
1990	Jean-Marie Lehn, Ulrich Koert, and Margaret Harding reported the synthesis of a new class of compounds, called nucleohelicates, that mimic the double helical structure of DNA, turned inside out.
1993	US chemists at the University of California and the Scripps Institute synthesized rapamycin, one of a group of complex, naturally occurring antibiotics and immunosuppressants that are being tested as anticancer agents.
1994	Elements 110 (ununnilium) and 111 (unununium) discovered at the GSI heavy-ion cyclotron, Darmstadt, Germany.
1995	German chemists built the largest ever wheel molecule made up of 154 molybdenum atoms surrounded by oxygen atoms. It has a relative molecular mass of 24,000 and is soluble in water.
1996	Element 112 discovered at the GSI heavy-ion cyclotron, Darmstadt, Germany.
1997	The International Union of Pure and Applied Chemistry (IUPAC) stated that elements 104–109 should be named rutherfordium (104), dubnium (105), seaborgium (106), bohrium (107), hassium (108), and meitnerium (109).

matter, the changes which matter may undergo and the phenomena which occur in the course of these changes.

Organic chemistry is the branch of chemistry that deals with carbon compounds. **Inorganic chemistry** deals with the description, properties, reactions, and preparation of all the elements and their compounds, with the exception of carbon compounds. **Physical chemistry** is concerned with the quantitative explanation of chemical phenomena and reactions, and the measurement of data required for such explanations. This branch studies in particular the movement of molecules and the effects of temperature and pressure, often with regard to gases and liquids.

All matter can exist in three states: gas, liquid, or solid. It is composed of minute particles termed **molecules**, which are constantly moving, and may be further divided into ◊**atoms**.

Molecules that contain atoms of one kind only are known as **elements**; those that contain atoms of different kinds are called **compounds**.

Chemical compounds are produced by a chemical action that alters the arrangement of the atoms in the reacting molecules. Heat, light, vibration, catalytic action, radiation, or pressure, as well as moisture (for ionization), may be necessary to produce a chemical change. Examination and possible breakdown of compounds to determine their components is **analysis**, and the building up

of compounds from their components is **synthesis**. When substances are brought together without changing their molecular structures they are said to be **mixtures**.

Symbols are used to denote the elements. The symbol is usually the first letter or letters of the English or Latin name of the element – for example, C for carbon; Ca for calcium; Fe for iron (*ferrum*). These symbols represent one atom of the element; molecules containing more than one atom of an element are denoted by a subscript figure – for example, water is H_2O.

In some substances a group of atoms acts as a single entity, and these are enclosed in parentheses in the symbol – for example $(NH_4)_2SO_4$ denotes ammonium sulphate. The symbolic representation of a molecule is known as a **formula**. A figure placed before a formula represents the number of molecules of a substance taking part in, or being produced by, a chemical reaction – for example, $2H_2O$ indicates two molecules of water. Chemical reactions are expressed by means of **equations** as in:

$$NaCl + H_2SO_4 \rightarrow NaHSO_4 + HCl$$

This equation states the fact that sodium chloride (NaCl) on being treated with sulphuric acid (H_2SO_4) is converted into sodium bisulphate (sodium hydrogensulphate, $NaHSO_4$) and hydrogen chloride (HCl). See also ◊chemical equation.

Chemistry: Chronology of Industrial Processes

c. AD 1100	Alcohol was first distilled.
1746	John Roebuck invented the lead-chamber process for the manufacture of sulphuric acid.
1790	Nicolas Leblanc developed a process for making sodium carbonate from sodium chloride (common salt).
1827	John Walker invented phosphorus matches.
1831	Peregrine Phillips developed the contact process for the production of sulphuric acid; it was first used on an industrial scale 1875.
1834	Justus von Liebig developed melamine.
1835	Tetrachloroethene (vinyl chloride) was first prepared.
1850	Ammonia was first produced from coal gas.
1855	A technique was patented for the production of cellulose nitrate (nitrocellulose) fibres, the first artificial fibres.
1856	Henry Bessemer developed the Bessemer converter for the production of steel.
1857	William Henry Perkin set up the first synthetic-dye factory, for the production of mauveine.
1861	Ernest Solvay patented a method for the production of sodium carbonate from sodium chloride and ammonia; the first production plant was established 1863.
1862	Alexander Parkes produced the first known synthetic plastic (Parkesine, or xylonite) from cellulose nitrate, vegetable oils, and camphor; it was the forerunner of celluloid.
1864	William Siemens and Pierre Emile Martin developed the Siemens–Martin process (open-hearth method) for the production of steel.
1868	Henry Deacon invented the Deacon process for the production of chlorine by the catalytic oxidation of hydrogen chloride.
1869	Celluloid was first produced from cellulose nitrate and camphor.
1880	The first laboratory preparation of polyacrylic substances.
1886	Charles M Hall and Paul-Louis-Toussaint Héroult developed, independently of each other, a method for producing aluminium by the electrolysis of aluminium oxide.
1891	Rayon was invented. Herman Frasch patented the Frasch process for the recovery of sulphur from underground deposits. Lindemann produced the first epoxy resins.
1894	Carl Kellner and Hamilton Castner developed, independently of each other, a method for the production of sodium hydroxide by the electrolysis of brine; collaboration gave rise to the Castner–Kellner process.
1895	The Thermit reaction for the reduction of metallic oxides to their molten metals was developed by Johann Goldschmidt.
1902	Friedrich Ostwald patented a process for the production of nitric acid by the catalytic oxidation of ammonia.
1908	Fritz Haber invented the Haber process for the production of ammonia from nitrogen and hydrogen. Heike Kamerlingh-Onnes prepared liquid helium.
1909	The first totally synthetic plastic (Bakelite) was produced by Leo Baekeland.
1912	I Ostromislensky patented the use of plasticizers, which rendered plastics mouldable.
1913	The thermal cracking of petroleum was established.
1919	Elwood Haynes patented non-rusting stainless steel.
1927	The commercial production of polyacrylic polymers began.
1930	Freons were first prepared and used in refrigeration plants. William Chalmers produced the polymer of methyl methacrylate (later marketed as Perspex).
1933	E W Fawcett and R O Gibson first produced polyethylene (polyethene) by the high-pressure polymerization of ethene.
1935	The catalytic cracking of petroleum was introduced. Triacetate film (used as base for photographic film) was developed.
1937	Wallace Carothers invented nylon. Polyurethanes were first produced.
1938	Roy Plunkett first produced polytetrafluoroethene (PTFE, marketed as Teflon).
1943	The industrial production of silicones was initiated. J R Whinfield invented Terylene.
1953	The German chemist Karl Zeigler produced high-density polyethylene.
1955	Artificial diamonds were first produced.
1959	The Du Pont company developed Lycra.
1963	Leslie Phillips and co-workers at the Royal Aircraft Establishment, Farnborough, England, invented carbon fibre.
1980	Nippon Oil patented the use of methyl-tert-butyl ether (MTBE) as a lead-free antiknock additive to petrol.
1984	About 2,500 people died in Bhopal, central India, when poisonous methyl isocyanate gas escaped from a chemical plant owned by US company Union Carbide.
1991	ICI began production of the hydrofluorocarbon HFA-134a, a substitute for CFCs in refrigerators and air-conditioning systems. Superconducting salts of buckminsterfullerene were discovered by researchers at AT & T Bell Laboratories, New Jersey, USA.
1993	Scientists at BP Chemicals built a pilot plant to convert plastic waste into an oil like naphtha, the crude oil fraction from which most plastics are derived. Chemists at the University of Cambridge, England, developed light-emitting diodes (LEDs) from the polymer poly(*p*-phenylenevinyl) that emit as much light as conventional, semiconductor-based LEDs and in a variety of colours.
1996	US scientists announced the invention of the all-plastic battery.

Elements are divided into **metals**, which have lustre and conduct heat and electricity, and **nonmetals**, which usually lack these properties.

The **periodic system**, developed by John Newlands in 1863 and established by Dmitri Mendeleyev in 1869, classified elements according to their relative atomic masses. Those elements that resemble each other in general properties were found to bear a relation to one another by weight, and these were placed in groups or families. Certain anomalies in this system were later removed by classifying the elements according to their atomic numbers. The latter is equivalent to the positive charge on the nucleus of the atom.

chemosynthesis method of making ◊protoplasm (contents of a cell) using the energy from chemical reactions, in contrast to the use of light energy employed for the same purpose in ◊photosynthesis. The process is used by certain bacteria, which can synthesize organic compounds from carbon dioxide and water using the energy from special methods of ◊respiration.

Nitrifying bacteria are a group of chemosynthetic organisms which change free nitrogen into a form that can be taken up by plants; nitrobacteria, for example, oxidize nitrites to nitrates. This is a vital part of the ◊nitrogen cycle. As chemosynthetic bacteria can survive without light energy, they can live in dark and inhospitable regions, including the hydrothermal vents of the Pacific ocean. Around these vents, where temperatures reach up to 350°C/662°F, the chemosynthetic bacteria are the basis of a food web supporting fishes and other marine life.

chemotaxis in biology, the property that certain cells have of attracting or repelling other cells. For example, white blood cells are attracted to the site of infection by the release of substances during certain types of immune response.

chemotherapy any medical treatment with chemicals. It usually refers to treatment of cancer with cytotoxic and other drugs. The term was coined by the German bacteriologist Paul Ehrlich for the use of synthetic chemicals against infectious diseases.

chemotropism movement by part of a plant in response to a chemical stimulus. The response by the plant is termed 'positive' if the growth is towards the stimulus or 'negative' if the growth is away from the stimulus.

Fertilization of flowers by pollen is achieved because the ovary releases chemicals that produce a positive chemotropic response from the developing pollen tube.

Chernobyl town in northern Ukraine; site of a nuclear power station. In April 1986 two huge explosions destroyed a central reactor, breaching the 1,000-tonne roof. In the immediate vicinity of Chernobyl, 31 people died (all firemen or workers at the plant) and 135,000 were permanently evacuated. It has been estimated that there will be an additional 20–40,000 deaths from cancer in the following 60 years; 600,000 are officially classified as at risk. According to WHO figures of 1995, the incidence of thyroid cancer in children has increased 200-fold in Belarus as a result of fallout from the disaster.

The resulting clouds of radioactive isotopes were traced all over Europe, from Ireland to Greece. Together with the fallout from nuclear weapons testing conducted in the past, the Chernobyl explosion currently contributes 0.4% to the annual average radiation dose in the UK, with the greatest effects on average concentration occurring in Scotland and Northern Ireland.

chickenpox or **varicella** common, usually mild disease, caused by a virus of the ◊herpes group and transmitted by airborne droplets. Chickenpox chiefly attacks children under the age of ten. The incubation period is two to three weeks. One attack normally gives immunity for life.

The temperature rises and spots (later inflamed blisters) develop on the torso, then on the face and limbs. The sufferer recovers within a week, but remains infectious until the last scab disappears.

The US Food and Drug Administration approved a chickenpox vaccine in March 1995. Based on a weakened form of the live virus, the vaccine is expected to be 70–90% effective. A vaccine is available in Europe, but is only used in children with an impaired immune system.

chilblain painful inflammation of the skin of the feet, hands, or ears, due to cold. The parts turn red, swell, itch violently, and are very tender. In bad cases, the skin cracks, blisters, or ulcerates.

childbirth the expulsion of a baby from its mother's body following ◊pregnancy. In a broader sense, it is the period of time involving labour and delivery of the baby.

chimera or **chimaera** in biology, an organism composed of tissues that are genetically different. Chimeras can develop naturally if a ◊mutation occurs in a cell of a developing embryo, but are more commonly produced artificially by implanting cells from one organism into the embryo of another.

china clay commercial name for ◊kaolin.

chinook warm dry wind that blows downhill on the E side of the Rocky Mountains of North America. It often occurs in winter and spring when it produces a rapid thaw, and so is important to the agriculture of the area.

chip or **silicon chip** another name for an ◊**integrated circuit**, a complete electronic circuit on a slice of silicon (or other semiconductor) crystal only a few millimetres square.

Chiron unusual Solar System object orbiting between Saturn and Uranus, discovered in 1977 by US astronomer Charles T Kowal (1940–).

Initially classified as an asteroid, it is now believed to be a giant cometary nucleus about 200 km/120 mi across, composed of ice with a dark crust of carbon dust. It has a 51-year orbit and a coma (cloud of gas and dust) caused by evaporation from its surface, resembling that of a comet. It is classified as a centaur.

chiropractic in alternative medicine, technique of manipulation of the spine and other parts of the body, based on the principle that physical disorders are attributable to aberrations in the functioning of the nervous system, which manipulation can correct.

Developed in the 1890s by US practitioner Daniel David Palmer, chiropractic is widely practised today by accredited therapists, although orthodox medicine remains sceptical of its efficacy except for the treatment of back problems.

chitin complex long-chain compound, or ◊polymer; a nitrogenous derivative of glucose. Chitin is widely found in invertebrates. It forms the ◊exoskeleton of insects and other arthropods. It combines with protein to form a covering that can be hard and tough, as in beetles, or soft and flexible, as in caterpillars and other insect larvae. It is insoluble in water and resistant to acids, alkalis, and many organic solvents. In crustaceans such as crabs, it is impregnated with calcium carbonate for extra strength.

Chitin also occurs in some ◊protozoans and coelenterates (such as certain jellyfishes), in the jaws of annelid worms, and as the cell-wall polymer of fungi. Its uses include coating apples (still fresh after six months), coating seeds, and dressing wounds. In 1993 chemists at North Carolina State University found that it can be used to filter pollutants from industrial waste water.

chlamydia virus-like bacteria which live parasitically in animal cells, and cause disease in humans and birds. Chlamydiae are thought to be descendants of bacteria that have lost certain metabolic processes. In humans, a strain of chlamydia causes ◊trachoma, a disease found mainly in the tropics (a leading cause of blindness); venereally transmitted chlamydiae cause genital and urinary infections.

Protein from *C. Pneumoniae* (which accounts for 10% of pneumonia cases) has been found in 79% of cases of atheroma (furring up of the arteries) in a US study, and it has also been cultured from a diseased coronary artery, providing a possible link between chlamydia infection and heart disease. A link has also been established between *C. Pneumoniae* infection and chronic high blood pressure.

chloral or *trichloroethanal* CCl₃CHO oily, colourless liquid with a characteristic pungent smell, produced by the action of chlorine on ethanol. It is soluble in water and its compound chloral hydrate is a powerful sleep-inducing agent.

chloramphenicol the first broad-spectrum antibiotic to be used commercially. It was discovered in 1947 in a Venezuelan soil sample containing the bacillus *Streptomyces venezuelae*, which produces the antibiotic substance C₁₁H₁₂Cl₂N₂O₅, now synthesized. Because of its toxicity, its use is mainly limited to the treatment of life-threatening infections, such as meningitis, legionnaire's disease, and typhoid fever.

chlorate any salt derived from an acid containing both chlorine and oxygen and possessing the negative ion ClO⁻, ClO₂⁻, ClO₃⁻, or ClO₄⁻. Common chlorates are those of sodium, potassium, and barium. Certain chlorates are used in weedkillers.

chloride Cl⁻ negative ion formed when hydrogen chloride dissolves in water, and any salt containing this ion, commonly formed by the action of hydrochloric acid (HCl) on various metals or by direct combination of a metal and chlorine. Sodium chloride (NaCl) is common table salt.

chlorinated solvent any liquid organic compound that contains chlorine atoms, often two or more. These compounds are very effective solvents for fats and greases, but many have toxic properties.

They include trichloromethane (chloroform, CHCl₃), tetrachloromethane (carbon tetrachloride, CCl₄), and trichloroethene (CH₂ClCHCl₂).

chlorination the treatment of water with chlorine in order to disinfect it; also, any chemical reaction in which a chlorine atom is introduced into a chemical compound.

chlorine greenish-yellow, gaseous, nonmetallic element with a pungent odour, symbol Cl, atomic number 17, relative atomic mass 35.453. It is a member of the ◊halogen group and is widely distributed, in combination with the ◊alkali metals, as chlorates or chlorides.

Chlorine was discovered in 1774 by the German chemist Karl Scheele, but English chemist Humphry Davy first proved it to be an element in 1810 and named it after its colour (chloros is Greek for green). In nature it is always found in the combined form, as in hydrochloric acid, produced in the mammalian stomach for digestion. Chlorine is obtained commercially by the electrolysis of concentrated brine and is an important bleaching agent and germicide, used for both drinking and swimming-pool water. As an oxidizing agent it finds many applications in organic chemistry. The pure gas (Cl₂) is a poison and was used in gas warfare in World War I, where its release seared the membranes of the nose, throat, and lungs, producing pneumonia.

Chlorine is a component of chlorofluorocarbons (CFCs) and is partially responsible for the depletion of the ◊ozone layer; it is released from the CFC molecule by the action of ultraviolet radiation in the upper atmosphere, making it available to react with and destroy the ozone. The concentration of chlorine in the atmosphere in 1997 reached just over 3 parts per billion. It is expected to reach its peak in 1999 and then start falling rapidly due to international action to curb ozone-destroying chemicals.

chlorofluorocarbon (CFC) synthetic chemical that is odourless, nontoxic, nonflammable, and chemically inert. The first CFC was synthesized in 1892, but no use was found for it until the 1920s. Since then their stability and apparently harmless properties have made CFCs popular in ◊aerosol cans, as refrigerants in refrigerators and air conditioners, and in the manufacture of foam packaging. They are partly responsible for the destruction of the ◊ozone layer. In June 1990 representatives of 93 nations, including the UK and the USA, agreed to phase out production of CFCs and various other ozone-depleting chemicals by the end of the 20th century.

When CFCs are released into the atmosphere, they drift up slowly into the stratosphere, where, under the influence of ultraviolet radiation from the Sun, they break down into chlorine atoms which destroy the ozone layer and allow harmful radiation from the Sun to reach the Earth's surface. CFCs can remain in the atmosphere for more than 100 years. Replacements for CFCs are being developed, and research into safe methods of destroying existing CFCs is being carried out. In Jan 1996 it was reported that US chemists at Yale University had developed a process for breaking down freons and other gases containing CFCs into nonhazardous compounds.

The European Union agreed to ban by the end of 1995 the five 'full hydrogenated' CFCs that are restricted under the ◊Montréal Protocol and a range of CFCs used as industrial solvents, refrigerants, and in fire extinguishers.

chloroform (technical name *trichloromethane*) CHCl₃ clear, colourless, toxic, carcinogenic liquid with a characteristic pungent, sickly sweet smell and taste, formerly used as an anaesthetic (now superseded by less harmful substances).

It is used as a solvent and in the synthesis of organic chemical compounds.

chlorophyll green pigment present in most plants; it is

responsible for the absorption of light energy during ◊photosynthesis. The pigment absorbs the red and blue-violet parts of sunlight but reflects the green, thus giving plants their characteristic colour. Chlorophyll is found within chloroplasts, present in large numbers in leaves. Cyanobacteria (blue-green algae) and other photosynthetic bacteria also have chlorophyll, though of a slightly different type. Chlorophyll is similar in structure to ◊haemoglobin, but with magnesium instead of iron as the reactive part of the molecule.

chloroplast structure (◊organelle) within a plant cell containing the green pigment chlorophyll. Chloroplasts occur in most cells of the green plant that are exposed to light, often in large numbers. Typically, they are flattened and disclike, with a double membrane enclosing the stroma, a gel-like matrix. Within the stroma are stacks of fluid-containing cavities, or vesicles, where ◊photosynthesis occurs.

It is thought that the chloroplasts were originally free-living cyanobacteria (blue-green algae) which invaded larger, non-photosynthetic cells and developed a symbiotic relationship with them. Like ◊mitochondria, they contain a small amount of DNA and divide by fission. Chloroplasts are a type of ◊plastid.

chlorosis abnormal condition of green plants in which the stems and leaves turn pale green or yellow. The yellowing is due to a reduction in the levels of the green chlorophyll pigments. It may be caused by a deficiency in essential elements (such as magnesium, iron, or manganese), a lack of light, genetic factors, or viral infection.

choke coil in physics, a coil employed to limit or suppress alternating current without stopping direct current, particularly the type used as a 'starter' in the circuit of fluorescent lighting.

choking in medicine, a process that results from an obstruction in the larynx that inhibits breathing. It is usually due to irritation produced by a piece of food or other substance that has been introduced via the mouth. The irritation of the sensitive mucus membrane lining the larynx provokes coughing as an attempt to expel the irritant. If the foreign body is large, the face may appear livid because of partial suffocation.

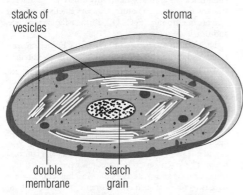

chloroplast Green chlorophyll molecules on the membranes of the vesicle stacks capture light energy to produce food by photosynthesis.

The choking individual should be encouraged to take slow, deep breaths that do not force the foreign body further into the respiratory tract. If the individual attempts to catch breath suddenly between coughs, the foreign body may penetrate further into the respiratory tract and stimulate further coughing. Blows with the palm of the hand over the shoulder blades, timed to coincide with the coughs, may assist the effect of the coughing. Medical attention is necessary if these methods do not result in the foreign body being expelled.

cholecalciferol or **vitamin D** fat-soluble chemical important in the uptake of calcium and phosphorous for bones. It is found in liver, fish oils, and margarine. It can be produced in the skin, provided that the skin is adequately exposed to sunlight. Lack of vitamin D leads to rickets and other bone diseases.

cholecystectomy surgical removal of the ◊gall bladder. It is carried out when gallstones or infection lead to inflammation of the gall bladder, which may then be removed either by conventional surgery or by a 'keyhole' procedure (see ◊endoscopy).

cholera disease caused by infection with various strains of the bacillus *Vibrio cholerae*, transmitted in contaminated water and characterized by violent diarrhoea and vomiting. It is prevalent in many tropical areas.

The formerly high death rate during epidemics has been much reduced by treatments to prevent dehydration and loss of body salts, together with the use of antibiotics. There is an effective vaccine that must be repeated at frequent intervals for people exposed to continuous risk of infection. The worst epidemic in the Western hemisphere for 70 years occurred in Peru 1991, with 55,000 confirmed cases and 258 deaths. It was believed to have been spread by the consumption of seafood contaminated by untreated sewage.

cholesterol white, crystalline ◊sterol found throughout the body, especially in fats, blood, nerve tissue, and bile; it is also provided in the diet by foods such as eggs, meat, and butter. A high level of cholesterol in the blood is thought to contribute to atherosclerosis (hardening of the arteries). Cholesterol is an integral part of all cell membranes and the starting point for steroid hormones, including the sex hormones. It is broken down by the liver into bile salts, which are involved in fat absorption in the digestive system, and it is an essential component of *lipoproteins*, which transport fats and fatty acids in the blood. *Low-density lipoprotein cholesterol* (LDL-cholesterol), when present in excess, can enter the tissues and become deposited on the surface of the arteries, causing ◊atherosclerosis. *High-density lipoprotein cholesterol* (HDL-cholesterol) acts as a scavenger, transporting fat and cholesterol from the tissues to the liver to be broken down. The composition of HDL-cholesterol can vary and some forms may not be as effective as others. Blood cholesterol levels can be altered by reducing the amount of alcohol and fat in the diet and by substituting some of the saturated fat for polyunsaturated fat, which gives a reduction in LDL-cholesterol. HDL-cholesterol can be increased by exercise.

choline in medicine, one of the members of the vitamin B complex. The daily requirement in adults, approximately 500 mg, is present in egg yolk, liver, and meat and sufficient choline can be obtained from a normal diet.

Figure labels:
stacks of vesicles
stroma
double membrane
starch grain

Choline can be synthesized by the body and deficiency results in a fatty liver.

cholinergic in biology, activity of nerve fibres to release the neurotransmitter ◊acetylcholine that mediates the transmission of nerve impulses between nerves or between nerves and muscles. Anticholinergic agents, such as pilocarpine, prolong the action of acetylcholine and have a role in the treatment of glaucoma.

chondrule in astronomy a small, round mass of silicate material found. Chondrites (stony ◊meteorites) are characterized by the presence of chondrules.

Chondrules are thought to be mineral grains that condensed from hot gas in the early Solar System, most of which were later incorporated into larger bodies from which the ◊planets formed.

chord in geometry, a straight line joining any two points on a curve. The chord that passes through the centre of a circle (its longest chord) is the diameter. The longest and shortest chords of an ellipse (a regular oval) are called the major and minor axes respectively.

chordate animal belonging to the phylum Chordata, which includes vertebrates, sea squirts, amphioxi, and others. All these animals, at some stage of their lives, have a supporting rod of tissue (notochord or backbone) running down their bodies.

Chordates are divided into three major groups: tunicates, cephalochordates, and craniates (including all vertebrates).

chorea condition featuring involuntary movements of the face muscles and limbs. It is seen in a number of neurological diseases, including ◊Huntington's chorea. See also ◊St Vitus' dance.

chorion outermost of the three membranes enclosing the embryo of reptiles, birds, and mammals; the amnion is the innermost membrane.

chorionic villus sampling (CVS) ◊biopsy of a small sample of placental tissue, carried out in early pregnancy at 10–12 weeks' gestation. Since the placenta forms from embryonic cells, the tissue obtained can be tested to reveal genetic abnormality in the fetus. The advantage of CVS over ◊amniocentesis is that it provides an earlier diagnosis, so that if any abnormality is discovered, and the parents opt for an abortion, it can be carried out more safely.

choroid layer found at the rear of the ◊eye beyond the retina. By absorbing light that has already passed through the retina, it stops back-reflection and so prevents blurred vision.

chromatography technique for separating or analysing a mixture of gases, liquids, or dissolved substances. This is brought about by means of two immiscible substances, one of which (*the mobile phase*) transports the sample mixture through the other (*the stationary phase*). The mobile phase may be a gas or a liquid; the stationary phase may be a liquid or a solid, and may be in a column, on paper, or in a thin layer on a glass or plastic support. The components of the mixture are absorbed or impeded by the stationary phase to different extents and therefore become separated. The technique is used for both qualitative and quantitive analyses in biology and chemistry.

In *paper chromatography*, the mixture separates because the components have differing solubilities in the solvent flowing through the paper and in the chemically bound water of the paper.

In *thin-layer chromatography*, a wafer-thin layer of adsorbent medium on a glass plate replaces the filter paper. The mixture separates because of the differing solubilities of the components in the solvent flowing up the solid layer, and their differing tendencies to stick to the solid (adsorption). The same principles apply in *column chromatography*.

In *gas–liquid chromatography*, a gaseous mixture is passed into a long, coiled tube (enclosed in an oven) filled with an inert powder coated in a liquid. A carrier gas flows through the tube. As the mixture proceeds along the tube it separates as the components dissolve in the liquid to differing extents or stay as a gas. A detector locates the different components as they emerge from the tube. The technique is very powerful, allowing tiny quantities of substances (fractions of parts per million) to be separated and analysed.

Preparative chromatography is carried out on a large scale for the purification and collection of one or more of a mixture's constituents; for example, in the recovery of protein from abattoir wastes.

Analytical chromatography is carried out on far smaller quantities, often as little as one microgram (one-millionth of a gram), in order to identify and quantify component parts of a mixture. It is used to determine the identities and amounts of amino acids in a protein, and the alcohol content of blood and urine samples. The technique was first used in the separation of coloured mixtures into their component pigments.

chromite $FeCr_2O_4$ iron chromium oxide, the main chromium ore. It is one of the ◊spinel group of minerals, and crystallizes in dark-coloured octahedra of the cubic system. Chromite is usually found in association with ultrabasic and basic rocks; in Cyprus, for example, it occurs with ◊serpentine, and in South Africa it forms continuous layers in a layered ◊intrusion.

chromium hard, brittle, grey-white, metallic element, symbol Cr, atomic number 24, relative atomic mass 51.996. It takes a high polish, has a high melting point, and is very resistant to corrosion. It is used in chromium electroplating, in the manufacture of stainless steel and other alloys, and as a catalyst. Its compounds are used for tanning leather and for ◊alums. In human nutrition it is a vital trace element. In nature, it occurs chiefly as chrome iron ore or chromite ($FeCr_2O_4$). Kazakhstan, Zimbabwe, and Brazil are sources.

The element was named in 1797 by the French chemist Louis Vauquelin (1763–1829) after its brightly coloured compounds.

chromium ore essentially the mineral chromite, $FeCr_2O_4$, from which chromium is extracted. South Africa and Zimbabwe are major producers.

chromosome structure in a cell nucleus that carries the ◊genes. Each chromosome consists of one very long strand of DNA, coiled and folded to produce a compact body. The point on a chromosome where a particular gene occurs is known as its locus. Most higher organisms have two copies of each chromosome (they are ◊diploid) but some have only one (they are ◊haploid). There are 46 chromosomes in a normal human cell. See also ◊mitosis and ◊meiosis.

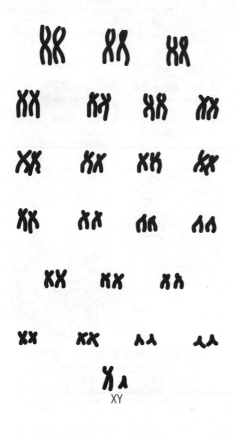

XY

chromosome The 23 pairs of chromosomes of a normal human male.

Chromosomes are only visible during cell division; at other times they exist in a less dense form called chromatin.

The first artificial human chromosome was built by US geneticists early in 1997. They constructed telomeres, centromeres, and DNA containing genetic information, which they removed from white blood cells, and inserted them into human cancer cells. The cells assembled the material into chromosomes. The artificial chromosome was successfully passed on to all daughter cells.

Chromosome

There are 46 chromosomes in the living cells of a human being. Other living things also have chromosomes but the number varies widely from species to species. For example, a garden pea has 14 chromosomes, a potato 48, and a crayfish 200.

chromosphere layer of mostly hydrogen gas about 10,000 km/6,000 mi deep above the visible surface of the Sun (the photosphere). It appears pinkish red during ◊eclipses of the Sun.

chronic in medicine, term used to describe a condition that is of slow onset and then runs a prolonged course, such as rheumatoid arthritis or chronic bronchitis. In contrast, an *acute* condition develops quickly and may be of relatively short duration.

chronic fatigue syndrome (CFS) a common debilitating condition also known as myalgic encephalomyelitis (ME) or postviral fatigue syndrome. It is characterized by a diffuse range of symptoms present for at least six months including extreme fatigue, muscular pain, weakness, depression, poor balance and coordination, joint pains, and gastric upset. It is usually diagnosed after exclusion of other diseases and frequently follows a flulike illness.

The cause of CFS remains unknown, but it is believed to have its origin in a combination of viral, social, and psychological factors. Theories based on one specific cause (such as Epstein-Barr virus) have been largely discredited. There is no definitive treatment, but with time the symptoms become less severe. Depression is treated if present, and recent research has demonstrated the effectiveness of ◊cognitive therapy.

chronometer instrument for measuring time precisely, originally used at sea. It is designed to remain accurate through all conditions of temperature and pressure. The first accurate marine chronometer, capable of an accuracy of half a minute a year, was made in 1761 by John Harrison in England.

chrysotile mineral in the ◊serpentine group, $Mg_3Si_2O_5$ $(OH)_4$. A soft fibrous silky mineral, the primary source of asbestos.

chyme general term for the stomach contents. Chyme resembles a thick creamy fluid and is made up of partly digested food, hydrochloric acid, and a range of enzymes.

The muscular activity of the stomach churns this fluid constantly, continuing the mechanical processes initiated by the mouth. By the time the chyme leaves the stomach for the duodenum, it is a smooth liquid ready for further digestion and absorption by the small intestine.

Cibachrome in photography, a process of printing directly from transparencies. It can be home-processed and the rich, saturated colours are highly resistant to fading. It was introduced in 1963.

cilia (singular *cilium*) small hairlike organs on the surface of some cells, particularly the cells lining the upper respiratory tract. Their wavelike movements waft particles of dust and debris towards the exterior. Some single-celled organisms move by means of cilia. In multicellular animals, they keep lubricated surfaces clear of debris. They also move food in the digestive tracts of some invertebrates.

ciliary body in biology, the part of the eye that connects the iris and the choroid. The ◊ciliary muscles contract and relax to change the shape of the lens during accommodation. The ciliary body is lined with cells that secrete aqueous humour into the anterior chamber of the eye.

ciliary muscle ring of muscle surrounding and controlling the lens inside the vertebrate eye, used in ◊accommodation (focusing). Suspensory ligaments, resembling spokes of a wheel, connect the lens to the ciliary muscle and pull the lens into a flatter shape when the muscle relaxes. When the muscle is relaxed the lens has its longest ◊focal length and focuses rays from distant objects. On contraction, the lens returns to its normal spherical state and therefore has a shorter focal length and focuses images of near objects.

cine camera camera that takes a rapid sequence of still photographs – 24 frames (pictures) each second. When the pictures are projected one after the other at the same speed on to a screen, they appear to show movement, because our eyes hold on to the image of one picture until the next one appears.

The cine camera differs from an ordinary still camera in having a motor that winds the film on. The film is held still by a claw mechanism while each frame is exposed. When the film is moved between frames, a semicircular disc slides between the lens and the film and prevents exposure.

cinnabar mercuric sulphide mineral, HgS, the only commercially useful ore of mercury. It is deposited in veins and impregnations near recent volcanic rocks and hot springs. The mineral itself is used as a red pigment, commonly known as ***vermilion***. Cinnabar is found in the USA (California), Spain (Almadén), Peru, Italy, and Slovenia.

circadian rhythm metabolic rhythm found in most organisms, which generally coincides with the 24-hour day. Its most obvious manifestation is the regular cycle of sleeping and waking, but body temperature and the concentration of ◊hormones that influence mood and behaviour also vary over the day. In humans, alteration of habits (such as rapid air travel round the world) may result in the circadian rhythm being out of phase with actual activity patterns, causing malaise until it has had time to adjust.

In mammals the circadian rhythm is controlled by the suprachiasmatic nucleus in the ◊hypothalamus. US researchers discovered a second circadian control mechanism in 1996; they found that cells within the retina also produced the hormone melatonin. In 1997, US geneticists identified a gene, *clock*, in chromosome 5 in mice, that regulated the circadian rhythm.

circle perfectly round shape, the path of a point that moves so as to keep a constant distance from a fixed point (the centre). Each circle has a ***radius*** (the distance from any point on the circle to the centre), a ***circumference*** (the boundary of the circle), ***diameters*** (straight lines crossing the circle through the centre), ***chords*** (lines joining two points on the circumference), ***tangents*** (lines that touch the circumference at one point only), ***sectors*** (regions inside the circle between two radii), and ***segments*** (regions between a chord and the circumference).

The ratio of the distance all around the circle (the circumference) to the diameter is an ◊irrational number called π (***pi***), roughly equal to 3.1416. A circle of radius r and diameter d has a circumference $C = \pi d$, or $C = 2\pi r$, and an area $A = \pi r^2$. The area of a circle can be shown by dividing it into very thin sectors and reassembling them to make an approximate rectangle. The proof of $A = \pi r^2$ can be done only by using ◊integral calculus.

circuit in physics or electrical engineering, an arrangement of electrical components through which a current can flow. There are two basic circuits, series and parallel. In a ◊series circuit, the components are connected end to end so that the current flows through all components one after the other. In a parallel circuit, components are connected side by side so that part of the current passes through each component. A circuit diagram shows in

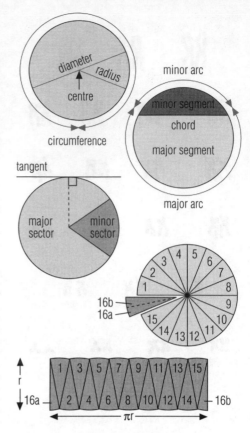

circle Technical terms used in the geometry of the circle; the area of a circle can be seen to equal πr^2 by dividing the circle into segments which form a rectangle.

graphical form how components are connected together, using standard symbols for the components.

circuit breaker switching device designed to protect an electric circuit from excessive current. It has the same action as a ◊fuse, and many houses now have a circuit breaker between the incoming mains supply and the domestic circuits. Circuit breakers usually work by means of ◊solenoids. Those at electricity-generating stations have to be specially designed to prevent dangerous arcing (the release of luminous discharge) when the high-voltage supply is switched off. They may use an air blast or oil immersion to quench the arc.

circulatory system system of vessels in an animal's body that transports essential substances (blood or other circulatory fluid) to and from the different parts of the body. Except for simple animals such as sponges and coelenterates (jellyfishes, sea anemones, corals), all animals have a circulatory system.

In fishes, blood passes once around the body before returning to a two-chambered heart (single circulation). In birds and mammals, blood passes to the lungs and back to the heart before circulating around the remainder of the body (double circulation). In all vertebrates, blood flows in one direction. Valves in the heart, large arteries, and

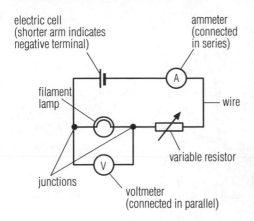

electric cell
(shorter arm indicates
negative terminal)

ammeter
(connected
in series)

filament
lamp

wire

variable resistor

junctions

voltmeter
(connected in parallel)

circuit diagram

veins prevent backflow, and the muscular walls of the arteries assist in pushing the blood around the body.

Although most animals have a heart or hearts to pump the blood, normal body movements circulate the fluid in some small invertebrates. In the *open system*, found in snails and other molluscs, the blood (more correctly called ◊haemolymph) passes from the arteries into a body cavity (haemocoel), and from here is gradually returned to the heart, via the gills, by other blood vessels. Insects and other arthropods have an open system with a heart. In the *closed system* of earthworms, blood flows directly from the main artery to the main vein, via smaller lateral vessels in each body segment. Vertebrates, too, have a closed system with a network of tiny ◊capillaries carrying the blood from arteries to veins. *See illustration overleaf.*

Circulatory System

There are about 96,500 km/60,000 mi of blood vessels in the human body – enough to go more than twice around the Earth. A blood cell makes one circuit of the blood system in about 60 seconds.

---

circumcision surgical removal of all or part of the foreskin (prepuce) of the penis, usually performed on the newborn; it is practised among Jews and Muslims. In some societies in Africa and the Middle East, female circumcision or clitoridectomy (removal of the labia minora and/or clitoris) is practised on adolescents as well as babies; it is illegal in the West. Female circumcision has no medical benefit and often causes disease and complications in childbirth. Male circumcision too is usually carried out for cultural reasons and not as a medical necessity, apart from cases where the opening of the prepuce is so small as to obstruct the flow of urine. Some evidence indicates that it protects against the development of cancer of the penis later in life and that women with circumcised partners are at less risk from cancer of the cervix.

circumference in geometry, the curved line that encloses a curved plane figure, for example a ◊circle or an ellipse. Its length varies according to the nature of the curve, and may be ascertained by the appropriate formula. The circumference of a circle is πd or $2\pi r$, where d

is the diameter of the circle, r is its radius, and π is the constant pi, approximately equal to 3.1416.

circumpolar in astronomy, a description applied to celestial objects that remain above the horizon at all times and do not set as seen from a given location. The amount of sky that is circumpolar depends on the observer's latitude on Earth. At the Earth's poles, all the visible sky is circumpolar, but at the Earth's equator none of it is circumpolar.

cirque french name for a ◊corrie, a steep-sided hollow in a mountainside.

cirrhosis any degenerative disease in an organ of the body, especially the liver, characterized by excessive development of connective tissue, causing scarring and painful swelling.

Cirrhosis of the liver may be caused by an infection such as viral hepatitis, chronic obstruction of the common bile duct, chronic alcoholism or drug use, blood disorder, heart failure, or malnutrition. However, often no cause is apparent. If cirrhosis is diagnosed early, it can be arrested by treating the cause; otherwise it will progress to coma and death.

CISC (acronym for *complex instruction-set computer*) in computing, a microprocessor (processor on a single chip) that can carry out a large number of ◊machine code instructions – for example, the Intel 80386. The term was introduced to distinguish them from the more rapid ◊RISC (reduced instruction-set computer) processors, which handle only a smaller set of instructions.

cistron in genetics, the segment of ◊DNA that is required to synthesize a complete polypeptide chain. It is the molecular equivalent of a ◊gene.

CITES (abbreviation for *Convention on International Trade in Endangered Species*) international agreement under the auspices of the ◊IUCN with the aim of regulating trade in ◊endangered species of animals and plants. The agreement came into force in 1975 and by 1997 had been signed by 138 states. It prohibits any trade in a category of 8,000 highly endangered species and controls trade in a further 30,000 species.

citizens' band (CB) short-range radio communication facility (around 27 MHz) used by members of the public in the USA and many European countries to talk to one another or call for emergency assistance.

citric acid HOOCCH$_2$C(OH)(COOH)CH$_2$COOH organic acid widely distributed in the plant kingdom; it is found in high concentrations in citrus fruits and has a sharp, sour taste. At one time it was commercially prepared from concentrated lemon juice, but now the main source is the fermentation of sugar with certain moulds.

civil aviation operation of passenger and freight transport by air. With increasing traffic, control of air space is a major problem, and in 1963 Eurocontrol was established by Belgium, France, West Germany, Luxembourg, the Netherlands, and the UK to supervise both military and civil movement in the air space over member countries. There is also a tendency to coordinate services and other facilities between national airlines; for example, the establishment of Air Union in 1963 by France (Air France), West Germany (Lufthansa), Italy (Alitalia), and Belgium (Sabena).

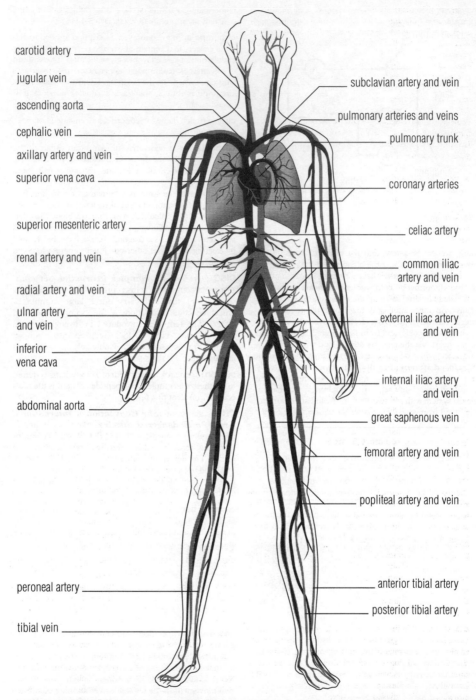

carotid artery

jugular vein

ascending aorta

cephalic vein

axillary artery and vein

superior vena cava

superior mesenteric artery

renal artery and vein

radial artery and vein

ulnar artery and vein

inferior vena cava

abdominal aorta

peroneal artery

tibial vein

subclavian artery and vein

pulmonary arteries and veins

pulmonary trunk

coronary arteries

celiac artery

common iliac artery and vein

external iliac artery and vein

internal iliac artery and vein

great saphenous vein

femoral artery and vein

popliteal artery and vein

anterior tibial artery

posterior tibial artery

circulatory system Blood flows through 96,500 km/60,000 mi of arteries and veins, supplying oxygen and nutrients to organs and limbs. Oxygen-poor blood (black) circulates from the heart to the lungs where oxygen is absorbed. Oxygen-rich blood (grey) flows back to the heart and is then pumped round the body through the aorta, the largest artery, to smaller arteries and capillaries. Here oxygen and nutrients are exchanged with carbon dioxide and waste products and the blood returns to the heart via the veins. Waste products are filtered by the liver, spleen, and kidneys and nutrients are absorbed from the stomach and small intestine.

In the UK there are about 170 airports. Heathrow, City, Gatwick, and Stansted (all serving London), Prestwick, and Edinburgh are managed by the British Airports Authority (founded in 1965). The British Airways Board supervises British Airways (BA), formerly British European Airways (BEA) and British Overseas Airways Corporation (BOAC); there are also independent companies. Close cooperation is maintained with authorities in other countries, including the Federal Aviation Agency, which is responsible for regulating development of aircraft, air navigation, traffic control, and communications in the USA. The Civil Aeronautics Board is the US authority prescribing safety regulations and investigating accidents. There are no state airlines in the USA, although many of the private airlines are large. The world's largest airline was the USSR's government-owned Aeroflot (split among the republics in 1992), which operated 1,300 aircraft over 1 million km/620,000 mi of routes.

civil engineering branch of engineering that is concerned with the construction of roads, bridges, airports, aqueducts, waterworks, tunnels, canals, irrigation works, and harbours.

The term is thought to have been used for the first time by British engineer John Smeaton in about 1750 to distinguish civilian from military engineering projects.

cladistics method of biological ◊classification (taxonomy) that uses a formal step-by-step procedure for objectively assessing the extent to which organisms share particular characters, and for assigning them to taxonomic groups. Taxonomic groups (for example, ◊species, ◊genus, family) are termed *clades*.

cladode in botany, a flattened stem that is leaflike in appearance and function. It is an adaptation to dry conditions because a stem contains fewer ◊stomata than a leaf, and water loss is thus minimized. The true leaves in such plants are usually reduced to spines or small scales. Examples of plants with cladodes are butcher's-broom *Ruscus aculeatus*, asparagus, and certain cacti. Cladodes may bear flowers or fruit on their surface, and this distinguishes them from leaves.

Clarke orbit alternative name for ◊*geostationary orbit*, an orbit 35,900 km/22,300 mi high, in which satellites circle at the same speed as the Earth turns. This orbit was first suggested by writer Arthur C Clarke in 1945.

class in biological classification, a group of related ◊orders. For example, all mammals belong to the class Mammalia and all birds to the class Aves. Among plants, all class names end in 'idae' (such as Asteridae) and among fungi in 'mycetes'; there are no equivalent conventions among animals. Related classes are grouped together in a ◊phylum.

classification in biology, the arrangement of organisms into a hierarchy of groups on the basis of their similarities in biochemical, anatomical, or physiological characters. The basic grouping is a ◊species, several of which may constitute a ◊genus, which in turn are grouped into families, and so on up through orders, classes, phyla (in plants, sometimes called divisions), to kingdoms.

class interval in statistics, the range of each class of data, used when dealing with large amounts of data. To obtain an idea of the distribution, the data are broken down into convenient classes, which must be mutually exclusive and

are usually equal. The class interval defines the range of each class; for example, if the class interval is five and the data begin at zero, the classes are 0–4, 5–9, 10–14, and so on.

clathrate compound formed when the small molecules of one substance fill in the holes in the structural lattice of another, solid, substance – for example, sulphur dioxide molecules in ice crystals. Clathrates are therefore intermediate between mixtures and true compounds (which are held together by ◊ionic or covalent chemical bonds).

clathration in chemistry, a method of removing water from an aqueous solution, and therefore increasing the solution's concentration, by trapping it in a matrix with inert gases such as freons.

clausius in engineering, a unit of ◊entropy (the loss of energy as heat in any physical process). It is defined as the ratio of energy to temperature above absolute zero.

claustrophobia ◊phobia involving fear of enclosed spaces.

clavicle the collar bone of many vertebrates. In humans it is vulnerable to fracture, since falls involving a sudden force on the arm may result in very high stresses passing into the chest region by way of the clavicle and other bones. It is connected at one end with the sternum (breastbone), and at the other end with the shoulder-blade, together with which it forms the arm socket. The wishbone of a chicken is composed of its two fused clavicles.

claw hard, hooked, pointed outgrowth of the digits of mammals, birds, and most reptiles. Claws are composed of the protein keratin, and grow continuously from a bundle of cells in the lower skin layer. Hooves and nails are modified structures with the same origin as claws.

clay very fine-grained ◊sedimentary deposit that has undergone a greater or lesser degree of consolidation. When moistened it is plastic, and it hardens on heating, which renders it impermeable. It may be white, grey, red, yellow, blue, or black, depending on its composition. Clay minerals consist largely of hydrous silicates of aluminium and magnesium together with iron, potassium, sodium, and organic substances. The crystals of clay minerals have a layered structure, capable of holding water, and are responsible for its plastic properties. According to international classification, in mechanical analysis of soil, clay has a grain size of less than 0.002 mm/0.00008 in.

clay mineral one of a group of hydrous silicate minerals that form most of the fine-grained particles in clays. Clay minerals are normally formed by weathering or alteration of other silicate minerals. Virtually all have sheet silicate structures similar to the ◊micas. They exhibit the following useful properties: loss of water on heating, swelling and shrinking in different conditions, cation exchange with other media, and plasticity when wet. Examples are kaolinite, illite, and montmorillonite.

Kaolinite $Al_2Si_2O_5(OH)_4$ is a common white clay mineral derived from alteration of aluminium silicates, especially feldspars. Illite contains the same constituents as kaolinite, plus potassium, and is the main mineral of clay sediments, mudstones, and shales; it is a weathering product of feldspars and other silicates. Montmorillonite contains the constituents of kaolinite plus sodium and magnesium; along with related magnesium- and iron-bearing clay minerals, it is derived from alteration and weathering of mafic igneous rocks. Kaolinite (the mineral name for kaolin or china clay) is economically important in the ceramic and paper industries. Illite, along with other clay minerals, may also be used in ceramics. Montmorillonite is the chief constituent of fuller's earth, and is also used in drilling muds (muds used to cool and lubricate drilling equipment). Vermiculite (similar to montmorillonite) will expand on heating to produce a material used in insulation.

cleavage in mineralogy, the tendency of a mineral to split along defined, parallel planes related to its internal structure. It is a useful distinguishing feature in mineral identification. Cleavage occurs where bonding between atoms is weakest, and cleavages may be perfect, good, or poor, depending on the bond strengths; a given mineral may possess one, two, three, or more orientations along which it will cleave. Some minerals have no cleavage, for example, quartz will fracture to give curved surfaces similar to those of broken glass. Some other minerals, such as apatite, have very poor cleavage that is sometimes known as a parting. Micas have one perfect cleavage and therefore split easily into very thin flakes. Pyroxenes have two good cleavages and break (less perfectly) into long prisms. Galena has three perfect cleavages parallel to the cube edges, and readily breaks into smaller and smaller cubes. Baryte has one perfect cleavage plus good cleavages in other orientations.

cleft palate fissure of the roof of the mouth, often accompanied by a harelip, the result of the two halves of the palate failing to join properly during embryonic development. It can be remedied by plastic surgery.

cleistogamy production of flowers that never fully open and that are automatically self-fertilized. Cleistogamous flowers are often formed late in the year, after the pro-

duction of normal flowers, or during a period of cold weather, as seen in several species of violet *Viola*.

client-server architecture in computing, a system in which the mechanics of looking after data are separated from the programs that use the data. For example, the 'server' might be a central database, typically located on a large computer that is reserved for this purpose. The 'client' would be an ordinary program that requests data from the server as needed. Most Internet services are examples of client-server applications, including the World Wide Web, FTP, Telnet, and Gopher.

climacteric period during the lifespan when an important physiological change occurs, usually referring to ◊menopause.

climate weather conditions at a particular place over a period of time. Climate encompasses all meteorological elements and the factors that influence them. The primary factors that determine the variations of climate over the surface of the Earth are: (a) the effect of latitude and the tilt of the Earth's axis to the plane of the orbit about the Sun (66.5°); (b) the large-scale movements of different wind belts over the Earth's surface; (c) the temperature difference between land and sea; (d) contours of the ground; and (e) location of the area in relation to ocean currents. Catastrophic variations to climate may be caused by the impact of another planetary body, or by clouds resulting from volcanic activity. The most important local or global meteorological changes brought about by human activity are those linked with ◊ozone depleters and the ◊greenhouse effect.

How much heat the Earth receives from the Sun varies in different latitudes and at different times of the year. In the equatorial region the mean daily temperature of the air near the ground has no large seasonal variation. In the polar regions the temperature in the long winter, when

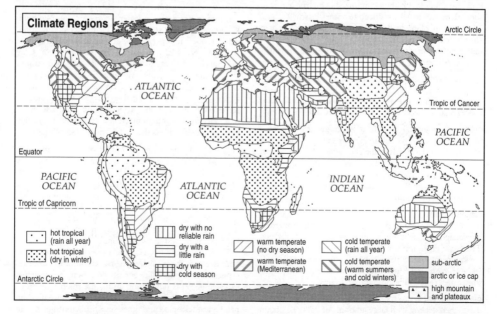

climate The world's climatic zones. There are many systems of classifying climate. One system, that of Wladimir Köppen, was based on temperature and plant type. Other systems take into account the distribution of global winds.

there is no incoming solar radiation, falls far below the summer value. Climate types were first classified by Wladimir Köppen (1846–1940) in 1884. The temperature of the sea, and of the air above it, varies little in the course of day or night, whereas the surface of the land is rapidly cooled by lack of solar radiation. In the same way the annual change of temperature is relatively small over the sea and great over the land. Continental areas are thus colder than the sea in winter and warmer in summer. Winds that blow from the sea are warm in winter and cool in summer, while winds from the central parts of continents are hot in summer and cold in winter. On average, air temperature drops with increasing land height at a rate of 1°C/1.8°F per 90 m/300 ft. Thus places situated above mean sea level usually have lower temperatures than places at or near sea level. Even in equatorial regions, high mountains are snow-covered during the whole year. Rainfall is produced by the condensation of water vapour in air. When winds blow against a range of mountains so that the air is forced to ascend, rain results, the amount depending on the height of the ground and the dampness of the air. The complexity of the distribution of land and sea, and the consequent complexity of the general circulation of the atmosphere, have a direct effect on the distribution of the climate. Centred on the Equator is a belt of tropical ◊rainforest, which may be either constantly wet or monsoonal (seasonal with wet and dry seasons in each year). On each side of this is a belt of savannah, with lighter seasonal rainfall and less dense vegetation, largely in the form of grasses. Usually there is then a transition through ◊steppe (semi-arid) to desert (arid), with a further transition through steppe to Mediterranean climate with dry summer, followed by the moist temperate climate of middle latitudes. Next comes a zone of cold climate with moist winter. Where the desert extends into middle latitudes, however, the zones of Mediterranean and moist temperate climates are missing, and the transition is from desert to a cold climate with moist winter. In the extreme east of Asia a cold climate with dry winters extends from about 70° N to 35° N. The polar caps have ◊tundra and glacial climates, with little or no ◊precipitation (rain or snow).

Climate changes over the last millennium can be detected by geophysicists using downhole measurement.

climatic change change in the climate of an area or of the whole world over an appreciable period of time. The geological record shows that climatic changes have taken place regularly, most notably during the ◊ice age. Modern climatic changes may be linked to increasing levels of pollution changing the composition of the atmosphere and producing a ◊greenhouse effect.

climatology study of climate, its global variations and causes.

Climatologists record mean daily, monthly, and annual temperatures and monthly and annual rainfall totals, as well as maximum and minimum values. Other data collected relate to pressure, humidity, sunshine, cold cover, and the frequency of days of frost, snow, hail, thunderstorms, and gales. The main facts are summarized in tables and climatological atlases published by nearly all the national meteorological services of the world.

climax community assemblage of plants and animals that is relatively stable in its environment. It is brought about by ecological ◊succession, and represents the point at which succession ceases to occur.

In temperate or tropical conditions, a typical climax community comprises woodland or forest and its associated fauna (for example, an oak wood in the UK). In essence, most land management is a series of interferences with the process of succession.

The theory, created by Frederic Clement in 1916, has been criticized for not explaining 'retrogressive' succession, when some areas revert naturally to pre-climax vegetation.

clinical ecology in medicine, ascertaining environmental factors involved in illnesses, particularly those manifesting nonspecific symptoms such as fatigue, depression, allergic reactions, and immune-system malfunctions, and prescribing means of avoiding or minimizing these effects.

clinical psychology branch of psychology dealing with the understanding and treatment of health problems, particularly mental disorders. The main problems dealt with include anxiety, phobias, depression, obsessions, sexual and marital problems, drug and alcohol dependence, childhood behavioural problems, psychoses (such as schizophrenia), mental disability, and brain disease (such as dementia) and damage. Other areas of work include forensic psychology (concerned with criminal behaviour) and health psychology.

Assessment procedures assess intelligence and cognition (for example, in detecting the effects of brain damage) by using psychometric tests. *Behavioural approaches* are methods of treatment that apply learning theories to clinical problems. *Behaviour therapy* helps clients change unwanted behaviours (such as phobias, obsessions, sexual problems) and to develop new skills (such as improving social interactions). *Behaviour modification* relies on operant conditioning, making selective use of rewards (such as praise) to change behaviour. This is helpful for children, the mentally disabled, and for patients in institutions, such as mental hospitals. *Cognitive therapy* is a new approach to treating emotional problems, such as anxiety and depression, by teaching clients how to deal with negative thoughts and attitudes. *Counselling*, developed by Carl Rogers, is widely used to help clients solve their own problems. *Psychoanalysis*, as developed by Sigmund Freud and Carl Jung, is little used by clinical psychologists today. It emphasizes childhood conflicts as a source of adult problems.

clinical trial in medicine, the evaluation of the effectiveness of medical treatment in a systematic fashion. Clinical trials compare new treatments with established treatments or placebos under standardized conditions. Such treatments may be drugs or surgical procedures. Ethical standards are maintained by ethics committees and they also ensure that the clinical trial procedure is explained to patients. The system was established with the development of new medicines in the 1940s and 1950s.

clinometer handheld surveying instrument for measuring angles of slope.

clint one of a number of flat-topped limestone blocks that make up a ◊limestone pavement. Clints are separated from each other by enlarged joints called grykes.

clo unit of thermal insulation of clothing. Standard

clothes have an insulation of about 1 clo; the warmest clothing is about 4 clo per 2.5 cm/1 in of thickness. See also ◊tog.

cloaca the common posterior chamber of most vertebrates into which the digestive, urinary, and reproductive tracts all enter; a cloaca is found in most reptiles, birds, and amphibians; many fishes; and, to a reduced degree, marsupial mammals. Placental mammals, however, have a separate digestive opening (the anus) and urinogenital opening. The cloaca forms a chamber in which products can be stored before being voided from the body via a muscular opening, the cloacal aperture.

clock any device that measures the passage of time, usually shown by means of pointers moving over a dial or by a digital display. Traditionally a timepiece consists of a train of wheels driven by a spring or weight controlled by a balance wheel or pendulum. Many clocks now run by batteries rather than clockwork. The watch is a portable clock.

In ancient Egypt the time during the day was measured by a shadow clock, a primitive form of ◊sundial, and at night the water clock was used. Up to the late 16th century the only clock available for use at sea was the sand clock, of which the most familiar form is the hourglass. During the Middle Ages various types of sundial were widely used, and portable sundials were in use from the 16th to the 18th century. Watches were invented in the 16th century – the first were made in Nürnberg, Germany, shortly after 1500 – but it was not until the 19th century that they became cheap enough to be widely available. The first known public clock was set up in Milan, Italy, in 1353. The timekeeping of both clocks and watches was revolutionized in the 17th century by the application of pendulums to clocks and of balance springs to watches.

The **marine chronometer** is a precision timepiece of special design, used at sea for giving Greenwich mean time (GMT). **Electric timepieces** were made possible by the discovery early in the 19th century of the magnetic effects of electric currents. One of the earliest and most satisfactory methods of electrical control of a clock was invented by Matthaeus Hipp in 1842. In one kind of electric clock, the place of the pendulum or spring-controlled balance wheel is taken by a small synchronous electric motor, which counts up the alternations (frequency) of the incoming electric supply and, by a suitable train of wheels, records the time by means of hands on a dial. The **quartz crystal clock** (made possible by the ◊piezoelectric effect of certain crystals) has great precision, with a short-term variation in accuracy of about one-thousandth of a second per day. More accurate still is the ◊**atomic clock**. This utilizes the natural resonance of certain atoms (for example, caesium) as a regulator controlling the frequency of a quartz crystal ◊oscillator. It is accurate to within one-millionth of a second per day.

clock rate the frequency of a computer's internal electronic clock. Every computer contains an electronic clock, which produces a sequence of regular electrical pulses used by the control unit to synchronize the components of the computer and regulate the ◊fetch-execute cycle by which program instructions are processed.

A fixed number of time pulses is required in order to execute each particular instruction. The speed at which a computer can process instructions therefore depends on the clock rate: increasing the clock rate will decrease the time required to complete each particular instruction.

Clock rates are measured in **megahertz** (MHz), or millions of pulses a second. Microcomputers commonly have a clock rate of 8–50 MHz.

clone an exact replica. In genetics, any one of a group of genetically identical cells or organisms. An identical ◊twin is a clone; so, too, are bacteria living in the same colony. The term 'clone' has also been adopted by computer technology to describe a (nonexistent) device that mimics an actual one to enable certain software programs to run correctly.

In Aug 1996, scientists in Oregon, USA, cloned two rhesus monkeys from embryo cells.

British scientists confirmed in Feb 1997 that they had cloned an adult sheep from a single cell to produce a lamb with the same genes as its mother. A cell was taken from the udder of the mother sheep, and its DNA combined with an unfertilized egg that had had its DNA removed. The fused cells were grown in the laboratory and then implanted into the uterus of a surrogate mother sheep. The resulting lamb, Dolly, came from an animal that was six years old, whose genes have already been damaged by environmental toxins and cosmic rays; the sheep could therefore develop cancers abnormally early.

It was the first time cloning had been achieved using cells other then reproductive cells. The cloning breakthrough has ethical implications, as the same principle could be used with human cells and eggs. The news was met with international calls to prevent the cloning of humans. The UK, Spain, Germany, Canada, and Denmark already have laws against cloning humans, as do some individual states in the USA. President Clinton announced in March 1997 a ban on using federal funds to support human cloning research, and called for a moratorium on this type of scientific research. He also asked the National Bioethics Advisory Commission to review and issue a report on the ramifications that cloning would have on humans. France and Portugal also have very restrictive laws on cloning.

closed in mathematics, descriptive of a set of data for which an operation (such as addition or multiplication) done on any members of the set gives a result that is also a member of the set.

For example, the set of even numbers is closed with respect to addition, because two even numbers added to each other always give another even number.

closed-circuit television (CCTV) localized television system in which programmes are sent over relatively short distances, the camera, receiver, and controls being linked by cable. Closed-circuit TV systems are used in department stores and large offices as a means of internal security, monitoring people's movements.

cloud water vapour condensed into minute water particles that float in masses in the atmosphere. Clouds, like fogs or mists, which occur at lower levels, are formed by the cooling of air containing water vapour, which generally condenses around tiny dust particles.

Clouds are classified according to the height at which they occur and their shape. **Cirrus** and **cirrostratus** clouds occur at around 10 km/33,000 ft. The former, sometimes called mares' tails, consist of minute specks of ice and appear as feathery white wisps, while cirrostratus

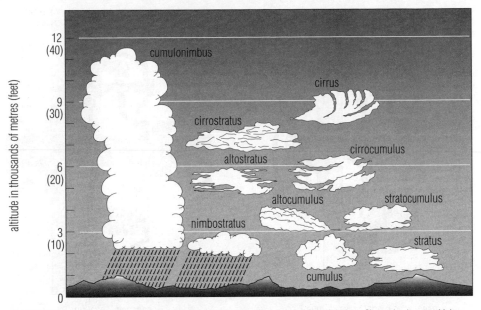

cloud Standard types of cloud. The height and nature of a cloud can be deduced from its name. Cirrus clouds are at high levels and have a wispy appearance. Stratus clouds form at low level and are layered. Middle-level clouds have names beginning with 'alto'. Cumulus clouds, ball or cottonwool clouds, occur over a range of heights.

clouds stretch across the sky as a thin white sheet. Three types of cloud are found at 3–7 km/10,000–23,000 ft: cirrocumulus, altocumulus, and altostratus. *Cirro-cumulus* clouds occur in small or large rounded tufts, sometimes arranged in the pattern called mackerel sky. *Altocumulus* clouds are similar, but larger, white clouds, also arranged in lines. *Altostratus* clouds are like heavy cirrostratus clouds and may stretch across the sky as a grey sheet. *Stratocumulus* clouds are generally lower, occurring at 2–6 km/6,500–20,000 ft. They are dull grey clouds that give rise to a leaden sky that may not yield rain. Two types of clouds, **cumulus** and **cumulo-nimbus**, are placed in a special category because they are produced by daily ascending air currents, which take moisture into the cooler regions of the atmosphere. Cumulus clouds have a flat base generally at 1.4 km/4,500 ft where condensation begins, while the upper part is dome-shaped and extends to about 1.8 km/6,000 ft. Cumulonimbus clouds have their base at much the same level, but extend much higher, often up to over 6 km/20,000 ft. Short heavy showers and sometimes thunder may accompany them. *Stratus* clouds, occurring below 1–2.5 km/3,000–8,000 ft, have the appearance of sheets parallel to the horizon and are like high fogs.

In addition to their essential role in the water cycle, clouds are important in the regulation of radiation in the Earth's atmosphere. They reflect short-wave radiation from the Sun, and absorb and re-emit long-wave radiation from the Earth's surface.

cloud chamber apparatus for tracking ionized particles. It consists of a vessel fitted with a piston and filled with air or other gas, saturated with water vapour. When the volume of the vessel is suddenly expanded by moving the piston outwards, the vapour cools and a cloud of tiny droplets forms on any nuclei, dust, or ions present. As fast-moving ionizing particles collide with the air or gas molecules, they show as visible tracks.

Much information about interactions between such particles and radiations has been obtained from photographs of these tracks.

The system has been improved upon in recent years by the use of liquid hydrogen or helium instead of air or gas (see ◊particle detector). The cloud chamber was devised in 1897 by Charles Thomson Rees Wilson (1869–1959) at Cambridge University.

Club of Rome informal international organization that aims to promote greater understanding of the interdependence of global economic, political, natural, and social systems. Members include industrialists, economists, and research scientists. Membership is limited to 100 people. It was established in 1968.

The organization, set up after a meeting at the Accademia dei Lincei, Rome, seeks to initiate new policies and take action to overcome some of the global problems facing humanity which traditional national organizations and short-term policies are unable to tackle effectively.

clusec unit for measuring the power of a vacuum pump.

Cluster ◊European Space Agency project to study the interaction of the solar wind with the Earth's ◊magnetosphere from an array of four identical satellites. Cluster works in conjunction with *SOHO* (Solar and Heliospheric Observatory).

clutch any device for disconnecting rotating shafts, used especially in a car's transmission system. In a car with a manual gearbox, the driver depresses the clutch when changing gear, thus disconnecting the engine from the gearbox.

The clutch consists of two main plates, a pressure plate

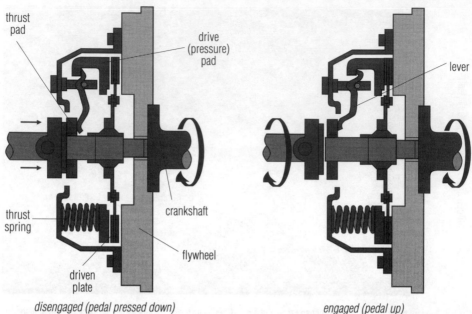

thrust pad

drive (pressure) pad

lever

thrust spring

crankshaft

driven plate

flywheel

disengaged (pedal pressed down)   *engaged (pedal up)*

clutch The clutch consists of two main plates: a drive plate connected to the engine crankshaft and a driven plate connected to the wheels. When the clutch is disengaged, the drive plate does not press against the driven plate. When the clutch is engaged, the two plates are pressed into contact and the rotation of the crankshaft is transmitted to the wheels.

and a driven plate, which is mounted on a shaft leading to the gearbox. When the clutch is engaged, the pressure plate presses the driven plate against the engine ◊fly-wheel, and drive goes to the gearbox. Depressing the clutch springs the pressure plate away, freeing the driven plate. Cars with **automatic transmission** have no clutch. Drive is transmitted from the flywheel to the automatic gearbox by a liquid coupling or torque converter.

cm symbol for **centimetre**.

coaching conveyance by coach – a horse-drawn passenger carriage on four wheels, sprung and roofed in. Public **stagecoaches** made their appearance in the middle of the 17th century; the first British mail coach began in 1784, and they continued until 1840 when railways began to take over the traffic.

The main roads were kept in good repair by turnpike trusts, and large numbers of inns – many of which still exist – catered for stagecoach passengers and horses. In the UK, coaches still in use on ceremonial occasions include those of the Lord Mayor of London in 1757 and the state coach built in 1761 for George III.

coagulation in biology, another term for ◊blood clotting, the process by which bleeding is stopped in the body.

coal black or blackish mineral substance formed from the compaction of ancient plant matter in tropical swamp conditions. It is used as a fuel and in the chemical industry. Coal is classified according to the proportion of carbon it contains. The main types are ◊**anthracite** (shiny, with about 90% carbon), **bituminous coal** (shiny and dull patches, about 75% carbon), and **lignite** (woody, grading into peat, about 50% carbon). Coal burning is one of the main causes of ◊acid rain.

coal gas gas produced when coal is destructively distilled or heated out of contact with the air. Its main constituents are methane, hydrogen, and carbon monoxide. Coal gas has been superseded by ◊natural gas for domestic purposes.

coal mining extraction of coal (a ◊sedimentary rock) from the Earth's crust. Coal mines may be opencast (see ◊opencast mining), adit, or deepcast. The least expensive is opencast but this may result in scars on the landscape.

coal tar black oily material resulting from the destructive distillation of bituminous coal.

Further distillation of coal tar yields a number of fractions: light oil, middle oil, heavy oil, and anthracene oil; the residue is called pitch. On further fractionation a large number of substances are obtained, about 200 of which have been isolated. They are used as dyes and in medicines.

coastal erosion the erosion of the land by the constant battering of the sea's waves. This produces two effects. The first is a hydraulic effect, in which the force of the wave compresses air pockets in coastal rocks and cliffs, and the air then expands explosively. The second is the effect of ◊corrasion, in which rocks and pebbles are flung against the cliffs, wearing them away. Frost shattering (or freeze-thaw), caused by the expansion of frozen seawater in cavities, and ◊biological weathering, caused by the burrowing of rock-boring molluscs, also result in the breakdown of the rock.

In areas where there are beaches, the waves cause longshore drift, in which sand and stone fragments are carried parallel to the shore, causing buildups (sandspits) in some areas and beach erosion in others.

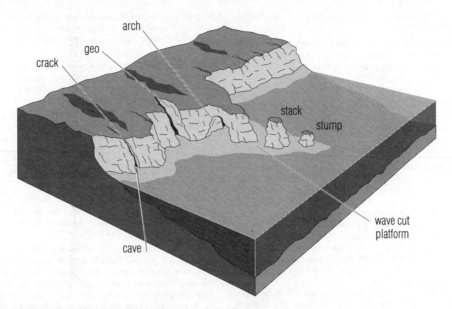

arch
geo
crack
stack
stump
cave
wave cut
platform

coastal erosion Typical features resulting from erosion of a headland.

coaxial cable electric cable that consists of a solid or stranded central conductor insulated from and surrounded by a solid or braided conducting tube or sheath. It can transmit the high-frequency signals used in television, telephone, and other telecommunications transmissions.

cobalt hard, lustrous, grey, metallic element, symbol Co, atomic number 27, relative atomic mass 58.933. It is found in various ores and occasionally as a free metal, sometimes in metallic meteorite fragments. It is used in the preparation of magnetic, wear-resistant, and high-strength alloys; its compounds are used in inks, paints, and varnishes.

The isotope Co-60 is radioactive (half-life 5.3 years) and is produced in large amounts for use as a source of gamma rays in industrial radiography, research, and cancer therapy. Cobalt was named in 1730 by Swedish chemist Georg Brandt (1694–1768); the name derives from the fact that miners considered its ore malevolent because it interfered with copper production (*Kobalt* is German for evil spirit).

cobalt-60 radioactive (half-life 5.3 years) isotope produced by neutron radiation of cobalt in heavy-water reactors, used in large amounts for gamma rays in cancer therapy, industrial radiography, and research, substituting for the much more costly radium.

cobalt chloride $CoCl_2$ compound that exists in two forms: the hydrated salt ($CoCl_2.6H_2O$), which is pink, and the anhydrous salt, which is blue. The anhydrous form is used as an indicator because it turns pink if water is present. When the hydrated salt is gently heated the blue anhydrous salt is reformed.

cobalt ore cobalt is extracted from a number of minerals, the main ones being *smaltite*, $(CoNi)As_3$; *linnaeite*, Co_3S_4; *cobaltite*, CoAsS; and *glaucodot*, $(CoFe)AsS$.

All commercial cobalt is obtained as a by-product of

other metals, usually associated with other ores, such as copper. Congo (formerly Zaire) is the largest producer of cobalt, and it is obtained there as a by-product of the copper industry. Other producers include Canada and Morocco. Cobalt is also found in the manganese nodules that occur on the ocean floor, and was successfully refined in 1988 from the Pacific Ocean nodules, although this process has yet to prove economic.

COBOL (acronym for *common business-oriented language*) high-level computer-programming language, designed in the late 1950s for commercial data-processing problems; it has become the major language in this field. COBOL features powerful facilities for file handling and business arithmetic. Program instructions written in this language make extensive use of words and look very much like English sentences. This makes COBOL one of the easiest languages to learn and understand.

cocaine alkaloid $C_{17}H_{21}NO_4$ extracted from the leaves of the coca tree. It has limited medical application, mainly as a local anaesthetic agent that is readily absorbed by mucous membranes (lining tissues) of the nose and throat. It is both toxic and addictive. Its use as a stimulant is illegal. ◊Crack is a derivative of cocaine.

Cocaine was first extracted from the coca plant in Germany in the 19th century. Most of the world's cocaine is produced from coca grown in Peru, Bolivia, Colombia, and Ecuador. Estimated annual production totals 215,000 tonnes, with most of the processing done in Colombia. Long-term use may cause mental and physical deterioration.

coccolithophore microscopic, planktonic marine alga, which secretes a calcite shell. The shells (coccoliths) of coccolithophores are a major component of deep sea ooze. Coccolithophores were particularly abundant during the late ◊Cretaceous period and their remains form the northern European chalk deposits, such as the white cliffs of Dover.

coccus (plural *cocci*) member of a group of globular bacteria, some of which are harmful to humans. The cocci contain the subgroups **streptococci**, where the bacteria associate in straight chains, and **staphylococci**, where the bacteria associate in branched chains.

cochlea part of the inner ◊ear. It is equipped with approximately 10,000 hair cells, which move in response to sound waves and thus stimulate nerve cells to send messages to the brain. In this way they turn vibrations of the air into electrical signals.

cocktail effect the effect of two toxic, or potentially toxic, chemicals when taken together rather than separately. Such effects are known to occur with some mixtures of drugs, with the active ingredient of one making the body more sensitive to the other.

This sometimes occurs because both drugs require the same ◊enzyme to break them down. Chemicals such as pesticides and food additives are only ever tested singly, not in combination with other chemicals that may be consumed at the same time, so no allowance is made for cocktail effects.

'Gulf War syndrome' may have resulted from the cocktail effect of an anti-nerve gas drug and two different insecticides.

codeine opium derivative that provides ◊analgesia in mild to moderate pain. It also suppresses the cough centre of the brain. It is an alkaloid, derived from morphine but less toxic and addictive.

codominance in genetics, the failure of a pair of alleles, controlling a particular characteristic, to show the classic recessive-dominant relationship. Instead, aspects of both alleles may show in the phenotype.

The snapdragon shows codominance in respect to colour. Two alleles, one for red petals and the other for white, will produce a pink colour if the alleles occur together as a heterozygous form.

codon in genetics, a triplet of bases (see ◊base pair) in a molecule of DNA or RNA that directs the placement of a particular amino acid during the process of protein (polypeptide) synthesis. There are 64 codons in the ◊genetic code.

coefficient the number part in front of an algebraic term, signifying multiplication. For example, in the expression $4x^2 + 2xy - x$, the coefficient of x^2 is 4 (because $4x^2$ means $4 \times x^2$), that of xy is 2, and that of x is -1 (because $-1 \times x = -x$).

In general algebraic expressions, coefficients are represented by letters that may stand for numbers; for example, in the equation $ax^2 + bx + c = 0$, a, b, and c are coefficients, which can take any number.

coefficient of relationship the probability that any two individuals share a given gene by virtue of being descended from a common ancestor. In sexual reproduction of diploid species, an individual shares half its genes with each parent, with its offspring, and (on average) with each sibling; but only a quarter (on average) with its grandchildren or its siblings' offspring; an eighth with its great-grandchildren, and so on.

In certain species of insects (for example honey bees), females have only one set of chromosomes (inherited from the mother), so that sisters are identical in genetic make-up; this produces a different set of coefficients. These coefficients are important in calculations of ◊inclusive fitness.

coelenterate any freshwater or marine organism of the phylum Coelenterata, having a body wall composed of two layers of cells. They also possess stinging cells. Examples are jellyfish, hydra, and coral.

coeliac disease disease in which the small intestine fails to digest and absorb food. The disease can appear at any age but has a peak incidence in the 30–50 age group; it is more common in women. It is caused by an intolerance to gluten (a constituent of wheat, rye, and barley) and characterized by diarrhoea and malnutrition. Treatment is by a gluten-free diet.

coelom in all but the simplest animals, the fluid-filled cavity that separates the body wall from the gut and associated organs, and allows the gut muscles to contract independently of the rest of the body.

coevolution evolution of those structures and behaviours within a species that can best be understood in relation to another species. For example, insects and flowering plants have evolved together: insects have produced mouthparts suitable for collecting pollen or drinking nectar, and plants have developed chemicals and flowers that will attract insects to them.

Coevolution occurs because both groups of organisms, over millions of years, benefit from a continuing association, and will evolve structures and behaviours that maintain this association.

cognition in psychology, a general term covering the functions involved in synthesizing information – for example, perception (seeing, hearing, and so on), attention, memory, and reasoning.

cognitive therapy or *cognitive behaviour therapy* treatment for emotional disorders such as ◊depression and ◊anxiety states. It encourages the patient to challenge the distorted and unhelpful thinking that is characteristic of depression, for example. The treatment may include ◊behaviour therapy.

cohesion in physics, a phenomenon in which interaction between two surfaces of the same material in contact makes them cling together (with two different materials the similar phenomenon is called adhesion). According to kinetic theory, cohesion is caused by attraction between particles at the atomic or molecular level. ◊Surface tension, which causes liquids to form spherical droplets, is caused by cohesion.

coil in medicine, another name for an ◊intrauterine device.

coke clean, light fuel produced, along with town gas, when coal is strongly heated in an airtight oven. Coke contains 90% carbon and makes a useful domestic and industrial fuel (used, for example in the iron and steel industries).

The process was patented in England in 1622, but it was only in 1709 that Abraham Darby devised a commercial method of production.

cold, common minor disease of the upper respiratory tract, caused by a variety of viruses. Symptoms are headache, chill, nasal discharge, sore throat, and occasionally cough. Research indicates that the virulence of a cold depends on psychological factors and either a reduction or an increase of social or work activity, as a result of stress, in the previous six months.

There is little immediate hope of an effective cure since the viruses transform themselves so rapidly.

cold-blooded of animals, dependent on the surrounding temperature; see ◊ *poikilothermy*.

cold dark matter theory in cosmology, a theory in which the bulk of the matter in the universe is in the form of dark, unseen material consisting of slow-moving particles. The gravitational clumping of this dark matter in the early universe may have lead to the formation of clusters and superclusters of ◊ galaxies.

cold fusion in nuclear physics, the fusion of atomic nuclei at room temperature. If cold fusion were possible it would provide a limitless, cheap, and pollution-free source of energy, and it has therefore been the subject of research around the world. In 1989, Martin Fleischmann (1927–   ) and Stanley Pons (1943–   ) of the University of Utah, USA, claimed that they had achieved cold fusion in the laboratory, but their results could not be substantiated.

cold-working method of shaping metal at or near atmospheric temperature.

coleoptile protective sheath that surrounds the young shoot tip of a grass during its passage through the soil to the surface. Although of relatively simple structure, most coleoptiles are very sensitive to light, ensuring that seedlings grow upwards.

colic spasmodic attack of pain in the abdomen, usually coming in waves. Colicky pains are caused by the painful muscular contraction and subsequent distension of a hollow organ; for example, the bowels, gall bladder (biliary colic), or ureter (renal colic).

Intestinal colic is due to partial or complete blockage of the intestine, or constipation; *infantile colic* is usually due to wind in the intestine.

colitis inflammation of the colon (large intestine) with diarrhoea (often bloody). It is usually due to infection or some types of bacterial dysentery.

collagen protein that is the main constituent of ◊ connective tissue. Collagen is present in skin, cartilage, tendons, and ligaments. Bones are made up of collagen, with the mineral calcium phosphate providing increased rigidity.

It was identified in a yeastlike fungus in 1996, the first time it has been found in a non-animal source.

collective farm farm in which a group of farmers pool their land, domestic animals, and agricultural implements, retaining as private property enough only for the members' own requirements. The profits of the farm are divided among its members. In cooperative farming, farmers retain private ownership of the land.

Collective farming was first developed in the USSR in 1917, where it became general after 1930. Stalin's collectivization drive (1929–33) wrecked a flourishing agricultural system and alienated the Soviet peasants from the land: 15 million people were left homeless, 1 million of whom were sent to labour camps and some 12 million deported to Siberia. In subsequent years, millions of those peasants forced into collectives died. Collective farming is practised in other countries; it was adopted from 1953 in China, and Israel has a large number of collective farms.

collective unconscious in psychology, a shared pool of memories, ideas, modes of thought, and so on, which, according to the Swiss psychiatrist Carl Jung, comes from the life experience of one's ancestors, indeed from the entire human race. It coexists with the personal ◊ unconscious, which contains the material of individual experience, and may be regarded as an immense depository of ancient wisdom.

Primal experiences are represented in the collective unconscious by archetypes, symbolic pictures, or personifications that appear in dreams and are the common element in myths, fairy tales, and the literature of the world's religions. Examples include the serpent, the sphinx, the Great Mother, the anima (representing the nature of woman), and the mandala (representing balanced wholeness, human or divine).

collenchyma plant tissue composed of relatively elongated cells with thickened cell walls, in particular at the corners where adjacent cells meet.

It is a supporting and strengthening tissue found in nonwoody plants, mainly in the stems and leaves.

collimator (1) small telescope attached to a larger optical instrument to fix its line of sight; (2) optical device for producing a nondivergent beam of light; (3) any device for limiting the size and angle of spread of a beam of radiation or particles.

collinear in mathematics, lying on the same straight line.

collision theory theory that explains how chemical reactions take place and why rates of reaction alter. For a reaction to occur the reactant particles must collide. Only a certain fraction of the total collisions cause chemical change; these are called *fruitful collisions*. The fruitful collisions have sufficient energy (activation energy) at the moment of impact to break the existing bonds and form new bonds, resulting in the products of the reaction. Increasing the concentration of the reactants and raising the temperature bring about more collisions and therefore more fruitful collisions, increasing the rate of reaction.

When a ◊ catalyst undergoes collision with the reactant molecules, less energy is required for the chemical change to take place, and hence more collisions have sufficient energy for reaction to occur. The reaction rate therefore increases.

colloid substance composed of extremely small particles of one material (the dispersed phase) evenly and stably distributed in another material (the continuous phase). The size of the dispersed particles (1–1,000 nanometres across) is less than that of particles in suspension but greater than that of molecules in true solution. Colloids involving gases include *aerosols* (dispersions of liquid or solid particles in a gas, as in fog or smoke) and *foams* (dispersions of gases in liquids).

Those involving liquids include *emulsions* (in which both the dispersed and the continuous phases are liquids) and *sols* (solid particles dispersed in a liquid). Sols in which both phases contribute to a molecular three-dimensional network have a jellylike form and are known as *gels*; gelatin, starch 'solution', and silica gel are common examples.

Milk is a natural emulsion of liquid fat in a watery liquid; synthetic emulsions such as some paints and cosmetic lotions have chemical emulsifying agents to stabilize the colloid and stop the two phases from separating out. Colloids were first studied thoroughly by the British

a fruitful collision

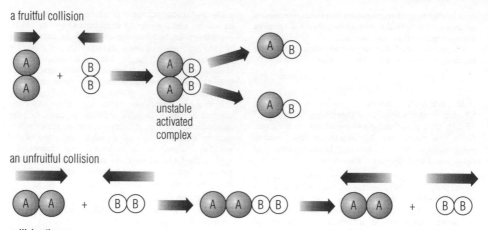

unstable
activated
complex

an unfruitful collision

collision theory

chemist Thomas Graham, who defined them as substances that will not diffuse through a semipermeable membrane (as opposed to what he termed crystalloids, solutions of inorganic salts, which will diffuse through).

colon in anatomy, the main part of the large intestine, between the caecum and rectum. Water and mineral salts are absorbed from undigested food in the colon, and the residue passes as faeces towards the rectum.

colonization in ecology, the spread of species into a new habitat, such as a freshly cleared field, a new motorway verge, or a recently flooded valley. The first species to move in are called **pioneers**, and may establish conditions that allow other animals and plants to move in (for example, by improving the condition of the soil or by providing shade). Over time a range of species arrives and the habitat matures; early colonizers will probably be replaced, so that the variety of animal and plant life present changes. This is known as ◊succession.

colour quality or wavelength of light emitted or reflected from an object. Visible white light consists of electromagnetic radiation of various wavelengths, and if a beam is refracted through a prism, it can be spread out into a spectrum, in which the various colours correspond to different wavelengths. From long to short wavelengths (from about 700 to 400 nanometres) the colours are red, orange, yellow, green, blue, indigo, and violet.

The light entering our eyes is either reflected from the objects we see, or emitted by hot or luminous objects.

Emitted light sources of light have a characteristic ◊spectrum or range of wavelengths. Hot solid objects emit light with a broad range of wavelengths, the maximum intensity being at a wavelength which depends on the temperature. The hotter the object, the shorter the wavelengths emitted, as described by ◊Wien's displacement law. Hot gases, such as the vapour of sodium street lights, emit light at discrete wavelengths. The pattern of wavelengths emitted is unique to each gas and can be used to identify the gas (see ◊spectroscopy).

Reflected light when an object is illuminated by white light, some of the wavelengths are absorbed and some are reflected to the eye of an observer. The object appears coloured because of the mixture of wavelengths in the reflected light. For instance, a red object absorbs all wavelengths falling on it except those in the red end of the spectrum. This process of subtraction also explains why certain mixtures of paints produce different colours. Blue and yellow paints when mixed together produce green because between them the yellow and blue pigments absorb all wavelengths except those around green. A suitable combination of three pigments – cyan (blue-green), magenta (blue-red), and yellow – can produce any colour when mixed. This fact is used in colour printing, although additional black pigment is also added.

Primary colours in the light-sensitive lining of our eyeball (the ◊retina), cells called cones are responsible for colour vision. There are three kinds of cones. Each type is sensitive to one colour only, either red, green, or blue. The brain combines the signals sent from the set of cones to produce a sensation of colour. When all cones are stimulated equally the sensation is of white light. The three colours to which the cones respond are called the primary colours. By mixing lights of these three colours, it is possible to produce any colour. This process is called colour mixing by addition, and is used to produce the colour on a television screen, where glowing phosphor dots of red, green, and blue combine.

Complementary colours pairs of colours that produce white light, such as yellow and blue, are called complementary colours.

Classifying colours many schemes have been proposed for classifying colours. The most widely used is the Munsell scheme, which classifies colours according to their hue (dominant wavelength), saturation (the degree of whiteness), and brightness (intensity).

colour blindness hereditary defect of vision that reduces the ability to discriminate certain colours, usually red and green. The condition is sex-linked, affecting men more than women.

colour index in astronomy, a measure of the colour of a star made by comparing its brightness through different coloured filters. It is defined as the difference between the ◊magnitude of the star measured through two standard photometric filters. Colour index is directly related to the surface temperature of a star and its spectral classification.

colouring food additive used to alter or improve the

colour of processed foods. Colourings include artificial colours, such as tartrazine and amaranth, which are made from petrochemicals, and the 'natural' colours such as chlorophyll, caramel, and carotene. Some of the natural colours are actually synthetic copies of the naturally occurring substances, and some of these, notably the synthetically produced caramels, may be injurious to health.

columbium (Cb) former name for the chemical element ◊niobium. The name is still used occasionally in metallurgy.

COM acronym for ◊*computer output on microfilm/microfiche*.

coma in astronomy, the hazy cloud of gas and dust that surrounds the nucleus of a ◊comet.

coma in medicine, a state of deep unconsciousness from which the subject cannot be roused. Possible causes include head injury, brain disease, liver failure, cerebral haemorrhage, and drug overdose.

coma in optics, one of the geometrical aberrations of a lens, whereby skew rays from an object make a comet-shaped spot on the image plane instead of meeting at a point.

combe or *coombe* steep-sided valley found on the scarp slope of a chalk ◊escarpment. The inclusion of 'combe' in a placename usually indicates that the underlying rock is chalk.

combination in mathematics, a selection of a number of objects from some larger number of objects when no account is taken of order within any one arrangement. For example, 123, 213, and 312 are regarded as the same combination of three digits from 1234. Combinatorial analysis is used in the study of ◊probability.

The number of ways of selecting r objects from a group of n is given by the formula:

$$n!/[r!(n-r)!]$$

(see ◊factorial). This is usually denoted by ^nC_r.

combine harvester or *combine* machine used for harvesting cereals and other crops, so called because it combines the actions of reaping (cutting the crop) and threshing (beating the ears so that the grain separates).

Combines, drawn by horses, were used in the Californian cornfields in the 1850s. Today's mechanical combine harvesters are capable of cutting a swath of up to 9 m/30 ft or more.

combustion burning, defined in chemical terms as the rapid combination of a substance with oxygen, accompanied by the evolution of heat and usually light. A slow-burning candle flame and the explosion of a mixture of petrol vapour and air are extreme examples of combustion.

comet small, icy body orbiting the Sun, usually on a highly elliptical path. A comet consists of a central nucleus a few kilometres across, and has been likened to a dirty snowball because it consists mostly of ice mixed with dust. As a comet approaches the Sun its nucleus heats up, releasing gas and dust which form a tenuous coma, up to 100,000 km/60,000 mi wide, around the nucleus. Gas and dust stream away from the coma to form one or more tails, which may extend for millions of kilometres. US astronomers concluded in 1996 that there are two distinct types of comet: one rich in methanol and one low in methanol. Evidence for this comes in part from observations of the spectrum of Comet Hyakutake.

Major Comets

Name	First recorded sighting	Orbital period (yrs)	Interesting facts
Halley's comet	240 BC	76	parent of Eta Aquarid and Orionid meteor showers
Comet Tempel-Tuttle	1366	33	parent of Leonid meteors
Biela's comet	1772	6.6	broke in half in 1846; not seen since 1852
Encke's comet	1786	3.3	parent of Taurid meteors
Comet Swift-Tuttle	1862	130	parent of Perseid meteors; reappeared 1992
Comet Ikeya-Seki	1965	880	so-called 'Sun-grazing' comet, passed 500,000 km/300,000 mi above surface of the Sun on 21 October 1965
Comet Kohoutek	1973	–	observed from space by *Skylab* astronauts
Comet West	1975	500,000	nucleus broke into four parts
Comet Bowell	1980	–	ejected from Solar System after close encounter with Jupiter
Comet IRAS-Araki-Alcock	1983	–	passed only 4.5 million km/2.8 million mi from the Earth on 11 May 1983
Comet Austin	1989	–	passed 32 million km/20 million mi from the Earth in 1990
Comet Shoemaker-Levy 9	1993	–	made up of 21 fragments; crashed into Jupiter in July 1994
Comet Hale-Bopp	1995	1,000	spitting out of gas and debris produced a coma with greater volume than the Sun; the bright coma is due to the outgassing of carbon monoxide; clearly visible with the naked eye in March 1997
Comet Hyakutake	1996	–	passed 15 million km/9,300,000 mi from the Earth

Comets are believed to have been formed at the birth of the Solar System. Billions of them may reside in a halo (the ◊**Oort cloud**) beyond Pluto. The gravitational effect of passing stars pushes some towards the Sun, when they eventually become visible from Earth. Most comets swing around the Sun and return to distant space, never to be seen again for thousands or millions of years, although some, called **periodic comets**, have their orbits altered by the gravitational pull of the planets so that they reappear every 200 years or less. Periodic comets are thought to come from the ◊Kuiper belt, a zone just beyond Neptune. Of the 800 or so comets whose orbits have been calculated, about 160 are periodic. The one with the shortest known period is ◊Encke's comet, which orbits the Sun every 3.3 years. A dozen or more comets are discovered every year, some by amateur astronomers.

Comet Hale-Bopp (C/1995 01) large and exceptionally active comet, which in March 1997 made its closest flyby to Earth since 2000 BC, coming within 190 million km/ 118 million mi. It has a diameter of approximately 40 km/ 25 mi and an extensive gas coma (when close to the Sun Hale-Bopp released 10 tonnes of gas every second). Unusually, Hale-Bopp has three tails: one consisting of dust particles, one of charged particles, and a third of sodium particles.

Comet Hale-Bopp was discovered independently in July 1995 by two amateur US astronomers, Alan Hale and Thomas Bopp.

Comet Shoemaker-Levy 9 comet that crashed into ◊Jupiter. The fragments crashed into Jupiter at 60 kps/37 mps over the period 16–22 July 1994. The impacts occurred on the far side of Jupiter, but the impact sites came into view of Earth about 25 minutes later. Analysis of the impacts shows that most of the pieces were solid bodies about 1 km/0.6 mi in diameter, but that at least three of them were clusters of smaller objects.

When first sighted in 24 March 1993 by US astronomers Carolyn and Eugene Shoemaker and David Levy, it was found to consist of at least 21 fragments in an unstable orbit around Jupiter. It is believed to have been captured by Jupiter in about 1930 and fragmented by tidal forces on passing within 21,000 km/ 13,050 mi of the planet in July 1992.

command language in computing, a set of commands and the rules governing their use, by which users control a program. For example, an ◊operating system may have commands such as SAVE and DELETE, or a payroll program may have commands for adding and amending staff records.

commensalism in biology, a relationship between two ◊species whereby one (the commensal) benefits from the association, whereas the other neither benefits nor suffers. For example, certain species of millipede and silverfish inhabit the nests of army ants and live by scavenging on the refuse of their hosts, but without affecting the ants.

commercial farming production of crops for sale and profit, although the farmers and their families may use a small amount of what they produce. Profits may be re-invested to improve the farm. Large-scale commercial farming is ◊agribusiness; see also ◊plantation and ◊ranching. The opposite of commercial farming is ◊subsistence farming where no food is produced for sale.

Committee on Safety of Medicines (CSM) UK authority processing licence applications for new drugs. The members are appointed by the secretary of state for health. Drugs are licensed on the basis of safety alone, according to the manufacturers' own data; usefulness is not considered.

The CSM operates a 'yellow card' reporting system: once a drug has entered clinical use doctors are expected to report any side effects. If these prove unacceptable the CSM can withdraw the drug's licence.

common denominator denominator that is a common multiple of, and hence exactly divisible by, all the denominators of a set of fractions, and which therefore enables their sums or differences to be found.

For example, 2/3 and 3/4 can both be converted to equivalent fractions of denominator 12, 2/3 being equal to 8/12 and 3/4 to 9/12. Hence their sum is 17/12 and their difference is 1/12. The **lowest common denominator** (lcd) is the smallest common multiple of the denominators of a given set of fractions.

common logarithm another name for a ◊logarithm to the base ten.

communication in biology, the signalling of information by one organism to another, usually with the intention of altering the recipient's behaviour. Signals used in communication may be **visual** (such as the human smile or the display of colourful plumage in birds), **auditory** (for example, the whines or barks of a dog), **olfactory** (such as the odours released by the scent glands of a deer), **electrical** (as in the pulses emitted by electric fish), or **tactile** (for example, the nuzzling of male and female elephants).

communications satellite relay station in space for sending telephone, television, telex, and other messages around the world. Messages are sent to and from the satellites via ground stations. Most communications satellites are in ◊geostationary orbit, appearing to hang fixed over one point on the Earth's surface.

The first satellite to carry TV signals across the Atlantic Ocean was *Telstar* in July 1962. The world is now linked by a system of communications satellites called Intelsat. Other satellites are used by individual countries for internal communications, or for business or military use. A new generation of satellites, called **direct broadcast satellites**, are powerful enough to transmit direct to small domestic aerials. The power for such satellites is produced by solar cells (see ◊solar energy). The total energy requirement of a satellite is small; a typical communications satellite needs about 2 kW of power, the same as an electric heater.

community in ecology, an assemblage of plants, animals, and other organisms living within a circumscribed area. Communities are usually named by reference to a dominant feature such as characteristic plant species (for example, a beech-wood community), or a prominent physical feature (for example, a freshwater-pond community).

commutative operation in mathematics, an operation that is independent of the order of the numbers or symbols concerned. For example, addition is commutative: the result of adding 4 + 2 is the same as that of adding 2 + 4; subtraction is not as 4 − 2 = 2, but 2 − 4 = −2. Compare ◊associative operation and ◊distributive operation.

commutator device in a DC (direct-current) electric motor that reverses the current flowing in the armature coils as the armature rotates.

A DC generator, or ◊dynamo, uses a commutator to convert the AC (alternating current) generated in the armature coils into DC. A commutator consists of opposite pairs of conductors insulated from one another, and contact to an external circuit is provided by carbon or metal brushes.

compact disc (CD) disc for storing digital information, about 12 cm/4.5 in across, mainly used for music, when it can have over an hour's playing time. The compact disc is made of aluminium with a transparent plastic coating; the metal disc underneath is etched by a ◊laser beam with microscopic pits that carry a digital code representing the sounds. During playback, a laser beam reads the code and produces signals that are changed into near-exact replicas of the original sounds.

CD-ROM, or **compact disc read-only memory**, is used to store written text, pictures, and video clips in addition to music. The discs are ideal for large works, such as catalogues and encyclopedias. CD-I, or **compact-disc interactive**, is a form of CD-ROM used with a computerized reader, which responds intelligently to the user's instructions. Recordable CDs, called **WORM**s ('write once, read many times'), are used as computer discs, but are rather expensive for home use. **Video CD**s, on sale in 1994, store an hour of video. High-density video discs, first publicly demonstrated in 1995, can hold full-length features. Erasable CDs, which can be erased and recorded many times, are also used by the computer industry. These are coated with a compound of cobalt and the rare earth metal gadolinium, which alters the polarization of light falling on it. In the reader, the light

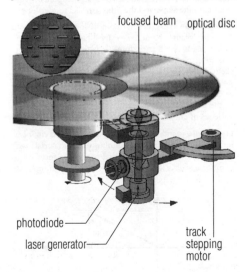

focused beam   optical disc

photodiode
laser generator
track stepping motor

compact disc The compact disc is a digital storage device; music is recorded as a series of etched pits representing numbers in digital code. During playing, a laser scans the pits and the pattern of reflected light reveals the numbers representing the sound recorded. The optical signal is converted to electrical form by a photocell and sent to the amplifiers and loudspeakers.

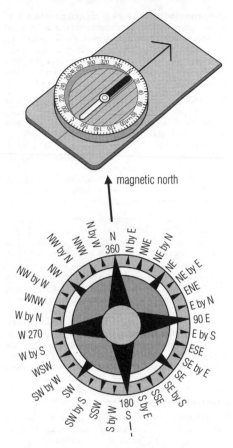

magnetic north

compass As early as 2500 BC, the Chinese were using pieces of magnetic rock, magnetite, as simple compasses. By the 12th century, European navigators were using compasses consisting of a needle-shaped magnet floating in a bowl of water.

reflected from the disc is passed through polarizing filters and the changes in polarization are converted into electrical signals.

Multi-layer CDs with increased storage capacity were developed in 1996. Two layers are enough to store a film 2 hours long, and up to 16 layers have been reliably read.

compass any instrument for finding direction. The most commonly used is a magnetic compass, consisting of a thin piece of magnetic material with the north-seeking pole indicated, free to rotate on a pivot and mounted on a compass card on which the points of the compass are marked. When the compass is properly adjusted and used, the north-seeking pole will point to the magnetic north, from which true north can be found from tables of magnetic corrections.

Compasses not dependent on the magnet are gyro-compasses, dependent on the ◊gyroscope, and radio-compasses, dependent on the use of radio. These are unaffected by the presence of iron and by magnetic anomalies of the Earth's magnetic field, and are widely used in ships and aircraft. See ◊navigation.

compensation point in biology, the point at which there is just enough light for a plant to survive. At this point all the food produced by ◊photosynthesis is used up by ◊respiration. For aquatic plants, the compensation point is the depth of water at which there is just enough light to sustain life (deeper water = less light = less photosynthesis).

competition in ecology, the interaction between two or more organisms, or groups of organisms (for example, species), that use a common resource which is in short supply. Competition invariably results in a reduction in the numbers of one or both competitors, and in ◊evolution contributes both to the decline of certain species and to the evolution of ◊adaptations.

Thus plants may compete with each other for sunlight, or nutrients from the soil, while animals may compete amongst themselves for food, water, or refuge.

compiler computer program that translates programs written in a ◊high-level language into machine code (the form in which they can be run by the computer). The compiler translates each high-level instruction into several machine-code instructions – in a process called *compilation* – and produces a complete independent program that can be run by the computer as often as required, without the original source program being present.

Different compilers are needed for different high-level languages and for different computers. In contrast to using an ◊interpreter, using a compiler adds slightly to the time needed to develop a new program because the machine-code program must be recompiled after each change or correction. Once compiled, however, the machine-code program will run much faster than an interpreted program.

complement in mathematics, the set of the elements within the universal set that are not contained in the designated set. For example, if the universal set is the set of all positive whole numbers and the designated set S is the set of all even numbers, then the complement of S (denoted S') is the set of all odd numbers.

complementary angles two angles that add up to 90°.

complementary medicine in medicine, systems of care based on methods of treatment or theories of disease that differ from those taught in most western medical schools. See ◊medicine, alternative.

complementary number in number theory, the number obtained by subtracting a number from its base. For example, the complement of 7 in numbers to base 10 is 3. Complementary numbers are necessary in computing, as the only mathematical operation of which digital computers (including pocket calculators) are directly capable is addition. Two numbers can be subtracted by adding one number to the complement of the other; two numbers can be divided by using successive subtraction (which, using complements, becomes successive addition); and multiplication can be performed by using successive addition.

complementation in genetics, the interaction that can occur between two different mutant alleles of a gene in a ◊diploid organism, to make up for each other's deficiencies and allow the organism to function normally.

complex in psychology, a group of ideas and feelings that have become repressed because they are distasteful to the person in whose mind they arose, but are still active in the depths of the person's unconscious mind, continuing to affect his or her life and actions, even though he or she is no longer fully aware of their existence. Typical examples include the ◊Oedipus complex and the ◊inferiority complex.

complex number in mathematics, a number written in the form $a + ib$, where a and b are ◊real numbers and i is the square root of -1 (that is, $i^2 = -1$); i used to be known as the 'imaginary' part of the complex number. Some equations in algebra, such as those of the form $x^2 + 5 = 0$, cannot be solved without recourse to complex numbers, because the real numbers do not include square roots of negative numbers.

The sum of two or more complex numbers is obtained by adding separately their real and imaginary parts, for example:

$$(a + bi) + (c + di) = (a + c) + (b + d)i$$

Complex numbers can be represented graphically on an Argand diagram, which uses rectangular ◊Cartesian coordinates in which the x-axis represents the real part of the number and the y-axis the imaginary part. Thus the number $z = a + bi$ is plotted as the point (a, b). Complex numbers have applications in various areas of science, such as the theory of alternating currents in electricity.

component in physics, one of two or more vectors, normally at right angles to each other, that add together to produce the same effect as a single resultant vector. Any ◊vector quantity, such as force, velocity, or electric field, can be resolved into components chosen for ease of calculation. For example, the weight of a body resting on a slope can be resolved into two force components; one normal to the slope and the other parallel to the slope.

Compositae daisy family, comprising dicotyledonous flowering plants characterized by flowers borne in composite heads (see ◊capitulum). It is the largest family of flowering plants, the majority being herbaceous. Birds seem to favour the family for use in nest 'decoration', possibly because many species either repel or kill insects. Species include the daisy and dandelion; food plants such as the artichoke, lettuce, and safflower; and the garden varieties of chrysanthemum, dahlia, and zinnia.

composite in industry, any purpose-designed engineering material created by combining single materials with complementary properties into a composite form. Most composites have a structure in which one component

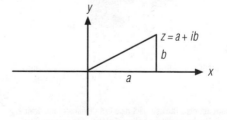

complex number A complex number can be represented graphically as a line whose end-point coordinates equal the real and imaginary parts of the complex number. This type of diagram is called an Argand diagram after the French mathematician Jean Argand (1768–1822) who devised it.

consists of discrete elements, such as fibres, dispersed in a continuous matrix. For example, lengths of asbestos, glass, or carbon steel, or 'whiskers' (specially grown crystals a few millimetres long) of substances such as silicon carbide may be dispersed in plastics, concrete, or steel.

compost organic material decomposed by bacteria under controlled conditions to make a nutrient-rich natural fertilizer for use in gardening or farming. A well-made compost heap reaches a high temperature during the composting process, killing most weed seeds that might be present.

compound chemical substance made up of two or more ◊elements bonded together, so that they cannot be separated by physical means. Compounds are held together by ionic or covalent bonds.

compressor machine that compresses a gas, usually air, commonly used to power pneumatic tools, such as road drills, paint sprayers, and dentist's drills. Compressed air expands when the pressure is removed, thus supplying the power to drive the tool.

Compressors are grouped into either positive displacement machines (which use a mechanical means of compression) or dynamic machines (which produce compression by creating high air speeds and thus high levels of kinetic energy, which is then converted into static pressure by a diffuser).

computer programmable electronic device that processes data and performs calculations and other symbol-manipulation tasks. There are three types: the ◊**digital computer**, which manipulates information coded as binary numbers (see ◊binary number system); the

◊**analogue computer**, which works with continuously varying quantities; and the **hybrid computer**, which has characteristics of both analogue and digital computers.

There are four types of digital computer, corresponding roughly to their size and intended use. **Microcomputers** are the smallest and most common, used in small businesses, at home, and in schools. They are usually single-user machines. **Minicomputers** are found in medium-sized businesses and university departments. They may support from 10 to 200 or so users at once. **Mainframes**, which can often service several hundred users simultaneously, are found in large organizations, such as national companies and government departments. **Supercomputers** are mostly used for highly complex scientific tasks, such as analysing the results of nuclear physics experiments and weather forecasting.

We do not need to have an infinity of different machines doing different jobs. A single one will suffice. The engineering problem of producing various machines for various jobs is replaced by the office work of 'programming' the universal machine to do these jobs.

On **computer** Alan Turing, quoted in
A Hodges *Alan Turing: The Enigma of Intelligence* 1985

Microcomputers now come in a range of sizes from battery-powered pocket PCs and electronic organizers, notebook and laptop PCs to floor-standing tower systems

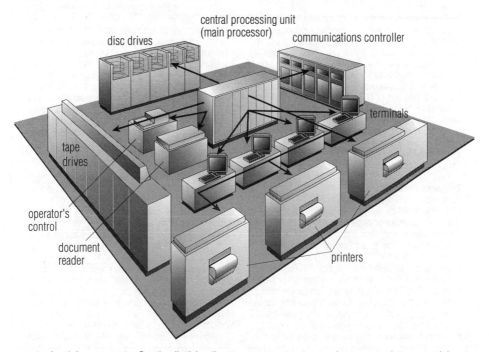

central processing unit
(main processor)

disc drives

communications controller

terminals

tape drives

operator's control

document reader

printers

computer A mainframe computer. Functionally, it has the same component parts as a microcomputer, but on a much larger scale. The central processing unit is at the hub, and controls all the attached devices.

Computing: the mechanics of an electronics industry

Computing is based mainly on three things: hardware, software, and economics. Hardware developments are generally predictable, but software availability has a significant effect on the adoption of technical advances, while economics affects its wide-scale deployment.

Moore's Law

For the past two decades, advances in computing have followed 'Moore's Law': the observation by Gordon Moore, a co-founder Intel, that the power of microchips doubles every 18 months or so. (Moore didn't actually say this, but it's a good rule of thumb.) The regular appearance of faster microprocessors and bigger memory chips provides ever more powerful computers at lower prices, while leading to the rapid obsolescence of existing models. This gives the industry its breakneck pace, because unsold computers diminish in value like perishable fruits.

A PC based on a new Intel processor of a certain speed will typically be a high-specification, high-priced machine. Six months to a year later, the same specification enters the mainstream at less than half the price. A year after that, it will have become a low cost 'entry-level' system, unless it has already been discontinued.

This process leads to rapid progress. The first mass market personal computers, launched at the end of the 1970s, ran at speeds of about 2 MHz and had 32 or 64 kilobytes of memory. Twenty years later, processors run at 200 MHz to 300 MHz or more, and new computers have 32 or 64 megabytes of memory (1024 times as much). The hard disc drives used for storing programs and data have also grown in capacity from about 5 megabytes to 2–5 gigabytes – and while today's drives store perhaps a thousand times more data, they cost less than one tenth of the price.

Software stability

The rapid changes in hardware technology are balanced by relative stability in software. If a new computer only has a working life of two to four years, there is no time to develop new software for it, and no one would buy it even if there were. (Developing a major program takes at least two years, while writing and testing an operating system can take four or five years.) A new computer must therefore be compatible with existing programs to provide users with a wide selection of applications, and to ensure that they can move the same programs and data on to the next new model, when it appears. The need for compatibility has helped create the dominance of the so-called 'Wintel' PC, based on Microsoft's DOS and Windows operating software and Intel's microprocessors. DOS – first introduced with the IBM PC in 1981 – is on the way out but is still used by some database and accounting programs, where the continuity of data is more important than it is with, say, word processors or games. Microsoft Windows took off with the launch of Windows 3 in 1990, and became dominant with the arrival of Windows 95 in 1995. It is now starting to be replaced by Windows NT (New Technology) that is superficially similar to Windows 95 but technically far superior. However, it is also less compatible with old DOS and Windows 3 programs and PC peripherals.

The PC's continuous upgrade cycle makes it unattractive to switch to different chips and operating systems. In theory, the market could change to a non-Microsoft operating system, but users are unwilling to give up access to thousands of popular Windows programs and games. They are not, after all, buying the operating system for its own sake, but as a 'platform' for applications. Compact cassette decks and VHS video recorders retain their popularity for the same reason, even though technically-superior systems have been available for many years.

In theory, the market could also change to non-Intel processors, and Microsoft allowed for this by designing Windows NT to run on a variety of chips from IBM/Motorola, Mips, and Digital as well as Intel. However, these alternatives were not successful: users decided that Intel compatibility was more important. Rival chips may be faster, but almost anyone with a PC can already buy a faster Intel-based machine, or simply wait a few months for the next advance from Intel.

Wintel's dominance is reinforced by both old and new economics. The 'old economics' is involved in the 'economies of scale' on which mass-market computing depends. It costs perhaps $200 million to write a new operating system, and about $2 billion to produce an advanced microprocessor, including the factory in which to make it. But once those tasks are completed, the cost of producing further copies is close to zero. If an operating system sells a million copies, then the cost of writing it is $200 per copy. However, if it sells 100 million copies, like Windows 95, the cost falls to $2 per copy. Economies of scale give Intel and Microsoft an almost unbeatable advantage: skulduggery is not required.

The virtuous circle

The second factor – an apparent contradiction of old economic principles – is known as the 'law of increasing returns'. In other words, the more people use a particular system, the more valuable it becomes. The more people use Windows, for example, the more software is written for it, which then attracts more people to use Windows. The same effect is apparent in hardware availability, the number of books, magazines, and training courses, and much else besides. This phenomenon tends to squeeze out proprietary systems (ones owned by a single company) such as the Acorn Archimedes, Apple Macintosh, Atari ST, and Commodore Amiga.

The 'virtuous circle' of increasing returns is not new: it has been part of the computer scene at least since the IBM 360 mainframe range appeared in 1964. However, the huge volume of PC sales, which surpassed annual sales of TV sets in 1996, has given the phenomenon unprecedented impact. Instead of just being limited to desktop computers, economies of scale and increasing returns have driven the use of Wintel-based computers into all areas of computing from laptops to departmental file servers, and they are now invading mainframe data centres. PCs running Windows NT are now competing with graphics workstations running UNIX, with network file-servers running Novel's NetWare operating system, with IBM minicomputers running OS/400, and so on, as well as with desktop machines like the Apple Macintosh.

The Java Virtual Machine

The future of computing is likely to resemble the recent past. Most PC users can be expected to upgrade from Windows 3 running on Intel 486 processors and Windows 95 running on Pentiums to Windows 98 and NT running on Pentium MMX and Pentium II processors, with the next generation – the IA-64 (64-bit Intel Architecture) chip, codenamed Merced, now being developed by Intel and Hewlett-Packard – appearing in high-end machines around the millennium. However, there is always the possibility of fundamental change. The Wintel hegemony may be challenged by new electronics devices equipped with a sort of software computer – a Java Virtual Machine or JVM – able to run the same programs regardless of the operating system and processor used. Java is intended to run on all kinds of devices from smartcards to supercomputers, including mobile phones, televisions set-top boxes (used to bring the Internet to TV sets), network computers and PCs. But whether Java is a real threat to Windows on the desktop, or wishful thinking, only time will tell.

Jack Schofield

---

that may serve local area ◊networks or work as minicomputers. Indeed, most minicomputers are now built using low-cost microprocessors, and large-scale computers built out of multiple microprocessors are starting to challenge traditional mainframe and supercomputer designs.

computer-aided design use of computers to create and modify design drawings; see ◊CAD.

computer-aided manufacturing use of computers to regulate production processes in industry; see ◊CAM.

computer art art produced with the help of a computer.

Since the 1950s the aesthetic use of computers has been increasingly evident in most artistic disciplines, including film animation, architecture, and music. ◊Computer graphics has been the most developed area, with the 'paint-box' computer liberating artists from the confines of the canvas. It is now also possible to programme computers in advance to generate graphics, music, and sculpture, according to 'instructions' which may include a preprogrammed element of unpredictability. In this last function, computer technology has been seen as a way of challenging the elitist nature of art by putting artistic creativity within relatively easy reach of anyone owning a computer.

computer-assisted learning use of computers in education and training; see ◊CAL.

computer game or **video game** any computer-controlled game in which the computer (sometimes) opposes the human player. Computer games typically employ fast, animated graphics on a ◊VDU (visual display unit) and synthesized sound.

Commercial computer games became possible with the advent of the ◊microprocessor in the mid-1970s and rapidly became popular as amusement-arcade games, using dedicated chips. Available games range from chess to fighter-plane simulations.

Some of the most popular computer games in the early 1990s were id Software's *Wolfenstein 3D* and *Doom*, which are designed to be played across networks including the Internet. A whole subculture built up around those particular games, as users took advantage of id's help to create their own additions to the game.

The computer games industry has been criticized for releasing many violent games with little intellectual content.

computer generation any of the five broad groups into which computers may be classified: **first generation** the earliest computers, developed in the 1940s and 1950s, made from valves and wire circuits; **second generation** from the early 1960s, based on transistors and printed circuits; **third generation** from the late 1960s, using integrated circuits and often sold as families of computers, such as the IBM 360 series; **fourth gener-**

Computer: Chronology

1614	John Napier invented logarithms.
1615	William Oughtred invented the slide rule.
1623	Wilhelm Schickard invented the mechanical calculating machine.
1645	Blaise Pascal produced a calculator.
1672–74	Gottfried Leibniz built his first calculator, the Stepped Reckoner.
1801	Joseph-Marie Jacquard developed an automatic loom controlled by punch cards.
1820	The first mass-produced calculator, the Arithometer, was developed by Charles Thomas de Colmar.
1822	Charles Babbage completed his first model for the difference engine.
1830s	Babbage created the first design for the analytical engine.
1890	Herman Hollerith developed the punched-card ruler for the US census.
1936	Alan Turing published the mathematical theory of computing.
1938	Konrad Zuse constructed the first binary calculator, using Boolean algebra.
1939	US mathematician and physicist J V Atanasoff became the first to use electronic means for mechanizing arithmetical operations.
1943	The Colossus electronic code-breaker was developed at Bletchley Park, England. The Harvard University Mark I or Automatic Sequence Controlled Calculator (partly financed by IBM) became the first program-controlled calculator.
1946	ENIAC (acronym for electronic numerator, integrator, analyser, and computer), the first general purpose, fully electronic digital computer, was completed at the University of Pennsylvania, USA.
1948	Manchester University (England) Mark I, the first stored-program computer, was completed. William Shockley of Bell Laboratories invented the transistor.
1951	Launch of Ferranti Mark I, the first commercially produced computer. Whirlwind, the first real-time computer, was built for the US air-defence system. Grace Murray Hopper of Remington Rand invented the compiler computer program.
1952	EDVAC (acronym for electronic discrete variable computer) was completed at the Institute for Advanced Study, Princeton, USA (by John Von Neumann and others).
1953	Magnetic core memory was developed.
1958	The first integrated circuit was constructed.
1963	The first minicomputer was built by Digital Equipment (DEC). The first electronic calculator was built by Bell Punch Company.
1964	Launch of IBM System/360, the first compatible family of computers. John Kemeny and Thomas Kurtz of Dartmouth College invented BASIC (Beginner's All-purpose Symbolic Instruction Code), a computer language similar to FORTRAN.
1965	The first supercomputer, the Control Data CD6600, was developed.
1971	The first microprocessor, the Intel 4004, was announced.
1974	CLIP–4, the first computer with a parallel architecture, was developed by John Backus at IBM.
1975	Altair 8800, the first personal computer (PC), or microcomputer, was launched.
1981	The Xerox Star system, the first WIMP system (acronym for windows, icons, menus, and pointing devices), was developed. IBM launched the IBM PC.
1984	Apple launched the Macintosh computer.
1985	The Inmos T414 transputer, the first 'off-the-shelf' microprocessor for building parallel computers, was announced.
1988	The first optical microprocessor, which uses light instead of electricity, was developed.
1989	Wafer-scale silicon memory chips, able to store 200 million characters, were launched.
1990	Microsoft released Windows 3, a popular windowing environment for PCs.
1992	Philips launched the CD-I (Compact-Disc Interactive) player, based on CD audio technology, to provide interactive multimedia programs for the home user.
1993	Intel launched the Pentium chip containing 3.1 million transistors and capable of 100 MIPs (millions of instructions per second). The Personal Digital Assistant (PDA), which recognizes user's handwriting, went on sale.
1995	Intel launched the Pentium Pro microprocessor (formerly codenamed P6).
1996	IBM's computer Deep Blue beat grand master Gary Kasparnov at chess, the first time a computer has beaten a human grand master.
1997	In the USA, an attempt to bring legislation to control the Internet, intended to prevent access to sexual material, was rejected as unconstitutional.

ation using ◊microprocessors, large-scale integration (LSI), and sophisticated programming languages, still in use in the 1990s; and **fifth generation** based on parallel processing and very large-scale integration, currently under development.

computer graphics use of computers to display and manipulate information in pictorial form. Input may be achieved by scanning an image, by drawing with a mouse or stylus on a graphics tablet, or by drawing directly on the screen with a light pen.

The output may be as simple as a pie chart, or as complex as an animated sequence in a science-fiction film, or a seemingly three-dimensional engineering blueprint. The drawing is stored in the computer as ◊raster graphics or ◊vector graphics. Computer graphics are increasingly used in computer-aided design (◊CAD), and to generate models and simulations in engineering, meteorology, medicine and surgery, and other fields of science.

computerized axial tomography medical technique, usually known as ◊CAT scan, for noninvasive investigation of disease or injury.

computer output on microfilm/microfiche (COM) technique for producing computer output in very compact, photographically reduced form (◊microform).

computer program coded instructions for a computer; see ◊program.

computer simulation representation of a real-life situation in a computer program. For example, the program might simulate the flow of customers arriving at a bank. The user can alter variables, such as the number of cashiers on duty, and see the effect.

More complex simulations can model the behaviour of chemical reactions or even nuclear explosions. The behaviour of solids and liquids at high temperatures can be simulated using quantum simulation. Computers also control the actions of machines – for example, a ◊flight simulator models the behaviour of real aircraft and allows training to take place in safety. Computer simulations are very useful when it is too dangerous, time consuming, or simply impossible to carry out a real experiment or test.

computer terminal the device whereby the operator communicates with the computer; see ◊terminal.

computing device any device built to perform or help perform computations, such as the ◊abacus, ◊slide rule, or ◊computer.

The earliest known example is the abacus. Mechanical devices with sliding scales (similar to the slide rule) date from ancient Greece. In 1642, French mathematician Blaise Pascal built a mechanical adding machine and, in 1671, German mathematician Gottfried Leibniz produced a machine to carry out multiplication. The first mechanical computer, the analytical engine, was designed by British mathematician Charles Babbage in 1835. For the subsequent history of computing, see ◊computer.

concave of a surface, curving inwards, or away from the eye. For example, a bowl appears concave when viewed from above. In geometry, a concave polygon is one that has an interior angle greater than 180°. Concave is the opposite of ◊convex.

concave lens lens that possesses at least one surface that curves inwards. It is a diverging lens, spreading out those light rays that have been refracted through it. A concave lens is thinner at its centre than at its edges, and is used to correct short-sightedness.

Common forms include the **biconcave** lens (with both surfaces curved inwards) and the **plano-concave** (with one flat surface and one concave). The whole lens may be further curved overall, making a **convexo-concave** or diverging meniscus lens, as in some lenses used for corrective purposes.

concave mirror curved mirror that reflects light from its inner surface. It may be either circular or parabolic in section. A concave mirror converges light rays to form a reduced, inverted, real image in front, or an enlarged, upright, virtual image seemingly behind it, depending on how close the object is to the mirror.

Only a parabolic concave mirror has a true, single-point ◊principal focus for parallel rays. For this reason, parabolic mirrors are used as reflectors to focus light in telescopes, or to focus microwaves in satellite communication systems. The reflector behind a spot lamp or car headlamp is parabolic.

concentration in chemistry, the amount of a substance (◊solute) present in a specified amount of a solution. Either amount may be specified as a mass or a volume (liquids only). Common units used are ◊moles per cubic decimetre, grams per cubic decimetre, grams per 100 cubic centimetres, and grams per 100 grams.

The term also refers to the process of increasing the concentration of a solution by removing some of the substance (◊solvent) in which the solute is dissolved. In a **concentrated solution**, the solute is present in large quantities. Concentrated brine is around 30% sodium chloride in water; concentrated caustic soda (caustic liquor) is around 40% sodium hydroxide; and concentrated sulphuric acid is 98% acid.

concentric circles two or more circles that share the same centre.

Concorde the only supersonic airliner, which cruises at Mach 2, or twice the speed of sound, about 2,170 kph/1,350 mph. Concorde, the result of Anglo-French co-operation, made its first flight in 1969 and entered commercial service seven years later. It is 62 m/202 ft long and has a wing span of nearly 26 m/84 ft. Developing Concorde cost French and British taxpayers £2 billion.

concrete building material composed of cement, stone, sand, and water. It has been used since Egyptian and Roman times. Since the late 19th century, it has been increasingly employed as an economical alternative to materials such as brick and wood, and has been combined with steel to increase its tension capacity.

Reinforced concrete and prestressed concrete are strengthened by combining concrete with another material, such as steel rods or glass fibres. The addition of carbon fibres to concrete increases its conductivity. The electrical resistance of the concrete changes with increased stress or fracture, so this 'smart concrete' can be used as an early indicator of structural damage.

concurrent lines two or more lines passing through a single point; for example, the diameters of a circle are all concurrent at the centre of the circle.

concussion temporary unconsciousness resulting from a blow to the head. It is often followed by amnesia for events immediately preceding the blow.

condensation in organic chemistry, a reaction in which two organic compounds combine to form a larger molecule, accompanied by the removal of a smaller molecule (usually water). This is also known as an addition–elimination reaction. Polyamides (such as nylon) and polyesters (such as Terylene) are made by condensation ◊polymerization.

condensation conversion of a vapour to a liquid. This is frequently achieved by letting the vapour come into contact with a cold surface. It is the process by which water vapour turns into fine water droplets to form ◊cloud.

Condensation in the atmosphere occurs when the air becomes completely saturated and is unable to hold any more water vapour. As air rises it cools and contracts – the cooler it becomes the less water it can hold. Rain is frequently associated with warm weather fronts because the air rises and cools, allowing the water vapour to condense as rain. The temperature at which the air becomes saturated is known as the ◊dew point. Water vapour will not condense in air if there are not enough condensation nuclei (particles of dust, smoke or salt) for the droplets to form on. It is then said to be supersaturated. Condensation is an important part of the ◊water cycle.

condensation polymerization ◊polymerization reaction in which one or more monomers, with more than one reactive functional group, combine to form a polymer with the elimination of water or another small molecule.

condenser laboratory apparatus used to condense vapours back to liquid so that the liquid can be recovered. It is used in ◊distillation and in reactions where the liquid mixture can be kept boiling without the loss of solvent.

condenser in electronic circuits, a former name for a ◊capacitor.

condenser in optics, a ◊lens or combination of lenses with a short focal length used for concentrating a light source onto a small area, as used in a slide projector or microscope substage lighting unit. A condenser can also be made using a concave mirror.

conditioning in psychology, two major principles of behaviour modification.

In *classical conditioning*, described by Russian psychologist Ivan Pavlov, a new stimulus can evoke an automatic response by being repeatedly associated with a stimulus that naturally provokes that response. For example, the sound of a bell repeatedly associated with food will eventually trigger salivation, even if sounded without food being presented. In *operant conditioning*, described by US psychologists Edward Lee Thorndike (1874–1949) and B F Skinner, the frequency of a voluntary response can be increased by following it with a reinforcer or reward.

condom or *sheath* or *prophylactic* barrier contraceptive, made of rubber, which fits over an erect penis and holds in the sperm produced by ejaculation. It is an effective means of preventing pregnancy if used carefully, preferably with a ◊spermicide. A condom with spermicide is 97% effective; one without spermicide is 85% effective as a contraceptive. Condoms can also give some protection against sexually transmitted diseases, including AIDS.

In 1996 the European Union agreed a standard for condoms, which is 17 cm/6.7 in long; although the width can be variable, a regular width was agreed as 5.2 cm/2 in.

conductance ability of a material to carry an electrical current, usually given the symbol G. For a direct current, it is the reciprocal of ◊resistance: a conductor of resistance R has a conductance of $1/R$. For an alternating current, conductance is the resistance R divided by the ◊impedance Z: $G = R/Z$. Conductance was formerly expressed in reciprocal ◊ohms (or mhos); the SI unit is the ◊siemens (S).

conduction, electrical flow of charged particles through a material giving rise to electric current. Conduction in metals involves the flow of negatively charged free ◊electrons. Conduction in gases and some liquids involves the flow of ◊ions that carry positive charges in one direction and negative charges in the other. Conduction in a ◊semiconductor such as silicon involves the flow of electrons and positive holes.

conduction, heat flow of heat energy through a material without the movement of any part of the material itself (compare ◊conduction, electrical). Heat energy is present in all materials in the form of the kinetic energy of their vibrating molecules, and may be conducted from one molecule to the next in the form of this mechanical vibration. In the case of metals, which are particularly good conductors of heat, the free electrons within the material carry heat around very quickly.

conductor any material that conducts heat or electricity (as opposed to an insulator, or nonconductor). A good conductor has a high electrical or heat conductivity, and is generally a substance rich in free electrons such as a metal. A poor conductor (such as the nonmetals glass and porcelain) has few free electrons. Carbon is exceptional in being nonmetallic and yet (in some of its forms) a relatively good conductor of heat and electricity. Substances such as silicon and germanium, with intermediate conductivities that are improved by heat, light, or voltage, are known as ◊semiconductors.

cone in botany, the reproductive structure of the conifers and cycads; also known as a ◊strobilus. It consists of a central axis surrounded by numerous, overlapping, scalelike, modified leaves (sporophylls) that bear the reproductive organs. Usually there are separate male and female cones, the former bearing pollen sacs containing pollen grains, and the larger female cones bearing the ovules that contain the ova or egg cells. The pollen is carried from male to female cones by the wind (◊anemophily). The seeds develop within the female cone and are released as the scales open in dry atmospheric conditions, which favour seed dispersal.

In some groups (for example, the pines) the cones take two or even three years to reach maturity. The cones of junipers have fleshy cone scales that fuse to form a berrylike structure. One group of ◊angiosperms, the alders, also bear conelike structures; these are the woody remains of the short female catkins, and they contain the alder ◊fruits. In zoology, cones are a type of light-sensitive cell found in the retina of the ◊eye.

cone in geometry, a solid or surface consisting of the set of all straight lines passing through a fixed point (the vertex) and the points of a circle or ellipse whose plane does not contain the vertex. A circular cone of perpendic-

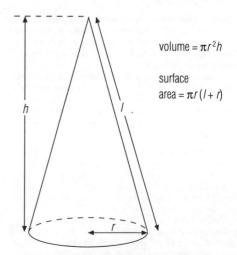

$$\text{volume} = \pi r^2 h$$

$$\text{surface area} = \pi r (l + r)$$

cone The volume and surface area of a cone are given by formulae involving a few simple dimensions.

ular height, with its apex above the centre of the circle, is known as a ***right circular cone***; it is generated by rotating an isosceles triangle or framework about its line of symmetry. A right circular cone of perpendicular height h and base of radius r has a volume $V = 1/3\pi r^2 h$.

The distance from the edge of the base of a cone to the vertex is called the slant height. In a right circular cone of slant height l, the curved surface area is $\pi r l$, and the area of the base is πr^2. Therefore the total surface area is

$$A = \pi r l + \pi r^2 = \pi r (l + r).$$

congenital disease in medicine, a disease that is present at birth. It is not necessarily genetic in origin; for example, congenital herpes may be acquired by the baby as it passes through the mother's birth canal.

congestion in traffic, the overcrowding of a route, leading to slow and inefficient flow. Congestion on the roads is a result of the large increase in car ownership. It may lead to traffic jams and long delays as well as pollution. Congestion within urban areas may also restrict accessibility.

conglomerate in geology, coarse-grained clastic ◊sedimentary rock, composed of rounded fragments (clasts) of pre-existing rocks cemented in a finer matrix, usually sand.

The fragments in conglomerates are pebble- to boulder-sized, and the rock can be regarded as the lithified equivalent of gravel. A ◊bed of conglomerate is often associated with a break in a sequence of rock beds (an unconformity), where it marks the advance of the sea over an old eroded landscape. An ***oligomict conglomerate*** contains one type of pebble; a ***polymict conglomerate*** has a mixture of pebble types. If the rock fragments are angular, it is called a ◊breccia.

congruent in geometry, having the same shape and size, as applied to two-dimensional or solid figures. With plane congruent figures, one figure will fit on top of the other exactly, though this may first require rotation and/or rotation of one of the figures.

conic section curve obtained when a conical surface is intersected by a plane. If the intersecting plane cuts both extensions of the cone, it yields a ◊parabola. Other intersecting planes produce ◊circles or ◊ellipses.

The Greek mathematician Apollonius wrote eight books with the title *Conic Sections*, which superseded previous work on the subject by Aristarchus and Euclid.

conjugate in mathematics, a term indicating that two elements are connected in some way; for example, $(a + ib)$ and $(a - ib)$ are conjugate complex numbers.

conjugate angles two angles that add up to 360°.

conjugation in biology, the bacterial equivalent of sexual reproduction. A fragment of the ◊DNA from one bacterium is passed along a thin tube, the pilus, into another bacterium.

conjugation in organic chemistry, the alternation of

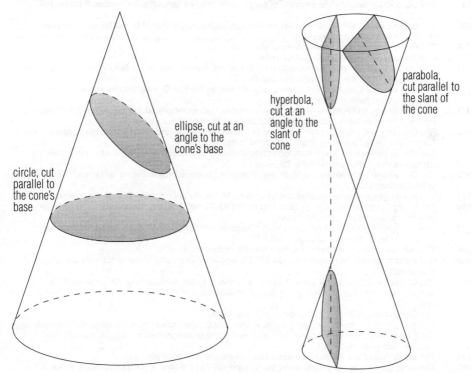

circle, cut parallel to the cone's base

ellipse, cut at an angle to the cone's base

hyperbola, cut at an angle to the slant of cone

parabola, cut parallel to the slant of the cone

conic section The four types of curve that may be obtained by cutting a single or double right-circular cone with a plane (two-dimensional surface).

double (or triple) and single carbon–carbon bonds in a molecule – for example, in penta-1,3-diene, $H_2C=CH–CH=CH–CH_3$. Conjugation imparts additional stability as the double bonds are less reactive than isolated double bonds.

conjunction in astronomy, the alignment of two celestial bodies as seen from Earth. A ◊superior planet (or other object) is in conjunction when it lies behind the Sun. An ◊inferior planet (or other object) comes to **inferior conjunction** when it passes between the Earth and the Sun; it is at **superior conjunction** when it passes behind the Sun.

  Planetary conjunction takes place when a planet is closely aligned with another celestial object, such as the Moon, a star, or another planet.

  Because the orbital planes of the inferior planets are tilted with respect to that of the Earth, they usually pass either above or below the Sun at inferior conjunction. If they line up exactly, a ◊transit will occur.

conjunctiva membrane covering the front of the vertebrate ◊eye. It is continuous with the epidermis of the eyelids, and lies on the surface of the cornea.

conjunctivitis inflammation of the conjunctiva, the delicate membrane that lines the inside of the eyelids and covers the front of the eye. Symptoms include redness, swelling, and a watery or pus-filled discharge. It may be caused by infection, allergy, or other irritant.

connective tissue in animals, tissue made up of a noncellular substance, the ◊extracellular matrix, in which some cells are embedded. Skin, bones, tendons, cartilage, and adipose tissue (fat) are the main connective tissues. There are also small amounts of connective tissue in organs such as the brain and liver, where they maintain shape and structure.

conservation in the life sciences, action taken to protect and preserve the natural world, usually from pollution, overexploitation, and other harmful features of human activity. The late 1980s saw a great increase in public concern for the environment, with membership of conservation groups, such as ◊Friends of the Earth, ◊Greenpeace, and the US Sierra Club, rising sharply. Globally the most important issues include the depletion of atmospheric ozone by the action of ◊chlorofluorocarbons (CFCs), the build-up of carbon dioxide in the atmosphere (thought to contribute to an intensification of the ◊greenhouse effect), and ◊deforestation. *See feature on page 160.*

Conservation: Chronology

1627	Last surviving aurochs, long-horned wild cattle that previously roamed Europe, southwest Asia, and North Africa, became extinct in Poland.
1664	A Dutch mandate drawn up to protect forest in Cape Colony, South Africa.
1681	The last dodo, a long-standing symbol of the need for species conservation, died on the island of Mauritius.
1764	The British established forest reserves on Tobago, after deforestation in Barbados and Jamaica resulted in widespread soil erosion.
1769	The French passed conservation laws in Mauritius.
1868	First laws passed in the UK to protect birds.
1948	The International Union for Conservation of Nature and Natural Resources (IUCN) was founded, with its sister organization, the World Wildlife Fund (WWF).
1970	The Man and the Biosphere Programme was initiated by UNESCO, providing for an international network of biosphere reserves.
1971	The Convention on Wetlands of International Importance (especially concerned with wildfowl habitat) signed in Ramsar, Iran starting a List of Wetlands of International Importance.
1972	The Convention Concerning the Protection of the World Cultural and Natural Heritage adopted in Paris, France, providing for the designation of World Heritage Sites.
1972	The UN Conference on the Human Environment held in Stockholm, Sweden, lead to the creation of the UN Environment Programme (UNEP).
1973	The Convention on International Trade in Endangered Species of Wild Fauna and Flora (CITES) signed in Washington, DC.
1974	The world's largest protected area, the Greenland National Park covering 97 million hectares, created.
1980	The World Conservation Strategy launched by the IUCN, with the WWF and UNEP, showed how conservation contributes to development.
1982	The first herd of 10 Arabian oryx bred from a 'captive breeding' programme released into the wild in Oman. The last wild oryx had been killed 1972.
1986	The first 'Red List' of endangered animal species compiled by IUCN.
1989	International trade in ivory banned under CITES legislation in an effort to protect the African elephant from poachers.
1992	The UN convened the 'Earth Summit' in Rio de Janeiro, Brazil, to discuss global planning for a sustainable future. The Convention on Biological Diversity and the Convention on Climate Change were opened for signing.
1993	The Convention on Biological Diversity came into force.
1995	The Arabian oryx conservation programme (began 1962), the most successful attempt at reintroducing zoo-bred animals to the wild, came to an end as the last seven animals were flown from the USA to join the 228-strong herd in Oman.
1996	The World Wide Fund for Nature had 5 million members in 28 countries.
1997	The world ban on the ivory trade was lifted in June 1997 at the tenth CITES convention. Trade is scheduled to resume in 1999 with Zimbabwe, Botswana, and Namibia the only countries allowed to export.

Conservation must come before recreation.

On **conservation** Charles, Prince of Wales
The Times 5 July 1989

conservation of energy in chemistry, the principle that states that in a chemical reaction, the total amount of energy in the system remains unchanged.

For each component there may be changes in energy due to change of physical state, changes in the nature of chemical bonds, and either an input or output of energy. However, there is no net gain or loss of energy.

conservation of mass in chemistry, the principle that states that in a chemical reaction the sum of all the masses of the substances involved in the reaction (reactants) is equal to the sum of all of the masses of the substances produced by the reaction (products) – that is, no matter is gained or lost.

conservation of momentum in mechanics, a law that states that total ◊momentum is conserved (remains constant) in all collisions, providing no external resultant force acts on the colliding bodies. The principle may be expressed as an equation used to solve numerical problems: total momentum before collision = total momentum after collision.

conservative margin or *passive margin* In plate tectonics, a region on the Earth's surface in which one plate slides past another. An example is the San Andreas Fault, California, where the movement of the plates is irregular and sometimes takes the form of sudden jerks, which cause the ◊earthquakes common in the San Francisco–Los Angeles area.

constant in mathematics, a fixed quantity or one that does not change its value in relation to ◊variables. For example, in the algebraic expression $y^2 = 5x - 3$, the numbers 3 and 5 are constants. In physics, certain quanti-ties are regarded as universal constants, such as the speed of light in a vacuum.

constantan or *eureka* high-resistance alloy of approximately 40% nickel and 60% copper with a very low coefficient of ◊thermal expansion (measure of expansion on heating). It is used in electrical resistors.

constant composition, law of in chemistry, the law that states that the proportions of the amounts of the elements in a pure compound are always the same and are independent of the method by which the compound was produced.

constellation one of the 88 areas into which the sky is divided for the purposes of identifying and naming celestial objects. The first constellations were simple, arbitrary patterns of stars in which early civilizations visualized gods, sacred beasts, and mythical heroes.

The constellations in use today are derived from a list of 48 known to the ancient Greeks, who inherited some from the Babylonians. The current list of 88 constellations was adopted by the International Astronomical Union, astronomy's governing body in 1930. *See table on page 161.*

constipation in medicine, the infrequent emptying of the bowel. The intestinal contents are propelled by peristaltic contractions of the intestine in the digestive process. The faecal residue collects in the rectum, distending it and promoting defecation. Constipation may be due to illness, alterations in food consumption, stress, or as an adverse effect of certain drugs. An increased intake of dietary fibre (see ◊fibre, dietary) can alleviate constipation. Laxatives may be used to relieve temporary constipation but they should not be used routinely.

constructive margin or *divergent margin* In plate tectonics, a region in which two plates are moving away from each other. Magma, or molten rock, escapes to the surface along this margin to form new crust, usually in the form of a ridge. Over time, as more and more magma reaches the surface, the sea floor spreads – for example, the upwelling of magma at the Mid-Atlantic Ridge causes

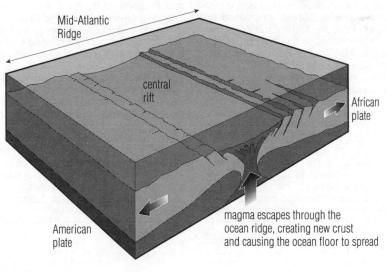

constructive margin

The race to prevent mass extinction

About a million different living species have been identified so far. Recent studies in tropical forests – where biodiversity is greatest – suggest the true figure is nearer 30 million. Most are animals, and most of those are insects. Because the tropical forests are threatened, at least half the animal species could become extinct within the next century. There have been at least five 'mass extinctions' in our planet's history; the last removed the dinosaurs 65 million years ago. The present wave of extinction is of similar scale, but hundreds of times faster.

There is conflict within the conservation movement. Some believe that *habitats* (the places where animals and plants live) should be conserved; others prefer to concentrate upon individual *species*. Both approaches have strengths and weaknesses, and they must operate in harmony.

Habitat protection

Habitat protection has obvious advantages. Many species benefit if land is preserved. Animals need somewhere to live; unless the habitat is preserved it may not be worth saving the individual animal. Habitat protection *seems* cheap; for example, tropical forest can often be purchased for only a few dollars per hectare. Only by habitat protection can we save more than a handful of the world's animals.

But there are difficulties. Even when a protected area is designated a 'national park', its animals may not be safe. All five remaining species of rhinoceros are heavily protected in the wild, but are threatened by poaching. Early in 1991 Zimbabwe had 1,500 black rhinos – the world's largest population. Patrols of game wardens shoot poachers on sight. Yet by late 1992, 1,000 of the 1,500 had been poached. In many national parks worldwide, the habitat is threatened by the local farmers' need to graze their cattle.

Computer models and field studies show that wild populations need several hundred individuals to be viable. Smaller populations will eventually become extinct in the wild, because of accidents to key breeding individuals, or epidemics. The big predators need vast areas. One tiger may command hundreds of square kilometres; a viable population needs an area as big as Wales or Holland. Only one of the world's five remaining subspecies of tiger – a population of Bengals in India – occupies an area large enough to be viable. All the rest (Indo-Chinese, Sumatran, Chinese, and Siberian) seem bound to die out. Three other subspecies have become extinct in the past 100 years – the latest, the Javan, in the 1970s.

Mosaic

Ecologists now emphasize the concept of *mosaic*. All animals need different things from their habitat; a failure of any one is disastrous. Giant pandas feed mainly on bamboo, but give birth in old hollow trees – of which there is a shortage. Birds commonly roost in one place, but feed in special areas far way. Nature reserves must either contain all essentials for an animal's life, or else allow access to such areas elsewhere. For many animals in a reserve, these conditions are not fulfilled. Hence year by year, after reserves are created, species go extinct: a process called *species relaxation*. The remaining fauna and flora may be a poor shadow of the original.

Captive breeding

Interest is increasing in *captive breeding*, carried out mainly by the world's 800 zoos. Their task is formidable; each captive species should include several hundred individuals. Zoos maintain such numbers through *cooperative breeding*, organized regionally and coordinated by the Captive Breeding Specialist Group or the World Conservation Union, based in Minneapolis, Minnesota. Each programme is underpinned by a studbook, showing which individuals are related to which.

Breeding for conservation is different from breeding for livestock improvement. Livestock breeders breed *uniform* creatures by selecting animals conforming to some prescribed ideal. Conservation breeders maintain *maximum genetic diversity* by encouraging every individual to breed, including those reluctant to breed in captivity; by equalizing family size, so one generation's genes are all represented in the next; and by swapping individuals between zoos, to prevent inbreeding.

Cooperative breeding programmes are rapidly diversifying; by the year 2000 there should be several hundred. They can only make a small impression on the 15 million endangered species, but they can contribute greatly to particular groups of animals, especially the land vertebrates – mammals, birds, reptiles, and amphibians. These include most of the world's largest animals, with the greatest impact on their habitats. There are 24,000 species of land vertebrate, of which 2,000 probably require captive breeding to survive. Zoos could save all 2,000; that would be a great contribution.

Some successes

Captive breeding is not intended to establish 'museum' populations, but to provide a temporary 'lifeboat'. Things are hard for wild animals, but over the next few decades, despite the growing human population, it should be possible to establish more, safe national parks. Arabian oryx, California condor, black-footed ferret, red wolf, and Mauritius kestrel are among the creatures saved from extinction by captive breeding and returned to the wild. In the future, we can expect to see many more.

Colin Tudge

Constellations

Constellation	Abbreviation	Popular name	Constellation	Abbreviation	Popular name
Andromeda	And	–	Leo	Leo	Lion
Antlia	Ant	Airpump	Leo Minor	LMi	Little Lion
Apus	Aps	Bird of Paradise	Lepus	Lep	Hare
Aquarius	Aqr	Water-bearer	Libra	Lib	Balance
Aquila	Aqi	Eagle	Lupus	Lup	Wolf
Ara	Ara	Altar	Lynx	Lyn	–
Aries	Ari	Ram	Lyra	Lyr	Lyre
Auriga	Aur	Charioteer	Mensa	Men	Table
Boötes	Boo	Herdsman	Microscopium	Mic	Microscope
Caelum	Cae	Chisel	Monoceros	Mon	Unicorn
Camelopardalis	Cam	Giraffe	Musca	Mus	Southern Fly
Cancer	Cnc	Crab	Norma	Nor	Rule
Canes Venatici	CVn	Hunting Dogs	Octans	Oct	Octant
Canis Major	CMa	Great Dog	Ophiuchus	Oph	Serpent-bearer
Canis Minor	CMi	Little Dog	Orion	Ori	–
Capricornus	Cap	Sea-goat	Pavo	Pav	Peacock
Carina	Car	Keel	Pegasus	Peg	Flying Horse
Cassiopeia	Cas	–	Perseus	Per	–
Centaurus	Cen	Centaur	Phoenix	Phe	Phoenix
Cepheus	Cep	–	Pictor	Pic	Painter
Cetus	Cet	Whale	Pisces	Psc	Fishes
Chamaeleon	Cha	Chameleon	Piscis Austrinus	PsA	Southern Fish
Circinus	Cir	Compasses	Puppis	Pup	Poop
Columba	Col	Dove	Pyxis	Pyx	Compass
Coma Berenices	Com	Berenice's Hair	Reticulum	Ret	Net
Corona Australis	CrA	Southern Crown	Sagitta	Sge	Arrow
Corona Borealis	CrB	Northern Crown	Sagittarius	Sgr	Archer
Corvus	Crv	Crow	Scorpius	Sco	Scorpion
Crater	Crt	Cup	Sculptor	Scl	–
Crux	Cru	Southern Cross	Scutum	Sct	Shield
Cygnus	Cyn	Swan	Serpens	Ser	Serpent
Delphinus	Del	Dolphin	Sextans	Sex	Sextant
Dorado	Dor	Goldfish	Taurus	Tau	Bull
Draco	Dra	Dragon	Telescopium	Tel	Telescope
Equuleus	Equ	Foal	Triangulum	Tri	Triangle
Eridanus	Eri	River	Triangulum		
Fornax	For	Furnace	Australe	TrA	Southern Triangle
Gemini	Gem	Twins	Tucana	Tuc	Toucan
Grus	Gru	Crane	Ursa Major	UMa	Great Bear
Hercules	Her	–	Ursa Minor	UMi	Little Bear
Horologium	Hor	Clock	Vela	Vel	Sails
Hydra	Hya	Watersnake	Virgo	Vir	Virgin
Hydrus	Hyi	Little Snake	Volans	Vol	Flying Fish
Indus	Ind	Indian	Vulpecula	Vul	Fox
Lacerta	Lac	Lizard			

the floor of the Atlantic Ocean to grow at a rate of about 5 cm/2 in a year.

◊Volcanoes can form along the ridge and islands may result (for example, Iceland was formed in this way). Eruptions at constructive plate margins tend to be relatively gentle; the lava produced cools to form ◊basalt.

consumption in medicine, former popular name for the disease ◊tuberculosis.

contact lens lens, made of soft or hard plastic, that is worn in contact with the cornea and conjunctiva of the eye, beneath the eyelid, to correct defective vision. In special circumstances, contact lenses may be used as protective shells or for cosmetic purposes, such as changing eye colour.

The earliest use of contact lenses in the late 19th century was protective, or in the correction of corneal malformation. It was not until the 1930s that simplification of fitting technique by taking eye impressions made general use possible. Recent developments are a type of soft lens that can be worn for lengthy periods without removal, and a disposable soft lens that needs no cleaning but should be discarded after a week of constant wear.

continent any one of the seven large land masses of the Earth, as distinct from the oceans. They are Asia, Africa, North America, South America, Europe, Australia, and Antarctica. Continents are constantly moving and evolving (see ◊plate tectonics). A continent does not end at the coastline; its boundary is the edge of the shallow continental shelf, which may extend several hundred kilometres out to sea.

At the centre of each continental mass lies a shield or ◊craton, a deformed mass of old ◊metamorphic rocks dating from Precambrian times. The shield is thick, compact, and solid (the Canadian Shield is an example), having undergone all the mountain-building activity it is ever likely to, and is usually worn flat. Around the shield is a concentric pattern of fold mountains, with older ranges, such as the Rockies, closest to the shield, and younger ranges, such as the coastal ranges of North America, farther away. This general concentric pattern is modified when two continental masses have drifted together and they become welded with a great mountain range along the join, the way Europe and N Asia are joined along the Urals. If a continent is torn apart, the new continental edges have no fold mountains; for instance, South America has fold mountains (the Andes) along its western flank, but none along the east where it tore away from Africa 200 million years ago.

continental drift in geology, the theory that, about 250–200 million years ago, the Earth consisted of a single large continent (◊Pangaea), which subsequently broke apart to form the continents known today. The theory was proposed in 1912 by German meteorologist Alfred Wegener, but such vast continental movements could not be satisfactorily explained until the study of ◊plate tectonics in the 1960s.

The term 'continental drift' is not strictly correct, since land masses do not drift through the oceans. The continents form part of a plate, and the amount of crust created at divergent plate margins must equal the amount of crust destroyed at subduction zones.

continental rise the portion of the ocean floor rising gently from the abyssal plain toward the steeper continen-

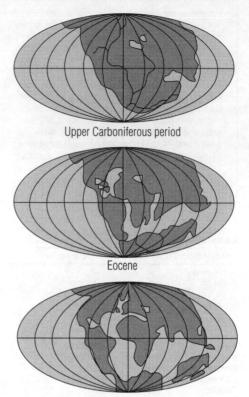

Upper Carboniferous period

Eocene

Lower Quaternary

continental drift The drifting continents. The continents are slowly shifting their positions, driven by fluid motion beneath the Earth's crust. Over 200 million years ago, there was a single large continent called Pangaea. By 200 million years ago, the continents had started to move apart. By 50 million years ago, the continents were approaching their present positions.

tal slope. The continental rise is a depositional feature formed from sediments transported down the slope mainly by turbidity currents. Much of the continental rise consists of coalescing submarine alluvial fans bordering the continental slope.

continental shelf the submerged edge of a continent, a gently sloping plain that extends into the ocean. It typically has a gradient of less than 1°. When the angle of the sea bed increases to 1°–5° (usually several hundred kilometres away from land), it becomes known as the **continental slope**.

continental slope sloping, submarine portion of a continent. It extends downward from the edge of the continental shelf. In some places, such as south of the Aleutian Islands of Alaska, continental slopes extend directly to the ocean deeps or abyssal plain. In others, such as the E coast of North America, they grade into the gentler continental rises that in turn grade into the abyssal plains.

continuous data data that can take any of an infinite number of values between whole numbers and so may not be measured completely accurately. This type of data contrasts with ◊discrete data, in which the variable can only take one of a finite set of values. For example, the sizes of apples on a tree form continuous data, whereas the numbers of apples form discrete data.

continuum in mathematics, a ◊set that is infinite and everywhere continuous, such as the set of points on a line.

contour on a map, a line drawn to join points of equal height. Contours are drawn at regular height intervals; for example, every 10 m. The closer together the lines are, the steeper the slope. Contour patterns can be used to interpret the relief of an area and to identify land forms.

contraceptive any drug, device, or technique that prevents pregnancy. The contraceptive pill (the ◊Pill) contains female hormones that interfere with egg production or the first stage of pregnancy. The 'morning-after' pill can be taken up to 72 hours after unprotected intercourse. Barrier contraceptives include ◊condoms (sheaths), ◊diaphragms, also called caps or Dutch caps, and sponges impregnated with spermicide; they prevent the sperm entering the cervix (neck of the womb).

◊Intrauterine devices, also known as IUDs or coils, cause a slight inflammation of the lining of the womb; this prevents the fertilized egg from becoming implanted. See also ◊family planning.

Other contraceptive methods include ◊sterilization (women) and ◊vasectomy (men); these are usually non-reversible. 'Natural' methods include withdrawal of the penis before ejaculation (coitus interruptus), and avoidance of intercourse at the time of ovulation (◊rhythm method). These methods are unreliable and normally only used on religious grounds. The use of any contraceptive (birth control) is part of family planning. The effectiveness of a contraceptive method is often given as a percentage. To say that a method has 95% effectiveness means that, on average, out of 100 healthy couples using that method for a year, 95 will not conceive.

contractile root in botany, a thickened root at the base of a corm, bulb, or other organ that helps position it at an appropriate level in the ground. Contractile roots are found, for example, on the corms of plants of the genus *Crocus*. After they have become anchored in the soil, the upper portion contracts, pulling the plant deeper into the ground.

control experiment essential part of a scientifically valid experiment, designed to show that the factor being tested is actually responsible for the effect observed. In the control experiment all factors, apart from the one under test, are exactly the same as in the test experiments, and all the same measurements are carried out. In drug trials, a placebo (a harmless substance) is given alongside the substance being tested in order to compare effects.

control total in computing, a ◊validation check in which an arithmetic total of a specific field from a group of records is calculated. This total is input together with the data to which it refers. The program recalculates the control total and compares it with the one entered to ensure that no entry errors have been made.

control unit the component of the ◊central processing unit that decodes, synchronizes, and executes program instructions.

convection heat energy transfer that involves the movement of a fluid (gas or liquid). According to ◊kinetic theory, molecules of fluid in contact with the source of heat expand and tend to rise within the bulk of the fluid. Less energetic, cooler molecules sink to take their place, setting up convection currents. This is the principle of natural convection in many domestic hot-water systems and space heaters.

convectional rainfall rainfall associated with hot climates, resulting from the uprising of convection currents of warm air. Air that has been warmed by the extreme heating of the ground surface rises to great heights and is abruptly cooled. The water vapour carried by the air condenses and rain falls heavily. Convectional rainfall is usually associated with a ◊thunderstorm.

convection current current caused by the expansion of a liquid or gas as its temperature rises. The expanded material, being less dense, rises above colder and therefore denser material. Convection currents arise in the atmosphere above warm land masses or seas, giving rise to sea breezes and land breezes respectively. In some heating systems, convection currents are used to carry hot water upwards in pipes.

Convection currents in the viscous rock of the Earth's mantle help to drive the movement of the rigid plates making up the Earth's surface (see ◊plate tectonics).

conventional current direction in which an electric current is considered to flow in a circuit. By convention, the direction is that in which positive-charge carriers would flow – from the positive terminal of a cell to its negative terminal. It is opposite in direction to the flow of electrons. In circuit diagrams, the arrows shown on symbols for components such as diodes and transistors point in the direction of conventional current flow.

convergence in mathematics, the property of a series of numbers in which the difference between consecutive terms gradually decreases. The sum of a converging series approaches a limit as the number of terms tends to ◊infinity.

convergent evolution in biology, the independent evolution of similar structures in species (or other taxonomic groups) that are not closely related, as a result of living in a similar way. Thus, birds and bats have wings, not because they are descended from a common winged ancestor, but because their respective ancestors independently evolved flight.

converging lens lens that converges or brings to a focus those light rays that have been refracted by it. It is a ◊convex lens, with at least one surface that curves outwards, and is thicker towards the centre than at the edge. Converging lenses are used to form real images in many ◊optical instruments, such as cameras and projectors. A converging lens that forms a virtual, magnified image may be used as a magnifying glass or to correct ◊long-sightedness.

converse in mathematics, the reversed order of a conditional statement; the converse of the statement 'if a, then b' is 'if b, then a'. The converse does not always hold true; for example, the converse of 'if $x = 3$, then $x^2 = 9$' is 'if $x^2 = 9$, then $x = 3$', which is not true, as x could also be -3.

convertiplane ◊vertical takeoff and landing craft (VTOL) with rotors on its wings that spin horizontally for

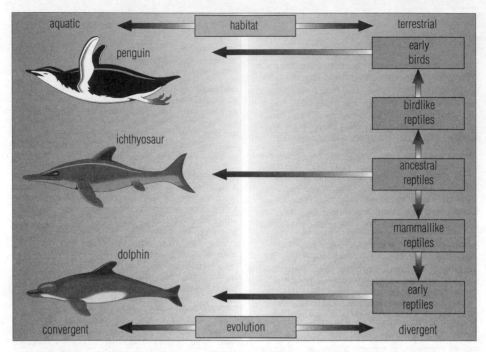

aquatic · habitat · terrestrial

penguin

early
birds

birdlike
reptiles

ichthyosaur

ancestral
reptiles

mammallike
reptiles

dolphin

early
reptiles

convergent · evolution · divergent

convergent evolution Convergent evolution produced the similarly-shaped streamlined bodies seen in the penguin and the dolphin today. Despite their very different evolutionary origins both have evolved and adapted to their aquatic environment.

takeoff, but tilt to spin in a vertical plane for forward flight.

At takeoff it looks like a two-rotor helicopter, with both rotors facing skywards. As forward speed is gained, the rotors tilt slowly forward until they are facing directly ahead.

convex of a surface, curving outwards, or towards the eye. For example, the outer surface of a ball appears convex. In geometry, the term is used to describe any polygon possessing no interior angle greater than 180°. Convex is the opposite of ◊concave.

convex lens lens that possesses at least one surface that curves outwards. It is a ◊converging lens, bringing rays of light to a focus. A convex lens is thicker at its centre than at its edges, and is used to correct long-sightedness.

Common forms include the **biconvex** lens (with both surfaces curved outwards) and the **plano-convex** (with one flat surface and one convex). The whole lens may be further curved overall, making a **concavo-convex** or converging meniscus lens, as in some lenses used in corrective eyewear.

convex mirror curved mirror that reflects light from its outer surface. It diverges reflected light rays to form a reduced, upright, virtual image. Convex mirrors give a wide field of view and are therefore particularly suitable for car wing mirrors and surveillance purposes in shops.

conveyor device used for transporting materials. Widely used throughout industry is the **conveyor belt**, usually a rubber or fabric belt running on rollers. Trough-shaped belts are used, for example in mines, for transporting ores

and coal. **Chain conveyors** are also used in coal mines to remove coal from the cutting machines. Overhead endless chain conveyors are used to carry components and bodies in car-assembly works. Other types include **bucket conveyors** and **screw conveyors**, powered versions of the ◊Archimedes screw.

convulsion series of violent contractions of the muscles over which the patient has no control. It may be associated with loss of consciousness. Convulsions may arise from any one of a number of causes, including brain disease (such as ◊epilepsy), injury, high fever, poisoning, and electrocution.

cooperative farming system in which individual farmers pool their resources (excluding land) to buy commodities such as seeds and fertilizers, and services such as marketing. It is a system of farming found throughout the world and is particularly widespread in Denmark and the ex-Soviet republics. In a ◊collective farm, land is also held in common.

coordinate in geometry, a number that defines the position of a point relative to a point or axis (reference line). ◊Cartesian coordinates define a point by its perpendicular distances from two or more axes drawn through a fixed point mutually at right angles to each other. ◊Polar coordinates define a point in a plane by its distance from a fixed point and direction from a fixed line.

coordinate geometry or **analytical geometry** system of geometry in which points, lines, shapes, and surfaces are represented by algebraic expressions. In plane (two-dimensional) coordinate geometry, the plane is usually

Cartesian coordinates

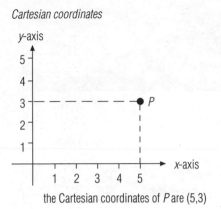

the Cartesian coordinates of *P* are (5,3)

Polar coordinates

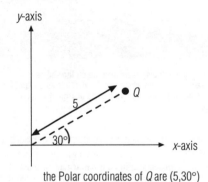

the Polar coordinates of *Q* are (5,30°)

coordinate Coordinates are numbers that define the position of points in a plane or in space. In the Cartesian coordinate system, a point in a plane is charted based upon its location along intersecting horizontal and vertical axes. In the polar coordinate system, a point in a plane is defined by its distance from a fixed point and direction from a fixed line.

defined by two axes at right angles to each other, the horizontal *x*-axis and the vertical *y*-axis, meeting at O, the origin. A point on the plane can be represented by a pair of ◊Cartesian coordinates, which define its position in terms of its distance along the *x*-axis and along the *y*-axis from O. These distances are respectively the *x* and *y* coordinates of the point.

Lines are represented as equations; for example, $y = 2x + 1$ gives a straight line, and $y = 3x^2 + 2x$ gives a ◊parabola (a curve). The graphs of varying equations can be drawn by plotting the coordinates of points that satisfy their equations, and joining up the points. One of the advantages of coordinate geometry is that geometrical solutions can be obtained without drawing but by manipulating algebraic expressions. For example, the coordinates of the point of intersection of two straight lines can be determined by finding the unique values of *x* and *y* that satisfy both of the equations for the lines, that is, by solving them as a pair of ◊simultaneous equations. The curves studied

in simple coordinate geometry are the ◊conic sections (circle, ellipse, parabola, and hyperbola), each of which has a characteristic equation.

copepod ◊crustacean of the subclass Copepoda, mainly microscopic and found in plankton.

coplanar in geometry, describing lines or points that all lie in the same plane.

copper orange-pink, very malleable and ductile, metallic element, symbol Cu (from Latin *cuprum*), atomic number 29, relative atomic mass 63.546. It is used for its durability, pliability, high thermal and electrical conductivity, and resistance to corrosion.

It was the first metal used systematically for tools by humans; when mined and worked into utensils it formed the technological basis for the Copper Age in prehistory. When alloyed with tin it forms bronze, which strengthens the copper, allowing it to hold a sharp edge; the systematic production and use of this was the basis for the prehistoric Bronze Age. Brass, another hard copper alloy, includes zinc. The element's name comes from the Greek for Cyprus (*Kyprios*), where copper was mined.

copper ore any mineral from which copper is extracted, including native copper, Cu; chalcocite, Cu_2S; chalcopyrite, $CuFeS_2$; bornite, Cu_5FeS_4; azurite, $Cu_3(CO_3)_2(OH)_2$; malachite, $Cu_2CO_3(OH)_2$; and chrysocolla, $CuSiO_3.2H_2O$.

Native copper and the copper sulphides are usually found in veins associated with igneous intrusions. Chrysocolla and the carbonates are products of the weathering of copper-bearing rocks. Copper was one of the first metals to be worked, because it occurred in native form and needed little refining. Today the main producers are the USA, Russia, Kazakhstan, Georgia, Uzbekistan, Armenia, Zambia, Chile, Peru, Canada, and Congo (formerly Zaire).

coppicing woodland management practice of severe pruning where trees are cut down to near ground level at regular intervals, typically every 3–20 years, to promote the growth of numerous shoots from the base.

This form of ◊forestry was once commonly practised in Europe, principally on hazel and chestnut, to produce large quantities of thin branches for firewood, fencing, and so on; alder, eucalyptus, maple, poplar, and willow were also coppiced. The resulting thicket was known as a coppice or copse. See also ◊pollarding.

copulation act of mating in animals with internal ◊fertilization. Male mammals have a ◊penis or other organ that is used to introduce spermatozoa into the reproductive tract of the female. Most birds transfer sperm by pressing their cloacas (the openings of their reproductive tracts) together.

coral marine invertebrate of the class Anthozoa in the phylum Cnidaria, which also includes sea anemones and jellyfish. It has a skeleton of lime (calcium carbonate) extracted from the surrounding water. Corals exist in warm seas, at moderate depths with sufficient light. Some coral is valued for decoration or jewellery, for example, Mediterranean red coral *Corallum rubrum*.

Corals live in a symbiotic relationship with microscopic algae (zooxanthellae), which are incorporated into the soft tissue. The algae obtain carbon dioxide from the coral polyps, and the polyps receive nutrients from the algae. Corals also have a relationship to the fish that rest or take

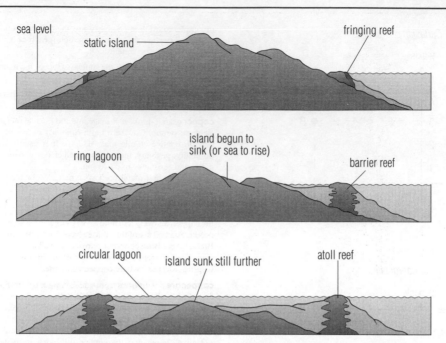

coral The formation of a coral atoll by the gradual sinking of a volcanic island. Fringing reefs build up on the shores of continents or islands. Barrier reefs are separated from the shore by a saltwater lagoon which may be as much as 30 km/20 mi wide. The coral reefs fringing the island build up as the island sinks, eventually producing a ring of coral around a lagoon (the spot where the island sank).

refuge within their branches, and which excrete nutrients that make the corals grow faster. The majority of corals form large colonies although there are species that live singly. Their accumulated skeletons make up large coral reefs and atolls. The Great Barrier Reef, to the NE of Australia, is about 1,600 km/1,000 mi long, has a total area of 20,000 sq km/7,700 sq mi, and adds 50 million tonnes of calcium to the reef each year. The world's reefs cover an estimated 620,000 sq km/240,000 sq mi.

Coral reefs provide a habitat for a diversity of living organisms. In 1997, some 93,000 species were identified. One third of the world's marine fishes live in reefs.

cord unit for measuring the volume of wood cut for fuel. One cord equals 128 cubic feet (3.456 cubic metres), or a stack 8 feet (2.4 m) long, 4 feet (1.2 m) wide, and 4 feet high.

cordierite silicate mineral, $(Mg,Fe)_2Al_4Si_5O_{18}$, blue to purplish in colour. It is characteristic of metamorphic rocks formed from clay sediments under conditions of low pressure but moderate temperature; it is the mineral that forms the spots in spotted slate and spotted hornfels.

cordillera group of mountain ranges and their valleys, all running in a specific direction, formed by the continued convergence of two tectonic plates (see ◊plate tectonics) along a line.

core in archaeology, a solid cylinder of sediment or soil collected with a coring device and used to evaluate the geological context and stratigraphy of archaeological material or to obtain palaeobotanical samples. Core can also mean the tool used to extract a core sample from the

ground, or a stone blank from which flakes or blades are removed.

core in earth science, the innermost part of Earth. It is divided into an outer core, which begins at a depth of 2,898 km/1,800 mi, and an inner core, which begins at a depth of 4,982 km/3,095 mi. Both parts are thought to consist of iron-nickel alloy. The outer core is liquid and the inner core is solid.

Evidence for the nature of the core comes from seismology (observation of the paths of earthquake waves through Earth), and calculations of Earth's density. The temperature of the core is estimated to be at least 4,000°C/7,232°F.

Coriolis effect the effect of the Earth's rotation on the atmosphere and on all objects on the Earth's surface. In the northern hemisphere it causes moving objects and currents to be deflected to the right; in the southern hemisphere it causes deflection to the left. The effect is named after its discoverer, French mathematician Gaspard de Coriolis (1792–1843).

cork light, waterproof outer layers of the bark covering the branches and roots of almost all trees and shrubs. The cork oak (*Quercus suber*), a native of S Europe and N Africa, is cultivated in Spain and Portugal; the exceptionally thick outer layers of its bark provide the cork that is used commercially.

corm short, swollen, underground plant stem, surrounded by protective scale leaves, as seen in the genus *Crocus*. It stores food, provides a means of ◊vegetative reproduction, and acts as a ◊perennating organ.

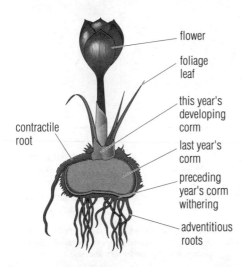

flower

foliage leaf

this year's developing corm

last year's corm

preceding year's corm withering

adventitious roots

contractile root

corm Corms, found in plants such as the gladiolus and crocus, are underground storage organs. They provide the food for growth during adverse conditions such as cold or drought.

During the year, the corm gradually withers as the food reserves are used for the production of leafy, flowering shoots formed from axillary buds. Several new corms are formed at the base of these shoots, above the old corm.

corn general term for the main ◊cereal crop of a region – for example, wheat in the UK, oats in Scotland and Ireland, maize in the USA. Also, another word for ◊maize.

cornea transparent front section of the vertebrate ◊eye. The cornea is curved and behaves as a fixed lens, so that light entering the eye is partly focused before it reaches the lens.

There are no blood vessels in the cornea and it relies on the fluid in the front chamber of the eye for nourishment. Further protection for the eye is provided by the ◊conjunctiva. In humans, diseased or opaque parts may be replaced by grafts of corneal tissue from a donor.

corneal graft in medicine, the replacement of a damaged or diseased cornea by a cornea from a human donor. Indications for corneal grafting include severe corneal scarring as a result of herpes virus infections, alkali burns, or eye injuries. Rejection of the corneal graft may occur due to stimulation of the immune system of the recipient by foreign proteins of the donor. Rejection results in oedema (fluid accumulation) and poor vision.

corolla collective name for the petals of a flower. In some plants the petal margins are partly or completely fused to form a **corolla tube**, for example in bindweed *Convolvulus arvensis*.

corona faint halo of hot (about 2,000,000°C/3,600,000°F) and tenuous gas around the Sun, which boils from the surface. It is visible at solar ◊eclipses or through a **coronagraph**, an instrument that blocks light from the Sun's brilliant disc. Gas flows away from the corona to form the ◊solar wind.

Corona Australis or **Southern Crown** small constellation of the southern hemisphere, located near the constellation ◊Sagittarius. It is similar in size and shape to ◊Corona Borealis but is not as bright.

Corona Borealis or **Northern Crown** small but easily recognizable constellation of the northern hemisphere, between ◊Hercules and ◊Boötes, traditionally identified with the jewelled crown of Ariadne that was cast into the sky by Bacchus in Greek mythology. Its brightest star is Alphecca (or Gemma), which is 78 light years from Earth.

It contains several variable stars. R Coronae Borealis is normally fairly constant in brightness but fades at irregular intervals and stays faint for a variable length of time. T Coronae Borealis is normally faint, but very occasionally blazes up and for a few days may be visible to the naked eye. It is a recurrent ◊nova.

coronary in biology, describing any of several structures in the body that encircle an organ in the manner of a crown. For example, the coronary arteries arise from the aorta and deliver blood to the heart muscle.

coronary artery disease condition in which the fatty deposits of ◊atherosclerosis form in the coronary arteries that supply the heart muscle, narrowing them and restricting the blood flow.

These arteries may already be hardened (arteriosclerosis). If the heart's oxygen requirements are increased, as during exercise, the blood supply through the narrowed arteries may be inadequate, and the pain of ◊angina results. A ◊heart attack occurs if the blood supply to an area of the heart is cut off, for example because a blood clot (thrombus) has blocked one of the coronary arteries. The subsequent lack of oxygen damages the heart muscle (infarct), and if a large area of the heart is affected, the attack may be fatal. Coronary artery disease tends to run in families and is linked to smoking, lack of exercise, and a diet high in saturated (mostly animal) fats, which tends to increase the level of blood ◊cholesterol. It is a common cause of death in many industrialized countries; older men are the most vulnerable group. The condition is treated with drugs or bypass surgery.

coronary thrombosis in medicine, acute manifestation of coronary heart disease causing a ◊heart attack.

corpuscle in biology, a small body. The cellular components of blood are sometimes referred to as corpuscles; see ◊red blood cell and ◊white blood cell.

corpus luteum glandular tissue formed in the mammalian ◊ovary after ovulation from the Graafian follicle, a group of cells associated with bringing the egg to maturity. It secretes the hormone progesterone in anticipation of pregnancy.

After the release of an egg the follicle enlarges under the action of luteinizing hormone, released from the pituitary. The corpus luteum secretes the hormone progesterone, which maintains the uterus wall ready for pregnancy. If pregnancy occurs, the corpus luteum continues to secrete progesterone until the fourth month; if pregnancy does not occur the corpus luteum breaks down.

corrasion the grinding away of solid rock surfaces by particles carried by water, ice, and wind. It is generally held to be the most significant form of ◊erosion. As the eroding particles are carried along they become eroded themselves due to the process of ◊attrition.

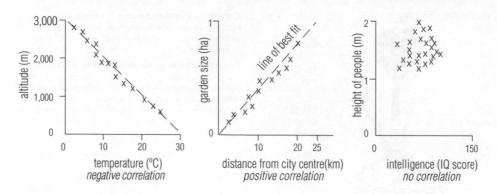

correlation Scattergraphs showing different kinds of correlation. In this way, a causal relationship between two variables may be proved or disproved, provided there are no hidden factors.

correlation the degree of relationship between two sets of information. If one set of data increases at the same time as the other, the relationship is said to be positive or direct. If one set of data increases as the other decreases, the relationship is negative or inverse. Correlation can be shown by plotting a best-fit line on a ◊scatter diagram.

In statistics, such relations are measured by the calculation of ◊coefficients of correlation. These generally measure correlation on a scale with 1 indicating perfect positive correlation, 0 no correlation at all, and −1 perfect inverse correlation. Correlation coefficients for assumed linear relations include the Pearson product moment correlation coefficient (known simply as the correlation coefficient), Kendall's tau correlation coefficient, or Spearman's rho correlation coefficient, which is used in nonparametric statistics (where the data are measured on ordinal rather than interval scales). A high correlation does not always indicate dependence between two variables; it may be that there is a third (unstated) variable upon which both depend.

correspondence in mathematics, the relation between two sets where an operation on the members of one set maps some or all of them onto one or more members of the other. For example, if A is the set of members of a family and B is the set of months in the year, A and B are in correspondence if the operation is: '... has a birthday in the month of ...'.

corresponding angles a pair of equal angles lying on the same side of a transversal (a line that cuts through two or more lines in the same plane), and making an interior and exterior angle with the intersected lines.

corrie (Welsh **cwm**; French, North American **cirque**) Scottish term for a steep-sided hollow in the mountainside of a glaciated area. The weight and movement of the ice has ground out the bottom and worn back the sides. A corrie is open at the front, and its sides and back are formed of ◊arêtes. There may be a lake in the bottom, called a tarn.

A corrie is formed as follows: (1) snow accumulates in a hillside hollow (enlarging the hollow by nivation), and turns to ice ; (2) the hollow is deepened by ◊abrasion and ◊plucking; (3) the ice in the corrie rotates under the influence of gravity, deepening the hollow still further; (4) since the ice is thinner and moves more slowly at the foot of the hollow, a rock lip forms; (5) when the ice melts, a lake or tarn may be formed in the corrie. The steep back wall may be severely weathered by freeze-thaw, weathering providing material for further abrasion.

corrosion the eating away and eventual destruction of metals and alloys by chemical attack. The rusting of ordinary iron and steel is the most common form of corrosion. Rusting takes place in moist air, when the iron combines with oxygen and water to form a brown-orange deposit of ◊rust (hydrated iron oxide). The rate of corrosion is increased where the atmosphere is polluted with sulphur dioxide. Salty road and air conditions accelerate the rusting of car bodies.

Corrosion is largely an electrochemical process, and acidic and salty conditions favour the establishment of electrolytic cells on the metal, which cause it to be eaten away. Other examples of corrosion include the green deposit that forms on copper and bronze, called verdigris, a basic copper carbonate. The tarnish on silver is a corrosion product, a film of silver sulphide.

corrosion in earth science, an alternative name for ◊solution, the process by which water dissolves rocks such as limestone.

corruption of data introduction or presence of errors in data. Most computers use a range of ◊verification and ◊validation routines to prevent corrupt data from entering the computer system or detect corrupt data that are already present.

cortex in biology, the outer part of a structure such as the brain, kidney, or adrenal gland. In botany the cortex includes non-specialized cells lying just beneath the surface cells of the root and stem.

corticosteroid any of several steroid hormones secreted by the cortex of the ◊adrenal glands; also synthetic forms with similar properties. Corticosteroids have anti-inflammatory and ◊immunosuppressive effects and may be used to treat a number of conditions, including rheumatoid arthritis, severe allergies, asthma, some skin diseases, and some cancers. Side effects can be serious, and therapy must be withdrawn very gradually.

The two main groups of corticosteroids include **glucocorticoids** (◊cortisone, hydrocortisone, prednisone, and dexamethasone), which are essential to carbohydrate, fat and protein metabolism, and to the body's response to stress; and **mineralocorticoids** (aldosterone, fluorocortisone), which control the balance of water and salt in the body.

cortisone natural corticosteroid produced by the ◊adrenal gland, now synthesized for its anti-inflammatory qualities and used in the treatment of rheumatoid arthritis.

Cortisone was discovered by Tadeus Reichstein of Basel, Switzerland, and put to practical clinical use for rheumatoid arthritis by Philip Hench (1896–1965) and Edward Kendall (1886–1972) in the USA (all three shared a Nobel prize 1950).

A product of the adrenal gland, it was first synthesized from a constituent of ox bile, and is now produced commercially from a Mexican yam and from a by-product of the sisal plant. It is used for treating allergies and certain cancers, as well as rheumatoid arthritis. The side effects of cortisone steroids include muscle wasting, fat redistribution, diabetes, bone thinning, and high blood pressure.

corundum native aluminium oxide, Al_2O_3, the hardest naturally occurring mineral known apart from diamond (corundum rates 9 on the Mohs' scale of hardness); lack of ◊cleavage also increases its durability. Its crystals are barrel-shaped prisms of the trigonal system. Varieties of gem-quality corundum are **ruby** (red) and **sapphire** (any colour other than red, usually blue). Poorer-quality and synthetic corundum is used in industry, for example as an ◊abrasive.

Corundum forms in silica-poor igneous and metamorphic rocks. It is a constituent of emery, which is metamorphosed bauxite.

cosecant in trigonometry, a ◊function of an angle in a right-angled triangle found by dividing the length of the hypotenuse (the longest side) by the length of the side opposite the angle. Thus the cosecant of an angle A, usually shortened to cosec A, is always greater than (or equal to) 1. It is the reciprocal of the sine of the angle, that is, cosec A = 1/sin A.

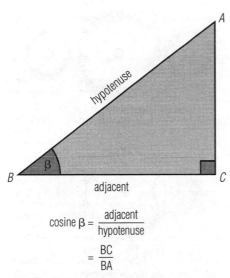

$$\text{cosine } \beta = \frac{\text{adjacent}}{\text{hypotenuse}}$$

$$= \frac{BC}{BA}$$

cosine The cotangent of angle β is equal to the ratio of the length of the adjacent side to the length of the hypotenuse (the longest side, opposite to the right angle).

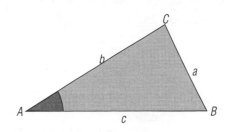

the cosine rule states that
$$a^2 = b^2 + c^2 - 2bc \cos A$$

cosine rule The cosine rule is a rule of trigonometry that relates the sides and angles of triangles. It can be used to find a missing length or angle in a triangle.

cosine in trigonometry, a ◊function of an angle in a right-angled triangle found by dividing the length of the side adjacent to the angle by the length of the hypotenuse (the longest side). It is usually shortened to **cos**.

cosine rule in trigonometry, a rule that relates the sides and angles of triangles. The rule has the formula:

$$a^2 = b^2 + c^2 - 2bc \cos A$$

Where a, b, and c are the sides of the triangle, and A is the angle opposite a.

cosmic background radiation or *3° radiation* electromagnetic radiation left over from the original formation of the universe in the Big Bang around 15 billion years ago. It corresponds to an overall background temperature of 3K (−270°C/−454°F), or 3°C above absolute zero. In 1992 the Cosmic Background Explorer satellite, COBE, detected slight 'ripples' in the strength of the background radiation that are believed to mark the first stage in the formation of galaxies.

Cosmic background radiation was first detected 1965 by US physicists Arno Penzias (1933–) and Robert Wilson (1936–), who in 1978 shared the Nobel Prize for Physics for their discovery.

cosmic radiation streams of high-energy particles from outer space, consisting of protons, alpha particles, and light nuclei, which collide with atomic nuclei in the Earth's atmosphere, and produce secondary nuclear particles (chiefly ◊mesons, such as pions and muons) that shower the Earth.

Those of low energy seem to be galactic in origin, and those of high energy of extragalactic origin. The galactic particles may come from ◊supernova explosions or ◊pulsars. At higher energies, other sources are necessary, possibly the giants jets of gas which are emitted from some galaxies.

cosmology branch of astronomy that deals with the structure and evolution of the universe as an ordered whole. Its method is to construct 'model universes' mathematically and compare their large-scale properties with those of the observed universe.

Modern cosmology began in the 1920s with the discovery that the universe is expanding, which suggested that it began in an explosion, the ◊Big Bang. An alternative – now discarded – view, the ◊steady-state theory, claimed that the universe has no origin, but is expanding because new matter is being continually created.

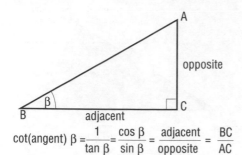

$$\text{cot(angent) } \beta = \frac{1}{\tan \beta} = \frac{\cos \beta}{\sin \beta} = \frac{\text{adjacent}}{\text{opposite}} = \frac{BC}{AC}$$

cotangent The cotangent of angle β is equal to the ratio of the length of the adjacent side to the length of the opposite side.

cotangent in trigonometry, a ◊function of an angle in a right-angled triangle found by dividing the length of the side adjacent to the angle by the length of the side opposite it. It is usually written as cotan, or cot and it is the reciprocal of the tangent of the angle, so that cot $A = 1/\tan A$, where A is the angle in question.

cot death or ***sudden infant death syndrome*** (SIDS) death of an apparently healthy baby, almost always during sleep. It is most common in the winter months, and strikes more boys than girls. The cause is not known but risk factors that have been identified include prematurity, respiratory infection, overheating and sleeping position.

There was a 60% reduction in the number of cot deaths in the UK in 1993, following a massive advertising campaign advising parents to put their babies to sleep on their backs, ensure they do not overheat, and avoid smoking near them. In New Zealand (1991), where the cot death rate had been the highest in the world, the rate was halved by a similar campaign.

cotyledon structure in the embryo of a seed plant that may form a 'leaf' after germination and is commonly known as a seed leaf. The number of cotyledons present in an embryo is an important character in the classification of flowering plants (◊angiosperms).

Monocotyledons (such as grasses, palms, and lilies) have a single cotyledon, whereas dicotyledons (the majority of plant species) have two. In seeds that also contain ◊endosperm (nutritive tissue), the cotyledons are thin, but where they are the primary food-storing tissue, as in peas and beans, they may be quite large. After germination the cotyledons either remain below ground (hypogeal) or, more commonly, spread out above soil level (epigeal) and become the first green leaves. In gymnosperms there may be up to a dozen cotyledons within each seed.

coulomb SI unit (symbol C) of electrical charge. One coulomb is the quantity of electricity conveyed by a current of one ◊ampere in one second.

courtship behaviour exhibited by animals as a prelude to mating. The behaviour patterns vary considerably from one species to another, but are often ritualized forms of behaviour not obviously related to courtship or mating (for example, courtship feeding in birds).

Courtship ensures that copulation occurs with a member of the opposite sex of the right species. It also synchronizes the partners' readiness to mate and allows each partner to assess the suitability of the other.

covalent bond chemical ◊bond produced when two atoms share one or more pairs of electrons (usually each atom contributes an electron). The bond is often represented by a single line drawn between the two atoms. Covalently bonded substances include hydrogen (H_2), water (H_2O), and most organic substances.

CP/M (abbreviation for ***control program/monitor*** or ***control program for microcomputers***) one of the earliest ◊operating systems for microcomputers. It was written by Gary Kildall, who founded Digital Research, and became a standard for microcomputers based on the Intel 8080 and Zilog Z80 8-bit microprocessors. In the 1980s it was superseded by Microsoft's ◊MS-DOS, written for 16-bit microprocessors.

CPU in computing, abbreviation for ◊central processing unit.

Crab nebula cloud of gas 6,000 light years from Earth, in the constellation ◊Taurus. It is the remains of a star that according to Chinese records exploded as a ◊supernova observed as a brilliant light on Earth on 4 July 1054.

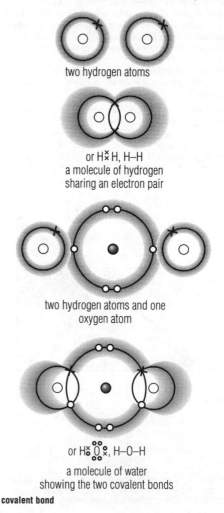

two hydrogen atoms

or $H\overset{\times}{\cdot} H$, H–H
a molecule of hydrogen
sharing an electron pair

two hydrogen atoms and one
oxygen atom

or $H\overset{\times}{\cdot}\overset{\circ\circ}{O}\overset{\circ}{\cdot}$, H–O–H
a molecule of water
showing the two covalent bonds

covalent bond

At its centre is a ◊pulsar that flashes 30 times a second. It was named by Lord Rosse after its crablike shape.

crack street name for a chemical derivative (bicarbonate) of ◊cocaine in hard, crystalline lumps; it is heated and inhaled (smoked) as a stimulant. Crack was first used in San Francisco in the early 1980s, and is highly addictive.

Its use has led to numerous deaths but it is the fastest-growing sector of the illegal drug trade, since it is less expensive than cocaine.

cracking reaction in which a large ◊alkane molecule is broken down by heat into a smaller alkane and a small ◊alkene molecule. The reaction is carried out at a high temperature (600°C or higher) and often in the presence of a catalyst. Cracking is a commonly used process in the ◊petrochemical industry.

It is the main method of preparation of alkenes and is also used to manufacture petrol from the higher-boiling-point ◊fractions that are obtained by fractional ◊distillation (fractionation) of crude oil.

crag in previously glaciated areas, a large lump of rock that a glacier has been unable to wear away. As the glacier passed up and over the crag, weaker rock on the far side was largely protected from erosion and formed a tapering ridge, or *tail*, of debris.

An example of a crag-and-tail feature is found in Edinburgh in Scotland; Edinburgh Castle was built on the crag (Castle Rock), which dominates the city beneath.

crane in engineering, a machine for raising, lowering, or placing in position heavy loads. The three main types are the jib crane, the overhead travelling crane, and the tower crane. Most cranes have the machinery mounted on a revolving turntable. This may be mounted on trucks or be self-propelled, often being fitted with ◊caterpillar tracks.

The main features of a *jib crane* are a power winch, a rope or cable, and a movable arm or jib. The cable, which carries a pulley block, hangs from the end of the jib and is wound up and down by the winch. The *overhead travelling crane*, chiefly used in workshops, consists of a fixed horizontal arm, along which runs a trolley carrying the pulley block. *Tower cranes*, seen on large building sites, have a long horizontal arm able to revolve on top of a tall tower. The arm carries the trolley.

Cranes can also be mounted on barges or large semi-submersibles for marine construction work.

cranium the dome-shaped area of the vertebrate skull that protects the brain. It consists of eight bony plates fused together by sutures (immovable joints). Fossil remains of the human cranium have aided the development of theories concerning human evolution.

The cranium has been studied as a possible indicator of intelligence or even of personality. The Victorian argument that a large cranium implies a large brain, which in turn implies a more profound intelligence, has been rejected.

crankshaft essential component of piston engines that converts the up-and-down (reciprocating) motion of the pistons into useful rotary motion. The car crankshaft carries a number of cranks. The pistons are connected to the cranks by connecting rods and ◊bearings; when the pistons move up and down, the connecting rods force the offset crank pins to describe a circle, thereby rotating the crankshaft.

crater bowl-shaped depression in the ground, usually round and with steep sides. Craters are formed by explosive events such as the eruption of a volcano or a bomb, or by the impact of a meteorite.

Crater

Hell is on the Moon. It is a crater 32 km/20 mi across and was named after 19th-century Hungarian astronomer Maximilian Hell. It has nothing to do with the flaming inferno. Utopia, on the other hand, is a smooth, low-lying region on Mars.

The Moon has more than 300,000 craters over 1 km/6 mi in diameter, formed by meteorite bombardment; similar craters on Earth have mostly been worn away by erosion. Craters are found on many other bodies in the Solar System.

Studies at the Jet Propulsion Laboratory in California, USA, have shown that craters produced by impact or by volcanic activity have distinctive shapes, enabling astronomers to distinguish likely methods of crater formation on planets in the Solar System. Unlike volcanic craters, impact craters have a raised rim and central peak and are almost always circular, irrespective of the meteorite's angle of incidence.

craton or *shield* core of a continent, a vast tract of highly deformed ◊metamorphic rock around which the continent has been built. Intense mountain-building periods shook these shield areas in Precambrian times before stable conditions set in.

Cratons exist in the hearts of all the continents, a typical example being the Canadian Shield.

creationism theory concerned with the origins of matter and life, claiming, as does the Bible in Genesis, that the world and humanity were created by a supernatural Creator, not more than 6,000 years ago. It was developed in response to Darwin's theory of ◊evolution; it is not recognized by most scientists as having a factual basis.

After a trial (1981–82) a US judge ruled unconstitutional an attempt in Arkansas schools to enforce equal treatment of creationism and evolutionary theory. In Alabama from 1996 all biology textbooks must contain a statement that evolution is a controversial theory and not a proven fact.

creep in civil and mechanical engineering, the property of a solid, typically a metal, under continuous stress that causes it to deform below its ◊yield point (the point at which any elastic solid normally stretches without any increase in load or stress). Lead, tin, and zinc, for example, exhibit creep at ordinary temperatures, as seen in the movement of the lead sheeting on the roofs of old buildings.

Copper, iron, nickel, and their alloys also show creep at high temperatures.

creosote black, oily liquid derived from coal tar, used as a wood preservative. Medicinal creosote, which is transparent and oily, is derived from wood tar.

crescent curved shape of the Moon when it appears less than half illuminated. It also refers to any object or symbol resembling the crescent Moon. Often associated with Islam, it was first used by the Turks on their standards after the capture of Constantinople in 1453, and appears on the flags of many Muslim countries. The *Red Crescent* is the Muslim equivalent of the Red Cross.

Cretaceous period of geological time approximately 144.2–65 million years ago. It is the last period of the Mesozoic era, during which angiosperm (seed-bearing) plants evolved, and dinosaurs reached a peak before their extinction at the end of the period. The north European chalk, which forms the white cliffs of Dover, was deposited during the latter half of the Cretaceous.

crevasse deep crack in the surface of a glacier; it can reach several metres in depth. Crevasses often occur where a glacier flows over the break of a slope, because the upper layers of ice are unable to stretch and cracks result. Crevasses may also form at the edges of glaciers owing to friction with the bedrock.

crith unit of mass used for weighing gases. One crith is the mass of one litre of hydrogen gas (H_2) at standard temperature and pressure.

critical density in cosmology, the minimum average density that the universe must have for it to stop expanding at some point in the future.

The precise value depends on ◊Hubble's constant and so is not fixed, but it is approximately 10–26 kg m^{-3}, equivalent to a few hydrogen atoms per cubic metre. The density parameter (symbol Ω) is the ratio of the actual density to the critical density. If $\Omega < 1$, the universe is open and will expand forever (see heat death). If $\Omega > 1$, the universe is closed and the expansion will eventually halt, to be followed by a contraction (see Big Crunch). Current estimates from visible matter in the universe indicate that Ω is about 0.01, well below critical density, but unseen dark matter may be sufficient to raise Ω to somewhere between 0.1 and 2.

critical mass in nuclear physics, the minimum mass of fissile material that can undergo a continuous ◊chain reaction. Below this mass, too many ◊neutrons escape from the surface for a chain reaction to carry on; above the critical mass, the reaction may accelerate into a nuclear explosion.

critical path analysis procedure used in the management of complex projects to minimize the amount of time taken. The analysis shows which subprojects can run in parallel with each other, and which have to be completed before other subprojects can follow on.

By identifying the time required for each separate subproject and the relationship between the subprojects, it is possible to produce a planning schedule showing when each subproject should be started and finished in order to complete the whole project most efficiently. Complex projects may involve hundreds of subprojects, and computer ◊applications packages for critical path analysis are widely used to help reduce the time and effort involved in their analysis.

Crohn's disease or *regional ileitis* chronic inflammatory bowel disease. It tends to flare up for a few days at a time, causing diarrhoea, abdominal cramps, loss of appetite, weight loss, and mild fever. The cause of Crohn's disease is unknown, although stress may be a factor.

Crohn's disease may occur in any part of the digestive system but usually affects the small intestine. It is characterized by ulceration, abscess formation, small perforations, and the development of adhesions binding the loops of the small intestine. Affected segments of intestine may constrict, causing obstruction, or may perforate. It is treated by surgical removal of badly affected segments, and by corticosteroids. Mild cases respond to rest, bland diet, and drug treatment. Crohn's disease first occurs most often in adults aged 20–40.

crop in birds, the thin-walled enlargement of the digestive tract between the oesophagus and stomach. It is an effective storage organ especially in seed-eating birds; a pigeon's crop can hold about 500 cereal grains. Digestion begins in the crop, by the moisturizing of food. A crop also occurs in insects and annelid worms.

Crop

A sandgrouse can hold about 9,000 seeds in its crop. This is useful, as ground-feeding is a risky business and the birds need to feed rapidly as they move about.

---

crop circle circular area of flattened grain found in fields especially in SE England, with increasing frequency every summer since 1980. More than 1,000 such formations were reported in the UK in 1991. The cause is unknown.

Most of the research into crop circles has been conducted by dedicated amateur investigators rather than scientists. Physicists who have studied the phenomenon have suggested that an electromagnetic whirlwind, or 'plasma vortex', can explain both the crop circles and some UFO sightings, but this does not account for the increasing geometrical complexity of crop circles, nor for the fact that until 1990 they were unknown outside the UK. Crop circles began to appear in the USA only after a US magazine published an article about them. A few people have confessed publicly to having made crop circles that were accepted as genuine by investigators.

crop rotation system of regularly changing the crops grown on a piece of land. The crops are grown in a particular order to utilize and add to the nutrients in the soil and to prevent the build-up of insect and fungal pests. Including a legume crop, such as peas or beans, in the rotation helps build up nitrate in the soil because the roots contain bacteria capable of fixing nitrogen from the air.

A simple seven-year rotation, for example, might include a three-year ley followed by two years of wheat and then two years of barley, before returning the land to temporary grass once more. In this way, the cereal crops can take advantage of the build-up of soil fertility that occurs during the period under grass. In the 18th century, a four-year rotation was widely adopted with autumn-sown cereal, followed by a root crop, then spring cereal, and ending with a leguminous crop. Since then, more elaborate rotations have been devised with two, three, or four successive cereal crops, and with the root crop replaced by a cash crop such as sugar beet or potatoes, or by a legume crop such as peas or beans.

cross-section the surface formed when a solid is cut through by a plane at right angles to its axis.

croup inflammation of the larynx in small children, with harsh, difficult breathing and hoarse coughing. Croup is most often associated with viral infection of the respiratory tract.

crude oil the unrefined form of ◊petroleum.

crust the outermost part of the structure of Earth, consisting of two distinct parts, the oceanic crust and the continental crust. The *oceanic* crust is on average about 10 km/6.2 mi thick and consists mostly of basaltic types of rock. By contrast, the *continental* crust is largely made

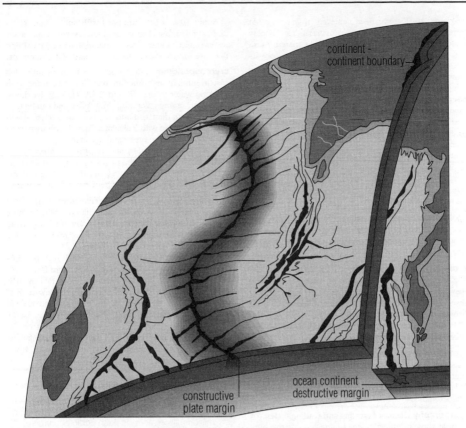

continent - continent boundary

constructive plate margin

ocean continent destructive margin

crust The crust of the Earth is made up of plates with different kinds of margins. In mid-ocean, there are constructive plate margins, where magma wells up from the Earth's interior, forming new crust. On continent–continent margins, mountain ranges are flung up by the collision of two continents. At an ocean–continent destructive margin, ocean crust is forced under the denser continental crust, forming an area of volcanic instability.

of granite and is more complex in its structure. Because of the movements of ◊plate tectonics, the oceanic crust is in no place older than about 200 million years. However, parts of the continental crust are over 3 billion years old.

Beneath a layer of surface sediment, the oceanic crust is made up of a layer of basalt, followed by a layer of gabbro. The composition of the oceanic crust overall shows a high proportion of silicon and magnesium oxides, hence named **sima** by geologists. The continental crust varies in thickness from about 40 km/25 mi to 70km/45 mi, being deeper beneath mountain ranges. The surface layer consists of many kinds of sedimentary and igneous rocks. Beneath lies a zone of metamorphic rocks built on a thick layer of granodiorite. Silicon and aluminium oxides dominate the composition and the name **sial** is given to continental crustal material.

crustacean one of the class of arthropods that includes crabs, lobsters, shrimps, woodlice, and barnacles. The external skeleton is made of protein and chitin hardened with lime. Each segment bears a pair of appendages that may be modified as sensory feelers (antennae), as mouthparts, or as swimming, walking, or grasping structures.

Crux constellation of the southern hemisphere, popularly known as the Southern Cross, the smallest of the 88

constellations but one of the brightest, and one of the best known as it is represented on the flags of Australia and New Zealand. Its brightest stars are Alpha Crucis (or Acrux), a ◊double star about 400 light years from Earth, and Beta Crucis (or Mimosa).

Near Beta Crucis lies a glittering star cluster known as the Jewel Box, named by John Herschel. The constellation also contains the Coalsack, a dark nebula silhouetted against the bright starry background of the Milky Way.

cryogenics science of very low temperatures (approaching ◊absolute zero), including the production of very low temperatures and the exploitation of special properties associated with them, such as the disappearance of electrical resistance (◊superconductivity).

Low temperatures can be produced by the Joule–Thomson effect (cooling a gas by making it do work as it expands). Gases such as oxygen, hydrogen, and helium may be liquefied in this way, and temperatures of 0.3K can be reached. Further cooling requires magnetic methods; a magnetic material, in contact with the substance to be cooled and with liquid helium, is magnetized by a strong magnetic field. The heat generated by the process is carried away by the helium. When the material is then demagnetized, its temperature falls; temperatures of around 10^{-3}K have been achieved in this way. At temperatures near absolute zero, materials can display unusual properties. Some metals, such as mercury and lead, exhibit superconductivity. Liquid helium loses its viscosity and becomes a 'superfluid' when cooled to below 2K; in this state it flows up the sides of its container. Cryogenics has several practical applications. *Cryotherapy* is a process used in eye surgery, in which a freezing probe is briefly applied to the outside of the eye to repair a break in the retina. Electronic components called ◊Josephson junctions, which could be used in very fast computers, need low temperatures to function. Magnetic levitation (◊maglev) systems must be maintained at low temperatures. Food can be frozen for years, and freezing eggs, sperm and pre-embryos is now routine. In Sept 1996 South African researchers resuscitated a rat's heart that had been frozen to $-196°$C.

cryolite rare granular crystalline mineral (sodium aluminium fluoride), Na_3AlF_6, used in the electrolytic reduction of ◊bauxite to aluminium. It is chiefly found in Greenland.

cryonics practice of freezing a body at the moment of clinical death with the aim of enabling eventual resuscitation. The body, drained of blood, is indefinitely preserved in a thermos-type container filled with liquid nitrogen at $-196°$C/$-321°$F.

The first human treated was James H Bedford, a lung-cancer patient of 74, in the USA in 1967.

cryptogam obsolete name applied to the lower plants. It included the algae, liverworts, mosses, and ferns (plus the fungi and bacteria in very early schemes of classification). In such classifications seed plants were known as ◊phanerogams.

cryptography science of creating and reading codes; for example, those produced by the Enigma coding machine used by the Germans in World War II and those used in commerce by banks encoding electronic fund-transfer messages, business firms sending computer-conveyed memos between headquarters, and in the growing field of electronic mail. The breaking and decipherment of such codes is known as 'cryptanalysis'. No method of encrypting is completely unbreakable, but decoding can be made extremely complex and time consuming.

cryptorchism or *cryptorchidism* condition marked by undescended testicles; failure of the testes to complete their descent into the scrotum before birth. When only one testicle has descended, the condition is known as monorchism.

About 10% of boys are born with one or both testes undescended. Usually the condition resolves within a few weeks of birth. Otherwise, an operation is needed to bring the testes down and ensure normal sexual development.

cryptosporidium waterborne parasite that causes disease in humans and other animals. It has been found in drinking water in the UK and USA, causing diarrhoea, abdominal cramps, vomiting, and fever, and can be fatal in people with damaged immune systems, such as AIDS sufferers or those with leukaemia. Just 30 cryptosporidia are enough to cause prolonged diarrhoea.

Conventional filtration and chlorine disinfection are ineffective at removing the parasite. Slow-sand filtration is the best method of removal, but the existing systems were dismantled in the 1970s because of their slowness.

crystal substance with an orderly three-dimensional arrangement of its atoms or molecules, thereby creating an external surface of clearly defined smooth faces having characteristic angles between them. Examples are table salt and quartz.

Each geometrical form, many of which may be combined in one crystal, consists of two or more faces – for example, dome, prism, and pyramid. A mineral can often be identified by the shape of its crystals and the system of crystallization determined. A single crystal can vary in size from a submicroscopic particle to a mass some 30 m/100 ft in length. Crystals fall into seven crystal systems or groups, classified on the basis of the relationship of three or four imaginary axes that intersect at the centre of any perfect, undistorted crystal.

Crystal

Crystals of salt (NaCl) are made up of small cubes of atoms. A crystal of rock salt contains 40,000 million million million atoms in each cubic centimetre, arranged in tiny cubes 0.00000002 cm/ 0.000000008 in across.

---

crystallography the scientific study of crystals. In 1912 it was found that the shape and size of the repeating atomic patterns (unit cells) in a crystal could be determined by passing X-rays through a sample. This method, known as ◊X-ray diffraction, opened up an entirely new way of 'seeing' atoms. It has been found that many substances have a unit cell that exhibits all the symmetry of the whole crystal; in table salt (sodium chloride, NaCl), for instance, the unit cell is an exact cube.

Many materials were not even suspected of being crystals until they were examined by X-ray crystallography. It

Crystal System

Crystal system	Minimum symmetry	Possible shape	Mineral examples
cubic	4 threefold axes	cube, octahedron, dodecahedron	diamond, garnet, pyrite
tetragonal	1 fourfold axis	square-based prism	zircon
orthorhombic	3 twofold axes or mirror planes	matchbox shape	baryte
monoclinic	1 twofold axis or mirror plane	matchbox distorted in one plane	gypsum
triclinic	no axes or mirror planes	matchbox distorted in three planes	plagioclase feldspar
trigonal	1 threefold axis	triangular prism, rhombohedron	calcite, quartz
hexagonal	1 sixfold axis	hexagonal prism	beryl

has been shown that purified biomolecules, such as proteins and DNA, can form crystals, and such compounds may now be studied by this method. Other applications include the study of metals and their alloys, and of rocks and soils.

crystal system all known crystalline substances crystallize in one of the seven crystal systems defined by symmetry. The elements of symmetry used for this purpose are: (1) planes of *mirror symmetry*, across which a mirror image is seen, and (2) axes of *rotational symmetry*, about which, in a 360° rotation of the crystal, equivalent faces are seen twice, three, four, or six times. To be assigned to a particular crystal system, a mineral must possess a certain minimum symmetry, but it may also possess additional symmetry elements. Since crystal symmetry is related to internal structure, a given mineral will always crystallize in the same system, although the crystals may not always grow into precisely the same shape. In cases where two minerals have the same chemical composition but different internal structures (for example graphite and diamond, or quartz and cristobalite), they will generally have different crystal systems.

CT scanner medical device used to obtain detailed X-ray pictures of the inside of a patient's body. See ◊CAT scan.

cu abbreviation for *cubic* (measure).

cube in geometry, a regular solid figure whose faces are all squares. It has six equal-area faces and 12 equal-length edges.

If the length of one edge is l, the volume V of the cube is given by:

$$V = l^3$$

and its surface area A by:

$$A = 6l^2$$

cube to multiply a number by itself and then by itself again. For example, 5 cubed = $5^3 = 5 \times 5 \times 5 = 125$. The term also refers to a number formed by cubing; for example, 1, 8, 27, 64 are the first four cubes.

cubic equation any equation in which the largest power of x is x^3. For example, $x^3 + 3x^2y + 4y^2 = 0$ is a cubic equation.

cubit earliest known unit of length, which originated between 2800 and 2300 BC. It is approximately 50.5 cm/20.6 in long, which is about the length of the human forearm measured from the tip of the middle finger to the elbow.

cuboid six-sided three-dimensional prism whose faces are all rectangles. A brick is a cuboid.

cultivar variety of a plant developed by horticultural or agricultural techniques. The term derives from '*cultivated variety*'.

culture in biology, the growing of living cells and tissues in laboratory conditions.

cumulative frequency in statistics, the total frequency of a given value up to and including a certain point in a set of data. It is used to draw the cumulative frequency curve, the ogive.

cuprite Cu_2O ore (copper(I) oxide), found in crystalline form or in earthy masses. It is red to black in colour, and is often called ruby copper.

cupronickel copper alloy (75% copper and 25% nickel), used in hardware products and for coinage.

curare black, resinous poison extracted from the bark and juices of various South American trees and plants. Originally used on arrowheads by Amazonian hunters to paralyse prey, it blocks nerve stimulation of the muscles. Alkaloid derivatives (called curarines) are used in medicine as muscle relaxants during surgery.

curing method of preserving meat by soaking it in salt (sodium chloride) solution, with saltpetre (sodium nitrate) added to give the meat its pink colour and characteristic taste. The nitrates in cured meats are converted to nitrites and nitrosamines by bacteria, and these are potentially carcinogenic to humans.

curium synthesized, radioactive, metallic element of the *actinide* series, symbol Cm, atomic number 96, relative atomic mass 247. It is produced by bombarding plutonium or americium with neutrons. Its longest-lived isotope has a half-life of 1.7×10^7 years.

current flow of a body of water or air, or of heat, moving in a definite direction. Ocean currents are fast-flowing currents of seawater generated by the wind or by variations in water density between two areas. They are partly responsible for transferring heat from the Equator to the poles and thereby evening out the global heat imbalance. There are three basic types of ocean current: *drift currents* are broad and slow-moving; *stream currents* are narrow and swift-moving; and *upwelling currents* bring cold, nutrient-rich water from the ocean bottom.

Stream currents include the ◊Gulf Stream and the ◊Japan (or Kuroshio) Current. Upwelling currents, such as the Gulf of Guinea Current and the Peru (Humboldt) current, provide food for plankton, which in turn supports fish and sea birds. At approximate five-to-eight-year intervals, the Peru Current that runs from the Antarctic up the W coast of South America, turns warm, with heavy rain and rough seas, and has disastrous results (as in 1982–83) for Peruvian wildlife and for the anchovy industry. The phenomenon is called *El Niño* (Spanish 'the Child') because it occurs towards Christmas.

current, electric see ◊electric current.

curve in geometry, the ◊locus of a point moving according to specified conditions. The circle is the locus of all points equidistant from a given point (the centre). Other common geometrical curves are the ◊ellipse, ◊parabola, and ◊hyperbola, which are also produced when a cone is cut by a plane at different angles.

Many curves have been invented for the solution of special problems in geometry and mechanics – for example, the cissoid (the inverse of a parabola) and the ◊cycloid.

Cushing's syndrome condition in which the body chemistry is upset by excessive production of ◊steroid hormones from the adrenal cortex.

Symptoms include weight gain in the face and trunk, raised blood pressure, excessive growth of facial and body hair (hirsutism), demineralization of bone, and, sometimes, diabeteslike effects. The underlying cause may be an adrenal or pituitary tumour, or prolonged high-dose therapy with ◊corticosteroid drugs.

cusp point where two branches of a curve meet and the tangents to each branch coincide.

cuticle the horny noncellular surface layer of many invertebrates such as insects; in botany, the waxy surface layer on those parts of plants that are exposed to the air, continuous except for ◊stomata and ◊lenticels. All types are secreted by the cells of the ◊epidermis. A cuticle reduces water loss and, in arthropods, acts as an ◊exoskeleton.

cutting technique of vegetative propagation involving taking a section of root, stem, or leaf and treating it so that it develops into a new plant.

cwt symbol for ◊ *hundredweight*, a unit of weight equal to 112 pounds (50.802 kg); 100 lb (45.36 kg) in the USA.

cyanide CN⁻ ion derived from hydrogen cyanide (HCN), and any salt containing this ion (produced when hydrogen cyanide is neutralized by alkalis), such as potassium cyanide (KCN). The principal cyanides are potassium, sodium, calcium, mercury, gold, and copper. Certain cyanides are poisons.

cyanobacteria (singular *cyanobacterium*) alternative name for ◊blue-green algae.

cyanocobalamin chemical name for vitamin B_{12}, which is normally produced by microorganisms in the gut. The richest sources are liver, fish, and eggs. It is essential to the replacement of cells, the maintenance of the myelin sheath which insulates nerve fibres, and the efficient use of folic acid, another vitamin in the B complex. Deficiency can result in pernicious anaemia (defective production of red blood cells), and possible degeneration of the nervous system.

cybernetics science concerned with how systems organize, regulate, and reproduce themselves, and also how they evolve and learn. In the laboratory, inanimate objects are created that behave like living systems. Applications range from the creation of electronic artificial limbs to the running of the fully automated factory where decisionmaking machines operate up to managerial level.

Cybernetics was founded and named in 1947 by US mathematician Norbert Wiener. Originally, it was the study of control systems using feedback to produce automatic processes.

cyberspace the imaginary, interactive 'worlds' created by networked computers; often used interchangeably with 'virtual world'. The invention of the word 'cyberspace' is generally credited to US science-fiction writer William Gibson (1948–) and his first novel *Neuromancer* published in 1984.

As well as meaning the interactive environment encountered in a virtual reality system, cyberspace is 'where' the global community of computer-linked individuals and groups lives. From the mid-1980s, the development of computer networks and telecommunications, both international (such as the ◊Internet) and local (such

as the services known as 'bulletin board' or conferencing systems), made possible the instant exchange of messages using ◊electronic mail and electronic conferencing systems directly from the individual's own home.

cycle in physics, a sequence of changes that moves a system away from, and then back to, its original state. An example is a vibration that moves a particle first in one direction and then in the opposite direction, with the particle returning to its original position at the end of the vibration.

cyclic compound any of a group of organic chemicals that have rings of atoms in their molecules, giving them a closed-chain structure.

cycloid in geometry, a curve resembling a series of arches traced out by a point on the circumference of a circle that rolls along a straight line. Its applications include the study of the motion of wheeled vehicles along roads and tracks.

cyclone alternative name for a ◊depression, an area of low atmospheric pressure. A severe cyclone that forms in the tropics is called a tropical cyclone or ◊hurricane.

cyclosporin ◊immunosuppressive drug derived from a fungus (*Tolypocladium inflatum*). In use by 1978, it revolutionized transplant surgery by reducing the incidence and severity of rejection of donor organs.

It suppresses the T-cells (see ◊lymphocyte) that reject foreign tissue without suppressing other cells that fight infection and cancer.

cyclotron circular type of particle ◊accelerator.

Cygnus large prominent constellation of the northern hemisphere, represented as a swan. Its brightest star is first-magnitude Alpha Cygni or ◊Deneb.

Beta Cygni (Albireo) is a yellow and blue ◊double star, visible through small telescopes. The constellation contains the North America nebula (named after its shape), the Veil nebula (the remains of a ◊supernova that exploded about 50,000 years ago), Cygnus A (apparently a double galaxy, a powerful radio source, and the first radio star to be discovered), and the X-ray source Cygnus X-1, thought to mark the position of a black hole. The area is rich in high luminosity objects, nebulae, and clouds of obscuring matter. Deneb marks the tail of the swan, which is depicted as flying along the Milky Way. Some of the brighter stars form the Northern Cross, the upright being defined by Alpha, Gamma, Eta, and Beta, and the crosspiece by Delta, Gamma, and Epsilon Cygni.

cylinder in geometry, a tubular solid figure with a circular base. In everyday use, the term applies to a *right cylinder*, the curved surface of which is at right angles to the base.

The volume V of a cylinder is given by the formula

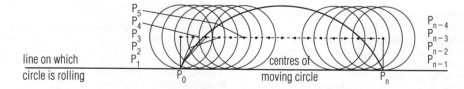

line on which circle is rolling

centres of moving circle

cycloid The cycloid is the curve traced out by a point on a circle as it rolls along a straight line. The teeth of gears are often cut with faces that are arcs of cycloids so that there is rolling contact when the gears are in use.

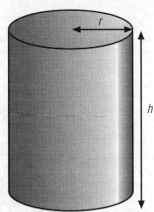

volume = $\pi r^2 h$
area or curved
surface = $2\pi rh$

total surface area
= $2\pi r(r + h)$

cylinder The volume and area of a cylinder are given by simple formulae relating the dimensions of the cylinder.

$V = \pi r^2 h$, where r is the radius of the base and h is the height of the cylinder. Its total surface area A has the formula $A = 2\pi r(h + r)$, where $2\pi rh$ is the curved surface area, and $2\pi r^2$ is the area of both circular ends.

cystic fibrosis hereditary disease involving defects of various tissues, including the sweat glands, the mucous glands of the bronchi (air passages), and the pancreas. The sufferer experiences repeated chest infections and digestive disorders and generally fails to thrive. In 1989 a gene for cystic fibrosis was identified by teams of researchers in Michigan, USA, and Toronto, Canada. This discovery enabled the development of a screening test for carriers; the disease can also be detected in the unborn child.

One person in 22 is a carrier of the disease. If two carriers have children, each child has a one-in-four chance of having the disease, so that it occurs in about one in 2,000 pregnancies. Around 10% of newborns with cystic fibrosis develop an intestinal blockage (meconium ileus) which requires surgery.

Cystic fibrosis was once universally fatal at an early age; now, although there is no definitive cure, treatments have raised both the quality and expectancy of life. Results in 1995 from a four-year US study showed that the painkiller ibuprofen, available over the counter, slowed lung deterioration in children by almost 90% when taken in large doses.

Management of cystic fibrosis is by diets and drugs, physiotherapy to keep the chest clear, and use of antibiotics to combat infection and minimize damage to the lungs. Some sufferers have benefited from heart-lung transplants.

In 1993, UK researchers (at the Imperial Cancer Research Fund, Oxford, and the Wellcome Trust, Cambridge) successfully introduced a corrective version of the gene for cystic fibrosis into the lungs of mice with induced cystic fibrosis, restoring normal function (in 1992, US researchers had introduced such a gene into the lungs of healthy laboratory rats, greatly improving the prospect of

a cure). Trials in human subjects began in 1993, and the cystic fibrosis defect in the nasal cavities of three patients in the USA was successfully corrected, though a later trial was halted after a patient became ill following a dose of the genetically altered virus. Patients treated by gene therapy administered in the form of a nasal spray were showing signs of improvement following a preliminary trial in 1996. Cystic fibrosis is seen as a promising test case for ◊gene therapy. It is the commonest fatal hereditary disease amongst white people.

cystitis inflammation of the bladder, usually caused by bacterial infection, and resulting in frequent and painful urination. It is more common in women. Treatment is by antibiotics and copious fluids with vitamin C.

cytochrome protein responsible for part of the process of ◊respiration by which food molecules are broken down in ◊aerobic organisms. Cytochromes are part of the electron transport chain, which uses energized electrons to reduce molecular oxygen (O_2) to oxygen ions (O^{2-}). These combine with hydrogen ions (H^+) to form water (H_2O), the end product of aerobic respiration. As electrons are passed from one cytochrome to another, energy is released and used to make ◊ATP.

cytokine in biology, chemical messenger that carries information from one cell to another, for example the ◊lymphokines.

cytokinin ◊plant hormone that stimulates cell division. Cytokinins affect several different aspects of plant growth and development, but only if ◊auxin is also present. They may delay the process of senescence, or ageing, break the dormancy of certain seeds and buds, and induce flowering.

cytology the study of ◊cells and their functions. Major advances have been made possible in this field by the development of ◊electron microscopes.

cytoplasm the part of the cell outside the ◊nucleus. Strictly speaking, this includes all the ◊organelles (mitochondria, chloroplasts, and so on), but often cytoplasm refers to the jellylike matter in which the organelles are embedded (correctly termed the cytosol). The cytoplasm is the site of protein synthesis.

In many cells, the cytoplasm is made up of two parts: the *ectoplasm* (or plasmagel), a dense gelatinous outer layer concerned with cell movement, and the *endoplasm* (or plasmasol), a more fluid inner part where most of the organelles are found.

cytoskeleton in a living cell, a matrix of protein filaments and tubules that occurs within the cytosol (the liquid part of the cytoplasm). It gives the cell a definite shape, transports vital substances around the cell, and may also be involved in cell movement.

cytotoxic drug any drug used to kill the cells of a malignant tumour; it may also damage healthy cells. Side effects include nausea, vomiting, hair loss, and bone-marrow damage. Some cytotoxic drugs are also used to treat other diseases and to suppress rejection in transplant patients.

because they are expensive and time-consuming to build. Other concrete dams are built in the shape of an arch, with the curve facing upstream: the **arch dam** derives its strength from the arch shape, just as an arch bridge does, and has been widely used in the 20th century. They require less construction material than other dams but are the strongest type.

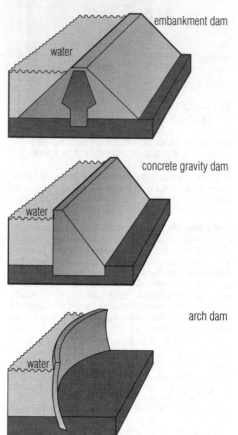

DAC abbreviation for ◊digital-to-analogue converter.

daguerreotype in photography, a single-image process using mercury vapour and an iodine-sensitized silvered plate; it was invented by Louis Daguerre 1838.

dairying the business of producing ◊milk and milk products.

In the UK and the USA, over 70% of the milk produced is consumed in its liquid form, whereas areas such as the French Alps and New Zealand rely on easily transportable milk products such as butter, cheese, and condensed and dried milk.

It is now usual for dairy farms to concentrate on the production of milk, and for factories to take over the handling, processing, and distribution of milk as well as the manufacture of dairy products.

daisywheel printing head in a computer printer or typewriter that consists of a small plastic or metal disc made up of many spokes (like the petals of a daisy). At the end of each spoke is a character in relief. The daisywheel is rotated until the spoke bearing the required character is facing an inked ribbon, then a hammer strikes the spoke against the ribbon, leaving the impression of the character on the paper beneath.

The daisywheel can be changed to provide different typefaces; however, daisywheel printers cannot print graphics nor can they print more than one typeface in the same document. For these reasons, they are rapidly becoming obsolete.

dam structure built to hold back water in order to prevent flooding, to provide water for irrigation and storage, and to provide hydroelectric power. The biggest dams are of the earth- and rock-fill type, also called **embankment dams**. Early dams in Britain, built before and about 1800, had a core made from puddled clay (clay which has been mixed with water to make it impermeable). Such dams are generally built on broad valley sites. Deep, narrow gorges dictate a **concrete dam**, where the strength of reinforced concrete can withstand the water pressures involved.

Concrete dams a valuable development in arid regions, as in parts of Brazil, is the **underground dam**, where water is stored on a solid rock base, with a wall to ground level, so avoiding rapid evaporation. Many concrete dams are triangular in cross section, with their vertical face pointing upstream. Their sheer weight holds them in position, and they are called **gravity dams**. They are no longer favoured for very large dams, however,

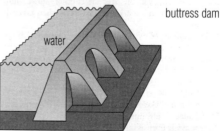

dam There are two basic types of dam: the gravity dam and the arch dam. The gravity dam relies upon the weight of its material to resist the forces imposed upon it; the arch dam uses an arch shape to take the forces in a horizontal direction into the sides of the river valley. The largest dams are usually embankment dams. Buttress dams are used to hold back very wide rivers or lakes.

Buttress dams are used when economy of construction is important or foundation conditions preclude any other type. The upstream portion of a buttress dam may comprise series of cantilevers, slabs, arches or domes supported from the back by a line of buttresses. They are usually made from reinforced and pre-stressed concrete.

Earth dams earth dams have a watertight core wall, formerly made of puddle clay but nowadays constructed of concrete. Their construction is very economical even for very large structures. **Rock-fill dams** are a variant of the earth dam in which dumped rock takes the place of compacted earth fill.

Major dams Rogun (Tajikistan) is the world's highest at 335 m/1,099 ft. New Cornelia Tailings (USA) is the world's biggest in volume, 209 million cu m/7.4 billion cu ft. Owen Falls (Uganda) has the world's largest reservoir capacity, 204.8 billion cu m/7.2 trillion cu ft. Itaipu (Brazil/Paraguay) is the world's most powerful, producing 12,700 megawatts of electricity. The Three Gorges Dam on the Chang Jiang was officially inaugurated in 1994. A treaty between Nepal and India, ratified by Nepal in 1996, included plans to construct the 315 m/1,035 ft Pancheshwar dam.

In 1997 there were approximately 40,000 large dams (more than 15 m in height) and 800,000 small ones worldwide.

damask textile of woven linen, cotton, wool, or silk, with a reversible figured pattern. It was first made in the city of Damascus, Syria.

daminozide (trade name **Alar**) chemical formerly used by fruit growers to make apples redder and crisper. In 1989 a report published in the USA found the consumption of daminozide to be linked with cancer, and the US Environment Protection Agency (EPA) called for an end to its use. The makers have now withdrawn it worldwide.

damper any device that deadens or lessens vibrations or oscillations; for example, one used to check vibrations in the strings of a piano. The term is also used for the movable plate in the flue of a stove or furnace for controlling the draught.

darcy c.g.s. unit (symbol D) of permeability, used mainly in geology to describe the permeability of rock (for example, to oil, gas, or water).

dark cloud in astronomy, a cloud of cold dust and gas seen in silhouette against background stars or an ◊HII region.

dark matter matter that, according to current theories of ◊cosmology, makes up 90–99% of the mass of the universe but so far remains undetected. Dark matter, if shown to exist, would explain many currently unexplained gravitational effects in the movement of galaxies.

Theories of the composition of dark matter include unknown atomic particles (cold dark matter) or fast-moving neutrinos (hot dark matter) or a combination of both.

In 1993 astronomers identified part of the dark matter in the form of stray planets and ◊brown dwarfs, and possibly, stars that have failed to light up. These objects are known as MACHOs (massive astrophysical compact halo objects) and, according to US astronomers in 1996, make up approximately half of the dark matter in the Milky Way's halo.

DAT abbreviation for ◊ **digital audio tape**.

data (singular **datum**) facts, figures, and symbols, especially as stored in computers. The term is often used to mean raw, unprocessed facts, as distinct from information, to which a meaning or interpretation has been applied.

database in computing, a structured collection of data, which may be manipulated to select and sort desired items of information. For example, an accounting system might be built around a database containing details of customers and suppliers. In larger computers, the database makes data available to the various programs that need it, without the need for those programs to be aware of how the data are stored. The term is also sometimes used for simple record-keeping systems, such as mailing lists, in which there are facilities for searching, sorting, and producing records.

There are three main types (or 'models'): hierarchical, network, and ◊relational, of which relational is the most widely used. A **free-text database** is one that holds the unstructured text of articles or books in a form that permits rapid searching.

A collection of databases is known as a **databank**. A database-management system (DBMS) program ensures that the integrity of the data is maintained by controlling the degree of access of the ◊application programs using the data. Databases are normally used by large organizations with mainframes or minicomputers.

A telephone directory stored as a database might allow all the people whose names start with the letter B to be selected by one program, and all those living in Chicago by another.

data bus in computing, the electrical pathway, or ◊bus, used to carry data between the components of the computer.

data communications sending and receiving data via any communications medium, such as a telephone line. The term usually implies that the data are digital (such as computer data) rather than analogue (such as voice messages). However, in the ISDN (◊Integrated Services Digital Network) system, all data – including voices and video images – are transmitted digitally. See ◊telecommunications.

data compression in computing, techniques for reducing the amount of storage needed for a given amount of data. They include word tokenization (in which frequently used words are stored as shorter codes), variable bit lengths (in which common characters are represented by fewer ◊bits than less common ones), and run-length encoding (in which a repeated value is stored once along with a count).

In **lossless compression** the original file is retrieved unchanged after decompression. Some types of data (sound and pictures) can be stored by **lossy compression** where some detail is lost during compression, but the loss is not noticeable. Lossy compression allows a greater level of compression. The most popular compression program is PKZIP, widely available as shareware.

data flow chart diagram illustrating the possible routes that data can take through a system or program; see ◊flow chart.

data processing (DP) or *electronic data processing* (EDP) use of computers for performing clerical tasks such as stock control, payroll, and dealing with orders. DP systems are typically ◊batch systems, running on mainframe computers.

data protection safeguarding of information about individuals stored on computers, to protect privacy.

data security in computing, precautions taken to prevent the loss or misuse of data, whether accidental or deliberate. These include measures that ensure that only authorized personnel can gain entry to a computer system or file, and regular procedures for storing and 'backing up' data, which enable files to be retrieved or recreated in the event of loss, theft, or damage.

A number of ◊verification and ◊validation techniques may also be used to prevent data from being lost or corrupted by misprocessing.

data terminator or *rogue value* in computing, a special value used to mark the end of a list of input data items. The computer must be able to detect that the data terminator is different from the input data in some way – for instance, a negative number might be used to signal the end of a list of positive numbers, or 'XXX' might be used to terminate the entry of a list of names.

dating science of determining the age of geological structures, rocks, and fossils, and placing them in the context of geological time. The techniques are of two types: relative dating and absolute dating.

Relative dating can be carried out by identifying fossils of creatures that lived only at certain times (marker fossils), and by looking at the physical relationships of rocks to other rocks of a known age.

Absolute dating is achieved by measuring how much of a rock's radioactive elements have changed since the rock was formed, using the process of ◊radiometric dating.

day time taken for the Earth to rotate once on its axis. The *solar day* is the time that the Earth takes to rotate once relative to the Sun. It is divided into 24 hours, and is the basis of our civil day. The *sidereal day* is the time that the Earth takes to rotate once relative to the stars. It is 3 minutes 56 seconds shorter than the solar day, because the Sun's position against the background of stars as seen from Earth changes as the Earth orbits it.

dBASE family of microcomputer programs used for manipulating large quantities of data; also, a related ◊fourth-generation language. The first version, dBASE II, appeared in 1981; it has since become the basis for a recognized standard for database applications, known as Xbase.

DC in physics, abbreviation for *direct current* (electricity).

DCC abbreviation for ◊ *digital compact cassette*.

DDT abbreviation for *dichloro-diphenyl-trichloro-ethane* ($ClC_6H_5)_2CHC(HCl_2)$ insecticide discovered in 1939 by Swiss chemist Paul Müller. It is useful in the control of insects that spread malaria, but resistant strains have developed. DDT is highly toxic and persists in the environment and in living tissue. Its use is now banned in most countries, but it continues to be used on food plants in Latin America.

deafness partial or total deficit of hearing in either ear.

Of assistance are hearing aids, lip-reading, a cochlear implant in the ear in combination with a special electronic processor, sign language, and 'cued speech' (manual clarification of ambiguous lip movement during speech).

Conductive deafness is due to faulty conduction of sound inwards from the external ear, usually due to infection (see ◊otitis), or a hereditary abnormality of the bones of the inner ear (see ◊otosclerosis).

Perceptive deafness may be inborn or caused by injury or disease of the cochlea, auditory nerve, or the hearing centres in the brain. It becomes more common with age.

Deafness

Alexander Graham Bell's invention of the telephone stemmed from his interest in hearing defects. His wife was profoundly deaf. On the day that Bell was buried in 1922, the entire telephone network in the USA was closed down for one minute in tribute.

---

deamination removal of the amino group (-NH_2) from an unwanted ◊amino acid. This is the nitrogen-containing part, and it is converted into ammonia, uric acid, or urea (depending on the type of animal) to be excreted in the urine. In vertebrates, deamination occurs in the ◊liver.

death cessation of all life functions, so that the molecules and structures associated with living things become disorganized and indistinguishable from similar molecules found in nonliving things. In medicine, a person is pronounced dead when the brain ceases to control the vital functions, even if breathing and heartbeat are maintained artificially.

debt-for-nature swap agreement under which a proportion of a country's debts are written off in exchange for a commitment by the debtor country to undertake projects for environmental protection. Debt-for-nature swaps were set up by environment groups in the 1980s in an attempt to reduce the debt problem of poor countries, while simultaneously promoting conservation.

Most debt-for-nature swaps have concentrated on setting aside areas of land, especially tropical rainforest, for protection and have involved private conservation foundations. The first swap took place in 1987, when a US conservation group bought $650,000 of Bolivia's national debt from a bank for $100,000, and persuaded the Bolivian government to set aside a large area of rainforest as a nature reserve in exchange for never having to pay back the money owed. Other countries participating in debt-for-nature swaps are the Philippines, Costa Rica, Ecuador, and Poland. However, the debtor country is expected to ensure that the area of land remains adequately protected, and in practice this does not always happen. The practice has also produced complaints of neocolonialism.

decagon in geometry, a ten-sided ◊polygon.

decay, radioactive see ◊radioactive decay.

decibel unit (symbol dB) of measure used originally to compare sound intensities and subsequently electrical or electronic power outputs; now also used to compare voltages. An increase of 10 dB is equivalent to a 10-fold increase in intensity or power, and a 20-fold increase in voltage. A whisper has an intensity of 20 dB; 140 dB (a jet aircraft taking off nearby) is the threshold of pain.

Decibel Scale

Decibels	Typical sound
0	threshold of hearing
10	rustle of leaves in gentle breeze
10	quiet whisper
20	average whisper
20–50	quiet conversation
40–45	hotel; theatre (between performances)
50–65	loud conversation
65–70	traffic on busy street
65–90	train
75–80	factory (light/medium work)
90	heavy traffic
90–100	thunder
110–140	jet aircraft at take-off
130	threshold of pain
140–190	space rocket at take-off

The decibel scale is used primarily to compare sound intensities although it can be used to compare voltages.

The difference in decibels between two levels of intensity (or power) L_1 and L_2 is $10 \log_{10}(L_1/L_2)$; a difference of 1 dB thus corresponds to a change of about 25%. For two voltages V_1 and V_2, the difference in decibels is $20 \log_{10}(V_1/V_2)$; 1 dB corresponding in this case to a change of about 12%.

Decibel

The low-frequency call of the humpback whale is the loudest noise made by a living creature. At 190 decibels, it is louder than Concorde taking off, and can be detected up to 800 km/500 mi away.

deciduous of trees and shrubs, that shed their leaves at the end of the growing season or during a dry season to reduce ◊transpiration (the loss of water by evaporation).

Most deciduous trees belong to the ◊angiosperms, plants in which the seeds are enclosed within an ovary, and the term 'deciduous tree' is sometimes used to mean 'angiosperm tree', despite the fact that many angiosperms are evergreen, especially in the tropics, and a few ◊gymnosperms, plants in which the seeds are exposed, are deciduous (for example, larches). The term **broadleaved** is now preferred to 'deciduous' for this reason.

decimal fraction fraction in which the denominator is any higher power of 10. Thus 3/10, 51/100, and 23/1,000 are decimal fractions and are normally expressed as 0.3, 0.51, 0.023. The use of decimals greatly simplifies addition and multiplication of fractions, though not all fractions can be expressed exactly as decimal fractions.

The regular use of the decimal point appears to have been introduced about 1585, but the occasional use of decimal fractions can be traced back as far as the 12th century.

decimal number system or *denary number system* the most commonly used number system, to the base ten. Decimal numbers do not necessarily contain a decimal point; 563, 5.63, and −563 are all decimal numbers. Other systems are mainly used in computing and include the ◊binary number system, ◊octal number system, and ◊hexadecimal number system.

Decimal numbers may be thought of as written under column headings based on the number ten. For example,

the number 2,567 stands for 2 thousands, 5 hundreds, 6 tens, and 7 ones. Large decimal numbers may also be expressed in floating-point notation.

declination in astronomy, the coordinate on the ◊celestial sphere (imaginary sphere surrounding the Earth) that corresponds to latitude on the Earth's surface. Declination runs from 0° at the celestial equator to 90° at the north and south celestial poles.

decoder in computing, an electronic circuit used to select one of several possible data pathways. Decoders are, for example, used to direct data to individual memory locations within a computer's immediate access memory.

decomposer in biology, any organism that breaks down dead matter. Decomposers play a vital role in the ◊ecosystem by freeing important chemical substances, such as nitrogen compounds, locked up in dead organisms or excrement. They feed on some of the released organic matter, but leave the rest to filter back into the soil as dissolved nutrients, or pass in gas form into the atmosphere, for example as nitrogen and carbon dioxide.

The principal decomposers are bacteria and fungi, but earthworms and many other invertebrates are often included in this group. The ◊nitrogen cycle relies on the actions of decomposers.

decomposition process whereby a chemical compound is reduced to its component substances. In biology, it is the destruction of dead organisms either by chemical reduction or by the action of decomposers, such as bacteria and fungi.

decompression sickness illness brought about by a sudden and substantial change in atmospheric pressure. It is caused by a too rapid release of nitrogen that has been dissolved into the bloodstream under pressure; when the nitrogen bubbles it causes the ◊bends. The condition causes breathing difficulties, joint and muscle pain, and cramps, and is experienced mostly by deep-sea divers who surface too quickly.

After a one-hour dive at 30 m/100 ft, 40 minutes of decompression are needed, according to US Navy tables.

decontamination factor in radiological protection, a measure of the effectiveness of a decontamination process. It is the ratio of the original contamination to the remaining radiation after decontamination: 1,000 and above is excellent; 10 and below is poor.

dedicated computer computer built into another device for the purpose of controlling or supplying information to it. Its use has increased dramatically since the advent of the ◊microprocessor: washing machines, digital watches, cars, and video recorders all now have their own processors.

A dedicated system is a general-purpose computer system confined to performing only one function for reasons of efficiency or convenience. A word processor is an example.

deep freezing method of preserving food by lowering its temperature to −18°/0°F or below; see ◊food technology. It stops almost all spoilage processes, although there may be some residual enzyme activity in uncooked vegetables, which is why these are blanched (dipped in hot water to destroy the enzymes) before freezing. Microorganisms cannot grow or divide while frozen, but most remain alive and can resume activity once defrosted.

Commercial freezing is usually done by one of the following methods: blast, the circulation of air at −40°C/ −40°F; contact, in which a refrigerant is circulated through hollow shelves; immersion, for example, fruit in a solution of sugar and glycerol; or cryogenic means, for example, by liquid nitrogen spray.

Accelerated ◊freeze-drying (AFD) involves rapid freezing followed by heat drying in a vacuum, for example, prawns for later rehydration. The product does not have to be stored in frozen conditions.

Deep-Sea Drilling Project research project initiated by the USA in 1968 to sample the rocks of the ocean ◊crust. In 1985 it became known as the ◊Ocean Drilling Program (ODP).

deep-sea trench another term for ◊ocean trench.

defibrillation use of electrical stimulation to restore a chaotic heartbeat to a rhythmical pattern. In fibrillation, which may occur in most kinds of heart disease, the heart muscle contracts irregularly; the heart is no longer working as an efficient pump. Paddles are applied to the chest wall, and one or more electric shocks are delivered to normalize the beat.

Implantable defibrillators are inserted into the chest with leads threading through veins into the right side of the heart in patients suffering with ◊arrhythmia. The first was implanted in 1980 and by 1996 around 50,000– 80,000 had been implanted worldwide.

deforestation destruction of forest for timber, fuel, charcoal burning, and clearing for agriculture and extractive industries, such as mining, without planting new trees to replace those lost (reafforestation) or working on a cycle that allows the natural forest to regenerate. Deforestation causes fertile soil to be blown away or washed into rivers, leading to ◊soil erosion, drought, flooding, and loss of wildlife. It may also increase the carbon dioxide content of the atmosphere and intensify the ◊greenhouse effect, because there are fewer trees absorbing carbon dioxide from the air for photosynthesis.

Many people are concerned about the rate of deforestation as great damage is being done to the habitats of plants and animals. Deforestation ultimately leads to famine, and is thought to be partially responsible for the flooding of lowland areas – for example, in Bangladesh – because trees help to slow down water movement.

degaussing neutralization of the magnetic field around a body by encircling it with a conductor through which a current is maintained. Ships were degaussed in World War II to prevent them from detonating magnetic mines.

degeneration in biology, a change in the structure or chemical composition of a tissue or organ that interferes with its normal functioning. Examples of degeneration include fatty degeneration, fibroid degeneration (cirrhosis), and calcareous degeneration, all of which are part of natural changes that occur in old age.

The causes of degeneration are often unknown. Heredity often has a role in the degeneration of organs; for example, fibroid changes in the kidney can be seen in successive generations. Defective nutrition and continued stress on particular organs can cause degenerative changes. Alcoholism can result in cirrhosis of the liver and tuberculosis causes degeneration of the lungs.

degree in mathematics, a unit (symbol °) of measurement of an angle or arc. A circle or complete rotation is divided into 360°. A degree may be subdivided into 60 minutes (symbol ′), and each minute may be subdivided in turn into 60 seconds (symbol ″).

Temperature is also measured in degrees, which are divided on a decimal scale. See also ◊Celsius, and ◊Fahrenheit.

A degree of latitude is the length along a meridian such that the difference between its north and south ends subtend an angle of 1° at the centre of the Earth. A degree of longitude is the length between two meridians making an angle of 1° at the centre of the Earth.

dehydration in chemistry, the removal of water from a substance to give a product with a new chemical formula; it is not the same as drying.

There are two types of dehydration. For substances such as hydrated copper sulphate ($CuSO_4.5H_2O$) that contain ◊water of crystallization, dehydration means removing this water to leave the anhydrous substance. This may be achieved by heating, and is reversible.

Some substances, such as ethanol, contain the elements of water (hydrogen and oxygen) joined in a different form. **Dehydrating agents** such as concentrated sulphuric acid will remove these elements in the ratio 2:1.

dehydration process to preserve food. Moisture content is reduced to 10–20% in fresh produce, and this provides good protection against moulds. Bacteria are not inhibited by drying, so the quality of raw materials is vital.

The process was developed commercially in France in about 1795 to preserve sliced vegetables, using a hot-air blast. The earliest large-scale application was to starch products such as pasta, but after 1945 it was extended to milk, potato, soups, instant coffee, and prepared baby and pet foods. A major benefit to food manufacturers is reduction of weight and volume of the food products, lowering distribution cost.

Deimos one of the two moons of Mars. It is irregularly shaped, $15 \times 12 \times 11$ km/$9 \times 7.5 \times 7$ mi, orbits at a height of 24,000 km/15,000 mi every 1.26 days, and is not as heavily cratered as the other moon, Phobos. Deimos was discovered in 1877 by US astronomer Asaph Hall (1829– 1907), and is thought to be an asteroid captured by Mars's gravity.

deliquescence phenomenon of a substance absorbing so much moisture from the air that it ultimately dissolves in it to form a solution.

Deliquescent substances make very good drying agents and are used in the bottom chambers of desiccators. Calcium chloride ($CaCl_2$) is one of the commonest.

delirium in medicine, a state of acute confusion in which the subject is incoherent, frenzied, and out of touch with reality. It is often accompanied by delusions or hallucinations.

Delirium may occur in feverish illness, some forms of mental illness, brain disease, and as a result of drug or alcohol intoxication. In chronic alcoholism, attacks of **delirium tremens** (DTs), marked by hallucinations, sweating, trembling, and anxiety, may persist for several days.

delta tract of land at a river's mouth, composed of silt deposited as the water slows on entering the sea. Familiar examples of large deltas are those of the Mississippi, Ganges and Brahmaputra, Rhône, Po, Danube, and Nile;

the shape of the Nile delta is like the Greek letter *delta* Δ, and thus gave rise to the name.

The *arcuate delta* of the Nile is only one form. Others are *birdfoot deltas*, like that of the Mississippi which is a seaward extension of the river's ◊levee system; and *tidal deltas*, like that of the Mekong, in which most of the material is swept to one side by sea currents.

Delta rocket US rocket used to launch many scientific and communications satellites since 1960, based on the Thor ballistic missile. Several increasingly powerful versions produced as satellites became larger and heavier. Solid-fuel boosters were attached to the first stage to increase lifting power.

delta wing aircraft wing shaped like the Greek letter *delta* Δ. Its design enables an aircraft to pass through the ◊sound barrier with little effect. The supersonic airliner ◊Concorde and the US ◊space shuttle have delta wings.

delusion in psychiatry, a false belief that is unshakeably held. Delusions are a prominent feature of schizophrenia and paranoia, but may also occur in other psychiatric states.

dementia mental deterioration as a result of physical changes in the brain. It may be due to degenerative change, circulatory disease, infection, injury, or chronic poisoning. *Senile dementia*, a progressive loss of mental faculties such as memory and orientation, is typically a disease process of old age, and can be accompanied by ◊depression.

Dementia is distinguished from amentia, or severe congenital mental insufficiency.

demodulation in radio, the technique of separating a transmitted audio frequency signal from its modulated radio carrier wave. At the transmitter the audio frequency signal (representing speech or music, for example) may be made to modulate the amplitude (AM broadcasting) or frequency (FM broadcasting) of a continuously transmitted radio-frequency carrier wave. At the receiver, the signal from the aerial is demodulated to extract the required speech or sound component. In early radio systems, this process was called detection. See ◊modulation.

dendrite part of a ◊nerve cell or neuron. The dendrites are slender filaments projecting from the cell body. They receive incoming messages from many other nerve cells and pass them on to the cell body.

If the combined effect of these messages is strong enough, the cell body will send an electrical impulse along the axon (the threadlike extension of a nerve cell). The tip of the axon passes its message to the dendrites of other nerve cells.

dendritic ulcer in medicine, a branching ulcer on the surface of the cornea that is caused by infection with the *Herpes simplex* virus. Such infections are painful and can result in loss of vision if they remain untreated. The infection is treated with eye drops or ointments that contain antiviral agents.

dendrochronology or *tree-ring dating* analysis of the ◊annual rings of trees to date past events by determining the age of timber. Since annual rings are formed by variations in the water-conducting cells produced by the plant during different seasons of the year, they also provide a means of establishing past climatic conditions in a given area.

Samples of wood are obtained by driving a narrow metal tube into a tree to remove a core extending from the bark to the centre. Samples taken from timbers at an archaeological site can be compared with a master core on file for that region or by taking cores from old living trees; the year when they were felled can be determined by locating the point where the rings of the two samples correspond and counting back from the present.

Moisture levels will affect growth, the annual rings being thin in dry years, thick in moist ones, although in Europe ring growth is most affected by temperature change and insect defoliation.

In North America studies are now extremely extensive, covering many wood types, including sequoia, juniper, and sagebrush. Sequences of tree rings extending back over 8,000 years have been obtained for the SW USA and N Mexico by using cores from the bristle-cone pine *Pinus aristata* of the White Mountains, California, which can live for over 4,000 years in that region. The dryness of the SW USA has preserved wood in its archaeological sites. In wet temperate regions, wood is usually absorbed by soil acidity so that this dating technique cannot be used.

Deneb or *Alpha Cygni* brightest star in the constellation ◊Cygnus, and the 19th brightest star in the night sky. It is one of the greatest supergiant stars known, with a true luminosity of about 60,000 times that of the Sun. Deneb is about 1,800 light years from Earth.

The name Deneb is derived from the Arabic word for tail.

dengue tropical viral fever transmitted by mosquitoes and accompanied by joint pains, headache, rash, and glandular swelling. The incubation time is a week and the fever also lasts about a week. A more virulent form, dengue haemorrhagic fever, is thought to be caused by a second infection on top of the first, and causes internal bleeding. It mainly affects children.

In Delhi between Aug–Sept 1996, 230 people died of dengue. A further 5,000 reported to hospital, indicating a much higher actual rate of infection.

denier unit used in measuring the fineness of yarns, equal to the mass in grams of 9,000 metres of yarn. Thus 9,000 metres of 15 denier nylon, used in nylon stockings, weighs 15 g/0.5 oz, and in this case the thickness of thread would be 0.00425 mm/0.0017 in. The term is derived from the French silk industry; the *denier* was an old French silver coin.

denitrification process occurring naturally in soil, where bacteria break down ◊nitrates to give nitrogen gas, which returns to the atmosphere.

denominator in mathematics, the bottom number of a fraction, so called because it names the family of the fraction. The top number, or numerator, specifies how many unit fractions are to be taken.

density measure of the compactness of a substance; it is equal to its mass per unit volume and is measured in kg per cubic metre/lb per cubic foot. Density is a ◊scalar quantity. The average density D of a mass m occupying a volume V is given by the formula:

$$D = m/V$$

Densities of Some Common Substances

Substance	Density in kg m⁻³
Solids	
balsa wood	200
oak	700
butter	900
ice	920
ebony	120
sand (dry)	1,600
concrete	2,400
aluminium	2,700
steel	7,800
copper	8,900
lead	11,300
uranium	19,000
Liquids	
water	1,000
petrol, paraffin	800
olive oil	900
milk	1,030
sea water	1,030
glycerine	1,260
Dead Sea brine	1,800
Gases	
(at standard temperature and pressure of 0°C and 1 atm)	
air	1.30
hydrogen	0.09
helium	0.18
methane	0.72
nitrogen	1.25
oxygen	1.43
carbon dioxide	1.98
propane	2.02
butane (iso)	2.60

◊Relative density is the ratio of the density of a substance to that of water at 4°C.

Density

The least-dense solid material is made from seaweed and is called Seagel. It was first made in 1992 at the US Government Lawrence Livermore National Laboratory in California. Seagel is lighter than air and would float away if it were not weighed down by the air trapped in its pores.

In photography, density refers to the degree of opacity of a negative; in population studies, it is the quantity or number per unit of area; in electricity, current density is the amount of current passing through a cross-sectional area of a conductor (usually given in amperes per sq in or per sq cm).

density wave in astrophysics, a concept proposed to account for the existence of spiral arms in ◊galaxies. In the density wave theory, stars in a spiral galaxy move in elliptical orbits in such a way that they crowd together in waves of temporarily enhanced density that appear as spiral arms. The idea was first proposed by Swedish astronomer Bertil Lindblad in the 1920s and developed by US astronomers C C Lin and Frank Shu in the 1960s.

dental caries in medicine, another name for ◊caries.

dental formula way of showing the number of teeth in an animal's mouth. The dental formula consists of eight numbers separated by a line into two rows. The four above the line represent the teeth in one side of the upper jaw, starting at the front. If this reads 2 1 2 3 (as for humans) it means two incisors, one canine, two premolars, and three molars (see ◊tooth). The numbers below the line represent the lower jaw. The total number of teeth can be calculated by adding up all the numbers and multiplying by two.

dentistry care and treatment of the teeth and gums. **Orthodontics** deals with the straightening of the teeth for aesthetic and clinical reasons, and **periodontics** with care of the supporting tissue (bone and gums).

The bacteria that start the process of dental decay are normal, nonpathogenic members of a large and varied group of microorganisms present in the mouth. They are strains of oral streptococci, and it is only in the presence of sucrose (from refined sugar) in the mouth that they become damaging to the teeth. ◊Fluoride in the water supply has been one attempted solution, and in 1979 a vaccine was developed from a modified form of the bacterium *Streptococcus mutans*.

dentition type and number of teeth in a species. Different kinds of teeth have different functions; a grass-eating animal will have large molars for grinding its food, whereas a meat-eater will need powerful canines for catching and killing its prey. The teeth that are less useful to an animal's lifestyle may be reduced in size or missing altogether. An animal's dentition is represented diagramatically by a ◊dental formula.

Young children have **deciduous dentition**, popularly known as 'milk teeth', the first ones erupting at about six months of age. **Mixed dentition** is present from the ages

herbivore (sheep)

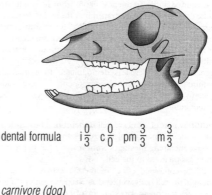

dental formula $i\frac{0}{3}$ $c\frac{0}{0}$ $pm\frac{3}{3}$ $m\frac{3}{3}$

carnivore (dog)

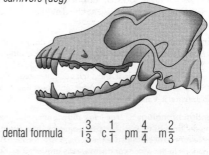

dental formula $i\frac{3}{3}$ $c\frac{1}{1}$ $pm\frac{4}{4}$ $m\frac{2}{3}$

dentition

of about six (when the first milk teeth are shed) to about 12. **Permanent dentition** (up to 32 teeth) is usually complete by the mid-teens, although the third molars (wisdom teeth) may not appear until around the age of 21.

denudation natural loss of soil and rock debris, blown away by wind or washed away by running water, that lays bare the rock below. Over millions of years, denudation causes a general lowering of the landscape.

deoxyribonucleic acid full name of ◊DNA.

depilatory any instrument or substance used to remove growing hair, usually for cosmetic reasons. Permanent eradication is by electrolysis, the destruction of each individual hair root by an electrolytic needle or an electrocautery, but there is a danger of some regrowth as well as scarring.

depolarizer oxidizing agent used in dry-cell batteries that converts hydrogen released at the negative electrode into water. This prevents the build-up of gas, which would otherwise impair the efficiency of the battery. ◊Manganese(IV) oxide is used for this purpose.

depression or *cyclone* or *low* in meteorology, a region of low atmospheric pressure. In mid latitudes a depression forms as warm, moist air from the tropics mixes with cold, dry polar air, producing warm and cold boundaries (◊fronts) and unstable weather – low cloud and drizzle, showers, or fierce storms. The warm air, being less dense, rises above the cold air to produce the area of low pressure on the ground. Air spirals in towards the centre of the depression in an anticlockwise direction in the northern hemisphere, clockwise in the southern hemisphere, generating winds up to gale force. Depressions tend to travel eastwards and can remain active for several days.

depression in medicine, an emotional state characterized by sadness, unhappy thoughts, apathy, and dejection. Sadness is a normal response to major losses such as bereavement or unemployment. After childbirth, ◊postnatal depression is common. However, clinical depression, which is prolonged or unduly severe, often requires treatment, such as antidepressant medication, ◊cognitive therapy, or, in very rare cases, electroconvulsive therapy (ECT), in which an electrical current is passed through the brain.

Periods of depression may alternate with periods of high optimism, over-enthusiasm, and confidence. This is the manic phase in a disorder known as **manic depression** or **bipolar disorder**. A manic depressive state is one in which a person switches repeatedly from one extreme to the other. Each mood can last for weeks or for months. Typically, the depressive state lasts longer than the manic phase.

derivative or **differential coefficient** in mathematics, the limit of the gradient of a chord linking two points on a curve as the distance between the points tends to zero; for a function of a single variable, $y = f'(x)$, it is denoted by $f'(x)$, $Df(x)$, or dy/dx, and is equal to the gradient of the curve.

dermatitis inflammation of the skin (see ◊eczema), usually related to an allergy. **Dermatosis** refers to any skin disorder and may be caused by contact or systemic problems.

dermatology medical speciality concerned with the diagnosis and treatment of skin disorders.

derrick simple lifting machine consisting of a pole carrying a block and tackle. Derricks are commonly used on ships that carry freight. In the oil industry the tower used for hoisting the drill pipes is known as a derrick.

desalination removal of salt, usually from sea water, to produce fresh water for irrigation or drinking. Distillation has usually been the method adopted, but in the 1970s a cheaper process, using certain polymer materials that

a typical depression showing low pressure at the centre

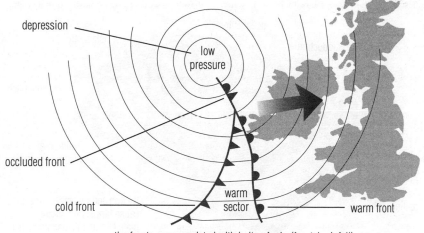

depression

low pressure

occluded front

cold front

warm sector

warm front

the fronts are associated with belts of rain (frontal rainfall)

depression

filter the molecules of salt from the water by reverse osmosis, was developed.

Desalination plants occur along the shores of the Middle East where fresh water is in short supply.

desert arid area without sufficient rainfall and, consequently, vegetation to support human life. The term includes the ice areas of the polar regions (known as cold deserts). Almost 33% of the Earth's land surface is desert, and this proportion is increasing.

The **tropical desert** belts of latitudes from 5° to 30° are caused by the descent of air that is heated over the warm land and therefore has lost its moisture. Other natural desert types are the **continental deserts**, such as the Gobi, that are too far from the sea to receive any moisture; **rain-shadow deserts**, such as California's Death Valley, that lie in the lee of mountain ranges, where the ascending air drops its rain only on the windward slopes; and **coastal deserts**, such as the Namib, where cold ocean currents cause local dry air masses to descend. Desert surfaces are usually rocky or gravelly, with only a small proportion being covered with sand. Deserts can be created by changes in climate, or by the human-aided process of desertification.

Characteristics common to all deserts include irregular rainfall of less than 250 mm per year, very high evaporation rates often 20 times the annual precipitation, and low relative humidity and cloud cover. Temperatures are more variable; tropical deserts have a big diurnal temperature range and very high daytime temperatures (58°C has been recorded at Azizia in Libya), whilst mid-latitude deserts have a wide annual range and much lower winter temperatures (in the Mongolian desert the mean temperature is below freezing point for half the year).

Desert soils are infertile, lacking in ◊humus and generally grey or red in colour. The few plants capable of surviving such conditions are widely spaced, scrubby and often thorny. Long-rooted plants (phreatophytes) such as the date palm and musquite commonly grow along dry stream channels. Salt-loving plants (◊halophytes) such as saltbushes grow in areas of highly saline soils and near the edges of ◊playas (dry saline lakes). Others, such as the ◊xerophytes are drought-resistant and survive by remaining leafless during the dry season or by reducing water

losses with small waxy leaves. They frequently have shallow and widely branching root systems and store water during the wet season (for example, succulents and cacti with pulpy stems).

desertification spread of deserts by changes in climate, or by human-aided processes. Desertification can sometimes be reversed by special planting (marram grass, trees) and by the use of water-absorbent plastic grains, which, added to the soil, enable crops to be grown.

The processes leading to desertification include overgrazing, destruction of forest belts, and exhaustion of the soil by intensive cultivation without restoration of fertility – all of which may be prompted by the pressures of an expanding population or by concentration in land ownership. About 135 million people are directly affected by desertification, mainly in Africa, the Indian subcontinent, and South America.

desktop publishing (DTP) use of microcomputers for small-scale typesetting and page makeup. DTP systems are capable of producing camera-ready pages (pages ready for photographing and printing), made up of text and graphics, with text set in different typefaces and sizes. The page can be previewed on the screen before final printing on a laser printer.

destructive margin or **convergent margin** in plate tectonics, a region on the Earth's crust in which two plates are moving towards one another. Usually one plate (the denser of the two) is forced to dive below the other into what is called the **subduction zone**. The descending plate melts to form a body of magma, which may then rise to the surface through cracks and faults to form volcanoes. If the two plates consist of more buoyant continental crust, subduction does not occur. Instead, the crust crumples gradually to form fold mountains, such as the Himalayas.

detergent surface-active cleansing agent. The common detergents are made from ◊fats (hydrocarbons) and sulphuric acid, and their long-chain molecules have a type of structure similar to that of ◊soap molecules: a salt group at one end attached to a long hydrocarbon 'tail'. They have the advantage over soap in that they do not produce scum

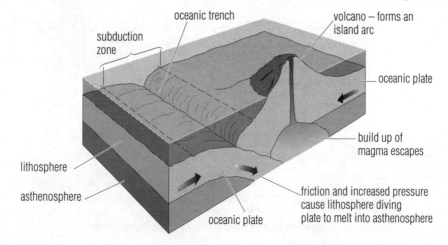

destructive margin

by forming insoluble salts with the calcium and magnesium ions present in hard water.

To remove dirt, which is generally attached to materials by means of oil or grease, the hydrocarbon 'tails' (soluble in oil or grease) penetrate the oil or grease drops, while the 'heads' (soluble in water but insoluble in grease) remain in the water and, being salts, become ionized. Consequently the oil drops become negatively charged and tend to repel one another; thus they remain in suspension and are washed away with the dirt.

Detergents were first developed from coal tar in Germany during World War I, and synthetic organic detergents were increasingly used after World War II.

Domestic powder detergents for use in hot water have alkyl benzene as their main base, and may also include bleaches and fluorescers as whiteners, perborates to free stain-removing oxygen, and water softeners. Environment-friendly detergents contain no phosphates or bleaches. Liquid detergents for washing dishes are based on epoxyethane (ethylene oxide). Cold-water detergents consist of a mixture of various alcohols, plus an ingredient for breaking down the surface tension of the water, so enabling the liquid to penetrate fibres and remove the dirt. When these surface-active agents (surfactants) escape the normal processing of sewage, they cause troublesome foam in rivers; phosphates in some detergents can also cause the excessive enrichment (◊eutrophication) of rivers and lakes.

determinant in mathematics, an array of elements written as a square, and denoted by two vertical lines enclosing the array. For a 2 × 2 matrix, the determinant is given by the difference between the products of the diagonal terms. Determinants are used to solve sets of ◊simultaneous equations by matrix methods.

When applied to transformational geometry, the determinant of a 2 × 2 matrix signifies the ratio of the area of the transformed shape to the original and its sign (plus or minus) denotes whether the image is direct (the same way round) or indirect (a mirror image).

For example, the determinant of the matrix:

$$\begin{pmatrix} a & b \\ c & d \end{pmatrix} = \begin{vmatrix} a & b \\ c & d \end{vmatrix} = ad - bc$$

detonator or **blasting cap** or **percussion cap** small explosive charge used to trigger off a main charge of high explosive. The relatively unstable compounds mercury fulminate and lead azide are often used in detonators, being set off by a lighted fuse or, more commonly, an electric current.

detritus in biology, the organic debris produced during the ◊decomposition of animals and plants.

deuterium naturally occurring heavy isotope of hydrogen, mass number 2 (one proton and one neutron), discovered by Harold Urey 1932. It is sometimes given the symbol D. In nature, about one in every 6,500 hydrogen atoms is deuterium. Combined with oxygen, it produces 'heavy water' (D_2O), used in the nuclear industry.

Deuterium

Deuterium is the most concentrated energy source known. When 1 kg of deuterium is converted to helium, it releases 3 million times more energy than burning the same weight of coal.

---

deuteron nucleus of an atom of deuterium (heavy hydrogen). It consists of one proton and one neutron, and is used in the bombardment of chemical elements to synthesize other elements.

developing in photography, the process that produces a visible image on exposed photographic ◊film, involving the treatment of the exposed film with a chemical developer.

The developing liquid consists of a reducing agent that changes the light-altered silver salts in the film into darker metallic silver. The developed image is made permanent with a fixer, which dissolves away any silver salts which were not affected by light. The developed image is a negative, or reverse image: darkest where the strongest light hit the film, lightest where the least light fell. To produce a positive image, the negative is itself photographed, and the development process reverses the shading, producing the final print. Colour and black-and-white film can be obtained as direct reversal, slide, or transparency material. Slides and transparencies are used for projection or printing with a positive-to-positive process such as Cibachrome.

development in biology, the process whereby a living thing transforms itself from a single cell into a vastly complicated multicellular organism, with structures, such as limbs, and functions, such as respiration, all able to work correctly in relation to each other. Most of the details of this process remain unknown, although some of the central features are becoming understood.

Apart from the sex cells (◊gametes), each cell within an organism contains exactly the same genetic code. Whether a cell develops into a liver cell or a brain cell depends therefore not on which ◊genes it contains, but on which genes are allowed to be expressed. The development of forms and patterns within an organism, and the production of different, highly specialized cells, is a problem of control, with genes being turned on and off according to the stage of development reached by the organism.

developmental psychology study of development of cognition and behaviour from birth to adulthood.

devil wind minor form of ◊tornado, usually occurring in fine weather; formed from rising thermals of warm air (as is a ◊cyclone). A fire creates a similar updraught.

A **fire devil** or **firestorm** may occur in oil-refinery fires, or in the firebombings of cities, for example Dresden, Germany, in World War II.

Devonian period of geological time 408–360 million years ago, the fourth period of the Palaeozoic era. Many desert sandstones from North America and Europe date from this time. The first land plants flourished in the Devonian period, corals were abundant in the seas, amphibians evolved from air-breathing fish, and insects developed on land.

The name comes from the county of Devon in SW England, where Devonian rocks were first studied.

dew precipitation in the form of moisture that collects on the ground. It forms after the temperature of the ground has fallen below the ◊dew point of the air in contact with it. As the temperature falls during the night, the air and its water vapour become chilled, and condensation takes place on the cooled surfaces.

dew point temperature at which the air becomes saturated with water vapour. At temperatures below the dew point, the water vapour condenses out of the air as droplets. If the droplets are large they become deposited on the ground as dew; if small they remain in suspension in the air and form mist or fog.

diabase alternative name for ◊dolerite (a form of basalt that contains very little silica), especially dolerite that has metamorphosed.

diabetes disease *diabetes mellitus* in which a disorder of the islets of Langerhans in the ◊pancreas prevents the body producing the hormone ◊insulin, so that sugars cannot be used properly. Treatment is by strict dietary control and oral or injected insulin, depending on the type of diabetes. There are two forms of diabetes: Type 1, or insulin-dependent diabetes, which usually begins in childhood (early onset) and is an autoimmune condition; and Type 2, or noninsulin-dependent diabetes, which occurs in later life (late onset).

diagenesis in geology, the physical and chemical changes by which a sediment becomes a ◊sedimentary rock. The main processes involved include compaction of the grains, and the cementing of the grains together by the growth of new minerals deposited by percolating groundwater.

dialysis technique for removing waste products from the blood in chronic or acute kidney failure. There are two main methods, haemodialysis and peritoneal dialysis.

In **haemodialysis**, the patient's blood is passed through a pump, where it is separated from sterile dialysis fluid by a semipermeable membrane. This allows any toxic substances which have built up in the bloodstream, and which would normally be filtered out by the kidneys, to diffuse out of the blood into the dialysis fluid. Haemo-

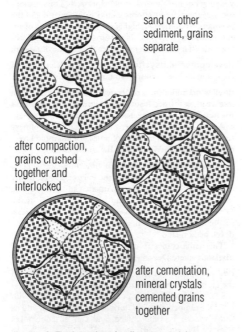

sand or other sediment, grains separate

after compaction, grains crushed together and interlocked

after cementation, mineral crystals cemented grains together

diagenesis The formation of sedimentary rock by diagenesis. Sand and other sediment grains are compacted and cemented together.

dialysis is very expensive and usually requires the patient to attend a specialized unit.

Peritoneal dialysis uses one of the body's natural semipermeable membranes for the same purpose. About two litres of dialysis fluid is slowly instilled into the peritoneal cavity of the abdomen, and drained out again, over about two hours. During that time toxins from the blood diffuse into the peritoneal cavity across the peritoneal membrane. The advantage of peritoneal dialysis is that the patient can remain active while the dialysis is proceeding. This is known as continuous ambulatory peritoneal dialysis (CAPD).

In the long term, dialysis is expensive and debilitating, and ◊transplants are now the treatment of choice for patients in chronic kidney failure.

diameter straight line joining two points on the circumference of a circle that passes through the centre of that circle. It divides a circle into two equal halves.

diamond generally colourless, transparent mineral, an ◊allotrope of carbon. It is regarded as a precious gemstone, and is the hardest substance known (10 on the ◊Mohs' scale). Industrial diamonds, which may be natural or synthetic, are used for cutting, grinding, and polishing.

Diamond crystallizes in the cubic system as octahedral crystals, some with curved faces and striations. The high refractive index of 2.42 and the high dispersion of light, or 'fire', account for the spectral displays seen in polished diamonds.

DIANE (acronym for *direct information access network for Europe*) collection of information suppliers, or 'hosts', for the European computer network.

diapause period of suspended development that occurs in some species of insects, characterized by greatly reduced metabolism. Periods of diapause are often timed to coincide with the winter months, and improve the insect's chances of surviving adverse conditions.

diaphragm in mammals, a thin muscular sheet separating the thorax from the abdomen. It is attached by way of the ribs at either side and the breastbone and backbone, and a central tendon. Arching upwards against the heart and lungs, the diaphragm is important in the mechanics of breathing. It contracts at each inhalation, moving downwards to increase the volume of the chest cavity, and relaxes at exhalation.

diaphragm or *cap* or *Dutch cap* barrier ◊contraceptive that is passed into the vagina to fit over the cervix (neck of the uterus), preventing sperm from entering the uterus. For a cap to be effective, a ◊spermicide must be used and the diaphragm left in place for 6–8 hours after intercourse. This method is 97% effective if practised correctly.

diarrhoea frequent or excessive action of the bowels so that the faeces are liquid or semiliquid. It is caused by intestinal irritants (including some drugs and poisons), infection with harmful organisms (as in dysentery, salmonella, or cholera), or allergies.

Diarrhoea is the biggest killer of children in the world. In 1996 the World Health Organization reported that 3.1 million deaths had been caused by diarrhoeal disease during 1995. The commonest cause of dehydrating diarrhoea is human rotavirus infection, responsible for about 870,000 infant deaths annually. Dehydration as a result of diarrhoeal disease can be treated by giving a solution of salt and glucose by mouth in large quantities (to restore the electrolyte balance in the blood). Since most diarrhoea is viral in origin, antibiotics are ineffective.

diastasis in biology, the separation of the end of a growing bone from the shaft. Although the condition resembles a fracture, it is more serious because the damage to the growing cartilage that occurs during the separation reduces the possibility of future bone growth.

diastole in biology, the relaxation of a hollow organ. In particular, the term is used to indicate the resting period between beats of the heart when blood is flowing into it.

diastolic pressure in medicine, measurement due to the pressure of blood against the arterial wall during diastole (relaxation of the heart). It is the lowest ◊blood pressure during the cardiac cycle. The average diastolic pressure in healthy young adults is about 80 mmHg. The variation of diastolic pressure due to changes in body position and mood is greater than that of systolic pressure. Diastolic pressure is also a more accurate predictor of hypertension (high blood pressure).

diathermy generation of heat in body tissues by the passage of high-frequency electric currents between two electrodes placed on the skin. The heat engendered helps to relieve rheumatic and arthritic pain.

The principle of diathermy is also applied in surgery. The high-frequency current produces, at the tip of a diathermy knife, sufficient heat to cut tissues, or to coagulate and kill cells, with a minimum of bleeding.

diatom microscopic alga found in all parts of the world. Diatoms consist of single cells, but are sometimes grouped in colonies. (Division Bacillariophyta.)

The cell wall of a diatom is made up of two overlapping valves known as *frustules*, which are usually impregnated with silica, and which fit together like the lid and body of a pillbox. Diatomaceous earths (diatomite) are made up of the valves of fossil diatoms, and are used in the manufacture of dynamite and in the rubber and plastics industries.

diatomic molecule molecule composed of two atoms joined together. In the case of an element such as oxygen (O_2), the atoms are identical.

dichloro-diphenyl-trichloroethane full name of the insecticide ◊DDT.

dicotyledon major subdivision of the ◊angiosperms, containing the great majority of flowering plants. Dicotyledons are characterized by the presence of two seed leaves, or ◊cotyledons, in the embryo, which is usually surrounded by an ◊endosperm. They generally have broad leaves with netlike veins.

diecasting form of ◊casting in which molten metal is injected into permanent metal moulds or dies.

dielectric an insulator or nonconductor of electricity, such as rubber, glass, and paraffin wax. An electric field in a dielectric material gives rise to no net flow of electricity. However, the applied field causes electrons within the material to be displaced, creating an electric charge on the surface of the material. This reduces the field strength within the material by a factor known as the dielectric constant (or relative permittivity) of the material. Dielec-

trics are used in capacitors, to reduce dangerously strong electric fields, and have optical applications.

diesel engine ◊internal-combustion engine that burns a lightweight fuel oil. The diesel engine operates by compressing air until it becomes sufficiently hot to ignite the fuel. It is a piston-in-cylinder engine, like the ◊petrol engine, but only air (rather than an air-and-fuel mixture) is taken into the cylinder on the first piston stroke (down). The piston moves up and compresses the air until it is at a very high temperature. The fuel oil is then injected into the hot air, where it burns, driving the piston down on its power stroke. For this reason the engine is called a compression-ignition engine.

Diesel engines have sometimes been marketed as 'cleaner' than petrol engines because they do not need lead additives and produce fewer gaseous pollutants. However, they do produce high levels of the tiny black carbon particles called particulates, which are believed to be carcinogenic and may exacerbate or even cause asthma.

The principle of the diesel engine was first explained in England by Herbert Akroyd (1864–1937) in 1890, and was applied practically by Rudolf Diesel in Germany in 1892.

diesel oil lightweight fuel oil used in diesel engines. Like petrol, it is a petroleum product. When used in vehicle engines, it is also known as *derv* (diesel-engine road vehicle).

diet the range of foods eaten by an animal, also a particular selection of food, or the overall intake and selection of food for a particular person or people. The basic components of any diet are a group of chemicals: proteins, carbohydrates, fats, vitamins, minerals, and water. Different animals require these substances in different proportions, but the necessity of finding and processing an appropriate diet is a very basic drive in animal evolution. For instance, all guts are adapted for digesting and absorbing food, but different guts have adapted to cope with particular diets.

For humans, an adequate diet is one that fulfils the body's nutritional requirements and gives an energy intake proportional to the person's activity level (the average daily requirement is 2,400 calories for men, less for women, more for active children). In the Third World and in famine or poverty areas some 450 million people in the world subsist on fewer than 1,500 calories per day, whereas in the developed countries the average daily intake is 3,300 calories.

Dietary requirements may vary over the lifespan of an organism, according to whether it is growing, reproducing, highly active, or approaching death. For instance, increased carbohydrate for additional energy, or increased minerals, may be necessary during periods of growth.

dietetics specialized branch of human nutrition, dealing with the promotion of health through the proper kinds and quantities of food.

Therapeutic dietetics has a large part to play in the treatment of certain illnesses, such as allergies, arthritis, and diabetes; it is sometimes used alone, but often in conjunction with drugs. The preventative or curative effects of specific diets, such as the 'grape cure' or raw vegetable diets sometimes prescribed for cancer patients, are disputed by orthodox medicine.

Recommended Daily Intake of Nutrients

Age ranges (years)	Energy	kcal	Protein (g)	Calcium (mg)	Iron (mg)	Vitamin A (µg)	Thiamine (retinol equivalent) (mg)	Riboflavin (mg)	Niacin (mg)	Vitamin C (mg)	Vitamin D[1] (µg)
Boys											
<1	3.25	780	19	600	6	450	0.3	0.4	5	20	7.5
1	5.0	1,200	30	600	7	300	0.5	0.6	7	20	10
2	5.75	1,400	35	600	7	300	0.6	0.7	8	20	10
3–4	6.5	1,560	39	600	8	300	0.6	0.8	9	20	10
5–6	7.25	1,740	43	600	10	300	0.7	0.9	10	20	–
7–8	8.25	1,980	49	600	10	400	0.8	1.0	11	20	–
9–11	9.5	2,280	56	700	12	575	0.9	1.2	14	25	–
12–14	11.0	2,640	66	700	12	725	1.1	1.4	16	25	–
15–17	12.0	2,880	72	600	12	750	1.2	1.7	19	30	–
Girls											
<1	3.0	720	18	600	6	450	0.3	0.4	5	20	7.5
1	4.5	1,100	27	600	7	300	0.4	0.6	7	20	10
2	5.5	1,300	32	600	7	300	0.5	0.7	8	20	10
3–4	6.25	1,500	37	600	8	300	0.6	0.8	9	20	10
5–6	7.0	1,680	42	600	10	300	0.7	0.9	10	20	–
7–8	8.0	1,900	48	600	10	400	0.8	1.0	11	20	–
9–11	8.5	2,050	51	700	12[2]	575	0.8	1.2	14	25	–
12–14	9.0	2,150	53	700	12[2]	725	0.9	1.4	16	25	–
15–17	9.0	2,150	53	600	12[2]	750	0.9	1.7	19	30	–
Men											
18–34											
sedentary	10.5	2,510	62	500	10	750	1.0	1.6	18	30	–
moderately active	12.0	2,900	72	500	10	750	1.2	1.6	18	30	–
very active	14.0	3,350	84	500	10	750	1.3	1.6	18	30	–
35–64											
sedentary	10.0	2,400	60	500	10	750	1.0	1.6	18	30	–
moderately active	11.5	2,750	69	500	10	750	1.1	1.6	18	30	–
very active	14.0	3,350	84	500	10	750	1.3	1.6	18	30	–
65–74	10.0	2,400	60	500	10	750	1.0	1.6	18	30	–
≥75	9.0	2,150	54	500	10	750	0.9	1.6	18	30	–
Women											
18–54											
most occupations	9.0	2,150	54	500	12[2]	750	0.9	1.3	15	30	–
very active	10.5	2,500	62	500	12[2]	750	1.0	1.3	15	30	–
55–74	8.0	1,900	47	500	10	750	0.8	1.3	15	30	–
≥75	7.0	1,680	42	500	10	750	0.7	1.3	15	30	–
pregnant	10.0	2,400	60	1,200	13	750	1.0	1.6	18	60	10
lactating	11.5	2,750	69	1,200	15	1,200	1.1	1.8	21	60	10

[1] Most people who go out in the sun need no dietary source of vitamin D, but children and adolescents in winter, and housebound adults, are recommended to take 10 µg vitamin D daily.
[2] These iron recommendations may not cover heavy menstrual losses

difference in mathematics, the result obtained when subtracting one number from another. Also, those elements of one ◊set that are not elements of another.

difference engine mechanical calculating machine designed (and partly built in 1822) by the British mathematician Charles Babbage to produce reliable tables of life expectancy. A precursor of the analytical engine, it was to calculate mathematical functions by solving the differences between values given to ◊variables within equations. Babbage designed the calculator so that once the initial values for the variables were set it would produce the next few thousand values without error.

differential arrangement of gears in the final drive of a vehicle's transmission system that allows the driving wheels to turn at different speeds when cornering. The differential consists of sets of bevel gears and pinions within a cage attached to the crown wheel. When cornering, the bevel pinions rotate to allow the outer wheel to turn faster than the inner.

differential calculus branch of ◊calculus involving applications such as the determination of maximum and minimum points and rates of change.

differentiation in embryology, the process by which cells become increasingly different and specialized, giving rise to more complex structures that have particular functions in the adult organism. For instance, embryonic cells may develop into nerve, muscle, or bone cells.

differentiation in mathematics, a procedure for determining the ◊derivative or gradient of the tangent to a curve f(x) at any point x.

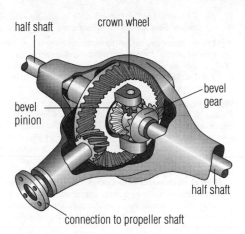

differential The differential lies midway between the driving wheels of a motorcar. When the car is turning, the bevel pinions spin, allowing the outer wheel to turn faster than the inner wheel.

diffraction the slight spreading of a light beam into a pattern of light and dark bands when it passes through a narrow slit or past the edge of an obstruction. A *diffraction grating* is a plate of glass or metal ruled with close, equidistant parallel lines used for separating a wave train such as a beam of incident light into its component frequencies (white light results in a spectrum).

The regular spacing of atoms in crystals are used to diffract X-rays, and in this way the structure of many substances has been elucidated, including recently that of proteins (see ◊X-ray diffraction). Sound waves can also be diffracted by a suitable array of solid objects.

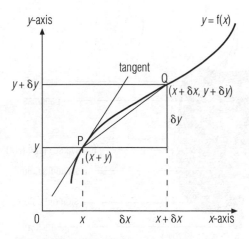

differentiation A mathematical procedure for determining the gradient, or slope, of the tangent to any curve f(x) at any point x. Part of a branch of mathematics called differential calculus, it is used to solve problems involving continuously varying quantities (such as the change in velocity or altitude of a rocket), to find the rates at which these variations occur and to obtain maximum and minimum values for the quantities.

diffusion spontaneous and random movement of molecules or particles in a fluid (gas or liquid) from a region in which they are at a high concentration to a region of lower concentration, until a uniform concentration is achieved throughout. No mechanical mixing or stirring is involved. For instance, if a drop of ink is added to water, its molecules will diffuse until their colour becomes evenly distributed throughout.

In biological systems, diffusion plays an essential role in the transport, over short distances, of molecules such as nutrients, respiratory gases, and neurotransmitters. It provides the means by which small molecules pass into and out of individual cells and microorganisms, such as amoebae, that possess no circulatory system. Diffusion of water across a semi-permeable membrane is termed ◊osmosis.

One application of diffusion is the separation of isotopes, particularly those of uranium. When uranium hexafluoride diffuses through a porous plate, the ratio of the 235 and 238 isotopes is changed slightly. With sufficient number of passages, the separation is nearly complete. There are large plants in the USA and UK for obtaining enriched fuel for fast nuclear reactors and the fissile uranium-235, originally required for the first atom bombs. Another application is the diffusion pump, used extensively in vacuum work, in which the gas to be evacuated diffuses into a chamber from which it is carried away by the vapour of a suitable medium, usually oil or mercury.

Laws of diffusion were formulated by Thomas Gra-

sugar and water molecules become evenly mixed

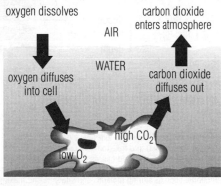

diffusion

ham in 1829 (for gases) and Adolph Fick between 1829–1901 (for solutions).

digestion process whereby food eaten by an animal is broken down mechanically, and chemically by ◊enzymes, mostly in the ◊stomach and ◊intestines, to make the nutrients available for absorption and cell metabolism.

In some single-celled organisms, such as amoebae, a food particle is engulfed by the cell and digested in a ◊vacuole within the cell.

digestive system mouth, stomach, intestine, and associated glands of animals, which are responsible for digesting food. The food is broken down by physical and chemical means in the ◊stomach; digestion is completed, and most nutrients are absorbed in the small intestine; what remains is stored and concentrated into faeces in the large intestine. In birds, additional digestive organs are the ◊crop and ◊gizzard.

In smaller, simpler animals such as jellyfish, the digestive system is simply a cavity (coelenteron or enteric cavity) with a 'mouth' into which food is taken; the digest-

ible portion is dissolved and absorbed in this cavity, and the remains are ejected back through the mouth.

digit in mathematics, any of the numbers from 0 to 9 in the decimal system. Different bases have different ranges of digits. For example, the ◊hexadecimal system has digits 0 to 9 and A to F, whereas the binary system has two digits (or ◊bits), 0 and 1.

digital in electronics and computing, a term meaning 'coded as numbers'. A digital system uses two-state, either on/off or high/low voltage pulses, to encode, receive, and transmit information. A **digital display** shows discrete values as numbers (as opposed to an analogue signal, such as the continuous sweep of a pointer on a dial).

Digital electronics is the technology that underlies digital techniques. Low-power, miniature, integrated circuits (chips) provide the means for the coding, storage, transmission, processing, and reconstruction of information of all kinds.

digital audio tape (DAT) digitally recorded audio tape produced in cassettes that can carry up to two hours of

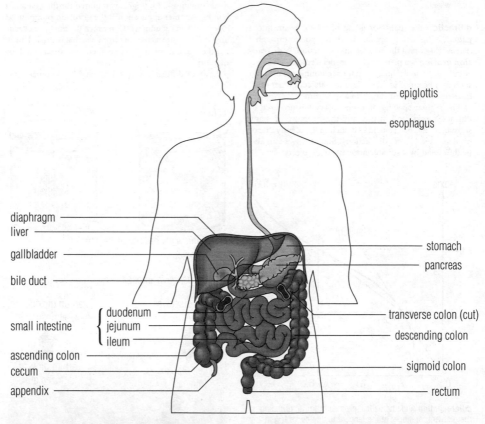

epiglottis

esophagus

diaphragm
liver

gallbladder

bile duct

small intestine { duodenum
                  jejunum
                  ileum

ascending colon
cecum
appendix

stomach

pancreas

transverse colon (cut)

descending colon

sigmoid colon

rectum

digestive system The human digestive system. When food is swallowed, it is moved down the oesophagus by the action of muscles (peristalsis) into the stomach. Digestion starts in the stomach as the food is mixed with enzymes and strong acid. After several hours, the food passes to the small intestine. Here more enzymes are added and digestion is completed. After all nutrients have been absorbed, the indigestible parts pass into the large intestine and thence to the rectum. The liver has many functions, such as storing minerals and vitamins and making bile, which is stored in the gall bladder until needed for the digestion of fats. The pancreas supplies enzymes. The appendix appears to have no function in human beings.

GREAT EXPERIMENTS AND DISCOVERIES

An army marches on its stomach

On 6 June 1822 at Fort Mackinac, Michigan, USA, an 18-year-old French Canadian was accidentally wounded in the abdomen by the discharge of a musket. He was brought to the army surgeon, US physician William Beaumont (1785–1853), who noted several serious wounds and, in particular, a hole in the abdominal wall and stomach. The surgeon observed that through this hole in the patient 'was pouring out the food he had taken for breakfast'.

The patient, Alexis St Martin, a trapper by profession, was serving with the army as a porter and general servant. Not surprisingly, St Martin was at first unable to keep food in his stomach. As the wound gradually healed, firm dressings were needed to retain the stomach contents. Beaumont tended his patient assiduously and tried during the ensuing months to close the hole in his stomach, without success. After 18 months, a small, protruding fleshy fold had grown to fill the aperture (fistula). This 'valve' could be opened simply by pressing it with a finger.

Digestion ... inside and outside

At this point, it occurred to Beaumont that here was an ideal opportunity to study the process of digestion. His patient must have been an extremely tough character to have survived the accident at all. For the next nine years he was the subject of a remarkable series of pioneering experiments, in which Beaumont was able to vary systematically the conditions under which digestion took place and discover the chemical principles involved.

Beaumont attacked the problem of digestion in two ways. He studied how various substances were actually digested in the stomach, and also how they were 'digested' outside the stomach in the digestive juices he extracted from St Martin. He found it was easy enough to drain out the digestive juices from his fortuitously wounded patient 'by placing the subject on his left side, depressing the valve within the aperture, introducing a gum elastic tube and then turning him ... on introducing the tube the fluid soon began to run.'

Beaumont was basically interested in the rate and temperature of digestion, and also the chemical conditions that favoured different stages of the process of digestion. He describes a typical experiment (he performed hundreds), where (a) digestion in the stomach is contrasted (b) with artificial digestion in glass containers kept at suitable temperatures, like this:

(a) 'At 9 o'clock he breakfasted on bread, sausage, and coffee, and kept exercising. 11 o'clock, 30 minutes, stomach two-thirds empty, aspects of weather similar, thermometer 298°F, temperature of stomach 1011/28 and 1003/48. The appearance of contraction and dilation and alternative piston motions were distinctly observed at this examination. 12 o'clock, 20 minutes, stomach empty.'

(b) 'February 7. At 8 o'clock, 30 minutes a.m. I put twenty grains of boiled codfish into three drachms of gastric juice and placed them on the bath.'

'At 1 o'clock, 30 minutes, p.m., the fish in the gastric juice on the bath was almost dissolved, four grains only remaining: fluid opaque, white, nearly the colour of milk. 2 o'clock, the fish in the vial all completely dissolved.'

All a matter of chemistry

Beaumont's research showed clearly for the first time just what happened during digestion and that digestion, as a process, could take place independently outside the body. He wrote that gastric juice: 'so far from being inert as water as some authors assert, is the most general solvent in nature of alimentary matter – even the hardest bone cannot withstand its action. It is capable, even out of the stomach, of effecting perfect digestion, with the aid of due and uniform degree of heat (100°Fahrenheit) and gentle agitation ... I am impelled by the weight of evidence ... to conclude that the change effected by it on the aliment, is purely chemical.'

Our modern understanding of the physiology of digestion as a process whereby foods are gradually broken down into their basic components follows logically from his work. An explanation of how the digestive juices flowed in the first place came in 1889, when Russian physiologist Ivan Pavlov (1849–1936) showed that their secretion in the stomach was controlled by the nervous system. By preventing the food eaten by a dog from actually entering the stomach, he found that the secretions of gastric juices began the moment the dog started eating, and continued as long as it did so. Since no food had entered the stomach, the secretions must be mediated by the nervous system.

Later it was found that the further digestion that takes place beyond the stomach was hormonally controlled. But it was Beaumont's careful scientific work, which was published in 1833 with the title *Experiments and Observations on the Gastric Juice and Physiology of Digestion*, that triggered subsequent research in this field.

Julian Rowe

sound on each side and are about half the size of standard cassettes. DAT players/recorders were developed 1987. The first DAT for computer data was introduced 1988.

DAT machines are constructed like video cassette recorders (though they use metal audio tape), with a movable playback head, the tape winding in a spiral around a rotating drum. The tape can also carry additional information; for example, a time code for instant location of any point on the track. The music industry delayed releasing prerecorded DAT cassettes because of fears of bootlegging, but a system has now been internationally agreed whereby it is not possible to make more than one copy of any prerecorded compact disc or DAT. DAT is mainly used in recording studios for making master tapes. The system was developed by Sony.

By 1990, DATs for computer data had been developed to a capacity of around 2.5 gigabytes per tape, achieved by using helical scan recording (in which the tape covers some 90% of the total head area of the rotating drum). This enables data from the tape to be read over 200 times faster than it can be written. Any file can be located within 60 seconds.

digital compact cassette (DCC) digitally recorded audio cassette that is roughly the same size as a standard cassette. It cannot be played on a normal tape recorder, though standard tapes can be played on a DCC machine; this is known as 'backwards compatibility'. The playing time is 90 minutes.

A DCC player has a stationary playback and recording head similar to that in ordinary tape decks, though the tape used is chrome video tape. The cassettes are copy-protected and can be individually programmed for playing order. Some DCC decks have a liquid-crystal digital-display screen, which can show track titles and other information encoded on the tape.

digital computer computing device that operates on a two-state system, using symbols that are internally coded as binary numbers (numbers made up of combinations of the digits 0 and 1); see ◊computer.

digital data transmission in computing, a way of sending data by converting all signals (whether pictures, sounds, or words) into numeric (normally binary) codes before transmission, then reconverting them on receipt. This virtually eliminates any distortion or degradation of the signal during transmission, storage, or processing.

digitalis drug that increases the efficiency of the heart by strengthening its muscle contractions and slowing its rate. It is derived from the leaves of the common European woodland plant *Digitalis purpurea* (foxglove).

It is purified to digoxin, digitoxin, and lanatoside C, which are effective in cardiac regulation but induce the side effects of nausea, vomiting, and pulse irregularities. Pioneered in the late 1700s by William Withering, an English physician and botanist, digitalis was the first cardiac drug.

digital recording technique whereby the pressure of sound waves is sampled more than 30,000 times a second and the values converted by computer into precise numerical values. These are recorded and, during playback, are reconverted to sound waves.

This technique gives very high-quality reproduction. The numerical values converted by computer represent the original sound-wave form exactly and are recorded on

compact disc. When this is played back by ◊laser, the exact values are retrieved.

When the signal is fed via an amplifier to a loudspeaker, sound waves exactly like the original ones are reproduced.

digital sampling electronic process used in ◊telecommunications for transforming a constantly varying (analogue) signal into one composed of discrete units, a digital signal. In the creation of recorded music, sampling enables the composer, producer, or remix engineer to borrow discrete vocal or instrumental parts from other recorded work (it is also possible to sample live sound).

A telephone microphone changes sound waves into an analogue signal that fluctuates up and down like a wave. In the digitizing process the waveform is sampled thousands of times a second and each part of the sampled wave is given a binary code number (made up of combinations of the digits 0 and 1) related to the height of the wave at that point, which is transmitted along the telephone line. Using digital signals, messages can be transmitted quickly, accurately, and economically.

digital-to-analogue converter electronic circuit that converts a digital signal into an ◊analogue (continuously varying) signal. Such a circuit is used to convert the digital output from a computer into the analogue voltage required to produce sound from a conventional loudspeaker.

digitizer in computing, a device that converts an analogue video signal into a digital format so that video images can be input, stored, displayed, and manipulated by a computer. The term is sometimes used to refer to a ◊graphics tablet.

dihybrid inheritance in genetics, a pattern of inheritance observed when two characteristics are studied in succeeding generations.

The first experiments of this type, as well as in ◊monohybrid inheritance, were carried out by Austrian biologist Gregor Mendel using pea plants.

dilatation and curettage (D and C) common gynaecological procedure in which the cervix (neck of the womb) is widened, or dilated, giving access so that the lining of the womb can be scraped away (curettage). It may be carried out to terminate a pregnancy, treat an incomplete miscarriage, discover the cause of heavy menstrual bleeding, or for biopsy.

dilution process of reducing the concentration of a solution by the addition of a solvent.

The extent of a dilution normally indicates the final volume of solution required. A five-fold dilution would mean the addition of sufficient solvent to make the final volume five times the original.

dimension in science, any directly measurable physical quantity such as mass (M), length (L), and time (T), and the derived units obtainable by multiplication or division from such quantities.

For example, acceleration (the rate of change of velocity) has dimensions (LT^{-2}), and is expressed in such units as km s^{-2}. A quantity that is a ratio, such as relative density or humidity, is dimensionless.

In geometry, the dimensions of a figure are the number of measures needed to specify its size. A point is considered to have zero dimension, a line to have one dimension, a plane figure to have two, and a solid body to have three.

dimethyl sulphoxide (DMSO) $(CH_3)_2SO$ colourless liquid used as an industrial solvent and an antifreeze. It is obtained as a by-product of the processing of wood to paper.

dinitrogen oxide alternative name for ◊nitrous oxide, or 'laughing gas', one of the nitrogen oxides.

dinosaur any of a group (sometimes considered as two separate orders) of extinct reptiles living between 205 million and 65 million years ago. Their closest living relations are crocodiles and birds. Many species of dinosaur evolved during the millions of years when they were the dominant land animals. Most were large (up to 27 m/90 ft), but some were as small as chickens. They disappeared 65 million years ago for reasons not fully understood, although many theories exist.

A currently popular theory of dinosaur extinction suggests that the Earth was struck by a giant meteorite or a swarm of comets 65 million years ago and this sent up such a cloud of debris and dust that climates were changed and the dinosaurs could not adapt quickly enough. The evidence for this includes a bed of rock rich in ◊iridium – an element rare on Earth but common in extraterrestrial bodies – dating from the time.

An alternative theory suggests that changes in geography brought about by the movements of continents and variations in sea level led to climate changes and the mixing of populations between previously isolated regions. This resulted in increased competition and the spread of disease.

The term dinosaur was coined 1842 by Richard Owen, although there were findings of dinosaur bones as far back as the 17th century. In 1822 G A Mantell (1790–1852) found teeth of Iguanodon in a quarry in Sussex, UK. The first dinosaur to be described in a scientific journal was in 1824, when William Buckland, professor of geology at Oxford University, published his finding of a 'megalosaurus or great fossil lizard' found at Stonesfield, a village northwest of Oxford, although a megalosaurus bone had been found in 1677.

An almost complete fossil of a dinosaur skeleton was found in 1969 in the Andean foothills, South America; it had been a two-legged carnivore 2 m/6 ft tall and weighed more than 100 kg/220 lb. More than 230 million years old, it is the oldest known dinosaur. In 1982 a number of nests and eggs were found in 'colonies' in Montana, suggesting that some bred together like modern seabirds. In 1987 finds were made in China that may add much to the traditional knowledge of dinosaurs, chiefly gleaned from North American specimens. In 1989 and 1990 an articulated *Tyrannosaurus rex* was unearthed by a palaeontological team in Montana, with a full skull, one of only six known. Short stretches of dinosaur DNA were extracted 1994 from unfossilized bone retrieved from coal deposits approximately 80 million years old.

The discovery of a small dinosaur was announced in China in 1996. Sinosauropteryx is about 120 million years old and was 0.5 m/1.6 ft tall. It had long hindlegs, short forelegs, a long tail, and short feathers, mainly on its neck and shoulders.

US scientists claimed in Feb 1997 that 65 million-year-old remains discovered in the Atlantic Ocean were proof that a massive asteroid impact on Earth killed the dinosaurs. A sea-drilling expedition discovered three samples that have the signature of an asteroid impact. Previous evidence from sediment suggested that the dinosaurs did not become extinct at exactly the same time as an impact occurred. The new evidence appeared to substantiate the theories of geologists such as Walter Alvarez, who championed the theory that the dinosaurs disappeared from fossil history because of such an impact.

diode combination of a cold anode and a heated cathode (or the semiconductor equivalent, which incorporates a p–n junction; see ◊semiconductor diode). Either device allows the passage of direct current in one direction only, and so is commonly used in a ◊rectifier to convert alternating current (AC) to direct current (DC).

dioecious of plants with male and female flowers borne on separate individuals of the same species. Dioecism occurs, for example, in the willows *Salix*. It is a way of avoiding self-fertilization.

dioptre optical unit in which the power of a ◊lens is expressed as the reciprocal of its focal length in metres. The usual convention is that convergent lenses are positive and divergent lenses negative. Short-sighted people need lenses of power about −0.7 dioptre; a typical value for long sight is about +1.5 dioptre.

diorite igneous rock intermediate in composition between mafic (consisting primarily of dark-coloured minerals) and felsic (consisting primarily of light-coloured minerals); the coarse-grained plutonic equivalent of ◊andesite.

dioxin any of a family of over 200 organic chemicals, all of which are heterocyclic hydrocarbons (see ◊cyclic compounds).

The term is commonly applied, however, to only one member of the family, 2,3,7,8-tetrachlorodibenzo-*p*-dioxin (2,3,7,8-TCDD), a highly toxic chemical that occurs, for example, as an impurity in the defoliant Agent Orange, used in the Vietnam War, and sometimes in the weedkiller 2,4,5-T. It has been associated with a disfiguring skin complaint (chloracne), birth defects, miscarriages, and cancer.

Disasters involving accidental release of large amounts of dioxin into the environment have occurred at Seveso, Italy, and Times Beach, Missouri, USA. Small amounts of dioxins are released by the burning of a wide range of chlorinated materials (treated wood, exhaust fumes from fuels treated with chlorinated additives, and plastics) and as a side-effect of some techniques of paper-making. The possibility of food becoming contaminated by dioxins in the environment has led the European Community to decrease significantly the allowed levels of dioxin emissions from incinerators. Dioxin may be produced as a by-product in the manufacture of the bactericide ◊hexachlorophene.

diphtheria acute infectious disease in which a membrane forms in the throat (threatening death by ◊asphyxia), along with the production of a powerful toxin that damages the heart and nerves. The organism responsible is a bacterium (*Corynebacterium diphtheriae*). It is treated with antitoxin and antibiotics.

Although its incidence has been reduced greatly by immunization, an epidemic in the former Soviet Union resulted in 47,802 cases and 1,746 deaths in 1994, and 1,500 deaths in 1995. In 1995 the World Health Organization (WHO) declared the epidemic 'an international public health emergency' after 20 linked cases were identified in other parts of Europe. The epidemic showed signs of abating by 1996, with a 59% decrease in the number of cases for the first three months, compared with the same period in 1995, but the WHO still considers it an international medical emergency.

diploblastic in biology, having a body wall composed of two layers. The outer layer is the **ectoderm**, the inner layer is the **endoderm**. This pattern of development is shown by ◊coelenterates.

diploid having paired ◊chromosomes in each cell. In sexually reproducing species, one set is derived from each parent, the ◊gametes, or sex cells, of each parent being ◊haploid (having only one set of chromosomes) due to ◊meiosis (reduction cell division).

diplopia double vision, usually due to a lack of coordination of the movements of the eyes. It may arise from disorder in, or damage to, the nerve supply or muscles of the eye, or from intoxication.

dip, magnetic angle at a particular point on the Earth's surface between the direction of the Earth's magnetic field and the horizontal. It is measured using a **dip circle**, which has a magnetized needle suspended so that it can turn freely in the vertical plane of the magnetic field. In the northern hemisphere the needle dips below the horizontal, pointing along the line of the magnetic field towards its north pole. At the magnetic north and south poles, the needle dips vertically and the angle of dip is 90°. See also ◊angle of declination.

dipole the uneven distribution of magnetic or electrical characteristics within a molecule or substance so that it behaves as though it possesses two equal but opposite poles or charges, a finite distance apart.

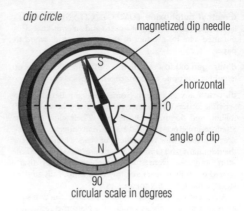

dip circle

magnetized dip needle

S

horizontal

0

angle of dip

N

90

circular scale in degrees

dip, magnetic

The uneven distribution of electrons within a molecule composed of atoms of different ◊electronegativities may result in an apparent concentration of electrons towards one end of the molecule and a deficiency towards the other, so that it forms a dipole consisting of apparently separated but equal positive and negative charges. The product of one charge and the distance between them is the **dipole moment**. A bar magnet behaves as though its magnetism were concentrated in separate north and south magnetic poles because of the uneven distribution of its magnetic field.

dipole in radio, a rod aerial, usually one half-wavelength or a whole wavelength long.

Exploring - Graphic

File Edit View Tools Help

Graphic

All Folders · Contents of 'Graphic'

Name	Size	Type	Modified
1945n014.art	14KB	ART Image	03/10/96 15:11
Ampersan.art	6KB	ART Image	15/08/95 09:56
Art_b.art	26KB	ART Image	15/08/95 10:06
Bluejay.art	4KB	ART Image	15/08/95 10:04
Demo_b.art	10KB	ART Image	15/08/95 11:21
Faq_b.art	12KB	ART Image	15/08/95 11:19
Fractal.art	23KB	ART Image	15/08/95 11:18
Franklin.art	15KB	ART Image	15/08/95 11:36
Happy.art	5KB	ART Image	15/08/95 11:31
Home_b.art	25KB	ART Image	15/08/95 11:35
Info_b.art	12KB	ART Image	15/08/95 11:33
Jg_b.art	16KB	ART Image	15/08/95 11:32
Lc_b.art	18KB	ART Image	15/08/95 11:22
Masks.art	9KB	ART Image	15/08/95 11:30
Modart.art	5KB	ART Image	14/08/95 11:47
New_b.art	19KB	ART Image	15/08/95 11:43
News_b.art	12KB	ART Image	15/08/95 11:42
Oa_b.art	7KB	ART Image	15/08/95 11:44
Prod_b.art	20KB	ART Image	15/08/95 11:53
Pub_b.art	8KB	ART Image	15/08/95 11:52
Recycle.art	3KB	ART Image	15/08/95 11:50
S_arrow2.art	2KB	ART Image	15/08/95 11:49
Scifi.art	22KB	ART Image	15/08/95 11:48
Sites_b.art	12KB	ART Image	15/08/95 11:46

Desktop
My Computer
3½ Floppy (A:)
Local1 (C:)
Local2 (D:)
Audio CD (E:)
Apple volume on 'Hythe' (O:)
CMI.screens
FROM HD
Artpress
AOLtest
Flags
Images
Graphic
Hutch
Photo
mapsart
Test Scripts
COMPUTING
Fractimg
Grinder
Utils
Tiffs
XIPHSIZED
u on 'random' (Q:)
Public on 'Hythe' (S:)

folders

files

24 object(s) 290KB (Disk free space: 849MB)

directory A graphical illustration of the directory filing system on a computer. On the left hand side of the screen are the sub-directories available from the root; on the right are the files contained within the active directory.

dipole, magnetic see ◊magnetic dipole.

direct access or **_random access_** type of ◊file access. A direct-access file contains records that can be accessed by the computer directly because each record has its own address on the storage disc.

direct current (DC) electric current that flows in one direction, and does not reverse its flow as ◊alternating current does. The electricity produced by a battery is direct current.

directed number a number with a positive (+) or negative (−) sign attached, for example +5 or −5. On a graph, a positive sign shows a movement to the right or upwards; a negative sign indicates movement downwards or to the left.

directory in computing, a list of file names, together with information that enables a computer to retrieve those files from ◊backing storage. The computer operating system will usually store and update a directory on the backing storage to which it refers. So, for example, on each ◊disc used by a computer a directory file will be created listing the disc's contents.

dirigible another name for ◊airship.

disability limitation of a person's ability to carry out the activities of daily living, to the extent that he or she may need help in doing so.

disaccharide sugar made up of two monosaccharides or simple sugars. Sucrose ($C_{12}H_{22}O_{11}$) or table sugar, is a disaccharide.

disc or **_disk_** in computing, a common medium for storing large volumes of data (an alternative is ◊magnetic tape). A **_magnetic disc_** is rotated at high speed in a disc-drive unit as a read/write (playback or record) head passes over its surfaces to record or read the magnetic variations that encode the data. Recently, **_optical discs_**, such as ◊CD-ROM (compact-disc read-only memory) and ◊WORM (write once, read many times), have been used to store computer data. Data are recorded on the disc surface as etched microscopic pits and are read by a laser-scanning device. Optical discs have an enormous capacity – about 550 megabytes (million ◊bytes) on a compact disc, and thousands of megabytes on a full-size optical disc.

Magnetic discs come in several forms: **_fixed hard discs_** are built into the disc-drive unit, occasionally stacked on top of one another. A fixed disc cannot be

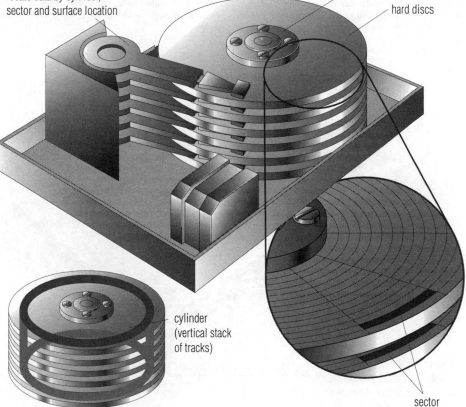

read-write heads locate data by cylinder, sector and surface location

drive spindle

hard discs

cylinder (vertical stack of tracks)

sector

disc A hard disc. Data is stored in sectors within cylinders and is read by a head which passes over the spinning surface of each disc.

removed: once it is full, data must be deleted in order to free space or a complete new disc drive must be added to the computer system in order to increase storage capacity. Large fixed discs, used with mainframe and minicomputers, provide up to 3,000 megabytes. Small fixed discs for use with microcomputers were introduced in the 1980s and typically hold 40–400 megabytes. **Removable hard discs** are common in minicomputer systems. The discs are contained, individually or as stacks (disc packs), in a protective plastic case, and can be taken out of the drive unit and kept for later use. By swapping such discs around, a single hard-disc drive can be made to provide a potentially infinite storage capacity. However, access speeds and capacities tend to be lower that those associated with large fixed hard discs. A **floppy disc** (or diskette) is the most common form of backing store for microcomputers. It is much smaller in size and capacity than a hard disc, normally holding 0.5–2 megabytes of data. The floppy disc is so called because it is manufactured from thin flexible plastic coated with a magnetic material. The earliest form of floppy disc was packaged in a card case and was easily damaged; more recent versions are contained in a smaller, rigid plastic case and are much more robust. All floppy discs can be removed from the drive unit.

disc in astronomy, the flat, roughly circular region of a spiral or lenticular ◊galaxy containing stars, ◊nebulas, and dust clouds orbiting about the nucleus. Discs contain predominantly young stars and regions of star formation.

The disc of our own Galaxy is seen from Earth as the band of the ◊Milky Way.

disc drive mechanical device that reads data from and writes data to a magnetic ◊disc.

disc formatting in computing, preparing a blank magnetic disc in order that data can be stored on it. Data are recorded on a disc's surface on circular tracks, each of which is divided into a number of sectors. In formatting a disc, the computer's operating system adds control information such as track and sector numbers, which enables the data stored to be accessed correctly by the disc-drive unit.

Some floppy discs, called **hard-sectored discs**, are sold already formatted. However, because different makes of computer use different disc formats, discs are also sold unformatted, or **soft-sectored**, and computers are provided with the necessary ◊utility program to format these discs correctly before they are used.

discharge tube device in which a gas conducting an electric current emits visible light. It is usually a glass tube from which virtually all the air has been removed (so that it 'contains' a near vacuum), with electrodes at each end. When a high-voltage current is passed between the electrodes, the few remaining gas atoms in the tube (or some deliberately introduced ones) ionize and emit coloured light as they conduct the current along the tube. The light originates as electrons change energy levels in the ionized atoms.

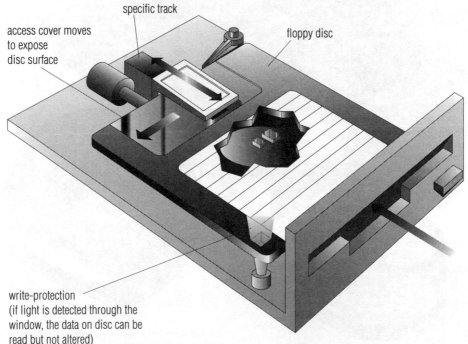

read-write head moves to locate specific track

access cover moves to expose disc surface

floppy disc

write-protection (if light is detected through the window, the data on disc can be read but not altered)

disc drive A floppy disc drive. As the disc is inserted into the drive, its surface is exposed to the read-write head, which moves over the spinning disc surface to locate a specific track.

By coating the inside of the tube with a phosphor, invisible emitted radiation (such as ultraviolet light) can produce visible light; this is the principle of the fluorescent lamp.

discrete data data that can take only whole-number or fractional values. The opposite is ◊continuous data, which can take all in-between values. Examples of discrete data include frequency and population data. However, measurements of time and other dimensions can give rise to continuous data.

disease any condition that impairs the normal state of an organism, and usually alters the functioning of one or more of its organs or systems. A disease is usually characterized by a set of specific symptoms and signs, although these may not always be apparent to the sufferer. Diseases may be inborn (see ◊congenital disease) or acquired through infection, injury, or other cause. Many diseases have unknown causes.

According to the World Health Organization's 1996 World Health Report, 17 million deaths (one-third of the total number) occur as a result of infectious diseases.

disinfectant agent that kills, or prevents the growth of, bacteria and other microorganisms. Chemical disinfectants include carbolic acid (phenol, used by Lister in surgery in the 1870s), ethanol, methanol, chlorine, and iodine.

dispersal phase of reproduction during which gametes, eggs, seeds, or offspring move away from the parents into other areas. The result is that overcrowding is avoided and parents do not find themselves in competition with their own offspring. The mechanisms are various, including a reliance on wind or water currents and, in the case of animals, locomotion. The ability of a species to spread widely through an area and to colonize new habitats has survival value in evolution.

dispersion in physics, the separation of waves of different frequencies by passage through a dispersive medium, in which the speed of the wave depends upon its frequency or wavelength. In optics, the splitting of white light into a spectrum; for example, when it passes through a prism or diffraction grating. It occurs because the prism or grating bends each component wavelength through a slightly different angle. A rainbow is formed when sunlight is dispersed by raindrops.

displacement activity in animal behaviour, an action that is performed out of its normal context, while the animal is in a state of stress, frustration, or uncertainty. Birds, for example, often peck at grass when uncertain whether to attack or flee from an opponent; similarly, humans scratch their heads when nervous.

dissection cutting apart of bodies to study their organization, or tissues to gain access to a site in surgery.

dissociation in chemistry, the process whereby a single compound splits into two or more smaller products, which may be capable of recombining to form the reactant.

Where dissociation is incomplete (not all the compound's molecules dissociate), a ◊chemical equilibrium exists between the compound and its dissociation products. The extent of incomplete dissociation is defined by a numerical value (dissociation constant).

distance modulus in astronomy, a method of finding the distance to an object in the universe, such as a star or ◊galaxy, using the difference between the actual and observed brightness of the object. The actual brightness is deduced from the object's type and its size. The apparent brightness is obtained by direct observation.

distance ratio in a machine, the distance moved by the input force, or effort, divided by the distance moved by the output force, or load. The ratio indicates the movement magnification achieved, and is equivalent to the machine's velocity ratio.

distemper any of several infectious diseases of animals characterized by catarrh, cough, and general weakness. Specifically, it refers to a virus disease in young dogs, also found in wild animals, which can now be prevented by vaccination. In 1988 an allied virus killed over 10,000 common seals in the Baltic and North seas.

distillation technique used to purify liquids or to separate mixtures of liquids possessing different boiling points. *Simple distillation* is used in the purification of liquids (or the separation of substances in solution from their solvents) – for example, in the production of pure water from a salt solution.

The solution is boiled and the vapours of the solvent rise into a separate piece of apparatus (the condenser) where they are cooled and condensed. The liquid produced (the distillate) is the pure solvent; the non-volatile solutes (now in solid form) remain in the distillation vessel to be discarded as impurities or recovered as required. Mixtures of liquids (such as ◊petroleum or aqueous ethanol) are separated by *fractional distillation*, or fractionation. When the mixture is boiled, the vapours of its most volatile component rise into a vertical ◊fractionating column where they condense to liquid form. However, as this liquid runs back down the column it is reheated to boiling point by the hot rising vapours of the next-most-volatile component and so its vapours ascend the column once more. This boiling-condensing process occurs repeatedly inside the column, eventually bringing about a temperature gradient along its length. The vapours of the

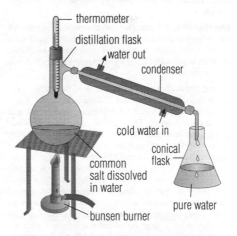

distillation Laboratory apparatus for simple distillation. Other forms of distillation include steam distillation, in which steam is passed into the mixture being distilled, and vacuum distillation, in which air is removed from above the mixture to be distilled.

more volatile components therefore reach the top of the column and enter the condenser for collection before those of the less volatile components. In the fractional distillation of petroleum, groups of compounds (fractions) possessing similar relative molecular masses and boiling points are tapped off at different points on the column. The earliest-known reference to the process is to the distillation of wine in the 12th century by Adelard of Bath. The chemical retort used for distillation was invented by Muslims, and was first seen in the West about 1570.

distributive operation in mathematics, an operation, such as multiplication, that bears a relationship to another operation, such as addition, such that

$$a \times (b + c) = (a \times b) + (a \times c).$$

For example,

$$3 \times (2 + 4) = (3 \times 2) + (3 \times 4) = 18.$$

Multiplication may be said to be distributive over addition. Addition is not, however, distributive over multiplication because $3 + (2 \times 4) = (3 + 2) \times (3 + 4)$.

distributor device in the ignition system of a piston engine that distributes pulses of high-voltage electricity to the ◊spark plugs in the cylinders. The electricity is passed to the plug leads by the tip of a rotor arm, driven by the engine camshaft, and current is fed to the rotor arm from the ignition coil. The distributor also houses the contact point or breaker, which opens and closes to interrupt the battery current to the coil, thus triggering the high-voltage pulses. With electronic ignition it is absent.

diuresis in medicine, increase in the production of urine. It may be due to increased fluid intake, decreased levels of antidiuretic hormone (vasopressin), renal disease, or the use of ◊diuretics. Diuretics are used in the treatment of high blood pressure (hypertension) and heart failure.

diuretic any drug that increases the output of urine by the kidneys. It may be used in the treatment of high blood pressure and to relieve ◊oedema associated with heart, lung, kidney or liver disease, and some endocrine disorders.

diverticulitis inflammation of diverticula (pockets of herniation) in the large intestine. It is usually triggered by infection and causes diarrhoea or constipation, and lower abdominal pain. Usually it can be controlled by diet and antibiotics.

diving apparatus any equipment used to enable a person to spend time underwater. Diving bells were in use in the 18th century, the diver breathing air trapped in a bell-shaped chamber. This was followed by cumbersome diving suits in the early 19th century. Complete freedom of movement came with the ◊aqualung, invented by Jacques Cousteau in the early 1940s. For work at greater depths the technique of saturation diving was developed in the 1970s by which divers live for a week or more breathing a mixture of helium and oxygen at the pressure existing on the seabed where they work (as in work on North Sea platforms and tunnel building). The first diving suit, with a large metal helmet supplied with air through a hose, was invented in the UK by the brothers John and Charles Deane in 1828. Saturation diving was developed for working in offshore oilfields. Working divers are ferried down to the work site by a lock-out ◊submersible. By this technique they avoid the need for

lengthy periods of decompression after every dive. Slow decompression is necessary to avoid the dangerous consequences of an attack of the bends, or ◊decompression sickness.

dizygotic twin in medicine, one of a pair of twins born to the same parents at the same time following the fertilization of two separate ova. They may be of different sexes and they are no more likely to resemble each other than any other sibling pair.

dizziness another word for ◊vertigo.

DNA (abbreviation for *deoxyribonucleic acid*) complex giant molecule that contains, in chemically coded form, the information needed for a cell to make proteins. DNA is a ladderlike double-stranded ◊nucleic acid which forms the basis of genetic inheritance in all organisms, except for a few viruses that have only ◊RNA. DNA is organized into ◊chromosomes and, in organisms other than bacteria, it is found only in the cell nucleus. *See illustration on page 202.*

DNA fingerprinting or *DNA profiling* another name for ◊genetic fingerprinting.

document in computing, data associated with a particular application. For example, a *text document* might be produced by a ◊word processor and a *graphics document* might be produced with a ◊CAD package. An ◊*OMR* or ◊*OCR* document is a paper document containing data that can be directly input to the computer using a ◊document reader.

documentation in computing, the written information associated with a computer program or applications package. Documentation is usually divided into two categories: program documentation and user documentation.

Program documentation is the complete technical description of a program, drawn up as the software is written and intended to support any later maintenance or development of that program. It typically includes details of when, where, and by whom the software was written; a general description of the purpose of the software, including recommended input, output, and storage methods; a detailed description of the way the software functions, including full program listings and ◊flow charts; and details of software testing, including sets of ◊test data with expected results.

User documentation explains how to operate the software. It typically includes a non-technical explanation of the purpose of the software; instructions for loading, running, and using the software; instructions for preparing any necessary input data; instructions for requesting and interpreting output data; and explanations of any error messages that the program may produce.

document reader in computing, an input device that reads marks or characters, usually on preprepared forms and documents. Such devices are used to capture data by ◊optical mark recognition (OMR), ◊optical character recognition (OCR), and ◊mark sensing.

dodecahedron regular solid with 12 pentagonal faces and 12 vertices. It is one of the five regular ◊polyhedra, or Platonic solids. *See illustration on page 202.*

doldrums area of low atmospheric pressure along the Equator, in the ◊intertropical convergence zone where the NE and SE trade winds converge. The doldrums are characterized by calm or very light winds, during which

GREAT EXPERIMENTS AND DISCOVERIES

Too pretty not to be true!

The first announcement
'We wish to suggest a structure for the salt of deoxyribose nucleic acid (DNA). This structure has novel features which are of considerable biological interest.'

So began a 900-word article that was published in the journal *Nature* in April 1953. Its authors were British molecular biologist Francis Crick and US biochemist James Watson. The article described the correct structure of DNA, a discovery that many scientists have called the most important since Austrian botanist and monk Gregor Mendel (1822–1884) laid the foundations of the science of genetics. DNA is the molecule of heredity, and by knowing its structure, scientists can see exactly how forms of life are transmitted from one generation to the next.

The problem of inheritance
The story of DNA really begins with British naturalist Charles Darwin (1809–1882). When, in Nov 1859, he published an abstract entitled 'On the Origin of Species by Means of Natural Selection' outlining his theory of evolution, he was unable to explain exactly how inheritance came about. For at that time it was believed that offspring inherited an average of the features of their parents. If this were so, as Darwin's critics pointed out, any remarkable features produced in a living organism by evolutionary processes would, in the natural course of events, soon disappear.

The work of Gregor Mendel, only rediscovered 20 years after Darwin's death, provided a clear demonstration that inheritance was not a 'blending' process at all. His description of the mathematical basis to genetics followed years of careful plant-breeding experiments. He concluded that each of the features he studied, such as colour or stem length, was determined by two 'factors' of inheritance, one coming from each parent. Each egg or sperm cell contained only one factor of each pair. In this way a particular factor, say for the colour red, would be preserved through subsequent generations.

Today, we call Mendel's factors *genes*. Through the work of many scientists, it came to be realized that genes are part of the chromosomes located in the nucleus of living cells and that DNA, rather than protein as was first thought, was a hereditary material.

The double helix
In the early 1960s, scientists realized that X-ray crystallography, a method of using X-rays to obtain an exact picture of the atoms in a molecule, could be successfully applied to the large and complex molecules found in living cells.

It had been known since 1946 that genes consist of DNA. At King's College, London, New Zealand–British biophysicist Maurice Wilkins had been using X-ray crystallography to examine the structure of DNA, together with his colleague, British X-ray crystallographer Rosalind Franklin (1920–1958), and had made considerable progress.

While in Copenhagen, James Watson had realized that one of the major unresolved problems of biology was the precise structure of DNA. In 1952, he came as a young postdoctoral student to join the Medical Research Council Unit at the Cavendish Laboratory, Cambridge, where Francis Crick was already working. Convinced that a gene must be some kind of molecule, the two scientists set to work on DNA.

Helped by the work of Wilkins, they were able to build an accurate model of DNA. They showed that DNA had a double helical structure, rather like a spiral staircase. Because the molecule of DNA was made from two strands, they envisaged that as a cell divides, the strands unravel, and each could serve as a template as new DNA was formed in the resulting daughter cells. Their model also explained how genetic information might be coded in the sequence of the simpler molecules of which DNA is comprised. Here for the first time was a complete insight into the basis of heredity. James Watson commented that this result was 'too pretty not to be true!'

Cracking the code
Later, working with South African–British molecular biologist Sidney Brenner (1927–), Crick went on to work out the genetic code, and so ascribe a precise function to each specific region of the molecule of DNA. These triumphant results created a tremendous flurry of scientific activity around the world. The pioneering work of Crick, Wilkins, and Watson was recognized in the award of the Nobel Prize for physiology or medicine in 1962.

The unravelling of the structure of DNA lead to a new scientific discipline, molecular biology, and laid the foundation stones for genetic engineering – a powerful new technique that will revolutionize biology, medicine, and food production through the purposeful adaptation of living organisms.

Julian Rowe

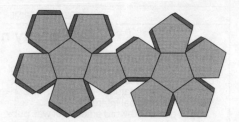

dodecahedron A dodecahedron is a solid figure which has 12 pentagonal faces and 12 vertices. It is one of the five regular polyhedra (with all faces the same size and shape).

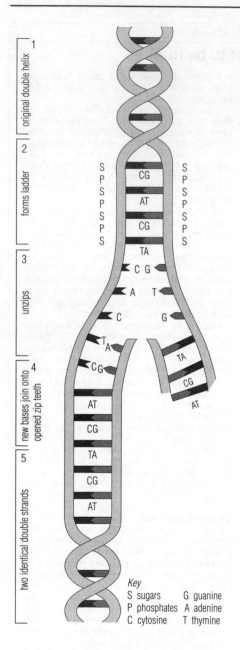

Key
S sugars   G guanine
P phosphates A adenine
C cytosine   T thymine

DNA How the DNA molecule divides. The DNA molecule consists of two strands wrapped around each other in a spiral or helix. The main strands consist of alternate sugar (S) and phosphate (P) groups, and attached to each sugar is a nitrogenous base – adenine (A), cytosine (C), guanine (G), or thymine (T). The sequence of bases carries the genetic code which specifies the characteristics of offspring. The strands are held together by weak bonds between the bases, cytosine to guanine, and adenine to thymine. The weak bonds allow the strands to split apart, allowing new bases to attach, forming another double strand.

there may be sudden squalls and stormy weather. For this reason the areas are avoided as far as possible by sailing ships.

dolerite igneous rock formed below the Earth's surface, a form of basalt, containing relatively little silica (mafic in composition). Dolerite is a medium-grained (hypabyssal) basalt and forms in shallow intrusions, such as dykes, which cut across the rock strata, and sills, which push between beds of sedimentary rock. When exposed at the surface, dolerite weathers into spherical lumps.

dolmen prehistoric ◊megalith in the form of a chamber built of three or more large upright stone slabs, capped by a horizontal flat stone. Dolmens are the burial chambers of Neolithic (New Stone Age) chambered tombs and passage graves, revealed by the removal of the covering burial mound.

dolomite in mineralogy, white mineral with a rhombohedral structure, calcium magnesium carbonate ($CaMg(CO_3)_2$). Dolomites are common in geological successions of all ages and are often formed when ◊limestone is changed by the replacement of the mineral calcite with the mineral dolomite.

domain small area in a magnetic material that behaves like a tiny magnet. The magnetism of the material is due to the movement of electrons in the atoms of the domain. In an unmagnetized sample of material, the domains point in random directions, or form closed loops, so that there is no overall magnetization of the sample. In a magnetized sample, the domains are aligned so that their magnetic effects combine to produce a strong overall magnetism.

dominance in genetics, the masking of one allele (an alternative form of a gene) by another allele. For example, if a ◊heterozygous person has one allele for blue eyes and one for brown eyes, his or her eye colour will be brown. The allele for blue eyes is described as ◊recessive and the allele for brown eyes as dominant.

dopa in medicine, a naturally occurring amino acid from which dopamine, a compound involved in the transmission of nervous impulses in the brain, is made. Dopamine deficiency is one of the causative factors of ◊Parkinson's disease.

dopamine neurotransmitter, hydroxytyramine $C_8H_{11}NO_2$, an intermediate in the formation of adrenaline. There are special nerve cells (neurons) in the brain that use dopamine for the transmission of nervous impulses. One such area of dopamine neurons lies in the basal ganglia, a region that controls movement. Patients suffering from the tremors of Parkinson's disease show nerve degeneration in this region. Another dopamine area lies in the limbic system, a region closely involved with emotional responses. It has been found that schizophrenic patients respond well to drugs that limit dopamine excess in this area.

Doppler effect change in the observed frequency (or wavelength) of waves due to relative motion between the wave source and the observer. The Doppler effect is responsible for the perceived change in pitch of a siren as it approaches and then recedes, and for the ◊red shift of light from distant galaxies. It is named after the Austrian physicist Christian Doppler.

Dorado constellation of the southern hemisphere, represented as a goldfish. It is easy to locate, since the Large ◊Magellanic Cloud marks its southern border. Its brightest star is Alpha Doradus, just under 200 light years from Earth.

One of the most conspicuous objects in the Large Magellanic Cloud is the 'Great Looped Nebula' that surrounds 30 Doradus.

dormancy in botany, a phase of reduced physiological activity exhibited by certain buds, seeds, and spores. Dormancy can help a plant to survive unfavourable conditions, as in annual plants that pass the cold winter season as dormant seeds, and plants that form dormant buds.

For various reasons many seeds exhibit a period of dormancy even when conditions are favourable for growth. Sometimes this dormancy can be broken by artificial methods, such as penetrating the seed coat to facilitate the uptake of water (chitting) or exposing the seed to light. Other seeds require a period of ◊after-ripening.

dorsal in vertebrates, the surface of the animal closest to the backbone. For most vertebrates and invertebrates this is the upper surface, or the surface furthest from the ground. For bipedal primates such as humans, where the dorsal surface faces backwards, then the word is 'back'.

Not all animals can be said to have a dorsal surface, just as many animals cannot be said to have a front; for example, jellyfish, anemones, and sponges.

DOS (acronym for **disc operating system**) computer ◊operating system specifically designed for use with disc storage; also used as an alternative name for a particular operating system, ◊MS-DOS.

dot matrix printer computer printer that produces each character individually by printing a pattern, or matrix, of very small dots. The printing head consists of a vertical line or block of either 9 or 24 printing pins. As the printing head is moved from side to side across the paper, the pins are pushed forwards selectively to strike an inked ribbon and build up the dot pattern for each character on the paper beneath.

A dot matrix printer is more flexible than a ◊daisywheel printer because it can print graphics and text in many different typefaces. It is cheaper to buy and maintain than a ◊laser printer or ◊ink-jet printer, and, because its pins physically strike the paper, is capable of producing carbon copies. However, it is noisy in operation and cannot produce the high-quality printing associated with the non-impact printers.

double blind trial in medicine, method often used in clinical trials of new medicines. Patients are placed in groups using a randomization code and each group receives a different treatment or a placebo. Neither the investigator nor the patient knows which treatment is being administered during the trial. The randomization code is broken at the end of the trial and the results are analysed statistically. Double blind trials eliminate bias but they must be rigorously controlled using recognized ethical standards.

double bond two covalent bonds between adjacent atoms, as in the ◊alkenes ($-C=C-$) and ◊ketones ($-C=O-$).

double decomposition reaction between two chemical substances (usually ◊salts in solution) that results in the exchange of a constituent from each compound to create two different compounds.

For example, if silver nitrate solution is added to a solution of sodium chloride, there is an exchange of ions yielding sodium nitrate and silver chloride.

double star two stars that appear close together. Many stars that appear single to the naked eye appear double when viewed through a telescope. Some double stars attract each other due to gravity, and orbit each other, forming a genuine ◊binary star, but other double stars are at different distances from Earth, and lie in the same line of sight only by chance. Through a telescope both types look the same.

Double stars of the second kind, which are of little astronomical interest, are referred to as 'optical pairs'; those of the first as 'physical pairs' or, more usually, 'visual binaries'. They are the principal source from which is derived our knowledge of stellar masses.

dough mixture consisting primarily of flour, water, and yeast, which is used in the manufacture of bread.

The preparation of dough involves thorough mixing (kneading) and standing in a warm place to 'prove' (increase in volume) so that the ◊enzymes in the dough can break down the starch from the flour into smaller sugar molecules, which are then fermented by the yeast. This releases carbon dioxide, which causes the dough to rise.

Down's syndrome condition caused by a chromosomal abnormality (the presence of an extra copy of chromosome 21), which in humans produces mental retardation; a flattened face; coarse, straight hair; and a fold of skin at the inner edge of the eye (hence the former name 'mongolism'). The condition can be detected by prenatal testing.

Those afflicted are usually born to mothers over 40 (one in 100), and in 1995 French researchers discovered a link between Down's syndrome incidence and paternal age, with men over 40 having an increased likelihood of fathering a Down's syndrome baby.

The syndrome is named after J L H Down (1828–1896), an English physician who studied it. All people with Down's syndrome who live long enough eventually develop early-onset ◊Alzheimer's disease, a form of dementia. This fact led to the discovery in 1991 that some forms of early-onset Alzheimer's disease are caused by a gene defect on chromosome 21.

dowsing ascertaining the presence of water or minerals beneath the ground with a forked twig or a pendulum. Unconscious muscular action by the dowser is thought to move the twig, usually held with one fork in each hand, possibly in response to a local change in the pattern of electrical forces. The ability has been known since at least the 16th century and, though not widely recognized by science, it has been used commercially and in archaeology.

Draco in astronomy, a large but faint constellation represented as a dragon coiled around the north celestial pole. Due to ◊precession the star Alpha Draconis (Thuban) was the pole star 4,800 years ago.

This star seems to have faded, for it is no longer the brightest star in the constellation as it was at the beginning of the 17th century. Gamma Draconis is more than a magnitude brighter. It was extensively observed by Bradley, who from its apparent changes in position discovered the ◊aberration of starlight and ◊nutation.

drag resistance to motion a body experiences when passing through a fluid, gas or liquid. The aerodynamic drag aircraft experience when travelling through the air represents a great waste of power, so they must be carefully shaped, or streamlined, to reduce drag to a minimum. Cars benefit from ◊streamlining, and aerodynamic drag is used to slow down spacecraft returning from space. Boats travelling through water experience hydrodynamic drag on their hulls, and the fastest vessels are ◊hydrofoils, whose hulls lift out of the water while cruising.

Drag

When spiny lobsters migrate, they do so by walking in queues of up to 65 along the sea-bed, each clinging with its claws to the rear of the one in front. Scientific experiments with dead lobsters, weights, and pulleys have shown that such an arrangement reduces drag and allows a 25% improvement in speed through the water.

DRAM (acronym for *dynamic random-access memory*) computer memory device in the form of a silicon chip commonly used to provide the ◊immediate-access memory of microcomputers. DRAM loses its contents unless they are read and rewritten every 2 milliseconds or so. This process is called **refreshing** the memory. DRAM is slower but cheaper than ◊SRAM, an alternative form of silicon-chip memory.

dream series of events or images perceived through the mind during sleep. Their function is unknown, but Sigmund Freud saw them as wish fulfilment (nightmares being failed dreams prompted by fears of 'repressed' impulses). Dreams occur in periods of rapid eye movement (REM) by the sleeper, when the cortex of the brain is approximately as active as in waking hours. Dreams occupy about a fifth of sleeping time.

If a high level of acetylcholine (chemical responsible for transmission of nerve impulses) is present, dreams occur too early in sleep, causing wakefulness, confusion, and ◊depression, which suggests that a form of memory search is involved. Prevention of dreaming, by taking sleeping pills, for example, has similar unpleasant results. For the purposes of (allegedly) foretelling the future, dreams fell into disrepute in the scientific atmosphere of the 18th century.

drilling common woodworking and metal machinery process that involves boring holes with a drill bit. The commonest kind of drill bit is the fluted drill, which has spiral grooves around it to allow the cut material to escape. In the oil industry, rotary drilling is used to bore oil wells. The drill bit usually consists of a number of toothed cutting wheels, which grind their way through the rock as the drill pipe is turned, and mud is pumped through the pipe to lubricate the bit and flush the ground-up rock to the surface.

In rotary drilling, a drill bit is fixed to the end of a length of drill pipe and rotated by a turning mechanism, the rotary table. More lengths of pipe are added as the hole deepens. The long drill pipes are handled by lifting gear in a steel tower or ◊derrick.

driver in computing, a program that controls a peripheral device. Every device connected to the computer needs a driver program. The driver ensures that communication between the computer and the device is successful.

For example, it is often possible to connect many different types of printer, each with its own special operating codes, to the same type of computer. This is because driver programs are supplied to translate the computer's standard printing commands into the special commands needed for each printer.

drug any of a range of substances, natural or synthetic, administered to humans and animals as therapeutic agents: to diagnose, prevent, or treat disease, or to assist recovery from injury. Traditionally many drugs were obtained from plants or animals; some minerals also had medicinal value. Today, increasing numbers of drugs are synthesized in the laboratory.

Drugs are administered in various ways, including: orally, by injection, as a lotion or ointment, as a ◊pessary, by inhalation, and by transdermal patch.

drug and alcohol dependence physical or psychological craving for addictive drugs such as alcohol, nicotine, narcotics, tranquillizers, or stimulants (for example, amphetamines). Such substances can alter mood or behaviour. When dependence is established, sudden withdrawal from the drug can cause unpleasant physical and/or psychological reactions, which may be dangerous. See also ◊addiction and ◊alcoholism.

drug, generic any drug produced without a brand name that is identical to a branded product. Usually generic drugs are produced when the patent on a branded drug has expired, and are cheaper than their branded equivalents.

drug misuse illegal use of drugs for nontherapeutic purposes. Under the UK Misuse of Drugs regulations drugs used illegally include: narcotics, such as heroin, morphine, and the synthetic opioids; barbiturates; amphetamines and related substances; ◊benzodiazepine tranquillizers; cocaine, LSD, and cannabis.

Designer drugs, for example ecstasy, are usually modifications of the amphetamine molecule, altered in order to evade the law as well as for different effects, and may be many times more powerful and dangerous. Crack, a highly toxic derivative of cocaine, became available to drug users in the 1980s. Some athletes misuse drugs such as ◊ephedrine and ◊anabolic steroids.

Sources of traditional drugs include the 'Golden Triangle' (where Myanmar, Laos, and Thailand meet), Mexico, Colombia, China, and the Middle East.

drupe fleshy ◊fruit containing one or more seeds which are surrounded by a hard, protective layer – for example

Illegal Drugs

	Street names	Description	Method of use	Positive effects	Negative effects	Long-term negative effects
amphetamines	speed, uppers, sulphate, whizz	synthetic drugs that stimulate the nervous system	sniffed, swallowed, injected, smoked	alertness, euphoria	paranoia, panic, anxiety, insomnia	addiction, mental illness, liver damage, depression, weight loss
cannabis	dope, pot, grass, marijuana, hash, ganja, spliff, skunk, puff, blow	dried leaves (marijuana) or resinous substance (hashish) produced from the plant *Cannabis sativa*	smoked, put into food	relaxation	paranoia, anxiety	forgetfulness, confusion
LSD	acid	synthetic hallucinogen derived from a fungus; odourless and colourless	swallowed on small squares of paper	hallucinations, intensified sensory perceptions	anxiety, paranoia, unpleasant hallucinations, impaired judgment	mental illness, depression
magic mushrooms	mush	hallucinogenic mushrooms	swallowed, cooked, brewed	euphoria	visual distortion, nausea, vomiting, stomach pains; risk of picking poisonous mushrooms	
cocaine	coke, charlie, crack, snow, rocks	white stimulant powder derived from the coca tree	sniffed, injected, smoked (crack)	exhilaration, insensitivity to pain and fatigue	restlessness, nausea, insomnia	addiction, tolerance, weight loss, respiratory and nasal damage
heroin	gear, H, smack, junk, skag, China white, Boy	opiate powder derived from morphine	injected, sniffed, smoked	relaxation, drowsiness, contentedness	depression of breathing; risk of overdose	addiction, tolerance, withdrawal symptoms; risk of infection with HIV or hepatitis from injecting
ecstasy	E, XTC, cloves, adam, disco burghers, diamonds, strawberries, apples, rhubarb and custard	tablets of the modified amphetamine MDMA with mild psychedelic effects	swallowed	exhilaration, euphoria	anxiety, panic, depression, visual disturbances, dry mouth, dehydration, kidney failure	brain damage
solvents		aerosols and glues with effects similar to alcohol	sniffed	hallucinations, lightheadedness, giddiness	confusion, headache, nausea, suffocation, impaired judgment, heart failure	addiction, skin irritation, brain and lung damage

cherry, almond, and plum. The wall of the fruit (◊pericarp) is differentiated into the outer skin (exocarp), the fleshy layer of tissues (mesocarp), and the hard layer surrounding the seed (endocarp).

The coconut is a drupe, but here the pericarp becomes dry and fibrous at maturity. Blackberries are an aggregate fruit composed of a cluster of small drupes.

dry-cleaning method of cleaning textiles based on the use of volatile solvents, such as trichloroethene (trichloroethylene), that dissolve grease. No water is used. Dry-cleaning was first developed in France in 1849.

Some solvents are known to damage the ozone layer and one, tetrachloroethene (perchloroethylene), is toxic in water and gives off toxic fumes when heated.

dry ice solid carbon dioxide (CO_2), used as a refrigerant. At temperatures above −79°C/−110.2°F, it sublimes (turns into vapour without passing through a liquid stage) to gaseous carbon dioxide.

DTP abbreviation for ◊*desktop publishing*.

dubnium synthesized, radioactive, metallic element of the ◊transactinide series, symbol Db, atomic number 105, relative atomic mass 261. Six isotopes have been synthesized, each with very short (fractions of a second) half-lives. Two institutions claim to have been the first to produce it: the Joint Institute for Nuclear Research in Dubna, Russia, in 1967; and the University of California at Berkeley, USA, who disputed the Soviet claim, in 1970. Its temporary name was unnilpentium.

ductless gland alternative name for an ◊endocrine gland.

dump in computing, the process of rapidly transferring data to external memory or to a printer. It is usually done to help with debugging (see ◊bug) or as part of an error-recovery procedure designed to provide ◊data security. A

◊screen dump makes a printed copy of the current screen display.

dune mound or ridge of wind-drifted sand common on coasts and in sandy deserts. Loose sand is blown and bounced along by the wind, up the windward side of a dune. The sand particles then fall to rest on the lee side, while more are blown up from the windward side. In this way a dune moves gradually downwind.

In sandy deserts, the typical crescent-shaped dune is called a **barchan**. **Seif dunes** are longitudinal and lie parallel to the wind direction, and **star-shaped dunes** are formed by irregular winds.

duodecimal system system of arithmetic notation using 12 as a base, at one time considered superior to the decimal number system in that 12 has more factors (2, 3, 4, 6) than 10 (2, 5).

It is now superseded by the universally accepted decimal system.

duodenum in vertebrates, a short length of ◊alimentary canal found between the stomach and the small intestine. Its role is in digesting carbohydrates, fats, and proteins. The smaller molecules formed are then absorbed, either by the duodenum or the ileum.

Entry of food to the duodenum is controlled by the pyloric sphincter, a muscular ring at the outlet of the stomach. Once food has passed into the duodenum it is mixed with bile released from the gall bladder and with a range of enzymes secreted from the pancreas, a digestive gland near the top of the intestine. The bile neutralizes the acidity of the gastric juices passing out of the stomach and aids fat digestion.

duralumin lightweight aluminium ◊alloy widely used in aircraft construction, containing copper, magnesium, and manganese.

dust bowl area in the Great Plains region of North America (Texas to Kansas) that suffered extensive wind erosion as the result of drought and poor farming practice in once-fertile soil. Much of the topsoil was blown away in the droughts of the 1930s and the 1980s.

Similar dust bowls are being formed in many areas today, noticeably across Africa, because of overcropping and overgrazing.

Dutch cap common name for a barrier method of contraception; see ◊diaphragm.

Dutch elm disease disease of elm trees *Ulmus*, principally Dutch, English, and American elm, caused by the fungus *Certocystis ulmi*. The fungus is usually spread from tree to tree by the elm-bark beetle, which lays its eggs beneath the bark. The disease has no cure, and control methods involve injecting insecticide into the trees annually to prevent infection, or the destruction of all elms in a broad band around an infected area, to keep the beetles out.

The disease was first described in the Netherlands and by the early 1930s had spread across Britain and continental Europe, as well as North America.

dye substance that, applied in solution to fabrics, imparts a colour resistant to washing. **Direct dyes** combine with the material of the fabric, yielding a coloured compound; **indirect dyes** require the presence of another substance (a mordant), with which the fabric must first be treated; **vat dyes** are colourless soluble substances that on exposure to air yield an insoluble coloured compound.

Naturally occurring dyes include indigo, madder (alizarin), logwood, and cochineal, but industrial dyes (introduced in the 19th century) are usually synthetic: acid green was developed 1835 and bright purple 1856.

Industrial dyes include ◊azo dyestuffs, ◊acridine, ◊anthracene, and ◊aniline.

barchans with weak wind

barchans with strong wind

star dunes with irregular winds

seif dunes on bare rock, parallel to wind direction

dune The shape of a dune indicates the prevailing wind pattern. Crescent-shaped dunes form in sandy desert with winds from a constant direction. Seif dunes form on bare rocks, parallel to the wind direction. Irregular star dunes are formed by variable winds.

dye-transfer print in photography, a print made by a relatively permanent colour process that uses red, yellow, and blue separation negatives printed together.

dyke in earth science, a sheet of ◊igneous rock created by the intrusion of magma (molten rock) across layers of pre-existing rock. (By contrast, a sill is intruded *between* layers of rock.) It may form a ridge when exposed on the surface if it is more resistant than the rock into which it intruded. A dyke is also a human-made embankment built along a coastline (for example, in the Netherlands) to prevent the flooding of lowland coastal regions.

dynamics or **kinetics** in mechanics, the mathematical and physical study of the behaviour of bodies under the action of forces that produce changes of motion in them.

dynamite explosive consisting of a mixture of nitroglycerine and diatomaceous earth (diatomite, an absorbent, chalk-like material). It was first devised by Alfred Nobel.

dynamo simple generator or machine for transforming mechanical energy into electrical energy. A dynamo in basic form consists of a powerful field magnet between the poles of which a suitable conductor, usually in the form of a coil (armature), is rotated. The mechanical energy of rotation is thus converted into an electric current in the armature.

Present-day dynamos work on the principles described by English physicist Michael Faraday in 1830, that an ◊electromotive force is developed in a conductor when it is moved in a magnetic field.

dyne c.g.s. unit (symbol dyn) of force. 10^5 dynes make one newton. The dyne is defined as the force that will accelerate a mass of one gram by one centimetre per second per second.

dyscalculia disability demonstrated by a poor aptitude with figures. A similar disability in reading and writing is called ◊dyslexia.

dysentery infection of the large intestine causing abdominal cramps and painful ◊diarrhoea with blood. There are two kinds of dysentery: **amoebic** (caused by a protozoan), common in the tropics, which may lead to liver damage; and **bacterial**, the kind most often seen in the temperate zones.

Both forms are successfully treated with antibacterials and fluids to prevent dehydration.

dyslexia malfunction in the brain's synthesis and interpretation of written information, popularly known as 'word blindness'.

Dyslexia may be described as specific or developmental to distinguish it from reading or writing difficulties which are acquired. It results in poor ability in reading and writing, though the person may excel in other areas, for example, in mathematics. A similar disability with figures is called **dyscalculia**. **Acquired dyslexia** may occur as a result of brain injury or disease.

Dyslexia

Hans Christian Andersen, the Danish writer of fairy tales, was a bad speller and was probably dyslexic.

---

dysmenorrhoea in medicine, pain during menstruation. Dysmenorrhoea is fairly common during adolescents and it may be due to excessive production of ◊prostaglandins which mediate pain. Psychological factors are sometimes important. In older women, inflammation of the womb, ovaries and Fallopian tubes can result in dysmenorrhoea. Symptoms of dysmenorrhoea include pain in the pelvis, groin, thighs, and lower back. Other symptoms are headache, backache, general lassitude, and irritability. Analgesics and rest can relieve the pain in the majority of women.

dyspepsia another word for indigestion.

dysphagia difficulty in swallowing. It may be due to infection, obstruction, spasm in the throat or oesophagus (gullet), or neurological disease or damage.

dyspnoea difficulty in breathing, or shortness of breath disproportionate to effort. It is mostly caused by circulatory or respiratory diseases.

dysprosium silver-white, metallic element of the ◊lanthanide series, symbol Dy, atomic number 66, relative atomic mass 162.50. It is among the most magnetic of all known substances and has a great capacity to absorb neutrons.

It was discovered in 1886 by French chemist Paul Lecoq de Boisbaudran (1838–1912).

cells that stimulate the auditory nerve connected to the brain. There are approximately 30,000 sensory hair cells (**stereocilia**). Exposure to loud noise and the process of ageing damages the stereocilia, with resulting hearing loss. Three fluid-filled canals of the inner ear detect changes of position; this mechanism, with other sensory inputs, is responsible for the sense of balance.

When a loud noise occurs, muscles behind the eardrum contract automatically, suppressing the noise to enhance perception of sound and prevent injury.

earth electrical connection between an appliance and the ground. In the event of a fault in an electrical appliance, for example, involving connection between the live part of the circuit and the outer casing, the current flows to earth, causing no harm to the user.

In most domestic installations, earthing is achieved by a connection to a metal water-supply pipe buried in the ground before it enters the premises.

Earth third planet from the Sun. It is almost spherical, flattened slightly at the poles, and is composed of three concentric layers: the ◊core, the ◊mantle, and the ◊crust. About 70% of the surface (including the north and south polar icecaps) is covered with water. The Earth is surrounded by a life-supporting atmosphere and is the only planet on which life is known to exist.

Mean distance from the Sun 149,500,000 km/ 92,860,000 mi.

Equatorial diameter 12,756 km/7,923 mi.

Circumference 40,070 km/24,900 mi.

E abbreviation for *east*.

ear organ of hearing in animals. It responds to the vibrations that constitute sound, and these are translated into nerve signals and passed to the brain. A mammal's ear consists of three parts: outer ear, middle ear, and inner ear. The *outer ear* is a funnel that collects sound, directing it down a tube to the *ear drum* (tympanic membrane), which separates the outer and *middle ear*s. Sounds vibrate this membrane, the mechanical movement of which is transferred to a smaller membrane leading to the *inner ear* by three small bones, the auditory ossicles. Vibrations of the inner ear membrane move fluid contained in the snail-shaped cochlea, which vibrates hair

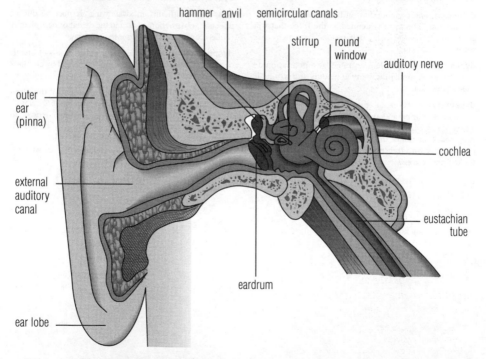

hammer anvil   semicircular canals

stirrup   round window

auditory nerve

outer ear (pinna)

external auditory canal

cochlea

eustachian tube

eardrum

ear lobe

ear The structure of the ear. The three bones of the middle ear – hammer, anvil, and stirrup – vibrate in unison and magnify sounds about 20 times. The spiral-shaped cochlea is the organ of hearing. As sound waves pass down the spiral tube, they vibrate fine hairs lining the tube, which activate the auditory nerve connected to the brain. The semicircular canals are the organs of balance, detecting movements of the head.

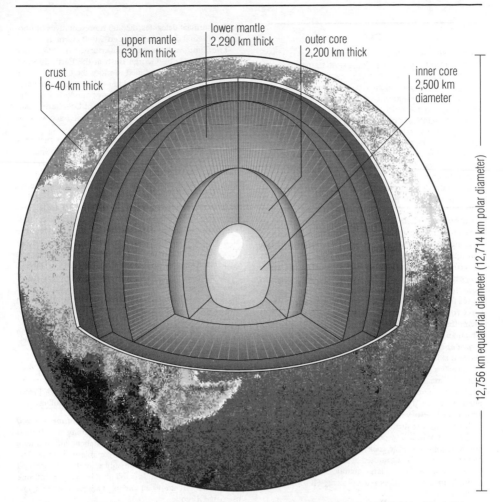

crust
6-40 km thick

upper mantle
630 km thick

lower mantle
2,290 km thick

outer core
2,200 km thick

inner core
2,500 km
diameter

12,756 km equatorial diameter (12,714 km polar diameter)

Earth Inside the Earth. The surface of the Earth is a thin crust about 6 km/4 mi thick under the sea and 40 km/25 mi thick under the continents. Under the crust lies the mantle about 2,900 km/1,800 mi thick and with a temperature of 1,500–3,000°C/2,700–5,400°F. The inner core is probably solid iron and nickel at about 5,000°C/9,000°F.

Rotation period 23 hr 56 min 4.1 sec.

Year (complete orbit, or sidereal period) 365 days 5 hr 48 min 46 sec. Earth's average speed around the Sun is 30 kps/18.5 mps; the plane of its orbit is inclined to its equatorial plane at an angle of 23.5°, the reason for the changing seasons.

> *How inappropriate to call this planet*
> *Earth when quite clearly it is an Ocean.*
> On **Earth** Arthur C Clarke
> *Nature* 1990

Atmosphere nitrogen 78.09%; oxygen 20.95%; argon 0.93%; carbon dioxide 0.03%; and less than 0.0001% neon, helium, krypton, hydrogen, xenon, ozone, radon.

Surface land surface 150,000,000 sq km/ 57,500,000 sq mi (greatest height above sea level 8,872 m/29,118 ft Mount Everest); water surface 361,000,000 sq km/139,400,000 sq mi (greatest depth 11,034 m/36,201 ft ◊Mariana Trench in the Pacific). The interior is thought to be an inner core about 2,600 km/ 1,600 mi in diameter, of solid iron and nickel; an outer core about 2,250 km/1,400 mi thick, of molten iron and nickel; and a mantle of mostly solid rock about 2,900 km/1,800 mi thick, separated by the ◊Mohorovičić discontinuity from the Earth's crust. The crust and the topmost layer of the mantle form about 12 major moving plates, some of which carry the continents. The plates are in constant, slow motion, called tectonic drift. US geophysicists announced in 1996 that they had detected a difference in the spinning time of the Earth's core and the rest of the planet; the core is spinning slightly faster.

Satellite the ◊Moon.

Age 4.6 billion years. The Earth was formed with the rest of the ◊Solar System by consolidation of interstellar dust. Life began 3.5–4 billion years ago.

Earth

If one year were to represent the entire span of life on Earth, then one day would represent 10 million years. On this scale, algae-like organisms appeared in August and the first soft-bodied animals appeared towards the end of September. The first fish developed in the middle of November, along with the first land plants. Amphibians and insects emerged at the end of November, and reptiles appeared in early December. The first dinosaur appeared in the middle of December. A week later, the first birds flew. By the beginning of the last week of the year the dinosaurs had became extinct. The earliest apes appeared halfway through that week. Humans did not come on the scene until the evening of 31 December.

earthquake shaking of the Earth's surface as a result of the sudden release of stresses built up in the Earth's crust. The study of earthquakes is called ◊seismology. Most earthquakes occur along ◊faults (fractures or breaks) in the crust. ◊Plate tectonic movements generate the major proportion: as two plates move past each other they can become jammed and deformed, and a series of shock waves (seismic waves) occur when they spring free. Their force (magnitude) is measured on the ◊Richter scale, and their effect (intensity) on the Mercalli scale. The point at which an earthquake originates is the *seismic focus* or *hypocentre*; the point on the Earth's surface directly above this is the *epicentre*.

In 1987 a California earthquake was successfully predicted by measurement of underground pressure waves; prediction attempts have also involved the study of such phenomena as the change in gases issuing from the ◊crust, the level of water in wells, slight deformation of the rock surface, a sequence of minor tremors, and the behaviour of animals. The possibility of earthquake prevention is remote. However, rock slippage might be slowed at movement points or promoted at stoppage points by the extraction or injection of large quantities of water underground, since water serves as a lubricant. This would ease overall pressure.

earth science scientific study of the planet Earth as a whole, a synthesis of several traditional subjects such as ◊geology, ◊meteorology, ◊oceanography, ◊geophysics, ◊geochemistry, and ◊palaeontology.

The mining and extraction of minerals and gems, the prediction of weather and earthquakes, the pollution of the atmosphere, and the forces that shape the physical world all fall within its scope of study. The emergence of the discipline reflects scientists' concern that an understanding of the global aspects of the Earth's structure and its past will hold the key to how humans affect its future, ensuring that its resources are used in a sustainable way.

Earth Summit (official name *United Nations Conference on Environment and Development*) international meetings aiming at drawing measures towards environmental protection of the world. The first summit took place in Rio de Janeiro, Brazil, in June 1992. Treaties were made to combat global warming and protect wildlife ('biodiversity') (the latter was not signed by the USA). The second Earth Summit took place in New York in June 1997.

The Rio summit, which cost $23 million to stage (of which 60% was spent on security), was attended by 10,000 official delegates, 12,000 representatives of non-governmental organizations, and 7,000 journalists.

The Clinton administration overturned in 1993 certain decisions made by George Bush at the Earth Summit. The USA, which had failed to ratify the Convention of Biological Diversity pact along with other nations in 1994, came under renewed pressure to endorse it in April 1995 after India threatened to prevent US pharmaceutical and cosmetic companies from gaining access to its natural resources.

By 1996 most wealthy nations estimated that they would exceed their emissions targets, including Spain by 24%, Australia by 25%, and the USA by 3%. Britain and Germany were expected to meet their targets.

In 1997, the second UN Earth Summit took place in New York. The summit failed to agree a new deal to address the world's escalating environmental crisis. Dramatic falls in aid to the so-called Third World countries, which the 1992 Earth Summit promised to increase, were at the heart of the breakdown. British Prime Minister Tony Blair condemned the USA, Japan, Canada and Australia for failing to deliver on commitments to stabilise rising emissions of climate-changing greenhouse gases. The European Community as a whole was on target to meet its stabilisation commitment because of cuts in emissions in Germany and the UK.

east one of the four cardinal points of the compass, indicating that part of the horizon where the Sun rises; when facing north, east is to the right.

eau de cologne refreshing toilet water (weaker than perfume), made of alcohol and aromatic oils. Its invention is ascribed to Giovanni Maria Farina (1685–1766), who moved from Italy to Cologne 1709 to manufacture it.

EBCDIC (abbreviation for *extended binary-coded decimal interchange code*) in computing, a code used for storing and communicating alphabetic and numeric characters. It is an 8-bit code, capable of holding 256 different characters, although only 85 of these are defined in the standard version. It is still used in many mainframe computers, but almost all mini-and microcomputers now use ◊ASCII code.

eccentricity in geometry, a property of a ◊conic section (circle, ellipse, parabola, or hyperbola). It is the distance of any point on the curve from a fixed point (the focus) divided by the distance of that point from a fixed line (the directrix). A circle has an eccentricity of zero; for an ellipse it is less than one; for a parabola it is equal to one; and for a hyperbola it is greater than one.

ecdysis periodic shedding of the ◊exoskeleton by insects and other arthropods to allow growth. Prior to shedding, a new soft and expandable layer is first laid down underneath the existing one. The old layer then splits, the animal moves free of it, and the new layer expands and hardens.

ECG abbreviation for ◊*electrocardiogram*.

echinoderm marine invertebrate of the phylum Echinodermata ('spiny-skinned'), characterized by a five-radial symmetry. Echinoderms have a water-vascular system which transports substances around the body. They include starfishes (or sea stars), brittle-stars, sea lilies, sea urchins, and sea cucumbers. The skeleton is external, made of a series of limy plates. Echinoderms generally

Earth Science: Chronology

1735	English lawyer George Hadley described the circulation of the atmosphere as large-scale convection currents centred on the Equator.
1743	Christopher Packe produced the first geological map, of southern England.
1744	The first map produced on modern surveying principles was produced by César-François Cassini in France.
1745	In Russia, Mikhail Vasilievich Lomonosov published a catalogue of over 3,000 minerals.
1746	A French expedition to Lapland proved the Earth to be flattened at the poles.
1760	Lomonosov explained the formation of icebergs. John Mitchell proposed that earthquakes are produced when one layer of rock rubs against another.
1766	The fossilized bones of a huge animal (later called *Mosasaurus*) were found in a quarry near the river Meuse, the Netherlands.
1776	James Keir suggested that some rocks, such as those making up the Giant's Causeway in Ireland, may have formed as molten material that cooled and then crystallized.
1779	French naturalist Comte George de Buffon speculated that the Earth may be much older than the 6,000 years suggested by the Bible.
1785	Scottish geologist James Hutton proposed the theory of uniformitarianism: all geological features are the result of processes that are at work today, acting over long periods of time.
1786	German–Swiss Johann von Carpentier described the European ice age.
1793	Jean Baptiste Lamarck argued that fossils are the remains of once-living animals and plants.
1794	William Smith produced the first large-scale geological maps of England.
1795	In France, Georges Cuvier identified the fossil bones discovered in the Netherlands in 1766 as being those of a reptile, now extinct.
1804	French physicists Jean Biot and Joseph Gay-Lussac studied the atmosphere from a hot-air balloon.
1809	The first geological survey of the eastern USA was produced by William Maclure.
1815	In England, William Smith showed how rock strata (layers) can be identified on the basis of the fossils found in them.
1822	Mary Ann Mantell discovered on the English coast the first fossil to be recognized as that of a dinosaur (an iguanodon). In Germany, Friedrich Mohs introduced a scale for specifying mineral hardness.
1825	Cuvier proposed his theory of catastrophes as the cause of the extinction of large groups of animals.
1830	Scottish geologist Charles Lyell published the first volume of *The Principles of Geology*, which described the Earth as being several hundred million years old.
1839	In the USA, Louis Agassiz described the motion and laying down of glaciers, confirming the reality of the ice ages.
1842	English palaeontologist Richard Owen coined the name 'dinosaur' for the reptiles, now extinct, that lived about 175 million years ago.
1846	Irish physicist William Thomson (Lord Kelvin) estimated, using the temperature of the Earth, that the Earth is 100 million years old.
1850	US naval officer Matthew Fontaine Maury mapped the Atlantic Ocean, noting that it is deeper near its edges than at the centre.
1852	Edward Sabine in Ireland showed a link between sunspot activity and changes in the Earth's magnetic field.
1853	James Coffin described the three major wind bands that girdle each hemisphere.
1854	English astronomer George Airy calculated the mass of the Earth by measuring gravity at the top and bottom of a coal mine.
1859	Edwin Drake drilled the world's first oil well at Titusville, Pennsylvania, USA.
1872	The beginning of the world's first major oceanographic expedition, the four-year voyage of the *Challenger*.
1882	Scottish physicist Balfour Stewart postulated the existence of the ionosphere (the ionized layer of the outer atmosphere) to account for differences in the Earth's magnetic field.
1884	German meteorologist Vladimir Köppen introduced a classification of the world's temperature zones.
1890	English geologist Arthur Holmes used radioactivity to date rocks, establishing the Earth to be 4.6 billion years old.
1895	In the USA, Jeanette Picard launched the first balloon to be used for stratospheric research.
1896	Swedish chemist Svante Arrhenius discovered a link between the amount of carbon dioxide in the atmosphere and the global temperature.
1897	Norwegian-US meteorologist Jacob Bjerknes and his father Vilhelm developed the mathematical theory of weather forecasting.
1902	British physicist Oliver Heaviside and US engineer Arthur Edwin Kennelly predicted the existence of an electrified layer in the atmosphere that reflects radio waves. In France, Léon Teisserenc discovered layers of different temperatures in the atmosphere, which he called the troposphere and stratosphere.
1906	Richard Dixon Oldham proved the Earth to have a molten core by studying seismic waves.
1909	Yugoslav physicist Andrija Mohorovičić discovered a discontinuity in the Earth's crust, about 30 km/18 mi below the surface, that forms the boundary between the crust and the mantle.
1912	In Germany, Alfred Wegener proposed the theory of continental drift and the existence of a supercontinent, Pangaea, in the distant past.
1913	French physicist Charles Fabry discovered the ozone layer in the upper atmosphere.
1914	German-US geologist Beno Gutenberg discovered the discontinuity that marks the boundary between the Earth's mantle and the outer core.

Earth Science: Chronology – *continued*

1922 British meteorologist Lewis Fry Richardson developed a method of numerical weather forecasting.
1925 A German expedition discovered the Mid-Atlantic Ridge by means of sonar. Edward Appleton discovered a layer of the atmosphere that reflects radio waves; it was later named after him.
1929 By studying the magnetism of rocks, Japanese geologist Motonori Matuyama showed that the Earth's magnetic field reverses direction from time to time.
1935 US seismologist Charles Francis Richter established a scale for measuring the magnitude of earthquakes.
1936 Danish seismologist Inge Lehmann postulated the existence of a solid inner core of the Earth from the study of seismic waves.
1939 In Germany, Walter Maurice Elsasser proposed that eddy currents in the molten iron core cause the Earth's magnetism.
1950 Hungarian-US mathematician John Von Neumann made the first 24-hour weather forecast by computer.
1956 US geologists Bruce Charles Heezen and Maurice Ewing discovered a global network of oceanic ridges and rifts that divide the Earth's surface into plates.
1958 Using rockets, US physicist James Van Allen discovered a belt of radiation (later named after him) around the Earth.
1960 The world's first weather satellite, *TIROS 1*, was launched. US geologist Harry Hammond Hess showed that the sea floor spreads out from ocean ridges and descends back into the mantle at deep-sea trenches.
1963 British geophysicists Fred Vine and Drummond Matthews analysed the magnetism of rocks in the Atlantic Ocean floor and found conclusive proof of seafloor spreading.
1985 A British expedition to the Antarctic discovered a hole in the ozone layer above the South Pole.
1991 A borehole in the Kola Peninsula in Arctic Russia, begun in the 1970s, reached a depth of 12,261 m/ 40,240 ft (where the temperature was found to be 210°C/410°F).
1996 US geophysicists detected a difference between the spinning time of the core and that of the rest of the Earth.

move by using tube-feet, small water-filled sacs that can be protruded or pulled back to the body.

echo repetition of a sound wave, or of a ◊radar or ◊sonar signal, by reflection from a surface. By accurately measuring the time taken for an echo to return to the transmitter, and by knowing the speed of a radar signal (the speed of light) or a sonar signal (the speed of sound in water), it is possible to calculate the range of the object causing the echo (◊echolocation).

echolocation or *biosonar* method used by certain animals, notably bats, whales, and dolphins, to detect the positions of objects by using sound. The animal emits a stream of high-pitched sounds, generally at ultrasonic frequencies (beyond the range of human hearing), and listens for the returning echoes reflected off objects to determine their exact location.

The location of an object can be established by the time difference between the emitted sound and its differential return as an echo to the two ears. Echolocation is of particular value under conditions when normal vision is poor (at night in the case of bats, in murky water for dolphins). A few species of bird can also echolocate, including cave-dwelling birds such as some species of swiftlets and the South American Oil Bird.

The frequency range of bats' echolocation calls is 20–215 kHz. Many species produce a specific and identifiable pattern of sound. Bats vary in the way they use echolocation: some emit pure sounds lasting up to 150 milliseconds, while others use a series of shorter 'chirps', and sounds may be emitted through the mouth or nostrils depending on species.

Echolocation was first described in the 1930s, though it was postulated by Italian biologist Lazzaro Spallanzani (1729–1799).

echo sounder or *sonar device* device that detects objects under water by means of ◊sonar – by using

reflected sound waves. Most boats are equipped with echo sounders to measure the water depth beneath them. An echo sounder consists of a transmitter, which emits an ultrasonic pulse, and a receiver, which detects the pulse after reflection from the seabed. The time between transmission and receipt of the reflected signal is a measure of the depth of water. Fishing boats use echo sounders to detect shoals of fish and navies use them to find enemy submarines.

eclampsia convulsions occurring during pregnancy following ◊pre-eclampsia.

eclipse passage of an astronomical body through the shadow of another.

The term is usually employed for solar and lunar eclipses, which may be either partial or total, but also, for example, for eclipses by Jupiter of its satellites. An eclipse of a star by a body in the Solar System is called an occultation.

A *solar eclipse* occurs when the Moon passes in front of the Sun as seen from Earth, and can happen only at new Moon. During a total eclipse the Sun's ◊corona can be seen. A total solar eclipse can last up to 7.5 minutes. When the Moon is at its farthest from Earth it does not completely cover the face of the Sun, leaving a ring of sunlight visible. This is an **annular eclipse** (from the Latin word *annulus* 'ring'). Between two and five solar eclipses occur each year.

A *lunar eclipse* occurs when the Moon passes into the shadow of the Earth, becoming dim until emerging from the shadow. Lunar eclipses may be partial or total, and they can happen only at full Moon. Total lunar eclipses last for up to 100 minutes; the maximum number each year is three.

eclipsing binary binary (double) star in which the two stars periodically pass in front of each other as seen from Earth.

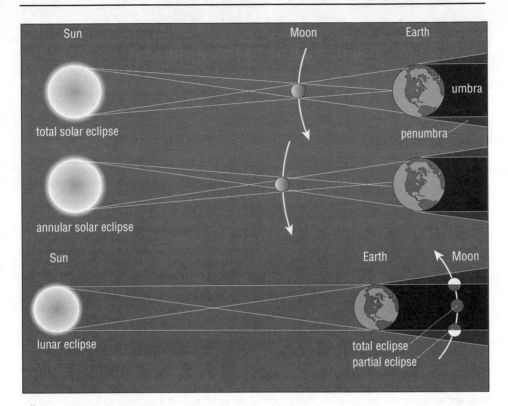

Sun | Moon | Earth
umbra
total solar eclipse | penumbra

annular solar eclipse

Sun | Earth | Moon

lunar eclipse | total eclipse
partial eclipse

eclipse

When one star crosses in front of the other the total light received on Earth from the two stars declines. The first eclipsing binary to be noticed was ◊Algol.

ecliptic path, against the background of stars, that the Sun appears to follow each year as the Earth orbits the Sun. It can be thought of as the plane of the Earth's orbit projected on to the ◊celestial sphere (imaginary sphere around the Earth).

The ecliptic is tilted at about 23.5° with respect to the celestial equator, a result of the tilt of the Earth's axis relative to the plane of its orbit around the Sun.

ecliptic coordinates in astronomy, a system for measuring the position of astronomical objects on the ◊celestial sphere with reference to the plane of the Earth's orbit, the ◊ecliptic.

Ecliptic latitude (symbol β) is measured in degrees from the ecliptic ($\beta = 0°$) to the north ($\beta = 90°$) and south ($\beta = -90°$) ecliptic poles.

Ecliptic longitude (symbol λ) is measured in degrees eastward along the ecliptic ($\lambda = 0°$ to 360°) from a fixed point known as the first point of ◊Aries or the ◊vernal equinox. Ecliptic coordinates are often used to measure the positions of the Sun and planets with respect to the Earth.

Ecliptic latitude and longitude are sometimes known as celestial latitude and longitude or ◊declination and ◊ascension. The ecliptic longitude of the Sun (solar longitude) is a convenient measure of the position of the Earth in its orbit.

E. coli abbreviation for ◊*Escherichia coli*.

ecology study of the relationship among organisms and the environments in which they live, including all living and nonliving components. The chief environmental factors governing the distribution of plants and animals are temperature, humidity, soil, light intensity, daylength, food supply, and interaction with other organisms. The term was coined by the biologist Ernst Haeckel in 1866.

Ecology may be concerned with individual organisms (for example, behavioural ecology, feeding strategies), with populations (for example, population dynamics), or with entire communities (for example, competition between species for access to resources in an ecosystem, or predator–prey relationships). Applied ecology is concerned with the management and conservation of habitats and the consequences and control of pollution.

ecosystem in ecology, an integrated unit consisting of the ◊community of living organisms and the nonliving, or physical, environment in a particular area. The relationships among species in an ecosystem are usually complex and finely balanced, and removal of any one species may be disastrous. The removal of a major predator, for example, can result in the destruction of the ecosystem through overgrazing by herbivores.

Ecosystems can be identified at different scales – for example, the global ecosystem consists of all the organisms living on Earth, the Earth itself (both land and sea), and the atmosphere above; a freshwater-pond ecosystem consists of the plants and animals living in the pond, the

Solar and Lunar Eclipses

Month	Day	Type of eclipse	Time of maximum eclipse	Main area of visibility
1998				
February	26	Sun total	17h 29	southern and eastern USA, Central America, northern South America
August	22	Sun annular	2h 07	Southeast Asia, Oceania, Australasia
1999				
February	16	Sun annular	6h 35	southern Indian Ocean, Antarctica, Australia
August	11	Sun total	11h 04	Europe, North Africa, Arabia, western Asia
2000				
January	21	Moon total	4h 44	the Americas, Europe, Africa, western Asia
February	5	Sun partial	12h 50	Antarctica
July	1	Sun partial	19h 34	southeastern Pacific Ocean
July	16	Moon total	13h 56	southeastern Asia, Australasia
July	31	Sun partial	2h 14	Arctic regions
December	25	Sun partial	17h 36	USA, eastern Canada, Central America, Caribbean
2001				
January	9	Moon total	20h 21	Africa, Europe, Asia
June	21	Sun total	12h 04	central and southern Africa
December	14	Sun annular	20h 52	Pacific Ocean

Table does not include partial eclipses of the Moon.

pond water and all the substances dissolved or suspended in that water, and the rocks, mud, and decaying matter that make up the pond bottom.

Energy and nutrients pass through organisms in an ecosystem in a particular sequence (see ◊food chain): energy is captured through ◊photosynthesis, and nutrients are taken up from the soil or water by plants; both are passed to herbivores that eat the plants and then to carnivores that feed on herbivores. These nutrients are returned to the soil through the ◊decomposition of excrement and dead organisms, thus completing a cycle that is crucial to the stability and survival of the ecosystem.

ecstasy or **MDMA** (3,4-methylenedioxymethamphetamine) illegal drug in increasing use from the 1980s. It is a modified ◊amphetamine with mild psychedelic effects, and works by depleting serotonin (a neurotransmitter) in the brain. Its long-term effects are unknown, but animal experiments have shown brain damage.

Ecstasy was first synthesized in 1914 by the Merck pharmaceutical company in Germany, and was one of eight psychedelics tested by the US army in 1953, but was otherwise largely forgotten until the mid-1970s. It can be synthesized from nutmeg oil. Since 1996 chemical recipes for making the drug have been circulated on the Internet.

ECT abbreviation for ◊*electroconvulsive therapy*.

ectoparasite ◊parasite that lives on the outer surface of its host.

ectopic in medicine, term applied to an anatomical feature that is displaced or found in an abnormal position. An ectopic pregnancy is one occurring outside the womb, usually in a Fallopian tube.

ectoplasm outer layer of a cell's ◊cytoplasm.

ectotherm 'cold-blooded' animal (see ◊poikilothermy), such as a lizard, that relies on external warmth (ultimately from the Sun) to raise its body temperature so that it can become active. To cool the body, ectotherms seek out a cooler environment.

eczema inflammatory skin condition, a form of dermatitis, marked by dryness, rashes, itching, the formation of blisters, and the exudation of fluid. It may be allergic in origin and is sometimes complicated by infection.

eddy current electric current induced, in accordance with ◊Faraday's laws, in a conductor located in a changing magnetic field. Eddy currents can cause much wasted energy in the cores of transformers and other electrical machines.

educational psychology the work of psychologists primarily in schools, including the assessment of children with achievement problems and advising on problem behaviour in the classroom.

EEG abbreviation for ◊*electroencephalogram*.

EEPROM (acronym for **electrically erasable programmable read-only memory**) computer memory that can record data and retain it indefinitely. The data can be erased with an electrical charge and new data recorded.

Some EEPROM must be removed from the computer and erased and reprogrammed using a special device. Other EEPROM, called **flash memory**, can be erased and reprogrammed without removal from the computer.

Effelsberg site, near Bonn, Germany, of the world's largest fully steerable radio telescope, the 100 m/328 ft radio dish of the Max Planck Institute for Radio Astronomy, opened in 1971.

efficiency in physics, a general term indicating the degree to which a process or device can convert energy from one form to another without loss. It is normally expressed as a fraction or percentage, where 100% indicates conversion with no loss. The efficiency of a machine, for example, is the ratio of the work done by the machine to the energy put into the machine; it is always less than 100% because of frictional heat losses. Certain electrical machines with no moving parts, such as transformers, can approach 100% efficiency.

Since the ◊mechanical advantage, or force ratio, is the ratio of the load (the output force) to the effort (the input force), and the velocity ratio is the distance moved by the effort divided by the distance moved by the load, for certain machines the efficiency can also be defined as the mechanical advantage divided by the velocity ratio.

EFTPOS acronym for *electronic funds transfer at point of sale*, a form of electronic funds transfer.

egg in animals, the ovum, or female ◊gamete (reproductive cell).

After fertilization by a sperm cell, it begins to divide to form an embryo. Eggs may be deposited by the female (◊oviparity) or they may develop within her body (◊vivip-

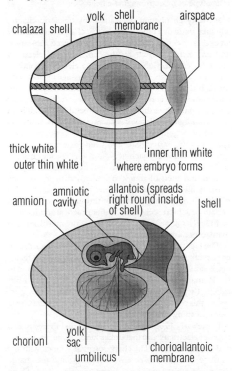

egg Section through a fertilized bird egg. Inside a bird's egg is a complex structure of liquids and membranes designed to meet the needs of the growing embryo. The yolk, which is rich in fat, is gradually absorbed by the embryo. The white of the egg provides protein and water. The chalaza is a twisted band of protein which holds the yolk in place and acts as a shock absorber. The airspace allows gases to be exchanged through the shell. The allantois contains many blood vessels which carry gases between the embryo and the outside.

ary and ◊ovoviviparity). In the oviparous reptiles and birds, the egg is protected by a shell, and well supplied with nutrients in the form of yolk.

Egg

The kiwi produces the largest egg, proportionally to size, of any bird. The female lays one or two eggs, each weighing a quarter of her body weight.

---

ego in psychology, the processes concerned with the self and a person's conception of himself or herself, encompassing values and attitudes. In Freudian psychology, the term refers specifically to the element of the human mind that represents the conscious processes concerned with reality, in conflict with the ◊id (the instinctual element) and the ◊superego (the ethically aware element).

einsteinium synthesized, radioactive, metallic element of the actinide series, symbol Es, atomic number 99, relative atomic mass 254.09.

It was produced by the first thermonuclear explosion, in 1952, and discovered in fallout debris in the form of the isotope Es-253 (half-life 20 days). Its longest-lived isotope, Es-254, with a half-life of 276 days, allowed the element to be studied at length. It is now synthesized by bombarding lower-numbered ◊transuranic elements in particle accelerators. It was first identified by A Ghiorso and his team who named it in 1955 after Albert Einstein, in honour of his theoretical studies of mass and energy.

ejecta in astronomy, any material thrown out of a ◊crater due to volcanic eruption or the impact of a ◊meteorite or other object. Ejecta from impact craters on the ◊Moon often form long bright streaks known as rays, which in some cases can be traced for thousands of kilometres across the lunar surface.

ejector seat device for propelling an aircraft pilot out of the plane to parachute to safety in an emergency, invented by the British engineer James Martin (1893–1981). The first seats of 1945 were powered by a compressed spring; later seats used an explosive charge. By the early 1980s, 35,000 seats had been fitted worldwide, and the lives of 5,000 pilots saved by their use.

Seats that can be ejected on takeoff and landing or at low altitude were a major breakthrough of the 1980s. They are as effective as those originally designed for parachuting from high altitudes.

Ekman spiral effect an application of the ◊Coriolis effect to ocean currents, whereby the currents flow at an angle to the winds that drive them. It derives its name from the Swedish oceanographer Vagn Ekman (1874–1954).

In the northern hemisphere, surface currents are deflected to the right of the wind direction. The surface current then drives the subsurface layer at an angle to its original deflection. Consequent subsurface layers are similarly affected, so that the effect decreases with increasing depth. The result is that most water is transported at about right-angles to the wind direction. Directions are reversed in the southern hemisphere.

elasticity in physics, the ability of a solid to recover its shape once deforming forces (stresses modifying its dimensions or shape) are removed. An elastic material obeys ◊Hooke's law: that is, its deformation is proportional to the applied stress up to a certain point, called the *elastic limit*, beyond which additional stress will deform it permanently. Elastic materials include metals and rubber; however, all materials have some degree of elasticity.

E layer (formerly called the Kennelly–Heaviside layer) the lower regions (90–120 km/56–75 mi) of the ◊ionosphere, which reflect radio waves, allowing their reception around the surface of the Earth. The E layer approaches the Earth by day and recedes from it at night.

electrical relay an electromagnetic switch; see ◊relay.

electric arc a continuous electric discharge of high current between two electrodes, giving out a brilliant light and heat. The phenomenon is exploited in the carbon-arc lamp, once widely used in film projectors. In the electric-arc furnace an arc struck between very large carbon electrodes and the metal charge provides the heating. In arc ◊welding an electric arc provides the heat to fuse the metal. The discharges in low-pressure gases, as in neon and sodium lights, can also be broadly considered as electric arcs.

electric charge property of some bodies that causes them to exert forces on each other. Two bodies both with positive or both with negative charges repel each other, whereas bodies with opposite or 'unlike' charges attract each other, since each is in the ◊electric field of the other. In atoms, ◊electrons possess a negative charge, and ◊protons an equal positive charge. The ◊SI unit of electric charge is the coulomb (symbol C).

A body can be charged by friction, induction, or chemical change and shows itself as an accumulation of electrons (negative charge) or loss of electrons (positive charge) on an atom or body. Atoms have no charge but can sometimes gain electrons to become negative ions or lose them to become positive ions. So-called ◊static electricity, seen in such phenomena as the charging of nylon shirts when they are pulled on or off, or in brushing hair, is in fact the gain or loss of electrons from the surface atoms. A flow of charge (such as electrons through a copper wire) constitutes an electric current; the flow of current is measured in amperes (symbol A).

electric current the flow of electrically charged particles through a conducting circuit due to the presence of a ◊potential difference. The current at any point in a circuit is the amount of charge flowing per second; its SI unit is the ampere (coulomb per second).

Current carries electrical energy from a power supply, such as a battery of electrical cells, to the components of the circuit, where it is converted into other forms of energy, such as heat, light, or motion. It may be either ◊direct current or ◊alternating current.

heating effect When current flows in a component possessing resistance, electrical energy is converted into heat energy. If the resistance of the component is R ohms and the current through it is I amperes, then the heat energy W (in joules) generated in a time t seconds is given by the formula:

$$W = I^2Rt$$

magnetic effect A ◊magnetic field is created around all conductors that carry a current. When a current-bearing conductor is made into a coil it forms an ◊electromagnet with a magnetic field that is similar to that of a bar magnet, but which disappears as soon as the current is

switched off. The strength of the magnetic field is directly proportional to the current in the conductor – a property that allows a small electromagnet to be used to produce a pattern of magnetism on recording tape or disc that accurately represents the sound or data to be stored. The direction of the field created around a conducting wire may be predicted by using ◊Maxwell's screw rule.

motor effect A conductor carrying current in a magnetic field experiences a force, and is impelled to move in a direction perpendicular to both the direction of the current and the direction of the magnetic field. The direction of motion may be predicted by Fleming's left-hand rule (see ◊Fleming's rules). The magnitude of the force experienced depends on the length of the conductor and on the strengths of the current and the magnetic field, and is greatest when the conductor is at right angles to the field. A conductor wound into a coil that can rotate between the poles of a magnet forms the basis of an ◊electric motor.

electric field in physics, a region in which a particle possessing electric charge experiences a force owing to the presence of another electric charge. The strength of an electric field, E, is measured in volts per metre ($V\,m^{-1}$). It is a type of ◊electromagnetic field.

electricity all phenomena caused by ◊electric charge, whether static or in motion. Electric charge is caused by an excess or deficit of electrons in the charged substance, and an electric current by the movement of electrons around a circuit. Substances may be electrical conductors, such as metals, which allow the passage of electricity through them, or insulators, such as rubber, which are extremely poor conductors. Substances with relatively poor conductivities that can be improved by the addition of heat or light are known as ◊semiconductors.

Electricity generated on a commercial scale was available from the early 1880s and used for electric motors driving all kinds of machinery, and for lighting, first by carbon arc, but later by incandescent filaments (first of carbon and then of tungsten), enclosed in glass bulbs partially filled with inert gas under vacuum. Light is also produced by passing electricity through a gas or metal vapour or a fluorescent lamp. Other practical applications include telephone, radio, television, X-ray machines, and many other applications in ◊electronics.

The fact that amber has the power, after being rubbed, of attracting light objects, such as bits of straw and feathers, is said to have been known to Thales of Miletus and to the Roman naturalist Pliny. William Gilbert, Queen Elizabeth I's physician, found that many substances possessed this power, and he called it 'electric' after the Greek word meaning 'amber'.

In the early 1700s, it was recognized that there are two types of electricity and that unlike kinds attract each other and like kinds repel. The charge on glass rubbed with silk came to be known as positive electricity, and the charge on amber rubbed with wool as negative electricity. These two charges were found to cancel each other when brought together.

In 1800 Alessandro Volta found that a series of cells containing brine, in which were dipped plates of zinc and copper, gave an electric current, which later in the same year was shown to evolve hydrogen and oxygen when passed through water (◊electrolysis). Humphry Davy, in 1807, decomposed soda and potash (both thought to be elements) and isolated the metals sodium and potassium, a discovery that led the way to ◊electroplating. Other properties of electric currents discovered were the heating effect, now used in lighting and central heating, and the deflection of a magnetic needle, described by Hans Oersted in 1820 and elaborated by André Ampère in 1825. This work made possible the electric telegraph.

One day, Sir, you may tax it.

Michael Faraday (1791–1867) to
Mr Gladstone, Chancellor of the Exchequer,
when asked about the usefulness of
electricity

For Michael Faraday, the fact that an electric current passing through a wire caused a magnet to move suggested that moving a wire or coil of wire rapidly between the poles of a magnet would induce an electric current. He demonstrated this in 1831, producing the first ◊dynamo, which became the basis of electrical engineering. The characteristics of currents were crystallized in about 1827 by Georg Ohm, who showed that the current passing along a wire was equal to the electromotive force (emf) across the wire multiplied by a constant, which was the conductivity of the wire. The unit of resistance (ohm) is named after Ohm, the unit of emf (volt) is named after Volta, and the unit of current (amp) after Ampère.

The work of the late 1800s indicated the wide interconnections of electricity (with magnetism, heat, and light), and in about 1855 James Clerk Maxwell formulated a single electromagnetic theory. The universal importance of electricity was decisively proved by the discovery that the atom, until then thought to be the ultimate particle of matter, is composed of a positively charged central core, the nucleus, about which negatively charged electrons rotate in various orbits.

Electricity is the most useful and most convenient form of energy, readily convertible into heat and light and used to power machines. Electricity can be generated in one place and distributed anywhere because it readily flows through wires. It is generated at power stations where a suitable energy source is harnessed to drive ◊turbines that spin electricity generators. Current energy sources are coal, oil, water power (hydroelectricity), natural gas, and ◊nuclear energy. Research is under way to increase the contribution of wind, tidal, solar, and geothermal power. Nuclear fuel has proved a more expensive source of electricity than initially anticipated and worldwide concern over radioactivity may limit its future development.

Electricity is generated at power stations at a voltage of about 25,000 volts, which is not a suitable voltage for long-distance transmission. For minimal power loss, transmission must take place at very high voltage (400,000 volts or more). The generated voltage is therefore increased ('stepped up') by a ◊transformer. The resulting high-voltage electricity is then fed into the main arteries of the ◊grid system, an interconnected network of power stations and distribution centres covering a large area. After transmission to a local substation, the line voltage is reduced by a step-down transformer and distributed to consumers.

Among specialized power units that convert energy directly to electrical energy without the intervention of any moving mechanisms, the most promising are thermionic converters. These use conventional fuels such as propane gas, as in portable military power packs, or, if refuelling is to be avoided, radioactive fuels, as in uncrewed navigational aids and spacecraft.

electric motor a machine that converts electrical energy into mechanical energy. There are various types, including direct-current and induction motors, most of which produce rotary motion. A linear induction motor produces linear (in a straight line) rather than rotary motion.

A simple **direct-current motor** consists of a horseshoe-shaped permanent ◊magnet with a wire-wound coil (◊armature) mounted so that it can rotate between the poles of the magnet. A ◊commutator reverses the current (from a battery) fed to the coil on each half-turn, which rotates because of the mechanical force exerted on a conductor carrying a current in a magnetic field.

An **induction motor** employs ◊alternating current. It comprises a stationary current-carrying coil (stator) surrounding another coil (rotor), which rotates because of

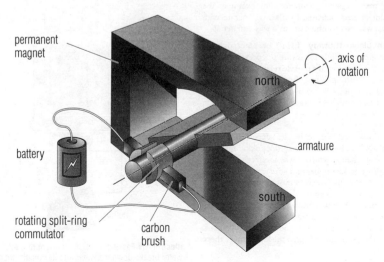

electric motor An example of a simple direct-current motor.

permanent magnet

axis of rotation

north

armature

battery

south

rotating split-ring commutator

carbon brush

the current induced in it by the magnetic field created by the stator; it thus requires no commutator.

electric power the rate at which an electrical machine uses electrical ◊energy or converts it into other forms of energy – for example, light, heat, mechanical energy. Usually measured in watts (equivalent to joules per second), it is equal to the product of the voltage and the current flowing.

An electric lamp that passes a current of 0.4 amps at 250 volts uses 100 watts of electrical power and converts it into light and heat – in ordinary terms it is a 100-watt lamp. An electric motor that requires 6 amps at the same voltage consumes 1,500 watts (1.5 kilowatts), equivalent to delivering about 2 horsepower of mechanical power.

electrocardiogram (ECG) graphic recording of the electrical activity of the heart, as detected by electrodes placed on the skin. Electrocardiography is used in the diagnosis of heart disease.

electrochemical series or *electromotive series* list of chemical elements arranged in descending order of the ease with which they can lose electrons to form cations (positive ions). An element can be displaced (displacement reaction) from a compound by any element above it in the series.

electrochemistry the branch of science that studies chemical reactions involving electricity. The use of electricity to produce chemical effects, ◊electrolysis, is employed in many industrial processes, such as the manufacture of chlorine and the extraction of aluminium. The use of chemical reactions to produce electricity is the basis of electrical ◊cells, such as the dry cell and the Leclanché cell.

Since all chemical reactions involve changes to the electronic structure of atoms, all reactions are now recognized as electrochemical in nature. Oxidation, for example, was once defined as a process in which oxygen was combined with a substance, or hydrogen was removed from a compound; it is now defined as a process in which electrons are lost.

Electrochemistry is also the basis of new methods of destroying toxic organic pollutants. For example, the development of electrochemical cells that operate with supercritical water to combust organic waste materials.

electroconvulsive therapy (ECT) or *electroshock therapy* treatment mainly for severe ◊depression, given under anaesthesia and with a muscle relaxant. An electric current is passed through one or both sides of the brain to induce alterations in its electrical activity. The treatment can cause distress and loss of concentration and memory, and so there is much controversy about its use and effectiveness.

ECT was first used in 1938 but its success in treating depression lead to its excessive use for a wide range of mental illnesses against which it was ineffective. Its side effects included broken bones and severe memory loss.

The procedure in use today is much improved, using the minimum shock necessary to produce a seizure, administered under general anaesthetic with muscle relaxants to prevent spasms and fractures. It is the seizure rather than the shock itself that produces improvement. The smaller the shock administered the less damage there is to memory.

electrocrystal diagnosis technique recently developed by British biologist Harry Oldfield, based on the finding that stimulated electromagnetic fields of the human body resonate at a particular frequency which varies with individuals, and that actual or incipient disease can be pinpointed by a scanning device responsive to local deviations from the person's norm.

electrode any terminal by which an electric current passes in or out of a conducting substance; for example, the anode or cathode in a battery or the carbons in an arc lamp. The terminals that emit and collect the flow of electrons in thermionic ◊valves (electron tubes) are also called electrodes: for example, cathodes, plates, and grids.

electrodynamics the branch of physics dealing with electric charges, electric currents and associated forces. ◊Quantum electrodynamics (QED) studies the interaction between charged particles and their emission and absorption of electromagnetic radiation. This field combines quantum theory and relativity theory, making accurate predictions about subatomic processes involving charged particles such as electrons and protons.

electroencephalogram (EEG) graphic record of the electrical discharges of the brain, as detected by electrodes placed on the scalp. The pattern of electrical activity revealed by electroencephalography is helpful in the diagnosis of some brain disorders, in particular epilepsy.

electrolysis in archaeological conservation, a cleaning process, especially of material from underwater archaeology, involving immersing the object in a chemical solution and passing a weak current between it and a surrounding metal grille. Corrosive salts move slowly from the object (cathode) to the grille (anode), leaving the artefact clean.

electrolysis in chemistry, the production of chemical changes by passing an electric current through a solution or molten salt (the electrolyte), resulting in the migration of ions to the electrodes: positive ions (cations) to the negative electrode (cathode) and negative ions (anions) to the positive electrode (anode).

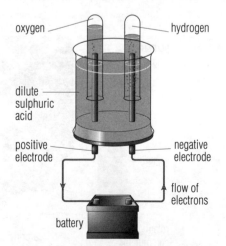

electrolysis Passing an electric current through acidified water breaks down the water into its constituent elements – hydrogen and oxygen.

During electrolysis, the ions react with the electrode, either receiving or giving up electrons. The resultant atoms may be liberated as a gas, or deposited as a solid on the electrode, in amounts that are proportional to the amount of current passed, as discovered by English chemist Michael Faraday. For instance, when acidified water is electrolysed, hydrogen ions (H^+) at the cathode receive electrons to form hydrogen gas; hydroxide ions (OH^-) at the anode give up electrons to form oxygen gas and water.

One application of electrolysis is *electroplating*, in which a solution of a salt, such as silver nitrate ($AgNO_3$), is used and the object to be plated acts as the negative electrode, thus attracting silver ions (Ag^+). Electrolysis is used in many industrial processes, such as coating metals for vehicles and ships, and refining bauxite into aluminium; it also forms the basis of a number of electrochemical analytical techniques, such as polarography.

electrolyte solution or molten substance in which an electric current is made to flow by the movement and discharge of ions in accordance with Faraday's laws of ◊electrolysis.

The term 'electrolyte' is frequently used to denote a substance that, when dissolved in a specified solvent, usually water, dissociates into ◊ions to produce an electrically conducting medium.

In medicine the term is often used for the ion itself (sodium or potassium, for example). Electrolyte balance may be severely disrupted in illness or injury.

electromagnet coil of wire wound around a soft iron core that acts as a magnet when an electric current flows through the wire. Electromagnets have many uses: in switches, electric bells, solenoids, and metal-lifting cranes.

electromagnetic field in physics, the region in which a particle with an ◊electric charge experiences a force. If it does so only when moving, it is in a pure *magnetic field*; if it does so when stationary, it is in an *electric field*. Both can be present simultaneously.

electromagnetic force one of the four fundamental ◊forces of nature, the other three being gravity, the strong nuclear force, and the weak nuclear force. The ◊elementary particle that is the carrier for the electromagnetic (em) force is the photon.

electromagnetic induction in electronics, the production of an ◊electromotive force (emf) in a circuit by a change of magnetic flux through the circuit or by relative motion of the circuit and the magnetic flux. In a closed circuit an ◊induced current will be produced. All dynamos and generators make use of this effect. When magnetic tape is driven past the playback head (a small coil) of a tape-recorder, the moving magnetic field induces an emf in the head, which is then amplified to reproduce the recorded sounds.

If the change of magnetic flux is due to a variation in the current flowing in the same circuit, the phenomenon is known as self-induction; if it is due to a change of current flowing in another circuit it is known as mutual induction.

electromagnetic spectrum the complete range, over all wavelengths, of ◊electromagnetic waves.

electromagnetic system of units former system of absolute electromagnetic units (emu) based on the ◊c.g.s.

system and having, as its primary electrical unit, the unit magnetic pole. It was replaced by ◊SI units.

electromagnetic waves oscillating electric and magnetic fields travelling together through space at a speed of nearly 300,000 km/186,000 mi per second.

The (limitless) range of possible wavelengths or ◊frequencies of electromagnetic waves, which can be thought of as making up the *electromagnetic spectrum*, includes radio waves, infrared radiation, visible light, ultraviolet radiation, X-rays, and gamma rays.

electromotive force (emf) loosely, the voltage produced by an electric battery or generator or, more precisely, the energy supplied by a source of electric power in driving a unit charge around an electrical circuit. The unit is the ◊volt.

electron stable, negatively charged ◊elementary particle; it is a constituent of all atoms, and a member of the class of particles known as ◊leptons. The electrons in each atom surround the nucleus in groupings called shells; in a neutral atom the number of electrons is equal to the number of protons in the nucleus. This electron structure is responsible for the chemical properties of the atom (see ◊atomic structure).

Electrons are the basic particles of electricity. Each carries a charge of 1.602192×10^{-19} coulomb, and all electrical charges are multiples of this quantity. A beam of electrons will undergo ◊diffraction (scattering) and produce interference patterns in the same way as ◊electromagnetic waves such as light; hence they may also be regarded as waves.

electronegativity the ease with which an atom can attract electrons to itself. Electronegative elements attract electrons, so forming negative ions.

Linus Pauling devised an electronegativity scale to indicate the relative power of attraction of elements for electrons. Fluorine, the most nonmetallic element, has a value of 4.0 on this scale; oxygen, the next most nonmetallic, has a value of 3.5.

In a covalent bond between two atoms of different electronegativities, the bonding electrons will be located close to the more electronegative atom, creating a ◊dipole.

electron gun a part in many electronic devices consisting of a series of ◊electrodes, including a cathode for producing an electron beam. It plays an essential role in ◊cathode-ray tubes (television tubes) and ◊electron microscopes.

electronic mail or *E-mail* private messages sent electronically from computer to computer via network connections such as Ethernet or the Internet, or via telephone lines to a host system. Messages once sent are stored on the network or by the host system until the recipient picks them up.

Subscribers to an electronic mail system type messages in ordinary letter form on a word processor, or microcomputer, and 'drop' the letters into a central computer's memory bank by means of a computer/telephone connector (a ◊modem). The recipient 'collects' the letter by calling up the central computer and feeding a unique password into the system. *See illustration on page 221.*

electronic point of sale (EPOS) system used in retailing in which a bar code on a product is scanned at the cash till and the information relayed to the store computer.

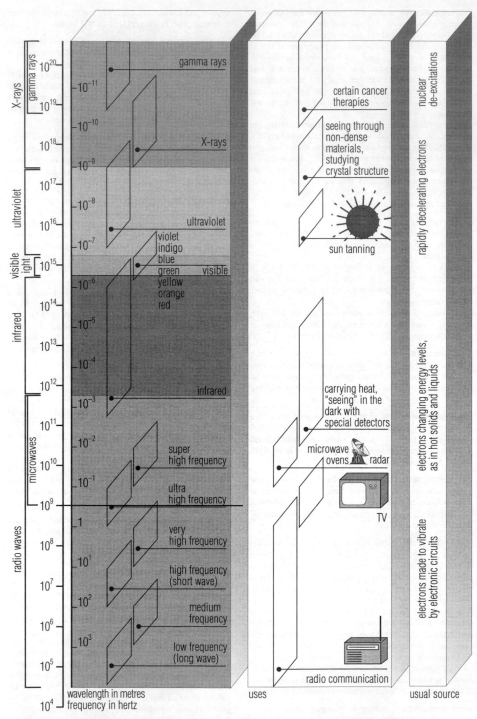

electromagnetic waves Radio waves have the lowest frequency. Infra-red radiation, visible light, ultraviolet radiation, X-rays, and gamma rays have progressively higher frequencies.

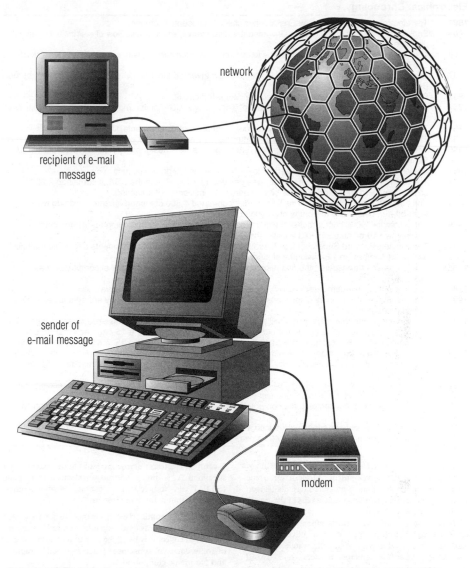

network

recipient of e-mail
message

sender of
e-mail message

modem

electronic mail The basic structure of an electronic mail system. A message is sent via a telephone line and stored in a
central computer. The message remains there until the recipient calls up the central computer and collects the message.

The computer will then relay back the price of the item to
the cash till. The customer can then be given an itemized
receipt while the computer removes the item from stock
figures.

EPOS enables efficient computer stock control and
reordering as well as giving a wealth of information about
turnover, profitability on different lines, stock ratios, and
other important financial indicators.

electronic publishing the distribution of information
using computer-based media such as ◊multimedia and
◊hypertext in the creation of electronic 'books'. Critical
technologies in the development of electronic publishing

were ◊CD-ROM, with its massive yet compact storage
capabilities, and the advent of computer networking with
its ability to deliver information instantaneously anywhere in the world.

electronics branch of science that deals with the emission of ◊electrons from conductors and ◊semiconduc-
tors, with the subsequent manipulation of these electrons,
and with the construction of electronic devices. The first
electronic device was the thermionic ◊valve, or vacuum
tube, in which electrons moved in a vacuum, and led to
such inventions as ◊radio, ◊television, ◊radar, and the
digital ◊computer. Replacement of valves with the com-

Electronics: Chronology

1897	The electron was discovered by English physicist John Joseph Thomson.
1904	English physicist Ambrose Fleming invented the diode valve, which allows flow of electricity in one direction only.
1906	The triode electron valve, the first device to control an electric current, was invented by US physicist Lee De Forest.
1947	John Bardeen, William Shockley, and Walter Brattain invented the junction germanium transistor at the Bell Laboratories, New Jersey, USA.
1952	British physicist G W A Dunner proposed the integrated circuit.
1953	Jay Forrester of the Massachusetts Institute of Technology, USA, built a magnetic memory smaller than existing vacuum-tube memories.
1954	The silicon transistor was developed by Gordon Teal of Texas Instruments, USA.
1958	The first integrated circuit, containing five components, was built by US electrical physicist Jack Kilby.
1959	The planar transistor, which is built up in layers, or planes, was designed by Robert Noyce of Fairchild Semiconductor Corporation, USA.
1961	Steven Hofstein designed the field-effect transistor used in integrated circuits.
1971	The first microprocessor, the Intel 4004, was designed by Ted Hoff in the USA; it contained 2,250 components and could add two four-bit numbers in 11-millionths of a second.
1974	The Intel 8080 microprocessor was launched; it contained 4,500 components and could add two eight-bit numbers in 2.5-millionths of a second.
1979	The Motorola 68000 microprocessor was introduced; it contained 70,000 components and could multiply two 16-bit numbers in 3.2-millionths of a second.
1981	The Hewlett-Packard Superchip was introduced; it contained 450,000 components and could multiply two 32-bit numbers in 1.8-millionths of a second.
1985	The Inmos T414 transputer, the first microprocessor designed for use in parallel computers, was launched.
1988	The first optical microprocessor, which uses light instead of electricity, was developed.
1989	Wafer-scale silicon memory chips were introduced: the size of a beer mat, they are able to store 200 million characters.
1990	Memory chips capable of holding 4 million bits of information began to be mass-produced in Japan. The chips can store the equivalent of 520,000 characters, or the contents of a 16-page newspaper. Each chip contains 9 million components packed on a piece of silicon less than 15 mm long by 5 mm wide.
1992	Transistors made from high-temperature superconducting ceramics rather than semiconductors were produced in Japan by Sanyo Electric. The new transistors are ten times faster than semiconductor transistors.
1993	US firm Intel launched the Pentium 64-bit microprocessor. With two separate integer processing units that can run in parallel, it was several times faster than earlier processors.
1996	Japanese researchers completed a computer able to perform 1.08 billion floating-point operations per second.
1997	The US firm Texas Instruments introduced a digital-signal microprocessor chip that can process 1.6 billion instructions a second – about 40 times more powerful than current chips.

paratively tiny and reliable transistor from 1948 revolutionized electronic development. Modern electronic devices are based on minute ◊integrated circuits (silicon chips), wafer-thin crystal slices holding tens of thousands of electronic components.

By using solid-state devices such as integrated circuits, extremely complex electronic circuits can be constructed, leading to the development of ◊digital watches, pocket calculators, powerful ◊microcomputers, and ◊word processors.

electron microscope instrument that produces a magnified image by using a beam of ◊electrons instead of light rays, as in an optical ◊microscope. An *electron lens* is an arrangement of electromagnetic coils that control and focus the beam. Electrons are not visible to the eye, so instead of an eyepiece there is a fluorescent screen or a photographic plate on which the electrons form an image. The wavelength of the electron beam is much shorter than that of light, so much greater magnification and resolution (ability to distinguish detail) can be achieved. The development of the electron microscope has made possible the observation of very minute organisms, viruses, and even large molecules.

A ◊transmission electron microscope passes the electron beam through a very thin slice of a specimen. A ◊scanning electron microscope looks at the exterior of a specimen. A ◊scanning transmission electron microscope (STEM) can produce a magnification of 90 million times. See also ◊atomic force microscope.

electrons, delocalized electrons that are not associated with individual atoms or identifiable chemical bonds, but are shared collectively by all the constituent atoms or ions of some chemical substances (such as metals, graphite, and ◊aromatic compounds).

A metallic solid consists of a three-dimensional arrangement of metal ions through which the delocalized electrons are free to travel. Aromatic compounds are characterized by the sharing of delocalized electrons by several atoms within the molecule.

electrons, localized a pair of electrons in a ◊covalent bond that are located in the vicinity of the nuclei of the two contributing atoms. Such electrons cannot move beyond this area.

electron spin resonance in archaeology, a nondestructive dating method applicable to teeth, bone, heat-treated flint, ceramics, sediments, and stalagmitic concretions. It enables electrons, displaced by natural radiation and then trapped in the structure, to be measured; their number indicates the age of the specimen.

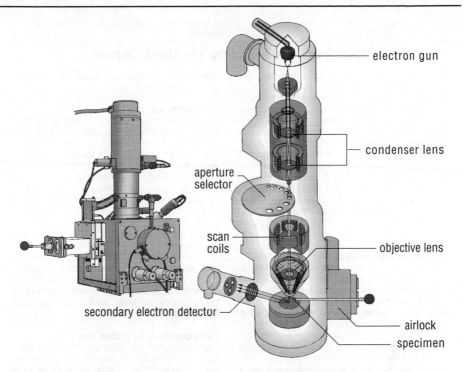

electron gun

condenser lens

aperture
selector

scan
coils

objective lens

secondary electron detector

airlock

specimen

electron microscope The scanning electron microscope. Electrons from the electron gun are focused to a fine point on the specimen surface by the lens systems. The beam is moved across the specimen by the scan coils. Secondary electrons are emitted by the specimen surface and pass through the detector, which produces an electrical signal. The signal is passed to an electronic console, and produces an image on a television-like screen.

electron volt unit (symbol eV) for measuring the energy of a charged particle (◊ion or ◊electron) in terms of the energy of motion an electron would gain from a potential difference of one volt. Because it is so small, more usual units are mega-(million) and giga-(billion) electron volts (MeV and GeV).

electrooculography in medicine, a method of recording movements of the eyes. It is used to assess the function of the ◊retina.

electrophoresis the ◊diffusion of charged particles through a fluid under the influence of an electric field. It can be used in the biological sciences to separate ◊molecules of different sizes, which diffuse at different rates. In industry, electrophoresis is used in paint-dipping operations to ensure that paint reaches awkward corners.

electroplating deposition of metals upon metallic surfaces by electrolysis for decorative and/or protective purposes. It is used in the preparation of printers' blocks, 'master' audio discs, and in many other processes.

A current is passed through a bath containing a solution of a salt of the plating metal, the object to be plated being the cathode (negative terminal); the anode (positive terminal) is either an inert substance or the plating metal. Among the metals most commonly used for plating are zinc, nickel, chromium, cadmium, copper, silver, and gold.

In *electropolishing*, the object to be polished is made the anode in an electrolytic solution and by carefully controlling conditions the high spots on the surface are

dissolved away, leaving a high-quality stain-free surface. This technique is useful in polishing irregular stainless-steel articles.

electropositivity in chemistry, a measure of the ability of elements (mainly metals) to donate electrons to form positive ions. The greater the metallic character, the more electropositive the element.

electroscope apparatus for detecting ◊electric charge. The simple gold-leaf electroscope consists of a vertical conducting (metal) rod ending in a pair of rectangular pieces of gold foil, mounted inside and insulated from an earthed metal case or glass jar. An electric charge applied to the end of the metal rod makes the gold leaves diverge, because they each receive a similar charge (positive or negative) and so repel each other.

The polarity of the charge can be found by bringing up another charge of known polarity and applying it to the metal rod. A like charge has no effect on the gold leaves, whereas an opposite charge neutralizes the charge on the leaves and causes them to collapse. *See illustration on page 225.*

electrostatics the study of stationary electric charges and their fields (not currents). See ◊static electricity.

electrovalent bond another name for an ◊ionic bond, a chemical bond in which the combining atoms lose or gain electrons to form ions.

electrum naturally occurring alloy of gold and silver used

Electron microscopy via the Internet

Researchers can now make use of the high voltage electron microscope at the Lawrence Berkeley Laboratory without travelling to the laboratory in California. The laboratory has created a set of interactive, online computing tools that allow scientists to manipulate the instrument, conduct experiments, and view images from their own office, via the Internet. In the first demonstration of the system, researchers 2,000 miles from Berkeley used a computer to take control of the microscope, heat an advanced alloy specimen, and observe the ensuing progression of structural changes on their computer monitor.

A fast-moving subject

Remote control of machines or scientific instruments is not new; the Hubble Space Telescope is a well-known example. Remote operation of an electron microscope, however, is especially difficult because of the slight but inevitable time delays that occur with Internet use. At the atomic level, things move fast; atoms move and shapes change quickly. This means that an electron microscope needs constant adjustment to keep it in focus and trained on the area being studied.

According to Bahram Parvin, the computer scientist who leads the team at Berkeley with Michael O'Keefe, deputy head of the Berkeley National Center for Electron Microscopy, normally an operator sitting next to the microscope has to constantly adjust dials just to keep the area of the specimen under study within the field of view. Focusing also requires continuous attention. Trying to do this over the Internet, which has a time lag analogous to the time lag that occurs during an international phone call, is impossible. By the time the remote operator's commands reach the microscope, they are too late.

Computer vision

To sidestep this limitation of computer networks, Parvin's team has automated the positioning and focusing of the Berkeley microscope using advanced computer vision techniques.

As Parvin explains, 'You start with what to a computer is an indiscriminate field. You then detect and lock onto objects of interest. This is computer vision. Very soon, from a remote location, computer vision will self-calibrate the microscope, automatically focus it, and compensate for thermal drift. Underlying this is a complex package of algorithms dealing with shape analysis, background measurements, and so on. Essentially though, we are making the microscope smarter, making it do automatically what users would normally have to do on their own'.

Developments in online tools

Parvin is also developing another set of computer vision tools in collaboration with Berkeley scientists Dan Callahan and Marcos Maestre. These not only detect objects of interest – for example, single DNA molecules – but lock onto and track them. The starting point for this technology is an algorithm that defines a shape. Parvin's team have constructed algorithms for a range of shapes, including circular, tubular, and complex objects. The team has also created a tool for drawing an outline around a uniquely-shaped object and then launching a process that finds and tracks these objects. Ultimately, Parvin hopes to integrate tracking software into a package of online tools.

Peter Lafferty

---

by early civilizations to make the first coins, about the 6th century BC.

element substance that cannot be split chemically into simpler substances. The atoms of a particular element all have the same number of protons in their nuclei (their ◊atomic number). Elements are classified in the ◊periodic table of the elements. Of the 112 known elements, 92 are known to occur in nature (those with atomic numbers 1–92). Those from 96 to 112 do not occur in nature and are synthesized only, produced in particle accelerators. Eighty-one of the elements are stable; all the others, which include atomic numbers 43, 61, and from 84 up, are radioactive.

Elements are classified as metals, nonmetals, or metalloids (weakly metallic elements) depending on a combination of their physical and chemical properties; about 75% are metallic. Some elements occur abundantly (oxygen, aluminium); others occur moderately or rarely (chromium, neon); some, in particular the radioactive ones, are found in minute (neptunium, plutonium) or very minute (technetium) amounts.

Symbols (devised by Swedish chemist Jöns Berzelius) are used to denote the elements; the symbol is usually the first letter or letters of the English or Latin name (for example, C for carbon, Ca for calcium, Fe for iron, *ferrum*). The symbol represents one atom of the element.

According to current theories, hydrogen and helium were produced in the ◊Big Bang at the beginning of the universe. Of the other elements, those up to atomic number 26 (iron) are made by nuclear fusion within the stars. The more massive elements, such as lead and uranium,

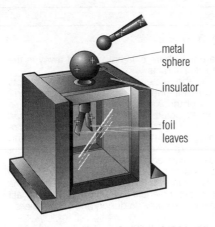

electroscope

are produced when an old star explodes; as its centre collapses, the gravitational energy squashes nuclei together to make new elements.

Element

Of the 92 naturally occurring chemical elements, only two (bromine and mercury) are liquids at ordinary temperatures, and only 11 are gases. All the other elements are solids, mostly metals.

---

elementary particle in physics, a subatomic particle that is not made up of smaller particles, and so can be considered one of the fundamental units of matter. There are three groups of elementary particles: quarks, leptons, and gauge bosons.

Quarks, of which there are 12 types (up, down, charm, strange, top, and bottom, plus the antiparticles of each), combine in groups of three to produce heavy particles called baryons, and in groups of two to produce intermediate-mass particles called mesons. They and their composite particles are influenced by the strong nuclear force.

Leptons are light particles. Again, there are 12 types: the electron, muon, tau; their neutrinos, the electron neutrino, muon neutrino, and tau neutrino; and the antiparticles of each. These particles are influenced by the weak nuclear force.

Gauge bosons carry forces between other particles. There are four types: the gluons, photon, weakons, and graviton. The gluon carries the strong nuclear force, the photon the electromagnetic force, the weakons the weak nuclear force, and the graviton the force of gravity (see ◊forces, fundamental).

elements, the four earth, air, fire, and water. The Greek philosopher Empedocles believed that these four elements made up the fundamental components of all matter and that they were destroyed and renewed through the action of love and discord.

This belief was shared by Aristotle who also claimed that the elements were mutable and contained specific qualities: cold and dry for earth, hot and wet for air, hot and dry for fire, and cold and wet for water. The transformation of the elements formed the basis of medieval

alchemy, and the belief that base metals could be turned into gold. The theory of the elements prevailed until the 17th century when Robert Boyle redefined an element as a substance 'simple or unmixed, not made of other bodies' and proposed the existence of a greater number than four.

elephantiasis in the human body, a condition of local enlargement and deformity, most often of a leg, though the scrotum, vulva, or breast may also be affected.

The commonest form of elephantiasis is the tropical variety (filariasis) caused by infestation by parasitic roundworms (filaria); the enlargement is due to damage of the lymphatic system and consequent impaired immunity. This left sufferers susceptible to infection from bacteria and fungi, entering through skin splits. The swelling reduces dramatically if the affected area is kept rigorously clean and treated with antibiotic cream, combined with rest, after drug treatment has killed all filarial worms.

ellipse curve joining all points (loci) around two fixed points (foci) such that the sum of the distances from those points is always constant. The diameter passing through the foci is the major axis, and the diameter bisecting this at right angles is the minor axis. An ellipse is one of a series of curves known as ◊conic sections. A slice across a cone that is not made parallel to, and does not pass through, the base will produce an ellipse.

elliptical galaxy in astronomy, one of the main classes of ◊galaxy in the Hubble classification and characterized by a featureless elliptical profile. Unlike spiral galaxies, elliptical galaxies have very little gas or dust and no stars have recently formed within them. They range greatly in size from giant ellipticals, which are often found at the centres of clusters of galaxies and may be strong radio sources, to tiny dwarf ellipticals, containing about a million stars, which are the most common galaxies of any type. More than 60% of known galaxies are elliptical.

El Niño Spanish 'the child' warm ocean surge of the ◊Peru Current, so called because it tends to occur at Christmas, recurring about every 5–8 years in the E Pacific off South America. It involves a change in the direction of ocean currents, which prevents the upwelling of cold, nutrient-rich waters along the coast of Ecuador and Peru, killing fish and plants. It is an important factor in global weather.

El Niño is believed to be caused by the failure of trade winds and, consequently, of the ocean currents normally driven by these winds. Warm surface waters then flow in from the east. The phenomenon can disrupt the climate

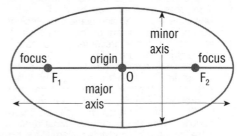

ellipse Technical terms used to describe the ellipse; for all points on the ellipse, the sum of the distances from the two foci, F_1 and F_2, is the same.

El Niño

El Niño is the name that climatologists give to a change in the currents and winds of the eastern Pacific ocean. The effect results in the warming of the surface waters of the area with subsequent damage to local fisheries. It usually happens about Christmas-time – hence the name, which means 'the Child' – and appears at irregular intervals of a few years. Attempts are now being made to analyse past patterns so that the event can be predicted. El Niño usually lasts for about 18 months, but the 1990 El Niño lasted until June 1995. No records exist of it having lasted five years before.

A small but significant event

In early 1991 climatologists at the National Weather Service's Climate Analysis Centre in Camp Springs, Maryland, at the Lamont–Doherty Geological Observatory in Palisades, New York, and at the Scripps Institution of Oceanography in La Jolla, California, independently predicted that there would be a small El Niño event beginning in summer 1991. The event did take place, with a number of alarming effects.

The effects

The northeasterly trade winds that bring moisture to Australia failed, giving rise to the worst drought on that continent for 200 years. Eucalyptus trees, normally well adapted to dry conditions, dried out and died, with some forests in Queensland recording a death rate of 50%. The level of the Darling and other rivers was so low that sunlight reaching the bed gave rise to a bloom of cyanobacteria, poisoning the water. Fertilizer runoff added to this problem. The subsequent effect was that many of Australia's unique animals, such as koalas and platypuses, died in large numbers.

Further north there was a change in the regular pattern of typhoons or hurricanes. A permanent high-pressure area over the western Pacific tends to deflect the typhoons from Japan, but in Sept and Oct 1991 this high-pressure area collapsed. El Niño, which causes a warming of the east Pacific, also causes a cooling of the west Pacific, and in this instance prevented the formation of the high-pressure area.

Japan was subjected to severe typhoons in Sept and Oct 1991, producing rainfall that was about 3.6 times the average and destroying the apple crop. Further south, in the Philippines, typhoon Thelma killed 6,000 people in Nov the same year. Most deaths were the result of mudslides. Illegal logging operations in the hills had left large areas of bare soil that washed away in the torrential rains.

In the eastern Pacific the El Niño effects were felt further north than usual. The warming of the waters produced heavy rains in British Columbia and southern Alaska in Jan 1992. In Feb the storms hit California, dropping 30 cm/12 in of rain in four days and causing extensive flooding. These conditions are only expected in the area once in 50 years or so. The unpredictability of El Niño and its effects is shown by the fact that rains failed to materialize along the Peruvian coast as they usually do during an El Niño event.

The current El Niño

The latest El Niño started in January 1997. Since March 1997, sea surface temperatures have continued to rise and its global effects are already being felt with wetter than normal conditions in the equatorial central and eastern Pacific, floods in Chile, and drier than normal conditions being experienced in Indonesia, Pakistan, parts of India, central America, southern Mexico, and more droughts in Australia.

Efforts to forecast El Niño

With all these side-effects, as well as the inevitable failure of the fish harvests, it is important that the effects should be accurately predicted in the future so that adequate preparations can be made. The recent history of the climates of the Pacific Ocean is being studied to try to detect the patterns.

A new tool in this study is the single-celled alga *Emiliana huxleyi*. This blooms in the summer and dies back in the winter, giving rise to bottom sediments that are alternately rich and poor in their remains. The layering can be used for accurate dating. Scientists at Stanford University in California and the University of Bristol have found that this alga synthesizes an organic molecule called an alkenone, and that the chemical bonding in this molecule depends on the temperature of the surrounding water. By studying the layering of the sediment, and the structure of the alkenone molecules, it has been possible to produce a very accurate profile of the changing sea temperatures of the eastern Pacific between 1915 and 1988. Up to now these studies have been carried out using the different isotopes of carbon and oxygen in carbonates formed at different times, but the alkenone molecules are much more stable.

Nigel Dudley

of the area disastrously, and has played a part in causing famine in Indonesia, drought and bush fires in the Galápagos Islands, rainstorms in California and South America, and the destruction of Peru's anchovy harvest and wildlife in 1982–1983. El Niño contributed to algal blooms in Australia's drought-stricken rivers and an unprecedented number of typhoons in Japan in 1991. It also caused the 1997 drought in Australia.

El Niño usually lasts for about 18 months, but the 1990 occurrence lasted until June 1995; US climatologists estimated this duration to be the longest in 2,000 years. The last prolonged El Niño of 1939–41 caused extensive drought and famine in Bengal. It is understood that there might be a link between El Niño and ◊global warming.

elongation in astronomy, the angular distance between the Sun and a planet or other solar-system object. This angle is 0° at ◊conjunction, 90° at quadrature, and 180° at ◊opposition.

embolism blockage of a blood vessel by an obstruction called an embolus (usually a blood clot, fat particle, or bubble of air).

embryo early developmental stage of an animal or a plant following fertilization of an ovum (egg cell), or activation of an ovum by ◊parthenogenesis. In humans, the term embryo describes the fertilized egg during its first seven weeks of existence; from the eighth week onwards it is referred to as a fetus.

In animals the embryo exists either within an egg (where it is nourished by food contained in the yolk), or in mammals, in the ◊uterus of the mother. In mammals (except marsupials) the embryo is fed through the ◊placenta. The plant embryo is found within the seed in higher plants. It sometimes consists of only a few cells, but usually includes a root, a shoot (or primary bud), and one or two ◊cotyledons, which nourish the growing seedling.

embryology study of the changes undergone by an organism from its conception as a fertilized ovum (egg) to its emergence into the world at hatching or birth. It is mainly concerned with the changes in cell organization in the embryo and the way in which these lead to the structures and organs of the adult (the process of ◊differentiation).

Applications of embryology include embryo transplants, both commercial (for example, in building up a prize dairy-cow herd quickly at low cost) and in obstetric medicine (as a method for helping couples with fertility problems to have children).

embryo research the study of human embryos at an early stage, in order to detect hereditary disease and genetic defects, and to investigate the problems of subfertility and infertility.

Eggs are fertilized in vitro (see ◊in vitro fertilization) and allowed to grow to the eight-cell stage. One or two cells are then removed for analysis. Diseases that can be tested for include cystic fibrosis, Duchenne's muscular dystrophy, Lesch-Nyhan syndrome, Tay-Sachs disease, and haemophilia A. If the embryo appears healthy it is transferred to the mother.

embryo sac large cell within the ovule of flowering plants that represents the female ◊gametophyte when fully developed. It typically contains eight nuclei. Fertilization

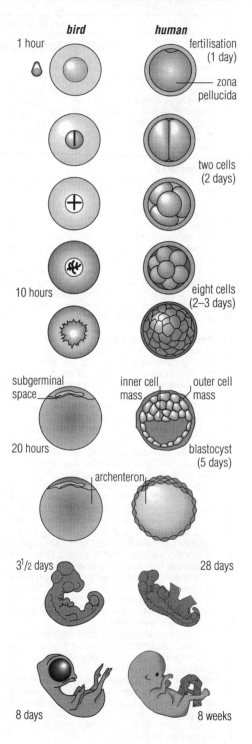

embryo The development of a bird and a human embryo.

Embryology: early observations

The essence of science is systematic observation of the natural world. This may seem obvious to us, but it is a hard-won insight. Some appreciation of the value of observation stirred in Greece about 2,000 years ago, when Aristotle began to use experiments and scientific methods in the study of biology.

Hippocrates

The origins of animals and plants had been considered by Greek thinkers before Aristotle, beginning with Hippocrates in the 5th century BC. Did an organism form stage by stage or did it come into existence small but fully formed and then grow? The Hippocratic writings contain a clear suggestion for resolving the question:

'Take twenty eggs or more, and set them for brooding under two or more hens. Then on each day of incubation from the second to the last, that of hatching, remove one egg and open it for examination.'

Aristotle's experiments

There is no sign that the writer bothered to carry out the experiment; that was left to Aristotle, and he wrote a detailed description of what he saw. It seems that he extended the work to other birds, and that he understood something of the process which unfolded as the chick grew. He wrote:

'Generation from the egg proceeds in an identical manner with all birds, but the full periods from conception to birth differ. With the common hen after three days and three nights there is the first indication of the embryo; with larger birds the interval being longer, with smaller birds shorter. Meanwhile the yolk comes into being, rising towards the sharp end, where the primal element of the egg is situated, and where the egg gets hatched; and the heart appears, like a speck of blood, in the white of the egg. This point beats and moves as if endowed with life ...'

On this Aristotle was wrong; the heart is not the first organ to develop. However, he correctly emphasizes that fetal development is closely linked to the development of blood vessels and membranes:

'... two vein-ducts with blood in them trend in a convoluted course ... and a membrane carrying bloody fibres now envelops the yolk, leading off from the vein ducts. A little afterwards the body is differentiated, at first very small and white. The head is clearly distinguished, and in it the eyes, swollen out to a great extent.'

He understands correctly the role of the egg white and yolk:

'The life-element of the chick is in the white of the egg, and the nutrient comes through the navel-string out of the yolk.'

And the functions of the membranes surrounding the chick:

'The disposition of the several constituent parts is as follows. First and outermost comes the membrane of the egg, not that of the shell, but underneath it. Inside this membrane is a white liquid; then comes the chick, and a membrane round about it, separating it off so as to keep the chick free from liquid; next after the chick comes the yolk, into which one of the two veins was described as leading, the other leading into the enveloping white substance.'

Day by day, Aristotle describes the development of the various organs:

'When the egg is now ten days old the chick and all its parts are distinctly visible. The head is still larger than the rest of its body, and the eyes are larger than the head, but still devoid of vision. ... At this time also the larger internal organs are visible, as also the stomach and the arrangement of the viscera; and the veins seen to proceed from the heart are now close to the navel.'

'About the twentieth day, if you open the egg and touch the chick, it moves inside and chirps; and it is already coming to be covered with down, when, after the twentieth day is past, the chick begins to break the shell. The head is situated over the right leg close to the flank, and the wing is placed over the head; and about this time is plain to be seen the membrane resembling an afterbirth that comes next after the outermost membrane of the shell, into which membrane the one of the navel strings was described as leading ...'

He details the way the various membranes change, shrivel and detach during birth, ending with the observation:

'By and by the yolk, diminishing gradually in size, at length becomes entirely used up and comprehended within the chick (so that, ten days after hatching, if you cut open the chick, a small remnant of the yolk is still left in connection with the gut) ...'

Modern embryology

Obviously, Aristotle believed in following up his experiments, seeking the unexpected. His description sounds like a modern scientist at work. For around 2,000 years, researchers could only refine the picture and correct a few small errors. Modern embryology is in the midst of an exciting revolution; we are beginning to understand the processes seen by Aristotle at molecular level – the level of genes. Genes control development and the same mechanisms are found in most animals. But the modern approach can still involve experiments on the embryo, and may mean opening chicken eggs and observing their development as Aristotle did.

Peter Lafferty

occurs when one of these nuclei, the egg nucleus, fuses with a male ◊gamete.

emerald a clear, green gemstone variety of the mineral ◊beryl. It occurs naturally in Colombia, the Ural Mountains, in Russia, Zimbabwe, and Australia.

emery greyish-black opaque metamorphic rock consisting of ◊corundum and magnetite, together with other minerals such as hematite. It is used as an ◊abrasive.

emetic any substance administered to induce vomiting. Emetics are used to empty the stomach in many cases of deliberate or accidental drug overdose. The most frequently used is ipecacuanha.

emf in physics, abbreviation for ◊ *electromotive force*.

emission line in astronomy, bright line in the spectrum of a luminous object caused by ◊atoms emitting light at sharply defined ◊wavelengths.

emission spectroscopy in analytical chemistry, a technique for determining the identity or amount present of a chemical substance by measuring the amount of electromagnetic radiation it emits at specific wavelengths; see ◊spectroscopy.

emotion in psychology, a powerful feeling; a complex state of body and mind involving, in its bodily aspect, changes in the viscera (main internal organs) and in facial expression and posture, and in its mental aspect, heightened perception, excitement and, sometimes, disturbance of thought and judgement. The urge to action is felt and impulsive behaviour may result.

emphysema incurable lung condition characterized by disabling breathlessness. Progressive loss of the thin walls dividing the air spaces (alveoli) in the lungs reduces the area available for the exchange of oxygen and carbon dioxide, causing the lung tissue to expand. The term 'emphysema' can also refer to the presence of air in other body tissues.

Emphysema is most often seen at an advanced stage of chronic ◊bronchitis, although it may develop in other long-standing diseases of the lungs. It destroys lung tissue, leaving behind scar tissue in the form of air blisters called bullae. As the disease progresses, the bullae occupy more and more space in the chest cavity, inflating the lungs and causing severe breathing difficulties. The bullae may be removed surgically, and since early 1994 US trials have achieved measured success using lasers to eliminate them in a procedure called lung-reduction pneumenoplasty (LRP). Lasers are particularly useful where the emphysema is diffuse and bullae are interspersed within healthy tissue. As LRP is a less invasive process, survival rates are improved (90% compared with 75% for conventional surgery) and patients are quicker to recover.

emulator in computing, an item of software or firmware that allows one device to imitate the functioning of another. Emulator software is commonly used to allow one make of computer to run programs written for a different make of computer. This allows a user to select from a wider range of ◊applications programs, and perhaps to save money by running programs designed for an expensive computer on a cheaper model.

Many printers contain emulator firmware that enables them to imitate Hewlett Packard and Epson printers, because so much software is written to work with these widely used machines.

emulsifier food additive used to keep oils dispersed and in suspension, in products such as mayonnaise and peanut butter. Egg yolk is a naturally occurring emulsifier, but most of the emulsifiers in commercial use today are synthetic chemicals.

emulsion a stable dispersion of a liquid in another liquid – for example, oil and water in some cosmetic lotions.

encephalin a naturally occurring chemical produced by nerve cells in the brain that has the same effect as morphine or other derivatives of opium, acting as a natural painkiller. Unlike morphine, encephalins are quickly degraded by the body, so there is no build-up of tolerance to them, and hence no addiction. Encephalins are a variety of ◊peptides, as are ◊endorphins, which have similar effects.

encephalitis inflammation of the brain, nearly always due to viral infection but it may also occur in bacterial and other infections. It varies widely in severity, from short-lived, relatively slight effects of headache, drowsiness, and fever to paralysis, coma, and death.

encephalomyelitis in medicine, inflammation of the brain and the spinal cord. It is caused by bacteria, viruses, fungi, malignant cells, and blood following subarachnoid haemorrhage. The disease is serious and requires urgent treatment in hospital.

Encke's comet comet with the shortest known orbital period, 3.3 years. It is named after German mathematician and astronomer Johann Franz Encke (1791–1865), who calculated its orbit in 1819 from earlier sightings.

It was first seen in 1786 by the French astronomer Pierre Méchain (1744–1804). It is the parent body of the Taurid meteor shower and a fragment of it may have hit the Earth in the ◊Tunguska Event in 1908.

In 1913, it became the first comet to be observed throughout its entire orbit when it was photographed near ◊aphelion (the point in its orbit furthest from the Sun) by astronomers at Mount Wilson Observatory in California, USA.

endangered species plant or animal species whose numbers are so few that it is at risk of becoming extinct. Officially designated endangered species are listed by the ◊International Union for the Conservation of Nature (IUCN).

Endangered species are not a new phenomenon; extinction is an integral part of evolution. The replacement of one species by another usually involves the eradication of the less successful form, and ensures the continuance and diversification of life in all forms. However, this only holds good for natural extinctions; those induced by humans are negative, representing an evolutionary dead-end, allowing for no succession. The great majority of recent extinctions have been directly or indirectly induced by humans; most often by the loss, modification, or pollution of the organism's habitat, but also by hunting for 'sport' or for commercial purposes.

According to the Red Data List of endangered species, published in 1996 by the IUCN, 25% of all mammal species (including 46% of primates, 36% of insectivores, and 33% of pigs and antelopes) and 11% of all bird species are threatened with extinction.

Endangered Species

Species	Observation
plants	a quarter of the world's plants are threatened with extinction by the year 2010
amphibians	worldwide decline in numbers; half of New Zealand's frog species are now extinct
birds	three-quarters of all bird species are declining
carnivores	almost all species of cats and bears are declining in numbers
fish	one-third of North American freshwater fish are rare or endangered; half the fish species in Lake Victoria, Africa's largest lake, are close to extinction due to predation by the introduced Nile perch
invertebrates	about 100 species are lost each day due to deforestation; half the freshwater snails in the southeastern USA are now extinct or threatened; a quarter of German invertebrates are threatened
mammals	half of Australia's mammals are threatened; 40% of mammals in France, the Netherlands, Germany, and Portugal are threatened
primates	two-thirds of primate species are threatened
reptiles	over 40% of reptile species are threatened; 20% with extinction

endocrine gland gland that secretes hormones into the bloodstream to regulate body processes. Endocrine glands are most highly developed in vertebrates, but are also found in other animals, notably insects. In humans the main endocrine glands are the pituitary, thyroid, parathyroid, adrenal, pancreas, ovary, and testis.

endogenous in medicine, arising within the body, for example the endogenous amines, serotonin, and noradrenaline, in the brain that are implicated in ◊depression.

endolymph fluid found in the inner ◊ear, filling the central passage of the cochlea as well as the semicircular canals.

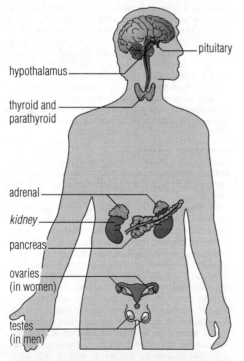

endocrine gland The main human endocrine glands. These glands produce hormones – chemical messengers – which travel in the bloodstream to stimulate certain cells.

Sound waves travelling into the ear pass eventually through the three small bones of the middle ear and set up vibrations in the endolymph. These vibrations are detected by receptors in the cochlea, which send nerve impulses to the hearing centres of the brain.

endoplasm inner, liquid part of a cell's ◊cytoplasm.

endoplasmic reticulum (ER) a membranous system of tubes, channels, and flattened sacs that form compartments within ◊eukaryotic cells. It stores and transports proteins within cells and also carries various enzymes needed for the synthesis of ◊fats. The ◊ribosomes, or the organelles that carry out protein synthesis, are attached to parts of the ER.

Under the electron microscope, ER looks like a series of channels and vesicles, but it is in fact a large, sealed, baglike structure crumpled and folded into a convoluted mass. The interior of the 'bag', the ER lumen, stores various proteins needed elsewhere in the cell, then organizes them into transport vesicles formed by a small piece of ER membrane budding from the main membrane.

endorphin natural substance (a polypeptide) that modifies the action of nerve cells. Endorphins are produced by the pituitary gland and hypothalamus of vertebrates. They lower the perception of pain by reducing the transmission of signals between nerve cells.

Endorphins not only regulate pain and hunger, but are also involved in the release of sex hormones from the pituitary gland. Opiates act in a similar way to endorphins, but are not rapidly degraded by the body, as natural endorphins are, and thus have a long-lasting effect on pain perception and mood. Endorphin release is stimulated by exercise.

endoscope in medicine, tubular instrument inserted into a cavity in the body to investigate and treat diseases during a process known as ◊endoscopy.

endoscopy examination of internal organs or tissues by an instrument allowing direct vision. An endoscope is equipped with an eyepiece, lenses, and its own light source to illuminate the field of vision. The endoscope used to examine the digestive tract is a flexible fibreoptic instrument swallowed by the patient.

There are various types of endoscope in use – some rigid, some flexible – with names prefixed by their site of application (for example, bronchoscope and laryngoscope). The value of endoscopy is in permitting diagnosis without the need for exploratory surgery. Biopsies (tissue samples) and photographs may be taken by way of the endoscope as an aid to diagnosis, or to monitor the effects of treatment. Some surgical procedures can be performed using fine instruments introduced through the endoscope. Keyhole surgery is increasingly popular as a cheaper, safer option for some conditions than conventional surgery.

endoskeleton the internal supporting structure of vertebrates, made up of cartilage or bone. It provides support, and acts as a system of levers to which muscles are attached to provide movement. Certain parts of the skeleton (the skull and ribs) give protection to vital body organs.

Sponges are supported by a network of rigid, or semi-rigid, spiky structures called spicules; a bath sponge is the proteinaceous skeleton of a sponge.

endosperm nutritive tissue in the seeds of most flowering plants. It surrounds the embryo and is produced by an unusual process that parallels the ◊fertilization of the ovum by a male gamete. A second male gamete from the pollen grain fuses with two female nuclei within the ◊embryo sac. Thus endosperm cells are triploid (having three sets of chromosomes); they contain food reserves such as starch, fat, and protein that are utilized by the developing seedling.

In 'non-endospermic' seeds, absorption of these food molecules by the embryo is completed early, so that the endosperm has disappeared by the time of germination.

endotherm 'warm-blooded', or homeothermic, animal. Endotherms have internal mechanisms for regulating their body temperatures to levels different from the environmental temperature. See ◊homeothermy.

endothermic reaction chemical reaction that requires an input of energy in the form of heat for it to proceed; the energy is absorbed from the surroundings by the reactants.

The dissolving of sodium chloride in water and the process of photosynthesis are both endothermic changes. See ◊energy of reaction.

endotoxin in biology, heat stable complex of protein and lipopolysaccharide that is produced following the death of certain bacteria. Endotoxins are typically produced by the Gram negative bacteria and can cause fever. They can also cause shock by rendering the walls of the blood vessels permeable so that fluid leaks into the tissues and blood pressure falls sharply.

end user the user of a computer program; in particular, someone who uses a program to perform a task (such as accounting or playing a computer game), rather than someone who writes programs (a programmer).

Energiya most powerful Soviet space rocket, first launched on 15 May 1987.

Used to launch the Soviet space shuttle, the Energiya ◊booster is capable, with the use of strap-on boosters, of launching payloads of up to 190 tonnes into Earth's orbit.

energy capacity for doing ◊work. Potential energy (PE) is energy deriving from position; thus a stretched spring has elastic PE, and an object raised to a height above the Earth's surface, or the water in an elevated reservoir, has gravitational PE. A lump of coal and a tank of petrol, together with the oxygen needed for their combustion, have chemical energy. Other sorts of energy include elec-

trical and nuclear energy, and light and sound. Moving bodies possess kinetic energy (KE). Energy can be converted from one form to another, but the total quantity stays the same (in accordance with the ◊conservation of energy principle). For example, as an apple falls, it loses gravitational PE but gains KE.

Although energy is never lost, after a number of conversions it tends to finish up as the kinetic energy of random motion of molecules (of the air, for example) at relatively low temperatures. This is 'degraded' energy that is difficult to convert back to other forms.

So-called energy resources are stores of convertible energy. Nonrenewable resources include the fossil fuels (coal, oil, and gas) and nuclear-fission 'fuels' – for example, uranium-235. Renewable resources, such as wind, tidal, and geothermal power, have so far been less exploited. Hydroelectric projects are well established, and wind turbines and tidal systems are being developed.

$E = mc^2$ Einstein's special theory of ◊relativity in 1905 correlates any gain, E, in energy with a gain, m, in mass, by the equation $E = mc^2$, in which c is the speed of light. The conversion of mass into energy in accordance with this equation applies universally, although it is only for nuclear reactions that the percentage change in mass is large enough to detect.

energy, alternative energy from sources that are renewable and ecologically safe, as opposed to sources that are nonrenewable with toxic by-products, such as coal, oil, or gas (fossil fuels), and uranium (for nuclear power). The most important alternative energy source is flowing water, harnessed as ◊hydroelectric power. Other sources include the oceans' tides and waves (see ◊tidal power station and ◊wave power), ◊wind power (harnessed by windmills and wind turbines), the Sun (◊solar energy), and the heat trapped in the Earth's crust (◊geothermal energy) (see also ◊cold fusion).

energy conservation methods of reducing energy use through insulation, increasing energy efficiency, and changes in patterns of use. Profligate energy use by industrialized countries contributes greatly to air pollution and the ◊greenhouse effect when it draws on nonrenewable energy sources.

It has been calculated that increasing energy efficiency alone could reduce carbon dioxide emissions in several high-income countries by 1–2% a year.

energy of reaction energy released or absorbed during a chemical reaction, also called **enthalpy of reaction** or **heat of reaction**. In a chemical reaction, the energy stored in the reacting molecules is rarely the same as that stored in the product molecules. Depending on which is the greater, energy is either released (an exothermic reaction) or absorbed (an endothermic reaction) from the surroundings (see ◊conservation of energy). The amount of energy released or absorbed by the quantities of substances represented by the chemical equation is the energy of reaction.

engine device for converting stored energy into useful work or movement. Most engines use a fuel as their energy store. The fuel is burnt to produce heat energy – hence the name 'heat engine' – which is then converted into movement. Heat engines can be classified according to the fuel they use (◊petrol engine or ◊diesel engine), or according to whether the fuel is burnt inside (◊internal combustion engine) or outside (◊steam engine) the engine, or according to whether they produce a reciprocating or rotary motion (◊turbine or ◊Wankel engine).

engineering the application of science to the design, construction, and maintenance of works, machinery, roads, railways, bridges, harbour installations, engines, ships, aircraft and airports, spacecraft and space stations, and the generation, transmission, and use of electrical power. The main divisions of engineering are aerospace, chemical, civil, electrical, electronic, gas, marine, materials, mechanical, mining, production, radio, and structural.

English Nature agency created in 1991 from the division of the ◊Nature Conservancy Council into English, Scottish, and Welsh sections.

enterovirus in medicine, one of a family of small polyhedral RNA-containing viruses that enter the body through the gut and are able to penetrate the central nervous system. They include the ◊polio virus; Coxsackie viruses that cause diseases such as herpangina and epidemic myalgia; and ECHO viruses that cause aseptic meningitis.

enthalpy in chemistry, alternative term for ◊energy of reaction, the heat energy associated with a chemical change.

entomology study of ◊insects.

entropy in ◊thermodynamics, a parameter representing the state of disorder of a system at the atomic, ionic, or molecular level; the greater the disorder, the higher the entropy. Thus the fast-moving disordered molecules of water vapour have higher entropy than those of more ordered liquid water, which in turn have more entropy than the molecules in solid crystalline ice.

In a closed system undergoing change, entropy is a measure of the amount of energy unavailable for useful work. At ◊absolute zero (−273.15°C/−459.67°F/0 K), when all molecular motion ceases and order is assumed to be complete, entropy is zero.

E number code number for additives that have been approved for use by the European Commission (EC). The E written before the number stands for European. E numbers do not have to be displayed on lists of ingredients, and the manufacturer may choose to list additives by their name instead. E numbers cover all categories of additives apart from flavourings.

Additives, other than flavourings, that are not approved by the European Commission, but are still used in Britain, are represented by a code number without an E.

envelope in geometry, a curve that touches all the members of a family of lines or curves. For example, a family of three equal circles all touching each other and forming a triangular pattern (like a clover leaf) has two envelopes: a small circle that fits in the space in the middle, and a large circle that encompasses all three circles.

environment in ecology, the sum of conditions affecting a particular organism, including physical surroundings, climate, and influences of other living organisms. See also ◊biosphere and ◊habitat.

In common usage, 'the environment' often means the total global environment, without reference to any particular organism. In genetics, it is the external influences that affect an organism's development, and thus its ◊phenotype.

environmental archaeology subfield of archaeology aimed at identifying processes, factors, and conditions of past biological and physical environmental systems and how they relate to cultural systems.

It is a field where archaeologists and natural scientists combine their skills to reconstruct the human uses of plants and animals and how societies adapted to changing environmental conditions.

environmental audit another name for ◊green audit, the inspection of a company to assess its environmental impact.

Environmentally Sensitive Area (ESA) scheme introduced by the UK Ministry of Agriculture 1984, as a result of EC legislation, to protect some of the most beautiful areas of the British countryside from the loss and damage caused by agricultural change. The first areas to be designated ESAs were in the Pennine Dales, the North Peak District, the Norfolk Broads, the Breckland, the Suffolk River Valleys, the Test Valley, the South Downs, the Somerset Levels and Moors, West Penwith, Cornwall, the Shropshire Borders, the Cambrian Mountains, and the Lleyn Peninsula.

The scheme is voluntary, with farmers being encouraged to adapt their practices so as to enhance or maintain the natural features of the landscape and conserve wildlife habitat. A farmer who joins the scheme agrees to manage the land in this way for at least five years. In return for this agreement, the Ministry of Agriculture pays the farmer a sum that reflects the financial losses incurred as a result of reconciling conservation with commercial farming.

Environmental Protection Agency (EPA) US agency set up in 1970 to control water and air quality, industrial and commercial wastes, pesticides, noise, and radiation. In its own words, it aims to protect 'the country from being degraded, and its health threatened, by a multitude of human activities initiated without regard to long-ranging effects upon the life-supporting properties, the economic uses, and the recreational value of air, land, and water'.

environment–heredity controversy see ◊nature–nurture controversy.

enzyme biological ◊catalyst produced in cells, and capable of speeding up the chemical reactions necessary for life. They are large, complex ◊proteins, and are highly specific, each chemical reaction requiring its own particular enzyme. The enzyme's specificity arises from its *active site*, an area with a shape corresponding to part of the molecule with which it reacts (the substrate). The enzyme and the substrate slot together forming an enzyme–substrate complex that allows the reaction to take place, after which the enzyme falls away unaltered.

The activity and efficiency of enzymes are influenced by various factors, including temperature and pH conditions. Temperatures above 60°C/140°F damage (denature) the intricate structure of enzymes, causing reactions to cease. Each enzyme operates best within a specific pH range, and is denatured by excessive acidity or alkalinity.

Digestive enzymes include amylases (which digest starch), lipases (which digest fats), and proteases (which digest protein). Other enzymes play a part in the conver-

E Numbers *A selection of food additives authorized by the European Commission*

Number	Name	Typical use	Number	Name	Typical use
Colours			**Preservatives**		
E102	tartrazine	soft drinks	E215	sodium ethyl para-hydroxy-benzoate	
E104	quinoline yellow				
E110	sunset yellow	biscuits	E216	propyl para-hydroxy-benzoate	
E120	cochineal	alcoholic drinks			
E122	carmoisine	jams and preserves	E217	sodium propyl para-hydroxy-benzoate	
E123	amaranth				
E124	ponceau 4R	dessert mixes	E218	methyl para-hydroxy-benzoate	
E127	erythrosine	glacé cherries			
E131	patent blue V		E220	sulphur dioxide	
E132	indigo carmine		E221	sodium sulphate	dried fruit, dehydrated vegetables, fruit juices and syrups, sausages, fruit-based dairy desserts, cider, beer, and wine; also used to prevent browning of peeled potatoes and to condition biscuit doughs
E142	green S	pastilles	E222	sodium bisulphite	
E150	caramel	beers, soft drinks, sauces, gravy browning	E223	sodium metabisulphite	
E151	black PN		E224	potassium metabisulphite	
E160 (b)	annatto; bixin; norbixin	crisps			
E180	pigment rubine (lithol rubine BK)		E226	calcium sulphite	
Anti-oxidants			E227	calcium bisulphite	
E310	propyl gallate	vegetable oils; chewing gum	E249	potassium nitrite	
E311	octyl gallate		E250	sodium nitrite	bacon, ham, cured meats, corned beef and some cheeses
E312	dodecyl gallate		E251	sodium nitrate	
E320	butylated hydroxynisole (BHA)	beef stock cubes; cheese spread	E252	potassium nitrate	
E321	butylated hydroxytoluene (BHT)	chewing gum	**Others**		
Emulsifiers and stabilizers			E450 (a)	disodium dihydrogen diphosphate trisodium diphosphate tetrasodium diphosphate tetrapotassium diphosphate	butters, sequestrants, emulsifying salts, stabilizers, texturizers
E407	carageenan	quick-setting jelly mixes; milk shakes			
E413	tragacanth	salad dressings; processed cheese	E450 (b)	pentasodium triphosphate pentapotassium triphosphate	raising agents, used in whipping cream, fish and meat products, bread, processed cheese, canned vegetables
Preservatives					
E210	benzoic acid				
E211	sodium benzoate	beer, jam, salad cream, soft drinks, fruit pulp			
E212	potassium benzoate	fruit-based pie fillings, marinated herring and mackerel			
E213	calcium benzoate				
E214	ethyl para-hydroxy-benzoate				

sion of food energy into ◊ATP; the manufacture of all the molecular components of the body; the replication of ◊DNA when a cell divides; the production of hormones; and the control of movement of substances into and out of cells.

Enzymes have many medical and industrial uses, from washing powders to drug production, and as research tools in molecular biology. They can be extracted from bacteria and moulds, and ◊genetic engineering now makes it possible to tailor an enzyme for a specific purpose.

enzyme-linked immunosorbant assay (ELISA) in medicine, technique for accurately determining the quantity of a substance in a sample, for example hormone levels in a blood sample. It involves the preparation of an antibody to the substance and the enzyme to which the antibody binds. The presence of a substance can be measured accurately as a result of colour changes that occur following reaction between the antibody and the enzyme in the presence of the substance. The ELISA technique is used in the detection of altered hormone levels in the urine in pregnancy testing kits.

Eocene second epoch of the Tertiary period of geological time, 56.5–35.5 million years ago. Originally considered the earliest division of the Tertiary, the name means 'early recent', referring to the early forms of mammals evolving at the time, following the extinction of the dinosaurs.

ephedrine drug that acts like adrenaline on the sympathetic ◊nervous system (sympathomimetic). Once used to relieve bronchospasm in ◊asthma, it has been superseded by safer, more specific drugs. It is contained in some cold remedies as a decongestant. Side effects include rapid heartbeat, tremor, dry mouth, and anxiety.

Ephedrine is an alkaloid, $C_{10}H_{15}NO_1$, derived from Asian gymnosperms (genus *Ephedra*) or synthesized. It is sometimes misused, and excess leads to mental confusion and increased confidence in one's own capabilities as they actually decline.

ephemeral plant plant with a very short life cycle, sometimes as little as six or eight weeks. It may complete several generations in one growing season.

epicentre the point on the Earth's surface immediately above the seismic focus of an ◊earthquake. Most damage usually takes place at an earthquake's epicentre. The term sometimes refers to a point directly above or below a nuclear explosion ('at ground zero').

epicyclic gear or *sun-and-planet gear* gear system that consists of one or more gear wheels moving around another. Epicyclic gears are found in bicycle hub gears and in automatic gearboxes.

epicycloid in geometry, a curve resembling a series of arches traced out by a point on the circumference of a circle that rolls around another circle of a different diameter. If the two circles have the same diameter, the curve is a ◊cardioid.

epidemic outbreak of infectious disease affecting large numbers of people at the same time. A widespread epidemic that sweeps across many countries (such as the ◊Black Death in the late Middle Ages) is known as a *pandemic*.

epicycloid A seven-cusped epicycloid, formed by a point on the circumference of a circle (of diameter d) that rolls around another circle (of diameter 7*d*/3).

epidermis outermost layer of ◊cells on an organism's body. In plants and many invertebrates such as insects, it consists of a single layer of cells. In vertebrates, it consists of several layers of cells.

The epidermis of plants and invertebrates often has an outer noncellular ◊cuticle that protects the organism from desiccation.

Epidermis

Each person sheds an average of 18 kg/40lb of skin in a lifetime. The outer layer of skin – the epidermis – consists entirely of dead cells that fall off and are replaced by the level below.

---

epidural anaesthetic in medicine, procedure that induces loss of sensation in the lower part of the body. The spinal nerves are blocked in the epidural space by injection of small volumes of local ◊anaesthetic into the tissues surrounding the spinal cord through a catheter. The catheter can be used to introduce further volumes of local anaesthetic if the block begins to wear off before the end of the procedure. Epidurals are commonly used as a method of pain relief during childbirth.

epigeal seed germination in which the ◊cotyledons (seed leaves) are borne above the soil.

epiglottis small flap located behind the root of the tongue in mammals. It closes off the end of the windpipe during swallowing to prevent food from passing into it and causing choking.

The action of the epiglottis is a highly complex reflex process involving two phases. During the first stage a mouthful of chewed food is lifted by the tongue towards the top and back of the mouth. This is accompanied by the cessation of breathing and by the blocking of the nasal areas from the mouth. The second phase involves the epiglottis moving over the larynx while the food passes down into the oesophagus.

epilepsy medical disorder characterized by a tendency to develop fits, which are convulsions or abnormal feelings caused by abnormal electrical discharges in the cerebral hemispheres of the ◊brain. Epilepsy can be controlled with a number of anticonvulsant drugs.

The term epilepsy covers a range of conditions from mild 'absences', involving momentary loss of awareness, to major convulsions. In some cases the abnormal electrical activity is focal (confined to one area of the brain); in others it is generalized throughout the cerebral cortex. Fits are classified according to their clinical type. They include: the *grand mal* seizure with convulsions and loss of consciousness; the fleeting absence of *petit mal*, almost exclusively a disorder of childhood; *Jacksonian* seizures, originating in the motor cortex; and *temporal-lobe* fits, which may be associated with visual hallucinations and bizarre disturbances of the sense of smell.

Epilepsy affects 1–3% of the world's population. It may arise spontaneously or may be a consequence of brain surgery, organic brain disease, head injury, metabolic disease, alcoholism or withdrawal from some drugs. Almost a third of patients have a family history of the condition.

Most epileptics have infrequent fits that have little impact on their daily lives. Epilepsy does not imply that the sufferer has any impairment of intellect, behaviour, or personality.

epiphyte any plant that grows on another plant or object above the surface of the ground, and has no roots in the soil. An epiphyte does not parasitize the plant it grows on but merely uses it for support. Its nutrients are obtained from rainwater, organic debris such as leaf litter, or from the air. The greatest diversity of epiphytes is found in tropical areas and includes many orchids.

episiotomy incision made in the perineum (the tissue bridging the vagina and rectum) to facilitate childbirth and prevent tearing of the vagina.

Episiotomy may be necessary, mainly for women giving birth for the first time, to widen the birth outlet and prevent perineal tearing. The incision is made in the second stage of labour, as the largest part of the baby's head begins to emerge from the birth canal.

An episiotomy is quickly repaired using absorbable stitches under local anaesthetic.

epithelium in animals, tissue of closely packed cells that forms a surface or lines a cavity or tube. Epithelium may be protective (as in the skin) or secretory (as in the cells lining the wall of the gut).

epoch subdivision of a geological period in the geological time scale. Epochs are sometimes given their own names (such as the Palaeocene, Eocene, Oligocene, Miocene, and Pliocene epochs comprising the Tertiary period), or they are referred to as the late, early, or middle portions of a given period (as the Late Cretaceous or the Middle Triassic epoch).

epoxy resin synthetic ◊resin used as an ◊adhesive and as an ingredient in paints. Household epoxy resin adhesives come in component form as two separate tubes of chemical, one tube containing resin, the other a curing agent (hardener). The two chemicals are mixed just before application, and the mix soon sets hard.

EPROM (acronym for *erasable programmable read-only memory*) computer memory device in the form of an ◊integrated circuit (chip) that can record data and retain it indefinitely. The data can be erased by exposure to ultraviolet light, and new data recorded. Other kinds of computer memory chips are ◊ROM (read-only memory), ◊PROM (programmable read-only memory), and ◊RAM (random-access memory).

Epsom salts $MgSO_4.7H_2O$ hydrated magnesium sulphate, used as a relaxant and laxative and added to baths to soothe the skin. The name is derived from a bitter saline spring at Epsom, Surrey, England, which contains the salt in solution.

equation in chemistry, representation of a chemical reaction by symbols and numbers; see ◊chemical equation.

equation in mathematics, expression that represents the equality of two expressions involving constants and/or variables, and thus usually includes an equals sign (=). For example, the equation $A = \pi r^2$ equates the area A of a circle of radius r to the product πr^2.

The algebraic equation $y = mx + c$ is the general one in coordinate geometry for a straight line.

If a mathematical equation is true for all variables in a given domain, it is sometimes called an identity and denoted by $\equiv$.

Thus $(x + y)^2 \equiv x^2 + 2xy + y^2$ for all $x, y \in R$.

An *indeterminate equation* is an equation for which there is an infinite set of solutions – for example, $2x = y$.

A *diophantine equation* is an indeterminate equation in which both the solution and the terms must be whole numbers (after Diophantus of Alexandria, c. AD 250).

Equator or *terrestrial equator* the ◊great circle whose plane is perpendicular to the Earth's axis (the line joining the poles). Its length is 40,092 km/24,901.8 mi, divided into 360 degrees of longitude. The Equator encircles the broadest part of the Earth, and represents 0° latitude. It divides the Earth into two halves, called the northern and the southern hemispheres.

The *celestial equator* is the circle in which the plane of the Earth's Equator intersects the ◊celestial sphere.

equatorial coordinates in astronomy, a system for measuring the position of astronomical objects on the ◊celestial sphere with reference to the plane of the Earth's equator.

◊Declination (symbol δ), analogous to latitude, is measured in degrees from the equator to the north ($\delta = 90°$) or south ($\delta = -90°$) celestial poles. Right ◊ascension (symbol α), analogous to longitude, is normally measured in hours of time ($\alpha = 0$ h to 24 h) eastward along the equator from a fixed point known as the first point of ◊Aries or the ◊vernal equinox.

equatorial mounting in astronomy, a method of mounting a ◊telescope to simplify the tracking of celestial objects. One axis (the polar axis) is mounted parallel to the rotation axis of the Earth so that the telescope can be turned about it to follow objects across the sky. The ◊declination axis moves the telescope in declination and is clamped before tracking begins. Another advantage over the simpler altazimuth mounting is that the orientation of the image is fixed, permitting long-exposure photography.

equilateral of a geometrical figure, having all sides of equal length.

equilibrium in physics, an unchanging condition in which an undisturbed system can remain indefinitely in a state of balance. In a *static equilibrium*, such as an object resting on the floor, there is no motion. In a *dynamic equilibrium*, in contrast, a steady state is maintained by constant, though opposing, changes. For example, in a sealed bottle half-full of water, the constancy of the water level is a result of molecules evaporating from the surface and condensing on to it at the same rate.

equinox the points in spring and autumn at which the Sun's path, the ◊ecliptic, crosses the celestial equator, so that the day and night are of approximately equal length. The *vernal equinox* occurs about 21 March and the *autumnal equinox*, 23 Sept.

era any of the major divisions of geological time, each including several periods, but smaller than an eon. The currently recognized eras all fall within the Phanerozoic eon – or the vast span of time, starting about 570 million years ago, when fossils are found to become abundant. The eras in ascending order are the Palaeozoic, Mesozoic, and Cenozoic. We are living in the Recent epoch of the Quaternary period of the Cenozoic era.

Eratosthenes' sieve a method for finding ◊prime numbers. It involves writing in sequence all numbers from 2. Then, starting with 2, cross out every second number (but not 2 itself), thus eliminating numbers that can be divided by 2. Next, starting with 3, cross out every third number (but not 3 itself), and continue the process for 5, 7, 11, 13, and so on. The numbers that remain are primes.

erbium soft, lustrous, greyish, metallic element of the ◊lanthanide series, symbol Er, atomic number 68, relative atomic mass 167.26. It occurs with the element yttrium or as a minute part of various minerals. It was discovered in 1843 by Carl Mosander (1797–1858), and named after the town of Ytterby, Sweden, near which the lanthanides (rare-earth elements) were first found.

Erbium has been used since 1987 to amplify data pulses in optical fibre, enabling faster transmission. Erbium ions in the fibreglass, charged with infrared light, emit energy by amplifying the data pulse as it moves along the fibre.

erg c.g.s. unit of work, replaced in the SI system by the ◊joule. One erg of work is done by a force of one ◊dyne moving through one centimetre.

ergonomics study of the relationship between people and the furniture, tools, and machinery they use at work. The object is to improve work performance by removing sources of muscular stress and general fatigue: for example, by presenting data and control panels in easy-to-view form, making office furniture comfortable, and creating a generally pleasant environment.

Good ergonomic design makes computer systems easier to use and minimizes the health hazards and physical stresses of working with computers for many hours a day: it helps data entry workers to avoid conditions like ◊repetitive strain injury (RSI), eyestrain, and back and muscle aches.

ergosterol substance that, under the action of the Sun's ultraviolet rays on the skin, gives rise to the production of vitamin D2 – a vitamin that helps in calcium and phosphorus metabolism, promotes bone formation, and in children prevents ◊rickets.

Ergosterol $C_{28}H_{43}OH$ is a ◊sterol that occurs in the fungus ergot (hence the name), yeast and other fungi, and some animal fats. The principal source of commercial ergosterol is yeast.

ergotamine ◊alkaloid $C_{33}H_{35}O_5N_5$ administered to treat migraine. Isolated from ergot, a fungus that colonizes rye, it relieves symptoms by causing the cranial arteries to constrict. Its use is limited by severe side effects, including nausea and abdominal pain.

Eridanus in astronomy, the sixth largest constellation, which meanders from the celestial equator (see ◊celestial sphere) deep into the southern hemisphere of the sky. Eridanus is represented as a river. Its brightest star is ◊Achernar, a corruption of the Arabic for 'the end of the river'.

ERNIE acronym for *electronic random number indicator*, machine designed and produced by the UK Post Office Research Station to select a series of random 9-figure numbers to indicate prizewinners among Premium Bond holders.

Eros in astronomy, an asteroid, discovered in 1898, that can pass 22 million km/14 million mi from the Earth, as observed in 1975. Eros was the first asteroid to be discovered that has an orbit coming within that of Mars. It is elongated, measures about 36×12 km/22×7 mi, rotates around its shortest axis every 5.3 hours, and orbits the Sun every 1.8 years.

The Near Earth Asteroid Rendezvous (NEAR) launched in Feb 1996 is estimated to take three years to reach Eros. It will spend a year circling the asteroid in an attempt to determine what it is made of.

erosion wearing away of the Earth's surface, caused by the breakdown and transportation of particles of rock or soil (by contrast, ◊weathering does not involve transportation). Agents of erosion include the sea, rivers, glaciers, and wind.

Water, consisting of sea waves and currents, rivers, and rain; ice, in the form of glaciers; and wind, hurling sand fragments against exposed rocks and moving dunes along, are the most potent forces of erosion.

People also contribute to erosion by bad farming practices and the cutting down of forests, which can lead to the formation of dust bowls.

There are several processes of erosion including ◊hydraulic action, ◊corrosion, ◊attrition, and ◊solution.

erratic in geology, a displaced rock that has been transported by a glacier or some other natural force to a site of different geological composition.

error in computing, a fault or mistake, either in the software or on the part of the user, that causes a program to stop running (crash) or produce unexpected results. Program errors, or bugs, are largely eliminated in the course of the programmer's initial testing procedure, but some will remain in most programs. All computer operating systems are designed to produce an *error message* (on the display screen, or in an error file or printout) whenever an error is detected, reporting that an error has taken place and, wherever possible, diagnosing its cause.

error detection in computing, the techniques that enable a program to detect incorrect data. A common method is to add a check digit to important codes, such as account numbers and product codes. The digit is chosen so that the code conforms to a rule that the program can verify. Another technique involves calculating the sum (called the ◊hash total) of each instance of a particular item of data, and storing it at the end of the data.

erythroblast in biology, a series of nucleated cells that go through various stages of development in the bone marrow until they form red blood cells (erythrocytes). This process is known as erythropoiesis. Erythroblasts can appear in the blood in people with blood cancers.

erythrocyte another name for ◊red blood cell.

erythrocyte sedimentation rate (ESR) in medicine, measure of the rate at which red blood cells (erythrocytes) settle out of suspension in blood plasma. This is affected by the amount of protein in the plasma which, in turn, is affected by the presence of disease. Determination of the ESR is useful in the diagnosis of certain diseases, such as infection and malignancy.

erythropoietin (EPO) in biology, a naturally occurring hormone, secreted mainly by the kidneys in adults and the liver in children, that stimulates production of red blood cells, which carry oxygen around the body. It is released in response to a lowered percentage of oxygen in the blood reaching the kidneys, such as in anaemic subjects. Recombinant human erythropoietin is used therapeutically to treat the anaemia associated with chronic kidney failure. A synthetic version is sometimes used illegally by athletes in endurance sports as it increases the oxygen-carrying capacity of the blood.

ESA abbreviation for ◊*European Space Agency*.

escalator automatic moving staircase that carries people between floors or levels. It consists of treads linked in an endless belt arranged to form strips (steps), powered by an electric motor that moves both steps and handrails at the same speed. Towards the top and bottom the steps flatten out for ease of passage. The first escalator was exhibited in Paris in 1900.

escape velocity in physics, minimum velocity with which an object must be projected for it to escape from the gravitational pull of a planetary body. In the case of the Earth, the escape velocity is 11.2 kps/6.9 mps; the Moon, 2.4 kps/1.5 mps; Mars, 5 kps/3.1 mps; and Jupiter, 59.6 kps/37 mps.

escarpment or *cuesta* large ridge created by the erosion of dipping sedimentary rocks. It has one steep side (scarp) and one gently sloping side (dip). Escarpments are common features of chalk landscapes, such as the Chiltern Hills and the North Downs in England. Certain features are associated with chalk escarpments, including dry valleys (formed on the dip slope), combes (steep-sided valleys on the scarp slope), and springs.

Escherichia coli rod-shaped Gram-negative bacterium (see ◊bacteria) that lives, usually harmlessly, in the colon of most warm-blooded animals. It is the commonest cause of urinary tract infections in humans. It is sometimes found in water or meat where faecal contamination has occurred and can cause severe gastric problems.

Escherichia coli is the only species in the bacterial family Enterobacteriaceae.

esker narrow, steep-walled ridge, often sinuous and sometimes branching, formed beneath a glacier. It is made of sands and gravels, and represents the course of a subglacial river channel. Eskers vary in height from 3–30 m/10–100 ft and can be up to 160 km/100 mi or so in length.

ESP abbreviation for ◊*extrasensory perception*.

essential amino acid water-soluble organic molecule vital to a healthy diet; see ◊amino acid.

essential fatty acid organic compound consisting of a hydrocarbon chain and important in the diet; see ◊fatty acid.

ester organic compound formed by the reaction between an alcohol and an acid, with the elimination of water. Unlike ◊salts, esters are covalent compounds.

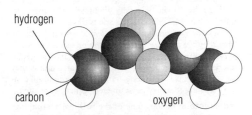

ester Molecular model of the ester ethyl ethanoate (ethyl acetate) $CH_3CH_2COOCH_3$.

estradiol alternative spelling of oestradiol, a type of ◊oestrogen (female sex hormone).

estuary river mouth widening into the sea, where fresh water mixes with salt water and tidal effects are felt.

ethanal common name *acetaldehyde* CH_3CHO one of the chief members of the group of organic compounds known as ◊aldehydes. It is a colourless inflammable liquid boiling at 20.8°C/69.6°F. Ethanal is formed by the oxidation of ethanol or ethene and is used to make many other organic chemical compounds.

ethanal trimer common name *paraldehyde* $(CH_3CHO)_3$ colourless liquid formed from ethanal. It is soluble in water.

ethane CH_3CH_3 colourless, odourless gas, the second member of the ◊alkane series of hydrocarbons (paraffins).

ethane-1,2-diol technical name for ◊glycol.

ethanoate common name *acetate* $CH_3CO_2^-$ negative ion derived from ethanoic (acetic) acid; any salt containing this ion. In textiles, acetate rayon is a synthetic fabric made from modified cellulose (wood pulp) treated with ethanoic acid; in photography, acetate film is a non-flammable film made of cellulose ethanoate.

ethanoic acid common name *acetic acid* CH_3CO_2H one of the simplest fatty acids (a series of organic acids). In the pure state it is a colourless liquid with an unpleasant pungent odour; it solidifies to an icelike mass of crystals at 16.7°C/62.4°F, and hence is often called glacial ethanoic acid. Vinegar contains 5% or more ethanoic acid, produced by fermentation.

Cellulose (derived from wood or other sources) may be treated with ethanoic acid to produce a cellulose ethanoate (acetate) solution, which can be used to make plastic items by injection moulding or extruded to form synthetic textile fibres.

ethanol common name *ethyl alcohol* C_2H_5OH alcohol found in beer, wine, cider, spirits, and other alcoholic drinks. When pure, it is a colourless liquid with a pleasant odour, miscible with water or ether; it burns in air with a pale blue flame. The vapour forms an explosive mixture with air and may be used in high-compression internal combustion engines.

It is produced naturally by the fermentation of carbohydrates by yeast cells. Industrially, it can be made by absorption of ethene and subsequent reaction with water, or by the reduction of ethanal in the presence of a catalyst, and is widely used as a solvent.

Ethanol is used as a raw material in the manufacture of ether, chloral, and iodoform. It can also be added to petrol, where it improves the performance of the engine, or be used as a fuel in its own right (as in Brazil). Crops such as sugar cane may be grown to provide ethanol (by fermentation) for this purpose.

ethene common name *ethylene* C_2H_4 colourless, flammable gas, the first member of the ◊alkene series of hydrocarbons. It is the most widely used synthetic organic chemical and is used to produce the plastics polyethene (polyethylene), polychloroethene, and polyvinyl chloride (PVC). It is obtained from natural gas or coal gas, or by the dehydration of ethanol.

Ethene is produced during plant metabolism and is classified as a plant hormone. It is important in the ripening of fruit and in ◊abscission. Small amounts of ethene are often added to the air surrounding fruit to artificially promote ripening. Tomato and marigold plants show distorted growth in concentrations as low as 0.01 parts per million. Plants also release ethene when they are under stress. German physicists invented a device in 1997 that measured stress levels in plants by measuring surrounding ethene levels.

ether in chemistry, any of a series of organic chemical compounds having an oxygen atom linking the carbon atoms of two hydrocarbon radical groups (general formula R-O-R′); also the common name for ethoxyethane $C_2H_5OC_2H_5$ (also called diethyl ether).

This is used as an anaesthetic and as an external cleansing agent before surgical operations. It is also used as a solvent, and in the extraction of oils, fats, waxes, resins, and alkaloids.

Ethoxyethane is a colourless, volatile, inflammable liquid, slightly soluble in water, and miscible with ethanol. It is prepared by treatment of ethanol with excess concentrated sulphuric acid at 140°C/284°F.

ether or **aether** in the history of science, a hypothetical medium permeating all of space. The concept originated with the Greeks, and has been revived on several occasions to explain the properties and propagation of light. It was supposed that light and other electromagnetic radiation – even in outer space – needed a medium, the ether, in which to travel. The idea was abandoned with the acceptance of ◊relativity.

ethics committee in medicine, review panel made up of both medical and lay people to facilitate medical research in the interests of society, to protect subjects from possible harm due to the research being undertaken, preserve the rights of research subjects, and to provide reassurances to the public on the safety of clinical research. Information provided by researchers should give sufficient detail to enable the ethics committee to evaluate the ethics as well as the scientific merit of the research proposal.

Research involving humans has been governed by protective codes since the 1960s. These codes include those of the World Medical Association (Declaration of Helsinki) and the Medical Research Council in the UK. The development of ethics committees parallels the development of controlled ◊clinical trials of new medicines and surgical procedures since the 1940s and 1950s.

ethnoarchaeology the study of human behaviour and the material culture of living societies, in order to see how materials enter the archaeological record, and thus to provide hypotheses explaining the production, use, and disposal patterns of ancient material culture.

ethnography study of living cultures, using anthropological techniques like participant observation (where the anthropologist lives in the society being studied) and a reliance on informants. Ethnography has provided much data of use to archaeologists as analogies.

ethnology study of contemporary peoples, concentrating on their geography and culture, as distinct from their social systems. Ethnologists make a comparative analysis of data from different cultures to understand how cultures work and why they change, with a view to deriving general principles about human society.

ethology comparative study of animal behaviour in its natural setting. Ethology is concerned with the causal mechanisms (both the stimuli that elicit behaviour and the physiological mechanisms controlling it), as well as the development of behaviour, its function, and its evolutionary history.

Ethology was pioneered during the 1930s by the Austrians Konrad Lorenz and Karl von Frisch who, with the Dutch zoologist Nikolaas Tinbergen, received the Nobel prize in 1973. Ethologists believe that the significance of an animal's behaviour can be understood only in its natural context, and emphasize the importance of field studies and an evolutionary perspective. A recent development within ethology is ◊sociobiology, the study of the evolutionary function of ◊social behaviour.

ethyl alcohol common name for ◊ethanol.

ethylene common name for ◊ethene.

ethylene glycol alternative name for ◊glycol.

ethyne common name acetylene CHCH colourless inflammable gas produced by mixing calcium carbide and water. It is the simplest member of the ◊alkyne series of hydrocarbons. It is used in the manufacture of the synthetic rubber neoprene, and in oxyacetylene welding and cutting.

Ethyne was discovered by Edmund Davy in 1836 and was used in early gas lamps, where it was produced by the reaction between water and calcium carbide. Its combustion provides more heat, relatively, than almost any other fuel known (its calorific value is five times that of hydrogen). This means that the gas gives an intensely hot flame; hence its use in oxyacetylene torches.

etiolation in botany, a form of growth seen in plants receiving insufficient light. It is characterized by long, weak stems, small leaves, and a pale yellowish colour (◊chlorosis) owing to a lack of chlorophyll. The rapid increase in height enables a plant that is surrounded by others to quickly reach a source of light, after which a return to normal growth usually occurs.

eugenics study of ways in which the physical and mental characteristics of the human race may be improved. The eugenic principle was abused by the Nazi Party in Germany during the 1930s and early 1940s to justify the attempted extermination of entire social and ethnic groups and the establishment of selective breeding programmes. Modern eugenics is concerned mainly with the elimination of genetic disease.

The term was coined by the English scientist Francis Galton in 1883, and the concept was originally developed in the late 19th century with a view to improving human intelligence and behaviour.

In 1986 Singapore became the first democratic country to adopt an openly eugenic policy by guaranteeing pay increases to female university graduates when they give birth to a child, while offering grants towards house purchases for non-graduate married women on condition that they are sterilized after the first or second child. In China in June 1995, a law was passed making it illegal for carriers of certain genetic diseases to marry unless they agree to sterilization or long-term contraception. All couples wishing to marry must undergo genetic screening.

eukaryote in biology, one of the two major groupings into which all organisms are divided. Included are all organisms, except bacteria and cyanobacteria (◊blue-green algae), which belong to the ◊prokaryote grouping.

The cells of eukaryotes possess a clearly defined nucleus, bounded by a membrane, within which DNA is formed into distinct chromosomes. Eukaryotic cells also contain mitochondria, chloroplasts, and other structures (organelles) that, together with a defined nucleus, are lacking in the cells of prokaryotes.

eureka in chemistry, alternative name for the copper–nickel alloy ◊constantan, which is used in electrical equipment.

Eurocodes series of codes giving design rules for all types of engineering structures, except certain very specialized forms, such as nuclear reactors. The codes will be given the status of ENs (European standards) and will be administered by CEN (European Committee for Standardization). ENs will eventually replace national codes, in Britain currently maintained by the BSI (British Standards Institute), and will include parameters to reflect local requirements.

Europa in astronomy, the fourth-largest moon of the planet Jupiter, diameter 3,140 km/1,950 mi, orbiting 671,000 km/417,000 mi from the planet every 3.55 days. It is covered by ice and criss-crossed by thousands of thin cracks, each some 50,000 km/30,000 mi long.

NASA's robot probe, *Galileo*, began circling Europa in Feb 1997 and is expected to send back around 800 images from 50 different sites by 1999. NASA plans to send a $250 million robot probe in 2001 or 2002 to circle 100 km/60 mi above the surface, using radar and lasers to establish whether water exists beneath the icy surface. If found, a lander would be launched by 2006 to release a robot that would melt the ice and release a submarine to take pictures. This operation is also billed at $250 million. One of the first discoveries was that what were thought to be cracks covering the surface of the moon are in fact low ridges. Further investigation is needed to determine their origin.

European Southern Observatory observatory operated jointly by Belgium, Denmark, France, Germany, Italy, the Netherlands, Sweden, and Switzerland with headquarters near Munich. Its telescopes, located at La Silla, Chile, include a 3.6 m/142 in reflector opened in 1976 and the 3.58 m/141 in New Technology Telescope opened in 1990. By 1988 work began on the Very Large Telescope, consisting of four 8 m/315 in reflectors mounted independently but capable of working in combination.

European Space Agency (ESA) organization of European countries (Austria, Belgium, Denmark, Finland, France, Germany, Ireland, Italy, the Netherlands, Norway, Spain, Sweden, Switzerland, and the UK) that engages in space research and technology. It was founded in 1975, with headquarters in Paris.

ESA has developed various scientific and communications satellites, the *Giotto* space probe, and the ◊*Ariane* rockets. ESA built ◊*Spacelab*, and plans to build its own space station, *Columbus*, for attachment to a US space station.

The ESA's earth-sensing satellite ERS-2 was launched successfully in April 1995. It will work in tandem with ERS-1 launched in 1991, and should improve measurements of global ozone.

europium soft, greyish, metallic element of the ◊lanthanide series, symbol Eu, atomic number 63, relative atomic mass 151.96. It is used in lasers and as the red phosphor in colour televisions; its compounds are used to make control rods for nuclear reactors. It was named in 1901 by French chemist Eugène Demarçay (1852–1904) after the continent of Europe, where it was first found.

eusociality form of social life found in insects such as honey bees and termites, in which the colony is made up of special castes (for example, workers, drones, and reproductives) whose membership is biologically determined. The worker castes do not usually reproduce. Only one mammal, the naked mole rat, has a social organization of this type. A eusocial shrimp was discovered in 1996 living in the coral reefs of Belize. *Synalpheus regalis* lives in colonies of up to 300 individuals, all the offspring of a single reproductive female. See also ◊social behaviour.

Eustachian tube small air-filled canal connecting the middle ◊ear with the back of the throat. It is found in all land vertebrates and equalizes the pressure on both sides of the eardrum.

eustatic change worldwide rise or fall in sea level caused by a change in the amount of water in the oceans (by contrast, ◊isostasy involves a rising or sinking of the land). During the last ice age, sea level fell because water became 'locked-up' in the form of ice and snow, and less water reached the oceans.

Eutelsat acronym for *European Telecommunications Satellite Organization*.

euthanasia in medicine, mercy killing of someone with a severe and incurable condition or illness. Euthanasia is an issue that creates much controversy on medical and ethical grounds.

In Australia, a bill legalizing voluntary euthanasia for terminally ill patients was passed by the Northern Territory state legislature in May 1995.

In the Netherlands, where approximately 2,700 patients formally request it each year, euthanasia is technically illegal. However, provided guidelines issued by the Royal Dutch Medical Association are followed, doctors are not prosecuted.

In the USA, the Supreme Court ruled in June 1997 that the terminally ill do not have the fundamental right to have doctors assist them to die. The Court upheld state laws in Washington and New York that forbid assisted suicides.

A patient's right to refuse life-prolonging treatment is recognized in several countries, including the UK.

eutrophication excessive enrichment of rivers, lakes, and shallow sea areas, primarily by nitrate fertilizers washed from the soil by rain, by phosphates from fertilizers, and from nutrients in municipal sewage, and by sewage itself. These encourage the growth of algae and bacteria which use up the oxygen in the water, thereby making it uninhabitable for fish and other animal life.

evaporation process in which a liquid turns to a vapour without its temperature reaching boiling point. A liquid left to stand in a saucer eventually evaporates because, at any time, a proportion of its molecules will be fast enough (have enough kinetic energy) to escape through the attractive intermolecular forces at the liquid surface into the atmosphere. The temperature of the liquid tends to fall because the evaporating molecules remove energy from the liquid. The rate of evaporation rises with increased temperature because as the mean kinetic energy of the liquid's molecules rises, so will the number possessing enough energy to escape.

A fall in the temperature of the liquid, known as the **cooling effect**, accompanies evaporation because as the faster molecules escape through the surface the mean energy of the remaining molecules falls. The effect may be noticed when wet clothes are worn, and sweat evaporates from the skin. ◊Refrigeration makes use of the cooling effect to extract heat from foodstuffs, and in the body it plays a part in temperature control.

evaporite sedimentary deposit precipitated on evaporation of salt water.

With a progressive evaporation of seawater, the most common salts are deposited in a definite sequence: calcite (calcium carbonate), gypsum (hydrous calcium sulphate), halite (sodium chloride), and finally salts of potassium and magnesium.

Calcite precipitates when seawater is reduced to half its original volume, gypsum precipitates when the seawater body is reduced to one-fifth, and halite when the volume is reduced to one-tenth. Thus the natural occurrence of chemically precipitated calcium carbonate is common, of gypsum fairly common, and of halite less common.

Because of the concentrations of different dissolved salts in seawater, halite accounts for about 95% of the chlorides precipitated if evaporation is complete. More unusual evaporite minerals include borates (for example borax, hydrous sodium borate) and sulphates (for example glauberite, a combined sulphate of sodium and calcium).

evening primrose any of a group of plants that typically have pale yellow flowers which open in the evening. About 50 species are native to North America, several of which now also grow in Europe. Some are cultivated for their oil, which is rich in gamma-linoleic acid (GLA). The body converts GLA into substances which resemble hormones, and **evening primrose oil** is beneficial in relieving the symptoms of ◊premenstrual tension. It is also used in treating eczema and chronic fatigue syndrome. (Genus *Oenothera*, family Onagraceae.)

evergreen in botany, a plant such as pine, spruce, or holly, that bears its leaves all year round. Most conifers are evergreen. Plants that shed their leaves in autumn or during a dry season are described as ◊deciduous.

evolution slow process of change from one form to another, as in the evolution of the universe from its formation in the ◊Big Bang to its present state, or in the evolution of life on Earth. Some Christians and Muslims deny the theory of evolution as conflicting with the belief that God created all things (see ◊creationism). English naturalist Charles Darwin assigned the main role in evolutionary change to ◊natural selection acting on randomly occurring variations (now known to be produced by spontaneous changes or ◊mutations in the genetic material of organisms).

Organic evolution traces the development of simple unicellular forms to more complex forms, ultimately to the flowering plants and vertebrate animals, including man. The Earth contains an immense diversity of living organisms: about a million different species of animals and half a million species of plants have so far been described. There is overwhelming evidence that this vast array arose by a gradual process of evolutionary divergence and not by individual acts of divine creation as described in the Book of Genesis. There are several lines of evidence: the fossil record, the existence of similarities or homologies between different groups of organisms, embryology, and geographical distribution.

evolutionary stable strategy (ESS) in ◊sociobiology, an assemblage of behavioural or physical characters (collectively termed a 'strategy') of a population that is resistant to replacement by any forms bearing new traits, because the new traits will not be capable of successful reproduction.

ESS analysis is based on ◊game theory and can be applied both to genetically determined physical characters (such as horn length), and to learned behavioural responses (for example, whether to fight or retreat from an opponent).

An ESS may be conditional on the context, as in the rule 'fight if the opponent is smaller, but retreat if the opponent is larger'.

excavation or *dig* in archaeology, the systematic recovery of data through the exposure of buried sites and artefacts. Excavation is destructive, and is therefore accompanied by a comprehensive recording of all material found and its three-dimensional locations (its context). As much material and information as possible must be recovered from any dig. A full record of all the techniques employed in the excavation itself must also be made, so that future archaeologists will be able to evaluate the results of the work accurately.

Besides being destructive, excavation is also costly. For both these reasons, it should be used only as a last resort. It can be partial, with only a sample of the site investigated, or total. Samples are chosen either intuitively, in which case excavators investigate those areas they feel will be most productive, or statistically, in which case the sample is drawn using various statistical techniques, so as to ensure that it is representative.

An important goal of excavation is a full understanding of a site's stratigraphy; that is, the vertical layering of a site.

These layers or levels can be defined naturally (for example, soil changes), culturally (for example, different occupation levels), or arbitrarily (for example, 10 cm/4 in levels). Excavation can also be done horizontally, to uncover larger areas of a particular layer and reveal the spatial relationships between artefacts and features in that layer. This is known as open-area excavation and is used especially where single-period deposits lie close to the surface, and the time dimension is represented by lateral movement rather than by the placing of one building on top of the preceding one.

Most excavators employ a flexible combination of vertical and horizontal digging adapting to the nature of their site and the questions they are seeking to answer.

excavator machine designed for digging in the ground, or for earth-moving in general. Diggers with hydraulically powered digging arms are widely used on building sites. They may run on wheels or on ◊caterpillar tracks. The largest excavators are the draglines used in mining to strip away earth covering the coal or mineral deposit.

exchange transfusion in medicine, blood ◊transfusion procedure carried out on a newborn infant. Blood is removed from the baby using the umbilical vein and is replaced with the same quantity of blood obtained from a donor. The procedure is repeated several times to remove damaged red blood cells while maintaining the blood volume and red blood cell count of the infant. Haemolytic disease of the newborn and severe jaundice may require an exchange transfusion.

exclusion principle in physics, a principle of atomic structure originated by Austrian–US physicist Wolfgang Pauli. It states that no two electrons in a single atom may have the same set of ◊quantum numbers.

Hence, it is impossible to pack together certain elementary particles, such as electrons, beyond a certain critical density, otherwise they would share the same location and quantum number. A white dwarf star is thus prevented from contracting further by the exclusion principle and never collapses.

excretion in biology, the removal of waste products from the cells of living organisms. In plants and simple animals, waste products are removed by diffusion, but in higher animals they are removed by specialized organs. In mammals the kidneys are the principle organs of excretion. Water and metabolic wastes are also excreted in the faeces and, in humans, through the sweat glands; carbon dioxide and water are removed via the lungs.

Excretion

Even the bowels of the sloth work slowly. It defecates on a weekly basis, descending discreetly to the ground to do so. For its safety the sloth needs to spend as little time on the ground as possible, but this strategy means it also avoids revealing its position by the give-away sounds of excrement crashing through the trees.

---

exfoliation in biology, the separation of pieces of dead bone or skin in layers.

exobiology study of life forms that may possibly exist elsewhere in the universe, and of the effects of extraterrestrial environments on Earth organisms. Techniques include space probe experiments designed to detect organic molecules, and the monitoring of radio waves from other star systems.

exocrine gland gland that discharges secretions, usually through a tube or a duct, on to a surface. Examples include sweat glands which release sweat on to the skin, and digestive glands which release digestive juices on to the walls of the intestine. Some animals also have ◊endocrine glands (ductless glands) that release hormones directly into the bloodstream.

exogenous in medicine, generated from outside the body. Drugs used to treat diseases may be described as exogenous. For example, exogenous insulin is used in the treatment of ◊diabetes mellitus.

exoskeleton the hardened external skeleton of insects, spiders, crabs, and other arthropods. It provides attachment for muscles and protection for the internal organs, as well as support. To permit growth it is periodically shed in a process called ◊ecdysis.

exosphere the uppermost layer of the ◊atmosphere. It is an ill-defined zone above the thermosphere, beginning at about 700 km/435 mi and fading off into the vacuum of space. The gases are extremely thin, with hydrogen as the main constituent.

exothermic reaction a chemical reaction during which heat is given out (see ◊energy of reaction).

expansion in physics, the increase in size of a constant mass of substance caused by, for example, increasing its temperature (◊thermal expansion) or its internal pressure. The *expansivity*, or coefficient of thermal expansion, of a material is its expansion (per unit volume, area, or length) per degree rise in temperature.

expectorant any substance, often added to cough mixture, to encourage secretion of mucus in the airways to make it easier to cough up. It is debatable whether expectorants have an effect on lung secretions.

experiment in science, a practical test designed with the intention that its results will be relevant to a particular theory or set of theories. Although some experiments may be used merely for gathering more information about a topic that is already well understood, others may be of crucial importance in confirming a new theory or in undermining long-held beliefs.

The manner in which experiments are performed, and the relation between the design of an experiment and its value, are therefore of central importance. In general an experiment is of most value when the factors that might affect the results (variables) are carefully controlled; for this reason most experiments take place in a well-managed environment such as a laboratory or clinic.

experimental archaeology the controlled replication of ancient technologies and behaviour in order to provide hypotheses that can be tested by actual archaeological data. Experiments can range in size from the reproduction of ancient tools in order to learn about their processes of manufacture and use, and their effectiveness, to the construction of whole villages and ancient subsistence practices in long-term experiments.

The most exciting phrase to hear in science, the one that heralds new discoveries, is not 'Eureka!' (I've found it!) but 'That's funny...'

On **experiment** Isaac Asimov

---

experimental psychology the application of scientific methods to the study of mental processes and behaviour.

This covers a wide range of fields of study, including: *human and animal learning*, in which learning theories describe how new behaviours are acquired and modified; *cognition*, the study of a number of functions, such as perception, attention, memory, and language; and *physiological psychology*, which relates the study of cognition to different regions of the brain. *Artificial intelligence* refers to the computer simulation of cognitive processes, such as language and problem-solving.

expert system computer program for giving advice (such as diagnosing an illness or interpreting the law) that incorporates knowledge derived from human expertise. It is a kind of ◊knowledge-based system containing rules that can be applied to find the solution to a problem. It is a form of ◊artificial intelligence.

explanation in science, an attempt to make clear the cause of any natural event by reference to physical laws and to observations.

The extent to which any explanation can be said to be

true is one of the chief concerns of philosophy, partly because observations may be wrongly interpreted, partly because explanations should help us predict how nature will behave. Although it may be reasonable to expect that a physical law will hold in the future, that expectation is problematic in that it relies on induction, a much criticized feature of human thought; in fact no explanation, however 'scientific', can be held to be true for all time, and thus the difference between a scientific and a common-sense explanation remains the subject of intense philosophical debate.

Explorer series of US scientific satellites. *Explorer 1*, launched in Jan 1958, was the first US satellite in orbit and discovered the Van Allen radiation belts around the Earth.

explosive any material capable of a sudden release of energy and the rapid formation of a large volume of gas, leading when compressed to the development of a high-pressure wave (blast).
types of explosive Combustion and explosion differ essentially only in rate of reaction, and many explosives (called *low explosives*) are capable of undergoing relatively slow combustion under suitable conditions. **High explosives** produce uncontrollable blasts. The first low explosive was ◊gunpowder; the first high explosive was ◊nitroglycerine.

exponent or *index* in mathematics, a number that indicates the number of times a term is multiplied by itself; for example $x^2 = x \times x$, $4^3 = 4 \times 4 \times 4$.
Exponents obey certain rules. Terms that contain them are multiplied together by adding the exponents; for example, $x^2 \times x^5 = x^7$. Division of such terms is done by subtracting the exponents; for example, $y^5 \div y^3 = y^2$. Any number with the exponent 0 is equal to 1; for example, $x^0 = 1$ and $99^0 = 1$.

exponential in mathematics, descriptive of a ◊function in which the variable quantity is an exponent (a number indicating the power to which another number or expression is raised).
Exponential functions and series involve the constant e = 2.71828.... . Scottish mathematician John Napier devised natural ◊logarithms in 1614 with e as the base.
Exponential functions are basic mathematical functions, written as e^x or exp x. The expression e^x has five definitions, two of which are: (i) e^x is the solution of the differential equation $dx/dt = x$ ($x = 1$ if $t = 0$); (ii) e^x is the limiting sum of the infinite series $1 + x + (x^2/2!) + (x^3/3!) + ... + (x^n/n!)$.

export file in computing, a file stored by the computer in a standard format so that it can be accessed by other programs, possibly running on different makes of computer.
For example, a word-processing program running on an Apple ◊Macintosh computer may have a facility to save a file on a floppy disc in a format that can be read by a word-processing program running on an IBM PC-compatible computer. When the file is being read by the second program or computer, it is often referred to as an *import file*.

exposure meter instrument used in photography for indicating the correct exposure – the length of time the camera shutter should be open under given light conditions. Meters use substances such as cadmium sulphide

and selenium as light sensors. These materials change electrically when light strikes them, the change being proportional to the intensity of the incident light. Many cameras have a built-in exposure meter that sets the camera controls automatically as the light conditions change.

extensive agriculture farming system where the area of the farm is large but there are low inputs (such as labour or fertilizers). Extensive farming generally gives rise to lower yields per hectare than ◊intensive agriculture. For example, in East Anglia, UK, intensive use of land may give wheat yields as high as 53 tonnes per hectare, whereas an extensive wheat farm on the Canadian prairies may produce an average of 8.8 tonnes per hectare.

extensor a muscle that straightens a limb.

extinction in biology, the complete disappearance of a species. In the past, extinctions are believed to have occurred because species were unable to adapt quickly enough to a naturally changing environment. Today, most extinctions are due to human activity. Some species, such as the dodo of Mauritius, the moas of New Zealand, and the passenger pigeon of North America, were exterminated by hunting. Others became extinct when their habitat was destroyed. ◊Endangered species are close to extinction.
Mass extinctions are episodes during which whole groups of species have become extinct, the best known being that of the dinosaurs, other large reptiles, and various marine invertebrates about 65 million years ago. Another mass extinction occurred about 10,000 years ago when many giant species of mammal died out. This is known as the 'Pleistocene overkill' because their disappearance was probably hastened by the hunting activities of prehistoric humans. The greatest mass extinction occurred about 250 million years ago marking the Permian-Triassic boundary (see ◊geological time), when up to 96% of all living species became extinct. It was proposed in 1982 that mass extinctions occur periodically, at intervals of approximately 26 million years.
The current mass extinction is largely due to human destruction of habitats, as in the tropical forests and coral reefs; it is far more serious and damaging than mass extinctions of the past because of the speed at which it occurs. Human-made climatic changes and pollution also make it less likely that the biosphere can recover and evolve new species to suit a changed environment.
The rate of extinction is difficult to estimate, since most losses occur in the rich environment of the tropical rainforest, where the total number of existent species is not known. Conservative estimates put the rate of loss due to deforestation alone at 4,000 to 6,000 species a year. Overall, the rate could be as high as one species an hour, with the loss of one species putting those dependent on it at risk. Globally, 168 mammal species and 168 birds are judged critically endangered. Australia has the worst record for extinction: 18 mammals have disappeared since Europeans settled there, and 40 more are threatened.

extracellular matrix strong material naturally occurring in animals and plants, made up of protein and long-chain sugars (polysaccharides) in which cells are embedded. It is often called a 'biological glue', and forms part of ◊connective tissues such as bone and skin.
The cell walls of plants and bacteria, and the ◊exo-

skeletons of insects and other arthropods, are also formed by types of extracellular matrix.

extrasensory perception (ESP) any form of perception beyond and distinct from the known sensory processes. The main forms of ESP are clairvoyance (intuitive

> *There are children playing in the street who could solve some of my top problems in physics, because they have modes of sensory perception that I lost long ago.*
>
> On **extrasensory perception**
> J Robert Oppenheimer

perception or vision of events and situations without using the senses); precognition (the ability to foresee events); and telepathy or thought transference (communication between people without using any known visible, tangible, or audible medium). Verification by scientific study has yet to be achieved.

extroversion or **extraversion** personality dimension described by the psychologists Carl Jung and, later, Hans Eysenck. The typical extrovert is sociable, impulsive, and carefree. The opposite of extroversion is ◊introversion.

extrusion common method of shaping metals, plastics, and other materials. The materials, usually hot, are forced through the hole in a metal die and take its cross-sectional shape. Rods, tubes, and sheets may be made in this way.

extrusive rock or **volcanic rock** ◊Igneous rock formed on the surface of the Earth; for example, basalt. It is usually fine-grained (having cooled quickly), unlike the more coarse-grained ◊intrusive rocks (igneous rocks formed under the surface). The magma (molten rock) that cools to form extrusive rock may reach the surface through a crack, such as the ◊constructive margin at the Mid-Atlantic Ridge, or through the vent of a ◊volcano.

Extrusive rock can be either **lava** (solidified from a flow) or a **pyroclastic deposit** (hot rocks or ash).

eye the organ of vision. In the human eye, the light is focused by the combined action of the curved **cornea**, the internal fluids, and the **lens**. The insect eye is compound – made up of many separate facets – known as ommatidia, each of which collects light and directs it separately to a receptor to build up an image. Invertebrates have much simpler eyes, with no lenses. Among molluscs, cephalopods have complex eyes similar to those of vertebrates.

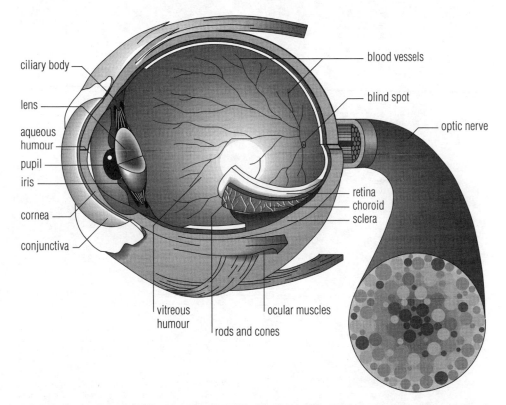

ciliary body
blood vessels
lens
blind spot
aqueous humour
optic nerve
pupil
iris
cornea
retina
choroid
sclera
conjunctiva
vitreous humour
ocular muscles
rods and cones

eye The human eye. The retina of the eye contains about 137 million light-sensitive cells in an area of about 650 sq mm/1 sq in. There are 130 million rod cells for black and white vision and 7 million cone cells for colour vision. The optic nerve contains about 1 million nerve fibres. The focusing muscles of the eye adjust about 100,000 times a day. To exercise the leg muscles to the same extent would need an 80 km/50 mi walk.

The mantis shrimp's eyes contain ten colour pigments with which to perceive colour; some flies and fish have five, while the human eye has only three.

The human eye is a roughly spherical structure contained in a bony socket. Light enters it through the cornea, and passes through the circular opening (***pupil***) in the iris (the coloured part of the eye).

The ciliary muscles act on the lens (the rounded transparent structure behind the iris) to change its shape, so that images of objects at different distances can be focused on the ◊***retina***. This is at the back of the eye, and is packed with light-sensitive cells (rods and cones), connected to the brain by the optic nerve.

Eye

The eyes of the giant squid – at a diameter of 37 cm/15 in – are the size of dustbin lids. They are the largest in the animal kingdom.

---

eye, defects of the abnormalities of the eye that impair vision. Glass or plastic lenses, in the form of spectacles or contact lenses, are the usual means of correction. Common optical defects are ◊short-sightedness or myopia; far-sightedness or hypermetropia; lack of ◊accommodation or presbyopia; and ◊astigmatism. Other eye defects include ◊colour blindness.

°F symbol for degrees ◊ *Fahrenheit*.

'f/64' group group of US photographers, including Edward Weston, Ansel Adams, and Imogen Cunningham, formed in 1932. The sharp focus and clarity of their black and white pictures was achieved by setting the lens aperture to f/64.

facies in geology, any assemblage of mineral, rock, or fossil features that reflect the environment in which rock was formed.

The set of characters that distinguishes one facies from another in a given time stratigraphic unit is used to interpret local changes in simultaneously existing environments. Thus one facies in a body of rock might consist of porous limestone containing fossil reef-building organisms in their living positions. This facies might laterally pass into a reef-flank facies of steeply dipping deposits of rubble from the reef, which in turn might grade into an inter-reef basin composed of fine, clayey limestone.

Ancient floods and migrations of the seashore up or down can also be traced by changes in facies.

facsimile transmission full name for ◊ *fax* or *telefax*.

factor a number that divides into another number exactly. For example, the factors of 64 are 1, 2, 4, 8, 16, 32, and 64. In algebra, certain kinds of polynomials (expressions consisting of several or many terms) can be factorized. For example, the factors of $x^2 + 3x + 2$ are $x + 1$ and $x + 2$, since $x^2 + 3x + 2 = (x + 1)(x + 2)$. This is called factorization. See also ◊ prime number.

factorial of a positive number, the product of all the whole numbers (integers) inclusive between 1 and the number itself. A factorial is indicated by the symbol '!'. Thus $6! = 1 \times 2 \times 3 \times 4 \times 5 \times 6 = 720$. Factorial zero, 0!, is defined as 1.

factory farming intensive rearing of poultry or animals for food, usually on high-protein foodstuffs in confined quarters. Chickens for eggs and meat, and calves for veal are commonly factory farmed. Some countries restrict the use of antibiotics and growth hormones as aids to factory farming, because they can persist in the flesh of the animals after they are slaughtered. The emphasis is on productive yield rather than animal welfare so that conditions for the animals are often very poor. For this reason, many people object to factory farming on moral as well as health grounds.

Egg-laying hens are housed in 'batteries' of cages arranged in long rows. If caged singly, they lay fewer eggs, so there are often four to a cage with a floor area of only 2,400 sq cm/372 sq in. In the course of a year, battery hens average 261 eggs each, whereas for free-range chickens the figure is 199.

faeces remains of food and other waste material eliminated from the digestive tract of animals by way of the anus. Faeces consist of quantities of fibrous material, bacteria and other microorganisms, rubbed-off lining of the digestive tract, bile fluids, undigested food, minerals, and water.

Fahrenheit scale temperature scale invented in 1714 by Gabriel Fahrenheit which was commonly used in English-speaking countries until the 1970s, after which the ◊ Celsius scale was generally adopted, in line with the rest of the world. In the Fahrenheit scale, intervals are measured in degrees (°F); °F = (°C × 9/5) + 32.

Fahrenheit took as the zero point the lowest temperature he could achieve anywhere in the laboratory, and as the other fixed point, body temperature, which he set at 96°F. On this scale, water freezes at 32°F and boils at 212°F.

fainting sudden, temporary loss of consciousness caused by reduced blood supply to the brain. It may be due to emotional shock or physical factors, such as pooling of blood in the legs from standing still for long periods.

Fallopian tube or *oviduct* in mammals, one of two tubes that carry eggs from the ovary to the uterus. An egg is fertilized by sperm in the Fallopian tubes, which are lined with cells whose ◊ cilia move the egg towards the uterus.

Fallot's tetralogy in medicine, a hereditary defect of the heart that is associated with defects in the septum, an increase in the size of the right ventricle (due to the demands placed upon it), narrowing of the pulmonary arteries, and displacement of the aorta. It is usually possible to correct these deficits by surgery during childhood.

fallout harmful radioactive material released into the atmosphere in the debris of a nuclear explosion and descending to the surface of the Earth. Such material can enter the food chain, cause ◊ radiation sickness, and last for hundreds of thousands of years (see ◊ half-life).

fallow land ploughed and tilled, but left unsown for a season to allow it to recuperate. In Europe, it is associated with the mediaeval three-field system. It is used in some modern ◊ crop rotations and in countries that do not have access to fertilizers to maintain soil fertility.

false-colour imagery graphic technique that displays images in false (not true-to-life) colours so as to enhance certain features. It is widely used in displaying electronic images taken by spacecraft; for example, Earth-survey satellites such as *Landsat*. Any colours can be selected by a computer processing the received data.

falsificationism in philosophy of science, the belief that a scientific theory must be under constant scrutiny and that its merit lies only in how well it stands up to rigorous testing. It was first expounded by philosopher Karl Popper in his *Logic of Scientific Discovery* published in 1934.

Such thinking also implies that a theory can be held to be scientific only if it makes predictions that are clearly testable. Critics of this belief acknowledge the strict logic of this process, but doubt whether the whole of scientific method can be subsumed into so narrow a programme.

Philosophers and historians such as Thomas Kuhn and Paul Feyerabend have attempted to use the history of science to show that scientific progress has resulted from a more complicated methodology than Popper suggests.

family in biological classification, a group of related genera (see ◊genus). Family names are not printed in italic (unlike genus and species names), and by convention they all have the ending -idae (animals) or -aceae (plants and fungi). For example, the genera of hummingbirds are grouped in the hummingbird family, Trochilidae. Related families are grouped together in an ◊order.

family planning spacing or preventing the birth of children. Access to family-planning services (see ◊contraceptive) is a significant factor in women's health as well as in limiting population growth. If all those women who wished to avoid further childbirth were able to do so, the number of births would be reduced by 27% in Africa, 33% in Asia, and 35% in Latin America; and the number of women who die during pregnancy or childbirth would be reduced by about 50%.

The average number of pregnancies per woman is two in the industrialized countries, where 71% use family planning, as compared with six or seven pregnancies per woman in the Third World. According to a World Bank estimate, doubling the annual $2 billion spent on family planning would avert the deaths of 5.6 million infants and 250,000 mothers each year.

farad SI unit (symbol F) of electrical capacitance (how much electric charge a ◊capacitor can store for a given voltage). One farad is a capacitance of one ◊coulomb per volt. For practical purposes the microfarad (one millionth of a farad, symbol μF) is more commonly used.

faraday unit of electrical charge equal to the charge on one mole of electrons. Its value is 9.648×10^4 coulombs.

Faraday's constant constant (symbol F) representing the electric charge carried on one mole of electrons. It is found by multiplying Avogadro's constant by the charge carried on a single electron, and is equal to 9.648×10^4 coulombs per mole.

One **faraday** is this constant used as a unit. The constant is used to calculate the electric charge needed to discharge a particular quantity of ions during ◊electrolysis.

Faraday's laws three laws of electromagnetic induction, and two laws of electrolysis, all proposed originally by English scientist Michael Faraday:

Induction (1) a changing magnetic field induces an electromagnetic force in a conductor; (2) the electromagnetic force is proportional to the rate of change of the field; (3) the direction of the induced electromagnetic force depends on the orientation of the field.

Electrolysis (1) the amount of chemical change during electrolysis is proportional to the charge passing through the liquid; (2) the amount of chemical change produced in a substance by a given amount of electricity is proportional to the electrochemical equivalent of that substance.

fast breeder or *breeder reactor* alternative names for ◊fast reactor, a type of nuclear reactor.

fasting the practice of voluntarily going without food. It can be undertaken as a religious observance, a sign of mourning, a political protest (hunger strike), or for slimming purposes.

Fasting or abstinence from certain types of food or beverages occurs in most religious traditions. It is seen as an act of self-discipline that increases spiritual awareness by lessening dependence on the material world. In the Roman Catholic church, fasting is seen as a penitential rite, a means to express repentance for sin. The most commonly observed Christian fasting is in Lent, from Ash Wednesday to Easter Sunday, and recalls the 40 days Jesus spent in the wilderness. Roman Catholics and Orthodox Christians usually fast before taking communion and monastic communities observe regular weekly fasts. Devout Muslims go without food or water between sunrise and sunset during the month of Ramadan. Jews fast for Yom Kippur and before several other festivals. Many devout Hindus observe a weekly day of partial or total fast.

Total abstinence from food for a limited period is prescribed by some ◊naturopaths to eliminate body toxins or make available for recuperative purposes the energy normally used by the digestive system. Prolonged fasting can be dangerous. The liver breaks up its fat stores, releasing harmful by-products called ketones, which results in a condition called ketosis, which develops within three days. An early symptom is a smell of pear drops on the breath. Other symptoms include nausea, vomiting, fatigue, dizziness, severe depression, and irritability. Eventually the muscles and other body tissues become wasted, and death results.

fast reactor or *fast breeder reactor* ◊nuclear reactor that makes use of fast neutrons to bring about fission. Unlike other reactors used by the nuclear-power industry, it has little or no ◊moderator, to slow down neutrons. The reactor core is surrounded by a 'blanket' of uranium carbide. During operation, some of this uranium is converted into plutonium, which can be extracted and later used as fuel.

Fast breeder reactors can extract about 60 times the amount of energy from uranium that thermal reactors do. In the 1950s, when uranium stocks were thought to be dwindling, the fast breeder was considered to be the reactor of the future. Now, however, as new uranium reserves have been found and because of various technical difficulties in their construction, development of the fast breeder has slowed.

fat in the broadest sense, a mixture of ◊lipids – chiefly triglycerides (lipids containing three ◊fatty acid molecules linked to a molecule of glycerol). More specifically, the term refers to a lipid mixture that is solid at room temperature (20°C); lipid mixtures that are liquid at room temperature are called *oils*. The higher the proportion of saturated fatty acids in a mixture, the harder the fat.

Boiling fats in strong alkali forms soaps (saponification). Fats are essential constituents of food for many animals, with a calorific value twice that of carbohydrates; however, eating too much fat, especially fat of animal origin, has been linked with heart disease in humans. In many animals and plants, excess carbohydrates and proteins are converted into fats for storage. Mammals and other vertebrates store fats in specialized connective tissues (◊adipose tissues), which not only act as energy reserves but also insulate the body and cushion its organs.

As a nutrient fat serves five purposes: it is a source of energy (9 kcal/g); makes the diet palatable; provides basic

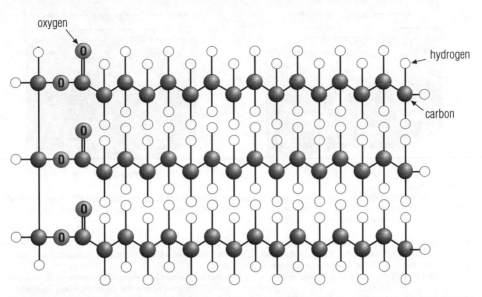

oxygen

hydrogen

carbon

fat The molecular structure of typical fat. The molecule consists of three fatty acid molecules linked to a molecule of glycerol.

building blocks for cell structure; provides essential fatty acids (linoleic and linolenic); and acts as a carrier for fat-soluble vitamins (A, D, E, and K). Foods rich in fat are butter, lard, margarine, and cooking oils. Products high in monounsaturated or polyunsaturated fats are thought to be less likely to contribute to cardiovascular disease.

fathom in mining, seafaring, and handling timber, a unit of depth measurement (1.83 m/6 ft) used prior to metrication; it approximates to the distance between an adult man's hands when the arms are outstretched.

fatigue in muscle, reduced response brought about by the accumulation of lactic acid in muscle tissue due to excessive cellular activity.

fatigue in medicine, general weariness, possibly due to lack of sleep, too little exercise, and poor nutrition. It is also a symptom of certain diseases, such as ◊anaemia, ◊hypothyroidism, or ◊myasthenia gravis. See also ◊chronic fatigue syndrome.

fatty acid or **carboxylic acid** organic compound consisting of a hydrocarbon chain, up to 24 carbon atoms long, with a carboxyl group (–COOH) at one end. The covalent bonds between the carbon atoms may be single or double; where a double bond occurs the carbon atoms concerned carry one instead of two hydrogen atoms. Chains with only single bonds have all the hydrogen they can carry, so they are said to be **saturated** with hydrogen. Chains with one or more double bonds are said to be **unsaturated** (see ◊polyunsaturate). Fatty acids are produced in the small intestine when fat is digested.

Saturated fatty acids include palmitic and stearic acids; unsaturated fatty acids include oleic (one double bond), linoleic (two double bonds), and linolenic (three double bonds). Linoleic acid accounts for more than one third of some margarines. Supermarket brands that say they are high in polyunsaturates may contain as much as 39%.

Fatty acids are generally found combined with glycerol in ◊lipids such as tryglycerides.

fault in geology, a fracture in the Earth's crust along which the two sides have moved as a result of differing strains in the adjacent rock bodies. Displacement of rock masses horizontally or vertically along a fault may be microscopic, or it may be massive, causing major ◊earthquakes.

If the movement has a major vertical component, the fault is termed a **normal fault**, where rocks on each side have moved apart, or a **reverse fault**, where one side has overridden the other (a low angle reverse fault is called a **thrust**). A **lateral fault**, or **tear fault**, occurs where the relative movement is sideways. A particular kind of fault found only in ocean ridges is the **transform fault** (a term coined by Canadian geophysicist John Tuzo Wilson in 1965). On a map an ocean ridge has a stepped appearance. The ridge crest is broken into sections, each section offset from the next. Between each section of the ridge crest the newly generated plates are moving past one another, forming a transform fault.

Faults produce lines of weakness that are often exploited by processes of ◊weathering and ◊erosion. Coastal caves and geos (narrow inlets) often form along faults and, on a larger scale, rivers may follow the line of a fault.

fax (common name for **facsimile transmission** or **telefax**) the transmission of images over a ◊telecommunications link, usually the telephone network. When placed on a fax machine, the original image is scanned by a transmitting device and converted into coded signals, which travel via the telephone lines to the receiving fax machine, where an image is created that is a copy of the original. Photographs as well as printed text and drawings can be sent. The standard transmission takes place at 4,800 or 9,600 bits of information per second.

The world's first fax machine, the *pantélégraphe*, was invented by Italian physicist Giovanni Caselli in 1866,

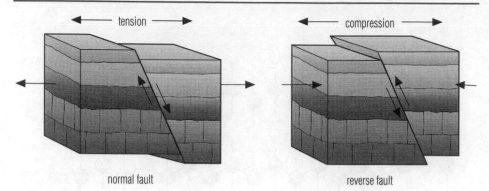

fault Faults are caused by the movement of rock layers, producing such features as block mountains and rift valleys. A normal fault is caused by a tension or stretching force acting in the rock layers. A reverse fault is caused by compression forces. Faults can continue to move for thousands or millions of years.

over a century before the first electronic model came on the market. Standing over 2 m/6.5 ft high, it transmitted by telegraph nearly 5,000 handwritten documents and drawings between Paris and Lyon in its first year.

feather rigid outgrowth of the outer layer of the skin of birds, made of the protein keratin. Feathers provide insulation and facilitate flight. There are several types, including long quill feathers on the wings and tail, fluffy down feathers for retaining body heat, and contour feathers covering the body. The colouring of feathers is often important in camouflage or in courtship and other displays. Feathers are normally replaced at least once a year.

Feathers generally consist of two main parts, axis and barbs, the former of which is divided into the quill, which is bare and hollow, and the shaft, which bears the barbs. The quill is embedded in the skin, and has at its base a small hole through which nourishment passes during the growth of the feather. The barbs which constitute the vane are lath-shaped and taper to a point, and each one supports a series of outgrowths known as barbules, so that

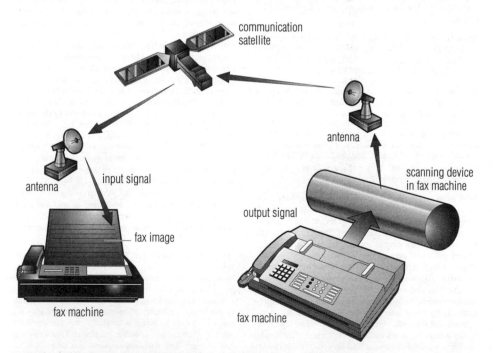

fax The fax machine scans a document as it is fed through the rollers. The photodiode array produces an electrical signal representing the dot-by-dot image of the page, which is transmitted over the telephone lines. At the receiving end, the process is reversed to reconstitute the original page.

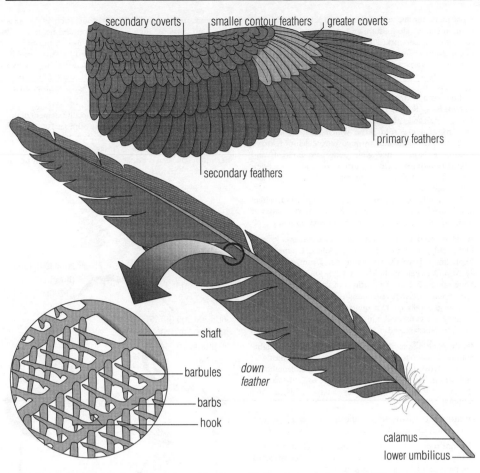

secondary coverts   smaller contour feathers   greater coverts

primary feathers

secondary feathers

shaft

barbules

down feather

barbs

hook

calamus

lower umbilicus

feather Types of feather. A bird's wing is made up of two types of feather: contour feathers and flight feathers. The primary and secondary feathers are flight feathers. The coverts are contour feathers, used to streamline the bird. Semi-plume and down feathers insulate the bird's body and provide colour.

each barb is like a tiny feather. Adjacent barbs are linked to each other by hooks on the barbules.

fecundity the rate at which an organism reproduces, as distinct from its ability to reproduce (◊fertility). In vertebrates, it is usually measured as the number of offspring produced by a female each year.

feedback general principle whereby the results produced in an ongoing reaction become factors in modifying or changing the reaction; it is the principle used in self-regulating control systems, from a simple ◊thermostat and steam-engine ◊governor to automatic computer-controlled machine tools. A fully computerized control system, in which there is no operator intervention, is called a ***closed-loop feedback*** system. A system that also responds to control signals from an operator is called an ***open-loop feedback*** system.

In self-regulating systems, information about what *is* happening in a system (such as level of temperature, engine speed, or size of workpiece) is fed back to a controlling device, which compares it with what *should* be happening. If the two are different, the device takes suitable action (such as switching on a heater, allowing more steam to the engine, or resetting the tools). The idea that the Earth is a self-regulating system, with feedback operating to keep nature in balance, is a central feature of the ◊Gaia hypothesis.

Fehling's test chemical test to determine whether an organic substance is a reducing agent (substance that donates electrons to other substances in a chemical reaction). It is usually used to detect reducing sugars (monosaccharides, such as glucose, and the disaccharides maltose and lactose) and aldehydes.

If the test substance is heated with a freshly prepared solution containing copper(II) sulphate, sodium hydroxide and sodium potassium tartrate, the production of a brick-red precipitate indicates the presence of a reducing agent.

feldspar one of a group of rock-forming minerals, the most abundant group in the Earth's crust. They are the chief constituents of ◊igneous rock and are present in most metamorphic and sedimentary rocks. All feldspars contain silicon, aluminium, and oxygen, linked together to form a framework; spaces within this structure are occupied by sodium, potassium, calcium, or occasionally barium, in various proportions. Feldspars form white, grey, or pink crystals and rank 6 on the ◊Mohs' scale of hardness.

The four extreme compositions of feldspar are **orthoclase**, $KAlSi_3O_8$; **albite**, $NaAlSi_3O_8$; **anorthite**, $CaAl_2Si_2O_8$; and **celsian**, $BaAl_2Si_2O_8$. These are grouped into **plagioclase feldspars**, which range from pure sodium feldspar (albite) through pure calcium feldspar (anorthite) with a negligible potassium content; and **alkali feldspars** (including orthoclase and microcline), which have a high potassium content, less sodium, and little calcium.

The type known as ◊**moonstone** has a pearl-like effect and is used in jewellery. Approximately 4,000 tonnes of feldspar are used in the ceramics industry annually.

feldspathoid any of a group of silicate minerals resembling feldspars but containing less silica. Examples are nepheline ($NaAlSiO_4$ with a little potassium) and leucite ($KAlSi_2O_6$). Feldspathoids occur in igneous rocks that have relatively high proportions of sodium and potassium. Such rocks may also contain alkali feldspar, but they do not generally contain quartz because any free silica would have combined with the feldspathoid to produce more feldspar instead.

felsic rock a ◊plutonic rock composed chiefly of light-coloured minerals, such as quartz, feldspar and mica. It is derived from **feldspar**, **lenad** (meaning feldspathoid), and **silica**. The term **felsic** also applies to light-coloured minerals as a group, especially quartz, feldspar, and feldspathoids.

femur the **thigh-bone**; also the upper bone in the hind limb of a four-limbed vertebrate.

Fermat's principle in physics, the principle that a ray of light, or other radiation, moves between two points along the path that takes the minimum time.

The principle is named after French mathematician Pierre de Fermat, who used it to deduce the laws of ◊reflection and ◊refraction.

fermentation the breakdown of sugars by bacteria and yeasts using a method of respiration without oxygen (◊anaerobic). Fermentation processes have long been utilized in baking bread, making beer and wine, and producing cheese, yoghurt, soy sauce, and many other foodstuffs.

In baking and brewing, yeasts ferment sugars to produce ◊ethanol and carbon dioxide; the latter makes bread rise and puts bubbles into beers and champagne. Many antibiotics are produced by fermentation; it is one of the processes that can cause food spoilage.

Fermilab (shortened form of **Fermi National Accelerator Laboratory**) US centre for ◊particle physics at Batavia, Illinois, near Chicago. It is named after Italian–US physicist Enrico Fermi. Fermilab was opened in 1972, and is the home of the Tevatron, the world's most powerful particle ◊accelerator. It is capable of boosting protons and antiprotons to speeds near that of light (to energies of 20 TeV).

fermion in physics, a subatomic particle whose spin can only take values that are half-integers, such as 1/2 or 3/2.

Fermions may be classified as leptons, such as the electron, and baryons, such as the proton and neutron. All elementary particles are either fermions or ◊bosons.

The exclusion principle, formulated by Austrian–US physicist Wolfgang Pauli in 1925, asserts that no two fermions in the same system (such as an atom) can possess the same position, energy state, spin, or other quantized property.

fermium synthesized, radioactive, metallic element of the ◊actinide series, symbol Fm, atomic number 100, relative atomic mass 257.10. Ten isotopes are known, the longest-lived of which, Fm-257, has a half-life of 80 days. Fermium has been produced only in minute quantities in particle accelerators.

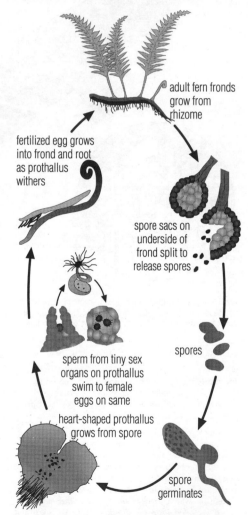

adult fern fronds grow from rhizome

fertilized egg grows into frond and root as prothallus withers

spore sacs on underside of frond split to release spores

spores

sperm from tiny sex organs on prothallus swim to female eggs on same

heart-shaped prothallus grows from spore

spore germinates

fern The life cycle of a fern. Ferns have two distinct forms that alternate during their life cycle. For the main part of its life, a fern consists of a short stem (or rhizome) from which roots and leaves grow. The other part of its life is spent as a small heart-shaped plant called a prothallus.

It was discovered in 1952 in the debris of the first thermonuclear explosion. The element was named in 1955 in honour of Italian–US physicist Enrico Fermi.

fern any of a group of plants related to horsetails and clubmosses. Ferns are spore-bearing, not flowering, plants and most are perennial, spreading by slow-growing roots. The leaves, known as fronds, vary widely in size and shape. Some taller types, such as tree ferns, grow in the tropics. There are over 7,000 species. (Order Filicales.)

ferrimagnetism form of ◊magnetism in which adjacent molecular magnets are aligned anti-parallel, but have unequal strength, producing a strong overall magnetization. Ferrimagnetism is found in certain inorganic substances, such as ◊ferrites.

ferrite ceramic ferrimagnetic material. Ferrites are iron oxides to which small quantities of ◊transition metal oxides (such as cobalt and nickel oxides) have been added. They are used in transformer cores, radio antennae, and, formerly, in computer memories.

ferro-alloy alloy of iron with a high proportion of elements such as manganese, silicon, chromium, and molybdenum. Ferro-alloys are used in the manufacture of alloy steels. Each alloy is generally named after the added metal – for example, ferrochromium.

ferroelectric material ceramic dielectric material that, like ferromagnetic materials, has a ◊domain structure that makes it exhibit magnetism and usually the ◊piezoelectric effect. An example is Rochelle salt (potassium sodium tartrate tetrahydrate, $KNaC_4H_4O_6.4H_2O$).

ferromagnetism form of ◊magnetism in which magnetism can be acquired in an external magnetic field and usually retained in its absence, so that ferromagnetic materials are used to make permanent magnets. A ferromagnetic material may therefore be said to have a high magnetic permeability and susceptibility (which depends upon temperature). Examples are iron, cobalt, nickel, and their alloys.

Ultimately, ferromagnetism is caused by spinning electrons in the atoms of the material, which act as tiny weak magnets. They align parallel to each other within small regions of the material to form ◊domains, or areas of stronger magnetism. In an unmagnetized material, the domains are aligned at random so there is no overall magnetic effect. If a magnetic field is applied to that material, the domains align to point in the same direction, producing a strong overall magnetic effect. Permanent magnetism arises if the domains remain aligned after the external field is removed. Ferromagnetic materials exhibit hysteresis.

fertility an organism's ability to reproduce, as distinct from the rate at which it reproduces (◊fecundity). Individuals become infertile (unable to reproduce) when they cannot generate gametes (eggs or sperm) or when their gametes cannot yield a viable ◊embryo after fertilization.

fertility drug any of a range of drugs taken to increase a female's fertility, developed in Sweden in the mid-1950s. They increase the chances of a multiple birth.

The most familiar is gonadotrophin, which is made from hormone extracts taken from the human pituitary gland: follicle-stimulating hormone and lutenizing hormone. It stimulates ovulation in women. As a result of a fertility drug, the first sextuplets to survive were born to Susan Rosenkowitz in 1974 of South Africa.

fertilization in ◊sexual reproduction, the union of two ◊gametes (sex cells, often called egg and sperm) to produce a ◊zygote, which combines the genetic material contributed by each parent. In self-fertilization the male and female gametes come from the same plant; in cross-fertilization they come from different plants. Self-fertilization rarely occurs in animals; usually even ◊hermaphrodite animals cross-fertilize each other.

In terrestrial insects, mammals, reptiles, and birds, fertilization occurs within the female's body. In humans it usually takes place in the ◊Fallopian tube. In the majority of fishes and amphibians, and most aquatic invertebrates, fertilization occurs externally, when both sexes release their gametes into the water. In most fungi, gametes are not released, but the hyphae of the two parents grow towards each other and fuse to achieve fertilization. In higher plants, ◊pollination precedes fertilization.

fertilizer substance containing some or all of a range of about 20 chemical elements necessary for healthy plant growth, used to compensate for the deficiencies of poor or depleted soil. Fertilizers may be ***organic***, for example

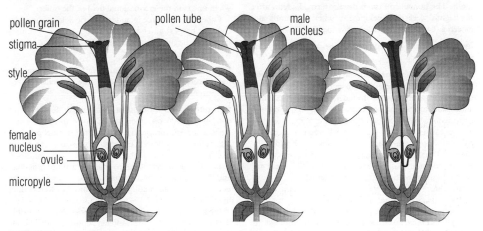

pollen grain
stigma
style
female nucleus
ovule
micropyle
pollen tube
male nucleus

fertilization

farmyard manure, composts, bonemeal, blood, and fish-meal; or *inorganic*, in the form of compounds, mainly of nitrogen, phosphate, and potash, which have been used on a very much increased scale since 1945.

Because externally applied fertilizers tend to be in excess of plant requirements and drain away to affect lakes and rivers (see ◊eutrophication), attention has turned to the modification of crop plants themselves. Plants of the legume family, including the bean, clover, and lupin, live in symbiosis with bacteria located in root nodules, which fix nitrogen from the atmosphere. Research is now directed to producing a similar relationship between these bacteria and crops such as wheat.

fetal therapy diagnosis and treatment of conditions aris-ing in the unborn child. While some anomalies can be diagnosed antenatally, fetal treatments are only appropri-ate in a few cases – mostly where the development of an organ is affected.

Fetal therapy was first used in 1963 with exchange transfusion for haemolytic disease of the newborn, once a serious problem (see also ◊rhesus factor). Today the use of fetal therapy remains limited. Most treatments involve 'needling': introducing fine instruments through the mother's abdominal and uterine walls under ultrasound guidance. Open-womb surgery (hysterotomy) remains controversial because of the risks involved. It is available only in some centres in the United States.

fetch–execute cycle or *processing cycle* in comput-ing, the two-phase cycle used by the computer's central processing unit to process the instructions in a program. During the *fetch phase*, the next program instruction is transferred from the computer's immediate-access mem-ory to the instruction register (memory location used to hold the instruction while it is being executed). During the *execute phase*, the instruction is decoded and obeyed. The process is repeated in a continuous loop.

fetishism in psychology, the transfer of erotic interest to an object, such as an item of clothing, whose real or fantasized presence is necessary for sexual gratification. The fetish may also be a part of the body not normally considered erogenous, such as the feet.

fetus or *foetus* stage in mammalian ◊embryo develop-ment. The human embryo is usually termed a fetus after the eighth week of development, when the limbs and external features of the head are recognizable.

fever condition of raised body temperature, usually due to infection.

fibre, dietary or *roughage* plant material that cannot be digested by human digestive enzymes; it consists largely of cellulose, a carbohydrate found in plant cell walls. Fibre adds bulk to the gut contents, assisting the muscular contractions that force food along the intestine. A diet low in fibre causes constipation and is believed to increase the risk of developing diverticulitis, diabetes, gall-bladder dis-ease, and cancer of the large bowel – conditions that are rare in nonindustrialized countries, where the diet con-tains a high proportion of unrefined cereals.

Soluble fibre consists of indigestible plant carbo-hydrates (such as pectins, hemicelluloses, and gums) that dissolve in water. A high proportion of the fibre in such foods as oat bran, pulses, and vegetables is of this sort. Its presence in the diet has been found to reduce the amount

of cholesterol in blood over the short term, although the mechanism for its effect is disputed.

fibreglass glass that has been formed into fine fibres, either as long continuous filaments or as a fluffy, short-fibred glass wool. Fibreglass is heat- and fire-resistant and a good electrical insulator. It has applications in the field of fibre optics and as a strengthener for plastics in ◊GRP (glass-reinforced plastics).

The long filament form is made by forcing molten glass through the holes of a spinneret, and is woven into tex-tiles. Glass wool is made by blowing streams of molten glass in a jet of high-pressure steam, and is used for electrical, sound, and thermal insulation, especially for the roof space in houses.

fibre optics branch of physics dealing with the trans-mission of light and images through glass or plastic fibres known as ◊optical fibres.

fibrin insoluble protein involved in blood clotting. When an injury occurs fibrin is deposited around the wound in the form of a mesh, which dries and hardens, so that bleeding stops. Fibrin is developed in the blood from a soluble protein, fibrinogen.

The conversion of fibrinogen to fibrin is the final stage in blood clotting. Platelets, a type of cell found in blood, release the enzyme thrombin when they come into con-tact with damaged tissue, and the formation of fibrin then occurs. Calcium, vitamin K, and a variety of enzymes called factors are also necessary for efficient blood clotting.

fibrositis inflammation and overgrowth of fibrous tissue, mainly of the muscle sheaths. It is also known as muscular rheumatism. Symptoms are sudden pain and stiffness, usually relieved by analgesics and rest.

fibula the rear lower bone in the hind leg of a vertebrate. It is paired and often fused with a smaller front bone, the tibia.

field in computing, a specific item of data. A field is usually part of a *record*, which in turn is part of a ◊file.

field in physics, a region of space in which an object exerts a force on another separate object because of cer-tain properties they both possess. For example, there is a force of attraction between any two objects that have mass when one is in the gravitational field of the other.

Other fields of force include ◊electric fields (caused by electric charges) and ◊magnetic fields (caused by circu-lating electric currents), both of which can involve attract-ive or repulsive forces.

field enclosed area of land used for farming. Traditionally fields were measured in ◊acres; the current unit of measurement is the hectare (2.47 acres).

fifth-generation computer anticipated new type of computer based on emerging microelectronic technol-ogies with high computing speeds and ◊parallel process-ing. The development of very large-scale integration (VLSI) technology, which can put many more circuits on to an integrated circuit (chip) than is currently possible, and developments in computer hardware and software design may produce computers far more powerful than those in current use.

It has been predicted that such a computer will be able to communicate in natural spoken language with its user; store vast knowledge databases; search rapidly through these databases, making intelligent inferences and drawing logical conclusions; and process images and 'see' objects in the way that humans do.

In 1981 Japan's Ministry of International Trade and Industry launched a ten-year project to build the first fifth-generation computer, the 'parallel inference machine', consisting of over a thousand microprocessors operating in parallel with each other. By 1992, however, the project was behind schedule and had only produced 256-processor modules. It has since been suggested that research into other technologies, such as ◊neural networks, may present more promising approaches to artificial intelligence. Compare earlier ◊computer generations.

filament in astronomy, a dark, winding feature occasionally seen on images of the Sun in hydrogen light. Filaments are clouds of relatively cool gas suspended above the Sun by magnetic fields and seen in silhouette against the hotter ◊photosphere below. During total ◊eclipses they can be seen as bright features against the sky at the edge of the Sun where they are known as ◊prominences.

filariasis collective term for several diseases, prevalent in tropical areas, caused by certain roundworm (nematode) parasites. About 80 million people worldwide are infected with filarial worms, mostly in India and Africa.

Symptoms include damaged and swollen lymph vessels leading to grotesque swellings of the legs and genitals (Bancroftian filariasis, ◊elephantiasis), blindness, and dry, scaly skin (◊onchocerciasis). The disease-causing worms are spread mainly by insects, notably mosquitoes and blackflies. Filariasis is treated by drugs to kill the worms, though this does not reverse any swelling.

file in computing, a collection of data or a program stored in a computer's external memory (for example, on ◊disc). It might include anything from information on a company's employees to a program for an adventure game. *Serial files* hold information as a sequence of characters, so that, to read any particular item of data, the program must read all those that precede it. *Random-access files* allow the required data to be reached directly.

file access in computing, the way in which the records in a file are stored, retrieved, or updated by computer. There are four main types of file organization, each of which allows a different form of access to the records.

Records in a *serial file* are not stored in any particular order, so a specific record can be accessed only by reading through all the previous records.

Records in a *sequential file* are sorted by reference to a key field (see ◊sorting) and the computer can use a searching technique, such as a binary search, to access a specific record.

An *indexed sequential file* possesses an index, which records the position of each block of records and is created and updated with that file. By consulting the index, the computer can obtain the address of the block containing the required record, and search just that block rather than the whole file.

A *direct-access* or *random-access file* contains records that can be accessed directly by the computer.

film, photographic strip of transparent material (usually cellulose acetate) coated with a light-sensitive emulsion, used in cameras to take pictures. The emulsion contains a mixture of light-sensitive silver halide salts (for example, bromide or iodide) in gelatin. When the emulsion is exposed to light, the silver salts are invisibly altered, giving a latent image, which is then made visible by the process of ◊developing. Films differ in their sensitivities to light, this being indicated by their speeds. Colour film consists of several layers of emulsion, each of which records a different colour in the light falling on it.

In *colour film* the front emulsion records blue light, then comes a yellow filter, followed by layers that record green and red light respectively. In the developing process the various images in the layers are dyed yellow, magenta (red), and cyan (blue), respectively.

When they are viewed, either as a transparency or as a colour print, the colours merge to produce the true colour of the original scene photographed.

filter in chemistry, a porous substance, such as blotting paper, through which a mixture can be passed to separate out its solid constituents.

filter in electronics, a circuit that transmits a signal of some frequencies better than others. A low-pass filter transmits signals of low frequency and direct current; a high-pass filter transmits high-frequency signals; a band-pass filter transmits signals in a band of frequencies.

filter in optics, a device that absorbs some parts of the visible ◊spectrum and transmits others. For example, a green filter will absorb or block all colours of the spectrum except green, which it allows to pass through. A yellow filter absorbs only light at the blue and violet end of the spectrum, transmitting red, orange, green, and yellow light.

filtration technique by which suspended solid particles in a fluid are removed by passing the mixture through a filter, usually porous paper, plastic, or cloth. The particles are retained by the filter to form a residue and the fluid passes through to make up the filtrate. For example, soot may be filtered from air, and suspended solids from water.

finsen unit unit (symbol FU) for measuring the intensity of ultraviolet (UV) light; for instance, UV light of 2 FUs causes sunburn in 15 minutes.

fiord alternative spelling of ◊fjord.

fireball in astronomy, a very bright ◊meteor, often bright enough to be seen in daylight and occasionally leading to the fall of a ◊meteorite. Some fireballs are caused by ◊satellites or other space debris burning up in the Earth's atmosphere.

fire clay a ◊clay with refractory characteristics (resistant to high temperatures), and hence suitable for lining furnaces (firebrick). Its chemical composition consists of a high percentage of silicon and aluminium oxides, and a low percentage of the oxides of sodium, potassium, iron, and calcium.

firedamp gas that occurs in coal mines and is explosive when mixed with air in certain proportions. It consists chiefly of methane (CH_4, natural gas or marsh gas) but always contains small quantities of other gases, such as nitrogen, carbon dioxide, and hydrogen, and sometimes ethane and carbon monoxide.

fire extinguisher device for putting out a fire. Fire extinguishers work by removing one of the three conditions necessary for fire to continue (heat, oxygen, and fuel), either by cooling the fire or by excluding oxygen.

The simplest fire extinguishers contain water, which

when propelled onto the fire cools it down. Water extinguishers cannot be used on electrical fires, as there is a danger of electrocution, or on burning oil, as the oil will float on the water and spread the blaze.

Many domestic extinguishers contain liquid carbon dioxide under pressure. When the handle is pressed, carbon dioxide is released as a gas that blankets the burning material and prevents oxygen from reaching it. Dry extinguishers spray powder, which then releases carbon dioxide gas. Wet extinguishers are often of the soda-acid type; when activated, sulphuric acid mixes with sodium bicarbonate, producing carbon dioxide. The gas pressure forces the solution out of a nozzle, and a foaming agent may be added.

Some extinguishers contain halons (hydrocarbons with one or more hydrogens substituted by a halogen such as chlorine, bromine, or fluorine). These are very effective at smothering fires, but cause damage to the ◊ozone layer, and their use is now restricted.

firewood the principal fuel for some 2 billion people, mainly in the Third World. In principle a renewable energy source, firewood is being cut far faster than the trees can regenerate in many areas of Africa and Asia, leading to ◊deforestation.

In Mali, for example, wood provides 97% of total energy consumption, and deforestation is running at an estimated 9,000 hectares a year. The heat efficiency of firewood can be increased by use of well-designed stoves, but for many people they are either unaffordable or unavailable. With wood for fuel becoming scarcer the UN Food and Agricultural Organization has estimated that by the year 2000, 3 billion people worldwide will face chronic problems in getting food cooked.

firmware computer program held permanently in a computer's ◊ROM (read-only memory) chips, as opposed to a program that is read in from external memory as it is needed.

first aid action taken immediately in a medical emergency in order to save a sick or injured person's life, prevent further damage, or facilitate later treatment. See also ◊resuscitation.

fish aquatic vertebrate that uses gills to obtain oxygen from fresh or sea water. There are three main groups: the bony fish or Osteichthyes (goldfish, cod, tuna); the cartilaginous fish or Chondrichthyes (sharks, rays); and the jawless fish or Agnatha (hagfishes, lampreys).

Fish of some form are found in virtually every body of water in the world except for the very salty water of the Dead Sea and some of the hot larval springs. Of the 30,000 fish species, approximately 2,500 are freshwater.

Bony fish These constitute the majority of living fish (about 20,000 species). The skeleton is bone, movement is controlled by mobile fins, and the body is usually covered with scales. The gills are covered by a single flap. Many have a ◊swim bladder with which the fish adjusts its buoyancy. Most lay eggs, sometimes in vast numbers; some cod can produce as many as 28 million. These are laid in the open sea, and probably no more than 28 of them will survive to become adults. Those species that produce small numbers of eggs very often protect them in nests, or brood them in their mouths. Some fish are internally fertilized and retain eggs until hatched inside the body, then giving birth to live young. Most bony fish are ray-finned fish, but a few, including lungfish and coelacanths, are fleshy-finned.

Cartilaginous fish These are efficient hunters. There are fewer than 600 known species of sharks and rays. The skeleton is cartilage, the mouth is generally beneath the head, the nose is large and sensitive, and there is a series of open gill slits along the neck region. They have no swim-bladder and, in order to remain buoyant, must keep swimming. They may lay eggs ('mermaid's purses') or bear live young. Some types of cartilaginous fish, such as sharks, retain the shape they had millions of years ago.

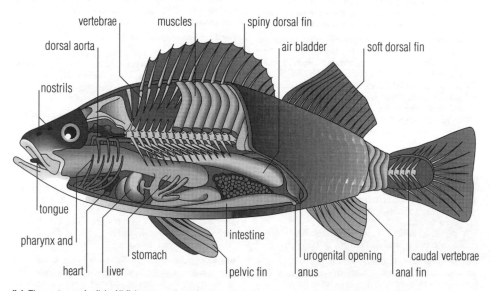

fish The anatomy of a fish. All fishes move through water using their fins for propulsion. The bony fish, like the specimen shown here, constitute the largest group of fish with about 20,000 species.

FEATURE FOCUS

Fishing: declining stocks

Declining fish stocks used to be seen as a regional issue affecting particular communities, but has developed into a crisis that threatens both the ecological balance of the oceans and the chief source of protein for many of the world's people.

In theory, commercial fisheries remove a surplus of each fish population – the fish that would naturally be killed by other predators or would fail to reproduce – so that the overall population level of a particular species does not suffer. This has some ecological implications, of course, but in general allows a balanced ecosystem to remain while providing humans with a valuable food source.

Overexploiting fish stocks

Over the last few decades, fish stocks have been exploited way beyond their natural limits. In the last 40 years, the world catch has increased four-fold. Gigantic drift nets trawl up vast populations of fish, including many unwanted species that are simply abandoned. The largest nets can now encompass the equivalent of 12 Boeing 747 jumbo jets, and even larger nets are under construction.

Far from taking a surplus, fishing fleets have been mining the core population of successive fish species, resulting in a series of population crashes. Populations of Atlantic cod, herring and mackerel have fallen dramatically. Many lesser species, caught by accident, are also endangered. Of the world's 15 leading oceanic fisheries, 13 are in decline. The survival of turtles, dolphins, and other sea creatures is under serious threat simply because they are inadvertently caught up in the world's fishing catch. Sea birds around the UK have suffered a rapid population collapse because there are insufficient fish to feed their young.

An unsustainable system

Such a wasteful system is demonstrably unsustainable. The catches of the world's vast fishing fleets threaten both the marine environment and the livelihoods of many smaller fishing communities throughout the world. About 200 million people in the developing world rely on fisheries for their livelihood.

The problem of declining fish stocks is likely to become one of the key environmental issues in the last years of the century. One of the main reasons for Norway refusing to join the European Union in 1995 was a fear that doing so would open up its rich fishing grounds to Spanish trawlers; Spanish fishermen had clashed with Canadian authorities earlier the same year.

Nigel Dudley

Jawless fish Jawless fish have a body plan like that of some of the earliest vertebrates that existed before true fish with jaws evolved. There is no true backbone but a ◊notochord. The lamprey, for example, attaches itself to the fish on which it feeds by a suckerlike rasping mouth.

The world's largest fish is the whale shark *Rhineodon typus*, more than 20 m/66 ft long; the smallest is the dwarf pygmy goby *Pandaka pygmaea*), 7.5–9.9 mm long. The study of fish is called ichthyology.

Fish

Spurdog, alewife, twaite shad, jollytail, tadpole, madtom, bummalow, walleye pollock, wrestling halfbeak, mummichog, jolthead porgy, sweetlip emperor, and slippery dick are all common names for species of fish.

---

fish farming or **aquaculture** raising fish (including molluscs and crustaceans) under controlled conditions in tanks and ponds, sometimes in offshore pens. It has been practiced for centuries in the Far East, where Japan today produces some 100,000 tonnes of fish a year; the US, Norway, and Canada are also big producers. In the 1980s 10% of the world's consumption of fish was farmed, notably carp, catfish, trout, Atlantic salmon, turbot, eel, mussels, clams, oysters, and shrimp.

Fish farms are environmentally controversial because of the risk of escapees that could spread disease and alter the genetic balance of wild populations.

fishing and fisheries fisheries can be classified by (1) type of water: freshwater (lake, river, pond); marine (inshore, midwater, deep sea); (2) catch: for example, salmon fishing; (3) fishing method: diving, stunning or poisoning, harpooning, trawling, drifting.

Marine fishing Most of the world's catch comes from the oceans, and marine fishing accounts for around 20% of the world's animal-based protein. A wide range of species is included in the landings of the world's marine fishing nations, but the majority belong to the herring and cod groups. The majority of the crustaceans landed are shrimps, and squid and bivalves, such as oysters, are dominant among the molluscs.

Almost all marine fishing takes place on or above the continental shelf, in the photic zone, the relatively thin surface layer (50 m/165 ft) of water that can be penetrated by light, allowing photosynthesis by plant ◊plankton to take place.

Pelagic fishing exploits not only large fish such as tuna, which live near the surface in the open sea and are caught in purse-seine nets, with an annual catch of over 30 million tonnes, but also small, shoaling and plankton-feeding fish that live in the main body of the water.

Examples are herring, sardines, anchovies, and mackerel, which are caught with drift nets, purse seines, and pelagic trawls. The fish are often used for fish meal rather than for direct human consumption.

Demersal fish, such as haddock, halibut, plaice, and cod, live primarily on or near the ocean floor, and feed on various invertebrate marine animals. Over 20 million tonnes of them are caught each year by trawling.

Freshwater fishing Such species as salmon and eels, which migrate to and from fresh and salt water for spawning, may be fished in either system. About a third of the total freshwater catch comes from ◊fish farming methods, which are better developed in freshwater than in marine systems. There is much demand for salmon, trout, carp, eel, bass, pike, perch, and catfish. These are caught in ponds, lakes, rivers, or swamps. In Africa, although marine fishing is generally more important, certain areas have significant freshwater fisheries; Lake Victoria annually yields a catch of 100,000 tonnes, which is four times the total catch from the whole E African seaboard. In W Europe there is very little food production from fresh water; instead the fish are usually exploited for recreational purposes or sport.

Methods The gear and methods used to catch fish are very varied and show much geographical and historical variation. The method chosen for a particular situation will depend on the species being hunted and the nature of the habitat (for example, the speed of the current, the depth of water, and the roughness of the sea bed). It is often useful to divide gear types into active (for example trawls, seines, harpoons, dredges) and passive (drift nets, traps, hooks and lines). Passive gear relies on the fish's own movements to bring them into contact with it, and may involve some method of artificial attraction such as baits or lights. Most fishing gear is operated from boats, ranging from one-person canoes to trawlers about 100 m/330 ft long.

Much of the world's fish catch is caught by trawls. These may be used on the sea bed (demersal) or in midwater (pelagic), but in all cases the equipment consists essentially of a tapered bag of netting which is towed through the water. The mouth of the net is kept open in the vertical plane by having floats on the headline and weights on the footrope. On bottom trawls these weights are usually hollow iron spheres that roll over the sea bed. In addition there may be tickler chains which help to dislodge or disturb fish from the sea bed in advance of the trawl so that they are more likely to be caught. There are three methods of keeping the net open in the horizontal plane: (1) pair trawling, in which the two trawl warps are towed by separate vessels; (2) beam trawling, in which the net is supported on a rigid frame consisting of a horizontal wooden or metal beam with a shoe at either end ;and (3) otter trawling, in which an otter board (a weighted board with lines and baited hooks attached) is incorporated into each warp and acts as a hydroplane to push the warp out sideways. Most modern trawlers are otter trawlers, and many haul the net up over the stern rather than the side. The main problem with pelagic trawls is to control the depth of fishing and to relate this to the concentrations of fish. The most effective means of tracking fish for this purpose is by using an ◊echo sounder.

Seine nets operate by trapping fish within encircling gear. The **Danish seine** resembles a light trawl with very long side pieces, or wings, but is operated differently. The method consists of dropping a large buoy and then paying out up to 4 km/2.5 mi of warp in dogleg shape, then the net, followed by a further length of rope warp in reverse dogleg, bringing the boat back to the buoy, which is then picked up before hauling in the rope warps. As the ropes straighten on the sea bed, they channel the fish into a narrow path between them. The fish are then swept up by the net as it is hauled towards the boat. This method requires smooth sandy ground. **Beach seines** are similar to Danish seines but may consist simply of a wall of netting. They are set in a line or semicircle parallel to a beach and can then be hauled onto the beach, trapping the fish between the net and the shore. Salmon are often caught this way in estuaries. **Purse seines**, nets that close like a purse, are used to catch pelagic fish such as herring, mackerel, and tuna, which form dense shoals near the surface. Once a shoal has been located, usually by echo sounder, the net is shot around it by one vessel and later hauled in towards another. The nets are large, often as long as 30 nautical miles, and are not usually hauled aboard like trawls. Instead the fish are scooped or pumped out of the net into the ship's hold. They have caused a crisis in the S Pacific where Japan, Taiwan, and South Korea fish illegally in other countries' fishing zones.

Gill nets passively depend on the fish entangling themselves in the meshes of the net, usually being held fast by their gill covers. An example is the drift net used for pelagic fish, but in many areas it is now superseded by purse seines and pelagic trawls. Drift nets are walls of netting suspended from floats on the surface. Those used in the East Anglian herring fishery were only 70 m/230 ft long but were set in fleets up to 4 km/2.5 mi long. Herring were caught in them as they came up to the surface at night to feed. Other types of gill net can be used near the sea bed, and one, the trammel, is still used quite commonly in inshore fisheries along the S coast of England. This typically consists of a curtain of large- and small-mesh netting into which the fish swim, forcing the small-mesh net into the large and becoming trapped in a net bag.

Netting panels can be arranged to form traps into which fish are guided or attracted; those used on the NE coast of England to catch salmon are good examples. Many crustaceans, such as lobsters and crabs, are normally taken in baited baskets set in strings of several hundred laid over suitable ground. Earthenware jars are used as octopus traps in the Mediterranean.

Although a distinct method of catching fish, lines and hooks can be regarded as a special type of trap. Natural or artificial baits are used and the gear may be fished anywhere from the sea bed to the surface. Hooks and lines fished off the sea bed may be towed from moving boats, which is called trolling. The largest lines, called long lines, are those used by the Japanese to catch tuna in ocean areas. These are up to 80 km/50 mi long and the baited hooks hang well below the surface from the buoyed lines.

Dredges act like small trawls to collect molluscs and other sluggish or sessile organisms; some are hydraulic and use jets of water to dislodge the molluscs from the bottom and wash them into the dredge bag or directly onto the boat via a conveyor belt.

Molluscs may also be gathered by hand, either on foot at low water or by divers below the shoreline. Rakes may be used to dig out cockles from within the sand. Other methods include dip, lift and cast nets, harpoons, and spears.

fission in physics, the splitting of a heavy atomic nucleus into two or more major fragments. It is accompanied by the emission of two or three neutrons and the release of large amounts of ◊nuclear energy.

Fission occurs spontaneously in nuclei of uranium-235, the main fuel used in nuclear reactors. However, the process can also be induced by bombarding nuclei with neutrons because a nucleus that has absorbed a neutron becomes unstable and soon splits. The neutrons released spontaneously by the fission of uranium nuclei may therefore be used in turn to induce further fissions, setting up a ◊chain reaction that must be controlled if it is not to result in a nuclear explosion.

fistula in medicine, an abnormal pathway developing between adjoining organs or tissues, or leading to the exterior of the body. A fistula developing between the bowel and the bladder, for instance, may give rise to urinary-tract infection by intestinal organisms.

fit in medicine, popular term for ◊convulsion.

fitness in genetic theory, a measure of the success with which a genetically determined character can spread in future generations. By convention, the normal character is assigned a fitness of one, and variants (determined by other ◊alleles) are then assigned fitness values relative to this. Those with fitness greater than one will spread more rapidly and will ultimately replace the normal allele; those with fitness less than one will gradually die out.

fixed point temperature that can be accurately reproduced and used as the basis of a temperature scale. In the Celsius scale, the fixed points are the temperature of melting ice, defined to be 0°C (32°F), and the temperature of boiling water (at standard atmospheric pressure), defined to be 100°C (212°F).

fjord or **fiord** narrow sea inlet enclosed by high cliffs. Fjords are found in Norway, New Zealand, and western parts of Scotland. They are formed when an over-deepened U-shaped glacial valley is drowned by a rise in sea-level. At the mouth of the fjord there is a characteristic lip causing a shallowing of the water. This is due to reduced glacial erosion and the deposition of moraine at this point.

flaccidity in botany, the loss of rigidity (turgor) in plant cells, caused by loss of water from the central vacuole so that the cytoplasm no longer pushes against the cellulose cell wall. If this condition occurs throughout the plant then wilting is seen.

Flaccidity can be induced in the laboratory by immersing the plant cell in a strong saline solution. Water leaves the cell by ◊osmosis causing the vacuole to shrink. In extreme cases the actual cytoplasm pulls away from the cell wall, a phenomenon known as plasmolysis.

flagellum small hairlike organ on the surface of certain cells. Flagella are the motile organs of certain protozoa and single-celled algae, and of the sperm cells of higher animals. Unlike ◊cilia, flagella usually occur singly or in pairs; they are also longer and have a more complex whiplike action.

Each flagellum consists of contractile filaments producing snakelike movements that propel cells through fluids, or fluids past cells. Water movement inside sponges is also produced by flagella.

flag of convenience national flag flown by a ship that has registered in that country in order to avoid legal or tax commitments (also known as offshore registry). Flags of convenience are common in the merchant fleets of Liberia and Panama; ships registered in these countries avoid legislation governing, for example, employment of sailors and minimum rates of pay.

In 1995 75 million gross tonnes of shipping was registered in Panama, making it the world's largest merchant fleet. Ships registered under flags of convenience accounted in 1993 for 34% of the world's fleet but 60% of all maritime accidents.

flame test in chemistry, the use of a flame to identify metal ◊cations present in a solid.

A nichrome or platinum wire is moistened with acid, dipped in a compound of the element, either powdered or in solution, and then held in a hot flame. The colour produced in the flame is characteristic of metals present; for example, sodium burns with an orange-yellow flame, and potassium with a lilac one. *See table overleaf.*

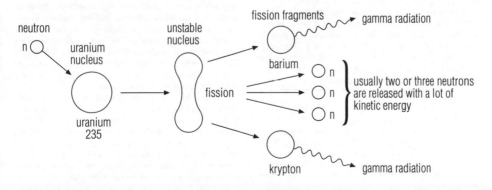

fission Nuclear fission is harnessed in the reactors of nuclear power stations to generate energy. When a nucleus of uranium-235 is struck by a neutron (a subatomic particle) it quickly becomes unstable and splits into two parts – a barium nucleus and a krypton nucleus – liberating a large amount of energy and ejecting two or three neutrons. The emitted neutrons may then be used to trigger the fission of more neutrons, setting up a chain reaction.

Flame Test

Element	Colour of flame
sodium	orange-yellow
potassium	lilac
calcium	red or yellow-red
strontium, lithium	crimson
barium, manganese (manganese chloride)	pale green
copper, thallium, boron (boric acid)	bright green
lead, arsenic, antimony	livid blue
copper (copper (II) chloride)	bright blue

flare, solar brilliant eruption on the Sun above a ◊sunspot, thought to be caused by release of magnetic energy. Flares reach maximum brightness within a few minutes, then fade away over about an hour. They eject a burst of atomic particles into space at up to 1,000 kps/600 mps. When these particles reach Earth they can cause radio blackouts, disruptions of the Earth's magnetic field, and ◊aurorae.

flash flood flood of water in a normally arid area brought on by a sudden downpour of rain. Flash floods are rare and usually occur in mountainous areas. They may travel many kilometres from the site of the rainfall.

Because of the suddenness of flash floods, little warning can be given of their occurrence.

flash memory type of ◊EEPROM memory that can be erased and reprogrammed without removal from the computer.

flatworm invertebrate of the phylum Platyhelminthes. Some are free-living, but many are parasitic (for example, tapeworms and flukes). The body is simple and bilaterally symmetrical, with one opening to the intestine. Many are hermaphroditic (with both male and female sex organs) and practise self-fertilization.

flax any of a group of plants including the cultivated *L. usitatissimum*; **linen** is produced from the fibre in its stems. The seeds yield **linseed oil**, used in paints and varnishes. The plant, of almost worldwide distribution, has a stem up to 60 cm/24 in high, small leaves, and bright blue flowers. (Genus *Linum*, family Linaceae.)

After extracting the oil, what is left of the seeds is fed to cattle. The stems are retted (soaked) in water after harvesting, and then dried, rolled, and scutched (pounded), separating the fibre from the central core of woody tissue. The long fibres are spun into linen thread, twice as strong as cotton, yet more delicate, and suitable for lace; shorter fibres are used to make string or paper.

Flax

Perhaps the most demanding demon in European mythology is the Pszepolnica. This is a Polish–German horse-footed witch who accosts passers-by and beheads them if they cannot talk for an hour on the subject of flax.

Fleming's rules memory aids used to recall the relative directions of the magnetic field, current, and motion in an electric generator or motor, using one's fingers. The three directions are represented by the thu*m*b (for *m*otion), *f*orefinger (for *f*ield), and se*c*ond finger (for *c*urrent), all

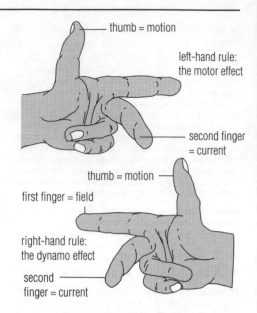

thumb = motion

left-hand rule: the motor effect

second finger = current

thumb = motion

first finger = field

right-hand rule: the dynamo effect

second finger = current

Fleming's rules Fleming's rules give the direction of the magnetic field, motion, and current in electrical machines. The left hand is used for motors, and the right hand for generators and dynamos.

held at right angles to each other. The right hand is used for generators and the left for motors.

The rules were devised by the English physicist John Fleming.

flexor any muscle that bends a limb. Flexors usually work in opposition to other muscles, the extensors, an arrangement known as antagonistic.

flight or *aviation* method of transport in which aircraft carry people and goods through the air. People first took to the air in ◊balloons and began powered flight in 1852 in ◊airships, but the history of flying, both for civilian and military use, is dominated by the ◊aeroplane. The earliest planes were designed for gliding; the advent of the petrol engine saw the first powered flight by the Wright brothers in 1903 in the USA. This inspired the development of aircraft throughout Europe. Biplanes were succeeded by monoplanes in the 1930s. The first jet plane (see ◊jet propulsion) was produced in 1939, and after the end of World War II the development of jetliners brought about a continuous expansion in passenger air travel. In 1969 came the supersonic aircraft ◊Concorde.

In the 14th century the English philosopher Roger Bacon spoke of constructing an aircraft by means of a hollow globe and liquid fire. He was followed in the 15th century by Albert of Saxony, who also spoke of balloon flight by means of fire in a light sphere. During the 16th and 17th centuries a number of fantastic ideas were put forward; one was that swans' eggs be filled with sulphur or mercury and thereby drawn up to the Sun.

In 1670 Francisco de Lana proposed that four hollow balls made of very thin brass should be emptied of air. To them should be attached a small boat and sail, and in that way a balloon would be contrived which could carry a person. The idea was not feasible, since the globes, made

Flight: Chronology

1783	First human flight, by Jean F Pilâtre de Rozier and the Marquis d'Arlandes, in Paris, using a hot-air balloon made by Joseph and Etienne Montgolfier; first ascent in a hydrogen-filled balloon by Jacques Charles and M N Robert in Paris.
1785	Jean-Pierre Blanchard and John J Jeffries made the first balloon crossing of the English Channel.
1852	Henri Giffard flew the first steam-powered airship over Paris.
1853	George Cayley flew the first true aeroplane, a model glider 1.5 m/5 ft long.
1891–96	Otto Lilienthal piloted a glider in flight.
1903	First powered and controlled flight of a heavier-than-air craft (aeroplane) by Orville Wright, at Kitty Hawk, North Carolina, USA.
1908	First powered flight in the UK by Samuel Cody.
1909	Louis Blériot flew across the English Channel in 36 minutes.
1914–18	World War I stimulated improvements in speed and power.
1919	First E–W flight across the Atlantic by Albert C Read, using a flying boat; first nonstop flight across the Atlantic E–W by John William Alcock and Arthur Whitten Brown in 16 hours 27 minutes; first complete flight from Britain to Australia by Ross Smith and Keith Smith.
1923	Juan de la Cieva flew the first autogiro with a rotating wing.
1927	Charles Lindbergh made the first W–E solo nonstop flight across the Atlantic.
1928	First transpacific flight, from San Francisco to Brisbane, by Charles Kinsford Smith and C T P Ulm.
1930	Frank Whittle patented the jet engine; Amy Johnson became the first woman to fly solo from England to Australia.
1937	The first fully pressurized aircraft, the Lockheed XC-35, came into service.
1939	Erich Warsitz flew the first Heinkel jet plane, in Germany; Igor Sikorsky designed the first helicopter, with a large main rotor and a smaller tail rotor.
1939–45	World War II – developments included the Hawker Hurricane and Supermarine Spitfire fighters, and Avro Lancaster and Boeing Flying Fortress bombers.
1947	A rocket-powered plane, the Bell X-1, was the first aircraft to fly faster than the speed of sound.
1949	The de Havilland Comet, the first jet airliner, entered service; James Gallagher made the first nonstop round-the-world flight, in a Boeing Superfortress.
1953	The first vertical takeoff aircraft, the Rolls-Royce 'Flying Bedstead', was tested.
1968	The world's first supersonic airliner, the Russian TU-144, flew for the first time.
1970	The Boeing 747 jumbo jet entered service, carrying 500 passengers.
1976	Anglo-French Concorde, making a transatlantic crossing in under three hours, came into commercial service. A Lockheed SR-17A, piloted by Eldon W Joersz and George T Morgan, set the world air-speed record of 3,529.56 kph/2,193.167 mph over Beale Air Force Base, California, USA.
1978	A US team made the first transatlantic crossing by balloon, in the helium-filled *Double Eagle II*.
1979	First crossing of the English Channel by a human-powered aircraft, *Gossamer Albatross*, piloted by Bryan Allen.
1981	The solar-powered *Solar Challenger* flew across the English Channel, from Paris to Kent, taking 5 hours for the 262 km/162.8 mi journey.
1986	Dick Rutan and Jeana Yeager made the first nonstop flight around the world without refuelling, piloting *Voyager*, which completed the flight in 9 days 3 minutes 44 seconds.
1987	Richard Branson and Per Lindstrand made the first transatlantic crossing by hot-air balloon, in *Virgin Atlantic Challenger*.
1988	*Daedelus*, a human-powered craft piloted by Kanellos Kanellopoulos, flew 118 km/74 mi across the Aegean Sea.
1991	Richard Branson and Per Lindstrand crossed the Pacific Ocean in the hot-air balloon *Virgin Otsouka Pacific Flyer* from the southern tip of Japan to northwest Canada in 46 hours 15 minutes.
1992	US engineers demonstrated a model radio-controlled ornithopter, the first aircraft to be successfully propelled and manoeuvred by flapping wings.
1993	The US Federal Aviation Authority made the use of an automatic on-board collision avoidance system (TCAS-2) mandatory in US airspace.
1994	The US Boeing 777 airliner made its first flight. A scale model scramjet (supersonic combustion ramjet) was tested and produced speeds of 9,000 kph/5,590 mph (Mach 8.2). The scramjet uses oxygen from the atmosphere to burn its fuel.
1996	Japan tested the first fire-fighting helicopter, designed to reach skyscrapers beyond the range of fire-engine ladders.
1997–98	In an attempt to circle the globe, Bertrand Piccard, Wim Verstraeten and Andy Elson overtook the 1986 World non-stop flight record in their *Breitling Orbiter 2* balloon after 9 days 17 hours and 55 minutes.

of brass only 0.1 mm thick, would have collapsed by reason of their own weight. But although de Lana saw this difficulty, he argued that their shape would prevent that.

It was not until the next century that the real ◊balloon was invented. The beginning of the development of the balloon was the work of two brothers, Joseph and Etienne Montgolfier, who came to the conclusion that a paper bag filled with a 'substance of a cloud-like nature' would float in the atmosphere. They made a number of experiments which attracted attention and further efforts from others. Progress was made gradually, and the first person-carrying ascent took place in October 1783, when Jean F Pilâtre de Rozier went up in a Montgolfier captive balloon. The first woman to ascend was Madame Thible, who went up from Lyons in 1784. In 1859 a flight of over 1,600 km/994 mi was made in the USA.

It had long been recognized that the difficulty with balloons was navigating through the air. Oars were tried, but were not successful. The first attempt to navigate the balloon by means of a small, light engine came in 1852, the experiment being made by Henri Giffard. From 1897 the development of the airship was the special work of Ferdinand Zeppelin. In 1900 he made his first flight with a dirigible balloon carrying five men. It was made of aluminium, supported by gas-bags, and driven by two motors, each of about 12 kW. His first experiment met with some success, a second, more powerful version was wrecked, and a third met with great success. This airship carried 11 passengers and attained a speed of about 55 kph/34 mph, travelling about 400 km/248 mi in 11 hours, but was wrecked by a storm in 1908, caught fire, and was completely destroyed.

In the late 19th century experiments were being made with soaring machines and hang gliders, chiefly by Otto Lilienthal, who, with an arrangement formed on the plan of birds' wings, attempted to imitate their 'soaring flight'. Following up Lilienthal's ideas, the Wright brothers produced their first aeroplane in 1903. Their first invention was simply an aeroplane that flew in a straight line, but this received many modifications; and in 1908 they went to France to carry on experiments, during which Wilbur Wright created a record by remaining in the air for over an hour while carrying a passenger. He also attained a speed of 60 kph/37 mph.

In Europe, at the beginning of the 20th century, France led in aeroplane design and Louis Blériot brought aviation much publicity by crossing the Channel 1909, as did the Reims air races of that year. The first powered flight in the UK was made by Samuel Franklin Cody 1908. In 1912 Sopwith and Bristol both built small biplanes. The first big twin-engined aeroplane was the Handley Page bomber 1917. The stimulus of World War I (1914–18) and rapid development of the petrol engine led to increased power, and speeds rose to 320 kph/200 mph. Streamlining the body of planes became imperative: the body, wings, and exposed parts were reshaped to reduce drag. Eventually the biplane was superseded by the internally braced monoplane structure, for example, the Hawker Hurricane and Supermarine Spitfire fighters and Avro Lancaster and Boeing Flying Fortress bombers of World War II (1939–45).

The German Heinkel 178, built 1939, was the first jet plane; it was driven, not by a ◊propeller as all planes before it, but by a jet of hot gases. The first British jet aircraft, the Gloster E.28/39, flew from Cranwell, Lincolnshire, on 15 May 1941, powered by a jet engine invented by British engineer Frank Whittle. Twin-jet Meteor fighters were in use by the end of the war. The rapid development of the jet plane led to enormous increases in power and speed until air-compressibility effects were felt near the speed of sound, which at first seemed to be a flight speed limit (the sound barrier). To attain ◊supersonic speed, streamlining the aircraft body became insufficient: wings were swept back, engines buried in wings and tail units, and bodies were even eliminated in all-wing delta designs. In the 1950s the first jet airliners, such as the Comet (first introduced 1949), were introduced into service. Today jet planes dominate both military and civilian aviation, although many light planes still use piston engines and propellers. The late

1960s saw the introduction of the ◊jumbo jet, and in 1976 the Anglo-French Concorde, which makes a transatlantic crossing in under three hours, came into commercial service.

During the 1950s and 1960s research was done on V/STOL (vertical and/or short takeoff and landing) aircraft. The British Harrier jet fighter has been the only VTOL aircraft to achieve commercial success, but STOL technology has fed into subsequent generations of aircraft. The 1960s and 1970s also saw the development of variable geometry ('swing-wing') aircraft, the wings of which can be swept back in flight to achieve higher speeds. In the 1980s much progress was made in 'fly-by-wire' aircraft with computer-aided controls. International partnerships have developed both civilian and military aircraft. The Panavia Tornado is a joint project of British, German, and Italian aircraft companies. It is an advanced swing-wing craft of multiple roles – interception, strike, ground support, and reconnaissance. The airbus is a wide-bodied airliner built jointly by companies from France, Germany, the UK, the Netherlands, and Spain.

flight simulator computer-controlled pilot-training device, consisting of an artificial cockpit mounted on hydraulic legs, that simulates the experience of flying a real aircraft. Inside the cockpit, the trainee pilot views a screen showing a computer-controlled projection of the view from a real aircraft, and makes appropriate adjustments to the controls. The computer monitors these adjustments, changing both the alignment of the cockpit on its hydraulic legs, and the projected view seen by the pilot. In this way a trainee pilot can progress to quite an advanced stage of training without leaving the ground.

flint compact, hard, brittle mineral (a variety of chert), brown, black, or grey in colour, found in nodules in limestone or shale deposits. It consists of fine-grained silica, SiO_2 (usually ◊quartz), in cryptocrystalline form. Flint implements were widely used in prehistory.

When chipped, the flint nodules show a shell-like fracture and a sharp cutting edge. The earliest flint implements, belonging to Palaeolithic cultures and made by striking one flint against another, are simple, whereas those of the Neolithic are expertly chipped and formed, and are often ground or polished.

The best flint, used for Neolithic tools, is **floorstone**, a shiny black flint that occurs deep within the chalk.

Because of their hardness (7 on the ◊Mohs' scale), flint splinters are used for abrasive purposes and, when ground into powder, added to clay during pottery manufacture. Flints have been used for making fire by striking the flint against steel, which produces a spark, and for discharging guns. Flints in cigarette lighters are made from cerium alloy.

flocculation in soils, the artificially induced coupling together of particles to improve aeration and drainage. Clay soils, which have very tiny particles and are difficult to work, are often treated in this way. The method involves adding more lime to the soil.

flooding the inundation of land that is not normally covered with water. Flooding from rivers commonly takes place after heavy rainfall or in the spring after winter snows have melted. The river's discharge (volume of water carried in a given period) becomes too great, and water spills over the banks onto the surrounding flood plain. Small floods may happen once a year – these are called **annual floods** and are said to have a one-year return period. Much larger floods may occur on average only once every 50 years.

Flooding is least likely to occur in an efficient channel that is semicircular in shape. Flooding can also occur at the coast in stormy conditions (see ◊storm surge) or when there is an exceptionally high tide. The Thames Flood Barrier was constructed in 1982 to prevent the flooding of London from the sea.

flood plain area of periodic flooding along the course of river valleys. When river discharge exceeds the capacity of the channel, water rises over the channel banks and floods the adjacent low-lying lands. As water spills out of the channel some alluvium (silty material) will be deposited on the banks to form ◊levees (raised river banks). This water will slowly seep into the flood plain, depositing a new layer of rich fertile alluvium as it does so.

Many important floodplains, such as the inner Niger delta in Mali, occur in arid areas where their exceptional productivity has great importance for the local economy.

A flood plain (sometimes called inner ◊delta) can be regarded as part of a river's natural domain, statistically certain to be claimed by the river at repeated intervals. By plotting floods that have occurred and extrapolating from these data we can speak of ten-year floods, 100-year floods, 500-year floods, and so forth, based on the statistical probability of flooding across certain parts of the flood plain.

Even the most energetic flood-control plans (such as dams, dredging, and channel modification) will sometimes fail, and using flood plains as the site of towns and villages is always laden with risk. It is more judicious to use flood plains in ways compatible with flooding, such as for agriculture or parks.

Floodplain features include ◊meanders and ◊oxbow lakes.

floppy disc in computing, a storage device consisting of a light, flexible disc enclosed in a cardboard or plastic jacket. The disc is placed in a disc drive, where it rotates at high speed. Data are recorded magnetically on one or both surfaces.

Floppy discs were invented by IBM in 1971 as a means of loading programs into the computer. They were originally 20 cm/8 in in diameter and typically held about 240 ◊kilobytes of data. Present-day floppy discs, widely used on ◊microcomputers, are usually either 13.13 cm/5.25 in or 8.8 cm/3.5 in in diameter, and generally hold 0.5–2 ◊megabytes, depending on the disc size, recording method, and whether one or both sides are used.

Floppy discs are inexpensive, and light enough to send through the post, but have slower access speeds and are more fragile than hard discs. (See also ◊disc.)

floral diagram diagram showing the arrangement and number of parts in a flower, drawn in cross-section. An ovary is drawn in the centre, surrounded by representations of the other floral parts, indicating the position of each at its base. If any parts such as the petals or sepals are fused, this is also indicated. Floral diagrams allow the structures of different flowers to be compared, and are usually shown with the floral formula.

floral formula symbolic representation of the structure of a flower. Each kind of floral part is represented by a letter (K for calyx, C for corolla, P for perianth, A for androecium, G for gynoecium) and a number to indicate the quantity of the part present, for example, C5 for a flower with five petals. The number is in brackets if the parts are fused. If the parts are arranged in distinct whorls within the flower, this is shown by two separate figures, such as A5 + 5, indicating two whorls of five stamens each. A flower with radial symmetry is known as **actinomorphic**; a flower with bilateral symmetry as **zygomorphic**.

floret small flower, usually making up part of a larger, composite flower head. There are often two different types present on one flower head: disc florets in the central area, and ray florets around the edge which usually have a single petal known as the ligule. In the common daisy, for example, the disc florets are yellow, while the ligules are white.

flotation, law of law stating that a floating object displaces its own weight of the fluid in which it floats. See ◊Archimedes principle.

flotation process common method of preparing mineral ores for subsequent processing by making use of the different wetting properties of various components. The ore is finely ground and then mixed with water and a specially selected wetting agent. Air is bubbled through the mixture, forming a froth; the desired ore particles attach themselves to the bubbles and are skimmed off, while unwanted dirt or other ores remain behind.

flow chart diagram, often used in computing, to show the possible paths that data can take through a system or program.

A **system flow chart**, or **data flow chart**, is used to describe the flow of data through a complete data-processing system. Different graphic symbols represent clerical operations involved and the different input, storage, and output equipment required. Although the flow chart may indicate the specific programs used, no details are given of how the programs process the data.

A **program flow chart** is used to describe the flow of data through a particular computer program, showing the exact sequence of operations performed by that program in order to process the data. Different graphic symbols are used to represent data input and output, decisions, branches, and ◊subroutines. *See illustrations on page 262.*

flower the reproductive unit of an angiosperm or flowering plant, typically consisting of four whorls of modified leaves: ◊sepals, ◊petals, ◊stamens, and ◊carpels. These are borne on a central axis or ◊receptacle. The many variations in size, colour, number, and arrangement of parts are closely related to the method of pollination. Flowers adapted for wind pollination typically have reduced or absent petals and sepals and long, feathery ◊stigmas that hang outside the flower to trap airborne pollen. In contrast, the petals of insect-pollinated flowers are usually conspicuous and brightly coloured.

structure The sepals and petals form the **calyx** and **corolla** respectively and together comprise the **perianth** with the function of protecting the reproductive organs and attracting pollinators. The stamens lie within the corolla, each having a slender stalk, or filament, bearing the pollen-containing anther at the top. Collectively they are known as the **androecium** (male organs). The inner whorl of the flower comprises the carpels, each usually consisting of an ovary in which are borne the ◊ovules, and a stigma borne at the top of a slender stalk, or style.

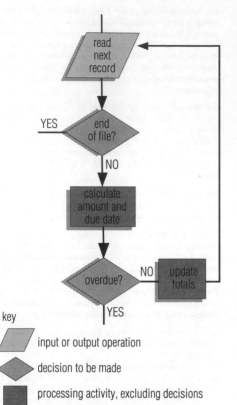

key

input or output operation

decision to be made

processing activity, excluding decisions

flow chart A program flow chart shows the sequence of operations needed to achieve a task, in this case reading customer accounts and calculating the amount due for each customer. After an account has been processed, the program loops back to process the next one.

Collectively the carpels are known as the **gynoecium** (female organs).

Flower

The world's largest flower smells of rotting flesh. The flower of the parasitic *Rafflesia* has a diameter of up to 1 m/3.3 ft and attracts pollinating insects by mimicking the smell of decomposing corpses. It flowers only once every ten years.

---

types of flower In size, flowers range from the tiny blooms of duckweeds scarcely visible to the naked eye to the gigantic flowers of the Malaysian *Rafflesia*, which can reach over 1 m/3.3 ft across. Flowers may either be borne singly or grouped together in ◊inflorescences. The stalk of the whole inflorescence is termed a **peduncle**, and the stalk of an individual flower is termed a **pedicel**. A flower is termed hermaphrodite when it contains both male and female reproductive organs. When male and female organs are carried in separate flowers, they are termed **monoecious**; when male and female flowers are on separate plants, the term **dioecious** is used.

flowering plant term generally used for ◊angiosperms, which bear flowers with various parts, including sepals, petals, stamens, and carpels. Sometimes the term is used more broadly, to include both angiosperms and ◊gymnosperms, in which case the ◊cones of conifers and cycads are referred to as 'flowers'. Usually, however, the angiosperms and gymnosperms are referred to collectively as ◊seed plants, or spermatophytes. In 1996 UK palaeontologists found fossils in S England of what may be the world's oldest flowering plant.

Bevhalstia pebja, a wetland herb about 25 cm/10 in high, has been dated as early Cretaceous, about 130 million years old.

flue-gas desulphurization process of removing harmful sulphur pollution from gases emerging from a boiler. Sulphur compounds such as sulphur dioxide are commonly produced by burning ◊fossil fuels, especially coal in power stations, and are the main cause of ◊acid rain.

The process is environmentally beneficial but expensive, adding about 10% to the cost of electricity generation.

fluid mechanics the study of the behaviour of fluids (liquids and gases) at rest and in motion. Fluid mechanics is important in the study of the weather, the design of aircraft and road vehicles, and in industries, such as the chemical industry, which deal with flowing liquids or gases.

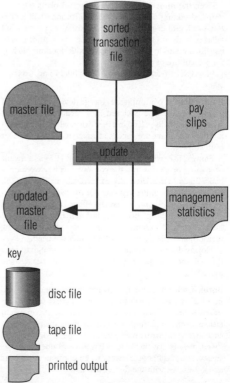

key

disc file

tape file

printed output

flow chart A system flow chart describes the flow of data through a data-processing system. This chart shows the data flow in a basic accounting system.

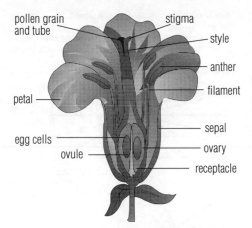

pollen grain and tube — stigma — style — anther — filament — petal — sepal — egg cells — ovary — ovule — receptacle

flower Cross-section of a typical flower showing its basic components: sepals, petals, stamens (anthers and filaments), and carpel (ovary and stigma). Flowers vary greatly in the size, shape, colour, and arrangement of these components.

fluke any of various parasitic flatworms of the classes Monogenea and Digenea, that as adults live in and destroy the livers of sheep, cattle, horses, dogs, and humans. Monogenetic flukes can complete their life cycle in one host; digenetic flukes require two or more hosts, for example a snail and a human being, to complete their life cycle.

fluorescence in scientific usage, very short-lived ◊luminescence (a glow not caused by high temperature). Generally, the term is used for any luminescence regardless of the persistence. ◊Phosphorescence lasts a little longer.

Fluorescence is used in strip and other lighting, and was developed rapidly during World War II because it was a more efficient means of illumination than the incandescent lamp. Recently, small bulb-size fluorescence lamps have reached the market. It is claimed that, if widely used, their greater efficiency could reduce demand for electricity. Other important applications are in fluorescent screens for television and cathode-ray tubes.

fluorescence microscopy technique for examining samples under a ◊microscope without slicing them into thin sections. Instead, fluorescent dyes are introduced into the tissue and used as a light source for imaging purposes. Fluorescent dyes can also be bonded to monoclonal antibodies and used to highlight areas where particular cell proteins occur.

fluoridation addition of small amounts of fluoride salts to drinking water by certain water authorities to help prevent tooth decay. Experiments in Britain, the USA, and elsewhere have indicated that a concentration of fluoride of 1 part per million in tap water retards the decay of children's teeth by more than 50%.

Much concern has been expressed about the risks of medicating the population at large by the addition of fluoride to the water supply, but the medical evidence demonstrates conclusively that there is no risk to the general health from additions of 1 part per million of fluoride to drinking water.

fluoride negative ion (F^-) formed when hydrogen fluor-ide dissolves in water; compound formed between fluorine and another element in which the fluorine is the more electronegative element (see ◊electronegativity, ◊halide).

In parts of India, the natural level of fluoride in water is 10 parts per million. This causes fluorosis, or chronic fluoride poisoning, mottling teeth and deforming bones.

fluorine pale yellow, gaseous, nonmetallic element, symbol F, atomic number 9, relative atomic mass 19. It is the first member of the halogen group of elements, and is pungent, poisonous, and highly reactive, uniting directly with nearly all the elements. It occurs naturally as the minerals fluorite (CaF_2) and cryolite (Na_3AlF_6). Hydrogen fluoride is used in etching glass, and the freons, which all contain fluorine, are widely used as refrigerants.

Fluorine was discovered by the Swedish chemist Karl Scheele in 1771 and isolated by the French chemist Henri Moissan in 1886.

Combined with uranium as UF_6, it is used in the separation of uranium isotopes.

fluorite or **fluorspar** a glassy, brittle halide mineral, calcium fluoride, CaF_2, forming cubes and octahedra; colourless when pure, otherwise violet, blue, yellow, brown, or green.

Fluorite is used as a flux in iron and steel making; colourless fluorite is used in the manufacture of microscope lenses. It is also used for the glaze on pottery, and as a source of fluorine in the manufacture of hydrofluoric acid.

fluorocarbon compound formed by replacing the hydrogen atoms of a hydrocarbon with fluorine. Fluorocarbons are used as inert coatings, refrigerants, synthetic resins, and as propellants in aerosols.

There is concern that the release of fluorocarbons – particularly those containing chlorine (chlorofluorocarbons, CFCs) – depletes the ◊ozone layer, allowing more ultraviolet light from the Sun to penetrate the Earth's atmosphere, and increasing the incidence of skin cancer in humans.

fluoroscopy in medicine, a technique in which X-rays are rendered visible after they have passed through the body by projecting them on to a film or screen containing a fluorescent material using a fluoroscope. The technique allows the beating of the heart or the movements in the intestine after a barium meal to be observed and it can assist diagnosis of disease in these organs.

fluvioglacial of a process or landform, associated with glacial meltwater. Meltwater, flowing beneath or ahead of a glacier, is capable of transporting rocky material and creating a variety of landscape features, including eskers, kames, and outwash plains.

flux in smelting, a substance that combines with the unwanted components of the ore to produce a fusible slag, which can be separated from the molten metal. For example, the mineral fluorite, CaF_2, is used as a flux in iron smelting; it has a low melting point and will form a fusible mixture with substances of higher melting point such as silicates and oxides.

flux in soldering, a substance that improves the bonding properties of solder by removing contamination from metal surfaces and preventing their oxidation, and by reducing the surface tension of the molten solder alloy. For example, with solder made of lead-tin alloys, the flux may be resin, borax, or zinc chloride.

flystrike or *blowfly strike* or *sheep strike* infestation of the flesh of living sheep by blowfly maggots, especially those of the blue blowfly. It is one of the most costly sheep diseases in Australia, affecting all the grazing areas of New South Wales. Control has mainly been by insecticide, but non-chemical means, such as docking of tails and mulesing, are increasingly being encouraged. Mulesing involves an operation to remove the wrinkles of skin which trap moisture and lay the sheep open to infestation.

flywheel heavy wheel in an engine that helps keep it running and smooths its motion. The ◊crankshaft in a petrol engine has a flywheel at one end, which keeps the crankshaft turning in between the intermittent power strokes of the pistons. It also comes into contact with the ◊clutch, serving as the connection between the engine and the car's transmission system.

FM in physics, abbreviation for ◊*frequency modulation*.

f-number or *f-stop* measure of the relative aperture of a telescope or camera lens; it indicates the light-gathering power of the lens. In photography, each successive f-number represents a halving of exposure speed.

focal length or *focal distance* the distance from the centre of a lens or curved mirror to the focal point. For a concave mirror or convex lens, it is the distance at which parallel rays of light are brought to a focus to form a real image (for a mirror, this is half the radius of curvature). For a convex mirror or concave lens, it is the distance from the centre to the point at which a virtual image (an image produced by diverging rays of light) is formed.

With lenses, the greater the power (measured in dioptres) of the lens, the shorter its focal length. The human eye has a lens of adjustable focal length to allow the light from objects of varying distance to be focused on the retina.

focus in astronomy, either of two points lying on the major axis of an elliptical ◊orbit on either side of the centre. One focus marks the centre of mass of the system and the other is empty. In a circular orbit the two foci coincide at the centre of the circle and in a parabolic orbit the second focus lies at infinity. See ◊Kepler's Laws.

focus or *focal point* in optics, the point at which light rays converge, or from which they appear to diverge, to form a sharp image. Other electromagnetic rays, such as microwaves, and sound waves may also be brought together at a focus. Rays parallel to the principal axis of a lens or mirror are converged at, or appear to diverge from, the ◊principal focus.

focus in photography, the distance that a lens must be moved in order to focus a sharp image on the light-sensitive film at the back of the camera. The lens is moved away from the film to focus the image of closer objects. The focusing distance is often marked on a scale around the lens; however, some cameras now have an automatic focusing (autofocus) mechanism that uses an electric motor to move the lens.

fog cloud that collects at the surface of the Earth, composed of water vapour that has condensed on particles of dust in the atmosphere. Cloud and fog are both caused by the air temperature falling below ◊dew point. The thickness of fog depends on the number of water particles it contains. Officially, fog refers to a condition when visi-

bility is reduced to 1 km/0.6 mi or less, and mist or haze to that giving a visibility of 1–2 km or about 1 mi.

There are two types of fog. An *advection fog* is formed by the meeting of two currents of air, one cooler

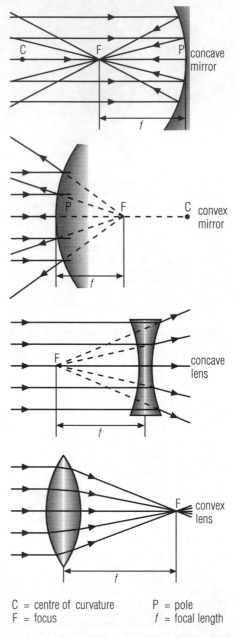

C = centre of curvature   P = pole
F = focus   f = focal length

focal length The distance from the pole (P), or optical centre, of a lens or spherical mirror to its principal focus (F). The focal length of a spherical mirror is equal to half the radius of curvature ($f = CP/2$). The focal length of a lens is inversely proportional to the power of that lens (the greater the power the shorter the focal length).

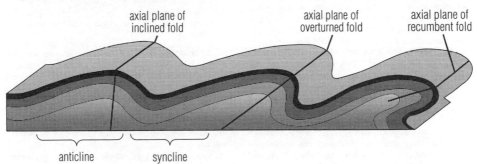

advection fog
warm moist air cools either as it passes over a cool sea or comes into contact with cold land surface

heat lost

heat lost

heat lost

heat lost

fog

fog

sea

radiation fog
during a clear night, heat is lost rapidly from the land. This cools the air which, if moist, becomes saturated – fog forms as it condenses

fog

than the other, or by warm air flowing over a cold surface. Sea fogs commonly occur where warm and cold currents meet and the air above them mixes. A *radiation fog* forms on clear, calm nights when the land surface loses heat rapidly (by radiation); the air above is cooled to below its dew point and condensation takes place. A *mist* is produced by condensed water particles, and a *haze* by smoke or dust.

In drought areas, for example, Baja California, Canary Islands, Cape Verde Islands, Namib Desert, Peru, and Chile, coastal fogs enable plant and animal life to survive without rain and are a potential source of water for human use (by means of water collectors exploiting the effect of condensation).

Industrial areas uncontrolled by pollution laws have a continual haze of smoke over them, and if the temperature falls suddenly, a dense yellow smog forms. At some airports since 1975 it has been possible for certain aircraft to land and take off blind in fog, using radar navigation.

föhn or **foehn** warm dry wind that blows down the leeward slopes of mountains.

The air heats up as it descends because of the increase in pressure, and it is dry because all the moisture was dropped on the windward side of the mountain. In the

valleys of Switzerland it is regarded as a health hazard, producing migraine and high blood pressure. A similar wind, chinook, is found on the eastern slopes of the Rocky Mountains in North America.

fold in geology, a bend in ◊beds or layers of rock. If the bend is arched in the middle it is called an **anticline**; if it sags downwards in the middle it is called a **syncline**. The line along which a bed of rock folds is called its axis. The axial plane is the plane joining the axes of successive beds.

fold mountain mountains formed from large-scale folding of the Earth's crust at a destructive margin (where two tectonic plates collide). The Himalayas are an example.

folic acid a ◊vitamin of the B complex. It is found in liver and green leafy vegetables, and is also synthesized by the intestinal bacteria. It is essential for growth, and plays many other roles in the body. Lack of folic acid causes anaemia because it is necessary for the synthesis of nucleic acids and the formation of red blood cells.

follicle in botany, a dry, usually many-seeded fruit that splits along one side only to release the seeds within. It is derived from a single ◊carpel, examples include the fruits of the larkspur *Delphinium* and columbine *Aquilegia*. It

axial plane of inclined fold

axial plane of overturned fold

axial plane of recumbent fold

anticline

syncline

fold The folding of rock strata occurs where compression causes them to buckle. Over time, folding can assume highly complicated forms, as can sometimes be seen in the rock layers of cliff faces or deep cuttings in the rock. Folding contributed to the formation of great mountain chains such as the Himalayas.

differs from a pod, which always splits open (dehisces) along both sides.

follicle in zoology, a small group of cells that surround and nourish a structure such as a hair (hair follicle) or a cell such as an egg (Graafian follicle; see ◊menstrual cycle).

follicle-stimulating hormone (FSH) a ◊hormone produced by the pituitary gland. It affects the ovaries in women, stimulating the production of an egg cell.

Luteinizing hormone is needed to complete the process. In men, FSH stimulates the testes to produce sperm. It is used to treat some forms of infertility.

Fomalhaut or *Alpha Piscis Austrini* the brightest star in the southern constellation ◊Piscis Austrinus and the 18th brightest star in the night sky. It is 22 light years from Earth, with a true luminosity 13 times that of the Sun.

Fomalhaut is one of a number of stars around which IRAS (the Infra-Red Astronomy Satellite) detected excess infrared radiation, presumably from a region of solid particles around the star. This material may be a planetary system in the process of formation.

font or *fount* complete set of printed or display characters of the same typeface, size, and style (bold, italic, underlined, and so on). In the UK, font sizes are measured in points, a point being approximately 0.3 mm.

Fonts used in computer setting are of two main types: bit-mapped and outline. *Bit-mapped fonts* are stored in the computer memory as the exact arrangement of ◊pixels or printed dots required to produce the characters in a particular size on a screen or printer. *Outline fonts* are stored in the computer memory as a set of instructions for drawing the circles, straight lines, and curves that make up the outline of each character. They require a powerful computer because each character is separately generated from a set of instructions and this requires considerable computation. Bit-mapped fonts become very ragged in appearance if they are enlarged and so a separate set of bit maps is required for each font size. In contrast, outline fonts can be scaled to any size and still maintain exactly the same appearance.

food anything eaten by human beings and other animals and plants to sustain life and health. The building blocks of food are nutrients, and humans can utilize the following nutrients: **carbohydrates**, as starches found in bread, potatoes, and pasta; as simple sugars in sucrose and honey; as fibres in cereals, fruit, and vegetables; **proteins** as from nuts, fish, meat, eggs, milk, and some vegetables; **fats** as found in most animal products (meat, lard, dairy products, fish), also in margarine, nuts and seeds, olives, and edible oils; **vitamins**, found in a wide variety of foods, except for vitamin B_{12}, which is mainly found in foods of animal origin; **minerals**, found in a wide variety of foods (for example, calcium from milk and broccoli, iodine from seafood, and iron from liver and green vegetables); **water** ubiquitous in nature; **alcohol**, found in fermented distilled beverages, from 40% in spirits to 0.01% in low-alcohol lagers and beers.

Food is needed both for energy, measured in ◊calories or kilojoules, and nutrients, which are converted to body tissues. Some nutrients, such as fat, carbohydrate, and alcohol, provide mainly energy; other nutrients are important in other ways; for example, fibre is an aid to metabolism. Proteins provide energy and are necessary for building cell and tissue structure.

Food and Agriculture Organization (FAO) United Nations agency that coordinates activities to improve food and timber production and levels of nutrition throughout the world. It is also concerned with investment in agriculture and dispersal of emergency food supplies. It's headquarters are in Rome, and it was founded in 1945.

food chain in ecology, a sequence showing the feeding relationships between organisms in a particular ◊ecosystem. Each organism depends on the next lowest member of the chain for its food.

Energy in the form of food is shown to be transferred from ◊autotrophs, or producers, which are principally plants and photosynthetic microorganisms, to a series of ◊heterotrophs, or consumers. The heterotrophs comprise the ◊herbivores, which feed on the producers; ◊carnivores, which feed on the herbivores; and ◊decomposers, which break down the dead bodies and waste products of all four groups (including their own), ready for recycling.

In reality, however, organisms have varied diets, relying on different kinds of foods, so that the food chain is an oversimplification. The more complex *food web* shows a greater variety of relationships, but again emphasizes that energy passes from plants to herbivores to carnivores.

Environmentalists have used the concept of the food chain to show how poisons and other forms of pollution can pass from one animal to another, threatening rare species. For example, the pesticide DDT has been found in lethal concentrations in the bodies of animals at the top of the food chain, such as the golden eagle *Aquila chrysaetos*.

food irradiation the exposure of food to low-level ◊irradiation to kill microorganisms; a technique used in ◊food technology. Irradiation is highly effective, and does not make the food any more radioactive than it is naturally. Irradiated food is used for astronauts and for immunocompromised patients in hospitals. Some vitamins are partially destroyed, such as vitamin C, and it would be unwise to eat only irradiated fruit and vegetables.

The main cause for concern is that it may be used by unscrupulous traders to 'clean up' consignments of food, particularly shellfish, with high bacterial counts. Bacterial toxins would remain in the food, so that it could still cause illness, although irradiation would have removed signs of live bacteria. Stringent regulations would be needed to prevent this happening. Other damaging changes may take place in the food, such as the creation of ◊free radicals, but research so far suggests that the process is relatively safe.

food poisoning any acute illness characterized by vomiting and diarrhoea and caused by eating food contaminated with harmful bacteria (for example, ◊listeriosis), poisonous food (for example, certain mushrooms, puffer fish), or poisoned food (such as lead or arsenic introduced accidentally during processing). A frequent cause of food poisoning is ◊Salmonella bacteria. Salmonella comes in many forms, and strains are found in cattle, pigs, poultry, and eggs.

Deep freezing of poultry before the birds are properly cooked is a common cause of food poisoning. Attacks of salmonella also come from contaminated eggs that have been eaten raw or cooked only lightly. Pork may carry the roundworm *Trichinella*, and rye the parasitic fungus

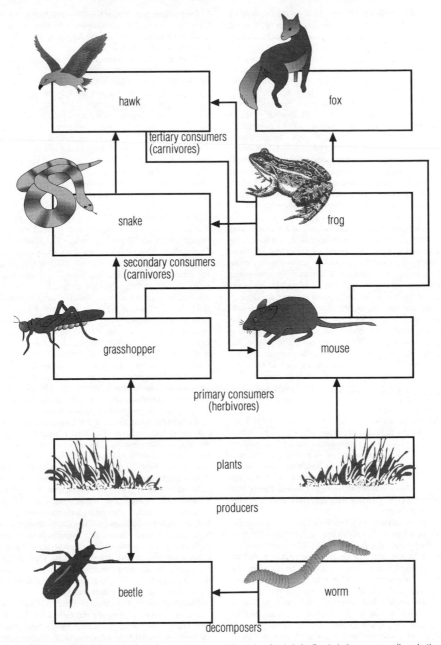

food chain The complex interrelationships between animals and plants in a food chain. Food chains are normally only three or four links long. This is because most of the energy at each link is lost in respiration, and so cannot be passed on to the next link.

ergot. The most dangerous food poison is the bacillus that causes ◊botulism. This is rare but leads to muscle paralysis and, often, death. ◊Food irradiation is intended to prevent food poisoning.

food technology the application of science to the com-mercial processing of foodstuffs. Food is processed to make it more palatable or digestible, for which the tra-ditional methods include boiling, frying, flour-milling, bread-, yoghurt-, and cheese-making, and brewing; or to prevent the growth of bacteria, moulds, yeasts, and other microorganisms; or to preserve it from spoilage caused by the action of ◊enzymes within the food that change its chemical composition, resulting in changes in flavour, odour, colour, and texture of the food. These changes are not always harmful or undesirable; examples of desirable changes are the ripening of cream in butter manufacture, flavour development of cheese, and the hanging of meat to tenderize the muscle fibres. Fatty or oily foods suffer oxidation of the fats, which makes them rancid.

Preservation enables foods that are seasonally produced to be available all the year. Traditional forms of **food preservation** include salting, smoking, pickling, drying, bottling, and preserving in sugar. Modern food technology also uses many novel processes and additives, which allow a wider range of foodstuffs to be preserved. All foods undergo some changes in quality and nutritional value when subjected to preservation processes. No preserved food is identical in quality to its fresh counterpart, hence only food of the highest quality should be preserved.

In order to grow, bacteria, yeasts, and moulds need moisture, oxygen, a suitable temperature, and food. The various methods of food preservation aim to destroy the microorganisms within the food, to remove one or more of the conditions essential for their growth, or to make the foods unsuitable for their growth. Adding large amounts of salt or sugar reduces the amount of water available to microorganisms, because the water tied up by these solutes cannot be used for microbial growth. This is the principle in salting meat and fish, and in the manufacture of jams and jellies. These conditions also inhibit the enzyme activity in food. Preservatives may also be developed in the food by the controlled growth of microorganisms to produce fermentation that may make alcohol, or acetic or lactic acid. Examples of food preserved in this way are vinegar, sour milk, yoghurt, sauerkraut, and alcoholic beverages.

Refrigeration below 5°C/41°F (or below 3°C/37°F for cooked foods) slows the processes of spoilage, but is less effective for foods with a high water content. This process cannot kill microorganisms, nor stop their growth completely, and a failure to realize its limitations causes many cases of food poisoning. Refrigerator temperatures should be checked as the efficiency of the machinery (see ◊refrigeration) can decline with age, and higher temperatures are dangerous.

Deep freezing (−18°C/−1°F or below) stops almost all spoilage processes, except residual enzyme activity in uncooked vegetables and most fruits, which are blanched (dipped in hot water to destroy the enzymes) before freezing. Preservation by freezing works by rendering the water in foodstuffs unavailable to microorganisms by converting it to ice. Microorganisms cannot grow or divide while frozen, but most remain alive and can resume activity once defrosted. Some foods are damaged by freezing, notably soft fruits and salad vegetables, the cells of which are punctured by ice crystals, leading to loss of crispness. Fatty foods such as cow's milk and cream tend to separate. Freezing has little effect on the nutritive value of foods, though a little vitamin C may be lost in the blanching process for fruit and vegetables. Various processes are used for ◊deep freezing foods commercially.

Pasteurization is used mainly for milk. By holding the milk at 72°C/161.6°F for 15 seconds, all disease-causing bacteria can be destroyed. Less harmful bacteria survive, so the milk will still go sour within a few days.

Ultra-heat treatment is used to produce UHT milk. This process uses higher temperatures than pasteurization, and kills all bacteria present, giving the milk a long shelf life but altering the flavour.

Drying is effective because both microorganisms and enzymes need water to be active. This is one of the oldest, simplest, and most effective ways of preserving foods. In addition, drying concentrates the soluble ingredients in foods, and this high concentration prevents the growth of bacteria, yeasts, and moulds. Dried food will deteriorate rapidly if allowed to become moist, but provided they are suitably packaged, products will have a long shelf life. Traditionally, foods were dried in the sun and wind, but commercially today, products such as dried milk and instant coffee are made by spraying the liquid into a rising column of dry, heated air; solid foods, such as fruit, are spread in layers on a heated surface.

Freeze-drying is carried out under vacuum. It is less damaging to food than straight dehydration in the sense that foods reconstitute better, and is used for quality instant coffee and dried vegetables. The foods are fast frozen, then dried by converting the ice to vapour under very low pressure. The foods lose much of their weight, but retain the original size and shape. They have a spongelike texture, and rapidly reabsorb liquid when reconstituted. Refrigeration is unnecessary during storage; the shelf life is similar to dried foods, provided the product is not allowed to become moist. The success of the method is dependent on a fast rate of freezing, and rapid conversion of the ice to vapour. Hence, the most acceptable results are obtained with thin pieces of food, and the method is not recommended for pieces thicker than 3 cm/1 in. Fruit, vegetables, meat, and fish have proved satisfactory. This method of preservation is commercially used but the products are most often used as constituents of composite dishes, such as packet meals.

Canning relies on high temperatures to destroy microorganisms and enzymes. The food is sealed in a can to prevent recontamination. The effect of heat processing on the nutritive value of food is variable. For instance, the vitamin-C content of green vegetables is much reduced, but, owing to greater acidity, in fruit juices vitamin C is quite well retained. There is also a loss of 25–50% of water-soluble vitamins if the liquor is not used. Vitamin B (thiamine) is easily destroyed by heat treatment, particularly in alkaline conditions. Acid products retain thiamine well, because they require only minimum heat during sterilization. The sterilisation process seems to have little effect on retention of vitamins A and B_2. During storage of canned foods, the proportion of vitamins B and C decreases gradually. Drinks may be canned to preserve the carbon dioxide that makes them fizzy.

Pickling utilizes the effect of acetic (ethanoic) acid, found in vinegar, in stopping the growth of moulds. In sauerkraut, lactic acid, produced by bacteria, has the same effect. Similar types of nonharmful, acid-generating bacteria are used to make yoghurt and cheese.

Curing of meat involves soaking in salt (sodium chloride) solution, with saltpetre (sodium nitrate) added to give the meat its pink colour and characteristic taste. Bacteria convert the nitrates in cured meats to nitrites and nitrosamines, which are potentially carcinogenic to humans.

Irradiation is a method of preserving food by subjecting it to low-level radiation (see ◊food irradiation).

Puffing is a method of processing cereal grains. They

are subjected to high pressures, then suddenly ejected into a normal atmospheric pressure, causing the grain to expand sharply. This is used to make puffed wheat cereals and puffed rice cakes.

Chemical treatments are widely used, for example in margarine manufacture, in which hydrogen is bubbled through vegetable oils in the presence of a ◊catalyst to produce a more solid, spreadable fat. The catalyst is later removed. Chemicals introduced in processing that remain in the food are known as *food additives* and include flavourings, preservatives, anti-oxidants, emulsifiers, and colourings.

food test any of several types of simple test, easily performed in the laboratory, used to identify the main classes of food.

Starch–iodine test Food is ground up in distilled water and iodine is added. A dense black colour indicates that starch is present.

Sugar–Benedict's test Food is ground up in distilled water and placed in a test tube with Benedict's reagent. The tube is then heated in a boiling water bath. If glucose is present the colour changes from blue to brick-red.

Protein–Biuret test Food is ground up in distilled water and a mixture of copper(II) sulphate and sodium hydroxide is added. If protein is present a mauve colour is seen.

foot imperial unit of length (symbol ft), equivalent to 0.3048 m, in use in Britain since Anglo-Saxon times. It originally represented the length of a human foot. One foot contains 12 inches and is one-third of a yard.

foot-and-mouth disease contagious eruptive viral disease of cloven-hoofed mammals, characterized by blisters in the mouth and around the hooves. In cattle it causes deterioration of milk yield and abortions. It is an airborne virus, which makes its eradication extremely difficult.

forage crop plant that is grown to feed livestock; for example, grass, clover, and kale (a form of cabbage). Forage crops cover a greater area of the world than food crops, and grass, which dominates this group, is the world's most abundant crop, though much of it is still in an unimproved state.

forbidden line in astronomy, emission line seen in the spectra of certain astronomical objects that are not seen under the conditions prevailing in laboratory experiments. They indicate that the hot gas emitting them is at extremely low density. Forbidden lines are seen, for example, in the tenuous gas of the Sun's ◊corona, in ◊HII regions, and in the nucleuses of certain active galaxies.

force any influence that tends to change the state of rest or the uniform motion in a straight line of a body. The action of an unbalanced or resultant force results in the acceleration of a body in the direction of action of the force, or it may, if the body is unable to move freely, result in its deformation (see ◊Hooke's law). Force is a vector quantity, possessing both magnitude and direction; its SI unit is the newton.

According to Newton's second law of motion the magnitude of a resultant force is equal to the rate of change of ◊momentum of the body on which it acts; the force F producing an acceleration a m s^{-2} on a body of mass m kilograms is therefore given by:

$$[F = ma]$$

See also ◊Newton's laws of motion.

forceps in medicine, twin-bladed metal instruments that are used to seize and hold objects firmly in surgical, obstetric, and dental procedures.

force ratio the magnification of a force by a machine; see ◊mechanical advantage.

forces, fundamental in physics, the four fundamental interactions believed to be at work in the physical universe. There are two long-range forces: *gravity*, which keeps the planets in orbit around the Sun, and acts between all particles that have mass; and *electromagnetic force*, which stops solids from falling apart, and acts between all particles with ◊electric charge. There are two very short-range forces which operate only inside the atomic nucleus: *weak nuclear force*, responsible for the reactions that fuel the Sun and for the emission of ◊beta particles from certain nuclei; and *strong nuclear force*, which binds together the protons and neutrons in the nuclei of atoms. The relative strengths of the four forces are: strong, 1; electromagnetic, 10^{-2}; weak, 10^{-6}; gravitational, 10^{-40}.

By 1971, US physicists Steven Weinberg and Sheldon Glashow, Pakistani physicist Abdus Salam, and others had developed a theory which suggested that the weak and electromagnetic forces were aspects of a single force called the *electroweak force*; experimental support came from observation at ◊CERN in the 1980s. Physicists are now working on theories to unify all four forces.

forensic medicine branch of medicine concerned with the resolution of crimes. Examples of forensic medicine include the determination of the cause of death in suspicious circumstances or the identification of a criminal by examining tissue found at the scene of a crime. Forensic psychology involves the establishment of a psychological profile of a criminal that can assist in identification.

forensic science the use of scientific techniques to solve criminal cases. A multidisciplinary field embracing chemistry, physics, botany, zoology, and medicine, forensic science includes the identification of human bodies or traces. Ballistics (the study of projectiles, such as bullets), another traditional forensic field, makes use of such tools as the comparison microscope and the electron microscope.

Forensic entomology

Cocaine speeds up the growth of certain insects. This enables forensic entomologists to determine if a corpse is that of a cocaine-user.

Traditional methods such as fingerprinting are still used, assisted by computers; in addition, blood analysis, forensic dentistry, voice and speech spectrograms, and ◊genetic fingerprinting are increasingly applied. Chemicals, such as poisons and drugs, are analysed by ◊chromatography. ESDA (electrostatic document analysis) is a technique used for revealing indentations on paper, which helps determine if documents have been tampered with. Forensic entomology is also a branch of forensic science.

forest area where trees have grown naturally for centuries, instead of being logged at maturity (about 150–200 years). A natural, or old-growth, forest has a multistorey canopy and includes young and very old trees (this gives the canopy its range of heights). There are also fallen trees contributing to the very complex ecosystem,

Forests: the hunt for sustainability

After years of neglect, forests are now higher on the international agenda than ever before. Alarm about deforestation in the tropics is now matched by fears concerning a parallel loss of forest quality in temperate and boreal countries. Recognition that forests are important for more than just timber production is also forcing a rethink about their management and protection.

A variety of different initiatives

Since the 1992 Earth Summit the global crisis has been addressed by a bewildering array of initiatives. The World Commission on Forests and Sustainable Development has held a series of public consultations around the world aimed at identifying the root causes of conflicts between different forest users. The International Tropical Timber Organisation, long a thorn in the side of environmental groups because of its support for timber exploitation, has now set a year 2000 deadline for introducing sustainable forest management within the tropics. In addition, forestry issues are being addressed by both the Convention on Biological Diversity and the Climate Change Convention, in the latter case because of the role that forest biomass has in storing carbon and thus reducing global warming.

Perhaps even more important are a series of regional, intergovernmental efforts to develop criteria and indicators of sustainable forest management. This is less esoteric than it seems at first sight. Many governments are agreeing on how to identify and measure the full range of goods and services that forests supply, with the aim of helping to design management policies that cater for all these needs. For the first time, forestry departments around the world are sitting down together and discussing how biodiversity conservation, the rights of indigenous peoples and even the spiritual and cultural values of forests can be incorporated into day-to-day management policies.

To date, the best developed of these schemes have been the Helsinki Process, involving most European countries, and the Montreal Process, catering for other temperate and boreal forest countries. Other initiatives are under way – including the Tarapoto Process for South American countries (named after the small Peruvian town which hosted the first meeting) – and discussions are taking place in Africa, Central America, and Asia. Governments will not only have to agree on far broader approaches to man-agement, but countries' performances will be measured and compared. The spotlight on countries failing to take forest management seriously will be further increased by the United Nations' *Forest Resource Assessment* which, from now on, will be measuring a wide range of issues connected to forest quality.

News from the forest floor

Unfortunately, the discussions in the boardrooms of international conferences have not, as yet, always been translated into similar changes in the forest. Since the Earth Summit, rates of destruction have continued to increase in many countries, and new forest ecosystems have come under threat in Surinam, Guyana, Bolivia, Peru, Zaire, Cameroon, Cambodia, the Solomon Islands, Canada, and elsewhere. Asian transnational logging companies have joined those from Europe and North America in a rapid and usually unsustainable harvest of many tropical forests. Worse still, there has been a concerted backlash against environmental regulations in some countries, including the USA where much of the 1–2% of old-growth forest that still remains intact has been opened up to logging by the government.

On the other hand, there have been some welcome improvements. Governments and some private forest owners are starting to change systems of forest management to take more account of nature. The so-called 'new forestry' includes, for example, leaving some older trees standing after cutting, retaining some dead timber to act as habitat for specialized plant and animal species, and using prescribed burning to mimic fire ecology. Timber certification is rapidly increasing; for example, buyers responsible for approximately one quarter of UK timber purchases are already committed to buying only certified timber when this becomes available.

Important new protected areas have also been created in many countries. Australia has committed to setting 15% of its forests aside as a 'Gift to the Earth' – an area twice the size of Germany.

Discussions regarding the Global Forest Convention are starting again in earnest, although the shape and remit of this remain unclear. Whatever happens, it seems likely that international interest in forests will remain high for years to come.

Nigel Dudley

which may support more than 150 species of mammals and many thousands of species of insects.

The Pacific forest of the west coast of North America is one of the few remaining old-growth forests in the temperate zone. It consists mainly of conifers and is threatened by logging – less than 10% of the original forest remains.

forestry the science of forest management. Recommended forestry practice aims at multipurpose crops, allowing the preservation of varied plant and animal species as well as human uses (lumbering, recreation). Forestry has often been confined to the planting of a single species, such as a rapid-growing conifer providing softwood for paper pulp and construction timber, for which world demand is greatest. In tropical countries, logging contributes to the destruction of ◊rainforests, causing global environmental problems. Small unplanned forests are ◊woodland.

The earliest planned forest dates from 1368 at Nürnberg, Germany; in Britain, planning of forests began in the 16th century. In the UK, Japan, and other countries, forestry practices have been criticized for concentration on softwood conifers to the neglect of native hardwoods.

forging one of the main methods of shaping metals, which involves hammering or a more gradual application of pressure. A blacksmith hammers red-hot metal into shape on an anvil, and the traditional place of work is called a forge. The blacksmith's mechanical equivalent is the drop forge. The metal is shaped by the blows from a falling hammer or ram, which is usually accelerated by steam or air pressure. Hydraulic presses forge by applying pressure gradually in a squeezing action.

formaldehyde common name for ◊methanal.

formalin aqueous solution of formaldehyde (methanal) used to preserve animal specimens.

formatting in computing, short for ◊disc formatting.

Formica trademark of the Formica Corporation for a heat-proof plastic laminate, widely used as a veneer on wipe-down kitchen surfaces and children's furniture. It is made from formaldehyde resins similar to ◊Bakelite. It was first put on the market in 1913.

formic acid common name for ◊methanoic acid.

formula in chemistry, a representation of a molecule, radical, or ion, in which the component chemical elements are represented by their symbols. An *empirical formula* indicates the simplest ratio of the elements in a compound, without indicating how many of them there are or how they are combined. A *molecular formula* gives the number of each type of element present in one molecule. A *structural formula* shows the relative positions of the atoms and the bonds between them. For example, for ethanoic acid, the empirical formula is CH_2O, the molecular formula is $C_2H_4O_2$, and the structural formula is CH_3COOH.

formula in mathematics, a set of symbols and numbers that expresses a fact or rule. $A = \pi r^2$ is the formula for calculating the area of a circle. Einstein's famous formula relating energy and mass is $E = mc^2$.

formulary in medicine, a list of drugs and other medical preparations that are available for the treatment of diseases. Formularies may contain general advice on pre-

scribing and notes on the various drugs available. These are usually classified according to disease states. Important information relating to drug usage may be listed separately. For example, interactions between drugs, the use of drugs during pregnancy and breast-feeding, and the treatment of poisoning are included in some formularies.

FORTRAN (or *fortran*, acronym for *formula translation*) high-level computer-programming language suited to mathematical and scientific computations. Developed in 1956, it is one of the earliest computer languages still in use. A version, Fortran 90, is now being used on advanced parallel computers. ◊BASIC was strongly influenced by FORTRAN and is similar in many ways.

fossil a cast, impression, or the actual remains of an animal or plant preserved in rock. Fossils were created during periods of rock formation, caused by the gradual accumulation of sediment over millions of years at the bottom of the sea bed or an inland lake. Fossils may include footprints, an internal cast, or external impression. A few fossils are preserved intact, as with mammoths fossilized in Siberian ice, or insects trapped in tree resin that is today amber. The study of fossils is called ◊palaeontology. Palaeontologists are able to deduce much of the geological history of a region from fossil remains.

About 250,000 fossil species have been discovered – a figure that is believed to represent less than 1 in 20,000 of the species that ever lived. *Microfossils* are so small they can only be seen with a microscope. They include the fossils of pollen, bone fragments, bacteria, and the remains of microscopic marine animals and plants, such as foraminifera and diatoms.

fossil fuel fuel, such as coal, oil, and natural gas, formed from the fossilized remains of plants that lived hundreds of millions of years ago. Fossil fuels are a ◊nonrenewable resource and will eventually run out. Extraction of coal and oil causes considerable environmental pollution, and burning coal contributes to problems of ◊acid rain and the ◊greenhouse effect.

four-colour process colour ◊printing using four printing plates, based on the principle that any colour is made up of differing proportions of the primary colours blue, red, and green. The first stage in preparing a colour picture for printing is to produce separation films, one each for the blue, red, and green respectively in the picture (colour separations). From these separations three printing plates are made, with a fourth plate for black (for shading or outlines and type). Ink colours complementary to those represented on the plates are used for printing – yellow for the blue plate, cyan for the red, and magenta for the green.

Fourdrinier machine papermaking machine patented by the Fourdrinier brothers Henry and Sealy in England in 1803. On the machine, liquid pulp flows onto a moving wire-mesh belt, and water drains and is sucked away, leaving a damp paper web. This is passed first through a series of steam-heated rollers, which dry it, and then between heavy calendar rollers, which give it a smooth finish.

Such machines can measure up to 90 m/300 ft in length, and are still in use.

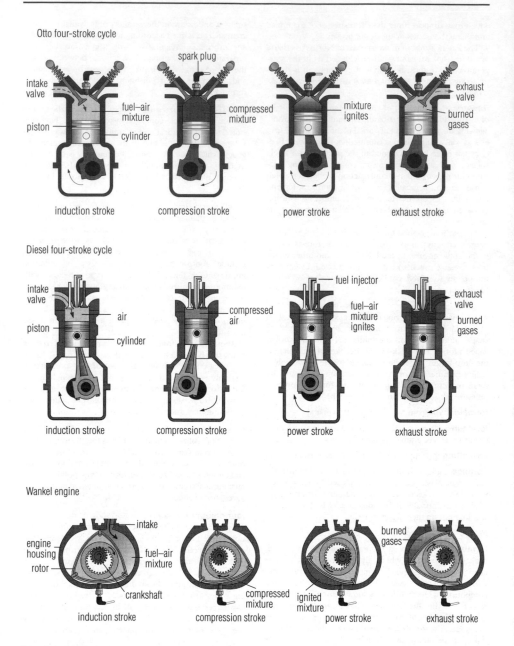

Otto four-stroke cycle

spark plug

intake valve

fuel–air mixture

piston

cylinder

compressed mixture

mixture ignites

exhaust valve

burned gases

induction stroke   compression stroke   power stroke   exhaust stroke

Diesel four-stroke cycle

intake valve

air

piston

cylinder

compressed air

fuel injector

fuel–air mixture ignites

exhaust valve

burned gases

induction stroke   compression stroke   power stroke   exhaust stroke

Wankel engine

intake

engine housing

rotor

fuel–air mixture

crankshaft

compressed mixture

ignited mixture

burned gases

induction stroke   compression stroke   power stroke   exhaust stroke

four stroke engine

four-stroke cycle the engine-operating cycle of most petrol and ◊diesel engines. The 'stroke' is an upward or downward movement of a piston in a cylinder. In a petrol engine the cycle begins with the induction of a fuel mixture as the piston goes down on its first stroke. On the second stroke (up) the piston compresses the mixture in the top of the cylinder. An electric spark then ignites the mixture, and the gases produced force the piston down on

its third, power, stroke. On the fourth stroke (up) the piston expels the burned gases from the cylinder into the exhaust.

fourth-generation language in computing, a type of programming language designed for the rapid programming of ◊applications but often lacking the ability to control the individual parts of the computer. Such a language typically provides easy ways of designing screens and reports, and of using databases. Other 'generations' (the term implies a class of language rather than a chronological sequence) are ◊machine code (first generation); ◊assembly languages, or low-level languages (second); and conventional high-level languages such as ◊BASIC and ◊PASCAL (third).

f.p.s. system system of units based on the foot, pound, and second as units of length, mass, and time, respectively. It has now been replaced for scientific work by the ◊SI system.

fractal irregular shape or surface produced by a procedure of repeated subdivision. Generated on a computer screen, fractals are used in creating models of geographical or biological processes (for example, the creation of a coastline by erosion or accretion, or the growth of plants).

Sets of curves with such discordant properties were developed in the 19th century in Germany by Georg Cantor and Karl Weierstrass. The name was coined by the French mathematician Benoit Mandelbrod. Fractals are also used for computer art.

fraction in chemistry, a group of similar compounds, the boiling points of which fall within a particular range and which are separated during fractional ◊distillation (fractionation).

fraction in mathematics, a number that indicates one or more equal parts of a whole. Usually, the number of equal parts into which the unit is divided (denominator) is written below a horizontal line, and the number of parts comprising the fraction (numerator) is written above; thus 2/3 or 3/4. Such fractions are called **vulgar** or **simple** fractions. The denominator can never be zero. A **proper fraction** is one in which the numerator is less than the denominator. An **improper fraction** has a numerator that is larger than the denominator, for example 3/2. It can therefore be expressed as a mixed number, for example, 11/2. A combination such as 5/0 is not regarded as a fraction (an object cannot be divided into zero equal parts), and mathematically any number divided by 0 is equal to infinity. A **decimal fraction** has as its denominator a power of 10, and these are omitted by use of the decimal point and notation, for example 0.04, indicate the numerators of vulgar fractions whose denominators are 10, 100, 1,000, and so on. Most fractions can be expressed exactly as decimal fractions (1/3 = 0.333...). Fractions are also known as the **rational numbers**; that is, numbers formed by a ratio. **Integers** may be expressed as fractions with a denominator of 1.

fractionating column device in which many separate ◊distillations can occur so that a liquid mixture can be separated into its components.

Various designs exist but the primary aim is to allow maximum contact between the hot rising vapours and the cooling descending liquid. As the mixture of vapours ascends the column it becomes progressively enriched in the lower-boiling-point components, so these separate out first.

fractionation or **fractional distillation** process used to split complex mixtures (such as crude oil) into their components, usually by repeated heating, boiling, and condensation; see ◊distillation.

francium radioactive metallic element, symbol Fr, atomic number 87, relative atomic mass 223. It is one of

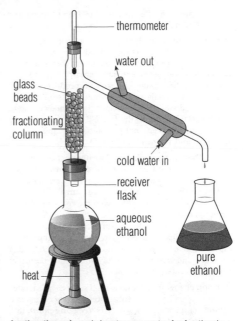

fractionating column Laboratory apparatus for fractional distillation. Fractional distillation is the main means of separating the components of crude oil.

the alkali metals and occurs in nature in small amounts as a decay product of actinium. Its longest-lived isotope has a half-life of only 21 minutes. Francium was discovered and named in 1939 by Marguérite Perey to honour her country.

Frasch process process used to extract underground deposits of sulphur. Superheated steam is piped into the sulphur deposit and melts it. Compressed air is then pumped down to force the molten sulphur to the surface. The process was developed in the USA, in 1891 by German-born Herman Frasch (1851–1914).

freckles in medicine, localized collections of brown spots that are often seen on the arms and faces of fair-skinned individuals. They are due to the excessive production of ◊melanin following exposure to sunlight and they fade during the winter months.

free fall the state in which a body is falling freely under the influence of ◊gravity, as in freefall parachuting (skydiving). The term **weightless** is normally used to describe a body in free fall in space.

In orbit, astronauts and spacecraft are still held by gravity and are in fact falling toward the Earth. Because of their speed (orbital velocity), the amount they fall towards the Earth just equals the amount the Earth's surface curves away; in effect they remain at the same height, apparently weightless.

free radical in chemistry, an atom or molecule that has an unpaired electron and is therefore highly reactive. Most free radicals are very short-lived. They are by-products of normal cell chemistry and rapidly oxidize other molecules they encounter. Free radicals are thought to do considerable damage. They are neutralized by protective enzymes.

Free radicals are often produced by high temperatures and are found in flames and explosions.

freeze-drying method of preserving food; see ◊food technology. The product to be dried is frozen and then put in a vacuum chamber that forces out the ice as water vapour, a process known as sublimation.

Many of the substances that give products such as coffee their typical flavour are volatile, and would be lost in a normal drying process because they would evaporate along with the water. In the freeze-drying process these volatile compounds do not pass into the ice that is to be sublimed, and are therefore largely retained.

freeze-thaw form of physical ◊weathering, common in mountains and glacial environments, caused by the expansion of water as it freezes. Water in a crack freezes and expands in volume by 9% as it turns to ice. This expansion exerts great pressure on the rock causing the crack to enlarge. After many cycles of freeze-thaw, rock fragments may break off to form ◊scree slopes.

For freeze-thaw to operate effectively the temperature must fluctuate regularly above and below 0°C/32°F. It is therefore uncommon in areas of extreme and perpetual cold, such as the polar regions.

freezing change from liquid to solid state, as when water becomes ice. For a given substance, freezing occurs at a definite temperature, known as the *freezing point*, that is invariable under similar conditions of pressure, and the temperature remains at this point until all the liquid is frozen. The amount of heat per unit mass that has to be removed to freeze a substance is a constant for any given substance, and is known as the latent heat of fusion.

freezing point, depression of lowering of a solution's freezing point below that of the pure solvent; it depends on the number of molecules of solute dissolved in it. For a single solvent, such as pure water, all solute substances in the same molar concentration produce the same lowering of freezing point. The depression d produced by the presence of a solute of molar concentration C is given by the equation $d = KC$, where K is a constant (called the cryoscopic constant) for the solvent concerned.

Antifreeze mixtures for car radiators and the use of salt to melt ice on roads are common applications of this principle. Animals in arctic conditions, for example insects or fish, cope with the extreme cold either by manufacturing natural 'antifreeze' and staying active, or by allowing themselves to freeze in a controlled fashion, that is, they manufacture proteins to act as nuclei for the formation of ice crystals in areas that will not produce cellular damage, and so enable themselves to thaw back to life again.

Measurement of freezing-point depression is a useful method of determining the molecular weights of solutes. It is also used to detect the illicit addition of water to milk.

frequency in physics, the number of periodic oscillations, vibrations, or waves occurring per unit of time. The SI unit of frequency is the hertz (Hz), one hertz being equivalent to one cycle per second.

Human beings can hear sounds from objects vibrating in the range 20–15,000 Hz. Ultrasonic frequencies well above 15,000 Hz can be detected by such mammals as bats. Infrasound (low frequency sound) can be detected by some animals and birds. Pigeons can detect sounds as low as 0.1 Hz; elephants communicate using sounds as low as 1 Hz.

frequency in statistics, the number of times an event occurs. For example, when two dice are thrown repeatedly and the two scores added together, each of the numbers 2 to 12 may have a frequency of occurrence. The set of data including the frequencies is called a *frequency distribution*, usually presented in a frequency table or shown diagramatically, by a frequency polygon.

frequency modulation (FM) method by which radio waves are altered for the transmission of broadcasting signals. FM is constant in amplitude and varies the frequency of the carrier wave in accordance with the signal being transmitted. Its advantage over AM (◊amplitude modulation) is its better signal-to-noise ratio. It was invented by the US engineer Edwin Armstrong.

Freudian therapy in medicine, the body of psychological knowledge derived by Sigmund Freud and his followers; see ◊psychoanalysis.

friction in physics, the force that opposes the relative motion of two bodies in contact. The *coefficient of friction* is the ratio of the force required to achieve this relative motion to the force pressing the two bodies together.

Friction is greatly reduced by the use of lubricants such as oil, grease, and graphite. Air bearings are now used to minimize friction in high-speed rotational machinery. In other instances friction is deliberately increased by making the surfaces rough – for example, brake linings, driving belts, soles of shoes, and tyres.

Friends of the Earth (FoE or FOE) environmental pressure group, established in the UK in 1971, that aims to protect the environment and to promote rational and sustainable use of the Earth's resources. It campaigns on such issues as acid rain; air, sea, river, and land pollution; recycling; disposal of toxic wastes; nuclear power and renewable energy; the destruction of rainforests; pesticides; and agriculture. FoE has branches in 30 countries.

frond large leaf or leaflike structure; in ferns it is often pinnately divided. The term is also applied to the leaves of palms and less commonly to the plant bodies of certain seaweeds, liverworts, and lichens.

front in meteorology, the boundary between two air masses of different temperature or humidity. A *cold front* marks the line of advance of a cold air mass from below, as it displaces a warm air mass; a *warm front* marks the advance of a warm air mass as it rises up over a cold one. Frontal systems define the weather of the mid-

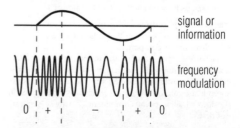

signal or information

frequency modulation

0   +   −   +   0

frequency modulation In FM radio transmission, the frequency of the carrier wave is modulated, rather than its amplitude (as in AM broadcasts). The FM system is not affected by the many types of interference which change the amplitude of the carrier wave, and so provides better quality reception than AM broadcasts.

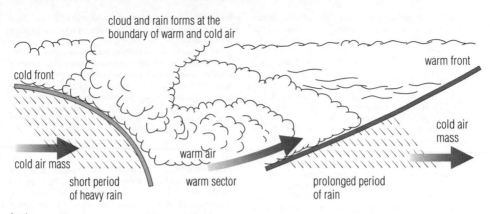

cloud and rain forms at the
boundary of warm and cold air

warm front

cold front

cold air
mass

cold air mass

warm air

short period
of heavy rain

warm sector

prolonged period
of rain

front

latitudes, where warm tropical air is constantly meeting cold air from the poles.

Warm air, being lighter, tends to rise above the cold; its moisture is carried upwards and usually falls as rain or snow, hence the changeable weather conditions at fronts. Fronts are rarely stable and move with the air mass. An *occluded front* is a composite form, where a cold front catches up with a warm front and merges with it.

frontal lobotomy operation on the brain. See ◊leucotomy.

front-end processor small computer used to coordinate and control the communications between a large mainframe computer and its input and output devices.

frost condition of the weather that occurs when the air temperature is below freezing (0°C/32°F). Water in the atmosphere is deposited as ice crystals on the ground or exposed objects. As cold air is heavier than warm, ground frost is more common than hoar frost, which is formed by the condensation of water particles in the same way that ◊dew collects.

frostbite the freezing of skin or flesh, with formation of ice crystals leading to tissue damage. The treatment is slow warming of the affected area; for example, by skin-to-skin contact or with lukewarm water. Frostbitten parts are extremely vulnerable to infection, with the risk of gangrene.

frost hollow depression or steep-sided valley in which cold air collects on calm, clear nights. Under clear skies, heat is lost rapidly from ground surfaces, causing the air above to cool and flow downhill (as katabatic wind) to collect in valley bottoms. Fog may form under these conditions and, in winter, temperatures may be low enough to cause frost.

frost shattering alternative name for ◊freeze-thaw.

FRS abbreviation for *Fellow of the ◊Royal Society*.

fructose $C_6H_{12}O_6$ a sugar that occurs naturally in honey, the nectar of flowers, and many sweet fruits; it is commercially prepared from glucose.

It is a monosaccharide, whereas the more familiar cane or beet sugar is a disaccharide, made up of two monosaccharide units: fructose and glucose. It is sweeter than cane sugar and can be used to sweeten foods for people with diabetes.

fruit in botany, the ripened ovary in flowering plants that develops from one or more seeds or carpels and encloses one or more seeds. Its function is to protect the seeds during their development and to aid in their dispersal.

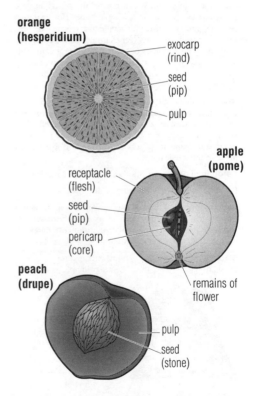

orange (hesperidium)

exocarp (rind)

seed (pip)

pulp

apple (pome)

receptacle (flesh)

seed (pip)

pericarp (core)

remains of flower

peach (drupe)

pulp

seed (stone)

fruit A fruit contains the seeds of a plant. Its outer wall is the exocarp, or epicarp; its inner layers are the mesocarp and endocarp. The orange is a hesperidium, a berry having a leathery rind and containing many seeds. The mango is a drupe, a fleshy fruit with a hard seed, or 'stone', at the centre. The apple is a pome, a fruit with a fleshy outer layer and a core containing the seeds.

Fruits are often edible, sweet, juicy, and colourful. When eaten they provide vitamins, minerals, and enzymes, but little protein. Most fruits are borne by perennial plants.

Fruits are divided into three agricultural categories on the basis of the climate in which they grow. **Temperate fruits** require a cold season for satisfactory growth; the principal temperate fruits are apples, pears, plums, peaches, apricots, cherries, and soft fruits, such as strawberries. **Subtropical fruits** require warm conditions but can survive light frosts; they include oranges and other citrus fruits, dates, pomegranates, and avocados. **Tropical fruits** cannot tolerate temperatures that drop close to freezing point; they include bananas, mangoes, pineapples, papayas, and litchis. Fruits can also be divided botanically into **dry** (such as the ◊capsule, ◊follicle, ◊schizocarp, ◊nut, ◊caryopsis, pod or legume, ◊lomentum, and ◊achene) and those that become **fleshy** (such as the ◊drupe and the ◊berry). The fruit structure consists of the ◊pericarp or fruit wall, which is usually divided into a number of distinct layers. Sometimes parts other than the ovary are incorporated into the fruit structure, resulting in a false fruit or ◊pseudocarp, such as the apple and strawberry. True fruits include the tomato, orange, melon, and banana. Fruits may be dehiscent, which open to shed their seeds, or indehiscent, which remain unopened and are dispersed as a single unit. Simple fruits (for example, peaches) are derived from a single ovary, whereas compositae or multiple fruits (for example, blackberries) are formed from the ovaries of a number of flowers. In ordinary usage, 'fruit' includes only sweet, fleshy items; it excludes many botanical fruits such as acorns, bean pods, thistledown, and cucumbers.

frustum in geometry, a 'slice' taken out of a solid figure by a pair of parallel planes. A conical frustum, for example, resembles a cone with the top cut off. The volume and area of a frustum are calculated by subtracting the volume or area of the 'missing' piece from those of the whole figure.

f-stop in photography, another name for ◊f-number.

ft symbol for ◊**foot**, a measure of distance.

fuel any source of heat or energy, embracing the entire range of materials that burn in air (combustibles). A **nuclear fuel** is any material that produces energy by nuclear fission in a nuclear reactor.

fuel cell cell converting chemical energy directly to electrical energy. It works on the same principle as a battery but is continually fed with fuel, usually hydrogen. Fuel cells are silent and reliable (no moving parts) but expensive to produce.

Hydrogen is passed over an ◊electrode (usually nickel or platinum) containing a ◊catalyst, which strips electrons off the atoms. These pass through an external circuit while hydrogen ions (charged atoms) pass through an ◊electrolyte to another electrode, over which oxygen is passed. Water is formed at this electrode (as a by-product) in a chemical reaction involving electrons, hydrogen ions, and oxygen atoms. If the spare heat also produced is used for hot water and space heating, 80% efficiency in fuel is achieved.

fuel injection injecting fuel directly into the cylinders of an internal combustion engine, instead of by way of a carburettor. It is the standard method used in ◊diesel engines, and is now becoming standard for petrol engines. In the diesel engine, oil is injected into the hot compressed air at the top of the second piston stroke and explodes to drive the piston down on its power stroke. In the petrol engine, fuel is injected into the cylinder at the start of the first induction stroke of the ◊four-stroke cycle.

fullerene form of carbon, discovered in 1985, based on closed cages of carbon atoms. The molecules of the most symmetrical of the fullerenes are called ◊buckminster-fullerenes (or buckyballs). They are perfect spheres made up of 60 carbon atoms linked together in 12 pentagons and 20 hexagons fitted together like those of a spherical football. Other fullerenes, with 28, 32, 50, 70, and 76 carbon atoms, have also been identified.

Fullerenes can be made by arcing electricity between carbon rods. They may also occur in candle flames and in clouds of interstellar gas. Fullerene chemistry may turn out to be as important as organic chemistry based on the benzene ring. Already, new molecules based on the buckyball enclosing a metal atom, and 'buckytubes' (cylinders of carbon atoms arranged in hexagons), have been made. Applications envisaged include using the new molecules as lubricants, semiconductors, and superconductors, and as the starting point for making new drugs.

fuller's earth soft, greenish-grey rock resembling clay, but without clay's plasticity. It is formed largely of clay minerals, rich in montmorillonite, but a great deal of silica is also present. Its absorbent properties make it suitable for removing oil and grease, and it was formerly used for cleaning fleeces ('fulling'). It is still used in the textile industry, but its chief application is in the purification of oils. Beds of fuller's earth are found in the southern USA, Germany, Japan, and the UK.

fulminate any salt of fulminic (cyanic) acid (HOCN), the chief ones being silver and mercury. The fulminates detonate (are exploded by a blow); see ◊detonator.

fumigation in medicine, the use of vapours, such as camphor, to destroy bacteria or vermin. The process was widely used in the past but it has been superseded by more effective disinfection procedures.

function in computing, a small part of a program that supplies a specific value – for example, the square root of a specified number, or the current date. Most programming languages incorporate a number of built-in functions; some allow programmers to write their own. A function may have one or more arguments (the values on which the function operates). A **function key** on a key-

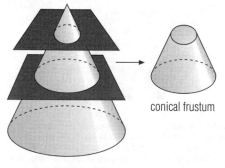

conical frustum

cone

frustum The frustum, a slice taken out of a cone.

board is one that, when pressed, performs a designated task, such as ending a program.

function in mathematics, a function f is a non-empty set of ordered pairs $(x, f(x))$ of which no two can have the same first element. Hence, if

$$[f(x) = x^2]$$

two ordered pairs are $(-2,4)$ and $(2,4)$. The set of all first elements in a function's ordered pairs is called the **domain**; the set of all second elements is the **range**. In the algebraic expression $y = 4x^3 + 2$, the dependent variable y is a function of the independent variable x, generally written as $f(x)$.

Functions are used in all branches of mathematics, physics, and science generally; for example, the formula

$$t = 2\pi\sqrt{(l/g)}$$

shows that for a simple pendulum the time of swing t is a function of its length l and of no other variable quantity (π and g, the acceleration due to gravity, are ◊constants).

functional group in chemistry, a small number of atoms in an arrangement that determines the chemical properties of the group and of the molecule to which it is attached (for example, the carboxyl group COOH, or the amine group NH_2). Organic compounds can be considered as structural skeletons, with a high carbon content, with functional groups attached.

fundamental constant physical quantity that is constant in all circumstances throughout the whole universe. Examples are the electric charge of an electron, the speed of light, Planck's constant, and the gravitational constant.

fundamental forces see ◊forces, fundamental.

fundamental particle another term for ◊ *elementary particle*.

fungal infection in medicine, infection with a fungus, such as the ◊ *Candida albicans*. Sites of infection are the mouth, skin folds, and vagina. These are treated by the use of suspensions, creams, and pessaries containing antifungal agents, such as nystatin and clotrimazole. More generalized candidiasis may occur in patients with damaged immune systems due to AIDS or to the use of immunosuppressive drugs. These infections can be severe and they are treated with intravenous infusions of amphotericin and other antifungal drugs.

Dermatophyte infections are due to fungi that infect the keratin tissues of the skin, hair, and nails. Such infections are usually superficial and they tend to occur in moist and warm situations. Examples are athlete's foot and ◊ringworm. They often respond to local treatment with clotrimazole or miconazole. Oral antifungal agents, such as terbinafine, may be required to treat persistent infections. Therapy may need to be prolonged over several weeks due to low concentrations of the drug reaching the site of infection because of poor blood supply. This is a particular problem with nail infections.

fungicide any chemical ◊pesticide used to prevent fungus diseases in plants and animals. Inorganic and organic compounds containing sulphur are widely used.

fungus (plural *fungi*) any of a unique group of organisms that includes moulds, yeasts, rusts, smuts, mildews, mushrooms, and toadstools. About 50,000 species have been identified. They are not considered to be plants for three main reasons: they have no leaves or roots; they

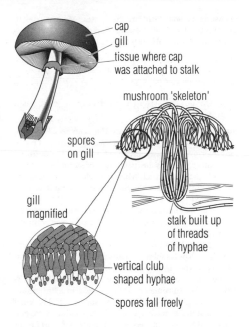

fungus

contain no chlorophyll (green colouring) and are therefore unable to make their own food by ◊photosynthesis; and they reproduce by ◊spores. Some fungi are edible but many are highly poisonous; they often cause damage and sometimes disease to the organic matter they live and feed on, but some fungi are exploited in the production of food and drink (for example, yeasts in baking and brewing) and in medicine (for example, penicillin). (Kingdom Fungi.)

Fungus

In 1992 an individual honey fungus, *Armallaria ostoyae*, in Washington State, USA, was identified as the world's largest living thing – estimated to be between 500 and 1,000 years old, its underground network of hyphae covers 600 hectares/1,480 acres.

Fungi are either ◊parasites, existing on living plants or animals, or ◊saprotrophs, living on dead matter. Many of the most serious plant diseases are caused by fungi, and several fungi attack humans and animals. Athlete's foot, ◊thrush, and ◊ringworm are fungal diseases.

Before the classification Fungi came into use, they were included within the division Thallophyta, along with algae and ◊bacteria. Two familiar fungi are bread mould, which illustrates the typical many-branched body (mycelium) of the organism, made up of threadlike chains of cells called hyphae; and mushrooms, which are the sexually reproductive fruiting bodies of an underground mycelium.

The mycelium of a true fungus is made up of many intertwined hyphae. When the fungus is ready to reproduce, the hyphae become closely packed into a solid mass called the fruiting body, which is usually small and inconspicuous but can be very large; mushrooms, toadstools, and bracket fungi are all examples of large fruiting bodies.

These carry and distribute the spores. Most species of fungi reproduce both asexually (on their own) and sexually (involving male and female parents).

funicular railway railway with two cars connected by a wire cable wound around a drum at the top of a steep incline. Funicular railways of up to 1.5 km/1 mi exist in Switzerland.

In Britain, the system is used only in seaside cliff railways.

fur the ◊hair of certain animals. Fur is an excellent insulating material and so has been used as clothing. This is, however, vociferously criticized by many groups on humane grounds, as the methods of breeding or trapping animals are often cruel. Mink, chinchilla, and sable are among the most valuable, the wild furs being finer than the farmed.

Fur such as mink is made up of a soft, thick, insulating layer called underfur and a top layer of longer, lustrous guard hairs.

Furs have been worn since prehistoric times and have long been associated with status and luxury (ermine traditionally worn by royalty, for example), except by certain ethnic groups like the Inuit. The fur trade had its origin in North America, where in the late 17th century the Hudson Bay Company was established. The chief centres of the fur trade are New York, London, St Petersburg, and Kastoria in Greece. It is illegal to import furs or skins of endangered species listed by ◊CITES (such as the leopard). Many synthetic fibres are widely used as substitutes.

furlong unit of measurement, originating in Anglo-Saxon England, equivalent to 220 yd (201.168 m).

A furlong consists of 40 rods, poles, or perches; 8 furlongs equal one statute ◊mile. Its literal meaning is 'furrow-long', and refers to the length of a furrow in the common field characteristic of medieval farming.

furnace structure in which fuel such as coal, coke, gas, or oil is burned to produce heat for various purposes. Furnaces are used in conjunction with ◊boilers for heating, to produce hot water, or steam for driving turbines – in ships for propulsion and in power stations for generating electricity. The largest furnaces are those used for smelting and refining metals, such as the ◊blast furnace, electric furnace, and ◊open-hearth furnace.

fuse in electricity, a wire or strip of metal designed to melt when excessive current passes through. It is a safety device to stop at that point in the circuit when surges of current would otherwise damage equipment and cause fires. In explosives, a fuse is a cord impregnated with chemicals so that it burns slowly at a predetermined rate. It is used to set off a main explosive charge, sufficient length of fuse being left to allow the person lighting it to get away to safety.

fusel oil liquid with a characteristic unpleasant smell, obtained as a by-product of the distillation of the product of any alcoholic fermentation, and used in paints, varnishes, essential oils, and plastics. It is a mixture of fatty acids, alcohols, and esters.

fusion in physics, the fusing of the nuclei of light elements, such as hydrogen, into those of a heavier element, such as helium. The resultant loss in their combined mass is converted into energy. Stars and thermonuclear weapons work on the principle of nuclear fusion.

fuzzy logic in mathematics and computing, a form of knowledge representation suitable for notions (such as 'hot' or 'loud') that cannot be defined precisely but depend on their context. For example, a jug of water may be described as too hot or too cold, depending on whether it is to be used to wash one's face or to make tea.

The central idea of fuzzy logic is *probability of set membership*. For instance, when referring to someone 175 cm/5 ft 9 in tall, the statement 'this person is tall' (or 'this person is a member of the set of tall people') might be about 70% true if that person is a man, and about 85% true if that person is a woman.

g symbol for ◊ *gram*.

g symbol for the ◊ gravitational field strength, the strength of the Earth's gravitational field at any point, and for acceleration due to gravity.

GABA (abbreviation for ***gamma aminobutyric acid***) in medicine, an amino acid that occurs in the central nervous system, mainly in the tissue of the brain. It is involved in the transmission of inhibitory impulses from nerve to nerve and between nerves and tissues in the brain. Imbalances in GABA concentrations in the brain may be implicated in a variety of disorders, including ◊ anxiety and ◊ epilepsy.

gabbro mafic (consisting primarily of dark-coloured crystals) igneous rock formed deep in the Earth's crust. It contains pyroxene and calcium-rich feldspar, and may contain small amounts of olivine and amphibole. Its coarse crystals of dull minerals give it a speckled appearance.

Gabbro is the plutonic version of basalt (that is, derived from magma that has solidified below the Earth's surface), and forms in large, slow-cooling intrusions.

gadolinium silvery-white metallic element of the lanthanide series, symbol Gd, atomic number 64, relative atomic mass 157.25. It is found in the products of nuclear fission and used in electronic components, alloys, and products needing to withstand high temperatures.

Gaia hypothesis theory that the Earth's living and non-living systems form an inseparable whole that is regulated and kept adapted for life by living organisms themselves. The planet therefore functions as a single organism, or a giant cell. The hypothesis was elaborated by British scientist James Lovelock and first published in 1968.

When I first introduced Gaia, I had vague hopes that it might be denounced from the pulpit and thus made acceptable to my scientific colleagues. As it was, Gaia was embraced by theologians and by a wide range of New Age writers and thinkers but denounced by biologists.

On **Gaia hypothesis** James Lovelock
Earthwatch 1992

---

gain in electronics, the ratio of the amplitude of the output signal produced by an amplifier to that of the input signal.

In a ◊ voltage amplifier the voltage gain is the ratio of the output voltage to the input voltage; in an inverting ◊ operational amplifier (op-amp) it is equal to the ratio of the resistance of the feedback resistor to that of the input resistor.

gait in medicine, the way that an individual walks as an important indicator of health and disease. Gait is affected by paralysis on one side of the body following stroke, locomotor ataxia, spastic paralysis, ◊ Parkinson's disease, and ◊ chorea.

Children usually learn to stand in their first year and they begin to walk between 12 and 18 months of age and an unusual gait may be an indication of an underlying problem, for example mental retardation, cerebral palsy, or malformation of the hip joints.

gal symbol for ◊ *gallon*, ◊ *galileo*.

galactic coordinates in astronomy, a system for measuring the position of astronomical objects on the ◊ celestial sphere with reference to the galactic equator (or ◊ great circle).

Galactic latitude (symbol b) is measured in degrees from the galactic equator (b = 0°) to the north (b = 90°) and south (b = – 90°) galactic poles.

Galactic longitude (symbol l) is measured in degrees eastward (l = 0° to 360°) from a fixed point in the constellation of ◊ Sagittarius that approximates to the centre of the Galaxy. Galactic coordinates are often used when astronomers are studying the distribution of material in the Galaxy.

galactic halo in astronomy, the outer, sparsely populated region of a galaxy, roughly spheroid in shape and extending far beyond the bulk of the visible stars. In our own Galaxy, the halo contains the globular clusters, and may harbour large quantities of ◊ dark matter.

galactic plane in astronomy, a plane passing through the ◊ Sun and the centre of the Galaxy defining the mid-plane of the galactic disc. Viewed from the Earth, the galactic plane is a ◊ great circle (galactic equator) marking the approximate centre line of the ◊ Milky Way.

galaxy congregation of millions or billions of stars, held together by gravity.

Spiral galaxies, such as the ◊ Milky Way, are flattened in shape, with a central bulge of old stars surrounded by a disc of younger stars, arranged in spiral arms like a Catherine wheel.

Barred spirals are spiral galaxies that have a straight bar of stars across their centre, from the ends of which the spiral arms emerge. The arms of spiral galaxies contain gas and dust from which new stars are still forming.

Elliptical galaxies contain old stars and very little gas. They include the most massive galaxies known, containing a trillion stars. At least some elliptical galaxies are thought to be formed by mergers between spiral galaxies. There are also irregular galaxies. Most galaxies occur in clusters, containing anything from a few to thousands of members.

Our own Galaxy, the Milky Way, is about 100,000 light years in diameter, and contains at least 100 billion stars. It is a member of a small cluster, the ◊ Local Group. The Sun lies in one of its spiral arms, about 25,000 light years from the centre.

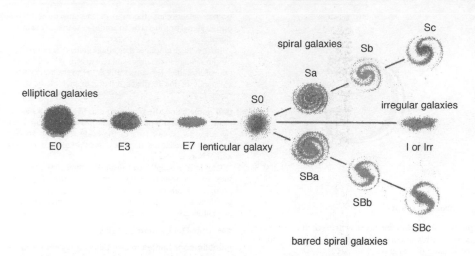

galaxy Galaxies were classified by US astronomer Edwin Hubble in 1925. He placed the galaxies in a 'tuning-fork' pattern, in which the two prongs correspond to the barred and non-barred spiral galaxies.

After a five-year study ending in 1995, US astronomers had identified 600 previously uncatalogued galaxies, mostly 200–400 million light years away, leading to the conclusion that there may be 30–100% more galaxies than previously estimated. Two galaxies were discovered obscured by galactic dust at the edge of the Milky Way. One, named MB1, is a spiral galaxy 17,000 light years across; the other, MB2, is an irregular-shaped dwarf galaxy about 4,000 light years across. In 1996 US astronomers discovered a further new galaxy 17 million light years away. The galaxy, NGC2915, is a blue compact dwarf galaxy and 95% of its mass is in the form of dark matter. Also in 1996, astronomers announced their discovery of the most distant galaxy ever detected; in the constellation Virgo, the galaxy is estimated to be around 14 billion light years from Earth.

Galaxy

An average galaxy contains 100,000 million stars. If you counted the stars in a galaxy at a rate of one every second, it would take 3,000 years to count them all.

---

galena mineral consisting of lead sulphide, PbS, the chief ore of lead. It is lead-grey in colour, has a high metallic lustre and breaks into cubes because of its perfect cubic cleavage. It may contain up to 1% silver, and so the ore is sometimes mined for both metals. Galena occurs mainly among limestone deposits in Australia, Mexico, Russia, Kazakhstan, the UK, and the USA.

galileo unit (symbol gal) of acceleration, used in geological surveying. One galileo is 10^{-2} metres per second per second. The Earth's gravitational field often differs by several milligals (thousandths of gals) in different places, because of the varying densities of the rocks beneath the surface.

Galileo spacecraft launched from the space shuttle *Atlantis* in Oct 1989, on a six-year journey to Jupiter. *Galileo's* probe entered the atmosphere of Jupiter in Dec 1995. It radioed information back to the orbiter for 57 minutes before it was destroyed by atmospheric pressure.

gall abnormal outgrowth on a plant that develops as a result of attack by insects or, less commonly, by bacteria, fungi, mites, or nematodes. The attack causes an increase in the number of cells or an enlargement of existing cells in the plant. Gall-forming insects generally pass the early stages of their life inside the gall.

Gall wasps are responsible for the conspicuous bud galls forming on oak trees, 2.5–4 cm/1–1.5 in across, known as 'oak apples'. The organisms that cause galls are host-specific. Thus, for example, gall wasps tend to parasitize oaks, and sawflies willows.

gall bladder small muscular sac, part of the digestive system of most, but not all, vertebrates. In humans, it is situated on the underside of the liver and connected to the small intestine by the bile duct. It stores bile from the liver.

galley ship powered by oars, and usually also equipped with sails. Galleys typically had a crew of hundreds of rowers arranged in banks. They were used in warfare in the Mediterranean from antiquity until the 18th century. France maintained a fleet of some 40 galleys, crewed by over 10,000 convicts, until 1748. The maximum speed of a galley is estimated to have been only four knots (7.5 kph/4.5 mph), because only 20% of the rower's effort was effective, and galleys could not be used in stormy weather because of their very low waterline.

gallium grey metallic element, symbol Ga, atomic number 31, relative atomic mass 69.72. It is liquid at room temperature. Gallium arsenide (GaAs) crystals are used in microelectronics, since electrons travel a thousand times faster through them than through silicon. The element was discovered in 1875 by Lecoq de Boisbaudran (1838–1912).

gallium arsenide GaAs compound of gallium and arsenic, used in lasers, photocells, and microwave generators. Its semiconducting properties make it a possible rival to ◊silicon for use in microprocessors. Chips made from gallium arsenide require less electric power and process data faster than those made from silicon.

gallon imperial liquid or dry measure, equal to 4.546 litres, and subdivided into four quarts or eight pints. The US gallon is equivalent to 3.785 litres.

gallstone pebblelike, insoluble accretion formed in the human gall bladder or bile ducts from cholesterol or calcium salts present in bile. Gallstones may be symptomless or they may cause pain, indigestion, or jaundice. They can be dissolved with medication or removed, either by means of an endoscope or, along with the gall bladder, in an operation known as cholecystectomy.

galvanizing process for rendering iron rust-proof, by plunging it into molten zinc (the dipping method), or by electroplating it with zinc.

galvanometer instrument for detecting small electric currents by their magnetic effect.

gamete cell that functions in sexual reproduction by merging with another gamete to form a ◊zygote. Examples of gametes include sperm and egg cells. In most organisms, the gametes are haploid (they contain half the number of chromosomes of the parent), owing to reduction division or ◊meiosis.

In higher organisms, gametes are of two distinct types: large immobile ones known as eggs or egg cells (see ◊ovum) and small ones known as ◊sperm. They come together at ◊fertilization. In some lower organisms the gametes are all the same, or they may belong to different mating strains but have no obvious differences in size or appearance.

game theory group of mathematical theories, developed in 1944 by Oscar Morgenstern (1902–1977) and John Von Neumann, that seeks to abstract from invented game-playing scenarios and their outcome the essence of situations of conflict and/or cooperation in the real political, business, and social world.

A feature of such games is that the rationality of a decision by one player will depend on what the others do; hence game theory has particular application to the study of oligopoly (a market largely controlled by a few producers).

gametophyte the ◊haploid generation in the life cycle of a plant that produces gametes; see ◊alternation of generations.

gamma radiation very high-frequency electromagnetic radiation, similar in nature to X-rays but of shorter wavelength, emitted by the nuclei of radioactive substances during decay or by the interactions of high-energy electrons with matter. Cosmic gamma rays have been identified as coming from pulsars, radio galaxies, and quasars, although they cannot penetrate the Earth's atmosphere.

Gamma rays are stopped only by direct collision with an atom and are therefore very penetrating; they can, however, be stopped by about 4 cm/1.5 in of lead or by a very thick concrete shield. They are less ionizing in their effect than alpha and beta particles, but are dangerous nevertheless because they can penetrate deeply into body tissues such as bone marrow. They are not deflected by either magnetic or electric fields.

Gamma radiation is used to kill bacteria and other microorganisms, sterilize medical devices, and change the molecular structure of plastics to modify their properties

(for example, to improve their resistance to heat and abrasion).

gamma-ray astronomy the study of gamma rays from space. Much of the radiation detected comes from collisions between hydrogen gas and cosmic rays in our Galaxy. Some sources have been identified, including the Crab nebula and the Vela pulsar (the most powerful gamma-ray source detected).

Gamma rays are difficult to detect and are generally studied by use of balloon-borne detectors and artificial satellites. The first gamma-ray satellites were *SAS II* (1972) and *COS B* (1975), although gamma-ray detectors were carried on the *Apollo 15* and *16* missions. *SAS II* failed after only a few months, but *COS B* continued working until 1982, carrying out a complete survey of the galactic disc.

ganglion (plural *ganglia*) solid cluster of nervous tissue containing many cell bodies and ◊synapses, usually enclosed in a tissue sheath; found in invertebrates and vertebrates.

In many invertebrates, the central nervous system consists mainly of ganglia connected by nerve cords. The ganglia in the head (cerebral ganglia) are usually well developed and are analogous to the brain in vertebrates. In vertebrates, most ganglia occur outside the central nervous system.

gangrene death and decay of body tissue (often of a limb) due to bacterial action; the affected part gradually turns black and causes blood poisoning.

Gangrene sets in as a result of loss of blood supply to the area. This may be due to disease (diabetes, atherosclerosis), an obstruction of a major blood vessel (as in ◊thrombosis), injury, or frostbite. Bacteria colonize the site unopposed, and a strong risk of blood poisoning often leads to surgical removal of the tissue or the affected part (amputation).

Gangrene

Wounds infested with maggots heal quickly and without spread of gangrene or other infection. This is because the maggots eat only the dead and suppurating flesh within a wound. This fact, first noted by World War I doctors in the trenches, is still used today in 'maggot therapy'.

---

gangue the part of an ore deposit that is not itself economically valuable; for example, calcite may occur as a gangue mineral with galena.

Ganymede in astronomy, the largest moon of the planet Jupiter, and the largest moon in the Solar System, 5,260 km/3,270 mi in diameter (larger than the planet Mercury). It orbits Jupiter every 7.2 days at a distance of 1.1 million km/700,000 mi. Its surface is a mixture of cratered and grooved terrain. Molecular oxygen was identified on Ganymede's surface in 1994.

The space probe *Galileo* detected a magnetic field around Ganymede in 1996; this suggests it may have a molten core. *Galileo* photographed Ganymede at a distance of 7,448 km/4,628 mi. The resulting images were 17 times clearer than those taken by *Voyager 2* in 1979, and show the surface to be extensively cratered and ridged, probably as a result of forces similar to those that create mountains on Earth. *Galileo* also detected molecules containing both carbon and nitrogen on the surface

March 1997. Their presence may indicate that Ganymede harboured life at some time.

garnet group of silicate minerals with the formula $X_3Y_2(SiO_4)_3$, when X is calcium, magnesium, iron, or manganese, and Y is iron, aluminium, or chromium. Garnets are used as semiprecious gems (usually pink to deep red) and as abrasives.

They occur in metamorphic rocks such as gneiss and schist.

gas in physics, a form of matter, such as air, in which the molecules move randomly in otherwise empty space, filling any size or shape of container into which the gas is put.

A sugar-lump sized cube of air at room temperature contains 30 trillion molecules moving at an average speed of 500 metres per second (1,800 kph/1,200 mph). Gases can be liquefied by cooling, which lowers the speed of the molecules and enables attractive forces between them to bind them together.

gas constant in physics, the constant R that appears in the equation $PV = nRT$, which describes how the pressure P, volume V, and temperature T of an ideal gas are related (n is the amount of gas in moles). This equation combines ◊Boyle's law and ◊Charles's law.

R has a value of 8.3145 joules per kelvin per mole.

gas engine internal-combustion engine in which a gas (coal gas, producer gas, natural gas, or gas from a blast furnace) is used as the fuel.

The first practical gas engine was built in 1860 by Jean Etienne Lenoir, and the type was subsequently developed by Nikolaus August Otto, who introduced the ◊four-stroke cycle.

gas exchange movement of gases between an organism and the atmosphere, principally oxygen and carbon dioxide. All aerobic organisms (most animals and plants) take in oxygen in order to burn food and manufacture ◊ATP. The resultant oxidation reactions release carbon dioxide as a waste product to be passed out into the environment. Green plants also absorb carbon dioxide during ◊photosynthesis, and release oxygen as a waste product.

Specialized respiratory surfaces have evolved during evolution to make gas exchange more efficient. In humans and other tetrapods (four-limbed vertebrates), gas exchange occurs in the ◊lungs, aided by the breathing movements of the ribs. Many adult amphibia and terrestrial invertebrates can absorb oxygen directly through the skin. The bodies of insects and some spiders contain a system of air-filled tubes known as ◊tracheae. Fish have ◊gills as their main respiratory surface. In plants, gas exchange generally takes place via the ◊stomata and the air-filled spaces between the cells in the interior of the leaf.

gas giant in astronomy, any of the four large outer ◊planets of the Solar System, ◊Jupiter, ◊ Saturn, ◊Uranus, and ◊Neptune, which consist largely of gas and have no solid surface.

gas laws physical laws concerning the behaviour of gases. They include ◊Boyle's law and ◊Charles's law, which are concerned with the relationships between the pressure, temperature, and volume of an ideal (hypothetical) gas. These two laws can be combined to give the *general* or *universal gas law*, which may be expressed as:

$$(\text{pressure} \times \text{volume})/\text{temperature} = \text{constant}$$

Van der Waals' law includes corrections for the non-ideal behaviour of real gases.

We have no right to assume that any physical laws exist, or if they have existed up to now, that they will continue to exist in a similar manner in the future.

On **gas laws** Max Planck
The Universe in the Light of Modern Physics

gasohol motor fuel that is 90% petrol and 10% ethanol (alcohol). The ethanol is usually obtained by fermentation, followed by distillation, using maize, wheat, potatoes, or sugar cane. It was used in early cars before petrol became economical, and its use was revived during the 1940s war shortage and the energy shortage of the 1970s, for example in Brazil.

gastritis inflammation of the lining of the stomach. The term is a vague one, applied to a range of conditions.

gastroenteritis inflammation of the stomach and intestines, giving rise to abdominal pain, vomiting, and diarrhoea. It may be caused by food or other poisoning, allergy, or infection. Dehydration may be severe and it is a particular risk in infants.

gastroenterology medical speciality concerned with disorders of the digestive tract and associated organs such as the liver, gall bladder and pancreas.

gastrolith stone that was once part of the digestive system of a dinosaur or other extinct animal. Rock fragments were swallowed to assist in the grinding process in the dinosaur digestive tract, much as some birds now swallow grit and pebbles to grind food in their crop. Once the animal has decayed, smooth round stones remain – often the only clue to their past use is the fact that they are geologically different from their surrounding strata.

gastroscope in medicine, instrument for viewing the interior of the stomach. It consists of a tube with a light and mirrors attached that is introduced into the stomach via the mouth and the oesophagus. A camera attachment allows photographs to be taken of the stomach interior.

Gastroscopy using fibre-optic technology allows the observation of images that are transmitted through a fibreoptic bundle or a small video camera. Flexibility of the instrument allows photographs of all areas of the stomach and tissue samples can be taken to assist in diagnosis when required.

gas turbine engine in which burning fuel supplies hot gas to spin a ◊turbine. The most widespread application of gas turbines has been in aviation. All jet engines (see ◊jet propulsion) are modified gas turbines, and some locomotives and ships also use gas turbines as a power source.

They are also used in industry for generating and pumping purposes.

In a typical gas turbine a multi-vaned compressor draws in and compresses air. The compressed air enters a combustion chamber at high pressure, and fuel is sprayed in and ignited. The hot gases produced escape through the blades of (typically) two turbines and spin them around. One of the turbines drives the compressor; the other provides the external power that can be harnessed.

gauge any scientific measuring instrument – for example, a wire gauge or a pressure gauge. The term is also applied to the width of a railway or tramway track.

gauge boson or **field particle** any of the particles that carry the four fundamental forces of nature (see ◊forces, fundamental).

Gauge bosons are ◊elementary particles that cannot be subdivided, and include the photon, the graviton, the gluons, and the weakons.

gauss c.g.s. unit (symbol Gs) of magnetic induction or magnetic flux density, replaced by the SI unit, the ◊tesla, but still commonly used. It is equal to one line of magnetic flux per square centimetre. The Earth's magnetic field is about 0.5 Gs, and changes to it over time are measured in gammas (one gamma equals 10^{-5} gauss).

gear toothed wheel that transmits the turning movement of one shaft to another shaft. Gear wheels may be used in pairs, or in threes if both shafts are to turn in the same direction. The gear ratio – the ratio of the number of teeth on the two wheels – determines the torque ratio, the turning force on the output shaft compared with the turning force on the input shaft. The ratio of the angular velocities of the shafts is the inverse of the gear ratio.

The most common type of gear for parallel shafts is the **spur gear**, with straight teeth parallel to the shaft axis. The **helical gear** has teeth cut along sections of a helix or corkscrew shape; the double form of the helix gear is the most efficient for energy transfer. **Bevel gears**, with tapering teeth set on the base of a cone, are used to connect intersecting shafts.

Geiger counter any of a number of devices used for detecting nuclear radiation and/or measuring its intensity by counting the number of ionizing particles produced (see ◊radioactivity). It detects the momentary current that passes between ◊electrodes in a suitable gas when a nuclear particle or a radiation pulse causes the ionization of that gas. The electrodes are connected to electronic devices that enable the number of particles passing to be measured. The increased frequency of measured particles indicates the intensity of radiation. The device is named after the German physicist Hans Geiger.

The Geiger–Müller, Geiger–Klemperer, and Rutherford–Geiger counters are all devices often referred to loosely as Geiger counters.

gel solid produced by the formation of a three-dimensional cage structure, commonly of linked large-molecular-mass polymers, in which a liquid is trapped. It is a form of ◊colloid. A gel may be a jellylike mass (pectin, gelatin) or have a more rigid structure (silica gel).

gelignite type of ◊dynamite.

gem mineral valuable by virtue of its durability (hardness), rarity, and beauty, cut and polished for ornamental use, or engraved. Of 120 minerals known to have been used as gemstones, only about 25 are in common use in jewellery today; of these, the diamond, emerald, ruby, and sapphire are classified as precious, and all the others semi-precious; for example, the topaz, amethyst, opal, and aquamarine. Pearls are not technically gems.

Gemini prominent zodiacal constellation in the northern hemisphere represented as the twins Castor and Pollux. Its brightest star is ◊Pollux; ◊Castor is a system of six stars. The Sun passes through Gemini from late June to late July. Each Dec, the Geminid meteors radiate from Gemini. In astrology, the dates for Gemini are between about 21 May and 21 June (see ◊precession).

Gemini project US space programme (1965–66) in which astronauts practised rendezvous and docking of spacecraft, and working outside their spacecraft, in preparation for the ◊Apollo Moon landings.

Gemini spacecraft carried two astronauts and were launched by Titan rockets.

gemma (plural **gemmae**) unit of ◊vegetative reproduction, consisting of a small group of undifferentiated green cells. Gemmae are found in certain mosses and liverworts, forming on the surface of the plant, often in cup-shaped structures, or gemmae cups. Gemmae are dispersed by splashes of rain and can then develop into new plants. In many species, gemmation is more common than reproduction by ◊spores.

gene unit of inherited material, encoded by a strand of ◊DNA and transcribed by ◊RNA. In higher organisms, genes are located on the ◊chromosomes. A gene consistently affects a particular character in an individual – for example, the gene for eye colour. Also termed a Mendelian gene, after Austrian biologist Gregor Mendel, it occurs at a particular point, or locus, on a particular chromosome and may have several variants, or ◊alleles, each specifying a particular form of that character – for example, the alleles for blue or brown eyes. Some alleles show ◊dominance. These mask the effect of other alleles, known as ◊recessive.

We are survival machines – robot vehicles blindly programmed to preserve the selfish molecules known as genes. This is a truth which still fills me with astonishment.

On the **gene** Richard Dawkins
The Selfish Gene Preface

In the 1940s, it was established that a gene could be identified with a particular length of DNA, which coded for a complete protein molecule, leading to the 'one gene, one enzyme' principle. Later it was realized that proteins can be made up of several ◊polypeptide chains, each with a separate gene, so this principle was modified to 'one gene, one polypeptide'. However, the fundamental idea remains the same, that genes produce their visible effects simply by coding for proteins; they control the structure of those proteins via the genetic code, as well as the amounts produced and the timing of production.

In modern genetics, the gene is identified either with the ◊cistron (a set of ◊codons that determines a complete polypeptide) or with the unit of selection (a Mendelian gene that determines a particular character in the organism on which ◊natural selection can act). Genes undergo ◊mutation and ◊recombination to produce the variation on which natural selection operates.

gene amplification technique by which selected DNA from a single cell can be duplicated indefinitely until there is a sufficient amount to analyse by conventional genetic techniques.

Gene amplification uses a procedure called the polymerase chain reaction. The sample of DNA is mixed with a solution of enzymes called polymerases, which enable it to replicate, and with a plentiful supply of nucleotides, the building blocks of DNA. The mixture is repeatedly heated and cooled. At each warming, the double-stranded DNA present separates into two single strands, and with each cooling the polymerase assembles a new paired strand for each single strand. Each cycle takes approximately 30 minutes to complete, so that after 10 hours there is one million times more DNA present than at the start.

The technique has been used to analyse DNA from a man who died in 1959, showing the presence of sequences from the HIV virus in his cells. It can also be used to test for genetic defects in a single cell taken from an embryo, before the embryo is reimplanted in ◊ in vitro fertilization.

gene bank collection of seeds or other forms of genetic material, such as tubers, spores, bacterial or yeast cultures, live animals and plants, frozen sperm and eggs, or frozen embryos. These are stored for possible future use in agriculture, plant and animal breeding, or in medicine, genetic engineering, or the restocking of wild habitats where species have become extinct. Gene banks will be increasingly used as the rate of extinction increases, depleting the Earth's genetic variety (biodiversity).

gene pool total sum of ◊ alleles (variants of ◊ genes) possessed by all the members of a given population or species alive at a particular time.

generate in mathematics, to produce a sequence of numbers from either the relationship between one number and the next or the relationship between a member of the sequence and its position. For example, $u_{n+1} = 2u_n$ generates the sequence 1, 2, 4, 8, ... ; $a_n = n(n+1)$ generates the sequence of numbers 2, 6, 12, 20, ...

generation in computing, stage of development in computer electronics (see ◊ computer generation) or a class of programming language (see ◊ fourth-generation language).

generator machine that produces electrical energy from mechanical energy, as opposed to an ◊ electric motor, which does the opposite. A simple generator (dynamo) consists of a wire-wound coil (◊ armature) that is rotated between the poles of a permanent magnet. The movement of the wire in the magnetic field induces a current in the coil by ◊ electromagnetic induction, which can be fed by means of a ◊ commutator as a continuous direct current into an external circuit. Slip rings instead of a commutator produce an alternating current, when the generator is called an alternator.

gene shears technique in ◊ genetic engineering which may have practical applications in the future. The gene shears are pieces of messenger ◊ RNA that can bind to other pieces of messenger RNA, recognizing specific sequences, and cut them at that point. If a piece of ◊ DNA which codes for the shears can be inserted in the chromosomes of a plant or animal cell, that cell will then destroy all messenger RNA of a particular type. Genetic shears may be used to protect plants against viruses which infect them and cause disease. They might also be useful against ◊ AIDS.

gene-splicing see ◊ genetic engineering.

gene therapy proposed medical technique for curing or alleviating inherited diseases or defects, certain infections, and several kinds of cancer in which affected cells from a sufferer would be removed from the body, the ◊ DNA repaired in the laboratory (◊ genetic engineering), and the functioning cells reintroduced. In 1990 a genetically engineered gene was used for the first time to treat a patient.

The first human being to undergo gene therapy, in Sept 1990, was one of the so-called 'bubble babies' – a four-year-old American girl suffering from a rare enzyme (ADA) deficiency that cripples the immune system. Unable to fight off infection, such children are nursed in a germ-free bubble; they usually die in early childhood. See ◊ severe combined immune deficiency.

Cystic fibrosis is the commonest inherited disorder and the one most keenly targeted by genetic engineers; it has been pioneered in patients in the USA and UK. Gene therapy is not the final answer to inherited disease; it may cure the patient but it cannot prevent him or her from passing on the genetic defect to any children. However, it does hold out the promise of a cure for various other conditions, including heart disease and some cancers; US researchers have successfully used a gene gun to target specific tumour cells. In 1995 tumour growth was halted in mice when DNA-coated gold bullets were fired into tumour cells.

By the end of 1995, although 600 people had been treated with gene therapy, nobody had actually been cured. Even in the ADA trials, the most successful to date, the children were still receiving injections of synthetic ADA, possibly the major factor in their improvement.

genetic code the way in which instructions for building proteins, the basic structural molecules of living matter, are 'written' in the genetic material ◊ DNA. This relationship between the sequence of bases (the subunits in a DNA molecule) and the sequence of ◊ amino acids (the subunits of a protein molecule) is the basis of heredity. The code employs ◊ codons of three bases each; it is the same in almost all organisms, except for a few minor differences recently discovered in some protozoa.

Only 2% of DNA is made up of base sequences, called *exons*, that code for proteins. The remaining DNA is known as 'junk' DNA or *introns*.

genetic counselling in medicine, the establishment of a detailed family history of any genetic disorders for individuals who are at risk of developing them, or concerned that they may pass them on to their children, and discussion as to likely incidence and available options. Prenatal diagnosis may be available to those at risk of having a child affected by a particular disorder. Genetic counselling of individuals at risk of developing a genetic disorder in adult life can result in the individual taking action to prevent the disorder developing; for example, removal of the ovaries in women at high risk of developing ovarian cancer.

There is a great deal of variation in the mode of inheritance of genetic disorders. Some disorders, such as Huntington's chorea, are inherited as dominant genes and there is a 50% chance of the children of an affected person and a normal person being affected by the disorder. Other disorders, such as cystic fibrosis, are inherited as recessive genes. The possession of a recessive gene does not result in overt disease but the children of an affected person and a normal person have a 25% chance of being affected by the disorder. Other genetic diseases, such as haemophilia, are linked to the X chromosome. These diseases are only found in male children but female children can carry the gene. Other diseases, such as diabetes, schizophrenia, and some forms of cancer have complex genetic and environmental components and risk can be determined from population studies.

Genes associated with some genetic disorders have been isolated in recent years. These have resulted in the development of tests for certain disorders in which inheritance is important, such as Alzheimer's disease and certain forms of breast cancer. After receiving genetic counselling, individuals at risk can be tested to determine if they are likely to develop such diseases later in life.

genetic disease any disorder caused at least partly by defective genes or chromosomes. In humans there are some 3,000 genetic diseases, including cystic fibrosis, Down's syndrome, haemophilia, Huntington's chorea, some forms of anaemia, spina bifida, and Tay–Sachs disease.

genetic engineering deliberate manipulation of genetic material by biochemical techniques. It is often achieved by the introduction of new ◊DNA, usually by means of a virus or ◊plasmid. This can be for pure research, ◊gene therapy, or to breed functionally specific plants, animals, or bacteria. These organisms with a foreign gene added are said to be transgenic. At the beginning of 1995 more than 60 plant species had been genetically engineered, and nearly 3,000 transgenic crops had been field-tested.

In genetic engineering, the splicing and reconciliation of genes is used to increase knowledge of cell function and reproduction, but it can also achieve practical ends. For example, plants grown for food could be given the ability to fix nitrogen, found in some bacteria, and so reduce the need for expensive fertilizers, or simple bacteria may be modified to produce rare drugs. A foreign gene can be inserted into laboratory cultures of bacteria to generate commercial biological products, such as synthetic insulin, hepatitis-B vaccine, and interferon. Gene splicing was invented in 1973 by the US scientists Stanley Cohen and Herbert Boyer, and patented in the USA in 1984.

Developments in genetic engineering have led to the production of growth hormone, and a number of other bone-marrow stimulating hormones. New strains of animals have also been produced; a new strain of mouse was patented in the USA in 1989 (the application was rejected in the European Patent Office). A ◊vaccine against a sheep parasite (a larval tapeworm) has been developed by genetic engineering; most existing vaccines protect against bacteria and viruses.

The first genetically engineered food went on sale in 1994; the 'Flavr Savr' tomato, produced by the US biotechnology company Calgene, was available in California and Chicago.

There is a risk that when transplanting genes between different types of bacteria (*Escherichia coli*, which lives in the human intestine, is often used) new and harmful strains might be produced. For this reason strict safety precautions are observed, and the altered bacteria are disabled in some way so they are unable to exist outside the laboratory.

genetic fingerprinting or *genetic profiling* technique used for determining the pattern of certain parts of the genetic material ◊DNA that is unique to each individual. Like conventional fingerprinting, it can accurately distinguish humans from one another, with the exception of identical siblings from multiple births. It can be applied to as little material as a single cell.

Genetic fingerprinting involves isolating DNA from cells, then comparing and contrasting the sequences of component chemicals between individuals. The DNA pattern can be ascertained from a sample of skin, hair, or semen. Although differences are minimal (only 0.1% between unrelated people), certain regions of DNA, known as *hypervariable regions*, are unique to individuals.

genetics the science of ◊heredity that attempts to explain how characteristics are passed on from one generation to the next. The founder of genetics was Austrian biologist Gregor Mendel, whose experiments with peas showed that inheritance of characteristics takes place by means of discrete 'particles', later called ◊genes.

Before Mendel, it had been assumed that the characteristics of the two parents were blended during inheritance, but Mendel showed that the genes remain intact, although their combinations change. Since Mendel, genetics has advanced greatly, first through ◊breeding experiments and light-microscope observations (classical genetics), later by means of biochemical and electron-microscope studies (molecular genetics).

genetic screening in medicine, the determination of the genetic make-up of an individual to determine if he or she is at risk of developing a hereditary disease later in life. Genetic screening can also be used to determine if an individual is a carrier for a particular genetic disease and, hence, can pass the disease on to any children. Genetic counselling should be undertaken at the same time as genetic screening of affected individuals. Diseases that can be screened for include cystic fibrosis, Huntington's chorea, and certain forms of cancer.

genitalia reproductive organs of sexually reproducing animals, particularly the external/visible organs of mammals: in males, the penis and the scrotum, which contains the testes, and in females, the clitoris and vulva.

genome the full complement of ◊genes carried by a single (haploid) set of ◊chromosomes. The term may be applied to the genetic information carried by an individual or to the range of genes found in a given species. The human genome is made up of 75,000 genes.

genotype the particular set of ◊alleles (variants of genes) possessed by a given organism. The term is usually used in conjunction with ◊phenotype, which is the product of the genotype and all environmental effects. See also ◊nature–nurture controversy.

genus (plural *genera*) group of ◊species with many characteristics in common.

Thus all doglike species (including dogs, wolves, and jackals) belong to the genus *Canis* (Latin 'dog').

Species of the same genus are thought to be descended from a common ancestor species. Related genera are grouped into ◊families.

geochemistry science of chemistry as it applies to geology. It deals with the relative and absolute abundances of the chemical elements and their ◊isotopes in the Earth, and also with the chemical changes that accompany geologic processes.

geochronology the branch of geology that deals with the dating of the Earth by studying its rocks and contained fossils. The ◊geological time chart is a result of these

Genetic engineering: a brave new world

The 1980s and 1990s have seen considerable advances in understanding of the structure of genes and how they work, thanks to the development of techniques for isolating and manipulating genes (*recombinant DNA technology*). We can determine the structure of the genetic material (*DNA sequencing*) and study minute quantities of DNA using PCR (*polymerase chain reaction*).

Large-scale production of pure proteins
Each gene is responsible for making a protein. By purifying genes and transferring them to bacterial cells, scientists can harvest large quantities of proteins that would not normally be made by bacteria. Most of the insulin for diabetics is now made in bacteria. Several hormones and enzymes, whose deficiency results in inherited diseases, are manufactured in microorganisms.

Vaccines produced by genetic engineering
Genetic engineering is being used to produce vaccines. These are normally made by growing *pathogens* (disease-causing organisms) in animals, or in organ or tissue culture, and then harvesting the live pathogen. The inactivated (attenuated) pathogens provoke our immune system, which reacts to the coat proteins covering the pathogen. The new method transfers the gene coding for the pathogen's coat protein into harmless bacteria or viruses which then make the protein themselves. The purified protein is used to stimulate the immune system. Vaccines against, for example, foot-and-mouth disease of cattle and hepatitis B in humans are proving much safer than attenuated vaccines.

Crop and livestock improvement
The transfer of 'foreign' genes into plants has led to the development of crops with desirable characteristics. A bacterial gene can be transferred, conferring resistance to caterpillars and other insects that can devastate crops, reducing the need for insecticides and their consequent environmental damage. Transgenic plants resistant to the alfalfa mosaic virus and the tobacco ringspot virus are also now available.

In 1997 the first successful clone of a mammal was achieved; Dolly, a Finn Dorset sheep was cloned from a tiny fleck of skin scraped from the udder of another sheep. The commercial and ethical implications of this are still being explored.

Genetic fingerprinting
Mini- and micro-satellite DNA are types of DNA that differ so much between individuals that each person is effectively unique. Forensic scientists can determine the origin of tissue samples like blood and semen with a certainty that was previously impossible. Genetic fingerprinting has rapidly become invaluable in convicting murderers and rapists, and in eliminating suspects. Genetic fingerprints also allow relationships between people to be established, and are increasingly being used in paternity and immigration disputes.

Screening and gene therapy
The fundamental genetic changes responsible for some inherited diseases have been identified, leading to deeper understanding of the causes of disease, and to diagnostic methods. In certain communities a particular inherited disease is so common that programmes of preventive medicine have been attempted. All pregnancies are screened and termination is offered to females shown to be carrying an affected fetus. Such community medicine has dramatically reduced the number of babies suffering from a blood disease in Sardinia, and from a type of inherited idiocy in US Jews.

Two recent projects promise exciting developments. One is human gene therapy – transferring normal genes to people with inherited diseases. Animals have been successfully treated, and after much discussion, medical approval has been granted for human trials. The other is the Human Genome Project, a huge international research effort to identify and study all 100,000 human genes. The project should lead to deeper understanding of development and disease, and may have unforeseen applications in many branches of biology.

Thomas Day

---

studies. Absolute dating methods involve the measurement of radioactive decay over time in certain chemical elements found in rocks, whereas relative dating methods establish the sequence of deposition of various rock layers by identifying and comparing their contained fossils.

geode in geology, a subspherical cavity into which crys-

tals have grown from the outer wall into the centre. Geodes often contain very well-formed crystals of quartz (including amethyst), calcite, or other minerals.

geodesic dome hemispherical dome, a type of spaceframe, whose surface is formed out of short rods arranged in triangles. The rods lie on geodesics (the shortest lines joining two points on a curved surface). This type of dome allows large spaces to be enclosed using the minimum of materials, and was patented by US engineer Buckminster Fuller in 1954.

geodesy methods of surveying the Earth for making maps and correlating geological, gravitational, and magnetic measurements. Geodesic surveys, formerly carried out by means of various measuring techniques on the surface, are now commonly made by using radio signals and laser beams from orbiting satellites.

geography the study of the Earth's surface, its topography, climate, and physical conditions, and how these factors affect people and society. It is usually divided into *physical geography*, dealing with landforms and climates, and *human geography*, dealing with the distribution and activities of peoples on Earth.
history Early preclassical geographers concentrated on map-making, surveying, and exploring. In classical Greece theoretical ideas first became a characteristic of geography. Aristotle and Pythagoras believed the Earth to be a sphere, Eratosthenes was the first to calculate the circumference of the world, and Herodotus investigated the origin of the Nile floods and the relationship between climate and human behaviour.

During the mediaeval period the study of geography progressed little in Europe, but the Muslim world retained much of the Greek tradition, embellishing the 2nd-century maps of Ptolemy. During the early Renaissance the role of the geographer as an explorer and surveyor became important once again.

The foundation of modern geography as an academic subject stems from the writings of Friedrich Humboldt and Johann Ritter, in the late 18th and early 19th centuries, who for the first time defined geography as a major branch of scientific inquiry.

geological time time scale embracing the history of the Earth from its physical origin to the present day. Geological time is traditionally divided into eons (Phanerozoic, Proterozoic, and Archaeozoic), which in turn are divided into eras, periods, epochs, ages, and finally chrons.

geology science of the Earth, its origin, composition, structure, and history. It is divided into several branches: *mineralogy* (the minerals of Earth), *petrology* (rocks), *stratigraphy* (the deposition of successive beds of sedimentary rocks), *palaeontology* (fossils), and *tectonics* (the deformation and movement of the Earth's crust).

Geology is regarded as part of earth science, a more widely embracing subject that brings in meteorology, oceanography, geophysics, and geochemistry.

geomagnetic reversal another term for ◊polar reversal.

geometric mean in mathematics, the *n*th root of the product of *n* positive numbers. The geometric mean *m* of two numbers *p* and *q* is such that $m = \sqrt{(p \times q)}$. For example, the mean of 2 and 8 is $\sqrt{(2 \times 8)} = \sqrt{16} = 4$.

geometric progression or *geometric sequence* in mathematics, a sequence of terms (progression) in which each term is a constant multiple (called the **common ratio**) of the one preceding it. For example, 3, 12, 48, 192, 768, ... is a geometric progression with a common ratio 4, since each term is equal to the previous term multiplied by 4. Compare ◊arithmetic progression.
The sum of *n* terms of a *geometric series*

$$1 + r + r^2 + r^3 + ... + r^{n-1}$$

is given by the formula

$$S_n = (1 - r^n)/(1 - r)$$

for all $r \neq 1$. For $r = 1$, the geometric series can be summed to infinity:

$$S\infty = 1/(1 - r).$$

In nature, many single-celled organisms reproduce by splitting in two so that one cell gives rise to 2, then 4, then 8 cells, and so on, forming a geometric sequence 1, 2, 4, 8, 16, 32, ..., in which the common ratio is 2.

geometry branch of mathematics concerned with the properties of space, usually in terms of plane (two-dimensional) and solid (three-dimensional) figures. The subject is usually divided into *pure geometry*, which embraces roughly the plane and solid geometry dealt with in Euclid's *Elements*, and *analytical* or ◊*coordinate geometry*, in which problems are solved using algebraic methods. A third, quite distinct, type includes the non-Euclidean geometries.

Geometry probably originated in ancient Egypt, in land measurements necessitated by the periodic inundations of the river Nile, and was soon extended into surveying and navigation. Early geometers were the Greek mathematicians Thales, Pythagoras, and Euclid. Analytical methods were introduced and developed by the French philosopher René Descartes in the 17th century. From the 19th century, various non-Euclidean geometries were devised by Carl Friedrich Gauss, János Bolyai, and Nikolai Lobachevsky. These were later generalized by Bernhard Riemann and found to have applications in the theory of relativity.

geomorphology branch of geology that deals with the nature and origin of surface landforms such as mountains, valleys, plains, and plateaus.

geophysics branch of earth science using physics to study the Earth's surface, interior, and atmosphere. Studies also include winds, weather, tides, earthquakes, volcanoes, and their effects.

geostationary orbit circular path 35,900 km/22,300 mi above the Earth's Equator on which a ◊satellite takes 24 hours, moving from west to east, to complete an orbit, thus appearing to hang stationary over one place on the Earth's surface. Geostationary orbits are used particularly for communications satellites and weather satellites. They were first thought of by the author Arthur C Clarke. A *geosynchronous orbit* lies at the same distance from Earth but is inclined to the Equator.

geothermal energy energy extracted for heating and electricity generation from natural steam, hot water, or hot dry rocks in the Earth's crust. Water is pumped down through an injection well where it passes through joints in the hot rocks. It rises to the surface through a recovery well and may be converted to steam or run through a heat exchanger. Dry steam may be directed through turbines to produce electricity. It is an important source of energy in volcanically active areas such as Iceland and New Zealand. *See illustration overleaf.*

geriatrics medical speciality concerned with diseases and problems of the elderly.

germ colloquial term for a microorganism that causes disease, such as certain ◊bacteria and ◊viruses. Formerly,

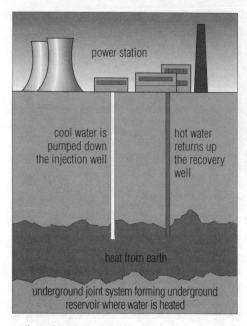

power station

cool water is pumped down the injection well

hot water returns up the recovery well

heat from earth

underground joint system forming underground reservoir where water is heated

geothermal energy

it was also used to mean something capable of developing into a complete organism (such as a fertilized egg, or the ◊embryo of a seed).

Germ

More germs are transmitted when shaking hands than when kissing. Louis Pasteur, the pioneer of hygienic methods, refused to shake hands with acquaintances for fear of infection.

---

germanium brittle, grey-white, weakly metallic (◊metalloid) element, symbol Ge, atomic number 32, relative atomic mass 72.6. It belongs to the silicon group, and has chemical and physical properties between those of silicon and tin. Germanium is a semiconductor material and is used in the manufacture of transistors and integrated circuits. The oxide is transparent to infrared radiation, and is used in military applications. It was discovered in 1886 by German chemist Clemens Winkler (1838–1904).

In parts of Asia, germanium and plants containing it are used to treat a variety of diseases, and it is sold in the West as a food supplement despite fears that it may cause kidney damage.

German measles or *rubella* mild, communicable virus disease, usually caught by children. It is marked by a sore throat, pinkish rash, and slight fever, and has an incubation period of two to three weeks. If a woman contracts it in the first three months of pregnancy, it may cause serious damage to the unborn child.

German silver or *nickel silver* silvery alloy of nickel, copper, and zinc. It is widely used for cheap jewellery and the base metal for silver plating. The letters EPNS on silverware stand for *e*lectro*p*lated *n*ickel *s*ilver.

germination in botany, the initial stages of growth in a seed, spore, or pollen grain. Seeds germinate when they are exposed to favourable external conditions of moisture, light, and temperature, and when any factors causing dormancy have been removed.

The process begins with the uptake of water by the seed. The embryonic root, or radicle, is normally the first organ to emerge, followed by the embryonic shoot, or plumule. Food reserves, either within the ◊endosperm or from the ◊cotyledons, are broken down to nourish the rapidly growing seedling. Germination is considered to have ended with the production of the first true leaves.

germ layer in ◊embryology, a layer of cells that can be distinguished during the development of a fertilized egg. Most animals have three such layers: the inner, middle, and outer. These differentiate to form the various body tissues.

The inner layer (**endoderm**) gives rise to the gut, the middle one (**mesoderm**) develops into most of the other organs, while the outer one (**ectoderm**) gives rise to the skin and nervous system. Simple animals, such as sponges, lack a mesoderm.

Gerson therapy in alternative medicine, radical nutritional therapy for degenerative diseases, particularly cancer, developed by German-born US physician Max Gerson (1881–1959).

Numerous cures of chronic cases were achieved by Gerson, but his opposition to orthodox cancer treatment, and the stringency of his alternative approach, have resulted in the therapy not being widely practised today, although the evidence for its effectiveness, both in relieving the suffering and drug-dependence of most patients and in achieving the recovery of many others, is substantial.

Gestalt concept of a unified whole that is greater than, or different from, the sum of its parts; that is, a complete structure whose nature is not explained simply by analysing its constituent elements. A chair, for example, will generally be recognized as a chair despite great variations between individual chairs in such attributes as size, shape, and colour.

Gestalt psychology regards all mental phenomena as being arranged in organized, structured wholes, as opposed to being composed of simple sensations. For example, learning is seen as a reorganizing of a whole situation (often involving insight), as opposed to the behaviourists' view that it consists of associations between stimuli and responses. Gestalt psychologists' experiments show that the brain is not a passive receiver of information, but that it structures all its input in order to make sense of it, a belief that is now generally accepted; however, other principles of Gestalt psychology have received considerable criticism.

The term 'Gestalt' was first used in psychology by the Austrian philosopher and psychologist Christian von Ehrenfels in 1890. Max Wertheimer, Wolfgang Köhler, and Kurt Koffka were cofounders of Gestalt psychology.

gestation in all mammals except the ◊monotremes (platypus and spiny anteaters), the period from the time of implantation of the embryo in the uterus to birth. This period varies among species; in humans it is about 266 days, in elephants 18–22 months, in cats about 60 days, and in some species of marsupial (such as opossum) as short as 12 days.

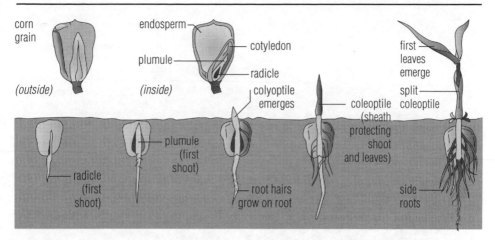

germination The germination of a corn grain. The plumule and radicle emerge from the seed coat and begin to grow into a new plant. The coleoptile protects the emerging bud and the first leaves.

geyser natural spring that intermittently discharges an explosive column of steam and hot water into the air due to the build-up of steam in underground chambers. One of the most remarkable geysers is Old Faithful, in Yellowstone National Park, Wyoming, USA. Geysers also occur in New Zealand and Iceland.

g-force force that pilots and astronauts experience when their craft accelerate or decelerate rapidly. One *g* is the ordinary pull of gravity.

Early astronauts were subjected to launch and reentry forces of up to six *g* or more; in the space shuttle, more than three *g* is experienced on liftoff. Pilots and astronauts wear *g*-suits that prevent their blood pooling too much under severe *g*-forces, which can lead to unconsciousness.

giant star in astronomy, a class of stars to the top right of the ◊Hertzsprung–Russell diagram characterized by great size and ◊luminosity. Giants have exhausted their supply of hydrogen fuel and derive their energy from the fusion of helium and heavier elements. They are roughly 10–300 times bigger than the Sun with 30–1,000 times the luminosity. The cooler giants are known as ◊red giants.

giardiasis in medicine, gastrointestinal infection usually characterized by severe diarrhoea, abdominal pain, and nausea, though in some cases it may be symptomless. It also produces a sulphurous smell on the breath. Infection is acquired through drinking water contaminated with the parasite protozoan *Giardia lamblia*, usually in in Middle Eastern and Asian countries. Giardiasis develops one or two weeks after exposure. Antibiotics such as metronidazole and tinidazole provide effective treatment.

gibberellin plant growth substance (see also ◊auxin) that promotes stem growth and may also affect the breaking of dormancy in certain buds and seeds, and the induction of flowering. Application of gibberellin can stimulate the stems of dwarf plants to additional growth, delay the ageing process in leaves, and promote the production of seedless fruit (◊parthenocarpy).

giga- prefix signifying multiplication by 10^9 (1,000,000,000 or 1 billion), as in **gigahertz**, a unit of frequency equivalent to 1 billion hertz.

gigabyte in computing, a measure of ◊memory capacity, equal to 1,024 ◊megabytes. It is also used, less precisely, to mean 1,000 million ◊bytes.

gill in biology, the main respiratory organ of most fishes and immature amphibians, and of many aquatic invertebrates. In all types, water passes over the gills, and oxygen diffuses across the gill membranes into the circulatory system, while carbon dioxide passes from the system out into the water.

In aquatic insects, these gases diffuse into and out of air-filled canals called tracheae.

gill imperial unit of volume for liquid measure, equal to one-quarter of a pint or 5 fluid ounces (0.142 litre). It is used in selling alcoholic drinks.

In S England it is also called a noggin, but in N England the large noggin is used, which is two gills.

Gilles de la Tourette syndrome in medicine, another name for ◊Tourette syndrome.

ginseng plant with a thick forked aromatic root used in alternative medicine as a tonic. (*Panax ginseng*, family Araliaceae.)

Giotto space probe built by the European Space Agency to study ◊Halley's comet. Launched by an Ariane rocket in July 1985, *Giotto* passed within 600 km/375 mi of the comet's nucleus on 13 March 1986. On 2 July 1990, it flew 23,000 km/14,000 mi from Earth, which diverted its path to encounter another comet, Grigg-Skjellerup, on 10 July 1992.

gizzard muscular grinding organ of the digestive tract, below the ◊crop of birds, earthworms, and some insects, and forming part of the ◊stomach. The gizzard of birds is lined with a hardened horny layer of the protein keratin, preventing damage to the muscle layer during the grinding process. Most birds swallow sharp grit which aids maceration of food in the gizzard.

glacial deposition the laying-down of rocky material once carried by a glacier. When ice melts, it deposits the material that it has been carrying.

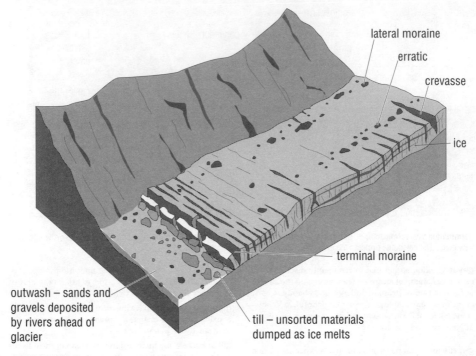

lateral moraine

erratic

crevasse

ice

terminal moraine

outwash – sands and
gravels deposited
by rivers ahead of
glacier

till – unsorted materials
dumped as ice melts

glacial deposition

The material dumped on the valley floor forms a deposit called till or boulder clay. It comprises angular particles of all sizes from boulders to clay. Till can be moulded by ice to form drumlins, egg-shaped hills. At the snout of the glacier, material piles up to form a ridge called a terminal moraine. Small depositional landforms may also result from glacial deposition, such as kames (small mounds) and kettle holes (small depressions, often filled with water).

Meltwater flowing away from a glacier will carry some of the till many kilometres away. This sediment will become rounded (by the water) and, when deposited, will form a gently sloping area called an outwash plain. Several landforms owe their existence to meltwater – these are called *fluvioglacial landforms* and include the long ridges called eskers. Meltwater may fill depressions eroded by the ice to form ribbon lakes.

glacial erosion the wearing-down and removal of rocks and soil by a glacier. Glacial erosion forms impressive landscape features, including the glacial trough (valley), arêtes (steep ridges), corries (enlarged hollows), and pyramidal peaks (high mountain peaks with concave faces).

Ice is a powerful agent of erosion. It can bulldoze its way down a valley, eroding away spurs to form truncated spurs. Rock fragments below the ice will abrade the valley floor, leading to overdeepening. Loose fragments will be plucked away from the bedrock to form jagged surfaces such as those on a roche moutonnée. Glacial erosion is made more effective by the process of freeze–thaw rock shattering (a form of physical weathering), which wea-

kens a rock surface and also provides material for abrasion.

glacial trough or *U-shaped valley* steep-sided, flat-bottomed valley formed by a glacier. The erosive action of the glacier and of the debris carried by it results in the formation not only of the trough itself but also of a number of associated features, such as truncated spurs (projections of rock that have been sheared off by the ice) and hanging valleys (smaller glacial valleys that enter the trough at a higher level than the trough floor). Features characteristic of glacial deposition, such as drumlins and eskers, are commonly found on the floor of the trough, together with linear lakes called ribbon lakes.

glacier tongue of ice, originating in mountains in snow-fields above the snowline, which moves slowly downhill and is constantly replenished from its source. The scenery produced by the erosive action of glaciers is characteristic and includes ◊glacial troughs (U-shaped valleys), ◊corries, and ◊arêtes. In lowlands, the laying down of ◊moraine (rocky debris once carried by glaciers) produces a variety of landscape features.

Glaciers form where annual snowfall exceeds annual melting and drainage. The snow compacts to ice under the weight of the layers above.

Under pressure the ice moves plastically (changing its shape permanently). When a glacier moves over an uneven surface, deep crevasses are formed in rigid upper layers of the ice mass; if it reaches the sea or a lake, it breaks up to form icebergs. A glacier that is formed by one or several valley glaciers at the base of a mountain is called a *piedmont* glacier. A body of ice that covers a large land surface or continent, for example Greenland or Antarctica, and flows outward in all directions is called an *ice sheet*.

In Oct 1996 a volcano erupted under Europe's largest glacier, Vatnajökull in Iceland, causing flooding.

Glacier

The longest glacier in the world is the Lambert Glacier in Australian Antarctic Territory. It is up to 64 km/40 mi wide and at least 400 km/250 mi long.

gland specialized organ of the body that manufactures and secretes enzymes, hormones, or other chemicals. In animals, glands vary in size from small (for example, tear glands) to large (for example, the pancreas), but in plants they are always small, and may consist of a single cell. Some glands discharge their products internally, ◊endocrine glands, and others, ◊exocrine glands, externally. Lymph nodes are sometimes wrongly called glands.

glandular fever or *infectious mononucleosis* viral disease characterized at onset by fever and painfully swollen lymph nodes; there may also be digestive upset, sore throat, and skin rashes. Lassitude persists for months and even years, and recovery can be slow. It is caused by the Epstein–Barr virus.

glass transparent or translucent substance that is physically neither a solid nor a liquid. Although glass is easily shattered, it is one of the strongest substances known. It is made by fusing certain types of sand (silica); this fusion occurs naturally in volcanic glass (see ◊obsidian).

In the industrial production of common types of glass, the type of sand used, the particular chemicals added to it (for example, lead, potassium, barium), and refinements of technique determine the type of glass produced. Types of glass include: soda glass; flint glass, used in cut-crystal ware; optical glass; stained glass; heat-resistant glass; and glasses that exclude certain ranges of the light spectrum. Blown glass is either blown individually from molten glass (using a tube up to 1.5 m/4.5 ft long), as in the making of expensive crafted glass, or blown automatically into a mould – for example, in the manufacture of light bulbs and bottles; pressed glass is simply pressed into moulds, for jam jars, cheap vases, and light fittings; while sheet glass, for windows, is made by putting the molten glass through rollers to form a 'ribbon', or by floating molten glass on molten tin in the 'float glass' process; ◊fibreglass is made from fine glass fibres. Metallic glass is produced by treating alloys so that they take on the properties of glass while retaining the malleability and conductivity characteristic of metals.

glass-reinforced plastic (GRP) a plastic material strengthened by glass fibres, sometimes erroneously called ◊fibreglass. Glass-reinforced plastic is a favoured material for boat hulls and for the bodies and some structural components of performance cars; it is also used in the manufacture of passenger cars.

Products are usually moulded, mats of glass fibre being sandwiched between layers of a polyester plastic, which sets hard when mixed with a curing agent.

Glauber's salt crystalline sodium sulphate decahydrate $Na_2SO_4.10H_2O$, produced by the action of sulphuric acid on common salt. It melts at 87.8°F/31°C; the latent heat stored as it solidifies makes it a convenient thermal energy store. It is used in medicine as a laxative.

glaucoma condition in which pressure inside the eye (intraocular pressure) is raised abnormally as excess fluid accumulates. It occurs when the normal outflow of fluid within the chamber of the eye (aqueous humour) is interrupted. As pressure rises, the optic nerve suffers irreversible damage, leading to a reduction in the field of vision and, ultimately, loss of eyesight.

The most common type, *chronic glaucoma*, usually affects people over the age of 40, when the trabecular meshwork (the filtering tissue at the margins of the eye) gradually becomes blocked and drainage slows down. The condition cannot be cured, but, in many cases, it is controlled by drug therapy. Laser treatment to the trabecular meshwork often improves drainage for a time; surgery to create an artificial channel for fluid to leave the eye offers more long-term relief. A tiny window may be cut in the iris during the same operation.

Acute glaucoma is a medical emergency. A precipitous rise in pressure occurs when the trabecular meshwork suddenly becomes occluded (blocked). This is treated surgically to remove the cause of the obstruction. Acute glaucoma is extremely painful. Treatment is required urgently since damage to the optic nerve begins within hours of onset.

global positioning system (GPS) US satellite-based navigation system, a network of satellites in six orbits, each circling the Earth once every 24 hours. Each satellite sends out a continuous time signal, plus an identifying signal. To fix position, a user needs to be within range of four satellites, one to provide a reference signal and three to provide directional bearings. The user's receiver can then calculate the position from the difference in time between receiving the signals from each satellite.

The position of the receiver can be calculated to within 10 m/33 ft, although only the US military can tap the full potential of the system. Other users can obtain a position to within 100 m/330 ft. This is accurate enough to be of use to boats, walkers, and motorists, and suitable receivers are on the market.

global warming climate change attributed to the ◊greenhouse effect. Greenhouse gases are warming the Earth's

Global warming

Marine algae help combat global warming by removing carbon dioxide from the atmosphere during photosynthesis. They remove 10 billion tonnes of carbon dioxide each year – more than all the land plants combined.

atmosphere and disastrous effects are predicted. These include a rise in global sea level, and resulting flooding of low-lying areas; fluctuations in temperature and precipitation, with droughts, heat waves, fires, and flooding; and the melting of mountain glaciers.

globular cluster spherical or near-spherical ◊star cluster containing from approximately 10,000 to millions of stars. More than a hundred globular clusters are distributed in a spherical halo around our Galaxy. They consist of old stars, formed early in the Galaxy's history. Globular clusters are also found around other galaxies.

glomerulus in the kidney, the cluster of blood capillaries at the threshold of the renal tubule, or nephron, responsible for filtering out the fluid that passes down the tubules and ultimately becomes urine. In the human kidney there are approximately one million tubules, each possessing its own glomerulus.

The structure of the glomerulus allows a wide range of substances including amino acids and sugar, as well as a large volume of water, to pass out of the blood. As the fluid moves through the tubules, most of the water and all of the sugars are reabsorbed, so that only waste remains, dissolved in a relatively small amount of water. This fluid collects in the bladder as urine.

glottis in medicine, narrow opening at the upper end of the larynx that contains the vocal cords.

glucagon in biology, a hormone secreted by the alpha cells of the islets of Langerhans in the ◊pancreas, which increases the concentration of glucose in the blood by promoting the breakdown of glycogen in the liver. Secretion occurs in response to a lowering of blood glucose concentrations.

Glucagon injections can be issued to close relatives of patients with ◊diabetes who are being treated with insulin. Hypoglycaemia may develop in such patients in the event of inadequate control of diabetes. An injection of glucagon can be used to reverse hypoglycaemia before serious symptoms, such as unconsciousness, develop.

glucose or *dextrose* or *grape sugar* $C_6H_{12}O_6$ sugar present in the blood and manufactured by green plants during ◊photosynthesis. The ◊respiration reactions inside cells involve the oxidation of glucose to produce ◊ATP, the 'energy molecule' used to drive many of the body's biochemical reactions.

In humans and other vertebrates optimum blood glucose levels are maintained by the hormone ◊insulin.

Glucose is prepared in syrup form by the hydrolysis of cane sugar or starch, and may be purified to a white crystalline powder. Glucose is a monosaccharide sugar (made up of a single sugar unit), unlike the more familiar sucrose (cane or beet sugar), which is a disaccharide (made up of two sugar units: glucose and fructose).

glucose tolerance test in medicine, a method of assessing the efficiency of the body to metabolize glucose, which is used in the diagnosis of ◊diabetes. It involves the starvation of the subject for several hours before giving a standard amount of glucose by mouth. Concentrations of glucose are measured in the blood and the urine over several hours.

glue type of ◊adhesive.

glue ear or *secretory otitis media* condition commonly affecting small children, in which the Eustachian tube, which normally drains and ventilates the middle ◊ear, becomes blocked with mucus. The resulting accumulation of mucus in the middle ear muffles hearing. It is the leading cause of deafness (usually transient) in children.

Glue ear resolves spontaneously after some months, but because the loss of hearing can interfere with a child's schooling the condition is often treated by a drainage procedure (myringotomy) and the surgical insertion of a small ventilating tube, or *grommet*, into the eardrum (tympanic membrane). This allows air to enter the middle ear, thereby enabling the mucus to drain freely once more along the Eustachian tube and into the back of the throat. The grommet is gradually extruded from the eardrum over several months, and the eardrum then heals naturally.

glue-sniffing or *solvent misuse* inhalation of the fumes from organic solvents of the type found in paints, lighter fuel, and glue, for their hallucinatory effects. As well as being addictive, solvents are dangerous for their effects on the user's liver, heart, and lungs. It is believed that solvents produce hallucinations by dissolving the cell membrane of brain cells, thus altering the way the cells conduct electrical impulses.

gluon in physics, a ◊gauge boson that carries the ◊strong nuclear force, responsible for binding quarks together to form the strongly interacting subatomic particles known as ◊hadrons. There are eight kinds of gluon.

Gluons cannot exist in isolation; they are believed to exist in balls ('glueballs') that behave as single particles.

gluten protein found in cereal grains, especially wheat and rye. Gluten enables dough to expand during rising. Sensitivity to gliadin, a type of gluten, gives rise to ◊coeliac disease.

glyceride ◊ester formed between one or more acids and glycerol (propan-1,2,3-triol). A glyceride is termed a mono-, di-, or triglyceride, depending on the number of hydroxyl groups from the glycerol that have reacted with the acids.

Glycerides, chiefly triglycerides, occur naturally as esters of ◊fatty acids in plant oils and animal fats.

glycerine another name for ◊glycerol.

glycerol or *glycerine* or *propan-1,2,3-triol* $HOCH_2CH(OH)CH_2OH$ thick, colourless, odourless, sweetish liquid. It is obtained from vegetable and animal oils and fats (by treatment with acid, alkali, superheated steam, or an enzyme), or by fermentation of glucose, and is used in the manufacture of high explosives, in antifreeze solutions, to maintain moist conditions in fruits and tobacco, and in cosmetics.

glycine $CH_2(NH_2)COOH$ the simplest amino acid, and one of the main components of proteins. When purified, it is a sweet, colourless crystalline compound.

Glycine was found in 1994 in the star-forming region Sagittarius B2. The discovery is important because of its bearing on the origins of life on Earth.

glycogen polymer (a polysaccharide) of the sugar ◊glucose made and retained in the liver as a carbohydrate store, for which reason it is sometimes called animal starch. It is a source of energy when needed by muscles, where it is converted back into glucose by the hormone ◊insulin and metabolized.

glycogen in biology, storage polysaccharide. Carbohydrates that are taken in as food are converted to glycogen and stored in the liver and muscles. Glycogen stores can be converted into glucose if the body requires additional sources of energy.

glycol or *ethylene glycol* or *ethane-1,2-diol* $HOCH_2CH_2OH$ thick, colourless, odourless, sweetish liquid. It is used in antifreeze solutions, in the preparation of ethers and esters (used for explosives), as a solvent, and as a substitute for glycerol.

glycoside in biology, compound containing a sugar and a non-sugar unit. Many glycosides occur naturally, for example, ◊digitalis is a preparation of dried and powdered foxglove leaves that contains a mixture of cardiac glycosides. One of its constituents, digoxin, is used in the treatment of congestive heart failure and cardiac arrhythmias.

GMT abbreviation for ◊ *Greenwich Mean Time*.

gneiss coarse-grained ◊metamorphic rock, formed under conditions of increasing temperature and pressure, and often occurring in association with schists and granites. It has a foliated, laminated structure, consisting of thin bands of micas and/or amphiboles alternating with granular bands of quartz and feldspar. Gneisses are formed during regional metamorphism; **paragneisses** are derived from sedimentary rocks and **orthogneisses** from igneous rocks. Garnets are often found in gneiss.

goblet cell in biology, cup-shaped cell present in the epithelium of the respiratory and gastrointestinal tracts. Goblet cells secrete mucin, the main constituent of mucous, which lubricates the mucous membranes of these tracts.

Goddard Space Flight Center NASA installation at Greenbelt, Maryland, USA, responsible for the operation of NASA's unmanned scientific satellites, including the ◊Hubble Space Telescope. It is also home of the National Space Science Data centre, a repository of data collected by satellites.

goitre enlargement of the thyroid gland seen as a swelling on the neck. It is most pronounced in simple goitre, which is caused by iodine deficiency. More common is toxic goitre or ◊hyperthyroidism, caused by overactivity of the thyroid gland.

gold heavy, precious, yellow, metallic element; symbol Au, atomic number 79, relative atomic mass 197.0. It is unaffected by temperature changes and is highly resistant to acids. For manufacture, gold is alloyed with another strengthening metal (such as copper or silver), its purity being measured in ◊carats on a scale of 24.

Gold occurs naturally in veins, but following erosion it can be transported and redeposited. It has long been valued for its durability, malleability, and ductility, and its uses include dentistry and jewellery. As it will not corrode, it is also used in the manufacture of electric contacts for computers and other electrical devices.

golden section visually satisfying ratio, first constructed by the Greek mathematician Euclid and used in art and architecture. It is found by dividing a line AB at a point O such that the rectangle produced by the whole line and one of the segments is equal to the square drawn on the other segment. The ratio of the two segments is about 8:13 or 1:1.618. A rectangle whose sides are in this ratio is called a **golden rectangle**. The ratio of consecutive Fibonacci numbers tends to the golden ratio.

In van Gogh's picture *Mother and Child*, for example, the Madonna's face fits perfectly into a golden rectangle.

gold rush large influx of gold prospectors to an area where gold deposits have recently been discovered. The result is a dramatic increase in population. Cities such as Johannesburg, Melbourne, and San Francisco either originated or were considerably enlarged by gold rushes. Melbourne's population trebled from 77,000 to some 200,000 between 1851 and 1853.

Golgi apparatus or *Golgi body* stack of flattened membranous sacs found in the cells of ◊eukaryotes. Many molecules travel through the Golgi apparatus on their way to other organelles or to the endoplasmic reticulum. Some

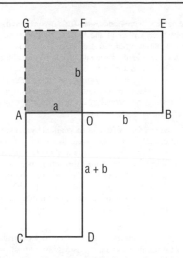

golden section The golden section is the ratio a:b, equal to 8:13. A golden rectangle is one, like that shaded in the picture, that has its length and breadth in this ratio. These rectangles are said to be pleasant to look at and have been used instinctively by artists in their pictures.

are modified or assembled inside the sacs. The Golgi apparatus is named after the Italian physician Camillo Golgi.

gonad the part of an animal's body that produces the sperm or egg cells (ova) required for sexual reproduction. The sperm-producing gonad is called a ◊testis, and the egg-producing gonad is called an ◊ovary.

gonadotrophin any hormone that supports and stimulates the function of the gonads (sex glands); some gonadotrophins are used as ◊fertility drugs.

Gondwanaland or *Gondwana* southern landmass formed 200 million years ago by the splitting of the single world continent ◊Pangaea. (The northern landmass was ◊Laurasia.) It later fragmented into the continents of South America, Africa, Australia, and Antarctica, which then drifted slowly to their present positions. The baobab tree found in both Africa and Australia is a relic of this ancient land mass.

A database of the entire geology of Gondwanaland has been constructed by geologists in South Africa. The database, known as Gondwana Geoscientific Indexing Database (GO-GEOID), displays information as a map of Gondwana 155 million years ago, before the continents drifted apart.

gonorrhoea common sexually transmitted disease arising from infection with the bacterium *Neisseria gonorrhoeae*, which causes inflammation of the genito-urinary tract. After an incubation period of two to ten days, infected men experience pain while urinating and a discharge from the penis; infected women often have no external symptoms.

Untreated gonorrhoea carries the threat of sterility to both sexes; there is also the risk of blindness in a baby born to an infected mother. The condition is treated with antibiotics, though ever-increasing doses are becoming necessary to combat resistant strains.

gorge narrow steep-sided valley (or canyon) that may or may not have a river at the bottom. A gorge may be formed as a ◊waterfall retreats upstream, eroding away the rock at the base of a river valley; or it may be caused by rejuvenation, when a river begins to cut downwards into its channel once again (for example, in response to a fall in sea level). Gorges are common in limestone country, where they may be formed by the collapse of the roofs of underground caverns.

gout hereditary form of ◊arthritis, marked by an excess of uric acid crystals in the tissues, causing pain and inflammation in one or more joints (usually the feet or hands). Acute attacks are treated with ◊anti-inflammatories.

The disease, ten times more common in men, poses a long-term threat to the blood vessels and the kidneys, so ongoing treatment may be needed to minimize the levels of uric acid in the bloodstream. It is aggravated by heavy drinking.

governor in engineering, any device that controls the speed of a machine or engine, usually by regulating the intake of fuel or steam.

Scottish inventor James Watt invented the steam-engine governor in 1788. It works by means of heavy balls, which rotate on the end of linkages and move in or out because of ◊centrifugal force according to the speed of rotation. The movement of the balls closes or opens the steam valve to the engine. When the engine speed increases too much, the balls fly out, and cause the steam valve to close, so the engine slows down. The opposite happens when the engine speed drops too much.

GP in medicine, abbreviation for *general practitioner*.

Graafian follicle fluid-filled capsule that surrounds and protects the developing egg cell inside the ovary during the ◊menstrual cycle. After the egg cell has been released, the follicle remains and is known as a corpus luteum.

graft in medicine, a piece of tissue that has been removed from one person and implanted either in another person or elsewhere on the same individual, to remedy a defect in that tissue. Skin grafts, for example, are used to treat patients with substantial burns. The skin is taken from an area of the body that has not been affected by the burn. An organ graft is often referred to as a ◊transplant.

Vein grafts can be used to replace parts of arteries in the heart and the lower limbs that have become blocked. Bone marrow grafting involves the transfusion of bone marrow from an unaffected individual to treat a patient with a haematological malignancy. Advances in chemotherapy have allowed the development of autologous bone marrow transplantation.

grafting in medicine, the operation by which an organ or other living tissue is removed from one organism and transplanted into the same or a different organism.

In horticulture, it is a technique widely used for propagating plants, especially woody species. A bud or shoot on one plant, termed the *scion*, is inserted into another, stock, so that they continue growing together, the tissues combining at the point of union. In this way some of the advantages of both plants are obtained.

Grafting is usually only successful between species that are closely related and is most commonly practised on roses and fruit trees. The grafting of nonwoody species is more difficult but it is sometimes used to propagate tomatoes and cacti. See also ◊transplant.

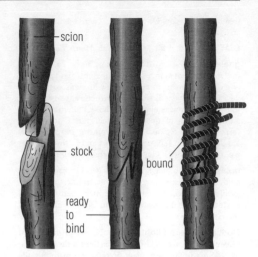

grafting Grafting, a method of artificial propagation in plants, is commonly used in the propagation of roses and fruit trees. A relatively small part, the scion, of one plant is attached to another plant so that growth continues. The plant receiving the transplanted material is called the stock.

grain the smallest unit of mass in the three English systems (avoirdupois, troy, and apothecaries' weights) used in the UK and USA, equal to 0.0648 g. It was reputedly the weight of a grain of wheat. One pound avoirdupois equals 7,000 grains; one pound troy or apothecaries' weight equals 5,760 grains.

gram metric unit of mass; one-thousandth of a kilogram.

grand unified theory in physics, a sought-for theory that would combine the theory of the strong nuclear force (called ◊quantum chromodynamics) with the theory of the weak nuclear and electromagnetic forces. The search for the grand unified theory is part of a larger programme seeking a ◊unified field theory, which would combine all the forces of nature (including gravity) within one framework.

granite coarse-grained intrusive ◊igneous rock, typically consisting of the minerals quartz, feldspar, and mica. It may be pink or grey, depending on the composition of the feldspar. Granites are chiefly used as building materials.

Granites often form large intrusions in the core of mountain ranges, and they are usually surrounded by zones of ◊metamorphic rock (rock that has been altered by heat or pressure). Granite areas have characteristic moorland scenery. In exposed areas the bedrock may be weathered along joints and cracks to produce a *tor*, consisting of rounded blocks that appear to have been stacked upon one another.

graph pictorial representation of numerical data, such as statistical data, or a method of showing the mathematical relationship between two or more variables by drawing a diagram.

There are often two axes, or reference lines, at right angles intersecting at the origin – the zero point, from which values of the variables (for example, distance and time for a moving object) are assigned along the axes. Pairs of simultaneous values (the distance moved after a

graph of a straight line

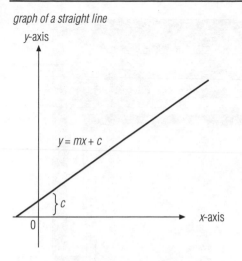

the equation of the straight-line graph takes the form $y = mx + c$, where m is the gradient (slope) of the line, and c is the y-intercept (the value of y where the line cuts the y-axis)for example, a graph of the equation $y = x4$ will have a gradient of -1 and will cut the y-axis at $y = 4$

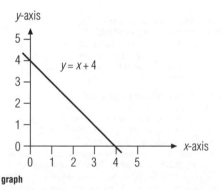

graph

particular time) are plotted as points in the area between the axes, and the points then joined by a smooth curve to produce a graph.

graphical user interface (GUI) or **WIMP** in computing, a type of ◊user interface in which programs and files appear as icons (small pictures), user options are selected from pull-down menus, and data are displayed in windows (rectangular areas), which the operator can manipulate in various ways. The operator uses a pointing device, typically a ◊mouse, to make selections and initiate actions. The concept of the graphical user interface was developed by the Xerox Corporation in the 1970s, was popularized with the Apple Macintosh computers in the 1980s, and is now available on many types of computer – most notably as Windows, an operating system for IBM PC-compatible microcomputers developed by the software company Microsoft. *See illustration overleaf.*

graphic equalizer control used in hi-fi systems that allows the distortions introduced by unequal amplification of different frequencies to be corrected.

The frequency range of the signal is divided into separate bands, usually third-octave bands. The amplification applied to each band is adjusted by a sliding contact; the position of the contact indicates the strength of the amplification applied to each frequency range.

graphics tablet or **bit pad** in computing, an input device in which a stylus or cursor is moved, by hand, over a flat surface. The computer can keep track of the position of the stylus, so enabling the operator to input drawings or diagrams into the computer.

A graphics tablet is often used with a form overlaid for users to mark boxes in positions that relate to specific registers in the computer, although recent developments in handwriting recognition may increase its future versatility.

graphite blackish-grey, laminar, crystalline form of ◊carbon. It is used as a lubricant and as the active component of pencil lead.

The carbon atoms are strongly bonded together in sheets, but the bonds between the sheets are weak, allowing other atoms to enter regions between the layers causing them to slide over one another. Graphite has a very high melting point (3,500°C/6,332°F), and is a good conductor of heat and electricity. It absorbs neutrons and is therefore used to moderate the chain reaction in nuclear reactors.

graph plotter alternative name for a ◊plotter.

grass any of a very large family of plants, many of which are economically important because they provide grazing for animals and food for humans in the form of cereals. There are about 9,000 species distributed worldwide

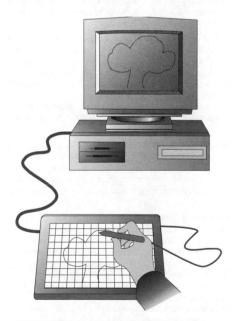

graphics tablet A graphics tablet enables images drawn freehand to be translated directly to the computer screen.

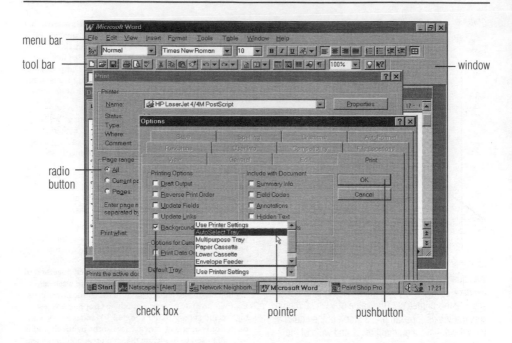

menu bar

tool bar

window

radio button

check box

pointer

pushbutton

graphical user interface A typical GUI, where the user is taken around the system by simply clicking on representative buttons or icons using the mouse.

except in the Arctic regions. Most are perennial, with long, narrow leaves and jointed, hollow stems; flowers with both male and female reproductive organs are borne on spikelets; the fruits are grainlike. Included in the family are bluegrass, wheat, rye, maize, sugarcane, and bamboo. (Family Gramineae.)

gravel coarse ◊sediment consisting of pebbles or small fragments of rock, originating in the beds of lakes and streams or on beaches. Gravel is quarried for use in road building, railway ballast, and for an aggregate in concrete. It is obtained from quarries known as gravel pits, where it is often found mixed with sand or clay.

Some gravel deposits also contain ◊placer deposits of metal ores (chiefly tin) or free metals (such as gold and silver).

gravimetric analysis in chemistry, a technique for determining, by weighing, the amount of a particular substance present in a sample. It usually involves the conversion of the test substance into a compound of known molecular weight that can be easily isolated and purified.

gravimetry study of the Earth's gravitational field. Small variations in the gravitational field (gravimetric anomalies) can be caused by varying densities of rocks and structure beneath the surface. Such variations are measured by a device called a gravimeter, which consists of a weighted spring that is pulled further downwards where the gravity is stronger (at a ◊Bouguer anomaly). Gravimetry is used by geologists to map the subsurface features of the Earth's crust, such as underground masses of heavy rock like granite, or light rock like salt.

gravitational field the region around a body in which

other bodies experience a force due to its gravitational attraction. The gravitational field of a massive object such as the Earth is very strong and easily recognized as the force of gravity, whereas that of an object of much smaller mass is very weak and difficult to detect. Gravitational fields produce only attractive forces.

gravitational field strength (symbol g) the strength of the Earth's gravitational field at a particular point. It is defined as the the gravitational force in newtons that acts on a mass of one kilogram. The value of g on the Earth's surface is taken to be 9.806 N kg^{-1}.

The symbol g is also used to represent the acceleration of a freely falling object in the Earth's gravitational field.

Near the Earth's surface and in the absence of friction due to the air, all objects fall with an acceleration of 9.806 m s^{-2}.

gravitational lensing bending of light by a gravitational field, predicted by Einstein's general theory of relativity. The effect was first detected in 1917 when the light from stars was found to be bent as it passed the totally eclipsed Sun. More remarkable is the splitting of light from distant quasars into two or more images by intervening galaxies. In 1979 the first double image of a quasar produced by gravitational lensing was discovered and a quadruple image of another quasar was later found.

graviton in physics, the ◊gauge boson that is the postulated carrier of the gravitational force.

gravity force of attraction that arises between objects by virtue of their masses. On Earth, gravity is the force of attraction between any object in the Earth's gravitational field and the Earth itself. It is regarded as one of the four

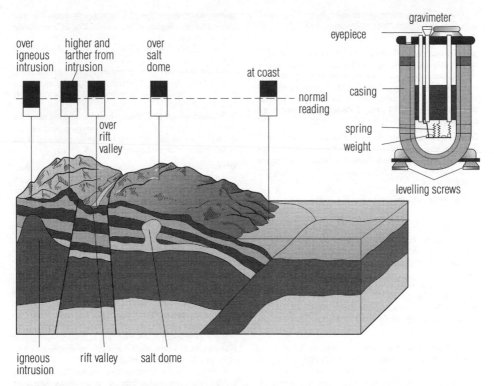

over igneous intrusion

higher and farther from intrusion

over salt dome

at coast

over rift valley

normal reading

gravimeter

eyepiece

casing

spring

weight

levelling screws

igneous intrusion

rift valley

salt dome

gravimetry The gravimeter is an instrument for measuring the force of gravity at a particular location. Variations in the force of gravity acting on a weight suspended by a spring cause the spring to stretch. The gravimeter is used in aerial surveys. Geological features such as intrusions and salt domes are revealed by the stretching of the spring.

Were it not for gravity one man might hurl another by a puff of his breath into the depths of space, beyond recall for all eternity.

On **gravity** Ruggerio Giuseppe Boscovich
Theoria

fundamental ◊forces of nature, the other three being the ◊electromagnetic force, the ◊strong nuclear force, and the ◊weak nuclear force. The gravitational force is the weakest of the four forces, but it acts over great distances. The particle that is postulated as the carrier of the gravitational force is the ◊graviton.

An experiment for determining the force of attraction between two masses was first planned in the mid-18th century by the Reverend J Mitchell, who did not live to work on the apparatus he had designed and completed. After Mitchell's death the apparatus came into the hands of Henry Cavendish, who largely reconstructed it but kept to Mitchell's original plan. The attracted masses consisted of two small balls, connected by a stiff wooden beam suspended at its middle point by a long, fine wire. The whole of this part of the apparatus was enclosed in a case, carefully coated with tinfoil to secure, as far as possible, a uniform temperature within the case. Irregular

distribution of temperature would have resulted in convection currents of air which would have had a serious disturbing effect on the suspended system. To the beam was attached a small mirror with its plane vertical. A small glazed window in the case allowed any motion of the mirror to be observed by the consequent deviations of a ray of light reflected from it. The attracting masses consisted of two equal, massive lead spheres. Using this apparatus, Cavendish, in 1797, obtained for the gravitational constant G the value 6.6×10^{-11} N m^2 kg^{-2}. The apparatus was refined by Charles Vernon Boys and he obtained the improved value 6.6576×10^{-11} N m^2 kg^{-2}. The value generally used today is 6.6720×10^{-11} N m^2 kg^{-2}.

Gravity

The maximum speed with which a falling raindrop can hit you is about 29 kph/18 mph. In a vacuum, the further an object falls, the more speed it gains, but in the real world, air resistance eventually balances out the accelerating effect of gravity.

gravure one of the three main ◊printing methods, in which printing is done from a plate etched with a pattern of recessed cells in which the ink is held. The greater the depth of a cell, the greater the strength of the printed ink.

Gravure plates are expensive to make, but the process is economical for high-volume printing and reproduces illustrations well.

gray SI unit (symbol Gy) of absorbed radiation dose. It replaces the rad (1 Gy equals 100 rad), and is defined as the dose absorbed when one kilogram of matter absorbs one joule of ionizing radiation. Different types of radiation cause different amounts of damage for the same absorbed dose; the SI unit of *dose equivalent* is the ◊sievert.

Great Artesian Basin the largest area of artesian water in the world. It underlies much of Queensland, New South Wales, and South Australia, and in prehistoric times formed a sea. It has an area of 1,750,000 sq km/ 676,250 sq mi.

Great Bear popular name for the constellation ◊Ursa Major.

great circle circle drawn on a sphere such that the diameter of the circle is a diameter of the sphere. On the Earth, all meridians of longitude are half great circles; among the parallels of latitude, only the Equator is a great circle.

The shortest route between two points on the Earth's surface is along the arc of a great circle. These are used extensively as air routes although on maps, owing to the distortion brought about by ◊projection, they do not appear as straight lines.

Great Red Spot prominent oval feature, 14,000 km/ 8,500 mi wide and some 30,000 km/20,000 mi long, in the atmosphere of the planet ◊Jupiter, S of the Equator. It was first observed in the 19th century. Space probes show it to be an anticlockwise vortex of cold clouds, coloured possibly by phosphorus.

green audit inspection of a company to assess the total environmental impact of its activities or of a particular product or process.

For example, a green audit of a manufactured product looks at the impact of production (including energy use and the extraction of raw materials used in manufacture), use (which may cause pollution and other hazards), and disposal (potential for recycling, and whether waste causes pollution).

Such 'cradle-to-grave' surveys allow a widening of the traditional scope of economics by ascribing costs to variables that are usually ignored, such as despoilation of the countryside or air pollution.

greenhouse effect phenomenon of the Earth's atmosphere by which solar radiation, trapped by the Earth and re-emitted from the surface, is prevented from escaping by various gases in the air. The result is a rise in the Earth's temperature (◊global warming. The main greenhouse gases are carbon dioxide, methane, and ◊chlorofluorocarbons (CFCs). Fossil-fuel consumption and forest fires are the main causes of carbon dioxide build-up; methane is a byproduct of agriculture (rice, cattle, sheep). Water vapour is another greenhouse gas.

The United Nations Environment Programme estimates that by the year 2025, average world temperatures will have risen by 1.5°C with a consequent rise of 20 cm in sea level. Low-lying areas and entire countries would be threatened by flooding and crops would be affected by the change in climate. However, predictions about global warming and its possible climatic effects are tentative and often conflict with each other.

At the 1992 Earth Summit it was agreed that by the year 2000 countries would stabilize carbon dioxide emissions at 1990 levels, but to halt global warming, emissions would probably need to be cut by 60%. Any increases in

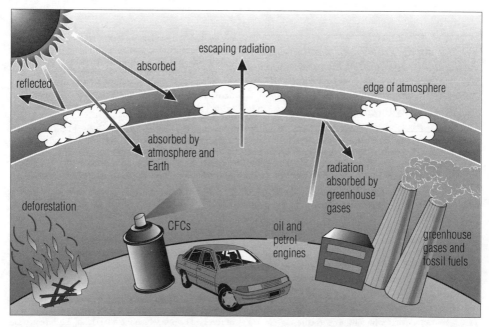

greenhouse effect

carbon dioxide emissions are expected to come from transport. The Berlin Mandate, agreed unanimously at the climate conference in Berlin in April 1995, committed industrial nations to the continuing reduction of greenhouse gas emissions after the year 2000, when the existing pact to stabilize emissions runs out. The stabilization of carbon dioxide emissions at 1990 levels by the year 2000 will not be achieved by a number of developed countries, including Spain, Australia, and the USA, according to 1997 estimates.

Dubbed the 'greenhouse effect' by Swedish scientist Svante Arrhenius, it was first predicted in 1827 by French mathematician Joseph Fourier.

green monkey disease another name for ◊Marburg disease, a virus originating in Central Africa.

green movement collective term for the individuals and organizations involved in efforts to protect the environment. The movement encompasses political parties such as the Green Party and organizations like ◊Friends of the Earth and ◊Greenpeace.

Despite a rapid growth of public support, and membership of environmental organizations running into many millions worldwide, political green groups have failed to win significant levels of the vote in democratic societies.

Greenpeace international environmental pressure group, founded in 1971, with a policy of nonviolent direct action backed by scientific research. During a protest against French atmospheric nuclear testing in the S Pacific in 1985, its ship *Rainbow Warrior* was sunk by French intelligence agents, killing a crew member.

green revolution in agriculture, the change in methods of arable farming instigated in the 1940s and 1950s in Third World countries. The intent was to provide more and better food for their populations, albeit with a heavy reliance on chemicals and machinery. It was abandoned by some countries in the 1980s.

The green revolution tended to benefit primarily those land-owners who could afford the investment necessary for such intensive agriculture. Without a dosage of 70–90 kg/154–198 lb of expensive nitrogen fertilizers per hectare, the high-yield varieties will not grow properly. Hence, rich farmers tend to obtain bigger yields while smallholders are unable to benefit from the new methods.

In terms of production, the green revolution was initially successful in SE Asia; India doubled its wheat yield in 15 years, and the rice yield in the Philippines rose by 75%. However, yields have levelled off in many areas; some countries that cannot afford the dams, fertilizers, and machinery required, have adopted ◊intermediate technologies.

Greenwich Mean Time (GMT) local time on the zero line of longitude (the *Greenwich meridian*), which passes through the Old Royal Observatory at Greenwich, London. It was replaced in 1986 by coordinated universal time (UTC), but continued to be used to measure longitudes and the world's standard time zones; see ◊time.

grey matter in biology, those parts of the brain and spinal cord that are made up of interconnected and tightly packed nerve cell nucleuses. The outer layers of the cerebellum contains most of the grey matter in the brain. It is the region of the brain that is responsible for advanced mental functions. Grey matter also constitutes the inner core of the spinal cord. This is in contrast to white matter, which is made of the axons of nerve cells.

grid network by which electricity is generated and distributed over a region or country. It contains many power stations and switching centres and allows, for example, high demand in one area to be met by surplus power generated in another.

The term is also used for any grating system, as in a cattle grid for controlling the movement of livestock across roads, and a conductor in a storage battery or electron gun.

grid reference on a map, numbers that are used to show location. The numbers at the bottom of the map (eastings) are given before those at the side (northings).

grooming in biology, the use by an animal of teeth, tongue, feet, or beak to clean fur or feathers. Grooming also helps to spread essential oils for waterproofing. In many social species, notably monkeys and apes, grooming of other individuals is used to reinforce social relationships.

ground water water collected underground in porous rock strata and soils; it emerges at the surface as springs and streams. The groundwater's upper level is called the *water table*. Sandy or other kinds of beds that are filled with groundwater are called *aquifers*. Recent estimates are that usable ground water amounts to more than 90% of all the fresh water on Earth; however, keeping such supplies free of pollutants entering the recharge areas is a critical environmental concern.

Most groundwater near the surface moves slowly through the ground while the water table stays in the same place. The depth of the water table reflects the balance between the rate of infiltration, called recharge, and the rate of discharge at springs or rivers or pumped water wells. The force of gravity makes underground water run 'downhill' underground just as it does above the surface. The greater the slope and the permeability, the greater the speed. Velocities vary from 100 cm/40 in per day to 0.5 cm/0.2 in.

group in chemistry, a vertical column of elements in the ◊periodic table. Elements in a group have similar physical and chemical properties; for example, the group I elements (the alkali metals: lithium, sodium, potassium, rubidium, caesium, and francium) are all highly reactive metals that form univalent ions. There is a gradation of properties down any group: in group I, melting and boiling points decrease, and density and reactivity increase.

group in mathematics, a finite or infinite set of elements that can be combined by an operation; formally, a group must satisfy certain conditions. For example, the set of all integers (positive or negative whole numbers) forms a group with regard to addition because: (1) addition is associative, that is, the sum of two or more integers is the same regardless of the order in which the integers are added; (2) adding two integers gives another integer; (3) the set includes an identity element 0, which has no effect on any integer to which it is added (for example, $0 + 3 = 3$); and (4) each integer has an inverse (for instance, 7 has the inverse -7), such that the sum of an integer and its inverse is 0. *Group theory* is the study of the properties of groups.

growth ring another name for ◊annual ring.

Green Movement: Chronology

1798	Thomas Malthus's *Essay on the Principle of Population* published, setting out the idea that humans are also bound by ecological constraints.
1824	Society for the Prevention of Cruelty to Animals founded.
1835–39	Droughts and famine in India resulted in the first connections being made between environmental damage (deforestation) and climate change.
1864	George Marsh's '*Man and Nature*' was the first comprehensive study of humans' impact on the environment.
1865	Commons Preservation Society founded, raising the issue of public access to the countryside, taken further by the mass trespasses of the 1930s.
1872	Yellowstone National Park created in the USA; a full system of national parks was established 40 years later.
1893	National Trust founded in the UK to buy land in order to preserve places of natural beauty and cultural landmarks.
1930	Chlorofluorocarbons (CFCs) invented; they were hailed as a boon for humanity as they were not only cheap and nonflammable but were also thought not to be harmful to the environment.
1934	Drought exacerbated soil erosion, causing the 'Dust Bowl Storm' in USA, during which some 350 million tons of topsoil were blown away.
1948	United Nations created special environmental agency, the International Union for the Conservation of Nature (IUCN).
1952	Air pollution caused massive smog in London, killing some 4,000 people and leading to clean-air legislation.
1968	Garret Hardin's essay '*The Tragedy of the Commons*' challenged individuals to recognize their personal reponsibility for environmental degradation as a result of lifestyle choices.
1969	Friends of the Earth launched in USA as a breakaway group from increasingly conservative Sierra Club; there was an upsurge of more radical active groups within the environmental movement over the following years.
1972	*Blueprint for Survival*, a detailed analysis of the human race's ecological predicament and proposed solutions, published in UK by Teddy Goldsmith and others from the *Ecologist* magazine.
1974	First scientific warning of serious depletion of protective ozone layer in upper atmosphere by CFCs.
1980	US president Jimmy Carter commissioned *Global 2000* report, reflecting entry of environmental concerns into mainstream of political issues.
1983	German Greens (Die Grünen) won 5% of vote, giving them 27 seats in the Bundestag.
1985	Greenpeace boat *Rainbow Warrior* sunk by French intelligence agents in a New Zealand harbour during a protest against French nuclear testing in the South Pacific. One crew member was killed.
1988	NASA scientist James Hansen warned US Congress of serious danger of global warming: 'The greenhouse effect is here'.
1989	European elections put green issues firmly on political agenda as Green parties across Europe attracted unprecedented support; especially in the UK, where the Greens received some 15% of votes cast (though not of seats).
1989	*The Green Consumer Guide* published in the UK, one of many such books worldwide advocating 'green consumerism'.
1991	The Gulf War had massive environmental consequences, primarily as a result of the huge quantity of oil discharged into the Persian Gulf from Kuwait's oilfields. Many felt this was just the start of a grim future of wars over ever decreasing resources.
1992	United Nations Earth Summit in Rio de Janeiro aroused great media interest but achieved little progress in tackling difficult global environmental issues because many nations feared possible effects on trade.
1994	Protests against roadbuilding in many parts of the UK; for example, in 'Battle of Wanstonia', green activists occupied buildings and trees in East London in attempt to halt construction of M11 motorway.
1995	Animal-rights activists campaigned against the export of live animals; activist Jill Phipps was killed 1 Feb during a protest at Coventry airport. In May 1995 Greenpeace's London headquarters were raided by the Ministry of Defence and files and computer discs were confiscated.
1996	A new political force, Real World, was formed from a coalition of 32 campaigning charities and pressure groups.

groyne wooden or concrete barrier built at right angles to a beach in order to block the movement of material along the beach by longshore drift. Groynes are usually successful in protecting individual beaches, but because they prevent beach material from passing along the coast they can mean that other beaches, starved of sand and shingle, are in danger of being eroded away by the waves. This happened, for example, at Barton-on-Sea in Hampshire, England, in the 1970s, following the construction of a large groyne at Bournemouth.

GRP abbreviation for ◊ *glass-reinforced plastic*.

g-scale scale for measuring force by comparing it with the force due to ◊gravity (*g*), often called ◊g-force.

guano dried excrement of fish-eating birds that builds up under nesting sites. It is a rich source of nitrogen and phosphorous, and is widely collected for use as fertilizer. Some 80% comes from the sea cliffs of Peru.

guard cell in plants, a specialized cell on the undersurface of leaves for controlling gas exchange and water loss. Guard cells occur in pairs and are shaped so that a pore, or ◊stomata, exists between them. They can change shape with the result that the pore disappears. During warm weather, when a plant is in danger of losing excessive water, the guard cells close, cutting down evaporation from the interior of the leaf.

Gulf Stream warm ocean ◊current that flows north from the warm waters of the Gulf of Mexico. Part of the current is diverted east across the Atlantic, where it is known as the *North Atlantic Drift*, and warms what would otherwise be a colder climate in the British Isles and NW Europe.

gum in botany, complex polysaccharides (carbohydrates) formed by many plants and trees, particularly by those from dry regions. They form four main groups: plant exudates (gum arabic); marine plant extracts (agar); seed extracts; and fruit and vegetable extracts. Some are made synthetically.

Gums are tasteless and odourless, insoluble in alcohol and ether but generally soluble in water. They are used for adhesives, fabric sizing, in confectionery, medicine, and calico printing.

gum in mammals, the soft tissues surrounding the base of the teeth. Gums are liable to inflammation (gingivitis) or to infection by microbes from food deposits (periodontal disease).

gun any kind of firearm or any instrument consisting of a metal tube from which a projectile is discharged; see also ◊pistol and ◊small arms.

gun metal type of ◊bronze, an alloy high in copper (88%), also containing tin and zinc, so-called because it was once used to cast cannons. It is tough, hard-wearing, and resists corrosion.

gunpowder or *black powder* the oldest known ◊explosive, a mixture of 75% potassium nitrate (saltpetre), 15% charcoal, and 10% sulphur. Sulphur ignites at a low temperature, charcoal burns readily, and the potassium nitrate provides oxygen for the explosion. As gunpowder produces lots of smoke and burns quite slowly, it has progressively been replaced since the late 19th century by high explosives, although it is still widely used for quarry blasting, fuses, and fireworks. Gunpowder has high ◊activation energy; a gun based on gunpowder alone requires igniting by a flint or a match.

Gunpowder is believed to have been invented in China in the 10th century, but may also have been independently discovered by the Arabs. Certainly the Arabs produced the first known working gun in 1304. Gunpowder was used in warfare from the 14th century but it was not generally adapted to civil purposes until the 17th century, when it began to be used in mining.

gut or *alimentary canal* In the ◊digestive system, the part of an animal responsible for processing food and preparing it for entry into the blood.

The gut consists of a tube divided into segments specialized to perform different functions. The front end (the mouth) is adapted for food intake and for the first stages of digestion. The stomach is a storage area, although digestion of protein by the enzyme pepsin starts here; in many herbivorous mammals this is also the site of cellulose digestion. The small intestine follows the stomach and is specialized for digestion and for absorption. The large intestine, consisting of the colon, caecum, and rectum, has a variety of functions, including cellulose digestion, water absorption, and storage of faeces. From the gut nutrients are carried to the liver via the hepatic portal vein, ready for assimilation by the cells.

guttation secretion of water on to the surface of leaves through specialized pores, or ◊hydathodes. The process occurs most frequently during conditions of high humidity when the rate of transpiration is low. Drops of water found on grass in early morning are often the result of guttation, rather than dew. Sometimes the water contains minerals in solution, such as calcium, which leaves a white crust on the leaf surface as it dries.

gymnosperm in botany, any plant whose seeds are exposed, as opposed to the structurally more advanced ◊angiosperms, where they are inside an ovary. The group includes conifers and related plants such as cycads and ginkgos, whose seeds develop in ◊cones. Fossil gymnosperms have been found in rocks about 350 million years old.

gynaecology medical speciality concerned with disorders of the female reproductive system.

gynaecomastia in medicine, an abnormal increase in the size of the male breast. It is caused by abnormalities in levels of the male sex hormones. Certain drugs, such as high doses of cimetidine, can cause gynaecomastia occasionally.

gynoecium or *gynaecium* collective term for the female reproductive organs of a flower, consisting of one or more ◊carpels, either free or fused together.

gypsum common sulphate ◊mineral, composed of hydrous calcium sulphate, $CaSO_4.2H_2O$. It ranks 2 on the Mohs' scale of hardness. Gypsum is used for making casts and moulds, and for blackboard chalk.

A fine-grained gypsum, called *alabaster*, is used for ornamental work. Burned gypsum is known as *plaster of Paris*, because for a long time it was obtained from the gypsum quarries of the Montmartre district of Paris.

gyre circular surface rotation of ocean water in each major sea (a type of ◊current). Gyres are large and permanent, and occupy the N and S halves of the three major oceans. Their movements are dictated by the prevailing winds and the ◊Coriolis effect. Gyres move clockwise in the northern hemisphere and anticlockwise in the southern hemisphere.

gyroscope mechanical instrument, used as a stabilizing device and consisting, in its simplest form, of a heavy wheel mounted on an axis fixed in a ring that can be rotated about another axis, which is also fixed in a ring capable of rotation about a third axis. Applications of the gyroscope principle include the gyrocompass, the gyropilot for automatic steering, and gyro-directed torpedoes.

The components of the gyroscope are arranged so that the three axes of rotation in any position pass through the wheel's centre of gravity. The wheel is thus capable of rotation about three mutually perpendicular axes, and its axis may take up any direction. If the axis of the spinning wheel is displaced, a restoring movement develops, returning it to its initial direction.

ha symbol for ◊*hectare*.

Haber process or *Haber–Bosch process* industrial process by which ammonia is manufactured by direct combination of its elements, nitrogen and hydrogen. The reaction is carried out at 400–500°C/752–932°F and at 200 atmospheres pressure. The two gases, in the proportions of 1:3 by volume, are passed over a ◊catalyst of finely divided iron.

Around 10% of the reactants combine, and the unused gases are recycled. The ammonia is separated either by being dissolved in water or by being cooled to liquid form.

$$N_2 + 3H_2 \rightleftharpoons 2NH_3$$

habitat localized ◊environment in which an organism lives, and which provides for all (or almost all) of its needs. The diversity of habitats found within the Earth's ecosystem is enormous, and they are changing all the time. Many can be considered inorganic or physical; for example, the Arctic ice cap, a cave, or a cliff face. Others are more complex; for instance, a woodland or a forest floor. Some habitats are so precise that they are called **microhabitats**, such as the area under a stone where a particular type of insect lives. Most habitats provide a home for many species.

hacking gaining unauthorized access to a computer, either for fun or for malicious or fraudulent purposes. Hackers generally use microcomputers and telephone lines to obtain access. In computing, the term is used in a wider sense to mean using software for enjoyment or self-education, not necessarily involving unauthorized access. The most destructive form of hacking is the introduction of a computer ◊virus.

Hacking can be divided into four main areas: ◊viruses, phreaking (tricking the phone system, eg. to get free calls), software piracy (stripping away the protective coding that should prevent the software being copied), and accessing operating systems.

A 1996 US survey co-sponsored by the FBI showed 41% of academic, corporate, and government organizations interviewed had had their computer systems hacked into during 1995.

hadal zone the deepest level of the ocean, below the abyssal zone, at depths of greater than 6,000 m/19,500 ft. The ocean trenches are in the hadal zone. There is no light in this zone and pressure is over 600 times greater than atmospheric pressure.

hadron in physics, a subatomic particle that experiences the strong nuclear force. Each is made up of two or three indivisible particles called ◊quarks. The hadrons are grouped into the ◊baryons (protons, neutrons, and hyperons) and the ◊mesons (particles with masses between those of electrons and protons).

haematology medical speciality concerned with disorders of the blood.

haemoglobin protein used by all vertebrates and some invertebrates for oxygen transport because the two substances combine reversibly. In vertebrates it occurs in red blood cells (erythrocytes), giving them their colour.

In the lungs or gills where the concentration of oxygen is high, oxygen attaches to haemoglobin to form **oxy-haemoglobin**. This process effectively increases the amount of oxygen that can be carried in the bloodstream. The oxygen is later released in the body tissues where it is at a low concentration, and the deoxygenated blood returned to the lungs or gills. Haemoglobin will also combine with carbon monoxide to form carboxyhaemoglobin, but in this case the reaction is irreversible.

Haemoglobin

A healthy person has 6,000 million million million haemoglobin molecules, with 400 million million being destroyed every second and replaced by new ones.

---

haemolymph circulatory fluid of those molluscs and insects that have an 'open' circulatory system. Haemolymph contains water, amino acids, sugars, salts, and white cells like those of blood. Circulated by a pulsating heart, its main functions are to transport digestive and excretory products around the body. In molluscs, it also transports oxygen and carbon dioxide.

haemolysis destruction of red blood cells. Aged cells are constantly being lysed (broken down), but increased wastage of red cells is seen in some infections and blood disorders. It may result in ◊jaundice (through the release of too much haemoglobin) and in ◊anaemia.

haemolytic anaemia in medicine, type of ◊anaemia caused by the excessive destruction of red blood cells. A form of hereditary haemolytic anaemia prevalent in populations originating from Africa, Asia, and Southern Europe occurs because of a lack of an enzyme known as glucose 6-phosphate dehydrogenase (G6PD). This results in the red blood cells being excessively fragile. Individuals with G6PD deficiency may develop acute haemolytic anaemia following the ingestion of fava beans and several types of antibiotics.

haemophilia any of several inherited diseases in which normal blood clotting is impaired. The sufferer experiences prolonged bleeding from the slightest wound, as well as painful internal bleeding without apparent cause.

Haemophilias are nearly always sex-linked, transmitted through the female line only to male infants; they have afflicted a number of European royal households. Males affected by the most common form are unable to synthesize Factor VIII, a protein involved in the clotting of blood. Treatment is primarily with Factor VIII (now mass-produced by recombinant techniques), but the haemophiliac remains at risk from the slightest incident of bleeding. The disease is a painful one that causes deformities of joints.

haemorrhage loss of blood from the circulatory system. It is 'manifest' when the blood can be seen, as when it flows from a wound, and 'occult' when the bleeding is internal, as from an ulcer or internal injury.

Rapid, profuse haemorrhage causes ◊shock and may prove fatal if the circulating volume cannot be replaced in time. Slow, sustained bleeding may lead to ◊anaemia. Arterial bleeding is potentially more serious than blood loss from a vein. It may be stemmed by applying pressure directly to the wound.

haemorrhagic fever any of several virus diseases of the tropics in which high temperatures over several days end in haemorrhage from nose, throat, and intestines, with up to 90% mortality.

Haemorrhagic fever viruses belong to a number of different families including flaviviruses (yellow fever and dengue), arenaviruses (Lassa fever), bunyaviruses (Rift Valley fever and Crimean–Congo fever), hantaviruses (haemorrhagic fever with renal syndrome), and filoviruses (Ebola virus disease and Marburg disease).

The causative organism of W African ◊Lassa fever lives in rats (which betray no symptoms), but in ◊Marburg disease and Ebola virus disease the host animal is the green monkey. See also ◊dengue fever.

haemorrhoids distended blood vessels (◊varicose veins) in the area of the anus, popularly called *piles*.

haemostasis natural or surgical stoppage of bleeding. In the natural mechanism, the damaged vessel contracts, restricting the flow, and blood ◊platelets plug the opening, releasing chemicals essential to clotting.

hafnium silvery, metallic element, symbol Hf, atomic number 72, relative atomic mass 178.49. It occurs in nature in ores of zirconium, the properties of which it resembles. Hafnium absorbs neutrons better than most metals, so it is used in the control rods of nuclear reactors; it is also used for light-bulb filaments.

It was named in 1923 by Dutch physicist Dirk Coster (1889–1950) and Hungarian chemist Georg von Hevesy after the city of Copenhagen, where the element was discovered.

hail precipitation in the form of pellets of ice (hailstones). It is caused by the circulation of moisture in strong convection currents, usually within cumulonimbus ◊clouds.

Water droplets freeze as they are carried upwards. As the circulation continues, layers of ice are deposited around the droplets until they become too heavy to be supported by the currents and they fall as a hailstorm.

Hail

Hailstones can kill. In the Gopalganji region of Bangladesh in 1988, 92 people died after being hit by huge hailstones weighing up to 1 kg/2.2 lb.

---

hair fine filament growing from mammalian skin. Each hair grows from a pit-shaped follicle embedded in the second layer of the skin, the dermis. It consists of dead cells impregnated with the protein keratin.

The average number of hairs on a human head varies from 98,000 (red-heads) to 120,000 (blondes). Each grows at the rate of 5–10 mm/0.2–0.4 in per month, lengthening for about three years before being replaced

by a new one. A coat of hair helps to insulate land mammals by trapping air next to the body. The thickness of this layer can be varied at will by raising or flattening the coat. In some mammals a really heavy coat may be so effective that it must be shed in summer and a thinner one grown. Hair also aids camouflage, as in the zebra and the white winter coats of Arctic animals; and protection, as in the porcupine and hedgehog; bluffing enemies by apparently increasing the size, as in the cat; sexual display, as in humans and the male lion; and its colouring or erection may be used for communication. In 1990 scientists succeeded for the first time in growing human hair in vitro.

Hale-Bopp, Comet see ◊Comet Hale-Bopp.

half-life during ◊radioactive decay, the time in which the strength of a radioactive source decays to half its original value. In theory, the decay process is never complete and there is always some residual radioactivity. For this reason, the half-life of a radioactive isotope is measured, rather than the total decay time. It may vary from millionths of a second to billions of years.

Radioactive substances decay exponentially; thus the time taken for the first 50% of the isotope to decay will be the same as the time taken by the next 25%, and by the 12.5% after that, and so on.

For example, carbon-14 takes about 5,730 years for half the material to decay; another 5,730 for half of the remaining half to decay; then 5,730 years for half of that remaining half to decay, and so on. Plutonium-239, one of the most toxic of all radioactive substances, has a half-life of about 24,000 years.

Half-Life

Isotope	Half-life
(least stable)	
lithium-5	4.4×10^{-22} sec
polonium-213	4.2×10^{-6} sec
lead-211	36 min
lead-209	3.3 hours
uranium-238	4.551×10^9 years
thorium-232	1.39×10^{10} years
tellurium-128	1.5×10^{24} years
(most stable)	

half-life in medicine, the time taken for the peak plasma concentration of a drug to decline by half. This is due to redistribution of the drug from the plasma to the tissues and to metabolism and excretion of the drug. The half-life of the drug is one of the factors that determines the frequency of administration required to achieve optimal therapeutic effects.

halftone process technique used in printing to reproduce the full range of tones in a photograph or other illustration. The intensity of the printed colour is varied from full strength to the lightest shades, even if one colour of ink is used. The picture to be reproduced is photographed through a screen ruled with a rectangular mesh of fine lines, which breaks up the tones of the original into areas of dots that vary in frequency according to the intensity of the tone. In the darker areas the dots run together; in the lighter areas they have more space between them.

halide any compound produced by the combination of a ◊halogen, such as chlorine or iodine, with a less electronegative element (see ◊electronegativity). Halides may be formed by ◊ionic bonds or by ◊covalent bonds.

halite mineral sodium chloride, NaCl, or common ◊salt. When pure it is colourless and transparent, but it is often pink, red, or yellow. It is soft and has a low density.

Halite occurs naturally in evaporite deposits that have precipitated on evaporation of bodies of salt water. As rock salt, it forms beds within a sedimentary sequence; it can also migrate upwards through surrounding rocks to form salt domes. It crystallizes in the cubic system.

halitosis bad breath. It may be caused by poor oral hygiene; disease of the mouth, throat, nose, or lungs; or disturbance of the digestion.

Hall effect production of a voltage across a conductor or semiconductor carrying a current at a right angle to a surrounding magnetic field. It was discovered in 1897 by the US physicist Edwin Hall (1855–1938). It is used in the **Hall probe** for measuring the strengths of magnetic fields and in magnetic switches.

Halley's comet comet that orbits the Sun about every 76 years, named after Edmond Halley who calculated its orbit. It is the brightest and most conspicuous of the periodic comets. Recorded sightings go back over 2,000 years. It travels around the Sun in the opposite direction to the planets. Its orbit is inclined at almost 20° to the main plane of the Solar System and ranges between the orbits of Venus and Neptune. It will next reappear in 2061.

The comet was studied by space probes at its last appearance in 1986. The European probe *Giotto* showed that the nucleus of Halley's comet is a tiny and irregularly shaped chunk of ice, measuring some 15 km/10 m long by 8 km/5 m wide, coated by a layer of very dark material, thought to be composed of carbon-rich compounds. This surface coating has a very low ◊albedo, reflecting just 4% of the light it receives from the Sun. Although the comet is one of the darkest objects known, it has a glowing head and tail produced by jets of gas from fissures in the outer dust layer. These vents cover 10% of the total surface area and become active only when exposed to the Sun. The force of these jets affects the speed of the comet's travel in its orbit.

Halley's comet

The *Anglo-Saxon Chronicle* records the 1066 visit of Halley's comet: 'Then there was seen all over England a sign such as no one had ever seen. Some said that the star was a comet, as some called the long-haired star. It had a tail streaming like smoke up to nearly half the sky.'

---

hallmark official mark stamped on British gold, silver, and (from 1913) platinum, instituted in 1327 (royal charter of London Goldsmiths) in order to prevent fraud. After 1363, personal marks of identification were added. Now tests of metal content are carried out at authorized assay offices in London, Birmingham, Sheffield, and Edinburgh; each assay office has its distinguishing mark, to which is added a maker's mark, date letter, and mark guaranteeing standard.

hallucinogen any substance that acts on the ◊central nervous system to produce changes in perception and mood and often hallucinations. Hallucinogens include ◊LSD, peyote, and ◊mescaline. Their effects are unpredictable and they are illegal in most countries.

In some circumstances hallucinogens may produce panic or even suicidal feelings, which can recur without warning several days or months after taking the drug. In rare cases they produce an irreversible psychotic state mimicking schizophrenia. Spiritual or religious experiences are common, hence the ritual use of hallucinogens in some cultures. They work by chemical interference with the normal action of neurotransmitters in the brain.

halogen any of a group of five nonmetallic elements with similar chemical bonding properties: fluorine, chlorine, bromine, iodine, and astatine. They form a linked group in the ◊periodic table of the elements, descending from fluorine, the most reactive, to astatine, the least reactive. They combine directly with most metals to form salts, such as common salt (NaCl). Each halogen has seven electrons in its valence shell, which accounts for the chemical similarities displayed by the group.

halon organic chemical compound containing one or two carbon atoms, together with ◊bromine and other ◊halogens. The most commonly used are halon 1211 (bromochlorodifluoromethane) and halon 1301 (bromotrifluoromethane). The halons are gases and are widely used in fire extinguishers. As destroyers of the ◊ozone layer, they are up to ten times more effective than ◊chlorofluorocarbons (CFCs), to which they are chemically related.

Levels in the atmosphere are rising by about 25% each year, mainly through the testing of fire-fighting equipment. The use of halons in fire extinguishers was banned Jan 1994.

halophyte plant adapted to live where there is a high concentration of salt in the soil, for example, in salt marshes and mud flats.

hand unit used in measuring the height of a horse from front hoof to shoulder (withers). One hand equals 10.2 cm/4 in.

hanging valley valley that joins a larger glacial trough at a higher level than the trough floor. During glaciation the ice in the smaller valley was unable to erode as deeply as the ice in the trough, and so the valley was left perched high on the side of the trough when the ice retreated. A river or stream flowing along the hanging valley often forms a waterfall as it enters the trough.

haploid having a single set of ◊chromosomes in each cell. Most higher organisms are ◊diploid – that is, they have two sets – but their gametes (sex cells) are haploid. Some plants, such as mosses, liverworts, and many seaweeds, are haploid, and male honey bees are haploid because they develop from eggs that have not been fertilized. See also ◊meiosis.

hard disc in computing, a storage device usually consisting of a rigid metal ◊disc coated with a magnetic material. Data are read from and written to the disc by means of a disc drive. The hard disc may be permanently fixed into the drive or in the form of a disc pack that can be removed and exchanged with a different pack. Hard discs vary from large units with capacities of more than 3,000 megabytes, intended for use with mainframe computers, to small units with capacities as low as 20 megabytes, intended for use with microcomputers.

hardening of oils transformation of liquid oils to solid products by ◊hydrogenation.

Vegetable oils contain double covalent carbon-to-carbon bonds and are therefore examples of ◊unsaturated compounds. When hydrogen is added to these double bonds, the oils become saturated. The more saturated oils are waxlike solids.

hardness physical property of materials that governs their use. Methods of heat treatment can increase the hardness of metals. A scale of hardness was devised by German–Austrian mineralogist Friedrich Mohs in the 1800s, based upon the hardness of certain minerals from soft talc (Mohs' hardness 1) to diamond (10), the hardest of all materials.

hardware the mechanical, electrical, and electronic components of a computer system, as opposed to the various programs, which constitute ◊software.

hard water water that does not lather easily with soap, and produces a deposit or 'scale' in kettles. It is caused by the presence of certain salts of calcium and magnesium.

Temporary hardness is caused by the presence of dissolved hydrogencarbonates (bicarbonates); when the water is boiled, they are converted to insoluble carbonates that precipitate as 'scale'. *Permanent hardness* is caused by sulphates and silicates, which are not affected by boiling. Water can be softened by ◊distillation, ◊ion exchange (the principle underlying commercial water softeners), addition of sodium carbonate or of large amounts of soap, or boiling (to remove temporary hardness).

harelip congenital facial deformity, a cleft in the upper lip and jaw, which may extend back into the palate (cleft palate). It can be remedied by surgery.

harmattan in meteorology, a dry and dusty NE wind that blows over W Africa.

Harrier the only truly successful vertical takeoff and landing fixed-wing aircraft, often called the *jump jet*. It was built in Britain and made its first flight in 1966. It has a single jet engine and a set of swivelling nozzles. These deflect the jet exhaust vertically downwards for takeoff and landing, and to the rear for normal flight. Designed to fly from confined spaces with minimal ground support, it refuels in midair.

harrow agricultural implement used to break up the furrows left by the ◊plough and reduce the soil to a fine consistency or tilth, and to cover the seeds after sowing. The traditional harrow consists of spikes set in a frame; modern harrows use sets of discs.

hashish drug made from the resin contained in the female flowering tops of hemp (◊cannabis).

hash total in computing, a ◊validation check in which an otherwise meaningless control total is calculated by adding together numbers (such as payroll or account numbers) associated with a set of records. The hash total is checked each time data are input, in order to ensure that no entry errors have been made.

hassium synthesized, radioactive element of the ◊transactinide series, symbol Hs, atomic number 108, relative atomic mass 265. It was first synthesized in 1984 by the Laboratory for Heavy Ion Research in Darmstadt, Germany. Its temporary name was unniloctium.

haustorium (plural *haustoria*) specialized organ produced by a parasitic plant or fungus that penetrates the cells of its host to absorb nutrients. It may be either an outgrowth of hyphae (see ◊hypha), as in the case of parasitic fungi, or of the stems of flowering parasitic plants, as in dodders (*Cuscuta*). The suckerlike haustoria of a dodder penetrate the vascular tissue of the host plant without killing the cells.

hay preserved grass used for winter livestock feed. The grass is cut and allowed to dry in the field before being removed for storage in a barn. The optimum period for cutting is when the grass has just come into flower and contains most feed value. During the natural drying process, the moisture content is reduced from 70–80% down to a safe level of 20%. In normal weather conditions, this takes from two to five days, during which time the hay is turned by machine to ensure even drying. Hay is normally baled before removal from the field.

Hayashi track in astronomy, a path on the ◊Hertzsprung–Russell diagram taken by protostars as they emerge from the clouds of dust and gas out of which they were born. A protostar appears on the right (cool) side of the Hertzsprung–Russell diagram and follows a Hayashi track until it arrives on the main sequence where hydrogen burning can start. It is named after the Japanese astrophysicist Chushiro Hayashi, who studied the theory of protostars in the 1960s.

hay fever allergic reaction to pollen, causing sneezing, with inflammation of the nasal membranes and conjunctiva of the eyes. Symptoms are due to the release of ◊histamine. Treatment is by antihistamine drugs.

hazardous waste waste substance, usually generated by industry, that represents a hazard to the environment or to people living or working nearby. Examples include radioactive wastes, acidic resins, arsenic residues, residual hardening salts, lead from car exhausts, mercury, nonferrous sludges, organic solvents, asbestos, chlorinated solvents, and pesticides. The cumulative effects of toxic waste can take some time to become apparent (anything from a few hours to many years), and pose a serious threat to the ecological stability of the planet; its economic disposal or recycling is the subject of research. *See illustration overleaf.*

HDTV abbreviation for ◊*high-definition television.*

headache pain felt within the skull. Most headaches are caused by stress or tension, but some may be symptoms of brain or ◊systemic disease, including ◊fever.

Chronic daily headache may be caused by painkiller misuse, according to the European Headache Foundation in 1996. People who take daily analgesics to treat chronic headaches may actually be causing the headaches by doing so. See also ◊migraine.

head louse parasitic insect living in human hair; see ◊louse.

headward erosion the backwards erosion of material at the source of a river or stream. Broken rock and soil at the source are carried away by the river, causing erosion to take place in the opposite direction to the river's flow. The resulting lowering of the land behind the source may, over time, cause the river to cut backwards into a neighbouring valley to 'capture' another river.

harmful/irritant

toxic

radioactive

explosive

flammable

corrosive

oxidizing/supports fire

hazard label Internationally recognized symbols indicating the potential dangers of handling certain substances.

health care implementation of a satisfactory regimen to ensure long-lasting good health. Life expectancy is determined by the overall efficiency of the body's vital organs and the rate at which these organs deteriorate. Fundamental health-care concerns include:

Smoking This is strongly linked to heart disease, stroke, bronchitis, lung cancer, and other serious diseases.

Exercise Regular physical exercise improves fitness, slows down the gradual decline in efficiency of the heart and lungs, and so helps to prolong life.

Diet A healthy diet contains plenty of vegetable fibre, complex carbohydrates, vitamins, minerals, and enzymes, and polyunsaturated fats (which keep the level of blood cholesterol low), not saturated (animal) fats (which contribute to cholesterol storage in blood vessels).

Weight Obesity (defined as generally being 20% or more above the desirable weight for age, sex, build, and height) is associated with many potentially dangerous conditions, such as coronary heart disease, diabetes, and stroke, as well as muscular and joint problems, and breathing difficulties.

Health Education Authority (HEA) authority established in 1987 to provide information and advice about health directly to the public; promote the development of health education in England and AIDS health education in the UK; and support and advise government departments, health authorities, local authorities, voluntary organizations, and other bodies or individuals concerned with health education. The HEA is also responsible for carrying out national campaigns and promoting research.

Following a review of the HEA's role and functions, in Dec 1994 a radical change in how it is to be funded was announced. Effective 1 April 1996, after a year of shadow operation 1995–96, the HEA will essentially be funded on a contract basis, seeking contracts from the Department of Health and other organizations to deliver health promotion programmes and projects for the supply of health promotion material, research, and expertise.

health screening testing large numbers of apparently healthy people for disease; see ◊screening.

health, world the health of people worldwide is monitored by the World Health Organization (WHO). Outside the industrialized world in particular, poverty and degraded environmental conditions mean that easily preventable diseases are widespread.

Vaccine-preventable diseases Every year, 46 million infants are not fully immunized; 2.8 million children die and 3 million are disabled due to vaccine-preventable diseases (polio, tetanus, diphtheria, whooping cough, tuberculosis, and measles).

Diarrhoea Every year, there are 750 million cases in children, causing 4 million deaths. Oral rehydration therapy can correct dehydration and prevent 65% of deaths due to diarrhoeal disease. The basis of therapy is prepackaged sugar and salt. Treatment to cure the disease costs less than 20 cents, but fewer than one-third of children are treated in this way.

Tuberculosis 1.6 billion people carry the bacteria, and there are 3 million deaths every year. Some 95% of all patients could be cured within six months using a specific antibiotic therapy which costs less than $30 per person.

Prevention and cure Increasing health spending in industrialized countries by only $2 per head would enable immunization of all children to be performed, polio to be eradicated, and drugs provided to cure all cases of diarrhoeal disease, acute respiratory infection, tuberculosis, malaria, schistosomiasis, and most sexually transmitted diseases.

hearing aid any device to improve the hearing of partially deaf people. Hearing aids usually consist of a battery-powered transistorized microphone/amplifier unit and earpiece. Some miniaturized aids are compact enough to fit in the ear or be concealed in the frame of eyeglasses.

heart muscular organ that rhythmically contracts to force blood around the body of an animal with a circulatory system. Annelid worms and some other invertebrates have simple hearts consisting of thickened sections of main blood vessels that pulse regularly. An earthworm has ten such hearts. Vertebrates have one heart. A fish heart has two chambers – the thin-walled *atrium* (once called the auricle) that expands to receive blood, and the thick-walled *ventricle* that pumps it out. Amphibians and most reptiles have two atria and one ventricle; birds and mammals have two atria and two ventricles. The beating of the heart is controlled by the autonomic nervous system and an internal control centre or pacemaker, the *sinoatrial node*.

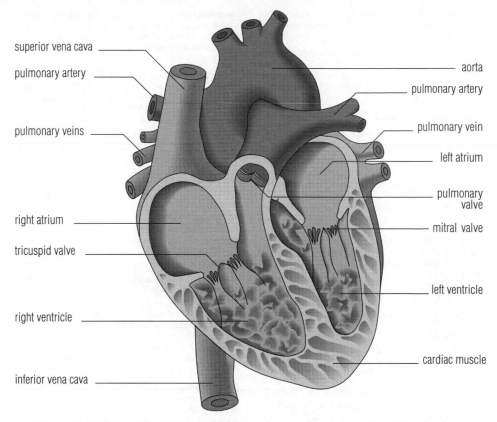

superior vena cava

pulmonary artery

pulmonary veins

right atrium

tricuspid valve

right ventricle

inferior vena cava

aorta

pulmonary artery

pulmonary vein

left atrium

pulmonary valve

mitral valve

left ventricle

cardiac muscle

heart The structure of the human heart. During an average lifetime, the human heart beats more than 2,000 million times and pumps 500 million l/110 million gal of blood. The average pulse rate is 70–72 beats per minute at rest for adult males, and 78–82 beats per minute for adult females.

The cardiac cycle is the sequence of events during one complete cycle of a heart beat. This consists of the simultaneous contraction of the two atria, a short pause, then the simultaneous contraction of the two ventricles, followed by a longer pause while the entire heart relaxes. The contraction phase is called 'systole' and the relaxation phase which follows is called 'diastole'. The whole cycle is repeated 70–80 times a minute under resting conditions.

When the atria contract, the blood in them enters the two relaxing ventricles, completely filling them. The mitral and tricuspid valves, which were open, now begin to shut and as they do so, they create vibrations in the heart walls and tendons, causing the first heart sound. The ventricles on contraction push open the pulmonary and aortic valves and eject blood into the respective vessels. The closed mitral and tricuspid valves prevent return of blood into the atria during this phase. As the ventricles start to relax, the aortic and pulmonary valves close to prevent backward flow of blood, and their closure causes the second heart sound. By now, the atria have filled once again and are ready to start contracting to begin the next cardiac cycle.

heart attack or *myocardial infarction* sudden onset of gripping central chest pain, often accompanied by sweat-

ing and vomiting, caused by death of a portion of the heart muscle following obstruction of a coronary artery by thrombosis (formation of a blood clot). Half of all heart attacks result in death within the first two hours, but in the remainder survival has improved following the widespread use of thrombolytic (clot-buster) drugs.

After a heart attack, most people remain in hospital for seven to ten days, and may make a gradual return to normal activity over the following months. How soon a patient is able to return to work depends on the physical and mental demands of their job.

heartbeat the regular contraction and relaxation of the heart, and the accompanying sounds. As blood passes through the heart a double beat is heard. The first is produced by the sudden closure of the valves between the atria and the ventricles. The second, slightly delayed sound, is caused by the closure of the valves found at the entrance to the major arteries leaving the heart. Diseased valves may make unusual sounds, known as heart murmurs.

heartburn burning sensation behind the breastbone (sternum). It results from irritation of the lower oesophagus (gullet) by excessively acid stomach contents, as sometimes happens during pregnancy and in cases of duodenal ulcer or obesity. It is often due to a weak valve at the entrance to the stomach that allows its contents to well up into the oesophagus.

heart disease disorder affecting the heart; for example, ◊ischaemic heart disease, in which the blood supply through the coronary arteries is reduced by ◊atherosclerosis; ◊valvular heart disease, in which a heart valve is damaged; and cardiomyopathy, where the heart muscle itself is diseased.

Heart disease

The Inuit people rarely suffer from heart disease. One explanation for this may be the high amount of fish, notably salmon, in their diet, which reduces the level of blood cholesterol.

---

heart–lung machine apparatus used during heart surgery to take over the functions of the heart and the lungs temporarily. It has a pump to circulate the blood around the body and is able to add oxygen to the blood and remove carbon dioxide from it. A heart–lung machine was first used for open-heart surgery in the USA in 1953.

heat form of energy possessed by a substance by virtue of the vibrating movement (kinetic energy) of its molecules or atoms. Heat energy is transferred by conduction, convection, and radiation. It always flows from a region of higher ◊temperature (heat intensity) to one of lower temperature. Its effect on a substance may be simply to raise its temperature, or to cause it to expand, melt (if a solid), vaporize (if a liquid), or increase its pressure (if a confined gas).

Quantities of heat are usually measured in units of energy, such as joules (J) or calories (cal). The specific heat of a substance is the ratio of the quantity of heat required to raise the temperature of a given mass of the substance through a given range of temperature to the heat required to raise the temperature of an equal mass of water through the same range. It is measured by a ◊calorimeter.

Conduction, convection, and radiation Conduction is the passing of heat along a medium to neighbouring parts with no visible motion accompanying the transfer of heat – for example, when the whole length of a metal rod is heated when one end is held in a fire. Convection is the transmission of heat through a fluid (liquid or gas) in currents – for example, when the air in a room is warmed by a fire or radiator. Radiation is heat transfer by infrared rays. It can pass through a vacuum, travels at the same speed as light, can be reflected and refracted, and does not affect the medium through which it passes. For example, heat reaches the Earth from the Sun by radiation.

For the transformation of heat, see ◊thermodynamics.

heat capacity in physics, the quantity of heat required to raise the temperature of an object by one degree. The *specific heat capacity* of a substance is the heat capacity per unit of mass, measured in joules per kilogram per kelvin ($J\ kg^{-1}\ K^{-1}$).

heat death in cosmology, a possible fate of the universe in which it continues expanding indefinitely while all the stars burn out and no new ones are formed. See ◊critical density.

heat of reaction alternative term for ◊energy of reaction.

heat pump machine, run by electricity or another power source, that cools the interior of a building by removing heat from interior air and pumping it out or, conversely, heats the inside by extracting energy from the atmosphere or from a hot-water source and pumping it in.

heat shield any heat-protecting coating or system, especially the coating (for example, tiles) used in spacecraft to protect the astronauts and equipment inside from the heat of re-entry when returning to Earth. Air friction can generate temperatures of up to 1,500°C/2,700°F on re-entry into the atmosphere.

heat storage any means of storing heat for release later. It is usually achieved by using materials that undergo phase changes, for example, Glauber's salt and sodium pyrophosphate, which melts at 70°C/158°F. The latter is used to store off-peak heat in the home: the salt is liquefied by cheap heat during the night and then freezes to give off heat during the day.

Other developments include the use of plastic crystals, which change their structure rather than melting when they are heated. They could be incorporated in curtains or clothing.

heatstroke or *sunstroke* rise in body temperature caused by excessive exposure to heat.

Mild heatstroke is experienced as feverish lassitude, sometimes with simple fainting; recovery is prompt following rest and replenishment of salt lost in sweat. Severe heatstroke causes collapse akin to that seen in acute ◊shock, and is potentially lethal without prompt treatment, including cooling the body carefully and giving fluids to relieve dehydration.

heat treatment in industry, the subjection of metals and alloys to controlled heating and cooling after fabrication to relieve internal stresses and improve their physical properties. Methods include ◊annealing, ◊quenching, and ◊tempering.

heavy water or *deuterium oxide* D_2O water containing the isotope deuterium instead of hydrogen (relative molecular mass 20 as opposed to 18 for ordinary water).

Its chemical properties are identical with those of ordinary water, but its physical properties differ slightly. It occurs in ordinary water in the ratio of about one part by mass of deuterium to 5,000 parts by mass of hydrogen, and can be concentrated by electrolysis, the ordinary water being more readily decomposed by this means than the heavy water. It has been used in the nuclear industry because it can slow down fast neutrons, thereby controlling the chain reaction.

hectare metric unit of area equal to 100 ares or 10,000 square metres (2.47 acres), symbol ha.

Trafalgar Square, London's only metric square, was laid out as one hectare.

hedge or *hedgerow* row of closely planted shrubs or low trees, generally acting as a land division and windbreak. Hedges also serve as a source of food and as a refuge for wildlife, and provide a ◊habitat not unlike the understorey of a natural forest.

height-weight chart in medicine, chart relating height to weight for either adults or children. The height-weight chart for children is used to indicate if the rate of growth falls within normal limits, whilst that for adults is used as an indication of a healthy weight.

There is a wide variation of height and weight in normal children, so charts are based on percentiles describing distribution in the population. They are obtained by measuring that characteristic in at least 1,000 boys and girls at each age. From these data, the fiftieth centile can be calculated. This represents the mean value for that characteristic and 50% of recordings will be above it and 50% below it. Investigation of the cause is usual when the height of a child falls below the third centile, indicating the child is shorter than 97% of other children of that age.

Height-weight charts for adults have been devised using statistics obtained from life insurance companies. These statistics have suggested that life expectancy is greatest if the average weight at age 25 to 30 years is maintained throughout the rest of the individual's life. A 10% range on either side of normal is allowed to account for the build of the individual.

helicopter powered aircraft that achieves both lift and propulsion by means of a rotary wing, or rotor, on top of the fuselage. It can take off and land vertically, move in any direction, or remain stationary in the air. It can be powered by piston or jet engine. The ◊autogiro was a precursor.

The rotor of a helicopter has two or more blades of aerofoil cross-section like an aeroplane's wings. Lift and propulsion are achieved by angling the blades as they rotate. Experiments using the concept of helicopter flight date from the early 1900s, with the first successful liftoff and short flight in 1907. Ukrainian–US engineer Igor Sikorsky built the first practical single-rotor craft in the USA in 1939.

A single-rotor helicopter must also have a small tail rotor to counter the torque, or tendency of the body to spin in the opposite direction to the main rotor. Twin-rotor helicopters, like the Boeing Chinook, have their rotors turning in opposite directions to prevent the body from spinning. Helicopters are now widely used in passenger service, rescue missions on land and sea, police pursuits and traffic control, firefighting, and agriculture. In war they carry troops and equipment into difficult terrain, make aerial reconnaissance and attacks, and carry the wounded to aid stations. A fire-fighting helicopter was tested in Japan in 1996, designed to reach skyscrapers beyond the reach of fire-engine ladders.

heliosphere region of space through which the ◊solar wind flows outwards from the Sun. The **heliopause** is the boundary of this region, believed to lie about 100 astronomical units from the Sun, where the flow of the solar wind merges with the interstellar gas.

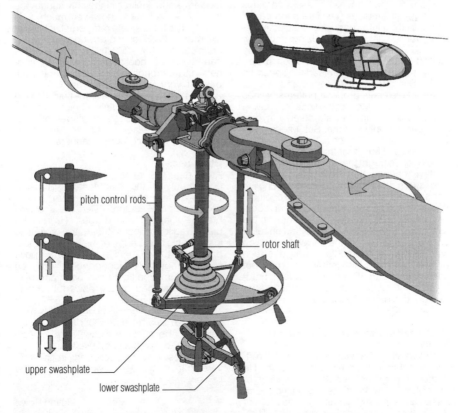

pitch control rods

rotor shaft

upper swashplate

lower swashplate

helicopter The helicopter is controlled by varying the rotor pitch (the angle of the rotor blade as it moves through the air). For backwards flight, the blades in front of the machine have greater pitch than those behind the craft. This means that the front blades produce more lift and a backwards thrust. For forwards flight, the situation is reversed. In level flight, the blades have unchanging pitch.

The second most abundant chemical

What is the second most abundant chemical element in the universe? A clue: nearly all of it was created only a few minutes after the Big Bang. This element was identified only 100 years ago, and, uniquely among the elements, was first discovered in the Sun. The answer, of course, is helium. Chemists somewhat dryly characterize helium as an inert gas. Indeed it is one of the most unreactive gases known, as well as being one of the lightest. It was these properties that spurred its first commercial production, when, in 1914, it was found that incendiary bullets fired into a Zeppelin airship filled with helium did not cause the airship to catch fire or explode.

Discovery in the Sun's spectrum
The story of helium begins with an eclipse of the Sun. For the 1868 eclipse, no fewer than four scientific expeditions set out to examine the solar prominences. These enormous eruptions of luminous gas flare hundreds of kilometres out into space and are clearly visible during an eclipse. Using the new method of spectrum analysis, invented by Robert Kirchhoff and Robert Bunsen in Germany, the scientists examined the spectrum of the Sun through telescopes and prisms. Different elements produce patterns of lines in the spectrum. Most of the lines they saw in the Sun's spectrum they attributed to hydrogen and possibly sodium. But it seemed to Pierre Janssen, a French astronomer on one of the expeditions, that the light of the solar flares was so bright that it could be examined in daylight and when there was no eclipse. The same idea occurred to two English astronomers who were not on the expedition, William Huggins and Norman Lockyer. Lockyer and Janssen published their idea on the same day in 1868. The new method enabled Lockyer to prove that the yellow line, which was thought to be sodium, was due to a different element altogether, and one not known on Earth. Professor Edward Frankland of Imperial College, London, suggested the name 'helium' from the Greek *helios* 'the Sun'.

Supposed extraterrestrial found on Earth
In 1895, William Ramsay, a Scottish chemist, showed that helium existed in Earth. He heated the mineral cleveite, a variety of pitchblende, with dilute sulphuric acid. Although he mostly got nitrogen gas (about 20%), there was small amount of helium, which he identified by its spectrum: it showed the same lines first seen during the 1868 eclipse expeditions.

Sources of helium
Helium is now known to be present in the Earth's atmosphere (0.0005%) and comes from gases issued by radioactive elements in the Earth's crust. Helium and other inert gases are evolved from hot springs that have their sources at great depths in the Earth. But the most important source of helium is in the natural gas from some petroleum springs in Texas, Utah, and Colorado in the USA and Medicine Hat in Canada.

Ultimate cool
Helium may be an inert gas – it has no known compounds – but its alternative family name, noble gas, is altogether more fitting, for its properties make it uniquely useful in the modern world. Helium is the only substance that never solidifies, however much it is cooled. It is the ultimate refrigerant. It boils at 4.2 K or –269°C, only a few degrees above absolute zero or –273.15°C. This is why it can be used to cool superconducting magnets and astronomical detectors that work best at very low temperatures. Cooled to 2.18 K, helium becomes a superfluid, flowing uphill and through the tiniest holes. In medicine, liquid helium cools the superconducting magnets of magnetic resonance-imaging body scanners.

Other uses
As a gas, helium is used in arc-welding equipment, in the manufacture of semiconductors, in powerful gas lasers, and to cool nuclear reactors. Deep-sea divers breathe it mixed with oxygen. Helium fills giant fragile balloons that carry scientific instruments to the edge of the Earth's atmosphere 35–40 km/20–25 mi high. The Apollo space programme depended critically on liquid helium. The liquid hydrogen and oxygen fuels needed by the powerful rockets were cooled by helium.

Today, annual consumption of helium is about 100 million cu m/3,500 million cu ft and rising by about 5% per year. There may be 20 years' accessible supply left.

The most abundant element in the universe? It is hydrogen, which is also the lightest.

Julian Rowe

helium colourless, odourless, gaseous, nonmetallic element, symbol He, atomic number 2, relative atomic mass 4.0026. It is grouped with the ◊inert gases, is non-reactive, and forms no compounds. It is the second-most abundant element (after hydrogen) in the universe, and has the lowest boiling ($-268.9°C/-452°F$) and melting points ($-272.2°C/-458°F$) of all the elements. It is present in small quantities in the Earth's atmosphere from gases issuing from radioactive elements (from ◊alpha decay) in the Earth's crust; after hydrogen it is the second-lightest element.

Helium is a component of most stars, including the Sun, where the nuclear-fusion process converts hydrogen into helium with the production of heat and light. It is obtained by compression and fractionation of naturally occurring gases. It is used for inflating balloons and as a dilutant for oxygen in deep-sea breathing systems. Liquid helium is used extensively in low-temperature physics (cryogenics).

helix in mathematics, a three-dimensional curve resembling a spring, corkscrew, or screw thread. It is generated by a line that encircles a cylinder or cone at a constant angle.

helminth in medicine, collective name to used to describe parasitic worms. There are several classes of helminth that can cause infections in humans, including ascarids (ascariasis), ◊tapeworms, and ◊threadworms.

hematite principal ore of iron, consisting mainly of iron(III) oxide, Fe_2O_3. It occurs as **specular hematite** (dark, metallic lustre), **kidney ore** (reddish radiating fibres terminating in smooth, rounded surfaces), and a red earthy deposit.

hemp annual plant originally from Asia, now cultivated in most temperate countries for the fibres produced in the outer layer of the stem, which are used in ropes, twines, and, occasionally, in a type of linen or lace. The drug ◊cannabis is obtained from certain varieties of hemp. (*Cannabis sativa*, family Cannabaceae.)

The name 'hemp' is also given to other similar types of fibre: **sisal hemp** and **henequen** obtained from the leaves of *Agave* species native to Yucatán and cultivated in many tropical countries, and **manila hemp** obtained from *Musa textilis*, a plant native to the Philippines and the Maluku Islands, Indonesia.

henry SI unit (symbol H) of ◊inductance (the reaction of an electric current against the magnetic field that surrounds it). One henry is the inductance of a circuit that produces an opposing voltage of one volt when the current changes at one ampere per second.

It is named after the US physicist Joseph Henry.

heparin anticoagulant substance produced by cells of the liver, lungs, and intestines. It normally inhibits the clotting of blood by interfering with the production of thrombin, which is necessary for clot formation. Heparin obtained from animals is administered after surgery to limit the risk of ◊thrombosis, or following pulmonary ◊embolism to ensure that no further clots form, and during haemodialysis.

hepatic of or pertaining to the liver.

hepatitis any inflammatory disease of the liver, usually caused by a virus. Other causes include alcohol, drugs, gallstones, ◊lupus erythematous, and amoebic ◊dysentery. Symptoms include weakness, nausea, and jaundice.

Five different hepatitis viruses have been identified; A, B, C, D, and E. The hepatitis A virus (HAV) is the commonest cause of viral hepatitis, responsible for up to 40% of cases worldwide. It is spread by contaminated food. Hepatitis B, or serum hepatitis, is a highly contagious disease spread by blood products or in body fluids. It often culminates in liver failure, and is also associated with liver cancer, although only 5% of those infected suffer chronic liver damage. During 1995, 1.1 million people died of hepatitis B. Around 300 million people are ◊carriers. Vaccines are available against hepatitis A and B.

Hepatitis C is mostly seen in people needing frequent transfusions. Hepatitis D, which only occurs in association with hepatitis B, is common in the Mediterranean region. Hepatitis E is endemic in India and South America.

Herb

The herb Catnip (*Nepeta cataria*) has a remarkable effect on cats. Members of the cat family roll over, extend their claws and twist in excitement when they smell the pungent odour of catnip. This reaction is believed to be caused by a chemical called trans-neptalacone, which is similar to a substance found in the female cat's urine.

---

herb any plant (usually a flowering plant) tasting sweet, bitter, aromatic, or pungent, used in cooking, medicine, or perfumery; technically, a herb is any plant in which the aerial parts do not remain above ground at the end of the growing season.

herbaceous plant plant with very little or no wood, dying back at the end of every summer. The herbaceous perennials survive winters as underground storage organs such as bulbs and tubers.

herbalism in alternative medicine, the prescription and use of plants and their derivatives for medication. Herbal products are favoured by alternative practitioners as 'natural medicine', as opposed to modern synthesized medicines and drugs, which are regarded with suspicion because of the dangers of side effects and dependence.

Many herbal remedies are of proven efficacy both in preventing and curing illness. Medical herbalists claim to be able to prescribe for virtually any condition, except those so advanced that surgery is the only option.

herbarium collection of dried, pressed plants used as an aid to identification of unknown plants and by taxonomists in the ◊classification of plants. The plant specimens are accompanied by information, such as the date and place of collection, by whom collected, details of habitat, flower colour, and local names.

Herbaria range from small collections containing plants of a limited region, to the large university and national herbaria (some at ◊botanical gardens) containing millions of specimens from all parts of the world.

herbicide any chemical used to destroy plants or check their growth; see ◊weedkiller.

herbivore animal that feeds on green plants (or photosynthetic single-celled organisms) or their products, including seeds, fruit, and nectar. The most numerous type of herbivore is thought to be the zooplankton, tiny invertebrates in the surface waters of the oceans that feed on small photosynthetic algae. Herbivores are more numerous than other animals because their food is the most abundant. They form a vital link in the food chain between plants and carnivores.

Hercules in astronomy, the fifth-largest constellation, lying in the northern hemisphere. Despite its size it contains no prominent stars. Its most important feature is the best example in the northern hemisphere of a ◊globular cluster of stars 22,500 light years from Earth, which lies between Eta and Zeta Herculis.

heredity in biology, the transmission of traits from parent to offspring. See also ◊genetics.

hermaphrodite organism that has both male and female sex organs. Hermaphroditism is the norm in such species as earthworms and snails, and is common in flowering plants. Cross-fertilization is the rule among hermaphrodites, with the parents functioning as male and female

simultaneously, or as one or the other sex at different stages in their development. Human hermaphrodites are extremely rare.

hernia or ***rupture*** protrusion of part of an internal organ through a weakness in the surrounding muscular wall, usually in the groin. The appearance is that of a rounded soft lump or swelling.

heroin or ***diamorphine*** powerful ◊opiate analgesic, an acetyl derivative of ◊morphine. It is more addictive than morphine but causes less nausea. It has an important place in the control of severe pain in terminal illness, severe injuries, and heart attacks, but is widely used illegally.

Heroin was discovered in Germany in 1898. The major regions of opium production, for conversion to heroin, are the 'Golden Crescent' of Afghanistan, Iran, and Pakistan, and the 'Golden Triangle' across parts of Myanmar (Burma), Laos, and Thailand.

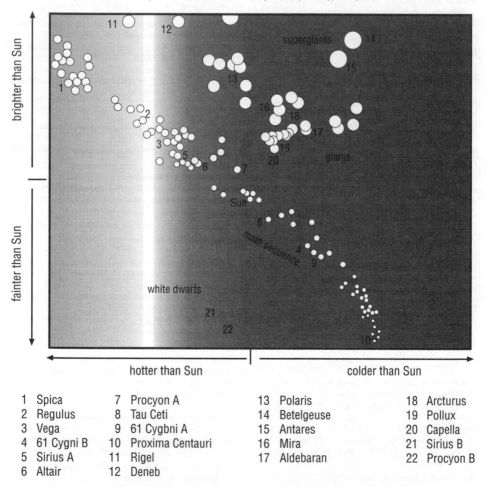

1	Spica	7	Procyon A	13	Polaris	18	Arcturus
2	Regulus	8	Tau Ceti	14	Betelgeuse	19	Pollux
3	Vega	9	61 Cygbni A	15	Antares	20	Capella
4	61 Cygni B	10	Proxima Centauri	16	Mira	21	Sirius B
5	Sirius A	11	Rigel	17	Aldebaran	22	Procyon B
6	Altair	12	Deneb				

Hertzsprung–Russell diagram The Hertzsprung–Russell diagram relates the brightness (or luminosity) of a star to its temperature. Most stars fall within a narrow diagonal band called the main sequence. A star moves off the main sequence when it grows old. The Hertzsprung–Russell diagram is one of the most important diagrams in astrophysics.

herpes any of several infectious diseases caused by viruses of the herpes group. ***Herpes simplex I*** is the causative agent of a common inflammation, the cold sore. ***Herpes simplex II*** is responsible for genital herpes, a highly contagious, sexually transmitted disease characterized by painful blisters in the genital area. It can be transmitted in the birth canal from mother to newborn. ***Herpes zoster*** causes ◊shingles; another herpes virus causes chickenpox.

A number of ◊antivirals treat these infections, which are particularly troublesome in patients whose immune systems have been suppressed medically; for example, after a transplant operation. The drug acyclovir, originally introduced for the treatment of genital herpes, has now been shown to modify the course of chickenpox and the related condition shingles, by reducing the duration of the illness.

herpes genitalis or ***genital herpes*** in medicine, an infection of the genital organs caused by the *Herpes simplex* virus. It is transmitted by sexual intercourse. Infection is marked by the formation of blisters on the genitals and pain. Treatment is with antiviral drugs, such as aciclovir, used orally or as creams.

hertz SI unit (symbol Hz) of frequency (the number of repetitions of a regular occurrence in one second). Radio waves are often measured in megahertz (MHz), millions of hertz, and the ◊clock rate of a computer is usually measured in megahertz. The unit is named after Heinrich Hertz.

Hertzsprung–Russell diagram in astronomy, a graph on which the surface temperatures of stars are plotted against their luminosities. Most stars, including the Sun, fall into a narrow band called the ◊***main sequence***. When a star grows old it moves from the main sequence to the upper right part of the graph, into the area of the giants and supergiants. At the end of its life, as the star shrinks to become a white dwarf, it moves again, to the bottom left area. It is named after the Dane Ejnar Hertzsprung (1873–1967) and the American Henry Norris Russell (1877–1957), who independently devised it in the years 1911–13.

heterogeneous reaction in chemistry, a reaction where there is an interface between the different components or reactants. Examples of heterogeneous reactions are those between a gas and a solid, a gas and a liquid, two immiscible liquids, or two different solids.

heterostyly in botany, having ◊styles of different lengths.

Certain flowers, such as primroses (*Primula vulgaris*), have different-sized ◊anthers and styles to ensure cross-fertilization (through ◊pollination) by visiting insects.

heterotroph any living organism that obtains its energy from organic substances produced by other organisms. All animals and fungi are heterotrophs, and they include herbivores, carnivores, and saprotrophs (those that feed on dead animal and plant material).

heterozygous in a living organism, having two different ◊alleles for a given trait. In ◊homozygous organisms, by contrast, both chromosomes carry the same allele. In an outbreeding population an individual organism will generally be heterozygous for some genes but homozygous for others.

For example, in humans, alleles for both blue-and brown-pigmented eyes exist, but the 'blue' allele is ◊recessive to the dominant 'brown' allele.

Only individuals with blue eyes are predictably homozygous for this trait; brown-eyed people can be either homozygous or heterozygous.

heuristics in computing, a process by which a program attempts to improve its performance by learning from its own experience.

hexachlorophene ($C_6HCl_3OH)_2CH_2$ white, odourless bactericide, used in minute quantities in soaps and surgical disinfectants.

Trichlorophenol is used in its preparation, and, without precise temperature control, the highly toxic TCDD (tetrachlorodibenzodioxin; see ◊dioxin) may form as a by-product.

hexadecimal number system or ***hex*** number system to the base 16, used in computing. In hex the decimal numbers 0–15 are represented by the characters 0, 1, 2, 3, 4, 5, 6, 7, 8, 9, A, B, C, D, E, F.

Each place in a number increases in value by a power of 16 going from right to left; for instance, 8F is equal to 15 + (8 × 16) = 143 in decimal. Hexadecimal numbers are often preferred by programmers writing in low-level languages because they are more easily converted to the computer's internal ◊binary (base-two) code than are decimal numbers, and because they are more compact than binary numbers and therefore more easily keyed, checked, and memorized.

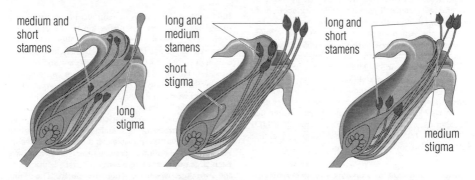

heterostyly Heterostyly, in which lengths of the stamens and stigma differ in flowers of different plants of the same species. This is a device to ensure cross-pollination by visiting insects.

HF in physics, abbreviation for **high ◊ frequency**. HF radio waves have frequencies in the range 3–30 MHz.

HGV abbreviation for **heavy goods vehicle**.

hiatus hernia in medicine, condition in which a portion of the stomach is displaced through the opening in the diaphragm, through which the oesophagus passes from the chest to the abdominal cavity.

hibernation state of dormancy in which certain animals spend the winter. It is associated with a dramatic reduction in all metabolic processes, including body temperature, breathing, and heart rate. It is a fallacy that animals sleep throughout the winter.

The body temperature of the Arctic ground squirrel falls to below 0°C/32°F during hibernation. Hibernating bats may breathe only once every 45 minutes, and can go for up to 2 hours without taking a breath.

hiccup sharp noise caused by a sudden spasm of the ◊ diaphragm with closing of the windpipe, commonly caused by digestive disorder. On rare occasions, hiccups may become continuous, when they are very debilitating; treatment with a muscle relaxant drug may be effective.

hi-fi (abbreviation for **high-fidelity**) faithful reproduction of sound from a machine that plays recorded music or speech. A typical hi-fi system includes a turntable for playing vinyl records, a cassette tape deck to play magnetic tape recordings, a tuner to pick up radio broadcasts, an amplifier to serve all the equipment, possibly a compact-disc player, and two or more loudspeakers.

Advances in mechanical equipment and electronics, such as digital recording techniques and compact discs, have made it possible to eliminate many distortions in sound-reproduction processes.

Higgs boson or **Higgs particle** postulated ◊ elementary particle whose existence would explain why particles have mass. The current theory of elementary particles, called the ◊ standard model, cannot explain how mass arises. To overcome this difficulty, Peter Higgs (1929–) of the University of Edinburgh and Thomas Kibble (1932–) of Imperial College, London proposed in 1964 a new particle that binds to other particles and gives them their mass. The Higgs boson has not yet been detected experimentally.

high-definition television (HDTV) ◊ television system offering a significantly greater number of scanning lines, and therefore a clearer picture, than that provided by conventional systems. Typically, HDTV has about twice the horizontal and vertical resolution of current 525-line (such as the American standard, NTSC) or 625-line standands (such as the British standard, PAL); a frame rate of at least 24 Hz; and a picture aspect ratio of 9:16 instead of the current 3:4. HDTV systems have been in development since the mid-1970s.

The Japanese HDTV system, or HiVision as it is tradenamed in Japan, uses 1,125 scanning lines and an aspect ratio of 16:9 instead of the squarish 4:3 that conventional television uses. A European HDTV system, called HD-MAC, using 1,250 lines, is under development. In the USA, a standard incorporating digital techniques is being discussed.

high-level language in computing, a programming language designed to suit the requirements of the programmer; it is independent of the internal machine code of any particular computer. High-level languages are used to solve problems and are often described as **problem-oriented languages** – for example, ◊ BASIC was designed to be easily learnt by first-time programmers; ◊ COBOL is used to write programs solving business problems; and ◊ FORTRAN is used for programs solving scientific and mathematical problems. In contrast, low-level languages, such as ◊ assembly languages, closely reflect the machine codes of specific computers, and are therefore described as **machine-oriented languages**.

Unlike low-level languages, high-level languages are relatively easy to learn because the instructions bear a close resemblance to everyday language, and because the programmer does not require a detailed knowledge of the internal workings of the computer. Each instruction in a high-level language is equivalent to several machine-code instructions. High-level programs are therefore more compact than equivalent low-level programs. However, each high-level instruction must be translated into machine code – by either a ◊ compiler or an ◊ interpreter program – before it can be executed by a computer. High-level languages are designed to be **portable** – programs written in a high-level language can be run on any computer that has a compiler or interpreter for that particular language.

high-yield variety crop that has been specially bred or selected to produce more than the natural varieties of the same species. During the 1950s and 1960s, new strains of wheat and maize were developed to reduce the food shortages in poor countries (the ◊ Green Revolution). Later, IR8, a new variety of rice that increased yields by up to six times, was developed in the Philippines. Strains of crops resistant to drought and disease were also developed. High-yield varieties require large amounts of expensive artificial fertilizers and sometimes pesticides for best results.

HII region in astronomy, a region of extremely hot ionized hydrogen, surrounding one or more hot stars, visible as a bright patch or emission ◊ nebula in the sky. The gas is ionized by the intense ultraviolet radiation from the stars within it. HII regions are often associated with interstellar clouds in which new stars are being born. An example is the ◊ Orion Nebula. It takes its name from a spectroscopic notation in which HI represents neutral hydrogen (H) and HII represents ionized hydrogen (H^+).

hill figure in Britain, any of a number of figures, usually of animals, cut from the turf to reveal the underlying chalk. Their origins are variously attributed to Celts, Romans, Saxons, Druids, or Benedictine monks, although most are of modern rather than ancient construction. Examples include 17 White Horses, and giants such as the Cerne Abbas Giant, near Dorchester, Dorset, associated with a prehistoric fertility cult.

Nearly 50 hill figures are known in Britain, of which all but four are on the southern chalk downs of England. Some are landmarks or memorials; others have a religious or ritual purpose. It is possible that the current figures are on the site of, or reinforce, previous ones. There may have been large numbers of figures dotted on the landscape in the Iron Age, which were not maintained. The White Horse at Uffington, on the Berkshire Downs, used to be annually 'scoured' in a folk ceremony.

hillfort European Iron Age site with massive banks and ditches for defence, used as both a military camp and a permanent settlement. Examples found across Europe, in particular France, central Germany, and the British Isles, include Heuneberg near Sigmaringen, Germany, Spinans Hill in County Wicklow, Ireland, and Maiden Castle, Dorset, England. Iron Age Germanic peoples spread the tradition of forts with massive defences, timberwork reinforcements, and sometimes elaborately defended gateways with guardrooms, the whole being overlooked from a rampart walk, as at Maiden Castle. The ramparts usually follow the natural line of a hilltop and are laid out to avoid areas of dead ground.

hinge joint in vertebrates, a joint where movement occurs in one plane only. Examples are the elbow and knee, which are controlled by pairs of muscles, the ◊flexors and ◊extensors.

Hipparcos (acronym for **high precision parallax collecting satellite**) satellite launched by the European Space Agency in Aug 1989. Named after the Greek astronomer Hipparchus, it is the world's first ◊astrometry satellite and is providing precise positions, distances, colours, brightnesses, and apparent motions for over 100,000 stars.

histamine inflammatory substance normally released in damaged tissues, which also accounts for many of the symptoms of ◊allergy. It is an amine, $C_5H_9N_3$. Substances that neutralize its activity are known as ◊antihistamines. Histamine was first described 1911 by British physiologist Henry Dale (1875–1968).

histogram in statistics, a graph showing frequency of data, in which the horizontal axis details discrete units or class boundaries, and the vertical axis represents the frequency. Blocks are drawn such that their areas (rather than their height as in a bar chart) are proportional to the frequencies within a class or across several class boundaries. There are no spaces between blocks.

histology study of plant and animal tissue by visual examination, usually with a ◊microscope.

◊Stains are often used to highlight structural characteristics such as the presence of starch or distribution of fats.

histology in medicine, the laboratory study of cells and tissues.

historical archaeology the archaeological study of historically documented cultures, especially in America and Australia, where it is directed at colonial and post-colonial settlements. The European equivalent is mediaeval and post-mediaeval archaeology.

HIV (abbreviation for **human immunodeficiency virus**) the infectious agent that is believed to cause ◊AIDS. It was first discovered in 1983 by Luc Montagnier of the Pasteur Institute in Paris, who called it lymphocyte-associated virus (LAV). Independently, US scientist Robert Gallo of the National Cancer Institute in Bethesda, Maryland, claimed its discovery in 1984 and named it human T-lymphocytotrophic virus 3 (HTLV-III).

Research during 1995 into how fast the HIV virus reproduces estimated that a billion new viruses are produced in the body daily, and that 2 billion white blood cells are destroyed every 24 hours. HIV is genetically very variable as a result of this rapid proliferation; an HIV carrier can harbour 1 million genetically distinct variants of the virus.

About 15% of babies born to HIV-positive mothers are themselves HIV-positive. A very small number of these babies (less than 3%) test negative for the virus some months later, a phenomenon yet to be explained.

hoard valuables or prized possessions that have been deliberately buried, often in times of conflict or war, and never reclaimed. Coins, objects in precious metals, and scrap metal are the most common objects found in hoards. In July 1991 the largest hoard found in Britain was discovered, consisting of 7,000 15th-century coins; it was declared treasure trove.

Hodgkin's disease or or *lymphadenoma* rare form of cancer mainly affecting the lymph nodes and spleen. It undermines the immune system, leaving the sufferer susceptible to infection.

However, it responds well to radiotherapy and ◊cytotoxic drugs, and long-term survival is usual.

Hoffman's voltameter in chemistry, an apparatus for collecting gases produced by the ◊electrolysis of a liquid.

It consists of a vertical E-shaped glass tube with taps at the upper ends of the outer limbs and a reservoir at the top of the central limb. Platinum electrodes fused into the lower ends of the outer limbs are connected to a source of direct current. At the beginning of an experiment, the outer limbs are completely filled with electrolyte by opening the taps. The taps are then closed and the current switched on. Gases evolved at the electrodes bubble up the outer limbs and collect at the top, where they can be measured.

hogback geological formation consisting of a ridge with a sharp crest and abruptly sloping sides, the outline of which resembles the back of a hog. Hogbacks are the result of differential erosion on steeply dipping rock strata composed of alternating resistant and soft beds. Exposed, almost vertical resistant beds provide the sharp crests.

holism in philosophy, the concept that the whole is greater than the sum of its parts.

A physician is obligated to consider more than a diseased organ, more even than the whole man – he must view the man in his world.

On **holistic medicine** Harvey Cushing, quoted in René Dubos *Man Adapting*

---

holistic medicine umbrella term for an approach that virtually all alternative therapies profess, which considers the overall health and lifestyle profile of a patient, and treats specific ailments not primarily as conditions to be alleviated but rather as symptoms of more fundamental disease.

holmium silvery, metallic element of the ◊lanthanide series, symbol Ho, atomic number 67, relative atomic mass 164.93. It occurs in combination with other rare-earth metals and in various minerals such as gadolinite. Its compounds are highly magnetic. The element was discovered in 1878, spectroscopically, by the Swiss chemists J L Soret and Delafontaine, and independently in 1879 by Swedish chemist Per Cleve (1840–1905), who named it after Stockholm, near which it was found.

Holocene epoch of geological time that began 10,000 years ago, the second and current epoch of the Quaternary period. During this epoch the glaciers retreated, the climate became warmer, and humans developed significantly.

holography method of producing three-dimensional (3-D) images, called holograms, by means of ◊laser light. Holography uses a photographic technique (involving the splitting of a laser beam into two beams) to produce a picture, or hologram, that contains 3-D information about the object photographed. Some holograms show meaningless patterns in ordinary light and produce a 3-D image only when laser light is projected through them, but reflection holograms produce images when ordinary light is reflected from them (as found on credit cards).

Although the possibility of holography was suggested as early as 1947 (by Hungarian-born British physicist Dennis Gabor), it could not be demonstrated until a pure coherent light source, the laser, became available 1963. The first laser-recorded holograms were created by Emmett Leith and Juris Upatnieks at the University of Michigan, USA, and Yuri Denisyuk in the Soviet Union. Research into holographic video and other spatial imaging techniques, led by Stephen Benton, is under way at the MIT Media Lab.

The technique of holography is also applicable to sound, and bats may navigate by ultrasonic holography.

Holographic techniques also have applications in storing dental records, detecting stresses and strains in construction and in retail goods, detecting forged paintings and documents, and producing three-dimensional body scans. The technique of detecting strains is of widespread application. It involves making two different holograms of an object on one plate, the object being stressed between exposures. If the object has distorted during stressing, the hologram will be greatly changed, and the distortion readily apparent.

Using holography, digital data can be recorded page by page in a crystal. In 1993 10,000 pages (100 megabytes) of digital data were stored in an iron-doped lithium nobate crystal measuring 1 cm^3.

homeopathy alternative spelling of ◊*homoeopathy*.

homeostasis maintenance of a constant internal state in an organism, particularly with regard to pH, salt concentration, temperature, and blood sugar levels. Stable conditions are important for the efficient functioning of the ◊enzyme reactions within the cells, which affect the performance of the entire organism.

homeothermy maintenance of a constant body temperature in endothermic (warm-blooded) animals, by the use of chemical processes to compensate for heat loss or gain when external temperatures change. Such processes include generation of heat by the breakdown of food and

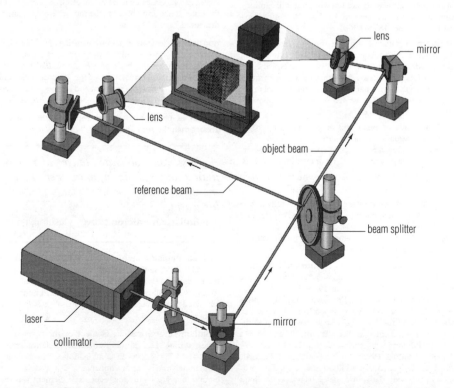

holography Recording a transmission hologram. Light from a laser is divided into two beams. One beam goes directly to the photographic plate. The other beam reflects off the object before hitting the photographic plate. The two beams combine to produce a pattern on the plate which contains information about the 3-D shape of the object. If the exposed and developed plate is illuminated by laser light, the pattern can be seen as a 3-D picture of the object.

the contraction of muscles, and loss of heat by sweating, panting, and other means.

Mammals and birds are homeotherms, whereas invertebrates, fish, amphibians, and reptiles are cold-blooded or poikilotherms. Homeotherms generally have a layer of insulating material to retain heat, such as fur, feathers, or fat (see ◊blubber). Their metabolism functions more efficiently due to homeothermy, enabling them to remain active under most climatic conditions.

homoeopathy or **homeopathy** system of alternative medicine based on the principle that symptoms of disease are part of the body's self-healing processes, and on the practice of administering extremely diluted doses of natural substances found to produce in a healthy person the symptoms manifest in the illness being treated. Developed by German physician Samuel Hahnemann (1755–1843), the system is widely practised today as an alternative to allopathic (orthodox) medicine, and many controlled tests and achieved cures testify its efficacy.

In 1992, the German health authority, the *Bundesgesundheitsamt*, banned 50 herbal and homeopathic remedies containing ◊alkaloids because they are toxic, and set dose limits on 550 other natural remedies.

homogeneous reaction in chemistry, a reaction where there is no interface between the components. The term applies to all reactions where only gases are involved or where all the components are in solution.

homologous in biology, a term describing an organ or structure possessed by members of different taxonomic groups (for example, species, genera, families, orders) that originally derived from the same structure in a common ancestor. The wing of a bat, the arm of a monkey, and the flipper of a seal are homologous because they all derive from the forelimb of an ancestral mammal.

homologous series any of a number of series of organic chemicals with similar chemical properties in which members differ by a constant relative molecular mass.

Alkanes (paraffins), alkenes (olefins), and alkynes (acetylenes) form such series in which members differ in mass by 14, 12, and 10 atomic mass units respectively. For example, the alkane homologous series begins with methane (CH_4), ethane (C_2H_6), propane (C_3H_8), butane (C_4H_{10}), and pentane (C_5H_{12}), each member differing from the previous one by a CH_2 group (or 14 atomic mass units).

homozygous in a living organism, having two identical ◊alleles for a given trait. Individuals homozygous for a trait always breed true; that is, they produce offspring that resemble them in appearance when bred with a genetically similar individual; inbred varieties or species are homozygous for almost all traits.

◊Recessive alleles are only expressed in the homozygous condition. See also ◊heterozygous.

honey sweet syrup produced by honey bees from the nectar of flowers. It is stored in honeycombs and made in excess of their needs as food for the winter. Honey comprises various sugars, mainly laevulose and dextrose, with enzymes, colouring matter, acids, and pollen grains. It has antibacterial properties and was widely used in ancient Egypt, Greece, and Rome as a wound salve. It is still popular for sore throats, in hot drinks or in lozenges.

honey guide in botany, line or spot on the petals of a flower that indicate to pollinating insects the position of the nectaries (see ◊nectar) within the flower. The orange dot on the lower lip of the toadflax flower (*Linaria vulgaris*) is an example. Sometimes the markings reflect only ultraviolet light, which can be seen by many insects although it is not visible to the human eye.

alkane	alcohol	aldehyde	ketone	carboxylic acid	alkene
CH_4 methane	CH_3OH methanol	HCHO methanal	—	HCOOH methanoic acid	—
CH_3CH_3 ethane	CH_3CH_2OH ethanol	CH_3CHO ethanal	—	CH_3COOH ethanoic acid	CH_2CH_2 ethene
$CH_3CH_2CH_3$ propane	$CH_3CH_2CH_2OH$ propanol	CH_3CH_2CHO propanal	CH_3COCH_3 propanone	CH_3CH_2COOH propanoic acid	CH_2CHCH_3 propene
methane	methanol	methanal	propanone	methanoic acid	ethene

homologous series The systematic naming of simple straight-chain organic molecules depends on two-part names. The first part of a name indicates the number of carbon atoms in the chain: one carbon, meth-; two carbons, eth-; three carbons, prop-; etc. The second part of each name indicates the kind of bonding between the carbon atoms, or the atomic group attached to the chain. The name of a molecule containing only single bonds ends in -ane. Molecules with double bonds have names ending with -ene. Molecules containing the OH group have names ending in -anol; those containing –CO– groups have names ending in -anone; those containing the carboxyl group –COOH– have names ending in -anoic acid.

Hooke's law law stating that the deformation of a body is proportional to the magnitude of the deforming force, provided that the body's elastic limit (see ◊elasticity) is not exceeded. If the elastic limit is not reached, the body will return to its original size once the force is removed. The law was discovered by Robert Hooke in 1676.

Hoover Dam highest concrete dam in the USA, 221 m/ 726 ft, on the Colorado River at the Arizona–Nevada border. It was built between 1931–36. Known as *Boulder Dam* 1933–47, its name was restored by President Truman as the reputation of the former president, Herbert Hoover, was revived. It impounds Lake Mead, and has a hydroelectric power capacity of 1,300 megawatts.

hops female fruit heads of the hop plant *Humulus lupulus*, family Cannabiaceae; these are dried and used as a tonic and in flavouring beer. In designated areas in Europe, no male hops may be grown, since seedless hops produced by the unpollinated female plant contain a greater proportion of the alpha acid that gives beer its bitter taste.

horizon the limit to which one can see across the surface of the sea or a level plain, that is, about 5 km/3 mi at 1.5 m/ 5 ft above sea level, and about 65 km/40 mi at 300 m/ 1,000 ft.

horizon in astronomy, the ◊great circle dividing the visible part of the sky from the part hidden by the Earth.

hormone secretion of the ◊endocrine glands, concerned with control of body functions. The major glands are the thyroid, parathyroid, pituitary, adrenal, pancreas, ovary, and testis. Hormones bring about changes in the functions of various organs according to the body's requirements. The ◊hypothalamus, which adjoins the pituitary gland, at the base of the brain, is a control centre for overall coordination of hormone secretion; the thyroid hormones determine the rate of general body chemistry; the adrenal hormones prepare the organism during stress for 'fight or flight'; and the sexual hormones such as oestrogen govern reproductive functions.

There are also hormone-secreting cells in the kidney, liver, gastrointestinal tract, thymus (in the neck), pineal (in the brain), and placenta. Many diseases due to hormone deficiency can be relieved with hormone preparations.

hormone-replacement therapy (HRT) use of ◊oestrogen and progesterone to help limit the unpleasant effects of the menopause in women. The treatment was first used in the 1970s.

At the menopause, the ovaries cease to secrete natural oestrogen. This results in a number of symptoms, including hot flushes, anxiety, and a change in the pattern of menstrual bleeding. It is also associated with osteoporosis, or a thinning of bone, leading to an increased incidence of fractures, frequently of the hip, in older women. Oestrogen preparations, taken to replace the decline in natural hormone levels, combined with regular exercise can help to maintain bone strength in women.

hornblende green or black rock-forming mineral, one of the ◊amphiboles; it is a hydrous silicate of calcium, iron, magnesium, and aluminium. Hornblende is found in both igneous and metamorphic rocks.

hornfels ◊metamorphic rock formed by rocks heated by contact with a hot igneous body. It is fine-grained and brittle, without foliation.

Hornfels may contain minerals only formed under conditions of great heat, such as andalusite, Al_2SiO_5, and cordierite, $(Mg,Fe)_2Al_4Si_5O_{18}$. This rock, originating from sedimentary rock strata, is found in contact with large igneous ◊intrusions where it represents the heat-altered equivalent of the surrounding clays. Its hardness makes it suitable for road building and railway ballast.

hornwort nonvascular plant (with no 'veins' to carry water and food), related to the ◊liverworts and ◊mosses. Hornworts are found in warm climates, growing on moist shaded soil. (Class Anthocerotae, order Bryophyta.)

The name is also given to a group of aquatic flowering plants which are found in slow-moving water. They have whorls of finely divided leaves and may grow up to 2 m/ 7 ft long. (Genus *Ceratophyllum*, family Ceratophyllaceae.)

Like liverworts and mosses, the bryophyte hornworts exist in two different reproductive forms, sexual and asexual, which appear alternately (see ◊alternation of generations). A leafy plant body, or gametophyte, produces gametes, or sex cells, and a small horned form, or sporophyte, which grows upwards from the gametophyte, produces spores. Unlike the sporophytes of mosses and liverworts, the hornwort sporophyte survives after the gametophyte has died.

horsepower imperial unit (abbreviation hp) of power, now replaced by the ◊watt. It was first used by the engineer James Watt, who employed it to compare the power of steam engines with that of horses.

horticulture art and science of growing flowers, fruit, and vegetables. Horticulture is practised in gardens and orchards, along with millions of acres of land devoted to vegetable farming. Some areas, like California, have specialized in horticulture because they have the mild climate and light fertile soil most suited to these crops.

hospice residential facility specializing in palliative care for terminally ill patients and their relatives.

hospital facility for the care of the sick, injured, and incapacitated. In ancient times, temples of deities such as Aesculapius offered facilities for treatment and by the 4th century, the Christian church had founded hospitals for lepers, cripples, the blind, the sick, and the poor. The oldest surviving hospital in Europe is the 7th-century Hôtel Dieu, Paris; in Britain, the most ancient are St Bartholomew's (1123) and St Thomas's (1200) in London; and in the Americas the Hospital of Jesus of Nazareth, Mexico, dates back to 1524. Medical knowledge advanced during the Renaissance, and hospitals became increasingly secularized after the Reformation. In the 19th century, further progress was made in hospital design, administration, and staffing (Florence Nightingale played a significant role in this). In the 20th century there has been an increasing trend towards specialization and the inclusion of maternity wards.

host in biology, an organism that is parasitized by another. In ◊commensalism, the partner that does not benefit may also be called the host.

hot spot in geology, isolated rising plume of molten mantle material that may escape to the surface of the Earth's crust creating features such as volcanoes, chains of ocean islands, and sea mounts. Hot spots occur mid-plate and so differ from areas of volcanic activity at plate margins (see ◊plate tectonics). Examples are the Hawaiian Islands, the Galápagos, and the Emperor Seamount chain in the Pacific Ocean.

A volcano forms on the ocean crust immediately above the hot spot, is carried away by ◊plate tectonic movement, and becomes extinct. A new volcano forms beside it, above the hot spot. The result is an active volcano and a chain of increasingly old and eroded volcanic stumps stretching away along the line of plate movement.

hour period of time comprising 60 minutes; 24 hours make one calendar day.

hovercraft vehicle that rides on a cushion of high-pressure air, free from all contact with the surface beneath, invented by British engineer Christopher Cockerell in 1959. Hovercraft need a smooth terrain when operating overland and are best adapted to use on waterways. They are useful in places where harbours have not been established.

Large hovercraft (SR-N4) operate a swift car-ferry service across the English Channel, taking only about 35 minutes between Dover and Calais. They are fitted with a flexible 'skirt' that helps maintain the air cushion.

A military version made of fibreglass, the M-10, is tough manoeuvrable, and less noisy.

hp abbreviation for ◊horsepower.

HTML (abbreviation for *Hypertext Markup Language*) standard for structuring and describing a docu-

ment on the World-Wide Web. The HTML standard provides labels for constituent parts of a document (eg. headings and paragraphs) and permits the inclusion of images, sounds, and 'hyperlinks' to other documents. A ◊browser program is then used to convert this information into a graphical document on-screen.

HTML is a specific example of SGML (the international standard for text encoding). As such it is not a rigid standard but is constantly being improved to incorporate new features and allow greater freedom of design. In 1995 the specifications for HTML version 3.0 were put forward, including provisions for display of such features as complex tabular information and captioned images.

Hubble classification in astronomy, a scheme for classifying ◊galaxies according to their shapes, originally devised by the US astronomer Edwin Hubble in the 1920s.

Elliptical galaxies are classed from type E0 to E7, where the figure denotes the degree of ellipticity. An E0 galaxy appears circular to an observer, while an E7 is highly elliptical (this is based on the apparent shape; the true shape, distorted by foreshortening, may be quite different). *Spiral galaxies* are classed as type Sa, Sb, or Sc, where Sa is a tightly wound spiral with a large central bulge and Sc is loosely wound with a small bulge. Intermediate types are denoted by Sab or Sbc. *Barred spiral galaxies*, which have a prominent bar across their centres, are similarly classed as type SBa, SBb, or SBc

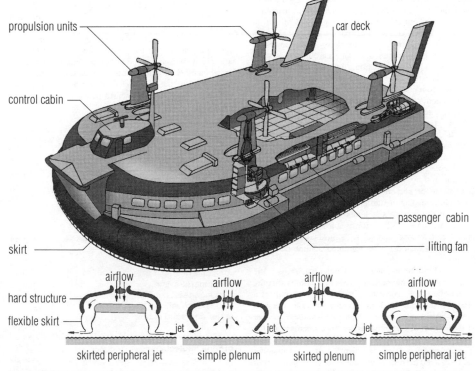

hovercraft There are several alternative ways of containing the cushion of air beneath the hull of a hovercraft. The passenger-carrying hovercraft that sails across the English Channel has a flexible skirt; other systems are the open plenum and the peripheral jet.

with intermediates SBab or SBbc. **Lenticular galaxies**, which have no spiral arms, are classed as type S0. **Irregular galaxies**, type Irr, can be subdivided into Irr I, which resemble poorly formed spirals, and Irr II which are otherwise

The Hubble classification was once believed to reveal an evolutionary sequence (from ellipticals to spirals) but this is now known not to be the case. Our own ◊Milky Way Galaxy is classified as type Sb or Sc, but may have a bar.

Hubble's constant in astronomy, a measure of the rate at which the universe is expanding, named after Edwin Hubble. Observations suggest that galaxies are moving apart at a rate of 50–100 kps/30–60 mps for every million ◊parsecs of distance. This means that the universe, which began at one point according to the ◊Big Bang theory, is between 10 billion and 20 billion years old (probably closer to 20). Observations by the Hubble Space Telescope in 1996 produced a revised constant of 73 kps/45 mps.

Hubble's law the law that relates a galaxy's distance from us to its speed of recession as the universe expands, announced in 1929 by Edwin Hubble. He found that galaxies are moving apart at speeds that increase in direct proportion to their distance apart. The rate of expansion is known as Hubble's constant.

Hubble Space Telescope (HST) space-based astronomical observing facility, orbiting the Earth at an altitude of 610 km/380 mi. It consists of a 2.4 m/94 in telescope and four complementary scientific instruments, is roughly cylindrical, 13 m/43 ft long, and 4 m/13 ft in diameter, with two large solar panels. HST produces a wealth of scientific data, and allows astronomers to observe the birth of stars, find planets around neighbouring stars, follow the expanding remnants of exploding stars, and search for black holes in the centre of galaxies. HST is a cooperative programme between the European Space Agency (ESA) and the US agency NASA, and is the first spacecraft specifically designed to be serviced in orbit as a permanent space-based observatory. It was launched in 1990.

By having a large telescope above Earth's atmosphere, astronomers are able to look at the universe with unprecedented clarity. Celestial observations by HST are unhampered by clouds and other atmospheric phenomena that distort and attenuate starlight. In particular, the apparent twinkling of starlight caused by density fluctuations in the atmosphere limits the clarity of ground-based telescopes. HST performs at least ten times better than such telescopes and can see almost back to the edge of the universe and to the beginning of time (see ◊Big Bang).

Before HST could reach its full potential, a flaw in the shape of its main mirror, discovered two months after the launch, had to be corrected. In Dec 1993, as part of a planned servicing and instrument upgrade mission, NASA astronauts aboard the space shuttle *Endeavor* installed a set of corrective lenses to compensate for the error in the mirror figure. COSTAR (corrective optics space telescope axial replacement), a device containing ten coin-sized mirrors, now feeds a corrected image from the main mirror to three of the HST's four scientific instruments. HST is also being used to detail the distribution of dust and stars in nearby galaxies, watch the

collisions of galaxies in detail, infer the evolution of galaxies, and measure the age of the universe.

In Dec 1995 HST was trained on an 'empty' area of sky near the Plough, now termed the **Hubble Deep Field**. Around 3,000 galaxies, mostly new discoveries, were photographed.

Two new instruments were added in Feb 1997. The Near Infared Camera and Multi-Object Spectrometer (NICMOS) will enable Hubble to see things even further away (and therefore older) than ever before. The Space Telescope Imaging Spectograph will work 30 times faster than its predecessor as it can gather information about different stars at the same time. Three new cameras had to be fitted shortly afterwards as one of the original ones was found to be faulty.

human body the physical structure of the human being. It develops from the single cell of the fertilized ovum, is born at 40 weeks, and usually reaches sexual maturity between 11 and 18 years of age. The bony framework (skeleton) consists of more than 200 bones, over half of which are in the hands and feet. Bones are held together by joints, some of which allow movement. The circulatory system supplies muscles and organs with blood, which provides oxygen and food and removes carbon dioxide and other waste products. Body functions are controlled by the nervous system and hormones. In the upper part of the trunk is the thorax, which contains the lungs and heart. Below this is the abdomen, containing the digestive system (stomach and intestines); the liver, spleen, and pancreas; the urinary system (kidneys, ureters, and bladder); and, in women, the reproductive organs (ovaries, uterus, and vagina). In men, the prostate gland and seminal vesicles only of the reproductive system are situated in the abdomen, the testes being in the scrotum, which, with the penis, is suspended in front of and below the abdomen. The bladder empties through a small channel (urethra); in the female this opens in the upper end of the vulval cleft, which also contains the opening of the vagina, or birth canal; in the male, the

Human Body: Composition

Chemical element or substance	Body weight (%)
pure elements	
oxygen	65
carbon	18
hydrogen	10
nitrogen	3
calcium	2
phosphorus	1.1
potassium	0.35
sulphur	0.25
sodium	0.15
chlorine	0.15
magnesium, iron, manganese, copper, iodine, cobalt, zinc	traces
water and solid matter	
water	60–80
total solid material	20–40
organic molecules	
protein	15–20
lipid	3–20
carbohydrate	1–15
small organic	0–1

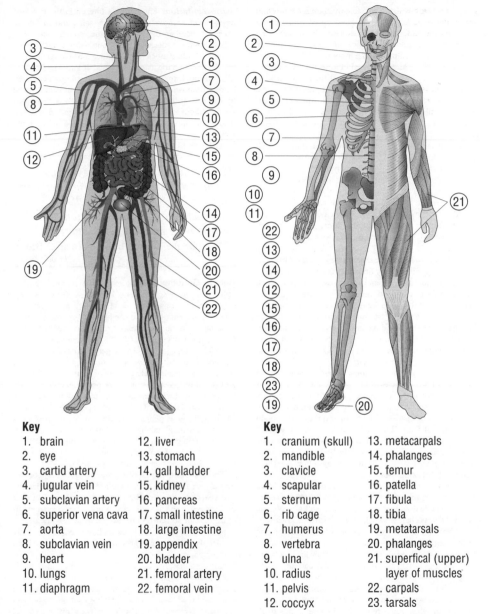

Key

1. brain
2. eye
3. cartid artery
4. jugular vein
5. subclavian artery
6. superior vena cava
7. aorta
8. subclavian vein
9. heart
10. lungs
11. diaphragm
12. liver
13. stomach
14. gall bladder
15. kidney
16. pancreas
17. small intestine
18. large intestine
19. appendix
20. bladder
21. femoral artery
22. femoral vein

Key

1. cranium (skull)
2. mandible
3. clavicle
4. scapular
5. sternum
6. rib cage
7. humerus
8. vertebra
9. ulna
10. radius
11. pelvis
12. coccyx
13. metacarpals
14. phalanges
15. femur
16. patella
17. fibula
18. tibia
19. metatarsals
20. phalanges
21. superfical (upper) layer of muscles
22. carpals
23. tarsals

human body The adult human body has approximately 650 muscles, 100 joints, 100,000 km/60,000 mi of blood vessels and 13,000 nerve cells. There are 206 bones in the adult body, nearly half of them in the hands and feet. The figure on the left shows the main organs and the circulatory system. The figure on the right illustrates the skeletal and muscular components.

urethra is continued into the penis. In both sexes, the lower bowel terminates in the anus, a ring of strong muscle situated between the buttocks.

The skull is mounted on the spinal column, or spine, a chain of 24 vertebrae. The ribs, 12 on each side, are articulated to the spinal column behind, and the upper seven meet the breastbone (sternum) in front. The lower end of the spine rests on the pelvic girdle, composed of the triangular sacrum, to which are attached the hipbones (ilia), which are fused in front. Below the sacrum is the tailbone (coccyx). The shoulder blades (scapulae) are held in place behind the upper ribs by muscles, and connected in front to the breastbone by the two collarbones (clavicles).

Each shoulder blade carries a cup (glenoid cavity) into which fits the upper end of the armbone (humerus). This articulates below with the two forearm bones (radius and ulna). These are articulated at the wrist (carpals) to the bones of the hand (metacarpals and phalanges). The upper end of each thighbone (femur) fits into a depression (acetabulum) in the hipbone; its lower end is articulated at the knee to the shinbone (tibia) and calf bone (fibula), which are articulated at the ankle (tarsals) to the bones of the foot (metatarsals and phalanges). At a moving joint, the end of each bone is formed of tough, smooth cartilage, lubricated by ◊synovial fluid. Points of special stress are reinforced by bands of fibrous tissue (ligaments).

Muscles are bundles of fibres wrapped in thin, tough layers of connective tissue (fascia); these are usually prolonged at the ends into strong, white cords (tendons, sinews) or sheets (aponeuroses), which connect the muscles to bones and organs, and by way of which the muscles do their work. Membranes of connective tissue also enfold the organs and line the interior cavities of the body. The thorax has a stout muscular floor, the diaphragm, which expands and contracts the lungs in the act of breathing.

The blood vessels of the *circulatory system*, branching into multitudes of very fine tubes (capillaries), supply all parts of the muscles and organs with blood, which carries oxygen and food necessary for life. The food passes out of the blood to the cells in a clear fluid (lymph); this is returned with waste matter through a system of lymphatic vessels that converge into collecting ducts that drain into large veins in the region of the lower neck. Capillaries join together to form veins which return blood, depleted of oxygen, to the heart.

A finely branching *nervous system* regulates the function of the muscles and organs, and makes their needs known to the controlling centres in the central nervous system, which consists of the brain and spinal cord. The inner spaces of the brain and the cord contain cerebrospinal fluid. The body processes are regulated both by the nervous system and by hormones secreted by the endocrine glands. Cavities of the body that open onto the surface are coated with mucous membranes, which secrete a lubricating fluid (mucus).

The exterior surface of the body is covered with *skin*. Within the skin are the sebaceous glands, which secrete sebum, an oily fluid that makes the skin soft and pliable, and the sweat glands, which secrete water and various salts. From the skin grow hairs, chiefly on the head, in the armpits, and around the sexual organs; and nails shielding the tips of the fingers and toes; both hair and nails are modifications of skin tissue. The skin also contains nerve receptors for sensations of touch, pain, heat, and cold.

The human *digestive system* is nonspecialized and can break down a wide variety of foodstuffs. Food is mixed with saliva in the mouth by chewing and is swallowed. It enters the stomach, where it is gently churned for some time and mixed with acidic gastric juice. It then passes into the small intestine. In the first part of this, the duodenum, it is broken down further by the juice of the pancreas and duodenal glands, and mixed with bile from the liver, which splits up the fat. The jejunum and ileum continue the work of digestion and absorb most of the nutritive substances from the food. The large intestine completes the process, reabsorbing water into the body, and ejecting the useless residue as faeces.

The body, to be healthy, must maintain water and various salts in the right proportions; the process is called *osmoregulation*. The blood is filtered in the two kidneys, which remove excess water, salts, and metabolic wastes. Together these form urine, which contains a yellow pigment derived from bile, and passes down through two fine tubes (ureters) into the bladder, a reservoir from which the urine is emptied at intervals (micturition) through the urethra. Heat is constantly generated by the combustion of food in the muscles and glands, and by the activity of nerve cells and fibres. It is dissipated through the skin by conduction and evaporation of sweat, through the lungs in the expired air, and in other excreted substances. Average body temperature is about 38°C/100°F (37°C/98.4°F in the mouth).

The human body is an energy system ... which is never a complete structure; never static; is in perpetual inner self-construction and self-destruction; we destroy in order to make it new.

On the **human body** Norman O Brown
Love's Body

Human Genome Project research scheme, begun in 1988, to map the complete nucleotide (see ◊nucleic acid) sequence of human ◊DNA. There are approximately 80,000 different ◊genes in the human genome, and one gene may contain more than 2 million nucleotides. The programme aims to collect 10-15,000 genetic specimens from 722 ethnic groups whose genetic make-up is to be preserved for future use and study. The knowledge gained is expected to help prevent or treat many crippling and lethal diseases, but there are potential ethical problems associated with knowledge of an individual's genetic make-up, and fears that it will lead to genetic engineering. Many indigenous people have condemned the project as 'bio- prospecting'–taking genetic material and exploiting it for economic gain–after attempts were made in 1993 to patent Human T-Lymphotropic Virus Type 2 taken from a Guayami woman with Leukaemia.

The Human Genome Organization (HUGO) coordinating the project expects to spend $1 billion over the first five years, making this the largest research project ever undertaken in the life sciences. Work is being carried out in more than 20 centres around the world. By the beginning of 1991, about 2,000 genes had been mapped. By late 1994 a genetic map of the complete genome had been completed.

Concern that, for example, knowledge of an individual's genes may make that person an unacceptable insurance risk has led to planned legislation on genome privacy in the USA, and 3% of HUGO's funds have been set aside for researching and reporting on the ethical implications of the project.

Each strand of DNA carries a sequence of chemical building blocks, the nucleotides. There are only four different types, but the number of possible combinations is immense. The different combinations of nucleotides produce different proteins in the cell, and thus determine the structure of the body and its individual variations. To establish the nucleotide sequence, DNA strands are broken into fragments, which are duplicated (by being introduced into cells of yeast or the bacterium *Escherichia coli*) and distributed to the research centres.

Genes account for only a small amount of the DNA

sequence. Over 90% of DNA appears not to have any function, although it is perfectly replicated each time the cell divides, and handed on to the next generation. Many higher organisms have large amounts of redundant DNA and it may be that this is an advantage, in that there is a pool of DNA available to form new genes if an old one is lost by mutation.

human reproduction an example of ◊sexual reproduction, where the male produces sperm and the female eggs. These gametes contain only half the normal number of chromosomes, 23 instead of 46, so that on fertilization the resulting cell has the correct genetic complement. Fertilization is internal, which increases the chances of conception; unusually for mammals, copulation and pregnancy can occur at any time of the year. Human beings are also remarkable for the length of childhood and for the highly complex systems of parental care found in society. The use of contraception and the development of laboratory methods of insemination and fertilization are issues that make human reproduction more than a merely biological phenomenon.

human species, origins of evolution of humans from ancestral ◊primates. The African apes (gorilla and chimpanzee) are shown by anatomical and molecular comparisons to be the closest living relatives of humans. The oldest known **hominids** (of the human group), the australopithecines, found in Africa, date from 3.5–4.4 million years ago. The first to use tools came 2 million years later, and the first humanoids to use fire and move out of Africa appeared 1.7 million years ago. Neanderthals were not our direct ancestors. Modern humans are all believed to descend from one African female who lived 200,000 years ago, although there is a rival theory that humans evolved in different parts of the world simultaneously.

Miocene apes Genetic studies indicate that the last common ancestor between chimpanzees and humans lived 5 to 10 million years ago. There are only fragmentary remains of ape and hominid fossils from this period. Dispute continues over the hominid status of *Ramapithecus*, the jaws and teeth of which have been found in India and Kenya in late Miocene deposits, dated between 14 and 10 million years. The lower jaw of a fossil ape found in the Otavi Mountains, Namibia, comes from deposits dated between 10 and 15 million years ago, and is similar to finds from E Africa and Turkey. It is thought to be close to the initial divergence of the great apes and humans.

Australopithecines Bones of the earliest known human ancestor, a hominid named *Australopithecus ramidus* were found in Ethiopia, in 1994 and dated as 4.4 million years old. *A. afarensis*, found in Ethiopia and Kenya, date from 3.9 to 4.4 million years ago. These hominids walked upright and they were either direct ancestors or an offshoot of the line that led to modern humans. They may have been the ancestors of *Homo habilis* (considered by some to be a species of *Australopithecus*), who appeared about 2 million years later, had slightly larger bodies and brains, and were probably the first to use stone tools. Also living in Africa at the same time was *A. africanus*, a gracile hominid thought to be a meat-eater, and *A.robustus*, a hominid with robust bones, large teeth, heavy jaws, and thought to be a vegetarian. They are not generally considered to be our ancestors.

Homo erectus Over 1.7 million years ago, *Homo erectus*, believed by some to be descended from *H. habilis*, appeared in Africa. *H. erectus* had prominent brow ridges, a flattened cranium, with the widest part of the skull low down, and jaws with a rounded tooth row, but the chin, characteristic of modern humans, was lacking. They also had much larger brains (900–1,200 cu cm), and were probably the first to use fire and the first to move out of Africa. Their remains are found as far afield as China, W Asia, Spain, and S Britain. Modern human *H. sapiens sapiens* and the Neanderthals *H. sapiens neanderthalensis* are probably descended from *H. erectus*.

Neanderthals Neanderthals were large-brained and heavily built, probably adapted to the cold conditions of the ice ages. They lived in Europe and the Middle East, and disappeared about 40,000 years ago, leaving *H. sapiens sapiens* as the only remaining species of the hominid group. Possible intermediate forms between Neanderthals and *H. sapiens sapiens* have been found at Mount Carmel in Israel and at Broken Hill in Zambia, but it seems that *H. sapiens sapiens* appeared in Europe quite rapidly and either wiped out the Neanderthals or interbred with them.

Modern humans There are currently two major views of human evolution: the **'out of Africa' model**, according to which *H. sapiens* emerged from *H. erectus*, or a descendant species, in Africa and then spread throughout the world; and the **multiregional model**, according to which selection pressures led to the emergence of similar advanced types of *H. sapiens* from *H. erectus* in different parts of the world at around the same time. Analysis of DNA in recent human populations suggests that *H. sapiens* originated about 200,000 years ago in Africa from a single female ancestor, 'Eve'. The oldest known fossils of *H.sapiens* also come from Africa, dating from 150,000–100,000 years ago. Separation of human populations would have occurred later, with separation of Asian, European, and Australian populations taking place between 100,000 and 50,000 years ago.

Humboldt Current former name of the ◊Peru Current.

hum, environmental disturbing sound of frequency about 40 Hz, heard by individuals sensitive to this range, but inaudible to the rest of the population. It may be caused by industrial noise pollution or have a more exotic origin, such as the jet stream, a fast-flowing high-altitude (about 15,000 m/50,000 ft) mass of air.

humerus the upper bone of the forelimb of tetrapods. In humans, the humerus is the bone above the elbow.

humidity the quantity of water vapour in a given volume of the atmosphere (absolute humidity), or the ratio of the amount of water vapour in the atmosphere to the saturation value at the same temperature (relative humidity). At ◊dew point the relative humidity is 100% and the air is said to be saturated. Condensation (the conversion of vapour to liquid) may then occur. Relative humidity is measured by various types of ◊hygrometer.

humour in biology, any fluid or semi-fluid tissue in the body, such as the vitreous and aqueous humours of the ◊eye.

humours, theory of theory prevalent in the West in classical and medieval times that the human body was composed of four kinds of fluid: phlegm, blood, choler or yellow bile, and melancholy or black bile. Physical and mental characteristics were explained by different proportions of humours in individuals.

An excess of phlegm produced a 'phlegmatic', or calm,

Out of Africa

Most palaeoanthropologists recognize the existence of two human species during the last million years – *Homo erectus*, now extinct, and *Homo sapiens* (including recent or 'modern' humans). In general, they believe that *Homo erectus* was the ancestor of *Homo sapiens*. How did the transition occur?

The multiregional model

There are two opposing views. The multiregional model says that *Homo erectus* gave rise to *Homo sapiens* across its whole range. About 700,000 years ago this included Africa, China, Java (Indonesia) and, probably, Europe. *Homo erectus*, following an African origin about 1.7 million years ago, dispersed around the Old World developing the regional variation that lies at the roots of modern 'racial' variation. Particular features in a given region persisted in the local descendant populations of today.

The multiregional model was first described in detail by Franz Weidenreich, the German palaeoanthropologist. It was developed further by American Carleton Coon, who tended to regard the regional lineages as genetically separate. Most recently, the model has become associated with researchers such as Milford Wolpoff (USA) and Alan Thorne (Australia), who have re-emphasized the importance of gene flow between the regional lines. In fact, they regard the continuity in time and space between the various forms of *Homo erectus* and their regional descendants to be so complete that they should be regarded as representing only one species – *Homo sapiens*.

The garden of Eden...

The opposing view is that *Homo sapiens* had a restricted origin in time and space. This is an old idea. Early in this century, workers such as Marcellin Boule (France) and Arthur Keith (UK) believed that the lineage of *Homo sapiens* was very ancient; it had developed in parallel with that of *Homo erectus* and the Neanderthals. However, much of the fossil evidence used to support their ideas has been re-evaluated, and few workers now support the idea of a very ancient and separate origin for modern *Homo sapiens*.

The modern equivalent of such ideas focuses on a recent and restricted origin for modern *Homo sapiens*. This was dubbed the 'Garden of Eden' or 'Noah's Ark' model by the US anthropologist William Howells in 1976 because of the idea that all modern human variation had a localized origin from one centre. Howells did not specify the centre of origin, but research since 1976 points to Africa as especially important in modern human origins.

The Out of Africa model

The consequent 'Out of Africa' model claims that *Homo erectus* evolved into modern *Homo sapiens* in Africa about 100,000–150,000 years ago. Part of the African stock of early modern humans spread from the continent into adjoining regions and eventually reached Australia, Europe, and the Americas (probably by 45,000, 40,000 and 15,000 years ago respectively). Regional ('racial') variation only developed during and after the dispersal, so that there is no continuity of regional features between *Homo erectus* and present counterparts in the same regions.

Like the multiregional model, this view accepts that *Homo erectus* evolved into new forms of human in inhabited regions outside Africa, but argues that these non-African lineages became extinct without evolving into modern humans. Some, such as the Neanderthals, were displaced and then replaced by the spread of modern humans into their regions.

An African Eve?

In 1987, research on the genetic material called mitochondrial DNA (mtDNA) in living humans led to the reconstruction of a hypothetical female ancestor for all present-day humanity. This 'Eve' was believed to have lived in Africa about 200,000 years ago. Recent re-examination of the Eve research has cast doubt on this hypothesis, but further support for an Out of Africa model has come from genetic studies of nuclear DNA, which also point to a recent African origin for present-day *Homo sapiens*.

Studies of fossil material from the last 50,000 years also seem to indicate that many 'racial' features in the human skeleton have only developed over the last 30,000 years, in line with the Out of Africa model, and at odds with the one-million-year timespan expected from the multiregional model.

Chris Stringer

temperament; of blood a 'sanguine', or passionate, one; of yellow bile a 'choleric', or irascible, temperament; and of black bile a 'melancholy', or depressive, one. The Greek physician Galen connected the theory to that of the four elements (see ◊elements, the four): the phlegmatic was associated with water, the sanguine with air, the choleric with fire, and the melancholic with earth. An imbalance of the humours could supposedly be treated by diet.

humus component of ◊soil consisting of decomposed or partly decomposed organic matter, dark in colour and usually richer towards the surface. It has a higher carbon content than the original material and a lower nitrogen content, and is an important source of minerals in soil fertility.

hundredweight imperial unit (abbreviation cwt) of mass, equal to 112 lb (50.8 kg). It is sometimes called the long hundredweight, to distinguish it from the short hundredweight or *cental*, equal to 100 lb (45.4 kg).

Huntington's chorea rare hereditary disease of the nervous system that mostly begins in middle age. It is characterized by involuntary movements (◊chorea), emotional disturbances, and rapid mental degeneration progressing to ◊dementia. There is no known cure but the genetic mutation giving rise to the disease was located 1993, making it easier to test individuals for the disease and increasing the chances of developing a cure.

hurricane revolving storm in tropical regions, called **typhoon** in the N Pacific. It originates at latitudes between 5° and 20° N or S of the Equator, when the surface temperature of the ocean is above 27°C/80°F. A central calm area, called the eye, is surrounded by inwardly spiralling winds (anticlockwise in the northern hemisphere) of up to 320 kph/200 mph. A hurricane is accompanied by lightning and torrential rain, and can cause extensive damage. In meteorology, a hurricane is a wind of force 12 or more on the ◊Beaufort scale.

hyaline membrane disease former name for ◊respiratory distress syndrome.

hybrid offspring from a cross between individuals of two different species, or two inbred lines within a species. In most cases, hybrids between species are infertile and unable to reproduce sexually. In plants, however, doubling of the chromosomes (see ◊polyploid) can restore the fertility of such hybrids.

hydathode specialized pore, or less commonly, a hair, through which water is secreted by hydrostatic pressure from the interior of a plant leaf onto the surface. Hydathodes are found on many different plants and are usually situated around the leaf margin at vein endings. Each pore is surrounded by two crescent-shaped cells and resembles an open ◊stoma, but the size of the opening cannot be varied as in a stoma. The process of water secretion through hydathodes is known as ◊guttation.

Hydra in astronomy, the largest constellation, winding across more than a quarter of the sky between ◊Cancer and ◊Libra in the southern hemisphere. Hydra is named after the multiheaded monster slain by Hercules. Despite its size, it is not prominent; its brightest star is second-magnitude Alphard.

hydrate chemical compound that has discrete water molecules combined with it. The water is known as *water of crystallization* and the number of water molecules

associated with one molecule of the compound is denoted in both its name and chemical formula: for example, $CuSO_4.5H_2O$ is copper(II) sulphate pentahydrate.

hydration in earth science, a form of ◊chemical weathering caused by the expansion of certain minerals as they absorb water. The expansion weakens the parent rock and may cause it to break up.

hydraulic action in earth science, the erosive force exerted by water (as distinct from the forces exerted by rocky particles carried by water). It can wear away the banks of a river, particularly at the outer curve of a meander (bend in the river), where the current flows most strongly.

Hydraulic action occurs as a river tumbles over a waterfall to crash onto the rocks below. It will lead to the formation of a plunge pool below the waterfall. The hydraulic action of ocean waves and turbulent currents forces air into rock cracks, and therefore brings about erosion by ◊cavitation.

hydraulic radius measure of a river's channel efficiency (its ability to discharge water), used by water engineers to assess the likelihood of flooding. The hydraulic radius of a channel is defined as the ratio of its cross-sectional area to its wetted perimeter (the part of the cross-section that is in contact with the water).

The greater the hydraulic radius, the greater the efficiency of the channel and the less likely the river is to flood. The highest values occur when channels are deep, narrow, and semi-circular in shape.

hydraulics field of study concerned with utilizing the properties of water and other liquids, in particular the way they flow and transmit pressure, and with the application of these properties in engineering. It applies the principles of ◊hydrostatics and hydrodynamics. The oldest type of

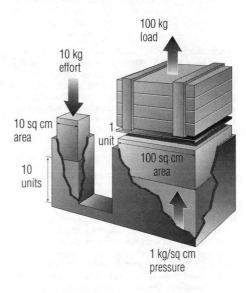

hydraulics The hydraulic jack transmits the pressure on a small piston to a larger one. A larger total force is developed by the larger piston but it moves a smaller distance than the small piston.

hydraulic machine is the **hydraulic press**, invented by Joseph Bramah in England in 1795. The hydraulic principle of pressurized liquid increasing mechanical efficiency is commonly used on vehicle braking systems, the forging press, and the hydraulic systems of aircraft and excavators.

A hydraulic press consists of two liquid-connected pistons in cylinders, one of narrow bore, one of large bore. A force applied to the narrow piston applies a certain pressure (force per unit area) to the liquid, which is transmitted to the larger piston. Because the area of this piston is larger, the force exerted on it is larger. Thus the original force has been magnified, although the smaller piston must move a great distance to move the larger piston only a little, hence mechanical efficiency is gained in force but lost in movement.

hydride chemical compound containing hydrogen and one other element, and in which the hydrogen is the more electronegative element (see ◊electronegativity).

Hydrides of the more reactive metals may be ionic compounds containing a hydride anion (H⁻).

hydrocarbon any of a class of chemical compounds containing only hydrogen and carbon (for example, the alkanes and alkenes). Hydrocarbons are obtained industrially principally from petroleum and coal tar.

hydrocephalus potentially serious increase in the volume of cerebrospinal fluid (CSF) within the ventricles of the brain. In infants, since their skull plates have not fused, it causes enlargement of the head, and there is a risk of brain damage from CSF pressure on the developing brain.

Hydrocephalus may be due to mechanical obstruction of the outflow of CSF from the ventricles or to faulty reabsorption. Treatment usually involves surgical placement of a shunt system to drain the fluid into the abdominal cavity. In infants, the condition is often seen in association with ◊spina bifida. Hydrocephalus may occur as a consequence of brain injury or disease.

hydrochloric acid HCl solution of hydrogen chloride (a colourless, acidic gas) in water. The concentrated acid is about 35% hydrogen chloride and is corrosive. The acid is a typical strong, monobasic acid forming only one series of salts, the chlorides. It has many industrial uses, including recovery of zinc from galvanized scrap iron and the production of chlorine. It is also produced in the stomachs of animals for the purposes of digestion.

hydrocyanic acid or **prussic acid** solution of hydrogen cyanide gas (HCN) in water. It is a colourless, highly poisonous, volatile liquid, smelling of bitter almonds.

hydrodynamics branch of physics dealing with fluids (liquids and gases) in motion.

hydroelectric power electricity generated by moving water. In a typical scheme, water stored in a reservoir, often created by damming a river, is piped into water ◊turbines, coupled to electricity generators. In ◊pumped storage plants, water flowing through the turbines is recycled. A ◊tidal power station exploits the rise and fall of the tides. About one-fifth of the world's electricity comes from hydroelectric power.

Hydroelectric plants have prodigious generating capacities. The Grand Coulee plant in Washington State, USA, has a power output of around 10,000 megawatts.

The Itaipu power station on the Paraná River (Brazil/Paraguay) has a potential capacity of 12,000 megawatts.

Work on the world's largest hydroelectric project, the Three Gorges Dam on the Chang Jiang, was officially inaugurated in Dec 1994. By 1996, around 600,000 sq km/231,660 sq mi of land had been flooded worldwide for hydroelectric reservoirs.

hydrofoil wing that develops lift in the water in much the same way that an aeroplane wing develops lift in the air. A hydrofoil boat is one whose hull rises out of the water owing to the lift, and the boat skims along on the hydrofoils. The first hydrofoil was fitted to a boat in 1906. The first commercial hydrofoil went into operation in 1956. One of the most advanced hydrofoil boats is the Boeing ◊jetfoil. Hydrofoils are now widely used for fast island ferries in calm seas.

hydrogen colourless, odourless, gaseous, nonmetallic element, symbol H, atomic number 1, relative atomic mass 1.00797. It is the lightest of all the elements and occurs on Earth chiefly in combination with oxygen as water. Hydrogen is the most abundant element in the universe, where it accounts for 93% of the total number of atoms and 76% of the total mass. It is a component of most stars, including the Sun, whose heat and light are produced through the nuclear-fusion process that converts hydrogen into helium. When subjected to a pressure 500,000 times greater than that of the Earth's atmosphere, hydrogen becomes a solid with metallic properties, as in one of the inner zones of Jupiter. Hydrogen's common and industrial uses include the hardening of oils and fats by hydrogenation, the creation of high-temperature flames for welding, and as rocket fuel. It has been proposed as a fuel for road vehicles.

Its isotopes ◊deuterium and ◊tritium (half-life 12.5 years) are used in nuclear weapons, and deuterons (deuterium nuclei) are used in synthesizing elements. The element's name, from the Greek for 'water generator', refers to the generation of water by the combustion of hydrogen, and was coined in 1787 by French chemist Louis Guyton de Morveau (1737–1816).

hydrogenation addition of hydrogen to an unsaturated organic molecule (one that contains ◊double bonds or ◊triple bonds). It is widely used in the manufacture of margarine and low-fat spreads by the addition of hydrogen to vegetable oils.

Vegetable oils contain double carbon-to-carbon bonds and are therefore examples of unsaturated compounds. When hydrogen is added to these double bonds, the oils become saturated and more solid in consistency.

hydrogen bomb bomb that works on the principle of nuclear ◊fusion. Large-scale explosion results from the thermonuclear release of energy when hydrogen nuclei are fused to form helium nuclei. The first hydrogen bomb was exploded at Enewetak Atoll in the Pacific Ocean by the USA in 1952.

hydrogen burning in astronomy, any of several processes by which hydrogen is converted to ◊helium by ◊nuclear fusion in the core of a star. In the Sun, the main process is the proton–proton chain, while in heavier stars the carbon cycle is more important. In both processes, four protons are converted to a helium nucleus with the emission of ◊positrons, ◊neutrinos, and gamma rays. The temperature must exceed several million K for hydrogen burning to start and the least massive stars (◊brown dwarfs) never become hot enough.

hydrogen carbonate or **bicarbonate** compound containing the ion HCO_3^-, an acid salt of carbonic acid (solution of carbon dioxide in water). When heated or treated with dilute acids, it gives off carbon dioxide. The most important compounds are ◊sodium hydrogen carbonate (bicarbonate of soda), and calcium hydrogen carbonate.

hydrogen cyanide HCN poisonous gas formed by the reaction of sodium cyanide with dilute sulphuric acid; it is used for fumigation.

The salts formed from it are cyanides – for example sodium cyanide, used in hardening steel and extracting gold and silver from their ores. If dissolved in water, hydrogen cyanide gives hydrocyanic acid.

hydrogen peroxide H_2O_2 in medicine, a liquid used, in diluted form, as an antiseptic. Oxygen is released when hydrogen peroxide is added to water and the froth helps to discharge dead tissue from wounds and ulcers. It is also used as a mouthwash and as a bleach.

hydrogen sulphide H_2S poisonous gas with the smell of rotten eggs. It is found in certain types of crude oil where it is formed by decomposition of sulphur compounds. It is removed from the oil at the refinery and converted to elemental sulphur.

hydrograph graph showing how the discharge of a river varies with time. By studying hydrographs, water engineers can predict when flooding is likely and take action to prevent its taking place.

A hydrograph shows the time lag, or delay, between peak rainfall and the resultant peak in discharge, and the length of time taken for that discharge to peak. The shorter the time lag and the higher the peak, the more likely it is that flooding will occur. Factors likely to give short time lags and high peaks include heavy rainstorms, steep slopes, deforestation, poor soil quality, and the covering of surfaces with impermeable substances such as tarmac and concrete. Actions taken by water engineers to increase time lags and lower peaks include planting trees in the drainage basin of a river.

hydrography study and charting of Earth's surface waters in seas, lakes, and rivers.

hydrological cycle alternative name for the ◊water cycle, by which water is circulated between the Earth's surface and its atmosphere.

hydrology study of the location and movement of inland water, both frozen and liquid, above and below ground. It is applied to major civil engineering projects such as irrigation schemes, dams, and hydroelectric power, and in planning water supply.

hydrolysis chemical reaction in which the action of water or its ions breaks down a substance into smaller molecules. Hydrolysis occurs in certain inorganic salts in solution, in nearly all nonmetallic chlorides, in esters, and in other organic substances. It is one of the mechanisms for the breakdown of food by the body, as in the conversion of starch to glucose.

hydrolysis in earth science, a form of ◊chemical weathering caused by the chemical alteration of certain minerals as they react with water. For example, the mineral feld-

spar in granite reacts with water to form a white clay called ◊china clay.

hydrometer in physics, an instrument used to measure the relative density of liquids (the density compared with that of water). A hydrometer consists of a thin glass tube ending in a sphere that leads into a smaller sphere, the latter being weighted so that the hydrometer floats upright, sinking deeper into less dense liquids than into denser liquids. Hydrometers are used in brewing and to test the strength of acid in car batteries.

The hydrometer is based on ◊Archimedes' principle.

hydrophilic in chemistry, a term describing ◊functional groups with a strong affinity for water, such as the carboxyl group (–COOH).

If a molecule contains both a hydrophilic and a ◊hydrophobic group (a group that repels water), it may have an affinity for both aqueous and nonaqueous molecules. Such compounds are used to stabilize ◊emulsions or as ◊detergents.

hydrophily type of ◊pollination where the pollen is transported by water. Water-pollinated plants occur in 31 genera in 11 different families. They are found in habitats as diverse as rainforests and seasonal desert pools. Pollen is either dispersed underwater or on the water's surface.

Pollen may be released directly onto the water's surface, as in the sea grass *Halodule pinifolia*, forming pollen rafts, or as in the freshwater plant *Vallisneria*, the pollen may be released within floating male flowers. In Caribbean turtle grass, *Thalassia testudinum*, pollen is released underwater embedded in strands of mucilage. Denser than water, it is carried by the current.

hydrophobia another name for the disease ◊rabies.

hydrophobic in chemistry, a term describing ◊functional groups that repel water (the opposite of ◊hydrophilic).

hydrophone underwater ◊microphone and ancillary equipment capable of picking up waterborne sounds. It was originally developed to detect enemy submarines but is now also used, for example, for listening to the sounds made by whales.

hydrophyte plant adapted to live in water, or in water-logged soil.

Hydrophytes may have leaves with a very reduced or absent ◊cuticle and no ◊stomata (since there is no need to conserve water), a reduced root and water-conducting system, and less supporting tissue since water buoys plants up. There are often numerous spaces between the cells in their stems and roots to make ◊gas exchange with all parts of the plant body possible. Many have highly divided leaves, which lessens resistance to flowing water; an example is spiked water milfoil *Myriophyllum spicatum*.

hydroplane on a submarine, a movable horizontal fin angled downwards or upwards when the vessel is descending or ascending. It is also a highly manoeuvrable motorboat with its bottom rising in steps to the stern, or a ◊hydrofoil boat that skims over the surface of the water when driven at high speed.

hydroponics cultivation of plants without soil, using specially prepared solutions of mineral salts. Beginning in the 1930s, large crops were grown by hydroponic methods, at first in California but since then in many other parts of the world.

Julius von Sachs (1832–1897) in 1860 and W Knop in 1865 developed a system of plant culture in water whereby the relation of mineral salts to plant growth could be determined, but it was not until about 1930 that large crops could be grown. The term was first coined by US scientist W F Gericke.

hydrosphere the water component of the Earth, usually encompassing the oceans, seas, rivers, streams, swamps, lakes, groundwater, and atmospheric water vapour.

hydrostatics in physics, the branch of ◊statics dealing with fluids in equilibrium – that is, in a static condition. Practical applications include shipbuilding and dam design.

hydrothermal in geology, pertaining to a fluid whose principal component is hot water, or to a mineral deposit believed to be precipitated from such a fluid.

hydrothermal vein crack in rock filled with minerals precipitated through the action of circulating high-temperature fluids. Igneous activity often gives rise to the circulation of heated fluids that migrate outwards and move through the surrounding rock. When such solutions carry metallic ions, ore-mineral deposition occurs in the new surroundings on cooling.

hydrothermal vent hot fissure in the ocean floor, known as a ◊smoker.

hydroxide any inorganic chemical compound containing one or more hydroxyl (OH) groups and generally combined with a metal. Hydroxides include sodium hydroxide (caustic soda, NaOH), potassium hydroxide (caustic potash, KOH), and calcium hydroxide (slaked lime, $Ca(OH)_2$).

hydroxyl group an atom of hydrogen and an atom of oxygen bonded together and covalently bonded to an organic molecule. Common compounds containing hydroxyl groups are alcohols and phenols.

In chemical reactions, the hydroxyl group (–OH) frequently behaves as a single entity.

hydroxypropanoic acid technical name for ◊lactic acid.

hygiene the science of the preservation of health and prevention of disease through the promotion of cleanliness. It is chiefly concerned with such factors as the purity of air and water; bodily cleanliness; cleanliness in the home and workplace; and sound practice in the preparation and distribution of food.

hygrometer in physics, any instrument for measuring the humidity, or water vapour content, of a gas (usually air). A wet and dry bulb hygrometer consists of two vertical thermometers, with one of the bulbs covered in absorbent cloth dipped into water. As the water evaporates, the bulb cools, producing a temperature difference between the two thermometers. The amount of evaporation, and hence cooling of the wet bulb, depends on the relative humidity of the air.

Other hygrometers work on the basis of a length of natural fibre, such as hair or a fine strand of gut, changing with variations in humidity. In a dew-point hygrometer, a polished metal mirror gradually cools until a fine mist of water (dew) forms on it. This gives a measure of the ◊dew point, from which the air's relative humidity can be calculated.

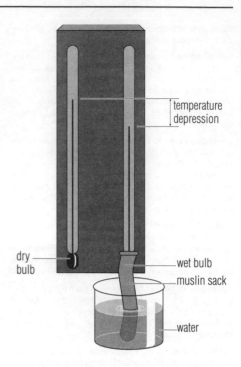

temperature depression

dry bulb

wet bulb

muslin sack

water

hygrometer The most common hygrometer, or instrument for measuring the humidity of a gas, is the wet and dry bulb hygrometer. The wet bulb records a lower temperature because water evaporates from the muslin, taking heat from the wet bulb. The degree of evaporation and hence cooling depends upon the humidity of the surrounding air or other gas.

hyoscine or **scopolamine** drug that acts on the autonomic nervous system and prevents muscle spasm. It is frequently included in ◊premedication to dry up lung secretions and as a postoperative sedative. It is also used to treat ulcers, to relax the womb in labour, for travel sickness, and to dilate the pupils before an eye examination. It is an alkaloid, $C_{17}H_{21}NO_2$, obtained from various plants of the nightshade family (such as ◊belladonna).

hyperactivity condition of excessive activity in young children, combined with restlessness, inability to concentrate, and difficulty in learning. There are various causes, ranging from temperamental predisposition to brain disease. In some cases food additives have come under suspicion; in such instances modification of the diet may help. Mostly there is improvement at puberty, but symptoms may persist in the small proportion diagnosed as having ◊attention-deficit hyperactivity disorder.

hyperalgesia in medicine, excessive sensitivity to ◊pain.

hyperbola in geometry, a curve formed by cutting a right circular cone with a plane so that the angle between the plane and the base is greater than the angle between the base and the side of the cone. All hyperbolae are bounded by two asymptotes (straight lines which the hyperbola moves closer and closer to but never reaches).

A hyperbola is a member of the family of curves known as ◊conic sections.

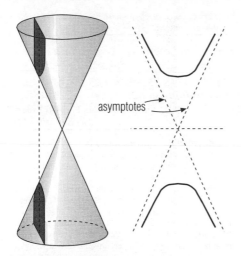

asymptotes

hyperbola The hyperbola is produced when a cone is cut by a plane. It is one of a family of curves called conic sections: the circle, ellipse, and parabola. These curves are produced when the plane cuts the cone at different angles and positions.

A hyperbola can also be defined as a path traced by a point that moves such that the ratio of its distance from a fixed point (focus) and a fixed straight line (directrix) is a constant and greater than 1; that is, it has an ◊eccentricity greater than 1.

hypercharge in physics, a property of certain ◊elementary particles, analogous to electric charge, that accounts for the absence of some expected behaviour (such as decay) in terms of the short-range strong nuclear force, which holds atomic nuclei together.

hyperkinetic syndrome in medicine, another name for ◊attention-deficit hyperactivity disorder.

hypertension abnormally high ◊blood pressure due to a variety of causes, leading to excessive contraction of the smooth muscle cells of the walls of the arteries. It increases the risk of kidney disease, stroke, and heart attack.

Hypertension is one of the major public health problems of the developed world, affecting 15–20% of adults in industrialized countries. It may be of unknown cause (*essential hypertension*), or it may occur in association with some other condition, such as kidney disease (*secondary or symptomatic hypertension*). It is controlled with a low-salt diet and drugs.

hypertext system for viewing information (both text and pictures) on a computer screen in such a way that related items of information can easily be reached. For example, the program might display a map of a country; if the user clicks (with a ◊mouse) on a particular city, the program will display information about that city.

hyperthyroidism or *thyrotoxicosis* overactivity of the thyroid gland due to enlargement or tumour. Symptoms include accelerated heart rate, sweating, anxiety, tremor, and weight loss. Treatment is by drugs or surgery.

hypertrophy abnormal increase in size of a body organ or tissue.

hyperventilation in medicine, an abnormally rapid resting respiratory rate. It occurs in patients with chest and heart diseases, such as severe chronic obstructive lung disease, because of a lack of oxygen or an increase in carbon dioxide in the blood.

hypha (plural *hyphae*) delicate, usually branching filament, many of which collectively form the mycelium and fruiting bodies of a ◊fungus. Food molecules and other substances are transported along hyphae by the movement of the cytoplasm, known as 'cytoplasmic streaming'.

Typically hyphae grow by increasing in length from the tips and by the formation of side branches. Hyphae of the higher fungi (the ascomycetes and basidiomycetes) are divided by cross walls or septa at intervals, whereas those of lower fungi (for example, bread mould) are undivided. However, even the higher fungi are not truly cellular, as each septum is pierced by a central pore, through which cytoplasm, and even nuclei, can flow. The hyphal walls contain ◊chitin, a polysaccharide.

hypnosis artificially induced state of relaxation or altered attention characterized by heightened suggestibility. There is evidence that, with susceptible persons, the sense of pain may be diminished, memory of past events enhanced, and illusions or hallucinations experienced. Posthypnotic amnesia (forgetting what happened during hypnosis) and posthypnotic suggestion (performing an action after hypnosis that had been suggested during it) have also been demonstrated.

Hypnosis has a number of uses in medicine. Hypnotically induced sleep, for example, may assist the healing process, and hypnotic suggestion (◊hypnotherapy) may help in dealing with the symptoms of emotional and psychosomatic disorders. The Austrian physician Friedrich Anton Mesmer is said to be the discoverer of hypnosis, but he called it 'animal magnetism', believing it to be a physical force or fluid. The term 'hypnosis' was coined by James Braid (1795–1860), a British physician and surgeon who was the first to regard it as a psychological phenomenon. The Scottish surgeon James Esdaile (1805–1859), working in India, performed hundreds of operations in which he used hypnosis to induce analgesia (insensitivity to pain) or general anaesthesia (total insensitivity).

hypnotherapy use of hypnotic trance and posthypnotic suggestions to relieve stress-related conditions such as insomnia and hypertension, or to break health-damaging habits or addictions.

Though it is an effective method of modifying behaviour, its effects are of short duration unless it is used as an adjunct to ◊psychotherapy.

hypnotic any substance (such as ◊barbiturate, ◊benzodiazepine, alcohol) that depresses brain function, inducing sleep. Prolonged use may lead to physical or psychological addiction.

hypo in photography, a term for sodium thiosulphate, discovered in 1819 by John Herschel, and used as a fixative for photographic images since 1837.

hypocycloid in geometry, a cusped curve traced by a point on the circumference of a circle that rolls around the inside of another larger circle. (Compare ◊epicycloid.)

hypodermic syringe instrument used for injecting fluids beneath the skin into either muscles or blood vessels. It consists of a small graduated tube with a close-fitting piston and a nozzle onto which a hollow needle can be fitted.

hypogeal term used to describe seed germination in which the ◊cotyledons remain below ground. It can refer to fruits that develop underground, such as peanuts *Arachis hypogea*.

hypoglycaemia condition of abnormally low level of sugar (glucose) in the blood (below 60 g/100 ml), which starves the brain. It causes weakness, sweating, and mental confusion, sometimes fainting.

Hypoglycaemia is most often seen in ◊diabetes. Low blood sugar occurs when the diabetic has taken too much insulin. It is treated by administering glucose.

hypotension in medicine, abnormally low blood pressure. Postural hypotension refers to the sudden fall in blood pressure that can occur on standing up suddenly. The patient experiences dizziness and may fall. It is most common in the elderly.

hypotenuse the longest side of a right-angled triangle, opposite the right angle. It is of particular application in Pythagoras' theorem (the square of the hypotenuse equals the sum of the squares of the other two sides), and in trigonometry where the ratios ◊sine and ◊cosine are defined as the ratios opposite/hypotenuse and adjacent/hypotenuse respectively.

hypothalamus region of the brain below the ◊cerebrum which regulates rhythmic activity and physiological stability within the body, including water balance and temperature. It regulates the production of the pituitary gland's hormones and controls that part of the ◊nervous system governing the involuntary muscles.

hypothermia condition in which the deep (core) temperature of the body falls below 35°C. If it is not discovered, coma and death ensue. Most at risk are the aged and babies (particularly if premature).

hypothesis in science, an idea concerning an event and its possible explanation. The term is one favoured by the followers of the philosopher Karl Popper, who argue that the merit of a scientific hypothesis lies in its ability to make testable predictions.

It is a good morning exercise for a research scientist to discard a pet hypothesis every day before breakfast. It keeps him young.

On **hypothesis** Konrad Lorenz
The So-Called Evil

hypothyroidism or *myxoedema* deficient functioning of the thyroid gland, causing slowed mental and physical performance, weight gain, sensitivity to cold, and susceptibility to infection.

This may be due to lack of iodine in the diet or a defect of the thyroid gland, both being productive of ◊goitre; or to the pituitary gland providing insufficient stimulus to the thyroid gland. Treatment of thyroid deficiency is by the hormone thyroxine. When present from birth, hypothyroidism can lead to cretinism if untreated.

hypoventilation in medicine, an abnormally slow resting respiratory rate characterized by shallow breathing. It can result in a lack of oxygen in the blood and can lead to organ damage and death if it is severe and remains untreated. Injury to the respiratory centre and some drugs, such as opioid analgesics, can cause hypoventilation.

hypsometer instrument for testing the accuracy of a thermometer at the boiling point of water. It was originally used for determining altitude by comparing changes in the boiling point with changes in atmospheric pressure.

hysterectomy surgical removal of all or part of the uterus (womb). The operation is performed to treat fibroids (benign tumours growing in the uterus) or cancer; also to relieve heavy menstrual bleeding. A woman who has had a hysterectomy will no longer menstruate and cannot bear children.

Hz in physics, the symbol for ◊*hertz*.

Mercury. In 1968 it passed 6 million km/4 million mi from the Earth.

ice solid formed by water when it freezes. It is colourless and its crystals are hexagonal. The water molecules are held together by hydrogen bonds.

The freezing point of ice, used as a standard for measuring temperature, is 0° for the Celsius and Réaumur scales and 32° for the Fahrenheit. Ice expands in the act of freezing (hence burst pipes), becoming less dense than water (0.9175 at 5°C/41°F).

ice form of methamphetamine that is smoked for its stimulating effect; its use has been illegal in the USA since 1989. Its use may be followed by a period of depression and psychosis.

Ice

The world's largest permanent ice sheet lies over the continent of Antarctica. It covers an area equivalent to the USA and Europe together, or twice the size of Australia, and contains 95% of the world's permanent ice.

Iapetus Ocean or **Proto-Atlantic** sea that existed in early ◊Palaeozoic times between the continent that was to become Europe and that which was to become North America. The continents moved together in the late Palaeozoic, obliterating the ocean. When they moved apart once more, they formed the Atlantic.

iatrogenic caused by medical treatment; the term 'iatrogenic disease' may be applied to any pathological condition or complication that is caused by the treatment, the facilities, or the staff.

IBM (abbreviation for **International Business Machines**) multinational company, the largest manufacturer of computers in the world. The company is a descendant of the Tabulating Machine Company, formed in 1896 by US inventor Herman Hollerith to exploit his punched-card machines. It adopted its present name in 1924. By 1991 it had an annual turnover of $64.8 billion and employed about 345,000 people, but in 1992 and 1993 it made considerable losses. The company acquired Lotus Development Corporation in June 1995. By 1996 IBM had, under new management, recovered financially, with an annual turnover of more than $70 billion, which means it is still a dominant industry player.

Its aquisition of the Lotus Development Corporation gave IBM access to its wide range of innovative software, including the 1–2–3 spreadsheet and Notes, a market leader in groupware.

ibuprofen in medicine, a non-steroidal anti-inflammatory drug (◊NSAID) that is used to relieve pain and inflammation. It acts by inhibiting the formation of ◊prostaglandins that mediate inflammation and so is used to treat conditions that are characterized by pain and inflammation, such as arthritic conditions, dental pain, period pain, sprains, and sports injuries. It is usually taken by mouth. Adverse effects include irritation of the gut, which can be reduced by taking ibuprofen at the same time as food. It should not be taken by patients with gastric ulceration and caution is required in patients with allergic disorders. Creams and gels are available that can be applied directly to the inflamed area.

IC abbreviation for ◊**integrated circuit**.

Icarus in astronomy, an ◊Apollo asteroid 1.5 km/1 mi in diameter, discovered 1949. It orbits the Sun every 409 days at a distance of 28–300 million km/18–186 million mi (0.19–2.0 astronomical units). It was the first asteroid known to approach the Sun closer than does the planet

ice age any period of glaciation occurring in the Earth's history, but particularly that in the Pleistocene epoch, immediately preceding historic times. On the North American continent, ◊glaciers reached as far south as the Great Lakes, and an ice sheet spread over N Europe, leaving its remains as far south as Switzerland.

There were several glacial advances separated by interglacial stages during which the ice melted and temperatures were higher than today.

Formerly there were thought to have been only three or four glacial advances, but recent research has shown about 20 major incidences. For example, ocean-bed cores record the absence or presence in their various layers of such cold-loving small marine animals as radiolaria, which indicate a fall in ocean temperature at regular intervals. Other ice ages have occurred throughout geological time: there were four in the Precambrian era, one in the Ordovician, and one at the end of the Carboniferous and beginning of the Permian. The occurrence of an ice age is governed by a combination of factors (the **Milankovitch hypothesis**): (1) the Earth's change of attitude in relation to the Sun, that is, the way it tilts in a 41,000-year cycle and at the same time wobbles on its axis in a 22,000-year cycle, making the time of its closest approach to the Sun come at different seasons; and (2) the 92,000-year cycle of eccentricity in its orbit round the Sun, changing it from an elliptical to a near circular orbit, the severest period of an ice age coinciding with the approach to circularity. There is a possibility that the Pleistocene ice age is not yet over. It may reach another maximum in another 60,000 years.

Ice Age: Major Ice Ages

Name	Date (years ago)
Pleistocene	1.64 million–10,000
Permo-Carboniferous	330–250 million
Ordovician	440–430 million
Verangian	615–570 million
Sturtian	820–770 million
Gnejso	940–880 million
Huronian	2,700–1,800 million

Ice Age, Little period of particularly severe winters that gripped N Europe between the 13th and 17th centuries. Contemporary writings and paintings show that Alpine glaciers were much more extensive than at present, and rivers such as the Thames, which do not ice over today, were so frozen that festivals could be held on them.

iceberg floating mass of ice, about 80% of which is submerged, rising sometimes to 100 m/300 ft above sea level. Glaciers that reach the coast become extended into a broad foot; as this enters the sea, masses break off and drift towards temperate latitudes, becoming a danger to shipping.

Iceland spar form of ◊calcite, $CaCO_3$, originally found in Iceland. In its pure form Iceland spar is transparent and exhibits the peculiar phenomenon of producing two images of anything seen through it. It is used in optical instruments. The crystals cleave into perfect rhombohedra.

iceman nickname given to the preserved body of a prehistoric man discovered in a glacier on the Austrian–Italian border in 1991. On the basis of the clothing and associated artefacts, the body was at first believed to be 4,000 years old, from the Bronze Age. Carbon dating established its age at about 5,300 years. The discovery led to a reappraisal of the boundary between the Bronze and the Stone Age.

icon in computing, a small picture on the computer screen, or ◊VDU, representing an object or function that the user may manipulate or otherwise use. It is a feature of ◊graphical user interface (GUI) systems. Icons make computers easier to use by allowing the user to point to and click with a ◊mouse on pictures, rather than type commands.

icosahedron (plural *icosahedra*) regular solid with 20 equilateral (equal-sided) triangular faces. It is one of the five regular ◊polyhedra, or Platonic solids.

id in Freudian psychology, the mass of motivational and instinctual elements of the human mind, whose activity is largely governed by the arousal of specific needs. It is regarded as the ◊unconscious element of the human psyche, and is said to be in conflict with the ◊ego and the ◊superego.

idiot savant person who has a specific mental skill that has developed at the expense of general intelligence. An idiot savant is educationally slow but may be able to calculate the day of the week for any date, or memorize a large quantity of text. Most idiot savants are male.

igneous rock rock formed from cooling magma or lava, and solidifying from a molten state. Igneous rocks are largely composed of silica (SiO_2) and they are classified according to their crystal size, texture, method of formation, or chemical composition, for example by the proportions of light and dark minerals.

ignis fatuus another name for ◊will-o'-the-wisp.

ignition coil ◊transformer that is an essential part of a petrol engine's ignition system. It consists of two wire coils wound around an iron core. The primary coil, which is connected to the car battery, has only a few turns. The secondary coil, connected via the ◊distributor to the ◊spark plugs, has many turns. The coil takes in a low voltage (usually 12 volts) from the battery and transforms it to a high voltage (about 20,000 volts) to ignite the engine.

When the engine is running, the battery current is periodically interrupted by means of the contact breaker in the distributor. The collapsing current in the primary coil induces a current in the secondary coil, a phenomenon known as ◊electromagnetic induction. The induced current in the secondary coil is at very high voltage, typically about 15,000–20,000 volts. This passes to the spark plugs to create sparks.

ignition temperature or *fire point* minimum temperature to which a substance must be heated before it will spontaneously burn independently of the source of heat; for example, ethanol has an ignition temperature of 425°C/798°F and a flash point of 12°C/54°F.

ileostomy in medicine, operation in which an artificial opening in the ◊ileum is brought through the abdominal wall and a stoma (opening) is created. The stoma acts as an artificial anus and a bag is attached to collect waste material from the remainder of the gut. The operation is performed most often when the rectum is removed in patients with rectal cancer.

ileum part of the small intestine of the ◊digestive system, between the duodenum and the colon, that absorbs digested food.

Its wall is muscular so that waves of contraction (peristalsis) can mix the food and push it forward. Numerous fingerlike projections, or villi, point inwards from the wall, increasing the surface area available for absorption. The ileum has an excellent blood supply, which receives the food molecules passing through the wall and transports them to the liver via the hepatic portal vein.

illumination or *illuminance* the brightness or intensity of light falling on a surface. It depends upon the brightness, distance, and angle of any nearby light sources. The SI unit is the ◊lux.

ilmenite oxide of iron and titanium, iron titanate ($FeTiO_3$); an ore of titanium. The mineral is black, with a metallic lustre. It is found as an accessory mineral in mafic igneous rocks and in sands.

ILS abbreviation for ◊*instrument landing system*, an automatic system for assisting aircraft landing at airports.

image picture or appearance of a real object, formed by light that passes through a lens or is reflected from a mirror. If rays of light actually pass through an image, it is called a *real image*. Real images, such as those produced by a camera or projector lens, can be projected onto a screen. An image that cannot be projected onto a screen, such as that seen in a flat mirror, is known as a *virtual image*.

imaginary number term often used to describe the non-real element of a ◊complex number. For the complex number $(a + ib)$, ib is the imaginary number where $i = \sqrt{-1}$, and b any real number.

immediate access memory in computing, ◊memory provided in the ◊central processing unit to store the programs and data in current use.

immiscible describing liquids that will not mix with each other, such as oil and water. When two immiscible liquids

FEATURE FOCUS

Immunity from the super-bugs

The recent flurry of deaths from antibiotic-resistant bacteria has pointed up the need for new drugs to fight infection. For the harsh fact is that, after half a century of widespread, often inappropriate use, antibiotics are beginning to lose their power. Microbiologists cite a growing number of bacterial strains – including some responsible for pneumonia, tuberculosis, and blood-poisoning – which defy all existing treatments. Most at risk from these so-called super-bugs are people whose immune systems are in some way impaired – the frail elderly, AIDS patients, transplant recipients, and patients receiving chemotherapy for cancer.

The search for new remedies

The phenomenon of super-resistance gives fresh urgency to the search for new remedies. Traditionally, researchers have mostly looked for antibiotics from organisms in the soil, and one promising new contender, rapamycin, has been isolated from bacteria and microscopic fungi contained in soil samples from Easter Island. Synthesized by workers in California, this experimental compound is thought also to have potential as an anti-cancer agent and immunosuppressant.

Meanwhile the hunt has been widened to other sources in the natural world – for instance the African clawed frog. In the late 1980s its nerve endings were found to contain a small peptide which not only kills bacteria but seems effective also against fungi, parasites, and some cancer cells. This compound, since named magainin (from the Hebrew word for a shield), is believed to have a similar mode of action to that of another new antibiotic substance, squalamine, discovered in tissue from the dogfish shark.

A new strategy

Other researchers have adopted a whole new strategy in the battle against infectious disease. Instead of targeting bacteria, as conventional antibiotics do, they are developing compounds to boost the body's natural defences. Among these is a genetically engineered carbohydrate that causes proliferation and increased bacterial activity of some white blood cells. One attraction of the immunological approach is that there is no likelihood of meeting the resistance seen with antibiotics since these new drugs – all still at the experimental stage – are merely stimulating a natural mechanism.

Paulette Pratt

are shaken together, a turbid mixture is produced. This normally forms separate layers on being left to stand.

immunity the protection that organisms have against foreign microorganisms, such as bacteria and viruses, and against cancerous cells (see ◊cancer). The cells that provide this protection are called white blood cells, or leucocytes, and make up the immune system. They include neutrophils and ◊macrophages, which can engulf invading organisms and other unwanted material, and natural killer cells that destroy cells infected by viruses and cancerous cells. Some of the most important immune cells are the ◊B cells and ◊T cells. Immune cells coordinate their activities by means of chemical messengers or ◊lymphokines, including the antiviral messenger ◊interferon. The lymph nodes play a major role in organizing the immune response.

Immunity is also provided by a range of physical barriers such as the skin, tear fluid, acid in the stomach, and mucus in the airways. ◊AIDS is one of many viral diseases in which the immune system is affected.

immunization conferring immunity to infectious disease by artificial methods. The most widely used technique is ◊vaccination.

Immunization is an important public health measure. If most of the population has been immunized against a particular disease, it is impossible for an epidemic to take hold.

Vaccination against smallpox was developed by Edward Jenner in 1796. In the late 19th century Louis Pasteur developed vaccines against cholera, typhoid, typhus, plague, and yellow fever.

When meditating over a disease, I never think of finding a remedy for it, but, instead, a means of preventing it.

On **immunization** Louis Pasteur, in address to the Fraternal Association of Former Students of the Ecole Centrale des Arts et Manufactures, Paris, 15 May 1884

immunization for travel in medicine, advised immunization for travellers. Although immunization against infectious diseases is rarely an essential legal requirement to enter a country, immunization is often advisable to protect the traveller from diseases not encountered at home. It is necessary to plan ahead for immunizations as some require more than one dose of vaccine and some cannot be given at the same time. General practitioners and travel health centres provide the most up-to-date advise on the immunizations recommended for travelling.

In general, travellers to all countries should have immunity to tetanus and poliomyelitis and childhood immunizations should be up to date. Additional immunizations are required in non-European areas surrounding

Immunization: Chronology

23–79	Pliny the Elder suggested using liver from mad dogs as protection against rabies.
1500s	Asian physicians immunized against smallpox using the crusts from pustules. This was only partially successful.
1720s	Lady Mary Wortley Montagu introduced smallpox immunization into Europe from Turkey.
1796	British physician Edward Jenner developed a safe smallpox vaccine using the cowpox virus.
1853	Vaccination against smallpox was made compulsory in Britain.
1885	French microbiologist Louis Pasteur developed a vaccine for rabies.
1894– 1904	German immunologist Emil von Behring and Japanese bacteriologist Kitasato Shibasaburo successfully tested vaccines for diphtheria and tetanus.
1896	A E Wright developed typhoid vaccine.
1914	Tetanus vaccine became available on a large scale.
1920s	Tuberculosis vaccine produced.
1937	South African-born US microbiologist Max Theiler developed vaccine 17-D, still the main form of protection against yellow fever.
1949	Whooping cough vaccine licensed.
1952	US microbiologist Jonas Salk developed the first successful vaccine for poliomyelitis.
1960s	Measles and rubella vaccines produced.
1961	The oral vaccine for poliomyelitis, developed by Russian-born virologist Albert Sabin, became widely available.
1967	The World Health Organization (WHO) began its global campaign against smallpox.
1970s	Vaccines produced for meningococcal diseases and chickenpox. The WHO began constructing its 'cold chain' to ensure adequate transport and refrigeration for vaccines, which can take up to two years to reach their target in a developing country.
1974	WHO launched the Expanded Programme on Immunization, as fewer than 5% of children in developing countries were immunized.
1978	Pulmonary disease vaccine was developed.
1980s	Vaccine for hepatitis B, and a combination vaccine for measles, mumps, and rubella (MMR) became available.
1980	Smallpox virtually eradicated.
1984	Leprosy vaccine developed.
1989–90	More than 100 children died when a measles epidemic swept through several large cities in the USA, highlighting the failure of the immunization programme. Three-quarters of the 45,000 children affected had not been vaccinated.
1990	Vaccine introduced for *Hemophilus influenzae*, a cause of meningitis.
1991	WHO estimated 80% of the world's children were immunized against diphtheria, whooping cough, tetanus, measles, polio, and tuberculosis. A US survey of nine major cities found that less than half of school age children had been fully vaccinated against infectious diseases by their second birthday.
1994	Fear of a measles epidemic in the UK led to a vaccination programme for all children aged 5–16.
1995	First human trials of a vaccine administered by eating genetically-engineered potatoes.
1996	Heat sensitive chemical monitors used on polio vaccine containers. If the monitor changes colour health workers will know that high temperatures have destroyed the vaccine. A vaccine for salmonella for use in poultry was approved in Australia.

the Mediterranean, in Africa, the Middle East, Asia, and South America.

Typhoid Typhoid vaccine is indicated for travellers to areas where typhoid is endemic, such as Asia and South America. However, vaccination is no substitution for good hygiene. Food should be freshly prepared and cooked but uncooked vegetables, including salads, should be avoided. Fruit should be peeled. Drinking water should be bottled, boiled or treated with water sterilising tablets.

Cholera Cholera vaccine provides little protection against this disease and good hygiene is particularly important in countries where cholera is endemic.

Meningitis There is a high incidence of meningococcal meningitis in some Asian countries and in southern sub-Saharan Africa and vaccination is recommended for these areas.

Hepatitis A Immunization against hepatitis A is advisable for people visiting Asia and Africa.

Yellow fever International certificates of vaccination against yellow fever are still required for travel to many countries in Africa and South America.

Malaria In addition to vaccination, travellers should take a course of antimalarial tablets if they are visiting areas in which malaria is endemic. Resistance to older antimalarial drugs has developed in some areas, such as central Africa, and it is essential to take this into account when requesting drugs for malaria prophylaxis.

The risks of being in contact with diseases such as hepatitis and typhoid are reduced for people staying for a short time in first-class accommodation. Immunization is especially important for people who are planning long backpacking trips or for those who are intending to work in rural areas of the countries visited. Additional immunizations, such as rabies vaccine and Japanese encephalitis vaccine, may be needed for people visiting very remote areas.

immunocompromised lacking a fully effective immune system. The term is most often used in connection with infections such as ◊AIDS where the virus interferes with the immune response (see ◊immunity).

Other factors that can impair the immune response are pregnancy, diabetes, old age, malnutrition and extreme stress, making someone susceptible to infections by microorganisms (such as listeria) that do not affect normal, healthy people. Some people are immunodeficient; others could be on ◊immunosuppressive drugs.

immunodeficient lacking one or more elements of a working immune system. Immune deficiency is the term generally used for patients who are born with such a defect, while those who acquire such a deficiency later in life are referred to as ◊ *immunocompromised* or immunosuppressed.

A serious impairment of the immune system is sometimes known as SCID, or Severe Combined Immune Deficiency. At one time children born with this condition would have died in infancy. They can now be kept alive in a germ-free environment, then treated with a bone-marrow transplant from a relative, to replace the missing immune cells. At present, the success rate for this type of treatment is still fairly low. See also ◊gene therapy.

immunoglobulin human globulin ◊protein that can be separated from blood and administered to confer immediate immunity on the recipient. It participates in the immune reaction as the antibody for a specific ◊antigen (disease-causing agent).

Normal immunoglobulin (gamma globulin) is the fraction of the blood serum that, in general, contains the most antibodies, and is obtained from plasma pooled from about a thousand donors. It is given for short-term (two to three months) protection when a person is at risk, mainly from hepatitis A (infectious hepatitis), or when a pregnant woman, not immunized against ◊German measles, is exposed to the rubella virus.

Specific immunoglobulins are injected when a susceptible (nonimmunized) person is at risk of infection from a potentially fatal disease, such as hepatitis B (serum hepatitis), rabies, or tetanus. These immunoglobulins are prepared from blood pooled from donors convalescing from the disease.

immunosuppressive any drug that suppresses the body's normal immune responses to infection or foreign tissue. It is used in the treatment of autoimmune disease (see ◊autoimmunity); as part of chemotherapy for leukaemias, lymphomas, and other cancers; and to help prevent rejection following organ transplantation.

Immunosuppressed patients are at greatly increased risk of infection.

impact printer computer printer that creates characters by striking an inked ribbon against the paper beneath. Examples of impact printers are dot-matrix printers, daisywheel printers, and most types of line printer.

Impact printers are noisier and slower than nonimpact printers, such as ink-jet and laser printers, but can be used to produce carbon copies.

impedance the total opposition of a circuit to the passage of alternating electric current. It has the symbol Z. For an ◊alternating current (AC) it includes the resistance R and the reactance X (caused by ◊capacitance or ◊inductance); the impedance can then be found using the equation $Z^2 = R^2 + X^2$.

imperial system traditional system of units developed in the UK, based largely on the foot, pound, and second (f.p.s.) system.

impetigo skin infection with either streptococcus or staphylococcus bacteria, characterized by encrusted yellow sores on the skin.

Particularly common in infants and small children, it is highly contagious but curable with antibiotics.

implantation in mammals, the process by which the developing ◊embryo attaches itself to the wall of the mother's uterus and stimulates the development of the ◊placenta. In humans it occurs 6–8 days after ovulation.

In some species, such as seals and bats, implantation is delayed for several months, during which time the embryo does not grow; thus the interval between mating and birth may be longer than the ◊gestation period.

import file in computing, a file that can be read by a program even though it was produced as an ◊export file by a different program or make of computer.

impotence in medicine, a physical inability to perform sexual intercourse (the term is not usually applied to women). Impotent men fail to achieve an erection, and this may be due to illness, the effects of certain drugs, or psychological factors.

The first prescription drug for treating the condition, alprostadil, was approved for sale in the USA in 1995. Self-administered by injection into the penis before intercourse, it functions as a muscle relaxant, increasing blood flow. Alprostadil is already marketed in more than 20 countries.

imprinting in ◊ethology, the process whereby a young animal learns to recognize both specific individuals (for example, its mother) and its own species.

Imprinting is characteristically an automatic response to specific stimuli at a time when the animal is especially sensitive to those stimuli (the **sensitive period**). Thus, goslings learn to recognize their mother by following the first moving object they see after hatching; as a result, they can easily become imprinted on other species, or even inanimate objects, if these happen to move near them at this time. In chicks, imprinting occurs only between 10 and 20 hours after hatching. In mammals, the mother's attachment to her infant may be a form of imprinting made possible by a sensitive period; this period may be as short as the first hour after giving birth.

impulse in mechanics, the product of a force and the time over which it acts. An impulse applied to a body causes its ◊momentum to change and is equal to that change in momentum. It is measured in newton seconds (N s).

For example, the impulse J given to a football when it is kicked is given by:

$$J = Ft$$

where F is the kick force in newtons and t is the time in seconds for which the boot is in contact with the ball.

in abbreviation for ◊ *inch*, a measure of distance.

Inbreeding

Three out of ten dalmatian dogs suffer from some form of hearing disability, and 8 % of all dalmations are entirely deaf. Their beautiful spotty coats are caused by intense inbreeding, which has the side-effect of increasing genetic disorders – such as deafness.

inbreeding in ◊genetics, the mating of closely related individuals. It is considered undesirable because it increases the risk that offspring will inherit copies of rare deleterious ◊recessive alleles (genes) from both parents and so suffer from disabilities.

incandescence emission of light from a substance in consequence of its high temperature. The colour of the emitted light from liquids or solids depends on their temperature, and for solids generally the higher the temperature the whiter the light. Gases may become incandescent through ◊ionizing radiation, as in the glowing vacuum ◊discharge tube.

The oxides of cerium and thorium are highly incandescent and for this reason are used in gas mantles. The light from an electric filament lamp is due to the incandescence of the filament, rendered white-hot when a current passes through it.

inch imperial unit of linear measure, a twelfth of a foot, equal to 2.54 centimetres.

It was defined in statute by Edward II of England as the length of three barley grains laid end to end.

incisor sharp tooth at the front of the mammalian mouth. Incisors are used for biting or nibbling, as when a rabbit or a sheep eats grass. Rodents, such as rats and squirrels, have large continually-growing incisors, adapted for gnawing. The elephant tusk is a greatly enlarged incisor. In humans, the incisors are the four teeth at the front centre of each jaw.

inclination angle between the ◊ecliptic and the plane of the orbit of a planet, asteroid, or comet. In the case of satellites orbiting a planet, it is the angle between the plane of orbit of the satellite and the equator of the planet.

inclusive fitness in ◊genetics, the success with which a given variant (or allele) of a ◊gene is passed on to future generations by a particular individual, after additional copies of the allele in the individual's relatives and their offspring have been taken into account.

The concept was formulated by W D Hamilton as a way of explaining the evolution of ◊altruism in terms of ◊natural selection. See also ◊fitness and ◊kin selection.

incontinence failure or inability to control evacuation of the bladder or bowel (or both in the case of double incontinence). It may arise as a result of injury, childbirth, disease, or senility.

indeterminacy principle alternative name for ◊uncertainty principle.

index (plural **indices**) in mathematics, another term for ◊exponent, the number that indicates the power to which a term should be raised.

indicator in chemistry, a compound that changes its structure and colour in response to its environment. The commonest chemical indicators detect changes in ◊pH (for example, ◊litmus), or in the oxidation state of a system (redox indicators).

indicator species plant or animal whose presence or absence in an area indicates certain environmental conditions, such as soil type, high levels of pollution, or, in rivers, low levels of dissolved oxygen. Many plants show a preference for either alkaline or acid soil conditions, while certain trees require aluminium, and are found only in soils where it is present. Some lichens are sensitive to sulphur dioxide in the air, and absence of these species indicates atmospheric pollution.

indium soft, ductile, silver-white, metallic element, symbol In, atomic number 49, relative atomic mass 114.82. It occurs in nature in some zinc ores, is resistant to abrasion, and is used as a coating on metal parts. It was discovered 1863 by German metallurgists Ferdinand Reich (1799–1882) and Hieronymus Richter (1824–1898), who named it after the two indigo lines of its spectrum.

induced current electric current that appears in a closed circuit when there is relative movement of its conductor in a magnetic field. The effect is known as the **dynamo effect**, and is used in all ◊dynamos and generators to produce electricity. See ◊electromagnetic induction.

There is no battery or other source of power in a circuit in which an induced current appears: the energy supply is provided by the relative motion of the conductor and the magnetic field. The magnitude of the induced current depends upon the rate at which the magnetic flux is cut by the conductor, and its direction is given by Fleming's right-hand rule (see ◊Fleming's rules).

inductance in physics, the phenomenon where a changing current in a circuit builds up a magnetic field which induces an ◊electromotive force either in the same circuit and opposing the current (self-inductance) or in another circuit (mutual inductance). The SI unit of inductance is the henry (symbol H).

A component designed to introduce inductance into a circuit is called an inductor (sometimes inductance) and is usually in the form of a coil of wire. The energy stored in the magnetic field of the coil is proportional to its inductance and the current flowing through it. See ◊electromagnetic induction.

induction in obstetrics, deliberate intervention to initiate labour before it starts naturally; then it usually proceeds normally.

Induction involves rupture of the fetal membranes (amniotomy) and the use of the hormone oxytocin to stimulate contractions of the womb. In biology, induction is a term used for various processes, including the pro-

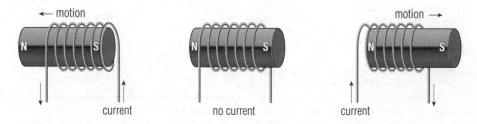

← motion   motion →

current   no current   current

induced current

duction of an ◊enzyme in response to a particular chemical in the cell, and the ◊differentiation of cells in an ◊embryo in response to the presence of neighbouring tissues.

In obstetrics, induction is recommended as a medical necessity where there is risk to the mother or baby in waiting for labour to begin of its own accord.

induction in physics, an alteration in the physical properties of a body that is brought about by the influence of a field. See ◊electromagnetic induction and ◊magnetic induction.

induction coil type of electrical transformer, similar to an ◊ignition coil, that produces an intermittent high-voltage alternating current from a low-voltage direct current supply.

It has a primary coil consisting of a few turns of thick wire wound around an iron core and passing a low voltage (usually from a battery). Wound on top of this is a secondary coil made up of many turns of thin wire. An iron armature and make-and-break mechanism (similar to that in an electric bell) repeatedly interrupts the current to the primary coil, producing a high, rapidly alternating current in the secondary circuit.

inductor device included in an electrical circuit because of its inductance.

industrial diseases in medicine, diseases connected with the workplace. Examples include lung diseases that develop following the chronic inhalation of dust, silica, and asbestos; back injuries due to heavy lifting; and dermatitis following the handling of irritant substances. Occupational diseases can also affect office workers. These include repetitive strain injury because of excessive keyboard use and the cluster of symptoms described by workers who occupy 'sick buildings' (see ◊sick building syndrome).

industry the extraction and conversion of raw materials, the manufacture of goods, and the provision of services. Industry can be either low technology, unspecialized, and labour-intensive, as in countries with a large unskilled labour force, or highly automated, mechanized, and specialized, using advanced technology, as in the industrialized countries. Major recent trends in industrial activity have been the growth of electronic, robotic, and microelectronic technologies, the expansion of the offshore oil industry, and the prominence of Japan and other Pacific-region countries in manufacturing and distributing electronics, computers, and motor vehicles.

inert gas or **noble gas** any of a group of six elements (helium, neon, argon, krypton, xenon, and radon), so named because they were originally thought not to enter into any chemical reactions. This is now known to be incorrect: in 1962, xenon was made to combine with fluorine, and since then, compounds of argon, krypton, and radon with fluorine and/or oxygen have been described.

The extreme unreactivity of the inert gases is due to the stability of their electronic structure. All the electron shells (energy levels) of inert gas atoms are full and, except for helium, they all have eight electrons in their outermost (◊valency) shell. The apparent stability of this electronic arrangement led to the formulation of the ◊octet rule to explain the different types of chemical bond found in simple compounds.

inertia in physics, the tendency of an object to remain in a state of rest or uniform motion until an external force is applied, as stated by Isaac Newton's first law of motion (see ◊Newton's laws of motion).

infantile paralysis former term for poliomyelitis. See ◊polio.

infant mortality rate measure of the number of infants dying under one year of age, usually expressed as the number of deaths per 1,000 live births. Improved sanitation, nutrition, and medical care have considerably lowered figures throughout much of the world; for example in the 18th century in the USA and UK infant mortality was about 500 per thousand, compared with under 10 per thousand in 1989. The lowest infant mortality rate is in Japan, at 4.5 per 1,000 live births. In much of the Third World, however, the infant mortality rate remains high.

infarct or **infarction** death and scarring of a portion of the tissue in an organ, as a result of congestion or blockage of a blood vessel serving it.

Myocardial infarction is the technical term for a heart attack.

infection invasion of the body by disease-causing organisms (pathogens, or germs) that become established, multiply, and produce symptoms. Bacteria and viruses cause most diseases, but diseases are also caused by other microorganisms, protozoans, and other parasites.

Most pathogens enter and leave the body through the digestive or respiratory tracts. Polio, dysentery, and typhoid are examples of diseases contracted by ingestion of contaminated foods or fluids. Organisms present in the saliva or nasal mucus are spread by airborne or droplet infection; fine droplets or dried particles are inhaled by others when the affected individual talks, coughs, or sneezes. Diseases such as measles, mumps, and tuberculosis are passed on in this way.

A less common route of entry is through the skin, either by contamination of an open wound (as in tetanus) or by penetration of the intact skin surface, as in a bite from a malaria-carrying mosquito. Relatively few diseases are transmissible by skin-to-skin contact. Glandular fever and herpes simplex (cold sore) may be passed on by kissing, and the group now officially bracketed as sexually transmitted diseases (◊STDs) are mostly spread by intimate contact.

inferiority complex in psychology, a ◊complex or cluster of repressed fears, described by Alfred Adler, based on physical inferiority. The term is popularly used to describe general feelings of inferiority and the overcompensation that often ensues.

inferior planet planet (Mercury or Venus) whose orbit lies within that of the Earth, best observed when at its greatest elongation from the Sun, either at eastern elongation in the evening (setting after the Sun) or at western elongation in the morning (rising before the Sun).

inferno in astrophysics, a unit for describing the temperature inside a star. One inferno is 1 billion K, or approximately 1 billion °C.

infertility in medicine, inability to reproduce. In women, this may be due to blockage in the Fallopian tubes, failure of ovulation, a deficiency in sex hormones, or general ill health. In men, impotence, an insufficient number of sperm or abnormal sperm may be the cause of infertility. Clinical investigation will reveal the cause of the infertility in about 75% of couples and assisted conception may then be appropriate.

infestation in medicine, occurrence of animal parasites in the intestine, hair, or clothing. Examples include head lice and scabies.

infinite series in mathematics, a series of numbers consisting of a denumerably infinite sequence of terms. The sequence $n, n^2, n^3, \ldots$ gives the series $n + n^2 + n^3 + \ldots$. For example, $1 + 2 + 3 + \ldots$ is a divergent infinite arithmetic series, and $8 + 4 + 2 + 1 + 1/2 + \ldots$ is a convergent infinite geometric series that has a sum to infinity of 16.

infinity mathematical quantity that is larger than any fixed assignable quantity; symbol ∞. By convention, the result of dividing any number by zero is regarded as infinity.

inflammation defensive reaction of the body tissues to disease or damage, including redness, swelling, and heat. Denoted by the suffix *-itis* (as in appendicitis), it may be acute or chronic, and may be accompanied by the formation of pus. This is an essential part of the healing process.

Inflammation occurs when damaged cells release a substance (◊histamine) that causes blood vessels to widen and leak into the surrounding tissues. This phenomenon accounts for the redness, swelling, and heat. Pain is due partly to the pressure of swelling and also to irritation of nerve endings. Defensive white blood cells congregate within an area of inflammation to engulf and remove foreign matter and dead tissue.

inflammatory bowel disease in medicine, one of a range of disorders that include ◊Crohn's disease and ulcerative ◊colitis. Dietary supplements to replace lost nutrients are necessary for patients with inflammatory bowel diseases.

inflation in cosmology, a phase of extremely fast expansion thought to have occurred within 10–32 seconds of the ◊Big Bang and in which almost all the matter and energy in the universe was created. The inflationary model based on this concept accounts for the density of the universe being very close to the ◊critical density, the smoothness of the ◊cosmic background radiation, and the homogeneous distribution of matter in the universe. Inflation was proposed by US astronomer Alan Guth in the early 1980s.

inflorescence in plants, a branch, or system of branches, bearing two or more individual flowers. Inflorescences can be divided into two main types: cymose (or definite) and racemose (or indefinite). In a **cymose inflorescence**, the tip of the main axis produces a single flower and subsequent flowers arise on lower side branches, as in forget-me-not *Myosotis* and chickweed *Stellaria*; the oldest flowers are, therefore, found at the tip. A **racemose inflorescence** has an active growing region at the tip of its main axis, and bears flowers along its length, as in hyacinth *Hyacinthus*; the oldest flowers are found near the base or, in cases where the inflorescence is flattened, towards the outside.

The stalk of the inflorescence is called a peduncle; the stalk of each individual flower is called a pedicel.

Types of racemose inflorescence include the **raceme**, a spike of similar, stalked flowers, as seen in lupin *Lupinus*. A **corymb**, seen in candytuft *Iberis amara*, is rounded or flat-topped because the pedicels of the flowers vary in length, the outer pedicels being longer than the inner ones. A **panicle** is a branched inflorescence made up of a number of racemes; such inflorescences are seen in many grasses, for example, the oat *Avena*. The pedicels of an **umbel**, seen in members of the carrot family (Umbelliferae), all arise from the same point on the main axis, like the spokes of an umbrella. Other types of racemose inflorescence include the ◊ **catkin**, a pendulous inflorescence, made up of many small stalkless flowers; the ◊ **spadix**, in which tiny flowers are borne on a fleshy axis; and the ◊ **capitulum**, in which the axis is flattened or rounded, bears many small flowers, and is surrounded by large petal-like bracts.

influenza any of various viral infections primarily affecting the air passages, accompanied by ◊systemic effects such as fever, chills, headache, joint and muscle pains, and lassitude. Treatment is with bed rest and analgesic drugs such as aspirin or paracetamol.

Depending on the virus strain, influenza varies in virulence and duration, and there is always the risk of secondary (bacterial) infection of the lungs (pneumonia). Vaccines are effective against known strains but will not give protection against newly evolving viruses. The 1918–19 influenza pandemic (see ◊epidemic) killed about 20 million people worldwide.

information technology (IT) collective term for the various technologies involved in processing and transmitting information. They include computing, telecommunications, and microelectronics.

Word processing, databases, and spreadsheets are just some of the computing ◊software packages that have revolutionized work in the office environment. Not only can work be done more quickly than before, but IT has given decisionmakers the opportunity to consider far more data when making decisions.

infrared absorption spectrometry technique used to determine the mineral or chemical composition of artefacts and organic substances, particularly amber. A sample is bombarded by infrared radiation, which causes the atoms in it to vibrate at frequencies characteristic of the substance present, and absorb energy at those frequencies from the infrared spectrum, thus forming the basis for identification.

infrared astronomy study of infrared radiation produced by relatively cool gas and dust in space, as in the areas around forming stars. In 1983, the Infra-Red Astronomy Satellite (IRAS) surveyed the entire sky at infrared wavelengths. It found five new comets, thousands of galaxies undergoing bursts of star formation, and the possibility of planetary systems forming around several dozen stars.

Planets and gas clouds emit their light in the far and mid-infrared region of the spectrum. The Infrared Space Observatory (ISO), launched in Nov 1995, observes a broad wavelength (3–200 micrometres) in this region. It is 10,000 times more sensitive than IRAS, and will search for ◊brown dwarfs (cool masses of gas smaller than the Sun).

infrared radiation invisible electromagnetic radiation of wavelength between about 0.75 micrometres and 1 millimetre – that is, between the limit of the red end of the visible spectrum and the shortest microwaves. All bodies above the ◊absolute zero of temperature absorb and radiate infrared radiation. Infrared radiation is used in medical photography and treatment, and in industry, astronomy, and criminology.

Infrared absorption spectra are used in chemical analysis, particularly for organic compounds. Objects that radiate infrared radiation can be photographed or made visible in the dark on specially sensitized emulsions. This is important for military purposes and in detecting people buried under rubble. The strong absorption by many substances of infrared radiation is a useful method of applying heat.

infrared telescope in astronomy, a ◊telescope designed to receive ◊electromagnetic waves in the infrared part of the spectrum. Infrared telescopes are always reflectors (glass lenses are opaque to infrared waves) and are normally of the ◊Cassegrain telescope type.

Since all objects at normal temperatures emit strongly in the infrared, careful design is required to ensure that the weak signals from the sky are not swamped by radiation from the telescope itself. Infrared telescopes are sited at high mountain observatories above the obscuring effects of water vapour in the atmosphere. Modern large telecopes are often designed to work equally well in both visible and infrared light.

ingestion process of taking food into the mouth. The method of food capture varies but may involve biting, sucking, or filtering. Many single-celled organisms have a region of their cell wall that acts as a mouth. In these cases surrounding tiny hairs (cilia) sweep food particles together, ready for ingestion.

inhalant in medicine, formulation of a drug that is designed to deliver it directly to the lungs. Traditional inhalants involve the dispersion of the drug in hot water and inhalation of the steam. Drugs can be inhaled from aerosols that have been developed for the treatment of asthma.

inhibition, neural in biology, the process in which activity in one ◊nerve cell suppresses activity in another. Neural inhibition in networks of nerve cells leading from sensory organs, or to muscles, plays an important role in allowing an animal to make fine sensory discriminations and to exercise fine control over movements.

ink-jet printer computer printer that creates characters and graphics by spraying very fine jets of quick-drying ink onto paper. Ink-jet printers range in size from small machines designed to work with microcomputers to very large machines designed for high-volume commercial printing.

Because they produce very high-quality printing and are virtually silent, small ink-jet printers (along with ◊laser printers) are replacing impact printers, such as dot-matrix and daisywheel printers, for use with microcomputers.

inoculation injection into the body of dead or weakened disease-carrying organisms or their toxins (◊vaccine) to produce immunity by inducing a mild form of a disease.

inorganic chemistry branch of chemistry dealing with the chemical properties of the elements and their compounds, excluding the more complex covalent compounds of carbon, which are considered in ◊organic chemistry.

The origins of inorganic chemistry lay in observing the characteristics and experimenting with the uses of the substances (compounds and elements) that could be extracted from mineral ores. These could be classified according to their chemical properties: elements could be classified as metals or nonmetals; compounds as acids or bases, oxidizing or reducing agents, ionic compounds (such as salts), or covalent compounds (such as gases). The arrangement of elements into groups possessing similar properties led to Mendeleyev's ◊periodic table of the elements, which prompted chemists to predict the properties of undiscovered elements that might occupy gaps in the table. This, in turn, led to the discovery of new elements, including a number of highly radioactive elements that do not occur naturally.

inorganic compound compound found in organisms that are not typically biological.

Water, sodium chloride, and potassium are inorganic compounds because they are widely found outside living cells. The term is also applied to those compounds that do not contain carbon and that are not manufactured by organisms. However, carbon dioxide is considered inorganic, contains carbon, and is manufactured by organisms during respiration. See ◊organic compound.

input device device for entering information into a computer. Input devices include keyboards, joysticks, mice, light pens, touch-sensitive screens, graphics tablets, speech-recognition devices, and vision systems. Compare ◊output device.

Input devices that are used commercially – for example, by banks, postal services, and supermarkets – must be able to read and capture large volumes of data very rapidly. Such devices include document readers for magnetic-ink character recognition (MICR), optical character recognition (OCR), and optical mark recognition (OMR); mark-sense readers; bar-code scanners; magnetic-strip readers; and point-of-sale (POS) terminals. Punched-card and paper-tape readers were used in earlier commercial applications but are now obsolete.

insanity in medicine and law, any mental disorder in which the patient cannot be held responsible for their actions. The term is no longer used to refer to psychosis.

insect any of a vast group of small invertebrate animals with hard, segmented bodies, three pairs of jointed legs, and, usually, two pairs of wings; they belong among the ◊arthropods and are distributed throughout the world. An insect's body is divided into three segments: head, thorax, and abdomen. On the head is a pair of feelers, or antennae. The legs and wings are attached to the thorax, or middle segment of the body. The abdomen, or end segment of the body, is where food is digested and excreted and where the reproductive organs are located.

Insects vary in size from 0.02 cm/0.007 in to 35 cm/13.5 in in length.

Many insects hatch out of their eggs as ◊larvae (an immature stage, usually in the form of a caterpillar, grub, or maggot) and have to pass through further major physical changes (◊metamorphosis) before reaching adulthood. An insect about to go through metamorphosis hides itself or makes a cocoon in which to hide, then rests while the changes take place; at this stage the insect is called a ◊pupa, or a chrysalis if it is a butterfly or moth. When the changes are complete, the adult insect emerges.

The *classification* of insects is largely based upon characteristics of the mouthparts, wings, and metamorphosis. Insects are divided into two subclasses (one with

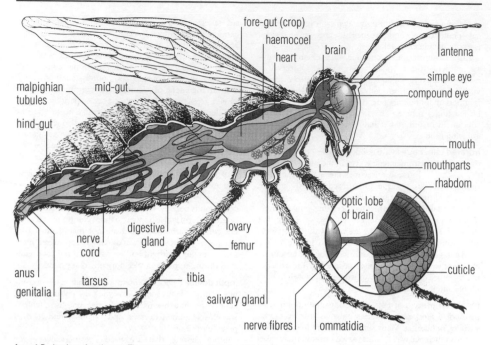

fore-gut (crop)
haemocoel
heart
brain
antenna
simple eye
compound eye
malpighian tubules
mid-gut
hind-gut
mouth
mouthparts
rhabdom
optic lobe of brain
digestive gland
ovary
femur
nerve cord
cuticle
anus
tarsus
tibia
genitalia
salivary gland
nerve fibres
ommatidia

insect Body plan of an insect. The general features of the insect body include a segmented body divided into head, thorax, and abdomen, jointed legs, feelers or antennae, and usually two pairs of wings. Insects often have compound eyes with a large field of vision.

two divisions) and 29 orders. More than 1 million species are known, and several thousand new ones are discovered each year.

The study of insects is called **entomology**.

Insect

The world's smallest winged insect is smaller than the eye of a house fly. It is the Tanzanian parasitic wasp, which has a wingspan of 0.2 mm/0.008 in.

insecticide any chemical pesticide used to kill insects. Among the most effective insecticides are synthetic organic chemicals such as ◊DDT and dieldrin, which are chlorinated hydrocarbons. These chemicals, however, have proved persistent in the environment and are also poisonous to all animal life, including humans, and are consequently banned in many countries. Other synthetic insecticides include organic phosphorus compounds such as malathion. Insecticides prepared from plants, such as derris and pyrethrum, are safer to use but need to be applied frequently and carefully.

insectivore any animal whose diet is made up largely or exclusively of ◊insects. In particular, the name is applied to mammals of the order Insectivora, which includes the shrews, hedgehogs, moles, and tenrecs.

insectivorous plant plant that can capture and digest live prey (normally insects), to obtain nitrogen compounds that are lacking in its usual marshy habitat. Some are passive traps, for example, the pitcher plants *Nepenthes* and *Sarracenia*. One pitcher-plant species has container-traps holding 1.6 l/3.5 pt of the liquid that 'digests' its food, mostly insects but occasionally even rodents.

Others, for example, sundews *Drosera*, butterworts *Pinguicula*, and Venus flytraps *Dionaea muscipula*, have an active trapping mechanism. Insectivorous plants have adapted to grow in poor soil conditions where the number of microorganisms recycling nitrogen compounds is very much reduced. In these circumstances other plants cannot gain enough nitrates to grow. See also ◊leaf.

inselberg or *kopje* prominent steep-sided hill of resistant solid rock, such as granite, rising out of a plain, usually in a tropical area. Its rounded appearance is caused by so-called onion-skin ◊weathering, in which the surface is eroded in successive layers.

The Sugar Loaf in Rio de Janeiro harbour in Brazil, and Ayers Rock in Northern Territory, Australia, are famous examples.

insemination, artificial see ◊artificial insemination.

instinct in ◊ethology, behaviour found in all equivalent members of a given species (for example, all the males, or all the females with young) that is presumed to be genetically determined.

Examples include a male robin's tendency to attack other male robins intruding on its territory and the tendency of many female mammals to care for their offspring. Instincts differ from ◊reflexes in that they involve very much more complex actions, and learning often plays an important part in their development.

instrument landing system (ILS) landing aid for aircraft that uses ◊radio beacons on the ground and instruments on the flight deck. One beacon (localizer) sends out a vertical radio beam along the centre line of the runway. Another beacon (glide slope) transmits a beam in the plane at right angles to the localizer beam at the ideal approach-path angle. The pilot can tell from the instruments how to manoeuvre to attain the correct approach path.

insulator any poor ◊conductor of heat, sound, or electricity. Most substances lacking free (mobile) ◊electrons, such as non-metals, are electrical or thermal insulators. Usually, devices of glass or porcelain, called insulators, are used for insulating and supporting overhead wires.

insulin protein ◊hormone, produced by specialized cells in the islets of Langerhans in the pancreas, that regulates the metabolism (rate of activity) of glucose, fats, and proteins. Insulin was discovered by Canadian physician Frederick Banting, who pioneered its use in treating ◊diabetes.

Normally, insulin is secreted in response to rising blood sugar levels (after a meal, for example), stimulating the body's cells to store the excess. Failure of this regulatory mechanism in diabetes mellitus requires treatment with insulin injections or capsules taken by mouth. Types vary from pig and beef insulins to synthetic and bioengineered ones. They may be combined with other substances to make them longer- or shorter-acting. Implanted, battery-powered insulin pumps deliver the hormone at a preset rate, to eliminate the unnatural rises and falls that result from conventional, subcutaneous (under the skin) delivery. Human insulin has now been produced from bacteria by ◊genetic engineering techniques, but may increase the chance of sudden, unpredictable ◊hypoglycaemia, or low blood sugar. In 1990 the Medical College of Ohio developed gelatin capsules and an aspirinlike drug which helps the insulin pass into the bloodstream.

integer any whole number. Integers may be positive or negative; 0 is an integer, and is often considered positive. Formally, integers are members of the set $Z = \{... -3, -2, -1, 0, 1, 2, 3,... \}$. Fractions, such as 1/2 and 0.35, are known as non-integral numbers ('not integers').

God made the integers,
man made the rest.

On **integer** Leopold Kronecker
Jahresberichte der deutschen Mathematiker
Vereinigung bk 2.
In F Cajori *A History of Mathematics* 1919

integral calculus branch of mathematics using the process of ◊integration. It is concerned with finding volumes and areas and summing infinitesimally small quantities.

integrated circuit (IC), popularly called *silicon chip*, A miniaturized electronic circuit produced on a single crystal, or chip, of a semiconducting material – usually silicon. It may contain many thousands of components and yet measure only 5 mm/0.2 in square and 1 mm/0.04 in thick. The IC is encapsulated within a plastic or ceramic case, and linked via gold wires to metal pins with which it is connected to a ◊printed circuit board and the other components that make up such electronic devices as computers and calculators.

Integrated Services Digital Network (ISDN) internationally developed telecommunications system for sending signals in ◊digital format along optical fibres and coaxial cable. It involves converting the 'local loop' – the link between the user's telephone (or private automatic

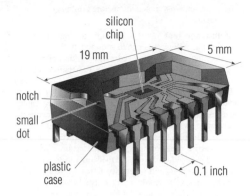

integrated circuit

branch exchange) and the digital telephone exchange – from an ◊analogue system into a digital system, thereby greatly increasing the amount of information that can be carried. The first large-scale use of ISDN began in Japan 1988.

ISDN has advantages in higher voice quality, better quality faxes, and the possibility of data transfer between computers faster than current modems. With ISDN's **Basic Rate Access**, a multiplexer divides one voice telephone line into three channels: two B bands and a D band. Each B band offers 64 kilobits per second and can carry one voice conversation or 50 simultaneous data calls at 1,200 bits per second. The D band is a data-signalling channel operating at 16 kilobits per second. With **Primary Rate Access**, ISDN provides 30 B channels.

integration in mathematics, a method in ◊calculus of determining the solutions of definite or indefinite integrals.

An example of a definite integral can be thought of as finding the area under a curve (as represented by an algebraic expression or function) between particular values of the function's variable. In practice, integral calculus provides scientists with a powerful tool for doing calculations that involve a continually varying quantity (such as determining the position at any given instant of a space rocket that is accelerating away from Earth). Its basic principles were discovered in the late 1660s independently by the German philosopher Leibniz and the British scientist Newton.

integument in seed-producing plants, the protective coat surrounding the ovule. In flowering plants there are two, in gymnosperms only one. A small hole at one end, the micropyle, allows a pollen tube to penetrate through to the egg during fertilization.

intelligence in psychology, a general concept that summarizes the abilities of an individual in reasoning and

It is not enough to have a good mind.
The main thing is to use it well.

On **intelligence** René Descartes
Discourse on Method

problem solving, particularly in novel situations. These consist of a wide range of verbal and nonverbal skills and therefore some psychologists dispute a unitary concept of intelligence.

intelligence test test that attempts to measure innate intellectual ability, rather than acquired ability.

It is now generally believed that a child's ability in an intelligence test can be affected by his or her environment, cultural background, and teaching. There is scepticism about the accuracy of intelligence tests, but they are still widely used as a diagnostic tool when children display learning difficulties. 'Sight and sound' intelligence tests, developed in 1981 by Christopher Brand, avoid cultural bias and the pitfalls of improvement by practice. Subjects are shown a series of lines being flashed on a screen at increasing speed, and are asked to identify in each case the shorter of a pair; and when two notes are relayed over headphones, they are asked to identify which is the higher. There is a close correlation between these results and other intelligence test scores.

intensity in physics, the power (or energy per second) per unit area carried by a form of radiation or wave motion. It is an indication of the concentration of energy present and, if measured at varying distances from the source, of the effect of distance on this. For example, the intensity of light is a measure of its brightness, and may be shown to diminish with distance from its source in accordance with the ◊inverse square law (its intensity is inversely proportional to the square of the distance).

intensive agriculture farming system where large quantities of inputs, such as labour or fertilizers, are involved over a small area of land. ◊Market gardening is an example. Yields are often much higher than those obtained from ◊extensive agriculture.

interactive computing in computing, a system for processing data in which the operator is in direct communication with the computer, receiving immediate responses to input data. In ◊batch processing, by contrast, the necessary data and instructions are prepared in advance and processed by the computer with little or no intervention from the operator.

interactive video (IV) computer-mediated system that enables the user to interact with and control information (including text, recorded speech, or moving images) stored on video disc. IV is most commonly used for training purposes, using analogue video discs, but has wider applications with digital video systems such as CD-I (Compact Disc Interactive, from Philips and Sony) which are based on the CD-ROM format derived from audio compact discs.

intercostal in biology, the nerves, blood vessels, and muscles that lie between the ribs.

interface in computing, the point of contact between two programs or pieces of equipment. The term is most often used for the physical connection between the computer and a peripheral device, which is used to compensate for differences in such operating characteristics as speed, data coding, voltage, and power consumption. For example, a *printer interface* is the cabling and circuitry used to transfer data from a computer to a printer, and to compensate for differences in speed and coding.

Common standard interfaces include the *Centronics interface*, used to connect parallel devices, and the *RS232 interface*, used to connect serial devices. For example, in many microcomputer systems, an RS232 interface is used to connect the microcomputer to a modem, and a Centronics device is used to connect it to a printer.

interference in physics, the phenomenon of two or more wave motions interacting and combining to produce a resultant wave of larger or smaller amplitude (depending on whether the combining waves are in or out of ◊phase with each other).

Interference of white light (multiwavelength) results in spectral coloured fringes; for example, the iridescent colours of oil films seen on water or soap bubbles (demonstrated by ◊Newton's rings). Interference of sound waves of similar frequency produces the phenomenon of beats, often used by musicians when tuning an instrument. With monochromatic light (of a single wavelength), interference produces patterns of light and dark bands. This is the basis of ◊holography, for example. Interferometry can also be applied to radio waves, and is a powerful tool in modern astronomy.

interferometer in physics, a device that splits a beam of light into two parts, the parts being recombined after travelling different paths to form an interference pattern of light and dark bands.

Interferometers are used in many branches of science and industry where accurate measurements of distances and angles are needed.

In the Michelson interferometer, a light beam is split into two by a semisilvered mirror. The two beams are then reflected off fully silvered mirrors and recombined. The pattern of dark and light bands is sensitive to small alterations in the placing of the mirrors, so the interferometer can detect changes in their position to within one ten-millionth of a metre. Using lasers, compact devices of this kind can be built to measure distances, for example to check the accuracy of machine tools.

In radio astronomy, interferometers consist of separate radio telescopes, each observing the same distant object, such as a galaxy, in the sky. The signal received by each telescope is fed into a computer. Because the telescopes are in different places, the distance travelled by the signal to reach each differs and the overall signal is akin to the interference pattern in the Michelson interferometer. Computer analysis of the overall signal can build up a detailed picture of the source of the radio waves.

In space technology, interferometers are used in radio and radar systems. These include space-vehicle guidance systems, in which the position of the spacecraft is determined by combining the signals received by two precisely spaced antennae mounted on it.

interferometry in astronomy, any of several techniques used in astronomy to obtain high-resolution images of astronomical objects. See ◊speckle interferometry and ◊VLBI (very long baseline interferometry).

interferon naturally occurring cellular protein that makes up part of the body's defences against viral disease. Three types (alpha, beta, and gamma) are produced by infected cells and enter the bloodstream and uninfected cells, making them immune to virus attack.

Interferon was discovered in 1957 by Scottish virologist Alick Isaacs. Interferons are cytokines, small molecules that carry signals from one cell to another. They can be divided into two main types: *type I* (alpha, beta, tau, and omega) interferons are more effective at bolstering cells' ability to resist infection; **type II** (gamma) interferon is more important to the normal functioning of the immune system. Alpha interferon may be used to treat some cancers; interferon beta 1b has been found useful in the treatment of ◊multiple sclerosis.

interleukin in medicine, polypeptide produced by an activated lymphocyte and involved in signalling between cells in the immune system as part of the immune response and in the formation of blood. There are several different interleukins with different activities and some have applications in the treatment of diseases, such as cancer.

intermediate technology application of mechanics, electrical engineering, and other technologies, based on inventions and designs developed in scientifically sophisticated cultures, but utilizing materials, assembly, and maintenance methods found in technologically less advanced regions (known as the Third World).

Intermediate technologies aim to allow developing countries to benefit from new techniques and inventions

windpump

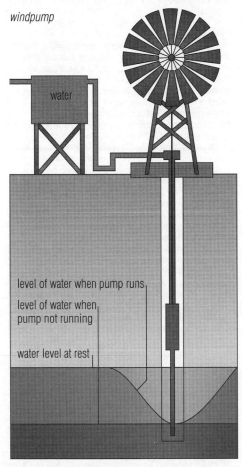

water

level of water when pump runs

level of water when pump not running

water level at rest

intermediate technology The simple windmill is an example of intermediate technology if it utilizes local materials and traditional design. In this way, there is no need for complex maintenance and repair, nor expensive spare parts.

of the 'First World', without the burdens of costly maintenance and supply of fuels and spare parts that in the Third World would represent an enormous and probably uneconomic overhead.

intermediate vector boson alternative name for **weakon**, the elementary particle responsible for carrying the ◊weak nuclear force.

intermolecular force or **van der Waals' force** force of attraction between molecules. Intermolecular forces are relatively weak; hence simple molecular compounds are gases, liquids, or low-melting-point solids.

internal-combustion engine heat engine in which fuel is burned inside the engine, contrasting with an external-combustion engine (such as the steam engine) in which fuel is burned in a separate unit. The ◊diesel engine and ◊petrol engine are both internal-combustion engines. Gas ◊turbines and ◊jet and ◊rocket engines are sometimes also considered to be internal-combustion engines because they burn their fuel inside their combustion chambers.

international classification of disease in medicine, classification system for all known diseases and syndromes devised by the World Health Organization. Diseases are subdivided according to the system they affect and to the type of disease and each is given a code to facilitate computerization. Classification of diseases allows the use of a uniform terminology and comparisons of data on an international basis.

International Date Line (IDL) imaginary line that approximately follows the 180° line of longitude. The date is put forward a day when crossing the line going west, and back a day when going east. The IDL was chosen at the International Meridian Conference in 1884.

International Organization for Standardization (ISO) international organization founded in 1947 to standardize technical terms, specifications, units, and so on. Its headquarters are in Geneva.

International Union for the Conservation of Nature (IUCN) organization established by the United Nations to promote the conservation of wildlife and habitats as part of the national policies of member states.

It has formulated guidelines and established research programmes (for example, International Biological Programme, IBP) and set up advisory bodies (such as Survival Commissions, SSC). In 1980, it launched the **World Conservation Strategy** to highlight particular problems, designating a small number of areas as **World Heritage Sites** to ensure their survival as unspoiled habitats (for example, Yosemite National Park in the USA, and the Simen Mountains in Ethiopia). It also compiles the **Red Data List of Threatened Animals**, classifying species according to their vulnerability to extinction.

Internet global computer network connecting governments, companies, universities, and many other networks and users. ◊Electronic mail, conferencing, and chat services are all supported across the network, as is the ability to access remote computers and send and retrieve files. By late 1994 it was estimated to have over 40 million users on 11,000 networks in 70 countries, with an estimated 1 million users joining each month. It was esti-

Censorship of the Internet

Most governments, and many citizens, are alarmed by the free flow of information, so it is hardly surprising if they are alarmed by the Internet, which is the largest free information system the world has ever seen. In the past, there have been dozens of attempts to limit the free flow of information – not just in 'restrictive' countries such as Singapore and China, but in more liberal ones including Germany, France, Australia, and the UK. Even in the USA, where information is most highly valued, the Communications Decency Act was passed to try to 'clean up' free speech, though after judicial examination, the Act's provisions were found to be unconstitutional.

Unstoppable information superhighway
The Internet is hard to censor for a number of pragmatic and practical reasons. The first being that is not a tangible thing, like a road, but a concept, like a transportation system. The Internet is simply a network of networks. While it is usually possible to close down a single computer network, just as it is possible to block a particular road, it is impossible to close down the whole system.

Also, in general terms, if users can connect to part of the Internet, they can find a route to any other part of the Internet. This is like saying that if a driver can get onto one road then in principle they can reach any other road anywhere in the world. The difference with the Internet is the speed at which information travels: a message can be sent to Antarctica and back in less than a second. The Internet is a global system, and does not stop at national boundaries.

A second problem with the Internet is that, like other transportation systems, it does not carry a single type of traffic. One type of Internet traffic, for example. consists of *World Wide Web pages*. These combine text and graphics in a way that is very similar to a magazine or newspaper. Another type of traffic is *electronic mail* (e-mail), which consists of (mostly) private messages of the sort that could otherwise be sent by post. A third type of traffic consists of *Internet Relay Chat* messages, which are typed in real time: this is the computer equivalent of CB radio. There are numerous other types of traffic including video conferencing, radio broadcasts, and computer file transfers.

The problem is that the same type and level of censorship cannot possibly be applied to all the different types of traffic. No sensible person would attempt to impose the same standards on, say, television broadcasts, private letters, and telephone conversations, and the same sensitivity must apply to the Internet. It might, for example, be reasonable to hold a publisher responsible for articles published on a Web site, because the publisher has control of the content, but not to messages posted on an open *bulletin board*.

Whose values?
It must also be observed that censoring the Internet is, in principle, dangerous: whose standards, whose laws, apply to something that is written in the US, held on a US-registered Internet server sited in Sweden, and read in Japan? This kind of problem has been highlighted by German attempts to block access to pro-white propaganda held on a server in Canada.

There is undoubtedly material on the Internet that is considered objectionable in the UK, for example, but almost every type of information is considered objectionable by someone somewhere. If one group can ban pro-Nazi material, another can ban pro-Capitalist tracts, or birth control information, or the discussion of evolution, or recipes for cooking and eating certain animals.

So far, the separate acts of individual governments to restrict information on the Internet have mainly resulted in increased publicity and the more widespread dissemination of whatever they have tried to ban. This seems likely to remain the case until there is international agreement about what can be controlled, and some means are devised to do it.

Jack Schofield

mated in 1997 that there would be 200 million users by the year 2000.

The technical underpinnings of the Internet were developed as a project funded by the Advanced Research Project Agency (ARPA) to research how to build a network that would withstand bomb damage. The Internet itself began in 1984 with funding from the US National Science Foundation as a means to allow US universities to share the resources of five regional supercomputing centres. The number of users grew quickly, and in the early 1990s access became cheap enough for domestic users to have their own links on home personal computers. As the amount of information available via the Internet grew, indexing and search services such as Gopher, Archie, Veronica, and WAIS were created by Internet users to help both themselves and others. The newer World Wide Web allows seamless browsing across the Internet via ◊hypertext.

interplanetary matter gas and dust thinly spread through the Solar System. The gas flows outwards from the Sun as the ◊solar wind.

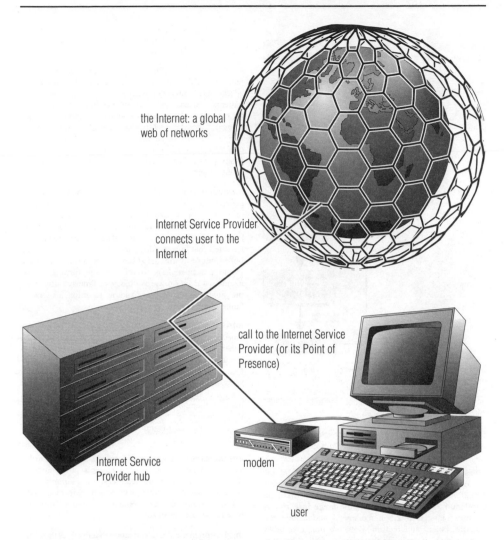

the Internet: a global
web of networks

Internet Service Provider
connects user to the
Internet

call to the Internet Service
Provider (or its Point of
Presence)

Internet Service
Provider hub

modem

user

Internet The Internet is accessed by users via a modem to the service provider's hub, which handles all connection requests. Once connected, the user can access a whole range of information from many different sources, including the World Wide Web.

Fine dust lies in the plane of the Solar System, scattering sunlight to cause the ◊zodiacal light. Swarms of dust shed by comets enter the Earth's atmosphere to cause ◊meteor showers.

interpreter computer program that translates and executes a program written in a high-level language. Unlike a ◊compiler, which produces a complete machine-code translation of the high-level program in one operation, an interpreter translates the source program, instruction by instruction, each time that program is run. *See illustration overleaf.*

intersex individual that is intermediate between a normal male and a normal female in its appearance (for example, a genetic male that lacks external genitalia and so resembles a female).

interstellar cirrus in astronomy, wispy cloud-like structures discovered in the mid-1980s by the Infrared Astronomy Satellite (IRAS) and believed to be the remains of dust shells blown into space from cool giant or supergiant stars.

interstitial in biology, undifferentiated tissue that is interspersed with the characteristic tissue of an organ. It is often formed of fibrous tissue and supports the organ. Interstitial fluid refers to the fluid present in small amounts in the tissues of an organ.

intertropical convergence zone (ITCZ) area of heavy rainfall found in the tropics and formed as the trade winds converge and rise to form cloud and rain. It moves a few degrees northwards during the northern summer and a few degrees southwards during the southern summer,

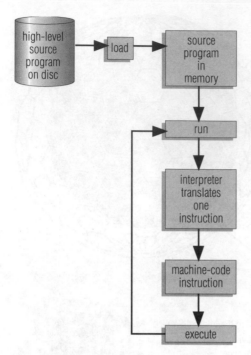

interpreter The sequence of events when running an interpreter on a high-level language program. Instructions are translated one at a time, making the process a slow one; however, interpreted programs do not need to be compiled and may be executed immediately.

following the apparent movement of the Sun. The ITCZ is responsible for most of the rain that falls in Africa. The ◊doldrums are also associated with this zone.

intestine in vertebrates, the digestive tract from the stomach outlet to the anus. The human **small intestine** is 6 m/20 ft long, 4 cm/1.5 in diameter, and consists of the duodenum, jejunum, and ileum; the **large intestine** is 1.5 m/5 ft long, 6 cm/2.5 in diameter, and includes the caecum, colon, and rectum. Both are muscular tubes comprising an inner lining that secretes alkaline digestive juice, a submucous coat containing fine blood vessels and nerves, a muscular coat, and a serous coat covering all, supported by a strong peritoneum, which carries the blood and lymph vessels, and the nerves. The contents are passed along slowly by ◊peristalsis (waves of involuntary muscular action). The term intestine is also applied to the lower digestive tract of invertebrates.

intrathecal space in medicine, space located between the arachnoid and the pia mater of the brain and spinal cord that contains the cerebrospinal fluid. Intrathecal injections of antibiotics may be given to treat infections of the central nervous system, such as meningitis.

intrauterine device (IUD) or **coil**, a contraceptive device that is inserted into the womb (uterus). It is a tiny plastic object, sometimes containing copper. By causing a mild inflammation of the lining of the uterus it prevents fertilized eggs from becoming implanted. IUDs are not usually given to women who have not had children. They are generally very reliable, as long as they stay in place,

with a success rate of about 98%. Some women experience heavier and more painful periods, and there is a very slight risk of a pelvic infection leading to infertility.

intravenous nutrition or **parenteral nutrition** in medicine, the provision of all essential nutrients in adequate amounts administered into the vein. It is used in patients with severe illnesses who are unable to take food by mouth, such as those who have undergone major surgery, have serious burns, or are suffering from renal failure. The risk of infection and the occurrence of biochemical abnormalities related to the compositions of the fluids infused are such that intravenous nutrition is reserved only for those who are severely ill.

introversion in psychology, preoccupation with the self, generally coupled with a lack of sociability. The opposite of introversion is ◊extroversion.

The term was introduced by the Swiss psychiatrist Carl Jung in 1924 in his description of ◊schizophrenia, where he noted that 'interest does not move towards the object but recedes towards the subject'. The term is also used within psychoanalysis to refer to the turning of the instinctual drives towards objects of fantasy rather than the pursuit of real objects. Another term for this sense is fantasy cathexis.

intrusion mass of ◊igneous rock that has formed by 'injection' of molten rock, or magma, into existing cracks beneath the surface of the Earth, as distinct from a volcanic rock mass which has erupted from the surface. Intrusion features include vertical cylindrical structures such as stocks, pipes, and necks; sheet structures such as dykes that cut across the strata and sills that push between them; laccoliths, which are blisters that push up the overlying rock; and batholiths, which represent chambers of solidified magma and contain vast volumes of rock.

intrusive rock ◊igneous rock formed beneath the Earth's surface. Magma, or molten rock, cools slowly at these depths to form coarse-grained rocks, such as granite, with large crystals. (◊Extrusive rocks, which are formed on the surface, are usually fine-grained.) A mass of intrusive rock is called an intrusion.

intubation in medicine, operation that involves the introduction of a tube through the mouth and into the larynx to keep the air passage open.

intuitionism in mathematics, the theory that propositions can be built up only from intuitive concepts that we all recognize easily, such as unity or plurality. The concept of ◊infinity, of which we have no intuitive experience, is thus not allowed.

inverse square law in physics, the statement that the magnitude of an effect (usually a force) at a point is inversely proportional to the square of the distance between that point and the object exerting the force.

Light, sound, electrostatic force (Coulomb's law), and gravitational force (Newton's law) all obey the inverse square law.

invertebrate animal without a backbone. The invertebrates comprise over 95% of the million or so existing animal species and include sponges, coelenterates, flatworms, nematodes, annelid worms, arthropods,

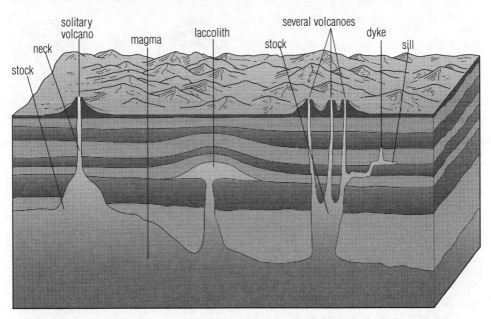

solitary volcano · neck · stock · magma · laccolith · stock · several volcanoes · dyke · sill

intrusion Igneous intrusions can be a variety of shapes and sizes. Laccoliths are domed circular shapes, and can be many miles across. Sills are intrusions that flow between rock layers. Pipes or necks connect the underlying magma chamber to surface volcanoes.

molluscs, echinoderms, and primitive aquatic chordates, such as sea squirts and lancelets.

Invertebrate

Every year, at least two species of British invertebrates become extinct.

---

in vitro fertilization (IVF; 'fertilization in glass') allowing eggs and sperm to unite in a laboratory to form embryos. The embryos (properly called pre-embryos in their two- to eight-celled state) are stored by cooling to the temperature of liquid air (cryopreservation) until they are implanted into the womb of the otherwise infertile mother (an extension of ◊ artificial insemination). The first baby to be produced by this method was born in 1978 in the UK. In cases where the Fallopian tubes are blocked, fertilization may be carried out by **intra-vaginal culture**, in which egg and sperm are incubated (in a plastic tube) in the mother's vagina, then transferred surgically into the uterus.

in vitro process biological experiment or technique carried out in a laboratory, outside the body of a living organism (literally 'in glass', for example in a test tube). By contrast, an in vivo process takes place within the body of an organism.

in vivo process biological experiment or technique carried out within a living organism; by contrast, an in vitro process takes place outside the organism, in an artificial environment such as a laboratory.

Io in astronomy, the third-largest moon of the planet Jupiter, 3,630 km/2,260 mi in diameter, orbiting in 1.77 days at a distance of 422,000 km/262,000 mi. It is the most volcanically active body in the Solar System,

covered by hundreds of vents that erupt not lava but sulphur, giving Io an orange-coloured surface.

In July 1995 the Hubble Space Telescope revealed the appearance of a 320 km/200 mi yellow spot on the surface of Io, located on the volcano Ra Patera. Though clearly volcanic in origin, astronomers are unclear as to the exact cause of the new spot.

Using data gathered by the spacecraft *Galileo*, US astronomers concluded in 1996 that Io has a large metallic core. The *Galileo* space probe also detected a 10-megawatt beam of electrons flowing between Jupiter and Io.

iodide compound formed between iodine and another element in which the iodine is the more electronegative element (see ◊ electronegativity, ◊ halide).

iodine greyish-black nonmetallic element, symbol I, atomic number 53, relative atomic mass 126.9044. It is a member of the ◊ halogen group. Its crystals give off, when heated, a violet vapour with an irritating odour resembling that of chlorine. It only occurs in combination with other elements. Its salts are known as iodides, and are found in sea water. As a mineral nutrient it is vital to the proper functioning of the thyroid gland, where it occurs in trace amounts as part of the hormone thyroxine. Absence of iodine from the diet leads to ◊ goitre. Iodine is used in photography, in medicine as an antiseptic, and in making dyes.

Its radioactive isotope ^{131}I (half-life of eight days) is a dangerous fission product from nuclear explosions and from the nuclear reactors in power plants, since, if ingested, it can be taken up by the thyroid and damage it. It was discovered in 1811 by French chemist B Courtois (1777–1838).

iodoform (chemical name **triiodomethane**) CHI_3, an antiseptic that crystallizes into yellow hexagonal plates. It is soluble in ether, alcohol, and chloroform, but not in water.

ion atom, or group of atoms, that is either positively charged (◊cation) or negatively charged (◊anion), as a result of the loss or gain of electrons during chemical reactions or exposure to certain forms of radiation.

ion engine rocket engine that uses ◊ions (charged particles) rather than hot gas for propulsion. Ion engines have been successfully tested in space, where they will eventually be used for gradual rather than sudden velocity changes. In an ion engine, atoms of mercury, for example, are ionized (given an electric charge by an electric field) and then accelerated at high speed by a more powerful electric field.

ion exchange process whereby an ion in one compound is replaced by a different ion, of the same charge, from another compound. It is the basis of a type of ◊chromatography in which the components of a mixture of ions in solution are separated according to the ease with which they will replace the ions on the polymer matrix through which they flow. The exchange of positively charged ions is called cation exchange; that of negatively charged ions is called anion exchange.

Ion-exchange is used in commercial water softeners to exchange the dissolved ions responsible for the water's hardness with others that do not have this effect. For example, when hard water is passed over an ion-exchange resin, the dissolved calcium and magnesium ions are replaced by either sodium or hydrogen ions, so the hardness is removed.

ion half equation equation that describes the reactions occurring at the electrodes of a chemical cell or in electrolysis. It indicates which ion is losing electrons (oxidation) or gaining electrons (reduction).

Examples are given from the electrolysis of dilute hydrochloric acid (HCl).

$$2Cl^- - 2e^- \rightarrow Cl_2 \text{ (positive electrode)}$$

$$2H^+ + 2e^- \rightarrow H_2 \text{ (negative electrode)}$$

ionic bond or **electrovalent bond** bond produced when atoms of one element donate electrons to atoms of another element, forming positively and negatively charged ◊ions respectively. The attraction between the oppositely charged ions constitutes the bond. Sodium chloride (Na^+Cl^-) is a typical ionic compound.

Each ion has the electronic structure of an inert gas (see ◊noble gas structure). The maximum number of electrons that can be gained is usually two.

ionic compound substance composed of oppositely charged ions. All salts, most bases, and some acids are examples of ionic compounds. They possess the following general properties: they are crystalline solids with a high melting point; are soluble in water and insoluble in organic solvents; and always conduct electricity when molten or in aqueous solution. A typical ionic compound is sodium chloride (Na^+Cl^-).

ionic equation equation showing only those ions in a chemical reaction that actually undergo a change, either by combining together to form an insoluble salt or by combining together to form one or more molecular compounds. Examples are the precipitation of insoluble barium sulphate when barium and sulphate ions are com-

electron transferred

electronic arrangement, 2.8.1 of a sodium atom

electronic arrangement, 2.8.7 of a chlorine atom

becomes a sodium ion, Na^+, with an electron arrangement 2.8

becomes a chloride ion, Cl^-, with an electron arrangement 2.8.8

ionic bond The formation of an ionic bond between a sodium atom and a chlorine atom to form a molecule of sodium chloride. The sodium atom transfers an electron from its outer electron shell (becoming the positive ion Na^+) to the chlorine atom (which becomes the negative chloride ion Cl^-). The opposite charges mean that the ions are strongly attracted to each other. The formation of the bond means that each atom becomes more stable, having a full quota of electrons in its outer shell.

bined in solution, and the production of ammonia and water from ammonium hydroxide.

$$Ba^{2+}_{(aq)} + SO_4^{2-}_{(aq)} \rightarrow BaSO_{4(s)}$$

$$NH_4^+_{(aq)} + OH^-_{(aq)} \rightarrow NH_{3(g)} + H_2O_{(l)}$$

The other ions in the mixtures do not take part and are called ◊spectator ions.

ionization process of ion formation. It can be achieved in two ways. The first way is by the loss or gain of electrons by atoms to form positive or negative ions.

$$Na - e^- \rightarrow Na^+$$

$$\tfrac{1}{2}Cl_2 + e^- \rightarrow Cl^-$$

In the second mechanism, ions are formed when a covalent bond breaks, as when hydrogen chloride gas is dissolved in water. One portion of the molecule retains both electrons, forming a negative ion, and the other portion becomes positively charged. This bond-fission process is sometimes called dissociation.

$$HCl_{(g)} + aq \rightleftharpoons H^+_{(aq)} + Cl^-_{(aq)}$$

ionization chamber device for measuring ◊ionizing radiation. The radiation ionizes the gas in the chamber and the ions formed are collected and measured as an electric charge. Ionization chambers are used for determining the intensity of X-rays or the disintegration rate of radioactive materials.

ionization potential measure of the energy required to remove an ◊electron from an ◊atom. Elements with a low ionization potential readily lose electrons to form ◊cations.

ionization therapy enhancement of the atmosphere of an environment by artificially boosting the negative ion content of the air.

Fumes, dust, cigarette smoke, and central heating cause negative ion deficiency, which particularly affects sufferers from respiratory disorders such as bronchitis, asthma, and sinusitis. Symptoms are alleviated by the use of ionizers in the home or workplace. In severe cases, ionization therapy is used as an adjunct to conventional treatment.

ionizing radiation radiation that knocks electrons from atoms during its passage, thereby leaving ions in its path. Alpha and beta particles are far more ionizing in their effect than are neutrons or gamma radiation.

ionosphere ionized layer of Earth's outer ◊atmosphere (60–1,000 km/38–620 mi) that contains sufficient free electrons to modify the way in which radio waves are propagated, for instance by reflecting them back to Earth. The ionosphere is thought to be produced by absorption of the Sun's ultraviolet radiation.

ion plating method of applying corrosion-resistant metal coatings. The article is placed in argon gas, together with some coating metal, which vaporizes on heating and becomes ionized (acquires charged atoms) as it diffuses through the gas to form the coating. It has important applications in the aerospace industry.

IR or *ir* in physics, abbreviation for *infrared*.

iridium hard, brittle, silver-white, metallic element, symbol Ir, atomic number 77, relative atomic mass 192.2. It is resistant to tarnish and corrosion. Iridium is one of the so-called platinum group of metals; it occurs in platinum ores and as a free metal (◊native metal) with osmium in osmiridium, a natural alloy that includes platinum, ruthenium, and rhodium.

It is alloyed with platinum for jewellery and used for watch bearings and in scientific instruments. It was named in 1804 by English chemist Smithson Tennant (1761–1815) for its iridescence in solution.

iridology diagnostic technique based on correspondences between specific areas of the iris and bodily functions and organs.

It was discovered in the 19th century independently by a Hungarian and a Swedish physician, and later refined and developed in the USA by Bernard Jensen. Iridology is of proven effectiveness in monitoring general wellbeing and indicating the presence of organic disorders, but cannot be as specific about the nature and extent of these as orthodox diagnostic techniques.

iris in anatomy, the coloured muscular diaphragm that controls the size of the pupil in the vertebrate eye. It contains radial muscle that increases the pupil diameter and circular muscle that constricts the pupil diameter. Both types of muscle respond involuntarily to light intensity.

iron hard, malleable and ductile, silver-grey, metallic element, symbol Fe (from Latin *ferrum*), atomic number 26, relative atomic mass 55.847. It is the fourth most abundant element (the second most abundant metal, after aluminium) in the Earth's crust. Iron occurs in concentrated deposits as the ores hematite (Fe_2O_3), spathic ore ($FeCO_3$), and magnetite (Fe_3O_4). It sometimes occurs as a free metal, occasionally as fragments of iron or iron–nickel meteorites.

Iron is the most common and most useful of all metals; it is strongly magnetic and is the basis for ◊steel, an alloy with carbon and other elements (see also ◊cast iron). In electrical equipment it is used in all permanent magnets and electromagnets, and forms the cores of transformers and magnetic amplifiers. In the human body, iron is an essential component of haemoglobin, the molecule in red blood cells that transports oxygen to all parts of the body. A deficiency in the diet causes a form of anaemia.

Iron Age developmental stage of human technology when weapons and tools were made from iron. Preceded by the Stone and Bronze ages, it is the last technological stage in the ◊Three Age System framework for prehistory. Iron was first produced in Thailand about 1600 BC, but was considered inferior in strength to bronze until about 1000 BC, when metallurgical techniques improved, and the alloy steel was produced by adding carbon during the smelting process.

iron ore any mineral from which iron is extracted. The chief iron ores are ◊*magnetite*, a black oxide; ◊*hematite*, or kidney ore, a reddish oxide; ◊*limonite*, brown, impure oxyhydroxides of iron; and *siderite*, a brownish carbonate.

Iron ores are found in a number of different forms, including distinct layers in igneous intrusions, as components of contact metamorphic rocks, and as sedimentary beds. Much of the world's iron is extracted in Russia, Kazakhstan, and the Ukraine. Other important producers are the USA, Australia, France, Brazil, and Canada; over 40 countries produce significant quantities of ore.

iron pyrites or *pyrite* FeS_2 common iron ore. Brassy yellow, and occurring in cubic crystals, it is often called 'fool's gold', since only those who have never seen gold would mistake it.

irradiation subjecting anything to radiation, including cancer tumours. See also ◊food irradiation.

irrational number a number that cannot be expressed as an exact ◊fraction. Irrational numbers include some square roots (for example, $\sqrt{2}$, $\sqrt{3}$, and $\sqrt{5}$ are irrational) and numbers such as π (the ratio of the circumference of a circle to its diameter, which is approximately equal to 3.14159) and e (the base of ◊natural logarithms, approximately 2.71828).

irregular galaxy in astronomy, a class of ◊galaxy with little structure, which does not conform to any of the standard shapes in the Hubble classification. The two satellite galaxies of the ◊Milky Way, the ◊Magellanic Clouds, are both irregulars. Some galaxies previously classified as irregulars are now known to be normal galaxies distorted by tidal effects or undergoing bursts of star formation (see ◊starburst galaxy).

irrigation artificial water supply for dry agricultural areas by means of dams and channels. Drawbacks are that it tends to concentrate salts at the surface, ultimately causing soil infertility, and that rich river silt is retained at dams, to the impoverishment of the land and fisheries below them.

Irrigation has been practised for thousands of years, in

Eurasia as well as the Americas. An example is the channelling of the annual Nile flood in Egypt, which has been done from earliest times to its present control by the Aswan High Dam.

ISBN (abbreviation for *International Standard Book Number*) code number used for ordering or classifying book titles. Every book printed now has a number on its back cover or jacket, preceded by the letters ISBN. It is a code to the country of origin and the publisher. The number is unique to the book, and will identify it anywhere in the world.

The final digit in each ISBN number is a check digit, which can be used by a computer program to validate the number each time it is input (see ◊validation).

ischaemic heart disease (IHD) disorder caused by reduced perfusion of the coronary arteries due to ◊atherosclerosis. It is the commonest cause of death in the Western world, leading to more than a million deaths each year in the USA and about 160,000 in the UK. See also ◊coronary artery disease.

Early symptoms of IHD include ◊angina or palpitations, but sometimes a heart attack is the first indication that a person is affected.

ISDN abbreviation for ◊*Integrated Services Digital Network*, a telecommunications system.

island area of land surrounded entirely by water. Australia is classed as a continent rather than an island, because of its size.

Islands can be formed in many ways. **Continental islands** were once part of the mainland, but became isolated (by tectonic movement, erosion, or a rise in sea level, for example). **Volcanic islands**, such as Japan, were formed by the explosion of underwater volcanoes. **Coral islands** consist mainly of ◊coral, built up over many years. An **atoll** is a circular coral reef surrounding a lagoon; atolls were formed when a coral reef grew up around a volcanic island that subsequently sank or was submerged by a rise in sea level. **Barrier islands** are found by the shore in shallow water, and are formed by the deposition of sediment eroded from the shoreline.

island arc curved chain of islands produced by volcanic activity at a ◊destructive margin (where one tectonic plate slides beneath another). Island arcs are common in the Pacific where they ring the ocean on both sides; the Aleutian Islands off Alaska are an example.

Such island arcs are often later incorporated into continental margins during mountain-building episodes.

islets of Langerhans groups of cells within the pancreas responsible for the secretion of the hormone insulin. They are sensitive to blood sugar, producing more hormone when glucose levels rise.

ISO in photography, a numbering system for rating the speed of films, devised by the International Standards Organization.

isobar line drawn on maps and weather charts linking all places with the same atmospheric pressure (usually measured in millibars).

When used in weather forecasting, the distance between the isobars is an indication of the barometric gradient (the rate of change in pressure).

Where the isobars are close together, cyclonic weather is indicated, bringing strong winds and a depression, and

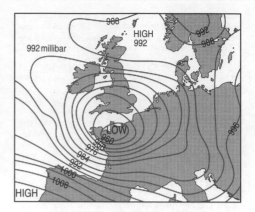

isobar The isobars around a low-pressure area or depression. In the northern hemisphere, winds blow anticlockwise around lows, approximately parallel to the isobars, and clockwise around highs. In the southern hemisphere, the winds blow in the opposite directions.

where far apart anticyclonic, bringing calmer, settled conditions.

isomer chemical compound having the same molecular composition and mass as another, but with different physical or chemical properties owing to the different structural arrangement of its constituent atoms. For example, the organic compounds butane ($CH_3(CH_2)_2$ CH_3) and methyl propane ($CH_3CH(CH_3)CH_3$) are iso-

butane $CH_3(CH_2)_2CH_3$

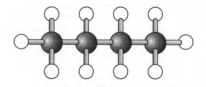

methyl propane $CH_3CH(CH_3)CH_3$

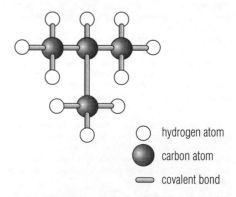

○ hydrogen atom
● carbon atom
⬭ covalent bond

isomer The chemicals butane and methyl propane are isomers. Each has the molecular formula C_4H_{10}, but with different spatial arrangements of atoms in their molecules.

mers, each possessing four carbon atoms and ten hydrogen atoms but differing in the way that these are arranged with respect to each other.

Structural isomers have obviously different constructions, but **geometrical** and **optical isomers** must be drawn or modelled in order to appreciate the difference in their three-dimensional arrangement. Geometrical isomers have a plane of symmetry and arise because of the restricted rotation of atoms around a bond; optical isomers are mirror images of each other. For instance, 1,1-dichloroethene ($CH_2=CCl_2$) and 1,2-dichloroethene ($CHCl=CHCl$) are structural isomers, but there are two possible geometric isomers of the latter (depending on whether the chlorine atoms are on the same side or on opposite sides of the plane of the carbon–carbon double bond).

isomorphism the existence of substances of different chemical composition but with similar crystalline form.

isoprene $CH_2CHC(CH_3)CH_2$ (technical name **methylbutadiene**) colourless, volatile fluid obtained from petroleum and coal, used to make synthetic rubber.

isostasy the theoretical balance in buoyancy of all parts of the Earth's ◊crust, as though they were floating on a denser layer beneath. There are two theories of the mechanism of isostasy, the Airy hypothesis and the Pratt hypothesis, both of which have validity. In the **Airy hypothesis** crustal blocks have the same density but different depths: like ice cubes floating in water, higher mountains have deeper roots. In the **Pratt hypothesis**, crustal blocks have different densities allowing the depth of crustal material to be the same.

There appears to be more geological evidence to support the Airy hypothesis of isostasy. During an ◊ice age the weight of the ice sheet pushes that continent into the Earth's mantle; once the ice has melted, the continent rises again. This accounts for shoreline features being found some way inland in regions that were heavily glaciated during the Pleistocene period.

isotonic solution in medicine, a solution that can be mixed with body fluids without affecting their constituents. Solutions which are isotonic with blood, such as sodium chloride 0.9%, have the same osmotic pressure as serum and they do not affect the membranes of the red blood cells.

isotope one of two or more atoms that have the same atomic number (same number of protons), but which contain a different number of neutrons, thus differing in their atomic masses. They may be stable or radioactive, naturally occurring or synthesized. The term was coined by English chemist Frederick Soddy, pioneer researcher in atomic disintegration.

iteroparity in biology, the repeated production of offspring at intervals throughout the life cycle. It is usually contrasted with ◊semelparity, where each individual reproduces only once during its life.

ITU abbreviation for **intensive therapy unit**, high-technology facility concerned with the care of patients with acute life-threatening conditions.

IUCN abbreviation for ◊ **International Union for the Conservation of Nature**.

IVF abbreviation for ◊ **in vitro fertilization**.

ivory hard white substance of which the teeth and tusks of certain mammals are made. Among the most valuable are elephants' tusks, which are of unusual hardness and density. Ivory is used in carving and other decorative work, and is so valuable that poachers continue to illegally destroy the remaining wild elephant herds in Africa to obtain it.

Poaching for ivory has led to the decline of the African elephant population from 2 million to approximately 600,000, with the species virtually extinct in some countries. Trade in ivory was halted by Kenya in 1989, but Zimbabwe continued its policy of controlled culling to enable the elephant population to thrive and to release ivory for export. China and Hong Kong have refused to obey an international ban on ivory trading. In 1997, the 138 member nations of the ◊CITES voted in Harare, amidst much controversy, to remove the ban on trade in ivory in Botswana, Namibia, and Zimbabwe. Trade is scheduled to resume in 1999 and would only be allowed from the existing stockpiles of ivory.

Vegetable ivory is used for buttons, toys, and cheap ivory goods. It consists of the hard albumen of the seeds of a tropical palm (*Phytelephas macrocarpa*), and is imported from Colombia.

J in physics, the symbol for *joule*, the SI unit of energy.

jack tool or machine for lifting, hoisting, or moving heavy weights, such as motor vehicles. A *screw jack* uses the principle of the screw to magnify an applied effort; in a car jack, for example, turning the handle many times causes the lifting screw to rise slightly, and the effort is magnified to lift heavy weights. A *hydraulic jack* uses a succession of piston strokes to increase pressure in a liquid and force up a lifting ram.

jade semiprecious stone consisting of either jadeite, NaAlSi$_2$O$_6$ (a pyroxene), or nephrite, Ca$_2$(Mg,Fe)$_5$Si$_8$O$_{22}$ (OH,F)$_2$ (an amphibole), ranging from colourless through shades of green to black according to the iron content. Jade ranks 5.5–6.5 on the Mohs' scale of hardness.

The early Chinese civilization discovered jade, bringing it from E Turkestan, and carried the art of jade-carving to its peak. The Olmecs, Aztecs, Maya, and the Maori have also used jade for ornaments, ceremony, and utensils.

jansky unit of radiation received from outer space, used in radio astronomy. It is equal to 10^{-26} watts per square metre per hertz, and is named after US engineer Karl Jansky.

Japan Current or *Kuroshio* warm ocean ◊current flowing from Japan to North America.

Japanese encephalitis in medicine, a disease carried by mosquitoes, which can be fatal. A vaccine is available but it is not given routinely as the disease is confined to remote areas of the world seldom visited by international travellers.

Jarvik 7 the first successful artificial heart intended for permanent implantation in a human being. Made from polyurethane plastic and aluminium, it is powered by compressed air. Barney Clark became the first person to receive a Jarvik 7, in Salt Lake City, Utah, USA, in Dec 1982; it kept him alive for 112 days.

jasper hard, compact variety of ◊chalcedony SiO$_2$, usually coloured red, brown, or yellow. Jasper can be used as a gem.

jaundice yellow discoloration of the skin and whites of the eyes caused by an excess of bile pigment in the bloodstream. Approximately 60% of newborn babies exhibit some degree of jaundice, which is treated by bathing in white, blue, or green light that converts the bile pigment

bilirubin into a water-soluble compound that can be excreted in urine. A serious form of jaundice occurs in rhesus disease (see ◊rhesus factor).

Bile pigment is normally produced by the liver from the breakdown of red blood cells, then excreted into the intestines. A build-up in the blood is due to abnormal destruction of red cells (as in some cases of ◊anaemia), impaired liver function (as in ◊hepatitis), or blockage in the excretory channels (as in gallstones or ◊cirrhosis). The jaundice gradually recedes following treatment of the underlying cause.

jaw one of two bony structures that form the framework of the mouth in all vertebrates except lampreys and hagfishes (the agnathous or jawless vertebrates). They consist of the upper jawbone (maxilla), which is fused to the skull, and the lower jawbone (mandible), which is hinged at each side to the bones of the temple by ◊ligaments.

Jeans mass in astronomy, the mass that a cloud (or part of a cloud) of interstellar gas must have before it can contract under its own weight to form a protostar. The Jeans mass is an expression of the *Jeans criterion*, which says that a cloud will contract when the gravitational force tending to draw material towards its centre is greater than the opposing force due to gas pressure. It is named after English mathematician James Hopwood Jeans, whose work focussed on the kinetic theory of gases and the origins of the cosmos.

jet hard, black variety of lignite, a type of coal. It is cut and polished for use in jewellery and ornaments. Articles made of jet have been found in Bronze Age tombs.

jet in astronomy, a narrow luminous feature seen protruding from a star or galaxy, and representing a rapid outflow of material. See ◊active galaxy.

JET (abbreviation for *Joint European Torus*) research facility in England that conducts experiments on nuclear fusion. It is the focus of the European effort to produce a safe and environmentally sound fusion-power reactor. On 9 November 1991, the JET ◊tokamak, operating with a mixture of deuterium and iritium, produced a 1.7 megawatt pulse of power in an experiment that lasted two seconds. This was the first time that a substantial amount of energy had been produced by nuclear power in a controlled experiment.

JET's replacement tokamak was given the go-ahead in May 1996. MAST (*Mega Amp Spherical Tokamak*) will be the largest spherical tokamak to date.

jetfoil advanced type of ◊hydrofoil boat built by Boeing, propelled by water jets. It features horizontal, fully submerged hydrofoils fore and aft and has a sophisticated computerized control system to maintain its stability in all waters

Jetfoils have been in service worldwide since 1975. A jetfoil service operates across the English Channel between Dover and Ostend, Belgium, with a passage time of about 1.5 hours. The cruising speed of the jetfoil is about 80 kph/50 mph.

jet lag the effect of a sudden switch of time zones in air travel, resulting in tiredness and feeling 'out of step' with day and night. In 1989 it was suggested that use of the hormone melatonin helped to lessen the effect of jet lag by resetting the body clock. See also ◊circadian rhythm.

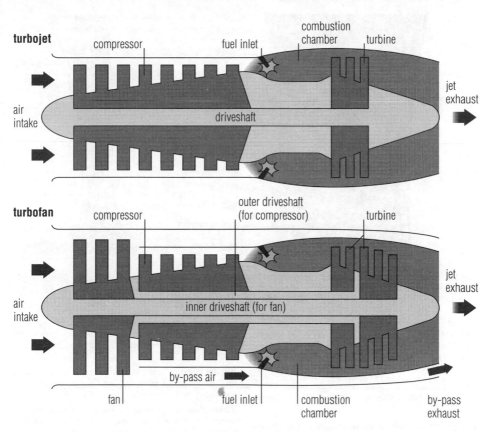

turbojet

compressor   fuel inlet   combustion chamber   turbine

air intake

driveshaft

jet exhaust

turbofan

compressor   outer driveshaft (for compressor)   turbine

air intake

inner driveshaft (for fan)

jet exhaust

by-pass air

fan   fuel inlet   combustion chamber   by-pass exhaust

jet propulsion Two forms of jet engine. In the turbojet, air passing into the air intake is compressed by the compressor and fed into the combustion chamber where fuel burns. The hot gases formed are expelled at high speed from the rear of the engine, driving the engine forwards and turning a turbine which drives the compressor. In the turbofan, some air flows around the combustion chamber and mixes with the exhaust gases. This arrangement is more efficient and quieter than the turbojet.

jet propulsion method of propulsion in which an object is propelled in one direction by a jet, or stream of gases, moving in the other. This follows from Isaac Newton's third law of motion: 'To every action, there is an equal and opposite reaction.' The most widespread application of the jet principle is in the jet engine, the most common kind of aircraft engine.

Jet Propulsion Laboratory NASA installation at Pasadena, California, operated by the California Institute of Technology. It is the command centre for NASA's deep-space probes such as the ◊Voyager, Magellan, and Galileo missions, with which it communicates via the Deep Space Network of radio telescopes at Goldstone, California; Madrid, Spain; and Canberra, Australia.

jet stream narrow band of very fast wind (velocities of over 150 kph/95 mph) found at altitudes of 10–16 km/ 6–10 mi in the upper troposphere or lower stratosphere. Jet streams usually occur about the latitudes of the Westerlies (35°–60°).

Jodrell Bank site in Cheshire, England, of the Nuffield Radio Astronomy Laboratories of the University of Manchester. Its largest instrument is the 76 m/250 ft radio dish

(the Lovell Telescope), completed in 1957 and modified in 1970. A 38 × 25 m/125 × 82 ft elliptical radio dish was introduced 1964, capable of working at shorter wave lengths.

These radio telescopes are used in conjunction with six smaller dishes up to 230 km/143 mi apart in an array called MERLIN (**m**ulti-**e**lement **r**adio-**l**inked **i**nterferometer **n**etwork) to produce detailed maps of radio sources.

joint in any animal with a skeleton, a point of movement or articulation. In vertebrates, it is the point where two bones meet. Some joints allow no motion (the sutures of the skull), others allow a very small motion (the sacroiliac joints in the lower back), but most allow a relatively free motion. Of these, some allow a gliding motion (one vertebra of the spine on another), some have a hinge action (elbow and knee), and others allow motion in all directions (hip and shoulder joints) by means of a ball-and-socket arrangement. The ends of the bones at a moving joint are covered with cartilage for greater elasticity and smoothness, and enclosed in an envelope (capsule) of tough white fibrous tissue lined with a membrane which

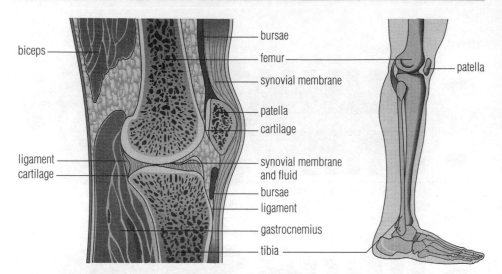

bursae
biceps
femur
patella
synovial membrane
patella
cartilage
ligament
cartilage
synovial membrane
and fluid
bursae
ligament
gastrocnemius
tibia

joint The knee joint is the largest and most complex in the human body. It is similar to a hinge joint but with a slight rotation which allows the leg to lock into a rigid position when extended. The ligaments surrounding the joint help prevent overextension. The ligament surrounding the patella or kneecap helps to protect the joint.

secretes a lubricating and cushioning ◊synovial fluid. The joint is further strengthened by ligaments. In invertebrates with an ◊exoskeleton, the joints are places where the exoskeleton is replaced by a more flexible outer covering, the arthrodial membrane, which allows the limb (or other body part) to bend at that point.

Joint European Torus experimental nuclear-fusion machine, known as ◊JET.

Josephson junction device used in 'superchips' (large and complex integrated circuits) to speed the passage of signals by a phenomenon called 'electron tunnelling'. Although these superchips respond a thousand times faster than the ◊silicon chip, they have the disadvantage that the components of the Josephson junctions operate only at temperatures close to ◊absolute zero. They are named after English theoretical physicist Brian Josephson.

joule SI unit (symbol J) of work and energy, replacing the ◊calorie (one joule equals 4.2 calories).

Joule

A bolt of lightning releases up to 3,000 million joules of energy. When a ton of TNT explodes, about 5,000 million joules of chemical energy are released. More than 100,000 million are needed to send a rocket into space. A tropical hurricane releases about 100,000 million million joules. In 1960 an earthquake in Chile released 10 million million million joules.

joystick in computing, an input device that signals to a computer the direction and extent of displacement of a hand-held lever. It is similar to the joystick used to control the flight of an aircraft.

Joysticks are sometimes used to control the movement of a cursor (marker) across a display screen, but are much more frequently used to provide fast and direct input for moving the characters and symbols that feature in computer games. Unlike a ◊mouse, which can move a pointer in any direction, simple games joysticks are often capable only of moving an object in one of eight different directions.

jugular vein one of two veins in the necks of vertebrates; they return blood from the head to the superior (or anterior) ◊vena cava and thence to the heart.

Julian date in astronomy, a measure of time used in astronomy in which days are numbered consecutively from noon ◊GMT on 1 January 4713 BC. It is useful where astronomers wish to compare observations made over long time intervals. The Julian date (JD) at noon on 1 January 2000 will be 2451545.0. The modified Julian date (MJD), defined as MJD = JD - 2400000.5, is more commonly used since the date starts at midnight GMT and the smaller numbers are more convenient.

jumbo jet popular name for a generation of huge wide-bodied airliners including the **Boeing 747**, which is 71 m/232 ft long, has a wingspan of 60 m/196 ft, a maximum takeoff weight of nearly 400 tonnes, and can carry more than 400 passengers.

jungle popular name for ◊rainforest.

Jupiter the fifth planet from the Sun, and the largest in the Solar System, with a mass equal to 70% of all the other planets combined, 318 times that of Earth's. It is largely composed of hydrogen and helium, liquefied by pressure in its interior, and probably with a rocky core larger than Earth. Its main feature is the Great Red Spot, a cloud of rising gases, 14,000 km/8,500 mi wide and 30,000 km/20,000 mi long, revolving anticlockwise.

Mean distance from the Sun 778 million km/484 million mi.

Equatorial diameter 142,800 km/88,700 mi.

Rotation period 9 hr 51 min.

Year (complete orbit) 11.86 Earth years.

Atmosphere consists of clouds of white ammonia crystals, drawn out into belts by the planet's high speed of

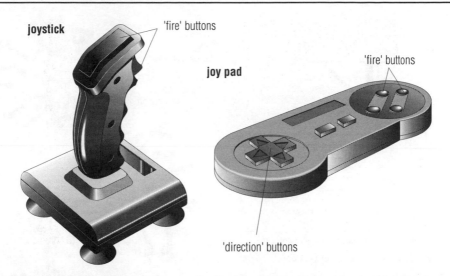

joystick   'fire' buttons

joy pad   'fire' buttons

'direction' buttons

joystick The directional and other controls on a conventional joystick may be translated to a joy pad, which enables all controls to be activated by buttons.

rotation (the fastest of any planet). Darker orange and brown clouds at lower levels may contain sulphur, as well as simple organic compounds. Further down still, temperatures are warm, a result of heat left over from Jupiter's formation, and it is this heat that drives the turbulent weather patterns of the planet.

In 1995, the *Galileo* probe revealed Jupiter's atmosphere to consist of 0.2% water, less than previously estimated.

Surface although largely composed of hydrogen and helium, Jupiter probably has a rocky core larger than Earth.

Satellites Jupiter has 16 moons. The four largest moons, Io, Europa (which is the size of our Moon), Ganymede, and Callisto, are the **Galilean satellites**, discovered in 1610 by Galileo (Ganymede, which is about the size of Mercury, is the largest moon in the Solar System). Three small moons were discovered in 1979 by the Voyager space probes, as was a faint ring of dust around Jupiter's equator 55,000 km/34,000 mi above the cloud tops.

The Great Red Spot was first observed in 1664. Its top is higher than the surrounding clouds; its colour is thought to be due to red phosphorus. Jupiter's strong magnetic field gives rise to a large surrounding magnetic 'shell', or magnetosphere, from which bursts of radio waves are detected. The Southern Equatorial Belt in which the Great Red Spot occurs is subject to unexplained fluctuation. In 1989 it sustained a dramatic and sudden fading.

Comet Shoemaker-Levy 9 crashed into Jupiter in July 1994. Impact zones were visible but are not likely to remain.

Jupiter

Jupiter is much smaller than the Sun. If the Sun were the size of a beach ball, 60 cm/2 ft across, Jupiter would be the size of a golf ball, 5 cm/2 in across. On this scale, the Earth would be less than 2.5 mm/0.1 in across.

Jurassic period of geological time 208–146 million years ago; the middle period of the Mesozoic era. Climates worldwide were equable, creating forests of conifers and ferns; dinosaurs were abundant, birds evolved, and limestones and iron ores were deposited.

The name comes from the Jura Mountains in France and Switzerland, where the rocks formed during this period were first studied.

justification in printing and word processing, the arrangement of text so that it is aligned with either the left or right margin, or both.

K

k symbol for *kilo-*, as in kg (kilogram) and km (kilometre).

K abbreviation for thousand, as in a salary of £30K.

K symbol for *kelvin*, a scale of temperature.

kale type of cabbage.

kame geological feature, usually in the form of a mound or ridge, formed by the deposition of rocky material carried by a stream of glacial meltwater. Kames are commonly laid down in front of or at the edge of a glacier (kame terrace), and are associated with the disintegration of glaciers at the end of an ice age.

Kames are made of well-sorted rocky material, usually sands and gravels. The rock particles tend to be rounded (by attrition) because they have been transported by water.

kaoliang variety of sorghum.

kaolin group of clay minerals, such as ◊kaolinite, $Al_2Si_2O_5(OH)_4$, derived from the alteration of aluminium silicate minerals, such as ◊feldspars and ◊mica. It is used in medicine to treat digestive upsets, and in poultices.

kaolinite white or greyish ◊clay mineral, hydrated aluminium silicate, $Al_2Si_2O_5(OH)_4$, formed mainly by the decomposition of feldspar in granite. China clay (kaolin) is derived from it. Kaolinite is economically important in the ceramic and paper industries. It is mined in the UK, the USA, France, and the Czech Republic. It is mined in France, the UK, Germany, China, and the USA.

karst landscape characterized by remarkable surface and underground forms, created as a result of the action of water on permeable limestone. The feature takes its name from the Karst region on the Adriatic coast in Slovenia and Croatia, but the name is applied to landscapes throughout the world, the most dramatic of which is found near the city of Guilin in the Guangxi province of China.

Limestone is soluble in the weak acid of rainwater. Erosion takes place most swiftly along cracks and joints in the limestone and these open up into gullies called grikes. The rounded blocks left upstanding between them are called clints.

karyotype in biology, the set of ◊chromosomes characteristic of a given species. It is described as the number, shape, and size of the chromosomes in a single cell of an organism. In humans for example, the karyotype consists

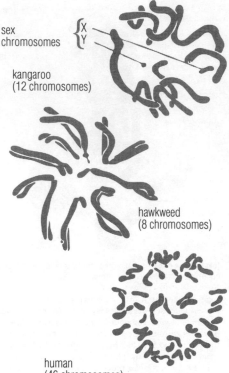

sex chromosomes { X Y

kangaroo (12 chromosomes)

hawkweed (8 chromosomes)

human (46 chromosomes)

karyotype The characteristics, or karyotype, of the chromosomes vary according to species. The kangaroo has 12 chromosomes, the hawkweed has 8, and a human being has 46.

of 46 chromosomes, in mice 40, crayfish 200, and in fruit flies 8.

The diagrammatic representation of a complete chromosome set is called a **karyogram**.

kayser unit of wave number (number of waves in a unit length), used in spectroscopy. It is expressed as waves per centimetre, and is the reciprocal of the wavelength. A wavelength of 0.1 cm has a wave number of 10 kaysers.

kcal symbol for *kilocalorie* (see ◊calorie).

Keck Telescope world's largest optical telescope, situated on Mauna Kea, Hawaii. It weighs 300 tonnes and has a primary mirror 10 m/33 ft in diameter, unique in that it consists of 36 hexagonal sections, each controlled and adjusted by a computer to generate single images of the objects observed. It received its first images in Nov 1990.

An identical telescope next to it, named Keck II, became operational in 1996. Both telescopes are jointly owned by the California Institute of Technology and the University of California.

keloid in medicine, overgrowth of fibrous tissue, usually produced at the site of a scar. Surgical removal is often unsuccessful, because the keloid returns.

kelvin scale temperature scale used by scientists. It begins at ◊absolute zero (−273.15°C) and increases by the same degree intervals as the Celsius scale; that is, 0°C is the same as 273.15 K and 100°C is 373.15 K.

Kennedy Space Center ◊NASA launch site on Merritt Island, near Cape Canaveral, Florida, used for Apollo and space-shuttle launches. The first flight to land on the Moon (1969) and *Skylab*, the first orbiting laboratory (1973), were launched here.

Kennelly–Heaviside layer former term for the ◊E layer of the ionosphere.

Kepler's laws in astronomy, three laws of planetary motion formulated in 1609 and 1619 by the German mathematician and astronomer Johannes Kepler: (1) the orbit of each planet is an ellipse with the Sun at one of the foci; (2) the radius vector of each planet sweeps out equal areas in equal times; (3) the squares of the periods of the planets are proportional to the cubes of their mean distances from the Sun.

Kepler derived the laws after exhaustive analysis of numerous observations of the planets, especially Mars, made by Tycho Brahe without telescopic aid. Isaac Newton later showed that Kepler's Laws were a consequence of the theory of universal gravitation.

keratin fibrous protein found in the ◊skin of vertebrates and also in hair, nails, claws, hooves, feathers, and the outer coating of horns.

If pressure is put on some parts of the skin, more keratin is produced, forming thick calluses that protect the layers of skin beneath.

keratitis in medicine, eye disease characterized by inflammation of the cornea. This may be the result of mechanical trauma, for example being hit in the eye, contact with chemicals, or infection with bacteria or viruses. The affected eye is red and painful and those affected should avoid exposure to light due to sensitivity. Treatment varies depending on the cause of the disorder.

kernel the inner, softer part of a ◊nut, or of a seed within a hard shell.

kernicterus in medicine, brain damage associated with haemolytic disease of the newborn. An incompatibility between the blood groups of mother and baby cause the baby's red blood cells to break down resulting in high concentrations of bilirubin (a pigment derived from haemoglobin) in the blood. Some of this bilirubin enters the brain causing toxic degeneration of the brain cells. Concentrations of bilirubin in the blood serum of affected infants with haemolytic disease should be monitored so that treatment can begin before dangerous concentrations are reached. Treatment is with blood transfusion, and can be administered before birth (see ◊fetal therapy).

kerosene thin oil obtained from the distillation of petroleum; a highly refined form is used in jet aircraft fuel. Kerosene is a mixture of hydrocarbons of the ◊paraffin series.

ketone member of the group of organic compounds containing the carbonyl group (C=O) bonded to two atoms of carbon (instead of one carbon and one hydrogen as in ◊aldehydes). Ketones are liquids or low-melting-point solids, slightly soluble in water.

An example is propanone (acetone, CH_3COCH_3), used as a solvent.

kettle hole pit or depression formed when a block of ice from a receding glacier becomes isolated and buried in glacial debris (till). As the block melts the till collapses to form a hollow, which may become filled with water to form a kettle lake or pond. Kettle holes range from 5 m/15 ft to 13 km/8 mi in diameter, and may exceed 33 m/100 ft in depth.

Kew Gardens popular name for the Royal Botanic Gardens, Kew, Surrey, England. They were founded in 1759 by Augusta of Saxe-Coburg (1719–1772), the mother of King George III, as a small garden and passed to the nation by Queen Victoria in 1840. By then they had expanded to almost their present size of 149 hectares/368 acres and since 1841 have been open daily to the public. They contain a collection of over 25,000 living plant species and many fine buildings. The gardens are also a centre for botanical research.

keyboard in computing, an input device resembling a typewriter keyboard, used to enter instructions and data. There are many variations on the layout and labelling of keys. Extra numeric keys may be added, as may special-purpose function keys, whose effects can be defined by programs in the computer.

kg symbol for ◊*kilogram*.

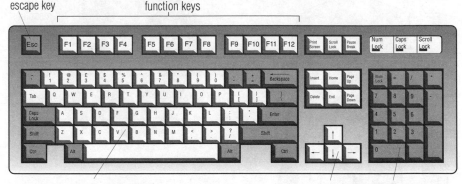

keyboard A standard 102-key keyboard. As well as providing a QWERTY typing keyboard, the function keys (labelled F1–F12) may be assigned tasks specific to a particular system.

khamsin hot southeasterly wind that blows from the Sahara desert over Egypt and parts of the Middle East from late March to May or June. It is called *sharav* in Israel.

kidney in vertebrates, one of a pair of organs responsible for fluid regulation, excretion of waste products, and maintaining the ionic composition of the blood. The kidneys are situated on the rear wall of the abdomen. Each one consists of a number of long tubules; the outer parts filter the aqueous components of blood, and the inner parts selectively reabsorb vital salts, leaving waste products in the remaining fluid (urine), which is passed through the ureter to the bladder.

The action of the kidneys is vital, although if one is removed, the other enlarges to take over its function. A patient with two defective kidneys may continue near-normal life with the aid of a kidney machine or continuous ambulatory peritoneal ◊dialysis (CAPD); or a kidney transplant may be recommended.

kidney machine medical equipment used in ◊dialysis.

kiln high-temperature furnace used commercially for drying timber, roasting metal ores, or for making cement, bricks, and pottery. Oil- or gas-fired kilns are used to bake ceramics at up to 1,760°C/3,200°F; electric kilns do not generally reach such high temperatures.

kilo- prefix denoting multiplication by 1,000, as in kilohertz, a unit of frequency equal to 1,000 hertz.

kilobyte (K or KB) in computing, a unit of memory equal to 1,024 ◊bytes. It is sometimes used, less precisely, to mean 1,000 bytes.

In the metric system, the prefix 'kilo-' denotes multiplication by 1,000 (as in kilometre, a unit equal to 1,000 metres). However, computer memory size is based on the ◊binary number system, and the most convenient binary equivalent of 1,000 is 2^{10}, or 1,024.

kilogram SI unit (symbol kg) of mass equal to 1,000 grams (2.24 lb). It is defined as a mass equal to that of the international prototype, a platinum-iridium cylinder held at the International Bureau of Weights and Measures in Sèvres, France.

kilometre unit of length (symbol km) equal to 1,000 metres, equivalent to 3,280.89 ft or 0.6214 (about 5/8) of a mile.

kilowatt unit (symbol kW) of power equal to 1,000 watts or about 1.34 horsepower.

kilowatt-hour commercial unit of electrical energy (symbol kWh), defined as the work done by a power of 1,000 watts in one hour and equal to 3,600 joules. It is used to calculate the cost of electrical energy taken from the domestic supply.

kimberlite an igneous rock that is ultramafic (containing very little silica); a type of alkaline ◊peridotite with a porphyritic texture (larger crystals in a fine-grained matrix), containing mica in addition to olivine and other minerals. Kimberlite represents the world's principal source of diamonds.

Kimberlite is found in carrot-shaped pipelike ◊intrusions called **diatremes**, where mobile material from very deep in the Earth's crust has forced itself upwards, expanding in its ascent. The material, brought upwards from near the boundary between crust and mantle, often altered and fragmented, includes diamonds. Diatremes are found principally near Kimberley, South Africa, from which the name of the rock is derived, and in the Yakut area of Siberia, Russia.

kinesis (plural **kineses**) in biology, a nondirectional movement in response to a stimulus; for example, woodlice move faster in drier surroundings. **Taxis** is a similar pattern of behaviour, but there the response is directional.

kinetic energy the energy of a body resulting from motion. It is contrasted with ◊potential energy.

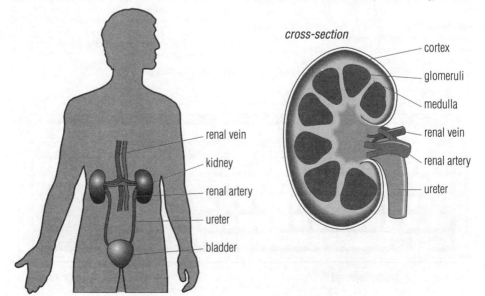

cross-section

cortex
glomeruli
medulla
renal vein
renal artery
ureter

renal vein
kidney
renal artery
ureter
bladder

kidney

kinetics the branch of chemistry that investigates the rates of chemical reactions.

kinetics branch of ◊dynamics dealing with the action of forces producing or changing the motion of a body; **kinematics** deals with motion without reference to force or mass.

kinetic theory theory describing the physical properties of matter in terms of the behaviour – principally movement – of its component atoms or molecules. The temperature of a substance is dependent on the velocity of movement of its constituent particles, increased temperature being accompanied by increased movement. A gas consists of rapidly moving atoms or molecules and, according to kinetic theory, it is their continual impact on the walls of the containing vessel that accounts for the pressure of the gas. The slowing of molecular motion as temperature falls, according to kinetic theory, accounts for the physical properties of liquids and solids, culminating in the concept of no molecular motion at ◊absolute zero (0K/−273°C).

By making various assumptions about the nature of gas molecules, it is possible to derive from the kinetic theory the various gas laws (such as ◊Avogadro's hypothesis, ◊Boyle's law, and ◊Charles's law).

kingdom the primary division in biological ◊classification. At one time, only two kingdoms were recognized: animals and plants. Today most biologists prefer a five-kingdom system, even though it still involves grouping together organisms that are probably unrelated. One widely accepted scheme is as follows: **Kingdom Animalia** (all multicellular animals); **Kingdom Plantae** (all plants, including seaweeds and other algae, except blue-green); **Kingdom Fungi** (all fungi, including the unicellular yeasts, but not slime moulds); **Kingdom Protista** or **Protoctista** (protozoa, diatoms, dinoflagellates, slime moulds, and various other lower organisms with eukaryotic cells); and **Kingdom Monera** (all prokaryotes – the bacteria and cyanobacteria, or ◊blue-green algae). The first four of these kingdoms make up the eukaryotes.

When only two kingdoms were recognized, any organism with a rigid cell wall was a plant, and so bacteria and fungi were considered plants, despite their many differences. Other organisms, such as the photosynthetic flagellates (euglenoids), were claimed by both kingdoms. The unsatisfactory nature of the two-kingdom system became evident during the 19th century, and the biologist Ernst Haeckel was among the first to try to reform it. High-power microscopes have revealed more about the structure of cells; it has become clear that there is a fundamental difference between cells without a nucleus (◊prokaryotes) and those with a nucleus (◊eukaryotes). However, these differences are larger than those between animals and higher plants, and are unsuitable for use as kingdoms. At present there is no agreement on how many kingdoms there are in the natural world.

Although the five-kingdom system is widely favoured, some schemes have as many as 20.

kin selection in biology, the idea that ◊altruism shown to genetic relatives can be worthwhile, because those relatives share some genes with the individual that is behaving altruistically, and may continue to reproduce. See ◊inclusive fitness.

Alarm-calling in response to predators is an example of a behaviour that may have evolved through kin selection: relatives that are warned of danger can escape and continue to breed, even if the alarm caller is caught.

Kirchhoff's laws two laws governing electric circuits devised by the German physicist Gustav Kirchhoff. **Kirchhoff's first law** states that the total current entering any junction in a circuit is the same as the total current leaving it. This is an expression of the conservation of electric charge. **Kirchhoff's second law** states that the sum of the potential drops across each resistance in any closed loop in a circuit is equal to the total electromotive force acting in that loop. The laws are equally applicable to DC and AC circuits.

Kirkwood gaps in astronomy, regions of the ◊asteroid belt, between ◊Mars, and ◊Jupiter, where there are relatively few asteroids.

The orbital periods of particles in the gaps correspond to simple fractions, especially 1/3, 2/5, 3/7, and 1/2, of the orbital period of Jupiter, indicating that they are caused by the gravitational influence of the larger planet. The gaps are named after Daniel Kirkwood, the 19th century US astronomer who first drew attention to them.

kiss of life (artificial ventilation) in first aid, another name for ◊artificial respiration.

kite quadrilateral with two pairs of adjacent equal sides. The geometry of this figure follows from the fact that it has one axis of symmetry.

Kitt Peak National Observatory observatory in the Quinlan Mountains near Tucson, Arizona, USA, operated by AURA (Association of Universities for Research into Astronomy). Its main telescopes are the 4 m/158 in Mayall reflector, opened in 1973, and the McMath Solar Telescope, opened in 1962, the world's largest of its type.

Among numerous other telescopes on the site is a 2.3 m/90 in reflector owned by the Steward Observatory of the University of Arizona.

kleptomania behavioural disorder characterized by an overpowering desire to possess articles for which one has no need. In kleptomania, as opposed to ordinary theft, there is no obvious need or use for what is stolen and sometimes the sufferer has no memory of the theft.

km symbol for ◊**kilometre**.

knocking in a spark-ignition petrol engine, a phenomenon that occurs when unburned fuel-air mixture explodes in the combustion chamber before being ignited by the spark. The resulting shock waves produce a metallic knocking sound. Loss of power occurs, which can be prevented by reducing the compression ratio, redesigning the geometry of the combustion chamber, or increasing the octane number of the petrol (usually by the use of tetraethyl lead anti-knock additives, or increasingly by MTBE – methyl tertiary butyl ether in unleaded petrol).

knot in navigation, unit by which a ship's speed is measured, equivalent to one ◊nautical mile per hour (one knot equals about 1.15 miles per hour). It is also sometimes used in aviation.

knowledge-based system (KBS) computer program that uses an encoding of human knowledge to help solve problems. It was discovered during research into ◊artificial intelligence that adding heuristics (rules of thumb) enabled programs to tackle problems that were otherwise difficult to solve by the usual techniques of computer science.

Chess-playing programs have been strengthened by including knowledge of what makes a good position, or of overall strategies, rather than relying solely on the computer's ability to calculate variations.

Kodak brand name of the US photographic company, Eastman Kodak Company of Rochester, New York. Together with Philips, they are the creators of the Photo CD system.

Koh-i-noor Persian 'mountain of light' diamond, originally part of the Aurangzeb treasure, seized in 1739 by the shah of Iran from the Moguls in India, taken back by Sikhs, and acquired by Britain in 1849 when the Punjab was annexed.

kolkhoz Russian term for a ◊collective farm, as opposed to a sovkhoz or state-owned farm.

Königsberg bridge problem long-standing puzzle that was solved by topology (the geometry of those properties of a figure which remain the same under distortion). In the city of Königsberg (now Kaliningrad in Russia), seven bridges connect the banks of the river Pregol'a and the islands in the river. For many years, people were challenged to cross each of the bridges in a single tour and return to their starting point. In 1736 Swiss mathematician Leonhard Euler converted the puzzle into a topological network, in which the islands and river banks were represented as nodes (junctions), and the connecting bridges as lines. By analysing this network he was able to show that it is not traversable – that is, it is impossible to cross each of the bridges once only and return to the point at which one started.

Kourou second-largest town of French Guiana, NW of Cayenne, site of the Guiana Space Centre of the European Space Agency; population (1996) 20,000 (20% of the total population of French Guiana).

kph or **km/h** symbol for **kilometres per hour**.

Krebs cycle or **citric acid cycle** or **tricarboxylic acid cycle** final part of the chain of biochemical reactions by which organisms break down food using oxygen to release energy (respiration). It takes place within structures called ◊mitochondria in the body's cells, and breaks down

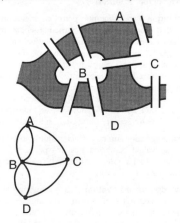

Königsberg bridge problem

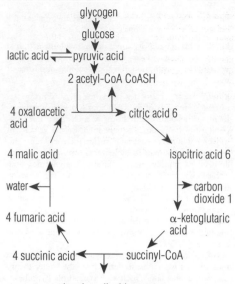

Krebs cycle The purpose of the Krebs (or citric acid) cycle is to complete the biochemical breakdown of food to produce energy-rich molecules, which the organism can use to fuel work. Acetyl coenzyme A (acetyl co-A) – produced by the breakdown of sugars, fatty acids, and some amino acids – reacts with oxaloacetic acid to produce citric acid, which is then converted in a series of enzyme-catalysed steps back to oxaloacetic acid. In the process, molecules of carbon dioxide and water are given off, and the precursors of the energy-rich molecules ATP are formed. (The numbers in the diagram indicate the number of carbon atoms in the principal compounds.)

food molecules in a series of small steps, producing energy-rich molecules of ◊ATP.

krypton colourless, odourless, gaseous, nonmetallic element, symbol Kr, atomic number 36, relative atomic mass 83.80. It is grouped with the inert gases and was long believed not to enter into reactions, but it is now known to combine with fluorine under certain conditions; it remains inert to all other reagents. It is present in very small quantities in the air (about 114 parts per million). It is used chiefly in fluorescent lamps, lasers, and gas-filled electronic valves.

Krypton was discovered in 1898 in the residue from liquid air by British chemists William Ramsay and Morris Travers; the name refers to their difficulty in isolating it. (*Kryptos* is Greek for 'hidden').

K-T boundary geologists' shorthand for the boundary between the rocks of the ◊Cretaceous and the ◊Tertiary periods 65 million years ago. It marks the extinction of the dinosaurs and in many places reveals a layer of iridium, possibly deposited by the meteorite that crashed into the Yucatá Peninsula, which may have caused the extinction by its impact. In 1996 US geologists discovered a small iridium-containing pebble believed to be a fragment of the meteorite.

Kuiper belt ring of small, icy bodies orbiting the Sun beyond the outermost planet. The Kuiper belt, named after US astronomer Gerard Kuiper who proposed its existence in 1951, is thought to be the source of comets that orbit the Sun with periods of less than 200 years. The first member of the Kuiper belt was seen in 1992. In 1995 the first comet-sized objects were discovered; previously the only objects found had diameters of at least 100 km/ 60 mi (comets generally have diameters of less than 10 km/6 mi).

Two new objects were discovered in the Kuiper belt in 1996. The first, 1996 TL66, is 500 km/300 mi in diameter and has an irregular orbit that takes it four to six times further from the Sun than Neptune. The second, 1996 RQ20, is slightly smaller, with an orbit that takes it about three times further from the Sun than Neptune. The orbits of both are at an angle of 20° to the plane of the Solar System.

Kuroshio or *Japan Current* warm ocean ◊ current flowing from Japan to North America.

kW symbol for ◊ *kilowatt*.

kwashiorkor severe protein deficiency in children under five years, resulting in retarded growth, lethargy, ◊ oedema, diarrhoea, and a swollen abdomen. It is common in Third World countries with a high incidence of malnutrition.

kyanite aluminium silicate, Al_2SiO_5, a pale-blue mineral occurring as blade-shaped crystals. It is an indicator of high-pressure conditions in metamorphic rocks formed from clay sediments. Andalusite, kyanite, and sillimanite are all polymorphs.

l symbol for ◊*litre*, a measure of liquid volume.

labelled compound or *tagged compound* chemical compound in which a radioactive isotope is substituted for a stable one. The path taken by such a compound through a system can be followed, for example by measuring the radiation emitted.

This powerful and sensitive technique is used in medicine, chemistry, biochemistry, and industry.

labellum lower petal of an orchid flower; it is a different shape from the two lateral petals and gives the orchid its characteristic appearance. The labellum is more elaborate and usually larger than the other petals. It often has distinctive patterning to encourage ◊pollination by insects; sometimes it is extended backwards to form a hollow spur containing nectar.

labyrinthitis inflammation of the part of the inner ear responsible for the sense of balance (the labyrinth). It results in dizziness, which may then cause nausea and vomiting. It is usually caused by a viral infection of the ear (◊otitis), which resolves in a few weeks. The nausea and vomiting may respond to antiemetic drugs.

laccolith intruded mass of igneous rock that forces apart two strata and forms a round lens-shaped mass many times wider than thick. The overlying layers are often pushed upward to form a dome. A classic development of laccoliths is illustrated in the Henry, La Sal, and Abajo mountains of SE Utah, USA, found on the Colorado plateau.

lacrimation in medicine, the production of tears by the lacrimal apparatus of the eye.

lactation secretion of milk in mammals, from the mammary glands. In late pregnancy, the cells lining the lobules inside the mammary glands begin extracting substances from the blood to produce milk. The supply of milk starts shortly after birth with the production of colostrum, a clear fluid consisting largely of water, protein, antibodies, and vitamins. The production of milk continues practically as long as the baby continues to suckle.

lacteal small vessel responsible for absorbing fat in the small intestine. Occurring in the fingerlike villi of the ◊ileum, lacteals have a milky appearance and drain into the lymphatic system.

Before fat can pass into the lacteal, bile from the liver causes its emulsification into droplets small enough for attack by the enzyme lipase. The products of this digestion form into even smaller droplets, which diffuse into the villi.

Large droplets re-form before entering the lacteal and this causes the milky appearance.

lactic acid or *2-hydroxypropanoic acid* CH₃CHOH-COOH organic acid, a colourless, almost odourless liquid, produced by certain bacteria during fermentation and by active muscle cells when they are exercised hard and are experiencing ◊oxygen debt. An accumulation of lactic acid in the muscles may cause cramp. It occurs in yoghurt, buttermilk, sour cream, poor wine, and certain plant extracts, and is used in food preservation and in the preparation of pharmaceuticals.

lactose white sugar, found in solution in milk; it forms 5% of cow's milk. It is commercially prepared from the whey obtained in cheese-making. Like table sugar (sucrose), it is a disaccharide, consisting of two basic sugar units (monosaccharides), in this case, glucose and galactose. Unlike sucrose, it is tasteless.

lagoon coastal body of shallow salt water, usually with limited access to the sea. The term is normally used to describe the shallow sea area cut off by a ◊coral reef or barrier islands.

Lagrangian points five locations in space where the centrifugal and gravitational forces of two bodies neutralize each other; a third, less massive body located at any one of these points will be held in equilibrium with respect to the other two. Three of the points, L1–L3, lie on a line joining the two large bodies. The other two points, L4 and L5, which are the most stable, lie on either side of this line. Their existence was predicted in 1772 by French mathematician Joseph Louis Lagrange.

The *Trojan asteroids* lie at Lagrangian points L4 and L5 in Jupiter's orbit around the Sun. Clouds of dust and debris may lie at the Lagrangian points of the Moon's orbit around the Earth.

lahar mudflow formed of a fluid mixture of water and volcanic ash. During a volcanic eruption, melting ice may combine with ash to form a powerful flow capable of causing great destruction. The lahars created by the eruption of Nevado del Ruiz in Colombia, South America, in 1985 buried 22,000 people in 8 m/26 ft of mud.

lake body of still water lying in depressed ground without direct communication with the sea. Lakes are common in formerly glaciated regions, along the courses of slow rivers, and in low land near the sea. The main classifications are by origin: *glacial lakes*, formed by glacial scouring; *barrier lakes*, formed by landslides and glacial moraines; *crater lakes*, found in volcanoes; and *tectonic lakes*, occurring in natural fissures.

Crater lakes form in the ◊calderas of extinct volcanoes, for example Crater Lake, Oregon. Subsidence of the roofs of limestone caves in ◊karst landscape exposes the subterranean stream network and provides a cavity in which a lake can develop. Tectonic lakes form during tectonic movement, as when a rift valley is formed. Lake Tanganyika was created in conjunction with the East African Great Rift Valley. Glaciers produce several distinct types of lake, such as the lochs of Scotland and the Great Lakes of North America.

Lakes are mainly freshwater, but salt and bitter lakes are found in areas of low annual rainfall and little surface runoff, so that the rate of evaporation exceeds the rate of inflow, allowing mineral salts to accumulate. The Dead Sea has a salinity of about 250 parts per 1,000 and the Great Salt Lake, Utah, about 220 parts per 1,000. Salinity can also be caused by volcanic gases or fluids, for example Lake Natron, Tanzania.

In the 20th century large artificial lakes have been created in connection with hydroelectric and other works. Some lakes have become polluted as a result of human activity. Sometimes ◊eutrophication (a state of overnourishment) occurs, when agricultural fertilizers leaching into lakes cause an explosion of aquatic life, which then depletes the lake's oxygen supply until it is no longer able to support life.

lake dwelling or *pile dwelling* prehistoric habitation built on piles driven into the bottom of a lake or at the edge of a lake or river. Such villages are found throughout Europe, in W Africa, South America, Borneo, and New Guinea.

Objects recovered from lake dwellings are often unusually well preserved by the mud or peat in which they are buried. Wooden items, wickerwork, woven fabrics, fruit, and pollen grains have been retrieved.

Lamarckism theory of evolution, now discredited, advocated during the early 19th century by French naturalist Jean Baptiste Lamarck.

Lamarckism is the theory that acquired characteristics were inherited. It differs from the Darwinian theory of evolution.

lambert unit of luminance (the light shining from a surface), equal to one ◊lumen per square centimetre. In scientific work the ◊candela per square metre is preferred.

lamina in flowering plants (◊angiosperms), the blade of the ◊leaf on either side of the midrib. The lamina is generally thin and flattened, and is usually the primary organ of ◊photosynthesis. It has a network of veins through which water and nutrients are conducted. More generally, a lamina is any thin, flat plant structure, such as the ◊thallus of many seaweeds.

lamp, electric device designed to convert electrical energy into light energy.

In a *filament lamp* such as a light bulb an electric current causes heating of a long thin coil of fine high-resistance wire enclosed at low pressure inside a glass bulb. In order to give out a good light the wire must glow white-hot and therefore must be made of a metal, such as tungsten, that has a high melting point. The efficiency of filament lamps is low because most of the electrical energy is converted to heat.

A *fluorescent lamp* uses an electrical discharge or spark inside a gas-filled tube to produce light. The inner surface of the tube is coated with a fluorescent material that converts the ultraviolet light generated by the discharge into visible light. Although a high voltage is needed to start the discharge, these lamps are far more efficient than filament lamps at producing light.

Landsat series of satellites used for monitoring Earth resources. The first was launched in 1972.

landslide sudden downward movement of a mass of soil or rocks from a cliff or steep slope. Landslides happen when a slope becomes unstable, usually because the base has been undercut or because materials within the mass have become wet and slippery.

A *mudflow* happens when soil or loose material is

mudflow landslide

slump landslide

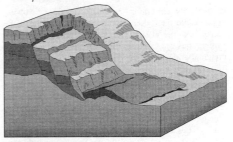

landslip landslide

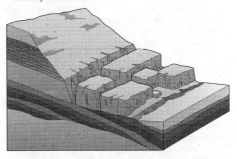

landslide Types of landslide. A mudflow is a tongue of mud that slides downhill. A slump is a fall of a large mass that stays together after the fall. A landslip occurs when beds of rock move along a lower bed.

soaked so that it no longer adheres to the slope; it forms a tongue of mud that reaches downhill from a semicircular hollow. A *slump* occurs when the material stays together as a large mass, or several smaller masses, and these may form a tilted steplike structure as they slide. A *landslip* is formed when ◊beds of rock dipping towards a cliff slide along a lower bed. Earthquakes may precipitate landslides.

lanolin sticky, purified wax obtained from sheep's wool and used in cosmetics, soap, and leather preparation.

lanthanide any of a series of 15 metallic elements (also known as rare earths) with atomic numbers 57 (lanthanum) to 71 (lutetium).

One of its members, promethium, is radioactive. All

occur in nature. Lanthanides are grouped because of their chemical similarities (most are trivalent, but some can be divalent or tetravalent), their properties differing only slightly with atomic number.

Lanthanides were called rare earths originally because they were not widespread and were difficult to identify and separate from their ores by their discoverers. The series is set out in a band in the periodic table of the elements, as are the ◊actinides.

lanthanum soft, silvery, ductile and malleable, metallic element, symbol La, atomic number 57, relative atomic mass 138.91, the first of the lanthanide series. It is used in making alloys. It was named in 1839 by Swedish chemist Carl Mosander (1797–1858).

laparoscope in medicine, another name for an ◊endoscope.

laparotomy exploratory surgical procedure involving incision into the abdomen, especially when done at the flanks. The use of laparotomy, as of other exploratory surgery, has decreased sharply with advances in medical imaging and the direct-viewing technique known as ◊endoscopy.

lapis lazuli rock containing the blue mineral lazurite in a matrix of white calcite with small amounts of other minerals. It occurs in silica-poor igneous rocks and metamorphic limestones found in Afghanistan, Siberia, Iran, and Chile. Lapis lazuli was a valuable pigment of the Middle Ages, also used as a gemstone and in inlaying and ornamental work.

laptop computer portable microcomputer, small enough to be used on the operator's lap. It consists of a single unit, incorporating a keyboard, ◊floppy disc and ◊hard disc drives, and a screen. The screen often forms a lid that folds back in use. It uses a liquid-crystal or gas-plasma display, rather than the bulkier and heavier cathode-ray tubes found in most display terminals. A typical laptop computer measures about 210×297 mm/8.3 $\times$ 11.7 in (A4), is 5 cm/2 in in depth, and weighs less than 3 kg/6 lb 9 oz.

Large Electron Positron Collider (LEP) the world's largest particle ◊accelerator, in operation from 1989 at the CERN laboratories near Geneva in Switzerland. It occupies a tunnel 3.8 m/12.5 ft wide and 27 km/16.7 mi long, which is buried 180 m/590 ft underground and forms a ring consisting of eight curved and eight straight sections.

In June 1996, LEP resumed operation after a £210 million upgrade. The upgraded machine will be known as LEP2, and can generate collision energy of 161 giga-electron volts.

Electrons and positrons enter the ring after passing through the Super Proton Synchrotron accelerator. They travel in opposite directions around the ring, guided by 3,328 bending magnets and kept within tight beams by 1,272 focusing magnets.

As they pass through the straight sections, the particles are accelerated by a pulse of radio energy. Once sufficient energy is accumulated, the beams are allowed to collide. Four giant detectors are used to study the resulting shower of particles.

In 1989 the LEP was used to measure the mass and lifetime of the Z particle, carrier of the weak nuclear force.

larva stage between hatching and adulthood in those species in which the young have a different appearance and way of life from the adults. Examples include tadpoles (frogs) and caterpillars (butterflies and moths). Larvae are typical of the invertebrates, some of which (for example, shrimps) have two or more distinct larval stages. Among vertebrates, it is only the amphibians and some fishes that have a larval stage.

The process whereby the larva changes into another stage, such as a pupa (chrysalis) or adult, is known as ◊metamorphosis.

Larva

Blowfly larvae develop in less than two weeks. During this time they can gain 5% of their final larval weight each hour.

laryngitis inflammation of the larynx, causing soreness of the throat, a dry cough, and hoarseness. The acute form is due to a virus or other infection, excessive use of the voice, or inhalation of irritating smoke, and may cause the voice to be completely lost. With rest, the inflammation usually subsides in a few days.

larynx in mammals, a cavity at the upper end of the trachea (windpipe) containing the vocal cords. It is stiffened with cartilage and lined with mucous membrane. Amphibians and reptiles have much simpler larynxes, with no vocal cords. Birds have a similar cavity, called the *syrinx*, found lower down the trachea, where it branches to form the bronchi. It is very complex, with well-developed vocal cords.

laser (acronym for *light amplification by stimulated emission of radiation*) device for producing a narrow beam of light, capable of travelling over vast distances without dispersion, and of being focused to give enormous power densities (10^8 watts per cm^2 for high-energy lasers). The laser operates on a principle similar to that of the ◊maser (a high-frequency microwave amplifier or oscillator). The uses of lasers include communications (a laser beam can carry much more information than can radio waves), cutting, drilling, welding, satellite tracking, medical and biological research, and surgery.

Any substance the majority of whose atoms or molecules can be put into an excited energy state can be used as laser material. Many solid, liquid, and gaseous substances have been used, including synthetic ruby crystal (used for the first extraction of laser light in 1960, and giving a high-power pulsed output) and a helium–neon gas mixture, capable of continuous operation, but at a lower power.

A blue shortwave laser was developed in Japan in 1988. Its expected application is in random access memory (◊RAM) and ◊compact disc recording, where its shorter wavelength will allow a greater concentration of digital information to be stored and read. A gallium arsenide chip, produced by IBM in 1989, contains the world's smallest lasers in the form of cylinders of ◊semiconductor roughly one-tenth of the thickness of a human hair; a million lasers can fit on a chip 1 cm/2.5 in square.

Sound wave vibrations from the window glass of a room can be picked up by a reflected laser beam. Lasers are also used as entertainment in theatres, concerts, and light shows.

laser printer computer printer in which the image to be printed is formed by the action of a laser on a light-sensitive drum, then transferred to paper by means of an

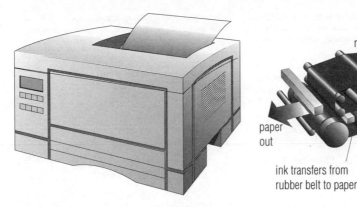

laser printer A laser printer works by transferring tiny ink particles contained in a toner cartridge to paper via a rubber belt. The image is produced by laser on a light-sensitive drum within the printer.

electrostatic charge. Laser printers are page printers, printing a complete page at a time. The printed image, which can take the form of text or pictures, is made up of tiny dots, or ink particles. The quality of the image generated depends on the fineness of these dots – most laser printers can print up to 120 dots per cm/300 dots per in across the page.

A typical desktop laser printer can print about 4–20 pages per minute. The first low-cost laser printer suitable for office use appeared in 1984.

laser surgery use of intense light sources to cut, coagulate, or vaporize tissue. Less invasive than normal surgery, it destroys diseased tissue gently and allows quicker, more natural healing. It can be used by way of a flexible endoscope to enable the surgeon to view the diseased area at which the laser needs to be aimed.

Lassa fever acute disease caused by an arenavirus, first detected in 1969, and spread by a species of rat found only in W Africa. It is classified as a ◊haemorrhagic fever and characterized by high fever, headache, muscle pain, and internal bleeding. There is no known cure, the survival rate being less than 50%.

lat abbreviation for ◊**latitude**.

latent heat in physics, the heat absorbed or released by a substance as it changes state (for example, from solid to liquid) at constant temperature and pressure.

lateral line system system of sense organs in fish and larval amphibians (tadpoles) that detects water movement. It usually consists of a row of interconnected pores on either side of the body that divide into a system of canals across the head.

lateral moraine linear ridge of rocky debris deposited near the edge of a ◊glacier. Much of the debris is material that has fallen from the valley side onto the glacier's edge, having been weathered by ◊freeze-thaw (the alternate freezing and thawing of ice in cracks); it will, therefore, tend to be angular in nature. Where two glaciers merge, two lateral moraines may join together to form a **medial moraine** running along the centre of the merged glacier.

laterite red residual soil characteristic of tropical rainforests. It is formed by the weathering of basalts, granites, and shales and contains a high percentage of aluminium and iron hydroxides. It may form an impermeable and infertile layer that hinders plant growth.

latex fluid of some plants (such as the rubber tree and poppy), an emulsion of resins, proteins, and other organic

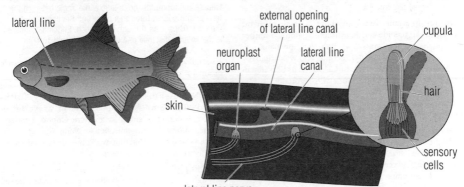

lateral line system In fish, the lateral line system detects water movement. Arranged along a line down the length of the body are two water-filled canals, just under the skin. The canals are open to the outside, and water movements cause water to move in the canals. Nerve endings detect the movements.

substances. It is used as the basis for making rubber. The name is also applied to a suspension in water of natural or synthetic rubber (or plastic) particles used in rubber goods, paints, and adhesives.

lathe machine tool, used for *turning*. The workpiece to be machined, usually wood or metal, is held and rotated while cutting tools are moved against it. Modern lathes are driven by electric motors, which can drive the spindle carrying the workpiece at various speeds.

latifundium in ancient Rome, a large agricultural estate designed to make maximum use of cheap labour, whether free workers or slaves.

In present-day Italy, Spain, and South America, the term *latifondo* refers to a large agricultural estate worked by low-paid casual or semiservile labour in the interests of absentee landlords.

latitude and longitude imaginary lines used to locate position on the globe. Lines of latitude are drawn parallel to the Equator, with 0° at the Equator and 90° at the north and south poles. Lines of longitude are drawn at right angles to these, with 0° (the Prime Meridian) passing through Greenwich, England.

laudanum alcoholic solution (tincture) of the drug ◊opium. Used formerly as a narcotic and painkiller, it was available in the 19th century from pharmacists on demand in most of Europe and the USA.

laughing gas popular name for ◊nitrous oxide, an anaesthetic.

Laurasia northern landmass formed 200 million years ago by the splitting of the single world continent ◊Pangaea. (The southern landmass was ◊Gondwanaland.) It consisted of what was to become North America, Greenland, Europe, and Asia, and is believed to have broken up about 100 million years ago with the separation of North America from Europe.

lava molten rock (usually 800–1,100°C/1,500–2,000°F) that erupts from a ◊volcano and cools to form extrusive ◊igneous rock. It differs from magma in that it is molten rock on the surface; magma is molten rock below the surface. Lava that is high in silica is viscous and sticky and does not flow far; it forms a steep-sided conical composite volcano. Low-silica lava can flow for long distances and forms a broad flat volcano.

lawrencium synthesized, radioactive, metallic element, the last of the actinide series, symbol Lr, atomic number 103, relative atomic mass 262. Its only known isotope, Lr-257, has a half-life of 4.3 seconds and was originally synthesized at the University of California at Berkeley in 1961 by bombarding californium with boron nuclei. The original symbol, Lw, was officially changed in 1963.

The element was named after Ernest Lawrence (1901–1958), the US inventor of the cyclotron.

laxative substance used to relieve constipation (infrequent bowel movement). Current medical opinion discourages regular or prolonged use. Regular exercise and a diet high in vegetable fibre is believed to be the best means of preventing and treating constipation.

lb symbol for ◊*pound* (weight).

LCD abbreviation for ◊*liquid-crystal display*.

L-dopa chemical, normally produced by the body, which is converted by an enzyme to dopamine in the brain. It is

Point X lies on longitude 60°W

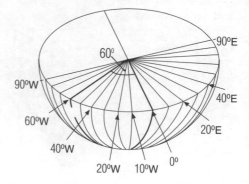

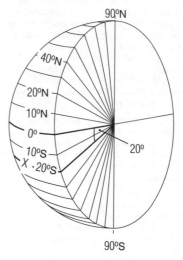
Point X lies on latitude 20°S

latitude and longitude Locating a point on a globe using latitude and longitude. Longitude is the angle between the terrestrial meridian through a place and the standard meridian 0° passing through Greenwich, England. Latitude is the angular distance of a place from the equator.

essential for integrated movement of individual muscle groups.

L-dopa is a left-handed isomer of an amino acid $C_9H_{11}NO_2$. As a treatment, it relieves the rigidity of ◊Parkinson's disease in 60% of sufferers, but may have significant side effects, such as extreme mood changes, hallucinations, and uncontrolled writhing movements. It is often given in combination with other drugs to improve its effectiveness at lower doses.

leaching process by which substances are washed through or out of the soil. Fertilizers leached out of the soil drain into rivers, lakes, and ponds and cause water pollution. In tropical areas, leaching of the soil after the destruction of forests removes scarce nutrients and can lead to a dramatic loss of soil fertility. The leaching of soluble minerals in soils can lead to the formation of distinct soil horizons as different minerals are deposited at successively lower levels.

lead heavy, soft, malleable, grey, metallic element, symbol Pb (from Latin *plumbum*), atomic number 82, relative atomic mass 207.19. Usually found as an ore (most often in galena), it occasionally occurs as a free metal (◊ native metal), and is the final stable product of the decay of uranium. Lead is the softest and weakest of the commonly used metals, with a low melting point; it is a poor conductor of electricity and resists acid corrosion. As a cumulative poison, lead enters the body from lead water pipes, lead-based paints, and leaded petrol. (In humans, exposure to lead shortly after birth is associated with impaired mental health between the ages of two and four.) The metal is an effective shield against radiation and is used in batteries, glass, ceramics, and alloys such as pewter and solder.

lead–acid cell type of ◊ accumulator (storage battery).

leaded petrol petrol that contains ◊ antiknock, a mixture of the chemicals tetraethyl lead and dibromoethane. The lead from the exhaust fumes enters the atmosphere, mostly as simple lead compounds, which are poisonous to the developing nervous systems of children.

lead ore any of several minerals from which lead is extracted. The primary ore is galena or lead sulphite PbS. This is unstable, and on prolonged exposure to the atmosphere it oxidizes into the minerals cerussite $PbCO_3$ and anglesite $PbSO_4$. Lead ores are usually associated with other metals, particularly silver – which can be mined at the same time – and zinc, which can cause problems during smelting.

Most commercial deposits of lead ore are in the form of veins, where hot fluids have leached the ore from cooling ◊ igneous masses and deposited it in cracks in the surrounding country rock, and in thermal ◊ metamorphic zones, where the heat of igneous intrusions has altered the minerals of surrounding rocks. Lead is mined in over 40 countries, but half of the world's output comes from the USA, Canada, Russia, Kazakhstan, Uzbekistan, Canada, and Australia.

lead poisoning in medicine, toxic condition due to the ingestion of lead. Symptoms include loss of appetite, nausea and vomiting, colic, anaemia, and nerve damage. It is treated by the removal of the source of lead and by chelating agents (see ◊ chelate) to remove the lead from the body. Most cases today are chronic and occur in workers at smelters or scrap yards, but lead poisoning used to occur more commonly because of the use of domestic lead-based paints and lead water pipes.

leaf lateral outgrowth on the stem of a plant, and in most species the primary organ of ◊ photosynthesis. The chief leaf types are cotyledons (seed leaves), scale leaves (on underground stems), foliage leaves, and bracts (in the axil of which a flower is produced).

Typically leaves are composed of three parts: the sheath or leaf base, the petiole or stalk, and the lamina or blade. The lamina has a network of veins through which water and nutrients are conducted. Structurally the leaf is made up of ◊ mesophyll cells surrounded by the epidermis and usually, in addition, a waxy layer, termed the cuticle, which prevents excessive evaporation of water from the leaf tissues by transpiration. The epidermis is interrupted by small pores, or stomata, through which gas exchange

between the plant and the atmosphere occurs. *See illustration overleaf.*

learning disability limitation in the ability to learn. Learning disabilities vary in degree and arise from a variety of origins, including dysfunction of the brain and childhood behavioural problems. Learning disabilities are often associated with other disabilities.

The major causes of brain dysfunction are due to the presence of abnormal chromosomes, such as in ◊ Down's syndrome, and to other genetic disorders, or it can be due to head injuries, such as epilepsy. Intellectual and social impairment results in the individual requiring additional care and supervision to live as normal a life as possible.

Profound learning disabilities result in the person having multiple disabilities and being unable to walk or talk and being totally dependent upon others for care.

learning theory in psychology, any theory or body of theories about how behaviour in animals and human beings is acquired or modified by experience. Two main theories are classical and operant ◊ conditioning.

least action, principle of in physics, an alternative expression of Newton's laws of motion that states that a particle moving between two points under the influence of a force will follow the path along which its total action is least. Action is a quantity related to the average difference between the kinetic energy and the potential energy of the particle along its path. The principle is only true where no energy is lost from the system; for example an object moving in free fall in a gravitational field. It is closely related to ◊ Fermat's principle of least time which governs the path taken by a ray of light.

leather material prepared from the hides and skins of animals, by tanning with vegetable tannins and chromium salts. Leather is a durable and water-resistant material, and is used for bags, shoes, clothing, and upholstery. There are three main stages in the process of converting animal skin into leather: cleaning, tanning, and dressing. Tanning is often a highly polluting process.

The skin, usually cattle hide, is dehydrated after removal to arrest decay. Soaking is necessary before tanning in order to replace the lost water with something that will bind the fibres together. The earliest practice, at least 7,000 years old, was to pound grease into the skin. In about 400 BC the Egyptians began to use vegetable extracts containing tannic acid, a method adopted in medieval Europe. Chemical tanning using mineral salts was introduced in the late 19th century.

Le Chatelier's principle or *Le Chatelier-Braun principle* in chemistry, principle that if a change in conditions is imposed on a system in equilibrium, the system will react to counteract that change and restore the equilibrium.

First stated in 1884 by French chemist Henri le Chatelier (1850–1936), it has been found to apply widely outside the field of chemistry.

lecithin lipid (fat), containing nitrogen and phosphorus, that forms a vital part of the cell membranes of plant and animal cells. The name is from the Greek *lekithos* 'egg yolk', eggs being a major source of lecithin.

LED abbreviation for ◊ *light-emitting diode*.

left-handedness using the left hand more skilfully and in preference to the right hand for most actions. It occurs

leaf margins

entire   serrate   dentate   incised   crenate   sinuate   scalloped   undulate

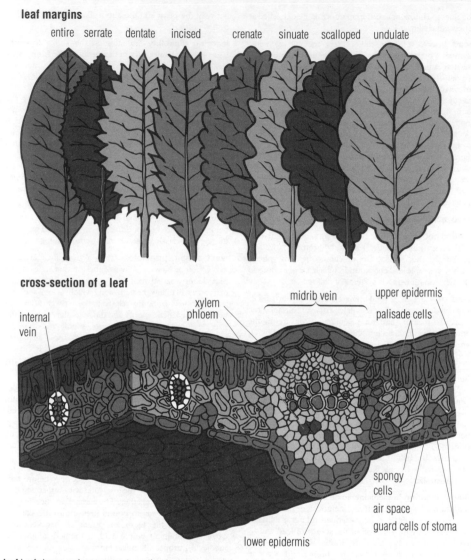

cross-section of a leaf

midrib vein   upper epidermis

xylem   palisade cells
phloem

internal
vein

spongy
cells

air space

guard cells of stoma

lower epidermis

leaf Leaf shapes and arrangements on the stem are many and varied; in cross-section, a leaf is a complex arrangement of cells surrounded by the epidermis. This is pierced by the stomata through which gases enter and leave.

in about 9% of the population, predominantly males. It is caused by dominance of the right side of the brain.

legionnaires' disease pneumonia-like disease, so called because it was first identified when it broke out at a convention of the American Legion in Philadelphia in 1976. Legionnaires' disease is caused by the bacterium *Legionella pneumophila*, which breeds in warm water (for example, in the cooling towers of air-conditioning systems). It is spread in minute water droplets, which may be inhaled. The disease can be treated successfully with antibiotics, though mortality can be high in elderly patients.

legume plant of the family Leguminosae, which has a pod containing dry seeds. The family includes peas, beans, lentils, clover, and alfalfa (lucerne). Legumes are important in agriculture because of their specialized roots, which have nodules containing bacteria capable of fixing nitrogen from the air and increasing the fertility of the soil. The edible seeds of legumes are called **pulses**.

leishmaniasis any of several parasitic diseases caused by microscopic protozoans of the genus *Leishmania*, identified by William Leishman (1865–1926), and transmitted by sandflies.

It occurs in two main forms: **visceral** (also called kala-azar), in which various internal organs are affected, and **cutaneous**, where the disease is apparent mainly in the skin. Leishmaniasis occurs in the Mediterranean region, Africa, Asia, and Central and S America. There are 12 million cases of leishmaniasis annually.

Indian researchers discovered in 1994 a cheap and effective way of keeping sandfly populations under control, by plastering the walls of houses and outbuildings with mud and lime. The plaster deprives flies of the moist crevices in which they breed, and the lime kills any existing larvae. In trials, sandfly numbers dropped by 90%.

lek in biology, a closely spaced set of very small ◊territories each occupied by a single male during the mating season. Leks are found in the mating systems of several ground-dwelling birds (such as grouse) and a few antelopes.

The lek is a traditional site where both males and females congregate during the breeding season. The males display to passing females in the hope of attracting them to mate. Once mated, the females go elsewhere to lay their eggs or to complete gestation.

lens in optics, a piece of a transparent material, such as glass, with two polished surfaces – one concave or convex,

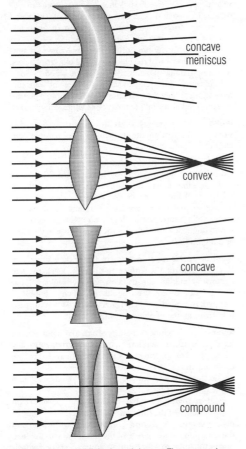

lens The passage of light through lenses. The concave lens diverges a beam of light from a distant source. The convex and compound lenses focus light from a distant source to a point. The distance between the focus and the lens is called the focal length. The shorter the focus, the more powerful the lens.

and the other plane, concave, or convex – that modifies rays of light. A convex lens brings rays of light together; a concave lens makes the rays diverge. Lenses are essential to spectacles, microscopes, telescopes, cameras, and almost all optical instruments.

The image formed by a single lens suffers from several defects or ◊aberrations, notably **spherical aberration** in which an image becomes blurred, and **chromatic aberration** in which an image in white light tends to have coloured edges. Aberrations are corrected by the use of compound lenses, which are built up from two or more lenses of different refractive index.

lensing, gravitational see ◊gravitational lensing.

lenticel small pore on the stems of woody plants or the trunks of trees. Lenticels are means of gas exchange between the stem interior and the atmosphere. They consist of loosely packed cells with many air spaces in between, and are easily seen on smooth-barked trees such as cherries, where they form horizontal lines on the trunk.

lenticular galaxy in astronomy, a lens-shaped ◊galaxy with a large central bulge and flat disc but no discernible spiral arms.

Lenz's law in physics, a law stating that the direction of an electromagnetically induced current (generated by moving a wire near a magnet or a wire in a magnetic field) will be such as to oppose the motion producing it. It is named after the German physicist Heinrich Friedrich Lenz (1804–1865), who announced it in 1833.

Leo zodiacal constellation in the northern hemisphere, represented as a lion. The Sun passes through Leo from mid-Aug to mid-Sept. Its brightest star is first-magnitude ◊Regulus at the base of a pattern of stars called the Sickle. In astrology, the dates for Leo are between about 23 July and 22 Aug (see ◊precession).

Lepenski Vir prehistoric settlement site in the valley of the river Danube, where it runs through the Iron Gates gorge, in former Yugoslavia on the Romanian border. One of Europe's oldest farming settlements, dating from the 6th millennium BC, it is possibly the earliest and best preserved late Mesolithic to Neolithic (Middle to New Stone Age) site in the Balkans. The site is now submerged by an artificial lake.

A series of unusual trapezoidal dwellings were found, dating from the mid-6th millennium BC, associated with large limestone sculptures of fishlike human beings, belonging to a preceding camp of hunter-fisher people. All the dwellings had their wide end pointing towards the river, with limestone plaster floors and probably a wooden structure over. Hearths comprised limestone blocks inside elongated pits; in some cases human burials were made close by. The roofs of the dwellings may have been covered in hides or reed thatch. A larger central house may indicate social stratification.

Diet depended on fish and the development of fishing and boat technology. The life style of the inhabitants was sedentary but had arisen out of adaptation from a previous hunting tradition. Resources included catfish, carp, deer, and wild pigs.

Lepidoptera order of insects, including butterflies and moths, which have overlapping scales on their wings; the order consists of about 165,000 species.

leprosy or **Hansen's disease** chronic, progressive

disease caused by a bacterium *Mycobacterium leprae* closely related to that of tuberculosis. The infection attacks the skin and nerves. Once common in many countries, leprosy is now confined almost entirely to the tropics. It is controlled with drugs.

According to a World Health Organization estimate there were 1.3 million people with leprosy in 1996; the year-2000 target for total elimination was declared unrealistic at an international leprosy conference in Delhi Oct 1996.

There are two principal manifestations. **Lepromatous leprosy** is a contagious, progressive form distinguished by the appearance of raised blotches and lumps on the skin and thickening of the skin and nerves, with numbness, weakness, paralysis, and ultimately deformity of the affected parts. In **tuberculoid leprosy**, sensation is lost in some areas of the skin; sometimes there is loss of pigmentation and hair. The visible effects of long-standing leprosy (joint damage, paralysis, loss of fingers or toes) are due to nerve damage and injuries of which the sufferer may be unaware. Damage to the nerves remains, and the technique of using the patient's muscle material to encourage nerve regrowth is being explored.

lepton any of a class of light ◊elementary particles that are not affected by the strong nuclear force; they do not interact strongly with other particles or nuclei. The leptons are comprised of the ◊electron, ◊muon, and ◊tau, and their ◊neutrinos (the electron neutrino, muon neutrino, and tau neutrino), plus their six ◊antiparticles.

leptospirosis any of several infectious diseases of domestic animals and humans caused by spirochetes of the genus *Leptospira*, found in sewage and natural waters. One such disease in cattle causes abortion; in humans, eyes, liver, and kidneys may be affected; meningitis (inflammation of the membrane surrounding the brain) is another symptom.

Leptospirosis can be transmitted to humans by rats and mice. The leptospira are carried in the kidneys, and transmitted via excreted urine in damp soils or water. For example, children may be infected if water from infected puddles splashes into their eyes. Many cases remain undiagnosed as symptoms may be no more serious than a mild dose of influenza. 1996 figures indicate that there are approximately 50 cases annually in the USA. If the symptoms become more serious, kidney and liver failure may result. The illness at this stage is called ◊Weil's disease.

lesion any change in a body tissue that is a manifestation of disease or injury.

lethargy in medicine, apathetic state that may arise from a variety of causes, including physical illnesses, such as glandular fever and other viral infections and anaemia; excessive fatigue due, for example, to insomnia; or as a symptom of mental illnesses, such as depression or anxiety. It can also be due to an adverse effect of some drugs, such as ◊beta-blockers. It is important to treat the underlying cause of lethargy.

letterpress method of printing from raised type, pioneered by Johann Gutenberg in Europe in the 1450s.

leucine one of the nine essential ◊amino acids.

leucite silicate mineral, $KAlSi_2O_6$, occurring frequently in some potassium-rich volcanic rocks. It is dull white to grey, and usually opaque. It is used as a source of potassium for fertilizer.

leucocyte another name for a ◊white blood cell.

leucotomy or *lobotomy* a brain operation to sever the connections between the frontal lobe and underlying structures. It was widely used in the 1940s and 1950s to treat severe psychotic or depressive illness. Though it achieved some success, it left patients dull and apathetic; there was also a considerable risk of epilepsy. It was largely replaced by the use of psychotropic drugs from the late 1950s.

Today, a limited amount of psychosurgery is performed in specialist centres under strict controls. It includes the creation of tiny, precise frontal lobe lesions to relieve severe conditions which have not responded to other treatments.

leukaemia any one of a group of cancers of the blood cells, with widespread involvement of the bone marrow and other blood-forming tissue. The central feature of leukaemia is runaway production of white blood cells that are immature or in some way abnormal. These rogue cells, which lack the defensive capacity of healthy white cells, overwhelm the normal ones, leaving the victim vulnerable to infection. Treatment is with radiotherapy and ◊cytotoxic drugs to suppress replication of abnormal cells, or by bone-marrow transplantation.

Abnormal functioning of the bone marrow also suppresses production of red blood cells and blood ◊platelets, resulting in ◊anaemia and a failure of the blood to clot.

Leukaemias are classified into acute or chronic, depending on their known rates of progression. They are also grouped according to the type of white cell involved.

leukotriene in biology, a group of naturally occurring substances that stimulate the activity, for example contraction, of smooth muscles.

levee naturally formed raised bank along the side of a river channel. When a river overflows its banks, the rate of flow is less than that in the channel, and silt is deposited on the banks. With each successive flood the levee increases in size so that eventually the river may be above the surface of the surrounding flood plain. Notable levees are found on the lower reaches of the Mississippi in the USA and the Po in Italy.

level or *spirit level* instrument for finding horizontal level, or adjusting a surface to an even level, used in surveying, building construction, and archaeology. It has a glass tube of coloured liquid, in which a bubble is trapped, mounted in an elongated frame. When the tube is horizontal, the bubble moves to the centre.

lever simple machine consisting of a rigid rod pivoted at a fixed point called the fulcrum, used for shifting or raising a heavy load or applying force in a similar way. Levers are classified into orders according to where the effort is applied, and the load-moving force developed, in relation to the position of the fulcrum.

A **first-order** lever has the load and the effort on opposite sides of the fulcrum – for example, a see-saw or pair of scissors. A **second-order** lever has the load and the effort on the same side of the fulcrum, with the load nearer the fulcrum – for example, nutcrackers or a wheelbarrow. A **third-order** lever has the effort nearer the fulcrum than the load, with both on the same side of it – for example, a pair of tweezers or tongs. The mechanical advantage of a lever is the ratio of load to effort, equal to the perpendicular distance of the effort's line of action

first-order lever

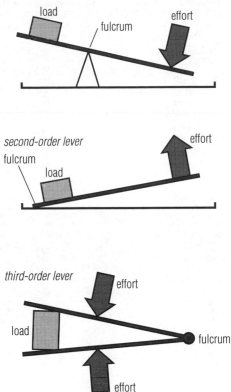

second-order lever

third-order lever

lever Types of lever. Practical applications of the first-order lever include the crowbar, seesaw, and scissors. The wheelbarrow and nutcrackers are second-order levers. A pair of tweezers, or tongs, is a third-order lever.

from the fulcrum divided by the distance to the load's line of action. Thus tweezers, for instance, have a mechanical advantage of less than one.

LF in physics, abbreviation for **low ◊frequency**. LF radio waves have frequencies in the range 30–300 kHz.

libido in Freudian psychology, the energy of the sex instinct, which is to be found even in a newborn child. The libido develops through a number of phases, described by Sigmund Freud in his theory of infantile sexuality. The source of the libido is the ◊id.

The phases of the libido are identified by Freud as the **oral stage**, when a child tests everything by mouth, the **anal stage**, when the child gets satisfaction from control of its body, and the **genital stage**, when the libido becomes concentrated in the sex organs.

Loss of adult libido is seen in some diseases; see also ◊impotence.

Libra faint zodiacal constellation on the celestial equator (see ◊celestial sphere) adjoining Scorpius, and represented as the scales of justice. The Sun passes through Libra during Nov. The constellation was once considered to be a part of Scorpius, seen as the scorpion's claws. In astrology, the dates for Libra are between about 23 Sept and 23 Oct (see ◊precession).

libration in astronomy, a slight, apparent wobble in the rotation of the Moon due to its variable speed of rotation and the tilt of its axis.

Generally, the Moon rotates on its axis in the same time as it takes to complete one orbit, causing it to keep one face turned permanently towards the Earth (see captured rotation). Its speed in orbit varies, however, because its orbit is not circular but elliptical, so at times the Moon's axial rotation appears to get either slightly ahead of or slightly behind its orbital motion, so that part of the 'dark side' of the Moon is visible around the east and west edges. This is known as **libration in longitude**.

Libration in latitude occurs because the Moon's axis is slightly tilted with respect to its orbital plane, so we can see over the north and south poles. In combination, these effects mean that a total of 59% of the Moon's surface is visible, rather than just 50% if libration did not occur.

lichen any organism of a unique group that consists of associations of a specific ◊fungus and a specific alga living together in a mutually beneficial relationship. Found as coloured patches or spongelike masses on trees, rocks, and other surfaces, lichens flourish in harsh conditions.

Some lichens are edible, for example, reindeer moss and Iceland moss; others are a source of colour dyes, such as litmus, or are used in medicine. They are sensitive to pollution in the air (see ◊indicator species).

Lick Observatory observatory of the University of California and Mount Hamilton, California. Its main instruments are the 3.04 m/120 in Shane reflector, opened in 1959, and a 91 cm/36 in refractor, opened in 1988, the second-largest refractor in the world.

lie detector instrument that records graphically certain body activities, such as thoracic and abdominal respiration, blood pressure, pulse rate, and galvanic skin response (changes in electrical resistance of the skin). Marked changes in these activities when a person answers a question may indicate that the person is lying.

life the ability to grow, reproduce, and respond to such stimuli as light, heat, and sound. It is thought that life on Earth began about 4 billion years ago. Over time, life has evolved from primitive single-celled organisms to complex multicellular ones. The earliest fossil evidence of life is threadlike chains of cells discovered in 1980 in deposits in NW Australia that have been dated as 3.5 billion years old.

Life exists in the universe only because the carbon atom possesses certain exceptional properties.

On **life** James Hopwood Jeans
Mysterious Universe

Life originated in the primitive oceans. The original atmosphere, 4,000 million years ago, consisted of carbon dioxide, nitrogen, and water. It has been shown in the laboratory that more complex organic molecules, such as ◊amino acids and ◊nucleotides, can be produced from these ingredients by passing electric sparks through a mixture. It has been suggested that lightning was extremely common in the early atmosphere, and that this combination of conditions could have resulted in the

Life Expectancy

Nation	Men	Women	Avrg.	Nation	Men	Women	Avrg.
Liechtenstein	78	83	81	Mongolia	63	67	65
Japan	76	82	79	Thailand	62	68	65
Switzerland	74	82	78	El Salvador	63	66	65
Australia	75	80	78	Turkey	63	66	65
Netherlands	74	81	78	Ecuador	62	66	64
Sweden	74	81	78	Vietnam	62	66	64
Iceland	74	80	77	Peru	61	66	64
Spain	74	80	77	Morocco	62	65	64
Italy	73	80	77	Colombia	61	66	64
Jamaica	75	78	77	Brazil	61	66	64
Norway	73	80	77	Dominican Republic	61	65	63
Canada	72	79	76	Iraq	62	63	63
USA	72	79	76	Nicaragua	61	63	62
Belgium	72	78	75	São Tome Príncipe	62	62	62
Denmark	72	78	75	Maldives	60	63	62
France	71	79	75	Tuvalu	60	63	62
New Zealand	72	78	75	Kenya	59	63	61
UK	72	78	75	Zimbabwe	59	63	61
Israel	73	76	75	Algeria	59	62	61
Luxembourg	71	78	75	Lesotho	59	62	61
Malta	72	77	75	Honduras	58	62	60
Portugal	71	78	75	Belize	60	60	60
Brunei	74	74	74	Botswana	59	59	59
Kuwait	72	76	74	Guatemala	57	61	59
Singapore	71	77	74	Cape Verde	57	61	59
Cyprus	72	76	74	Egypt	57	60	59
Finland	70	78	74	Dominica	57	59	58
Greece	72	76	74	Iran	57	57	57
Costa Rica	71	76	74	Oman	55	58	57
Cuba	72	75	74	Zambia	54	57	56
San Marino	70	77	74	India	56	55	56
Austria	70	77	74	Togo	53	57	55
Panama	71	75	73	Pakistan	54	55	55
Ireland	70	76	73	Liberia	53	56	55
Taiwan	70	75	73	Myanmar	53	56	55
Barbados	70	75	73	Ivory Coast	52	55	54
Yugoslavia (former)	69	75	72	Papua New Guinea	53	54	54
Tonga	69	74	72	Indonesia	52	55	54
Uruguay	68	75	72	Somalia	53	53	53
Czech Republic	68	75	72	Sudan	51	55	53
Slovakia	68	75	72	Senegal	51	54	53
Bulgaria	69	74	72	Congo (formerly Zaire)	51	54	53
St Vincent and the Grenadines	69	74	72	Bolivia	51	54	53
Albania	69	73	71	Haiti	51	54	53
Germany	68	74	71	Ghana	50	54	52
St Lucia	68	73	71	Tanzania	49	54	52
St Christopher-Nevis	69	72	71	Madagascar	50	53	52
Hungary	67	74	71	Rwanda	49	53	51
United Arab Emirates	68	72	70	Cameroon	49	53	51
Qatar	68	72	70	Bangladesh	50	52	51
Venezuela	67	73	70	Swaziland	47	54	51
Mexico	67	73	70	Djibouti	50	50	50
Korea, North	67	73	70	Uganda	49	51	50
Trinidad and Tobago	68	72	70	Comoros	48	52	50
Antigua and Barbuda	70	70	70	Nepal	50	49	50
Poland	66	74	70	Laos	48	51	50
Romania	67	73	70	Niger	48	50	49
Paraguay	67	72	70	Gabon	47	51	49
Tunisia	68	71	70	Yemen	47	50	49
Sri Lanka	67	72	70	Nigeria	47	49	48
Argentina	66	73	70	Malawi	46	50	48
Korea, South	66	73	70	Burundi	45	48	47
Bahrain	67	71	69	Mozambique	45	48	47
Jordan	67	71	69	Congo	45	48	47
Vanuatu	67	71	69	Equatorial Guinea	44	48	46
Fiji	67	71	69	Mauritania	43	48	46
Grenada	69	69	69	Mali	44	47	46
Solomon Islands	66	71	69	Burkina Faso	44	47	46
Guyana	66	71	69	Benin	42	46	44
Chile	64	73	69	Sierra Leone	41	47	44
Surinam	66	71	69	Chad	42	45	44
Syria	67	69	68	Cambodia	42	45	44
China	67	69	68	Bhutan	44	43	44
Malaysia	65	70	68	Central African Republic	41	45	43
Mauritius	64	71	68	Angola	40	44	42
Lebanon	65	70	68	Guinea-Bissau	42	42	42
Libya	64	69	67	Gambia	42	42	42
Samoa, Western	64	69	67	Afghanistan	43	41	42
Philippines	63	69	66	Guinea	39	42	41
Seychelles	66	66	66	Ethiopia	38	38	38
Saudi Arabia	64	67	66				

where separate figures are given for male and female life expectancy, the ranking is based on a simple average of the two

oceans becoming rich in organic molecules, the so-called 'primeval soup'. These molecules may then have organized into clusters capable of reproducing and of developing eventually into simple cells.

Once the atmosphere changed to its present composition, life could only be created by living organisms (a process called ◊biogenesis).

life cycle in biology, the sequence of developmental stages through which members of a given species pass. Most vertebrates have a simple life cycle consisting of ◊fertilization of sex cells or ◊gametes, a period of development as an ◊embryo, a period of juvenile growth after hatching or birth, an adulthood including ◊sexual reproduction, and finally death. Invertebrate life cycles are generally more complex and may involve major reconstitution of the individual's appearance (◊metamorphosis) and completely different styles of life. Plants have a special type of life cycle with two distinct phases, known as ◊alternation of generations. Many insects such as cicadas, dragonflies, and mayflies have a long larvae or pupae phase and a short adult phase. Dragonflies live an aquatic life as larvae and an aerial life during the adult phase. In many invertebrates and protozoa there is a sequence of stages in the life cycle, and in parasites different stages often occur in different host organisms.

life expectancy average lifespan that can be presumed of a person at birth. It depends on nutrition, disease control, environmental contaminants, war, stress, and living standards in general.

There is a marked difference between industrialized countries, which generally have an ageing population, and the poorest countries, where life expectancy is much shorter.

life sciences scientific study of the living world as a whole, a new synthesis of several traditional scientific disciplines including ◊biology, ◊zoology, and ◊botany, and newer, more specialized areas of study such as ◊biophysics and ◊sociobiology.

This approach has led to many new ideas and discoveries as well as to an emphasis on ◊ecology, the study of living organisms in their natural environments.

lift (US **elevator**) device for lifting passengers and goods vertically between the floors of a building. US inventor Elisha Graves Otis developed the first passenger lift, installed in 1857. The invention of the lift allowed the development of the skyscraper from the 1880s.

A lift usually consists of a platform or boxlike structure suspended by motor-driven cables with safety ratchets along the sides of the shaft. At first steam powered the movement, but hydraulic and then electric lifts were common from the early 1900s. Lift operators worked controls and gates until lifts became automatic.

ligament strong, flexible connective tissue, made of the protein ◊collagen, which joins bone to bone at moveable joints and sometimes encloses the joints. Ligaments prevent bone dislocation (under normal circumstances) but allow joint flexion. The ligaments around the joints are composed of white fibrous tissue. Other ligaments are composed of yellow elastic tissue, which is adapted to support a continuous but varying stress, as in the ligament connecting the various cartilages of the ◊larynx (voice box).

ligature any surgical device (nylon, gut, wire) used for tying a blood vessel to stop it bleeding, or to tie round the base of a growth to constrict its blood supply.

light electromagnetic waves in the visible range, having a wavelength from about 400 nanometres in the extreme violet to about 770 nanometres in the extreme red. Light is considered to exhibit particle and wave properties, and the fundamental particle, or quantum, of light is called the photon. The speed of light (and of all electromagnetic radiation) in a vacuum is approximately 300,000 km/186,000 mi per second, and is a universal constant denoted by c.

Isaac Newton was the first to discover, in 1666, that sunlight is composed of a mixture of light of different colours in certain proportions and that it could be separated into its components by dispersion. Before his time it was supposed that dispersion of light produced colour instead of separating already existing colours.

The ancients believed that light travelled at infinite speed; its speed was first measured accurately by Danish astronomer Ole Römer in 1676.

light bulb incandescent filament lamp, first demonstrated by Joseph Swan in the UK in 1878 and Thomas Edison in the USA in 1879. The present-day light bulb is a thin glass bulb filled with an inert mixture of nitrogen and argon gas. It contains a filament made of fine tungsten wire. When electricity is passed through the wire, it glows white hot, producing light.

Light bulb

Thomas Edison's invention of the light bulb did not meet with universal approval. When a British Parliamentary Committee was set up in the late 1870s to investigate its potential value, the report described it as 'good enough for our trans-Atlantic friends ... but unworthy of the attention of practical or scientific men'.

light curve in astronomy, a graph showing how the brightness of an astronomical object varies with time. Analysis of the light curves of variable stars, for example, gives information about the physical processes causing the variation.

light-emitting diode (LED) means of displaying symbols in electronic instruments and devices. An LED is made of ◊semiconductor material, such as gallium arsenide phosphide, that glows when electricity is passed through it. The first digital watches and calculators had LED displays, but many later models use ◊liquid-crystal displays.

A new generation of LEDs that can produce light in the mid-infrared range (3–10 micrometres) safely and cheaply were developed by British researchers in 1995, using thin alternating layers of indium arsenide and indium arsenide antimonide.

lighthouse structure carrying a powerful light to warn ships or aeroplanes that they are approaching a place (usually land) dangerous or important to navigation. The light is magnified and directed out to the horizon or up to the zenith by a series of mirrors or prisms. Increasingly lighthouses are powered by electricity and automated rather than staffed; the more recent models also emit radio signals.

Lights may be either flashing (the dark period exceeding the light) or rotating (the dark period being equal or

The nature of light

Alexander Pope, the poet, wrote of Isaac Newton's work:
'Nature, and Nature's Laws lay hid in Night: God said, *Let Newton be!* and All was *Light*.'

Newton discovers the spectrum

Isaac Newton, the English physicist and mathematician (1642–1727), would have been remembered as a great scientist for any one of his many discoveries. He made outstanding contributions to mathematics, astronomy, mechanics, and to understanding gravitation and the nature of light.

The first microscopes revealed the world of the very small to scientists for the very first time. They saw in great detail what microorganisms were really like. In the same way, at the other end of the scale, the first telescopes revealed to astronomers vast numbers of stars and the beauty of the Earth's planets. But there was a problem. Any image formed by the combination of the lenses then available to make microscopes or telescopes was surrounded by a colour fringe. This blurred the outline – an effect which became worse at higher magnifications.

From prisms and rainbows ...

People had long been familiar with the rainbow colours produced when light shone through a chandelier. Now, in order to improve their instruments, scientists needed to know why these colours were formed. Newton ground his own glass lenses and had, since he was an undergraduate student at Cambridge, been interested in the effect of a glass prism on sunlight.

Newton wrote: 'In the year 1666 (at which time I applied myself to the grinding of optick glass or other figures than spherical) I procured me a triangular glass prism, to try the celebrated phaenomena of colours.'

Newton made a small hole in the shutter covering the window in his room to let in a beam of sunlight. He placed the prism in front of this hole and viewed the vivid and intense spectrum of rainbow colours cast on the opposite wall.

Newton now examined these colours individually, using a second prism. He took two boards, each with a small hole in it. The first he placed behind the prism at the window. He positioned the second board so that only a single colour produced by the first prism fell on the hole in it. The second prism was placed so as to cast the light passing through the hole in the second board on to the wall. Newton saw at once that the coloured beam of light passing through the second prism was unchanged. This proved to be true for all the rainbow colours produced by the first prism.

Newton had performed the crucial experiment, because it had been assumed previously that light was basically white, and that colours could be added to it. Now it was clear that white light was a mixture of the colours of the rainbow; the prism simply split them up or refracted the light. A second prism could not 'split' them up further.

... to modern telescopes

In his book *Opticks*, published in 1704, Newton described a further experiment in which he used the second prism to recombine the rainbow colours of the spectrum to produce white light. Here for the first time was a simple explanation of the nature of light. The book had a great impact on 18th-century writers. What Newton had to say about what happened when white light passed through a prism could be immediately applied to rainbows, and this captured the imagination of artists and poets.

Because Newton suspected that the colour fringes or chromatic aberration produced by lenses in telescopes could not be avoided, in 1668 he designed a telescope that depended instead on the use of a curved mirror. Light reflected from a surface produces no colour effects, unlike light passing through a lens. Nowadays all large astronomical telescopes are of the reflecting type.

As a result of his elegant experiments with prisms and light, Newton also speculated about the ultimate nature of light itself. He proposed that light consisted of small 'corpuscles' (small particles) which are shot out from the source of light, rather in the same way that pellets are ejected from a shot gun. He was able to explain many of the known properties of light with this theory, which was widely accepted. Light has proved remarkably difficult to understand; nowadays it is regarded as having both particlelike and wavelike properties.

Julian Rowe

less); fixed lights are liable to cause confusion. The pattern of lighting is individually varied so that ships or aircraft can identify the lighthouse. In fog, sound signals are made (horns, sirens, explosives), and in the case of lightbuoys, fog bells and whistles are operated by the movement of the waves.

Among early lighthouses were the Pharos of Alexandria (about 280 BC) and those built by the Romans at Ostia, Ravenna, Boulogne, and Dover. In England beacons burning in church towers served as lighthouses until the 17th century, and in the earliest lighthouses, such as the Eddystone, first built in 1698, open fires or candles were used.

Where reefs or sandbanks made erection of a lighthouse impossible, lightships were often installed; increasingly these are being replaced by fixed, small, automated lighthouses. Where it is impossible to install a fixed structure, unattended lightbuoys equipped for up to a year's service may be used. In the UK, these are gradually being converted from acetylene gas in cylinders to solar power.

lightning high-voltage electrical discharge between two charged rainclouds or between a cloud and the Earth, caused by the build-up of electrical charges. Air in the path of lightning ionizes (becomes conducting), and expands; the accompanying noise is heard as thunder. Currents of 20,000 amperes and temperatures of 30,000°C/54,000°F are common.

Lightning

Around the world there are a total of 70 to 100 lightning flashes every second.

---

lightning conductor device that protects a tall building from lightning strike by providing an easier path for current to flow to earth than through the building. It consists of a thick copper strip of very low resistance connected to the ground below. A good connection to the ground is essential and is made by burying a large metal plate deep in the damp earth. In the event of a direct lightning strike, the current in the conductor may be so great as to melt or even vaporize the metal, but the damage to the building will nevertheless be limited.

light pen in computing, a device resembling an ordinary pen, used to indicate locations on a computer screen. With certain computer-aided design (◊CAD) programs, the light pen can be used to instruct the computer to change the shape, size, position, and colours of sections of a screen image.

The pen has a photoreceptor at its tip that emits signals when light from the screen passes beneath it. From the timing of this signal and a gridlike representation of the screen in the computer memory, a computer program can calculate the position of the light pen.

light second unit of length, equal to the distance travelled by light in one second. It is equal to 2.997925×10^8 m/9.835592×10^8 ft. See ◊light year.

light watt unit of radiant power (brightness of light). One light watt is the power required to produce a perceived brightness equal to that of light at a wavelength of 550 nanometres and 680 lumens.

light year in astronomy, the distance travelled by a beam of light in a vacuum in one year, approximately 9.46×10^{12} km/5.88×10^{12} mi.

lignin naturally occurring substance produced by plants to strengthen their tissues. It is difficult for ◊enzymes to attack lignin, so living organisms cannot digest wood, with the exception of a few specialized fungi and bacteria. Lignin is the essential ingredient of all wood and is, therefore, of great commercial importance.

Chemically, lignin is made up of thousands of rings of carbon atoms joined together in a long chain. The way in which they are linked up varies along the chain.

lignite type of ◊coal that is brown and fibrous, with a relatively low carbon content. As a fuel it is less efficient because more of it must be burned to produce the same amount of energy generated by bituminous coal. Lignite also has a high sulphur content and is more polluting. It is burned to generate power in Scandinavia and some former eastern block countries because it is the only fuel resource available without importing.

lignocaine short-term local anaesthetic injected into tissues or applied to skin. It is effective for brief, invasive procedures such as dental care or insertion of a cannula (small tube) into a vein. Temporary paralysis (to prevent involuntary movement during eye surgery, for example) can be achieved by injection directly into the nerve serving the region.

Rapidly absorbed by mucous membranes (lining tissues), lignocaine may be sprayed into the nose or throat to allow comfortable insertion of a viewing instrument during ◊endoscopy. Its action makes it a potent antiarrhythmia drug as well: given intravenously during or following a heart attack, it reduces the risk of cardiac arrest.

lime or **quicklime** CaO (technical name **calcium oxide**) white powdery substance used in making mortar and cement. It is made commercially by heating calcium carbonate ($CaCO_3$), obtained from limestone or chalk, in a ◊lime kiln. Quicklime readily absorbs water to become calcium hydroxide $Ca(OH)_2$, known as slaked lime, which is used to reduce soil acidity.

lime kiln oven used to make quicklime (calcium oxide, CaO) by heating limestone (calcium carbonate, $CaCO_3$) in the absence of air. The carbon dioxide is carried away to heat other kilns and to ensure that the reversible reaction proceeds in the right direction.

$$CaCO_3 \rightleftharpoons CaO + CO_2$$

limestone sedimentary rock composed chiefly of calcium carbonate $CaCO_3$, either derived from the shells of marine organisms or precipitated from solution, mostly in the ocean. Various types of limestone are used as building stone.

◊Marble is metamorphosed limestone. Certain so-called marbles are not in fact marbles but fine-grained fossiliferous limestones that take an attractive polish. Caves commonly occur in limestone. ◊Karst is a type of limestone landscape. *See illustration overleaf.*

limestone pavement bare rock surface resembling a block of chocolate, found on limestone plateaus. It is formed by the weathering of limestone into individual upstanding blocks, called clints, separated from each other by joints, called grykes. The weathering process is thought to entail a combination of freeze-thaw (the alter-

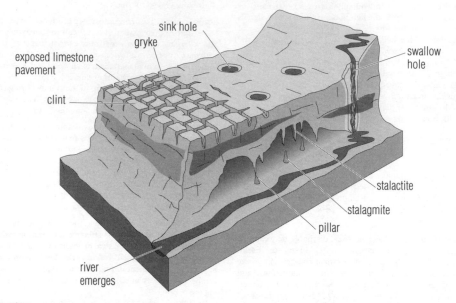

sink hole

gryke

exposed limestone
pavement

clint

swallow
hole

stalactite

stalagmite

pillar

river
emerges

limestone

nate freezing and thawing of ice in cracks) and carbonation (the dissolving of minerals in the limestone by weakly
acidic rainwater). Malham Tarn in North Yorkshire is an
example of a limestone pavement.

limewater common name for a dilute solution of slaked
lime (calcium hydroxide, $Ca(OH)_2$). In chemistry, it is
used to detect the presence of carbon dioxide.

If a gas containing carbon dioxide is bubbled through
limewater, the solution turns milky owing to the formation of calcium carbonate ($CaCO_3$). Continued bubbling
of the gas causes the limewater to clear again as the
calcium carbonate is converted to the more soluble calcium hydrogencarbonate ($Ca(HCO_3)_2$).

limiting factor in biology, any factor affecting the rate of
a metabolic reaction. Levels of light or of carbon dioxide
are limiting factors in ◊photosynthesis because both are
necessary for the production of carbohydrates. In experiments, photosynthesis is observed to slow down and
eventually stop as the levels of light decrease.

It is believed that the concentrations of carbon dioxide
building up in the atmosphere through the burning of
fossil fuels will allow faster plant growth.

limnology study of lakes and other bodies of open fresh
water, in terms of their plant and animal biology, chemistry, and physical properties.

limonite iron ore, mostly poorly crystalline iron oxy-
hydroxide, but usually mixed with ◊hematite and other
iron oxides. Also known as brown iron ore, it is often
found in bog deposits.

Lindow Man remains of an Iron Age man discovered in a
peat bog at Lindow Marsh, Cheshire, UK, in 1984. The
chemicals in the bog had kept the body in an excellent
state of preservation.

'Pete Marsh', as the archaeologists nicknamed him,

had been knocked unconscious, strangled, and then had
his throat cut before being thrown into the bog. He may
have been a sacrificial victim, as Celtic peoples often
threw offerings to the gods into rivers and marshes. His
stomach contained part of an unleavened barley 'ban-
nock' that might have been given as a sacrificial offering.
His well-cared-for nails indicate that he might have been a
Druid prince who became a willing sacrifice.

linear accelerator or **linac** in physics, a type of particle
◊accelerator in which the particles move along a straight
tube. Particles pass through a linear accelerator only once
– unlike those in a cyclotron (a ring-shaped accelerator),
which make many revolutions, gaining energy each time.

The world's longest linac is the Stanford Linear Collider, in which electrons and positrons are accelerated
along a straight track 3.2 km/2 mi long and then steered
into a head-on collision with other particles.

The first linear accelerator was built in 1928 by Norwegian engineer Ralph Widerøe to investigate the behaviour
of heavy ions (large atoms with one or more electrons
removed), but devices capable of accelerating smaller
particles such as protons and electrons could not be built
until after World War II and the development of high-
power radio-and microwave-frequency generators.

linear equation in mathematics, a relationship between
two variables that, when plotted on Cartesian axes produces a straight-line graph; the equation has the general
form $y = mx + c$, where m is the slope of the line represented by the equation and c is the y-intercept, or the
value of y where the line crosses the y-axis in the ◊Car-
tesian coordinate system. Sets of linear equations can be
used to describe the behaviour of buildings, bridges,
trusses, and other static structures.

linear motor type of electric motor, an induction motor
in which the fixed stator and moving armature are straight

and parallel to each other (rather than being circular and one inside the other as in an ordinary induction motor). Linear motors are used, for example, to power sliding doors. There is a magnetic force between the stator and armature; this force has been used to support a vehicle, as in the experimental ◊maglev linear motor train.

linear programming in mathematics and economics, a set of techniques for finding the maxima or minima of certain variables governed by linear equations or inequalities. These maxima and minima are used to represent 'best' solutions in terms of goals such as maximizing profit or minimizing cost.

linen yarn spun and the textile woven from the fibres of the stem of the ◊flax plant. Used by the ancient Egyptians, linen was introduced by the Romans to N Europe, where production became widespread. Religious refugees from the Low Countries in the 16th century helped to establish the linen industry in England, but here and elsewhere it began to decline in competition with cotton in the 18th century.

To get the longest possible fibres, flax is pulled, rather than cut by hand or machine, just as the ripened fruits, or bolls, are beginning to set. After preliminary drying, it is steeped in water so that the fibre can be more easily separated from the wood of the stem, then hackled (combed), classified, drawn into continuous fibres, and spun. Bleaching, weaving, and finishing processes vary according to the final product, which can be sailcloth, canvas, sacking, cambric, or lawn. Because of the length of its fibre, linen yarn has twice the strength of cotton, and yet is superior in delicacy, so that it is suitable for lace making. It mixes well with synthetic fibres.

line of force in physics, an imaginary line representing the direction of force at any point in a magnetic, gravitational, or electrical field.

line printer computer ◊printer that prints a complete line of characters at a time. Line printers can achieve very high printing speeds of up to 2,500 lines a minute, but can print in only one typeface, cannot print graphics, and are very noisy. Today, most users prefer ◊laser printers.

linkage in genetics, the association between two or more genes that tend to be inherited together because they are on the same chromosome.

The closer together they are on the chromosome, the less likely they are to be separated by crossing over (one of the processes of ◊recombination) and they are then described as being 'tightly linked'.

linoleum floor covering made from linseed oil, tall oil, rosin, cork, woodflour, chalk, clay, and pigments, pressed into sheets with a jute backing. Oxidation of the oil is accelerated by heating, so that the oil mixture solidifies into a tough, resilient material. Linoleum tiles have a backing made of polyester and glass.

Linoleum was invented in England in 1860 by Frederick Walton. In the early 20th century he invented a straight-line inlay machine which was able to produce patterned linoleum. Today, the manufacture of linoleum still follows the basic principles of Walton's process although production is much faster. Synthetic floor coverings are now popular and the use of linoleum has declined.

Linotype trademark for a typesetting machine once universally used for newspaper work, which sets complete lines (slugs) of hot-metal type as operators type the copy at a keyboard. It was invented in the USA in 1884 by German-born Ottmar Mergenthaler. It has been replaced by phototypesetting.

lipase enzyme responsible for breaking down fats into fatty acids and glycerol. It is produced by the ◊pancreas and requires a slightly alkaline environment. The products of fat digestion are absorbed by the intestinal wall.

lipid any of a large number of esters of fatty acids, commonly formed by the reaction of a fatty acid with glycerol (see ◊glyceride). They are soluble in alcohol but not in water. Lipids are the chief constituents of plant and animal waxes, fats, and oils.

Phospholipids are lipids that also contain a phosphate group, usually linked to an organic base; they are major components of biological cell membranes.

lipophilic in chemistry, a term describing ◊functional groups with an affinity for fats and oils.

lipophobic in chemistry, a term describing ◊functional groups that tend to repel fats and oils.

liposome in medicine, a minute droplet of oil that is separated from a medium containing water by a ◊phospholipid layer. Drugs, such as cytotoxic agents, can be incorporated into liposomes and given by injection or by mouth. The liposomes allow the drug to reach the site of action, such as a tumour, without being broken down in the body.

liquefaction in chemistry, the process of converting a gas to a liquid, normally associated with low temperatures and high pressures (see ◊condensation).

liquefaction in earth science, the conversion of a soft deposit, such as clay, to a jellylike state by severe shaking. During an earthquake buildings and lines of communication built on materials prone to liquefaction will sink and topple. In the Alaskan earthquake of 1964 liquefaction led to the destruction of much of the city of Anchorage.

liquefied petroleum gas (LPG) liquid form of butane, propane, or pentane, produced by the distillation of petroleum during oil refining. At room temperature these substances are gases, although they can be easily liquefied and stored under pressure in metal containers. They are used for heating and cooking where other fuels are not available: camping stoves and cigarette lighters, for instance, often use liquefied butane as fuel.

liquid state of matter between a ◊solid and a ◊gas. A liquid forms a level surface and assumes the shape of its container. Its atoms do not occupy fixed positions as in a crystalline solid, nor do they have freedom of movement as in a gas. Unlike a gas, a liquid is difficult to compress since pressure applied at one point is equally transmitted throughout (Pascal's principle). ◊Hydraulics makes use of this property.

liquid air air that has been cooled so much that it has liquefied. This happens at temperatures below about −196°C/−321°F. The various constituent gases, including nitrogen, oxygen, argon, and neon, can be separated from liquid air by the technique of ◊fractionation.

Air is liquefied by the **Linde process**, in which air is alternately compressed, cooled, and expanded, the expansion resulting each time in a considerable reduction in temperature. With the lower temperature the molecules move more slowly and occupy less space, so the air changes phase to become liquid.

liquid-crystal display (LCD) display of numbers (for example, in a calculator) or pictures (such as on a pocket television screen) produced by molecules of a substance in a semiliquid state with some crystalline properties, so that clusters of molecules align in parallel formations. The display is a blank until the application of an electric field, which 'twists' the molecules so that they reflect or transmit light falling on them. The two main types of LCD are **passive matrix** and **active matrix**.

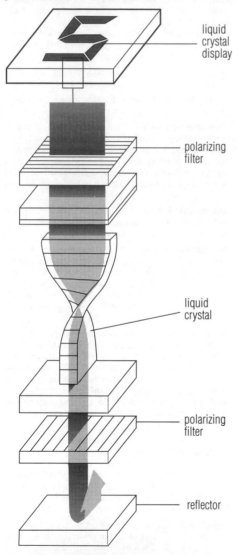

liquid crystal display

polarizing filter

liquid crystal

polarizing filter

reflector

liquid-crystal display A liquid-crystal display consists of a liquid crystal sandwiched between polarizing filters similar to polaroid sunglasses. When a segment of the seven-segment display is electrified, the liquid crystal twists the polarized light from the front filter, allowing the light to bounce off the rear reflector and illuminate the segment.

LISP (acronym for **list processing**) high-level computer-programming language designed for manipulating lists of data items. It is used primarily in research into ◊artificial intelligence (AI).

Developed in the late 1950s, and until recently common only in university laboratories, LISP is used more in the USA than in Europe, where the language ◊PROLOG is often preferred for AI work.

listeriosis disease of animals that may occasionally infect humans, caused by the bacterium *Listeria monocytogenes*. The bacteria multiply at temperatures close to 0°C/32°F, which means they may flourish in precooked frozen meals if the cooking has not been thorough. Listeriosis causes flulike symptoms and inflammation of the brain and its surrounding membranes. It can be treated with penicillin.

lithification the conversion of an unconsolidated newly deposited sediment into a coherent solid rock through processes such as compaction and cementation. The term ◊diagenesis includes many of the processes of lithification.

lithium soft, ductile, silver-white, metallic element, symbol Li, atomic number 3, relative atomic mass 6.941. It is one of the ◊alkali metals, has a very low density (far less than most woods), and floats on water (specific gravity 0.57); it is the lightest of all metals. Lithium is used to harden alloys, and in batteries; its compounds are used in medicine to treat manic depression.

Lithium was named in 1818 by Swedish chemist Jöns Berzelius, having been discovered the previous year by his student Johan A Arfwedson (1792–1841). Berzelius named it after 'stone' because it is found in most igneous rocks and many mineral springs.

lithography printmaking technique invented in 1798 by Aloys Senefelder, based on the mutual repulsion of grease and water. A drawing is made with greasy crayon on an absorbent stone, which is then wetted. The wet stone repels ink (which is greasy) applied to the surface and the crayon absorbs it, so that the drawing can be printed. Lithographic printing is used in book production, posters, and prints, and this basic principle has developed into complex processes.

Many artists have made brilliant use of the process since the early 19th century, including Delacroix, Goya, Isabey, Bonington, Daumier, Gavarni, Whistler, Toulouse-Lautrec (who devised colour effects of the most striking and original kind), Bonnard, and Vuillard.

Any sufficiently advanced technology is indistinguishable from magic.

On **new technology** Arthur C Clarke
The Lost Worlds of 2001

lithosphere topmost layer of the Earth's structure, forming the jigsaw of plates that take part in the movements of ◊plate tectonics. The lithosphere comprises the ◊crust and a portion of the upper ◊mantle. It is regarded as being rigid and moves about on the semi-molten ◊asthenosphere. The lithosphere is about 75 km/47 mi thick.

litmus dye obtained from various ◊lichens and used in chemistry as an indicator to test the acidic or alkaline nature of aqueous solutions; it turns red in the presence of acid, and blue in the presence of alkali.

litre metric unit of volume (symbol l), equal to one cubic decimetre (1.76 imperial pints/2.11 US pints). It was formerly defined as the volume occupied by one kilogram of pure water at 4°C at standard pressure, but this is slightly larger than one cubic decimetre.

Little Dipper another name for the most distinctive part of the constellation ◊Ursa Minor, the Little Bear.

liver large organ of vertebrates, which has many regulatory and storage functions.

The human liver is situated in the upper abdomen, and weighs about 2 kg/4.5 lb. It is divided into four lobes. The liver receives the products of digestion, converts glucose to glycogen (a long-chain carbohydrate used for storage), and breaks down fats. It removes excess amino acids from the blood, converting them to urea, which is excreted by the kidneys. The liver also synthesizes vitamins, produces bile and blood-clotting factors, and removes damaged red cells and toxins such as alcohol from the blood.

liverwort nonvascular plant (with no 'veins' to carry water and food), related to ◊hornworts and mosses; it is found growing in damp places. (Class Hepaticae, order Bryophyta.)

The plant exists in two different reproductive forms, sexual and asexual, which appear alternately (see ◊alternation of generations). The main sexual form consists of a plant body, or ◊thallus, which may be flat, green, and lobed like a small leaf, or leafy and mosslike. The asexual, spore-bearing form is smaller, typically parasitic on the thallus, and produces a capsule from which spores are scattered.

loam type of fertile soil, a mixture of sand, silt, clay, and organic material. It is porous, which allows for good air circulation and retention of moisture.

lobotomy another name for the former brain operation, ◊leucotomy.

local anaesthetic in medicine, an ◊anaesthetic that is applied only to the specific area under treatment.

local area network (LAN) in computing, a ◊network restricted to a single room or building. Local area networks enable around 500 devices to be connected together.

We do not need to have an infinity of different machines doing different jobs. A single one will suffice. The engineering problem of producing various machines for various jobs is replaced by the office work of 'programming' the universal machine to do these jobs.

Alan Turing
Quoted in A Hodges
Alan Turing: The Enigma of Intelligence 1985

Local Group in astronomy, a cluster of about 30 galaxies that includes our own, the Milky Way. Like other groups of galaxies, the Local Group is held together by the gravitational attraction among its members, and does not expand with the expanding universe. Its two largest galaxies are the Milky Way and the Andromeda galaxy; most of the others are small and faint.

lock construction installed in waterways to allow boats or ships to travel from one level to another. The earliest form, the **flash lock**, was first seen in the East in 1st-century-AD China and in the West in 11th-century Holland. By this method barriers temporarily dammed a river and when removed allowed the flash flood to propel the waiting boat through any obstacle. This was followed in 12th-century China and 14th-century Holland by the **pound lock**. In this system the lock has gates at each end. Boats enter through one gate when the levels are the same both outside and inside. Water is then allowed in (or out of) the lock until the level rises (or falls) to the new level outside the other gate.

Locks are important to shipping where canals link oceans of differing levels, such as the Panama Canal, or where falls or rapids are replaced by these adjustable water 'steps'. *See illustration overleaf.*

lock and key devices that provide security, usually fitted to a door of some kind. In 1778 English locksmith Robert Barron made the forerunner of the **mortise lock**, which contains levers that the key must raise to an exact height before the bolt can be moved. The **Yale lock**, a pin-tumbler cylinder design, was invented by US locksmith Linus Yale, Jr, in 1865. More secure locks include **combination locks**, with a dial mechanism that must be turned certain distances backwards and forwards to open, and **time locks**, which are set to be opened only at specific times.

lockjaw former name for ◊tetanus, a type of bacterial infection.

locomotive engine for hauling railway trains. In 1804 Richard Trevithick built the first steam engine to run on rails. Locomotive design did not radically improve until British engineer George Stephenson built the *Rocket* in 1829, which featured a multitube boiler and blastpipe, standard in all following **steam locomotives**. Today most locomotives are diesel or electric: **diesel locomotives** have a powerful diesel engine, and **electric locomotives** draw their power from either an overhead cable or a third rail alongside the ordinary track.

In a steam locomotive, fuel (usually coal, sometimes wood) is burned in a furnace. The hot gases and flames produced are drawn through tubes running through a huge water-filled boiler and heat up the water to steam. The steam is then fed to the cylinders, where it forces the pistons back and forth. Movement of the pistons is conveyed to the wheels by cranks and connecting rods. Diesel locomotives have a powerful diesel engine, burning oil.

The engine may drive a generator to produce electricity to power electric motors that turn the wheels, or the engine drives the wheels mechanically or through a hydraulic link. A number of **gas-turbine locomotives** are in use, in which a turbine spun by hot gases provides the power to drive the wheels.

locus in mathematics, traditionally the path traced out by a moving point, but now defined as the set of all points on a curve satisfying given conditions. For example, the locus of a point that moves so that it is always at the same distance from another fixed point is a circle; the locus of a

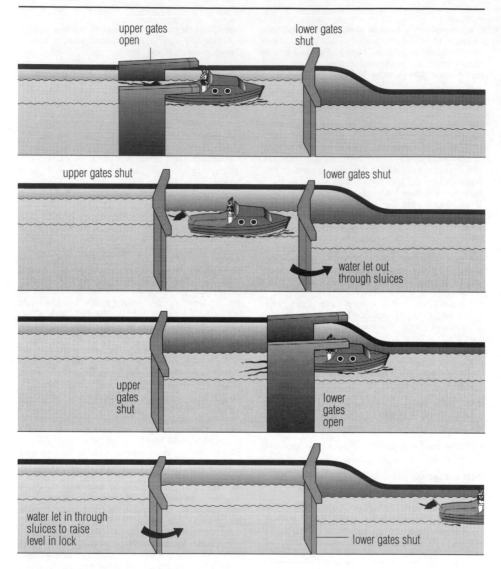

upper gates open

lower gates shut

upper gates shut

lower gates shut

water let out through sluices

upper gates shut

lower gates open

water let in through sluices to raise level in lock

lower gates shut

lock Travelling downstream, a boat enters the lock with the lower gates closed. The upper gates are then shut and the water level lowered by draining through sluices. When the water level in the lock reaches the downstream level, the lower gates are opened.

point that is always at the same distance from two fixed points is a straight line that perpendicularly bisects the line joining them.

lode geological deposit rich in certain minerals, generally consisting of a large vein or set of veins containing ore minerals. A system of veins that can be mined directly forms a lode, for example the mother lode of the California gold rush.

Lodes form because hot hydrothermal liquids and gases from magmas penetrate surrounding rocks, especially when these are limestones; on cooling, veins of ores formed from the magma then extend from the igneous mass into the local rock.

lodestar or **loadstar** a star used in navigation or astronomy, often ◊Polaris, the Pole Star.

loess yellow loam, derived from glacial meltwater deposits and accumulated by wind in periglacial regions during the ◊ice ages. Loess usually attains considerable depths, and the soil derived from it is very fertile. There are large deposits in central Europe (Hungary), China, and North America. It was first described in 1821 in the Rhine area, and takes its name from a village in Alsace.

log in mathematics, abbreviation for ◊*logarithm*.

log any apparatus for measuring the speed of a ship; also the daily record of events on board a ship or aircraft.

The log originally consisted of a piece of weighted wood attached to a line with knots at equal intervals that was cast from the rear of a ship. The vessel's speed was estimated by timing the passage of the knots with a sandglass (like an egg timer). Today logs use electromagnetism and sonar.

logarithm or *log* the ◊exponent or index of a number to a specified base – usually 10. For example, the logarithm to the base 10 of 1,000 is 3 because $10^3 = 1,000$; the logarithm of 2 is 0.3010 because $2 = 10^{0.3010}$. Before the advent of cheap electronic calculators, multiplication and division could be simplified by being replaced with the addition and subtraction of logarithms.

For any two numbers x and y (where $x = b^a$ and $y = b^c$) $x \times y = b^a \times b^c = b^{a+c}$; hence we would add the logarithms of x and y, and look up this answer in antilogarithm tables.

Tables of logarithms and antilogarithms are available that show conversions of numbers into logarithms, and vice versa.

For example, to multiply 6,560 by 980, one looks up their logarithms (3.8169 and 2.9912), adds them together (6.8081), then looks up the antilogarithm of this to get the answer (6,428,800). *Natural* or *Napierian logarithms* are to the base e, an ◊irrational number equal to approximately 2.7183.

The principle of logarithms is also the basis of the slide rule. With the general availability of the electronic pocket calculator, the need for logarithms has been reduced. The first log tables (to base e) were published by the Scottish mathematician John Napier in 1614. Base-ten logs were introduced by the Englishman Henry Briggs (1561–1631) and Dutch mathematician Adriaen Vlacq (1600–1667).

logic gate or *logic circuit* in electronics, one of the basic components used in building ◊integrated circuits. The five basic types of gate make logical decisions based on functions NOT, AND, OR, NAND (NOT AND), and NOR (NOT OR). With the exception of the NOT gate, each has two or more inputs.

Information is fed to a gate in the form of binary-coded input signals (logic value 0 stands for 'off' or 'low-voltage pulse', logic 1 for 'on' or 'high-voltage'), and each combination of input signals yields a specific output (logic 0 or 1). An *OR* gate will give a logic 1 output if one or more of its inputs receives a logic 1 signal; however, an *AND* gate will yield a logic 1 output only if it receives a logic 1 signal through both its inputs. The output of a *NOT* or *inverter* gate is the opposite of the signal received through its single input, and a *NOR* or *NAND* gate produces an output signal that is the opposite of the signal that would have been produced by an OR or AND gate respectively. The properties of a logic gate, or of a combination of gates, may be defined and presented in the form of a diagram called a **truth table**, which lists the output that will be triggered by each of the possible combinations of input signals. The process has close parallels in computer programming, where it forms the basis of binary logic.

LOGO high-level computer programming language designed to teach mathematical concepts. Developed in about 1970 at the Massachusetts Institute of Technology, it became popular in schools and with home computer users because of its 'turtle graphics' feature. This allows the user to write programs that create line drawings on a computer screen, or drive a small mobile robot (a 'turtle' or 'buggy') around the floor.

LOGO encourages the use of languages in a logical and structured way, leading to 'microworlds', in which problems can be solved by using a few standard solutions.

lomentum fruit similar to a pod but constricted between the seeds. When ripe, it splits into one-seeded units, as seen, for example, in the fruit of sainfoin *Onobrychis viciifolia* and radish *Raphanus raphanistrum*. It is a type of ◊schizocarp.

lone pair in chemistry, a pair of electrons in the outermost shell of an atom that are not used in bonding. In certain circumstances, they will allow the atom to bond with atoms, ions, or molecules (such as boron trifluoride, BF_3) that are deficient in electrons, forming coordinate

circuit symbols

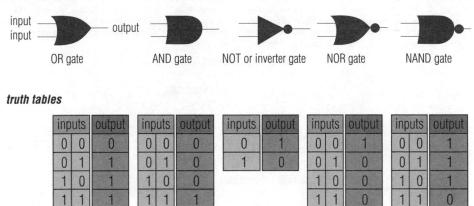

OR gate	AND gate	NOT or inverter gate	NOR gate	NAND gate

truth tables

inputs		output	inputs		output	inputs	output	inputs		output	inputs		output
0	0	0	0	0	0	0	1	0	0	1	0	0	1
0	1	1	0	1	0	1	0	0	1	0	0	1	1
1	0	1	1	0	0			1	0	0	1	0	1
1	1	1	1	1	1			1	1	0	1	1	0

OR gate	AND gate	NOT gate	NOR gate	NAND gate

logic gate The circuit symbols for the five basic types of logic gate: OR, AND, NOT, NOR, and NAND. The truth table displays the output results of each possible combination of input signal.

covalent (dative) bonds in which they provide both of the bonding electrons.

long. abbreviation for *longitude*; see ◊latitude and longitude.

longitude see ◊latitude and longitude.

longitudinal wave ◊wave in which the displacement of the medium's particles is in line with or parallel to the direction of travel of the wave motion.

long-sightedness or *hypermetropia* defect of vision in which a person is able to focus on objects in the distance, but not on close objects. It is caused by the failure of the lens to return to its normal rounded shape, or by the eyeball being too short, with the result that the image is focused on a point behind the retina. Long-sightedness is corrected by wearing spectacles fitted with ◊converging lenses, each of which acts like a magnifying glass.

loom any machine for weaving yarn or thread into cloth. The first looms were used to weave sheep's wool about 5000 BC. A loom is a frame on which a set of lengthwise threads (warp) is strung.

A second set of threads (weft), carried in a shuttle, is inserted at right angles over and under the warp.

In most looms the warp threads are separated by a device called a treddle to create a gap, or shed, through which the shuttle can be passed in a straight line. A kind of comb called a reed presses each new line of weave tight against the previous ones. All looms have similar features, but on the power loom, weaving takes place automatically at great speed. Mechanization of weaving began in 1733 when British inventor John Kay invented the flying shuttle. In 1785 British inventor Edmund Cartwright introduced a steam-powered loom. Among recent developments are shuttleless looms, which work at very high speed, passing the weft through the warp by means of 'rapiers', and jets of air or water.

loop in computing, short for ◊program loop.

Loran navigation system LOng-RAnge radio-based aid to Navigation. The current system (Loran C) consists of a master radio transmitter and two to four secondary transmitters situated within 1,000–2,000 km/600–1,250 mi. Signals from the transmitters are detected by the ship or aircraft's navigation receiver, and slight differences (phase differences) between the master and secondary transmitter signals indicate the position of the receiver.

The system is accurate to within 500 m/1,640 ft at a range of 2,000 km/1,250 mi, and covers the N Atlantic and N Pacific oceans, and adjacent areas such as the Mediterranean and Arabian Gulf.

The original Loran stem (Loran A) was used to guide bombers on night raids in Europe and N Africa in 1944.

Lorenzo's oil in medicine, a liquid containing glycerol trioleate oil and glycerol trierucate oil. It is used in the treatment of a muscle disease known as adrenoleukodystrophy.

Lotus 1–2–3 ◊spreadsheet computer program, produced by Lotus Development Corporation. It first appeared in 1982 and its combination of spreadsheet, graphics display, and data management contributed to the rapid acceptance of the IBM Personal Computer in businesses.

loudness subjective judgement of the level or power of sound reaching the ear. The human ear cannot give an absolute value to the loudness of a single sound, but can only make comparisons between two different sounds. The precise measure of the power of a sound wave at a particular point is called its ◊intensity.

Accurate comparisons of sound levels may be made using sound-level meters, which are calibrated in units called ◊decibels.

loudspeaker electromechanical device that converts electrical signals into sound waves, which are radiated into the air. The most common type of loudspeaker is the *moving-coil speaker*. Electrical signals from, for example, a radio are fed to a coil of fine wire wound around the top of a cone. The coil is surrounded by a magnet. When signals pass through it, the coil becomes an electromagnet, which by moving causes the cone to vibrate, setting up sound waves.

louse parasitic insect that lives on mammals. It has a flat, segmented body without wings, and a tube attached to the head, used for sucking blood from its host. (Order Anoplura.)

Some lice occur on humans, including the **head louse** (*Pediculus capitis*) and the **body louse** (*P. corporis*), a typhus carrier. Pediculosis is a skin disease caused by infestation of lice. Most mammals have a species of lice adapted to living on them. Biting lice belong to a different order of insects (Mallophaga) and feed on the skin, feathers, or hair.

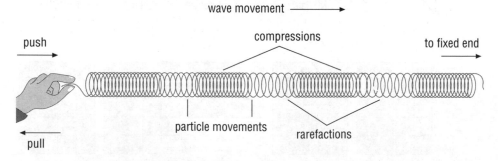

longitudinal wave The diagram illustrates the motion of a longitudinal wave. Sound, for example, travels through air in longitudinal waves: the waves vibrate back and forth in the direction of travel. In the compressions the particles are pushed together, and in the rarefactions they are pulled apart.

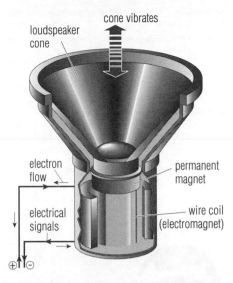

loudspeaker cone

cone vibrates

electron flow

electrical signals

permanent magnet

wire coil (electromagnet)

⊕ ⊖

loudspeaker A moving-coil loudspeaker. Electrical signals flowing through the wire coil turn it into an electromagnet, which moves as the signals vary. The attached cone vibrates, producing sound waves.

Louse

In Sweden in the Middle Ages, a mayor was once elected by a louse. The candidates rested their beards on a table and the louse was placed in the middle. The louse's chosen host was elected mayor.

Lowell Observatory US astronomical observatory founded by Percival Lowell at Flagstaff, Arizona, with a 61 cm/24 in refractor opened in 1896. The observatory now operates other telescopes at a nearby site on Anderson Mesa including the 1.83 m/72 in Perkins reflector of Ohio State and Ohio Wesleyan universities.

low-level language in computing, a programming language designed for a particular computer and reflecting its internal ◊machine code; low-level languages are therefore often described as *machine-oriented* languages. They cannot easily be converted to run on a computer with a different central processing unit, and they are relatively difficult to learn because a detailed knowledge of the internal working of the computer is required. Since they must be translated into machine code by an ◊assembler program, low-level languages are also called ◊assembly languages.

A mnemonic-based low-level language replaces binary machine-code instructions, which are very hard to remember, write down, or correct, with short codes chosen to remind the programmer of the instructions they represent. For example, the binary-code instruction that means '**sto**re the contents of the **a**ccumulator' may be represented with the mnemonic STA.

In contrast, ◊high-level languages are designed to solve particular problems and are therefore described as *problem-oriented languages*.

LSD (abbreviation for *lysergic acid diethylamide*) psychedelic drug, an ◊hallucinogen. Colourless, odour-

less, and easily synthesized, it is nonaddictive and non-toxic, but its effects are unpredictable. Its use is illegal in most countries.

The initials are from the German lyserg-säure-diäthylamid; the drug was first synthesized by a German chemist, Albert Hofmann, in 1943. In 1947 the US Central Intelligence Agency began experiments with LSD, often on unsuspecting victims. Many psychiatrists in North America used it in treatment in the 1950s. Its use as a means to increased awareness and enhanced perception was popularized in the 1960s by US psychologist Timothy Leary (1920–1997), novelist Ken Kesey, and chemist Augustus Owsley Stanley III. A series of laws to ban LSD were passed in the USA from 1965 (by which time 4 million Americans were estimated to have taken it) and in the UK in 1966; other countries followed suit. The drug had great influence on the hippie movement.

LSI (abbreviation for *large-scale integration*) the technology that enables whole electrical circuits to be etched into a piece of semiconducting material just a few millimetres square.

By the late 1960s a complete computer processor could be integrated on a single chip, or ◊integrated circuit, and in 1971 the US electronics company Intel produced the first commercially available ◊microprocessor. Very large-scale integration (VLSI) results in even smaller chips.

lubricant substance used between moving surfaces to reduce friction. Carbon-based (organic) lubricants, commonly called grease and oil, are recovered from petroleum distillation.

Extensive research has been carried out on chemical additives to lubricants, which can reduce corrosive wear, prevent the accumulation of 'cold sludge' (often the result of stop-start driving in city traffic jams), keep pace with the higher working temperatures of aviation gas turbines, or provide radiation-resistant greases for nuclear power plants. Silicon-based spray-on lubricants are also used; they tend to attract dust and dirt less than carbon-based ones.

A solid lubricant is graphite, an allotropic form of carbon, either flaked or emulsified (colloidal) in water or oil.

lucerne another name for the plant alfalfa.

lumbago pain in the lower region of the back, usually due to strain or faulty posture. If it occurs with ◊sciatica, it may be due to pressure on spinal nerves by a slipped disc. Treatment includes rest, application of heat, and skilled manipulation. Surgery may be needed in rare cases.

lumbar puncture or *spinal tap* insertion of a hollow needle between two lumbar (lower back) vertebrae to withdraw a sample of cerebrospinal fluid (CSF) for testing. Normally clear and colourless, the CSF acts as a fluid buffer around the brain and spinal cord. Changes in its quantity, colour, or composition may indicate neurological damage or disease.

lumen SI unit (symbol lm) of luminous flux (the amount of light passing through an area per second).

The lumen is defined in terms of the light falling on a unit area at a unit distance from a light source of luminous intensity of one ◊candela. One lumen at a wavelength of 5,550 angstroms equals 0.0014706 watts.

lumen in biology, the space enclosed by an organ, such as the bladder, or a tubular structure, such as the gastrointestinal tract.

luminescence emission of light from a body when its atoms are excited by means other than raising its temperature. Short-lived luminescence is called fluorescence; longer-lived luminescence is called phosphorescence.

When exposed to an external source of energy, the outer electrons in atoms of a luminescent substance absorb energy and 'jump' to a higher energy level. When these electrons 'jump' back to their former level they emit their excess energy as light. Many different exciting mechanisms are possible: visible light or other forms of electromagnetic radiation (ultraviolet rays or X-rays), electron bombardment, chemical reactions, friction, and ◊radioactivity. Certain living organisms produce ◊bioluminescence.

luminosity or *brightness* in astronomy, the amount of light emitted by a star, measured in ◊magnitudes. The apparent brightness of an object decreases in proportion to the square of its distance from the observer. The luminosity of a star or other body can be expressed in relation to that of the Sun.

luminous paint preparation containing a mixture of pigment, oil, and a phosphorescent sulphide, usually calcium or barium. After exposure to light it appears luminous in

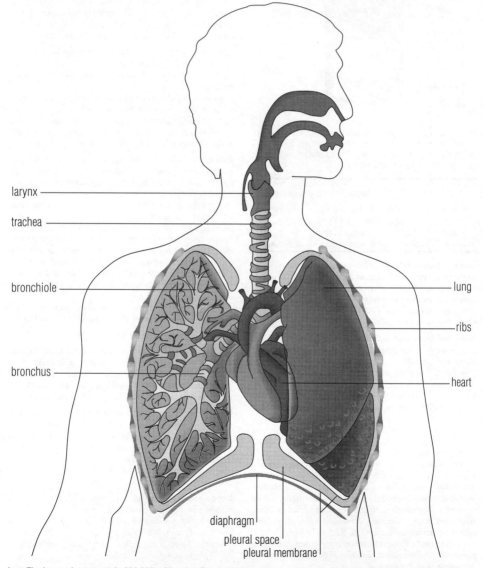

larynx
trachea
bronchiole
bronchus
lung
ribs
heart
diaphragm
pleural space
pleural membrane

lung The human lungs contain 300,000 million tiny blood vessels which would stretch for 2,400 km/1,500 mi if laid end to end. A healthy adult at rest breathes 12 times a minute; a baby breathes at twice this rate. Each breath brings 350 millilitres of fresh air into the lungs, and expels 150 millilitres of stale air from the nose and throat.

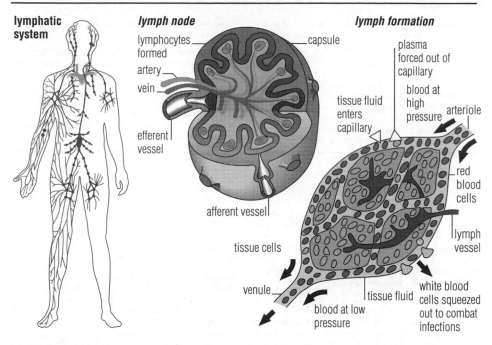

lymphatic system

lymphocytes formed

artery

vein

efferent vessel

lymph node

capsule

afferent vessel

tissue cells

venule

lymph formation

plasma forced out of capillary

blood at high pressure

tissue fluid enters capillary

arteriole

red blood cells

lymph vessel

white blood cells squeezed out to combat infections

tissue fluid

blood at low pressure

lymph Lymph is the fluid that carries nutrients and white blood cells to the tissues. Lymph enters the tissue from the capillaries (right) and is drained from the tissues by lymph vessels. The lymph vessels form a network (left) called the lymphatic system. At various points in the lymphatic system, lymph nodes (centre) filter and clean the lymph.

the dark. The luminous paint used on watch faces contains radium, is radioactive and therefore does not require exposure to light.

lung large cavity of the body, used for ◊gas exchange. It is essentially a sheet of thin, moist membrane that is folded so as to occupy less space. Most tetrapod (four-limbed) vertebrates have a pair of lungs occupying the thorax. The lung tissue, consisting of multitudes of air sacs and blood vessels, is very light and spongy, and functions by bringing inhaled air into close contact with the blood so that oxygen can pass into the organism and waste carbon dioxide can be passed out. The efficiency of lungs is enhanced by ◊breathing movements, by the thinness and moistness of their surfaces, and by a constant supply of circulating blood.

In humans, the principal diseases of the lungs are tuberculosis, pneumonia, bronchitis, emphysema, and cancer.

lupus in medicine, any of various diseases characterized by lesions of the skin. One form (lupus vulgaris) is caused by the tubercle bacillus (see ◊tuberculosis). The organism produces ulcers that spread and eat away the underlying tissues. Treatment is primarily with standard antituberculous drugs, but ultraviolet light may also be used.

Lupus erythematous (LE) has two forms: **discoid** LE, seen as red, scaly patches on the skin, especially the face; and **disseminated** or **systemic** LE, which may affect connective tissue anywhere in the body, often involving the internal organs. The latter is much more serious. Treatment is with ◊corticosteroids. LE is an ◊autoimmune disease.

luteinizing hormone ◊hormone produced by the pitu-

itary gland. In males, it stimulates the testes to produce androgens (male sex hormones). In females, it works together with follicle-stimulating hormone to initiate production of egg cells by the ovary. If fertilization occurs, it plays a part in maintaining the pregnancy by controlling the levels of the hormones oestrogen and progesterone in the body.

lutetium silver-white, metallic element, the last of the ◊lanthanide series, symbol Lu, atomic number 71, relative atomic mass 174.97. It is used in the 'cracking', or breakdown, of petroleum and in other chemical processes. It was named by its discoverer, French chemist Georges Urbain, (1872–1938) after his native city.

lux SI unit (symbol lx) of illuminance or illumination (the light falling on an object). It is equivalent to one ◊lumen per square metre or to the illuminance of a surface one metre distant from a point source of one ◊candela.

LW abbreviation for *long wave*, a radio wave with a wavelength of over 1,000 m/3,300 ft; one of the main wavebands into which radio frequency transmissions are divided.

Lyme disease disease transmitted by tick bites that affects all the systems of the body. It has a 10% mortality rate. First described in 1975 following an outbreak in children living around Lyme, Connecticut, USA, it is caused by the microorganism *Borrelia burgdorferi*, isolated by Burgdorfer and Barbour in the USA in 1982. Untreated, the disease attacks the joints, nervous system, heart, liver, kidneys, and eyes, but responds to penicillin or tetracycline.

The tick that carries the disease, *Ixodes dammini*, lives

on deer, while *B. burgdorferi* relies on mice during its life cycle. In the UK the sheep tick is the main carrier.

lymph fluid found in the lymphatic system of vertebrates.

Lymph is drained from the tissues by lymph capillaries, which empty into larger lymph vessels (lymphatics). These lead to lymph nodes (small, round bodies chiefly situated in the neck, armpit, groin, thorax, and abdomen), which process the ◊lymphocytes produced by the bone marrow, and filter out harmful substances and bacteria. From the lymph nodes, vessels carry the lymph to the thoracic duct and the right lymphatic duct, which drain into the large veins in the neck. Some vertebrates, such as amphibians, have a lymph heart, which pumps lymph through the lymph vessels. *See illustration on page 385.*

lymph nodes small masses of lymphatic tissue in the body that occur at various points along the major lymphatic vessels. Tonsils and adenoids are large lymph nodes. As the lymph passes through them it is filtered, and bacteria and other microorganisms are engulfed by cells known as macrophages.

Lymph nodes are sometimes mistakenly called lymph 'glands', and the term 'swollen glands' refers to swelling of the lymph nodes caused by infection.

lymphocyte type of white blood cell with a large nucleus, produced in the bone marrow. Most occur in the ◊lymph and blood, and around sites of infection. *B lymphocytes* or ◊B cells are responsible for producing ◊antibodies. *T*

lymphocytes or ◊T cells have several roles in the mechanism of ◊immunity.

lymphokines chemical messengers produced by lymphocytes that carry messages between the cells of the immune system (see ◊immunity). Examples include interferon, which initiates defensive reactions to viruses, and the interleukins, which activate specific immune cells.

lyophilization technical term for the ◊freeze-drying process used for foods and drugs and in the preservation of organic archaeological remains.

Lyra small but prominent constellation of the northern hemisphere, represented as the lyre of Orpheus. Its brightest star is ◊Vega.

Epsilon Lyrae is a system of four gravitationally linked stars. Beta Lyrae is an eclipsing binary. The Ring nebula, M57, is a ◊planetary nebula.

lysine one of the nine essential ◊amino acids.

lysis in biology, any process that destroys a cell by rupturing its membrane or cell wall (see ◊lysosome).

lysosome membrane-enclosed structure, or organelle, inside a ◊cell, principally found in animal cells. Lysosomes contain enzymes that can break down proteins and other biological substances. They play a part in digestion, and in the white blood cells known as phagocytes the lysosome enzymes attack ingested bacteria.

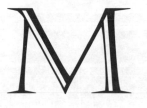

m symbol for ◊*metre*.

MA abbreviation for ◊*mechanical advantage*.

maceration in biology, a process of softening solids by soaking them in liquids. Food is macerated in the mouth as part of the process of ◊digestion.

machine device that allows a small force (the effort) to overcome a larger one (the load). There are three basic machines: the inclined plane (ramp), the lever, and the wheel and axle. All other machines are combinations of these three basic types. Simple machines derived from the inclined plane include the wedge, the gear, and the screw; the spanner is derived from the lever; the pulley from the wheel.

The principal features of a machine are its ◊mechanical advantage, which is the ratio of load to effort, its velocity ratio, and its ◊efficiency, which is the work done by the load divided by the work done by the effort; the latter is expressed as a percentage. In a perfect machine, with no friction, the efficiency would be 100%. All practical machines have efficiencies of less than 100%, otherwise perpetual motion would be possible.

machine code in computing, a set of instructions that a computer's central processing unit (CPU) can understand and obey directly, without any translation. Each type of CPU has its own machine code. Because machine-code programs consist entirely of binary digits (bits), most programmers write their programs in an easy-to-use ◊high-level language. A high-level program must be translated into machine code – by means of a ◊compiler or ◊interpreter program – before it can be executed by a computer.

Where no suitable high-level language exists or where very efficient machine code is required, programmers may choose to write programs in a low-level, or assembly, language, which is eventually translated into machine code by means of an ◊assembler program.

Microprocessors (CPUs based on a single integrated circuit) may be classified according to the number of machine-code instructions that they are capable of obeying: ◊CISC (complex instruction set computer) microprocessors support up to 200 instructions, whereas ◊RISC (reduced instruction set computer) microprocessors support far fewer instructions but execute programs more rapidly.

machine tool automatic or semi-automatic power-driven machine for cutting and shaping metals. Machine tools have powerful electric motors to force cutting tools into the metal: these are made from hardened steel containing heat-resistant metals such as tungsten and chromium. The use of precision machine tools in mass-production assembly methods ensures that all duplicate parts produced are virtually identical.

Many machine tools now work under computer control and are employed in factory ◊automation. The most common machine tool is the ◊lathe, which shapes shafts and similar objects. A milling machine cuts metal with a rotary toothed cutting wheel. Other machine tools cut, plane, grind, drill, and polish.

Mach number ratio of the speed of a body to the speed of sound in the undisturbed medium through which the body travels. Mach 1 is reached when a body (such as an aircraft) has a velocity greater than that of sound ('passes the sound barrier'), namely 331 m/1,087 ft per second at sea level. It is named after Austrian physicist Ernst Mach (1838–1916).

Macintosh range of microcomputers originally produced by Apple Computers. The Apple Macintosh, introduced in 1984, was the first popular microcomputer with a ◊graphical user interface. The success of the Macintosh prompted other manufacturers and software companies to create their own graphical user interfaces. Most notable of these are Microsoft Windows, which runs on IBM PC-compatible microcomputers, and OSF/Motif, from the Open Software Foundation, which is used with many UNIX systems.

Apple have faced major problems in the 1990s. Microsoft's introduction of Windows 95 eroded the competitive advantage of Macintosh's **System 7**.

macro in computer programming, a new command created by combining a number of existing ones. For example, a word processing macro might create a letterhead or fax cover sheet, inserting words, fonts, and logos with a single keystroke or mouse click. Macros are also useful to automate computer communications – for example, users can write a macro to ask their computer to dial an **Internet Service Provider** (ISP), retrieve e-mail and USENET articles, and then disconnect. A **macro key** on the keyboard combines the effects of pressing several individual keys.

macro- a prefix meaning on a very large scale, as opposed to micro.

macromolecule in chemistry, a very large molecule, generally a ◊polymer.

macrophage type of ◊white blood cell, or leucocyte, found in all vertebrate animals. Macrophages specialize in the removal of bacteria and other microorganisms, or of cell debris after injury. Like phagocytes, they engulf foreign matter, but they are larger than phagocytes and have a longer life span. They are found throughout the body, but mainly in the lymph and connective tissues, and especially the lungs, where they ingest dust, fibres, and other inhaled particles.

MAFF abbreviation for **Ministry of Agriculture, Fisheries and Food**.

mafic rock plutonic rock composed chiefly of dark-coloured minerals containing abundant magnesium and iron, such as olivine and pyroxene. It is derived from **magnesium** and **ferric** (iron). The term **mafic** also applies to dark-coloured minerals rich in iron and magnesium as a group.

Magdalenian final cultural phase of the Palaeolithic (Old Stone Age) in W Europe, best known for its art, and lasting from c. 16,000–10,000 BC. It was named after the rock-shelter of La Madeleine in SW France.

Magellan NASA space probe to ◊Venus, launched in May 1989; it went into orbit around Venus in Aug 1990 to make a detailed map of the planet by radar. It revealed volcanoes, meteorite craters, and fold mountains on the planet's surface. Magellan mapped 98% of Venus.

In Oct 1994 Magellan was instructed to self-destruct by entering the atmosphere around Venus where it burned up.

Magellanic Clouds in astronomy, the two galaxies nearest to our own galaxy. They are irregularly shaped, and appear as detached parts of the ◊Milky Way, in the southern constellations ◊Dorado, ◊Tucana, and Mensa.

The Large Magellanic Cloud spreads over the constellations of Dorado and Mensa. The Small Magellanic Cloud is in Tucana. The Large Magellanic Cloud is 169,000 light years from Earth, and about a third the diameter of our Galaxy; the Small Magellanic Cloud, 180,000 light years away, is about a fifth the diameter of our Galaxy. They are named after the navigator Ferdinand Magellan, who first described them.

magic bullet term sometimes used for a drug that is specifically targeted on certain cells or tissues in the body, such as a small collection of cancerous cells (see ◊cancer) or cells that have been invaded by a virus. Such drugs can be made in various ways, but ◊monoclonal antibodies are increasingly being used to direct the drug to a specific target.

The term was originally associated with the German chemist Paul Ehrlich (1854–1915) who discovered the first specific against ◊syphilis.

magic square in mathematics, a square array of numbers in which the rows, columns, and diagonals add up to the same total. A simple example employing the numbers 1 to 9, with a total of 15, has a first row of 6, 7, 2, a second row of 1, 5, 9, and a third row of 8, 3, 4.

maglev (acronym for *magnetic levitation*) high-speed surface transport using the repellent force of superconductive magnets (see ◊superconductivity) to propel and support, for example, a train above a track.

Technical trials on a maglev train track began in Japan in the 1970s, and a speed of 500 kph/310 mph has been reached, with a cruising altitude of 10 cm/4 in. The train is levitated by electromagnets and forward thrust is provided by linear motors aboard the cars, propelling the train along a reaction plate.

A small low-power maglev monorail system is in use at Birmingham Airport, England. A subway line using maglev carriages began operating in Osaka, Japan, in 1990.

magma molten rock material beneath the Earth's surface from which ◊igneous rocks are formed. ◊Lava is magma that has reached the surface and solidified, losing some of its components on the way.

magnesia common name for ◊magnesium oxide.

magnesium lightweight, very ductile and malleable, silver-white, metallic element, symbol Mg, atomic number

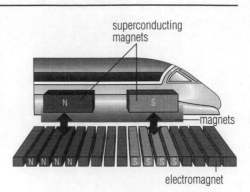

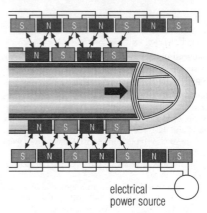

maglev The repulsion of superconducting magnets and electromagnets in the track keeps a maglev train suspended above the track. By varying the strength and polarity of the track electromagnets, the train can be driven forward.

12, relative atomic mass 24.305. It is one of the ◊alkaline-earth metals, and the lightest of the commonly used metals. Magnesium silicate, carbonate, and chloride are widely distributed in nature. The metal is used in alloys and flash photography. It is a necessary trace element in the human diet, and green plants cannot grow without it since it is an essential constituent of the photosynthetic pigment ◊chlorophyll ($C_{55}H_{72}MgN_4O_5$).

It was named after the ancient Greek city of Magnesia, near where it was first found. It was first recognized as an element by Scottish chemist Joseph Black in 1755 and discovered in its oxide by English chemist Humphry Davy in 1808. Pure magnesium was isolated in 1828 by French chemist Antoine-Alexandre-Brutus Bussy.

magnesium oxide or *magnesia* MgO white powder or colourless crystals, formed when magnesium is burned in air or oxygen; a typical basic oxide. It is used to treat acidity of the stomach, and in some industrial processes; for example, as a lining brick in furnaces, because it is very stable when heated (refractory oxide).

magnet any object that forms a magnetic field (displays ◊magnetism), either permanently or temporarily through induction, causing it to attract materials such as iron, cobalt, nickel, and alloys of these. It always has two ◊magnetic poles, called north and south.

magnetic compass device for determining the direction of the horizontal component of the Earth's magnetic field. It consists of a magnetized needle with its north-seeking pole clearly indicated, pivoted so that it can turn freely in a plane parallel to the surface of the Earth (in a horizontal circle).

The needle will turn so that its north-seeking pole points towards the Earth's magnetic north pole. See also ◊compass.

Walkers, sailors, and other travellers use a magnetic compass to find their direction. The direction of the geographic, or true, North Pole is, however, slightly different from that of the magnetic north pole, and so the readings obtained from a compass of this sort must be adjusted using tables of magnetic corrections or information marked on local maps.

magnetic declination see ◊angle of declination.

magnetic dip see ◊dip, magnetic and ◊angle of dip.

magnetic dipole the pair of north and south magnetic poles, separated by a short distance, that makes up all magnets. Individual magnets are often called 'magnetic dipoles'. Single magnetic poles, or monopoles, have never been observed despite being searched for. See also magnetic ◊domain.

magnetic field region around a permanent magnet, or around a conductor carrying an electric current, in which a force acts on a moving charge or on a magnet placed in the field. The field can be represented by lines of force, which by convention link north and south poles and are parallel to the directions of a small compass needle placed on them. A magnetic field's magnitude and direction are given by the ◊magnetic flux density, expressed in ◊teslas. See also ◊polar reversal.

magnetic flux measurement of the strength of the magnetic field around electric currents and magnets. Its SI unit is the ◊weber; one weber per square metre is equal to one tesla.

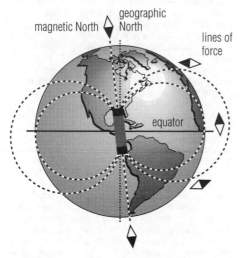

magnetic field The Earth's magnetic field is similar to that of a bar magnet with poles near, but not exactly at, the geographic poles. Compass needles align themselves with the magnetic field, which is horizontal near the equator and vertical at the magnetic poles.

The amount of magnetic flux through an area equals the product of the area and the magnetic field strength at a point within that area. It is a measure of the number of magnetic field lines passing through the area.

magnetic induction the production of magnetic properties in unmagnetized iron or other ferromagnetic material when it is brought close to a magnet. The material is influenced by the magnet's magnetic field and the two are attracted. The induced magnetism may be temporary, disappearing as soon as the magnet is removed, or permanent depending on the nature of the iron and the strength of the magnet.

◊Electromagnets make use of temporary induced magnetism to lift sheets of steel: the magnetism induced in the steel by the approach of the electromagnet enables it to be picked up and transported. To release the sheet, the current supplying the electromagnet is temporarily switched off and the induced magnetism disappears.

magnetic-ink character recognition (MICR) in computing, a technique that enables special characters printed in magnetic ink to be read and input rapidly to a computer. MICR is used extensively in banking because magnetic-ink characters are difficult to forge and are therefore ideal for marking and identifying cheques.

magnetic material one of a number of substances that are strongly attracted by magnets and can be magnetized. These include iron, nickel, and cobalt, and all those ◊alloys that contain a proportion of these metals.

Soft magnetic materials can be magnetized very easily, but the magnetism induced in them (see ◊magnetic induction) is only temporary. They include Stalloy, an alloy of iron with 4% silicon used to make the cores of electromagnets and transformers, and the materials used to make 'iron' nails and paper clips.

Hard magnetic materials can be permanently magnetized by a strong magnetic field. Steel and special alloys such as Alcomax, Alnico, and Ticonal, which contain various amounts of aluminium, nickel, cobalt, and copper, are used to make permanent magnets. The strongest permanent magnets are ceramic, made under high pressure and at high temperature from powders of various metal oxides.

magnetic pole region of a magnet in which its magnetic properties are strongest. Every magnet has two poles, called north and south.

The north (or north-seeking) pole is so named because a freely suspended magnet will turn so that this pole points towards the Earth's magnetic north pole. The north pole of one magnet will be attracted to the south pole of another, but will be repelled by its north pole.

Unlike poles may therefore be said to attract, like poles to repel.

magnetic resonance imaging (MRI) diagnostic scanning system based on the principles of nuclear magnetic resonance. MRI yields finely detailed three-dimensional images of structures within the body without exposing the patient to harmful radiation. The technique is invaluable for imaging the soft tissues of the body, in particular the brain and the spinal cord.

Claimed as the biggest breakthrough in diagnostic imaging since the discovery of X-rays, MRI is a non-invasive technique based on a magnet which is many thousands of times stronger than the Earth's magnetic field. It causes nuclei within the atoms of the body to align

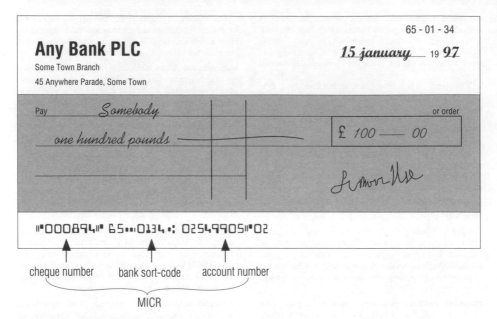

	65 - 01 - 34
Any Bank PLC	*15 january* 19 **97**
Some Town Branch	
45 Anywhere Parade, Some Town	

Pay   *Somebody*   or order

one hundred pounds   £ 100 —— 00

‖"000894‖" ⑃5‖‖‖0134⑈: 025499⑃05‖"02

cheque number   bank sort-code   account number

MICR

magnetic-ink character recognition An example of one of the uses of magnetic ink in automatic character recognition. Because of the difficulties in forging magnetic-ink characters, and the speed with which they can be read by computer systems, MICR is used extensively in banking.

themselves in one direction. When a brief radio pulse is beamed at the body the nuclei spin, emitting weak radio signals as they realign themselves to the magnet. These signals, which are characteristic for each type of tissue, are converted electronically into images on a viewing screen.

Also developed around magnetic technology, **magnetic resonance spectroscopy (MRS)** is a technique for investigating conditions in which there is a disturbance of the body's energy metabolism, including ischaemia and toxic damage due to drugs or other chemicals. MRS is also of value in diagnosing some cancers.

magnetic storm in meteorology, a sudden disturbance affecting the Earth's magnetic field, causing anomalies in radio transmissions and magnetic compasses. It is probably caused by ◊sunspot activity.

magnetic strip or **magnetic stripe** thin strip of magnetic material attached to a plastic card (such as a credit card) and used for recording data.

magnetic tape narrow plastic ribbon coated with an easily magnetizable material on which data can be recorded. It is used in sound recording, audiovisual systems (videotape), and computing. For mass storage on commercial mainframe computers, large reel-to-reel tapes are still used, but cartridges are coming in. Various types of cartridge are now standard on minis and PCs, while audio cassettes are sometimes used with home computers.

Magnetic tape was first used in **sound recording** in 1947, and made overdubbing possible, unlike the direct-to-disc system it replaced. Two-track tape was introduced in the 1950s and four-track in the early 1960s; today, studios use 16-, 24-, or 32-track tape, from which the tracks are mixed down to a stereo master tape.

In computing, magnetic tape was first used to record data and programs in 1951 as part of the UNIVAC 1 system. It was very popular as a storage medium for external memory in the 1950s and 1960s. Since then it has been largely replaced by magnetic ◊discs as a working medium, although tape is still used to make backup copies of important data. Information is recorded on the tape in binary form, with two different strengths of signal representing 1 and 0.

magnetism phenomena associated with ◊magnetic fields. Magnetic fields are produced by moving charged particles: in electromagnets, electrons flow through a coil of wire connected to a battery; in permanent magnets, spinning electrons within the atoms generate the field.

Substances differ in the extent to which they can be magnetized by an external field (susceptibility). Materials that can be strongly magnetized, such as iron, cobalt, and nickel, are said to be **ferromagnetic**; this is due to the formation of areas called ◊domains in which atoms, weakly magnetic because of their spinning electrons, align to form areas of strong magnetism. Magnetic materials lose their magnetism if heated to the Curie temperature. Most other materials are **paramagnetic**, being only weakly pulled towards a strong magnet. This is because their atoms have a low level of magnetism and do not form domains. **Diamagnetic** materials are weakly repelled by a magnet since electrons within their atoms act as electromagnets and oppose the applied magnetic force. **Anti-ferromagnetic** materials have a very low susceptibility that increases with temperature; a similar phenomenon in materials such as ferrites is called **ferrimagnetism**.

Apart from its universal application in dynamos, electric motors, and switch gears, magnetism is of considerable importance in advanced technology – for example, in particle ◊accelerators for nuclear research, memory stores for computers, tape recorders, and ◊cryogenics.

magnetite black, strongly magnetic opaque mineral, Fe_3O_4, of the spinel group, an important ore of iron. Widely distributed, magnetite is found in nearly all igneous and metamorphic rocks. Some deposits, called lodestone, are permanently magnetized. Lodestone has been used as a compass since the first millennium BC. Today the orientations of magnetite grains in rocks are used in the study of the Earth's magnetic field (see ◊palaeomagnetism).

magneto simple electric generator, often used to provide the electricity for the ignition system of motorcycles and used in early cars.

It consists of a rotating magnet that sets up an electric current in a coil, providing the spark.

magnetohydrodynamics (MHD) field of science concerned with the behaviour of ionized gases or liquid in a magnetic field. Systems have been developed that use MHD to generate electrical power.

MND-driven ships have been tested in Japan. In 1991 two cylindrical thrusters with electrodes and niobium-titanium superconducting coils, soaked in liquid helium, were placed under the passenger boat *Yamato 1*. The boat, 30 m/100 ft long, was designed to travel at 8 knots. An electric current passed through the electrodes accelerates water through the thrusters, like air through a jet engine, propelling the boat forward.

magnetometer device for measuring the intensity and orientation of the magnetic field of a particular rock or of a certain region. In archaeology, distortions of the magnetic field occur when structures, such as kilns or hearths, are present. A magnetometer allows for such features to be located without disturbing the ground, and for excavation to be concentrated in the most likely area.

magnetosphere volume of space, surrounding a planet, controlled by the planet's magnetic field, and acting as a magnetic 'shell'. The Earth's magnetosphere extends 64,000 km/40,000 mi towards the Sun, but many times this distance on the side away from the Sun.

The extension away from the Sun is called the *magnetotail*. The outer edge of the magnetosphere is the *magnetopause*. Beyond this is a turbulent region, the *magnetosheath*, where the ◊solar wind is deflected around the magnetosphere. Inside the magnetosphere, atomic particles follow the Earth's lines of magnetic force. The magnetosphere contains the ◊Van Allen radiation belts. Other planets have magnetospheres, notably Jupiter.

magnetron thermionic ◊valve (electron tube) for generating very high-frequency oscillations, used in radar and to produce microwaves in a microwave oven. The flow of electrons from the tube's cathode to one or more anodes is controlled by an applied magnetic field.

magnet therapy in alternative medicine, use of applied magnetic fields to regulate potentially pathogenic disorders in the electrical charges of body cells and structures.

Physicians Hippocrates and Galen in the ancient world, Paracelsus in the Middle Ages, and more recently the founder of homoeopathy, Samuel Hahnemann, practised magnet therapy. Today practitioners apply it as a stimulant of the body's self-healing processes, to enhance tissue repair and the healing of bone fractures, as an adjunct of structural therapies such as chiropractic, and as an alternative to needling in acupuncture.

magnification measure of the enlargement or reduction of an object in an imaging optical system. *Linear magnification* is the ratio of the size (height) of the image to that of the object.

Angular magnification is the ratio of the angle subtended at the observer's eye by the image to the angle subtended by the object when viewed directly.

magnitude in astronomy, measure of the brightness of a star or other celestial object. The larger the number denoting the magnitude, the fainter the object. Zero or first magnitude indicates some of the brightest stars. Still brighter are those of negative magnitude, such as Sirius, whose magnitude is −1.46. *Apparent magnitude* is the brightness of an object as seen from Earth; *absolute magnitude* is the brightness at a standard distance of 10 parsecs (32.6 light years).

Each magnitude step is equal to a brightness difference of 2.512 times. Thus a star of magnitude 1 is $(2.512)^5$ or 100 times brighter than a sixth-magnitude star just visible to the naked eye. The apparent magnitude of the Sun is −26.8, its absolute magnitude +4.8.

mail merge in computing, a feature offered by some word-processing packages that enables a list of personal details, such as names and addresses, to be combined with a general document outline to produce individualized documents.

For example, a club secretary might create a file containing a mailing list of the names and addresses of the club members. Whenever a letter is to be sent to all club members, a general letter outline is prepared with indications as to where individual names and addresses need to be added. The mail-merge feature then combines the file of names and addresses with the letter outline to produce and print individual letters addressed to each club member.

mainframe large computer used for commercial data processing and other large-scale operations. Because of the general increase in computing power, the differences between the mainframe, ◊supercomputer, ◊minicomputer, and ◊microcomputer (personal computer) are becoming less marked.

Mainframe manufacturers include IBM, Amdahl, Fujitsu, and Hitachi. Typical mainframes have from 32 to 256 MB of memory and tens of gigabytes of disc storage.

main sequence in astronomy, the part of the ◊Hertzsprung–Russell diagram that contains most of the stars, including the Sun. It runs diagonally from the top left of the diagram to the lower right. The most massive (and hence brightest) stars are at the top left, with the least massive (coolest) stars at the bottom right.

maize (North American *corn*) tall annual ◊cereal plant that produces spikes of yellow grains which are widely used as an animal feed. Grown extensively in all subtropical and warm temperate regions, its range has been extended to colder zones by hardy varieties developed in the 1960s. (*Zea mays.*)

Sweetcorn, a variety of maize in which the sugar is not converted to starch, is a familiar vegetable, known as corn on the cob; other varieties are made into hominy, polenta, popcorn, and corn bread. Sweetcorn is used in corn oil and fermented to make alcohol; its stalks are made into paper and hardboard.

malabsorption syndrome in medicine, a range of diseases that are characterized by impairment of absorption of essential nutrients from the gastrointestinal tract, for example ◊coeliac disease and ◊cystic fibrosis. Symptoms of malabsorption syndrome include anaemia, diarrhoea, and weight loss.

malachite common ◊copper ore, basic copper carbonate, $Cu_2CO_3(OH)_2$. It is a source of green pigment and is used as an antifungal agent in fish farming, as well as being polished for use in jewellery, ornaments, and art objects.

malaria infectious parasitic disease of the tropics transmitted by mosquitoes, marked by periodic fever and an enlarged spleen. When a female mosquito of the *Anopheles* genus bites a human who has malaria, it takes in with the human blood one of four malaria protozoa of the genus *Plasmodium*. This matures within the insect and is then transferred when the mosquito bites a new victim. Malaria affects about 267 million people in 103 countries, and in 1995 around 2.1 million people died of the disease. In sub-Saharan Africa alone between 1.5 and 2 million children die from malaria and its consequences each year.

Inside the human body the parasite settles first in the liver, then multiplies to attack the red blood cells. Within the red blood cells the parasites multiply, eventually causing the cells to rupture and other cells to become infected. The cell rupture tends to be synchronized, occurring every 2–3 days, when the symptoms of malaria become evident.

In Brazil a malaria epidemic broke out among new settlers in the Amazon region, with 287,000 cases in 1983 and 500,000 cases in 1988.

◊Quinine, the first drug used against malaria, has now been replaced by synthetics, such as chloroquine, used to prevent or treat the disease. However, chloroquine-resistant strains of the main malaria parasite, *Plasmodium falciparum*, are spreading rapidly in many parts of the world.

The drug mefloquine (Lariam) is widely prescribed for use in areas where chloroquine-resistant malaria prevails. It is surrounded by controversy, however, as it has been linked to unpleasant side effects, including psychiatric disturbances such as anxiety and hallucinations, epileptic seizures, and memory loss.

Another drug, artemether, derived from the shrub wormwood, was found in 1996 trials to be as effective as quinine in the treatment of cerebral malaria.

An experimental malaria vaccine SPf66, developed by Colombian scientist Manuel Patarroyo, was trialled in 1994 in rural Tanzania, where villagers are bitten an average of 300 times a year by infected mosquitoes. It reduced the incidence of malaria by one third. However, further trials of SPf66 in the Gambia concluded in 1995 that the vaccine provided only 8% protection for young children. A further trial in Thailand in 1996 failed to provide any evidence of its effectiveness.

Research has also been carried out into using chelating agents, which remove surplus iron from the blood, to treat malaria.

malic acid $HOOCCH_2CH(OH)COOH$ organic crystalline acid that can be extracted from apples, plums, cherries, grapes, and other fruits, but occurs in all living cells in smaller amounts, being one of the intermediates of the ◊Krebs cycle.

malignant hyperpyrexia or **malignant hyperthermia** in medicine, a rare complication of general anaesthesia. It is thought to be due to a reaction to volatile anaesthetics, such as halothane, and muscle relaxants. Symptoms include a rapid rise in temperature, muscle rigidity, abnormalities in heart rate and rhythm, and acidosis. The condition can be fatal. It is treated with the muscle relaxant dantrolene.

malignant hypertension or **accelerated hypertension** in medicine, very severe ◊hypertension (high blood pressure) that can become fatal and requires urgent admission to hospital. It may develop in patients with untreated essential hypertension.

malnutrition condition resulting from a defective diet where certain important food nutrients (such as protein, vitamins, or carbohydrates) are absent. It can lead to deficiency diseases.

A related problem is ◊undernourishment.

In a report released at the World Food Summit in Rome, Nov 1996, the World Bank warned of an impending international food crisis. In 1996, more than 800 million people were unable to get enough food to meet their basic needs. Eighty-two countries, half of them in Africa, did not grow enough food for their own people; nor could they afford to import it. The World Bank calculated that food production would have to double over the next 30 years as the world population increases. Contrary to earlier predictions, food stocks and particularly grain stocks have fallen during the 1990s.

By 1996, 20 countries did not have enough water to meet people's needs, and water tables were falling in crucial regions such as the American Midwest, India, and China. Meanwhile, 90 million people are added to the planet's population every year. The increasingly prosperous and populous nations of East Asia are placing pressure on supplies; as people become wealthier they often eat more meat, which is a less efficient use of resources.

malt in brewing, grain (barley, oats, or wheat) artificially germinated and then dried in a kiln. Malts are fermented to make beers or lagers, or fermented and then distilled to produce spirits such as whisky.

maltase enzyme found in plants and animals that breaks down the disaccharide maltose into glucose.

maltose $C_{12}H_{22}O_{11}$ a ◊disaccharide sugar in which both monosaccharide units are glucose.

It is produced by the enzymic hydrolysis of starch and is a major constituent of malt, produced in the early stages of beer and whisky manufacture.

mammal any of a large group of warm-blooded vertebrate animals characterized by having ◊mammary glands in the female; these are used for suckling the young. Other features of mammals are ◊hair (very reduced in some species, such as whales); a middle ear formed of three small bones (ossicles); a lower jaw consisting of two bones only; seven vertebrae in the neck; and no nucleus in the red blood cells. (Class Mammalia.)

Mammals are divided into three groups:

Placental mammals, where the young develop inside the mother's body, in the ◊uterus, receiving nourishment from the blood of the mother via the ◊placenta; **Marsupials**, where the young are born at an early stage of development and develop further in a pouch on the mother's body where they are attached to and fed from a nipple; and **Monotremes**, where the young hatch from an egg outside the mother's body and are then nourished with milk.

The monotremes are the least evolved and have been largely displaced by more sophisticated marsupials and placentals, so that there are only a few types surviving (platypus and echidna). Placentals have spread to all parts of the globe, where placentals have competed with marsupials, the placentals have in general displaced marsupial types. However, marsupials occupy many specialized niches in South America and, especially, Australasia.

The theory that marsupials succeed only where they do not compete with placentals was shaken in 1992, when a tooth, 55 million years old and belonging to a placental mammal, was found in Murgon, Australia, indicating that placental animals appeared in Australia at the same time as the marsupials. The marsupials, however, still prevailed.

There are over 4,000 species of mammals, adapted to almost every way of life. The smallest shrew weighs only 2 g/0.07 oz, the largest whale up to 140 tonnes.

mammary gland in female mammals, a milk-producing gland derived from epithelial cells underlying the skin, active only after the production of young. In all but monotremes (egg-laying mammals), the mammary glands terminate in teats which aid infant suckling. The number of glands and their position vary between species. In humans there are 2, in cows 4, and in pigs between 10 and 14.

The hatched young of monotremes simply lick milk from a specialized area of skin on the mother's abdomen.

mammography X-ray procedure used to screen for breast cancer. It can detect abnormal growths at an early stage, before they can be seen or felt.

manganese hard, brittle, grey-white metallic element, symbol Mn, atomic number 25, relative atomic mass 54.9380. It resembles iron (and rusts), but it is not magnetic and is softer. It is used chiefly in making steel alloys, also alloys with aluminium and copper.

It is used in fertilizers, paints, and industrial chemicals. It is a necessary trace element in human nutrition. The name is old, deriving from the French and Italian forms of Latin for *magnesia* (MgO), the white tasteless powder used as an antacid from ancient times.

manganese ore any mineral from which manganese is produced. The main ores are the oxides, such as **pyrolusite**, MnO_2; **hausmannite**, Mn_3O_4; and **manganite**, $MnO(OH)$.

Manganese ores may accumulate in metamorphic rocks or as sedimentary deposits, frequently forming nodules on the sea floor (since the 1970s many schemes have been put forward to harvest deep-sea manganese nodules). The world's main producers are Georgia, Ukraine, South Africa, Brazil, Gabon, and India.

manganese(IV) oxide or **manganese dioxide** MnO_2 brown solid that acts as a ◊depolarizer in dry batteries by oxidizing the hydrogen gas produced to water; without this process, the performance of the battery is impaired.

mangelwurzel or **mangold** variety of the common beet *Beta vulgaris* used chiefly as feed for cattle and sheep.

mangold another name for ◊mangelwurzel.

mania in psychiatry, term used to describe high mood. The affected individual can appear cheerful and optimistic or irritable and angry. Sleep is reduced and the sufferer can be overactive to the point of physical exhaustion. Speech is rapid and can convey grandiose delusions. Mania often occurs as part of ◊manic depression, which is characterized by mood swings from high mood to the low mood of depression. It is treated with an antipsychotic drug to control the mood and, in the long term, with ◊lithium.

manic depression or **bipolar disorder** mental disorder characterized by recurring periods of either ◊depression or mania (inappropriate elation, agitation, and rapid thought and speech) or both.

Sufferers may be genetically predisposed to the condition. Some cases have been improved by taking prescribed doses of ◊lithium.

manometer instrument for measuring the pressure of liquids (including human blood pressure) or gases. In its basic form, it is a U-tube partly filled with coloured liquid. Greater pressure on the liquid surface in one arm will force the level of the liquid in the other arm to rise. A difference between the pressures in the spaces in the two arms is therefore registered as a difference in the heights of the liquid in the arms.

mantissa in mathematics, the decimal part of a ◊logarithm. For example, the logarithm of 347.6 is 2.5411; in this case, the 0.5411 is the mantissa, and the integral (whole number) part of the logarithm, the 2, is the ◊characteristic.

mantle intermediate zone of the Earth between the ◊crust and the ◊core, accounting for 82% of Earth's volume. The boundary between the mantle and the crust above is the ◊Mohorovičić discontinuity, located at an average depth of 32 km/20 mi. The lower boundary with the core is the Gutenburg discontinuity at an average depth of 2,900 km/ 1813 mi.

The mantle is subdivided into **upper mantle, transition zone**, and **lower mantle**, based upon the different velocities with which seismic waves travel through these regions. The upper mantle includes a zone characterized by low velocities of seismic waves, called the **low**

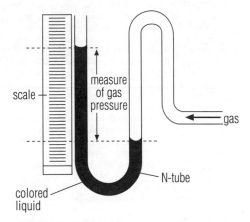

manometer The manometer indicates gas pressure by the rise of liquid in the tube.

velocity zone, at 72 km/45 mi to 250 km/155 mi depth. This zone corresponds to the ◊asthenosphere upon which Earth's tectonic plates of ◊lithosphere glide. Seismic velocities in the upper mantle are overall less than those in the transition zone and those of the transition zone are in turn less than those of the lower mantle. Faster propagation of seismic waves in the lower mantle implies that the lower mantle is more dense than the upper mantle.

The mantle is composed primarily of magnesium, silicon, and oxygen in the form of ◊silicate minerals. In the upper mantle, the silicon in silicate minerals, such as olivine, is surrounded by four oxygen atoms. Deeper in the transition zone greater pressures promote denser packing of oxygen such that some silicon is surrounded by six oxygen atoms, resulting in magnesium silicates with garnet and pyroxene structures. Deeper still, all silicon is surrounded by six oxygen atoms so that the new mineral $MgSiO_3$-perovskite predominates.

Mantoux test in medicine, test used to diagnose ◊tuberculosis. A small dose of tuberculin, a product obtained from killed *Mycobacterium tuberculosis* cells, is injected into the superficial layers of the skin. Reddening and inflammation of the skin will occur if the individual has been exposed to *M. tuberculosis*. This reaction does not confirm that the individual is suffering from active tuberculosis.

map diagrammatic representation of an area – for example, part of the Earth's surface or the distribution of the stars. Modern maps of the Earth are made using satellites in low orbit to take a series of overlapping stereoscopic photographs from which a three-dimensional image can be prepared. The earliest accurate large-scale maps appeared in about 1580 (see ◊atlas).

Conventional aerial photography, laser beams, microwaves, and infrared equipment are also used for land surveying. Many different kinds of ◊map projection (the means by which a three-dimensional body is shown in two dimensions) are used in map-making. Detailed maps requiring constant updating are kept in digital form on computer so that minor revisions can be made without redrafting.

map projection ways of depicting the spherical surface of the Earth on a flat piece of paper. Traditional projections include the **conic**, **azimuthal**, and **cylindrical**. The most famous cylindrical projection is the **Mercator projection**, which dates from 1569. The weakness of these systems is that countries in different latitudes are disproportionately large, and lines of longitude and latitude appear distorted.

In 1973 German historian Arno Peters devised the **Peters projection** in which the countries of the world retain their relative areas. In 1992 the US physicist Mitchell Feigenbaum devised the **optimal conformal** projection, using a computer program designed to take data about the boundary of a given area and calculate the projection that produces the minimum of inaccuracies.

The theory behind traditional map projection is that, if a light were placed at the centre of a transparent Earth, the surface features could be thrown as shadows on a piece of paper close to the surface.

This paper may be flat and placed on a pole (azimuthal or zenithal), or may be rolled around the equator (cylindrical), or may be in the form of a tall cone resting on the equator (conical). The resulting maps differ from one another, distorting either area or direction, and each is suitable for a particular purpose. For example, projections distorting area the least are used for distribution maps, and those with least distortion of direction are used for navigation charts.

marble rock formed by metamorphosis of sedimentary ◊limestone. It takes and retains a good polish, and is used in building and sculpture. In its pure form it is white and consists almost entirely of calcite $CaCO_3$. Mineral impurities give it various colours and patterns. Carrara, Italy, is known for white marble.

Marburg disease or **green monkey disease** viral disease of central Africa, first occurring in Europe in 1967 among research workers in Germany working with African green monkeys. Caused by a filovirus, it is characterized by haemorrhage of the mucous membranes, fever, vomiting, headache, and diarrhoea; mortality is high. It is a ◊haemorrhagic fever similar to Ebola virus disease.

mare (plural **maria**) dark lowland plain on the Moon. The name comes from Latin 'sea', because these areas were once wrongly thought to be water.

marginal land in farming, poor-quality land that is likely to yield a poor return. It is the last land to be brought into production and the first land to be abandoned. Examples are desert fringes in Africa and mountain areas in the UK.

Mariana Trench lowest region on the Earth's surface; the deepest part of the sea floor. The trench is 2,400 km/1,500 mi long and is situated 300 km/200 mi E of the Mariana Islands, in the NW Pacific Ocean. Its deepest part is the gorge known as the Challenger Deep, which extends 11,034 m/36,210 ft below sea level.

Mariana Trench

A heavy object dropped into the ocean over the 11 km/6.7 mi deep Mariana Trench in the Pacific would take over one hour to sink to the bottom.

---

marijuana dried leaves and flowers of the hemp plant ◊cannabis, used as a drug; it is illegal in most countries. Mexico is the world's largest producer.

Mariner spacecraft series of US space probes that explored the planets Mercury, Venus, and Mars between 1962 and 1975.

Mariner 1 (to Venus) had a failed launch. *Mariner 2* (1962) made the first fly-by of Venus, at 34,000 km/21,000 mi, confirmed the existence of ◊solar wind, and measured Venusian temperature. *Mariner 3* did not achieve its intended trajectory to Mars. *Mariner 4* (1965) passed Mars at a distance of 9,800 km/6,100 mi, and took photographs, revealing a dry, cratered surface. *Mariner 5* (1967) passed Venus at 4,000 km/2,500 mi, and measured Venusian temperature, atmosphere, mass, and diameter. *Mariner 6 and 7* (1969) photographed Mars' equator and southern hemisphere respectively, and also measured temperature, atmospheric pressure and composition, and diameter. *Mariner 8* (to Mars) had a failed launch. *Mariner 9* (1971) mapped the entire Martian surface, and photographed Mars' moons. Its photographs revealed the changing of the polar caps, and the extent of volcanism, canyons, and features, which suggested that there might once have been water on Mars.

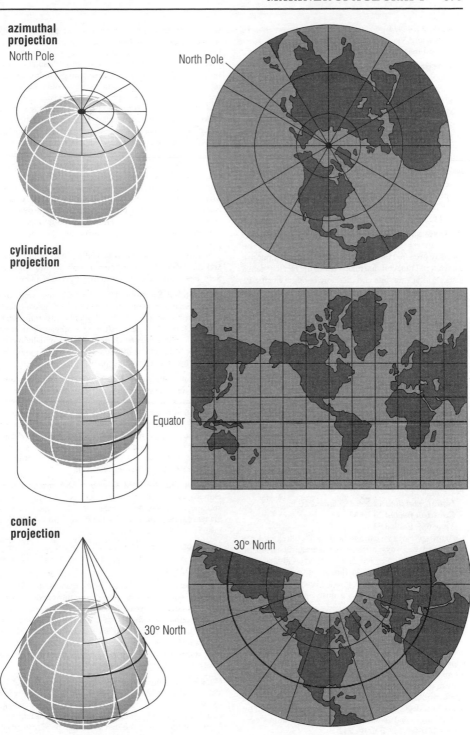

azimuthal projection

North Pole

North Pole

cylindrical projection

Equator

conic projection

30° North

30° North

map projection Three widely used map projections. If a light were placed at the centre of a transparent Earth, the shapes of the countries would be thrown as shadows on a sheet of paper. If the paper is flat, the azimuthal projection results; if it is wrapped around a cylinder or in the form of a cone, the cylindrical or conic projections result.

Mariner 10 (1974–75) took close-up photographs of Mercury and Venus, and measured temperature, radiation, and magnetic fields.

market gardening farming system that specializes in the commercial growing of vegetables, fruit, or flowers. It is an ◊intensive agriculture with crops often being grown inside greenhouses on small farms.

Market gardens may be located within easy access of markets, on the fringes of urban areas; for example, in the Home Counties for the London market. Such areas as the Channel Islands, where early crops can be grown outside because of a mild climate, are especially suitable.

Markov chain in statistics, an ordered sequence of discrete states (random variables) $x_1, x_2, ..., x_i, ..., x_n$ such that the probability of x_i depends only on n and/or the state x_{i-1} which has preceded it. If independent of n, the chain is said to be homogeneous.

mark sensing in computing, a technique that enables pencil marks made in predetermined positions on specially prepared forms to be rapidly read and input to a computer. The technique makes use of the fact that pencil marks contain graphite and therefore conduct electricity. A *mark sense reader* scans the form by passing small metal brushes over the paper surface. Whenever a brush touches a pencil mark a circuit is completed and the mark is detected.

marl crumbling sedimentary rock, sometimes called *clayey limestone*, including various types of calcareous ◊clays and fine-grained ◊limestones. Marls are often laid down in freshwater lakes and are usually soft, earthy, and of a white, grey, or brownish colour. They are used in cement-making and as fertilizer.

Mars fourth planet from the Sun. It is much smaller than Venus or Earth, with a mass 0.11 that of Earth. Mars is slightly pear-shaped, with a low, level northern hemisphere, which is comparatively uncratered and geologically 'young', and a heavily cratered 'ancient' southern hemisphere.

   Mean distance from the Sun 227.9 million km/141.6 million mi.

   Equatorial diameter 6,780 km/4,210 mi.

   Rotation period 24 hr 37 min.

   Year 687 Earth days.

   Atmosphere 95% carbon dioxide, 3% nitrogen, 1.5% argon, and 0.15% oxygen. Red atmospheric dust from the surface whipped up by winds of up to 450 kph/280 mph accounts for the light pink sky. The surface pressure is less than 1% of the Earth's atmospheric pressure at sea level.

   Surface The landscape is a dusty, red, eroded lava plain. Mars has white polar caps (water ice and frozen carbon dioxide) that advance and retreat with the seasons.

   Satellites two small satellites: ◊Phobos and Deimos.

There are four enormous volcanoes near the equator, of which the largest is Olympus Mons 24 km/15 mi high, with a base 600 km/375 mi across, and a crater 65 km/40 mi wide. To the east of the four volcanoes lies a high plateau cut by a system of valleys, Valles Marineris, some 4,000 km/2,500 mi long, up to 200 km/120 mi wide and 6 km/4 mi deep; these features are apparently caused by faulting and wind erosion. Recorded temperatures vary from −100°C/−148°F to 0°C/32°F.

Mars may approach Earth to within 54.7 million km/34 million mi. The first human-made object to orbit another planet was *Mariner 9*. *Viking 1 and 2*, which

landed, also provided much information. Studies in 1985 showed that enough water might exist to sustain prolonged missions by space crews.

A meteorite from Mars and found in Antarctica was revealed by NASA scientists to contain possible evidence of life in Aug 1996. The 15 million-year-old rock, which entered the Earth's atmosphere in the form of a meteorite 12,000 years ago, was probably broken from the surface of Mars when the planet collided with a large object in space. The meteorite was found to contain carbonate globules, on the surface of which are minute round or long and thin microorganisms. These microfossils are a hundred times smaller than the smallest known Earth microfossils. The globules also contain magnetite and iron sulphide particles comparable in shape and quantity to those produced by Earth microorganisms. Critics claim the meteorite became contaminated on arrival and by Nov 1996, two new papers had been published offering non-biological explanations for the meteorite particles.

NASA's *Mars Pathfinder* landed on Mars in July 1997 and the *Global Surveyor* entered orbit around Mars in Sept 1997. Their mission is to map the entire planet.

Mars Global Surveyor US spacecraft that went into orbit around ◊Mars on 12 September 1997 to conduct a detailed photographic survey of the planet commencing March 1998. The spacecraft used a previously untried technique called *aerobraking* to turn its initially highly elongated orbit into a 400 km/249 mi circular orbit by dipping into the outer atmosphere of the planet.

marsh low-lying wetland. Freshwater marshes are common wherever groundwater, surface springs, streams, or run-off cause frequent flooding or more or less permanent shallow water. A marsh is alkaline whereas a ◊bog is acid. Marshes develop on inorganic silt or clay soils. Rushes are typical marsh plants. Large marshes dominated by papyrus, cattail, and reeds, with standing water throughout the year, are commonly called ◊swamps. Near the sea, ◊salt marshes may form.

Marshall Space Flight Center NASA installation at Huntsville, Alabama, where the series of ◊Saturn rockets and the space-shuttle engines were developed. It also manages various payloads for the space shuttle, including the ◊Spacelab space station.

marsh gas gas consisting mostly of ◊methane. It is produced in swamps and marshes by the action of bacteria on dead vegetation.

Mars Observer NASA space probe launched in 1992 to orbit Mars and survey the planet, its atmosphere, and the polar caps over two years. The probe was also scheduled to communicate information from the robot vehicles delivered by Russia's Mars 94 mission. The $1 billion project miscarried, however, when the probe unaccountably stopped transmitting in Aug 1993, three days before it was due to drop into orbit.

Mars Pathfinder US spacecraft that landed in the Ares Vallis region of ◊Mars on 4 July 1997. It carried a small six-wheeled roving vehicle called *Sojourner* which examined rock and soil samples around the landing site. Mars Pathfinder was the first to use air bags instead of retro-rockets to cushion the landing

marsupial mammal in which the female has a pouch where she carries her young (born tiny and immature) for

Life on Mars?

For nearly a century, since the US astronomer Percival Lowell mapped what he claimed was a network of artificial canals criss-crossing the surface of Mars, the possibility of life on other planets has seized the imagination of public and scientists alike. Lowell's vision of an inhabited Mars spawned a genre of science-fiction stories, most notably H G Wells's *War of the Worlds*. Other astronomers, though, failed to see the canals and dismissed them as optical illusions.

More recently, the twin *Viking* space probes that landed on Mars in 1976 analysed the soil and reported it lifeless. Conditions on the planet are too harsh for life to exist today. So there was considerable surprise when, in August 1996, a team of American scientists announced that they had at last found what had eluded all previous searches – evidence of life on Mars, albeit long-dead.

Rocks in ice

The evidence lay in a 1.9 kg/4.2 lb rock from space which was found in the Allan Hills region of Antarctica in 1984; it was catalogued as sample ALH84001. Antarctica is a rich hunting ground for meteorites, because the extraterrestrial rocks lie preserved in deep freeze on the surface of the ice for many thousands of years after they fall.

Most meteorites are fragments from asteroids, but a dozen of them found at various places around the world are now identified as having come from Mars, on the basis of their composition. This one showed evidence of having been blasted off the surface of Mars by a giant impact 16 million years ago, and aimlessly circling the Sun before it encountered the Earth 13,000 years ago.

Meteorite ALH84001 was once part of the original crust of Mars. Detailed analysis shows that it formed 4,500 million years ago and was subsequently immersed under water when the climate on Mars was warmer and wetter than it is today. Hence it is a prime sample in which to find evidence of ancient life on Mars, should there have been any.

PAHs and microfossils

A team headed by David S McKay of NASA's Johnson Space Center in Houston, analysed the meteorite and found that it contained certain compounds, particularly complex molecules known as polycyclic aromatic hydrocarbons (PAHs), which are associated with life. On their own, these compounds would not be strong evidence for life on Mars, since they can be synthesized by normal chemical processes, too.

What tipped the balance was the discovery of wormlike structures, only a few millionths of an inch long, that the team interpreted as fossils of microscopic bacteria that once lived on Mars, probably about 3,600 million years ago.

However, the NASA team were unable to analyse the chemical composition of these supposed microfossils because they were so small, and so the team could not rule out the possibility that they are in fact composed of minerals that happen to have taken up a bacterium-like shape. One argument against the microfossil interpretation is that the structures are 'too' small – 100 times smaller than microfossils found in Earth rocks.

Too soon to say

Not surprisingly, the announcement of the discovery caused controversy in scientific circles. Even an optimist about the possibility of life on other worlds, Carl Sagan, called it 'not proven', while John F Kerridge, a planetary scientist at the University of California, San Diego, said, 'The conclusion is at best premature and more probably wrong'.

'We don't claim that we have conclusively proven it,' said team member Dr Everett Gibson. 'We are putting this evidence out to the scientific community for other investigators to verify, enhance, attack – disprove if they can – as part of the scientific process. Then, within a year or two, we hope to resolve the question one way or the other.'

Back to the source

The findings came at a good time for NASA, which was renewing its Mars exploration programme with the launch of two probes in late 1996: *Mars Global Surveyor*, an orbiter, and *Mars Pathfinder*, which landed a small rover on the surface in July 1997. Ultimately there will be a mission to bring back samples automatically from Mars, perhaps ten years from now. Possible target areas would be two unnamed craters, photographed 20 years ago by the *Viking* orbiter spacecraft, where meteorite ALH84001 could have been ejected from Mars. One crater is 20 km/12.4 mi across and the other is half that size, both in the planet's southern highlands. Both craters have the right age to be the source of ALH84001.

Ian Ridpath

a considerable time after birth. Marsupials include omnivorous, herbivorous, and carnivorous species, among them the kangaroo, wombat, opossum, phalanger, bandicoot, dasyure, and wallaby.

The Australian marsupial anteater known as the numbat is an exception to the rule in that it has no pouch.

maser (acronym for *microwave amplification by stimulated emission of radiation*) in physics, a high-frequency microwave amplifier or oscillator in which the signal to be amplified is used to stimulate excited atoms into emitting energy at the same frequency. Atoms or molecules are raised to a higher energy level and then allowed to lose this energy by radiation emitted at a precise frequency. The principle has been extended to other parts of the electromagnetic spectrum as, for example, in the ◊laser.

The two-level ammonia-gas maser was first suggested in 1954 by US physicist Charles Townes at Columbia University, New York, and independently the same year by Nikolai Basov and Aleksandr Prokhorov in Russia. The solid-state three-level maser, the most sensitive amplifier known, was envisaged by Nicolaas Bloembergen (1920–) at Harvard in 1956. The ammonia maser is used as a frequency standard oscillator (see ◊clock), and the three-level maser as a receiver for satellite communications and radio astronomy.

masochism desire to subject oneself to physical or mental pain, humiliation, or punishment, for erotic pleasure, to alleviate guilt, or out of destructive impulses turned inward. The term is derived from Leopold von Sacher-Masoch.

mass in physics, the quantity of matter in a body as measured by its inertia. Mass determines the acceleration produced in a body by a given force acting on it, the acceleration being inversely proportional to the mass of the body. The mass also determines the force exerted on a body by ◊gravity on Earth, although this attraction varies slightly from place to place. In the SI system, the base unit of mass is the kilogram.

At a given place, equal masses experience equal gravitational forces, which are known as the weights of the bodies. Masses may, therefore, be compared by comparing the weights of bodies at the same place.

The standard unit of mass to which all other masses are compared is a platinum-iridium cylinder of 1 kg, which is kept at the International Bureau of Weights and Measures in Sèvres, France.

mass action, law of in chemistry, a law stating that at a given temperature the rate at which a chemical reaction takes place is proportional to the product of the active masses of the reactants. The active mass is taken to be the molar concentration of the each reactant.

massage manipulation of the soft tissues of the body, the muscles, ligaments, and tendons, either to encourage the healing of specific injuries or to produce the general beneficial effects of relaxing muscular tension, stimulating blood circulation, and improving the tone and strength of the skin and muscles.

The benefits of massage were known to the ancient Chinese, Egyptian, and Greek cultures. The techniques most widely practised today were developed by the Swedish physician Per Henrik Ling (1776–1838).

mass–energy equation Einstein's equation $E = mc^2$, denoting the equivalence of mass and energy, where E is the energy in joules, m is the mass in kilograms, and c is the speed of light, in a vacuum, in metres per second.

mass number or *nucleon number* sum (symbol A) of the numbers of protons and neutrons in the nucleus of an atom. It is used along with the ◊atomic number (the number of protons) in ◊nuclear notation: in symbols that represent nuclear isotopes, such as $^{14}_{6}C$, the lower number is the atomic number, and the upper number is the mass number.

mass production manufacture of goods on a large scale, a technique that aims for low unit cost and high output. In factories mass production is achieved by a variety of means, such as division and specialization of labour and mechanization. These speed up production and allow the manufacture of near-identical, interchangeable parts. Such parts can then be assembled quickly into a finished product on an assembly line.

Division of labour means that a job is divided into a number of steps, and then groups of workers are employed to carry each step out, specializing and therefore doing the job in a routine way, producing more than if each individually had to carry out all the stages of manufacture. However, the system has been criticized for neglecting the skills of workers and removing their involvement with the end product.

Many of the machines now used in factories are robots (for example, on car-assembly lines): they work automatically under computer control. Such automation further streamlines production and raises output.

mass spectrometer in physics, an apparatus for analysing chemical composition. Positive ions (charged particles) of a substance are separated by an electromagnetic system, which permits accurate measurement of the relative concentrations of the various ionic masses present, particularly isotopes.

mast cell in medicine, one of the histamine-containing cells involved in the production of inflammatory reactions, in response to an external trigger factor. For example, cells in the airways or in the bronchial epithelium release ◊histamine and other mediators in response to allergic stimuli, such as allergens, infection, stress, and exercise.

match small strip of wood or paper, tipped with combustible material for producing fire. Friction matches containing phosphorus were first made in 1816 in France by François Derosne.

A *safety match* is one in which the oxidizing agent and the combustible body are kept apart, the former being incorporated into the striking surface on the side of the box, the latter into the match. Safety matches were patented by a Swede, J E Lundström, in 1855. Book matches were invented in the USA in 1892 by Joshua Pusey.

mathematical induction formal method of proof in which the proposition $P(n + 1)$ is proved true on the hypothesis that the proposition $P(n)$ is true. The proposition is then shown to be true for a particular value of n, say k, and therefore by induction the proposition must be true for $n = k + 1, k + 2, k + 3, ...$. In many cases $k = 1$, so then the proposition is true for all positive integers.

mathematics science of relationships between numbers, between spatial configurations, and abstract structures.

Mathematics: Chronology

c. 2500 BC	The people of Mesopotamia (now Iraq) developed a positional numbering (place-value) system, in which the value of a digit depends on its position in a number.
c. 2000	Mesopotamian mathematicians solved quadratic equations (algebraic equations in which the highest power of a variable is 2).
876	A symbol for zero was used for the first time, in India.
c. 550	Greek mathematician Pythagoras formulated a theorem relating the lengths of the sides of a right-angled triangle. The theorem was already known by earlier mathematicians in China, Mesopotamia, and Egypt.
c. 450	Hipparcos of Metapontum discovered that some numbers are irrational (cannot be expressed as the ratio of two integers).
300	Euclid laid out the laws of geometry in his book *Elements*, which was to remain a standard text for 2,000 years.
c. 230	Eratosthenes developed a method for finding all prime numbers.
c. 100	Chinese mathematicians began using negative numbers.
c. 190	Chinese mathematicians used powers of 10 to express magnitudes.
AD c. 210	Diophantus of Alexandria wrote the first book on algebra.
c. 600	A decimal number system was developed in India.
829	Persian mathematician Muhammad ibn-Mūsā al-Khwārizmī published a work on algebra that made use of the decimal number system.
1202	Italian mathematician Leonardo Fibonacci studied the sequence of numbers (1, 1, 2, 3, 5, 8, 13, 21, ...) in which each number is the sum of the two preceding ones.
1550	In Germany, Rheticus published trigonometrical tables that simplified calculations involving triangles.
1614	Scottish mathematician John Napier invented logarithms, which enable lengthy calculations involving multiplication and division to be carried out by addition and subtraction.
1623	Wilhelm Schickard invented the mechanical calculating machine.
1637	French mathematician and philosopher René Descartes introduced coordinate geometry.
1654	In France, Blaise Pascal and Pierre de Fermat developed probability theory.
1666	Isaac Newton developed differential calculus, a method of calculating rates of change.
1675	German mathematician Gottfried Wilhelm Leibniz introduced the modern notation for integral calculus, a method of calculating volumes.
1679	Leibniz introduced binary arithmetic, in which only two symbols are used to represent all numbers.
1684	Leibniz published the first account of differential calculus.
1718	Jakob Bernoulli in Switzerland published his work on the calculus of variations (the study of functions that are close to their minimum or maximum values).
1742	German mathematician Christian Goldbach conjectured that every even number greater than two can be written as the sum of two prime numbers. Goldbach's conjecture has still not been proven.
1746	In France, Jean le Rond d'Alembert developed the theory of complex numbers.
1747	D'Alembert used partial differential equations in mathematical physics.
1798	Norwegian mathematician Caspar Wessel introduced the vector representation of complex numbers.
1799	Karl Friedrich Gauss of Germany proved the fundamental theorem of algebra: the number of solutions of an algebraic equation is the same as the exponent of the highest term.
1810	In France, Jean Baptiste Joseph Fourier published his method of representing functions by a series of trigonometric functions.
1812	French mathematician Pierre Simon Laplace published the first complete account of probability theory.
1822	In the UK, Charles Babbage began construction of the first mechanical computer, the difference machine, a device for calculating logarithms and trigonometric functions.
1827	Gauss introduced differential geometry, in which small features of curves are described by analytical methods.
1829	In Russia, Nikolai Ivanonvich Lobachevsky developed hyperbolic geometry, in which a plane is regarded as part of a hyperbolic surface, shaped like a saddle. In France, Evariste Galois introduced the theory of groups (collections whose members obey certain simple rules of addition and multiplication).
1844	French mathematician Joseph Liouville found the first transcendental number, which cannot be expressed as an algebraic equation with rational coefficients. In Germany, Hermann Grassmann studied vectors with more than three dimensions.
1854	George Boole in the UK published his system of symbolic logic, now called Boolean algebra.
1858	English mathematician Arthur Cayley developed calculations using ordered tables called matrices.
1865	August Ferdinand Möbius in Germany described how a strip of paper can have only one side and one edge.
1892	German mathematician Georg Cantor showed that there are different kinds of infinity and studied transfinite numbers.
1895	Jules Henri Poincaré published the first paper on topology, often called 'the geometry of rubber sheets'.
1931	In the USA, Austrian-born mathematician Kurt Gödel proved that any formal system strong enough to include the laws of arithmetic is either incomplete or inconsistent.
1937	English mathematician Alan Turing published the mathematical theory of computing.
1944	John Von Neumann and Oscar Morgenstern developed game theory in the USA.

Mathematics: Chronology – *continued*

1945	The first general purpose, fully electronic digital computer, ENIAC (electronic numerator, integrator, analyser, and computer), was built at the University of Pennsylvania, USA.
1961	Meteorologist Edward Lorenz at the Massachusetts Institute of Technology, USA, discovered a mathematical system with chaotic behaviour, leading to a new branch of mathematics – chaos theory.
1962	Benoit Mandelbrot in the USA invented fractal images, using a computer that repeats the same mathematical pattern over and over again.
1975	US mathematician Mitchell Feigenbaum discovered a new fundamental constant (approximately 4.669201609103), which plays an important role in chaos theory.
1980	Mathematicians worldwide completed the classification of all finite and simple groups, a task that took over a hundred mathematicians more than 35 years to complete and whose results took up more than 14,000 pages in mathematical journals.
1989	A team of US computer mathematicians at Amdahl Corporation, California, discovered the highest known prime number (it contains 65,087 digits).
1993	British mathematician Andrew Wiles published a 1,000-page proof of Fermat's last theorem, one of the most baffling challenges in pure mathematics.
1996	Wiles's proof was accepted after revision.
1997	The largest number to be fractorized to date has 167 digits: $(3^{349}-1)/2$ was split into its 80 and 87-digit factors by a team of US mathematicians after 100,000 hours of computing.

The main divisions of **pure mathematics** include geometry, arithmetic, algebra, calculus, and trigonometry. Mechanics, statistics, numerical analysis, computing, the mathematical theories of astronomy, electricity, optics, thermodynamics, and atomic studies come under the heading of **applied mathematics**.

Prehistoric human beings probably learned to count at least up to ten on their fingers. The Chinese, Hindus, Babylonians, and Egyptians all devised methods of counting and measuring that were of practical importance in their everyday lives. The first theoretical mathematician is held to be Thales of Miletus (c. 580 BC) who is believed to have proposed the first theorems in plane geometry. His disciple Pythagoras established geometry as a recognized science among the Greeks.

The later school of Alexandrian geometers (4th and 3rd centuries BC) included Euclid and Archimedes. Our present decimal numerals are based on a Hindu–Arabic system that reached Europe about AD 100 from Arab mathematicians of the Middle East such as Khwārizmī.

Western mathematics began to develop from the 15th century. Geometry was revitalized by the invention of coordinate geometry by René Descartes in 1637; Blaise Pascal and Pierre de Fermat developed probability theory; John Napier invented logarithms; and Isaac Newton and Gottfried Leibniz invented calculus, later put on a more rigorous footing by Augustin Cauchy. In Russia, Nikolai Lobachevsky rejected Euclid's parallelism and developed a non-Euclidean geometry; this was subsequently generalized by Bernhard Riemann and later utilized by Einstein in his theory of relativity. In the mid-19th century a new major theme emerged: investigation of the logical foundations of mathematics. George Boole showed how logical arguments could be expressed in algebraic symbolism. Friedrich Ludwig Gottlob Frege and Giuseppe Peano considerably developed this symbolic logic.

In the 20th century, mathematics has become much more diversified. Each specialist subject is being studied in far greater depth and advanced work in some fields may be unintelligible to researchers in other fields. Mathematicians working in universities have had the economic freedom to pursue the subject for its own sake. Nevertheless, new branches of mathematics have been developed which are of great practical importance and which have basic ideas simple enough to be taught in secondary schools. Probably the most important of these is the mathematical theory of statistics in which much pioneering work was done by Karl Pearson. Another new development is operations research, which is concerned with finding optimum courses of action in practical situations, particularly in economics and management. As in the late medieval period, commerce began to emerge again as a major impetus for the development of mathematics.

Higher mathematics has a powerful tool in the high-speed electronic computer, which can create and manipulate mathematical 'models' of various systems in science, technology, and commerce.

Modern additions to school syllabuses such as sets, group theory, matrices, and graph theory are sometimes referred to as 'new' or 'modern' mathematics.

Et harum scientarum porta et clavis est Mathematica.

Mathematics is the door and the key to the sciences.

On **mathematics** Roger Bacon
Opus Majus part 4 *Distinctia Prima* cap 1,
1267 transl Robert Belle Burke, 1928

---

matrix in biology, usually refers to the ◊extracellular matrix.

matrix in mathematics, a square ($n \times n$) or rectangular ($m \times n$) array of elements (numbers or algebraic variables). They are a means of condensing information about mathematical systems and can be used for, among other things, solving simultaneous linear equations (see ◊simultaneous equations and ◊transformation).

Much early matrix theory was developed by the British mathematician Arthur Cayley, although the term was coined by his contemporary James Sylvester (1814–1897).

matter in physics, anything that has mass and can be detected and measured. All matter is made up of ◊atoms, which in turn are made up of ◊elementary particles; it exists ordinarily as a solid, liquid, or gas. The history of science and philosophy is largely taken up with accounts of theories of matter, ranging from the hard 'atoms' of Democritus to the 'waves' of modern quantum theory.

There are living systems; there is no 'living matter'.

On **matter** Jacques Monod,
in a lecture in Nov 1967

Mauna Kea astronomical observatory in Hawaii, USA, built on a dormant volcano at 4,200 m/13,784 ft above sea level. Because of its elevation high above clouds, atmospheric moisture, and artificial lighting, Mauna Kea is ideal for infrared astronomy. The first telescope on the site was installed in 1970.

Telescopes include the 2.24 m/88 in University of Hawaii reflector (1970). In 1979 three telescopes were erected: the 3.8 m/150 in United Kingdom Infrared Telescope (UKIRT) (also used for optical observations); the 3 m/120 in NASA Infrared Telescope Facility (IRTF); and the 3.6 m/142 in Canada–France–Hawaii Telescope (CFHT), designed for optical and infrared work. The 15 m/50 ft diameter UK/Netherlands James Clerk Maxwell Telescope (JCMT) is the world's largest telescope specifically designed to observe millimetre wave radiation from nebulae, stars, and galaxies. The JCMT is operated via satellite links by astronomers in Europe. The world's largest optical telescope, the ◊Keck Telescope, is also situated on Mauna Kea.

In 1996 the capacity of the JCMT was enhanced by the addition of SCUBA (Submillimetre Common-user Bolometer Array). SCUBA is a camera comprised of numerous detectors cooled to within a tenth of a degree of absolute zero (0 K) and is the world's most sensitive instrument at the 0.3–1.0 mm wavelength.

Maunder minimum in astronomy, the period between 1645–1715 when ◊sunspots were rarely seen and no ◊aurorae ('northern lights') were recorded. The Maunder minimum coincided with a time of unusually low temperature on Earth, known as the Little Ice Age, and is often taken as evidence that changes in solar activity can affect Earth's climate. The Maunder minimum is named after the English astronomer E W Maunder, who drew attention to it.

maximum and minimum in ◊coordinate geometry, points at which the slope of a curve representing a ◊function changes from positive to negative (maximum), or from negative to positive (minimum). A tangent to the curve at a maximum or minimum has zero gradient.

Maxima and minima can be found by differentiating the function for the curve and setting the differential to zero (the value of the slope at the turning point). For example, differentiating the function for the ◊parabola $y = 2x^2 - 8x$ gives $dy/dx = 4x - 8$. Setting this equal to zero gives $x = 2$, so that $y = -8$ (found by substituting $x = 2$ into the parabola equation). Thus the function has a minimum at the point (2, −8).

maxwell c.g.s. unit (symbol Mx) of magnetic flux (the strength of a ◊magnetic field in an area multiplied by the area). It is now replaced by the SI unit, the ◊weber (one maxwell equals 10^{-8} weber).

The maxwell is a very small unit. It is equal to the flux through one square centimetre normal to a magnetic field with an intensity of one gauss.

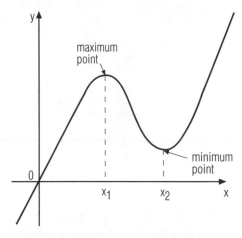

maximum and minimum A maximum point on a curve is higher than the points immediately on either side of it; it is not necessarily the highest point on the curve. Similarly, a minimum point is lower than the points immediately on either side.

Maxwell–Boltzmann distribution in physics, a statistical equation describing the distribution of velocities among the molecules of a gas. It is named after James Maxwell and Ludwig Boltzmann, who derived the equation, independently of each other, in the 1860s.

One form of the distribution is $n = Ne(-E/RT)$, where N is the total number of molecules, n is the number of molecules with energy in excess of E, T is the absolute temperature (temperature in kelvin), R is the ◊gas constant, and e is the exponential constant.

Maxwell's screw rule in physics, a rule formulated by Scottish physicist James Maxwell that predicts the direction of the magnetic field produced around a wire carrying electric current. It states that if a right-handed screw is turned so that it moves forwards in the same direction as the current, its direction of rotation will give the direction of the magnetic field. *See illustration overleaf.*

MD abbreviation for *Doctor of Medicine*.

MDMA (3,4-methylenedio-xymethamphetamine) psychedelic drug, also known as ◊ecstasy.

ME abbreviation for *myalgic encephalomyelitis*, a popular name for ◊chronic fatigue syndrome.

mean in mathematics, a measure of the average of a number of terms or quantities. The simple **arithmetic mean** is the average value of the quantities, that is, the sum of the quantities divided by their number. The **weighted mean** takes into account the frequency of the terms that are summed; it is calculated by multiplying each term by the number of times it occurs, summing the results and dividing this total by the total number of occurrences. The **geometric mean** of n quantities is the nth root of their product. In statistics, it is a measure of central tendency of a set of data.

meander loop-shaped curve in a river flowing across flat country. As a river flows, any curve in its course is accentuated by the current. The current is fastest on the outside of the curve where it cuts into the bank; on the curve's

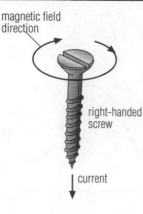

magnetic field
direction

right-handed
screw

current

Maxwell's screw rule Maxwell's screw rule, named after
the physicist James Maxwell, predicts the direction of the
magnetic field produced around a wire carrying electric
current. If a right-handed screw is turned so that it moves
forward in the same direction as the current, its direction of
rotation will give the direction of the magnetic field.

inside the current is slow and deposits any transported
material. In this way the river changes its course across the
flood plain.

A loop in a river's flow may become so accentuated that
it becomes cut off from the normal course and forms an
◊oxbow lake. The word comes from the river Menderes
in Turkey.

mean deviation in statistics, a measure of the spread of a
population from the ◊mean.

mean free path in physics, the average distance travelled
by a particle, atom, or molecule between successive
collisions. It is of importance in the ◊kinetic theory of
gases.

mean life in nuclear physics, the average lifetime of a
nucleus of a radioactive isotope equal to 1.44 times the
◊half-life. See ◊radioactivity.

measles acute virus disease (rubeola), spread by airborne infection.

Symptoms are fever, severe catarrh, small spots inside
the mouth, and a raised, blotchy red rash appearing for
about a week after two weeks' incubation. Prevention is
by vaccination.

In industrialized countries it is not usually a serious
disease, though serious complications may develop. More
than 1 million children a year die of measles (1995);
a high percentage of them are Third World children.
The North and South American Indians died by the
thousands in epidemics in the 17th, 18th, and 19th
centuries.

mechanical advantage (MA) in physics, the number
of times the load moved by a machine is greater than
the effort applied to that machine. In equation terms:
MA = load/effort.

The exact value of a working machine's MA is always
less than its predicted value because there will always be
some frictional resistance that increases the effort necessary to do the work.

mechanical equivalent of heat in physics, a constant
factor relating the calorie (the c.g.s. unit of heat) to the

joule (the unit of mechanical energy), equal to 4.1868
joules per calorie. It is redundant in the SI system of units,
which measures heat and all forms of energy in ◊joules (so
that the mechanical equivalent of heat is 1).

mechanics branch of physics dealing with the motions
of bodies and the forces causing these motions, and also
with the forces acting on bodies in ◊equilibrium. It is
usually divided into ◊dynamics and ◊statics.

Quantum mechanics is the system based on the
◊quantum theory, which has superseded Newtonian
mechanics in the interpretation of physical phenomena
on the atomic scale.

mechanization the use of machines in place of manual
labour or the use of animals. Until the 1700s there were
few machines available to help people in the home, on the
land, or in industry. There were no factories, only cottage
industries, in which people carried out work, such as
weaving, in their own homes for other people. The 1700s
saw a long series of inventions, initially in the textile
industry, that ushered in a machine age and brought
about the Industrial Revolution.

Among the first inventions in the textile industry were
those made by John Kay (flying shuttle, 1773), James
Hargreaves (spinning jenny, 1764), and Richard Arkwright (water frame, 1769). Arkwright pioneered the
mechanized factory system by installing many of his spinning machines in one building and employing people to
work them.

media (singular *medium*) in computing, the collective
name for materials on which data can be recorded. For
example, paper is a medium that can be used to record
printed data; a floppy disc is a medium for recording
magnetic data.

median in mathematics and statistics, the middle number
of an ordered group of numbers. If there is no middle
number (because there is an even number of terms),
the median is the ◊mean (average) of the two middle
numbers. For example, the median of the group 2, 3, 7,
11, 12 is 7; that of 3, 4, 7, 9, 11, 13 is 8 (the average of
7 and 9).

In geometry, the term refers to a line from the vertex of
a triangle to the midpoint of the opposite side.

medical ethics moral guidelines for doctors governing
good professional conduct. The basic aims are considered to be doing good, avoiding harm, preserving the
patient's autonomy, telling the truth, and pursuing justice. Ethical issues provoke the most discussion in medicine where these five aims cannot be simultaneously

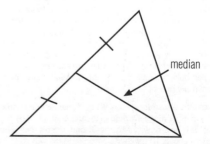

median

median The median is the name given to a line from the
vertex (corner) of a triangle to the mid-point of the opposite
side.

achieved – for example, what is 'good' for a child may clash with his or her autonomy or that of the parents.

Traditionally these principles have been set out in the Hippocratic Oath (introduced by Greek physician Hippocrates and including such injunctions as the command to preserve confidentiality, to help the sick to the best of one's ability, and to refuse fatal draughts), but in the late 20th century rapidly advancing technology has raised the question of how far medicine should intervene in natural processes.

medicine the practice of preventing, diagnosing, and treating disease, both physical and mental; also any substance used in the treatment of disease. The basis of medicine is anatomy (the structure and form of the body) and physiology (the study of the body's functions).

In the West, medicine increasingly relies on new drugs and sophisticated surgical techniques, while diagnosis of disease is more and more by noninvasive procedures. The time and cost of Western-type medical training makes it inaccessible to many parts of the Third World; where health care of this kind is provided it is often by auxiliary medical helpers trained in hygiene and the administration of a limited number of standard drugs for the prevalent diseases of a particular region.

medicine, alternative forms of medical treatment that do not use synthetic drugs or surgery in response to the symptoms of a disease, but aim to treat the patient as a whole (◊holism). The emphasis is on maintaining health (with diet and exercise) and on dealing with the underlying causes rather than just the symptoms of illness. It may involve the use of herbal remedies and techniques like ◊acupuncture, ◊homoeopathy, and ◊chiropractic. Some

Medicine, Western: Chronology

c. 400 BC	Hippocrates recognized that disease had natural causes.
c. AD 200	Galen consolidated the work of the Alexandrian doctors.
1543	Andreas Vesalius gave the first accurate account of the human body.
1628	William Harvey discovered the circulation of the blood.
1768	John Hunter began the foundation of experimental and surgical pathology.
1785	Digitalis was used to treat heart disease; the active ingredient was isolated 1904.
1798	Edward Jenner published his work on vaccination.
1877	Patrick Manson studied animal carriers of infectious diseases.
1882	Robert Koch isolated the bacillus responsible for tuberculosis.
1884	Edwin Klebs isolated the diphtheria bacillus.
1885	Louis Pasteur produced a vaccine against rabies.
1890	Joseph Lister demonstrated antiseptic surgery.
1895	Wilhelm Röntgen discovered X-rays.
1897	Martinus Beijerinck discovered viruses.
1899	Felix Hoffman developed aspirin; Sigmund Freud founded psychiatry.
1900	Karl Landsteiner identified the first three blood groups, later designated A, B, and O.
1910	Paul Ehrlich developed the first specific antibacterial agent, Salvarsan, a cure for syphilis.
1922	Insulin was first used to treat diabetes.
1928	Alexander Fleming discovered penicillin.
1932	Gerhard Domagk discovered the first antibacterial sulphonamide drug, Prontosil.
1937	Electro-convulsive therapy (ECT) was developed.
1940s	Lithium treatment for manic-depressive illness was developed.
1950s	Antidepressant drugs and beta-blockers for heart disease were developed. Manipulation of the molecules of synthetic chemicals became the main source of new drugs. Peter Medawar studied the body's tolerance of transplanted organs and skin grafts.
1950	Proof of a link between cigarette smoking and lung cancer was established.
1953	Francis Crick and James Watson announced the structure of DNA. Jonas Salk developed a vaccine against polio.
1958	Ian Donald pioneered diagnostic ultrasound.
1960s	A new generation of minor tranquillizers called benzodiazepines was developed.
1967	Christiaan Barnard performed the first human heart-transplant operation.
1971	Viroids, disease-causing organisms even smaller than viruses, were isolated outside the living body.
1972	The CAT scan, pioneered by Godfrey Hounsfield, was first used to image the human brain.
1975	César Milstein developed monoclonal antibodies.
1978	World's first 'test-tube baby' was born in the UK.
1980s	AIDS (acquired immune-deficiency syndrome) was first recognized in the USA. Barbara McClintock's discovery of the transposable gene was recognized.
1980	The World Health Organization reported the eradication of smallpox.
1983	The virus responsible for AIDS, now known as human immunodeficiency virus (HIV), was identified by Luc Montagnier at the Institut Pasteur, Paris; Robert Gallo at the National Cancer Institute, Maryland, USA discovered the virus independently 1984.
1984	The first vaccine against leprosy was developed.
1987	The world's longest-surviving heart-transplant patient died in France, 18 years after his operation.
1989	Grafts of fetal brain tissue were first used to treat Parkinson's disease.
1990	Gene for maleness discovered by UK researchers.
1991	First successful use of gene therapy (to treat severe combined immune deficiency) was reported in the USA.
1993	First trials of gene therapy against cystic fibrosis took place in the USA.
1996	An Australian man, Ben Dent, was the first person to end his life by legally sanctioned euthanasia.

Ethics in medicine

With advances in medical science, doctors often face agonizing moral decisions. What, for instance, should doctors do about a baby born with its brain outside its head? Many people would agree that it would be wrong to prolong the life of such a baby, but if a doctor lets the baby die, is this the same as killing it?

The moral dilemma
With less severe forms of handicap, the issue is even less clear. The quality of the lives of Down's syndrome children is usually not unbearably awful, and they can live enjoyable and serene lives. But what about a Down's Syndrome baby born with a life-threatening condition, such as a serious intestinal blockage?

Some years ago, the parents of just such a child – baby Alexandra – wanted her to die, believing that nature had 'made its own arrangements to terminate a life which could not be fruitful'. The doctor would not operate without their consent. Eventually, the local social services department succeeded in persuading the courts to overturn the parents' decision.

The case of Dr Arthur
A rather different judgement was given in the case of Dr Arthur, a senior consultant paediatrician in Derby. He was asked to examine a newly born Down's Syndrome baby, whose parents were very distressed. Dr Arthur noted: 'Parents do not wish it to survive. Nursing care only.' He prescribed a sedative to alleviate the baby's distress, as and when it arose. Three days later, the baby died of broncho-pneumonia.

Dr Arthur was charged with murder; but the prosecution's case fell apart when it emerged that a post-mortem examination had shown that the baby had potentially fatal heart, lung, and brain defects. Dr Arthur was then charged with attempted murder, but was acquitted. The judge observed that it was lawful to treat a baby with sedation and nursing care only, if certain criteria were met – that the child was rejected by its parents and was irreversibly disabled.

Dominant themes
There are, of course, many other issues in medical ethics. Here, there is space to mention only the main themes:

1. *respect for life*: Since we generally accept that we may kill other people in war and in self-defence, is life absolutely sacred or not? Is it right to preserve life if the quality of that life is likely to be poor? Should doctors preserve life at all costs and by extraordinary means?

And if not, where should doctors draw the line? Is switching off a life-support machine the same as murder?

2. *patient autonomy*: This theme arises in a variety of medical contexts – such as medical confidentiality, voluntary euthanasia, or experimentation on humans. Since patients must give informed consent to an experimental procedure, how much information is the patient to be given, and how detailed should it be? The problem is particularly acute in the case of randomly controlled trials which require the patient's agreement to receiving at random alternative forms of treatment.

3. *resource allocation*: What are the ethical and rational principles of distribution for medical resources? Where expensive but life-saving treatments are available, how should they be rationed? When kidney dialysis machines were novel and scarce, some doctors treated patients on a 'first come, first served' basis, while others gave precedence to younger patients or those with dependents.

4. *reproductive technologies*: Should a doctor prescribe the Pill to a 12-year-old girl? Should post-menopausal women have fertility treatment? Should parents be able to choose the sex of their babies? As genetic engineering and screening become available, what sort of people should there be? As we enter the 21st century, this is perhaps the most challenging ethical issue of all.

Moral philosophy in medical ethics
Medical ethics has been described, rather unfairly, as an adventure playground for moral philosophers. To anyone unfamiliar with philosophy, the fact that different philosophers come to different conclusions about the same problems in medical ethics is frustrating. But, then, philosophers can often do little more than clarify the questions to be answered or considered by busy doctors who frequently have to make difficult decisions under stress.

Moreover, in society at large, no single moral outlook currently prevails on matters of birth, of death, of suffering and of sexuality: there is moral pluralism, and a lack of moral consensus. However, by clarifying the issues, philosophers can help to build a consensus in medical ethics.

P M Rowntree

alternative treatments are increasingly accepted by ortho-dox medicine, but the absence of enforceable standards in some fields has led to the proliferation of eccentric or untrained practitioners.

medulla central part of an organ. In the mammalian kidney, the medulla lies beneath the outer cortex and is responsible for the reabsorption of water from the filtrate. In plants, it is a region of packing tissue in the centre of the stem. In the vertebrate brain, the medulla is the posterior region responsible for the coordination of basic activities, such as breathing and temperature control.

medusa the free-swimming phase in the life cycle of a coelenterate, such as a jellyfish or ◊coral. The other phase is the sedentary **polyp**.

meerschaum aggregate of minerals, usually the soft white clay mineral **sepiolite**, hydrous magnesium sili-cate. It floats on water and is used for making pipe bowls.

mega- prefix denoting multiplication by a million. For example, a megawatt (MW) is equivalent to a million watts.

megabyte (MB) in computing, a unit of memory equal to 1,024 ◊kilobytes.

It is sometimes used, less precisely, to mean 1 million bytes.

megalith prehistoric stone monument of the late Neo-lithic (New Stone Age) or early Bronze Age. Most com-mon in Europe, megaliths include single large uprights or ◊menhirs (for example, the Five Kings, Northumber-land, England); rows or **alignments** (for example, Car-nac, Brittany, France); stone circles; and the hut-like remains of burial chambers after the covering earth has disappeared, known as ◊dolmen (for example, Kits Coty, Kent, England, where only the entrance survives).

A number of explanations have been put forward for the building of megaliths during the Neolithic period in areas including Denmark, Ireland, NE Scotland, England, W France, and Spain. These range from econ-omic reasons to expressions of dominance (neo-Marxist) and symbolism. The great stone monuments at ◊Carnac in W Brittany, France; in Jersey, such as La Hougue Bie; and in W Britain and Ireland, suggest possible cultural links through trade among megalith builders whose rural economy encompassed arable farming, stockrearing, and the development of pottery and weaving.

In the later Neolithic, in Wessex, S England, the con-struction of stone monuments such as ◊Avebury and ◊Stonehenge involved large numbers of working hours and considerable organization; possibly the stone was transported over a great distance, as has been suggested in the case of the bluestone at Stonehenge, although glacial deposition is another explanation.

Changes in social structure and diversification of labour probably caused the practice of megalith building to be abandoned.

megaton one million (10^6) tons. Used with reference to the explosive power of a nuclear weapon, it is equivalent to the explosive force of one million tons of trinitrotoluene (TNT).

megavitamin therapy the administration of large doses of vitamins to combat conditions considered wholly or in part due to their deficiency.

Developed by US chemist Linus Pauling in the 1960s, and alternatively known as 'orthomolecular psychiatry', the treatment has proved effective with addicts, schizo-phrenics, alcoholics, and depressives.

meiosis in biology, a process of cell division in which the number of ◊chromosomes in the cell is halved. It only occurs in ◊eukaryotic cells, and is part of a life cycle that involves sexual reproduction because it allows the genes of two parents to be combined without the total number of chromosomes increasing.

In sexually reproducing ◊diploid animals (having two sets of chromosomes per cell), meiosis occurs during formation of the ◊gametes (sex cells, sperm and egg), so that the gametes are ◊haploid (having only one set of chromosomes). When the gametes unite during ◊fertil-ization, the diploid condition is restored. In plants, mei-osis occurs just before spore formation. Thus the spores are haploid and in lower plants such as mosses they develop into a haploid plant called a gametophyte which produces the gametes (see ◊alternation of generations). See also ◊mitosis. *See illustration overleaf.*

meitnerium synthesized radioactive element of the ◊transactinide series, symbol Mt, atomic number 109, relative atomic mass 266. It was first produced in 1982 at the Laboratory for Heavy Ion Research in Darmstadt, Germany, by fusing bismuth and iron nuclei; it took a week to obtain a single new, fused nucleus. It was named in 1997 after the Austrian-born Swedish physicist Lise Meitner. Its temporary name was unnilennium.

melamine $C_3H_6N_6$ ◊thermosetting ◊polymer based on urea–formaldehyde. It is extremely resistant to heat and is also scratch-resistant. Its uses include synthetic resins.

melanin brown pigment that gives colour to the eyes, skin, hair, feathers, and scales of many vertebrates. In humans, melanin helps protect the skin against ultraviolet radiation from sunlight. Both genetic and environmental factors determine the amount of melanin in the skin.

melanism black coloration of animal bodies caused by large amounts of the pigment melanin. Melanin is of significance in insects, because melanic ones warm more rapidly in sunshine than do pale ones, and can be more active in cool weather. A fall in temperature may stimulate such insects to produce more melanin. In industrial areas, dark insects and pigeons match sooty backgrounds and escape predation, but they are at a disadvantage in rural areas where they do not match their backgrounds. This is known as **industrial melanism**.

melanoma highly malignant tumour of the melanin-forming cells (melanocytes) of the skin. It develops from an existing mole in up to two thirds of cases, but can also arise in the eye or mucous membranes.

Malignant melanoma is the most dangerous of the skin cancers; it is associated with brief but excessive exposure to sunlight. It is easily treated if caught early but deadly once it has spread. There is a genetic factor in some cases.

Once rare, this disease is increasing at the rate of 7% in most countries with a predominantly fair-skinned popu-lation, owing to the increasing popularity of holidays in the sun. Most at risk are those with fair hair and light skin, and those who have had a severe sunburn in childhood. Cases of melanoma are increasing by 4% a year worldwide.

meltdown the melting of the core of a nuclear reactor, due to overheating.

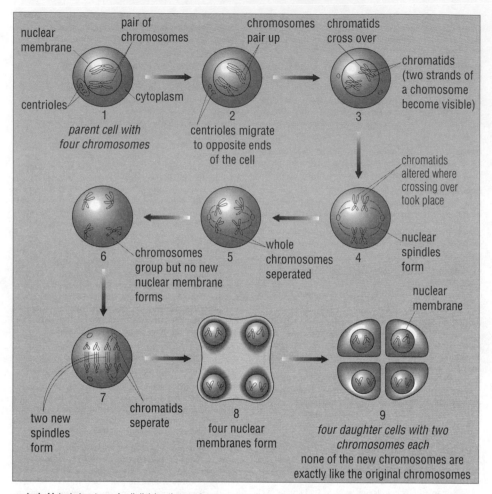

nuclear membrane

pair of chromosomes

chromosomes pair up

chromatids cross over

cytoplasm

centrioles

1

chromatids (two strands of a chromosome become visible)

2

3

parent cell with four chromosomes

centrioles migrate to opposite ends of the cell

chromatids altered where crossing over took place

6

chromosomes group but no new nuclear membrane forms

whole chromosomes seperated

5

4

nuclear spindles form

nuclear membrane

two new spindles form

chromatids seperate

7

8

four nuclear membranes form

9

four daughter cells with two chromosomes each

none of the new chromosomes are exactly like the original chromosomes

meiosis Meiosis is a type of cell division that produces gametes (sex cells, sperm and egg). This sequence shows an animal cell but only four chromosomes are present in the parent cell (1). There are two stages in the division process. In the first stage (2–6), the chromosomes come together in pairs and exchange genetic material. This is called crossing over. In the second stage (7–9), the cell divides to produce four gamete cells, each with only one copy of each chromosome from the parent cell.

To prevent such accidents all reactors have equipment intended to flood the core with water in an emergency. The reactor is housed in a strong containment vessel, designed to prevent radiation escaping into the atmosphere. The result of a meltdown would be an area radioactively contaminated for 25,000 years or more.

At Three Mile Island, Pennsylvania, USA, in March 1979, a partial meltdown occurred caused by a combination of equipment failure and operator error, and some radiation was released into the air. In April 1986, a reactor at Chernobyl, near Kiev, Ukraine, exploded, causing a partial meltdown of the core. Radioactive ◊fallout was detected as far away as Canada and Japan.

melting point temperature at which a substance melts, or changes from solid to liquid form. A pure substance under standard conditions of pressure (usually one atmosphere) has a definite melting point. If heat is sup-

plied to a solid at its melting point, the temperature does not change until the melting process is complete. The melting point of ice is 0°C/32°F.

membrane in living things, a continuous layer, made up principally of fat molecules, that encloses a ◊cell or ◊organelles within a cell. Small molecules, such as water and sugars, can pass through the cell membrane by ◊diffusion. Large molecules, such as proteins, are transported across the membrane via special channels, a process often involving energy input. The ◊Golgi apparatus within the cell is thought to produce certain membranes.

In cell organelles, enzymes may be attached to the membrane at specific positions, often alongside other enzymes involved in the same process, like workers at a conveyor belt. Thus membranes help to make cellular processes more efficient.

memory in computing, the part of a system used to store

data and programs either permanently or temporarily. There are two main types: immediate access memory and backing storage. Memory capacity is measured in ◊bytes or, more conveniently, in kilobytes (units of 1,024 bytes) or megabytes (units of 1,024 kilobytes).

Immediate access memory, or *internal memory*, describes the memory locations that can be addressed directly and individually by the central processing unit. It is either read-only (stored in ROM, PROM, and EPROM chips) or read/write (stored in RAM chips). Read-only memory stores information that must be constantly available and is unlikely to be changed. It is nonvolatile – that is, it is not lost when the computer is switched off. Read/write memory is volatile – it stores programs and data only while the computer is switched on.

Backing storage, or *external memory*, is nonvolatile memory, located outside the central processing unit, used to store programs and data that are not in current use. Backing storage is provided by such devices as magnetic ◊discs (floppy and hard discs), ◊magnetic tape (tape streamers and cassettes), optical discs (such as ◊CD-ROM), and ◊bubble memory. By rapidly switching blocks of information between the backing storage and the immediate-access memory, the limited size of the immediate-access memory may be increased artificially. When this technique is used to give the appearance of a larger internal memory than physically exists, the additional capacity is referred to as ◊virtual memory.

memory ability to store and recall observations and sensations. Memory does not seem to be based in any particular part of the brain; it may depend on changes to the pathways followed by nerve impulses as they move through the brain. Memory can be improved by regular use as the connections between ◊nerve cells (neurons) become 'well-worn paths' in the brain. Events stored in *short-term memory* are forgotten quickly, whereas those in *long-term memory* can last for many years, enabling recall of information and recognition of people and places over long periods of time.

Short-term memory is the most likely to be impaired by illness or drugs whereas long-term memory is very resistant to such damage. Memory changes with age and otherwise healthy people may experience a natural decline after the age of about 40. Research is just beginning to uncover the biochemical and electrical bases of the human memory.

mendelevium synthesized, radioactive metallic element of the ◊actinide series, symbol Md, atomic number 101, relative atomic mass 258. It was first produced by bombardment of Es-253 with helium nuclei. Its longest-lived isotope, Md-258, has a half-life of about two months. The element is chemically similar to thulium. It was named by the US physicists at the University of California at Berkeley who first synthesized it in 1955 after the Russian chemist Mendeleyev, who in 1869 devised the basis for the periodic table of the elements.

Mendelism in genetics, the theory of inheritance originally outlined by Austrian biologist Gregor Mendel. He suggested that, in sexually reproducing species, all characteristics are inherited through indivisible 'factors' (now identified with ◊genes) contributed by each parent to its offspring.

menhir prehistoric tall, upright stone monument or

◊megalith. Menhirs may be found singly as ◊monoliths or in groups. They have a wide geographical distribution in the Americas (mainly as monoliths), and in Europe, Asia, and Africa, and belong to many different periods. Most European examples were erected in the late Neolithic (New Stone Age) or early Bronze Age.

Ménière's disease or *Ménière's syndrome* recurring condition of the inner ear affecting mechanisms of both hearing and balance. It usually develops in the middle or later years. Symptoms, which include deafness, ringing in the ears, nausea, vertigo, and loss of balance, may be eased by drugs, but there is no cure.

meningitis inflammation of the meninges (membranes) surrounding the brain, caused by bacterial or viral infection. Bacterial meningitis, though treatable by antibiotics, is the more serious threat. Diagnosis is by ◊lumbar puncture.

Bacterial meningitis is caused by *Neisseria meningitidis*, a bacterium that colonizes the lining of the throat and is carried by 2–10% of the healthy population. Illness results if the bacteria enters the bloodstream, but normally the epithelial lining of the throat is a sufficient barrier.

Many common viruses can cause the occasional case of meningitis, although not usually in its more severe form. The treatment for viral meningitis is rest.

There are three strains of meningitis: serogroups A, B, and C. Vaccines exist only for A and C. However, they do not provide long-term protection nor are they suitable for children under the age of two. B is the most prevalent of the groups, causing over 50% of cases in Europe and the USA.

The severity of the disease varies from mild to rapidly lethal, and symptoms include fever, headache, nausea, neck stiffness, delirium, and (rarely) convulsions.

meningococcus in medicine, the causative organism of meningococcal ◊meningitis, *Neisseria meningitidis*.

meningoencephalitis in medicine, inflammation of the membrane (meninges) covering the brain with considerable involvement of the underlying brain tissue, as in ◊meningitis, for example.

meniscus in physics, the curved shape of the surface of a liquid in a thin tube, caused by the cohesive effects of ◊surface tension (capillary action). When the walls of the container are made wet by the liquid, the meniscus is concave, but with highly viscous liquids (such as

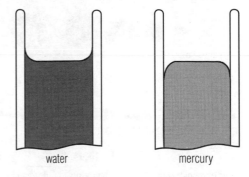

water   mercury

meniscus The curved shape, or meniscus, of a liquid surface is caused by the attraction or repulsion between liquid and container molecules.

mercury) the meniscus is convex. Meniscus is also the name of a concavo-convex or convexo-concave ◊lens.

meniscus in biology, the fibro-cartilage in joints, such as the knee joint.

menopause in women, the cessation of reproductive ability, characterized by menstruation (see ◊menstrual cycle) becoming irregular and eventually ceasing. The onset is at about the age of 50, but varies greatly. Menopause is usually uneventful, but some women suffer from complications such as flushing, excessive bleeding, and nervous disorders. Since the 1950s, ◊hormone-replacement therapy (HRT), using ◊oestrogen alone or with progestogen, a synthetic form of ◊progesterone, has been developed to counteract such effects.

Long-term use of HRT was previously associated with an increased risk of cancer of the uterus, and of clot formation in the blood vessels, but newer formulations using natural oestrogens are not associated with these risks. Without HRT there is increased risk of ◊osteoporosis (thinning of the bones) leading to broken bones, which may be indirectly fatal, particularly in the elderly.

The menopause is also known as the 'change of life'.

menstrual cycle cycle that occurs in female mammals of reproductive age, in which the body is prepared for pregnancy. At the beginning of the cycle, a Graafian (egg) follicle develops in the ovary, and the inner wall of the uterus forms a soft spongy lining. The egg is released from the ovary, and the uterus lining (endometrium) becomes vascularized (filled with blood vessels). If fertilization does not occur, the corpus luteum (remains of the Graafian follicle) degenerates, and the uterine lining breaks down, and is shed. This is what causes the loss of blood that marks menstruation. The cycle then begins

again. Human menstruation takes place from puberty to menopause, except during pregnancy, occurring about every 28 days.

Menstrual cycle

The menstrual cycles of women living in close proximity to each other become synchronized. This could be due to pheromones released in the sweat and may have evolved as means of thwarting male infidelity.

The cycle is controlled by a number of ◊hormones, including ◊oestrogen and ◊progesterone. If fertilization occurs, the corpus luteum persists and goes on producing progesterone.

mental disability arrested or incomplete development of mental capacities. It can be very mild, but in more severe cases is associated with social problems and difficulties in living independently. A person may be born with a mental disability (for example, ◊Down's syndrome) or may acquire it through brain damage. Between 90 and 130 million people in the world suffer from such disabilities.

Clinically, mental disability is graded as profound, severe, moderate, or mild, roughly according to IQ and the sufferer's ability to cope with everyday tasks. Among its many causes are genetic defect (◊phenylketonuria), chromosomal errors (Down's syndrome), infection before birth (◊rubella) or in infancy (◊meningitis), trauma (brain damage at birth or later), respiratory deficiency at the time of birth, toxins (lead poisoning), physical deprivation (lack of, or defective, ◊thyroid tissue, as in cretinism), and gross psychological deprivation. No clear cause of disability can be established for more than half of individuals with an IQ of less than 70.

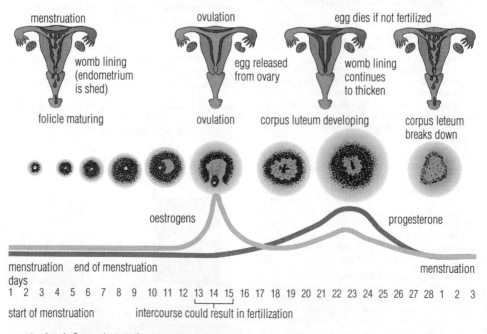

menstrual cycle From puberty to the menopause, most women produce a regular rhythm of hormones that stimulate the various stages of the menstrual cycle. The change in hormone levels may cause premenstrual tension.

mental illness disordered functioning of the mind. Since normal working cannot easily be defined, the borderline between mild mental illness and normality is a matter of opinion (not to be confused with normative behaviour). It is broadly divided into two categories: ◊neurosis, in which the patient remains in touch with reality; and ◊psychosis, in which perception, thought, and belief are disordered.

menu in computing, a list of options, displayed on screen, from which the user may make a choice – for example, the choice of services offered to the customer by a bank cash dispenser: withdrawal, deposit, balance, or statement. Menus are used extensively in ◊graphical user interface (GUI) systems, where the menu options are often selected using a pointing device called a ◊mouse.

Mercalli scale scale used to measure the intensity of an ◊earthquake. It differs from the ◊Richter scale, which measures *magnitude*. It is named after the Italian seismologist Giuseppe Mercalli (1850–1914).

Intensity is a subjective value, based on observed phenomena, and varies from place to place with the same earthquake.

merchant navy the passenger and cargo ships of a country. Most are owned by private companies. To avoid strict regulations on safety, union rules on crew wages, and so on, many ships are today registered under 'flags of convenience', that is, flags of countries that do not have such rules.

During wartime, merchant shipping may be drafted by the national government for military purposes.

mercury or *quicksilver* heavy, silver-grey, metallic element, symbol Hg (from Latin *hydrargyrum*), atomic number 80, relative atomic mass 200.59. It is a dense, mobile liquid with a low melting point (−38.87°C/ −37.96°F). Its chief source is the mineral cinnabar, HgS, but it sometimes occurs in nature as a free metal.

Its alloys with other metals are called amalgams (a silver-mercury amalgam is used in dentistry for filling cavities in teeth). Industrial uses include drugs and chemicals, mercury-vapour lamps, arc rectifiers, power-control switches, barometers, and thermometers.

Mercury is a cumulative poison that can contaminate the food chain, and cause intestinal disturbance, kidney and brain damage, and birth defects in humans. (The World Health Organization's 'safe' limit for mercury is 0.5 milligrams of mercury per kilogram of muscle tissue). The discharge into the sea by industry of organic mercury compounds such as dimethylmercury is the chief cause of mercury poisoning in the latter half of the 20th century. Between 1953 and 1975, 684 people in the Japanese fishing village of Minamata were poisoned (115 fatally) by organic mercury wastes that had been dumped into the bay and had accumulated in the bodies of fish and shellfish.

In a landmark settlement, a British multinational chemical company in April 1997 agreed to pay £1.3 million in compensation to 20 South African workers who were poisoned by mercury. Four of the black workers had died and a number of others were suffering severe brain and other neurological damage. The workers had accused Thor Chemical Holdings of adopting working practices in South Africa which would not have been allowed in Britain. The claimants had all worked at Thor's mercury plant at Cato Ridge in Natal. Thor had operated a mercury plant at Margate, in Kent, which during the 1980s was repeatedly criticised by the Health and Safety

Mercalli Scale

Intensity value	Description
I	not felt except by a very few under especially favourable conditions
II	felt only by a few persons at rest, especially on upper floors of buildings
III	felt quite noticeably by persons indoors, especially on upper floors of buildings; many people do not recognize it as an earthquake; standing motor cars may rock slightly; vibrations similar to the passing of a truck; duration estimated
IV	felt indoors by many, outdoors by few during the day; at night, some awakened; dishes, windows, doors disturbed; walls make cracking sound; sensation like heavy truck striking building; standing motor cars rock noticeably
V	felt by nearly everyone; many awakened; some dishes, windows broken; unstable objects overturned; pendulum clocks may stop
VI	felt by all, many frightened; some heavy furniture moved; a few instances of fallen plaster; damage slight
VII	damage negligible in buildings of good design and construction; slight to moderate in well-built ordinary structures; considerable damage in poorly-built or badly-designed structures; some chimneys broken
VIII	damage slight in specially-designed structures; considerable damage in ordinary substantial buildings with partial collapse; damage great in poorly-built structures; fall of chimneys, factory stacks, columns, monuments, walls; heavy furniture overturned
IX	damage considerable in specially-designed structures; well-designed frame structures thrown out of plumb; damage great in substantial buildings, with partial collapse; buildings shifted off foundations
X	some well-built wooden structures destroyed; most masonry and frame structures destroyed with foundations; rails bent
XI	few, if any (masonry) structures remain standing; bridges destroyed; rails bent greatly
XII	damage total; lines of sight and level are distorted; objects thrown into the air

The scale shown here is the Modified Mercalli Intensity Scale, developed in 1931 by US seismologists Harry Wood and Frank Neumann.

Executive (HSE) for bad working practices and over-exposure of British workers to mercury. Under pressure from the HSE, Thor closed down its mercury operations in Britain in 1987 and expanded them in South Africa.

The element was known to the ancient Chinese and Hindus, and is found in Egyptian tombs of about 1500 BC. It was named by the alchemists after the fast-moving god, for its fluidity.

Mercury in astronomy, the closest planet to the Sun. Its mass is 0.056 that of Earth. On its sunward side the surface temperature reaches over 400°C/752°F, but on the 'night' side it falls to −170°C/−274°F.

>   ***Mean distance from the Sun*** 58 million km/36 million mi.
>   ***Equatorial diameter*** 4,880 km/3,030 mi.
>   ***Rotation period*** 59 Earth days.
>   ***Year*** 88 Earth days.
>   ***Atmosphere*** Mercury has an atmosphere with minute traces of argon and helium.
>   ***Surface*** composed of silicate rock often in the form of lava flows. In 1974 the US space probe *Mariner 10* showed that Mercury's surface is cratered by meteorite impacts.
>   ***Satellites*** none.

mercury fulminate highly explosive compound used in detonators and percussion caps. It is a grey, sandy powder and extremely poisonous.

Mercury project US project to put a human in space in the one-seat Mercury spacecraft between 1961 and 1963.

The first two Mercury flights, on Redstone rockets, were short flights to the edge of space and back. The orbital flights, beginning with the third in the series (made by John Glenn), were launched by Atlas rockets.

meridian half a ◊great circle drawn on the Earth's surface passing through both poles and thus through all places with the same longitude. Terrestrial longitudes are usually measured from the Greenwich Meridian.

An astronomical meridian is a great circle passing through the celestial pole and the zenith (the point immediately overhead).

meristem region of plant tissue containing cells that are actively dividing to produce new tissues (or have the potential to do so). Meristems found in the tip of roots and stems, the apical meristems, are responsible for the growth in length of these organs.

The ◊cambium is a lateral meristem that is responsible for increase in girth in perennial plants. Some plants also have intercalary meristems, as in the stems of grasses, for example. These are responsible for their continued growth after cutting or grazing has removed the apical meristems of the shoots.

Meristem culture involves growing meristems taken from shoots on a nutrient-containing medium, and using them to grow new plants.

It is used to propagate infertile plants or hybrids that do not breed true from seed and to generate virus-free stock, since viruses rarely infect apical meristems.

MERLIN array radiotelescope network centred on ◊Jodrell Bank, N England.

mesa flat-topped, steep-sided plateau, consisting of horizontal weak layers of rock topped by a resistant formation; in particular, those found in the desert areas of the USA and Mexico. A small mesa is called a butte.

mescaline psychedelic drug derived from a small, spineless cactus *Lophophora williamsii* of N Mexico and the SW USA, known as peyote. The tops (called mescal buttons), which scarcely appear above ground, are dried and chewed, or added to alcoholic drinks. Mescaline is a crystalline alkaloid $C_{11}H_{17}NO_3$. It is used by some North American Indians in religious rites.

mesmerism former term for ◊hypnosis, after Austrian physician Friedrich Mesmer.

mesoglea layer of jelly-like noncellular tissue that separates the endoderm and ectoderm in jellyfish and other ◊coelenterates.

meson in physics, a group of unstable subatomic particles made up of two indivisible elementary particles called ◊quarks. It has a mass intermediate between that of the electron and that of the proton, is found in cosmic radiation, and is emitted by nuclei under bombardment by very high-energy particles.

The mesons form a subclass of the hadrons and include the kaons and pions. Their existence was predicted in 1935 by Japanese physicist Hideki Yukawa.

mesophyll the tissue between the upper and lower epidermis of a leaf blade (◊lamina), consisting of parenchyma-like cells containing numerous ◊chloroplasts.

In many plants, mesophyll is divided into two distinct layers.

The ***palisade mesophyll*** is usually just below the upper epidermis and is composed of regular layers of elongated cells. Lying below them is the ***spongy mesophyll***, composed of loosely arranged cells of irregular shape. This layer contains fewer chloroplasts and has many intercellular spaces for the diffusion of gases (required for ◊respiration and ◊photosynthesis), linked to the outside by means of ◊stomata.

mesosphere layer in the Earth's ◊atmosphere above the stratosphere and below the thermosphere. It lies between about 50 km/31 mi and 80 km/50 mi above the ground.

Mesozoic era of geological time 245–65 million years ago, consisting of the Triassic, Jurassic, and Cretaceous periods. At the beginning of the era, the continents were joined together as Pangaea; dinosaurs and other giant reptiles dominated the sea and air; and ferns, horsetails, and cycads thrived in a warm climate worldwide. By the end of the Mesozoic era, the continents had begun to assume their present positions, flowering plants were dominant, and many of the large reptiles and marine fauna were becoming extinct.

Messier catalogue in astronomy, a catalogue of 103 ◊galaxies, ◊nebulas, and star clusters (the Messier objects) published in 1784 by French astronomer Charles Messier. Catalogue entries are denoted by the prefix 'M'. Well known examples include M31 (the ◊Andromeda galaxy), M42 (the ◊Orion Nebula), and M45 (the ◊Pleiades star cluster).

Messier compiled the catalogue to identify fuzzy objects that could be mistaken for comets. The list was later extended to 109.

metabolism the chemical processes of living organisms enabling them to grow and to function. It involves a constant alternation of building up (**anabolism**) and breaking down (**catabolism**). For example, green plants build up complex organic substances from water, carbon dioxide, and mineral salts (photosynthesis); by digestion animals partially break down complex organic substances, ingested as food, and subsequently resynthesize them for use in their own bodies.

metal any of a class of chemical elements with certain chemical characteristics (◊metallic character) and physical properties: they are good conductors of heat and electricity; opaque but reflect light well; malleable, which enables them to be cold-worked and rolled into sheets; and ductile, which permits them to be drawn into thin wires.

Metallic elements compose about 75% of the 112 elements shown in the ◊periodic table of the elements. They form alloys with each other, ◊bases with the hydroxyl radical (OH), and replace the hydrogen in an ◊acid to form a salt. The majority are found in nature in the combined form only, as compounds or mineral ores; about 16 of them also occur in the elemental form, as ◊native metals. Their chemical properties are largely determined by the extent to which their atoms can lose one or more electrons and form positive ions (cations).

Metals have been put to many uses, both structural and decorative, since prehistoric times, and the Copper Age, Bronze Age, and Iron Age are named after the metal that formed the technological base for that stage of human evolution.

metal detector electronic device for detecting metal, usually below ground, developed from the wartime mine detector. In the head of the metal detector is a coil, which is part of an electronic circuit. The presence of metal causes the frequency of the signal in the circuit to change, setting up an audible note in the headphones worn by the user.

They are used to survey areas for buried metallic objects, occasionally by archaeologists. However, their indiscriminate use by treasure hunters has led to their being banned on recognized archaeological sites in some countries.

metal fatigue condition in which metals fail or fracture under relatively light loads, when these loads are applied repeatedly. Structures that are subject to flexing, such as the airframes of aircraft, are prone to metal fatigue.

metallic bond the force of attraction operating in a metal that holds the atoms together. In the metal the ◊valency electrons are able to move within the crystal and these electrons are said to be delocalized (see ◊electrons, delocalized). Their movement creates short-lived, positively charged ions. The electrostatic attraction between the delocalized electrons and the ceaselessly forming ions constitutes the metallic bond.

metallic character chemical properties associated with those elements classed as metals. These properties, which arise from the element's ability to lose electrons, are: the displacement of hydrogen from dilute acids; the formation of basic oxides; the formation of ionic chlorides; and their reducing reaction, as in the ◊thermite process (see ◊reduction).

In the periodic table of the elements, metallic character increases down any group and across a period from right to left.

metallic glass substance produced from metallic materials (non-corrosive alloys rather than simple metals) in a liquid state which, by very rapid cooling, are prevented from reverting to their regular metallic structure. Instead they take on the properties of glass, while retaining the metallic properties of malleability and relatively good electrical conductivity.

metallographic examination technique used by archaeologists for analysing the manufacturing techniques of metal artefacts. A cross-sectional slice of an artefact is polished, etched to highlight internal structures, and examined under a metallurgical microscope. The reflected light of the microscope enhances uneven surfaces, revealing grain size, shape, and boundaries, inclusions, fabric, defects, and other detail.

metalloid or **semimetal** any chemical element having some of but not all the properties of metals; metalloids are thus usually electrically semiconducting. They comprise the elements germanium, arsenic, antimony, and tellurium.

metallurgy the science and technology of producing metals, which includes extraction, alloying, and hardening. Extractive, or **process, metallurgy** is concerned with the extraction of metals from their ◊ores and refining and adapting them for use. **Physical metallurgy** is concerned with their properties and application. **Metallography** establishes the microscopic structures that contribute to hardness, ductility, and strength.

Metals can be extracted from their ores in three main ways: **dry processes**, such as smelting, volatilization, or amalgamation (treatment with mercury); **wet processes**, involving chemical reactions; and **electrolytic processes**, which work on the principle of ◊electrolysis.

The foundations of metallurgical science were laid about 3500 BC in Egypt, Mesopotamia, China, and India, where the art of ◊smelting metals from ores was discovered, starting with the natural alloy bronze. Later, gold, silver, copper, lead, and tin were worked in various ways, although they had been cold-hammered as native metals for thousands of years. The smelting of iron was discovered about 1500 BC. The Romans hardened and tempered iron into steel, using ◊heat treatment. From then until about AD 1400, advances in metallurgy came into Europe by way of Arabian chemists. ◊Cast iron began to be made in the 14th century in a crude blast furnace. The demands of the Industrial Revolution led to an enormous increase in ◊wrought iron production. The invention by British civil engineer Henry Bessemer of the ◊Bessemer process in 1856 made cheap steel available for the first time, leading to its present widespread use and the industrial development of many specialized steel alloys.

metamorphic rock rock altered in structure and composition by pressure, heat, or chemically active fluids after original formation. (If heat is sufficient to melt the original rock, technically it becomes an igneous rock upon cooling.) The term was coined in 1833 by Scottish geologist Charles Lyell (1797–1875).

metamorphism geological term referring to the changes in rocks of the Earth's crust caused by increasing pressure and temperature. The resulting rocks are metamorphic rocks. All metamorphic changes take place in solid rocks. If the rocks melt and then harden, they become ◊igneous rocks.

metamorphosis period during the life cycle of many invertebrates, most amphibians, and some fish, during which the individual's body changes from one form to another through a major reconstitution of its tissues. For example, adult frogs are produced by metamorphosis from tadpoles, and butterflies are produced from caterpillars following metamorphosis within a pupa.

metazoa another name for animals. It reflects an earlier system of classification, in which there were two main divisions within the animal kingdom, the multicellular animals, or metazoa, and the single-celled 'animals' or protozoa. The ◊protozoa are no longer included in the animal kingdom, so only the metazoa remain.

meteor flash of light in the sky, popularly known as a *shooting* or *falling star*, caused by a particle of dust, a *meteoroid*, entering the atmosphere at speeds up to 70 kps/45 mps and burning up by friction at a height of around 100 km/60 mi. On any clear night, several *sporadic meteors* can be seen each hour.

Several times each year the Earth encounters swarms of dust shed by comets, which give rise to a *meteor shower*.

This appears to radiate from one particular point in the sky, after which the shower is named; the Perseid meteor shower in August appears in the constellation Perseus. A brilliant meteor is termed a *fireball*. Most meteoroids are smaller than grains of sand. The Earth sweeps up an estimated 16,000 tonnes of meteoric material every year.

meteor-burst communications technique for sending messages by bouncing radio waves off the trails of ◊meteors. High-speed computer-controlled equipment is used to sense the presence of a meteor and to broadcast a signal during the short time that the meteor races across the sky.

The system, first suggested in the late 1920s, remained impracticable until data-compression techniques were developed, enabling messages to be sent in automatic high-speed bursts each time a meteor trail appeared. There are usually enough meteor trails in the sky at any time to permit continuous transmission of a message. The technique offers a communications link that is difficult to jam, undisturbed by storms on the Sun, and would not be affected by nuclear war.

meteorite piece of rock or metal from space that reaches the surface of the Earth, Moon, or other body. Most meteorites are thought to be fragments from asteroids, although some may be pieces from the heads of comets. Most are stony, although some are made of iron and a few have a mixed rock-iron composition.

Stony meteorites can be divided into two kinds: *chondrites* and *achondrites*. Chondrites contain chondrules, small spheres of the silicate minerals olivine and orthopyroxene, and comprise 85% of meteorites. Achondrites do not contain chondrules. Meteorites provide evidence for the nature of the Solar System and may be similar to the Earth's core and mantle, neither of which can be observed directly.

meteorology scientific observation and study of the ◊atmosphere, so that ◊weather can be accurately forecast.

Data from meteorological stations and weather satellites are collated by computer at central agencies, and forecast and weather maps based on current readings are issued at regular intervals. Modern analysis, employing some of the most powerful computers, can give useful forecasts for up to six days ahead.

At meteorological stations readings are taken of the factors determining weather conditions: atmospheric pressure, temperature, humidity, wind (using the ◊Beaufort scale), cloud cover (measuring both type of cloud and coverage), and precipitation such as rain, snow, and hail (measured at 12-hour intervals). ◊Satellites are used either to relay information transmitted from the Earth-based stations, or to send pictures of cloud development, indicating wind patterns, and snow and ice cover.

Apart from some observations included by Aristotle in his book *Meteorologia*, meteorology did not become a precise science until the end of the 16th century, when Galileo and the Florentine academicians constructed the first thermometer of any importance, and when Evangelista Torricelli in 1643 discovered the principle of the barometer. Robert Boyle's work on gases, and that of his assistant, Robert Hooke, on barometers, advanced the physics necessary for the understanding of the weather. Gabriel Fahrenheit's invention of a superior mercury thermometer provided further means for temperature recording.

In the early 19th century a chain of meteorological stations was established in France, and weather maps were constructed from the data collected. The first weather map in England, showing the trade winds and monsoons, was made in 1688, and the first telegraphic weather report appeared on 31 Aug 1848. The first daily telegraphic weather map was prepared at the Great Exhibition in 1851, but the Meteorological Office was not established in London until 1855. The first regular daily collections of weather observations by telegraph and the first British daily weather reports were made in 1860, and the first daily printed maps appeared in 1868.

Observations can be collected not only from land stations, but also from weather ships, aircraft, and self-recording and automatic transmitting stations, such as the ◊radiosonde. ◊Radar may be used to map clouds and storms. Satellites have played an important role in televising pictures of global cloud distribution.

meter any instrument used for measurement. The term is often compounded with a prefix to denote a specific type of meter: for example, ammeter, voltmeter, flowmeter, or pedometer.

methanal (common name *formaldehyde*) HCHO gas at ordinary temperatures, condensing to a liquid at $-21°C/-5.8°F$. It has a powerful, penetrating smell. Dissolved in water, it is used as a biological preservative. It is used in the manufacture of plastics, dyes, foam (for example urea-formaldehyde foam, used in insulation), and in medicine.

methane CH_4 the simplest hydrocarbon of the paraffin series. Colourless, odourless, and lighter than air, it burns with a bluish flame and explodes when mixed with air or oxygen. It is the chief constituent of natural gas and also occurs in the explosive firedamp of coal mines. Methane emitted by rotting vegetation forms marsh gas, which may ignite by spontaneous combustion to produce the pale flame seen over marshland and known as ◊will-o'-the-wisp.

Methane causes about 38% of the warming of the globe through the ◊greenhouse effect; weight for weight it is 60–70 times more potent than carbon dioxide at trapping solar radiation in the atmosphere and so heating the planet. The amount of methane in the air is predicted to double over the next 60 years. An estimated 15% of all methane gas into the atmosphere is produced by cows and other cud-chewing animals, and 20% is produced by termites that feed on soil.

methanogenic bacteria one of a group of primitive microorganisms, the ◊Archaea. They give off methane gas as a by-product of their metabolism, and are common in sewage treatment plants and hot springs, where the temperature is high and oxygen is absent. Archaeons were originally classified as bacteria, but were found to be unique in 1996 following the gene sequencing of the deep-sea vent *Methanococcus jannaschii*.

methanoic acid (common name **formic acid**) HCOOH, a colourless, slightly fuming liquid that freezes at 8°C/46.4°F and boils at 101°C/213.8°F. It occurs in stinging ants, nettles, sweat, and pine needles, and is used in dyeing, tanning, and electroplating.

methanol (common name **methyl alcohol**) CH_3OH the simplest of the alcohols. It can be made by the dry distillation of wood (hence it is also known as wood alcohol), but is usually made from coal or natural gas. When pure, it is a colourless, flammable liquid with a pleasant odour, and is highly poisonous.

Methanol is used to produce formaldehyde (from which resins and plastics can be made), methyl-ter-butyl ether (MTB, a replacement for lead as an octane-booster in petrol), vinyl acetate (largely used in paint manufacture), and petrol. In 1993 Japanese engineers built an engine, made largely of ceramics, that runs on methanol. The prototype is lighter and has a cleaner exhaust than comparable metal, petrol-powered engines.

methionine one of the nine essential ◊amino acids. It is also used as an antidote to paracetamol poisoning.

methyl alcohol common name for ◊methanol.

methylated spirit alcohol that has been rendered undrinkable, and is used for industrial purposes, as a fuel for spirit burners or a solvent.

It is nevertheless drunk by some individuals, resulting eventually in death. One of the poisonous substances in it is ◊methanol, or methyl alcohol, and this gives it its name. (The 'alcohol' of alcoholic drinks is ethanol.)

methyl benzene alternative name for ◊toluene.

methyl orange $C_{14}H_{14}N_3NaO_3S$ orange-yellow powder used as an acid–base indicator in chemical tests, and as a stain in the preparation of slides of biological material. Its colour changes with pH; below pH 3.1 it is red, above pH 4.4 it is yellow.

metre SI unit (symbol m) of length, equivalent to 1.093 yards. It is defined by scientists as the length of the path travelled by light in a vacuum during a time interval of 1/299,792,458 of a second.

metric system system of weights and measures developed in France in the 18th century and recognized by other countries in the 19th century.

In 1960 an international conference on weights and measures recommended the universal adoption of a revised International System (Système International d'Unités, or SI), with seven prescribed 'base units': the metre (m) for length, kilogram (kg) for mass, second (s) for time, ampere (A) for electric current, kelvin (K) for thermodynamic temperature, candela (cd) for luminous intensity, and mole (mol) for quantity of matter.

Two supplementary units are included in the SI system

– the radian (rad) and steradian (sr) – used to measure plane and solid angles. In addition, there are recognized derived units that can be expressed as simple products or divisions of powers of the basic units, with no other integers appearing in the expression; for example, the watt.

Some non-SI units, well established and internationally recognized, remain in use in conjunction with SI: minute, hour, and day in measuring time; multiples or submultiples of base or derived units which have long-established names, such as tonne for mass, the litre for volume; and specialist measures such as the metric carat for gemstones.

Prefixes used with metric units are tera (T) million million times; giga (G) billion (thousand million) times; mega (M) million times; kilo (k) thousand times; hecto (h) hundred times; deca (da) ten times; deci (d) tenth part; centi (c) hundredth part; milli (m) thousandth part; micro (µ) millionth part; nano (n) billionth part; pico (p) trillionth part; femto (f) quadrillionth part; atto (a) quintillionth part.

mg symbol for **milligram**.

MHD abbreviation for ◊**magnetohydrodynamics**.

mho SI unit of electrical conductance, now called the ◊siemens; equivalent to a reciprocal ohm.

mi symbol for ◊**mile**.

mica group of silicate minerals that split easily into thin flakes along lines of weakness in their crystal structure (perfect basal cleavage). They are glossy, have a pearly lustre, and are found in many igneous and metamorphic rocks. Their good thermal and electrical insulation qualities make them valuable in industry.

Their chemical composition is complicated, but they are silicates with silicon-oxygen tetrahedra arranged in continuous sheets, with weak bonding between the layers, resulting in perfect cleavage.

A common example of mica is muscovite (white mica), $KAl_2Si_3AlO_{10}(OH,F)_2$.

MICR abbreviation for ◊magnetic-ink character recognition.

micro- prefix (symbol µ) denoting a one-millionth part (10^{-6}). For example, a micrometre, µm, is one-millionth of a metre.

microbe another name for ◊microorganism.

microbiology the study of microorganisms, mostly viruses and single-celled organisms such as bacteria, protozoa, and yeasts. The practical applications of microbiology are in medicine (since many microorganisms cause disease); in brewing, baking, and other food and beverage processes, where the microorganisms carry out fermentation; and in genetic engineering, which is creating increasing interest in the field of microbiology.

microchip popular name for the silicon chip, or ◊integrated circuit.

microclimate the climate of a small area, such as a woodland, lake, or even a hedgerow. Significant differences can exist between the climates of two neighbouring areas – for example, a town is usually warmer than the surrounding countryside (forming a heat island), and a woodland cooler, darker, and less windy than an area of open land.

Microclimates play a significant role in agriculture and horticulture, as different crops require different growing conditions.

microcomputer or *micro* or *personal computer* small desktop or portable computer, typically designed to be used by one person at a time, although individual computers can be linked in a network so that users can share data and programs.

Its central processing unit is a ◊microprocessor, contained on a single integrated circuit.

Microcomputers are the smallest of the four classes of computer (the others are ◊supercomputer, ◊mainframe, and ◊minicomputer). Since the appearance in 1975 of the first commercially available microcomputer, the Altair 8800, micros have become ubiquitous in commerce, industry, and education.

microfiche sheet of film on which printed text is photographically reduced. See ◊microform.

microform generic name for media on which text or images are photographically reduced. The main examples are *microfilm* (similar to the film in an ordinary camera) and *microfiche* (flat sheets of film, generally 105 mm/4 in × 148 mm/6 in, holding the equivalent of 420 standard pages). Microform has the advantage of low reproduction and storage costs, but it requires special devices for reading the text. It is widely used for archiving and for storing large volumes of text, such as library catalogues.

monitor

CD-ROM drive

3¹/₂" disc drive

5¹/₄" disc drive

keyboard

system unit

floppy discs

mouse mat

mouse

microcomputer The component parts of the microcomputer: the system unit contains the hub of the system, including the central processing unit (CPU), information on all of the computer's peripheral devices, and often a fixed disc drive. The monitor (or visual display unit) displays text and graphics, the keyboard and mouse are used to input data, and the floppy disc and CD-ROM drives read data stored on discs.

Computer data may be output directly and quickly in microform by means of COM (computer output on microfilm/microfiche) techniques.

micrometer instrument for measuring minute lengths or angles with great accuracy; different types of micrometer are used in astronomical and engineering work.

The type of micrometer used in astronomy consists of two fine wires, one fixed and the other movable, placed in the focal plane of a telescope; the movable wire is fixed on a sliding plate and can be positioned parallel to the other until the object appears between the wires.

The movement is then indicated by a scale on the adjusting screw.

micrometre one-millionth of a ◊metre (symbol μm).

microorganism or *microbe* living organism invisible to the naked eye but visible under a microscope. Microorganisms include viruses and single-celled organisms such as bacteria, protozoa, yeasts, and some algae. The term

Microorganism

There are more living organisms on the skin of a single human than there are human beings on the surface of the Earth.

has no taxonomic significance in biology. The study of microorganisms is known as microbiology.

microphone primary component in a sound-reproducing system, whereby the mechanical energy of sound waves is converted into electrical signals by means of a ◊transducer. One of the simplest is the telephone receiver mouthpiece, invented by Scottish–US inventor Alexander Graham Bell in 1876; other types of microphone are used with broadcasting and sound-film apparatus.

Telephones have a *carbon microphone*, which reproduces only a narrow range of frequencies. For live music, a *moving-coil microphone* is often used. In it, a diaphragm that vibrates with sound waves moves a coil through a magnetic field, thus generating an electric current. The *ribbon microphone* combines the diaphragm and coil. The *condenser microphone* is most commonly used in recording and works by a ◊capacitor.

microprocessor complete computer ◊central processing unit contained on a single ◊integrated circuit, or chip. The appearance of the first microprocessor in 1971 designed by Intel for a pocket calculator manufacturer heralded the introduction of the microcomputer. The microprocessor has led to a dramatic fall in the size and cost of computers, and ◊dedicated computers can now be found in washing machines, cars, and so on. Examples of microprocessors are the Intel Pentium family and the IBM/Apple Power PC.

In Jan 1997 Texas Instruments introduced a digital-signal microprocessor chip that can process 1.6 billion instructions a second. This is about 40 times more powerful than a chip now found in today's computer modem. The new chip can reduce the time needed to download a file from the Internet from ten minutes to less than five seconds.

micropropagation the mass production of plants by placing tiny pieces of plant tissue in sterile glass containers along with nutrients. Perfect clones of superplants are produced in sterile cabinets, with filtered air and carefully controlled light, temperature, and humidity.

The system is used for the house-plant industry and for forestry – micropropagation gives immediate results, whereas obtaining genetically homogenous tree seed by traditional means would take over 100 years.

micropyle in flowering plants, a small hole towards one end of the ovule. At pollination the pollen tube growing down from the ◊stigma eventually passes through this pore. The male gamete is contained within the tube and is able to travel to the egg in the interior of the ovule. Fertilization can then take place, with subsequent seed formation and dispersal.

microscope instrument for magnification with high resolution for detail. Optical and electron microscopes are the ones chiefly in use; other types include acoustic, ◊scanning tunnelling, and ◊atomic force microscopes. In 1988 a scanning tunnelling microscope was used to photograph a single protein molecule for the first time.

The *optical microscope* usually has two sets of glass lenses and an eyepiece. It was invented in 1609 in the Netherlands by Zacharias Janssen (1580–c. 1638). *Fluorescence microscopy* makes use of fluorescent dyes to illuminate samples, or to highlight the presence of particular substances within a sample. Various illumination systems are also used to highlight details.

The ◊ *transmission electron microscope*, developed from 1932, passes a beam of electrons, instead of a beam of light, through a specimen. Since electrons are not visible, the eyepiece is replaced with a fluorescent screen or photographic plate; far higher magnification and resolution are possible than with the optical microscope.

The ◊*scanning electron microscope* (SEM), developed in the mid-1960s, moves a fine beam of electrons over the surface of a specimen, the reflected electrons being collected to form the image. The specimen has to be in a vacuum chamber.

The *acoustic microscope* passes an ultrasonic (ultra-high-frequency sound) wave through the specimen, the transmitted sound being used to form an image on a computer screen.

The *scanned-probe microscope*, developed in the late 1980s, runs a probe, with a tip so fine that it may consist only of a single atom, across the surface of the specimen, which requires no special preparation. In the *scanning tunnelling microscope*, an electric current that flows through the probe is used to construct an image of the specimen. In the *atomic force microscope*, the force felt by the probe is measured and used to form the image. These instruments can magnify a million times and give images of single atoms.

microsurgery part or all of an intricate surgical operation – rejoining a severed limb, for example – performed with the aid of a binocular microscope, using miniaturized instruments. Sewing of the nerves and blood vessels is done with a nylon thread so fine that it is only just visible to the naked eye.

The technique permits treatment of previously inaccessible lesions in the eye or brain.

microtubules tiny tubes found in almost all cells with a nucleus. They help to define the shape of a cell by forming scaffolding for cilia and they also form the fibres of the mitotic spindle (see ◊mitosis).

microwave ◊electromagnetic wave with a wavelength in the range 0. 3–30 cm/0.1 in–12 in, or 300–300,000 megahertz (between radio waves and ◊infrared radiation).

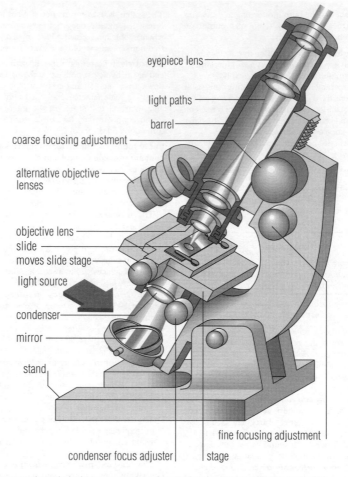

eyepiece lens

light paths

barrel

coarse focusing adjustment

alternative objective lenses

objective lens
slide
moves slide stage
light source
condenser
mirror
stand

fine focusing adjustment

condenser focus adjuster   stage

microscope In essence, the optical microscope consists of an eyepiece lens and an objective lens, which are used to produce an enlarged image of a small object by focusing light from a light source. Optical microscopes can achieve magnifications of up to 1,500–2,000. Higher magnifications and resolutions are obtained by electron microscopes.

Microwaves are used in radar, in radio broadcasting, and in microwave heating and cooking.

microwave heating heating by means of microwaves. Microwave ovens use this form of heating for the rapid cooking or reheating of foods, where heat is generated throughout the food simultaneously. If food is not heated completely, there is a danger of bacterial growth that may lead to food poisoning. Industrially, microwave heating is used for destroying insects in grain and enzymes in processed food, pasteurizing and sterilizing liquids, and drying timber and paper.

Mid-Atlantic Ridge ◊ocean ridge, formed by the movement of plates described by ◊plate tectonics, that runs along the centre of the Atlantic Ocean, parallel to its edges, for some 14,000 km/8,800 mi – almost from the Arctic to the Antarctic.

The Mid-Atlantic Ridge is central because the ocean crust beneath the Atlantic Ocean has continually grown outwards from the ridge at a steady rate during the past 200 million years. Iceland straddles the ridge and was formed by volcanic outpourings.

MIDI (acronym for *musical instrument digital interface*) manufacturer's standard allowing different pieces of digital music equipment used in composing and recording to be freely connected.

The information-sending device (any electronic instrument) is called a controller, and the reading device (such as a computer) the sequencer. Pitch, dynamics, decay rate, and stereo position can all be transmitted via the interface. A computer with a MIDI interface can input and store the sounds produced by the connected instruments, and can then manipulate these sounds in many different ways. For example, a single keystroke may change the key of an entire composition. Even a full written score for the composition may be automatically produced.

midnight sun the constant appearance of the Sun

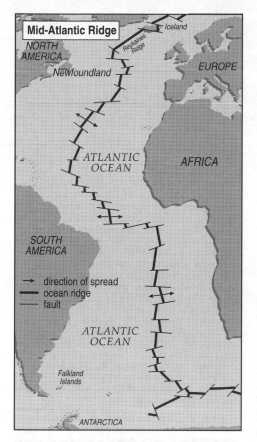

Mid-Atlantic Ridge

NORTH AMERICA
Iceland
Reykjanes Ridge
Newfoundland
EUROPE

ATLANTIC OCEAN
AFRICA

SOUTH AMERICA

→ direction of spread
— ocean ridge
— fault

ATLANTIC OCEAN

Falkland Islands

ANTARCTICA

Mid-Atlantic Ridge The Mid-Atlantic Ridge is the boundary between the crustal plates that form America, and Europe and Africa. An oceanic ridge cannot be curved since the material welling up to form the ridge flows at a right angle to the ridge. The ridge takes the shape of small straight sections offset by fractures transverse to the main ridge.

(within the Arctic and Antarctic circles) above the ◊horizon during the summer.

midwifery assistance of women in childbirth. Traditionally, it was undertaken by experienced specialists; in modern medical training it is a nursing speciality for practitioners called midwives.

The English physician William Harvey's 1653 work on generation contained an influential chapter on labour. Dr Peter Chamberlen II (1560–1631) made an important contribution to midwifery, particularly with his development of the first midwifery forceps.

mifepristone (formerly *RU486*) anti-progesterone drug used, in combination with a ◊prostaglandin, to procure early ◊abortion (up to the tenth week in pregnancy). It is administered only in hospitals or recognized clinics and a success rate of 95% is claimed. It was developed and first used in France in 1989. It was licensed in the UK in 1991 by which time 60,000 abortions had been carried out in France by this method.

Research is currently underway into using low doses of

mifepristone as a contraceptive pill both to be administered regularly or as emergency ('morning after') contraception.

migraine acute, sometimes incapacitating headache (generally only on one side), accompanied by nausea, that recurs, often with advance symptoms such as flashing lights. No cure has been discovered, but ◊ergotamine normally relieves the symptoms. Some sufferers learn to avoid certain foods, such as chocolate, which suggests an allergic factor.

In 1990 Hammersmith Hospital in London, England, reported successful treatment with goggles that depress beta waves in the brain (associated with stress) and stimulate alpha waves (whose effect is calming).

migration the movement, either seasonal or as part of a single life cycle, of certain animals, chiefly birds and fish, to distant breeding or feeding grounds.

The precise methods by which animals navigate and know where to go are still obscure. Birds have much sharper eyesight and better visual memory of ground clues than humans, but in long-distance flights appear to navigate by the Sun and stars, possibly in combination with a 'reading' of the Earth's magnetic field through an inbuilt 'magnetic compass', which is a tiny mass of tissue between the eye and brain in birds. Similar cells occur in 'homing' honeybees and in certain bacteria that use it to determine which way is 'down'. Leatherback turtles use the contours of underwater mountains and valleys to navigate by. Most striking, however, is the migration of young birds that have never flown a route before and are unaccompanied by adults. It is postulated that they may inherit as part of their genetic code an overall 'sky chart' of their journey that is triggered into use when they become aware of how the local sky pattern, above the place in which they hatch, fits into it. Similar theories have been advanced in the case of fish, such as eels and salmon, with whom vision obviously plays a lesser role, but for whom currents and changes in the composition and temperature of the sea in particular locations may play a part – for example, in enabling salmon to return to the precise river in which they were spawned. Migration also occurs with land animals; for example, lemmings and antelope.

Migration

The grey whale migrates further than any other mammal. It makes a round trip of 20,400 km/ 12,500 mi between its summer feeding grounds in the Arctic and its winter breeding lagoons off the western coast of Mexico.

Species migration is the spread of the home range of a species over many years, for example the spread of the collared dove (*Streptopelia decaocto*) from Turkey to Britain over the period 1920–52. Any journey which takes an animal outside of its normal home range is called **individual migration**; when the animal does not return to its home range it is called **removal migration**. An example of **return migration** is the movement of birds that fly south for the winter and return to their home ranges in the spring. Many types of whale also make return migrations. In **remigration**, the return leg of the migration is completed by a subsequent generation, for example locust swarms migrate, but each part of the circuit is completed by a different generation.

Related to migration is the homing ability of pigeons, bees, and other creatures.

Milankovitch hypothesis the combination of factors governing the occurrence of ◊ice ages proposed in 1930 by the Yugoslav geophysicist M Milankovitch (1879–1958). These include the variation in the angle of the Earth's axis, and the geometry of the Earth's orbit around the Sun.

mile imperial unit of linear measure. A statute mile is equal to 1,760 yards (1.60934 km), and an international nautical mile is equal to 2,026 yards (1,852 m).

milk secretion of the ◊mammary glands of female mammals, with which they suckle their young (during ◊lactation). Over 85% is water, the remainder comprising protein, fat, lactose (a sugar), calcium, phosphorus, iron, and vitamins. The milk of cows, goats, and sheep is often consumed by humans, but regular drinking of milk after infancy is principally a Western practice.

milking machine machine that uses suction to milk cows. The first milking machine was invented in the USA by L O Colvin in 1860. Later it was improved so that the suction was regularly released by a pulsating device, since it was found that continuous suction is harmful to cows.

milk teeth or **deciduous teeth** teeth that erupt in childhood between the ages of 6 and 30 months. They are replaced by the permanent teeth, which erupt between the ages of 6 and 21 years. See also ◊dentition and ◊tooth.

Milky Way faint band of light crossing the night sky, consisting of stars in the plane of our Galaxy. The name Milky Way is often used for the Galaxy itself. It is a spiral ◊galaxy, 100,000 light years in diameter and 2,000 light years thick, containing at least 100 billion ◊stars. The Sun is in one of its spiral arms, about 25,000 light years from the centre, not far from its central plane.

The densest parts of the Milky Way, towards the Galaxy's centre, lie in the constellation ◊Sagittarius. In places, the Milky Way is interrupted by lanes of dark dust that obscure light from the stars beyond, such as the Coalsack ◊nebula in ◊Crux (the Southern Cross). It is because of these that the Milky Way is irregular in width and appears to be divided into two between Centaurus and Cygnus.

milli- prefix (symbol m) denoting a one-thousandth part (10^{-3}). For example, a millimetre, mm, is one thousandth of a metre.

millibar unit of pressure, equal to one-thousandth of a ◊bar.

millilitre one-thousandth of a litre (ml), equivalent to one cubic centimetre (cc).

millimetre of mercury unit of pressure (symbol mmHg), used in medicine for measuring blood pressure defined as the pressure exerted by a column of mercury one millimetre high, under the action of gravity.

mimicry imitation of one species (or group of species) by another. The most common form is **Batesian mimicry** (named after English naturalist H W Bates), where the mimic resembles a model that is poisonous or unpleasant to eat, and has aposematic, or warning, coloration; the mimic thus benefits from the fact that predators have learned to avoid the model. Hoverflies that resemble bees or wasps are an example. Appearance is usually the basis for mimicry, but calls, songs, scents, and other signals can also be mimicked.

In **Mullerian mimicry**, two or more equally poisonous or distasteful species have a similar colour pattern, thereby reinforcing the warning each gives to predators. In some cases, mimicry is not for protection, but allows the mimic to prey on, or parasitize, the model.

mineral naturally formed inorganic substance with a particular chemical composition and a regularly repeating internal structure. Either in their perfect crystalline form or otherwise, minerals are the constituents of ◊rocks. In more general usage, a mineral is any substance economically valuable for mining (including coal and oil, despite their organic origins).

Mineral forming processes include: melting of pre-existing rock and subsequent crystallization of a mineral to form magmatic or volcanic rocks; weathering of rocks exposed at the land surface, with subsequent transport and grading by surface waters, ice or wind to form sediments; and recrystallization through increasing temperature and pressure with depth to form metamorphic rocks.

Minerals are usually classified as magmatic, sedimentary, or metamorphic. The magmatic minerals include the feldspars, quartz, pyroxenes, amphiboles, micas, and olivines that crystallize from silica-rich rock melts within the crust or from extruded lavas.

The most commonly occurring sedimentary minerals are either pure concentrates or mixtures of sand, clay minerals, and carbonates (chiefly calcite, aragonite, and dolomite).

Minerals typical of metamorphism include andalusite, cordierite, garnet, tremolite, lawsonite, pumpellyite, glaucophane, wollastonite, chlorite, micas, hornblende, staurolite, kyanite, and diopside.

mineral extraction recovery of valuable ores from the Earth's crust. The processes used include open-cast mining, shaft mining, and quarrying, as well as more specialized processes such as those used for oil and sulphur (see, for example, ◊Frasch process).

mineralogy study of minerals. The classification of minerals is based chiefly on their chemical composition and the kind of chemical bonding that holds these atoms together. The mineralogist also studies their crystallographic and physical characters, occurrence, and mode of formation.

The systematic study of minerals began in the 18th century, with the division of minerals into four classes: earths, metals, salts, and bituminous substances, distinguished by their reactions to heat and water.

mineral oil oil obtained from mineral sources, for example coal or petroleum, as distinct from oil obtained from vegetable or animal sources.

mineral salt in nutrition, a simple inorganic chemical that is required by living organisms. Plants usually obtain their mineral salts from the soil, while animals get theirs from their food. Important mineral salts include iron salts (needed by both plants and animals), magnesium salts (needed mainly by plants, to make chlorophyll), and calcium salts (needed by animals to make bone or shell). A ◊trace element is required only in tiny amounts.

minicomputer multiuser computer with a size and processing power between those of a ◊mainframe and a

Main Dietary Minerals

Mineral	Main dietary sources	Major functions in the body	Deficiency symptoms
calcium	milk, cheese, green vegetables, dried legumes	constituent of bones and teeth; essential for nerve transmission, muscle contraction, and clotting	tetany
chromium	vegetable oils, meat	involved in energy metabolism	impaired glucose metabolism
copper	drinking water, meat	associated with iron metabolism	anaemia
fluoride	drinking water, tea, seafoods	helps to keep bones and teeth healthy	increased rate of tooth decay
iodine	seafoods, dairy products, many vegetables, iodized table salt	essential for healthy growth and development	goitre
iron	meat (especially liver), legumes, green vegetables, whole grains, eggs	constituent of haemoglobin; involved in energy metabolism	anaemia
magnesium	whole grains, green vegetables	involved in protein synthesis	growth failure, weakness, behavioural disturbances
manganese	widely distributed in foods	involved in fat synthesis	not known in humans
molybdenum	legumes, cereals, offal	constituent of some enzymes	not known in humans
phosphorus	milk, cheese, meat, legumes, cereals	formation of bones and teeth, maintenance of acid–base balance	weakness, demineralization of bone
potassium	milk, meat, fruits	maintenance of acid–base balance, fluid balance, nerve transmission	muscular weakness, paralysis
selenium	seafoods, meat, cereals, egg yolk	role associated with that of vitamin E	not known in humans
sodium	widely distributed in foods	as for potassium	cramp, loss of appetite, apathy
zinc	widely distributed in foods	involved in digestion	growth failure, underdevelopment of reproductive organs

◊microcomputer. Nowadays almost all minicomputers are based on ◊microprocessors.

Minicomputers are often used in medium-sized businesses and in university departments handling ◊database or other commercial programs and running scientific or graphical applications.

mining extraction of minerals from under the land or sea for industrial or domestic uses. Exhaustion of traditionally accessible resources has led to development of new mining techniques; for example, extraction of oil from offshore deposits and from land shale reserves. Technology is also under development for the exploitation of minerals from entirely new sources such as mud deposits and mineral nodules from the sea bed.

Mud deposits are laid down by hot springs (about 350°C/660°F): sea water penetrates beneath the ocean floor and carries copper, silver, and zinc with it on its return. Such springs occur along the midocean ridges of the Atlantic and Pacific and in the geological rift between Africa and Arabia under the Red Sea.

Mineral nodules form on the ocean bed and contain manganese, cobalt, copper, molybdenum, and nickel; they stand out on the surface, and 'grow' by only a few millimetres every 100,000 years.

The deepest mine in Europe is a 1,100 m/3,630 ft deep working salt and potash mine at Boulby near Whitby on the northeast coast of England.

minor planet another name for an ◊asteroid.

minute unit of time consisting of 60 seconds; also a unit of angle equal to one sixtieth of a degree.

Miocene fourth epoch of the Tertiary period of geological time, 23.5–5.2 million years ago. At this time grasslands spread over the interior of continents, and hoofed mammals rapidly evolved.

mips (acronym for *million instructions per second*) in computing, a measure of the speed of a processor. It does not equal the computer power in all cases.

The original IBM PC had a speed of one-quarter mips, but now 50 mips PCs and 100 mips workstations are available.

Mir Russian 'peace' or 'world' Soviet space station, the core of which was launched on 20 Feb 1986. It is intended to be a permanently occupied space station.

A small wheat crop was harvested aboard the *Mir* space station on 6 Dec 1996. It was the first successful cultivation of a plant from seed in space.

Mira or *Omicron Ceti* brightest long-period pulsating ◊variable star, located in the constellation ◊Cetus. Mira was the first star discovered to vary periodically in brightness.

In 1596 Dutch astronomer David Fabricus noticed Mira as a third-magnitude object. Because it did not appear on any of the star charts available at the time, he mistook it for a ◊nova. The German astronomer Johann Bayer included it on his star atlas in 1603 and designated it Omicron Ceti. The star vanished from view again, only to reappear within a year. It was named 'Stella Mira', 'the wonderful star', by Hevelius, who observed it between 1659 and 1682.

It has a periodic variation between third or fourth magnitude and ninth magnitude over an average period of 331 days. It can sometimes reach second magnitude and once almost attained first magnitude in 1779. At times it is easily visible to the naked eye, being the brightest star in that part of the sky, while at others it cannot be seen without a telescope.

mirage illusion seen in hot weather of water on the horizon, or of distant objects being enlarged. The effect is caused by the ◊refraction, or bending, of light.

Light rays from the sky bend as they pass through the hot layers of air near the ground, so that they appear to come from the horizon.

Because the light is from a blue sky, the horizon appears blue and watery. If, during the night, cold air collects near the ground, light can be bent in the opposite direction, so that objects below the horizon appear to float above it. In the same way, objects such as trees or rocks near the horizon can appear enlarged.

mirror any polished surface that reflects light; often made from 'silvered' glass (in practice, a mercury-alloy coating of glass). A plane (flat) mirror produces a same-size, erect 'virtual' image located behind the mirror at the same distance from it as the object is in front of it. A spherical concave mirror produces a reduced, inverted real image in front or an enlarged, erect virtual image behind it (as in a shaving mirror), depending on how close the object is to the mirror. A spherical convex mirror produces a reduced, erect virtual image behind it (as in a car's rearview mirror).

In a plane mirror the light rays appear to come from behind the mirror but do not actually do so. The inverted real image from a spherical concave mirror is an image in which the rays of light pass through it. The ◊focal length f of a spherical mirror is half the radius of curvature; it is related to the image distance v and object distance u by the equation $1/v + 1/u = 1/f$.

miscarriage spontaneous expulsion of a fetus from the womb before it is capable of independent survival. Often, miscarriages are due to an abnormality in the developing fetus.

missile rocket-propelled weapon, which may be nuclear-armed. Modern missiles are often classified as surface-to-surface missiles (SSM), air-to-air missiles (AAM), surface-to-air missiles (SAM), or air-to-surface missiles (ASM). A **cruise missile** is in effect a pilotless, computer-guided aircraft; it can be sea-launched from submarines or surface ships, or launched from the air or the ground.

Rocket-propelled weapons were first used by the Chinese about AD 1100, and were encountered in the 18th century by the British forces. The rocket missile was then re-invented by William Congreve in England around 1805, and remained in use with various armies in the 19th century. The first wartime use of a long-range missile was against England in World War II, by the jet-powered German V1 (*Vergeltungswaffe*, 'revenge weapon' or Flying Bomb), a monoplane (wingspan about 6 m/18 ft, length 8.5 m/26 ft); the first rocket-propelled missile with a preset guidance system was the German V2, also launched by Germany against Britain in World War II.

Modern missiles are also classified as strategic or tactical: strategic missiles are the large, long-range **intercontinental ballistic missiles** (ICBMs, capable of reaching targets over 5,500 km/3,400 mi), and tactical

missiles are the short-range weapons intended for use in limited warfare (with a range under 1,100 km/680 mi).

Not all missiles are large. There are many missiles that are small enough to be carried by one person. The Stinger, for example, is an anti-aircraft missile fired by a single soldier from a shoulder-held tube. Most fighter aircraft are equipped with missiles to use against enemy aircraft or against ground targets. Other small missiles are launched from a type of truck, called a MLRS (multiple-launch rocket system), that can move around a battlefield. Ship-to-ship missiles like the Exocet have proved very effective in naval battles.

The vast majority of missiles have systems that guide them to their target. The guidance system may consist of radar and computers, either in the missile or on the ground. These devices track the missile and determine the correct direction and distance required for it to hit its target. In the radio-guidance system, the computer is on the ground, and guidance signals are radio-transmitted to the missile. In the inertial guidance system, the computer is on board the missile. Some small missiles have heat-seeking devices fitted to their noses to seek out the engines of enemy aircraft, or are guided by laser light reflected from the target. Others (called TOW missiles) are guided by signals sent along wires that trail behind the missile in flight.

Mississippian US term for the Lower or Early ◊Carboniferous period of geological time, 363–323 million years ago. It is named after the state of Mississippi.

mistral cold, dry, northerly wind that occasionally blows during the winter on the Mediterranean coast of France, particularly concentrated along the Rhône valley. It has been known to reach a velocity of 145 kph/90 mph.

mitochondria (singular **mitochondrion**) membrane-enclosed organelles within ◊eukaryotic cells, containing enzymes responsible for energy production during ◊aerobic respiration. These rodlike or spherical bodies are thought to be derived from free-living bacteria that, at a very early stage in the history of life, invaded larger cells and took up a symbiotic way of life inside. Each still contains its own small loop of DNA called mitochondrial DNA, and new mitochondria arise by division of existing ones.

mitosis in biology, the process of cell division by which identical daughter cells are produced. During mitosis the DNA is duplicated and the chromosome number doubled, so new cells contain the same amount of DNA as the original cell.

The genetic material of ◊eukaryotic cells is carried on a number of ◊chromosomes. To control movements of chromosomes during cell division so that both new cells get the correct number, a system of protein tubules, known as the spindle, organizes the chromosomes into position in the middle of the cell before they replicate. The spindle then controls the movement of chromosomes as the cell goes through the stages of division: **interphase**, **prophase**, **metaphase**, **anaphase**, and **telophase**. See also ◊meiosis.

mixed farming farming system where both arable and pastoral farming is carried out. Mixed farming is a lower-risk strategy than ◊monoculture. If climate, pests, or market prices are unfavourable for one crop or type of livestock, another may be more successful and the risk is

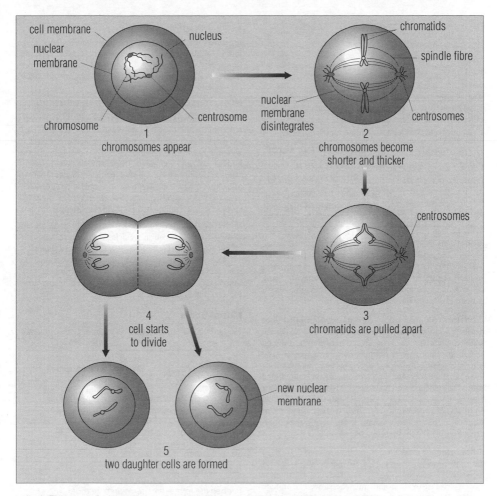

mitosis The stages of mitosis, the process of cell division that takes place when a plant or animal cell divides for growth or repair. The two daughter cells each receive the same number of chromosomes as were in the original cell.

shared. Animals provide manure for the fields and help to maintain soil fertility.

mixture in chemistry, a substance containing two or more compounds that still retain their separate physical and chemical properties. There is no chemical bonding between them and they can be separated from each other by physical means (compare ◊compound).

m.k.s. system system of units in which the base units metre, kilogram, and second replace the centimetre, gram, and second of the ◊c.g.s. system. From it developed the SI system (see ◊SI units).

It simplifies the incorporation of electrical units into the metric system, and was incorporated in SI. For application to electrical and magnetic phenomena, the ampere was added, creating what is called the m.k.s.a. system.

ml symbol for *millilitre*.

mm symbol for *millimetre*.

mmHg symbol for ◊*millimetre of mercury*.

mnemonic in computing, a short sequence of letters used in low-level programming languages (see ◊low-level language) to represent a ◊machine code instruction.

mobile ion in chemistry, an ion that is free to move; such ions are only found in the aqueous solutions or melts (molten masses) of an ◊electrolyte. The mobility of the ions in an electrolyte is what allows it to conduct electricity.

Möbius strip structure made by giving a half twist to a flat strip of paper and joining the ends together. It has certain remarkable properties, arising from the fact that it has only one edge and one side. If cut down the centre of the strip, instead of two new strips of paper, only one long strip is produced. It was invented by the German mathematician August Möbius. *See illustration overleaf.*

mode in mathematics, the element that appears most frequently in a given set of data. For example, the mode for the data 0, 0, 9, 9, 9, 12, 87, 87 is 9.

Möbius strip The Möbius strip has only one side and one edge. It consists of a strip of paper connected at its ends with a half-twist in the middle.

A theory has only the alternative of being right or wrong. A model has a third possibility: it may be right, but irrelevant.

On **model** Manfred Eigen, quoted in Jagdish Mehra (ed) *The Physicist's Conception of Nature* 1973

model simplified version of some aspect of the real world. Models are produced to show the relationships between two or more factors, such as land use and the distance from the centre of a town (for example, concentric-ring theory). Because models are idealized, they give only a general guide to what may happen.

modem (acronym for *modulator/demodulator*) device for transmitting computer data over telephone lines. Such a device is necessary because the ◊digital signals produced by computers cannot, at present, be transmitted directly over the telephone network, which uses ◊analogue signals. The modem converts the digital signals to analogue, and back again. Modems are used for linking remote terminals to central computers and enable computers to communicate with each other anywhere in the world. The fastest modems transmit at a maximum rate of about 28,000 bps (bits per second) (1996).

moderator in a ◊nuclear reactor, a material such as graphite or heavy water used to reduce the speed of high-energy neutrons. Neutrons produced by nuclear fission are fast-moving and must be slowed to initiate further fission so that nuclear energy continues to be released at a controlled rate.

Slow neutrons are much more likely to cause ◊fission in a uranium-235 nucleus than to be captured in a U-238 (nonfissile uranium) nucleus. By using a moderator, a reactor can thus be made to work with fuel containing only a small proportion of U-235.

modulation in radio transmission, the variation of frequency, or amplitude, of a radio carrier wave, in accordance with the audio characteristics of the speaking voice, music, or other signal being transmitted. See ◊pulse-code modulation, ◊AM (amplitude modulation), and ◊FM (frequency modulation).

module in construction, a standard or unit that governs the form of the rest. For example, Japanese room sizes are traditionally governed by multiples of standard tatami floor mats; today prefabricated buildings are mass-produced in a similar way. The components of a spacecraft are designed in coordination; for example, for the Apollo Moon landings the craft comprised a command module (for working, eating, sleeping), service module (electricity generators, oxygen supplies, manoeuvring rocket), and lunar module (to land and return the astronauts).

modulus in mathematics, a number that divides exactly into the difference between two given numbers. Also, the multiplication factor used to convert a logarithm of one base to a logarithm of another base. Also, another name for ◊absolute value.

MOH abbreviation for *Medical Officer of Health*.

Mohorovičić discontinuity also *Moho* or *M-discontinuity* boundary that separates the Earth's crust and mantle, marked by a rapid increase in the speed of earthquake waves. It follows the variations in the thickness of the crust and is found approximately 32 km/20 mi below the continents and about 10 km/6 mi below the oceans. It is named after the Yugoslav geophysicist Andrija Mohorovičić, (1857–1936), who suspected its presence after analysing seismic waves from the Kulpa Valley earthquake in 1909.

Mohs' scale scale of hardness for minerals (in ascending order): 1 talc; 2 gypsum; 3 calcite; 4 fluorite; 5 apatite; 6 orthoclase; 7 quartz; 8 topaz; 9 corundum; 10 diamond.

The scale is useful in mineral identification because any mineral will scratch any other mineral lower on the scale than itself, and similarly it will be scratched by any other mineral higher on the scale.

Mohs' Scale

Number	Defining mineral	Other Substances compared
1	talc	
2	gypsum	2 1/2 fingernail
3	calcite	3 1/2 copper coin
4	fluorite	
5	apatite	5 1/2 steel blade
6	orthoclase	5 3/4 glass
7	quartz	7 steel file
8	topaz	
9	corundum	
10	diamond	

note that the scale is not regular; diamond, at number 10 the hardest natural substance, is 90 times harder in absolute terms than corundum, number 9

molar one of the large teeth found towards the back of the mammalian mouth. The structure of the jaw, and the relation of the muscles, allows a massive force to be applied to molars. In herbivores the molars are flat with sharp ridges of enamel and are used for grinding, an adaptation to a diet of tough plant material. Carnivores have sharp powerful molars called carnassials, which are adapted for cutting meat.

molarity in chemistry, ◊concentration of a solution expressed as the number of ◊moles in grams of solute per cubic decimetre of solution.

molar solution in chemistry, solution that contains one ◊mole of a substance per litre of solvent.

molar volume volume occupied by one ◊mole (the molecular mass in grams) of any gas at standard temperature and pressure, equal to 2.24136×10^{-2} m^3.

mole SI unit (symbol mol) of the amount of a substance. It is defined as the amount of a substance that contains as many elementary entities (atoms, molecules, and so on) as there are atoms in 12 g of the ◊isotope carbon-12.

external modem

**external modem
for a notebook computer**

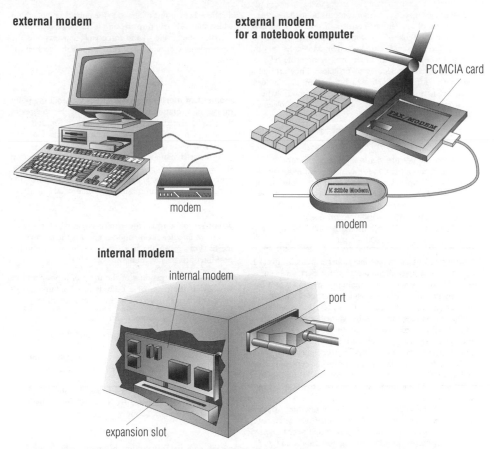

PCMCIA card

modem

internal modem

internal modem

port

expansion slot

modem

modem Modems are available in various forms: microcomputers may use an external device connected through a communications port, or an internal device, which takes the form of an expansion board inside the computer. Notebook computers use an external modem connected via a special interface card.

One mole of an element that exists as single atoms weighs as many grams as its ◊atomic number (so one mole of carbon weighs 12 g), and it contains 6.022045 × 10^{23} atoms, which is ◊Avogadro's number.

molecular biology study of the molecular basis of life, including the biochemistry of molecules such as DNA, RNA, and proteins, and the molecular structure and function of the various parts of living cells.

molecular clock use of rates of ◊mutation in genetic material to calculate the length of time elapsed since two related species diverged from each other during evolution. The method can be based on comparisons of the DNA or of widely occurring proteins, such as haemoglobin.

Since mutations are thought to occur at a constant rate, the length of time that must have elapsed in order to produce the difference between two species can be estimated. This information can be compared with the evidence obtained from palaeontology to reconstruct evolutionary events.

molecular cloud in astronomy, an enormous cloud of cool interstellar dust and gas containing hydrogen molecules and more complex molecular species. Giant molecular clouds (GMCs), about a million times as massive as the Sun and up to 300 light years in diameter, are regions in which stars are being born. The ◊Orion Nebula is part of a GMC.

molecular formula in chemistry, formula indicating the actual number of atoms of each element present in a single molecule of a chemical compound. This is determined by two pieces of information: the empirical ◊formula and the ◊relative molecular mass, which is determined experimentally.

molecular mass (also known as relative molecular mass) the mass of a molecule, calculated relative to one-twelfth the mass of an atom of carbon-12. It is found by adding the relative atomic masses of the atoms that make up the molecule.

molecular solid in chemistry, solid composed of molecules that are held together by relatively weak ◊intermolecular forces. Such solids are low-melting and tend to dissolve in organic solvents. Examples of molecular solids are sulphur, ice, sucrose, and solid carbon dioxide.

molecule group of two or more ◊atoms bonded together. A molecule of an element consists of one or more like ◊atoms; a molecule of a compound consists of two or more different atoms bonded together. Molecules vary in size and complexity from the hydrogen molecule (H_2) to the large ◊macromolecules of proteins. They are held together by ionic bonds, in which the atoms gain or lose electrons to form ◊ions, or by covalent bonds, where electrons from each atom are shared in a new molecular orbital.

Molecule

Most molecules are made up of small numbers of atoms but some contain rather more. For example, a molecule of aspirin contains 21 atoms. A molecule of haemoglobin, a substance found in blood, contains 758 carbon atoms, 1,203 hydrogen atoms, 195 oxygen atoms, 218 nitrogen atoms, 1 iron atom, and 3 sulphur atoms. Rubber molecules may have up to 65,000 atoms.

---

mollusc any of a group of invertebrate animals, most of which have a body divided into three parts: a head, a central mass containing the main organs, and a foot for movement; the more sophisticated octopuses and related molluscs have arms to capture their prey. The majority of molluscs are marine animals, but some live in fresh water, and a few live on dry land. They include clams, mussels, and oysters (bivalves), snails and slugs (gastropods), and cuttlefish, squids, and octopuses (cephalopods). The body is soft, without limbs (except for the cephalopods), and cold-blooded. There is no internal skeleton, but many species have a hard shell covering the body. (Phylum Mollusca.)

Molluscs have varying diets, the carnivorous species feeding mainly on other molluscs. Some are vegetarian. Reproduction is by means of eggs and is sexual; many species are hermaphrodite (having both male and female reproductive organs). The shells of molluscs take a variety of forms: single or univalve (like the snail), double or bivalve (like the clam), chambered (like the nautilus), and many other variations. In some cases (for example cuttlefish and squid), the shell is internal. Every mollusc has a fold of skin, the mantle, which covers either the whole body or only the back, and secretes the chalky substance that forms the shell. The lower ventral surface (belly area) of the body forms the foot, which enables the mollusc to move about.

molybdenite molybdenum sulphide, MoS_2, the chief ore mineral of molybdenum. It possesses a hexagonal crystal structure similar to graphite, has a blue metallic lustre, and is very soft (1–1.5 on Mohs' scale).

molybdenum heavy, hard, lustrous, silver-white, metallic element, symbol Mo, atomic number 42, relative atomic mass 95.94. The chief ore is the mineral molybdenite. The element is highly resistant to heat and conducts electricity easily. It is used in alloys, often to harden steels. It is a necessary trace element in human nutrition. It was named in 1781 by Swedish chemist Karl Scheele, after its isolation by P J Hjelm (1746–1813), for its resemblance to lead ore.

moment of a force in physics, measure of the turning effect, or torque, produced by a force acting on a body. It is equal to the product of the force and the perpendicular distance from its line of action to the point, or pivot, about which the body will turn. Its unit is the newton metre.

If the magnitude of the force is F newtons and the perpendicular distance is d metres then the moment is given by:

$$\text{moment} = Fd$$

moment of inertia in physics, the sum of all the point masses of a rotating object multiplied by the squares of their respective distances from the axis of rotation. It is analogous to the ◊mass of a stationary object or one moving in a straight line.

In linear dynamics, Newton's second law of motion states that the force F on a moving object equals the products of its mass m and acceleration a ($F = ma$); the analogous equation in rotational dynamics is:

$$T = I\alpha$$

where T is the torque (the turning effect of a force) that causes an angular acceleration α and I is the moment of inertia. For a given object, I depends on its shape and the position of its axis of rotation.

momentum the product of the mass of a body and its velocity. If the mass of a body is m kilograms and its velocity is v m s^{-1}, then its momentum is given by:

$$\text{Momentum} = mv$$

Its unit is the kilogram metre-per-second (kg m s^{-1}) or the newton second.

The momentum of a body does not change unless a resultant or unbalanced force acts on that body (see ◊Newton's laws of motion).

According to Newton's second law of motion, the magnitude of a resultant force F equals the rate of change of momentum brought about by its action, or:

$$F = (mv - mu)/t$$

where mu is the initial momentum of the body, mv is its final momentum, and t is the time in seconds over which the force acts. The change in momentum, or ◊impulse, produced can therefore be expressed as:

$$\text{impulse} = mv - mu = Ft$$

The law of conservation of momentum is one of the fundamental concepts of classical physics. It states that the total momentum of all bodies in a closed system is constant and unaffected by processes occurring within the system.

The **angular momentum** of an orbiting or rotating body of mass m travelling at a velocity v in a circular orbit of radius R is expressed as mvR. Angular momentum is conserved, and should any of the values alter (such as the radius of orbit), the other values (such as the velocity) will compensate to preserve the value of angular momentum, and that lost by one component is passed to another.

monazite mineral, $(Ce,La,Th)PO_4$, yellow to red, valued as a source of ◊lanthanides or rare earths, including cerium and europium; generally found in placer deposit (alluvial) sands.

mongolism former name (now considered offensive) for ◊Down's syndrome.

monocarpic or *hapaxanthic* describing plants that flower and produce fruit only once during their life cycle, after which they die. Most ◊annual plants and ◊biennial plants are monocarpic, but there are also a small number of monocarpic ◊perennial plants that flower just once, sometimes after as long as 90 years, dying shortly afterwards, for example, century plant *Agave* and some species of bamboo *Bambusa*. The general biological term related to organisms that reproduce only once during their lifetime is ◊semelparity.

monoclonal antibody (MAB) antibody produced by fusing an antibody-producing lymphocyte with a cancerous myeloma (bone-marrow) cell. The resulting fused cell, called a hybridoma, is immortal and can be used to produce large quantities of a single, specific antibody. By choosing antibodies that are directed against antigens found on cancer cells, and combining them with cytotoxic drugs, it is hoped to make so-called magic bullets that will be able to pick out and kill cancers.

It is the antigens on the outer cell walls of germs entering the body that provoke the production of antibodies as a first line of defence against disease. Antibodies 'recognize' these foreign antigens, and, in locking on to them, cause the release of chemical signals in the bloodstream to alert the immune system for further action. MABs are copies of these natural antibodies, with the same ability to recognize specific antigens. Introduced into the body, they can be targeted at disease sites.

The full potential of these biological missiles, developed by César Milstein and others at Cambridge University, England, in 1975, is still under investigation. However, they are already in use in blood-grouping, in pinpointing viruses and other sources of disease, in tracing cancer sites, and in developing vaccines.

monocotyledon angiosperm (flowering plant) having an embryo with a single cotyledon, or seed leaf (as opposed to ◊dicotyledons, which have two). Monocotyledons usually have narrow leaves with parallel veins and smooth edges, and hollow or soft stems. Their flower parts are arranged in threes. Most are small plants such as orchids, grasses, and lilies, but some are trees such as palms.

monoculture farming system where only one crop is grown. In Third World countries this is often a ◊cash crop, grown on ◊plantations, for example, sugar and coffee. Cereal crops in the industrialized world are also frequently grown on a monoculture basis; for example, wheat in the Canadian prairies.

Monoculture allows the farmer to tailor production methods to the requirements of one crop, but it is a high-risk strategy since the crop may fail (because of pests, disease, or bad weather) and world prices for the crop may fall. Monoculture without ◊crop rotation is likely to result in reduced soil quality despite the addition of artificial fertilizers, and it contributes to ◊soil erosion.

monocyte in biology, type of white blood cell. They are found in the tissues, the lymphatic and circulatory systems where their purpose is to remove foreign particles, such as bacteria and tissue debris, by ingesting them.

monoecious having separate male and female flowers on the same plant. Maize (*Zea mays*), for example, has a tassel of male flowers at the top of the stalk and a group of female flowers (on the ear, or cob) lower down. Monoecism is a way of avoiding self-fertilization.

◊Dioecious plants have male and female flowers on separate plants.

monohybrid inheritance pattern of inheritance seen in simple ◊genetics experiments, where the two animals (or two plants) being crossed are genetically identical except for one gene.

This gene may code for some obvious external features such as seed colour, with one parent having green seeds and the other having yellow seeds. The offspring are monohybrids, that is, hybrids for one gene only, having received one copy of the gene from each parent. Known as the F1 generation, they are all identical, and usually resemble one parent, whose version of the gene (the dominant ◊allele) masks the effect of the other version (the recessive allele). Although the characteristic coded for by the recessive allele (for example, green seeds) completely disappears in this generation, it can reappear in offspring of the next generation if they have two recessive alleles. On average, this will occur in one out of four offspring from a cross between two of the monohybrids. The next generation (called F2) show a 3:1 ratio for the characteristic in question, 75% being like the original parent with the recessive allele. Austrian biologist Gregor Mendel first carried out experiments of this type (crossing varieties of artificially bred plants, such as peas) and they revealed the principles of genetics. The same basic mechanism underlies all inheritance, but in most plants and animals there are so many genetic differences interacting to produce the external appearance (phenotype) that such simple, clear-cut patterns of inheritance are not evident.

monolith single isolated stone or column, usually standing and of great size, used as a form of monument. Some are natural features, such as the Buck Stone in the Forest of Dean, England. Other monoliths may be quarried, resited, finished, or carved; those in Egypt of about 3000 BC take the form of obelisks. They have a wide distribution including Europe, South America, N Africa, and the Middle East.

Apart from their ritual or memorial function, monoliths have been used as sundials and calendars in the civilizations of the Aztecs, Egyptians, and Chaldeans (ancient peoples of southern Babylonia). In landscape archaeology, monoliths are interpreted in a wider context, possibly as boundary markers. The largest cut stone, weighing about 1,500 tonnes, is sited in the ancient Syrian city of Baalbek.

monomer chemical compound composed of simple molecules from which ◊polymers can be made. Under certain conditions the simple molecules (of the monomer) join together (polymerize) to form a very long chain molecule (macromolecule) called a polymer. For example, the polymerization of ethene (ethylene) monomers produces the polymer polyethene (polyethylene).

$$2n\mathrm{CH}_2 = \mathrm{CH}_2 \rightarrow (\mathrm{CH}_2\text{-}\mathrm{CH}_2\text{-}\mathrm{CH}_2\text{-}\mathrm{CH}_2)_n$$

monorail railway that runs on a single rail; the cars can be balanced on it or suspended from it. It was invented 1882 to carry light loads, and when run by electricity was called a *telpher*.

The Wuppertal Schwebebahn, which has been running in Germany since 1901, is a suspension monorail, where the passenger cars hang from an arm fixed to a trolley that runs along the rail. Today most monorails are of the straddle type, where the passenger cars run on top of the rail. They are used to transport passengers between terminals at some airports; Japan has a monorail (1964) running from the city centre to Hameda airport.

monosaccharide or *simple sugar* ◊carbohydrate that cannot be hydrolysed (split) into smaller carbohydrate units. Examples are glucose and fructose, both of which have the molecular formula $C_6H_{12}O_6$.

monosodium glutamate (MSG) $NaC_5H_8NO_4$ a white, crystalline powder, the sodium salt of glutamic acid (an ◊amino acid found in proteins that plays a role in the metabolism of plants and animals). It has no flavour of its own, but enhances the flavour of foods such as meat and fish. It is used to enhance the flavour of many packaged and 'fast foods', and in Chinese cooking. Ill effects may arise from its overconsumption, and some people are very sensitive to it, even in small amounts. It is commercially derived from vegetable protein. It occurs naturally in soybeans and seaweed.

monotreme any of a small group of primitive egg-laying mammals, found in Australasia. They include the echidnas (spiny anteaters) and the platypus. (Order Monotremata.)

In 1995 Australian palaeontologists announced a new (extinct) family of monotreme, the Kollikodontidae, following the discovery of a 120-million-year-old jawbone in New South Wales.

monozygotic twin one of a set of ◊twins born to the same parents at the same time following the fertilization of a single ovum. They are of the same sex and virtually identical in appearance.

monsoon wind pattern that brings seasonally heavy rain to S Asia; it blows towards the sea in winter and towards the land in summer. The monsoon may cause destructive flooding all over India and SE Asia from April to Sept, leaving thousands of people homeless each year.

The monsoon cycle is believed to have started about 12 million years ago with the uplift of the Himalayas.

month unit of time based on the motion of the Moon around the Earth.

The time from one new or full Moon to the next (the **synodic** or **lunar month**) is 29.53 days. The time for the Moon to complete one orbit around the Earth relative to the stars (the **sidereal month**) is 27.32 days. The **solar month** equals 30.44 days, and is exactly one-twelfth of the solar or tropical year, the time taken for the Earth to orbit the Sun. The **calendar month** is a human invention, devised to fit the calendar year.

Montréal Protocol international agreement, signed in 1987, to stop the production of chemicals that are ◊ozone depleters by the year 2000.

Originally the agreement was to reduce the production of ozone depleters by 35% by 1999. The green movement criticized the agreement as inadequate, arguing that an 85% reduction in ozone depleters would be necessary just to stabilize the ozone layer at 1987 levels. The protocol (under the Vienna Convention for the Protection of the Ozone Layer) was reviewed in 1992. Amendments added another 11 chemicals to the original list of eight chemicals suspected of harming the ozone layer. A controversial amendment concerns a fund established to pay for the transfer of ozone-safe technology to poor countries.

moon in astronomy, any natural ◊satellite that orbits a planet. Mercury and Venus are the only planets in the Solar System that do not have moons.

Moon natural satellite of Earth, 3,476 km/2,160 mi in diameter, with a mass 0.012 (approximately one-eightieth) that of Earth.

Its surface gravity is only 0.16 (one-sixth) that of Earth. Its average distance from Earth is 384,400 km/238,855 mi, and it orbits in a west-to-east direction every 27.32 days (the **sidereal month**). It spins on its axis with one side permanently turned towards Earth. The Moon has no atmosphere or water.

The Moon is illuminated by sunlight, and goes through a cycle of phases of shadow, waxing from **new** (dark) via **first quarter** (half Moon) to **full**, and waning back again to new every 29.53 days (the **synodic month**, also known as a **lunation**). On its sunlit side, temperatures reach 110°C/230°F, but during the two-week lunar night the surface temperature drops to −170°C/−274°F.

The origin of the Moon is still open to debate. Scientists suggest the following theories: that it split from the Earth; that it was a separate body captured by Earth's gravity; that it formed in orbit around Earth; or that it was formed from debris thrown off when a body the size of Mars struck Earth. Future exploration of the Moon may detect water permafrost, which could be located at the permanently shadowed lunar poles.

The far side of the Moon was first photographed from the Soviet *Lunik 3* in Oct 1959. Much of our information about the Moon has been derived from this and other photographs and measurements taken by US and Soviet Moon probes, from geological samples brought back by US Apollo astronauts and by Soviet Luna probes, and from experiments set up by the US astronauts in 1969–72.

The Moon's composition is rocky, with a surface heavily scarred by ◊meteorite impacts that have formed craters up to 240 km/150 mi across. Seismic observations indicate that the Moon's surface extends downwards for tens of kilometres; below this crust is a solid mantle about 1,100 km/688 mi thick, and below that a silicate core, part of which may be molten. Rocks brought back by astronauts show the Moon is 4.6 billion years old, the same age as Earth. It is made up of the same chemical elements as Earth, but in different proportions, and differs from Earth in that most of the Moon's surface features were formed within the first billion years of its history when it was hit repeatedly by meteorites. The youngest craters are surrounded by bright rays of ejected rock. The largest scars have been filled by dark lava to produce the lowland plains called seas, or **maria** (plural of ◊mare). These dark patches form the so-called 'man-in-the-Moon' pattern. One of the Moon's easiest features to observe is the mare **Plato**, which is about 100 km/62 mi in diameter and 2,700 m/8,860 ft deep, and at times is visible with the naked eye alone.

Moon probe crewless spacecraft used to investigate the Moon. Early probes flew past the Moon or crash-landed on it, but later ones achieved soft landings or went into orbit. Soviet probes included the Luna/Lunik series. US probes (Ranger, Surveyor, Lunar Orbiter) prepared the way for the Apollo crewed flights.

moonstone translucent, pearly variety of potassium sodium ◊feldspar, found in Sri Lanka and Myanmar, and distinguished by a blue, silvery, or red opalescent tint. It is valued as a gem.

moor in earth science, a stretch of land, usually at a height, which is characterized by a vegetation of heather, coarse grass, and bracken. A moor may be poorly drained and contain boggy hollows.

moped lightweight motorcycle with pedals. Early mopeds (like the autocycle) were like motorized bicycles, using the pedals to start the bike and assist propulsion uphill. The pedals have little function in many mopeds today.

moraine rocky debris or ◊till carried along and deposited by a ◊glacier. Material eroded from the side of a glaciated valley and carried along the glacier's edge is called a *lateral moraine*; that worn from the valley floor and carried along the base of the glacier is called a *ground moraine*. Rubble dropped at the snout of a melting glacier is called a *terminal moraine*.

When two glaciers converge their lateral moraines unite to form a *medial moraine*. Debris that has fallen down crevasses and becomes embedded in the ice is termed an *englacial moraine*; when this is exposed at the surface due to partial melting it becomes ablation moraine.

morbidity in medicine, a term used to describe the condition of being diseased. The number of times a disease occurs within a particular number of the population is the morbidity rate of that disease.

moribund in medicine, term used to describe a patient who is dying.

morphine narcotic alkaloid $C_{17}H_{19}NO_3$ derived from ◊opium and prescribed only to alleviate severe pain. Its use produces serious side effects, including nausea, constipation, tolerance, and addiction, but it is highly valued for the relief of the terminally ill.

The risk of addiction arising from the use of morphine for pain relief is much lower than for recreational use (about 1 in 3,000) as the drug is processed differently by the body when pain is present.

morphing the metamorphosis of one shape or object into another by computer-generated animation. First used in film-making in 1990, it has transformed cinema special effects. Conventional animation is limited to two dimensions; morphing enables the creation of three-dimensional transformations.

To create such effects, the start and end of the transformation must be specified on screen using a wire-frame model that mathematically defines the object. To make the object three-dimensional, the wire can be extruded from a cross-section or turned as on a lathe to produce an evenly turned surface. This is then rendered, or filled in and shaded. Once the beginning and end objects have been created, the computer can calculate the morphing process.

morphogen in medicine, one of a class of substances believed to be present in the growing embryo, controlling its growth pattern.

It is thought that variations in the concentration of morphogens in different parts of the embryo cause them to grow at different rates.

morphology in biology, the study of the physical structure and form of organisms, in particular their soft tissues.

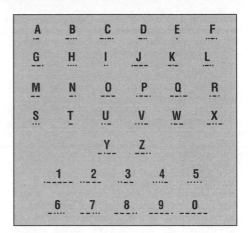

Morse code

Morse code international code for transmitting messages by wire or radio using signals of short (dots) and long (dashes) duration, originated by US inventor Samuel Morse for use on his invention, the telegraph (see ◊telegraphy).

The letters SOS (3 short, 3 long, 3 short) form the international distress signal, being distinctive and easily transmitted (popularly but erroneously 'save our souls'). By radio telephone the distress call is 'Mayday', for similar reasons (popularly alleged to derive from French *m'aidez*, help me).

mosquito any of a group of flies in which the female has needlelike mouthparts and sucks blood before laying eggs. The males feed on plant juices. Some mosquitoes carry diseases such as ◊malaria. (Family Culicidae, order Diptera.)

Human odour in general is attractive to mosquitos, as well as the lactic acid in sweat and the heat of the human body at close range. The varying reactions to mosquito bites depend on the individual's general allergic reaction and not on the degree of the bite; the allergic reaction is caused by the saliva injected from the mosquito's salivary glands to prevent the host's blood from clotting. The mosquito consumes 4 microlitres of blood when it feeds. Natural mosquito repellents include lavender oil, citronella (from lemon grass), thyme, and eucalyptus oils.

moss small nonflowering plant of the class Musci (10,000 species), forming with the ◊liverworts and the ◊hornworts the order Bryophyta. The stem of each plant bears ◊rhizoids that anchor it; there are no true roots. Leaves spirally arranged on its lower portion have sexual organs at their tips. Most mosses flourish best in damp conditions where other vegetation is thin.

The peat or bog moss *Sphagnum* was formerly used for surgical dressings. The smallest moss is the Cape pygmy moss *Ephemerum capensi*, only slightly larger than a pin head.

Mössbauer effect the recoil-free emission of gamma rays from atomic nuclei under certain conditions. The effect was discovered in 1958 by German physicist Rudolf Mössbauer, and used in 1960 to provide the first laboratory test of Einstein's general theory of relativity.

The absorption and subsequent re-emission of a

gamma ray by an atomic nucleus usually causes it to recoil, so affecting the wavelength of the emitted ray. Mössbauer found that at low temperatures, crystals will absorb gamma rays of a specific wavelength and resonate so that the crystal as a whole recoils while individual nuclei do not. The wavelength of the re-emitted gamma rays is therefore virtually unaltered by recoil and may be measured to a high degree of accuracy. Changes in the wavelength may therefore be studied as evidence of the effect of, say, neighbouring electrons or gravity. For example, the effect provided the first verification of the general theory of relativity by showing that gamma-ray wavelengths become longer in a gravitational field, as predicted by Einstein.

motherboard ◊printed circuit board that contains the main components of a microcomputer. The power, memory capacity, and capability of the microcomputer may be enhanced by adding expansion boards to the motherboard.

motility the ability to move spontaneously. The term is often restricted to those cells that are capable of independent locomotion, such as spermatozoa. Many single-celled organisms are motile, for example, the amoeba. Research has shown that cells capable of movement, including vertebrate muscle cells, have certain biochemical features in common. Filaments of the proteins actin and myosin are associated with motility, as are the metabolic processes needed for breaking down the energy-rich compound ◊ATP (adenosine triphosphate).

motor anything that produces or imparts motion; a machine that provides mechanical power – for example, an ◊electric motor. Machines that burn fuel (petrol, diesel) are usually called engines, but the internal-combustion engine that propels vehicles has long been called a motor, hence 'motoring' and 'motorcar'. Actually the motor is a part of the car engine.

motorboat small, waterborne craft for pleasure cruising or racing, powered by a petrol, diesel, or gas-turbine engine. A boat not equipped as a motorboat may be converted by a detachable outboard motor. For increased speed, such as in racing, motorboat hulls are designed to skim the water (aquaplane) and reduce frictional resistance. Plastics, steel, and light alloys are now used in construction as well as the traditional wood.

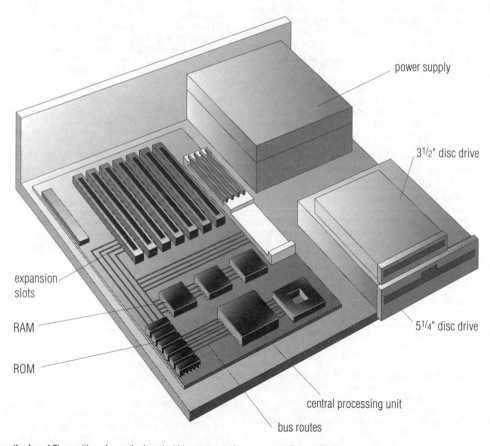

power supply

3¹/₂" disc drive

expansion slots

RAM

ROM

5¹/₄" disc drive

central processing unit

bus routes

motherboard The position of a motherboard within a computer's system unit. The motherboard contains the central processing unit, Random Access Memory (RAM) chips, Read-Only Memory (ROM), and a number of expansion slots.

In recent designs, drag is further reduced with hydro-fins and ◊hydrofoils, which enable the hull to rise clear of the water at normal speeds. Notable events in motorboat or 'powerboat' racing include the American Gold Cup in 1947 (over a 145 km/90 mi course) and the Round-Britain race in 1969.

motorcar another term for ◊car.

motorcycle or *motorbike* two-wheeled vehicle pro-pelled by a ◊petrol engine. The first successful motorized bicycle was built in France in 1901, and British and US manufacturers first produced motorbikes in 1903.

In 1868 Ernest and Pierre Michaux in France experi-mented with a steam-powered bicycle, but the steam-power unit was too heavy and cumbersome. Gottlieb Daimler, a German engineer, created the first motorcycle when he installed his lightweight petrol engine in a bicycle frame in 1885. Daimler soon lost interest in two wheels in favour of four and went on to pioneer the ◊car.

The first really successful two-wheel design was devised by Michael and Eugene Werner in France in 1901. They adopted the classic motorcycle layout with the engine low down between the wheels.

Harley Davidson in the USA and Triumph in the UK began manufacture in 1903. Road races like the Isle of Man TT (Tourist Trophy), established in 1907, helped improve motorcycle design and it soon evolved into more or less its present form. British bikes included the Vin-cent, BSA, and Norton.

In the 1970s British manufacturers were overtaken by Japanese ones, and such motorcycles as Honda, Kawa-saki, Suzuki, and Yamaha now dominate the world mar-ket. They make a wide variety of machines, from ◊mopeds (lightweights with pedal assistance) to stream-lined superbikes capable of speeds up to 250 kph/ 160 mph. There is still a smaller but thriving Italian motorcycle industry, making more specialist bikes. Laverda, Moto Guzzi, and Ducati continue to manufac-ture in Italy.

The lightweight bikes are generally powered by a two-stroke petrol engine (see ◊two-stroke cycle), while bikes with an engine capacity of 250 cc or more are generally four-strokes (see ◊four-stroke cycle). However, many special-use larger bikes (such as those developed for off-road riding and racing) are two-stroke. Most motorcycles are air-cooled – their engines are surrounded by metal fins to offer a large surface area – although some have a water-cooling system similar to that of a car. Most small bikes have single-cylinder engines, but larger machines can have as many as six. The single-cylinder engine is economical and was popular in British manufacture, then the Japanese developed multiple-cylinder models, but there has recently been some return to single-cylinder engines. In the majority of bikes a chain carries the drive from the engine to the rear wheel, though some machines are now fitted with shaft drive.

motor effect tendency of a wire carrying an electric current in a magnetic field to move. The direction of the movement is given by the left-hand rule (see ◊Fleming's rules). This effect is used in the ◊electric motor. It also explains why streams of electrons produced, for instance, in a television tube can be directed by electromagnets.

motor nerve in anatomy, any nerve that transmits impulses from the central nervous system to muscles or organs. Motor nerves cause voluntary and involuntary muscle contractions, and stimulate glands to secrete hormones.

motor neurone disease (MND) or *amyotrophic lat-eral sclerosis* chronic disease in which there is progress-ive degeneration of the nerve cells which instigate movement. It leads to weakness, wasting, and loss of muscle function and usually proves fatal within two to three years of onset. Motor neurone disease occurs in both hereditary and sporadic forms but its causes remain unclear. A gene believed to be implicated in familial cases was discovered in 1993. In Britain some 1,200 new cases are diagnosed each year.

Results of a US trial in 1995 showed that the drug Myotrophin, a genetically engineered version of a chemi-cal produced in the muscles, slowed deterioration in suf-ferers of MND by 25%.

motorway main road for fast motor traffic, with two or more lanes in each direction, and with special access points (junctions) fed by slip roads. The first motorway (85 km/53 mi) ran from Milan to Varese, Italy, and was completed in 1924; by 1939 some 500 km/300 mi of motorway (*autostrada*) had been built, although these did not attain the standards of later express highways. In Germany some 2,100 km/1,310 mi of *Autobahnen* had been completed by 1942. After World War II motorways were built in a growing number of countries, for example the USA, France, and the UK. The most ambitious build-ing programme was in the USA, which by 1974 had 70,800 km/44,000 mi of 'expressway'. Construction of new motorways causes much environmental concern.

mould furlike growth caused by any of a group of fungi (see ◊fungus) living on foodstuffs and other organic mat-ter; a few are parasitic on plants, animals, or each other. Many moulds are of medical or industrial importance; for example, the antibiotic penicillin comes from a type of mould.

moulding use of a pattern, hollow form, or matrix to give a specific shape to something in a plastic or molten state. It is commonly used for shaping plastics, clays, and glass. When metals are used, the process is called ◊casting.

In *injection moulding*, molten plastic, for example, is injected into a water-cooled mould and takes the shape of the mould when it solidifies. In *blow moulding*, air is blown into a blob of molten plastic inside a hollow mould. In *compression moulding*, synthetic resin powder is simultaneously heated and pressed into a mould.

moulting periodic shedding of the hair or fur of mam-mals, feathers of birds, or skin of reptiles. In mammals and birds, moulting is usually seasonal and is triggered by changes of day length.

The term is also often applied to the shedding of the ◊exoskeleton of arthropods, but this is more correctly called ◊ecdysis.

mountain natural upward projection of the Earth's sur-face, higher and steeper than a hill. The process of moun-tain building (orogeny) consists of volcanism, folding, faulting, and thrusting, resulting from the collision and welding together of two tectonic plates (see ◊plate tecton-ics). This process deforms the rock and compresses it between the two plates into mountain chains.

Mount Palomar astronomical observatory, 80 km/50 mi NE of San Diego, California, USA. It has a 5 m/200 in diameter reflector called the Hale. Completed in 1948, it was the world's premier observatory during the 1950s.

Mount Stromlo Observatory astronomical observatory established in Canberra, Australia, in 1923. Important observations have been made there on the Magellanic Clouds, which can be seen clearly from southern Australia.

Mount Wilson site near Los Angeles, California, of the 2.5 m/100 in Hooker telescope, opened in 1917, with which Edwin Hubble discovered the expansion of the universe. Two solar telescopes in towers 18.3 m/60 ft and 45.7 m/ 150 ft tall, and a 1.5 m/60 in reflector opened in 1908, also operate there.

mouse in computing, an input device used to control a pointer on a computer screen. It is a feature of ◊graphical user interface (GUI) systems. The mouse, about the size of a pack of playing cards, is connected to the computer by a wire, and incorporates one or more buttons that can be pressed. Moving the mouse across a flat surface causes a corresponding movement of the pointer. In this way, the operator can manipulate objects on the screen and make menu selections.

The mouse was invented in 1963 at the Stanford Research Institute, USA, by Douglas Engelbart, and developed by the Xerox Corporation in the 1970s. The first was made of wood; the Microsoft mouse was introduced in 1983, and the Apple Macintosh mouse in 1984. Mice work either mechanically (with electrical contacts to sense the movement in two planes of a ball on a level surface), or optically (photocells detecting movement by recording light reflected from a grid on which the mouse is moved).

mouth cavity forming the entrance to the digestive tract. In land vertebrates, air from the nostrils enters the mouth cavity to pass down the trachea. The mouth in mammals is enclosed by the jaws, cheeks, and palate.

moving-coil meter instrument used to detect and measure electrical current. A coil of wire pivoted between the poles of a permanent magnet is turned by the motor effect of an electric current (by which a force acts on a wire carrying a current in a magnetic field). The extent to which the coil turns can then be related to the magnitude of the current.

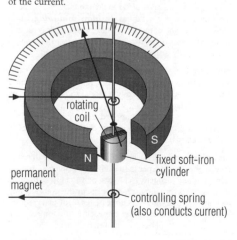

moving-coil meter

The sensitivity of the instrument depends directly upon the strength of the permanent magnet used, the number of turns making up the moving coil, and the coil's area. It depends inversely upon the strength of the controlling springs used to restrain the rotation of the coil. By the addition of a suitable resistor, a moving-coil meter can be adapted to read potential difference in volts.

mp in chemistry, abbreviation for **melting point**.

MS-DOS (abbreviation for **Microsoft Disc Operating System**) computer ◊operating system produced by Microsoft Corporation, widely used on ◊microcomputers with Intel x 86 and Pentium family microprocessors. A version called PC-DOS is sold by IBM specifically for its personal computers. MS-DOS and PC-DOS are usually referred to as DOS. MS-DOS first appeared in 1981, and was similar to an earlier system from Digital Research called CP/M.

MTBF (abbreviation for **mean time between failures**), the statistically average time a component can be used before it goes wrong. The MTBF of a computer hard disc, for example, is around 150,000 hours.

mucous membrane thin skin lining all animal body cavities and canals that come into contact with the air (for example, eyelids, breathing and digestive passages, genital tract). It secretes mucus, a moistening, lubricating, and protective fluid.

mucus lubricating and protective fluid, secreted by mucous membranes in many different parts of the body. In the gut, mucus smooths the passage of food and keeps potentially damaging digestive enzymes away from the gut lining. In the lungs, it traps airborne particles so that they can be expelled.

mudstone fine-grained sedimentary rock made up of clay-to silt-sized particles (up to 0.0625 mm/0.0025 in).

Mullard Radio Astronomy Observatory radio observatory of the University of Cambridge, England. Its main instrument is the Ryle Telescope, eight dishes 12.8 m/42 ft wide in a line of 5 km/3 mi long, opened in 1972.

multimedia computerized method of presenting information by combining audio and video components using text, sound, and graphics (still, animated, and video sequences). For example, a multimedia database of musical instruments may allow a user not only to search and retrieve text about a particular instrument but also to see pictures of it and hear it play a piece of music. Multimedia applications emphasize interactivity between the computer and the user.

As graphics, video, and audio are extremely demanding of storage space, multimedia PCs are usually fitted with ◊CD-ROM drives because of the high storage capacity of CD-ROM discs.

In the mid-1990s developments in compression techniques and software made it possible to incorporate multimedia elements into Internet Web sites.

multiple birth in humans, the production of more than two babies from one pregnancy. Multiple births can be caused by more than two eggs being produced and fertilized (often as the result of hormone therapy to assist pregnancy), or by a single fertilized egg dividing more than once before implantation. See also ◊twin.

Multiple Mirror Telescope telescope on Mount

Hopkins, Arizona, USA, opened in 1979, consisting of six 1.83 m/72 in mirrors mounted in a hexagon, the light-collecting area of which equals that of a single mirror of 4.5 m/176 in diameter. It is planned to replace the six mirrors with a single mirror 6.5 m/256 in wide.

multiple proportions, law of in chemistry, the principle that if two elements combine with each other to form more than one compound, then the ratio of the masses of one of them that combine with a particular mass of the other is a small whole number.

multiple sclerosis (MS) or *disseminated sclerosis* incurable chronic disease of the central nervous system, occurring in young or middle adulthood. Most prevalent in temperate zones, it affects more women than men. It is characterized by degeneration of the myelin sheath that surrounds nerves in the brain and spinal cord.

Depending on where the demyelination occurs – which nerves are affected – the symptoms of MS can mimic almost any neurological disorder. Typically seen are unsteadiness, ataxia (loss of muscular coordination), weakness, speech difficulties, and rapid involuntary movements of the eyes. The course of the disease is episodic, with frequent intervals of ◊remission. Its cause is unknown, but it may be initiated in childhood by some environmental factor, such as infection, in genetically susceptible people. It has been shown that there is a genetic component: identical twins of MS sufferers have a 1 in 4 chance of developing the disease, compared to the 1 in 1,000 chance for the general population.

In 1993 interferon beta 1b became the first drug to be approved in the United States for treating MS. It reduces the number and severity of relapses, and slows the formation of brain lesions giving hope that it may slow down the progression of the disease.

multiplexer in telecommunications, a device that allows a transmission medium to carry a number of separate signals at the same time – enabling, for example, several telephone conversations to be carried by one telephone line, and radio signals to be transmitted in stereo.

In *frequency-division multiplexing*, signals of different frequency, each carrying a different message, are transmitted.

Electrical frequency filters separate the message at the receiving station. In *time-division multiplexing*, the messages are broken into sections and the sections of several messages interleaved during transmission. ◊Pulse-code modulation allows hundreds of messages to be sent simultaneously over a single link.

multistage rocket rocket launch vehicle made up of several rocket stages (often three) joined end to end. The bottom stage fires first, boosting the vehicle to high speed, then it falls away. The next stage fires, thrusting the now lighter vehicle even faster. The remaining stages fire and fall away in turn, boosting the vehicle's payload (cargo) to an orbital speed that can reach 28,000 kph/17,500 mph.

multitasking or *multiprogramming* in computing, a system in which one processor appears to run several different programs (or different parts of the same program) at the same time. All the programs are held in memory together and each is allowed to run for a certain period.

For example, one program may run while other programs are waiting for a peripheral device to work or for input from an operator.

The ability to multitask depends on the ◊operating system rather than the type of computer. UNIX is one of the commonest.

multiuser system or *multiaccess system* in computing, an operating system that enables several users to access centrally-stored data and programs simultaneously over a network. Each user has a terminal, which may be local (connected directly to the computer) or remote (connected to the computer via a modem and a telephone line).

Multiaccess is usually achieved by *time-sharing*: the computer switches very rapidly between terminals and programs so that each user has sole use of the computer for only a fraction of a second but can work as if she or he had continuous access.

Multiuser systems are becoming increasingly common in the workplace, and have many advantages – such as enabling employees to refer to and update a shared corporate database.

mummy any dead body, human or animal, that has been naturally or artificially preserved. Natural mummification can occur through freezing (for example, mammoths in glacial ice from 25,000 years ago), drying, or preservation in bogs or oil seeps. Artificial mummification may be achieved by embalming (for example, the mummies of ancient Egypt) or by freeze-drying (see ◊cryonics).

mumps or *infectious parotitis* virus infection marked by fever, pain, and swelling of one or both parotid salivary glands (situated in front of the ears). It is usually short-lived in children, although meningitis is a possible complication. In adults the symptoms are more serious and it may cause sterility in men.

Mumps is the most common cause of ◊meningitis in children, but it follows a much milder course than bacterial meningitis, and a complete recovery is usual. Rarely, mumps meningitis may lead to deafness. An effective vaccine against mumps, measles, and rubella (MMR vaccine) is now offered for children aged 18 months.

Münchhausen's syndrome emotional disorder in which a patient feigns or invents symptoms to secure medical treatment. It is the chronic form of factitious disorder, which is more common, and probably underdiagnosed. In some cases the patient will secretly ingest substances to produce real symptoms. It was named after the exaggerated tales of Baron Münchhausen. Some patients invent symptoms for their children, a phenomenon known as *Münchhausen's by proxy*.

muon an ◊elementary particle similar to the electron except for its mass which is 207 times greater than that of the electron. It has a half-life of 2 millionths of a second, decaying into electrons and ◊neutrinos. The muon was originally thought to be a ◊meson and is thus sometimes called a mu meson, although current opinion is that it is a ◊lepton.

murmur in medicine, the characteristic uneven sound that is heard when examining the diseased heart and various blood vessels using a stethoscope. Murmurs may be an indication of disease of the heart valves.

muscle contractile animal tissue that produces locomotion and power, and maintains the movement of body substances. Muscle is made of long cells that can contract to between one-half and one-third of their relaxed length.

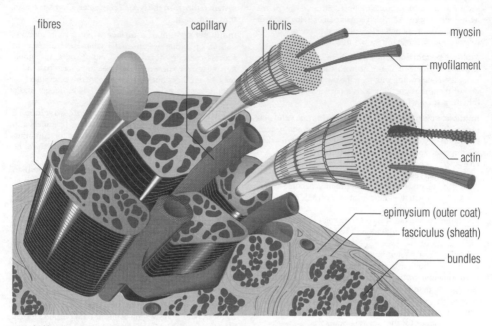

fibres   capillary   fibrils   myosin

myofilament

actin

epimysium (outer coat)

fasciculus (sheath)

bundles

muscle Muscles make up 35–45% of our body weight; there are over 650 skeletal muscles. Muscle cells may be up to 20 cm/0.8 in long. They are arranged in bundles, fibres, fibrils, and myofilaments.

Striped (or striated) muscles are activated by ◊motor nerves under voluntary control; their ends are usually attached via tendons to bones.

Involuntary or **smooth** muscles are controlled by motor nerves of the ◊autonomic nervous system, and are located in the gut, blood vessels, iris, and various ducts.

Cardiac muscle occurs only in the heart, and is also controlled by the autonomic nervous system.

The function of muscle is to pull and not to push, except in the case of the genitals and the tongue.

On **muscle** Leonardo da Vinci
The Notebooks of Leonardo da Vinci vol 1, ch 3

---

muscle relaxant in medicine, one of a number of drugs used to relax skeletal muscles. Muscle relaxants are used in patients undergoing surgery under general anaesthesia and in patients with disorders affecting the musculo-skeletal system, such as chronic muscle spasm or spasticity.

Muscle relaxants used in general anaesthesia are also known as neuromuscular blocking drugs. They block the transmission of nerve impulses between the nerve and the skeletal muscle, which results in relaxation of the abdomen and diaphragm. The vocal cords are also relaxed to allow the passage of an endotracheal tube.

muscovite white mica, $KAl_2Si_3AlO_{10}(OH,F)_2$, a common silicate mineral. It is colourless to silvery white with shiny surfaces, and like all micas it splits into thin flakes along its one perfect cleavage. Muscovite is a metamorphic mineral occurring mainly in schists; it is also found in some granites, and appears as shiny flakes on bedding planes of some sandstones.

muscular dystrophy any of a group of inherited chronic muscle disorders marked by weakening and wasting of muscle. Muscle fibres degenerate, to be replaced by fatty tissue, although the nerve supply remains unimpaired. Death usually occurs in early adult life.

The commonest form, Duchenne muscular dystrophy, strikes boys (1 in 3,000), usually before the age of four. The child develops a waddling gait and an inward curvature (lordosis) of the lumbar spine. The muscles affected by dystrophy and the rate of progress vary. There is no cure, but physical treatments can minimize disability. Death usually occurs before the age of 20.

mushroom fruiting body of certain fungi (see ◊fungus), consisting of an upright stem and a spore-producing cap with radiating gills on the undersurface. There are many edible species belonging to the genus *Agaricus*, including the field mushroom (*A. campestris*). See also ◊toadstool.

musical instrument digital interface manufacturer's standard for digital music equipment; see ◊MIDI.

music therapy use of music as an adjunct to ◊relaxation therapy, or in ◊psychotherapy to elicit expressions of suppressed emotions by prompting patients to dance, shout, laugh, cry, or whatever, in response.

Music therapists are most frequently called upon to help the mentally or physically disabled; for instance, patients suffering from speech difficulties or autism may be enabled to express themselves more effectively by making musical sounds, and music can help the physically disabled to develop better motor control.

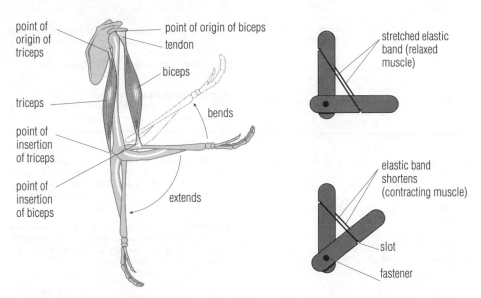

point of origin of triceps

point of origin of biceps

tendon

biceps

triceps

bends

point of insertion of triceps

point of insertion of biceps

extends

stretched elastic band (relaxed muscle)

elastic band shortens (contracting muscle)

slot

fastener

muscle The movements of the arm depend on two muscles, the biceps and the triceps. To lift the arm, the biceps shortens and the triceps lengthens. To lower the arm, the opposite occurs: the biceps lengthens and the triceps shortens.

mutagen any substance that increases the rate of gene ◊mutation. A mutagen may also act as a ◊carcinogen.

mutation in biology, a change in the genes produced by a change in the ◊DNA that makes up the hereditary material of all living organisms. Mutations, the raw material of evolution, result from mistakes during replication (copying) of DNA molecules. Only a few improve the organism's performance and are therefore favoured by ◊natural selection. Mutation rates are increased by certain chemicals and by radiation.

Common mutations include the omission or insertion of a base (one of the chemical subunits of DNA); these are known as *point mutations*. Larger-scale mutations include removal of a whole segment of DNA or its inversion within the DNA strand. Not all mutations affect the organism, because there is a certain amount of redundancy in the genetic information. If a mutation is 'translated' from DNA into the protein that makes up the organism's structure, it may be in a nonfunctional part of the protein and thus have no detectable effect. This is known as a *neutral mutation*, and is of importance in ◊molecular clock studies because such mutations tend to accumulate gradually as time passes. Some mutations do affect genes that control protein production or functional parts of protein, and most of these are lethal to the organism.

mutual induction in physics, the production of an electromotive force (emf) or voltage in an electric circuit caused by a changing ◊magnetic flux in a neighbouring circuit. The two circuits are often coils of wire, as in a ◊transformer, and the size of the induced emf depends largely on the numbers of turns of wire in each of the coils.

mutualism or ◊*symbiosis* an association between two organisms of different species whereby both profit from the relationship.

myasthenia gravis in medicine, an uncommon condition characterized by loss of muscle power, especially in the face and neck. The muscles tire rapidly and fail to respond to repeated nervous stimulation. ◊Autoimmunity is the cause.

mycelium interwoven mass of threadlike filaments or ◊hyphae, forming the main body of most fungi. The reproductive structures, or 'fruiting bodies', grow from the mycelium.

mycorrhiza mutually beneficial (mutualistic) association occurring between plant roots and a soil fungus. Mycorrhizal roots take up nutrients more efficiently than non-mycorrhizal roots, and the fungus benefits from obtaining carbohydrates from the plant or tree.

An *ectotrophic mycorrhiza* occurs on many tree species, which usually grow much better, most noticeably in the seeding stage, as a result. Typically the roots become repeatedly branched and coral-like, penetrated by hyphae of a surrounding fungal ◊mycelium. In an *endotrophic mycorrhiza*, the growth of the fungus is mainly inside the root, as in orchids. Such plants do not usually grow properly, and may not even germinate, unless the appropriate fungus is present.

Research by UK ecologists in 1996 showed that mycorrhizal fungi provides protection against ◊nematode worms that feed on plant roots, as well as pathogenic fungi.

myelin sheath insulating layer that surrounds nerve cells in vertebrate animals. It serves to speed up the passage of nerve impulses.

Myelin is made up of fats and proteins and is formed from up to a hundred layers, laid down by special cells, the *Schwann cells*.

myelitis in medicine, inflammation of the spinal cord. It occurs in diseases such as ◊meningitis.

myoglobin globular protein, closely related to ◊haemoglobin and located in vertebrate muscle. Oxygen binds to myoglobin and is released only when the haemoglobin can no longer supply adequate oxygen to muscle cells.

myopia or **short-sightedness** defect of the eye in which a person can see clearly only those objects that are close up. It is caused either by the eyeball being too long or by the cornea and lens system of the eye being too powerful, both of which cause the images of distant objects to be formed in front of the retina instead of on it. Nearby objects are sharply perceived. Myopia can be corrected by suitable glasses or contact lenses.

myopia, low-luminance poor night vision. About 20% of people have poor vision in twilight and nearly 50% in the dark. Low-luminance myopia does not show up in normal optical tests, but the degree of blurring can be measured by projecting images on a screen using a weak laser beam.

myrmecophyte plant that lives in association with a colony of ants and possesses specialized organs in which the ants live. For example, *Myrmecodia,* an epiphytic plant from Malaysia, develops root tubers containing a network of cavities inhabited by ants.

Several species of *Acacia* from tropical America have specialized hollow thorns for the same purpose. This is probably a mutualistic (mutually beneficial) relationship, with the ants helping to protect the plant from other insect pests and in return receiving shelter.

mystery reproductive syndrome (MRS) viral disease of pigs that causes sows to lose up to 10% of their litter. It was first seen in the USA in 1987 and in Europe in 1991. The symptoms are flulike.

myxoedema thyroid-deficiency disease developing in adult life, most commonly in middle-aged women. The symptoms include loss of energy and appetite, weight gain, inability to keep warm, mental dullness, and dry, puffy skin. It is reversed by giving the thyroid hormone thyroxine. See also ◊hypothyroidism.

myxomatosis contagious, usually fatal, virus infection of rabbits which causes much suffering. It has been deliberately introduced in the UK and Australia since the 1950s to reduce the rabbit population.

N abbreviation for **north**; **newton**; the chemical symbol for **nitrogen**.

nadir the point on the celestial sphere vertically below the observer and hence diametrically opposite the **zenith**. The term is used metaphorically to mean the low point of a person's fortunes.

naevus mole, or patch of discoloration on the skin which has been present from birth. There are many different types of naevi, including those composed of a cluster of small blood vessels, such as the 'strawberry mark' (which usually disappears early in life), and the 'port-wine stain'.

A naevus of moderate size is harmless, and such marks can usually be disguised cosmetically unless they are extremely disfiguring, when they can sometimes be removed by cutting out, burning with an electric needle, freezing with carbon dioxide snow, or by argon laser treatment. In rare cases a mole may be a precursor of a malignant ◊melanoma. Any changes in a mole, such as enlargement, itching, soreness, or bleeding, should be reported to a doctor.

nagana ◊sleeping sickness of African livestock, caused by a trypanosome and spread by the tsetse fly.

nail in biology, a hard, flat, flexible outgrowth of the digits of primates (humans, monkeys, and apes). Nails are composed of ◊keratin.

NAND gate type of ◊logic gate.

nano- prefix used in ◊SI units of measurement, equivalent to a one-billionth part (10^{-9}). For example, a nanosecond is one-billionth of a second.

nanotechnology experimental technology using individual atoms or molecules as the components of minute machines, measured by the nanometre, or millionth of a millimetre. Nanotechnology research in the 1990s focused on testing molecular structures and refining ways to manipulate atoms using a scanning tunnelling microscope. The ultimate aim is to create very small computers and molecular machines which could perform vital engineering or medical tasks.

The ◊scanning electron microscope can be used to see and position single atoms and molecules, and to drill holes a nanometre (billionth of a metre) across in a variety of materials. The instrument can be used for ultrafine etching; the entire 28 volumes of the *Encyclopedia Britannica* could be engraved on the head of a pin. In the USA a complete electric motor has been built, which is less than 0.1 mm across with a top speed of 600,000 rpm. It is etched out of silicon, using the ordinary methods of chip manufacturers.

naphtha the mixtures of hydrocarbons obtained by destructive distillation of petroleum, coal tar, and shale oil. It is a raw material for the petrochemical and plastics industries. The term was originally applied to naturally occurring liquid hydrocarbons.

naphthalene $C_{10}H_8$ solid, white, shiny, aromatic hydrocarbon obtained from coal tar. The smell of moth-balls is due to their naphthalene content. It is used in making indigo and certain azo dyes, as a mild disinfectant, and as an insecticide.

nappy rash in medicine, an inflamed skin condition that occurs on the buttocks of infants. It is usually due to changing nappies too infrequently and the use of tightly fitting plastic pants. Nappy rash can be avoided by careful washing and drying of skin creases and folds in the area, followed by the application of a barrier cream. If nappy rash does develop, it may clear if it is left exposed to the air. Creams containing dimethicone or zinc oxide can be used and, if the rash is associated with a fungal infection, creams containing the fungicide clotrimazole can be applied. It appears to occur less frequently in breast-fed infants and when disposable nappies are used.

narcissism in psychology, an exaggeration of normal self-respect and self-involvement which may amount to mental disorder when it precludes relationships with other people.

narcosis in medicine, condition that can be produced by opioid drugs. It is characterized by deep sleep from which the affected individual can be roused only with great difficulty.

narcotic pain-relieving and sleep-inducing drug. The term is usually applied to heroin, morphine, and other opium derivatives, but may also be used for other drugs which depress brain activity, including anaesthetic agents and ◊hypnotics.

NASA acronym for **National Aeronautics and Space Administration**, US government agency for spaceflight and aeronautical research, founded in 1958 by the National Aeronautics and Space Act. Its headquarters are in Washington DC and its main installation is at the ◊Kennedy Space Center in Florida. NASA's early planetary and lunar programs included Pioneer spacecraft from 1958, which gathered data for the later crewed missions, the most famous of which took the first people to the Moon in *Apollo 11* on 16–24 July 1969.

In the early 1990s, NASA moved towards lower-budget 'Discovery missions', which should not exceed a budget of $150 million (excluding launch costs), nor a development period of three years.

nastic movement plant movement that is caused by an external stimulus, such as light or temperature, but is directionally independent of its source, unlike ◊tropisms. Nastic movements occur as a result of changes in water pressure within specialized cells or differing rates of growth in parts of the plant.

Examples include the opening and closing of crocus flowers following an increase or decrease in temperature (**thermonasty**), and the opening and closing of evening-primrose *Oenothera* flowers on exposure to dark and light (**photonasty**).

The leaf movements of the Venus flytrap *Dionaea muscipula* following a tactile stimulus, and the rapid collapse of the leaflets of the sensitive plant *Mimosa pudica* are examples of **haptonasty**. Sleep movements, where the leaves or flowers of some plants adopt a different position at night, are described as **nyctinasty**. Other movement types include **hydronasty**, in response to a change in the atmospheric humidity, and **chemonasty**, in response to a chemical stimulus.

national park land set aside and conserved for public enjoyment. The first was Yellowstone National Park, USA, established in 1872. National parks include not only the most scenic places, but also places distinguished for their historic, prehistoric, or scientific interest, or for their superior recreational assets. They range from areas the size of small countries to pockets of just a few hectares.

National Rivers Authority (NRA) UK government agency launched in 1989. It is responsible for managing water resources, investigating and regulating pollution, and taking over flood controls and land drainage from the former ten regional water authorities of England and Wales. Following a judicial review of the authority in 1991 for allegedly failing to carry out its statutory duty to protect rivers and seas from pollution, river quality improved by 26% between 1993 and 1996. The NRA has granted licences to many UK companies to discharge polluting materials (a total of 12,000 licensed pipelines), and has in most cases failed to prosecute when the companies have exceeded their discharge limit.

In April 1996 the NRA was replaced by the Environment Agency, having begun to establish a reputation for being supportive to wildlife projects and tough on polluters.

National Security Agency (NSA) largest and most secret of US intelligence agencies. Established in 1952 to intercept foreign communications as well as to safeguard US transmissions, the NSA collects and analyses computer communications, telephone signals, and other electronic data, and gathers intelligence. Known as the Puzzle Palace, its headquarters are at Fort Meade, Maryland (with a major facility at Menwith Hill, England).

The NSA was set up by a classified presidential memorandum and its very existence was not acknowledged until 1962. It operates outside normal channels of government accountability, and its budget (also secret) is thought to exceed several billion dollars. Fort Meade has several Cray supercomputers.

In 1976, NSA's Harvest computer system intercepted 75 million individual messages, of which 1.8 million received further analysis.

native metal or **free metal** any of the metallic elements that occur in nature in the chemically uncombined or elemental form (in addition to any combined form). They include bismuth, cobalt, copper, gold, iridium, iron, lead, mercury, nickel, osmium, palladium, platinum, ruthenium, rhodium, tin, and silver. Some are commonly found in the free state, such as gold; others occur almost exclusively in the combined state, but under unusual conditions do occur as native metals, such as mercury. Examples of native nonmetals are carbon and sulphur.

Natural Environment Research Council (NERC) UK organization established by royal charter in 1965 to undertake and support research in the earth sciences, to give advice both on exploiting natural resources and on

protecting the environment, and to support education and training of scientists in these fields of study.

Research areas include geothermal energy, industrial pollution, waste disposal, satellite surveying, acid rain, biotechnology, atmospheric circulation, and climate. Research is carried out principally within the UK but also in Antarctica and in many Third World countries. It comprises 13 research bodies.

natural frequency the frequency at which a mechanical system will vibrate freely. A pendulum, for example, always oscillates at the same frequency when set in motion. More complicated systems, such as bridges, also vibrate with a fixed natural frequency. If a varying force with a frequency equal to the natural frequency is applied to such an object the vibrations can become violent, a phenomenon known as ◊resonance.

natural gas mixture of flammable gases found in the Earth's crust (often in association with petroleum). It is one of the world's three main fossil fuels (with coal and oil). Natural gas is a mixture of ◊hydrocarbons, chiefly methane (80%), with ethane, butane, and propane. Natural gas is usually transported from its source by pipeline, although it may be liquefied for transport and storage and is, therefore, often used in remote areas where other fuels are scarce and expensive. Prior to transportation, butane and propane are removed and liquefied to form 'bottled gas'.

natural logarithm in mathematics, the ◊exponent of a number expressed to base e, where e represents the ◊irrational number 2.71828... .

Natural ◊logarithms are also called Napierian logarithms, after their inventor, the Scottish mathematician John Napier.

natural radioactivity radioactivity generated by those radioactive elements that exist in the Earth's crust. All the elements from polonium (atomic number 84) to uranium (atomic number 92) are radioactive.

◊Radioisotopes of some lighter elements are also found in nature (for example potassium-40). See ◊background radiation.

natural selection the process whereby gene frequencies in a population change through certain individuals producing more descendants than others because they are better able to survive and reproduce in their environment.

The accumulated effect of natural selection is to produce ◊adaptations such as the insulating coat of a polar bear or the spadelike forelimbs of a mole. The process is slow, relying firstly on random variation in the genes of an organism being produced by ◊mutation and secondly on the genetic ◊recombination of sexual reproduction. It was recognized by Charles Darwin and English naturalist Alfred Russel Wallace as the main process driving ◊evolution.

I have called this principle, by which each slight variation, if useful, is preserved, by the term of Natural Selection.

On **natural selection** Charles Darwin
On the Origin of Species 1859

Nature Conservancy Council (NCC) former name of UK government agency divided in 1991 into English

Nature, Scottish Natural Heritage, and the Countryside Council for Wales.

The NCC was established by act of Parliament in 1973 (Nature Conservancy created by royal charter in 1949) with the aims of designating and managing national nature reserves and other conservation areas; identifying Sites of Special Scientific Interest; advising government ministers on policies; providing advice and information; and commissioning or undertaking relevant scientific research. In 1991 the Nature Conservancy Council was dissolved and its three regional bodies became autonomous agencies.

nature–nurture controversy or **environment–heredity controversy** long-standing dispute among philosophers and psychologists over the relative importance of environment, that is, upbringing, experience, and learning ('nurture'), and heredity, that is, genetic inheritance ('nature'), in determining the make-up of an organism, as related to human personality and intelligence.

One area of contention is the reason for differences between individuals; for example, in performing intelligence tests. The environmentalist position assumes that individuals do not differ significantly in their inherited mental abilities and that subsequent differences are due to learning, or to differences in early experiences. Opponents insist that certain differences in the capacities of individuals (and hence their behaviour) can be attributed to inherited differences in their genetic make-up.

nature reserve area set aside to protect a habitat and the wildlife that lives within it, with only restricted admission for the public. A nature reserve often provides a sanctuary for rare species. The world's largest is Etosha Reserve, Namibia; area 99,520 sq km/38,415 sq mi.

naturopathy in alternative medicine, facilitating of the natural self-healing processes of the body. Naturopaths are the general practitioners (GPs) of alternative medicine and often refer clients to other specialists, particularly in manipulative therapies, to complement their own work of seeking, through diet, the prescription of natural medicines and supplements, and lifestyle counselling, to restore or augment the vitality of the body and thereby its optimum health.

nausea in medicine, a feeling of sickness and a desire to vomit. It is a symptom of a wide range of diseases, such as gastrointestinal infections and motion sickness. Nausea can occur as an adverse reaction to some drugs that irritate the gastrointestinal tract, such as ibuprofen, and can be very severe in patients receiving chemotherapy or radiotherapy to treat cancer. Nausea often occurs during the first three months of pregnancy when it should not be treated on a routine basis.

nautical mile unit of distance used in navigation, an internationally agreed-on standard (since 1959) equalling the average length of one minute of arc on a great circle of the Earth, or 1,852 m/6,076.12 ft. The term was formerly applied to various units of distance used in navigation. In the UK the nautical mile was formerly defined as 6,082 ft.

navel or **umbilicus** small indentation in the centre of the abdomen of mammals, marking the site of attachment of the ◊umbilical cord, which connects the fetus to the ◊placenta.

navigation the science and technology of finding the position, course, and distance travelled by a ship, plane, or other craft. Traditional methods include the magnetic ◊compass and ◊sextant. Today the gyrocompass is usually used, together with highly sophisticated electronic methods, employing beacons of radio signals, such as Decca, ◊Loran, and Omega. Satellite navigation uses satellites that broadcast time and position signals.

The US ◊global positioning system (GPS) was introduced in 1992. When complete, it will feature 24 Navstar satellites that will enable users (including eventually motorists and walkers) to triangulate their position (from any three satellites) to within 15 m/50 ft.

In 1992, 85 nations agreed to take part in trials of a new navigation system which makes use of surplus military space technology left over from the Cold War. The new system, known as FANS or Future Navigation System, will make use of the 24 Russian Glonass satellites and the 24 US GPS satellites. Small computers will gradually be fitted to civil aircraft to process the signals from the satellite, allowing aircraft to navigate with pinpoint accuracy anywhere in the world. The signals from at least three satellites will guide the craft to within a few metres of accuracy. FANS will be used in conjunction with four Inmarsat satellites to provide worldwide communications between pilots and air-traffic controllers.

navigation, biological the ability of animals or insects to navigate. Although many animals navigate by following established routes or known landmarks, many animals can navigate without such aids; for example, birds can fly several thousand miles back to their nest site, over unknown terrain.

Such feats may be based on compass information derived from the position of the Sun, Moon, or stars, or on the characteristic patterns of Earth's magnetic field.

Biological navigation refers to the ability to navigate both in long-distance ◊migrations and over shorter distances when foraging (for example, the honey bee finding its way from the hive to a nectar site and back). Where reliant on known landmarks, birds may home on features that can be seen from very great distances (such as the cloud caps that often form above isolated mid-ocean islands). Even smells can act as a landmark. Aquatic species like salmon are believed to learn the characteristic taste of the river where they hatch and return to it, often many years later. Brain cells in some birds have been found to contain ◊magnetite and may therefore be sensitive to the Earth's magnetic field.

nebula cloud of gas and dust in space. Nebulae are the birthplaces of stars, but some nebulae are produced by gas thrown off from dying stars (see ◊planetary nebula; ◊supernova). Nebulae are classified depending on whether they emit, reflect, or absorb light.

An **emission nebula**, such as the ◊Orion nebula, glows brightly because its gas is energized by stars that have formed within it. In a **reflection nebula**, starlight reflects off grains of dust in the nebula, such as surround the stars of the ◊Pleiades cluster. A **dark nebula** is a dense cloud, composed of molecular hydrogen, which partially or completely absorbs light behind it. Examples include the Coalsack nebula in ◊Crux and the Horsehead nebula in Orion.

neck structure between the head and the trunk in animals. In the back of the neck are the upper seven vertebrae of the spinal column, and there are many powerful muscles that support and move the head. In front, the neck region contains the ◊pharynx and ◊trachea, and behind these the oesophagus. The large arteries (carotid, temporal, maxillary) and veins (jugular) that supply the brain and head are also located in the neck. The ◊larynx (voice box) occupies a position where the trachea connects with the pharynx, and one of its cartilages produces the projection known as Adam's apple. The ◊thyroid gland lies just below the larynx and in front of the upper part of the trachea.

necrosis death or decay of tissue in a particular part of the body, usually due to bacterial poisoning or loss of local blood supply.

nectar sugary liquid secreted by some plants from a nectary, a specialized gland usually situated near the base of the flower. Nectar often accumulates in special pouches or spurs, not always in the same location as the nectary. Nectar attracts insects, birds, bats, and other animals to the flower for ◊pollination and is the raw material used by bees in the production of honey.

negative/positive in photography, a reverse image, which when printed is again reversed, restoring the original scene. It was invented by Fox Talbot in about 1834.

nematode any of a group of unsegmented ◊worms that are pointed at both ends, with a tough, smooth outer skin. They include many free-living species found in soil and water, including the sea, but a large number are parasites, such as the roundworms and pinworms that live in humans, or the eelworms that attack plant roots. They differ from ◊flatworms in that they have two openings to the gut (a mouth and an anus). (Phylum Nematoda.)

Most nematode species are found in deep-sea sediment. Around 13,000 species are known, but a 1995 study by the Natural History Museum, London, based on the analysis of sediment from 17 seabed sites worldwide, estimated that nematodes may make up as much as 75% of all species, with there being an estimated 100 million species. Some are anhydrobiotic, which means they can survive becoming dehydrated, entering a state of suspended animation until they are rehydrated.

neo-Darwinism the modern theory of ◊evolution, built up since the 1930s by integrating the 19th-century English scientist Charles Darwin's theory of evolution through natural selection with the theory of genetic inheritance founded on the work of the Austrian biologist Gregor Mendel.

Neo-Darwinism asserts that evolution takes place because the environment is slowly changing, exerting a selection pressure on the individuals within a population. Those with characteristics that happen to adapt to the new environment are more likely to survive and have offspring and hence pass on these favourable characteristics. Over time the genetic make-up of the population changes and ultimately a new species is formed.

neodymium yellowish metallic element of the ◊lanthanide series, symbol Nd, atomic number 60, relative atomic mass 144.24. Its rose-coloured salts are used in colouring glass, and neodymium is used in lasers.

It was named in 1885 by Austrian chemist Carl von Welsbach (1858–1929), who fractionated it away from didymium (originally thought to be an element but actually a mixture of rare-earth metals consisting largely of neodymium, praesodymium, and cerium).

neon colourless, odourless, nonmetallic, gaseous element, symbol Ne, atomic number 10, relative atomic mass 20.183. It is grouped with the ◊inert gases, is non-reactive, and forms no compounds. It occurs in small quantities in the Earth's atmosphere.

Tubes containing neon are used in electric advertising signs, giving off a fiery red glow; it is also used in lasers. Neon was discovered by Scottish chemist William Ramsay and the Englishman Morris Travers.

neoplasm any lump or tumour, which may be benign or malignant (cancerous).

neoprene synthetic rubber, developed in the USA in 1931 from the polymerization of chloroprene. It is much more resistant to heat, light, oxidation, and petroleum than is ordinary rubber.

neoteny in biology, the retention of some juvenile characteristics in an animal that seems otherwise mature. An example is provided by the axolotl, a salamander that can reproduce sexually although still in its larval form.

neper unit used in telecommunications to express a ratio of powers and currents. It gives the attenuation of amplitudes as the natural logarithm of the ratio.

nephritis or *Bright's disease* general term used to describe inflammation of the kidney. The degree of illness varies, and it may be acute (often following a recent streptococcal infection), or chronic, requiring a range of treatments from antibiotics to ◊dialysis or transplant.

nephron microscopic unit in vertebrate kidneys that forms *urine*. A human kidney is composed of over a million nephrons. Each nephron consists of a knot of blood capillaries called a glomerulus, contained in the Bowman's capsule, and a long narrow tubule enmeshed with yet more capillaries. Waste materials and water pass from the bloodstream into the tubule, and essential minerals and some water are reabsorbed from the tubule back into the bloodstream. The remaining filtrate (urine) is passed out from the body.

Neptune in astronomy, the eighth planet in average distance from the Sun. It is a giant gas (hydrogen, helium, methane) planet, with a mass 17.2 times that of Earth. It has the highest winds in the Solar System.

Mean distance from the Sun 4.4 billion km/2.794 billion mi.

Equatorial diameter 48,600 km/30,200 mi.

Rotation period 16 hr 7 min.

Year 164.8 Earth years.

Atmosphere Methane in its atmosphere absorbs red light and gives the planet a blue colouring. Consists primarily of hydrogen (85%), with helium (13%) and methane (1–2%).

Surface hydrogen, helium and methane. Its interior is believed to have a central rocky core covered by a layer of ice.

Satellites Of Neptune's eight moons, two (◊Triton and Nereid) are visible from Earth. Six were discovered by the *Voyager 2* probe in 1989, of which Proteus (diameter 415 km/260 mi) is larger than Nereid (300 km/200 mi).

Rings There are four faint rings: Galle, Le Verrier, Arago, and Adams (in order from Neptune). Galle is the widest at 1,700 km/1,060 mi. Le Verrier and Arago are divided by a wide diffuse particle band called the plateau.

Neptune was located in 1846 by German astronomers J G Galle and Heinrich d'Arrest (1822–1875) after calculations by English astronomer John Couch Adams and

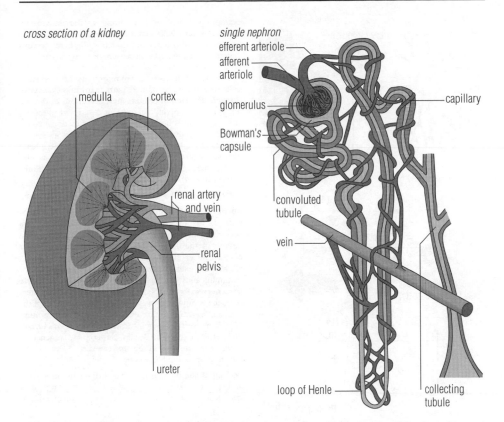

cross section of a kidney

medulla · cortex

renal artery and vein

renal pelvis

ureter

single nephron

efferent arteriole
afferent arteriole
glomerulus
Bowman's capsule
convoluted tubule
vein
capillary
loop of Henle
collecting tubule

nephron The kidney (left) contains more than a million filtering units, or nephrons (right), consisting of the glomerulus, Bowman's capsule, and the loop of Henle. Blood flows through the glomerulus – a tight knot of fine blood vessels from which water and metabolic wastes filter into the tubule. This filtrate flows through the convoluted tubule and loop of Henle where most of the water and useful molecules are reabsorbed into the blood capillaries. The waste materials are passed to the collecting tubule as urine.

French mathematician Urbain Leverrier had predicted its existence from disturbances in the movement of Uranus. *Voyager 2*, which passed Neptune in Aug 1989, revealed various cloud features, notably an Earth-sized oval storm cloud, the Great Dark Spot, similar to the Great Red Spot on Jupiter, but images taken by the Hubble Space Telescope in 1994 showed that the Great Dark Spot has disappeared. A smaller dark spot DS2 has also gone.

neptunium silvery, radioactive metallic element of the ◊actinide series, symbol Np, atomic number 93, relative atomic mass 237.048. It occurs in nature in minute amounts in ◊pitchblende and other uranium ores, where it is produced from the decay of neutron-bombarded uranium in these ores. The longest-lived isotope, Np-237, has a half-life of 2.2 million years. The element can be produced by bombardment of U-238 with neutrons and is chemically highly reactive. It was first synthesized in 1940 by US physicists E McMillan and P Abelson, who named it after the planet Neptune (since it comes after uranium as the planet Neptune comes after Uranus).

Neptunium was the first ◊transuranic element to be synthesized.

NERC abbreviation for ◊*Natural Environment Research Council*.

nerve bundle of nerve cells enclosed in a sheath of connective tissue and transmitting nerve impulses to and from the brain and spinal cord. A single nerve may contain both ◊motor and sensory nerve cells, but they function independently.

nerve cell or *neuron* elongated cell, the basic functional unit of the ◊nervous system that transmits information rapidly between different parts of the body. Each nerve cell has a cell body, containing the nucleus, from which trail processes called dendrites, responsible for receiving incoming signals. The unit of information is the *nerve impulse*, a travelling wave of chemical and electrical changes involving the membrane of the nerve cell. The cell's longest process, the ◊axon, carries impulses away from the cell body.

The impulse involves the passage of sodium and potassium ions across the nerve-cell membrane. Sequential changes in the permeability of the membrane to positive sodium (Na^+) ions and potassium (K^+) ions produce electrical signals called action potentials. Impulses are

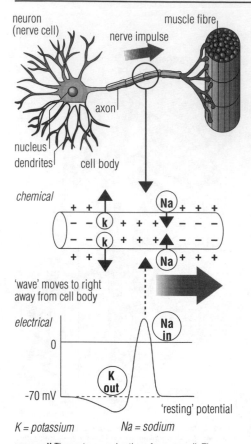

neuron (nerve cell)

nerve impulse

muscle fibre

axon

nucleus
dendrites   cell body

chemical

+ + (Na) + + +
– – (k) + + + – – –
– – (k) + + + – – –
+ + (Na) + + +

'wave' moves to right
away from cell body

electrical

0

Na
in

K
out

-70 mV

'resting' potential

K = potassium   *Na = sodium*

nerve cell The anatomy and action of a nerve cell. The nerve cell or neuron consists of a cell body with the nucleus and projections called dendrites which pick up messages. An extension of the cell, the axon, connects one cell to the dendrites of the next. When a nerve cell is stimulated, waves of sodium (Na⁺) and potassium (K⁺) ions carry an electrical impulse down the axon.

received by the cell body and passed, as a pulse of electric charge, along the axon. The axon terminates at the ◊synapse, a specialized area closely linked to the next cell (which may be another nerve cell or a specialized effector cell such as a muscle). On reaching the synapse, the impulse releases a chemical ◊neurotransmitter, which diffuses across to the neighbouring cell and there stimulates another impulse or the action of the effector cell.

Nerve impulses travel quickly – in humans, they may reach speeds of 160 m/525 ft per second.

nervous breakdown popular term for a reaction to overwhelming psychological stress. There is no equivalent medical term. People said to be suffering from a nervous breakdown may be suffering from a neurotic illness, such as depression or anxiety, or a psychotic illness, such as schizophrenia.

nervous system the system of interconnected ◊nerve cells of most invertebrates and all vertebrates. It is composed of the ◊central and ◊autonomic nervous systems. It may be as simple as the nerve net of coelenterates (for

example, jellyfish) or as complex as the mammalian nervous system, with a central nervous system comprising ◊brain and ◊spinal chord and a peripheral nervous system connecting up with sensory organs, muscles, and glands.

The human nervous system represents the product of millions of years of evolution, particularly in the degree of **encephalization** or brain complexity. It can be divided into central and peripheral parts for descriptive purposes, although there is both anatomical and functional continuity between the two parts. The central nervous system consists of the brain and the spinal cord. The peripheral nervous system is not so clearly subdivided, but its anatomical parts are: (1) the spinal nerves; (2) the cranial nerves; and (3) the autonomic nervous system.

nettle rash popular name for the skin disorder ◊urticaria.

network in computing, a method of connecting computers so that they can share data and peripheral devices, such as printers. The main types are classified by the pattern of the connections – star or ring network, for example – or by the degree of geographical spread allowed; for example, *local area networks* (LANs) for communication within a room or building, and *wide area networks* (WANs) for more remote systems. Internet is the computer network that connects major English-speaking institutions throughout the world, with around 12 million users. Janet (joint academic network), a variant of Internet, is used in Britain. SuperJanet, launched in 1992, is an extension of this that can carry 1,000 million bits of information per second.

neuralgia sharp or burning pain originating in a nerve and spreading over its area of distribution. Trigeminal neuralgia, a common form, is a severe pain on one side of the face.

neural network artificial network of processors that attempts to mimic the structure of nerve cells (neurons) in the human brain. Neural networks may be electronic, optical, or simulated by computer software.

A basic network has three layers of processors: an input layer, an output layer, and a 'hidden' layer in between. Each processor is connected to every other in the network by a system of 'synapses'; every processor in the top layer connects to every one in the hidden layer, and each of these connects to every processor in the output layer. This means that each nerve cell in the middle and bottom layers receives input from several different sources; only when the amount of input exceeds a critical level does the cell fire an output signal.

The chief characteristic of neural networks is their ability to sum up large amounts of imprecise data and decide whether they match a pattern or not. Networks of this type may be used in developing robot vision, matching fingerprints, and analysing fluctuations in stock-market prices. However, it is thought unlikely by scientists that such networks will ever be able accurately to imitate the human brain, which is very much more complicated; it contains around 10 billion nerve cells, whereas current artificial networks contain only a few hundred processors.

neurasthenia obsolete term for nervous exhaustion, covering mild ◊depression and various symptoms of ◊neurosis. Formerly thought to be a bodily malfunction, it is now generally considered to be mental in origin. Dating from the mid-19th century, the term became widely used

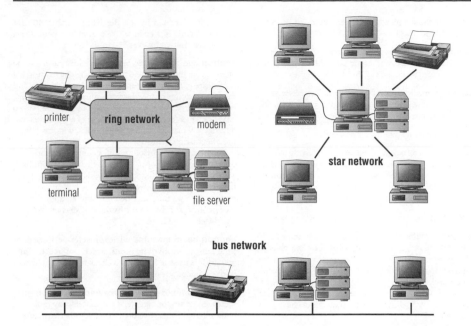

network Local area networks can be connected together via a ring circuit or in a star arrangement. In the ring arrangement, signals from a terminal or peripheral circulate around the ring to reach the terminal or peripheral addressed. In the star arrangement, signals travel via a central controller.

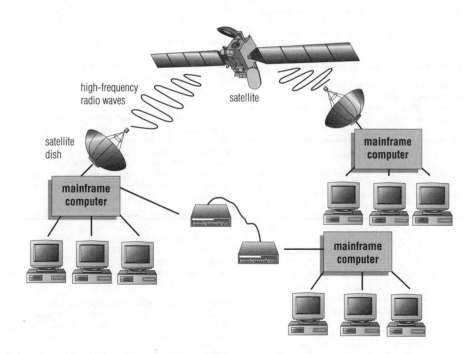

network A wide area network is used to connect remote computers via telephone lines or satellite links. The ISDN (Integrated Services Digital Network) telecommunications network allows high-speed transfer of digital data.

to describe the symptoms of soldiers returning from the front in World War I.

neuritis peripheral nerve inflammation caused by injury, poisoning, or disease, and accompanied by sensory and motor changes in the area of the affected nerve.

neurology medical speciality concerned with the study and treatment of disorders of the brain, spinal cord, and peripheral nerves.

neuron another name for a ◊nerve cell.

neurosis in psychology, a general term referring to emotional disorders, such as anxiety, depression, and phobias. The main disturbance tends to be one of mood; contact with reality is relatively unaffected, in contrast to ◊psychosis.

neurotoxin any substance that destroys nerve tissue.

neurotransmitter chemical that diffuses across a ◊synapse, and thus transmits impulses between ◊nerve cells, or between nerve cells and effector organs (for example, muscles). Common neurotransmitters are noradrenaline (which also acts as a hormone) and acetylcholine, the latter being most frequent at junctions between nerve and muscle. Nearly 50 different neurotransmitters have been identified.

neutralization in chemistry, a process occurring when the excess acid (or excess base) in a substance is reacted with added base (or added acid) so that the resulting substance is neither acidic nor basic.

In theory neutralization involves adding acid or base as required to achieve ◊pH 7. When the colour of an ◊indicator is used to test for neutralization, the final pH may differ from pH 7 depending upon the indicator used. It will also differ from 7 in reactions between strong acids and weak bases and weak acids and strong bases, as the salt formed will have acid or basic properties respectively.

neutral oxide oxide that has neither acidic nor basic properties (see ◊oxide). Neutral oxides are only formed by ◊nonmetals. Examples are carbon monoxide, water, and nitrogen(I) oxide.

Neutrino

A neutrino could pass through a solid piece of lead 965 million million km/600 million million mi thick without being absorbed. An estimated 10 million neutrinos from the Sun will pass through your body while you read this sentence. Neutrinos travel at the speed of light: by the time you have read this paragraph, the neutrinos that passed through your body will be further away than the Moon.

Neutrino in physics, any of three uncharged ◊elementary particles (and their antiparticles) of the ◊lepton class, having a mass too close to zero to be measured. The most familiar type, the antiparticle of the electron neutrino, is emitted in the beta decay of a nucleus. The other two are the muon and tau neutrinos.

US physicists at Los Alamos National Laboratory announced in 1995 that neutrinos have a mass of up to 5 electronvolts (one hundred-thousandth of the mass of an electron).

Neutrinos created by the Big Bang number 300 for every 1 cu cm/0.061 cu in of space. A million neutrinos a second are emitted by every 1 cm^2/0.155 in^2.

neutron one of the three main subatomic particles, the others being the proton and the electron. The neutron is a composite particle, being made up of three ◊quarks, and therefore belongs to the ◊baryon group of the ◊hadrons. Neutrons have about the same mass as protons but no electric charge, and occur in the nuclei of all atoms except hydrogen. They contribute to the mass of atoms but do not affect their chemistry.

For instance, the ◊isotopes of a single element differ only in the number of neutrons in their nuclei but have identical chemical properties. Outside a nucleus, a free neutron is unstable, decaying with a half-life of 11.6 minutes into a proton, an electron, and an antineutrino. The neutron was discovered by the British chemist James Chadwick in 1932.

neutron beam machine nuclear reactor or accelerator producing a stream of neutrons, which can 'see' through metals. It is used in industry to check molecular changes in metal under stress.

neutron bomb or *enhanced radiation weapon* small hydrogen bomb for battlefield use that kills by radiation, with minimal damage to buildings and other structures.

neutron star very small, 'superdense' star composed mostly of ◊neutrons. They are thought to form when massive stars explode as ◊supernovae, during which the protons and electrons of the star's atoms merge, owing to intense gravitational collapse, to make neutrons. A neutron star may have the mass of up to three Suns, compressed into a globe only 20 km/12 mi in diameter.

If its mass is any greater, its gravity will be so strong that it will shrink even further to become a ◊black hole. Being so small, neutron stars can spin very quickly. The rapidly flashing radio stars called ◊pulsars are believed to be neutron stars. The flashing is caused by a rotating beam of radio energy similar in behaviour to a lighthouse beam of light.

New Madrid seismic fault zone the largest system of geological faults in the eastern USA, centred on New Madrid, Missouri. Geologists estimate that there is a 50% chance of a magnitude 6 earthquake in the area by the year 2000. This would cause much damage because the solid continental rocks would transmit the vibrations over a wide area, and buildings in the region have not been designed with earthquakes in mind.

New Technology Telescope optical telescope that forms part of the ◊European Southern Observatory, La Silla, Chile; it came into operation in 1991. It has a thin, lightweight mirror, 3.38 m/141 in across, which is kept in shape by computer-adjustable supports to produce a sharper image than is possible with conventional mirrors. Such a system is termed *active optics*.

newton SI unit (symbol N) of ◊force. One newton is the force needed to accelerate an object with mass of one kilogram by one metre per second per second. The weight of a medium size (100 g/3 oz) apple is one newton.

Newtonian physics physics based on the concepts of the English scientist Isaac Newton, before the formulation of quantum theory or relativity theory.

GREAT EXPERIMENTS AND DISCOVERIES

The little neutral one

Most fundamental particles are discovered experimentally. Some are predicted by theory, but are often not taken seriously until detected. The neutrino is an exception. Postulated in 1930, its existence was taken as an act of faith for 25 years.

There were good reasons for predicting the neutrino's existence. Unexplained energy losses in certain radioactive decays (called beta decay) suggested the laws of conservation of energy and angular momentum were invalid at subatomic level. The great Danish theoretical physicist Niels Bohr maintained these laws held only on average, not for each individual process. Austrian physicist Wolfgang Pauli explained it more simply: in beta decay, there was a previously undiscovered neutral particle carrying away the missing energy and momentum. He called it a 'neutron'. Italian physicist Enrico Fermi introduced the alternative term 'neutrino': Italian for 'little neutral one'. He constructed a successful theory of beta decay based on the neutrino's existence.

An explosive suggestion

Experimental observation of neutrinos proved difficult. They pass easily through solid material, penetrating lead 965 million million km/600 million million mi thick without being absorbed. This makes them extremely hard to detect. In the 1950s, US physicists Frederick Reines and Clyde Cowan put forward a far-fetched scheme using a nuclear weapon as a neutrino source. Cowan described the proposed scheme several years later:

'We would dig a shaft near the centre of the explosion about 10 ft in diameter and about 150 ft deep. We would put a tank, 10 ft in diameter and 75 ft long on end at the bottom of the shaft. We would then suspend a detector from the top of the tank, along with its recording apparatus, and back-fill the shaft about the tank.'

'As the time for the explosion approached, we would start vacuum pumps and evacuate the tank as highly as possible. Then, when the countdown reached zero, we would break the suspension with a small explosive, allowing the detector to fall freely in the vacuum. For about two seconds, the falling detector would be seeing neutrinos and recording the pulses from them while the earth shock from the blast passed harmlessly by, rattling the tank mightily but not disturbing

our falling detector. When all was relatively quiet, the detector would reach the bottom of the tank, landing on a thick pile of foam rubber and feathers.'

'We would return to the site of the shaft in a few days (when surface radioactivity had died away sufficiently), dig down to the tank, recover the detector and discover the truth about neutrinos.'

The crucial experiment

Before the spectacular experiment could be set up, Reines and Cowan realized that, with suitable modifications, they could utilize the much smaller neutrino flux from a nuclear reactor. They set up their experiment next to the Savannah River nuclear plant's reactor. The detector comprised a tank containing cadmium chloride dissolved in water. A neutrino passing through the tank reacts with protons in the water, producing a positron and a neutron. The positron combines with electrons, producing gamma rays. The neutron also produces gamma rays by reacting with the cadmium. When a gamma ray falls on a scintillation detector it produces a flash of light, detected using a photomultiplier and showing that a neutrino has passed through the tank.

At first, there were about 1.63 'events' (or signals) per hour with the reactor operating and 0.4 with it shut down. Subsequent experiments with various adjustments raised the rate to nearly three events per hour. The next task was to confirm that these signals were produced by neutrinos. This involved eliminating all other possible sources: principally neutrons, gamma rays, electrons, and protons from the reactor.

The water was in two plastic tanks, holding 200 l/40 gal, between large scintillation counters, and surrounded by 110 photomultipliers, with lead walls and floor 10 cm/4 in thick, and roof and doors 20 cm/8 in thick to eliminate radiation from the reactor. To prove the lead's effectiveness, tests were done with additional shields. One experiment involved packing wet sawdust in bags round the detectors to reduce spurious neutron flow. There was no appreciable change in signal, and so it was concluded that it was not due to radiation produced outside the detectors. The neutrino had indeed been detected.

Peter Lafferty

Newtonian telescope in astronomy, a simple reflecting ◊telescope in which light collected by a parabolic primary mirror is directed to a focus at the side of the tube by a flat secondary mirror placed at 45 degrees to the optical axis. It is named after Isaac Newton, who constructed such a telescope in 1668.

Newton's laws of motion in physics, three laws that form the basis of Newtonian mechanics. (1) Unless acted upon by an unbalanced force, a body at rest stays at rest, and a moving body continues moving at the same speed in the same straight line. (2) An unbalanced force applied to a body gives it an acceleration proportional to the force and in the direction of the force. (3) When a body A exerts a force on a body B, B exerts an equal and opposite force on A; that is, to every action there is an equal and opposite reaction.

Newton's rings in optics, an ◊interference phenomenon seen (using white light) as concentric rings of spectral colours where light passes through a thin film of transparent medium, such as the wedge of air between a large-radius convex lens and a flat glass plate. With monochromatic light (light of a single wavelength), the rings take the form of alternate light and dark bands. They are caused by interference (interaction) between light rays reflected from the plate and those reflected from the curved surface of the lens.

NHS abbreviation for *National Health Service*, the UK state-financed health service.

niacin one of the 'B group' vitamins; see ◊nicotinic acid.

niche in ecology, the 'place' occupied by a species in its habitat, including all chemical, physical, and biological components, such as what it eats, the time of day at which the species feeds, temperature, moisture, the parts of the habitat that it uses (for example, trees or open grassland), the way it reproduces, and how it behaves.

It is believed that no two species can occupy exactly the same niche, because they would be in direct competition for the same resources at every stage of their life cycle.

Nichrome trade name for a series of alloys containing mainly nickel and chromium, with small amounts of other substances such as iron, magnesium, silicon, and carbon. Nichrome has a high melting point and is resistant to corrosion. It is therefore used in electrical heating elements and as a substitute for platinum in the ◊flame test.

nickel hard, malleable and ductile, silver-white metallic element, symbol Ni, atomic number 28, relative atomic mass 58.71. It occurs in igneous rocks and as a free metal (◊native metal), occasionally in fragments of iron-nickel meteorites. It is a component of the Earth's core, which is held to consist principally of iron with some nickel. It has a high melting point, low electrical and thermal conductivity, and can be magnetized. It does not tarnish and therefore is much used for alloys, electroplating, and for coinage.

It was discovered in 1751 by Swedish mineralogist Axel Cronstedt (1722–1765) and the name given as an abbreviated form of *kopparnickel*, Swedish 'false copper', since the ore in which it is found resembles copper but yields none.

nickel ore any mineral ore from which nickel is obtained. The main minerals are arsenides such as chloanthite ($NiAs_2$), and the sulphides millerite (NiS) and pentland-ite ($(Ni,Fe)_9S_8$), the commonest ore. The chief nickel-producing countries are Canada, Russia, Kazakhstan, Cuba, and Australia.

nicotine $C_{10}H_{14}N_2$ ◊alkaloid (nitrogenous compound) obtained from the dried leaves of the tobacco plant *Nicotiana tabacum* and used as an insecticide. A colourless oil, soluble in water, it turns brown on exposure to the air.

Nicotine in its pure form is one of the most powerful poisons known. It is the component of cigarette smoke that causes physical addiction. It is named after a 16th-century French diplomat, Jacques Nicot, who introduced tobacco to France.

nicotine patch plastic patch impregnated with nicotine that is stuck on to the skin to help the wearer give up smoking tobacco. Nicotine seeps out at a controlled rate onto the skin and is absorbed into the blood, thereby relieving the wearer's physical craving for the drug. The amount administered is reduced over time as the wearer proceeds from high-dose to low-dose patches.

Nicotine patches are more effective when combined with counselling because, although they can alleviate physical dependence on nicotine, they do not alter the habits that can make the wearer yearn for a cigarette. Very high doses of nicotine, such as those produced when wearing more than one patch at a time, can produce side effects such as vomiting, disturbed vision, and – in extreme cases – heart attacks.

nicotinic acid or *niacin* water-soluble ◊vitamin ($C_5H_5N.COOH$) of the B complex, found in meat, fish, and cereals; it can also be formed in small amounts in the body from the essential ◊amino acid tryptophan. Absence of nicotinic acid from the diet leads to the disease ◊pellagra.

niobium soft, grey-white, somewhat ductile and malleable, metallic element, symbol Nb, atomic number 41, relative atomic mass 92.906. It occurs in nature with tantalum, which it resembles in chemical properties. It is used in making stainless steel and other alloys for jet engines and rockets and for making superconductor magnets.

Niobium was discovered in 1801 by English chemist Charles Hatchett (1765–1847), who named it columbium (symbol Cb), a name that is still used in metallurgy. In 1844 it was renamed after Niobe by German chemist Heinrich Rose (1795–1864) because of its similarity to tantalum (Niobe is the daughter of Tantalus in Greek mythology).

nitrate salt or ester of nitric acid, containing the NO_3^- ion. Nitrates are used in explosives, in the chemical and pharmaceutical industries, in curing meat (see ◊nitre), and as fertilizers. They are the most water-soluble salts known and play a major part in the nitrogen cycle. Nitrates in the soil, whether naturally occurring or from inorganic or organic fertilizers, can be used by plants to make proteins and nucleic acids. However, runoff from fields can result in ◊nitrate pollution.

nitrate pollution the contamination of water by nitrates. Increased use of artificial fertilizers and land cultivation means that higher levels of nitrates are being washed from the soil into rivers, lakes, and aquifers. There they cause an excessive enrichment of the water (◊eutrophication), leading to a rapid growth of algae, which in turn darkens the water and reduces its oxygen content. The water is expensive to purify and many plants and animals die. High levels are now found in drinking water in arable areas. These may be harmful to newborn babies, and it is possible that they contribute to stomach cancer, although the evidence for this is unproven.

nitre or ***saltpetre*** potassium nitrate, KNO_3, a mineral found on and just under the ground in desert regions; used in explosives. Nitre occurs in Bihar, India, Iran, and Cape Province, South Africa. The salt was formerly used for the manufacture of gunpowder, but the supply of nitre for explosives is today largely met by making the salt from nitratine (also called Chile saltpetre, $NaNO_3$). Saltpetre is a ◊preservative and is widely used for curing meats.

nitric acid or ***aqua fortis*** HNO_3 fuming acid obtained by the oxidation of ammonia or the action of sulphuric acid on potassium nitrate. It is a highly corrosive acid, dissolving most metals, and a strong oxidizing agent. It is used in the nitration and esterification of organic substances, and in the making of sulphuric acid, nitrates, explosives, plastics, and dyes.

nitric oxide or ***nitrogen monoxide*** (NO) colourless gas released when metallic copper reacts with nitric acid and when nitrogen and oxygen combine at high temperatures. It is oxidized to nitrogen dioxide on contact with air. Nitric oxide has a wide range of functions in the body. It is involved the transmission of nerve impulses and the protection of nerve cells against stress. It is released by macrophages in the immune system in response to viral and bacterial infection or to the proliferation of cancer cells. It is also important in the control of blood pressure.

nitrification process that takes place in soil when bacteria oxidize ammonia, turning it into nitrates. Nitrates can be absorbed by the roots of plants, so this is a vital stage in the ◊nitrogen cycle.

nitrite salt or ester of nitrous acid, containing the nitrite ion (NO_2^-). Nitrites are used as preservatives (for example, to prevent the growth of botulism spores) and as colouring agents in cured meats such as bacon and sausages.

nitrocellulose alternative name for ◊cellulose nitrate.

nitrogen sodium or potassium nitrate Colourless, odourless, tasteless, gaseous, nonmetallic element, symbol N, atomic number 7, relative atomic mass 14.0067. It forms almost 80% of the Earth's atmosphere by volume and is a constituent of all plant and animal tissues (in proteins and nucleic acids). Nitrogen is obtained for industrial use by the liquefaction and fractional distillation of air. Its compounds are used in the manufacture of foods, drugs, fertilizers, dyes, and explosives.

Nitrogen has been recognized as a plant nutrient, found in manures and other organic matter, from early times, long before the complex cycle of ◊nitrogen fixation was understood. It was isolated in 1772 by English chemist Daniel Rutherford (1749–1819) and named in 1790 by French chemist Jean Chaptal (1756–1832).

nitrogen cycle the process of nitrogen passing through the ecosystem. Nitrogen, in the form of inorganic compounds (such as nitrates) in the soil, is absorbed by plants and turned into organic compounds (such as proteins) in plant tissue. A proportion of this nitrogen is eaten by ◊herbivores, with some of this in turn being passed on to the carnivores, which feed on the herbivores. The nitrogen is ultimately returned to the soil as excrement and when organisms die and are converted back to inorganic form by ◊decomposers.

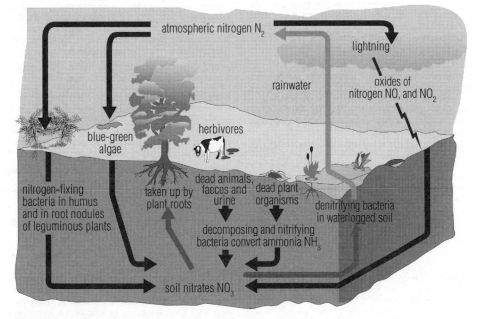

nitrogen cycle The nitrogen cycle is one of a number of cycles during which the chemicals necessary for life are recycled. The carbon, sulphur, and phosphorus cycles are others. Since there is only a limited amount of these chemicals in the Earth and its atmosphere, the chemicals must be continuously recycled if life is to go on.

Although about 78% of the atmosphere is nitrogen, this cannot be used directly by most organisms. However, certain bacteria and cyanobacteria (see ◊blue-green algae) are capable of nitrogen fixation. Some nitrogen-fixing bacteria live mutually with leguminous plants (peas and beans) or other plants (for example, alder), where they form characteristic nodules on the roots. The presence of such plants increases the nitrate content, and hence the fertility, of the soil.

nitrogen fixation the process by which nitrogen in the atmosphere is converted into nitrogenous compounds by the action of microorganisms, such as cyanobacteria (see ◊blue-green algae) and bacteria, in conjunction with certain ◊legumes. Several chemical processes duplicate nitrogen fixation to produce fertilizers; see ◊nitrogen cycle.

nitrogen oxide any chemical compound that contains only nitrogen and oxygen. All nitrogen oxides are gases. Nitrogen monoxide and nitrogen dioxide contribute to air pollution. See also ◊nitrous oxide.

Nitrogen oxide acts as a chemical messenger in small quantities within the human body, despite being toxic at higher concentrations, and its rapid reaction with oxygen. The medical condition of septic shock is linked to overproduction by the body of nitrogen oxide. Nitrogen oxide has an unpaired electron, which can be removed to produce the nitrosyl ion, NO^+.

Nitrogen monoxide NO, or nitric oxide, is a colourless gas released when metallic copper reacts with concentrated ◊nitric acid. It is also produced when nitrogen and oxygen combine at high temperature. On contact with air it is oxidized to nitrogen dioxide.

Nitrogen dioxide nitrogen(IV) oxide, NO_2, is a brown, acidic, pungent gas that is harmful if inhaled and contributes to the formation of ◊acid rain, as it dissolves in water to form nitric acid. It is the most common of the nitrogen oxides and is obtained by heating most nitrate salts (for example lead(II) nitrate, $Pb(NO_3)_2$). If liquefied, it gives a colourless solution (N_2O_4). It has been used in rocket fuels.

nitroglycerine $C_3H_5(ONO_2)_3$ flammable, explosive oil produced by the action of nitric and sulphuric acids on glycerol. Although poisonous, it is used in cardiac medicine. It explodes with great violence if heated in a confined space and is used in the preparation of dynamite, cordite, and other high explosives.

It was invented by the Italian Ascanio Soberro in 1846, and is unusual among explosives in that it is a liquid. Nitroglycerine is an effective explosive because it has low ◊activation energy, and produces little smoke when burned. However, it was initially so reactive it was virtually unusable. Alfred Nobel's innovation was to purify nitroglycerine (using water, with which it is immiscible, to dissolve the impurities), and thereby make it more stable.

nitrous acid HNO_2 weak acid that, in solution with water, decomposes quickly to form nitric acid and nitrogen dioxide.

nitrous oxide or *dinitrogen oxide* N_2O colourless, nonflammable gas that, used in conjunction with oxygen, reduces sensitivity to pain. In higher doses it is an anaesthetic. Well tolerated, it is often combined with other anaesthetic gases to enable them to be used in lower doses. It may be self-administered; for example, in childbirth. It is a greenhouse gas; about 10% of nitrous oxide

released into the atmosphere comes from the manufacture of nylon. It used to be known as 'laughing gas'.

nobelium synthesized, radioactive, metallic element of the ◊actinide series, symbol No, atomic number 102, relative atomic mass 259. It is synthesized by bombarding curium with carbon nuclei.

It was named in 1957 after the Nobel Institute in Stockholm, Sweden, where it was claimed to have been first synthesized. Later evaluations determined that this was in fact not so, as the successful 1958 synthesis at the University of California at Berkeley produced a different set of data. The name was not, however, challenged. In 1992 the International Unions for Pure and Applied Chemistry and Physics (IUPAC and IUPAP) gave credit to Russian scientists in Dubna for the discovery of nobelium.

Nobel prize annual international prize, first awarded in 1901 under the will of Alfred Nobel, Swedish chemist, who invented dynamite. The interest on the Nobel endowment fund is divided annually among the persons who have made the greatest contributions in the fields of physics, chemistry, medicine, literature, and world peace. The first four are awarded by academic committees based in Sweden, while the peace prize is awarded by a committee of the Norwegian parliament. A sixth prize, for economics, financed by the Swedish National Bank, was first awarded in 1969. The prizes have a large cash award and are given to organizations – such as the United Nations peacekeeping forces, which received the Nobel Peace Prize in 1988 – as well as individuals.

noble gas alternative name for ◊inert gas.

noble gas structure the configuration of electrons in noble or ◊inert gases (helium, neon, argon, krypton, xenon, and radon).

This is characterized by full electron shells around the nucleus of an atom, which render the element stable. Any ion, produced by the gain or loss of electrons, that achieves an electronic configuration similar to one of the inert gases is said to have a noble gas structure.

nocturnal enuresis in medicine, the involuntary passing of urine during sleep. See ◊bedwetting.

node in physics, a position in a ◊standing wave pattern at which there is no vibration. Points at which there is maximum vibration are called *antinodes*. Stretched strings, for example, can show nodes when they vibrate. Guitarists can produce special effects (harmonics) by touching a sounding string lightly to produce a node.

nodule in geology, a lump of mineral or other matter found within rocks or formed on the seabed surface; ◊mining technology is being developed to exploit them.

noise unwanted sound. Permanent, incurable loss of hearing can be caused by prolonged exposure to high noise levels (above 85 decibels). Over 55 decibels on a daily outdoor basis is regarded as an unacceptable level.

In scientific and engineering terms, a noise is any random, unpredictable signal.

Noise is a recognized form of pollution, but is difficult to measure because the annoyance or discomfort caused varies between individuals. If the noise is in a narrow frequency band, temporary hearing loss can occur even though the level is below 85 decibels or exposure is only for short periods. Lower levels of noise are an irritant, but seem not to increase fatigue or affect efficiency to any great extent. Loss of hearing is a common complaint of people working on factory production lines or in the construction and road industry. Minor psychiatric disease, stress-related ailments including high blood pressure, and disturbed sleep patterns are regularly linked to noise, although the causal links are in most cases hard to establish. Loud noise is a major pollutant in towns and cities.

Electronic noise takes the form of unwanted signals generated in electronic circuits and in recording processes by stray electrical or magnetic fields, or by temperature variations. In electronic recording and communication systems, **white noise** frequently appears in the form of high frequencies, or hiss. The main advantages of digital systems are their relative freedom from such noise and their ability to recover and improve noise-affected signals.

nomadic pastoralism farming system where animals (cattle, goats, camels) are taken to different locations in order to find fresh pastures. It is practised in the developing world; out of an estimated 30–40 million (1990) nomadic pastoralists worldwide, most are in central Asia and the Sahel region of W Africa. Increasing numbers of cattle may lead to overgrazing of the area and ◊desertification.

The increasing enclosure of land has reduced the area available for nomadic pastoralism and, as a result, this system of farming is under threat. The movement of farmers in this way contrasts with sedentary agriculture.

nonmetal one of a set of elements (around 20 in total) with certain physical and chemical properties opposite to those of metals. Nonmetals accept electrons (see ◊electronegativity) and are sometimes called electronegative elements.

nonrenewable resource natural resource, such as coal or oil, that takes thousands or millions of years to form naturally and can therefore not be replaced once it is consumed. The main energy sources used by humans are nonrenewable; ◊renewable resources, such as solar, tidal, and geothermal power, have so far been less exploited.

Nonrenewable resources have a high carbon content because their origin lies in the photosynthetic activity of plants millions of years ago. The fuels release this carbon back into the atmosphere as carbon dioxide. The rate at which such fuels are being burnt is thus resulting in rise in the concentration of carbon dioxide in the atmosphere.

nonsteroidal anti-inflammatory drug full name of ◊NSAID.

nonvolatile memory in computing, ◊memory that does not lose its contents when the power supply to the computer is disconnected.

noradrenaline in the body, a ◊catecholamine that acts directly on specific receptors to stimulate the sympathetic nervous system. Released by nerve stimulation or by drugs, it slows the heart rate mainly by constricting arterioles (small arteries) and so raising blood pressure. It is used therapeutically to treat ◊shock.

NOR gate in electronics, a type of ◊logic gate.

normal distribution curve the bell-shaped curve that results when a normal distribution is represented graphically by plotting the distribution $f(x)$ against x. The curve is symmetrical about the mean value.

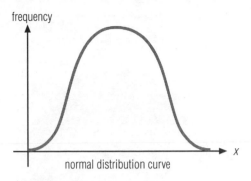

normal distribution curve

normal distribution

North Atlantic Drift warm ◊ocean current in the N Atlantic Ocean; an extension of the ◊Gulf Stream. It flows east across the Atlantic and has a mellowing effect on the climate of NW Europe, particularly the British Isles and Scandinavia.

northern lights common name for the ◊aurora borealis.

nose in humans, the upper entrance of the respiratory tract; the organ of the sense of smell. The external part is divided down the middle by a septum of ◊cartilage. The nostrils contain plates of cartilage that can be moved by muscles and have a growth of stiff hairs at the margin to prevent foreign objects from entering. The whole nasal cavity is lined with a ◊mucous membrane that warms and moistens the air as it enters and ejects dirt. In the upper parts of the cavity the membrane contains 50 million olfactory receptor cells (cells sensitive to smell). *See illustration overleaf.*

nosebleed or *epistaxis* bleeding from the nose. It may be caused by injury, infection, high blood pressure, or some disorders of the blood. Although usually minor and easily controlled, the loss of blood may occasionally be so rapid as to be life-threatening, particularly in small children. Most nosebleeds can be stopped by simply squeezing the nose for a few minutes with the head held forwards, but in exceptional cases transfusion may be required and the nose may need to be packed with ribbon gauze or cauterized.

nosocomial description of any infection acquired in a hospital or other medical facility, whether its effects are seen during the patient's stay or after discharge. Widely prevalent in some hospitals, nosocomial infections threaten patients who are seriously ill or whose immune systems have been suppressed. The threat is compounded by the prevalence of drug-resistant ◊pathogens endemic to the hospital environment.

nosology branch of medicine concerned with the classification of diseases. It can also be applied to the collection of evidence relating to whether a particular condition should be regarded as a disease entity.

nostril in vertebrates, the opening of the nasal cavity, in which cells sensitive to smell are located. (In fish, these cells detect water-borne chemicals, so they are effectively organs of taste.) In vertebrates with lungs, the nostrils also take in air. In humans, and most other mammals, the nostrils are located on the ◊nose.

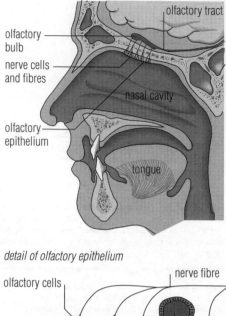

olfactory tract
olfactory bulb
nerve cells and fibres
nasal cavity
olfactory epithelium
tongue

detail of olfactory epithelium

nerve fibre
olfactory cells
supporting cells
mucous gland

nose The structure of the nose. The organs of smell are confined to a small area in the roof of the nasal cavity. The olfactory cells are stimulated when certain molecules reach them. Smell is one of our most subtle senses: tens of thousands of smells can be distinguished. By comparison, taste, although closely related to smell, is a crude sensation. All the subtleties of taste depend upon smell.

notebook computer small ◊laptop computer. Notebook computers became available in the early 1990s and, even complete with screen and hard disc drive, are no larger than a standard notebook.

NOT gate or **inverter gate** in electronics, a type of ◊logic gate.

notifiable disease any disease of humans or animals that must be notified, by law, to a health officer or local authority. The majority of these diseases are serious infectious diseases, such as meningitis, tetanus, and whooping cough in humans and anthrax and foot and mouth in sheep and cows. Some industrial diseases, such as lung diseases related to the inhalation of asbestos, are also notifiable.

notochord the stiff but flexible rod that lies between the gut and the nerve cord of all embryonic and larval chordates, including the vertebrates. It forms the supporting structure of the adult lancelet, but in vertebrates it is replaced by the vertebral column, or spine.

nova (plural **novae**) faint star that suddenly erupts in brightness by 10,000 times or more, remains bright for a few days, and then fades away and is not seen again for very many years, if at all. Novae are believed to occur in close ◊binary star systems, where gas from one star flows to a companion ◊white dwarf. The gas ignites and is thrown off in an explosion at speeds of 1,500 kps/930 mps or more. Unlike a ◊supernova, the star is not completely disrupted by the outburst. After a few weeks or months it subsides to its previous state; it may erupt many more times.

Although the name comes from the Latin 'new', photographic records show that such stars are not really new, but faint stars undergoing an outburst of radiation that temporarily gives them an absolute magnitude in the range ⁻6–⁻10, at least 100,000 times brighter than the Sun. They fade away, rapidly at first and then more slowly over several years. Two or three such stars are detected in our Galaxy each year, but on average one is sufficiently close to us to become a conspicuous naked-eye object only about once in ten years. Novae very similar to those appearing in our own Galaxy have also been observed in other galaxies.

Novocaine trade name of **procaine**, the first synthetic local anaesthetic, invented in 1905. It is now seldom used, having been replaced by agents such as ◊lignocaine. It is as effective as cocaine, formerly used as a painkiller, when injected, but only one third as toxic and not habit-forming.

NSAID (abbreviation for **nonsteroidal anti-inflammatory drug**) any of a class of drugs used in the long-term treatment of rheumatoid ◊arthritis and osteoarthritis; they act to reduce swelling and pain in soft tissues. Bleeding into the digestive tract is a serious side effect: NSAIDs should not be taken by persons with peptic ulcers.

NTP abbreviation for **normal temperature and pressure**, the former name for **STP** (◊**standard temperature and pressure**).

nuclear energy or **atomic energy** energy released from the inner core, or ◊nucleus, of the atom. Energy produced by **nuclear** ◊**fission** (the splitting of uranium or plutonium nuclei) has been harnessed since the 1950s to generate electricity, and research continues into the possible controlled use of ◊**nuclear fusion** (the fusing, or combining, of atomic nuclei).

In nuclear power stations, fission takes place in a ◊nuclear reactor. The nuclei of uranium or, more rarely, plutonium are induced to split, releasing large amounts of heat energy. The heat is then removed from the core of the reactor by circulating gas or water, and used to produce the steam that drives alternators and turbines to generate electrical power. Unlike fossil fuels, such as coal and oil, which must be burned in large quantities to produce energy, nuclear fuels are used in very small amounts and supplies are therefore unlikely to be exhausted in the foreseeable future. However, the use of nuclear energy has given rise to concern over safety such as the one at Chernobyl, Ukraine, in 1986. There has also been mounting concern about the production and disposal of toxic nuclear waste, which may have an active life of several thousand years, and the cost of maintaining nuclear power stations and decommissioning them at the end of their lives. *See illustration on page 450.*

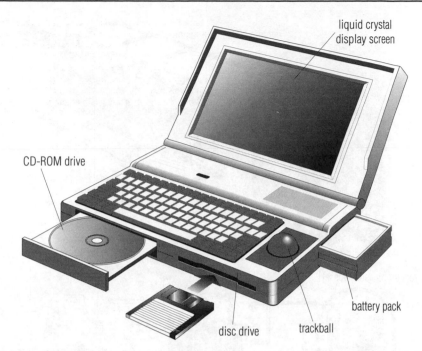

liquid crystal display screen

CD-ROM drive

battery pack

disc drive   trackball

notebook computer The component parts of a notebook computer. Although as powerful as a microcomputer, the battery pack enables the notebook to be used away from any source of power.

nuclear fusion process whereby two atomic nuclei are fused, with the release of a large amount of energy. Very high temperatures and pressures are thought to be required in order for the process to happen. Under these conditions the atoms involved are stripped of all their electrons so that the remaining particles, which together make up a *plasma*, can come close together at very high speeds and overcome the mutual repulsion of the positive charges on the atomic nuclei. At very close range the strong nuclear force will come into play, fusing the particles together to form a larger nucleus. As fusion is accompanied by the release of large amounts of energy, the process might one day be harnessed to form the basis of commercial energy production. Methods of achieving controlled fusion are therefore the subject of research around the world.

Fusion is the process by which the Sun and the other stars produce their energy.

nuclear magnetic resonance (NMR) in medicine, scanning technique used to produce images of the organs in the body. It is another term for ◊magnetic resonance imaging (MRI).

nuclear medicine in medicine, the use of radioactive isotopes in the diagnosis, investigation, and treatment of disease. See ◊radioisotope scanning.

nuclear notation method used for labelling an atom according to the composition of its nucleus. The atoms or isotopes of a particular element are represented by the symbol $^A_Z X$ where A is the mass number of their nuclei, Z is their atomic number, and X is the chemical symbol for that element.

nuclear physics study of the properties of the nucleus of

the ◊atom, including the structure of nuclei; nuclear forces; the interactions between particles and nuclei; and the study of radioactive decay. The study of elementary particles is ◊particle physics.

nuclear reaction reaction involving the nuclei of atoms. Atomic nuclei can undergo changes either as a result of radioactive decay, as in the decay of radium to radon (with the emission of an alpha particle) or as a result of particle bombardment in a machine or device, as in the production of cobalt-60 by the bombardment of cobalt-59 with neutrons.

$$^{226}_{88}Ra \rightarrow {}^{222}_{86}Rn + {}^4_2He$$

$$^{59}_{37}Co + {}^1_0n \rightarrow {}^{60}_{37}Co + \gamma$$

Nuclear ◊fission and nuclear ◊fusion are examples of nuclear reactions. The enormous amounts of energy released arise from the mass–energy relation put forward by Einstein, stating that $E = mc^2$ (where E is energy, m is mass, and c is the velocity of light).

In nuclear reactions the sum of the masses of all the products (on the atomic mass unit scale) is less than the sum of the masses of the reacting particles. This lost mass is converted to energy according to Einstein's equation.

nuclear reactor device for producing ◊nuclear energy in a controlled manner. There are various types of reactor in use, all using nuclear fission. In a *gas-cooled reactor*, a circulating gas under pressure (such as carbon dioxide) removes heat from the core of the reactor, which usually contains natural uranium. The efficiency of the fission process is increased by slowing neutrons in the core by using a ◊moderator such as carbon. The reaction is controlled with neutron-absorbing rods made of boron. An *advanced gas-cooled reactor* (AGR) generally has

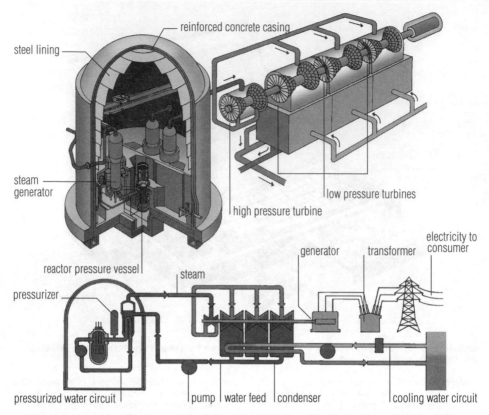

nuclear energy A pressurized water nuclear power station. Water at high pressure is circulated around the reactor vessel where it is heated. The hot water is pumped to the steam generator where it boils in a separate circuit; the steam drives the turbines coupled to the electricity generator. This is the most widely used type of reactor. More than 20 countries have pressurized water reactors.

enriched uranium as its fuel. In the US the **direct cycle gas turbine** is favoured. It has modular construction, enriched uranium fuel, and is cooled by helium. A **water-cooled reactor**, such as the steam-generating heavy water (deuterium oxide) reactor, has water circulating through the hot core. The water is converted to steam, which drives turbo-alternators for generating electricity. The most widely used reactor is the **pressurized-water reactor** (PWR), which contains a sealed system of pressurized water that is heated to form steam in heat exchangers in an external circuit. The **fast reactor** has no moderator and uses fast neutrons to bring about fission. It uses a mixture of plutonium and uranium oxide as fuel. When operating, uranium is converted to plutonium, which can be extracted and used later as fuel. It is also called the fast breeder because it produces more plutonium than it consumes. Heat is removed from the reactor by a coolant of liquid sodium.

Public concern over the safety of nuclear reactors has been intensified by explosions and accidental release of radioactive materials. The safest system allows for the emergency cooling of a reactor by automatically flooding an overheated core with water. Other concerns about nuclear power centre on the difficulties of reprocessing nuclear fuel and disposing safely of nuclear waste, and the

cost of maintaining nuclear power stations and of decommissioning them at the end of their lives. The break up of the former Soviet Union raised concerns about the ability of the new nation states to safely manage ageing reactors. In 1989, the UK government decided to postpone the construction of new nuclear power stations; in the USA, no new stations have been commissioned in over a decade. Rancho Seco, near Sacramento, California, was the first nuclear power station to be closed, by popular vote, in 1989. Sweden is committed to decommissioning its reactors. Some countries, such as France, are pressing ahead with their nuclear programmes.

nuclear safety measures to avoid accidents in the operation of nuclear reactors and in the production and disposal of nuclear weapons and of ◊nuclear waste. There are no guarantees of the safety of any of the various methods of disposal.

nuclear accidents

Windscale (now Sellafield), Cumbria, England. In 1957, fire destroyed the core of a reactor, releasing large quantities of radioactive fumes into the atmosphere.

Ticonderoga, 130 km/80 mi off the coast of Japan. In 1965 a US Navy Skyhawk jet bomber fell off the deck of this ship, sinking in 4,900 m/16,000 ft of water. It carried

Nuclear Energy: Chronology

1896	French physicist Henri Becquerel discovered radioactivity.
1905	In Switzerland, Albert Einstein showed that mass can be converted into energy.
1911	New Zealand physicist Ernest Rutherford proposed the nuclear model of the atom.
1919	Rutherford split the atom, by bombarding a nitrogen nucleus with alpha particles.
1939	Otto Hahn, Fritz Strassman, and Lise Meitner announced the discovery of nuclear fission.
1942	Enrico Fermi built the first nuclear reactor, in a squash court at the University of Chicago, USA.
1946	The first fast reactor, called Clementine, was built at Los Alamos, New Mexico.
1951	The Experimental Breeder Reactor, Idaho, USA, produced the first electricity to be generated by nuclear energy.
1954	The first reactor for generating electricity was built in the USSR, at Obninsk.
1956	The world's first commercial nuclear power station, Calder Hall, came into operation in the UK.
1957	The release of radiation from Windscale (now Sellafield) nuclear power station, Cumbria, England, caused 39 deaths to 1977. In Kyshym, USSR, the venting of plutonium waste caused high but undisclosed casualties (30 small communities were deleted from maps produced in 1958).
1959	Experimental fast reactor built in Dounreay, northern Scotland.
1979	Nuclear-reactor accident at Three Mile Island, Pennsylvania, USA.
1986	An explosive leak from a reactor at Chernobyl, the Ukraine, resulted in clouds of radioactive material spreading as far as Sweden; 31 people were killed and thousands of square kilometres were contaminated.
1991	The first controlled and substantial production of nuclear-fusion energy (a two-second pulse of 1.7 MW) was achieved at JET, the Joint European Torus, at Culham, Oxfordshire, England.
1994	The world record for producing nuclear fusion energy was broken at the Tokamak Fusion Test Reactor at Princeton University, New Jersey. The Princeton tokamak produced 9 megawatts of power.
1995	Sizewell B, the UK's first pressurized-water nuclear reactor and the most advanced nuclear power station in the world, begins operating in Suffolk, England.
1996	British Nuclear Fuels was fined £25,000 after safety failures that resulted in a worker becoming contaminated with radioactivity.
1997	English physicists at JET (Joint European Torus) fused isotopes of deuterium and tritium to produce a record 12 megawatts of nuclear-fusion power.

a one-megaton hydrogen bomb. The accident was only revealed in 1989.

Three Mile Island, Harrisburg, Pennsylvania, USA. In 1979, a combination of mechanical and electrical failure, as well as operator error, caused a pressurized water reactor to leak radioactive matter.

Church Rock, New Mexico, USA. In July 1979, 380 million litres/100 million gallons of radioactive water containing uranium leaked from a pond into the Rio Purco, causing the water to become over 6,500 times as radioactive as safety standards allow for drinking water.

Chernobyl, Ukraine. In April 1986 there was an explosive leak, caused by overheating, from a nonpressurized boiling-water reactor, one of the largest in Europe. The resulting clouds of radioactive material spread as far as the UK. 31 people were killed in the explosion, and thousands of square kilometres of land were contaminated by fallout. By June 1992, seven times as many children in the Ukraine and Belarus were contracting thyroid cancer as before the accident, the incidence of leukaemia was rising, and it was estimated that more than 6,000 people had died as a result of the accident, and that the death toll in the Ukraine alone would eventually reach 40,000.

Tomsk, Siberia. In April 1993 a tank exploded at a uranium reprocessing plant, sending a cloud of radioactive particles into the air.

nuclear waste the radioactive and toxic by-products of the nuclear-energy and nuclear-weapons industries. Nuclear waste may have an active life of several thousand years. Reactor waste is of three types: **high-level** spent fuel, or the residue when nuclear fuel has been removed from a reactor and reprocessed; **intermediate**, which may be long- or short-lived; and **low-level**, but bulky, waste from reactors, which has only short-lived radio-

activity. Disposal, by burial on land or at sea, has raised problems of safety, environmental pollution, and security. In absolute terms, nuclear waste cannot be safely relocated or disposed of.

The issue of nuclear waste is becoming the central controversy threatening the future of generating electricity by nuclear energy. The dumping of nuclear waste at sea officially stopped in 1983, when a moratorium was agreed by the members of the London Dumping Convention (a United Nations body that controls disposal of wastes at sea). Covertly, the USSR continued dumping, and deposited thousands of tonnes of nuclear waste and three faulty reactors into the sea between 1964–86. The USSR and the Russian Federation between them dumped an estimated 12 trillion becquerels of radioactivity in the sea between 1959–93. Russia has no way of treating nuclear waste and in 1993 announced its intention of continuing to dump it in the sea, in violation of international conventions, until 1997. Twenty reactors from Soviet nuclear-powered ships were dumped off the Arctic and Pacific coasts between 1965–93, and some are leaking. Fish-spawning grounds off Norway are threatened by plutonium from abandoned Soviet nuclear warheads.

nucleic acid complex organic acid made up of a long chain of ◊nucleotides, present in the nucleus and sometimes the cytoplasm of the living cell. The two types, known as ◊DNA (deoxyribonucleic acid) and ◊RNA (ribonucleic acid), form the basis of heredity. The nucleotides are made up of a sugar (deoxyribose or ribose), a phosphate group, and one of four purine or pyrimidine bases. The order of the bases along the nucleic acid strand contains the genetic code.

nucleolus in biology, a structure found in the nucleus of eukaryotic cells. It produces the RNA that makes up the ◊ribosomes, from instructions in the DNA.

The self-sustaining atomic pile

The world's first nuclear reactor was built in a squash court under Stagg Field football ground at the University of Chicago. On 2 Dec 1942 a team led by Italian physicist Enrico Fermi demonstrated a controlled self-sustaining nuclear chain reaction, opening the way to modern nuclear power and nuclear weapons.

The reactor – or 'atomic pile' – comprised 400 tonnes of graphite made into 45,000 bricks and built in layers into a structure about 6 m/20 ft high. Sixty tonnes of uranium was pressed into 22,000 small cylinders fitted into slots in the graphite, which slowed the neutrons produced by the splitting of the uranium nuclei. Cadmium control rods in the pile soaked up excess neutrons to control the reaction.

The pile took less than a month to build, following intensive research starting with the discovery of nuclear fission in 1939. Fermi had worked on the design for six months, calculating the amount of uranium needed, and the best arrangement of the fuel and graphite. They worked non-stop, knowing that the Nazis were also working to master nuclear power. Fermi, however, believed in regular routine. He always stopped at midday for lunch, and always went home at 5 p.m. Team member Albert Wattenberg recalls: 'He made everything look very logical and straightforward and he never made mistakes. It was eight years before I remember him making a mathematical mistake.'

Time for lunch

During the morning of 2 Dec 1942, Fermi began fine-tuning the pile, measuring the neutrons being produced. A single cadmium control rod regulated the neutron flow. The task was to determine how far to withdraw the rod for a chain reaction to proceed. Shortly before noon, there was a loud bang. Alarm followed – until the team realized that the safety control rod had operated automatically to shut the reactor down. The cutoff had been set too low. Fermi decided it was time for a break. 'I'm hungry; let's go to lunch', he said. Lunch was quiet; no-one talked about the experiment. Fermi sat silent and preoccupied. Back at work at 2 p.m., they eased the control rod out further. The instruments showing the reaction rate went off the scale and had to be reset; but the reaction was still not self-sustaining. Finally at 3.25 p.m., Fermi asked for the control rod

to be withdrawn another foot. 'This is going to do it. Now it will become self-sustaining. The trace will climb and continue to climb – it will not level off.' The rod was withdrawn and the reaction rate climbed. Team member Harold Agnew remembers: 'Hearing the counters going faster and faster and seeing the pen recorders continue to rise was scary.' Fermi did rapid calculations on his sliderule. The reaction counters clicked increasingly rapidly until they became a continuous buzz. After a few minutes, Fermi looked up from his calculations with a broad smile. 'The reaction is self-sustaining', he announced.

The reactor was allowed to operate for 28 minutes; at 3.53 p.m. Fermi told Wally Zinn to insert the control rod. The neutron rate dropped immediately. The first controlled nuclear chain reaction had stopped. Agnew recalls: 'No smoke, no smell. It was over and we were back where we were that morning. And the outside world was unaware of what had happened.'

The scientists applauded softly, aware of their work's implications. They fell silent and turned towards Fermi. Hungarian scientist Leo Szilard, an important team member, said, 'This will go down as a black day in the history of mankind.' Fermi's reply is not recorded.

Enrico sinks the admiral

The breakthrough was not announced publicly. Wartime secrecy prevailed; even Fermi's wife Laura did not learn of the success until after the war. At a party held after the successful experiment, she asked Leona Woods, the only woman in the team, what was being celebrated. 'Enrico has sunk a Japanese admiral', was the cryptic reply. The need for secrecy was paramount when Arthur Compton, in charge of the US nuclear weapons project, telephoned a colleague at Harvard with the news: 'The Italian navigator has landed in the New World', he said. 'How were the natives?' came the reply. 'Very friendly.' Compton answered.

The reactor did not operate again. In Feb 1943 it was dismantled and some of its parts used to make another reactor. The graphite blocks were eventually ground down to make ink to print invitations to the 40th anniversary celebrations, and to make into pencils given away as souvenirs.

Peter Lafferty

Nuclear Energy: Reactors

In 1992 a dozen nations depended on nuclear energy for at least a quarter of their electricity. France, with 72.7%, and Belgium and Sweden with well over 50%, top the list of 30 countries operating a total of 420 nuclear reactors around the world. Hungary and Korea are close behind with 48.4% and 47.5% respectively. The UK has one building, Sizewell B, to add to its current total of 37 operating reactors. The United States has 111 operating reactors and three more under construction. The 612.6 billion units (terawatt-hours or TWh) of nuclear electricity currently produced in the USA is just under a third of the world total and provides just over a fifth of the country's needs.

Country	Reactors in operation	Reactors being built	Nuclear electricity (% of total)
Argentina	2	1	19.1
Belgium	7		59.3
Brazil	1	1	0.6
Bulgaria	6		34.0
Canada	20	2	16.4
China	1	2	*
Cuba		2	
CSFR	8	6	28.6
Finland	4		33.3
France	56	5	72.7
Germany	21		27.6
Hungary	4		48.4
India	7	7	1.8
Iran		2	
Japan	42	10	23.8
Korea	9	3	47.5
Mexico	1	1	3.6
Netherlands	2		4.9
Pakistan	1		0.8
Romania		5	
South Africa	2		5.9
Spain	9		35.9
Sweden	12		51.6
Switzerland	5		40.0
Taiwan	6		37.8
UK	37	1	20.6
USA	111	3	21.7
ex-USSR	45	25	12.6
ex-Yugoslavia	1		6.3
total	420	76	

* Percentage figures for China unavailable.

nucleon in particle physics, either a ◊proton or a ◊neutron, when present in the atomic nucleus. **Nucleon number** is an alternative name for the ◊mass number of an atom.

nucleotide organic compound consisting of a purine (adenine or guanine) or a pyrimidine (thymine, uracil, or cytosine) base linked to a sugar (deoxyribose or ribose) and a phosphate group. ◊DNA and ◊RNA are made up of long chains of nucleotides.

nucleus in astronomy, the compact central core of a ◊galaxy, often containing powerful radio, X-ray and infrared sources. Active galaxies have extremely energetic nucleuses.

nucleus in biology, the central, membrane-enclosed part of a eukaryotic cell, containing the chromosomes.

nucleus in physics, the positively charged central part of an ◊atom, which constitutes almost all its mass. Except for hydrogen nuclei, which have only protons, nuclei are composed of both protons and neutrons. Surrounding the nuclei are electrons, of equal and opposite charge to that of the protons, thus giving the atom a neutral charge.

The nucleus was discovered by New Zealand physicist Ernest Rutherford in 1911 as a result of experiments in passing alpha particles through very thin gold foil.

When we have found how the nuclei of atoms are built up we shall have found the greatest secret of all – except life. We shall have found the basis of everything – of the earth we walk on, of the air we breathe, of the sunshine, of our physical body itself, of everything in the world, however great or however small – except life.

On the **nucleus** Ernest Rutherford
Passing Show

nuclide in physics, a species of atomic nucleus characterized by the number of protons (Z) and the number of neutrons (N). Nuclides with identical ◊proton number but differing neutron number are called ◊isotopes.

number symbol used in counting or measuring. In mathematics, there are various kinds of numbers. The everyday number system is the decimal ('proceeding by tens') system, using the base ten. ◊**Real numbers** include all rational numbers (integers, or whole numbers, and fractions) and irrational numbers (those not expressible as fractions). ◊**Complex numbers** include the real and imaginary numbers (real-number multiples of the square root of −1). The ◊binary number system, used in computers, has two as its base. The natural numbers, 0, 1, 2, 3, 4, 5, 6, 7, 8, and 9, give a counting system that, in the decimal system, continues 10, 11, 12, 13, and so on. These are whole numbers (integers), with fractions represented as, for example, 1/4, 1/2, 3/4, or as decimal

Number Systems

Binary (Base 2)	Octal (Base 8)	Decimal (Base 10)	Hexadecimal (Base 16)
0	0	0	0
1	1	1	1
10	2	2	2
11	3	3	3
100	4	4	4
101	5	5	5
110	6	6	6
111	7	7	7
1000	10	8	8
1001	11	9	9
1010	12	10	A
1011	13	11	B
1100	14	12	C
1101	15	13	D
1110	16	14	E
1111	17	15	F
10000	20	16	10
11111111	377	255	FF
11111010001	3721	2001	7D1

fractions (0.25, 0.5, 0.75). They are also *rational numbers*. *Irrational numbers* cannot be represented in this way and require symbols, such as $\sqrt{2}$, π, and e. They can be expressed numerically only as the (inexact) approximations 1.414, 3.142, and 2.718 (to three places of decimals) respectively. The symbols π and e are also examples of *transcendental numbers*, because they (unlike $\sqrt{2}$) cannot be derived by solving a ◊polynomial equation (an equation with one ◊variable quantity) with rational ◊coefficients (multiplying factors). Complex numbers, which include the real numbers as well as imaginary numbers, take the general form $a + bi$, where $i = \sqrt{-1}$ (that is, $i^2 = -1$), and a is the real part and bi the imaginary part.

The ancient Egyptians, Greeks, Romans, and Babylonians all evolved number systems, although none had a zero, which was introduced from India by way of Arab mathematicians in about the 8th century AD and allowed a place-value system to be devised on which the decimal system is based. Other number systems have since evolved and have found applications. For example, numbers to base two (binary numbers), using only 0 and 1, are commonly used in digital computers to represent the two-state 'on' or 'off' pulses of electricity. Binary numbers were first developed by German mathematician Gottfried Leibniz in the late 17th century.

numerator the number or symbol that appears above the line in a vulgar fraction. For example, the numerator of 5/6 is 5. The numerator represents the fraction's dividend and indicates how many of the equal parts indicated by the denominator (number or symbol below the line) comprise the fraction.

nursing care of the sick, the very young, the very old, and the disabled. Organized training originated in 1836 in Germany, and was developed in Britain by the work of Florence Nightingale, who, during the Crimean War, established standards of scientific, humanitarian care in military hospitals. Nurses give day-to-day care and carry out routine medical and surgical duties under the supervision of medical staff.

In ancient times very limited care was associated with some temples, and in Christian times nursing became associated with the religious orders until the Reformation brought it into secular hands in Protestant countries. Many specialities and qualifications now exist in Western countries, standards being maintained by professional bodies and boards.

nut any dry, single-seeded fruit that does not split open to release the seed, such as the chestnut. A nut is formed from more than one carpel, but only one seed becomes fully formed, the remainder aborting. The wall of the fruit, the pericarp, becomes hard and woody, forming the outer shell.

Examples of true nuts are the acorn and hazelnut. The term also describes various hard-shelled fruits and seeds, including almonds and walnuts, which are really the stones of ◊drupes, and brazil nuts and shelled peanuts, which are seeds. The kernels of most nuts provide a concentrated, nutritious food, containing vitamins, minerals, and enzymes, about 50% fat, and 10–20% protein, although a few, such as chestnuts, are high in carbohydrates and have only a moderate protein content of 5%. Nuts also provide edible and industrial oils. Most nuts are produced by perennial trees and shrubs. Whereas the majority of nuts are obtained from plantations, considerable quantities of pecans and brazil nuts are still collected from the wild. World production in the mid-1980s was about 4 million tonnes per year.

nut and bolt common method of fastening pieces of metal or wood together. The nut consists of a small block (usually metal) with a threaded hole in the centre for screwing on to a threaded rod or pin (bolt or screw). The method came into use at the turn of the 19th century, following Henry Maudslay's invention of a precision screw-cutting ◊lathe.

nutation in astronomy, a slight 'nodding' of the Earth in space, caused by the varying gravitational pulls of the Sun and Moon. Nutation changes the angle of the Earth's axial tilt (average 23.5°) by about 9 seconds of arc to either side of its mean position, a complete cycle taking just over 18.5 years.

nutation in botany, the spiral movement exhibited by the tips of certain stems during growth; it enables a climbing plant to find a suitable support. Nutation sometimes also occurs in tendrils and flower stalks.

The direction of the movements, clockwise or anticlockwise, is usually characteristic for particular species.

nutrition the strategy adopted by an organism to obtain the chemicals it needs to live, grow, and reproduce. Also, the science of food, and its effect on human and animal life, health, and disease. Nutrition involves the study of the basic nutrients required to sustain life, their bioavailability in foods and overall diet, and the effects upon them of cooking and storage. It is also concerned with dietary deficiency diseases.

There are six classes of nutrients: water, carbohydrates, proteins, fats, vitamins, and minerals.

Water is involved in nearly every body process. Animals and humans will succumb to water deprivation sooner than to starvation.

Carbohydrates are composed of carbon, hydrogen and oxygen. The major groups are starches, sugars, and cellulose and related material (or 'roughage'). The prime function of the carbohydrates is to provide energy for the body; they also serve as efficient sources of glucose, which the body requires for brain functioning, utilization of foods, maintenance of body temperature. Roughage includes the stiff structural materials of vegetables, fruits, and cereal products.

Proteins are made up of smaller units, amino acids. The primary function of dietary protein is to provide the amino acids required for growth and maintenance of body tissues. Both vegetable and animal foods are protein sources.

Fats serve as concentrated sources of energy, and protect vital organs such as the kidneys and skeleton. Saturated fats derive primarily from animal sources; unsaturated fats from vegetable sources such as nuts and seeds.

Vitamins are essential for normal growth, and are either fat-soluble or water-soluble. Fat-soluble vitamins include A, essential to the maintenance of mucous membranes, particularly the conjunctiva of the eyes; D, important to the absorption of calcium; E, an anti-oxidant; and K, which aids blood clotting. Water-soluble vitamins are the B complex, essential to metabolic reactions, and C, for maintaining connective tissue.

Minerals are vital to normal development; calcium and iron are particularly important as they are required in relatively large amounts. Minerals required by the body in trace amounts include chromium, copper, fluoride, iodine, iron, magnesium, manganese, molybdenum, phosphorus, potassium, selenium, sodium, and zinc.

nylon synthetic long-chain polymer similar in chemical structure to protein. Nylon was the first all-synthesized fibre, made from petroleum, natural gas, air, and water by the Du Pont firm in 1938. It is used in the manufacture of moulded articles, textiles, and medical sutures. Nylon fibres are stronger and more elastic than silk and are relatively insensitive to moisture and mildew. Nylon is used for hosiery and woven goods, simulating other materials such as silks and furs; it is also used for carpets.

nymph in entomology, the immature form of insects that do not have a pupal stage; for example, grasshoppers and dragonflies. Nymphs generally resemble the adult (unlike larvae), but do not have fully formed reproductive organs or wings.

oasis area of land made fertile by the presence of water near the surface in an otherwise arid region. The occurrence of oases affects the distribution of plants, animals, and people in the desert regions of the world.

oat type of annual grass, a ◊cereal crop. The plant has long narrow leaves and a stiff straw stem; the panicles of flowers (clusters around the main stem), and later of grain, hang downwards. The cultivated oat (*Avena sativa*) is produced for human and animal food.

obesity condition of being overweight (generally, 20% or more above the desirable weight for one's sex, build, and height). Obesity increases susceptibility to disease, strains the vital organs, and reduces life expectancy; it is usually remedied by controlled weight loss, healthy diet, and exercise.

In Dec 1994 US researchers discovered a gene in mice, *ob*, which controls the production of leptin, a protein involved in appetite control; defects in this gene cause obesity. Later research showed that obese mice lost weight dramatically after a 4-week treatment with the protein. An almost identical gene is found in humans, but obese humans have been found to have too much leptin rather than too little, as was the case in mice, so leptin injections will not cure obesity in humans.

object program in computing, the ◊machine code translation of a program written in a ◊source language.

observatory site or facility for observing astronomical or meteorological phenomena. The earliest recorded observatory was in Alexandria, N Africa, built by Ptolemy Soter in about 300 BC. The modern observatory dates from the invention of the telescope. Observatories may be ground-based, carried on aircraft, or sent into orbit as satellites, in space stations, and on the space shuttle.

The erection of observatories was revived in W Asia about AD 1000, and extended to Europe. The observatory built on the island of Hven (now Ven) in Denmark in 1576 for Tycho Brahe (1546–1601) was elaborate, but survived only to 1597. It was followed by those in Paris in 1667, Greenwich (the ◊Royal Greenwich Observatory) in 1675, and Kew, England. Most early observatories were near towns, but with the advent of big telescopes, clear skies with little background light, and hence high, remote sites, became essential.

The most powerful optical telescopes covering the sky are at ◊Mauna Kea, Hawaii; ◊Mount Palomar, California; ◊Kitt Peak National Observatory, Arizona; La Palma, Canary Islands; Cerro Tololo Inter-American

Observatory, and the ◊European Southern Observatory, Chile; ◊Siding Spring Mountain, Australia; and ◊Zelenchukskaya in the Caucasus.

Radio astronomy observatories include ◊Jodrell Bank, Cheshire, England; the ◊Mullard Radio Astronomy Observatory, Cambridge, England; ◊Arecibo, Puerto Rico; ◊Effelsberg, Germany; and ◊Parkes, Australia. The ◊Hubble Space Telescope was launched into orbit in April 1990. The Very Large Telescope is under construction by the European Southern Observatory in the mountains of N Chile.

obsession persistently intruding thought, emotion, or impulse, often recognized by the sufferer as irrational, but nevertheless causing distress. It may be a brooding on destiny or death, or chronic doubts interfering with everyday life (such as fearing the gas is not turned off and repeatedly checking), or an impulse leading to repetitive action, such as continually washing one's hands.

In obsessive-compulsive neurosis, these intrusions compel the patient to perform rituals or ceremonies, albeit reluctantly, no matter how absurd or distasteful they may seem.

obsidian black or dark-coloured glassy volcanic rock, chemically similar to ◊granite, but formed by cooling rapidly on the Earth's surface at low pressure.

The glassy texture is the result of rapid cooling, which inhibits the growth of crystals. Obsidian was valued by the early civilizations of Mexico for making sharp-edged tools and ceremonial sculptures.

obstetrics medical speciality concerned with the management of pregnancy, childbirth, and the immediate postnatal period.

obtuse angle an angle greater than 90° but less than 180°.

occluded front weather ◊front formed when a cold front catches up with a warm front. It brings clouds and rain as air is forced to rise upwards along the front, cooling and condensing as it does so.

occult in medicine, a substance that is present in small amounts but is not easily detected. Occult blood in the faeces can only be detected by a chemical test or by microscopic examination.

occultation in astronomy, the temporary obscuring of a star by a body in the Solar System. Occultations are used to provide information about changes in an orbit, and the structure of objects in space, such as radio sources.

The exact shapes and sizes of planets and asteroids can be found when they occult stars. The rings of Uranus were discovered when that planet occulted a star in 1977.

occupational psychology study of human behaviour at work. It includes dealing with problems in organizations, advising on management difficulties, and investigating the relationship between humans and machines (as in the design of aircraft controls; see also ◊ergonomics). Another area is ◊psychometrics and the use of assessment to assist in selection of personnel.

ocean great mass of salt water. Strictly speaking three oceans exist – the Atlantic, Indian, and Pacific – to which the Arctic is often added. They cover approximately 70% or 363,000,000 sq km/140,000,000 sq mi of the total surface area of the Earth. Water levels recorded in the

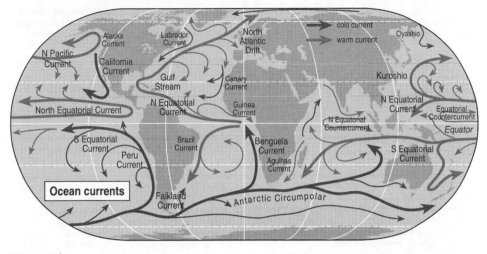

ocean currents

world's oceans have shown an increase of 10–15 cm/ 4–6 in over the past 100 years.

Depth (average) 3,660 m/12,000 ft, but shallow ledges (continental shelves) 180 m/600 ft run out from the continents, beyond which the continental slope reaches down to the ◊abyssal zone, the largest area, ranging from 2,000– 6,000 m/6,500–19,500 ft. Only the ◊deep-sea trenches go deeper, the deepest recorded being 11,034 m/36,201 ft (by the *Vityaz*, USSR) in the Mariana Trench of the W Pacific in 1957.

Features deep trenches (off E and SE Asia, and western South America), volcanic belts (in the W Pacific and E Indian Ocean), and ocean ridges (in the mid-Atlantic, E Pacific, and Indian Ocean).

Temperature varies on the surface with latitude (−2°C to +29°C); decreases rapidly to 370 m/1,200 ft, then more slowly to 2,200 m/7,200 ft; and hardly at all beyond that.

Water contents salinity averages about 3%; minerals commercially extracted include bromine, magnesium, potassium, salt; those potentially recoverable include aluminium, calcium, copper, gold, manganese, silver.

Pollution oceans have always been used as a dumping area for human waste, but as the quantity of waste increases, and land areas for dumping it diminish, the problem is exacerbated. Today ocean pollutants include airborne emissions from land (33% by weight of total marine pollution); oil from both shipping and land-based sources; toxins from industrial, agricultural, and domestic uses; sewage; sediments from mining, forestry, and farming; plastic litter; and radioactive isotopes. Thermal pollution by cooling water from power plants or other industry is also a problem, killing coral and other temperature-sensitive sedentary species.

oceanarium large display tank in which aquatic animals and plants live together much as they would in their natural environment. The first oceanarium was created by the explorer and naturalist W Douglas Burden in 1938 in Florida, USA.

ocean current fast-flowing ◊current of seawater generated by the wind or by variations in water density between two areas. Ocean currents are partly responsible for trans-ferring heat from the Equator to the poles and thereby evening out the global heat imbalance.

Ocean Drilling Program (ODP, formerly the *Deep-Sea Drilling Project* 1968–1985) research project initiated in the USA in 1968 to sample the rocks of the ocean ◊crust. Initially under the direction of Scripps Institution of Oceanography, the project was planned and administered by the Joint Oceanographic Institutions for Deep Earth Sampling (JOIDES). The operation became international in 1975, when Britain, France, West Germany, Japan, and the USSR also became involved.

Boreholes were drilled in all the oceans using the JOIDES ships *Glomar Challenger* and *Resolution*. Knowledge of the nature and history of the ocean basins was increased dramatically. The technical difficulty of drilling the seabed to a depth of 2,000 m/6,500 ft was overcome by keeping the ship in position with side-thrusting propellers and satellite navigation, and by guiding the drill using a radiolocation system. The project is intended to continue until 2005.

oceanography study of the oceans. Its subdivisions deal with each ocean's extent and depth, the water's evolution and composition, its physics and chemistry, the bottom topography, currents and wind effects, tidal ranges, the biology, and the various aspects of human use.

Oceanography involves the study of water movements – currents, waves, and tides – and the chemical and physical properties of the seawater. It deals with the origin and topography of the ocean floor – ocean trenches and ridges formed by ◊plate tectonics, and continental shelves from the submerged portions of the continents. Computer simulations are widely used in oceanography to plot the possible movements of the waters, and many studies are carried out by remote sensing.

ocean ridge mountain range on the seabed indicating the presence of a constructive plate margin (where tectonic plates are moving apart and magma rises to the surface; see ◊plate tectonics). Ocean ridges, such as the ◊Mid-Atlantic Ridge, consist of many segments offset along transform ◊faults, and can rise thousands of metres above the surrounding seabed.

Ocean ridges usually have a ◊rift valley along their crests, indicating where the flanks are being pulled apart by the growth of the plates of the ◊lithosphere beneath. The crests are generally free of sediment; increasing depths of sediment are found with increasing distance down the flanks.

ocean trench deep trench in the seabed indicating the presence of a destructive margin (produced by the movements of ◊plate tectonics). The subduction or dragging downwards of one plate of the ◊lithosphere beneath another means that the ocean floor is pulled down. Ocean trenches are found around the edge of the Pacific Ocean and the NE Indian Ocean; minor ones occur in the Caribbean and near the Falkland Islands.

Ocean trenches represent the deepest parts of the ocean floor, the deepest being the ◊Mariana Trench which has a depth of 11,034 m/36,201 ft. At depths of below 6 km/3.6 mi there is no light and very high pressure; ocean trenches are inhabited by crustaceans, coelenterates (for example, sea anemones), polychaetes (a type of worm), molluscs, and echinoderms.

OCR abbreviation for ◊ *optical character recognition*.

octahedron, regular regular solid with eight faces, each of which is an equilateral triangle. It is one of the five regular polyhedra or Platonic solids. The figure made by joining the midpoints of the faces is a perfect cube and the vertices of the octahedron are themselves the midpoints of the faces of a surrounding cube. For this reason, the cube and the octahedron are called dual solids.

octal number system number system to the base eight, used in computing. The highest digit that can appear in the octal system is seven. Whereas normal decimal, or base-ten, numbers may be considered to be written under column headings based on the number ten, octal, or base-eight, numbers can be thought of as written under column headings based on the number eight. The octal number 567 is therefore equivalent to the decimal number 375, since $(5 \times 64) + (6 \times 8) + (7 \times 1) = 375$.

The octal number system is sometimes used by computer programmers as an alternative to the ◊hexadecimal number system.

octane rating numerical classification of petroleum fuels indicating their combustion characteristics.

The efficient running of an ◊internal combustion engine depends on the ignition of a petrol–air mixture at the correct time during the cycle of the engine. Higher-rated petrol burns faster than lower-rated fuels. The use of the correct grade must be matched to the engine.

Octans faint constellation in the southern hemisphere, represented as an octant. It contains the southern celestial pole. The closest naked-eye star to the south celestial pole is fifth-magnitude Sigma Octantis.

octet rule in chemistry, rule stating that elements combine in a way that gives them the electronic structure of the nearest ◊inert gas. All the inert gases except helium have eight electrons in their outermost shell, hence the term octet.

oedema any abnormal accumulation of fluid in tissues or cavities of the body; waterlogging of the tissues due to excessive loss of ◊plasma through the capillary walls.

It may be generalized (the condition once known as dropsy) or confined to one area, such as the ankles.

Oedema may be mechanical – the result of obstructed veins or heart failure – or it may be due to increased permeability of the capillary walls, as in liver or kidney disease or malnutrition. Accumulation of fluid in the abdomen, a complication of cirrhosis, is known as *ascites*.

Oedipus complex in psychology, the unconscious antagonism of a son to his father, whom he sees as a rival for his mother's affection. For a girl antagonistic to her mother, as a rival for her father's affection, the term is *Electra complex*. The terms were coined by Sigmund Freud.

Freud saw this as a universal part of childhood development, which in most children is resolved during late childhood. Contemporary theory places less importance on the Oedipus/Electra complex than did Freud and his followers.

oersted c.g.s. unit (symbol Oe) of ◊magnetic field strength, now replaced by the SI unit ampere per metre. The Earth's magnetic field is about 0.5 oersted; the field near the poles of a small bar magnet is several hundred oersteds; and a powerful ◊electromagnet can have a field strength of 30,000 oersteds.

oesophagus muscular tube by which food travels from mouth to stomach. The human oesophagus is about 23 cm/9 in long. Its extends downwards from the ◊pharynx, immediately behind the windpipe. It is lined with a mucous membrane which secretes lubricant fluid to assist the downward movement of food (◊peristalsis).

oestrogen any of a group of hormones produced by the ◊ovaries of vertebrates; the term is also used for various synthetic hormones that mimic their effects. The principal oestrogen in mammals is oestradiol. Oestrogens control female sexual development, promote the growth of female secondary sexual characteristics, stimulate egg production, and, in mammals, prepare the lining of the uterus for pregnancy.

Oestrogens are used therapeutically for some hormone disorders and to inhibit lactation; they also form the basis of oral contraceptives.

US researchers in 1995 observed that oestrogen plays a role in the healing of damaged blood vessels. It has also been found that women recover more quickly from strokes if given a low oestrogen dose.

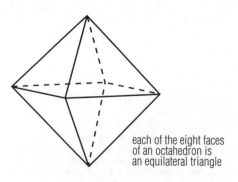

each of the eight faces of an octahedron is an equilateral triangle

octahedron, regular An octahedron is a solid figure which has eight faces, each of which is an equilateral triangle.

oestrus in mammals, the period during a female's reproductive cycle (also known as the oestrus cycle or ◊menstrual cycle) when mating is most likely to occur. It usually coincides with ovulation.

Office of Population Census and Surveys (OPCS) UK government department responsible for the compilation of statistics on local and national populations. It also conducts demographic surveys of births, marriages, and deaths. The OPCS carries out a census every 10 years and the statistics obtained are used by other government departments and local authorities to plan public services.

offset printing the most common method of ◊printing, which uses smooth (often rubber) printing plates. It works on the principle of ◊lithography: that grease and water repel one another.

The printing plate is prepared using a photographic technique, resulting in a type image that attracts greasy printing ink. On the printing press the plate is wrapped around a cylinder and wetted and inked in turn. The ink adheres only to the type area, and this image is then transferred via an intermediate blanket cylinder to the paper.

ohm SI unit (symbol Ω) of electrical ◊resistance (the property of a conductor that restricts the flow of electrons through it).

It was originally defined with reference to the resistance of a column of mercury, but is now taken as the resistance between two points when a potential difference of one volt between them produces a current of one ampere.

Ohm's law law that states that the current flowing in a metallic conductor maintained at constant temperature is directly proportional to the potential difference (voltage) between its ends. The law was discovered by German physicist Georg Ohm in 1827.

If a current of I amperes flows between two points in a conductor across which the potential difference is V volts, then V/I is a constant called the ◊resistance R ohms between those two points. Hence:

$$V/I = R$$

or

$$V = IR$$

Not all conductors obey Ohm's law; those that do are called **ohmic conductors**.

oil flammable substance, usually insoluble in water, and composed chiefly of carbon and hydrogen. Oils may be solids (fats and waxes) or liquids. The three main types are: **essential oils**, obtained from plants; **fixed oils**, obtained from animals and plants; and **mineral oils**, obtained chiefly from the refining of ◊petroleum.

Essential oils are volatile liquids that have the odour of their plant source and are used in perfumes, flavouring essences, and in ◊aromatherapy. Fixed oils are mixtures of ◊lipids, of varying consistency, found in both animals (for example, fish oils) and plants (in nuts and seeds); they are used as foods and as lubricants, and in the making of soaps, paints, and varnishes. Mineral oils are composed of a mixture of hydrocarbons, and are used as fuels and lubricants.

oil, cooking fat that is liquid at room temperature, extracted from the seeds or fruits of certain plants and used for frying, salad dressings, and sauces and condiments such as mayonnaise and mustard. Plants used for cooking oil include sunflower, olive, maize (corn), soya, peanut, and rape. Vegetable oil is a blend of more than one type of oil. Most oils are hot-pressed and refined, a process that leaves them without smell or flavour. Cold-pressed, unrefined oils keep their flavour. Oils are generally low in cholesterol and contain a high proportion of polyunsaturated or monounsaturated fatty acids, although all except soya and corn oil become saturated when heated.

oil crop plant from whose seeds vegetable oils are pressed. Cool temperate areas grow rapeseed and linseed; warm temperate regions produce sunflowers, olives, and soya beans; tropical regions produce groundnuts (peanuts), palm oil, and coconuts.

Some of the major vegetable oils, such as soya bean oil, peanut oil, and cottonseed oil, are derived from crops grown primarily for other purposes. Most vegetable oils are used as both edible oils and as ingredients in industrial products such as soaps, varnishes, printing inks, and paints.

oil spill oil released by damage to or discharge from a tanker or oil installation. An oil spill kills all shore life, clogging up the feathers of birds and suffocating other creatures. At sea, toxic chemicals leach into the water below, poisoning sea life. Mixed with dust, the oil forms globules that sink to the seabed, poisoning sea life there as well. Oil spills are broken up by the use of detergents but such chemicals can themselves damage wildlife. The annual spillage of oil is 8 million barrels a year. At any given time tankers are carrying 500 million barrels.

In March 1989 the *Exxon Valdez* (belonging to the Exxon Corporation) spilled oil in Alaska's Prince William Sound, covering 12,400 sq km/4,800 sq mi and killing at least 34,400 sea birds, 10,000 sea otters, and up to 16 whales. The incident led to the US Oil Pollution Act of 1990, which requires tankers operating in US waters to have double hulls.

The world's largest oil spill was in the Persian Gulf in Feb 1991 as a direct result of hostilities during the Gulf War. Around 6–8 million barrels of oil were spilled, polluting 675 km/420 mi of Saudi coastline. In some places, the oil was 30 cm/12 in deep in the sand.

old age later years of life. The causes of progressive degeneration of bodily and mental processes associated with it are still not precisely known, but every one of the phenomena of ◊ageing can occur at almost any age, and the process does not take place throughout the body at an equal rate. **Geriatrics** is the branch of medicine dealing with the diagnosis and treatment of disease in the elderly.

Normally, ageing begins after about 30. The arteries start to lose their elasticity, so that a greater strain is placed on the heart. The resulting gradual impairment of the blood supply is responsible for many of the changes, but between 30 and 60 there is a period of maturity in which ageing usually makes little progress. Research into the process of ageing (**gerontology**) includes study of dietary factors and the mechanisms behind structural changes in the tissues. Old age in itself is not a disease.

olefin common name for ◊alkene.

Oligocene third epoch of the Tertiary period of geological time, 35.5–3.25 million years ago. The name, from Greek, means 'a little recent', referring to the presence of the remains of some modern types of animals existing at that time.

oligosaccharide ◊carbohydrate comprising a few ◊monosaccharide units linked together. It is a general term used to indicate that a carbohydrate is larger than a simple di- or trisaccharide but not as large as a polysaccharide.

olivenite basic copper arsenate, $Cu_2AsO_4(OH)$, occurring as a mineral in olive-green prisms.

olivine greenish mineral, magnesium iron silicate, $(Mg,Fe)_2SiO_4$. It is a rock-forming mineral, present in, for example, peridotite, gabbro, and basalt. Olivine is called *peridot* when pale green and transparent, and used in jewellery.

omnivore animal that feeds on both plant and animal material. Omnivores have digestive adaptations intermediate between those of ◊herbivores and ◊carnivores, with relatively unspecialized digestive systems and gut microorganisms that can digest a variety of foodstuffs.

OMR abbreviation for ◊ *optical mark recognition*.

onchocerciasis or *river blindness* disease found in tropical Africa and Central America. It is transmitted by bloodsucking black flies, which infect the victim with parasitic filarial worms (genus *Onchocerca*), producing skin disorders and intense itching; some invade the eyes and may cause blindness. It is treated with antiparasitic drugs known as filaricides.

oncogene gene carried by a virus that induces a cell to divide abnormally, giving rise to a cancer. Oncogenes arise from mutations in genes (proto-oncogenes) found in all normal cells. They are usually also found in viruses that are capable of transforming normal cells to tumour cells. Such viruses are able to insert their oncogenes into the host cell's DNA, causing it to divide uncontrollably. More than one oncogene may be necessary to transform a cell in this way. In 1989 US scientists J Michael Bishop and Harold Varmus were jointly awarded the Nobel Prize for Physiology or Medicine for their concept of oncogenes, although credit for the discovery was claimed by a French cancer specialist, Dominique Stehelin.

oncology medical speciality concerned with the diagnosis and treatment of ◊neoplasms, especially cancer.

onco-mouse mouse that has a human ◊oncogene (gene that can cause certain cancers) implanted into its cells by genetic engineering. Such mice are used to test anticancer treatments and were patented within the USA by Harvard University in 1988, thereby protecting its exclusive rights to produce the animal and profit from its research.

on-line system in computing, originally a system that allows the computer to work interactively with its users, responding to each instruction as it is given and prompting users for information when necessary. Since almost all the computers used now work this way, 'on-line system' is now used to refer to large database, electronic mail, and conferencing systems accessed via a dial-up modem. These often have tens or hundreds of users from different places – sometimes from different countries – 'on line' at the same time.

ontogeny process of development of a living organism, including the part of development that takes place after hatching or birth. The idea that 'ontogeny recapitulates phylogeny' (the development of an organism goes through the same stages as its evolutionary history),

proposed by the German scientist Ernst Heinrich Haeckel, is now discredited.

onyx semiprecious variety of chalcedonic ◊silica (SiO_2) in which the crystals are too fine to be detected under a microscope, a state known as cryptocrystalline. It has straight parallel bands of different colours: milk-white, black, and red.

Sardonyx, an onyx variety, has layers of brown or red carnelian alternating with lighter layers of onyx. It can be carved into cameos.

oocyte in medicine, an immature ovum. Only a fraction of the oocytes produced in the ovary survive until puberty and not all of these undergo meiosis to become an ovum that can be fertilized by a sperm.

oolite limestone made up of tiny spherical carbonate particles, called *ooliths*, cemented together. Ooliths have a concentric structure with a diameter up to 2 mm/0.08 in. They were formed by chemical precipitation and accumulation on ancient sea floors.

The surface texture of oolites is rather like that of fish roe. The late Jurassic limestones of the British Isles are mostly oolitic in nature.

Oort cloud spherical cloud of comets beyond Pluto, extending out to about 100,000 astronomical units (1.5 light years) from the Sun. The gravitational effect of passing stars and the rest of our Galaxy disturbs comets from the cloud so that they fall in towards the Sun on highly elongated orbits, becoming visible from Earth. As many as 10 trillion comets may reside in the Oort cloud, named after Dutch astronomer Jan Oort who postulated it in 1950.

oosphere another name for the female gamete, or ◊ovum, of certain plants such as algae.

ooze sediment of fine texture consisting mainly of organic matter found on the ocean floor at depths greater than 2,000 m/6,600 ft. Several kinds of ooze exist, each named after its constituents.

Siliceous ooze is composed of the ◊silica shells of tiny marine plants (diatoms) and animals (radiolarians). *Calcareous ooze* is formed from the ◊calcite shells of microscopic animals (foraminifera) and floating algae (coccoliths).

opal form of hydrous ◊silica $(SiO_2.nH_2O)$, often occurring as stalactites and found in many types of rock. The common opal is translucent, milk-white, yellow, red, blue, or green, and lustrous. Precious opal is opalescent, the characteristic play of colours being caused by close-packed silica spheres diffracting light rays within the stone.

Opal is cryptocrystalline, that is, the crystals are too fine to be detected under an optical microscope. Opals are found in Hungary; New South Wales, Australia (black opals were first discovered there in 1905); and Mexico (red fire opals).

opencast mining or *open-pit mining* or *strip mining* mining at the surface rather than underground. Coal, iron ore, and phosphates are often extracted by opencast mining. Often the mineral deposit is covered by soil, which must first be stripped off, usually by large machines such as walking draglines and bucket-wheel excavators. The ore deposit is then broken up by explosives.

One of the largest excavations in the world has been

made by opencast mining at the Bingham Canyon copper mine in Utah, USA, measuring 790 m/2,590 ft deep and 3.7 km/2.3 mi across.

open cluster or *galactic cluster* in astronomy, a loose cluster of young stars. More than 1,200 open clusters have been catalogued, each containing between a dozen and several thousand stars. They are of interest to astronomers because they represent samples of stars that have been formed at the same time from similar material. Examples include the ◊Pleiades and the Hyades. See also ◊globular cluster and ◊star cluster.

open-hearth furnace method of steelmaking, now largely superseded by the ◊basic-oxygen process. It was developed in 1864 in England by German-born William and Friedrich Siemens, and improved by Pierre and Emile Martin in France in the same year. In the furnace, which has a wide, saucer-shaped hearth and a low roof, molten pig iron and scrap are packed into the shallow hearth and heated by overhead gas burners using preheated air.

operating system (OS) in computing, a program that controls the basic operation of a computer. A typical OS controls the peripheral devices such as printers, organizes the filing system, provides a means of communicating with the operator, and runs other programs.

Some operating systems were written for specific computers, but some are accepted standards. These include CP/M (by Digital Research) and MS-DOS (by Microsoft) for microcomputers. UNIX (developed at AT&T's Bell Laboratories) is the standard on workstations, minicomputers, and supercomputers; it is also used on desktop PCs and mainframes.

operational amplifier or *op-amp* electronic circuit that is used as a basic building block in electronic design.

Operational amplifiers are used in a wide range of electronic measuring instruments. The name arose because they were originally designed to carry out mathematical operations and solve equations.

operon group of genes that are found next to each other on a chromosome, and are turned on and off as an integrated unit. They usually produce enzymes that control different steps in the same biochemical pathway. Operons were discovered in 1961 (by the French biochemists François Jacob and Jacques Monod) in bacteria.

They are less common in higher organisms where the control of metabolism is a more complex process.

Ophiuchus large constellation along the celestial equator (see ◊celestial sphere), known as the serpent bearer because the constellation ◊Serpens is wrapped around it. The Sun passes through Ophiuchus each Dec, but the constellation is not part of the zodiac. Ophiuchus contains ◊Barnard's star.

ophthalmia neonatorum form of ◊conjunctivitis mostly contracted during delivery by an infant whose mother is infected with ◊gonorrhoea. It can lead to blindness unless promptly treated.

ophthalmology medical speciality concerned with diseases of the eye and its surrounding tissues.

opiate, endogenous naturally produced chemical in the body that has effects similar to morphine and other opiate drugs; a type of neurotransmitter.

Examples include ◊endorphins and ◊encephalins.

opioid analgesic or *narcotic analgesic* in medicine, pain-relieving agent that acts at receptor sites in the central nervous system. Opiates are used to relieve moderate to severe pain. Drugs in this class include morphine, pethidine, and diamorphine (heroin). Dependence and tolerance to these drugs can develop.

opioid antagonist in medicine, drug used to reverse the respiratory depression caused by opioid analgesics due to their effects on the mechanisms controlling respiration in the central nervous system. They are used to counteract respiratory depression postoperatively and in individuals who have taken overdoses of opioid analgesics.

opium drug extracted from the unripe seeds of the opium poppy (*Papaver somniferum*) of SW Asia. An addictive ◊narcotic, it contains several alkaloids, including *morphine*, one of the most powerful natural painkillers and addictive narcotics known, and *codeine*, a milder painkiller.

Heroin is a synthetic derivative of morphine and even more powerful as a drug. Opium is still sometimes given as a tincture, dissolved in alcohol and known as *laudanum*. Opium also contains the highly poisonous alkaloid *thebaine*.

Opium

Coleridge began to write his poem 'Kubla Khan' under the influence of opium. After writing 54 lines in a trance-like state, he was interrupted by 'a person on business from Porlock'. He never completed the poem.

---

opposition in astronomy, the moment at which a body in the Solar System lies opposite the Sun in the sky as seen from the Earth and crosses the ◊meridian at about midnight.

Although the ◊inferior planets cannot come to opposition, it is the best time for observation of the superior planets as they can then be seen all night.

optical aberration see ◊aberration, optical.

optical activity in chemistry, the ability of certain crystals, liquids, and solutions to rotate the plane of ◊polarized light as it passes through them. The phenomenon is related to the three-dimensional arrangement of the atoms making up the molecules concerned. Only substances that lack any form of structural symmetry exhibit optical activity.

optical character recognition (OCR) in computing, a technique for inputting text to a computer by means of a document reader. First, a ◊scanner produces a digital image of the text; then character-recognition software makes use of stored knowledge about the shapes of individual characters to convert the digital image to a set of internal codes that can be stored and processed by computer.

OCR originally required specially designed characters but current devices can recognize most standard typefaces and even handwriting. OCR is used, for example, by gas and electricity companies to input data collected on meter-reading cards, and by personal digital assistants to recognize users' handwriting.

optical disc in computing, a storage medium in which laser technology is used to record and read large volumes of digital data. Types include ◊CD-ROM, ◊WORM, and erasable optical disc.

optical fibre very fine, optically pure glass fibre through which light can be reflected to transmit images or data from one end to the other. Although expensive to produce and install, optical fibres can carry more data than traditional cables, and are less susceptible to interference.

Optical fibres are increasingly being used to replace metal communications cables, the messages being encoded as digital pulses of light rather than as fluctuating electric current. Current research is investigating how optical fibres could replace wiring inside computers.

Bundles of optical fibres are also used in endoscopes to inspect otherwise inaccessible parts of machines or of the living body (see ◊endoscopy).

optical illusion scene or picture that fools the eye. An example of a natural optical illusion is that the Moon appears bigger when it is on the horizon than when it is high in the sky, owing to the ◊refraction of light rays by the Earth's atmosphere.

optical instrument instrument that makes use of one or more lenses or mirrors, or of a combination of these, in order to change the path of light rays and produce an image. Optical instruments such as magnifying glasses, ◊microscopes, and ◊telescopes are used to provide a clear, magnified image of the very small or the very distant. Others, such as ◊cameras, photographic enlargers, and film ◊projectors, may be used to store or reproduce images.

optical mark recognition (OMR) in computing, a technique that enables marks made in predeter-mined positions on computer-input forms to be detected optically and input to a computer. An **optical mark reader** shines a light beam onto the input document and is able to detect the marks because less light is reflected back from them than from the paler, unmarked paper.

optic nerve large nerve passing from the eye to the brain, carrying visual information. In mammals, it may contain up to a million nerve fibres, connecting the sensory cells of the retina to the optical centres in the brain. Embryologically, the optic nerve develops as an outgrowth of the brain.

optics branch of physics that deals with the study of ◊light and vision – for example, shadows and mirror images, lenses, microscopes, telescopes, and cameras. For all practical purposes light rays travel in straight lines, although Albert Einstein demonstrated that they may be 'bent' by a gravitational field. On striking a surface they are reflected or refracted with some absorption of energy, and the study of this is known as geometrical optics.

optoelectronics branch of electronics concerned with the development of devices (based on the ◊semiconductor gallium arsenide) that respond not only to the ◊electrons of electronic data transmission, but also to ◊photons.

In 1989, scientists at IBM in the USA built a gallium arsenide microprocessor ('chip') containing 8,000 transistors and four photodetectors. The densest optoelectronic chip yet produced, this can detect and process data at a speed of 1 billion bits per second.

oral contraceptive in medicine, synthetic female

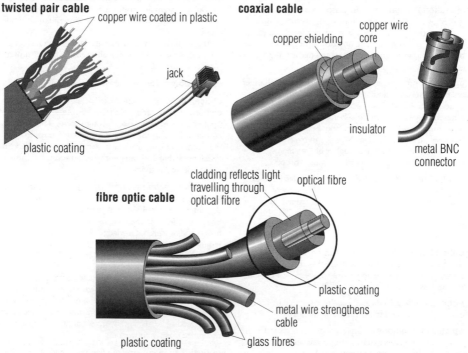

twisted pair cable

copper wire coated in plastic

jack

plastic coating

coaxial cable

copper wire core

copper shielding

insulator

metal BNC connector

fibre optic cable

cladding reflects light travelling through optical fibre

optical fibre

plastic coating

metal wire strengthens cable

plastic coating

glass fibres

optical fibre The major differences in construction between twisted pair (telephone), coaxial (Ethernet), and fibre optic cable.

hormones that are taken by mouth for contraception and commonly referred to as the ◊pill.

orbit path of one body in space around another, such as the orbit of Earth around the Sun, or the Moon around Earth. When the two bodies are similar in mass, as in a ◊binary star, both bodies move around their common centre of mass. The movement of objects in orbit follows Johann Kepler's laws, which apply to artificial satellites as well as to natural bodies.

As stated by the laws, the orbit of one body around another is an ellipse. The ellipse can be highly elongated, as are comet orbits around the Sun, or it may be almost circular, as are those of some planets. The closest point of a planet's orbit to the Sun is called **perihelion**; the most distant point is **aphelion**. (For a body orbiting the Earth, the closest and furthest points of the orbit are called **perigee** and **apogee**.)

orbit in mammals, one of two bony cavities on either side of the nose in which the eyes are situated. They protect the eyes from injury because of the thickness of the bone at the front of the orbit. The optic canal, which is the route between the optic nerve and the brain, opens into the rear of the orbits and nerves and blood vessels pass through the optical fissures. The eyes are held within the orbits by the extra-ocular muscles and the orbital septum.

orbital, atomic region around the nucleus of an atom (or, in a molecule, around several nuclei) in which an ◊electron is most likely to be found. According to ◊quantum theory, the position of an electron is uncertain; it may be found at any point. However, it is more likely to be found in some places than in others, and it is these that make up the orbital.

An atom or molecule has numerous orbitals, each of which has a fixed size and shape. An orbital is characterized by three numbers, called ◊quantum numbers, representing its energy (and hence size), its angular momentum (and hence shape), and its orientation. Each orbital can be occupied by one or (if their spins are aligned in opposite directions) two electrons.

order in biological classification, a group of related ◊families. For example, the horse, rhinoceros, and tapir famil-

ies are grouped in the order Perissodactyla, the odd-toed ungulates, because they all have either one or three toes on each foot. The names of orders are not shown in italic (unlike genus and species names) and by convention they have the ending '-formes' in birds and fish; '-a' in mammals, amphibians, reptiles, and other animals; and '-ales' in fungi and plants. Related orders are grouped together in a ◊class.

ordinal number in mathematics, one of the series first, second, third, fourth, ... Ordinal numbers relate to order, whereas ◊cardinal numbers (1, 2, 3, 4, ...) relate to quantity, or count.

ordinate in ◊coordinate geometry, the y coordinate of a point; that is, the vertical distance of the point from the horizontal or x-axis. For example, a point with the coordinates (3,4) has an ordinate of 4. See ◊abscissa.

Ordnance Survey (OS) official body responsible for the mapping of Britain. It was established in 1791 as the **Trigonometrical Survey** to continue work initiated in 1784 by Scottish military surveyor General William Roy (1726–1790). Its first accurate maps appeared in 1830, drawn to a scale of 1 in to the mile (1:63,000). In 1858 the OS settled on a scale of 1:2,500 for the mapping of Great Britain and Ireland (higher for urban areas, lower for uncultivated areas).

Subsequent revisions and editions include the 1:50,000 Landranger series (1971–86). In 1989, the OS began using a computerized system for the creation and continuous revision of maps. Customers can now have maps drafted to their own specifications, choosing from over 50 features (such as houses, roads, and vegetation). Since 1988 the OS has had a target imposed by the government to recover all its costs from sales.

Ordovician period of geological time 510–439 million years ago; the second period of the ◊Palaeozoic era. Animal life was confined to the sea: reef-building algae and the first jawless fish are characteristic.

The period is named after the Ordovices, an ancient Welsh people, because the system of rocks formed in the Ordovician period was first studied in Wales.

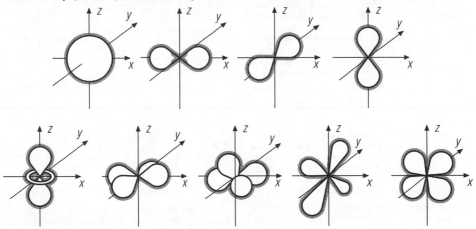

orbital, atomic The shapes of atomic orbitals. An atomic orbital is a picture of the 'electron cloud' that surrounds the nucleus of an atom. There are four basic shapes for atomic orbitals: spherical, dumbbell, clover-leaf, and complex (shown at bottom left).

ore body of rock, a vein within it, or a deposit of sediment, worth mining for the economically valuable mineral it contains. The term is usually applied to sources of metals. Occasionally metals are found uncombined (native metals), but more often they occur as compounds such as carbonates, sulphides, or oxides. The ores often contain unwanted impurities that must be removed when the metal is extracted.

Commercially valuable ores include bauxite (aluminium oxide, Al_2O_3) hematite (iron(III) oxide, Fe_2O_3), zinc blende (zinc sulphide, ZnS), and rutile (titanium dioxide, TiO_2).

Hydrothermal ore deposits are formed from fluids such as saline water passing through fissures in the host rock at an elevated temperature. Examples are the 'porphyry copper' deposits of Chile and Bolivia, the submarine copper–zinc–iron sulphide deposits recently discovered on the East Pacific Rise, and the limestone lead–zinc deposits that occur in the southern USA and in the Pennines of Britain.

Other ores are concentrated by igneous processes, causing the ore metals to become segregated from a magma – for example, the chromite and platinum-metal-rich bands within the Bushveld, South Africa. Erosion and transportation in rivers of material from an existing rock source can lead to further concentration of heavy minerals in a deposit – for example, Malaysian tin deposits.

Weathering of rocks in situ can result in residual metal-rich soils, such as the nickel-bearing laterites of New Caledonia.

organ in biology, part of a living body, such as the liver or brain, that has a distinctive function or set of functions.

organelle discrete and specialized structure in a living cell; organelles include mitochondria, chloroplasts, lysosomes, ribosomes, and the nucleus.

organic chemistry branch of chemistry that deals with carbon compounds. Organic compounds form the chemical basis of life and are more abundant than inorganic compounds. In a typical organic compound, each carbon atom forms bonds covalently with each of its neighbouring carbon atoms in a chain or ring, and additionally with other atoms, commonly hydrogen, oxygen, nitrogen, or sulphur.

The basis of organic chemistry is the ability of carbon to form long chains of atoms, branching chains, rings, and other complex structures. Compounds containing only carbon and hydrogen are known as *hydrocarbons*.

organic compound in chemistry, a class of compounds that contain carbon. The original distinction between organic and inorganic compounds was based on the belief that the molecules of living systems were unique, and could not be synthesized in the laboratory. Today it is routine to manufacture thousands of organic chemicals both in research and in the drug industry. Certain simple compounds of carbon, such as carbonates, oxides of carbon, carbon disulphide, and carbides are usually treated in inorganic chemistry.

organic farming farming without the use of synthetic fertilizers (such as ◊nitrates and phosphates) or ◊pesticides (herbicides, insecticides, and fungicides) or other agrochemicals (such as hormones, growth stimulants, or fruit regulators). Food produced by genetic engineering cannot be described as organic.

In place of artificial fertilizers, compost, manure, seaweed, or other substances derived from living things are used (hence the name 'organic'). Growing a crop of a nitrogen-fixing plant such as lucerne, then ploughing it back into the soil, also fertilizes the ground. Some organic farmers use naturally occurring chemicals such as nicotine or pyrethrum to kill pests, but control by non-chemical methods is preferred. Those methods include

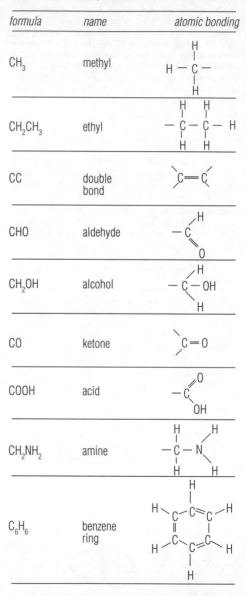

formula	name	atomic bonding
CH_3	methyl	
CH_2CH_3	ethyl	
CC	double bond	
CHO	aldehyde	
CH_2OH	alcohol	
CO	ketone	
COOH	acid	
CH_2NH_2	amine	
C_6H_6	benzene ring	

organic chemistry Common organic-molecule groupings. Organic chemistry is the study of carbon compounds, which make up over 90% of all chemical compounds. This diversity arises because carbon atoms can combine in many different ways with other atoms, forming a wide variety of loops and chains.

removal by hand, intercropping (planting with companion plants which deter pests), mechanical barriers to infestation, crop rotation, better cultivation methods, and ◊biological control. Weeds can be controlled by hoeing, mulching (covering with manure, straw, or black plastic), or burning off.

Organic farming methods produce food with minimal pesticide residues and greatly reduce pollution of the environment. They are more labour intensive, and therefore more expensive, but use less fossil fuel. Soil structure is greatly improved by organic methods, and recent studies show that a conventional farm can lose four times as much soil through erosion as an organic farm, although the loss may not be immediately obvious.

organizer in embryology, a part of the embryo that causes changes to occur in another part, through ◊induction, thus 'organizing' development and ◊differentiation.

organophosphorus insecticide in medicine, insecticidal compounds that cause the irreversible inhibition of the cholinesterase enzymes that break down acetylcholine.

As this mechanism of action is very toxic to humans, they should be used with great care. Malathion and permethrin may be used to control lice in humans and have many applications in veterinary medicine and in agriculture.

OR gate in electronics, a type of ◊logic gate.

orimulsion fuel made by mixing ◊bitumen and water that can be burnt in the same way as heavy oil. It is cheap to make but the smoke produced has a high sulphur content.

Orion in astronomy, a very prominent constellation in the equatorial region of the sky (see ◊celestial sphere), identified with the hunter of Greek mythology.

The bright stars Alpha (◊Betelgeuse), Gamma (Bellatrix), Beta (◊Rigel), and Kappa Orionis mark the shoulders and legs of Orion. Between them the belt is formed by Delta, Epsilon, and Zeta, three second-magnitude stars equally spaced in a straight line. Beneath the belt is a line of fainter stars marking Orion's sword. One of these, Theta, is not really a star but the brightest part of the ◊Orion nebula. Nearby is one of the most distinctive dark nebulae, the Horsehead.

Orion nebula luminous cloud of gas and dust 1,500 light years away, in the constellation Orion, from which stars are forming. It is about 15 light years in diameter, and contains enough gas to make a cluster of thousands of stars.

At the nebula's centre is a group of hot young stars, called the **Trapezium**, which make the surrounding gas glow. The nebula is visible to the naked eye as a misty patch below the belt of Orion.

Orion nebula

If the distance between the Earth and the Sun were 30 cm/1 ft, the Orion nebula would be about 190 km/120 mi across.

---

ormolu alloy of copper, zinc, and sometimes tin, used for furniture decoration.

ornithology study of birds. It covers scientific aspects relating to their structure and classification, and their habits, song, flight, and value to agriculture as destroyers of insect pests. Worldwide scientific banding (or the fitting of coded rings to captured specimens) has resulted in accurate information on bird movements and distribution. There is an International Council for Bird Preservation with its headquarters at the Natural History Museum, London.

ornithophily ◊pollination of flowers by birds. Ornithophilous flowers are typically brightly coloured, often red or orange. They produce large quantities of thin, watery nectar, and are scentless because most birds do not respond well to smell. They are found mostly in tropical areas, with hummingbirds being important pollinators in North and South America, and the sunbirds in Africa and Asia.

orogeny or **orogenesis** the formation of mountains. It is brought about by the movements of the rigid plates making up the Earth's crust (described by ◊plate tectonics). Where two plates collide at a destructive margin rocks become folded and lifted to form chains of fold mountains (such as the ◊young fold mountains of the Himalayas).

orrery mechanical device for demonstrating the motions of the heavenly bodies. Invented in about 1710 by George Graham, it was named after his patron, the 4th Earl of Orrery. It is the forerunner of the planetarium.

orthochromatic photographic film or paper of decreased sensitivity, which can be processed with a red safelight. Using it, blue objects appear lighter and red ones darker because of increased blue sensitivity.

orthodontics branch of ◊dentistry concerned with ◊dentition, and with treatment of any irregularities, such as correction of malocclusion (faulty position of teeth).

orthopaedics medical speciality concerned with the correction of disease or damage in bones and joints.

The first orthopaedic hospital was founded in 1780 at Orbe in Switzerland by Jean Venel. The first in England was founded in 1817.

OS/2 single-user computer ◊operating system produced jointly by Microsoft Corporation and IBM for use on large microcomputers. Its main features are ◊multitasking and the ability to access large amounts of internal ◊memory. OS/2 was announced in 1987. Microsoft abandoned it in 1992 to concentrate on ◊DOS, but it continues to be marketed by IBM. **OS/2 Warp** is a variation optimized to run Windows programs.

oscillating universe in astronomy, a theory that states that the gravitational attraction of the mass within the universe will eventually slow down and stop the expansion of the universe. The outward motions of the galaxies will then be reversed, eventually resulting in a 'Big Crunch' where all the matter in the universe would be contracted into a small volume of high density. This could undergo a further ◊Big Bang, thereby creating another expansion phase. The theory suggests that the universe would alternately expand and collapse through alternate Big Bangs and Big Crunches.

oscillation one complete to-and-fro movement of a vibrating object or system. For any particular vibration, the time for one oscillation is called its ◊period and the number of oscillations in one second is called its ◊frequency. The maximum displacement of the vibrating object from its rest position is called the ◊amplitude of the oscillation. *See illustration overleaf.*

oscillator any device producing a desired oscillation (vibration). There are many types of oscillator for different purposes, involving various arrangements of thermionic ◊valves or components such as ◊transistors, ◊inductors, ◊capacitors, and ◊resistors.

An oscillator is an essential part of a radio transmitter, generating the high-frequency carrier signal necessary for radio communication. The ◊frequency is often controlled by the vibrations set up in a crystal (such as quartz).

oscillograph instrument for displaying or recording the values of rapidly changing oscillations, electrical or mechanical.

oscilloscope or *cathode-ray oscilloscope* (CRO) instrument used to measure electrical voltages that vary over time and to display the waveforms of electrical oscillations or signals, by means of the deflection of a beam of ◊electrons. Readings are displayed graphically on the screen of a ◊cathode-ray tube.

osmium hard, heavy, bluish-white, metallic element, symbol Os, atomic number 76, relative atomic mass 190.2. It is the densest of the elements, and is resistant to tarnish and corrosion. It occurs in platinum ores and as a free metal (see ◊native metal) with iridium in a natural alloy called osmiridium, containing traces of platinum, ruthenium, and rhodium. Its uses include pen points and light-bulb filaments; like platinum, it is a useful catalyst.

It was discovered in 1803 and named in 1804 by English chemist Smithson Tennant (1761–1815).

osmoregulation process whereby the water content of living organisms is maintained at a constant level. If the water balance is disrupted, the concentration of salts will be too high or too low, and vital functions, such as nerve conduction, will be adversely affected.

In mammals, loss of water by evaporation is counteracted by increased intake and by mechanisms in the kidneys that enhance the rate at which water is resorbed before urine production. Both these responses are mediated by hormones, primarily those of the adrenal cortex (see ◊adrenal gland).

osmosis movement of solvent (usually water) through a semipermeable membrane separating solutions of different concentrations. The solvent passes from a less concentrated solution to a more concentrated solution until the two concentrations are equal. Applying external pressure to the solution on the more concentrated side arrests osmosis, and is a measure of the osmotic pressure of the solution.

Many cell membranes behave as semipermeable membranes, and osmosis is a vital mechanism in the transport of fluids in living organisms – for example, in the transport of water from the roots up the stems of plants.

ossification or *osteogenesis* process whereby bone is formed in vertebrate animals by special cells (*osteoblasts*) that secrete layers of ◊extracellular matrix on the surface of the existing ◊cartilage. Conversion to bone occurs through the deposition of calcium phosphate crystals within the matrix.

osteology part of the science of ◊anatomy, dealing with the structure, function, and development of bones.

osteomyelitis infection of bone, with spread of pus along the marrow cavity.

Now quite rare, it may follow from a compound fracture (where broken bone protrudes through the skin), or from infectious disease elsewhere in the body. It is more common in children whose bones are not yet fully grown.

The symptoms are high fever, severe illness, and pain over the limb. If the infection is at the surface of the bone it may quickly form an abscess; if it is deep in the bone marrow it may spread into the circulation and lead to blood poisoning. Most cases can be treated with immobilization, antibiotics, and surgical drainage.

osteopathy system of alternative medical practice that relies on physical manipulation to treat mechanical stress. It was developed over a century ago by US physician Andrew Taylor Still, who maintained that most ailments can be prevented or cured by techniques of spinal manipulation.

Osteopaths are generally consulted to treat problems of the musculo-skeletal structure such as back pain, and many doctors refer patients to them for such treatments. Although in the UK the wider applicability of their skills is not generally recognized, osteopathic doctors in the USA are also fully licensed to practice conventional medicine.

or

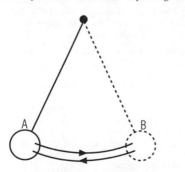

one complete oscillation or cycle is from A to B and back to A

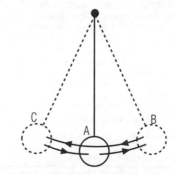

one complete oscillation or cycle is from A to B to C and back to A, moving in the same direction again

oscillation The repeated motion back and forth past a central neutral position is known as oscillation. For any particular vibration, the time for one complete oscillation or cycle is called its period, and the number of oscillations in one second is called its frequency.

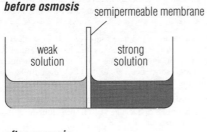

before osmosis   semipermeable membrane

weak
solution

strong
solution

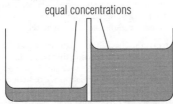

after osmosis

equal concentrations

osmosis Apparatus for measuring osmotic pressure. In 1877 German physicist Wilhelm Pfeffer used this apparatus to make the first ever measurement of osmotic pressure and show that osmotic pressure varies according to temperature and the strength of the solute (dissolved substance).

osteoporosis disease in which the bone substance becomes porous and brittle. It is common in older people, affecting more women than men. It may be treated with calcium supplements and etidronate. Approximately 1.7 million people worldwide, mostly women, suffer hip fractures, mainly due to osteoporosis. A single gene was discovered in 1993 to have a major influence on bone thinning.

Osteoporosis may occur in women whose ovaries have been removed, unless hormone-replacement therapy (HRT) is instituted; it may also occur in ◊Cushing's syndrome and as a side effect of long-term treatment with ◊corticosteroids. Early menopause in women, childlessness, small body build, lack of exercise, heavy drinking, smoking, and hereditary factors may be contributory factors.

otitis inflammation of the ear. *Otitis externa*, occurring in the outer ear canal, is easily treated with antibiotics. Inflamed conditions of the middle ear (*otitis media*) or inner ear (*otitis interna*) are more serious, carrying the risk of deafness and infection of the brain. Treatment is with antibiotics or, more rarely, surgery.

otosclerosis overgrowth of bone in the middle ear causing progressive deafness. This inherited condition is gradual in onset, developing usually before middle age. It is twice as common in women as in men.

The middle ear cavity houses the sound-conduction mechanism called the ossicular chain, consisting of three tiny bones (ossicles) that magnify vibrations received at the eardrum for onward transmission to the inner ear. In otosclerosis, extraneous growth of spongy bone immobilizes the chain, preventing the conduction of sound. Surgery is necessary to remove the diseased bone and reconstruct the ossicular chain.

Otto cycle alternative name for the ◊four-stroke cycle,

introduced by the German engineer Nikolaus Otto (1832–1891) in 1876. It improved on existing piston engines by compressing the fuel mixture in the cylinder before it was ignited.

ounce unit of mass, one-sixteenth of a pound ◊avoirdupois, equal to 437.5 grains (28.35 g); also one-twelfth of a pound troy, equal to 480 grains.

outback the inland region of Australia. Its main inhabitants are Aborigines, miners (including opal miners), and cattle ranchers.

output device in computing, any device for displaying, in a form intelligible to the user, the results of processing carried out by a computer.

ovary in female animals, the organ that generates the ◊ovum. In humans, the ovaries are two whitish rounded bodies about 25 mm/1 in by 35 mm/1.5 in, located in the lower abdomen to either side of the uterus. Every month, from puberty to the onset of the menopause, an ovum is released from the ovary. This is called ovulation, and forms part of the ◊menstrual cycle.

In botany, an ovary is the expanded basal portion of the ◊carpel of flowering plants, containing one or more ◊ovules. It is hollow with a thick wall to protect the ovules. Following fertilization of the ovum, it develops into the fruit wall or pericarp.

The ovaries of female animals secrete the hormones responsible for the secondary sexual characteristics of the female, such as smooth, hairless facial skin and enlarged breasts. An ovary in a half-grown human fetus contains 5 million eggs, and so the unborn baby already possesses the female genetic information for the next generation.

In botany, the relative position of the ovary to the other floral parts is often a distinguishing character in classification; it may be either inferior or superior, depending on whether the petals and sepals are inserted above or below.

overfishing fishing at rates that exceed the ◊sustained-yield cropping of fish species, resulting in a net population decline. For example, in the North Atlantic, herring has been fished to the verge of extinction and the cod and haddock populations are severely depleted. In the Third World, use of huge factory ships, often by fisheries from industrialized countries, has depleted stocks for local people who cannot obtain protein in any other way. See also ◊fishing and fisheries.

Ecologists have long been concerned at the wider implications of overfishing, in particular the devastation wrought on oceanic ◊food chains. The United Nations Food and Agriculture Organization estimates that worldwide overfishing has damaged oceanic ecosystems to such an extent that potential catches are on average reduced by 20%. With better management of fishing programmes the fishing catch could in principle be increased; it is estimated that, annually, 20 million tonnes of fish are discarded from fishing vessels at sea, because they are not the species sought.

According to an estimate by the Food and Agriculture Organization in 1993, nine of the world's 17 main fishing grounds were suffering a potentially catastrophic decline in some species. In Dec 1994 approximately 17% of fishing waters off the coast of New England, USA, was closed in an attempt to restore dwindling stocks. The area affected covered 17,000 sq km/6,600 sq mi and lay within the Georges Bank region of the Atlantic Ocean.

overtone note that has a frequency or pitch that is a multiple of the fundamental frequency, the sounding body's ◊natural frequency. Each sound source produces a unique set of overtones, which gives the source its quality or timbre.

oviparous method of animal reproduction in which eggs are laid by the female and develop outside her body, in contrast to ovoviviparous and viviparous. It is the most common form of reproduction.

ovoviviparous method of animal reproduction in which fertilized eggs develop within the female (unlike oviparous), and the embryo gains no nutritional substances from the female (unlike viviparous). It occurs in some invertebrates, fishes, and reptiles.

ovulation in female animals, the process of releasing egg cells (ova) from the ◊ovary. In mammals it occurs as part of the ◊menstrual cycle.

ovule structure found in seed plants that develops into a seed after fertilization. It consists of an ◊embryo sac containing the female gamete (◊ovum or egg cell), surrounded by nutritive tissue, the nucellus. Outside this there are one or two coverings that provide protection, developing into the testa, or seed coat, following fertilization.

In ◊angiosperms (flowering plants) the ovule is within an ◊ovary, but in ◊gymnosperms (conifers and their allies) the ovules are borne on the surface of an ovuliferous (ovule-bearing) scale, usually within a ◊cone, and are not enclosed by an ovary.

ovum (plural **ova**) female gamete (sex cell) before fertilization. In animals it is called an egg, and is produced in the ovaries. In plants, where it is also known as an egg cell or oosphere, the ovum is produced in an ovule. The ovum is nonmotile. It must be fertilized by a male gamete before it can develop further, except in cases of ◊parthenogenesis.

oxalic acid $(COOH)_2.2H_2O$ white, poisonous solid, soluble in water, alcohol, and ether. Oxalic acid is found in rhubarb, and its salts (oxalates) occur in wood sorrel (genus *Oxalis*, family Oxalidaceae) and other plants. It also occurs naturally in human body cells. It is used in the leather and textile industries, in dyeing and bleaching, ink manufacture, metal polishes, and for removing rust and ink stains.

oxbow lake curved lake found on the flood plain of a river. Oxbows are caused by the loops of ◊meanders being cut off at times of flood and the river subsequently adopting a shorter course. In the USA, the term bayou is often used.

oxidation in chemistry, the loss of ◊electrons, gain of oxygen, or loss of hydrogen by an atom, ion, or molecule during a chemical reaction.

Oxidation may be brought about by reaction with another compound (oxidizing agent), which simultaneously undergoes ◊reduction, or electrically at the anode (positive electrode) of an electrolytic cell.

oxidation in earth science, a form of ◊chemical weathering caused by the chemical reaction that takes place between certain iron-rich minerals in rock and the oxygen in water. It tends to result in the formation of a red-coloured soil or deposit. The inside walls of canal tunnels and bridges often have deposits formed in this way.

oxidation number roman numeral often seen in a chemical name, indicating the ◊valency of the element immediately before the number. Examples are lead(II) nitrate, manganese(IV) oxide, and potassium manganate(VII).

oxide compound of oxygen and another element, frequently produced by burning the element or a compound of it in air or oxygen.

Oxides of metals are normally ◊bases and will react with an acid to produce a ◊salt in which the metal forms the cation (positive ion). Some of them will also react with a strong alkali to produce a salt in which the metal is part of a complex anion (negative ion; see ◊amphoteric). Most oxides of nonmetals are acidic (dissolve in water to form an ◊acid). Some oxides display no pronounced acidic or basic properties.

oxyacetylene torch gas torch that burns ethyne (acetylene) in pure oxygen, producing a high-temperature (3,000°C/5,400°F) flame. It is widely used in welding to

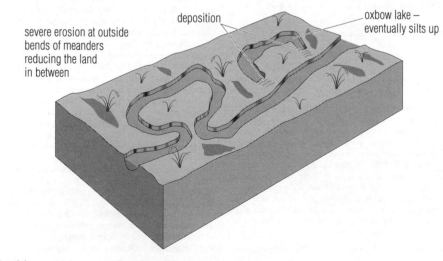

severe erosion at outside bends of meanders reducing the land in between

deposition

oxbow lake – eventually silts up

oxbow lake

fuse metals. In the cutting torch, a jet of oxygen burns through metal already melted by the flame.

oxygen colourless, odourless, tasteless, nonmetallic, gaseous element, symbol O, atomic number 8, relative atomic mass 15.9994. It is the most abundant element in the Earth's crust (almost 50% by mass), forms about 21% by volume of the atmosphere, and is present in combined form in water and many other substances. Oxygen is a by-product of ◊photosynthesis and the basis for ◊respiration in plants and animals.

Oxygen is very reactive and combines with all other elements except the ◊inert gases and fluorine. It is present in carbon dioxide, silicon dioxide (quartz), iron ore, calcium carbonate (limestone). In nature it exists as a molecule composed of two atoms (O_2); single atoms of oxygen are very short-lived owing to their reactivity. They can be produced in electric sparks and by the Sun's ultraviolet radiation in space, where they rapidly combine with molecular oxygen to form ozone (an allotrope of oxygen).

Oxygen is obtained for industrial use by the fractional distillation of liquid air, by the electrolysis of water, or by heating manganese (IV) oxide with potassium chlorate. It is essential for combustion, and is used with ethyne (acetylene) in high-temperature oxyacetylene welding and cutting torches.

The element was first identified by English chemist Joseph Priestley in 1774 and independently in the same year by Swedish chemist Karl Scheel. It was named by French chemist Antoine Lavoisier in 1777.

oxygen debt physiological state produced by vigorous exercise, in which the lungs cannot supply all the oxygen that the muscles need.

Oxygen is required for the release of energy from food molecules (aerobic ◊respiration). Instead of breaking food molecules down fully, muscle cells switch to a form of partial breakdown that does not require oxygen (anaerobic respiration) so that they can continue to generate energy. This partial breakdown produces ◊lactic acid, which results in a sensation of fatigue when it reaches certain levels in the muscles and the blood. Once the vigorous muscle movements cease, the body breaks down the lactic acid, using up extra oxygen to do so. Panting after exercise is an automatic mechanism to 'pay off' the oxygen debt.

oxygen toxicity in medicine, the ability of excess oxygen to cause cell and organ damage. The concentration and length of administration of oxygen for medical reasons must be controlled to avoid this. In adults, an excess of oxygen results in the accumulation of fluid in the lungs (oedema) and pulmonary hypertension. In neonates, it can cause convulsions that result in damage to the central nervous system. Free radical formation appears to be responsible for this damage.

oxytocin hormone that stimulates the uterus in late pregnancy to initiate and sustain labour. After birth, it stimulates the uterine muscles to contract, reducing bleeding at the site where the placenta was attached.

Intravenous injections of oxytocin may be given to induce labour, improve contractions, or control haemor-rhage after birth. It is also secreted during lactation. Oxytocin sprayed in the nose a few minutes before nursing improves milk production. It is secreted by the ◊pituitary gland.

oz abbreviation for ◊**ounce**.

Ozalid process trademarked copying process used to produce positive prints from drawn or printed materials or film, such as printing proofs from film images. The film is placed on top of chemically treated paper and then exposed to ultraviolet light. The image is developed dry using ammonia vapour.

ozone O_3 highly reactive pale-blue gas with a penetrating odour. Ozone is an allotrope of oxygen (see ◊allotropy), made up of three atoms of oxygen. It is formed when the molecule of the stable form of oxygen (O_2) is split by ultraviolet radiation or electrical discharge. It forms the ◊ozone layer in the upper atmosphere, which protects life on Earth from ultraviolet rays, a cause of skin cancer.

ozone depleter any chemical that destroys the ozone in the stratosphere. Most ozone depleters are chemically stable compounds containing chlorine or bromine, which remain unchanged for long enough to drift up to the upper atmosphere. The best known are ◊chlorofluorocarbons (CFCs), but many other ozone depleters are known, including halons, used in some fire extinguishers; methyl chloroform and carbon tetrachloride, both solvents; some CFC substitutes; and the pesticide methyl bromide.

CFCs accounted for approximately 75% of ozone depletion in 1995, whereas methyl chloroform (atmospheric concentrations of which had markedly decreased 1990–94) accounted for an estimated 12.5%. The ozone depletion rate overall is now decreasing as international agreements to curb the use of ozone-depleting chemicals begin to take effect. In 1996 there was a decrease in ozone depleters in the lower atmosphere. This trend is expected to continue into the stratosphere over the next few years.

ozone layer thin layer of the gas ◊ozone in the upper atmosphere that shields the Earth from harmful ultraviolet rays. A continent-sized hole has formed over Antarctica as a result of damage to the ozone layer. This has been caused in part by ◊chlorofluorocarbons (CFCs), but many reactions destroy ozone in the stratosphere: nitric oxide, chlorine, and bromine atoms are implicated.

It is believed that the ozone layer is depleting at a rate of about 5% every 10 years over northern Europe, with depletion extending south to the Mediterranean and southern USA. However, ozone depletion over the polar regions is the most dramatic manifestation of a general global effect. Ozone levels over the Arctic in spring 1997 fell over 10 % since 1987, despite the reduction in the concentration of CFCs and other industrial compounds which destroy the ozone when exposed to sunlight. It is thought that this may be because of an expanding vortex of cold air forming in the lower stratosphere above the Arctic, leading to increased ozone loss. It is expected that an Arctic hole as large as that over Antarctica could remain a threat to the northern hemisphere for several decades.

What was phlogiston?

Not many scientists are credited with founding a major branch of science, but Antoine Lavoisier, the French aristocrat (1743–1794), is universally regarded as the founder of modern chemistry.

Before Lavoisier, chemistry had been dominated for 100 years by an influential theory of matter proposed by the German chemist Georg Stahl (1660–1743). This theory held that a combustible material burned because it contained a substance called *phlogiston*. Charcoal was a prime example of a phlogiston-rich material. The fact that metallurgists obtained some metals from their ores by heating them with charcoal seemed to lend support to the phlogiston theory of combustion. With hindsight, the reasons why such a theory persisted are obvious enough. Chemists had no clear idea of what a chemical element was, nor any understanding of the nature of gases.

Lavoisier's contribution

Lavoisier restructured chemistry and gave it its modern form. He provided a firm foundation for the atomic theory proposed by British chemist and physicist John Dalton (1766–1844), and his chemical elements were later classified in the periodic table of the elements.

The key experiments Lavoisier did seem simple enough. He burned phosphorus, lead and other solid elements in a sealed container. He noted that although the weight of the container and its contents did not increase, the weights of the solids in it did. He showed that the increase in weight of the solid was exactly compensated by the diminished weight of air present.

In 1772, he recorded the observation that when phosphorus and sulphur burn, their weight increases. He deduced that they combine with air. It followed that there should be a way of releasing this 'fixed' air. It should be explained that in the 18th century, because of the imperfect understanding scientists had of the nature of gases, they generally and confusingly referred to them as different sorts of 'air'.

Then in 1774, Lavoisier showed that when litharge (lead oxide, a compound of lead containing oxygen) is roasted with charcoal, an enormous volume of 'air' was indeed liberated. Although other scientists had already made similar experiments, it was Lavoisier who finally drew all the evidence together, and discovered the oxygen theory of combustion.

In 1774, Lavoisier was visited in Paris by the English chemist, Joseph Priestley (1773–1804). Priestley had heated a calx (he used the oxide of mercury) in a closed apparatus, and collected the gas that was liberated in the process. Priestley had found that this gas supported combustion better than air: a candle flame burned with a marvellous brilliance when placed in a container full of it.

The acid former

Lavoisier repeated Priestley's experiments, and convinced himself of the presence in air of a gas which combined with substances when they burn, and that it was the same gas given off when the oxide of mercury was heated. He named this gas 'oxygine', or 'acid former' (from the Greek), because he believed all acids contained oxygen.

Lavoisier had in the meantime identified the other main component of air, nitrogen, which he named 'azote', from the Greek for 'no life'. He also demonstrated that when hydrogen, which chemists of the day called 'inflammable air', was burned with oxygen, water was formed.

Lavoisier could now put the final nail in the coffin of the phlogiston theory because he could now explain that when metals dissolved in an acid, the hydrogen evolved came from the water, and not from the metal, as the phlogistonists asserted.

From combustion to respiration

The oxygen theory of combustion was now complete: when a substance burned, it combined with the oxygen in the air.

Lavoisier also realized that respiration was like a slow form of combustion, in which the oxygen breathed in 'burned' the carbon in foodstuffs, which was then exhaled. As with the experiments on combustion, Lavoisier and his colleagues carried out exact measurements, using a guinea-pig. This is how the phrase 'to be a guinea-pig' originated.

Lavoisier was guillotined on 8 May 1794, executed on trumped-up charges during the Age of Terror. The mathematician Joseph Lagrange (1736–1813) said: 'It required only a moment to sever his head, and probably one hundred years will not suffice to produce another like it.'

Julian Rowe

intensity in either direction. The main type of pain transmitter is known simply as 'substance P', a neuropeptide concentrated in a certain area of the spinal cord. Substance P has been found in fish, and there is also evidence that substances that cause pain in humans (for example, bee venom) cause a similar reaction in insects and arachnids (for instance, spiders).

Since the sensation of pain is transmitted by separate nerves from those of fine touch, it is possible in diseases such as syringomyelia to have no sense of pain in a limb, yet maintain a normal sense of touch. Such a desensitized limb is at great risk of infection from unnoticed cuts and abrasions.

painkiller agent for relieving pain. Types of painkiller include analgesics such as ◊aspirin and aspirin substitutes, ◊morphine, ◊codeine, paracetamol, and synthetic versions of the natural inhibitors, the encephalins and endorphins, which avoid the side effects of the others.

Topical nerve irritants are also used in salves, such as camphor and eucalyptus; they cause the nerve endings to react to them, bringing increased blood flow to the areas and alleviating localized and joint pain.

paint any of various materials used to give a protective and decorative finish to surfaces or for making pictures. A paint consists of a pigment suspended in a vehicle, or binder, usually with added solvents. It is the vehicle that dries and hardens to form an adhesive film of paint. Among the most common kinds are cellulose paints (or lacquers), oil-based paints, emulsion (water-based) paints, and special types such as enamels and primers. **Lacquers** consist of a synthetic resin (such as an acrylic resin or cellulose acetate) dissolved in a volatile organic solvent, which evaporates rapidly to give a very quick-drying paint. A typical **oil-based paint** has a vehicle of a natural drying oil (such as linseed oil), containing a prime pigment of iron, lead, titanium, or zinc oxide, to which coloured pigments may be added. The finish – gloss, semimatt, or matt – depends on the amount of inert pigment (such as clay or silicates). Oil-based paints can be thinned with, and brushes cleaned in, a solvent such as turpentine or white spirit (a petroleum product). **Emulsion paints**, sometimes called latex paints, consist of pigments dispersed in a water-based emulsion of a polymer (such as polyvinyl chloride [PVC] or acrylic resin). They can be thinned with water, which can also be used to wash the paint out of brushes and rollers. **Enamels** have little pigment, and they dry to an extremely hard, high-gloss film. **Primers** for the first coat on wood or metal, on the other hand, have a high pigment content (as do undercoat paints). Aluminium or bronze powder may be used for priming or finishing objects made of metal.

Palaeocene first epoch of the Tertiary period of geological time, 65–56.5 million years ago. Many types of mammals spread rapidly after the disappearance of the great reptiles of the Mesozoic. Flying mammals replaced the flying reptiles, swimming mammals replaced the swimming reptiles, and all the ecological niches vacated by the reptiles were adopted by mammals.

palaeomagnetism science of the reconstruction of the Earth's ancient magnetic field and the former positions of the continents from the evidence of **remnant magnetization** in ancient rocks; that is, traces left by the Earth's magnetic field in ◊igneous rocks before they cool. Palaeomagnetism shows that the Earth's magnetic field has

P

pacemaker or **sinoatrial node** (SA node) in vertebrates, a group of muscle cells in the wall of the heart that contracts spontaneously and rhythmically, setting the pace for the contractions of the rest of the heart. The pacemaker's intrinsic rate of contraction is increased or decreased, according to the needs of the body, by stimulation from the ◊autonomic nervous system. The term also refers to a medical device implanted under the skin of a patient whose heart beats inefficiently. It delivers minute electric shocks to stimulate the heart muscles at regular intervals and restores normal heartbeat.

The latest pacemakers are powered by radioactive isotopes for long life and weigh no more than 15 g/0.5 oz.

Pacific Ocean world's largest ocean, extending from Antarctica to the Bering Strait; area 166,242,500 sq km/ 64,170,000 sq mi; average depth 4,188 m/13,749 ft; greatest depth of any ocean 11,034 m/36,210 ft in the ◊Mariana Trench.

packet switching in computing, a method of transmitting data between computers connected in a ◊network. A complete packet consists of the data being transmitted and information about which computer is to receive the data. The packet travels around the network until it reaches the correct destination.

paediatrics or **pediatrics** medical speciality concerned with the care of children.

page printer computer ◊printer that prints a complete page of text and graphics at a time. Page printers use electrostatic techniques, very similar to those used by photocopiers, to form images of pages, and range in size from small ◊laser printers designed to work with microcomputers to very large machines designed for high-volume commercial printing.

paging method of increasing a computer's apparent memory capacity. See ◊virtual memory.

pain sense that gives an awareness of harmful effects on or in the body. It may be triggered by stimuli such as trauma, inflammation, and heat. Pain is transmitted by specialized nerves and also has psychological components controlled by higher centres in the brain. Drugs that control pain are known as painkillers or ◊analgesics.

A pain message to the brain travels along the sensory nerves as electrical impulses. When these reach the gap between one nerve and another, biochemistry governs whether this gap is bridged and may also either increase or decrease the attention the message receives or modify its

reversed itself – the magnetic north pole becoming the magnetic south pole, and vice versa – at approximate half-million-year intervals, with shorter reversal periods in between the major spans.

Starting in the 1960s, this known pattern of magnetic reversals was used to demonstrate seafloor spreading or the formation of new ocean crust on either side of mid-oceanic ridges. As new material hardened on either side of a ridge, it would retain the imprint of the magnetic field, furnishing datable proof that material was spreading steadily outward. Palaeomagnetism is also used to demonstrate ◊continental drift by determining the direction of the magnetic field of dated rocks from different continents.

palaeontology in geology, the study of ancient life, encompassing the structure of ancient organisms and their environment, evolution, and ecology, as revealed by their ◊fossils. The practical aspects of palaeontology are based on using the presence of different fossils to date particular rock strata and to identify rocks that were laid down under particular conditions; for instance, giving rise to the formation of oil. The use of fossils to trace the age of rocks was pioneered in Germany by Johann Friedrich Blumenbach (1752–1830) at Göttingen, followed by Georges Cuvier and Alexandre Brongniart in France in 1811. The term palaeontology was first used in 1834, during the period when the first ◊dinosaur remains were discovered.

Palaeozoic era of geological time 570–245 million years ago. It comprises the Cambrian, Ordovician, Silurian, Devonian, Carboniferous, and Permian periods. The Cambrian, Ordovician, and Silurian constitute the Lower or Early Palaeozoic; the Devonian, Carboniferous, and Permian make up the Upper or Late Palaeozoic. The era includes the evolution of hard-shelled multicellular life forms in the sea; the invasion of land by plants and animals; and the evolution of fish, amphibians, and early reptiles. The earliest identifiable fossils date from this era.

The climate at this time was mostly warm with short ice ages. The continents were very different from the present ones but, towards the end of the era, all were joined together as a single world continent called ◊Pangaea.

palate in mammals, the roof of the mouth. The bony front part is the hard palate, the muscular rear part the soft palate. Incomplete fusion of the two lateral halves of the palate (◊cleft palate) causes interference with speech.

palindromic in medicine, a term used to describe symptoms or diseases that recur.

palisade cell cylindrical cell lying immediately beneath the upper epidermis of a leaf. Palisade cells normally exist as one closely packed row and contain many chloroplasts. During the hours of daylight palisade cells are photosynthetic, using the energy of the sun to create carbohydrates from water and carbon dioxide.

palladium lightweight, ductile and malleable, silver-white, metallic element, symbol Pd, atomic number 46, relative atomic mass 106.4.

It is one of the so-called platinum group of metals, and is resistant to tarnish and corrosion. It often occurs in nature as a free metal (see ◊native metal) in a natural alloy with platinum. Palladium is used as a catalyst, in alloys of

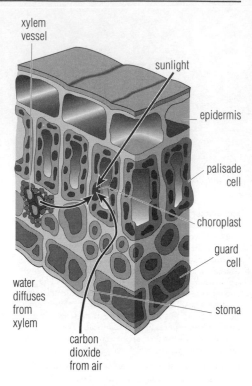

xylem vessel

sunlight

epidermis

palisade cell

choroplast

guard cell

stoma

water diffuses from xylem

carbon dioxide from air

palisade cell Palisade cells are closely packed, columnar cells, lying in the upper surfaces of leaves. They contain many chloroplasts (the structures responsible for photosynthesis) and are well adapted to receive and process the components necessary for photosynthesis – carbon dioxide, water, and sunlight. For instance, their vertical arrangement means that there are fewer cross-walls to interfere with the passage of sunlight.

gold (to make white gold) and silver, in electroplating, and in dentistry.

It was discovered in 1803 by British physicist William Wollaston (1766–1828), and named after the then recently discovered asteroid Pallas (found 1802).

palliative in medicine, any treatment given to relieve symptoms rather than to cure the underlying cause. In conditions that will resolve of their own accord (for instance, the common cold) or that are incurable, the entire treatment may be palliative.

palliative care in medicine, care aimed at reducing the suffering of those patients with terminal illnesses. Treatment centres on keeping patients as comfortable, alert, and free of pain as possible. This care also involves consideration of any emotional, social, or family problems that the patient is experiencing.

palpation in medicine, a method of examining the size, shape, and movement of the internal organs by laying the flat of the hand on the skin of the patient.

palsy in medicine, another term for ◊paralysis.

panchromatic in photography, a term describing highly sensitive black-and-white film made to render all visible spectral colours in correct grey tones. Panchromatic film is always developed in total darkness.

pancreas in vertebrates, an accessory gland of the digestive system located close to the duodenum. When stimulated by the hormone secretin, it releases enzymes into the duodenum that digest starches, proteins, and fats. In humans, it is about 18 cm/7 in long, and lies behind and below the stomach. It contains groups of cells called the *islets of Langerhans*, which secrete the hormones insulin and glucagon that regulate the blood sugar level.

Pangaea or *Pangea* single land mass, made up of all the present continents, believed to have existed between 300 and 200 million years ago; the rest of the Earth was covered by the Panthalassa ocean. Pangaea split into two land masses – ◊Laurasia in the north and ◊Gondwanaland in the south – which subsequently broke up into several continents. These then drifted slowly to their present positions (see ◊continental drift). The existence of a single 'supercontinent' was proposed by German meteorologist Alfred Wegener in 1912.

Panthalassa ocean that covered the surface of the Earth not occupied by the world continent ◊Pangaea between 300 and 200 million years ago.

pantothenic acid water-soluble ◊vitamin ($C_9H_{17}NO_5$) of the B complex, found in a wide variety of foods. Its absence from the diet can lead to dermatitis, and it is known to be involved in the breakdown of fats and carbohydrates.

Papanicolaou test in medicine, another name for ◊cervical smear.

paper thin, flexible material made in sheets from vegetable fibres (such as wood pulp) or rags and used for writing, drawing, printing, packaging, and various household needs. The name comes from papyrus, a form of writing material made from water reed, used in ancient Egypt. The invention of true paper, originally made of pulped fishing nets and rags, is credited to Tsai Lun, Chinese Minister of Agriculture, AD 105.

Paper came to the West with Arabs who had learned the secret from Chinese prisoners of war in Samarkand in 768. It spread from Morocco to Moorish Spain and to Byzantium in the 11th century, then to the rest of Europe. All early paper was handmade within frames.

With the spread of literacy there was a great increase in the demand for paper. Production by hand of single sheets could not keep pace with this demand, which led to the invention, by Louis Robert (1761–1828) in 1799, of a machine to produce a continuous reel of paper. The process was developed and patented in 1801 by François Didot, Robert's employer. Today most paper is made from wood pulp on a Fourdrinier machine, then cut to size; some high grade paper is still made from esparto or rag. Paper products absorb 35% of the world's annual commercial wood harvest; recycling avoids some of the enormous waste of trees, and most papermakers plant and replant their own forests of fast-growing stock.

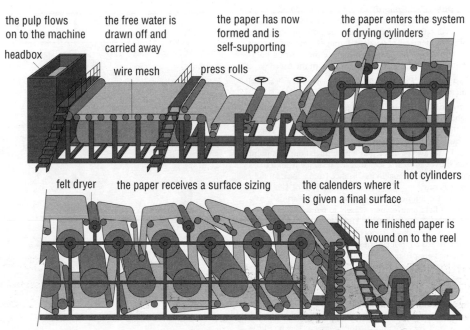

the pulp flows on to the machine
headbox

the free water is drawn off and carried away
wire mesh

the paper has now formed and is self-supporting
press rolls

the paper enters the system of drying cylinders

hot cylinders

felt dryer   the paper receives a surface sizing   the calenders where it is given a final surface

the finished paper is wound on to the reel

calender stacks

paper Today's fully automatic papermaking machines can be 200 m/640 ft long and produce over 1,000 m/3,200 ft of paper in a minute. The most common type of papermaking machine is the Fourdrinier, named after two British stationer brothers who invented it in 1803. Their original machine deposited the paper on pieces of felt, after which it was finished by hand.

papovavirus in medicine, group of viruses that includes the causative agents of papilloma (warts and verrucas) in humans.

pappus (plural **pappi**) in botany, a modified ♭calyx comprising a ring of fine, silky hairs, or sometimes scales or small teeth, that persists after fertilization. Pappi are found in members of the daisy family (Compositae) such as the dandelions *Taraxacum*, where they form a parachute-like structure that aids dispersal of the fruit.

Pap test or **Pap smear** common name for ♭cervical smear.

parabola in mathematics, a curve formed by cutting a right circular cone with a plane parallel to the sloping side of the cone. A parabola is one of the family of curves known as ♭conic sections. The graph of $y = x^2$ is a parabola. It can also be defined as a path traced out by a point that moves in such a way that the distance from a fixed point (focus) is equal to its distance from a fixed straight line (directrix); it thus has an ♭eccentricity of 1.

The trajectories of missiles within the Earth's gravitational field approximate closely to parabolas (ignoring the effect of air resistance). The corresponding solid figure, the paraboloid, is formed by rotating a parabola about its axis. It is a common shape for headlight reflectors, dish-shaped microwave and radar aerials, and radio-telescopes, since a source of radiation placed at the focus of a paraboloidal reflector is propagated as a parallel beam.

paracetamol analgesic, particularly effective for musculoskeletal pain. It is as effective as aspirin in reducing fever, and less irritating to the stomach, but has little anti-inflammatory action. An overdose can cause severe, often irreversible or even fatal, liver and kidney damage.

parachute any canopied fabric device strapped to a person or a package, used to slow down descent from a high altitude, or to return spent missiles or parts to a safe speed for landing, or sometimes to aid (through braking) the landing of a plane or missile.

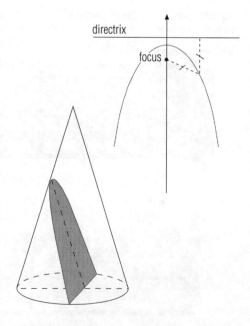

parabola The parabola is a curve produced when a cone is cut by a plane. It is one of a family of curves called conic sections, which also includes the circle, ellipse, and hyperbole. These curves are produced when the plane cuts the cone at different angles and positions.

paradigm all those factors, both scientific and sociological, that influence the research of the scientist. The term, first used by the US historian of science T S Kuhn, has subsequently spread to social studies and politics.

paraffin common name for ◊alkane, any member of the series of hydrocarbons with the general formula C_nH_{2n+2}. The lower members are gases, such as methane (marsh or natural gas). The middle ones (mainly liquid) form the basis of petrol, kerosene, and lubricating oils, while the higher ones (paraffin waxes) are used in ointment and cosmetic bases.

paraldehyde common name for ◊ethanal trimer.

parallax the change in the apparent position of an object against its background when viewed from two different positions. In astronomy, nearby stars show a shift owing to parallax when viewed from different positions on the Earth's orbit around the Sun. A star's parallax is used to deduce its distance.

Nearer bodies such as the Moon, Sun, and planets also show a parallax caused by the motion of the Earth. The **diurnal parallax** is caused by the Earth's rotation.

parallel circuit electrical circuit in which the components are connected side by side. The current flowing in the circuit is shared by the components. The division of the current across each conductor is in the ratio of their resistances.

If the currents across two conductors of resistance R_1 and R_2, connected in parallel, are I_1 and I_2 respec-

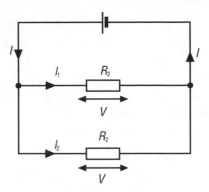

parallel circuit Two resistors, R_1 and R_2, connected in a parallel circuit. The potential difference V across each resistor is the same, but the current passing through each is divided up in the ratio of their resistance (I_1:I_2 is the same as R_1:R_2). The advantage of such an arrangement is that if one of the components fails, the other parallel components will continue to receive current.

tively, then the ratio of those currents is given by the equation:

$$I_1/I_2 = R_2/R_1$$

The total resistance R of those conductors is given by:

$$1/R = 1/R_1 + 1/R_2$$

Compare ◊series circuit.

parallel device in computing, a device that communicates binary data by sending the bits that represent each character simultaneously along a set of separate data lines, unlike a ◊serial device.

parallel lines and parallel planes in mathematics, straight lines or planes that always remain a constant distance from one another no matter how far they are extended. This is a principle of Euclidean geometry. Some non-Euclidean geometries, such as elliptical and hyperbolic geometry, however, reject Euclid's parallel axiom.

parallelogram in mathematics, a quadrilateral (four-sided plane figure) with opposite pairs of sides equal in length and parallel, and with opposite angles equal. The diagonals of a parallelogram bisect each other. Its area is the product of the length of one side and the perpendicular distance between this and the opposite side. In the special case when all four sides are equal in length, the parallelogram is known as a rhombus, and when the internal angles are right angles, it is a rectangle or square.

parallelogram of forces in physics and applied mathematics, a method of calculating the resultant (combined effect) of two different forces acting together on an object. Because a force has both magnitude and direction it is a ◊vector quantity and can be represented by a straight line. A second force acting at the same point in a different direction can be represented by another line drawn at an angle to the first. By completing the parallelogram (of which the two lines are sides) a diagonal may be drawn from the original angle to the opposite corner to represent the resultant force vector.

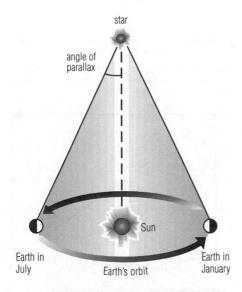

parallax The parallax of a star, the apparent change of its position during the year, can be used to find the star's distance from the Earth. The star appears to change its position because it is viewed at a different angle in July and January. By measuring the angle of parallax, and knowing the diameter of the Earth's orbit, simple geometry can be used to calculate the distance to the star.

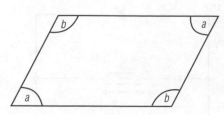

(i) opposite sides and angles are equal

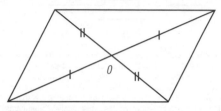

(ii) diagonals bisect each other at O

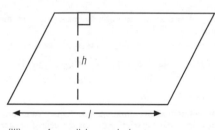

(iii) area of a parallelogram $l \times h$

parallelogram Some properties of a parallelogram.

parallel processing computer technology that allows more than one computation at the same time. Although in the 1980s this technology enabled only a small number of computer processor units to work in parallel, in theory thousands or millions of processors could be used at the same time.

Parallel processing, which involves breaking down computations into small parts and performing thousands of them simultaneously, rather than in a linear sequence, offers the prospect of a vast improvement in working speed for certain repetitive applications.

paralysis loss of voluntary movement due to failure of nerve impulses to reach the muscles involved. It may result from almost any disorder of the nervous system, including brain or spinal cord injury, poliomyelitis, stroke, and progressive conditions such as a tumour or multiple sclerosis. Paralysis may also involve loss of sensation due to sensory nerve disturbance.

parameter variable factor or characteristic. For example, length is one parameter of a rectangle; its height is another. In computing, it is frequently useful to describe a program or object with a set of variable parameters rather than fixed values.

For example, if a programmer writes a routine for drawing a rectangle using general parameters for the length, height, line thickness, and so on, any rectangle can be drawn by this routine by giving different values to the parameters.

Similarly, in a word-processing application that stores parameters for font, page layout, type of ◊justification, and so on, these can be changed by the user.

paranoia mental disorder marked by ◊delusions of grandeur or persecution. In popular usage, paranoia means baseless or exaggerated fear and suspicion.

In *chronic paranoia*, patients exhibit a rigid system of false beliefs and opinions, believing themselves, for example, to be followed by the secret police, to be loved by someone at a distance, or to be of great importance or in special relation to God. There are no hallucinations and patients are in other respects normal.

In *paranoid states*, the delusions of persecution or grandeur are present but not systematized.

In *paranoid ◊schizophrenia*, the patient suffers from many unsystematized and incoherent delusions, is extremely suspicious, and experiences hallucinations and the feeling that external reality has altered.

paranormal not within the range of, or explicable by, established science. Paranormal phenomena include ◊extrasensory perception (ESP) which takes in clairvoyance, precognition, and telepathy; *telekinesis*, the movement of objects from one position to another by human mental concentration; and *mediumship*, supposed contact with the spirits of the dead, usually via an intermediate 'guide' in the other world. ◊Parapsychology is the study of such phenomena.

Paranormal phenomena are usually attributed to the action of an unknown factor, ◊psi.

There have been many reports of sporadic paranormal phenomena, the most remarkable being reports by one person, or occasionally more, of apparitions or hallucinatory experiences associated with another person's death.

paraplegia paralysis of the lower limbs, involving loss of both movement and sensation; it is usually due to spinal injury.

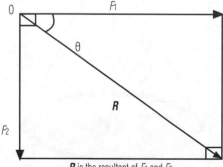

R is the resultant of F_1 and F_2

parallelogram of forces The diagram shows how the parallelogram of forces can be used to calculate the resultant (combined effect) of two different forces acting together on an object. The two forces are represented by two lines drawn at an angle to each other. By completing the parallelogram (of which the two lines are sides), a diagonal may be drawn from the original angle to the opposite corner to represent the resultant force vector.

parapsychology study of ◊paranormal phenomena, which are generally subdivided into two types: ◊extrasensory perception (ESP), or the paracognitive; and psychokinesis (PK), telekinesis, or the paraphysical – movement of an object without the use of physical force or energy.

Most research into parapsychology has been experimental. The first Society for Psychical Research was established in London 1882 by scientists, philosophers, classical scholars, and spiritualists. Despite continued scepticism within the scientific establishment, a chair of parapsychology was established 1984 at Edinburgh University, endowed by the Hungarian author Arthur Koestler.

paraquat $CH_3(C_5H_4N)_2CH_3 .2CH_3SO_4$ (technical name *1,1-dimethyl-4,4-dipyridylium*) nonselective herbicide (weedkiller). Although quickly degraded by soil microorganisms, it is deadly to human beings if ingested.

parasite organism that lives on or in another organism (called the host) and depends on it for nutrition, often at the expense of the host's welfare. Parasites that live inside the host, such as liver flukes and tapeworms, are called *endoparasites*; those that live on the exterior, such as fleas and lice, are called *ectoparasites*.

Great fleas have little fleas upon their backs to bite 'em, | And little fleas have lesser fleas, and so ad infinitum.

On **parasites**, Augustus De Morgan
Budget of Paradoxes

---

parasympathetic nervous system in medicine, a division of the ◊autonomic nervous system.

parathyroid one of a pair of small ◊endocrine glands. Most tetrapod vertebrates, including humans, possess two such pairs, located behind the ◊thyroid gland. They secrete parathyroid hormone, which regulates the amount of calcium in the blood.

paratyphoid fever infectious disease of the intestinal tract, similar to ◊typhoid fever but milder and less dangerous. It is caused by bacteria of the genus *Salmonella* and is treated with antibiotics.

parenchyma plant tissue composed of loosely packed, more or less spherical cells, with thin cellulose walls. Although parenchyma often has no specialized function, it is usually present in large amounts, forming a packing or ground tissue. It usually has many intercellular spaces.

parental care in biology, the time and energy spent by a parent in order to rear its offspring to maturity. Among animals, it ranges from the simple provision of a food supply for the hatching young at the time the eggs are laid (for example, many wasps) to feeding and protection of the young after hatching or birth, as in birds and mammals. In the more social species, parental care may include the teaching of skills – for example, female cats teach their kittens to hunt.

parity of a number, the state of being either even or odd. In computing, the term refers to the number of 1s in the binary codes used to represent data. A binary representation has *even parity* if it contains an even number of 1s and *odd parity* if it contains an odd number of 1s.

For example, the binary code 1000001, commonly used to represent the character 'A', has even parity

Parity

Character	Binary code	Parity	Base-ten representation
A	1000001	even	65
B	1000010	even	66
C	1000011	odd	67
D	1000100	even	68

because it contains two 1s, and the binary code 1000011, commonly used to represent the character 'C', has odd parity because it contains three 1s. A *parity bit* (0 or 1) is sometimes added to each binary representation to give them all the same parity so that a ◊validation check can be carried out each time data are transferred from one part of the computer to another. So, for example, the codes 1000001 and 1000011 could have parity bits added and become *0*1000001 and *1*1000011, both with even parity. If any bit in these codes should be altered in the course of processing the parity would change and the error would be quickly detected.

Parkes site in New South Wales of the Australian National Radio Astronomy Observatory, featuring a radio telescope of 64 m/210 ft aperture, run by the Commonwealth Scientific and Industrial Research Organization. It received a NASA-funded upgrade in 1996 to enable it to track *Galileo*.

Parkinson's disease or *parkinsonism* or *paralysis agitans* degenerative disease of the brain characterized by a progressive loss of mobility, muscular rigidity, tremor, and speech difficulties. The condition is mainly seen in people over the age of 50.

Parkinson's disease destroys a group of cells called the *substantia nigra* ('black substance') in the upper part of the ◊brainstem. These cells are concerned with the production of a neurotransmitter known as dopamine, which is essential to the control of voluntary movement. The almost total loss of these cells, and of their chemical product, produces the disabling effects. A defective gene responsible for 1 in 20 cases was identified in 1992.

The disease occurs in two forms: *multiple system atrophy (MSA)*, which is a failure of the central nervous system and accounts for 1 in 5 cases; and *pure autonomic failure (PAF)*, a deficit in the peripheral nerves. Symptoms, particularly in the early stages, can be identical. In 1996, an estimated 1.5 million Americans and 120,000 Britons were suffering with Parkinson's.

The introduction of the drug ◊L-dopa in the 1960s seemed at first the answer to Parkinson's disease. However, it became evident that long-term use brings considerable problems. At best, it postpones the terminal phase of the disease. Brain grafts with dopamine-producing cells were pioneered in the early 1980s, and attempts to graft Parkinson's patients with fetal brain tissue have been made. This experimental surgery brought considerable improvement to some PAF patients, but is ineffective in the MSA form. In 1989 a large US study showed that the drug deprenyl may slow the rate at which disability progresses in patients with early Parkinson's disease.

parsec in astronomy, a unit (symbol pc) used for distances to stars and galaxies. One parsec is equal to 3.2616 ◊light years, 2.063×10^5 ◊astronomical units, and 3.086×10^{13} km.

It is the distance at which a star would have a ◊parallax (apparent shift in position) of one second of arc when viewed from two points the same distance apart as the Earth's distance from the Sun; or the distance at which one astronomical unit subtends an angle of one second of arc.

parthenocarpy in botany, the formation of fruits without seeds. This phenomenon, of no obvious benefit to the plant, occurs naturally in some plants, such as bananas. It can also be induced in some fruit crops, either by breeding or by applying certain plant hormones.

parthenogenesis development of an ovum (egg) without any genetic contribution from a male. Parthenogenesis is the normal means of reproduction in a few plants (for example, dandelions) and animals (for example, certain fish). Some sexually reproducing species, such as aphids, show parthenogenesis at some stage in their life cycle to accelerate reproduction to take advantage of good conditions.

particle detector one of a number of instruments designed to detect subatomic particles and track their paths; they include the ◊cloud chamber, ◊bubble chamber, ◊spark chamber, and multiwire chamber.

The earliest particle detector was the cloud chamber, which contains a super-saturated vapour in which particles leave a trail of droplets, in much the same way that a jet aircraft leaves a trail of vapour in the sky. A bubble chamber contains a superheated liquid in which a particle leaves a trail of bubbles. A spark chamber contains a series of closely-packed parallel metal plates, each at a high voltage. As particles pass through the chamber, they leave a visible spark between the plates. A modern multiwire chamber consists of an array of fine, closely-packed wires, each at a high voltage. As a particle passes through the chamber, it produces an electrical signal in the wires. A computer analyses the signal and reconstructs the path of the particles. Multiwire detectors can be used to detect X-ray and gamma rays, and are used as detectors in ◊positron emission tomography (PET).

particle physics study of the particles that make up all atoms, and of their interactions. More than 300 subatomic particles have now been identified by physicists, categorized into several classes according to their mass, electric charge, spin, magnetic moment, and interaction. Subatomic particles include the ◊elementary particles (◊quarks, ◊leptons, and ◊gauge bosons), which are believed to be indivisible and so may be considered the fundamental units of matter; and the ◊hadrons (baryons, such as the proton and neutron, and mesons), which are composite particles, made up of two or three quarks. The proton, electron, and neutrino are the only stable particles (the neutron being stable only when in the atomic nucleus). The unstable particles decay rapidly into other particles, and are known from experiments with particle accelerators and cosmic radiation. See ◊atomic structure.

Pioneering research took place at the Cavendish laboratory, Cambridge, England. In 1897 English physicist Joseph John Thomson discovered that all atoms contain identical, negatively charged particles (◊electrons), which can easily be freed. By 1911 New Zealand physicist Ernest Rutherford had shown that the electrons surround a very small, positively-charged ◊nucleus. In the case of hydrogen, this was found to consist of a single positively charged particle, a ◊proton. The nuclei of other elements are made up of protons and uncharged particles called ◊neutrons.

1932 also saw the discovery of a particle (whose existence had been predicted by British theoretical physicist Paul Dirac in 1928) with the mass of an electron, but an equal and opposite charge – the ◊positron. This was the first example of ◊antimatter; it is now believed that almost all particles have corresponding antiparticles. In 1934 Italian–US physicist Enrico Fermi argued that a hitherto unsuspected particle, the ◊neutrino, must accompany electrons in beta-emission.

By the mid-1930s, four types of fundamental ◊force interacting between particles had been identified. The ◊electromagnetic force acts between all particles with electric charge, and is thought to be related to the exchange between these particles of ◊gauge bosons called ◊photons, packets of electromagnetic radiation. In 1935 Japanese physicist Hideki Yukawa suggested that the ◊strong nuclear force (binding protons and neutrons together in the nucleus) was transmitted by the exchange of particles with a mass about one-tenth of that of a proton; these particles, called ◊pions (originally pi mesons), were found by British physicist Cecil Powell in 1946.

Yukawa's theory was largely superseded from 1973 by the theory of ◊quantum chromodynamics, which postulates that the strong nuclear force is transmitted by the exchange of gauge bosons called ◊gluons between the quarks and antiquarks making up protons and neutrons. Theoretical work on the ◊weak nuclear force began with Enrico Fermi in the 1930s. The existence of the gauge bosons that carry this force, the ◊weakons (W and Z particles), was confirmed in 1983 at CERN, the European nuclear research organization. The fourth fundamental force, ◊gravity, is experienced by all matter; the postulated carrier of this force has been named the ◊graviton.

The electron, muon, tau, and their neutrinos comprise the ◊leptons – particles with half-integral spin that 'feel' the weak nuclear force but not the strong force. The muon (found by US physicist Carl Anderson in cosmic radiation in 1937) produces the muon neutrino when it decays; the tau, a surprise discovery of the 1970s, produces the tau neutrino when it decays.

The hadrons (particles that 'feel' the strong nuclear force) were found in the 1950s and 1960s. They are classified into ◊mesons, with whole-number or zero spins, and ◊baryons (which include protons and neutrons), with half-integral spins. It was shown in the early 1960s that if hadrons of the same spin are represented as points on suitable charts, simple patterns are formed. This symmetry enabled a hitherto unknown baryon, the omega-minus, to be predicted from a gap in one of the patterns; it duly turned up in experiments.

The whole history of particle physics, or of physics, is one of getting down the number of concepts to as few as possible.

On **particle physics** Abdus Salam, quoted in L Wolpert and A Richards *A Passion for Science* 1988

In 1964, US physicists Murray Gell-Mann and George Zweig suggested that all hadrons were built from three 'flavours' of a new particle with half-integral spin and a charge of magnitude either 1/3 or 2/3 that of an electron; Gell-Mann named the particle the *quark*. Mesons are quark–antiquark pairs (spins either add to one or cancel to zero), and baryons are quark triplets. To account for new mesons such as the psi (J) particle the number of quark flavours had risen to six by 1985.

particle, subatomic in physics, a particle that is smaller than an atom; see ◊particle physics.

parturition in medicine, another term for labour.

parvovirus in medicine, one of a group of viruses responsible for epidemic nausea and vomiting. Outbreaks can affect whole communities and are more common in the winter. Symptoms include nausea, vomiting, diarrhoea, and giddiness but the disease is not serious and usually lasts for less than three days.

PASCAL (French acronym for *program appliqué à la selection et la compilation automatique de la littérature*) a high-level computer-programming language. Designed by Niklaus Wirth in the 1960s as an aid to teaching programming, it is still widely used as such in universities, and as a good general-purpose programming language. Most professional programmers, however, now use ◊C or C++. PASCAL was named after 17th-century French mathematician Blaise Pascal.

pascal SI unit (symbol Pa) of pressure, equal to one newton per square metre. It replaces ◊bars and millibars (10^5 Pa equals one bar). It is named after the French mathematician Blaise Pascal.

Pascal's triangle triangular array of numbers (with 1 at the apex), in which each number is the sum of the pair of numbers above it. It is named after French mathematician Blaise Pascal, who used it in his study of probability. When plotted at equal distances along a horizontal axis, the numbers in the rows give the binomial probability distribution (with equal probability of success and failure) of an event, such as the result of tossing a coin.

passive smoking inhalation of tobacco smoke from other people's cigarettes; see ◊smoking.

pasteurization treatment of food to reduce the number of microorganisms it contains and so protect consumers from disease. Harmful bacteria are killed and the development of others is delayed. For milk, the method involves heating it to 72°C/161°F for 15 seconds followed by rapid cooling to 10°C/50°F or lower. The process also kills beneficial bacteria and reduces the nutritive property of milk.

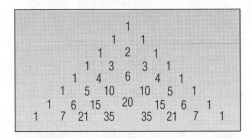

Pascal's triangle In Pascal's triangle, each number is the sum of the two numbers immediately above it, left and right – for example, 2 is the sum of 1 and 1, and 4 is the sum of 3 and 1. Furthermore, the sum of each row equals a power of 2 – for example, the sum of the 3rd row is $4 = 2^2$; the sum of the 4th row is $8 = 2^3$.

The experiments of Louis Pasteur on wine and beer in the 1850s and 1860s showed how heat treatment slowed the multiplication of bacteria and thereby the process of souring. Pasteurization of milk made headway in the dairy industries of Scandinavia and the USA before 1900 because of the realization that it also killed off bacteria associated with the diseases of tuberculosis, typhoid, diphtheria, and dysentery.

pastoral farming the rearing or keeping of animals in order to obtain meat or other products, such as milk, skins, and hair. Animals can be kept in one place or periodically moved (◊nomadic pastoralism).

patch test in medicine, the placing of small amounts of possible allergens on to the skin, usually on the arm or back, to identify the cause of an allergic reaction. A red flare and swelling will appear in the area of application if the individual is allergic to that substance. Patch tests can achieve very quick results, but in some cases the patch must remain in place for several days.

patella or *kneecap* flat bone embedded in the knee tendon of birds and mammals, which protects the joint from injury.

pathogen in medicine, any microorganism that causes disease. Most pathogens are ◊parasites, and the diseases they cause are incidental to their search for food or shelter inside the host. Nonparasitic organisms, such as soil bacteria or those living in the human gut and feeding on waste foodstuffs, can also become pathogenic to a person whose immune system or liver is damaged. The larger parasites that can cause disease, such as nematode worms, are not usually described as pathogens.

pathology medical speciality concerned with the study of disease processes and how these provoke structural and functional changes in the body.

PCP abbreviation for *phencyclidine hydrochloride*, a drug popularly known as ◊angel dust.

pd abbreviation for ◊potential difference.

peat fibrous organic substance found in bogs and formed by the incomplete decomposition of plants such as sphagnum moss. N Asia, Canada, Finland, Ireland, and other places have large deposits, which have been dried and used as fuel from ancient times. Peat can also be used as a soil additive. Peat bogs began to be formed when glaciers retreated, about 9,000 years ago. They grow at the rate of only a millimetre a year, and large-scale digging can result in destruction both of the bog and of specialized plants growing there.

pectoral relating to the upper area of the thorax associated with the muscles and bones used in moving the arms or forelimbs, in vertebrates. In birds, the *pectoralis major* is the very large muscle used to produce a powerful downbeat of the wing during flight.

pedicel the stalk of an individual flower, which attaches it to the main floral axis, often developing in the axil of a bract.

Pegasus in astronomy, a constellation of the northern hemisphere, near Cygnus, and represented as the winged horse of Greek mythology.

pegmatite extremely coarse-grained ◊igneous rock of any composition found in veins; pegmatites are usually associated with large granite masses.

Peking man Chinese representative of an early species of human, found as fossils, 500,000–750,000 years old, in the cave of Choukoutien in 1927 near Beijing (Peking). Peking man used chipped stone tools, hunted game, and used fire. Similar varieties of early human have been found in Java and E Africa.

Their classification is disputed: some anthropologists classify them as *Homo erectus*, others as *Homo sapiens pithecanthropus*.

pellagra chronic disease mostly seen in subtropical countries in which the staple food is maize. It is caused by deficiency of ◊nicotinic acid (one of the B vitamins), which is contained in protein foods, beans and peas, and yeast. Symptoms include diarrhoea, skin eruptions, and mental disturbances.

Peltier effect in physics, a change in temperature at the junction of two different metals produced when an electric current flows through them. The extent of the change depends on what the conducting metals are, and the nature of change (rise or fall in temperature) depends on the direction of current flow. It is the reverse of the ◊Seebeck effect. It is named after the French physicist Jean Charles Peltier (1785–1845) who discovered it in 1834.

pelvis in vertebrates, the lower area of the abdomen featuring the bones and muscles used to move the legs or hindlimbs. The *pelvic girdle* is a set of bones that allows movement of the legs in relation to the rest of the body and provides sites for the attachment of relevant muscles.

pendulum weight (called a 'bob') swinging at the end of a rod or cord. The regularity of a pendulum's swing was used in making the first really accurate clocks in the 17th century. Pendulums can be used for measuring the acceleration due to gravity (an important constant in physics).

Specialized pendulums are used to measure velocities (ballistic pendulum) and to demonstrate the Earth's rotation (Foucault's pendulum).

penicillin any of a group of ◊antibiotic (bacteria killing) compounds obtained from filtrates of moulds of the genus *Penicillium* (especially *P. notatum*) or produced synthetically. Penicillin was the first antibiotic to be discovered (by Alexander Fleming); it kills a broad spectrum of bacteria, many of which cause disease in humans.

The use of the original type of penicillin is limited by the increasing resistance of ◊pathogens and by allergic reactions in patients. Since 1941, numerous other antibiotics of the penicillin family have been discovered which are more selective against, or resistant to, specific microorganisms.

peninsula land surrounded on three sides by water but still attached to a larger landmass. Florida, USA, is an example.

penis male reproductive organ containing the ◊urethra, the channel through which urine and ◊semen are voided. It transfers sperm to the female reproductive tract to fertilize the ovum. In mammals, the penis is made erect by vessels that fill with blood, and in most mammals (but not humans) is stiffened by a bone.

Snakes and lizards have a paired structure that serves as a penis; other reptiles have a single organ. A few birds, mainly ducks and geese, also have a type of penis, as do snails, barnacles, and some other invertebrates. Many insects have a rigid, nonerectile male organ, usually referred to as an intromittent organ.

Pennsylvanian US term for the Upper or Late ◊Carboniferous period of geological time, 323–290 million years ago; it is named after the US state.

pentadactyl limb typical limb of the mammals, birds, reptiles, and amphibians. These vertebrates (animals with backbone) are all descended from primitive amphibians whose immediate ancestors were fleshy-finned fish. The limb which evolved in those amphibians had three parts: a 'hand/foot' with five digits (fingers/toes), a lower limb containing two bones, and an upper limb containing one bone.

This basic pattern has persisted in all the terrestrial vertebrates, and those aquatic vertebrates (such as seals) which are descended from these amphibians. Natural selection has modified the pattern to fit different ways of life. In flying animals (birds and bats) it is greatly altered and in some vertebrates, such as whales and snakes, the limbs are greatly reduced or lost. Pentadactyl limbs of different species are an example of ◊homologous organs.

pentagon five-sided plane figure. The regular pentagon has ◊golden section proportions between its sides and diagonals. The five-pointed star formed by drawing all the diagonals of a regular pentagon is called a *penta-gram*. This star has further golden sections.

pentanol $C_5H_{11}OH$ (common name *amyl alcohol*) clear, colourless, oily liquid, usually having a characteristic choking odour. It is obtained by the fermentation of starches and from the distillation of petroleum.

penumbra the region of partial shade between the totally dark part (umbra) of a ◊shadow and the fully illuminated region outside. It occurs when a source of light is only partially obscured by a shadow-casting object. The darkness of a penumbra varies gradually from total darkness at one edge to full brightness at the other. In astronomy, a penumbra is a region of the Earth from which only a partial ◊eclipse of the Sun or Moon can be seen.

pepsin enzyme that breaks down proteins during digestion. It requires a strongly acidic environment and is present in the stomach.

peptide molecule comprising two or more ◊amino acid molecules (not necessarily different) joined by *peptide bonds*, whereby the group of one acid is linked to the carboxyl amino group of the other (–CO–NH–). The number of amino acid molecules in the peptide is indicated by referring to it as a di-, tri-, or polypeptide (two, three, or many amino acids).

Proteins are built up of interacting polypeptide chains with various types of bonds occurring between the chains. Incomplete hydrolysis (splitting up) of a protein yields a mixture of peptides, examination of which helps to determine the sequence in which the amino acids occur within the protein.

percentage way of representing a number as a ◊fraction of 100. Thus 45 percent (45%) equals 45/100, and 45% of 20 is 45/100 × 20 = 9.

In general, if a quantity x changes to y, the percentage change is $100(x - y)/x$. Thus, if the number of people in a room changes from 40 to 50, the percentage increase is $(100 × 10)/40 = 25\%$. To express a fraction as a percentage, its denominator must first be converted to 100 – for

GREAT EXPERIMENTS AND DISCOVERIES

The luckiest accident

'I think the discovery and development of penicillin may be looked on as quite one of the luckiest accidents that have occurred in medicine.' With these words, Professor Howard Florey (1898–1968), an Australian pathologist, concluded a lecture he gave in 1943 at the Royal Institution. The topic was penicillin: the first, and still perhaps overall the best, of a range of natural chemo-therapeutic agents called antibiotics.

Together with Ernst Chain (1906–1979), Howard Florey had made the practical exploitation of penicillin possible on a worldwide scale.

A chance discovery

The story of the discovery of penicillin started with a chance observation made by British bacteriologist Alexander Fleming (1881–1955), while he was working in his laboratory at St Mary's Hospital, London in 1928.

Fleming was doing research on staphylococcus, a species of bacterium that can cause disease in humans and animals. In the laboratory such bacteria are grown in dishes containing a culture medium – a jelly-like substance which contains their food. Fleming noticed that one of his dishes had been accidentally contaminated with a mould. The mould could have entered the laboratory through an open window. All around where the mould was growing, the staphylococci had disappeared.

Fleming investigates the mould

Intrigued, because he had correctly concluded that the mould must contain a substance that killed the bacterium, Fleming isolated the mould and grew more of it in a culture broth. He found that the broth acquired a high antibacterial activity. He tested the action of the broth with a wide variety of pathogenic bacteria, and found that many of them were quickly destroyed.

Fleming was also able to demonstrate that the white corpuscles in human blood were not sensitive to the broth. This suggested to him that other human cells in the body would not be affected by it.

Penicillin is discovered

The mould that Fleming's acute powers of observation had noted in a single culture dish, was found to be *Penicillium notatum*. Fleming named the drug penicillin after this mould.

There the matter rested for 15 years. At the time that penicillin was discovered, it would have been extremely difficult to isolate and purify the drug using the chemical techniques then available. It would also have been quite easily destroyed by them. Without any of the pure substance, Fleming had no way of knowing its extraordinarily high antibacterial activity and its almost negligible toxicity. In fact, penicillin diluted 80,000,000 times will still inhibit the growth of staphylococcus. To translate this number of noughts into something tangible: if one drop of whisky is diluted 80,000,000 times it would fill over 6,000 whisky bottles.

The first practical application

In 1935, German–British biochemist Ernst Chain joined Professor Florey at Oxford. There, in 1939, he began a survey of anti-microbial substances. One of the first he looked at was Fleming's mould. It was chosen because it was already known to be active against staphylococcus, and because, since it was difficult to purify, it represented a biochemical challenge.

A method of purification using primitive apparatus was discovered, and an experiment using penicillin to treat infected mice was carried out with remarkable success. Eventually, enough penicillin was made to treat the first human patient in the terminal stage of a generalized infection. The patient showed an astonishing, although temporary, recovery. Five more seriously ill patients were successfully treated.

Because these patients had already failed to respond to sulphonamide drugs, the value of penicillin was now clearly apparent. For although sulphonamides are much more toxic to bacteria than to leucocytes (white blood cells), they do have some poisonous action on the whole human organism.

The new lifesaver

England, then in the midst of World War II, had insufficient resources to manufacture the new wonder drug on a sufficiently large scale. Florey went to the USA to try to interest the large pharmaceutical companies in penicillin. As a result of an outstanding effort, enough penicillin was produced in time to treat all the battle casualties of the Normandy landings in 1944. The use of penicillin in wartime saved countless lives, as it has continued to do ever since.

Julian Rowe

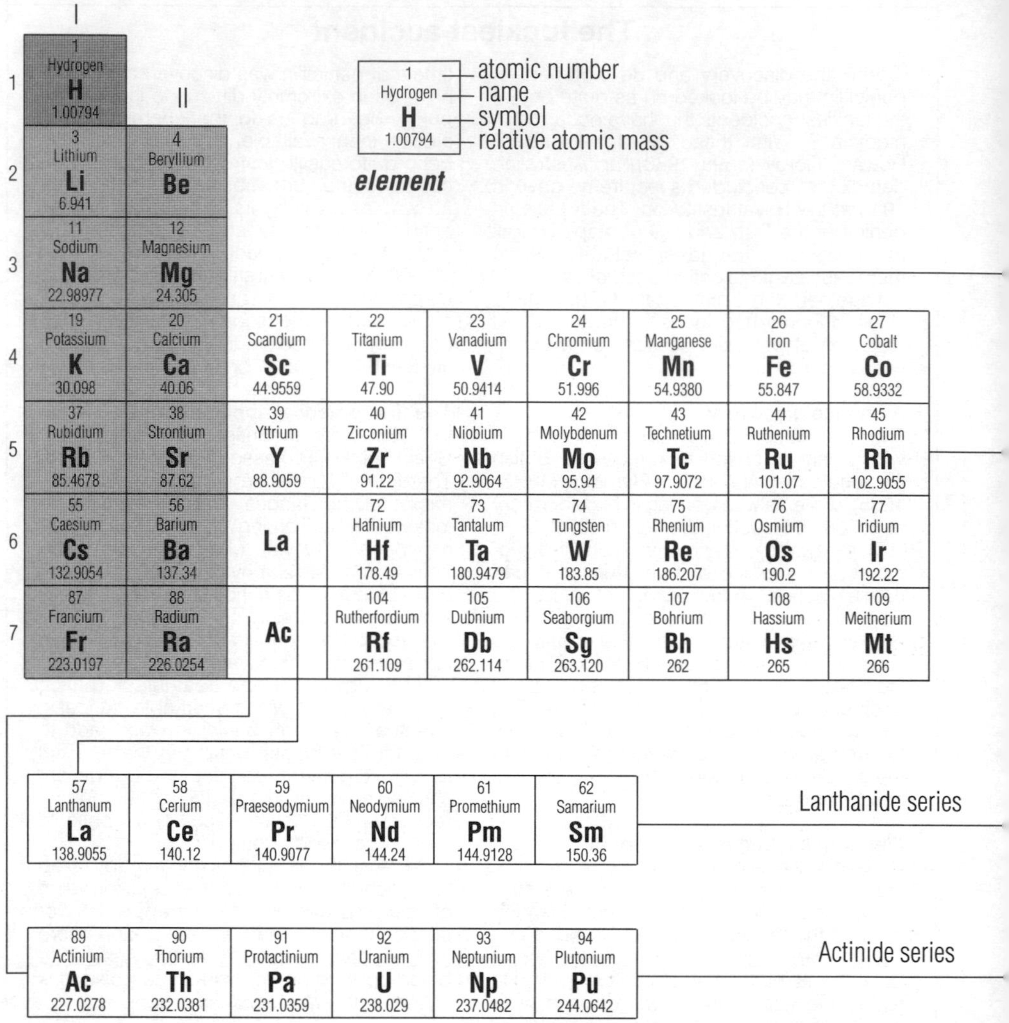

	I	II								
1	1 Hydrogen **H** 1.00794									
2	3 Lithium **Li** 6.941	4 Beryllium **Be**								
3	11 Sodium **Na** 22.98977	12 Magnesium **Mg** 24.305								
4	19 Potassium **K** 30.098	20 Calcium **Ca** 40.06	21 Scandium **Sc** 44.9559	22 Titanium **Ti** 47.90	23 Vanadium **V** 50.9414	24 Chromium **Cr** 51.996	25 Manganese **Mn** 54.9380	26 Iron **Fe** 55.847	27 Cobalt **Co** 58.9332	
5	37 Rubidium **Rb** 85.4678	38 Strontium **Sr** 87.62	39 Yttrium **Y** 88.9059	40 Zirconium **Zr** 91.22	41 Niobium **Nb** 92.9064	42 Molybdenum **Mo** 95.94	43 Technetium **Tc** 97.9072	44 Ruthenium **Ru** 101.07	45 Rhodium **Rh** 102.9055	
6	55 Caesium **Cs** 132.9054	56 Barium **Ba** 137.34	**La**	72 Hafnium **Hf** 178.49	73 Tantalum **Ta** 180.9479	74 Tungsten **W** 183.85	75 Rhenium **Re** 186.207	76 Osmium **Os** 190.2	77 Iridium **Ir** 192.22	
7	87 Francium **Fr** 223.0197	88 Radium **Ra** 226.0254	**Ac**	104 Rutherfordium **Rf** 261.109	105 Dubnium **Db** 262.114	106 Seaborgium **Sg** 263.120	107 Bohrium **Bh** 262	108 Hassium **Hs** 265	109 Meitnerium **Mt** 266	

1 — atomic number
Hydrogen — name
H — symbol
1.00794 — relative atomic mass

element

Lanthanide series

57 Lanthanum **La** 138.9055	58 Cerium **Ce** 140.12	59 Praeseodymium **Pr** 140.9077	60 Neodymium **Nd** 144.24	61 Promethium **Pm** 144.9128	62 Samarium **Sm** 150.36

Actinide series

89 Actinium **Ac** 227.0278	90 Thorium **Th** 232.0381	91 Protactinium **Pa** 231.0359	92 Uranium **U** 238.029	93 Neptunium **Np** 237.0482	94 Plutonium **Pu** 244.0642

periodic table of the elements The periodic table of the elements arranges the elements into horizontal rows (called periods) and vertical columns (called groups) according to their atomic numbers. The elements in a group or column all have similar properties – for example, all the elements in the far right-hand column are inert gases. Nonmetals are shown in mid grey, metals in light grey, and the metalloid (metal-like) elements in dark grey. The elements in white are called transition elements.

example, $1/8 = 12.5/100 = 12.5\%$. The use of percentages often makes it easier to compare fractions that do not have a common denominator.

The percentage sign is thought to have been derived as an economy measure when recording in the old counting houses; writing in the numeric symbol for 25/100 of a cargo would take two lines of parchment, and hence the '100' denominator was put alongside the 25 and rearranged to '%'.

perennating organ in plants, that part of a ◊biennial plant or herbaceous perennial that allows it to survive the winter; usually a root, tuber, rhizome, bulb, or corm.

perennial plant plant that lives for more than two years. Herbaceous perennials have aerial stems and leaves that die each autumn. They survive the winter by means of an underground storage (perennating) organ, such as a bulb or rhizome. Trees and shrubs or woody perennials have stems that persist above ground throughout the year, and may be either ◊deciduous or ◊evergreen. See also ◊annual plant, ◊biennial plant.

perforation in medicine, piercing of an organ through mechanical injury or disease. Ulcers, large tumours, or penetrating injuries may result in the perforation of the abdominal organs. If the intestinal contents enter the

0
2 Helium **He** 4002.60

III	IV	V	VI	VII
5 Boron **B** 10.81	6 Carbon **C** 12.011	7 Nitrogen **N** 14.0067	8 Oxygen **O** 15.9994	9 Fluorine **F** 18.99840
13 Aluminium **Al** 26.98154	14 Silicon **Si** 28.086	15 Phosphorus **P** 30.9738	16 Sulphur **S** 32.06	17 Chlorine **Cl** 35.453

10 Neon **Ne** 20.179								
18 Argon **Ar** 39.948								

28 Nickel **Ni** 58.70	29 Copper **Cu** 63.546	30 Zinc **Zn** 65.38	31 Gallium **Ga** 69.72	32 Germanium **Ge** 72.59	33 Arsenic **As** 74.9216	34 Selenium **Se** 78.96	35 Bromine **Br** 79.904	36 Krypton **Kr** 83.80
46 Palladium **Pd** 106.4	47 Silver **Ag** 107.868	48 Cadmium **Cd** 112.40	49 Indium **In** 114.82	50 Tin **Sn** 118.69	51 Antimony **Sb** 121.75	52 Tellurium **Te** 127.75	53 Iodine **I** 126.9045	54 Xenon **Xe** 131.30
78 Platinum **Pt** 195.09	79 Gold **Au** 196.9665	80 Mercury **Hg** 200.59	81 Thallium **Tl** 204.37	82 Lead **Pb** 207.37	83 Bismuth **Bi** 207.2	84 Polonium **Po** 210	85 Astatine **At** 211	86 Radon **Rn** 222.0176
110 Ununnilium **Uun** 269	111 Unununiun **Uuu** 272							

63 Europium **Eu** 151.96	64 Gadolinium **Gd** 157.25	65 Terbium **Tb** 158.9254	66 Dysprosium **Dy** 162.50	67 Holmium **Ho** 164.9304	68 Erbium **Er** 167.26	69 Thulium **Tm** 168.9342	70 Ytterbium **Yb** 173.04	71 Lutetium **Lu** 174.97
95 Americium **Am** 243.0614	96 Curium **Cm** 247.0703	97 Berkelium **Bk** 247	98 Californium **Cf** 251.0786	99 Einsteinium **Es** 252.0828	100 Fermium **Fm** 257.0951	101 Mendelevium **Md** 258.0986	102 Nobelium **No** 259.1009	103 Lawrencium **Lr** 260.1054

abdominal cavity, the large numbers of bacteria that they contain can cause peritonitis and septicaemia (blood poisoning), which can be fatal. Surgery is required to wash away the contamination and repair the perforation. Perforation of major blood vessels results in haemorrhage and requires urgent surgical treatment.

perfume fragrant essence used to scent the body, cosmetics, and candles. More than 100 natural aromatic chemicals may be blended from a range of 60,000 flowers, leaves, fruits, seeds, woods, barks, resins, and roots, combined with natural animal fixatives and various synthetics. Favoured ingredients include balsam, civet (from the African civet cat) hyacinth, jasmine, lily of the valley, musk (from the musk deer), orange blossom, rose, and tuberose.

Culture of the cells of fragrant plants, on membranes that are constantly bathed in a solution to carry the essen-

tial oils away for separation, is now being adopted to reduce costs.

perfusion in biology, the passage of fluid through tissues. For example, oxygen contained in the moist air that enters the lungs perfuses into the blood through the pulmonary tissues.

Pergau Dam hydroelectric dam on the Pergau River in Malaysia, near the Thai border. Building work began in 1991 with money from the UK foreign aid budget. Concurrently, the Malaysian government bought around £1 billion worth of arms from the UK. The suggested linkage of arms deals to aid become the subject of a UK government enquiry in March 1994. In Nov 1994 a High Court ruled as illegal British foreign secretary Douglas Hurd's allocation of £234 million towards the funding of the dam, on the grounds that it was not of economic or humanitarian benefit to the Malaysian people.

perianth in botany, a collective term for the outer whorls of the flower, which protect the reproductive parts during development.

In most ◊dicotyledons the perianth is composed of two distinct whorls, the calyx of ◊sepals and the corolla of ◊petals, whereas in many ◊monocotyledons the sepals and petals are indistinguishable and the segments of the perianth are then known individually as tepals.

pericarp wall of a ◊fruit. It encloses the seeds and is derived from the ◊ovary wall. In fruits such as the acorn, the pericarp becomes dry and hard, forming a shell around the seed.

In fleshy fruits the pericarp is typically made up of three distinct layers. The **epicarp**, or **exocarp**, forms the tough outer skin of the fruit, while the **mesocarp** is often fleshy and forms the middle layers. The innermost layer or **endocarp**, which surrounds the seeds, may be membranous or thick and hard, as in the ◊drupe (stone) of cherries, plums, and apricots.

peridot pale-green, transparent gem variety of the mineral ◊olivine.

peridotite rock consisting largely of the mineral olivine; pyroxene and other minerals may also be present. Peridotite is an ultramafic rock containing less than 45% silica by weight. It is believed to be one of the rock types making up the Earth's upper mantle, and is sometimes brought from the depths to the surface by major movements, or as inclusions in lavas.

perigee the point at which an object, travelling in an elliptical orbit around the Earth, is at its closest to the Earth.

The point at which it is furthest from the Earth is the apogee.

periglacial bordering a glacial area but not actually covered by ice, or having similar climatic and environmental characteristics, such as mountainous areas. Periglacial areas today include parts of Siberia, Greenland, and North America. The rock and soil in these areas is frozen to a depth of several metres (◊permafrost) with only the top few centimetres thawing during the brief summer. The vegetation is characteristic of ◊tundra.

During the last ice age all of southern England was periglacial. Weathering by ◊freeze-thaw (the alternate freezing and thawing of ice in rock cracks) would have been severe, and ◊solifluction would have taken place on a large scale, causing wet topsoil to slip from valley sides.

perihelion the point at which an object, travelling in an elliptical orbit around the Sun, is at its closest to the Sun. The point at which it is furthest from the Sun is the aphelion.

perimeter or **boundary** line drawn around the edge of an area or shape. For example, the perimeter of a rectangle is the sum of its four sides; the perimeter of a circle is known as its **circumference**.

period another name for menstruation; see ◊menstrual cycle.

period in physics, the time taken for one complete cycle of a repeated sequence of events. For example, the time taken for a pendulum to swing from side to side and back again is the period of the pendulum. The period is the reciprocal of the ◊frequency.

periodic table of the elements in chemistry, a table in which the elements are arranged in order of their atomic number. The table summarizes the major properties of the elements and enables predictions to be made about their behaviour.

There are striking similarities in the chemical properties of the elements in each of the vertical columns (called **groups**), which are numbered I–VII and 0 to reflect the number of electrons in the outermost unfilled shell and hence the maximum ◊valency. A gradation of properties may be traced along the horizontal rows (called **periods**). Metallic character increases across a period from right to left, and down a group. A large block of elements, between groups II and III, contains the transition elements, characterized by displaying more than one valency state.

These features are a direct consequence of the electronic (and nuclear) structure of the atoms of the elements. The relationships established between the positions of elements in the periodic table and their major properties has enabled scientists to predict the properties of other elements – for example, technetium, atomic number 43, first synthesized in 1937.

The first periodic table was devised by Russian chemist Dmitri Mendeleyev in 1869, the elements being arranged by atomic mass (rather than atomic number) in accordance with Mendeleyev's statement that 'the properties of elements are in periodic dependence upon their atomic weight'. *See illustration on pp. 482–483.*

periodontal disease (formerly known as **pyorrhoea**) disease of the gums and bone supporting the teeth, caused by the accumulation of plaque and microorganisms; the gums recede, and the teeth eventually become loose and may drop out unless treatment is sought.

Bacteria can eventually erode the bone that supports the teeth, so that surgery becomes necessary.

peripheral device in computing, any item connected to a computer's ◊central processing unit (CPU). Typical peripherals include keyboard, mouse, monitor and printer. Users who enjoy playing games might add a ◊joystick or a trackball; others might connect a ◊modem, ◊scanner, or ◊integrated services digital network (ISDN) terminal to their machines.

periscope optical instrument designed for observation from a concealed position such as from a submerged submarine. In its basic form it consists of a tube with parallel mirrors at each end, inclined at 45° to its axis. The periscope attained prominence in naval and military operations of World War I.

Although most often thought of as a submarine observation device, periscopes were widely used in the trenches during World War I to allow observation without exposing the observer, and special versions were also developed to be attached to rifles.

peristalsis wavelike contractions, produced by the contraction of smooth muscle, that pass along tubular organs, such as the intestines. The same term describes the wavelike motion of earthworms and other invertebrates, in which part of the body contracts as another part elongates.

peritoneum membrane lining the abdominal cavity and digestive organs of vertebrates. **Peritonitis**, inflammation within the peritoneum, can occur due to infection or other irritation. It is sometimes seen following a burst appendix and quickly proves fatal if not treated.

peripheral device Some of the many peripheral devices that can be connected to a computer.

permafrost condition in which a deep layer of soil does not thaw out during the summer. Permafrost occurs under ◊periglacial conditions. It is claimed that 26% of the world's land surface is permafrost.

Permafrost gives rise to a poorly drained form of grassland typical of N Canada, Siberia, and Alaska known as ◊tundra.

Permian period of geological time 290–245 million years ago, the last period of the Palaeozoic era. Its end was marked by a significant change in marine life, including the extinction of many corals and trilobites. Deserts were widespread, terrestrial amphibians and mammal-like reptiles flourished, and cone-bearing plants (gymnosperms) came to prominence. In the oceans, 49% of families and 72% of genera vanished in the late Permian. On land, 78% of reptile families and 67% of amphibian families disappeared.

permutation in mathematics, a specified arrangement of a group of objects.

It is the arrangement of a distinct objects taken b at a time in all possible orders. It is given by $a!/(a − b)!$, where '!' stands for ◊factorial. For example, the number of permutations of four letters taken from any group of six

different letters is $6!/2! = (1 × 2 × 3 × 4 × 5 × 6)/(1 × 2) = 360$. The theoretical number of four-letter 'words' that can be made from an alphabet of 26 letters is $26!/22! = 358,800$.

perpendicular in mathematics, at a right angle; also, a line at right angles to another or to a plane. For a pair of skew lines (lines in three dimensions that do not meet), there is just one common perpendicular, which is at right angles to both lines; the nearest points on the two lines are the feet of this perpendicular.

perpetual motion the idea that a machine can be designed and constructed in such a way that, once started, it will continue in motion indefinitely without requiring any further input of energy (motive power). Such a device contradicts the two laws of thermodynamics that state that (1) energy can neither be created nor destroyed (the law of conservation of energy) and (2) heat cannot by itself flow from a cooler to a hotter object. As a result, all practical (real) machines require a continuous supply of energy, and no heat engine is able to convert all the heat into useful work.

Perseus in astronomy, a bright constellation of the northern hemisphere, near ◊Cassiopeia. It is represented as the mythological hero; the head of the decapitated Gorgon, Medusa, is marked by ◊Algol (Beta Persei), the best known of the eclipsing binary stars.

Perseus lies in the Milky Way and contains the Double Cluster, a twin cluster of stars called h and Chi Persei. They are just visible to the naked eye as two hazy patches of light close to one another.

personal computer (PC) another name for ◊micro-computer. The term is also used, more specifically, to mean the IBM Personal Computer and computers compatible with it.

The first IBM PC was introduced in 1981; it had 64 kilobytes of random access memory (RAM) and one floppy-disc drive. It was followed in 1983 by the XT (with a hard-disc drive) and in 1984 by the AT (based on a more powerful ◊microprocessor). Many manufacturers have copied the basic design, which is now regarded as a standard for business microcomputers. Computers designed to function like an IBM PC are *IBM-compatible computers*.

personality individual's characteristic way of behaving across a wide range of situations. Two broad dimensions of personality are ◊extroversion and neuroticism. A number of more specific personal traits have also been described, including ◊psychopathy (antisocial behaviour).

perspective the realistic representation of a three-dimensional object in two dimensions. In a perspective drawing, vertical lines are drawn in parallel from the top of the page to the bottom. Horizontal lines, however, are represented by straight lines which meet at one of two perspective points. These perspective points lie to the right and left of the drawing at a distance which depends on the view being taken of the object.

Perspex trade name for a clear, lightweight, tough plastic first produced in 1930. It is widely used for watch glasses, advertising signs, domestic baths, motorboat windscreens, aircraft canopies, and protective shields. Its chemical name is polymethylmethacrylate (PMMA). It is manufactured under other names: Lucite, Acrylite, and Rhoplex (in the USA), and Oroglas (in Europe).

perspiration excretion of water and dissolved substances from the ◊sweat glands of the skin of mammals. Perspiration has two main functions: body cooling by the evaporation of water from the skin surface, and excretion of waste products such as salts.

pertussis medical name for ◊whooping cough, an infectious disease mainly seen in children.

Peru Current formerly known as *Humboldt Current* cold ocean ◊current flowing north from the Antarctic along the W coast of South America to S Ecuador, then west. It reduces the coastal temperature, making the W slopes of the Andes arid because winds are already chilled and dry when they meet the coast.

pessary medical device designed to be inserted into the vagina either to support a displaced womb or as a contraceptive. The word is also used for a vaginal suppository used for administering drugs locally, made from glycerol or oil of theobromine, which melts within the vagina to release the contained substance – for example, a contraceptive, antibiotic, antifungal agent, or ◊prostaglandin (to induce labour).

pest in biology, any insect, fungus, rodent, or other living organism that has a harmful effect on human beings, other than those that directly cause human diseases. Most pests damage crops or livestock, but the term also covers those that damage buildings, destroy food stores, and spread disease.

pesticide any chemical used in farming, gardening, or indoors to combat pests. Pesticides are of three main types: *insecticides* (to kill insects), *fungicides* (to kill fungal diseases), and *herbicides* (to kill plants, mainly those considered weeds). Pesticides cause a number of pollution problems through spray drift onto surrounding areas, direct contamination of users or the public, and as residues on food.

The safest pesticides include those made from plants, such as the insecticides pyrethrum and derris.

Pyrethrins are safe and insects do not develop resistance to them. Their impact on the environment is very small as the ingredients break down harmlessly.

More potent are synthetic products, such as chlorinated hydrocarbons. These products, including DDT and dieldrin, are highly toxic to wildlife and often to human beings, so their use is now restricted by law in some areas and is declining. Safer pesticides such as malathion are based on organic phosphorus compounds, but they still present hazards to health. The aid organization Oxfam estimates that pesticides cause about 10,000 deaths worldwide every year.

Pesticides were used to deforest SE Asia during the Vietnam War, causing death and destruction to the area's ecology and lasting health and agricultural problems.

petal part of a flower whose function is to attract pollinators such as insects or birds. Petals are frequently large and brightly coloured and may also be scented. Some have a nectary at the base and markings on the petal surface, known as honey guides, to direct pollinators to the source of the nectar. In wind-pollinated plants, however, the petals are usually small and insignificant, and sometimes absent altogether. Petals are derived from modified leaves, and are known collectively as a corolla.

Some insect-pollinated plants also have inconspicuous petals, with large colourful ◊bracts (leaflike structures) or ◊sepals taking over their role, or strong scents that attract pollinators such as flies.

petiole in botany, the stalk attaching the leaf blade, or ◊lamina, to the stem. Typically it is continuous with the midrib of the leaf and attached to the base of the lamina, but occasionally it is attached to the lower surface of the lamina, as in the nasturtium (a peltate leaf). Petioles that are flattened and leaflike are termed phyllodes. Leaves that lack a petiole are said to be sessile.

petrochemical chemical derived from the processing of petroleum (crude oil).

Petrochemical industries are those that obtain their raw materials from the processing of petroleum and natural gas. Polymers, detergents, solvents, and nitrogen fertilizers are all major products of the petrochemical industries. Inorganic chemical products include carbon black, sulphur, ammonia, and hydrogen peroxide.

petrol mixture of hydrocarbons derived from petroleum, mainly used as a fuel for internal-combustion engines. It is colourless and highly volatile. *Leaded petrol* contains antiknock (a mixture of tetraethyl lead and dibromoethane), which improves the combustion of petrol and the performance of a car engine. The lead from the exhaust fumes enters the atmosphere, mostly as simple lead compounds. There is strong evidence that it can act as a nerve poison on young children and cause mental impairment. This has prompted a gradual switch to the use of **unleaded petrol** in the UK.

The change-over from leaded petrol gained momentum from 1989 owing to a change in the tax on petrol, making it cheaper to buy unleaded fuel. Unleaded petrol contains a different mixture of hydrocarbons, and has a lower ◊octane rating than leaded petrol. Leaded petrol cannot be used in cars fitted with a ◊catalytic converter.

In the USA, petrol is called gasoline, and unleaded petrol has been used for many years.

petrol engine the most commonly used source of power for motor vehicles, introduced by the German engineers Gottlieb Daimler and Karl Benz in 1885. The petrol engine is a complex piece of machinery made up of about 150 moving parts. It is a reciprocating piston engine, in which a number of pistons move up and down in cylinders. The motion of the pistons rotate a crankshaft, at the end of which is a heavy flywheel. From the flywheel the power is transferred to the car's driving wheels via the transmission system of clutch, gearbox, and final drive.

The parts of the petrol engine can be subdivided into a number of systems. The *fuel system* pumps fuel from the petrol tank into the carburettor. There it mixes with air and is sucked into the engine cylinders. (With electronic fuel injection, it goes directly from the tank into the cylinders by way of an electronic monitor.) The *ignition system* supplies the sparks to ignite the fuel mixture in the cylinders. By means of an ignition coil and contact breaker, it boosts the 12-volt battery voltage to pulses of 18,000 volts or more. These go via a distributor to the spark plugs in the cylinders, where they create the sparks. (Electronic ignitions replace these parts.) Ignition of the fuel in the cylinders produces temperatures of 700°C/1,300°F or more, and the engine must be cooled to prevent overheating.

Most engines have a *water-cooling system*, in which water circulates through channels in the cylinder block, thus extracting the heat. It flows through pipes in a radiator, which are cooled by fan-blown air. A few cars and most motorbikes are air-cooled, the cylinders being surrounded by many fins to present a large surface area to the air. The *lubrication system* also reduces some heat, but its main job is to keep the moving parts coated with oil, which is pumped under pressure to the camshaft, crankshaft, and valve-operating gear.

petroleum or *crude oil* natural mineral oil, a thick greenish-brown flammable liquid found underground in permeable rocks. Petroleum consists of hydrocarbons mixed with oxygen, sulphur, nitrogen, and other elements in varying proportions. It is thought to be derived from ancient organic material that has been converted first by bacterial action, then heat and pressure (but its origin may also be chemical).

Various products are made from crude petroleum by distillation and other processes; for example, fuel oil, petrol, kerosene, diesel, and lubricating oil. Petroleum products and chemicals are used in large quantities in the manufacture of detergents, artificial fibres, plastics, insecticides, fertilizers, pharmaceuticals, toiletries, and synthetic rubber.

Petroleum was formed from the remains of marine plant and animal life which existed many millions of years ago (hence it is known as a fossil fuel). Some of these

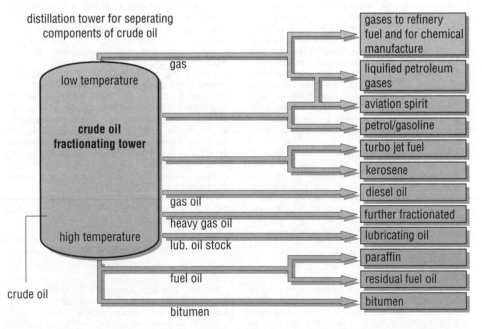

distillation tower for seperating components of crude oil

petroleum Refining petroleum using a distillation column. The crude petroleum is fed in at the bottom of the column where the temperature is high. The gases produced rise up the column, cooling as they travel. At different heights up the column, different gases condense to liquids called fractions, and are drawn off.

Petroleum: Chronology

1859	Edwin Drake drilled the world's first successful oil well in Titusville, Pennsylvania, USA, to a depth of 18 m/70 ft.
1865	The first oil pipeline, 9,750 m/32,000 ft long, was constructed at Oil Creek, Pennsylvania, USA, to carry oil from the well to a nearby coalfield.
1896	The first offshore wells were drilled from piers off the California coast.
1899	The first gravity meter was produced.
1914	Reginald Fessenden patented the seismograph.
1939	The aircraft-borne magnetometer was developed to measure magnetism of rocks.
1966	Oil was discovered beneath the North Sea.
1967	The worst oil spill in British waters from the *Torrey Canyon*, which struck rocks at Lands End. Over 108,000 tonnes of oil were spilled.
1974	The world's deepest oil well, 10,941 m/31,441 ft, was drilled in Oklahoma, USA.
1979	The oil rig *Ixtoc I* in the Gulf of Mexico accidentally released 545,000 tonnes of oil into the sea. The slick spread for 400 miles. Later the same year, the worst tanker spillage occurred. Two tankers, the *Atlantic Empress* and the *Aegean Captain*, collided off the island of Tobago, in the Caribbean Sea. More than 230,000 tonnes of oil were spilled.
1984	An exploratory well was drilled off the coast of New England, USA, in water of depth 2,116 m/6,942 ft – a world record.
1988	The *Piper Alpha* drilling rig in the North Sea caught fire in July, killing 167 people.
1989	The worst spill in American waters occurred when 55,000 tonnes of oil escaped from the *Exxon Valdez* off the Alaskan coast, near Prince William Sound.
1991	The worst spill to date was a consequence of the Gulf War when Iraqi forces opened the pipeline into the Persian Gulf and coalition forces caused further damage during Operation Desert Storm.
1993	Wrecked tanker *Braer* discharged 85,000 tonnes of oil to the Scottish coast.
1996	The supertanker *Sea Empress* ran aground off the southwest coast of Wales, spilling 72,000 tonnes of oil.

remains were deposited along with rock-forming sediments under the sea where they were decomposed anaerobically (without oxygen) by bacteria which changed the fats of the sediments into fatty acids which were then changed into an asphaltic material called kerogen. This was then converted over millions of years into petroleum by the combined action of heat and pressure. At an early stage the organic material was squeezed out of its original sedimentary mud into adjacent sandstones. Small globules of oil collected together in the pores of the rock and eventually migrated upwards through layers of porous rock by the action of the oil's own surface tension (capillary action), by the force of water movement within the rock, and by gas pressure. This migration ended either when the petroleum emerged through a fissure as a seepage of gas or oil on to the Earth's surface, or when it was trapped in porous reservoir rocks, such as sandstone or limestone, in anticlines and other traps below impervious rock layers.

The modern oil industry originates from the discovery of oil in western Ontario in 1857 followed by Edwin Drake's discovery in Pennsylvania in 1859. Drake used a steam engine to drive a punching tool to 18 m below the surface where he struck oil and started an oil boom. Rapid development followed in other parts of the USA, Canada, Mexico, and then Venezuela where commercial production began in 1878. Oil was found in Romania in 1860, Iran in 1908, Iraq in 1923, Bahrain in 1932, and Saudi Arabia and Kuwait in 1938.

The USA led in production until the 1960s, when the Middle East outproduced other areas, their immense reserves leading to a worldwide dependence on cheap oil for transport and industry. In 1961 the Organization of the Petroleum Exporting Countries (OPEC) was established to avoid exploitation of member countries; after OPEC's price rises in 1973, the International Energy Agency (IEA) was established in 1974 to protect the interests of oil-consuming countries. New technologies

were introduced to pump oil from offshore and from the Arctic (the Alaska pipeline) in an effort to avoid a monopoly by OPEC.

As shallow-water oil reserves dwindle, multinational companies have been developing deep-water oilfields at the edge of the continental shelf in the Gulf of Mexico. Shell has developed Mars, a 500-million barrel oilfield, in 2,940 ft of water, and the oil companies now have the technology to drill wells at up to 10,000 ft under the sea. It is estimated that the deep waters of Mexico could yield 8–15 million barrels in total; it could overtake the North Sea in importance as an oil source.

The burning of petroleum fuel is one cause of air pollution. The transport of oil can lead to catastrophes – for example, the *Torrey Canyon* tanker lost off SW England in 1967, which led to an agreement by the international oil companies in 1968 to pay compensation for massive shore pollution. The 1989 oil spill in Alaska from the *Exxon Valdez* damaged the area's fragile environment, despite clean-up efforts. Drilling for oil involves the risks of accidental spillage and drilling-rig accidents. The problems associated with oil have led to various alternative ◊energy technologies.

A new kind of bacterium was developed during the 1970s in the USA, capable of 'eating' oil as a means of countering oil spills.

petrology branch of geology that deals with the study of rocks, their mineral compositions, and their origins.

Peugeot France's second-largest car manufacturer, founded in 1885 when Armand Peugeot (1849–1915) began making bicycles; the company bought the rival firm Citroën in 1974 and the European operations of the American Chrysler Company in 1978.

In 1889 Armand Peugeot produced his first steam car and in 1890 his first petrol-driven car, with a Daimler engine. Peugeot's cars did well in races and were in demand from the public, and by 1900 he was producing a range of models. In the 1930s Peugeot sporting family cars sold widely. In 1978, after the acquisition of Chrysler in Europe, the Talbot marque was reintroduced.

pewter any of various alloys of mostly tin with varying amounts of lead, copper, or antimony. Pewter has been known for centuries and was once widely used for domestic utensils but is now used mainly for ornamental ware.

pH scale from 0 to 14 for measuring acidity or alkalinity. A pH of 7.0 indicates neutrality, below 7 is acid, while above 7 is alkaline. Strong acids, such as those used in car batteries, have a pH of about 2; strong alkalis such as sodium hydroxide are pH 13.

Acidic fruits such as citrus fruits are about pH 4. Fertile soils have a pH of about 6.5 to 7.0, while weak alkalis such as soap are 9 to 10.

phage another name for a ◊bacteriophage, a virus that attacks bacteria.

phagocyte type of white blood cell, or ◊leucocyte, that can engulf a bacterium or other invading microorganism. Phagocytes are found in blood, lymph, and other body tissues, where they also ingest foreign matter and dead tissue. A ◊macrophage differs in size and life span.

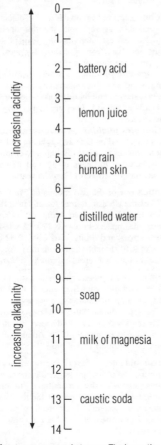

pH The pHs of some common substances. The lower the pH the more acid the substance.

phanerogam obsolete term for a plant that bears flowers or cones and reproduces by means of seeds, that is an ◊angiosperm and ◊gymnosperm, or a ◊seed plant. Plants such as mosses, fungi, and ferns were known as **cryptogams**.

Phanerozoic eon in Earth history, consisting of the most recent 570 million years. It comprises the Palaeozoic, Mesozoic, and Cenozoic eras. The vast majority of fossils come from this eon, owing to the evolution of hard shells and internal skeletons. The name means 'interval of well-displayed life'.

pharmacokinetics in medicine, the relationship between the absorption, distribution, metabolism, and excretion of a drug in mathematical terms. The effects and the duration of action of the drug are also taken into account. Clinical pharmacokinetics is the application of pharmacokinetic studies to clinical practice and to the safe and effective therapeutic management of the individual patient.

pharmacology study of the properties of drugs and their effects on the human body.

pharyngitis in medicine, an inflammatory condition that affects the wall of the ◊pharynx and the throat. It is usually due to a viral infection. Symptoms include irritation and discomfort of the throat, which may be relieved by throat pastilles and gargles. Infections due to bacteria can be treated by antibiotics.

pharynx muscular cavity behind the nose and mouth, extending downwards from the base of the skull. Its walls are made of muscle strengthened with a fibrous layer and lined with mucous membrane. The internal nostrils lead backwards into the pharynx, which continues downwards into the oesophagus and (through the epiglottis) into the windpipe. On each side, a Eustachian tube enters the pharynx from the middle ear cavity.

The upper part (nasopharynx) is an airway, but the remainder is a passage for food. Inflammation of the pharynx is named pharyngitis.

phase in astronomy, the apparent shape of the Moon or a planet when all or part of its illuminated hemisphere is facing the Earth.

The Moon undergoes a full cycle of phases from new (when between the Earth and the Sun) through first quarter (when at 90° eastern elongation from the Sun), full (when opposite the Sun), and last quarter (when at 90° western elongation from the Sun).

The Moon is gibbous (more than half but less than fully illuminated) when between first quarter and full or full and last quarter. Mars can appear gibbous at quadrature (when it is at right angles to the Sun in the sky). The gibbous appearance of Jupiter is barely noticeable.

The planets whose orbits lie within that of the Earth can also undergo a full cycle of phases, as can an asteroid passing inside the Earth's orbit.

Phase

February is the only month during which there may be no full or new Moon. This is because there are 29 days between new moons, but only 28 days in February (29 in leap years).

phase in chemistry, a physical state of matter: for example, ice and liquid water are different phases of water; a mixture of the two is termed a two-phase system.

phase in physics, a stage in an oscillatory motion, such as a wave motion: two waves are in phase when their peaks and their troughs coincide. Otherwise, there is a *phase difference*, which has consequences in ◊interference phenomena and ◊alternating current electricity.

phenol member of a group of aromatic chemical compounds with weakly acidic properties, which are characterized by a hydroxyl (OH) group attached directly to an aromatic ring. The simplest of the phenols, derived from benzene, is also known as phenol and has the formula C_6H_5OH. It is sometimes called *carbolic acid* and can be extracted from coal tar.

Pure phenol consists of colourless, needle-shaped crystals, which take up moisture from the atmosphere. It has a strong and characteristic smell and was once used as an antiseptic. It is, however, toxic by absorption through the skin.

phenotype in genetics, visible traits, those actually displayed by an organism. The phenotype is not a direct reflection of the ◊genotype because some alleles are masked by the presence of other, dominant alleles (see ◊dominance). The phenotype is further modified by the effects of the environment (for example, poor nutrition stunts growth).

phenylalanine one of the nine essential amino acids. ◊Phenylketonuria is a rare genetic disease which results from the inability to metabolize the phenylalanine present in food. It causes severe mental retardation if the phenylalanine content of the diet is not restricted. Routine testing of babies soon after birth can prevent the development of phenylketonuria and mental retardation.

phenylketonuria inherited metabolic condition in which the liver of a child cannot control the level of phenylalanine (an ◊amino acid derived from protein food) in the bloodstream. The condition must be detected promptly and a special diet started in the first few weeks of life if brain damage is to be avoided. Untreated, it causes stunted growth, epilepsy, and severe mental disability.

pheromone chemical signal (such as an odour) that is emitted by one animal and affects the behaviour of others. Pheromones are used by many animal species to attract mates.

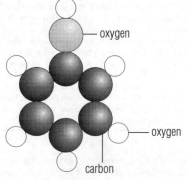

phenol The phenol molecule with its ring of six carbon atoms and a hydroxyl (OH) group attached. Phenol was first extracted from coal tar in 1834. It is used to make phenolic and epoxy resins, explosives, pharmaceuticals, perfumes, and nylon.

phlebitis inflammation of the wall of a vein. It is sometimes associated with ◊varicose veins or with a blockage by a blood clot (◊thrombosis), in which case it is more accurately described as thrombophlebitis.

Phlebitis may occur as a result of the hormonal changes associated with pregnancy, or due to long-term use of contraceptive pills, or following prolonged immobility (which is why patients are mobilized as soon as possible after surgery). If a major vein is involved, nearly always in a leg, the part beyond the blockage swells and may remain engorged for weeks. It is very painful. Treatment is with ◊anticoagulant drugs and sometimes surgery, depending on the cause.

phloem tissue found in vascular plants whose main function is to conduct sugars and other food materials from the leaves, where they are produced, to all other parts of the plant.

Phloem is composed of sieve elements and their associated companion cells, together with some ◊sclerenchyma and ◊parenchyma cell types. Sieve elements are long, thin-walled cells joined end to end, forming sieve tubes; large pores in the end walls allow the continuous passage of nutrients. Phloem is usually found in association with ◊xylem, the water-conducting tissue, but unlike the latter it is a living tissue.

phobia excessive irrational fear of an object or situation – for example, agoraphobia (fear of open spaces and crowded places), acrophobia (fear of heights), and claustrophobia (fear of enclosed places). ◊Behaviour therapy is one form of treatment.

A *specific phobia* is a severe dislike of a particular thing, including objects, animals or situations. Specific phobias start in childhood (particularly animal phobias) and early adulthood. They are more common in women than men.

Complex phobias have more complicated contributing factors and include agoraphobia and social phobia. These phobias are more disabling. Agoraphobia typically starts between the ages of 18 and 28. Social phobia usually onsets between 11 and 16 years.

Phobos one of the two moons of Mars, discovered in 1877 by the US astronomer Asaph Hall (1829–1907). It is an irregularly shaped lump of rock, cratered by ◊meteorite impacts. Phobos is $27 \times 22 \times 19$ km/$17 \times 13 \times 12$ mi across, and orbits Mars every 0.32 days at a distance of 9,400 km/5,840 mi from the planet's centre. It is thought to be an asteroid captured by Mars' gravity.

phon unit of loudness, equal to the value in decibels of an equally loud tone with frequency 1,000 Hz. The higher the frequency, the louder a noise sounds for the same decibel value; thus an 80-decibel tone with a frequency of 20 Hz sounds as loud as 20 decibels at 1,000 Hz, and the phon value of both tones is 20. An aircraft engine has a loudness of around 140 phons.

phosphate salt or ester of ◊phosphoric acid. Incomplete neutralization of phosphoric acid gives rise to acid phosphates (see ◊acid salts and ◊buffer). Phosphates are used as fertilizers, and are required for the development of healthy root systems. They are involved in many biochemical processes, often as part of complex molecules, such as ◊ATP.

phospholipid any ◊lipid consisting of a glycerol backbone, a phosphate group, and two long chains.

Phospholipids are found everywhere in living systems as the basis for biological membranes.

One of the long chains tends to be hydrophobic and the other hydrophilic (that is, they interrelate with water in opposite ways). This means that phospholipids will line up the same way round when in solution.

phosphor any substance that is phosphorescent, that is, gives out visible light when it is illuminated by a beam of electrons or ultraviolet light. The television screen is coated on the inside with phosphors that glow when beams of electrons strike them. Fluorescent lamp tubes are also phosphor-coated. Phosphors are also used in Day-Glo paints, and as optical brighteners in detergents.

phosphorescence in physics, the emission of light by certain substances after they have absorbed energy, whether from visible light, other electromagnetic radiation such as ultraviolet rays or X-rays, or cathode rays (a beam of electrons). When the stimulating energy is removed phosphorescence ceases, although it may persist for a short time after (unlike ◊fluorescence, which stops immediately).

phosphoric acid acid derived from phosphorus and oxygen. Its commonest form (H_3PO_4) is also known as orthophosphoric acid, and is produced by the action of phosphorus pentoxide (P_2O_5) on water. It is used in rust removers and for rust-proofing iron and steel.

phosphorus highly reactive, nonmetallic element, symbol P, atomic number 15, relative atomic mass 30.9738. It occurs in nature as phosphates (commonly in the form of the mineral ◊apatite), and is essential to plant and animal life. Compounds of phosphorus are used in fertilizers, various organic chemicals, for matches and fireworks, and in glass and steel.

Phosphorus was first identified in 1674 by German alchemist Hennig Brand, who prepared it from urine. The element has three allotropic forms: a black powder; a white-yellow, waxy solid that ignites spontaneously in air to form the poisonous gas phosphorus pentoxide; and a red-brown powder that neither ignites spontaneously nor is poisonous.

photocell or **photoelectric cell** device for measuring or detecting light or other electromagnetic radiation, since its electrical state is altered by the effect of light. In a **photoemissive** cell, the radiation causes electrons to be emitted and a current to flow (◊photoelectric effect); a **photovoltaic** cell causes an ◊electromotive force to be generated in the presence of light across the boundary of two substances. A **photoconductive** cell, which contains a semiconductor, increases its conductivity when exposed to electromagnetic radiation.

Photocells are used for photographers' exposure meters, burglar and fire alarms, automatic doors, and in solar energy arrays.

photochemical reaction any chemical reaction in which light is produced or light initiates the reaction. Light can initiate reactions by exciting atoms or molecules and making them more reactive: the light energy becomes converted to chemical energy. Many photochemical reactions set up a ◊chain reaction and produce ◊free radicals.

This type of reaction is seen in the bleaching of dyes or the yellowing of paper by sunlight. It is harnessed by plants in ◊photosynthesis and by humans in ◊photography.

Chemical reactions that produce light are most commonly seen when materials are burned. Light-emitting reactions are used by living organisms in ◊bioluminescence. One photochemical reaction is the action of sunlight on car exhaust fumes, which results in the production of ◊ozone. Some large cities, such as Los Angeles, and Santiago, Chile, now suffer serious pollution due to photochemical smog.

photocopier machine that uses some form of photographic process to reproduce copies of documents or illustrations. Most modern photocopiers, as pioneered by the Xerox Corporation, use electrostatic photocopying, or ◊xerography ('dry writing').

Additional functions of photocopiers include enlargement and reduction, copying on both sides of the sheet of paper, copying in colour, collating, and stapling.

photodiode semiconductor ◊p–n junction diode used to detect light or measure its intensity. The photodiode is encapsulated in a transparent plastic case that allows light to fall onto the junction. When this occurs, the reverse-bias resistance (high resistance in the opposite direction to normal current-flow) drops and allows a larger reverse-biased current to flow through the device. The increase in current can then be related to the amount of light falling on the junction.

Photodiodes that can detect small changes in light level are used in alarm systems, camera exposure controls, and optical communication links.

photoelectric cell alternative name for ◊photocell.

photoelectric effect in physics, the emission of ◊electrons from a substance (usually a metallic surface) when it is struck by ◊photons (quanta of electromagnetic radiation), usually those of visible light or ultraviolet radiation.

photography process for reproducing images on sensitized materials by various forms of radiant energy, including visible light, ultraviolet, infrared, X-rays, atomic radiations, and electron beams.

Photography was developed in the 19th century; among the pioneers were L J M Daguerre in France and Fox Talbot in the UK. Colour photography dates from the early 20th century.

The most familiar photographic process depends upon the fact that certain silver compounds (called ◊halides) are sensitive to light. A photographic film is coated with these compounds and, in a camera, is exposed to light. An image, or picture, of the scene before the camera is formed on the film because the silver halides become activated (light-altered) where light falls but not where light does not fall. The image is made visible by the process of ◊developing, made permanent by fixing, and, finally, is usually printed on paper. Motion-picture photography uses a camera that exposes a roll of film to a rapid succession of views that, when developed, are projected in equally rapid succession to provide a moving image.

photogravure ◊printing process that uses a plate prepared photographically, covered with a pattern of recessed cells in which the ink is held. See ◊gravure.

photolysis chemical reaction that is driven by light or ultraviolet radiation. For example, the light reaction of

Photography: Chronology

1515	Leonardo da Vinci described the camera obscura.
1790	Thomas Wedgwood in England made photograms – placing objects on leather, sensitized using silver nitrate.
1826	Nicéphore Niepce, a French doctor, produced the world's first photograph from nature on pewter plates with a camera obscura and an eight-hour exposure.
1838	As a result of his earlier collaboration with Niepce, L J M Daguerre produced the first daguerreotype camera photograph.
1840	Invention of the Petzval lens, which reduced exposure time by 90%. Herschel discovered sodium thiosulphate as a fixer for silver halides.
1841	Fox Talbot's calotype process was patented – the first multicopy method of photography using a negative/positive process, sensitized with silver iodide.
1851	Fox Talbot used a one-thousandth of a second exposure to demonstrate high-speed photography. Invention of the wet-collodion-on-glass process and the waxed-paper negative.
1855	Roger Fenton made documentary photographs of the Crimean War from a specially constructed caravan with a portable darkroom.
1858	Nadar took the first aerial photographs from a balloon.
1859	Nadar made photographs underground in Paris using battery-powered arc lights.
1861	The single-lens reflex plate camera was patented by Thomas Sutton. The principles of three-colour photography were demonstrated by Scottish physicist James Clerk Maxwell.
1871	Gelatin-silver bromide was developed.
1878	In the USA Eadweard Muybridge analysed the movements of animals through sequential photographs, using a series of cameras.
1879	The photogravure process was invented.
1880	A silver bromide emulsion was fixed with hypo. Photographs were first reproduced in newspapers in New York using the half-tone engraving process. The first twin-lens reflex camera was produced in London. Gelatin-silver chloride paper was introduced.
1884	George Eastman produced flexible negative film.
1889	The Eastman Company in the USA produced the Kodak No 1 camera and roll film, facilitating universal, hand-held snapshots.
1891	The first telephoto lens. The interference process of colour photography was developed by the French doctor Gabriel Lippmann.
1902	In Germany, Deckel invented a prototype leaf shutter and Zeiss introduced the Tessar lens.
1904	The autochrome colour process was patented by the Lumière brothers.
1907	The autochrome process began to be factory-produced.
1914	Oskar Barnack designed a prototype Leica camera for Leitz in Germany.
1924	Leitz launched the first 35mm camera, the Leica, delayed because of World War I. It became very popular with photojournalists because it was quiet, small, dependable, and had a range of lenses and accessories.
1929	Rolleiflex produced a twin-lens reflex camera in Germany.
1935	In the USA, Mannes and Godowsky invented Kodachrome transparency film, which produced sharp images and rich colour quality. Electronic flash was invented in the USA.
1940	Multigrade enlarging paper by Ilford was made available in the UK.
1942	Kodacolour negative film was introduced.
1947	Polaroid black and white instant process film was invented by Dr Edwin Land, who set up the Polaroid corporation in Boston, Massachusetts. The principles of holography were demonstrated in England by Dennis Gabor.
1955	Kodak introduced Tri-X, a black and white 200 ASA film.
1959	The zoom lens was invented by the Austrian firm of Voigtlander.
1960	The laser was invented in the USA, making holography possible. Polacolor, a self-processing colour film, was introduced by Polaroid, using a 60-second colour film and dye diffusion technique.
1963	Cibachrome, paper and chemicals for printing directly from transparencies, was made available by Ciba-Geigy of Switzerland. One of the most permanent processes, it is marketed by Ilford in the UK.
1969	Photographs were taken on the Moon by US astronauts.
1970	A charge-coupled device was invented at Bell Laboratories in New Jersey, USA, to record very faint images (for example in astronomy).
1972	The SX70 system, a single-lens reflex camera with instant prints, was produced by Polaroid.
1980	*Voyager 1* sent photographs of Saturn back to Earth across space.
1985	The Minolta Corporation in Japan introduced the Minolta 7000 – the world's first body-integral autofocus single-lens reflex camera.
1988	The electronic camera, which stores pictures on magnetic disc instead of on film, was introduced in Japan.
1990	Kodak introduced PhotoCD which converts 35mm camera pictures (on film) into digital form and stores them on compact disc (CD) for viewing on TV.
1992	Japanese company Canon introduced a camera with autofocus controlled by the user's eye. The camera focuses on whatever the user is looking at.
1996	Corbis, a company owned by Bill Gates, bought the exclusive rights to 2,500 photographs by Ansel Adams.

◊photosynthesis (the process by which green plants manufacture carbohydrates from carbon dioxide and water) is a photolytic reaction.

photometer instrument that measures luminous intensity, usually by comparing relative intensities from different sources. Bunsen's grease-spot photometer of 1844 compares the intensity of a light source with a known source by each illuminating one half of a translucent area. Modern photometers use ◊photocells, as in a photographer's exposure meter. A ◊photomultiplier can also be used as a photometer.

photomultiplier instrument that detects low levels of electromagnetic radiation (usually visible light or ◊infrared radiation) and amplifies it to produce a detectable signal.

One type resembles a ◊photocell with an additional series of coated ◊electrodes (dynodes) between the ◊cathode and ◊anode. Radiation striking the cathode releases electrons (primary emission) which hit the first dynode, producing yet more electrons (◊secondary emission), which strike the second dynode. Eventually this produces a measurable signal up to 100 million times larger than the original signal by the time it leaves the anode. Similar devices, called image intensifiers, are used in television camera tubes that 'see' in the dark.

photon in physics, the ◊elementary particle or 'package' (quantum) of energy in which light and other forms of electromagnetic radiation are emitted. The photon has both particle and wave properties; it has no charge, is considered massless but possesses momentum and energy. It is one of the ◊gauge bosons, a particle that cannot be subdivided, and is the carrier of the ◊electromagnetic force, one of the fundamental forces of nature.

According to ◊quantum theory the energy of a photon is given by the formula $E = hf$, where h is Planck's constant and f is the frequency of the radiation emitted.

photoperiodism biological mechanism that determines the timing of certain activities by responding to changes in day length. The flowering of many plants is initiated in this way. Photoperiodism in plants is regulated by a light-sensitive pigment, **phytochrome**. The breeding seasons of many temperate-zone animals are also triggered by increasing or declining day length, as part of their ◊biorhythms.

Autumn-flowering plants (for example, chrysanthemum and soya bean) and autumn-breeding mammals (such as goats and deer) require days that are shorter than a critical length; spring-flowering and spring-breeding ones (such as radish and lettuce, and birds) are triggered by longer days.

photosensitivity in medicine, skin sensitivity to sunlight characterized by rashes occurring on exposure. Abnormal skin sensitivity may result from the use of some cosmetics or medicines that contain photosensitizers.

photosphere visible surface of the Sun, which emits light and heat. About 300 km/200 mi deep, it consists of incandescent gas at a temperature of 5,800K (5,530°C/9,980°F).

Rising cells of hot gas produce a mottling of the photosphere known as **granulation**, each granule being about 1,000 km/620 mi in diameter. The photosphere is often marked by large, dark patches called ◊sunspots.

photosynthesis process by which green plants trap light energy and use it to drive a series of chemical reactions, leading to the formation of carbohydrates. All animals ultimately depend on photosynthesis because it is the method by which the basic food (sugar) is created. For photosynthesis to occur, the plant must possess ◊chlorophyll and must have a supply of carbon dioxide and water. Actively photosynthesizing green plants store excess sugar as starch (this can be tested for in the laboratory using iodine).

The chemical reactions of photosynthesis occur in two stages. During the **light reaction** sunlight is used to split water (H_2O) into oxygen (O_2), protons (hydrogen ions, H^+), and electrons, and oxygen is given off as a by-product. In the **dark reaction**, for which sunlight is not required, the protons and electrons are used to convert carbon dioxide (CO_2) into carbohydrates ($C_m(H_2O)_n$). Photosynthesis depends on the ability of chlorophyll to capture the energy of sunlight and to use it to split water molecules. The initial charge separation occurs in less than a billionth of a second, a speed that compares with that of current computers.

Other pigments, such as ◊carotenoids, are also involved in capturing light energy and passing it on to chlorophyll. Photosynthesis by cyanobacteria was responsible for the appearance of oxygen in the Earth's atmosphere 2 billion years ago, and photosynthesis by plants maintains the oxygen level today.

phototropism movement of part of a plant toward or away from a source of light. Leaves are positively phototropic, detecting the source of light and orientating themselves to receive the maximum amount.

phrenology study of the shape and protuberances of the skull, based on the now discredited theory of the Austrian physician Franz Josef Gall that such features revealed measurable psychological and intellectual traits.

phyllite ◊metamorphic rock produced under increasing temperature and pressure, in which minute mica crystals are aligned so that the rock splits along their plane of orientation, the resulting break being shiny and smooth. Intermediate between slate and schist, its silky sheen is an identifying characteristic.

phyllotaxis the arrangement of leaves on a plant stem. Leaves are nearly always arranged in a regular pattern and in the majority of plants they are inserted singly, either in a **spiral** arrangement up the stem, or on **alternate** sides. Other principal forms are opposite leaves, where two arise from the same node, and whorled, where three or more arise from the same node.

phylogeny historical sequence of changes that occurs in a given species during the course of its evolution. It was once erroneously associated with ontogeny (the process of development of a living organism).

phylum (plural **phyla**) major grouping in biological classification. Mammals, birds, reptiles, amphibians, fishes, and tunicates belong to the phylum Chordata; the phylum Mollusca consists of snails, slugs, mussels, clams, squid, and octopuses; the phylum Porifera contains sponges; and the phylum Echinodermata includes starfish, sea urchins, and sea cucumbers. In classifying plants (where the term 'division' often takes the place of

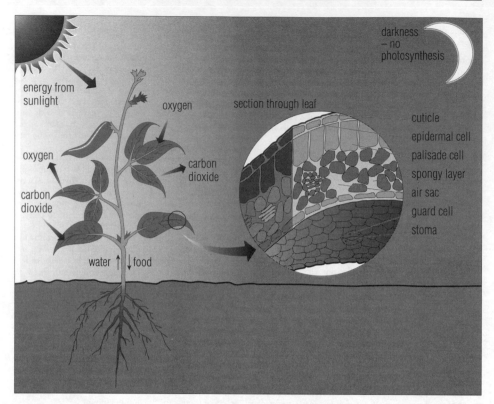

photosynthesis Process by which green plants and some bacteria manufacture carbohydrates from water and atmospheric carbon dioxide, using the energy of sunlight. Photosynthesis depends on the ability of chlorophyll molecules within plant cells to trap the energy of light to split water molecules, giving off oxygen as a by-product. The hydrogen of the water molecules is then used to reduce carbon dioxide to simple carbohydrates.

'phylum'), there are between four and nine phyla depending on the criteria used; all flowering plants belong to a single phylum, Angiospermata, and all conifers to another, Gymnospermata. Related phyla are grouped together in a ◊kingdom; phyla are subdivided into ◊classes.

There are 36 different phyla. The most recently identified is the Cycliophora described in Dec 1995. It contains a single known species, *Symbion pandora*, that lives on lobsters.

physical chemistry branch of chemistry concerned with examining the relationships between the chemical compositions of substances and the physical properties that they display. Most chemical reactions exhibit some physical phenomenon (change of state, temperature, pressure, or volume, or the use or production of electricity), and the measurement and study of such phenomena has led to many chemical theories and laws.

physics branch of science concerned with the laws that govern the structure of the universe, and the investigation of the properties of matter and energy and their interactions. For convenience, physics is often divided into branches such as atomic physics, nuclear physics, particle physics, solid-state physics, molecular physics, electricity and magnetism, optics, acoustics, heat, thermodynamics,

quantum theory, and relativity. Before the 20th century, physics was known as **natural philosophy**.

Our task is not to penetrate into the essence of things, the meaning of which we don't know anyway, but rather to develop concepts which allow us to talk in a productive way about phenomena in nature.

On **physics** Niels Bohr
in a letter to H P E Hansen 20 July 1935

physiology branch of biology that deals with the functioning of living organisms, as opposed to anatomy, which studies their structures.

physiotherapy treatment of injury and disease by physical means such as exercise, heat, manipulation, massage, and electrical stimulation.

phytomenadione one form of vitamin K, a fat-soluble chemical found in green vegetables. It is involved in the production of prothrombin, which is essential in blood clotting. It is given to newborns to prevent potentially fatal brain haemorrhages.

Physics: Chronology

c. 400 BC	The first 'atomic' theory was put forward by Democritus.
c. 250	Archimedes' principle of buoyancy was established.
AD 1600	Magnetism was described by William Gilbert.
1608	Hans Lippershey invented the refracting telescope.
c. 1610	The principle of falling bodies descending to earth at the same speed was established by Galileo.
1642	The principles of hydraulics were put forward by Blaise Pascal.
1643	The mercury barometer was invented by Evangelista Torricelli.
1656	The pendulum clock was invented by Christiaan Huygens.
1662	Boyle's law concerning the behaviour of gases was established by Robert Boyle.
c. 1665	Isaac Newton put forward the law of gravity.
1704	The corpuscular theory of light was put forward by Isaac Newton.
1714	The mercury thermometer was invented by Daniel Fahrenheit.
1764	Specific and latent heats were described by Joseph Black.
c. 1787	Charles's law relating the pressure, volume, and temperature of a gas was established.
1798	The link between heat and friction was discovered by Benjamin Rumford.
1800	Alessandro Volta invented the Voltaic cell.
1801	Interference of light was discovered by Thomas Young.
1808	The 'modern' atomic theory was propounded by John Dalton.
1811	Avogadro's hypothesis relating volumes and numbers of molecules of gases was proposed.
1814	Fraunhofer lines in the solar spectrum were mapped by Joseph von Fraunhofer.
1815	Refraction of light was explained by Augustin Fresnel.
1820	The discovery of electromagnetism was made by Hans Oersted.
1821	The thermocouple was discovered by Thomas Seebeck.
1822	The laws of electrodynamics were established by André Ampère.
1827	Ohm's law of electrical resistance was established by Georg Ohm; Brownian movement resulting from molecular vibrations was observed by Robert Brown.
1829	The law of gaseous diffusion was established by Thomas Graham.
1831	Electromagnetic induction was discovered by Faraday.
c. 1847	The mechanical equivalent of heat was described by James Joule.
1849	A measurement of speed of light was put forward by French physicist Armand Fizeau.
1851	The rotation of the Earth was demonstrated by Jean Foucault.
1859	Spectrographic analysis was made by Robert Bunsen and Gustav Kirchhoff.
1861	Osmosis was discovered.
1873	Light was conceived of as electromagnetic radiation by James Maxwell.
1880	Piezoelectricity was discovered by Pierre Curie.
1887	The existence of radio waves was predicted by Heinrich Hertz.
1895	X-Rays were discovered by Wilhelm Röntgen.
1896	The discovery of radioactivity was made by Antoine Becquerel.
1897	Joseph Thomson discovered the electron.
1899	Ernest Rutherford discovered alpha and beta rays.
1900	Quantum theory was propounded by Max Planck; the discovery of gamma rays was made by French physicist Paul-Ulrich Villard.
1904	The theory of radioactivity was put forward by Rutherford and Frederick Soddy.
1908	The Geiger counter was invented by Hans Geiger and Rutherford.
1911	The discovery of the atomic nucleus was made by Rutherford.
1915	X-Ray crystallography was discovered by William and Lawrence Bragg.
1916	Einstein put forward his general theory of relativity; mass spectrography was discovered by William Aston.
1926	Wave mechanics was introduced by Erwin Schrödinger.
1927	The uncertainty principle of quantum physics was established by Werner Heisenberg.
1931	The cyclotron was developed by Ernest Lawrence.
1932	The discovery of the neutron was made by James Chadwick; the electron microscope was developed by Vladimir Zworykin.
1933	The positron, the antiparticle of the electron, was discovered by Carl Anderson.
1939	The discovery of nuclear fission was made by Otto Hahn and Fritz Strassmann.
1942	The first controlled nuclear chain reaction was achieved by Enrico Fermi.
1956	The neutrino, an elementary particle, was discovered by Clyde Cowan and Fred Reines.
1960	The Mössbauer effect of atom emissions was discovered by Rudolf Mössbauer; the first laser and the first maser were developed by US physicist Theodore Maiman.
1964	Murray Gell-Mann and George Zweig discovered the quark.
1967	Jocelyn Bell (now Bell Burnell) and Antony Hewish discovered pulsars.
1983	Evidence of the existence of weakons (W and Z particles) was confirmed at CERN.
1995	Top quark discovered at Fermilab, the US particle-physics laboratory, near Chicago, USA.
1996	CERN physicists created the first atoms of antimatter (nine atoms of antihydrogen).
1997	A new subatomic particle, an exotic meson, was possibly discovered at Brookhaven National Laboratory, Upton, New York, USA. US physicists displayed the first atomic laser. It emitted atoms that act like lightwaves.

Lemons, crocodiles and wild yams

Birth control, in one form or another, has a long history. That well-known lover Casanova (1725–1798) is reputed to have used a lemon as a spermicide, with less than complete success. The ancient Egyptians used crocodile faeces. The sheath was probably introduced by Gabriel Fallopius, the famous Italian anatomist (1523–1562), but mainly as a prophylactic against venereal disease. Knowledge of human genital systems and their physiology, particularly the female system, remained rudimentary until the beginning of the 20th century. The most reliable form of birth control up to this point was simple abstinence.

The population of the world in 1830 was 1 billion; by 1929 it had increased to 2 billion. However, in the period following the carnage of World War I, people were more concerned about repopulation than the advancement of new methods of birth control. There were legal constraints on birth control as well. US social reformer Margaret Sanger (1883–1966) opened the first birth control advisory clinic in 1916 in Brooklyn, New York, USA, but it was closed by the police, and Sanger was imprisoned. So, in addition to the necessary scientific advances, before successful birth control could be practised there had to be changes both in the law and in social awareness.

Preventing ovulation

The key step forward in scientific understanding came in the 1950s when US physiologist Gregory Pincus discovered that the steroid hormone progesterone, found in greater concentrations during pregnancy, is responsible for the prevention of ovulation at that time.

The way had been paved in the 1920s when German physiologist Ludwig Haberlandt (1885–1932) conducted a crucial experiment. Starting from the idea that pregnancy somehow triggered the secretion of anti-ovulatory substances, he demonstrated that by transplanting the ovaries from pregnant rabbits and rats to non-pregnant females, he could induce a temporary sterility. He also found that simple ovarian extracts could achieve the same result. He suggested that specific hormones secreted during pregnancy were responsible.

The modern study of sex hormones began with US endocrinologist Edgar Allen (1892–1943) who discovered oestrogen and showed experimentally that it induced physiological changes normally found in the oestrous cycle. Allen and his colleagues extracted and crystallized a few milligrams of a hormone from the corpus luteum (part of the ovary) of the sow. To do this, they had to process the carcasses of several hundred sows. They suggested the name progesterone for the hormone.

A commercial source of progesterone

The story now moves to the Mexican jungle. In 1941, the US Bureau of Plant Industry, a branch of the Federal Bureau of Agronomy, received reports of the contraceptive properties of infusions used by the Indian women of Nevada. Like many indigenous tribes in different parts of the world, they made use of plant extracts. A US scientist, Russell Marker, found it difficult to pursue contraceptive studies in the prevailing climate of public opinion, and decided he would be better off in Mexico. Benefiting from local folklore, he investigated the wild yam as a possible source of progesterone.

After collecting considerable quantities of this tuber in the jungle, Marker succeeded in a remarkably short space of time in synthesizing 2 kg/4.4 lb of progesterone. It is related that he appeared one day in the offices of a small Mexican pharmaceutical firm and, placing four full 1 lb jars of progesterone on the manager's desk, enquired if he was interested in the industrial manufacture of the hormone!

The pill arrives

There the matter rested, until the work of Gregory Pincus made it possible to conduct a series of trials, one among the women of Port-au-Prince in Haiti, and two in Puerto Rico. The trials gave conclusive results, and the firm of Searle, in Chicago, USA, was then able to produce the first contraceptive pill, Enovid.

Progesterone is now known to play a vital role in maintaining the normal course of pregnancy. In high doses, however, it acts to prevent ovulation. By taking a contraceptive pill, in which the amounts of progesterone and oestrogen activity are carefully balanced, the female body behaves as if conception has occurred. Other types of pill operate only through progesterone activity.

The population of the world in 1961 was 3 billion; in 1975 4 billion and in the year 2000 it will top 6 billion. Most of the staggering 2 billion increase in the 25 years preceding the millenium will happen in countries other than North America and Europe. The population is relatively stable in the developed countries where, in some instances, the birth rate has been cut by up to 40%. This is attributable both to advances in the scientific knowledge of birth and its control made during this century and, equally, to an increase in overall wealth.

Julian Rowe

pi symbol π, the ratio of the circumference of a circle to its diameter. The value of pi is 3.1415926, correct to seven decimal places. Common approximations of pi are 22/7 and 3.14, although the value 3 can be used as a rough estimation.

Pi

In 1853, the English mathematician William Shanks published the value of pi to 707 decimal places. The calculation had taken him 15 years and was surpassed only in 1945, when computations made on an early desk calculator showed that the last 180 decimal places he had calculated were incorrect.

---

pickling method of preserving food by soaking it in acetic acid (found in vinegar), which stops the growth of moulds. In sauerkraut, lactic acid, produced by bacteria, has the same effect.

PID (abbreviation for *pelvic inflammatory disease*) serious gynaecological condition characterized by lower abdominal pain, malaise, and fever; menstruation may be disrupted; infertility may result. Treatment is with antibiotics. The incidence of the disease is twice as high in women using intrauterine contraceptive devices (IUDs).

PID is potentially life-threatening, and, while mild episodes usually respond to antibiotics, surgery may be necessary in cases of severe or recurrent pelvic infection. The bacterium *Chlamydia trachomatis* (see ◊chlamydia) has been implicated in a high proportion of cases. The condition is increasingly common.

piezoelectric effect property of some crystals (for example, quartz) to develop an electromotive force or voltage across opposite faces when subjected to tension or compression, and, conversely, to expand or contract in size when subjected to an electromotive force. Piezoelectric crystal ◊oscillators are used as frequency standards (for example, replacing balance wheels in watches), and for producing ultrasound.

pig iron or *cast iron* the quality of iron produced in a ◊blast furnace. It contains around 4% carbon plus some other impurities.

Pill, the commonly used term for the contraceptive pill, based on female hormones. The combined pill, which contains synthetic hormones similar to oestrogen and progesterone, stops the production of eggs, and makes the mucus produced by the cervix hostile to sperm. It is the most effective form of contraception apart from sterilization, being more than 99% effective.

The *minipill* or progesterone-only pill prevents implantation of a fertilized egg into the wall of the uterus. The minipill has a slightly higher failure rate, especially if not taken at the same time each day, but has fewer side effects and is considered safer for long-term use. Possible side effects of the Pill include migraine or headache and high blood pressure. More seriously, oestrogen-containing pills can slightly increase the risk of a clot forming in the blood vessels. This risk is increased in women over 35 if they smoke. Controversy surrounds other possible health effects of taking the Pill. The evidence for a link with cancer is slight (and the Pill may protect women from some forms of cancer). Once a woman ceases to take it, there is an increase in the likelihood of conceiving identical twins.

PIN (acronym for *personal identification number*) in banking, a unique number used as a password to establish the identity of a customer using an automatic cash dispenser. The PIN is normally encoded into the magnetic strip of the customer's bank card and is known only to the customer and to the bank's computer. Before a cash dispenser will issue money or information, the customer must insert the card into a slot in the machine (so that the PIN can be read from the magnetic strip) and enter the PIN correctly at a keyboard. This helps to prevent stolen cards from being used to obtain money from cash dispensers.

pineal body or *pineal gland* a cone-shaped outgrowth of the vertebrate brain. In some lower vertebrates, it develops a rudimentary lens and retina, which show it to be derived from an eye, or pair of eyes, situated on the top of the head in ancestral vertebrates. In fishes that can change colour to match their background, the pineal body perceives the light level and controls the colour change. In birds, the pineal body detects changes in daylight and stimulates breeding behaviour as spring approaches. Mammals also have a pineal gland, but it is located deeper within the brain. It secretes a hormone, melatonin, thought to influence rhythms of activity. In humans, it is a small piece of tissue attached by a stalk to the rear wall of the third ventricle of the brain.

pinhole camera the simplest type of camera, in which a pinhole rather than a lens is used to form an image. Light passes through the pinhole at one end of a box to form a sharp inverted image on the inside surface of the opposite end. The image is equally sharp for objects placed at different distances from the camera because only one ray from a particular distance or direction can enter through the tiny pinhole, and so only one corresponding point of light will be produced on the image. A photographic film or plate fitted inside the box will, if exposed for a long time, record the image.

pinna in botany, the primary division of a ◊pinnate leaf. In mammals, the pinna is the external part of the ear.

pinnate leaf leaf that is divided up into many small leaflets, arranged in rows along either side of a midrib, as in ash trees (*Fraxinus*). It is a type of compound leaf. Each leaflet is known as a *pinna*, and where the pinnae are themselves divided, the secondary divisions are known as pinnules.

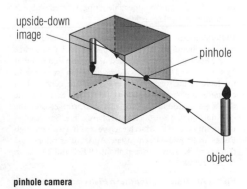

upside-down image

pinhole

object

pinhole camera

pins and needles or *paresthesiae* in medicine, numbness and disturbed sensation of the limbs. It may be experienced following long periods of immobility or as a symptom of some neurological diseases.

pint imperial dry or liquid measure of capacity equal to 20 fluid ounces, half a quart, one-eighth of a gallon, or 0.568 litre. In the US, a liquid pint is equal to 0.473 litre, while a dry pint is equal to 0.550 litre.

pion or *pi meson* in physics, a subatomic particle with a neutral form (mass 135 MeV) and a positively charged form (mass 139 MeV). The charged pion decays into muons and neutrinos and the neutral form decays into gamma-ray photons. They belong to the ◊hadron class of ◊elementary particles.

The mass of a positive pion is 273 times that of an electron; the mass of a neutral pion is 264 times that of an electron.

Pioneer probe any of a series of US Solar-System space probes between 1958 and 1978. The probes *Pioneer 4–9* went into solar orbit to monitor the Sun's activity during the 1960s and early 1970s. *Pioneer 5*, launched in 1960, was the first of a series to study the solar wind between the planets. *Pioneer 10*, launched in March 1972, was the first probe to reach Jupiter (Dec 1973) and to leave the Solar System (1983). *Pioneer 11*, launched in April 1973, passed Jupiter in Dec 1974, and was the first probe to reach Saturn (Sept 1979), before also leaving the Solar System.

Pioneer 10 and *11* carry plaques containing messages from Earth in case they are found by other civilizations among the stars. Pioneer Venus probes were launched in May and Aug 1978. One orbited Venus, and the other dropped three probes onto the surface. The orbiter finally burned up in the atmosphere of Venus in 1992. In 1992 *Pioneer 10* was more than 8 billion km/4.4 billion mi from the Sun. Both it and *Pioneer 11* were still returning data measurements of starlight intensity to Earth.

Pioneer 1, *2*, and *3*, launched in 1958, were intended Moon probes, but *Pioneer 2*'s launch failed, and *1* and *3* failed to reach their target, although they did measure the ◊Van Allen radiation belts. *Pioneer 4* began to orbit the Sun after passing the Moon.

pipette device for the accurate measurement of a known volume of liquid, usually for transfer from one container to another, used in chemistry and biology laboratories.

A pipette is a glass tube, often with an enlarged bulb, which is calibrated in one or more positions, or it may be a plastic device with an adjustable plunger, fitted with one or more disposable plastic tips.

Pisces inconspicuous zodiac constellation, mainly in the northern hemisphere between ◊Aries and ◊Aquarius, near ◊Pegasus. It is represented as two fish tied together by their tails. The Circlet, a delicate ring of stars, marks the head of the western fish in Pisces. The constellation contains the **vernal equinox**, the point at which the Sun's path around the sky (the *ecliptic*) crosses the celestial equator (see ◊celestial sphere). The Sun reaches this point around 21 March each year as it passes through Pisces from mid-March to late April. In astrology, the dates for Pisces are between about 19 Feb and 20 March (see ◊precession).

Piscis Austrinus or *Southern Fish* constellation of the southern hemisphere near ◊Capricornus. Its brightest star is the first-magnitude ◊Fomalhaut.

pistil general term for the female part of a flower, either referring to one single ◊carpel or a group of several fused carpels.

pistol any small firearm designed to be fired with one hand. Pistols have been in use from the early 15th century.

The problem of firing more than once without reloading was tackled by using many combinations of multiple barrels, both stationary and revolving. A breech-loading, multichambered revolver from as early as 1650 still survives; the first practical solution, however, was Samuel Colt's six-gun in 1847. Behind a single barrel, a short six-chambered cylinder was rotated by cocking the hammer to bring a fresh round of ammunition into the firing position. The automatic pistol, operated by gas or recoil, was introduced in Germany in the 1890s. Both revolvers and automatics remain in widespread military use.

piston barrel-shaped device used in reciprocating engines (steam, petrol, diesel oil) to harness power. Pistons are driven up and down in cylinders by expanding steam or hot gases. They pass on their motion via a connecting rod and crank to a crankshaft, which turns the driving wheels. In a pump or compressor, the role of the piston is reversed, being used to move gases and liquids. See also ◊internal-combustion engine.

pitch in chemistry, a black, sticky substance, hard when cold, but liquid when hot, used for waterproofing, roofing, and paving. It is made by the destructive distillation of wood or coal tar, and has been used since antiquity for caulking wooden ships.

pitch in mechanics, the distance between the adjacent threads of a screw or bolt. When a screw is turned through one full turn it moves a distance equal to the pitch of its thread. A screw thread is a simple type of machine, acting like a rolled-up inclined plane, or ramp (as may be illustrated by rolling a long paper triangle around a pencil). A screw has a ◊mechanical advantage greater than one.

pitchblende or *uraninite* brownish-black mineral, the major constituent of uranium ore, consisting mainly of uranium oxide (UO_2). It also contains some lead (the final, stable product of uranium decay) and variable amounts of most of the naturally occurring radioactive elements, which are products of either the decay or the fissioning of uranium isotopes. The uranium yield is 50–80%; it is also a source of radium, polonium, and actinium. Pitchblende was first studied by Pierre and Marie Curie, who found radium and polonium in its residues in 1898.

Pitot tube instrument that measures fluid (gas and liquid) flow. It is used to measure the speed of aircraft, and works by sensing pressure differences in different directions in the airstream.

It was invented in the 1730s by the French scientist Henri Pitot (1695–1771).

pituitary gland major ◊endocrine gland of vertebrates, situated in the centre of the brain. It is attached to the ◊hypothalamus by a stalk. The pituitary consists of two lobes. The posterior lobe is an extension of the hypothalamus, and is in effect nervous tissue. It stores two hormones synthesized in the hypothalamus: ◊ADH and

◊oxytocin. The anterior lobe secretes six hormones, some of which control the activities of other glands (thyroid, gonads, and adrenal cortex); others are direct-acting hormones affecting milk secretion and controlling growth.

pixel (derived from *picture element*) single dot on a computer screen. All screen images are made up of a collection of pixels, with each pixel being either off (dark) or on (illuminated, possibly in colour). The number of pixels available determines the screen's resolution. Typical resolutions of microcomputer screens vary from 320 × 200 pixels to 640 × 480 pixels, but screens with 1,024 × 768 pixels are now common for high-quality graphic (pictorial) displays.

placebo any harmless substance, often called a 'sugar pill', that has no active ingredient, but may nevertheless bring about improvement in the patient's condition.

The use of placebos in medicine is limited to drug trials, where a placebo is given alongside the substance being tested, to compare effects. The 'placebo effect', first named in 1945, demonstrates the control mind exerts over matter, bringing changes in blood pressure, perceived pain, and rates of healing. Recent research points to the release of certain neurotransmitters in the production of the placebo effect.

placenta organ that attaches the developing ◊embryo or ◊fetus to the ◊uterus in placental mammals (mammals other than marsupials, platypuses, and echidnas). Composed of maternal and embryonic tissue, it links the blood supply of the embryo to the blood supply of the mother, allowing the exchange of oxygen, nutrients, and waste products. The two blood systems are not in direct contact, but are separated by thin membranes, with materials diffusing across from one system to the other. The placenta also produces hormones that maintain and regulate pregnancy. It is shed as part of the afterbirth.

It is now understood that a variety of materials, including drugs and viruses, can pass across the placental membrane. HIV, the virus that causes ◊AIDS, can be transmitted in this way.

The tissue in plants that joins the ovary to the ovules is also called a placenta.

placer deposit detrital concentration of an economically important mineral, such as gold, but also other minerals

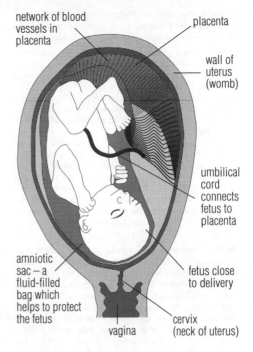

placenta The placenta is a disc-shaped organ about 25 cm/10 in in diameter and 3 cm/1 in thick. It is connected to the fetus by the umbilical cord.

such as cassiterite, chromite, and platinum metals. The mineral grains become concentrated during transport by water or wind because they are more dense than other detrital minerals such as quartz, and (like quartz) they are relatively resistant to chemical breakdown. Examples are the Witwatersrand gold deposits of South Africa, which are gold- and uranium-bearing conglomerates laid down by ancient rivers, and the placer tin deposits of the Malay Peninsula.

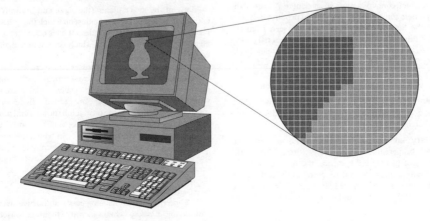

pixel

plage in astronomy, a bright patch in the ◊chromosphere above a group of ◊sunspots, occasionally seen on images of the Sun in hydrogen light.

plague term applied to any epidemic disease with a high mortality rate, but it usually refers to the bubonic plague. This is a disease transmitted by fleas (carried by the black rat) which infect the sufferer with the bacillus *Yersinia pestis*. An early symptom is swelling of lymph nodes, usually in the armpit and groin; such swellings are called 'buboes'. It causes virulent blood poisoning and the death rate is high.

Rarer but more virulent forms of plague are **septicaemic** and **pneumonic**; both still exert a formidable mortality. Outbreaks of plague still occur, mostly in poor countries, but never to the extent seen in the late Middle Ages. According to a World Health Organization report published in 1996, the incidence of plague is on the increase. It was reported in 13 states of the USA between 1984 and 1994, in comparison with just 3 states in the 1940s.

plain or **grassland** land, usually flat, upon which grass predominates. The plains cover large areas of the Earth's surface, especially between the deserts of the tropics and the rainforests of the Equator, and have rain in one season only. In such regions the climate belts move north and south during the year, bringing rainforest conditions at one time and desert conditions at another. Temperate plains include the North European Plain, the High Plains of the USA and Canada, and the Russian Plain, also known as the steppe.

Planck's constant in physics, a fundamental constant (symbol *h*) that relates the energy (*E*) of one quantum of electromagnetic radiation (the smallest possible 'packet' of energy; see ◊quantum theory) to the frequency (*f*) of its radiation by $E = hf$.

Its value is 6.6261×10^{-34} joule seconds.

planet large celestial body in orbit around a star, composed of rock, metal, or gas. There are nine planets in the ◊Solar System: Mercury, Venus, Earth, Mars, Jupiter, Saturn, Neptune, Uranus, and Pluto. The inner four, called the **terrestrial planets**, are small and rocky, and include the planet Earth. The outer planets, with the exception of Pluto, are called the **major planets**, and consist of large balls of rock, liquid, and gas; the largest is Jupiter, which contains a mass equivalent to 70% of all the other planets combined. Planets do not produce light, but reflect the light of their parent star. As seen from the Earth, all the historic planets are conspicuous naked-eye objects moving in looped paths against the stellar background. The size of these loops, which are caused by the Earth's own motion round the Sun, are inversely proportional to the planet's distance from the Earth.

planetarium optical projection device by means of which the motions of stars and planets are reproduced on a domed ceiling representing the sky.

planetary embryo in astronomy, one of numerous massive bodies thought to have formed from the accretion of planetesimals during the formation of the Solar System. Embryos in the region of the Earth's orbit would have been about 1,023 kg/ 2,251 lb in mass, and about 10–100 of them would have coalesced to make the Earth.

planetary nebula shell of gas thrown off by a star at the end of its life. Planetary nebulae have nothing to do with planets. They were named by William Herschel, who thought their rounded shape resembled the disc of a planet. After a star such as the Sun has expanded to become a ◊red giant, its outer layers are ejected into space to form a planetary nebula, leaving the core as a ◊white dwarf at the centre.

planimeter simple integrating instrument for measuring the area of a regular or irregular plane surface. It consists of two hinged arms: one is kept fixed and the other is traced around the boundary of the area. This actuates a small graduated wheel; the area is calculated from the wheel's change in position.

planisphere in astronomy, a graphical device for determining the ◊aspect of the sky for any date and time in the year. It consists of two discs mounted concentrically so that the upper disc, which has an aperture corresponding to the horizon of the observer, can rotate over the lower disc, which is printed with a map of the sky centred on the north or south celestial pole. In use, the observer aligns the time of day marked around the edge of the upper disc with the date marked around the edge of the lower disc. The aperture then shows which stars are above the horizon.

plankton small, often microscopic, forms of plant and animal life that live in the upper layers of fresh and salt water, and are an important source of food for larger animals. Marine plankton is concentrated in areas where rising currents bring mineral salts to the surface.

Pfiesteria piscicida is the only predatory phytoplankton. It stuns its prey by producing a powerful toxin that also causes the deaths of nearby fish and may be harmful to humans exposed to it.

plant organism that carries out ◊photosynthesis, has cellulose cell walls and complex cells, and is immobile. A few parasitic plants have lost the ability to photosynthesize but are still considered to be plants.

Plants are ◊autotrophs, that is, they make carbohydrates from water and carbon dioxide, and are the primary producers in all food chains, so that all animal life is dependent on them. They play a vital part in the carbon cycle, removing carbon dioxide from the atmosphere and generating oxygen. The study of plants is known as ◊botany.

Many of the lower plants (the algae and bryophytes) consist of a simple body, or thallus, on which the organs of reproduction are borne. Simplest of all are the threadlike algae, for example *Spirogyra*, which consist of a chain of cells.

The seaweeds (algae) and mosses and liverworts (bryophytes) represent a further development, with simple, multicellular bodies that have specially modified areas in which the reproductive organs are carried. Higher in the morphological scale are the ferns, club mosses, and horsetails (pteridophytes). Ferns produce leaflike fronds bearing sporangia on their undersurface in which the spores are carried. The spores are freed and germinate to produce small independent bodies carrying the sexual organs; thus the fern, like other pteridophytes and some seaweeds, has two quite separate generations in its life cycle (see ◊alternation of generations).

The pteridophytes have special supportive water-conducting tissues, which identify them as vascular plants, a group which includes all seed plants, that is the

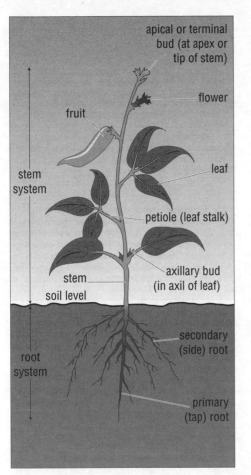

apical or terminal bud (at apex or tip of stem)

flower

fruit

leaf

stem system

petiole (leaf stalk)

stem

axillary bud (in axil of leaf)

soil level

secondary (side) root

root system

primary (tap) root

plant

diatoms, dinoflagellates, fungi, and slime moulds, but these are not now thought of as plants. The groups that are always classified as plants are the bryophytes (mosses and liverworts), pteridophytes (ferns, horsetails, and club mosses), gymnosperms (conifers, yews, cycads, and ginkgos), and angiosperms (flowering plants). The angiosperms are split into monocotyledons (for example, orchids, grasses, lilies) and dicotyledons (for example, oak, buttercup, geranium, and daisy).

The basis of plant classification was established by the Swedish naturalist Carolus Linnaeus. Among the angiosperms, it is largely based on the number and arrangement of the flower parts.

The unicellular algae, such as *Chlamydomonas*, are often now put with the protists (single-celled organisms) instead of the plants. Some classification schemes even classify the multicellular algae (seaweeds and freshwater weeds) in a new kingdom, the Protoctista, along with the protists.

plant hormone substance produced by a plant that has a marked effect on its growth, flowering, leaf fall, fruit ripening, or some other process. Examples include ◊auxin, ◊gibberellin, ◊ethylene, and ◊cytokinin.

Unlike animal hormones, these substances are not produced by a particular area of the plant body, and they may be less specific in their effects. It has therefore been suggested that they should not be described as hormones at all.

plant propagation production of plants. Botanists and horticulturalists can use a wide variety of means for propagating plants. There are the natural techniques of ◊vegetative reproduction, together with ◊cuttings, ◊grafting, and ◊micropropagation. The range is wide because most plant tissue, unlike animal tissue, can give rise to the complete range of tissue types within a particular species.

plaque any abnormal deposit on a body surface, especially the thin, transparent film of sticky protein (called mucin) and bacteria on tooth surfaces. If not removed, this film forms tartar (calculus), promotes tooth decay, and leads to gum disease. Another form of plaque is a deposit of fatty or fibrous material in the walls of blood vessels causing ◊atheroma.

plasma in biology, the liquid component of the ◊blood.

plasma in physics, an ionized gas produced at extremely high temperatures, as in the Sun and other stars, which contains positive and negative charges in equal numbers. It is a good electrical conductor. In thermonuclear reactions the plasma produced is confined through the use of magnetic fields.

plasmapheresis technique for acquiring plasma from blood. Blood is withdrawn from the patient and separated into its components (plasma and blood cells) by centrifugal force in a continuous-flow cell separator. Once separated, the plasma is available for specific treatments. The blood cells are transfused back into the patient.

plasmid small, mobile piece of ◊DNA found in bacteria and used in ◊genetic engineering. Plasmids are separate from the bacterial chromosome but still multiply during cell growth. Their size ranges from 3% to 20% of the size of the chromosome. There is usually only one copy of a single plasmid per cell, but occasionally several are found. Some plasmids carry 'fertility genes' that enable them to move from one bacterium to another and transfer genetic

gymnosperms (conifers, yews, cycads, and ginkgos) and the angiosperms (flowering plants).

The seed plants are the largest group, and structurally the most complex. They are usually divided into three parts: root, stem, and leaves. Stems grow above or below ground. Their cellular structure is designed to carry water and salts from the roots to the leaves in the ◊xylem, and sugars from the leaves to the roots in the ◊phloem. The leaves manufacture the food of the plant by means of photosynthesis, which occurs in the ◊chloroplasts they contain. Flowers and cones are modified leaves arranged in groups, enclosing the reproductive organs from which the fruits and seeds result.

plantation large farm or estate where commercial production of one crop – such as rubber (in Malaysia), palm oil (in Nigeria), or tea (in Sri Lanka) – is carried out. Plantations are usually owned by large companies, often multinational corporations, and run by an estate manager. Many plantations were established in countries under colonial rule, using slave labour.

plant classification taxonomy or classification of plants. Originally the plant kingdom included bacteria,

information between strains. Plasmid genes determine a wide variety of bacterial properties including resistance to antibiotics and the ability to produce toxins.

plaster of Paris form of calcium sulphate, obtained from gypsum; it is mixed with water for making casts and moulds.

plastic any of the stable synthetic materials that are fluid at some stage in their manufacture, when they can be shaped, and that later set to rigid or semi-rigid solids. Plastics today are chiefly derived from petroleum. Most are polymers, made up of long chains of identical molecules.

Since plastics have afforded an economical replacement for ivory in the manufacture of piano keys and billiard balls, the industrial chemist may well have been responsible for the survival of the elephant.

Most plastics cannot be broken down by microorganisms, so cannot easily be disposed of. Incineration leads to the release of toxic fumes, unless carried out at very high temperatures.

plastic surgery surgical speciality concerned with the repair of congenital defects and the reconstruction of tissues damaged by disease or injury, including burns. If a procedure is undertaken solely for reasons of appearance, for example, the removal of bags under the eyes or a double chin, it is called *cosmetic surgery*.

plastid general name for a cell ◊organelle of plants that is enclosed by a double membrane and contains a series of internal membranes and vesicles. Plastids contain ◊DNA and are produced by division of existing plastids. They can be classified into two main groups: the *chromoplasts*, which contain pigments such as carotenes and chlorophyll, and the *leucoplasts*, which are colourless; however, the distinction between the two is not always clear-cut.

◊Chloroplasts are the major type of chromoplast. They contain chlorophyll, are responsible for the green coloration of most plants, and perform ◊photosynthesis. Other chromoplasts give flower petals and fruits their distinctive colour. Leucoplasts are food-storage bodies and include amyloplasts, found in the roots of many plants, which store large amounts of starch.

plate or tectonic plate, one of several sections of ◊lithosphere approximately 100 km/60 mi thick and at least 200 km/120 mi across which together comprise the outermost layer of the Earth like the pieces of the cracked surface of a hard-boiled egg.

The plates are made up of two types of crustal material: oceanic crust (sima) and continental crust (sial), both of which are underlain by a solid layer of ◊mantle. Dense *oceanic crust* lies beneath Earth's oceans and consists largely of ◊basalt. *Continental crust*, which underlies the continents and their continental shelves, is thicker, less dense and consists of rocks rich in silica and aluminium.

Due to convection in the Earth's mantle (see ◊plate tectonics) these pieces of lithosphere are in motion, riding on a more plastic layer of the mantle, called the aesthenosphere. Mountains, volcanoes, earthquakes, and other geological features and phenomena all come about as a result of interaction between the plates.

plateau elevated area of fairly flat land, or a mountainous region in which the peaks are at the same height. An *intermontane plateau* is one surrounded by mountains. A *piedmont plateau* is one that lies between the mountains and low-lying land. A *continental plateau* rises abruptly from low-lying lands or the sea. Examples are the Tibetan Plateau and the Massif Central in France.

platelet tiny disc-shaped structure found in the blood, which helps it to clot. Platelets are not true cells, but membrane-bound cell fragments without nuclei that bud off from large cells in the bone marrow.

They play a vital role in blood clotting as they release blood clotting factors at the site of a cut. Over twelve clotting factors have been discovered and they produce a complex series of reactions which ultimately leads to fibrinogen, the inactive blood sealant always found in the plasma, being converted into fibrin. Fibrin aggregates into threads which form the fabric of a blood clot.

plate tectonics theory formulated in the 1960s to explain the phenomena of ◊continental drift and seafloor spreading, and the formation of the major physical features of the Earth's surface. The Earth's outermost layer is regarded as a jigsaw of rigid major and minor plates up to 100 km/62 mi thick, which move relative to each other, probably under the influence of convection currents in the mantle beneath. Major landforms occur at the margins of the plates, where plates are colliding or moving apart – for example, volcanoes, fold mountains, ocean trenches, and ocean ridges. The rate of drift is at most 15 cm/6 in per year.

Constructive margin Where two plates are moving apart from each other, molten rock from the mantle wells up in the space left between the plates and hardens to form new crust, usually in the form of an ocean ridge (such as the ◊Mid-Atlantic Ridge). The newly formed crust accumulates on either side of the ocean ridge, causing the seafloor to spread – for example, the floor of the Atlantic Ocean is growing by 5 cm/2 in each year because of the upwelling of new material at the Mid-Atlantic Ridge.

Destructive margin Where two plates are moving towards each other, the denser of the two plates may be forced under the other into a region called the subduction zone. The descending plate melts to form a body of magma, which may then rise to the surface through cracks and faults to form volcanoes. If the two plates consist of more buoyant continental crust, subduction does not occur. Instead, the crust crumples gradually to form ranges of young fold mountains, such as the Himalayas in Asia, the Andes in South America, and the Rockies in North America.

Conservative margin Sometimes two plates will slide past each other – an example is the San Andreas Fault, California, where the movement of the plates sometimes takes the form of sudden jerks, causing the earthquakes common in the San Francisco–Los Angeles area. Most of the earthquake and volcano zones of the world are, in fact, found in regions where two plates meet or are moving apart.

The concept of continental drift was first put forward in 1915 by the German geophysicist Alfred Wegener; plate tectonics was formulated by Canadian geophysicist John Tuzo Wilson in 1965 and has gained widespread acceptance among earth scientists.

In 1995 US and French geophysicists produced the first direct evidence that the Indo-Australian plate has split in two (in the middle of the Indian Ocean, just south

sea floor spreading

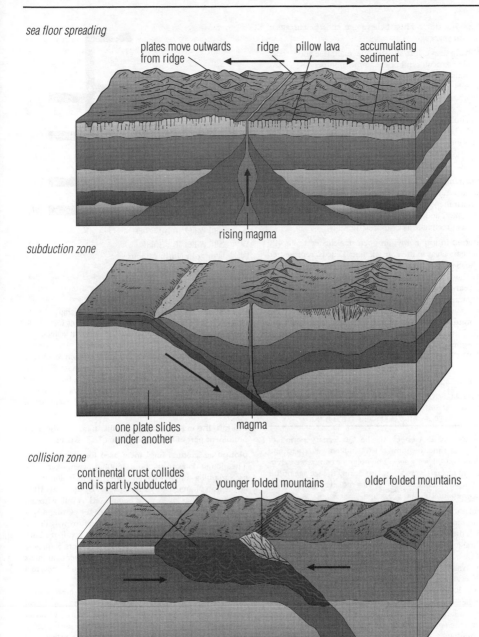

plates move outwards from ridge   ridge   pillow lava   accumulating sediment

rising magma

subduction zone

one plate slides under another   magma

collision zone

continental crust collides and is partly subducted   younger folded mountains   older folded mountains

plate tectonics The three main types of action in plate tectonics. (top) Sea-floor spreading. The up-welling of magma forces apart the crust plates, producing new crust at the joint. Rapid extrusion of magma produces a domed ridge; more gentle spreading produces a central valley. (middle) The drawing downwards of an oceanic plate beneath a continent produces a range of volcanic fold mountains parallel to the plate edge. (bottom) Collision of continental plates produces immense fold mountains, such as the Himalayas. Younger mountains are found near the coast with older ranges inland. The plates of the Earth's lithosphere are always changing in size and shape as material is added at constructive margins and removed at destructive margins. The process is extremely slow, but it means that the tectonic history of the Earth cannot be traced back further than about 200 million years.

of the Equator). They believe the split began about 8 million years ago.

platinum heavy, soft, silver-white, malleable and ductile, metallic element, symbol Pt, atomic number 78, relative atomic mass 195.09. It is the first of a group of six metallic elements (platinum, osmium, iridium, rhodium, ruthenium, and palladium) that possess similar traits, such as resistance to tarnish, corrosion, and attack by acid, and that often occur as free metals (◊native metals). They often occur in natural alloys with each other, the commonest of which is osmiridium. Both pure and as an alloy, platinum is used in dentistry, jewellery, and as a catalyst.

playa temporary lake in a region of interior drainage. Such lakes are common features in arid desert basins fed by intermittent streams. The streams bring dissolved salts to the lakes, and when the lakes shrink during dry spells, the salts precipitate as evaporite deposits.

Pleiades in astronomy, an open star cluster about 400 light years away in the constellation Taurus, represented as the Seven Sisters of Greek mythology. Its brightest stars (highly luminous, blue-white giants only a few million years old) are visible to the naked eye, but there are many fainter ones.

It is a young cluster, and the stars of the Pleiades are still surrounded by traces of the reflection ◊nebula from which they formed, visible on long-exposure photographs.

Pleiades

The Pleiades, or Seven Sisters, cluster of stars seems to contain seven stars, but with the aid of a telescope about 400 stars can be seen. There may indeed be as many as 3,000 stars in the group.

Pleistocene first epoch of the Quaternary period of geological time, beginning 1.64 million years ago and ending 10,000 years ago. The polar ice caps were extensive and glaciers were abundant during the ice age of this period, and humans evolved into modern *Homo sapiens sapiens* about 100,000 years ago.

pleurisy inflammation of the pleura, the thin, secretory membrane that covers the lungs and lines the space in which they rest. Pleurisy is nearly always due to bacterial or viral infection, but may also be a complication of other diseases.

Normally the two lung surfaces move easily on one another, lubricated by small quantities of fluid. When the pleura is inflamed, the surfaces may dry up or stick together, making breathing difficult and painful. Alternatively, a large volume of fluid may collect in the pleural cavity, the space between the two surfaces, and pus may accumulate.

Plimsoll line loading mark painted on the hull of merchant ships, first suggested by English politician Samuel Plimsoll. It shows the depth to which a vessel may be safely (and legally) loaded.

Pliocene fifth and last epoch of the Tertiary period of geological time, 5.2–1.64 million years ago. The earliest hominid, the humanlike ape *Australopithecines*, evolved in Africa.

plotter or *graph plotter* device that draws pictures or diagrams under computer control.

TF — Tropical fresh water
F — Fresh water
T — Tropical salt water
S — Salt water in Summer
W — Salt water in Winter
WNA — Winter in North Atlantic
LR — Lloyd's Register

Plimsoll line The Plimsoll line on the hull of a ship indicates maximum safe loading levels for sea or fresh water, winter or summer, in tropical or northern waters.

Plotters are often used for producing business charts, architectural plans, and engineering drawings. *Flatbed plotters* move a pen up and down across a flat drawing surface, whereas *roller plotters* roll the drawing paper past the pen as it moves from side to side. *See illustration opposite.*

Plough, the in astronomy, a popular name for the most prominent part of the constellation ◊Ursa Major.

plough agricultural implement used for tilling the soil. The plough dates from about 3500 BC, when oxen were used to pull a simple wooden blade, or ard. In about 500 BC the iron ploughshare came into use. By about AD 1000 horses as well as oxen were being used to pull wheeled ploughs, equipped with a ploughshare for cutting a furrow, a blade for forming the walls of the furrow (called a coulter), and a mouldboard to turn the furrow. In the 18th century an innovation introduced by Robert Ransome (1753–1830), led to a reduction in the number of animals used to draw a plough: from 8–12 oxen, or 6 horses, to a 2- or 4-horse plough.

Steam ploughs came into use in some areas in the 1860s, superseded half a century later by tractor-drawn ploughs. The modern plough consists of many 'bottoms', each comprising a curved ploughshare and angled mouldboard. The bottom is designed so that it slices into the ground and turns the soil over.

plucking in earth science, a process of glacial erosion. Water beneath a glacier will freeze fragments of loose rock to the base of the ice. When the ice moves, the rock fragment is 'plucked' away from the underlying bedrock. Plucking is thought to be responsible for the formation of steep, jagged slopes such as the backwall of the corrie and the downslope-side of the roche moutonnée.

plumbago alternative name for the mineral ◊graphite.

plumule part of a seed embryo that develops into the

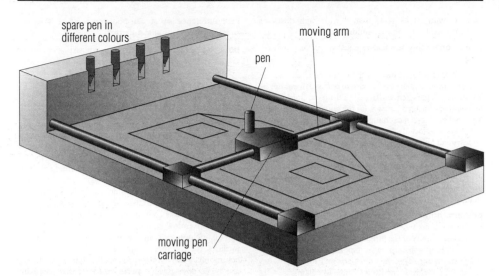

plotter A flatbed plotter, which may be used to produce plans, graphs, and other drawings. The moving arm, holding a pen of the appropriate colour, travels over the surface of the paper.

shoot, bearing the first true leaves of the plant. In most seeds, for example the sunflower, the plumule is a small conical structure without any leaf structure. Growth of the plumule does not occur until the ◊cotyledons have grown above ground. This is ◊epigeal germination. However, in seeds such as the broad bean, a leaf structure is visible on the plumule in the seed. These seeds develop by the plumule growing up through the soil with the cotyledons remaining below the surface. This is known as ◊hypogeal germination.

Pluto in astronomy, the smallest and, usually, outermost planet of the Solar System. The existence of Pluto was predicted by calculation by Percival Lowell and the planet was located by Clyde Tombaugh in 1930. Its highly elliptical orbit occasionally takes it within the orbit of Neptune, as in 1979–99. Pluto has a mass about 0.002 of that of Earth.
   Mean distance from the Sun 5.8 billion km/3.6 billion mi.
   Equatorial diameter 2,300 km/1,438 mi.
   Rotation period 6.39 Earth days.
   Year 248.5 Earth years.
   Atmosphere thin atmosphere with small amounts of methane gas.
   Surface low density, composed of rock and ice, primarily frozen methane. There is an ice cap at Pluto's north pole.
   Satellites one moon, Charon.

plutonic rock igneous rock derived from magma that has cooled and solidified deep in the crust of the Earth; granites and gabbros are examples of plutonic rocks.

plutonium silvery-white, radioactive, metallic element of the ◊actinide series, symbol Pu, atomic number 94, relative atomic mass 239.13. It occurs in nature in minute quantities in ◊pitchblende and other ores, but is produced in quantity only synthetically. It has six allotropic forms (see ◊allotropy) and is one of three fissile elements (elements capable of splitting into other elements – the

others are thorium and uranium). The element has awkward physical properties and is the most toxic substance known.

plywood manufactured panel of wood widely used in building. It consists of several thin sheets, or plies, of wood, glued together with the grain (direction of the wood fibres) of one sheet at right angles to the grain of the adjacent plies. This construction gives plywood equal strength in every direction.

Plywood

Many people know that Alfred Nobel, founder of the prizes that bear his name, was the inventor of dynamite. However, it is less well known that his father, Immanuel Nobel, was the inventor of plywood.

pneumatic drill drill operated by compressed air, used in mining and tunnelling, for drilling shot holes (for explosives), and in road repairs for breaking up pavements. It contains an air-operated piston that delivers hammer blows to the drill bit many times a second. The French engineer Germain Sommeiller (1815–1871) developed the pneumatic drill in 1861 for tunnelling in the Alps.

pneumatophore erect root that rises up above the soil or water and promotes ◊gas exchange. Pneumatophores, or breathing roots, are formed by certain swamp-dwelling trees, such as mangroves, since there is little oxygen available to the roots in waterlogged conditions. They have numerous pores or ◊lenticels over their surface, allowing gas exchange.

pneumoconiosis disease of the lungs caused by an accumulation of dust, especially from coal, asbestos, or silica. Inhaled particles make the lungs gradually fibrous and the victim has difficulty breathing.

pneumocystis pneumonia in medicine, type of ◊pneumonia due to infection with *Pneumocystis pneumoniae*. It occurs in immunosuppressed or severely debilitated patients and it is the most common cause of pneumonia in patients with AIDS. It is treated with high doses of co-trimoxazole or with pentamidine.

pneumonectomy surgical removal of all or part of a lung.

pneumonia inflammation of the lungs, generally due to bacterial or viral infection but also to particulate matter or gases. It is characterized by a build-up of fluid in the alveoli, the clustered air sacs (at the ends of the air passages) where oxygen exchange takes place.

Symptoms include fever and pain in the chest. With widespread availability of antibiotics, infectious pneumonia is much less common than it was. However, it remains a dire threat to patients whose immune systems are suppressed (including transplant recipients and AIDS and cancer victims) and to those who are critically ill or injured.

pneumothorax the presence of air in the pleural cavity, between a lung and the chest wall. It may be due to a penetrating injury of the lung or to lung disease, or it may occur without apparent cause (spontaneous pneumothorax) in an otherwise healthy person. Prevented from expanding normally, the lung is liable to collapse.

p–n junction diode in electronics, another name for ◊semiconductor diode.

pod in botany, a type of fruit that is characteristic of legumes (plants belonging to the Leguminosae family), such as peas and beans. It develops from a single ◊carpel and splits down both sides when ripe to release the seeds.

In certain species the seeds may be ejected explosively due to uneven drying of the fruit wall, which sets up tensions within the fruit. In agriculture, the name 'legume' is used for the crops of the pea and bean family. 'Grain legume' refers to those that are grown mainly for their dried seeds, such as lentils, chick peas, and soya beans.

podzol or **podsol** type of light-coloured soil found predominantly under coniferous forests and on moorlands in cool regions where rainfall exceeds evaporation. The constant downward movement of water leaches nutrients from the upper layers, making podzols poor agricultural soils.

The leaching of minerals such as iron, lime, and alumina leads to the formation of a bleached zone, often also depleted of clay.

These minerals can accumulate lower down the soil profile to form a hard, impermeable layer which restricts the drainage of water through the soil.

poikilothermy the condition in which an animal's body temperature is largely dependent on the temperature of the air or water in which it lives. It is characteristic of all animals except birds and mammals, which maintain their body temperatures by ◊homeothermy (they are 'warm-blooded').

Poikilotherms have behavioural means of temperature control; they can warm themselves up by basking in the sun, or shivering, and can cool themselves down by sheltering from the Sun under a rock or by bathing in water.

Poikilotherms are often referred to as 'cold-blooded animals', but this is not really correct: their internal temperatures, regulated by behavioural means, are often as high as those of birds and mammals during the times they need to be active for feeding and reproductive purposes, and may be higher, for example in very hot climates. The main difference is that their body temperatures fluctuate more than those of homeotherms.

point in geometry, a basic element, whose position in the Cartesian system may be determined by its ◊coordinates.

Mathematicians have had great difficulty in defining the point, as it has no size, and is only the place where two lines meet. According to the Greek mathematician Euclid, (i) a point is that which has no part; (ii) the straight line is the shortest distance between two points.

point-of-sale terminal (POS terminal) computer terminal used in shops to input and output data at the point where a sale is transacted; for example, at a supermarket checkout. The POS terminal inputs information about the identity of each item sold, retrieves the price and other details from a central computer, and prints out a fully itemized receipt for the customer. It may also input sales data for the shop's computerized stock-control system.

A POS terminal typically has all the facilities of a normal till, including a cash drawer and a sales register, plus facilities for the direct capture of sales information – commonly, a laser scanner for reading bar codes. It may also be equipped with a device to read customers' bank cards, so that payment can be transferred electronically from the customers' bank accounts to the shop's (see ◊EFTPOS).

poise c.g.s. unit (symbol P) of dynamic ◊viscosity (the property of liquids that determines how readily they flow). It is equal to one dyne-second per square centimetre. For most liquids the centipoise (one hundredth of a poise) is used. Water at 20°C/68°F has a viscosity of 1.002 centipoise.

poison or **toxin** any chemical substance that, when introduced into or applied to the body, is capable of injuring health or destroying life.

The liver removes some poisons from the blood. The majority of poisons may be divided into **corrosives**, such as sulphuric, nitric, and hydrochloric acids; **irritants**, including arsenic and copper sulphate; **narcotics** such as opium, and carbon monoxide; and **narcotico-irritants** from any substances of plant origin including carbolic acid and tobacco.

Corrosives all burn and destroy the parts of the body with which they come into contact; irritants have an irritating effect on the stomach and bowels; narcotics affect the brainstem and spinal cord, inducing a stupor; and narcotico-irritants can cause intense irritations and finally act as narcotics.

polar coordinates in mathematics, a way of defining the position of a point in terms of its distance r from a fixed point (the origin) and its angle θ to a fixed line or axis. The coordinates of the point are (r, θ).

Often the angle is measured in ◊radians, rather than degrees. The system is useful for defining positions on a plane in programming the operations of, for example, computer-controlled cloth-and metal-cutting machines.

polarimetry in astronomy, any technique for measuring the degree of polarization of radiation from stars, galaxies, and other objects.

Polaris or **Pole Star** or **North Star** bright star closest to the north celestial pole, and the brightest star in the constellation ◊Ursa Minor. Its position is indicated by the 'pointers' in ◊Ursa Major. Polaris is a yellow ◊supergiant about 500 light years away. It is also known as **Alpha Ursae Minoris**.

It currently lies within 1° of the north celestial pole; ◊precession (Earth's axial wobble) will bring Polaris closest to the celestial pole (less than 0.5° away) about AD 2100. Then its distance will start to increase, reaching 1° AD 2205 and 47° AD 28000. Other bright stars that have been, or will be close to the north celestial pole are Alpha Draconis (3000 BC), Gamma Cephei (AD 4000), Alpha Cephei (AD 7000), and ◊Vega (AD 14000).

polarized light light in which the electromagnetic vibrations take place in one particular plane. In ordinary (unpolarized) light, the electric fields vibrate in all planes perpendicular to the direction of propagation. After reflection from a polished surface or transmission through certain materials (such as Polaroid), the electric fields are confined to one direction, and the light is said to be **linearly polarized**. In **circularly polarized** and **elliptically polarized** light, the electric fields are confined to one direction, but the direction rotates as the light propagates. Polarized light is used to test the strength of sugar solutions, to measure stresses in transparent materials, and to prevent glare.

polarography electrochemical technique for the analysis of oxidizable and reducible compounds in solution. It involves the diffusion of the substance to be analyzed onto the surface of a small electrode, usually a bead of mercury, where oxidation or reduction occurs at an electric potential characteristic of that substance.

Polaroid camera instant-picture camera, invented by Edwin Land in the USA in 1947. The original camera produced black-and-white prints in about one minute. Modern cameras can produce black-and-white prints in a few seconds, and colour prints in less than a minute. An advanced model has automatic focusing and exposure.

It ejects a piece of film on paper immediately after the picture has been taken.

The film consists of layers of emulsion and colour dyes together with a pod of chemical developer. When the film is ejected the pod bursts and processing occurs in the light, producing a paper-backed print.

polar reversal change in polarity of the Earth's magnetic field. Like all magnets, the Earth's magnetic field has two opposing regions, or poles, one of attraction and one of repulsion, positioned approximately near the geographical North and South Poles. During a period of normal polarity the region of attraction corresponds with the North Pole. Today, a compass needle, like other magnetic materials, aligns itself parallel to the magnetizing force and points to the North pole. During a period of reversed polarity, the region of attraction would change to the South Pole and the needle of a compass would point south.

Studies of the magnetism retained in rocks at the time of their formation (like little compasses frozen in time) have shown that the polarity of the magnetic field has reversed repeatedly throughout geological time.

Polar reversals are a random process. Although the average time between reversals over the last ten million years has been 250,000 years, the rate of reversal has changed continuously over geological time. The most recent reversal was 700,000 years ago; scientists have no way of predicting when the next reversal will occur. The reversal process takes about a thousand years. Move-

ments of Earth's molten ◊core are thought to be responsible for the magnetic field and its polar reversals. Dating rocks using distinctive sequences of magnetic reversals is called palaeomagnetic stratigraphy.

polder area of flat reclaimed land that used to be covered by a river, lake, or the sea. Polders have been artificially drained and protected from flooding by building dykes. They are common in the Netherlands, where the total land area has been increased by nearly one-fifth since AD 1200. Such schemes as the Zuider Zee project have provided some of the best agricultural land in the country.

pole either of the geographic north and south points of the axis about which the Earth rotates. The geographic poles differ from the magnetic poles, which are the points towards which a freely suspended magnetic needle will point.

In 1985 the magnetic north pole was some 350 km/ 218 mi NW of Resolute Bay, Northwest Territories, Canada. It moves northwards about 10 km/6 mi each year, although it can vary in a day by about 80 km/50 mi from its average position. It is relocated every decade in order to update navigational charts.

It is thought that periodic changes in the Earth's core cause a reversal of the magnetic poles (see ◊polar reversal, ◊magnetic field). Many animals, including migrating birds and fish, are believed to orient themselves partly using the Earth's magnetic field. A permanent scientific base collects data at the South Pole.

Pole Star another name for ◊Polaris, the northern pole star. There is no bright star near the southern celestial pole.

polio (**poliomyelitis**) viral infection of the central nervous system affecting nerves that activate muscles. The disease used to be known as infantile paralysis. Two kinds of vaccine are available, one injected and one given by mouth. The Americas were declared to be polio-free by the Pan American Health Organization in Oct 1994. In 1997 the World Health Organization reported that cases of polio had dropped by nearly 90% since 1988 when the organization began its programme to eradicate the disease by the year 2000. Most cases remain in Africa and southeast Asia.

The polio virus is a common one and its effects are mostly confined to the throat and intestine, as with influenza or a mild digestive upset. There may also be muscle stiffness in the neck and back. Paralysis is seen in about 1% of cases, and the disease is life-threatening only if the muscles of the throat and chest are affected. Cases of this kind, once entombed in an 'iron lung', are today maintained on a respirator.

pollarding type of pruning whereby the young branches of a tree are severely cut back, about 2–4 m/6–12 ft above the ground, to produce a stumplike trunk with a rounded, bushy head of thin new branches. It is similar to ◊coppicing.

pollen the grains of ◊seed plants that contain the male gametes. In ◊angiosperms (flowering plants) pollen is produced within ◊anthers; in most ◊gymnosperms (cone-bearing plants) it is produced in male cones. A pollen grain is typically yellow and, when mature, has a hard outer wall. Pollen of insect-pollinated plants (see ◊pollination) is often sticky and spiny and larger than the

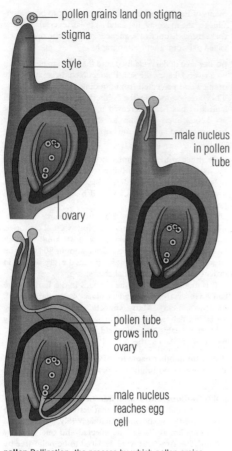

pollen grains land on stigma
stigma
style
male nucleus in pollen tube
ovary
pollen tube grows into ovary
male nucleus reaches egg cell

pollen Pollination, the process by which pollen grains transfer their male nuclei (gametes) to the ovary of a flower. (1) The pollen grains land on the stigma and (2) form a pollen tube that (3) grows down into the ovary. The male nuclei travel along the pollen tube.

smooth, light grains produced by wind-pollinated species.

The outer wall of pollen grains from both insect-pollinated and wind-pollinated plants is often elaborately sculptured with ridges or spines so distinctive that individual species or genera of plants can be recognized from their pollen. Since pollen is extremely resistant to decay, useful information on the vegetation of earlier times can be gained from the study of fossil pollen. The study of pollen grains is known as palynology.

pollen tube outgrowth from a pollen grain that grows towards the ◊ovule, following germination of the grain on the ◊stigma. In ◊angiosperms (flowering plants) the pollen tube reaches the ovule by growing down through the ◊style, carrying the male gametes inside. The gametes are discharged into the ovule and one fertilizes the egg cell.

pollination the process by which pollen is transferred from one plant to another. The male ◊gametes are contained in pollen grains, which must be transferred from the anther to the stigma in ◊angiosperms (flowering

plants), and from the male cone to the female cone in ◊gymnosperms (cone-bearing plants). Fertilization (not the same as pollination) occurs after the growth of the pollen tube to the ovary. Self-pollination occurs when pollen is transferred to a stigma of the same flower, or to another flower on the same plant; cross-pollination occurs when pollen is transferred to another plant. This involves external pollen-carrying agents, such as wind (see ◊anemophily), water (see ◊hydrophily), insects, birds (see ◊ornithophily), bats, and other small mammals.

Animal pollinators carry the pollen on their bodies and are attracted to the flower by scent, or by the sight of the petals. Most flowers are adapted for pollination by one particular agent only. Bat-pollinated flowers tend to smell of garlic, rotting vegetation, or fungus. Those that rely on animals generally produce nectar, a sugary liquid, or surplus pollen, or both, on which the pollinator feeds. Thus the relationship between pollinator and plant is an example of mutualism, in which both benefit. However, in some plants the pollinator receives no benefit (as in ◊pseudocopulation), while in others, nectar may be removed by animals that do not effect pollination.

polluter-pays principle the idea that whoever causes pollution is responsible for the cost of repairing any damage. The principle is accepted in British law but has in practice often been ignored; for example, farmers causing the death of fish through slurry pollution have not been fined the full costs of restocking the river.

pollution the harmful effect on the environment of by-products of human activity, principally industrial and agricultural processes – for example, noise, smoke, car emissions, chemical and radioactive effluents in air, seas, and rivers, pesticides, radiation, sewage (see ◊sewage disposal), and household waste. Pollution contributes to the ◊greenhouse effect. See also ◊air pollution.

Pollution control involves higher production costs for the industries concerned, but failure to implement adequate controls may result in irreversible environmental damage and an increase in the incidence of diseases such as cancer. Radioactive pollution results from inadequate ◊nuclear safety.

Transboundary pollution is when the pollution generated in one country affects another, for example as occurs with ◊acid rain. Natural disasters may also cause pollution; volcanic eruptions, for example, cause ash to be ejected into the atmosphere and deposited on land surfaces.

Pollux or *Beta Geminorum* the brightest star in the constellation ◊Gemini, and the 17th-brightest star in the sky. Pollux is a yellow star with a true luminosity 45 times that of the Sun. It is 35 light years away.

The first-magnitude Pollux and the second-magnitude ◊Castor, Alpha Geminorum, mark the heads of the Gemini twins. It is thought that the two stars may have changed their relative brightness, as Alpha is usually assigned to the brightest star in a constellation.

polonium radioactive, metallic element, symbol Po, atomic number 84, relative atomic mass 210. Polonium occurs in nature in small amounts and was isolated from ◊pitchblende. It is the element having the largest number of isotopes (27) and is 5,000 times as radioactive as radium, liberating considerable amounts of heat. It was the first element to have its radioactive properties recognized and investigated.

Pollution: Chronology

13th century	In York 58% of the population suffered from sinusitis as a result of the smoke and fumes.
1709	English industrialist Abraham Darby first smelted iron ore with coke, massively increasing industry's use of fossil fuels thereafter.
1852	Scottish chemist Robert Angus Smith first recorded unusually acidic rainfall in Manchester.
1856	The 'Year of the Great Stink' in London; the River Thames smelled so badly that parliament was evacuated.
1952	London was afflicted by severe smogs ('pea-soupers') which killed 4,000.
1962	US natural scientist Rachel Carson published *Silent Spring*, a treatise against the ecological dangers of pesticides.
1971	First international pollution compensation fund, the International Maritime Organization Oil Pollution Fund, established.
1984	A leak from a US-owned pesticide plant in Bhopal, India, killed 3,000, with another 30,000 seriously injured.
1985	A 'hole' in the stratospheric ozone layer over the Antarctic was first reported.
1986	Accident in the nuclear power plant at Chernobyl, Ukraine, contaminated large area of USSR. Fire at a chemical plant in Basel, Switzerland, released chemicals into the River Rhine, killing almost all living organisms in a 320 km stretch between Basel and Mainz.
1989	The tanker *Exxon Valdez* was wrecked off Alaska and discharged 37,000 tonnes of crude oil, causing substantial damage to bird and aquatic life and the Alaskan coast.
1991	Iraqi military discharged 7–8 million barrels of oil in Kuwait, nearly twice the size of the previous largest oil spill.
1992	UN Convention on Climate Change opened for signing.
1993	Wrecked tanker *Braer* discharged 85,000 tonnes of oil to the Scottish coast.
1994	First international deal in which states agreed to reduce pollutants by differing degrees according to the amount of damage they cause signed.
1995	US researchers studying voles around the devastated Chernobyl reactor have found mutation rates at much higher levels than in voles from outside the 'hot zone'. The Berlin Mandate, agreed unanimously at the climate conference April, committed industrial nations to the continuing reduction of greenhouse gas emissions after the year 2000, when the existing pact to stabilize emissions runs out.
1996	The supertanker *Sea Empress* ran aground off the southwest coast of Wales, spilling 72,000 tonnes of oil.
1997	Following forest clearance fires, smoke pollution in the city of Palangkaraya, Indonesia, reached levels of 7.5 mg per cu m (nearly 3 mg more than the London smog of 1952).

Polonium was isolated in 1898 from the pitchblende residues analyzed by French scientists Pierre and Marie Curie, and named after Marie Curie's native Poland.

polychlorinated biphenyl (PCB) any of a group of chlorinated isomers of biphenyl $(C_6H_5)_2$. They are dangerous industrial chemicals, valuable for their fire-resisting qualities. They constitute an environmental hazard because of their persistent toxicity. Since 1973 their use has been limited by international agreement.

polycystic kidney disease in medicine, an inherited disease in which the kidney contains many cysts. Continuing growth of the cysts results in the destruction of normal kidney tissue and kidney failure. Dialysis or kidney transplantation are the only methods of treating polycystic kidney disease

polyester synthetic resin formed by the ◊condensation of polyhydric alcohols (alcohols containing more than one hydroxyl group) with dibasic acids (acids containing two replaceable hydrogen atoms). Polyesters are thermosetting ◊plastics, used in making synthetic fibres, such as Dacron and Terylene, and constructional plastics. With glass fibre added as reinforcement, polyesters are used in car bodies and boat hulls.

polyethylene or **polyethene** polymer of the gas ethylene (technically called ethene, C_2H_4). It is a tough, white, translucent, waxy thermoplastic (which means it can be repeatedly softened by heating). It is used for packaging, bottles, toys, wood preservation, electric cable, pipes and tubing.

Polyethylene is produced in two forms: low-density polyethylene, made by high-pressure polymerization of ethylene gas, and high-density polyethylene, which is made at lower pressure by using catalysts. This form, first made in 1953 by German chemist Karl Ziegler, is more rigid at low temperatures and softer at higher temperatures than the low-density type. Polyethylene was first made in the 1930s at very high temperatures by ICI.

polygon in geometry, a plane (two-dimensional) figure with three or more straight-line sides. Common polygons have names which define the number of sides (for example, triangle, quadrilateral, pentagon).

These are all convex polygons, having no interior angle greater than 180°. The sum of the internal angles of a polygon having n sides is given by the formula $(2n - 4) \times 90°$; therefore, the more sides a polygon has, the larger the sum of its internal angles and, in the case of a convex polygon, the more closely it approximates to a circle.

polygraph technical name for a ◊lie detector.

polyhedron in geometry, a solid figure with four or more plane faces. The more faces there are on a polyhedron, the more closely it approximates to a sphere. Knowledge of the properties of polyhedra is needed in crystallography and stereochemistry to determine the shapes of crystals and molecules.

There are only five types of regular polyhedron (with all faces the same size and shape), as was deduced by early Greek mathematicians; they are the tetrahedron (four equilateral triangular faces), cube (six square faces), octahedron (eight equilateral triangles), dodecahedron (12

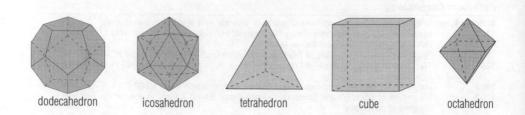

dodecahedron   icosahedron   tetrahedron   cube   octahedron

polyhedron The five regular polyhedra or Platonic solids.

regular pentagons) and icosahedron (20 equilateral triangles).

polymer compound made up of a large long-chain or branching matrix composed of many repeated simple units (**monomers**). There are many polymers, both natural (cellulose, chitin, lignin) and synthetic (polyethylene and nylon, types of plastic). Synthetic polymers belong to two groups: thermosoftening and thermosetting (see ◊plastic).

The size of the polymer matrix is determined by the amount of monomer used; it therefore does not form a molecule of constant molecular size or mass.

polymerization chemical union of two or more (usually small) molecules of the same kind to form a new compound. **Addition polymerization** produces simple multiples of the same compound. **Condensation polymerization** joins molecules together with the elimination of water or another small molecule. Addition polymerization uses only a single monomer (basic molecule); condensation polymerization may involve two or more different monomers (**co-polymerization**).

polymorph in biology, white blood cell that has a nucleus of irregular shape. Polymorphs constitute about 70% of the white blood cells and are involved in the immune response.

polymorphism in genetics, the coexistence of several distinctly different types in a ◊population (groups of animals of one species). Examples include the different blood groups in humans, different colour forms in some butterflies, and snail shell size, length, shape, colour, and stripiness.

polymorphism in mineralogy, the ability of a substance to adopt different internal structures and external forms, in response to different conditions of temperature and/or pressure. For example, diamond and graphite are both forms of the element carbon, but they have very different properties and appearance. Silica (SiO_2) also has several polymorphs, including quartz, tridymite, cristobalite, and stishovite (the latter a very high pressure form found in meteoritic impact craters).

polymytosis in medicine, a disease that usually affects the muscles in the shoulder or hip areas. The disease can be acute or chronic. The muscles weaken and may become tender and inflamed.

polynomial in mathematics, an algebraic expression that has one or more ◊variables (denoted by letters). A polynomial of degree one, that is, whose highest ◊power of x is 1, as in $2x + 1$, is called a linear polynomial; $3x^2 + 2x + 1$ is quadratic; $4x^3 + 3x^2 + 2x + 1$ is cubic.

polyp or **polypus** small 'stalked' benign tumour, usually found on mucous membrane of the nose or bowels. Intestinal polyps are usually removed, since some have been found to be precursors of cancer.

polyp in zoology, the sedentary stage in the life cycle of a coelenterate (such as a ◊coral or jellyfish), the other being the free-swimming **medusa**.

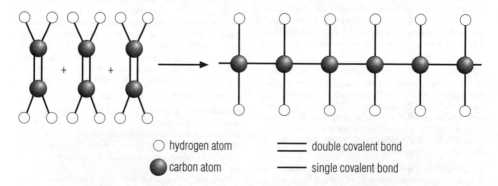

○ hydrogen atom   ▬▬ double covalent bond

● carbon atom   — single covalent bond

polymerization In polymerization, small molecules (monomers) join together to make large molecules (polymers). In the polymerization of ethene to polyethene, electrons are transferred from the carbon=carbon double bond of the ethene molecule, allowing the molecules to join together as a long chain of carbon=carbon single bonds.

polypeptide long-chain ◊peptide.

polyploid in genetics, possessing three or more sets of chromosomes in cases where the normal complement is two sets (◊diploid). Polyploidy arises spontaneously and is common in plants (mainly among flowering plants), but rare in animals. Many crop plants are natural polyploids, including wheat, which has four sets of chromosomes per cell (durum wheat) or six sets (common wheat).

Plant breeders can induce the formation of polyploids by treatment with a chemical, colchicine.

polypropylene plastic made by the polymerization, or linking together, of ◊propene molecules (CH_2=CH-CH_3). It is used as a moulding material.

polysaccharide long-chain ◊carbohydrate made up of hundreds or thousands of linked simple sugars (monosaccharides) such as glucose and closely related molecules.

The polysaccharides are natural polymers. They either act as energy-rich food stores in plants (starch) and animals (glycogen), or have structural roles in the plant cell wall (cellulose, pectin) or the tough outer skeleton of insects and similar creatures (chitin). See also ◊carbohydrate.

polystyrene type of ◊plastic used in kitchen utensils or, in an expanded form, in insulation and ceiling tiles. CFCs are used to produce expanded polystyrene so alternatives are being sought.

polytetrafluoroethene (PTFE) polymer made from the monomer tetrafluoroethene (CF_2CF_2). It is a thermosetting plastic with a high melting point that is used to produce 'non-stick' surfaces on pans and to coat bearings. Its trade name is Teflon.

Polythene trade name for a variety of ◊polyethylene.

polyunsaturate type of ◊fat or oil containing a high proportion of triglyceride molecules whose ◊fatty acid chains contain several double bonds. By contrast, the fatty-acid chains of the triglycerides in saturated fats (such as lard) contain only single bonds. Medical evidence suggests that polyunsaturated fats, used widely in margarines and cooking fats, are less likely to contribute to cardiovascular disease than saturated fats, but there is also some evidence that they may have adverse effects on health.

The more double bonds the fatty-acid chains contain, the lower the melting point of the fat. Unsaturated chains with several double bonds produce oils, such as vegetable

and fish oils, which are liquids at room temperature. Saturated fats, with no double bonds, are solids at room temperature. The polyunsaturated fats used for margarines are produced by taking a vegetable or fish oil and turning some of the double bonds to single bonds, so that the product is semi-solid at room temperature. This is done by bubbling hydrogen through the oil in the presence of a catalyst, such as platinum. The catalyst is later removed.

Mono-unsaturated oils, such as olive oil, whose fatty-acid chains contain a single double bond, are probably healthier than either saturated or polyunsaturated fats. Butter contains both saturated and unsaturated fats, together with ◊cholesterol, which also plays a role in heart disease.

polyurethane polymer made from the monomer urethane. It is a thermoset ◊plastic, used in liquid form as a paint or varnish, and in foam form for upholstery and in lining materials (where it may be a fire hazard).

polyvinyl chloride (PVC) a type of ◊plastic used for drainpipes, floor tiles, audio discs, shoes, and handbags.

pome type of ◊pseudocarp, or false fruit, typical of certain plants belonging to the Rosaceae family. The outer skin and fleshy tissues are developed from the ◊receptacle (the enlarged end of the flower stalk) after fertilization, and the five ◊carpels (the true fruit) form the pome's core, which surrounds the seeds. Examples of pomes are apples, pears, and quinces.

population in biology and ecology, a group of animals of one species, living in a certain area and able to interbreed; the members of a given species in a ◊community of living things.

population control measures taken by some governments to limit the growth of their countries' populations by trying to reduce birth rates. Propaganda, freely available contraception, and tax disincentives for large families are some of the measures that have been tried.

The population-control policies introduced by the Chinese government are the best known. In 1979 the government introduced a 'one-child policy' that encouraged ◊family planning and penalized couples who have more than one child. It has been only partially successful since it has been difficult to administer, especially in rural areas, and has in some cases led to the killing of girls in favour of sons as heirs.

population cycle in biology, regular fluctuations in the size of a population, as seen in lemmings, for example. Such cycles are often caused by density-dependent

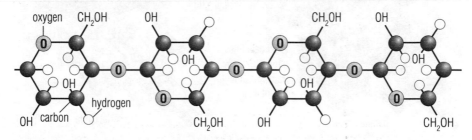

polysaccharide A typical polysaccharide molecule, glycogen (animal starch), is formed from linked glucose ($C_6H_{12}O_6$) molecules. A glycogen molecule has 100–1,000 linked glucose units.

mortality: high mortality due to overcrowding causes a sudden decline in the population, which then gradually builds up again. Population cycles may also result from an interaction between a predator and its prey.

pore in biology, a pore is a small opening in the skin that releases sweat and sebum. Sebum acts as a natural lubricant and protects the skin from the effects of moisture or excessive dryness.

porphyria group of rare genetic disorders caused by an enzyme defect. Porphyria affects the digestive tract, causing abdominal distress; the nervous system, causing psychotic disorder, epilepsy, and weakness; the circulatory system, causing high blood pressure; and the skin, causing extreme sensitivity to light. No specific treatments exist.

In porphyria the body accumulates and excretes (rather than utilizes) one or more porphyrins, the pigments that combine with iron to form part of the oxygen-carrying proteins haemoglobin and myoglobin; because of this urine turns reddish brown on standing. It is known as the 'royal disease' because sufferers are believed to have included Mary Queen of Scots, James I, and George III.

porphyry any ◊igneous rock containing large crystals in a finer matrix.

port in computing, a socket that enables a computer processor to communicate with an external device. It may be an *input port* (such as a joystick port), or an *output port* (such as a printer port), or both (an *i/o port*).

Microcomputers may provide ports for cartridges, televisions and/or monitors, printers, and modems, and sometimes for hard discs and musical instruments (MIDI, the musical-instrument digital interface). Ports may be serial or parallel.

port point where goods are loaded or unloaded from a water-based to a land-based form of transport. Most ports are coastal, though inland ports on rivers also exist. Ports often have specialized equipment to handle cargo in large quantities (for example, container or roll-on/roll-off facilities).

positron in physics, the antiparticle of the electron; an ◊elementary particle having the same mass as an electron but exhibiting a positive charge. The positron was discovered in 1932 by US physicist Carl Anderson at Caltech, USA, its existence having been predicted by the British physicist Paul Dirac in 1928.

positron emission tomography (PET) an imaging technique which enables doctors to observe the metabolic activity of the human body by following the progress of a radioactive chemical that has been inhaled or injected. PET detects ◊gamma radiation given out when ◊positrons emitted by the chemical are annihilated. The technique has been used to study a wide range of conditions, including schizophrenia, Alzheimer's disease and Parkinson's disease.

post-coital contraception in medicine, contraception used to prevent conception after intercourse has taken place. Hormonal contraceptives ('morning-after pills') containing a high dose of an oestrogen and a progestogen are effective if they are taken within three days of unprotected intercourse. Insertion of an intrauterine device (IUD) is more effective than hormonal post-coital contraceptives. It can be inserted up to five days after unprotected intercourse has taken place.

postmortem or *autopsy* dissection of a dead body to determine the cause of death.

postnatal depression mood change occurring in many mothers a few days after the birth of a baby, also known as 'baby blues'. It is usually a shortlived condition but can sometimes persist; one in five women suffer a lasting depression after giving birth. The most severe form of post-natal depressive illness, *puerperal psychosis*, requires hospital treatment.

In mild cases, antidepressant drugs and hormone treatment may help, although no link has been established between hormonal levels and postnatal depression. Research by UK psychologists 1996 showed that the mourning of a lost lifestyle may be a contributory factor.

postpartum in medicine, any events occurring immediately after childbirth.

parallel port

serial port

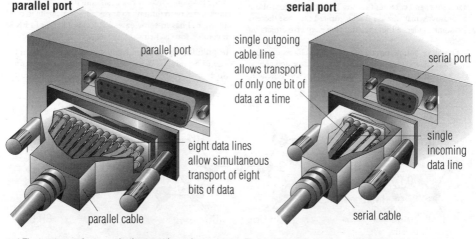

parallel port

single outgoing cable line allows transport of only one bit of data at a time

serial port

eight data lines allow simultaneous transport of eight bits of data

single incoming data line

parallel cable

serial cable

port The two types of communications port in a microcomputer. The parallel port enables up to eight bits of data to travel through it at any one time; the serial port enables only one.

post-viral fatigue syndrome in medicine, another name for ◊chronic fatigue syndrome.

potash general name for any potassium-containing mineral, most often applied to potassium carbonate (K_2CO_3) or potassium hydroxide (KOH). Potassium carbonate, originally made by roasting plants to ashes in earthenware pots, is commercially produced from the mineral sylvite (potassium chloride, KCl) and is used mainly in making artificial fertilizers, glass, and soap.

The potassium content of soils and fertilizers is also commonly expressed as potash, although in this case it usually refers to potassium oxide (K_2O).

potassium soft, waxlike, silver-white, metallic element, symbol K, atomic number 19, relative atomic mass 39.0983. It is one of the ◊alkali metals and has a very low density – it floats on water, and is the second lightest metal (after lithium). It oxidizes rapidly when exposed to air and reacts violently with water. Of great abundance in the Earth's crust, it is widely distributed with other elements and found in salt and mineral deposits in the form of potassium aluminium silicates.

Potassium is the main base ion of the fluid in the body's cells. Along with ◊sodium, it is important to the electrical potential of the nervous system and, therefore, for the efficient functioning of nerve and muscle. Shortage, which may occur with excessive fluid loss (prolonged diarrhoea, vomiting), may lead to muscular paralysis; potassium overload may result in cardiac arrest. It is also required by plants for growth. The element was discovered and named in 1807 by English chemist Humphry Davy, who isolated it from potash in the first instance of a metal being isolated by electric current.

potential difference (pd) difference in the electrical potential (see ◊potential, electric) of two points, being equal to the electrical energy converted by a unit electric charge moving from one point to the other. The SI unit of potential difference is the volt (V). The potential difference between two points in a circuit is commonly referred to as voltage. See also ◊Ohm's law.

potential, electric in physics, the potential at a point is equal to the energy required to bring a unit electric charge from infinity to the point. The SI unit of potential is the volt (V). Positive electric charges will flow 'downhill' from a region of high potential to a region of low potential. See ◊potential difference.

potential energy ◊energy possessed by an object by virtue of its relative position or state (for example, as in a compressed spring). It is contrasted with kinetic energy, the form of energy possessed by moving bodies.

potentiometer in physics, an electrical ◊resistor that can be divided so as to compare, measure, or control voltages. In radio circuits, any rotary variable resistance (such as volume control) is referred to as a potentiometer.

A simple type of potentiometer consists of a length of uniform resistance wire (about 1 m/3 ft long) carrying a constant current provided by a battery connected across the ends of the wire. The source of potential difference (voltage) to be measured is connected (to oppose the cell) between one end of the wire, through a ◊galvanometer (instrument for measuring small currents), to a contact free to slide along the wire. The sliding contact is moved until the galvanometer shows no deflection. The ratio of the length of potentiometer wire in the galvanometer

circuit to the total length of wire is then equal to the ratio of the unknown potential difference to that of the battery.

pound imperial unit (abbreviation lb) of mass. The commonly used avoirdupois pound, also called the *imperial standard pound* (7,000 grains/0.45 kg), differs from the *pound troy* (5,760 grains/0.37 kg), which is used for weighing precious metals. It derives from the Roman *libra*, which weighed 0.327 kg.

poundal imperial unit (abbreviation pdl) of force, now replaced in the SI system by the ◊newton. One poundal equals 0.1383 newtons.

It is defined as the force necessary to accelerate a mass of one pound by one foot per second per second.

powder metallurgy method of shaping heat-resistant metals such as tungsten. Metal is pressed into a mould in powdered form and then sintered (heated to very high temperatures).

power in mathematics, that which is represented by an ◊exponent or index, denoted by a superior small numeral. A number or symbol raised to the power of 2 – that is, multiplied by itself – is said to be squared (for example, 3^2, x^2), and when raised to the power of 3, it is said to be cubed (for example, 2^3, y^3).

power in optics, a measure of the amount by which a lens will deviate light rays. A powerful converging lens will converge parallel rays steeply, bringing them to a focus at a short distance from the lens. The unit of power is the *dioptre*, which is equal to the reciprocal of focal length in metres. By convention, the power of a converging (or convex) lens is positive and that of a diverging (or concave) lens negative.

power in physics, the rate of doing work or consuming energy. It is measured in watts (joules per second) or other units of work per unit time.

power station building where electrical energy is generated from a fuel or from another form of energy. Fuels used include fossil fuels such as coal, gas, and oil, and the nuclear fuel uranium. Renewable sources of energy include gravitational potential energy, used to produce ◊hydroelectric power, and ◊wind power.

The energy supply is used to turn ◊turbines either directly by means of water or wind pressure, or indirectly by steam pressure, steam being generated by burning fossil fuels or from the heat released by the fission of uranium nuclei. The turbines in their turn spin alternators, which generate electricity at very high voltage.

The world's largest power station is Turukhansk, on the Lower Tunguska river, Russia, with a capacity of 20,000 megawatts.

My personal desire would be to prohibit entirely the use of alternating currents. They are unnecessary as they are dangerous ... I can therefore see no justification for the introduction of a system which has no element of permanency and every element of danger to life and property.

On **electrical power** Thomas Edison quoted in R L Weber, *A Randon Walk in Science*

prairie the central North American plain, formerly grass-covered, extending over most of the region between the Rocky Mountains on the west and the Great Lakes and Ohio River on the east.

praseodymium silver-white, malleable, metallic element of the ◊lanthanide series, symbol Pr, atomic number 59, relative atomic mass 140.907. It occurs in nature in the minerals monzanite and bastnaesite, and its green salts are used to colour glass and ceramics. It was named in 1885 by Austrian chemist Carl von Welsbach (1858–1929).

He fractionated it from dydymium (originally thought to be an element but actually a mixture of rare-earth metals consisting largely of neodymium, praseodymium, and cerium), and named it for its green salts and spectroscopic line.

Precambrian in geology, the time from the formation of Earth (4.6 billion years ago) up to 570 million years ago. Its boundary with the succeeding Cambrian period marks the time when animals first developed hard outer parts (exoskeletons) and so left abundant fossil remains. It comprises about 85% of geological time and is divided into two periods: the Archaean, in which no life existed, and the Proterozoic, in which there was life in some form.

precession slow wobble of the Earth on its axis, like that of a spinning top. The gravitational pulls of the Sun and Moon on the Earth's equatorial bulge cause the Earth's axis to trace out a circle on the sky every 25,800 years. The position of the celestial poles (see ◊celestial sphere) is constantly changing owing to precession, as are the positions of the equinoxes (the points at which the celestial equator intersects the Sun's path around the sky). The *precession of the equinoxes* means that there is a gradual westward drift in the ecliptic – the path that the Sun appears to follow – and in the coordinates of objects on the celestial sphere.

This is why the dates of the astrological signs of the zodiac no longer correspond to the times of year when the Sun actually passes through the constellations. For example, the Sun passes through Leo from mid-Aug to mid-Sept, but the astrological dates for Leo are between about 23 July and 22 Aug.

Precession also occurs in other planets. Uranus has the Solar System's fastest known precession (264 days) determined in 1995.

precipitation in chemistry, the formation of an insoluble solid in a liquid as a result of a reaction within the liquid between two or more soluble substances. If the solid settles, it forms a *precipitate*; if the particles of solid are very small, they will remain in suspension, forming a *colloidal precipitate* (see ◊colloid).

precipitation in meteorology, water that falls to the Earth from the atmosphere. It includes rain, snow, sleet, hail, dew, and frost.

pre-eclampsia or *toxaemia of pregnancy* potentially serious condition developing in the third trimester and marked by high blood pressure and fluid retention. Arising from unknown causes, it disappears when pregnancy is over. It may progress to ◊eclampsia if untreated.

pregnancy in humans, the period during which an embryo grows within the womb. It begins at conception and ends at birth, and the normal length is 40 weeks. Menstruation usually stops on conception. About one in five pregnancies fails, but most of these failures occur very early on, so the woman may notice only that her period is late. After the second month, the breasts become tense and tender, and the areas round the nipples become darker. Enlargement of the uterus can be felt at about the end of the third month, and thereafter the abdomen enlarges progressively. Foetal movement can be felt at about 18 weeks; a heart-beat may be heard during the sixth month. Pregnancy in animals is called ◊gestation.

Occasionally the fertilized egg implants not in the womb but in the ◊Fallopian tube (the tube between the ovary and the uterus), leading to an ectopic ('out of place') pregnancy. This will cause the woman severe abdominal pain and vaginal bleeding. If the growing fetus ruptures the tube, life-threatening shock may ensue. Toxaemia is characterized by rising blood pressure, and if left untreated, can result in convulsions leading to coma.

prehistoric life the diverse organisms that inhabited Earth from the origin of life about 3.5 billion years ago to the time when humans began to keep written records, about 3500 BC. During the course of evolution, new forms of life developed and many other forms, such as the dinosaurs, became extinct. Prehistoric life evolved over this vast timespan from simple bacteria-like cells in the oceans to algae and protozoans and complex multicellular forms such as worms, molluscs, crustaceans, fishes, insects, land plants, amphibians, reptiles, birds, and mammals. On a geological timescale human beings evolved relatively recently, about 4 million years ago, although the exact dating is a matter of some debate. See also ◊geological time.

prehistory human cultures before the use of writing. The study of prehistory is mainly dependent on archaeology. General chronological dividing lines between prehistoric eras, or history and prehistory, are difficult to determine because communities have developed at differing rates. The ◊Three Age System of classification (published in 1836 by the Danish archaeologist Christian Thomsen) is based on the predominant materials used by early humans for tools and weapons: ◊Stone Age, ◊Bronze Age, and ◊Iron Age.

Human prehistory begins with the emergence of early modern hominids (see ◊human species, origins of). *Homo habilis*, the first tool user, was in evidence around 2 million years ago, and found at such sites as Koobi Fora, Kenya and Olduvai Gorge, Tanzania.

Stone Age Stone was the main material used for tools and weapons. The Stone Age is divided into:

Old Stone Age (Palaeolithic) 3,500,000–8500 BC. Stone and bone tools were chipped into shape by early humans or hominids from Africa, Asia, the Middle East, and Europe, as well as later Neanderthal and Cro-Magnon people; the only domesticated animals were dogs. Some Asians crossed the Bering land bridge to inhabit the Americas.

Middle Stone Age (Mesolithic) and *New Stone Age* (Neolithic). Bone tools and stone or flint implements were used. In Neolithic times, agriculture and the domestication of goats, sheep, and cattle began. Stone Age cultures survived in the Americas, Asia, Africa, Oceania, and Australia until the 19th and 20th centuries.

Bronze Age Bronze tools and weapons began to be used approximately 5000 BC in the Far East, and this continued in the Middle East until about 1200 BC; in Europe this period lasted from about 2000 to 500 BC.

Iron Age Iron was hardened (alloyed) by the addition

of carbon, so that it superseded bronze for tools and weapons; in the Old World this generally occurred from about 1000 BC.

prematurity the condition of an infant born before the full term. In obstetrics, an infant born before 37 weeks' gestation is described as premature.

Premature babies are often at risk. They lose heat quickly because they lack an insulating layer of fat beneath the skin; there may also be breathing difficulties. In hospitals with advanced technology, specialized neo-natal units can save some babies born as early as 23 weeks.

premedication combination of drugs given before surgery to prepare a patient for general anaesthesia.

One component (an anticholinergic) dries excess secretions produced by the airways when a tube is passed down the trachea, and during inhalation of anaesthetic gases. Other substances act to relax muscles, reduce anxiety, and relieve pain.

premenstrual tension (PMT) or **premenstrual syndrome** medical condition caused by hormone changes and comprising a number of physical and emotional features that occur cyclically before menstruation and disappear with its onset. Symptoms include mood changes, breast tenderness, a feeling of bloatedness, and headache.

premolar in mammals, one of the large teeth toward the back of the mouth. In herbivores they are adapted for grinding. In carnivores they may be carnassials. Premolars are present in milk ◊dentition as well as permanent dentition.

presbyopia vision defect, an increasing inability with advancing age to focus on near objects. It is caused by thickening and loss of elasticity in the lens, which is no longer able to relax to the near-spherical shape required for near vision.

prescription in medicine, an order written in a recognized form by a practitioner of medicine, dentistry, or veterinary surgery to a pharmacist for a preparation of medications to be used in treatment.

By tradition it used to be written in Latin, except for the directions addressed to the patient. It consists of (1) the superscription *recipe* ('take'), contracted to Rx; (2) the inscription or body, containing the names and quantities of the drugs to be dispensed; (3) the subscription, or directions to the pharmacist; (4) the signature, followed by directions to the patient; and (5) the patient's name, the date, and the practitioner's name.

presenile dementia in medicine, a term used to describe the symptoms of ◊dementia in younger people; it is a condition of premature ageing.

pregnancy The development of a human embryo. Division of the fertilized egg, or ovum, begins within hours of conception. Within a week a ball of cells –a blastocyst – has developed. After the third week, the embryo has changed from a mass of cells into a recognizable shape. At four weeks, the embryo is 3 mm/0.1 in long, with a large bulge for the heart and small pits for the ears. At six weeks, the embryo is 1.5 cm/0.6 in with a pulsating heart and ear flaps. At the eighth week, the embryo is 2.5 cm/1 in long and recognizably human, with eyelids, small fingers, and toes. From the end of the second month, the embryo is almost fully formed and further development is mainly by growth. After this stage, the embryo is termed a fetus.

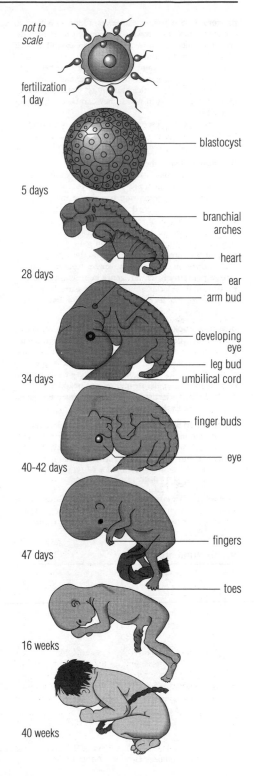

not to scale

fertilization 1 day

blastocyst

5 days

branchial arches

heart

28 days

ear

arm bud

developing eye

leg bud

34 days — umbilical cord

finger buds

eye

40-42 days

fingers

47 days

toes

16 weeks

40 weeks

preservative substance (additive) added to a food in order to inhibit the growth of bacteria, yeasts, moulds, and other microorganisms, and therefore extend its shelf life. The term sometimes refers to ◊anti-oxidants (substances added to oils and fats to prevent their becoming rancid) as well. All preservatives are potentially damaging to health if eaten in sufficient quantity. Both the amount used, and the foods in which they can be used, are restricted by law.

Alternatives to preservatives include faster turnover of food stocks, refrigeration, better hygiene in preparation, sterilization, and pasteurization (see ◊food technology).

pressure in a fluid, the force that would act normally (at right angles) per unit surface area of a body immersed in the fluid. The SI unit of pressure is the pascal (Pa), equal to a pressure of one newton per square metre. In the atmosphere, the pressure declines with height from about 100 kPa at sea level to zero where the atmosphere fades into space. Pressure is commonly measured with a ◊barometer, ◊manometer, or ◊Bourdon gauge. Other common units of pressure are the bar and the torr.

Absolute pressure is measured from a vacuum; *gauge pressure* is the difference between the absolute pressure and the local ◊atmospheric pressure. In a liquid, the pressure at a depth h is given by ρgh where ρ is the density and g is the acceleration of free fall.

pressure cooker closed pot in which food is cooked in water under pressure, where water boils at a higher temperature than normal boiling point (100°C/212°F) and therefore cooks food quickly. The modern pressure cooker has a quick-sealing lid and a safety valve that can be adjusted to vary the steam pressure inside.

The French scientist Denis Papin invented the pressure cooker in England in 1679.

pressurized water reactor (PWR) a ◊nuclear reactor design used in nuclear power stations in many countries, and in nuclear-powered submarines. In the PWR, water under pressure is the coolant and ◊moderator. It circulates through a steam generator, where its heat boils water to provide steam to drive power ◊turbines.

Prestel the ◊viewdata service provided by British Telecom (BT), which provides information on the television screen via the telephone network. BT pioneered the service in 1975.

prevalence in medicine, the proportion of a defined population having a particular condition at any one time. It is used to provide statistics on the occurrence of chronic diseases.

prickly heat acute skin condition characterized by small white or red itchy blisters (miliaria), resulting from inflammation of the sweat glands in conditions of heat and humidity.

primary sexual characteristic the endocrine gland producing maleness and femaleness. In males, the primary sexual characteristic is the ◊testis; in females it is the ◊ovary. Both are endocrine glands that produce hormones responsible for secondary sexual characteristics, such as facial hair and a deep voice in males and breasts in females.

primate in zoology, any member of the order of mammals that includes monkeys, apes, and humans (together called *anthropoids*), as well as lemurs, bushbabies, lorises, and tarsiers (together called *prosimians*).

Generally, they have forward-directed eyes, gripping hands and feet, opposable thumbs, and big toes. They tend to have nails rather than claws, with gripping pads on the ends of the digits, all adaptations to the arboreal, climbing mode of life.

In 1996 a new primate genus (probably extinct) was identified by a US anthropologist from a collection of bones believed to belong to a potto. The animal has been named *Pseudopotto martini*.

There are one hundred and ninety-three living species of monkeys and apes. One hundred and ninety-two are covered with hair. The exception is the naked ape self-named Homo sapiens.

On **primate** Desmond Morris
The Naked Ape 1967

prime number number that can be divided only by 1 or itself, that is, having no other factors. There is an infinite number of primes, the first ten of which are 2, 3, 5, 7, 11, 13, 17, 19, 23, and 29 (by definition, the number 1 is excluded from the set of prime numbers). The number 2

Prime Numbers

2	3	5	7	11	13	17	19	23	29
31	37	41	43	47	53	59	61	67	71
73	79	83	89	97	101	103	107	109	113
127	131	137	139	149	151	157	163	167	173
179	181	191	193	197	199	211	223	227	229
233	239	241	251	257	263	269	271	277	281
283	293	307	311	313	317	331	337	347	349
353	359	367	373	379	383	389	397	401	409
419	421	431	433	439	443	449	457	461	463
467	479	487	491	499	503	509	521	523	541
547	557	563	569	571	577	587	593	599	601
607	613	617	619	631	641	643	647	653	659
661	673	677	683	691	701	709	719	727	733
739	743	751	757	761	769	773	787	797	809
811	821	823	827	829	839	853	857	859	863
877	881	883	887	907	911	919	929	937	941
947	953	967	971	977	983	991	997		

All the Prime Numbers between 1 and 1,000

is the only even prime because all other even numbers have 2 as a factor.

Over the centuries mathematicians have sought general methods (algorithms) for calculating primes, from ◊Eratosthenes' sieve to programs on powerful computers.

The largest prime, $2^{859433}-1$ (258,716 digits long) was discovered in 1993. It is the thirty-third **Mersenne prime**. All Mersenne primes are in the form 2^q-1, where q is also a prime.

Prime number

The number formed by writing 1,031 ones in a row is a prime.

---

principal focus in optics, the point at which incident rays parallel to the principal axis of a ◊lens converge, or appear to diverge, after refraction. The distance from the lens to its principal focus is its ◊focal length.

printed circuit board (PCB) electrical circuit created by laying (printing) 'tracks' of a conductor such as copper on one or both sides of an insulating board. The PCB was invented in 1936 by Austrian scientist Paul Eisler, and was first used on a large scale in 1948.

Components such as integrated circuits (chips), resistors and capacitors can be soldered to the surface of the board (surface-mounted) or, more commonly, attached by inserting their connecting pins or wires into holes drilled in the board. PCBs include ◊motherboards, expansion boards, and adaptors.

printer in computing, an output device for producing printed copies of text or graphics. Types include the ◊**daisywheel printer**, which produces good-quality text but no graphics; the ◊**dot matrix printer**, which produces text and graphics by printing a pattern of small dots; the ◊**ink-jet printer**, which creates text and graphics by spraying a fine jet of quick-drying ink onto the paper; and the ◊**laser printer**, which uses electrostatic technology very similar to that used by a photocopier to produce high-quality text and graphics.

Printers may be classified as **impact printers** (such as daisywheel and dot-matrix printers), which form characters by striking an inked ribbon against the paper, and

non-impact printers (such as ink-jet and laser printers), which use a variety of techniques to produce characters without physical impact on the paper.

A further classification is based on the basic unit of printing, and categorizes printers as character printers, line printers, or page printers, according to whether they print one character, one line, or a complete page at a time.

printing reproduction of multiple copies of text or illustrative material on paper, as in books or newspapers, or on an increasing variety of materials; for example, on plastic containers. The first printing used woodblocks, followed by carved wood type and moulded metal type and hand-operated presses. Modern printing is effected by electronically controlled machinery. Current printing processes include electronic phototypesetting with ◊offset printing, and ◊gravure print.

In China the art of printing from a single wooden block was known by the 6th century AD, and movable type was being used by the 11th century. In Europe printing was unknown for another three centuries, and it was only in the 15th century that movable type was reinvented, traditionally by Johannes Gutenberg in Germany. From there printing spread to Italy, France, and England, where it was introduced by William Caxton.

There was no further substantial advance until, in the 19th century, steam power replaced hand-operation of printing presses, making possible long 'runs'; hand-composition of type (each tiny metal letter was taken from the case and placed individually in the narrow stick that carried one line of text) was replaced by machines operated by a keyboard. **Linotype**, a hot-metal process (it produced a line of type in a solid slug) used in newspapers, magazines, and books, was invented by Ottmar Mergenthaler 1886 and commonly used until the 1980s. The **Monotype**, used in bookwork (it produced a series of individual characters, which could be hand-corrected), was invented by Tolbert Lanston (1844–1913) in the USA 1889.

Important as these developments were, they represented no fundamental change but simply a faster method of carrying out the same basic typesetting operations. The actual printing process still involved pressing the inked type on to paper, by ◊**letterpress**.

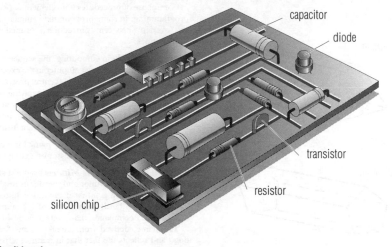

printed circuit board

In the 1960s, letterpress began to face increasing competition from ◊ *offset printing*, a method that prints from an inked flat surface, and from the ◊ *gravure* method (used for high-circulation magazines), which uses recessed plates. The introduction of electronic phototypesetting machines, also in the 1960s, allowed the entire process of setting and correction to be done in the same way that a typist operates, thus eliminating the hot-metal composing room (with its hazardous fumes, lead scraps, and noise) and leaving only the making of plates and the running of the presses to be done traditionally.

By the 1970s some final steps were taken to plateless printing, using various processes, such as a computer-controlled laser beam, or continuous jets of ink acoustically broken up into tiny equal-sized drops, which are electrostatically charged under computer control. Pictures can be fed into computer typesetting systems by optical ◊ scanners.

prion (acronym for *proteinaceous infectious particle*) exceptionally small microorganism, a hundred times smaller than a virus. Composed of protein, and without any detectable amount of nucleic acid (genetic material), it is thought to cause diseases such as scrapie in sheep, and certain degenerative diseases of the nervous system in humans. How it can operate without nucleic acid is not yet known.

prism in mathematics, a solid figure whose cross-section is constant in planes drawn perpendicular to its axis. A

triangular prism

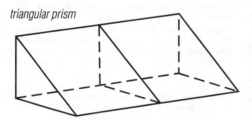

cross-section is the same throughout
the prism's length

trapezoidal prism

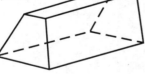

pentagonal prism

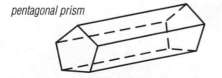

prism

cube, for example, is a rectangular prism with all faces (bases and sides) the same shape and size.

prism in optics, a triangular block of transparent material (plastic, glass, silica) commonly used to 'bend' a ray of light or split a beam into its spectral colours. Prisms are used as mirrors to define the optical path in binoculars, camera viewfinders, and periscopes. The dispersive property of prisms is used in the ◊ spectroscope.

probability likelihood, or chance, that an event will occur, often expressed as odds, or in mathematics, numerically as a fraction or decimal.

In general, the probability that n particular events will happen out of a total of m possible events is n/m. A certainty has a probability of 1; an impossibility has a probability of 0. Empirical probability is defined as the number of successful events divided by the total possible number of events.

In tossing a coin, the chance that it will land 'heads' is the same as the chance that it will land 'tails', that is, 1 to 1 or even; mathematically, this probability is expressed as 1/2 or 0.5. The odds against any chosen number coming up on the roll of a fair die are 5 to 1; the probability is 1/6 or 0.1666... . If two dice are rolled there are $6 \times 6 = 36$ different possible combinations. The probability of a double (two numbers the same) is 6/36 or 1/6 since there are six doubles in the 36 events: (1,1), (2,2), (3,3), (4,4), (5,5), and (6,6).

Probability theory was developed by the French mathematicians Blaise Pascal and Pierre de Fermat in the 17th century, initially in response to a request to calculate the odds of being dealt various hands at cards. Today probability plays a major part in the mathematics of atomic theory and finds application in insurance and statistical studies.

procedure in computing, a small part of a computer program that performs a specific task, such as clearing the screen or sorting a file. A *procedural language*, such as BASIC, is one in which the programmer describes a task in terms of how it is to be done, as opposed to a *declarative language*, such as PROLOG, in which it is described in terms of the required result. See ◊ programming.

Careful use of procedures is an element of ◊ structured programming. In some programming languages there is an overlap between procedures, ◊ functions, and ◊ subroutines.

processing cycle in computing, the sequence of steps performed repeatedly by a computer in the execution of a program. The computer's CPU (central processing unit) continuously works through a loop, involving fetching a program instruction from memory, fetching any data it needs, operating on the data, and storing the result in the memory, before fetching another program instruction.

processor in computing, another name for the ◊ central processing unit or ◊ microprocessor of a computer.

Procyon or *Alpha Canis Minoris* brightest star in the constellation ◊ Canis Minor and the eighth-brightest star in the sky. Procyon is a first-magnitude white star 11.4 light years from Earth, with a mass of 1.7 Suns. It has a ◊ white dwarf companion that orbits it every 40 years.

The name, derived from Greek, means 'before the dog', and reflects the fact that in midnorthern latitudes Procyon rises shortly before ◊ Sirius, the Dog Star.

Procyon and Sirius are sometimes called 'the Dog Stars'. Both are relatively close to us and have white dwarf companions.

productivity, biological in an ecosystem, the amount of material in the food chain produced by the primary producers (plants) that is available for consumption by animals. Plants turn carbon dioxide and water into sugars and other complex carbon compounds by means of photosynthesis. Their net productivity is defined as the quantity of carbon compounds formed, less the quantity used up by the respiration of the plant itself.

progesterone ◊steroid hormone that occurs in vertebrates. In mammals, it regulates the menstrual cycle and pregnancy. Progesterone is secreted by the corpus luteum (the ruptured Graafian follicle of a discharged ovum).

progestogen in medicine, drug closely related to progesterone, a female sex hormone that stimulates growth and secretion of the endometrial glands of the uterus in the first half of the menstrual cycle. Oral contraceptives contain progestogens, either alone or in combination with oestrogens. The contraceptive action of progestogens is due to their disruption of the menstrual cycle by their effect on the pituitary gland and because they make fertilization of an ovum less likely because of their effects on uterine tissue.

prognosis in medicine, prediction of the course or outcome of illness or injury, particularly the chance of recovery.

prograde or **direct motion** in astronomy, the orbit or rotation of a planet, or ◊satellite if the sense of rotation is the same as the general sense of rotation of the Solar System. On the ◊celestial sphere, it refers to motion from west to east against the background of stars.

program in computing, a set of instructions that controls the operation of a computer. There are two main kinds: ◊applications programs, which carry out tasks for the benefit of the user – for example, word processing; and ◊systems programs, which control the internal workings of the computer. A ◊utility program is a systems program that carries out specific tasks for the user. Programs can be written in any of a number of ◊programming languages but are always translated into machine code before they can be executed by the computer.

program counter in computing, an alternative name for ◊sequence-control register.

program loop part of a computer program that is repeated several times. The loop may be repeated a fixed number of times (**counter-controlled loop**) or until a certain condition is satisfied (**condition-controlled loop**). For example, a counter-controlled loop might be used to repeat an input routine until exactly ten numbers have been input; a condition-controlled loop might be used to repeat an input routine until the ◊data terminator 'XXX' is entered.

programming writing instructions in a programming language for the control of a computer. **Applications programming** is for end-user programs, such as accounts programs or word-processing packages. **Systems programming** is for operating systems and the like, which are concerned more with the internal workings of the computer.

There are several programming styles: **procedural**

programming, in which programs are written as lists of instructions for the computer to obey in sequence, is by far the most popular. It is the 'natural' style, closely matching the computer's own sequential operation; **declarative programming**, as used in the programming language PROLOG, does not describe how to solve a problem, but rather describes the logical structure of the problem. Running such a program is more like proving an assertion than following a procedure; **functional programming** is a style based largely on the definition of functions. There are very few functional programming languages, HOPE and ML being the most widely used, though many more conventional languages (for example C) make extensive use of functions; **object-oriented programming**, the most recently developed style, involves viewing a program as a collection of objects that behave in certain ways when they are passed certain 'messages'. For example, an object might be defined to represent a table of figures, which will be displayed on screen when a 'display' message is received.

programming language in computing, a special notation in which instructions for controlling a computer are written. Programming languages are designed to be easy for people to write and read, but must be capable of being mechanically translated (by a ◊compiler or an ◊interpreter) into the ◊machine code that the computer can execute. Programming languages may be classified as ◊high-level languages or ◊low-level languages. See also ◊source language. *See table overleaf.*

program trading buying and selling a group of shares using a computer program to generate orders automatically whenever there is an appreciable movement in prices.

progression sequence of numbers each occurring in a specific relationship to its predecessor. An **arithmetic progression** has numbers that increase or decrease by a common sum or difference (for example, 2, 4, 6, 8); a **geometric progression** has numbers each bearing a fixed ratio to its predecessor (for example, 3, 6, 12, 24); and a **harmonic progression** has numbers whose ◊reciprocals are in arithmetical progression, for example 1, 1/2, 1/3, 1/4.

projectile particle that travels with both horizontal and vertical motion in the Earth's gravitational field. If the frictional forces of air resistance are ignored, the two components of its motion can be analyzed separately: its vertical motion will be accelerated due to its weight in the gravitational field; its horizontal motion may be assumed to be at constant velocity. In a uniform gravitational field and in the absence of frictional forces the path of a projectile is a parabola.

projection of the earth on paper, see ◊map projection.

projector any apparatus that projects a picture on to a screen. In a **slide projector**, a lamp shines a light through the photographic slide or transparency, and a projection ◊lens throws an enlarged image of the slide onto the screen. A **film projector** has similar optics, but incorporates a mechanism that holds the film still while light is transmitted through each frame (picture). A shutter covers the film when it moves between frames.

prokaryote in biology, an organism whose cells lack organelles (specialized segregated structures such as nuclei, mitochondria, and chloroplasts). Prokaryote

Programming Languages

Language	Main uses	Description
Ada	defence applications	high level
assembler languages	jobs needing detailed control of the hardware, fast execution, and small program size	fast and efficient but require considerable effort and skill
BASIC (**b**eginners' **a**ll-purpose **s**ymbolic **i**nstruction **c**ode)	mainly in education, business, and the home, and among nonprofessional programmers, such as engineers	easy to learn; early versions lacked the features of other languages
C	systems programming; general programming	fast and efficient; widely used as a general-purpose language; especially popular among professional programmers
C++	systems and general programming; commercial software development	developed from C, adding the advantages of object-oriented programming
COBOL (**co**mmon **b**usiness-**o**riented **l**anguage)	business programming	strongly oriented towards commercial work; easy to learn but very verbose; widely used on mainframes
FORTH	control applications	reverse Polish notation language
FORTRAN (**for**mula **tran**slation)	scientific and computational work	based on mathematical formulae; popular among engineers, scientists, and mathematicians
Java	developed for consumer electronics; used for many interactive Web sites	multipurpose cross-platform object-oriented language with similar features to C and C++ but simpler
LISP (**lis**t **p**rocessing)	artificial intelligence	symbolic language with a reputation for being hard to learn; popular in the academic and research communities
Modula-2	systems and real-time programming; general programming	highly structured; intended to replace PASCAL for 'real-world' applications
OBERON	general programming	small, compact language incorporating many of the features of PASCAL and Modula-2
PASCAL (**p**rogram **a**ppliqué à la **s**élection et la **c**ompilation automatique de la **l**ittérature)	general-purpose language	highly structured; widely used for teaching programming in universities
Perl	systems programming and Web development	easy manipulation of text, files, and processes, especially in UNIX environment
PROLOG (**pro**gramming in **log**ic)	artificial intelligence	symbolic-logic programming system, originally intended for theorem solving but now used more generally in artificial intelligence

DNA is not arranged in chromosomes but forms a coiled structure called a *nucleoid*. The prokaryotes comprise only the *bacteria* and *cyanobacteria* (see ◊blue-green algae); all other organisms are eukaryotes.

prolapse displacement of an organ due to the effects of strain in weakening the supporting tissues. The term is most often used with regard to the rectum (due to chronic bowel problems) or the uterus (following several pregnancies).

PROLOG (acronym for *programming in logic*) high-level computer-programming language based on logic. Invented in 1971 at the University of Marseille, France, it did not achieve widespread use until more than ten years later. It is used mainly for ◊artificial intelligence programming.

PROM (acronym for *programmable read-only memory*) in computing, a memory device in the form of an integrated circuit (chip) that can be programmed after manufacture to hold information permanently. PROM chips are empty of information when manufactured, unlike ROM (read-only memory) chips, which have information built into them. Other memory devices are ◊EPROM (erasable programmable read-only memory) and ◊RAM (random-access memory).

promethium radioactive, metallic element of the ◊lanthanide series, symbol Pm, atomic number 61, relative atomic mass 145.

It occurs in nature only in minute amounts, produced as a fission product/by-product of uranium in ◊pitchblende and other uranium ores; for a long time it was considered not to occur in nature. The longest-lived isotope has a half-life of slightly more than 20 years.

Promethium is synthesized by neutron bombardment of neodymium, and is a product of the fission of uranium, thorium, or plutonium; it can be isolated in large amounts from the fission-product debris of uranium fuel in nuclear reactors. It is used in phosphorescent paints and as an X-ray source.

prominence bright cloud of gas projecting from the Sun into space 100,000 km/60,000 mi or more. *Quiescent prominences* last for months, and are held in place by magnetic fields in the Sun's corona. *Surge prominences* shoot gas into space at speeds of 1,000 kps/600 mps. *Loop prominences* are gases falling back to the Sun's surface after a solar ◊flare.

proof spirit numerical scale used to indicate the alcohol content of an alcoholic drink. Proof spirit (or 100% proof spirit) acquired its name from a solution of alcohol in water which, when used to moisten gunpowder, contained just enough alcohol to permit it to burn.

propane C_3H_8 gaseous hydrocarbon of the ◊alkane series, found in petroleum and used as fuel.

propanol or *propyl alcohol* third member of the homologous series of ◊alcohols. Propanol is usually a mixture of two isomeric compounds (see ◊isomer): propan-1-ol ($CH_3CH_2CH_2OH$) and propan-2-ol ($CH_3CHOHCH_3$). Both are colourless liquids that can be mixed with water and are used in perfumery.

propanone CH_3COCH_3 (common name **acetone**) colourless flammable liquid used extensively as a solvent, as in nail-varnish remover. It boils at 56.5°C/133.7°F, mixes with water in all proportions, and has a characteristic odour.

propellant substance burned in a rocket for propulsion. Two propellants are used: oxidizer and fuel are stored in separate tanks and pumped independently into the combustion chamber. Liquid oxygen (oxidizer) and liquid hydrogen (fuel) are common propellants, used, for example, in the space-shuttle main engines. The explosive charge that propels a projectile from a gun is also called a propellant.

propeller screwlike device used to propel some ships and aeroplanes. A propeller has a number of curved blades that describe a helical path as they rotate with the hub, and accelerate fluid (liquid or gas) backwards during rotation. Reaction to this backward movement of fluid sets up a propulsive thrust forwards. The marine screw propeller was developed by Francis Pettit Smith in the UK and Swedish-born John Ericson in the USA and was first used in 1839.

propene $CH_3CH=CH_2$ (common name **propylene**) second member of the alkene series of hydrocarbons. A colourless, flammable gas, it is widely used by industry to make organic chemicals, including polypropylene plastics.

propenoic acid $H_2C=CHCOOH$ (common name **acrylic acid**) acid obtained from the aldehyde propenal (acrolein) derived from glycerol or fats. Glasslike thermoplastic resins are made by polymerizing ◊esters of propenoic acid or methyl propenoic acid and used for transparent components, lenses, and dentures. Other

acrylic compounds are used for adhesives, artificial fibres, and artists' acrylic paint.

proper motion gradual change in the position of a star that results from its motion in orbit around our Galaxy, the Milky Way. Proper motions are slight and undetectable to the naked eye, but can be accurately measured on telescopic photographs taken many years apart.

Barnard's Star is the star with the largest proper motion, 10.3 arc seconds per year.

properties in chemistry, the characteristics a substance possesses by virtue of its composition.

Physical properties of a substance can be measured by physical means, for example boiling point, melting point, hardness, elasticity, colour, and physical state.

Chemical properties are the way it reacts with other substances; whether it is acidic or basic, an oxidizing or a reducing agent, a salt, or stable to heat, for example.

prophylaxis any measure taken to prevent disease, including exercise and ◊vaccination. Prophylactic (preventive) medicine is an aspect of public-health provision that is receiving increasing attention.

proportion the relation of a part to the whole (usually expressed as a fraction or percentage). In mathematics two variable quantities x and y are proportional if, for all values of x, $y = kx$, where k is a constant. This means that if x increases, y increases in a linear fashion.

A graph of x against y would be a straight line passing through the origin (the point $x = 0$, $y = 0$). y is inversely proportional to x if the graph of y against $1/x$ is a straight line through the origin. The corresponding equation is

$$y = k/x.$$

Many laws of science relate quantities that are proportional (for example, ◊Boyle's law).

proprioceptor in biology, one of the sensory nerve endings that are located in muscles, tendons, and joints. They relay information on the position of the body and the state of muscle contraction.

prop root or *stilt root* modified root that grows from the lower part of a stem or trunk down to the ground, providing a plant with extra support. Prop roots are common on some woody plants, such as mangroves, and also occur on a few herbaceous plants, such as maize. *Buttress roots* are a type of prop root found at the base of tree trunks, extended and flattened along the upper edge to form massive triangular buttresses; they are common on tropical trees.

propyl alcohol common name for ◊propanol.

propylene common name for ◊propene.

prosimian or *primitive primate* in zoology, any animal belonging to the suborder Strepsirhin or ◊primates. Prosimians are characterized by a wet nose with slitlike nostrils, the tip of the nose having a prominent vertical groove. Examples are lemurs, pottos, tarsiers, and the aye-aye.

prostacyclin in medicine, a ◊prostaglandin that is produced by the endothelial lining of the blood vessels. It inhibits the aggregation of platelets in the cardiovascular system and, consequently, the formation of blood clots. It is also a potent vasodilator, causing blood vessels to dilate.

Prostacyclin, either alone or in combination with ◊heparin, is used to inhibit platelet aggregation in patients undergoing renal dialysis. It has to be given by continuous intravenous infusion because its half-life is only three minutes. Adverse effects include flushing, hypotension, and headache because of its activity as a vasodilator.

prostaglandin any of a group of complex fatty acids present in the body that act as messenger substances between cells. Effects include stimulating the contraction of smooth muscle (for example, of the womb during birth), regulating the production of stomach acid, and modifying hormonal activity. In excess, prostaglandins may produce inflammatory disorders such as arthritis. Synthetic prostaglandins are used to induce labour in humans and domestic animals.

The analgesic actions of substances such as aspirin are due to inhibition of prostaglandin synthesis.

prostatectomy surgical removal of the ◊prostate gland. In many men over the age of 60 the prostate gland enlarges, causing obstruction to the urethra. This causes the bladder to swell with retained urine, leaving the sufferer more prone to infection of the urinary tract.

The treatment of choice is transurethral resection of the prostate, in which the gland is removed by passing an endoscope (slender optical instrument) up the urethra and using ◊diathermy to burn away the prostatic tissue. This procedure has now largely replaced 'open' prostatectomy, in which the prostate is removed via an incision in the abdomen.

prostate gland gland surrounding and opening into the ◊urethra at the base of the ◊bladder in male mammals.

The prostate gland produces an alkaline fluid that is released during ejaculation; this fluid activates sperm, and prevents their clumping together. Older men may develop **benign prostatic hyperplasia** (BHP), a painful condition in which the prostate becomes enlarged and restricts urine flow. This can cause further problems of the bladder and kidneys. It is treated by ◊prostatectomy.

prosthesis artificial device used to substitute for a body part which is defective or missing. Prostheses include artificial limbs, hearing aids, false teeth and eyes, heart ◊pacemakers and plastic heart valves and blood vessels.

Prostheses in the form of artificial limbs, such as wooden legs and metal hooks for hands, have been used for centuries, although artificial limbs are now more natural-looking and comfortable to wear. The comparatively new field of ◊bionics has developed myoelectric, or bionic, arms, which are electronically operated and worked by minute electrical impulses from body muscles.

protactinium silver–grey, radioactive, metallic element of the ◊actinide series, symbol Pa, atomic number 91, relative atomic mass 231.036. It occurs in nature in very small quantities, in ◊pitchblende and other uranium ores. It has 14 known isotopes; the longest-lived, Pa-231, has a half-life of 32,480 years.

protandry in a flower, the state where the male reproductive organs reach maturity before those of the female. This is a common method of avoiding self-fertilization. See also ◊protogyny.

protease general term for a digestive enzyme capable of splitting proteins. Examples include pepsin, found in the stomach, and trypsin, found in the small intestine.

protein complex, biologically important substance composed of amino acids joined by ◊peptide bonds. Proteins are essential to all living organisms. As **enzymes** they regulate all aspects of metabolism. Structural proteins such as **keratin** and **collagen** make up the skin, claws, bones, tendons, and ligaments; **muscle** proteins produce movement; **haemoglobin** transports oxygen; and **membrane** proteins regulate the movement of substances into and out of cells. For humans, protein is an essential part of the diet, and is found in greatest quantity in soya beans and other grain legumes, meat, eggs, and cheese.

Other types of bond, such as sulphur–sulphur bonds, hydrogen bonds, and cation bridges between acid sites, are responsible for creating the protein's characteristic three-dimensional structure, which may be fibrous, globular, or pleated. Protein provides 4 kcal of energy per gram (60 g per day is required).

protein engineering the creation of synthetic proteins designed to carry out specific tasks. For example, an enzyme may be designed to remove grease from soiled

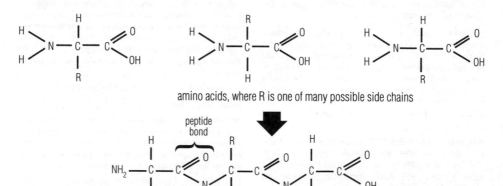

amino acids, where R is one of many possible side chains

peptide bond

protein A protein molecule is a long chain of amino acids linked by peptide bonds. The properties of a protein are determined by the order, or sequence, of amino acids in its molecule, and by the three-dimensional structure of the molecular chain. The chain folds and twists, often forming a spiral shape.

clothes and remain stable at the high temperatures in a washing machine.

protein synthesis manufacture, within the cytoplasm of the cell, of the proteins an organism needs. The building blocks of proteins are ◊amino acids, of which there are 20 types. The pattern in which the amino acids are linked decides what kind of protein is produced. In turn it is the genetic code, contained within ◊DNA, that determines the precise order in which the amino acids are linked up during protein manufacture.

Interestingly, DNA is found only in the nucleus, yet protein synthesis occurs only in the cytoplasm. The information necessary for making the proteins is carried from the nucleus to the cytoplasm by another nucleic acid, ◊RNA.

Proterozoic eon of geological time, possible 3.5 billion to 570 million years ago, the second division of the Precambrian. It is defined as the time of simple life, since many rocks dating from this eon show traces of biological activity, and some contain the fossils of bacteria and algae.

prothallus in botany, a short-lived gametophyte of many ferns and other ◊pteridophytes (such as horsetails or clubmosses). It bears either the male or female sex organs, or both. Typically it is a small, green, flattened structure that is anchored in the soil by several ◊rhizoids (slender, hairlike structures, acting as roots) and needs damp conditions to survive. The reproductive organs are borne on the lower surface close to the soil. See also ◊alternation of generations.

protist in biology, a single-celled organism which has a eukaryotic cell, but which is not a member of the plant, fungal, or animal kingdoms. The main protists are ◊protozoa.

Single-celled photosynthetic organisms, such as diatoms and dinoflagellates, are classified as protists or algae. Recently the term has also been used for members of the kingdom Protista, which features in certain five-kingdom classifications of the living world (see also ◊plant classification). This kingdom may include slime moulds, all algae (seaweeds as well as unicellular forms), and protozoa.

protocol in computing, an agreed set of **standards** for the transfer of data between different devices. They cover transmission speed, format of data, and the signals required to synchronize the transfer. See also ◊interface.

protogyny in a flower, the state where the female reproductive organs reach maturity before those of the male. Like protandry, in which the male organs reach maturity first, this is a method of avoiding self-fertilization, but it is much less common.

proton in physics, a positively charged subatomic particle, a constituent of the nucleus of all atoms. It belongs to the ◊baryon subclass of the ◊hadrons. A proton is extremely long-lived, with a lifespan of at least 10^{32} years. It carries a unit positive charge equal to the negative charge of an ◊electron. Its mass is almost 1,836 times that of an electron, or 1.67×10^{-27} kg. Protons are composed of two up ◊quarks and one down quark held together by ◊gluons. The number of protons in the atom of an element is equal to the atomic number of that element.

proton number alternative name for ◊atomic number.

protoplasm contents of a living cell. Strictly speaking it includes all the discrete structures (organelles) in a cell, but it is often used simply to mean the jellylike material in which these float. The contents of a cell outside the nucleus are called ◊cytoplasm.

protostar in astronomy, an early formation of a star that has recently condensed out of an interstellar cloud and which is not yet hot enough for hydrogen burning to start. Protostars derive their energy from gravitational contraction. See ◊Hayashi track.

prototype in technology, any of the first few machines of a new design. Prototypes are tested for performance, reliability, economy, and safety; then the main design can be modified before full-scale production begins.

protozoa group of single-celled organisms without rigid cell walls. Some, such as amoeba, ingest other cells, but most are ◊saprotrophs or parasites. The group is polyphyletic (containing organisms which have different evolutionary origins).

protractor instrument used to measure a flat ◊angle.

provitamin any precursor substance of a vitamin. Provitamins are ingested substances that become converted to active vitamins within the organism. One example is ergosterol (provitamin D_2), which through the action of sunlight is converted to calciferol (vitamin D_2); another example is beta-carotene, which is hydrolysed in the liver to vitamin A.

Proxima Centauri the closest star to the Sun, 4.2 light years away. It is a faint ◊red dwarf, visible only with a telescope, and is a member of the Alpha Centauri triple-star system.

It is called Proxima because it is about 0.1 light years closer to us than its two partners.

prussic acid former name for ◊hydrocyanic acid.

pseudocarp in botany, a fruitlike structure that incorporates tissue that is not derived from the ovary wall. The additional tissues may be derived from floral parts such as the ◊receptacle and ◊calyx. For example, the coloured, fleshy part of a strawberry develops from the receptacle and the true fruits are small ◊achenes – the 'pips' embedded in its outer surface. Rose hips are a type of pseudocarp that consists of a hollow, fleshy receptacle containing a number of achenes within. Different types of pseudocarp include pineapples, figs, apples, and pears.

A **coenocarpium** is a fleshy, multiple pseudocarp derived from an ◊inflorescence rather than a single flower. The pineapple has a thickened central axis surrounded by fleshy tissues derived from the receptacles and floral parts of many flowers. A fig is a type of pseudocarp called a **syconium**, formed from a hollow receptacle with small flowers attached to the inner wall. After fertilization the ovaries of the female flowers develop into one-seeded achenes. Apples and pears are ◊pomes, another type of pseudocarp.

pseudocopulation attempted copulation by a male insect with a flower. It results in ◊pollination of the flower and is common in the orchid family, where the flowers of many species resemble a particular species of female bee. When a male bee attempts to mate with a flower, the pollinia (groups of pollen grains) stick to its body. They are transferred to the stigma of another flower when the insect attempts copulation again. *See illustration overleaf.*

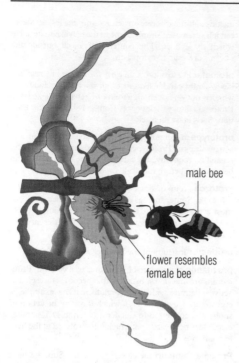

male bee

flower resembles
female bee

pseudocopulation The male bee, attracted to the orchid because of its resemblance to a female bee, attempts to mate with the flower. The bee's efforts cover its body with pollen, which is carried to the next flower it visits.

pseudomorph mineral that has replaced another in situ and has retained the external crystal shape of the original mineral.

psi in parapsychology, a hypothetical faculty common to humans and other animals, said to be responsible for ◊extrasensory perception, telekinesis, and other paranormal phenomena.

psittacosis infectious acute or chronic disease, contracted by humans from birds (especially parrots), which produces pneumonia-like symptoms. It is caused by a bacterium (*Chlamydia psittaci*, see ◊chlamydia) and treated with tetracycline.

psoriasis chronic, recurring skin disease characterized by raised, red, scaly patches, on the scalp, elbows, knees, and elsewhere. Tar preparations, steroid creams, and ultraviolet light are used to treat it, and sometimes it disappears spontaneously. Psoriasis may be accompanied by a form of arthritis (inflammation of the joints).

psychedelic drug any drug that produces hallucinations or altered states of consciousness. Such sensory experiences may be in the auditory, visual, tactile, olfactory, or gustatory fields or in any combination. Among drugs known to have psychedelic effects are LSD (lysergic acid diethylamide), mescaline, and, to a mild degree, marijuana, along with a number of other plant-derived or synthetically prepared substances.

psychiatry branch of medicine dealing with the diagnosis and treatment of mental disorder, normally divided

into the areas of **neurotic conditions**, including anxiety, depression, and hysteria, and **psychotic disorders**, such as schizophrenia. Psychiatric treatment consists of drugs, analysis, or electroconvulsive therapy.

In practice there is considerable overlap between psychiatry and ◊clinical psychology, the fundamental difference being that psychiatrists are trained medical doctors (holding an MD degree) and may therefore prescribe drugs, whereas psychologists may hold a PhD but do not need a medical qualification to practise. See also ◊psychoanalysis.

psychoanalysis theory and treatment method for neuroses, developed by Sigmund Freud in the 1890s. Psychoanalysis asserts that the impact of early childhood sexuality and experiences, stored in the ◊unconscious, can lead to the development of adult emotional problems. The main treatment method involves the free association of ideas, and their interpretation by patient and analyst, in order to discover these long-buried events and to grasp their significance to the patient, linking aspects of the patient's historical past with the present relationship to the analyst. Psychoanalytic treatment aims to free the patient from specific symptoms and from irrational inhibitions and anxieties.

As a theoretical system, psychoanalysis rests on three basic concepts. The central concept is that of the **unconscious**, a reservoir within one's mental state which contains elements and experiences of which one is unaware, but which may to some extent be brought into preconscious and conscious awareness, or inferred from aspects of behaviour. The second and related basic concept is that of **resistance**, a process by which unconscious elements are forcibly kept out of the conscious awareness by an active repressive force. Freud came to experience the third basic concept in his work, known as **transference**, with his earliest patients, who transferred to him aspects of their past relationships with others, so that their relationship with him was coloured by their previous feelings. The analysis of the transference in all its manifestations has become a vital aspect of current psychoanalytic practice.

Freud proposed a model of human psychology based on the concept of the conflicting ◊id, ◊ego, and ◊superego. The id is the mind's instinctual element which demands pleasure and satisfaction; the ego is the conscious mind which deals with the demands of the id and superego; the superego is the ethical element which acts as a conscience and may produce feelings of guilt. The conflicts between these three elements can be used to explain a range of neurotic symptoms.

In the early 1900s a group of psychoanalysts gathered around Freud. Some of these later broke away and formed their own schools, notably Alfred Adler in 1911 and Carl Jung in 1913. The significance of early infantile experience has been further elaborated in the field of child analysis, particularly in the work of Melanie Klein and her students, who pay particular attention to the development of the infant in the first six to eight months of life.

psychology systematic study of human and animal behaviour. The first psychology laboratory was founded in 1879 by Wilhelm Wundt at Leipzig, Germany. The subject includes diverse areas of study and application, among them the roles of instinct, heredity, environment, and culture; the processes of sensation, perception, learning, and memory; the bases of motivation and emotion; and the functioning of thought, intelligence, and language.

Significant psychologists have included Gustav Fechner, founder of psychophysics; Wolfgang Köhler, one of the ◊Gestalt or 'whole' psychologists; Sigmund Freud and his associates Carl Jung and Alfred Adler; William James, Jean Piaget; Carl Rogers; Hans Eysenck; J B Watson; and B F Skinner.

Experimental psychology emphasizes the application of rigorous and objective scientific methods to the study of a wide range of mental processes and behaviour, whereas *social psychology* concerns the study of individuals within their social environment; for example, within groups and organizations. This has led to the development of related fields such as *occupational psychology*, which studies human behaviour at work, and *educational psychology*. *Clinical psychology* concerns the understanding and treatment of mental health disorders, such as anxiety, phobias, or depression; treatment may include behaviour therapy, cognitive therapy, counselling, psychoanalysis, or some combination of these.

Modern studies have been diverse; for example, the psychological causes of obesity; the nature of religious experience; and the underachievement of women seen as resulting from social pressures. Other related subjects are the nature of sleep and dreams, and the possible extensions of the senses, which leads to the more contentious ground of ◊parapsychology.

psychometrics measurement of mental processes. This includes intelligence and aptitude testing to help in job selection and in the clinical assessment of cognitive deficiencies resulting from brain damage.

psychopathy personality disorder characterized by chronic antisocial behaviour (violating the rights of others, often violently) and an absence of feelings of guilt about the behaviour.

Because the term 'psychopathy' has been misused to refer to any severe mental disorder, many psychologists now prefer the term 'antisocial personality disorder', though this also includes cases in which absence or a lesser degree of guilt is not a characteristic feature.

psychosis or *psychotic disorder* general term for a serious mental disorder where the individual commonly loses contact with reality and may experience hallucinations (seeing or hearing things that do not exist) or delusions (fixed false beliefs). For example, in a paranoid psychosis, an individual may believe that others are plotting against him or her. A major type of psychosis is ◊schizophrenia.

psychosomatic of a physical symptom or disease thought to arise from emotional or mental factors.

The term 'psychosomatic' has been applied to many conditions, including asthma, migraine, hypertension, and peptic ulcers. Whereas it is unlikely that these and other conditions are wholly due to psychological factors, emotional states such as anxiety or depression do have a distinct influence on the frequency and severity of illness.

psychosurgery operation to relieve severe mental illness. See ◊leucotomy.

psychotherapy any treatment for psychological problems that involves talking rather than surgery or drugs.

Examples include ◊cognitive therapy and ◊psychoanalysis.

psychotic disorder another name for ◊psychosis.

pt symbol for ◊pint.

pteridophyte simple type of ◊vascular plant. The pteridophytes comprise four classes: the Psilosida, including the most primitive vascular plants, found mainly in the tropics; the Lycopsida, including the club mosses; the Sphenopsida, including the horsetails; and the Pteropsida, including the ferns. They do not produce seeds.

They are mainly terrestrial, non-flowering plants characterized by the presence of a vascular system; the possession of true stems, roots, and leaves; and by a marked ◊alternation of generations, with the sporophyte forming the dominant generation in the life cycle. The pteridophytes formed a large and dominant flora during the Carboniferous period, but many are now known only from fossils.

ptomaine any of a group of toxic chemical substances (alkaloids) produced as a result of decomposition by bacterial action on proteins.

puberty stage in human development when the individual becomes sexually mature. It may occur from the age of ten upwards. The sexual organs take on their adult form and pubic hair grows. In girls, menstruation begins, and the breasts develop; in boys, the voice breaks and becomes deeper, and facial hair develops.

pubes lowest part of the front of the human trunk, the region where the external generative organs are situated. The underlying bony structure, the pubic arch, is formed by the union in the midline of the two pubic bones, which are the front portions of the hip bones. In women this is more prominent than in men, to allow more room for the passage of the child's head at birth, and it carries a pad of fat and connective tissue, the *mons veneris* (mount of Venus), for its protection.

puddle clay clay, with sand or gravel, that has had water added and mixed thoroughly so that it becomes watertight. The term was coined in 1762 by the canal builder James Brindley, although the use of such clay in dams goes back to Roman times.

puerperal fever infection of the genital tract of the mother after childbirth, due to lack of aseptic conditions. Formerly often fatal, it is now rare and treated with antibiotics.

pull-down menu in computing, a list of options provided as part of a ◊graphical user interface. The presence of pull-down menus is normally indicated by a row of single words at the top of the screen. When the user points at a word with a ◊mouse, a full menu appears (is pulled down) and the user can then select the required option. Compare with pop-up menu.

In some graphical user interfaces the menus appear from the bottom of the screen and in others they may appear at any point on the screen when a special menu button is pressed on the mouse.

pulley simple machine consisting of a fixed, grooved wheel, sometimes in a block, around which a rope or chain can be run. A simple pulley serves only to change the direction of the applied effort (as in a simple hoist for raising loads). The use of more than one pulley results in a

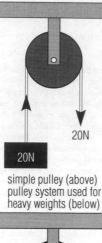

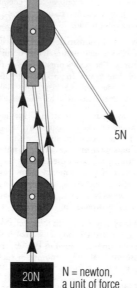

20N

20N

simple pulley (above)
pulley system used for
heavy weights (below)

5N

20N   N = newton,
a unit of force

pulley The mechanical advantage of a pulley increases with
the number of rope strands. If a pulley system has four
ropes supporting the load, the mechanical advantage is
four, and a 5 Newton force will lift a 20 Newton load.

mechanical advantage, so that a given effort can raise a
heavier load.

The mechanical advantage depends on the arrangement of the pulleys. For instance, a block and tackle
arrangement with three ropes supporting the load will lift
it with one-third of the effort needed to lift it directly (if
friction is ignored), giving a mechanical advantage of 3.

pulmonary pertaining to the ◊lungs.

pulmonary embolism in medicine, condition in which a
blood clot becomes lodged in the lungs. The clot usually
migrates in the circulatory system from a vein in the lower
abdomen or legs. The condition is characterized by a
sudden pain in the chest. It can be fatal if the clot is large.

pulmonary hypertension in medicine, high blood
pressure in the lungs due to a resistance to blood flow
caused by constriction of the smaller arteries. Lung diseases, such as chronic bronchitis and emphysema, bring
about pulmonary hypertension.

pulmonary oedema in medicine, the accumulation of
fluids in the lung. It is caused by heart diseases, such as left
ventricular failure or mitral stenosis (narrowing of the
◊bicuspid valve).

pulmonary surfactant in medicine, substance that lowers the surface tension in the finer branches of the lungs,
protecting them from collapse. Pulmonary surfactants are
used to treat ◊respiratory distress syndrome in premature
babies.

pulsar celestial source that emits pulses of energy at
regular intervals, ranging from a few seconds to a few
thousandths of a second. Pulsars are thought to be rapidly
rotating ◊neutron stars, which flash at radio and other
wavelengths as they spin. They were discovered in 1967
by Jocelyn Bell Burnell and Antony Hewish at the Mullard
Radio Astronomy Observatory, Cambridge, England.
Over 500 radio pulsars are now known in our Galaxy,
although a million or so may exist.

pulse impulse transmitted by the heartbeat throughout
the arterial systems of vertebrates. When the heart muscle
contracts, it forces blood into the ◊aorta (the chief artery).
Because the arteries are elastic, the sudden rise of pressure
causes a throb or sudden swelling through them. The
actual flow of the blood is about 60 cm/2 ft a second in
humans. The average adult pulse rate is generally about
70 per minute. The pulse can be felt where an artery is
near the surface, for example in the wrist or the neck.

pulse crop such as peas and beans. Pulses are grown
primarily for their seeds, which provide a concentrated
source of vegetable protein, and make a vital contribution
to human diets in poor countries where meat is scarce,
and among vegetarians. Soya beans are the major temperate protein crop in the West; most are used for oil production or for animal feed. In Asia, most are processed
into soya milk and beancurd. Peanuts dominate pulse
production in the tropical world and are generally consumed as human food. Pulses play a useful role in ◊crop
rotation as they help to raise soil nitrogen levels as well as
acting as break crops. In the mid-1980s, world production was about 50 million tonnes a year.

pulse-code modulation (PCM) in physics, a form of
digital ◊modulation in which microwaves or light waves
(the carrier waves) are switched on and off in pulses of
varying length according to a binary code. It is a relatively
simple matter to transmit data that are already in binary
code, such as those used by computer, by these means.
However, if an analogue audio signal is to be transmitted,
it must first be converted to a ***pulse-amplitude modulated*** signal (PAM) by regular sampling of its amplitude.
The value of the amplitude is then converted into a binary
code for transmission on the carrier wave. *See illustration
on page 528.*

pulsed high frequency (PHF) instrumental application
of high-frequency radio waves in short bursts to damaged
tissue to relieve pain, reduce bruising and swelling, and
speed healing.

An untidy bunch of squiggles

Making a major scientific discovery is not as many people imagine; often, there is no 'eureka moment' or single instant of discovery. Discovery is more often a process of checking and rechecking, of gradually eliminating spurious effects, until the truth is apparent. The discovery of the first pulsar shows this process in action.

Luck played a part in the discovery. In 1967 Antony Hewish at the Mullard Radio Astronomy Laboratory, Cambridge, constructed a new type of radio telescope – a large array of 2,048 aerials covering an area of 1.8 hectares/4.4 acres – to study the 'twinkling' or scintillation of radio galaxies. This is caused by clouds of ionized gas ejected from the Sun. It is most noticeable at metre wavelengths, so the new telescope had to be sensitive to radiation of this wavelength. Most radio telescopes achieve high sensitivity by averaging incoming signals for several seconds, and so are unsuitable for studying rapidly varying signals. They collect radiation with wavelengths of around a centimetre. The new telescope's ability to detect rapidly varying signals of metre wavelength was just what was needed to detect a pulsar.

Twinkle, twinkle, little galaxy

The telescope began work in July 1967. Its first task was to locate all radio galaxies twinkling in the area of sky accessible to it. Each day, as the Earth rotated, the telescope swept its radio eye across a band of sky. A complete scan took four days. Initially a graduate student, Jocelyn Bell, ran the survey and analysed the results, output on about 30 m/100 ft of chart paper each day.

After a few weeks, Bell noticed an unusual signal: not a single blip, but an untidy bunch of squiggles on the chart. It was nothing like the signals she was looking for, so she marked a query on the chart and did not investigate further. Later, when she saw the same signal on another chart, she realized it merited closer attention. The signal came from a part of the sky where scintillations were normally weak. It occurred at night, and scintillations are strongest during the day. A faster chart recorder was installed for a more detailed look at the signal. It would stretch out the signal over a longer chart, like a photographic enlargement.

What was going on?

For a while, Bell and Hewish's efforts were frustrated – the signal weakened and vanished. For a month, there was no sign of it. The researchers feared that they had seen a one-off event: possibly a star flaring brightly for a short time. If so, they had missed the chance to study it in detail. However, on 28 Nov, it returned. This time, the new recorder revealed the true nature of the signal. It was a series of short pulses about 1.3 seconds apart. Timing the pulses more accurately showed that they were in step to within one-millionth of a second. Their short duration indicated that they were coming from a very small object. Something peculiar was going on, but what?

The task now was to rule out spurious effects. Did the signal result from a machine malfunction? Was it caused by a satellite signal? Or a radar echo from the Moon? The fact that the signal appeared at regular intervals hinted that it was not a machine malfunction. Having calculated when the signal would next appear, at the appointed time the research team stood around the recorder. Nothing! Hewish and the others began to wander away. But before they reached the door, they were called back. 'Here it is!' said a student. They had miscalculated when the signal would be picked up. They knew now that it came from outside the laboratory.

Is there anybody there?

Further study established that the signal rotated with the stars, and came from beyond the Solar System, but from within our galaxy. At one stage, Hewish thought that the signal might be a message from an extraterrestrial civilization. It was given the name 'LGM', for 'little green men'. However, this possibility was ruled out. Other beings must live on a planet circling round a star. The planetary motion would show up as a slight variation in the pulse rate. After carefully timing the pulses for several weeks, the idea of a planetary origin was given up.

On 21 Dec 1967 Bell discovered a second signal elsewhere in the sky. This clinched the reality of the phenomenon, and the name 'pulsar' was quickly coined to describe the object radiating the pulsation. The results were published in Feb 1968; within a year it was generally accepted that the signals came from rapidly spinning neutron stars. Since then more than 500 pulsars have been discovered. Antony Hewish shared the 1974 Nobel Prize for physics with British radio astronomer Martin Ryle for their work in radio astronomy and, in particular, the discovery of pulsars.

Peter Lafferty

PCM microwave

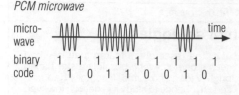

an analogue signal

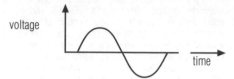

pulse-amplitude-modulated signal (PAM)

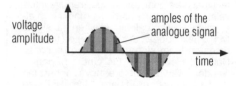

pulse-code modulation

pumice light volcanic rock produced by the frothing action of expanding gases during the solidification of lava. It has the texture of a hard sponge and is used as an abrasive.

pump any device for moving liquids and gases, or compressing gases.

Some pumps, such as the traditional *lift pump* used to raise water from wells, work by a reciprocating (up-and-down) action. Movement of a piston in a cylinder with a one-way valve creates a partial vacuum in the cylinder, thereby sucking water into it.

Gear pumps, used to pump oil in a car's lubrication system, have two meshing gears that rotate inside a housing, and the teeth move the oil. *Rotary pumps* contain a rotor with vanes projecting from it inside a casing, sweeping the oil round as they move.

pumped storage hydroelectric plant that uses surplus electricity to pump water back into a high-level reservoir. In normal working conditions the water flows from this reservoir through the ◊turbines to generate power for feeding into the grid. At times of low power demand, electricity is taken from the grid to turn the turbines into pumps that then pump the water back again. This ensures that there is always a maximum 'head' of water in the reservoir to give the maximum output when required.

punched card in computing, an early form of data storage and input, now almost obsolete. The 80-column card widely used in the 1960s and 1970s was a thin card, measuring 190 mm × 84 mm/7.5 in × 3.33 in, holding up to 80 characters of data encoded as small rectangular holes.

The punched card was invented by French textile manufacturer Joseph-Marie Jacquard (1752–1834) about 1801 to control weaving looms. The first data-processing machine using punched cards was developed by US inventor Herman Hollerith in the 1880s for the US census.

punctuated equilibrium model evolutionary theory developed by Niles Eldredge and US palaeontologist Stephen Jay Gould in 1972 to explain discontinuities in the fossil record. It claims that periods of rapid change alternate with periods of relative stability (stasis), and that the appearance of new lineages is a separate process from the gradual evolution of adaptive changes within a species.

The pattern of stasis and more rapid change is now widely accepted, but the second part of the theory remains unsubstantiated.

The *turnover pulse hypothesis* of US biologist Elisabeth Vrba postulates that the periods of rapid evolutionary change are triggered by environmental changes, particularly changes in climate.

pupa nonfeeding, largely immobile stage of some insect life cycles, in which larval tissues are broken down, and adult tissues and structures are formed.

In many insects, the pupa is *exarate*, with the appendages (legs, antennae, wings) visible outside the pupal case; in butterflies and moths, it is called a chrysalis, and is *obtect*, with the appendages developing inside the case.

purgative substance used to ease constipation. See ◊laxative.

purpura condition marked by purplish patches on the skin or mucous membranes due to localized spontaneous bleeding. It may be harmless, as sometimes with the elderly, or linked with disease, allergy, or drug reactions.

pus yellowish fluid that forms in the body as a result of bacterial infection; it includes white blood cells (leucocytes), living and dead bacteria, dead tissue, and serum. An enclosed collection of pus is called an abscess.

putrefaction decomposition of organic matter by microorganisms.

PUVA in medicine, treatment used for psoriasis that has failed to respond to other forms of treatment. The method involves the administration of a psoralen (P) to sensitize the skin to the effects of irradiation. About two hours later, the skin is irradiated with long-wave ultraviolet radiation (UVA). A course of PUVA is usually given twice weekly for four to six weeks. PUVA is only available in specialist centres and has to be regulated carefully because of the short-term hazard of severe burning and the long-term hazards of cataract formation, premature ageing, and the development of skin cancer.

PVC abbreviation for *polyvinylchloride*, a type of ◊plastic derived from vinyl chloride (CH_2=CHCl).

P-wave (abbreviation of *primary wave*) in seismology, a class of seismic wave that passes through the Earth in the form of longitudinal pressure waves at speeds of 6–7 kps/3.7–4.4 mps in the crust and up to 13 kps/8 mps in deeper layers. P–waves from an earthquake travel faster than S–waves and are the first to arrive at monitoring stations (hence primary waves). They can travel both through solid rock and the liquid outer core of the Earth.

PWR abbreviation for ◊*pressurized water reactor*, a type of nuclear reactor.

pyelitis inflammation of the renal pelvis, the central part of the kidney where urine accumulates before discharge. It is caused by bacterial infection and is more common in women than in men.

pylorus in biology, the lower opening of the stomach. Food that has been softened and partially digested in the mouth and the stomach passes into the intestine through the pylorus.

pyorrhoea former name for gum disease, now known as ◊periodontal disease.

pyramid in geometry, a three-dimensional figure with triangular side-faces meeting at a common vertex (point) and with a ◊polygon as its base. The volume V of a pyramid is given by $V = 1/3Bh$, where B is the area of the base and h is the perpendicular height.

Pyramids are generally classified by their bases. For example, the Egyptian pyramids have square bases, and are therefore called square pyramids. Triangular pyramids are also known as tetrahedra ('four sides').

pyramidal peak angular mountain peak with concave faces found in glaciated areas; for example, the Matterhorn in Switzerland. It is formed when three or four ◊corries (steep-sided hollows) are eroded, back-to-back, around the sides of a mountain, leaving an isolated peak in the middle.

pyridine C_5H_5N a heterocyclic compound (see ◊cyclic compounds). It is a liquid with a sickly smell and occurs in coal tar. It is soluble in water, acts as a strong ◊base, and is used as a solvent, mainly in the manufacture of plastics.

pyridoxine or *vitamin B_6* $C_8H_{11}NO_3$ water-soluble ◊vitamin of the B complex. There is no clearly identifiable disease associated with deficiency but its absence from the diet can give rise to malfunction of the central nervous system and general skin disorders. Good sources are liver, meat, milk, and cereal grains. Related compounds may also show vitamin B_6 activity.

pyrite iron sulphide FeS_2; also called *fool's gold* because of its yellow metallic lustre. Pyrite has a hardness of 6–6.5 on the Mohs scale. It is used in the production of sulphuric acid.

pyroclastic in geology, pertaining to fragments of solidified volcanic magma, ranging in size from fine ash to large boulders, that are extruded during an explosive volcanic eruption; also the rocks that are formed by consolidation of such material. Pyroclastic rocks include tuff (ash deposit) and agglomerate (volcanic breccia).

pyroclastic deposit deposit made up of fragments of rock, ranging in size from fine ash to large boulders, ejected during an explosive volcanic eruption.

pyrogallol $C_6H_3(OH)_3$ (technical name *trihydroxybenzene*) derivative of benzene, prepared from gallic acid. It is used in gas analysis for the measurement of oxygen because its alkaline solution turns black as it rapidly absorbs oxygen. It is also used as a developer in photography.

pyrolysis decomposition of a substance by heating it to a high temperature in the absence of air. The process is used to burn and dispose of old tyres, for example, without contaminating the atmosphere.

pyrometer in physics, any instrument used for measuring high temperatures by means of the thermal radiation emitted by a hot object. In a *radiation pyrometer* the emitted radiation is detected by a sensor such as a thermocouple. In an *optical pyrometer* the brightness of an electrically heated filament is matched visually to that of the emitted radiation. Pyrometers are especially useful for measuring the temperature of distant, moving or inaccessible objects.

pyroxene any one of a group of minerals, silicates of calcium, iron, and magnesium with a general formula X,YSi_2O_6, found in igneous and metamorphic rocks. The internal structure is based on single chains of silicon and oxygen. Diopside (X = Ca, Y = Mg) and augite (X = Ca, Y = Mg,Fe,Al) are common pyroxenes.

Jadeite ($NaAlSi_2O_6$) is also a pyroxene.

Pythagoras' theorem in geometry, a theorem stating that in a right-angled triangle, the area of the square on the hypotenuse (the longest side) is equal to the sum of the areas of the squares drawn on the other two sides. If the hypotenuse is h units long and the lengths of the other sides are a and b, then $h^2 = a^2 + b^2$.

The theorem provides a way of calculating the length of any side of a right-angled triangle if the lengths of the other two sides are known. It is also used to determine certain trigonometrical relationships such as:
$$\sin^2 \theta + \cos^2 \theta = 1.$$

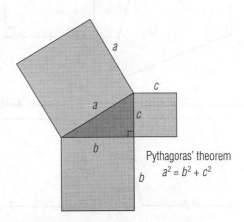

Pythagoras' theorem
$$a^2 = b^2 + c^2$$

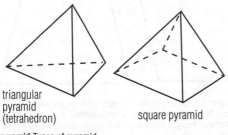

triangular
pyramid
(tetrahedron)   square pyramid

pyramid Types of pyramid.

Pythagoras' theorem Pythagoras' theorem for right-angled triangles is likely to have been known long before the time of Pythagoras. It was probably used by the ancient Egyptians to lay out the pyramids.

In ◊coordinate geometry, a quadratic function represents a ◊parabola.

Some quadratic equations can be solved by factorization, or the values of x can be found by using the formula for the general solution

$$x = [-b \pm \sqrt{(b^2 - 4ac)}]/2a$$

Depending on the value of the discriminant $b^2 - 4ac$, a quadratic equation has two real, two equal, or two complex roots (solutions). When $b^2 - 4ac > 0$, there are two distinct real roots. When $b^2 - 4ac = 0$, there are two equal real roots. When $b^2 - 4ac < 0$, there are two distinct complex roots.

quadrilateral plane (two-dimensional) figure with four straight sides. The following are all quadrilaterals, each with distinguishing properties: ***square*** with four equal angles and sides, four axes of symmetry; ***rectangle*** with four equal angles, opposite sides equal, two axes of symmetry; ***rhombus*** with four equal sides, two axes of symmetry; ***parallelogram*** with two pairs of parallel sides, rotational symmetry; and ***trapezium*** one pair of parallel sides.

qualitative analysis in chemistry, a procedure for determining the identity of the component(s) of a single substance or mixture. A series of simple reactions and tests can be carried out on a compound to determine the elements present.

quantitative analysis in chemistry, a procedure for determining the precise amount of a known component present in a single substance or mixture. A known amount of the substance is subjected to particular procedures.

QALY (acronym for ***quality adjusted life year***) in medicine, a measure derived by health economists to indicate the likely effectiveness of a particular treatment. It can contribute to the assessment of treatment and care that a patient receives. Values for the QALY are sometimes given in reports of clinical trials. For example, intense chemotherapy and radiotherapy may be used on patients with advanced cancer to prolong life for a few months but the quality of that life may be very poor.

quadratic equation in mathematics, a polynomial equation of second degree (that is, an equation containing as its highest power the square of a variable, such as x^2). The general formula of such equations is $ax^2 + bx + c = 0$, in which a, b, and c are real numbers, and only the coefficient a cannot equal 0.

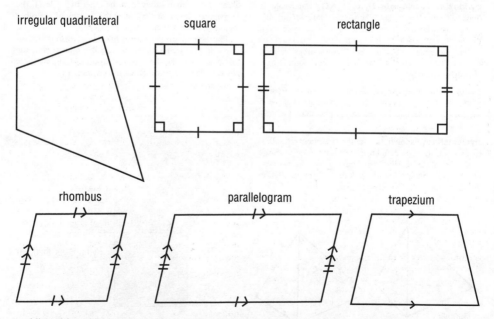

quadrilateral A quadrilateral is a plane figure with four straight sides. The diagram shows different types of quadrilaterals, each with distinguishing properties. A square has four equal angles, four axes of symmetry; a rectangle has four equal angles, two axes of symmetry; a rhombus has four equal sides, two axes of symmetry; a parallelogram has two pairs of parallel sides, rotational symmetry; and a trapezium has one pair of parallel sides.

Gravimetric analysis determines the mass of each constituent present; ◊*volumetric analysis* determines the concentration of a solution by ◊titration against a solution of known concentration.

quantum chromodynamics (QCD) in physics, a theory describing the interactions of ◊quarks, the ◊elementary particles that make up all ◊hadrons (subatomic particles such as protons and neutrons). In quantum chromodynamics, quarks are considered to interact by exchanging particles called gluons, which carry the ◊strong nuclear force, and whose role is to 'glue' quarks together.

The mathematics involved in the theory is complex, and, although a number of successful predictions have been made, the theory does not compare in accuracy with ◊quantum electrodynamics, upon which it is modelled. See ◊forces, fundamental.

quantum electrodynamics (QED) in physics, a theory describing the interaction of charged subatomic particles within electric and magnetic fields. It combines ◊quantum theory and ◊relativity, and considers charged particles to interact by the exchange of photons. QED is remarkable for the accuracy of its predictions; for example, it has been used to calculate the value of some physical quantities to an accuracy of ten decimal places, a feat equivalent to calculating the distance between New York and Los Angeles to within the thickness of a hair. The theory was developed by US physicists Richard Feynman and Julian Schwinger, and by Japanese physicist Sin-Itiro Tomonaga in 1948.

quantum mechanics branch of physics dealing with the interaction of ◊matter and ◊radiation, the structure of the ◊atom, the motion of atomic particles, and with related phenomena (see ◊elementary particle and ◊quantum theory).

quantum number in physics, one of a set of four numbers that uniquely characterize an ◊electron and its state in an ◊atom. The *principal quantum number* n defines the electron's main energy level. The *orbital quantum number* l relates to its angular momentum. The *magnetic quantum number* m describes the energies of electrons in a magnetic field. The *spin quantum number* m_s gives the spin direction of the electron.

The principal quantum number, defining the electron's energy level, corresponds to shells (energy levels) also known by their spectroscopic designations K, L, M, and so on. The orbital quantum number gives rise to a series of subshells designated s, p, d, f, and so on, of slightly different energy levels. The magnetic quantum number allows further subdivision of the subshells (making three subdivisions p_x, p_y, and p_z in the p subshell, for example, of the same energy level). No two electrons in an atom can have the same set of quantum numbers (the Pauli exclusion principle).

quantum theory or *quantum mechanics* in physics, the theory that ◊energy does not have a continuous range of values, but is, instead, absorbed or radiated discontinuously, in multiples of definite, indivisible units called quanta. Just as earlier theory showed how light, generally seen as a wave motion, could also in some ways be seen as being composed of discrete particles (◊photons), quantum theory shows how atomic particles such as electrons may also be seen as having wavelike properties. Quantum theory is the basis of particle physics, modern theoretical chemistry, and the solid-state physics that describes the behaviour of the silicon chips used in computers.

The theory began with the work of Max Planck in 1900 on radiated energy, and was extended by Albert Einstein to electromagnetic radiation generally, including light. Danish physicist Niels Bohr used it to explain the ◊spectrum of light emitted by excited hydrogen atoms. Later work by Erwin Schrödinger, Werner Heisenberg, Paul Dirac, and others elaborated the theory to what is called quantum mechanics (or wave mechanics).

This problem of getting the interpretation proved to be rather more difficult than just working out the equations.

On **quantum theory** Paul Dirac in Hungarian Academy of Sciences Report

---

quarantine any period for which people, animals, plants, or vessels may be detained in isolation to prevent the spread of contagious disease.

quark in physics, the ◊elementary particle that is the fundamental constituent of all ◊hadrons (baryons, such as neutrons and protons, and mesons). There are six types, or 'flavours': up, down, top, bottom, strange, and charmed, each of which has three varieties, or 'colours': red, green, and blue (visual colour is not meant, although the analogy is useful in many ways). To each quark there is an antiparticle, called an antiquark. See ◊quantum chromodynamics.

quart imperial liquid or dry measure, equal to two pints or 1.136 litres. In the USA, a liquid quart is equal to 0.946 litre, while a dry quart is equal to 1.101 litres.

quartz crystalline form of ◊silica SiO_2, one of the most abundant minerals of the Earth's crust (12% by volume). Quartz occurs in many different kinds of rock, including sandstone and granite. It ranks 7 on the Mohs' scale of hardness and is resistant to chemical or mechanical breakdown. Quartzes vary according to the size and purity of their crystals. Crystals of pure quartz are coarse, colourless, transparent, show no cleavage, and fracture unevenly; this form is usually called rock crystal. Impure coloured varieties, often used as gemstones, include ◊agate, citrine quartz, and ◊amethyst. Quartz is also used as a general name for the cryptocrystalline and noncrystalline varieties of silica, such as chalcedony, chert, and opal.

Quartz is used in ornamental work and industry, where its reaction to electricity makes it valuable in electronic instruments (see ◊piezoelectric effect). Quartz can also be made synthetically.

Crystals that would take millions of years to form naturally can now be 'grown' in pressure vessels to a standard that allows them to be used in optical and scientific instruments and in electronics, such as for quartz wristwatches.

quartzite ◊metamorphic rock consisting of pure quartz sandstone that has recrystallized under increasing heat and pressure.

quasar (from *quas*i-stell*ar* object or QSO) one of the most distant extragalactic objects known, discovered in 1963. Quasars appear starlike, but each emits more energy than 100 giant galaxies. They are thought to be at the centre of galaxies, their brilliance emanating from the

Testing quantum theory

In the 1930s German physicist Albert Einstein and his colleagues Boris Podolsky and Nathan Rosen proposed an experiment which they thought would expose a flaw in quantum theory. The experiment was based on the behaviour of pairs of particles that were once close together but had become separated. Such pairs are formed when an atom gives out two particles of light, called photons. According to quantum theory, if we observe one photon of such a pair, the other photon instantly knows that we have made an observation and adjusts its behaviour in certain ways. This would happen even if we waited until the photons were billions of kilometres apart before making our observation. According to Albert Einstein's theory of relativity, no information can travel faster than the speed of light, so how can such widely separated particles react instantly after the observation? Quantum theory must be wrong, said Einstein.

Einstein was never comfortable with quantum theory. He could not accept the idea that in the subatomic world events happen almost by chance, and particles do not have exact positions. Instead there is a range of positions where the particle might be found and all we can do is calculate the probability of it being at any particular point. There is a similar uncertainty about other physical quantities, such as energy and momentum. Einstein felt that behind the uncertainty of quantum theory there must be an exact reality. Paradoxically, when his suggested experiment was eventually performed, it proved him wrong.

Sunglasses for photons

It took a long time to convert Einstein's idea into a practicable experiment. However, in Paris in the early 1980s a series of experiments was carried out by a team of French scientists led by Alain Aspect, Jean Dalibard, and Gerard Roger. The apparatus used consisted of a source of light – calcium atoms – halfway along a long tube. Pairs of photons simultaneously emitted by the calcium atoms split so that each photon travelled towards opposite ends of the tube. At each end of the tube there were devices which would detect photons. Just before the detectors were further devices, called a light switch and polarizer, that could change a certain property – the polarization – of a photon within 0.000,000,01 second of the photon reaching it.

The light switches were ingenious devices. Each consisted of a small cell of water in which two vibrating piezoelectric crystals set up an ultrasonic wave. Depending upon the exact state of the ultrasonic vibration when a photon arrived, the photon either passed through the cell, or was deflected at right angles. The photons then passed through the polarizer, which worked just like a pair of sunglasses to change the polarization of the photon. There were two polarizers behind each light switch, set at different angles, so that the photons could be polarized in two different ways.

Why were such high-speed switches needed? Because the speed of the polarization change had to be faster than the time that it took a photon to travel to the end of the tube. This meant that, when the polarization was changed, there must be no possibility of a 'message' passing down the tube to the second photon (because by then the second photon would have reached the end of the tube and entered the detector). As the experiment proceeded, the light switch operated to continually change the polarization of the photons passing through it. The scientists measured the polarization of the photons arriving at each end of the tube. Photons which arrived simultaneously at the tube ends were obviously emitted simultaneously by the same calcium atom and therefore, according to quantum theory, should be linked in some mysterious way.

There were two possible results of this experiment. If quantum theory was correct, the polarization of photons arriving simultaneously at the detectors would always be the same. If quantum theory was incorrect, then the polarization of the photons would not always be the same.

Faster than light?

The results went against Einstein. Aspect and his team found that if one of a pair of photons was polarized in a certain way, its twin at the other end of the apparatus would always be polarized in the same way. It is as if the photons know what is happening to each other, even though there can be no possible communication between them. So, there is a mysterious instantaneous faster-than-light 'action at a distance' between once-linked photons, and presumably between once-linked particles, too. Scientists and philosophers are still examining the implications of this result. According to the Big Bang theory, all particles now in existence originate from a common point at the birth of the universe. Does this mean that there is a hidden connection between all the particles in the universe? How does this web of connections manifest itself?

Peter Lafferty

stars and gas falling towards an immense ◊black hole at their nucleus.

Quasar light shows a large ◊red shift, indicating that they are very distant. Some quasars emit radio waves (see ◊radio astronomy), which is how they were first identified in 1963, but most are radio-quiet. The furthest are over 10 billion light years away.

Quasar

From the Earth, quasars appear about as bright as a candle on the Moon. Astronomers need to amplify the light from quasars by 10 million times in order to study them. If a radio telescope collected the energy from a quasar for 10,000 years, there would only be enough energy to light a small bulb for a fraction of a second.

Quaternary period of geological time that began 1.64 million years ago and is still in process. It is divided into the ◊Pleistocene and ◊Holocene epochs.

quenching ◊heat treatment used to harden metals. The metals are heated to a certain temperature and then quickly plunged into cold water or oil.

quicksilver another name for the element ◊mercury.

quinine antimalarial drug extracted from the bark of the cinchona tree. Peruvian Indians taught French missionaries how to use the bark in 1630, but quinine was not isolated until 1820. It is a bitter alkaloid $C_{20}H_{24}N_2O_2$.

Other drugs against malaria have since been developed with fewer side effects, but quinine derivatives are still valuable in the treatment of unusually resistant strains.

rabies or *hydrophobia* viral disease of the central nervous system that can afflict all warm-blooded creatures. It is caused by a lyssavirus. It is almost invariably fatal once symptoms have developed. Its transmission to humans is generally by a bite from an infected animal. Rabies continues to kill hundreds of thousands of people every year; almost all these deaths occur in Asia, Africa, and South America.

After an incubation period, which may vary from ten days to more than a year, symptoms of fever, muscle spasm, and delirium develop. As the disease progresses, the mere sight of water is enough to provoke convulsions and paralysis. Death is usual within four or five days from the onset of symptoms. Injections of rabies vaccine and antiserum may save those bitten by a rabid animal from developing the disease. Louis Pasteur was the first to produce a preventive vaccine, and the Pasteur Institute was founded to treat the disease.

As a control measure for foxes and other wild animals, vaccination (by bait) is recommended. In France, Germany, and the border areas of Austria and the Czech Republic, foxes are now vaccinated against rabies with capsules distributed by helicopter; as a result, rabies has been virtually eradicated in Western Europe.

raceme in botany, a type of ◊inflorescence.

rack railway railway, used in mountainous regions, that uses a toothed pinion running in a toothed rack to provide traction. The rack usually runs between the rails. Ordinary wheels lose their grip even on quite shallow gradients, but rack railways, like that on Mount Pilatus in Switzerland, can climb slopes as steep as 50% (1 in 2).

rad unit of absorbed radiation dose, now replaced in the SI system by the ◊gray (one rad equals 0.01 gray), but still commonly used. It is defined as the dose when one kilogram of matter absorbs 0.01 joule of radiation energy (formerly, as the dose when one gram absorbs 100 ergs).

radar (acronym for *radio direction and ranging*) device for locating objects in space, direction finding, and navigating by means of transmitted and reflected high-frequency radio waves.

The direction of an object is ascertained by transmitting a beam of short-wavelength (1–100 cm/½–40 in), short-pulse radio waves, and picking up the reflected beam. Distance is determined by timing the journey of the radio waves (travelling at the speed of light) to the object and back again. Radar is also used to detect objects underground, for example service pipes, and in archaeology.

Contours of remains of ancient buildings can be detected down to 20 m/66 ft below ground. Radar is essential to navigation in darkness, cloud, and fog, and is widely used in warfare to detect enemy aircraft and missiles. To avoid detection, various devices, such as modified shapes (to reduce their radar cross-section), radar-absorbent paints and electronic jamming are used. To pinpoint small targets ◊laser 'radar' instead of microwaves, has been developed.

Developed independently in Britain, France, Germany, and the USA in the 1930s, it was first put to practical use for aircraft detection by the British, who had a complete coastal chain of radar sets installed by Sept 1938. It proved invaluable in the Battle of Britain in 1940, when the ability to spot incoming German aircraft did away with the need to fly standing patrols. Chains of ground radar stations are used to warn of enemy attack – for example, North Warning System (1985), consisting of 52 stations across the Canadian Arctic and N Alaska. Radar is also used in ◊meteorology and ◊astronomy.

radar astronomy bouncing of radio waves off objects in the Solar System, with reception and analysis of the 'echoes'. Radar contact with the Moon was first made in 1945 and with Venus in 1961. The travel time for radio reflections allows the distances of objects to be determined accurately. Analysis of the reflected beam reveals the rotation period and allows the object's surface to be mapped. The rotation periods of Venus and Mercury were first determined by radar. Radar maps of Venus were first obtained by Earth-based radar and subsequently by orbiting space probes.

radial artery in biology, artery that passes down the forearm and supplies blood to the hand and the fingers. The brachial artery, a large artery supplying blood to the arm, divides at the elbow to form the radial and ulnar arteries. The pulsation of blood through the radial artery can be felt at the wrist. This is generally known as the pulse.

radial nerve in biology, the nerve in the upper arm. Nervous impulses to regulate the function of the muscles which extend the arm, the wrist and some fingers pass along these nerves. They also relay sensation to parts of the arm and hand. The radial nerve arises from the brachial plexus (network of nerves supplying the arm) in the armpit and descends through the upper arm before dividing into the superficial radial and interosseous nerves.

radial velocity in astronomy, the velocity of an object, such as a star or galaxy, along the line of sight, moving towards or away from an observer. The amount of ◊Doppler shift (apparent change in wavelength) of the light reveals the object's velocity. If the object is approaching, the Doppler effect causes a ◊blue shift in its light. That is, the wavelengths of light coming from the object appear to be shorter, tending toward the blue end of the ◊spectrum. If the object is receding, there is a ◊red shift, meaning the wavelengths appear to be longer, toward the red end of the spectrum.

radian SI unit (symbol rad) of plane angles, an alternative unit to the ◊degree. It is the angle at the centre of a circle when the centre is joined to the two ends of an arc (part of the circumference) equal in length to the radius of the circle. There are 2π (approximately 6.284) radians in a full circle (360°).

One radian is approximately 57°, and 1° is $\pi/180$ or

approximately 0.0175 radians. Radians are commonly used to specify angles in ◊polar coordinates.

radiant heat energy that is radiated by all warm or hot bodies. It belongs to the ◊infrared part of the electromagnetic ◊spectrum and causes heating when absorbed. Radiant heat is invisible and should not be confused with the red glow associated with very hot objects, which belongs to the visible part of the spectrum.

Infrared radiation can travel through a vacuum and it is in this form that the radiant heat of the Sun travels through space. It is the trapping of this radiation by carbon dioxide and water vapour in the atmosphere that gives rise to the ◊greenhouse effect.

radiation in physics, emission of radiant ◊energy as particles or waves – for example, heat, light, alpha particles, and beta particles (see ◊electromagnetic waves and ◊radioactivity). See also ◊atomic radiation.

Of the radiation given off by the Sun, only a tiny fraction of it, called insolation, reaches the Earth's surface; much of it is absorbed and scattered as it passes through the ◊atmosphere. The radiation given off by the Earth itself is called (*ground radiation*).

radiation biology study of how living things are affected by radioactive (ionizing) emissions (see ◊radioactivity) and by electromagnetic (nonionizing) radiation (◊electromagnetic waves). Both are potentially harmful and can cause mutations as well as leukaemia and other cancers; even low levels of radioactivity are very dangerous. Both are, however, used therapeutically, for example to treat cancer, when the radiation dose is very carefully controlled (◊*radiotherapy* or X-ray treatment).

Radioactive emissions are more harmful. Exposure to high levels produces radiation burns and radiation sickness, plus genetic damage (resulting in birth defects) and cancers in the longer term. Exposure to low-level ionizing radiation can also cause genetic damage and cancers, particularly leukaemia.

Electromagnetic radiation is usually harmful only if exposure is to high-energy emissions, for example close to powerful radio transmitters or near radar-wave sources. Such exposure can cause organ damage, cataracts, loss of hearing, leukaemia and other cancers, or premature ageing. It may also affect the nervous system and brain, distorting their electrical nerve signals and leading to depression, disorientiation, headaches, and other symptoms. Individual sensitivity varies and some people are affected by electrical equipment, such as televisions, computers, and refrigerators.

Background radiation is the natural radiation produced by cosmic rays and radioactive rocks such as granite, and this must be taken into account when calculating the effects of nuclear accidents or contamination from power stations.

radiation sickness sickness resulting from exposure to radiation, including X-rays, gamma rays, neutrons, and other nuclear radiation, such as from weapons and fallout. Such radiation ionizes atoms in the body and causes nausea, vomiting, diarrhoea, and other symptoms. The body cells themselves may be damaged even by very small doses, causing leukaemia.

radiation units units of measurement for radioactivity and radiation doses. In SI units, the activity of a radioactive source is measured in becquerels (symbol Bq) where one becquerel is equal to one nuclear disintegration

per second (an older unit is the curie). The exposure is measured in coulombs per kilogram (C kg^{-1}); the amount of ionizing radiation (X-rays or gamma-rays) which produces one coulomb of charge in one kilogram of dry air (replacing the roentgen). The absorbed dose of ionizing radiation is measured in grays (symbol Gy) where one gray is equal to one joule of energy being imparted to one kilogram of matter (the rad is the previously used unit). The dose equivalent, which is a measure of the effects of radiation on living organisms, is the absorbed dose multiplied by a suitable factor which depends upon the type of radiation. It is measured in sieverts (symbol Sv), where one sievert is a dose equivalent of one joule per kilogram (an older unit is the rem).

radical in chemistry, a group of atoms forming part of a molecule, which acts as a unit and takes part in chemical reactions without disintegration, yet often cannot exist alone; for example, the methyl radical $-CH_3$, or the carboxyl radical $-COOH$.

radicle part of a plant embryo that develops into the primary root. Usually it emerges from the seed before the embryonic shoot, or ◊plumule, its tip protected by a root cap, or calyptra, as it pushes through the soil. The radicle may form the basis of the entire root system, or it may be replaced by adventitious roots (positioned on the stem).

radio transmission and reception of radio waves. In radio transmission a microphone converts sound waves (pressure variations in the air) into ◊electromagnetic waves that are then picked up by a receiving aerial and fed to a loudspeaker, which converts them back into sound waves.

The theory of electromagnetic waves was first developed by Scottish physicist James Clerk Maxwell in 1864, given practical confirmation in the laboratory in 1888 by German physicist Heinrich Hertz, and put to practical use by Italian inventor Guglielmo Marconi, who in 1901 achieved reception of a signal in Newfoundland transmitted from Cornwall, England.

To carry the transmitted electrical signal, an ◊oscillator produces a carrier wave of high frequency; different stations are allocated different transmitting carrier frequencies. A modulator superimposes the audiofrequency signal on the carrier. There are two main ways of doing this: ◊*amplitude modulation* (AM), used for long- and medium-wave broadcasts, in which the strength of the carrier is made to fluctuate in time with the audio signal; and ◊*frequency modulation* (FM), as used for VHF broadcasts, in which the frequency of the carrier is made to fluctuate. The transmitting aerial emits the modulated electromagnetic waves, which travel outwards from it.

In radio reception a receiving aerial picks up minute voltages in response to the waves sent out by a transmitter. A tuned circuit selects a particular frequency, usually by means of a variable ◊capacitor connected across a coil of wire. A demodulator disentangles the audio signal from the carrier, which is now discarded, having served its purpose. An amplifier boosts the audio signal for feeding to the loudspeaker. In a ◊superheterodyne receiver, the incoming signal is mixed with an internally-generated signal of fixed frequency so that the amplifier circuits can operate near their optimum frequency.

radioactive decay process of continuous disintegration undergone by the nuclei of radioactive elements, such as radium and various isotopes of uranium and the transuranic elements. This changes the element's atomic

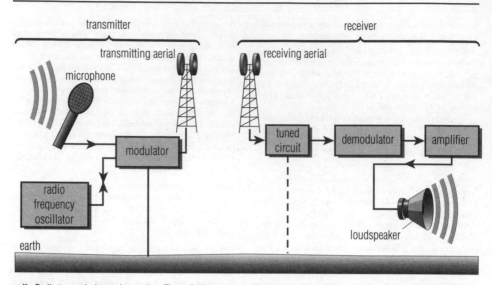

transmitter

receiver

transmitting aerial

receiving aerial

microphone

modulator

tuned circuit

demodulator

amplifier

radio frequency oscillator

earth

loudspeaker

radio Radio transmission and reception. The radio frequency oscillator generates rapidly varying electrical signals, which are sent to the transmitting aerial. In the aerial, the signals produce radio waves (the carrier wave), which spread out at the speed of light. The sound signal is added to the carrier wave by the modulator. When the radio waves fall on the receiving aerial, they induce an electrical current in the aerial. The electrical current is sent to the tuning circuit, which picks out the signal from the particular transmitting station desired. The demodulator separates the sound signal from the carrier wave and sends it, after amplification, to the loudspeaker.

number, thus transmuting one element into another, and is accompanied by the emission of radiation. Alpha and beta decay are the most common forms.

In **alpha decay** (the loss of a helium nucleus – two protons and two neutrons) the atomic number decreases by two; in **beta decay** (the loss of an electron) the atomic number increases by one. Certain lighter artificially created isotopes also undergo radioactive decay. The associated radiation consists of alpha rays, beta rays, or gamma rays (or a combination of these), and it takes place at a constant rate expressed as a specific half-life, which is the time taken for half of any mass of that particular isotope to decay completely. Less commonly occurring decay forms include heavy-ion emission, electron capture, and spontaneous fission (in each of these the atomic number decreases).

The original nuclide is known as the parent substance, and the product is a daughter nuclide (which may or may not be radioactive). The final product in all modes of decay is a stable element.

radioactive tracer any of various radioactive ◊isotopes used in labelled compounds; see ◊tracer.

radioactive waste any waste that emits radiation in excess of the background level. See ◊nuclear waste.

radioactivity spontaneous alteration of the nuclei of radioactive atoms, accompanied by the emission of radiation. It is the property exhibited by the radioactive ◊isotopes of stable elements and all isotopes of radioactive elements, and can be either natural or induced. See ◊radioactive decay.

Radioactivity establishes an equilibrium in parts of the nuclei of unstable radioactive substances, ultimately to form a stable arrangement of nucleons (protons and neutrons); that is, a non-radioactive (stable) element. This is

most frequently accomplished by the emission of ◊alpha particles (helium nuclei); ◊beta particles (electrons and positrons); or ◊gamma radiation (electromagnetic waves of very high frequency). Alpha, beta, and gamma radiation are ionizing in their effect and are therefore dangerous to body tissues, especially if a radioactive substance is ingested or inhaled. It takes place either directly, through a one-step decay, or indirectly, through a number of decays that transmute one element into another. This is called a decay series or chain, and sometimes produces an element more radioactive than its predecessor.

The instability of the particle arrangements in the nucleus of a radioactive atom (the ratio of neutrons to protons and/or the total number of both) determines the lengths of the ◊half-lives of the isotopes of that atom, which can range from fractions of a second to billions of years. All isotopes of relative atomic mass 210 and greater are radioactive.

Radioactivity

Our bodies contain tiny amounts of the radioactive isotopes potassium-40 and carbon-14. Around 38,000 atoms of potassium-40 and 1,200 of carbon-14 explode in your body each second. However, it has been calculated that the chance of this causing any serious damage to the body is negligible.

radio astronomy study of radio waves emitted naturally by objects in space, by means of a ◊radio telescope. Radio emission comes from hot gases (**thermal radiation**); electrons spiralling in magnetic fields (**synchrotron radiation**); and specific wavelengths (**lines**) emitted by atoms and molecules in space, such as the 21 cm/8 in line emitted by hydrogen gas.

Radio astronomy began in 1932 when US astronomer Karl Jansky detected radio waves from the centre of our Galaxy, but the subject did not develop until after World War II. Radio astronomy has greatly improved our understanding of the evolution of stars, the structure of galaxies, and the origin of the universe. Astronomers have mapped the spiral structure of the Milky Way from the radio waves given out by interstellar gas, and they have detected many individual radio sources within our Galaxy and beyond.

Among radio sources in our Galaxy are the remains of ◊supernova explosions, such as the ◊Crab nebula and ◊pulsars. Short-wavelength radio waves have been detected from complex molecules in dense clouds of gas where stars are forming. Searches have been undertaken for signals from other civilizations in the Galaxy, so far without success.

Strong sources of radio waves beyond our Galaxy include ◊radio galaxies and ◊quasars. Their existence far off in the universe demonstrates how the universe has evolved with time. Radio astronomers have also detected weak *background radiation* thought to be from the ◊Big Bang explosion that marked the birth of the universe.

radio beacon radio transmitter in a fixed location, used in marine and aerial ◊navigation. Ships and aircraft pinpoint their positions by reference to continuous signals given out by two or more beacons.

radiocarbon dating or *carbon dating* method of dating organic materials (for example, bone or wood), used in archaeology. Plants take up carbon dioxide gas from the atmosphere and incorporate it into their tissues, and some of that carbon dioxide contains the radioactive isotope of carbon, ^{14}C or carbon-14. As this decays at a known rate (half of it decays every 5,730 years), the time elapsed since the plant died can be measured in a laboratory. Animals take carbon-14 into their bodies from eating plant tissues and their remains can be similarly dated. After 120,000 years so little carbon-14 is left that no measure is possible (see ◊half-life).

Radiocarbon dating was first developed in 1949 by the US chemist Willard Libby. The method yields reliable ages back to about 50,000 years, but its results require correction since Libby's assumption that the concentration of carbon-14 in the atmosphere was constant through time has subsequently been proved wrong. Discrepancies were noted between carbon-14 dates for Egyptian tomb artefacts and construction dates recorded in early local texts. Radiocarbon dates from tree rings (see ◊dendrochronology) showed that material before 1000 BC had been exposed to greater concentrations of carbon-14. Now radiocarbon dates are calibrated against calendar dates obtained from tree rings, or, for earlier periods, against uranium/thorium dates obtained from coral. The carbon-14 content is determined by counting beta particles with either a proportional gas or a liquid scintillation counter for a period of time. A new advance, accelerator mass spectrometry, requires only tiny samples and counts the atoms of carbon-14 directly, disregarding their decay.

radio, cellular portable telephone system; see ◊cellular phone.

radiochemistry chemical study of radioactive isotopes and their compounds (whether produced from naturally

radioactive or irradiated materials) and their use in the study of other chemical processes.

When such isotopes are used in labelled compounds, they enable the biochemical and physiological functioning of parts of the living body to be observed. They can help in the testing of new drugs, showing where the drug goes in the body and how long it stays there. They are also useful in diagnosis – for example cancer, fetal abnormalities, and heart disease.

radio frequencies and wavelengths classification of, see ◊electromagnetic waves.

radio galaxy galaxy that is a strong source of electromagnetic waves of radio wavelengths. All galaxies, including our own, emit some radio waves, but radio galaxies are up to a million times more powerful.

In many cases the strongest radio emission comes not from the visible galaxy but from two clouds, invisible through an optical telescope, that can extend for millions of light years either side of the galaxy. This double structure at radio wavelengths is also shown by some ◊quasars, suggesting a close relationship between the two types of object. In both cases, the source of energy is thought to be a massive black hole at the centre. Some radio galaxies are thought to result from two galaxies in collision or recently merged.

radiography branch of science concerned with the use of radiation (particularly ◊X-rays) to produce images on photographic film or fluorescent screens. X-rays penetrate matter according to its nature, density, and thickness. In doing so they can cast shadows on photographic film, producing a radiograph. Radiography is widely used in medicine for examining bones and tissues and in industry for examining solid materials, for example to check welded seams in pipelines.

radioimmunoassay (RIA) in medicine, technique for measuring small quantities of circulating hormones. The assay depends upon the ability of a hormone to inhibit the binding of the same hormone (which has been labelled with a radioactive isotope) to a specific antibody by competition for the binding sites. The sensitivity of radioimmunoassays ensures their usefulness in the diagnosis of hormonal disorders. The technique was developed by US physicist Rosalyn Yalow.

radioisotope (contraction of *radioactive isotope*) in physics, a naturally occurring or synthetic radioactive form of an element. Most radioisotopes are made by bombarding a stable element with neutrons in the core of a nuclear reactor. The radiations given off by radioisotopes are easy to detect (hence their use as ◊tracers), can in some instances penetrate substantial thicknesses of materials, and have profound effects (such as genetic ◊mutation) on living matter. Although dangerous, radioisotopes are used in the fields of medicine, industry, agriculture, and research.

radioisotope scanning use of radioactive materials (radioisotopes or radionucleides) to pinpoint disease. It reveals the size and shape of the target organ and whether any part of it is failing to take up radioactive material, usually an indication of disease.

The speciality known as nuclear medicine makes use of the affinity of different chemical elements for certain parts of the body. Iodine, for instance, always makes its way to the thyroid gland. After being made radioactive, these materials can be given by mouth or injected, and then traced on scanners working on the Geiger-counter principle. The diagnostic record gained from radioisotope scanning is known as a *scintigram*.

radiology medical speciality concerned with the use of radiation, including X-rays, and radioactive materials in the diagnosis and treatment of injury and disease.

radiometric dating method of dating rock by assessing the amount ◊radioactive decay of naturally occurring ◊isotopes. The dating of rocks may be based on the gradual decay of uranium into lead. The ratio of the amounts of 'parent' to 'daughter' isotopes in a sample gives a measure of the time it has been decaying, that is, of its age. Different elements and isotopes are used depending on the isotopes present and the age of the rocks to be dated. Once-living matter can often be dated by ◊radiocarbon dating, employing the half-life of the isotope carbon-14, which is naturally present in organic tissue.

Radiometric methods have been applied to the decay of long-lived isotopes, such as potassium-40, rubidium-87, thorium-232, and uranium-238 which are found in rocks. These isotopes decay very slowly and this has enabled rocks as old as 3,800 million years to be dated accurately. Carbon dating can be used for material between 100,000 and 1,000 years old. **Potassium** dating is used for material more than 100,000 years old, **rubidium** for rocks more than 10 million years old, and **uranium** and **thorium** dating are suitable for rocks older than 20 million years.

radiosonde balloon carrying a compact package of meteorological instruments and a radio transmitter, used to 'sound', or measure, conditions in the atmosphere. The instruments measure temperature, pressure, and humidity, and the information gathered is transmitted back to observers on the ground. A radar target is often attached, allowing the balloon to be tracked.

radio telescope instrument for detecting radio waves from the universe in ◊radio astronomy. Radio telescopes usually consist of a metal bowl that collects and focuses radio waves the way a concave mirror collects and focuses light waves. Radio telescopes are much larger than optical telescopes, because the wavelengths they are detecting are much longer than the wavelength of light. The largest single dish is 305 m/1,000 ft across, at Arecibo, Puerto Rico.

A large dish such as that at ◊Jodrell Bank, England, can see the radio sky less clearly than a small optical telescope sees the visible sky. *Interferometry* is a technique in which the output from two dishes is combined to give better resolution of detail than with a single dish. *Very long baseline interferometry* (VBLI) uses radio telescopes spread across the world to resolve minute details of radio sources.

In *aperture synthesis*, several dishes are linked together to simulate the performance of a very large single dish. This technique was pioneered by Martin Ryle at Cambridge, England, site of a radio telescope consisting of eight dishes in a line 5 km/3 mi long. The ◊Very Large Array in New Mexico consists of 27 dishes arranged in a Y-shape, which simulates the performance of a single dish 27 km/17 mi in diameter. Other radio telescopes are shaped like long troughs, and some consist of simple rod-shaped aerials.

radiotherapy treatment of disease by ◊radiation from X-ray machines or radioactive sources. Radiation, which reduces the activity of dividing cells, is of special value for its effect on malignant tissues, certain nonmalignant tumours, and some diseases of the skin.

Generally speaking, the rays of the diagnostic X-ray machine are not penetrating enough to be efficient in treatment, so for this purpose more powerful machines are required, operating from 10,000 to over 30 million volts. The lower-voltage machines are similar to conventional X-ray machines; the higher-voltage ones may be of special design; for example, linear accelerators and betatrons. Modern radiotherapy is associated with fewer side effects than formerly, but radiotherapy to the head can cause temporary hair loss, and if the treatment involves the gut, diarrhoea and vomiting may occur. Much radiation now given uses synthesized ◊radioisotopes. Radioactive cobalt is the most useful, since it produces gamma rays, which are highly penetrating, and it is used instead of very high-energy X-rays.

radio wave electromagnetic wave possessing a long wavelength (ranging from about 10^{-3} to 10^4 m) and a low frequency (from about 10^5 to 10^{11} Hz). Included in the radio-wave part of the spectrum are ◊microwaves, used for both communications and for cooking; ultra high-and very high-frequency waves, used for television and FM (◊frequency modulation) radio communications; and short, medium, and long waves, used for AM (◊amplitude modulation) radio communications. Radio waves that are used for communications have all been modulated (see ◊modulation) to carry information. Certain astronomical objects emit radio waves, which may be detected and studied using ◊radio telescopes.

radium white, radioactive, metallic element, symbol Ra, atomic number 88, relative atomic mass 226.02. It is one of the ◊alkaline-earth metals, found in nature in ◊pitchblende and other uranium ores. Of the 16 isotopes, the commonest, Ra-226, has a half-life of 1,620 years. The element was discovered and named in 1898 by Pierre and Marie Curie, who were investigating the residues of pitchblende.

Radium decays in successive steps to produce radon (a gas), polonium, and finally a stable isotope of lead. The isotope Ra-223 decays through the uncommon mode of heavy-ion emission, giving off carbon-14 and transmuting directly to lead. Because radium luminesces, it was formerly used in paints that glowed in the dark; when the hazards of radioactivity became known its use was abandoned, but factory and dump sites remain contaminated and many former workers and neighbours contracted fatal cancers.

radius in biology, one of the two bones in the lower forearm of tetrapod (four-limbed) vertebrates.

radius a straight line from the centre of a circle to its circumference, or from the centre to the surface of a sphere.

radon colourless, odourless, gaseous, radioactive, non-metallic element, symbol Rn, atomic number 86, relative atomic mass 222. It is grouped with the ◊inert gases and was formerly considered to be non-reactive, but is now known to form some compounds with fluorine. Of the 20 known isotopes, only three occur in nature; the longest half-life is 3.82 days (Rn-222).

Radon is the densest gas known and occurs in small amounts in spring water, streams, and the air, being formed from the natural radioactive decay of radium. Ernest Rutherford discovered the isotope Rn-220 in 1899, and Friedrich Dorn (1848–1916) in 1900; after several other chemists discovered additional isotopes, William Ramsay and R W Whytlaw-Gray isolated the element, which they named niton in 1908. The name radon was adopted in the 1920s.

railway method of transport in which trains convey passengers and goods along a twin rail track. Following the work of British steam pioneers such as Scottish engineer James Watt, English engineer George Stephenson built the first public steam railway, from Stockton to Darlington, England, in 1825. This heralded extensive railway building in Britain, continental Europe, and North America, providing a fast and economical means of transport and communication. After World War II, steam engines were replaced by electric and diesel engines. At the same time, the growth of road building, air services, and car ownership destroyed the supremacy of the railways.

rain form of ◊precipitation in which separate drops of water fall to the Earth's surface from clouds. The drops are formed by the accumulation of fine droplets that condense from water vapour in the air. The condensation is usually brought about by rising and subsequent cooling of air.

Rain can form in three main ways – frontal (or cyclonic) rainfall, orographic (or relief) rainfall, and convectional rainfall. **Frontal rainfall** takes place at the boundary, or ◊front, between a mass of warm air from the tropics and a mass of cold air from the poles. The water vapour in the warm air is chilled and condenses to form clouds and rain.

Orographic rainfall occurs when an airstream is forced to rise over a mountain range. The air becomes cooled and precipitation takes place. In the UK, the Pennine hills, which extend southwards from Northumbria to Derbyshire in N England, interrupt the path of the prevailing southwesterly winds, causing orographic rainfall. Their presence is partly responsible for the west of the UK being wetter than the east. **Convectional rainfall**, associated with hot climates, is brought about by rising and abrupt cooling of air that has been warmed by the extreme heat of the ground surface. The water vapour carried by the air condenses and so rain falls heavily.

Railways: Chronology

1500s	Tramways – wooden tracks along which trolleys ran – were in use in mines.
1789	Flanged wheels running on cast-iron rails were first introduced; cars were still horse-drawn.
1804	Richard Trevithick built the first steam locomotive, and ran it on the track at the Pen-y-darren ironworks in South Wales.
1825	George Stephenson in England built the first public railway to carry steam trains – the Stockton and Darlington line – using his engine *Locomotion*.
1829	Stephenson designed his locomotive *Rocket*.
1830	Stephenson completed the Liverpool and Manchester Railway, the first steam passenger line. The first US-built locomotive, *Best Friend of Charleston*, went into service on the South Carolina Railroad.
1835	Germany pioneered steam railways in Europe, using *Der Adler*, a locomotive built by Stephenson.
1863	Robert Fairlie, a Scot, patented a locomotive with pivoting driving bogies, allowing tight curves in the track (this was later applied in the Garratt locomotives). London opened the world's first underground railway, powered by steam.
1869	The first US transcontinental railway was completed at Promontory, Utah, when the Union Pacific and the Central Pacific railroads met. George Westinghouse of the USA invented the compressed-air brake.
1879	Werner von Siemens demonstrated an electric train in Germany. Volk's Electric Railway along the Brighton seafront in England was the world's first public electric railway.
1883	Charles Lartique built the first monorail, in Ireland.
1885	The trans-Canada continental railway was completed, from Montréal in the east to Port Moody, British Columbia, in the west.
1890	The first electric underground railway opened in London.
1901	The world's longest-established monorail, the Wuppertal Schwebebahn, went into service in Germany.
1912	The first diesel locomotive took to the rails in Germany.
1938	The British steam locomotive *Mallard* set a steam-rail speed record of 203 kph/126 mph.
1941	Swiss Federal Railways introduced a gas-turbine locomotive.
1964	Japan National Railways inaugurated the 515 km/320 mi New Tokaido line between Osaka and Tokyo, on which the 210 kph/130 mph 'bullet' trains run.
1973	British Rail's High Speed Train (HST) set a diesel-rail speed record of 229 kph/142 mph.
1979	Japan National Railways' maglev test vehicle ML-500 attained a speed of 517 kph/321 mph.
1981	France's Train à Grande Vitesse (TGV) superfast trains began operation between Paris and Lyons, regularly attaining a peak speed of 270 kph/168 mph.
1987	British Rail set a new diesel-traction speed record of 238.9 kph/148.5 mph, on a test run between Darlington and York; France and the UK began work on the Channel Tunnel, a railway link connecting the two countries, running beneath the English Channel.
1988	The West German Intercity Experimental train reached 405 kph/252 mph on a test run between Würzburg and Fulda.
1990	A new rail-speed record of 515 kph/320 mph was established by a French TGV train, on a stretch of line between Tours and Paris.
1991	The British and French twin tunnels met 23 km/14 mi out to sea to form the Channel Tunnel.
1993	British Rail privatization plans announced; government investment further reduced.
1994	Rail services started through the Channel Tunnel.

Convectional rainfall is usually accompanied by a thunderstorm, and it can be intensified over urban areas due to higher temperatures.

rainbow arch in the sky displaying the colours of the ◊spectrum formed by the refraction and reflection of the Sun's rays through rain or mist. Its cause was discovered by Theodoric of Freiburg in the 14th century.

rainforest dense forest usually found on or near the ◊Equator where the climate is hot and wet. Heavy rainfall results as the moist air brought by the converging tradewinds rises because of the heat. Over half the tropical rainforests are in Central and South America, the rest in SE Asia and Africa. They provide the bulk of the oxygen needed for plant and animal respiration. Tropical rainforests once covered 14% of the Earth's land surface, but are now being destroyed at an increasing rate as their valuable timber is harvested and the land cleared for agriculture, causing problems of ◊deforestation. Although by 1991 over 50% of the world's rainforests had been removed, they still comprise about 50% of all growing wood on the planet, and harbour at least 40% of the Earth's species (plants and animals).

Tropical rainforests are characterized by a great diversity of species, usually of tall broad-leafed evergreen trees, with many climbing vines and ferns, some of which are a main source of raw materials for medicines. A tropical forest, if properly preserved, can yield medicinal plants, oils (from cedar, juniper, cinnamon, sandalwood), spices, gums, resins (used in inks, lacquers, linoleum), tanning and dyeing materials, forage for animals, beverages, poisons, green manure, rubber, and animal products (feathers, hides, honey). Other types of rainforests include montane, upper montane or cloud, mangrove, and subtropical.

Rainforests comprise some of the most complex and diverse ecosystems on the planet and help to regulate global weather patterns. When deforestation occurs, the microclimate of the mature forest disappears; soil erosion and flooding then become major problems since rainforests protect the shallow tropical soils. Once an area is cleared it is very difficult for shrubs and bushes to re-establish because soils are poor in nutrients. This causes problems for plans to convert rainforests into agricultural land – after two or three years the crops fail and the land is left bare. Clearing of the rainforests may lead to a global warming of the atmosphere, and contribute to the ◊greenhouse effect.

RAM (acronym for *random-access memory*) in computing, a memory device in the form of a collection of integrated circuits (chips), frequently used in microcomputers. Unlike ◊ROM (read-only memory) chips, RAM chips can be both read from and written to by the computer, but their contents are lost when the power is switched off. Many modern commercial programs require a great deal of RAM to work efficiently.

ramjet simple jet engine (see under ◊jet propulsion) used in some guided missiles. It only comes into operation at high speeds. Air is then 'rammed' into the combustion chamber, into which fuel is sprayed and ignited.

ranching commercial form of ◊pastoral farming that involves extensive use of large areas of land (◊extensive agriculture) for grazing cattle or sheep.

Ranches may be very large, especially where the soil quality is poor; for example, the estancias on the Pampas grasslands in Argentina. Cattle have in the past been allowed to graze freely but more are now enclosed. In the Amazon some deforested areas have been given over to beef-cattle ranching.

random access in computing, an alternative term for ◊direct access.

random number one of a series of numbers having no detectable pattern. Random numbers are used in ◊computer simulation and ◊computer games. It is impossible for an ordinary computer to generate true random numbers, but various techniques are available for obtaining pseudo-random numbers – close enough to true randomness for most purposes.

rangefinder instrument for determining the range or distance of an object from the observer; used to focus a camera or to sight a gun accurately. A *rangefinder camera* has a rotating mirror or prism that alters the image seen through the viewfinder, and a secondary window. When the two images are brought together into one, the lens is sharply focused.

rare-earth element alternative name for ◊lanthanide.

rarefaction in medicine, the reduction in bone density due to the depletion of calcium stores, caused by an infection of the bone, for example.

rare gas alternative name for ◊inert gas.

rash in medicine, eruption on the surface of the skin. It is usually raised and red or it may contain vesicles filled with fluid. It may also be scaly or crusty. Characteristic rashes are produced by infectious diseases, such as chickenpox, measles, German measles, and scarlet fever. The severity of the rash usually reflects the severity of the disease. Rashes are also produced as an allergic response to stings from insects and plants. These are often alleviated by antihistamines and they usually resolve within a few days.

Chronic rashes are present in patients with some skin diseases, such as ◊eczema and ◊psoriasis.

raster graphics computer graphics that are stored in the computer memory by using a map to record data (such as colour and intensity) for every ◊pixel that makes up the image. When transformed (enlarged, rotated, stretched, and so on), raster graphics become ragged and suffer loss of picture resolution, unlike ◊vector graphics. Raster graphics are typically used for painting applications, which allow the user to create artwork on a computer screen much as if they were painting on paper or canvas.

rate of reaction the speed at which a chemical reaction proceeds. It is usually expressed in terms of the concentration (usually in ◊moles per litre) of a reactant consumed, or product formed, in unit time; so the units would be moles per litre per second ($mol\,l^{-1}\,s^{-1}$). The rate of a reaction may be affected by the concentration of the reactants, the temperature of the reactants, and the presence of a ◊catalyst. If the reaction is entirely in the gas state, the rate is affected by pressure, and, for solids, it is affected by the particle size.

During a reaction at constant temperature the concentration of the reactants decreases and so the rate of reaction decreases.

These changes can be represented by drawing graphs.

The rate of reaction is at its greatest at the beginning of the reaction and it gradually slows down. For an ◊endothermic reaction (one that absorbs heat) increasing the

(a) *rate of reaction decreases with time*

(b) *concentration of reactant decreases with time*

(c) *concentration of product increases with time*

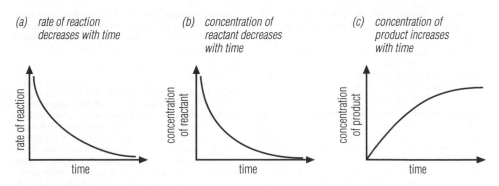

rate of reaction

temperature may produce large increases in the rate of reaction. A 10°C rise can double the rate while a 40°C rise can produce a 50–100-fold increase in the rate.

◊Collision theory is used to explain these effects. Increasing the concentration or the pressure of a gas means there are more particles per unit volume, therefore there are more collisions and more fruitful collisions. Increasing the temperature makes the particles move much faster, resulting in more collisions per unit time and more fruitful collisions; consequently the rate increases.

ratio measure of the relative size of two quantities or of two measurements (in similar units), expressed as a proportion. For example, the ratio of vowels to consonants in the alphabet is 5:21; the ratio of 500 m to 2 km is 500:2,000, or 1:4.

rational number in mathematics, any number that can be expressed as an exact fraction (with a denominator not equal to 0), that is, as $a \div b$ where a and b are integers. For example, 2/1, 1/4, 15/4, −3/5 are all rational numbers, whereas π (which represents the constant 3.141592 ...) is not. Numbers such as π are called ◊irrational numbers.

Raunkiaer system method of classification devised by the Danish ecologist Christen Raunkiaer (1860–1938) whereby plants are divided into groups according to the position of their ◊perennating (overwintering) buds in relation to the soil surface. For example, plants in cold areas, such as the tundra, generally have their buds protected below ground, whereas in hot, tropical areas they are above ground and freely exposed. This scheme is useful for comparing vegetation types in different parts of the world.

The main divisions are **phanerophytes** with buds situated well above the ground; **chamaephytes** with buds borne within 25 cm/10 in of the soil surface; **hemi-cryptophytes** with buds at or immediately below the soil surface; and **cryptophytes** with their buds either beneath the soil (**geophyte**) or below water (**hydrophyte**).

Raynaud's disease chronic condition in which the blood supply to the extremities is reduced by periodic spasm of the blood vessels on exposure to cold. It is most often seen in young women.

Attacks are usually brought on by cold or by emotional factors. Typically, the hands and/or feet take on a corpse-like pallor, changing to blue as the circulation begins to return; initial numbness is replaced by a tingling or burn-ing sensation. Drugs may be necessary to control the condition, which in severe cases can give rise to ulceration or gangrene.

rayon any of various shiny textile fibres and fabrics made from ◊cellulose. It is produced by pressing whatever cellulose solution is used through very small holes and solidifying the resulting filaments. A common type is ◊viscose, which consists of regenerated filaments of pure cellulose. Acetate and triacetate are kinds of rayon consisting of filaments of cellulose acetate and triacetate.

reactance property of an alternating current circuit that together with any ◊resistance makes up the ◊impedance (the total opposition of the circuit to the passage of a current).

The reactance of an inductance L is wL and that of capacitance is $1/wC$. Reactance is measured in ◊ohms.

reaction in chemistry, the coming together of two or more atoms, ions, or molecules with the result that a ◊chemical change takes place. The nature of the reaction is portrayed by a chemical equation.

reaction force in physics, the equal and opposite force described by Newton's third law of motion that arises whenever one object applies a force (**action force**) to another. For example, if a magnet attracts a piece of iron, then that piece of iron will also attract the magnet with a force that is equal in magnitude but opposite in direction. When any object rests on the ground the downwards contact force applied to the ground always produces an equal, upwards reaction force.

reactivity series chemical series produced by arranging the metals in order of their ease of reaction with reagents such as oxygen, water, and acids. This arrangement aids the understanding of the properties of metals, helps to explain differences between them, and enables predictions to be made about a metal's behaviour, based on a knowledge of its position or properties.

real number in mathematics, any of the ◊rational numbers (which include the integers) or ◊irrational numbers. Real numbers exclude ◊imaginary numbers, found in ◊complex numbers of the general form $a + bi$ where $i = \sqrt{-1}$, although these do include a real component a.

real-time system in computing, a program that responds to events in the world as they happen. For example, an automatic-pilot program in an aircraft must respond instantly in order to correct deviations from its course. Process control, robotics, games, and many military applications are examples of real-time systems.

receptacle the enlarged end of a flower stalk to which the floral parts are attached. Normally the receptacle is rounded, but in some plants it is flattened or cup-shaped. The term is also used for the region on that part of some seaweeds which becomes swollen at certain times of the year and bears the reproductive organs.

receptor in biology, receptors are discrete areas of cell membranes or areas within cells with which neurotransmitters, hormones, and drugs interact. Such interactions control the activities of the body. For example, adrenaline transmits nervous impulses to receptors in the sympathetic nervous system which initiates the characteristic response to excitement and fear in an individual.

Other types of receptors, such as the proprioceptors, are located in muscles, tendons, and joints. They relay information on the position of the body and the state of muscle contraction to the brain.

recessive gene in genetics, an ◊allele (alternative form of a gene) that will show in the ◊phenotype (observed characteristics of an organism) only if its partner allele on the paired chromosome is similarly recessive. Such an allele will not show if its partner is dominant, that is if the organism is ◊heterozygous for a particular characteristic. Alleles for blue eyes in humans, and for shortness in pea plants are recessive. Most mutant alleles are recessive and therefore are only rarely expressed (see ◊haemophilia and ◊sickle-cell disease).

reciprocal in mathematics, the result of dividing a given quantity into 1. Thus the reciprocal of 2 is 1/2; of 2/3 is 3/2; of x^2 is $1/x^2$ or x^{-2}. Reciprocals are used to replace division by multiplication, since multiplying by the reciprocal of a number is the same as dividing by that number.

recombinant DNA in genetic engineering, ◊DNA formed by splicing together genes from different sources into new combinations.

recombination in genetics, any process that recombines, or 'shuffles', the genetic material, thus increasing genetic variation in the offspring. The two main processes of recombination both occur during meiosis (reduction division of cells). One is **crossing over**, in which chromosome pairs exchange segments; the other is the random reassortment of chromosomes that occurs when each gamete (sperm or egg) receives only one of each chromosome pair.

recording any of a variety of techniques used to capture, store and reproduce music, speech and other information carried by sound waves. A microphone first converts the sound waves into an electrical signal which varies in proportion to the loudness of the sound. The signal can be stored in digital or analogue form, or on magnetic tape.

In an **analogue recording**, the pattern of the signal is copied into another form. In a **vinyl** gramophone record, for example, a continuous spiral groove is cut into a plastic disc by a vibrating needle. The recording is replayed by a stylus which follows the undulations in the groove, so reproducing the vibrations which are amplified and turned back into sound. In a **magnetic tape** recording, the signal is recorded as a pattern of magnetization on a plastic tape coated with a magnetic powder. When the tape is played back, the magnetic patterns

create an electrical signal which, as with the gramophone record, is used to recreate the original sound. All analogue recording techniques suffer from background noise and the quality of reproduction is gradually degraded as the disc or tape wears.

In **digital recording**, the signals picked up by the microphone are converted into a stream of numbers which can then be stored in several ways. The most well-known of these is the ◊compact disc, in which numbers are coded as a string of tiny pits pressed into a 12 cm plastic disc. When the recording is played back, using a laser, the exact values are retrieved and converted into a varying electrical signal and then back into sound. Digital recording is relatively immune to noise and interference and gives a very high quality of reproduction. It is also suitable for storing information to be processed by computers.

record player device for reproducing recorded sound stored as a spiral groove on a vinyl disc. A motor-driven turntable rotates the record at a constant speed, and a stylus or needle on the head of a pick-up is made to vibrate by the undulations in the record groove. These vibrations are then converted to electrical signals by a ◊transducer in the head (often a ◊piezoelectric crystal). After amplification, the signals pass to one or more loudspeakers, which convert them into sound. Alternative formats are ◊compact disc and magnetic ◊tape recording.

The pioneers of the record player were Thomas Edison, with his phonograph, and Emile Berliner (1851–1929), who invented the predecessor of the vinyl record in 1896. More recent developments are stereophonic sound and digital recording on compact disc.

rectangle quadrilateral (four-sided plane figure) with opposite sides equal and parallel and with each interior angle a right angle (90°). Its area A is the product of the length l and height h; that is, $A = l \times h$. A rectangle with all four sides equal is a ◊square.

A rectangle is a special case of a ◊parallelogram. The diagonals of a rectangle are equal and bisect each other.

rectifier in electrical engineering, a device used for obtaining one-directional current (DC) from an alternating source of supply (AC). (The process is necessary because almost all electrical power is generated, transmitted, and supplied as alternating current, but many devices, from television sets to electric motors, require direct current.) Types include plate rectifiers, thermionic ◊diodes, and ◊semiconductor diodes.

rectum lowest part of the large intestine of animals, which stores faeces prior to elimination (defecation).

recursion in computing and mathematics, a technique whereby a ◊function or ◊procedure calls itself into use in order to enable a complex problem to be broken down into simpler steps. For example, a function that finds the factorial of a number n (calculates the product of all the whole numbers between 1 and n) would obtain its result by multiplying n by the factorial of $n - 1$.

recycling processing of industrial and household waste (such as paper, glass, and some metals and plastics) so that the materials can be reused. This saves expenditure on scarce raw materials, slows down the depletion of ◊nonrenewable resources, and helps to reduce pollution.

Aluminium is frequently recycled because of its value and special properties that allow it to be melted down and re-pressed without loss of quality, unlike paper and glass, which deteriorate when recycled.

red blood cell or **erythrocyte** the most common type of blood cell, responsible for transporting oxygen around the body. It contains haemoglobin, which combines with oxygen from the lungs to form oxyhaemoglobin. When transported to the tissues, these cells are able to release the oxygen because the oxyhaemoglobin splits into its original constituents.

Mammalian erythrocytes are disc-shaped with a depression in the centre and no nucleus; they are manufactured in the bone marrow and, in humans, last for only four months before being destroyed in the liver and spleen. Those of other vertebrates are oval and nucleated.

red dwarf any star that is cool, faint, and small (about one-tenth the mass and diameter of the Sun). Red dwarfs burn slowly, and have estimated lifetimes of 100 billion years. They may be the most abundant type of star, but are difficult to see because they are so faint. Two of the closest stars to the Sun, ◊Proxima Centauri and ◊Barnard's Star, are red dwarfs.

red giant any large bright star with a cool surface. It is thought to represent a late stage in the evolution of a star like the Sun, as it runs out of hydrogen fuel at its centre. Red giants have diameters between 10 and 100 times that of the Sun. They are very bright because they are so large, although their surface temperature is lower than that of the Sun, about 2,000–3,000K (1,700–2,700°C/ 3,000°–5,000°F). See also Red ◊supergiants.

redox reaction chemical change where one reactant is reduced and the other reactant oxidized. The reaction can only occur if both reactants are present and each changes simultaneously. For example, hydrogen reduces copper (II) oxide to copper while it is itself oxidized to water. The corrosion of iron and the reactions taking place in electric and electrolytic cells are just a few instances of redox reactions.

red shift in astronomy, the lengthening of the wavelengths of light from an object as a result of the object's motion away from us. It is an example of the ◊Doppler effect. The red shift in light from galaxies is evidence that the universe is expanding.

Lengthening of wavelengths causes the light to move or shift towards the red end of the ◊spectrum, hence the name. The amount of red shift can be measured by the displacement of lines in an object's spectrum. By measuring the amount of red shift in light from stars and galaxies, astronomers can tell how quickly these objects are moving away from us. A strong gravitational field can also produce a red shift in light; this is termed **gravitational red shift**.

Redstone rocket short-range US military missile, modified for use as a space launcher. Redstone rockets launched the first two flights of the ◊Mercury project. A modified Redstone, *Juno 1*, launched the first US satellite, *Explorer 1*, in 1958.

reduction in chemistry, the gain of electrons, loss of oxygen, or gain of hydrogen by an atom, ion, or molecule during a chemical reaction.

Reduction may be brought about by reaction with another compound, which is simultaneously oxidized (reducing agent), or electrically at the cathode (negative electrode) of an electric cell.

reflecting telescope in astronomy, a ◊telescope in which light is collected and brought to a focus by a concave mirror. The ◊Cassegrain telescope and ◊Newtonian telescope are examples.

refining any process that purifies or converts something into a more useful form. Metals usually need refining after they have been extracted from their ores by such processes as The ◊smelting. Petroleum, or crude oil, needs refining before it can be used; the process involves fractional ◊distillation, the separation of the substance into separate components or 'fractions'.

Electrolytic metal-refining methods use the principle of ◊electrolysis to obtain pure metals. When refining petroleum, or crude oil, further refinery processes after fractionation convert the heavier fractions into more useful lighter products. The most important of these processes is

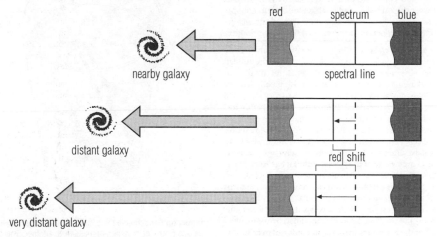

red shift The red shift causes lines in the spectra of galaxies to be shifted towards the red end of the spectrum. More distant galaxies have greater red shifts than closer galaxies. The red shift indicates that distant galaxies are moving apart rapidly, as the universe expands.

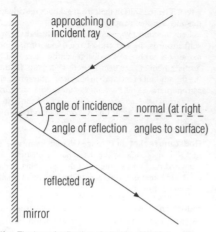

reflection The law of reflection: the angle of incidence of a light beam equals the angle of reflection of the beam.

◊cracking; others include ◊polymerization, hydrogenation, and reforming.

reflection the throwing back or deflection of waves, such as ◊light or sound waves, when they hit a surface. The *law of reflection* states that the angle of incidence (the angle between the ray and a perpendicular line drawn to the surface) is equal to the angle of reflection (the angle between the reflected ray and a perpendicular to the surface).

reflex in animals, a very rapid involuntary response to a particular stimulus. It is controlled by the ◊nervous system. A reflex involves only a few nerve cells, unlike the slower but more complex responses produced by the many processing nerve cells of the brain.

Reflex

When threatened by a predator, many tree-frog species urinate before running away. This is not just a fear-induced reflex, but enables the frog to escape more quickly by jumping further. Tree-frogs with empty bladders can jump an extra two body lengths.

A *simple reflex* is entirely automatic and involves no learning. Examples of such reflexes include the sudden withdrawal of a hand in response to a painful stimulus, or the jerking of a leg when the kneecap is tapped. Sensory cells (receptors) in the knee send signals to the spinal cord along a sensory nerve cell. Within the spine a *reflex arc* switches the signals straight back to the muscles of the leg (effectors) via an intermediate nerve cell and then a motor nerve cell; contraction of the leg occurs, and the leg kicks upwards. Only three nerve cells are involved, and the brain is only aware of the response after it has taken place. Such reflex arcs are particularly common in lower animals, and have a high survival value, enabling organisms to take rapid action to avoid potential danger. In higher animals (those with a well-developed ◊central nervous system) the simple reflex can be modified by the involvement of the brain – for instance, humans can override the automatic reflex to withdraw a hand from a source of pain.

A *conditioned reflex* involves the modification of a reflex action in response to experience (learning). A stimulus that produces a simple reflex response becomes linked with another, possibly unrelated, stimulus. For example, a dog may salivate (a reflex action) when it sees its owner remove a tin-opener from a drawer because it has learned to associate that stimulus with the stimulus of being fed.

reflex angle an angle greater than 180° but less than 360°.

reflex camera camera that uses a mirror and prisms to reflect light passing through the lens into the viewfinder, showing the photographer the exact scene that is being shot. When the shutter button is released the mirror springs out of the way, allowing light to reach the film. The most common type is the single-lens reflex (◊SLR) camera. The twin-lens reflex (◊TLR) camera has two lenses: one has a mirror for viewing, the other is used for exposing the film.

reflexology in alternative medicine, manipulation and massage of the feet to ascertain and treat disease or dysfunction elsewhere in the body.

Correspondence between reflex points on the feet and remote organic and physical functions was discovered early in the 20th century by US physician William Fitzgerald, who also found that pressure and massage applied to these reflex points beneficially affect the related organ or function.

refraction the bending of a wave when it passes from one medium to another. Refraction occurs because waves travel at different velocities in different media.

refractive index measure of the refraction of a ray of light as it passes from one transparent medium to another.

refraction Refraction is the bending of a light beam when it passes from one transparent medium to another. This is why a spoon appears bent when standing in a glass of water and pools of water appear shallower than they really are.

If the angle of incidence is i and the angle of refraction is r, the ratio of the two refractive indices is given by

$$n_1/n_2 = \sin i/\sin r$$

It is also equal to the speed of light in the first medium divided by the speed of light in the second, and it varies with the wavelength of the light.

refractor in astronomy, ◊telescope in which light is collected and brought to a focus by a convex lens (the object lens or objective).

refractory (of a material) able to resist high temperature, for example ceramics made from clay, minerals, or other earthy materials. Furnaces are lined with refractory materials such as silica and dolomite.

Alumina (aluminium oxide) is an excellent refractory, often used for the bodies of spark plugs. Titanium and tungsten are often called refractory metals because they are temperature resistant. ◊Cermets are refractory materials made up of ceramics and metals.

refrigeration use of technology to transfer heat from cold to warm, against the normal temperature gradient, so that a body can remain substantially colder than its surroundings. Refrigeration equipment is used for the chilling and deep-freezing of food in ◊food technology, and in air conditioners and industrial processes.

Refrigeration is commonly achieved by a vapour-compression cycle, in which a suitable chemical (the refrigerant) travels through a long circuit of tubing, during which it changes from a vapour to a liquid and back again. A compression chamber makes it condense, and thus give out heat. In another part of the circuit, called the evaporator coils, the pressure is much lower, so the refrigerant evaporates, absorbing heat as it does so. The evaporation process takes place near the central part of the refrigerator, which therefore becomes colder, while the com-

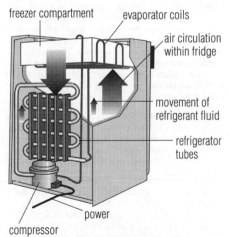

freezer compartment   evaporator coils
air circulation within fridge
movement of refrigerant fluid
refrigerator tubes
power
compressor

refrigeration The circulating refrigerant fluid moves heat from the inside to the outside of a refrigerator. The refrigerant is pumped around a circuit of tubing which wraps around the freezer compartment and forms the radiator coil at the rear. The refrigerant evaporates when travelling around the freezer, extracting heat in the process. The compressor condenses the refrigerant back into liquid while it is travelling through the radiator coils, releasing heat.

pression process takes place near a ventilation grille, transferring the heat to the air outside. The most commonly used refrigerants in modern systems were ◊chlorofluorocarbons, but these are now being replaced by coolants that do not damage the ozone layer.

Refrigeration

The philosopher Francis Bacon was probably the first person whose death was caused by frozen food. Experimenting with the idea of preserving meat by freezing, he stuffed a chicken with snow. Unfortunately the experiment resulted in his catching a chill, then bronchitis, which led to his death.

regelation phenomenon in which water refreezes to ice after it has been melted by pressure at a temperature below the freezing point of water. Pressure makes an ice skate, for example, form a film of water that freezes once again after the skater has passed.

regeneration in biology, regrowth of a new organ or tissue after the loss or removal of the original. It is common in plants, where a new individual can often be produced from a 'cutting' of the original. In animals, regeneration of major structures is limited to lower organisms; certain lizards can regrow their tails if these are lost, and new flatworms can grow from a tiny fragment of an old one. In mammals, regeneration is limited to the repair of tissue in wound healing and the regrowth of peripheral nerves following damage.

regimen in medicine, a course of treatment that is designed to improve the health of a patient. Many regimens consist of combinations of drugs but other elements, such as diet, exercise, and physiotherapy, may be incorporated.

register in computing, a memory location that can be accessed rapidly; it is often built into the computer's central processing unit.

Some registers are reserved for special tasks – for example, an *instruction register* is used to hold the machine-code command that the computer is currently executing, while a *sequence-control register* keeps track of the next command to be executed. Other registers are used for holding frequently used data and for storing intermediate results.

regolith the surface layer of loose material that covers most bedrock. It consists of eroded rocky material, volcanic ash, river alluvium, vegetable matter, or a mixture of these known as ◊soil.

regular of geometric figures, having all angles and sides equal. Also, of solids, having bases comprised of regular ◊polygons.

Regulus or *Alpha Leonis* the brightest star in the constellation Leo, and the 21st brightest star in the sky. First-magnitude Regulus has a true luminosity 100 times that of the Sun, and is 69 light years from Earth.

Regulus was one of the four royal stars of ancient Persia marking the approximate positions of the Sun at the equinoxes and solstices. The other three were ◊Aldebaran, ◊Antares, and ◊Fomalhaut.

regurgitation in medicine, the return of food that has already been swallowed from the oesophagus or the stomach to the mouth.

Regurgitation of blood in the heart results from

defective heart valves. These inhibit the passage of blood from the atria to the ventricles or from the ventricles to the arteries.

rehabilitation in medicine, the period required to restore general health and the capacity to work following recovery from the immediate effects of disease, mental illness, or addiction.

Reichian therapy in alternative medicine, any of a group of body therapies based on the theory, propounded in the 1930s by Austrian doctor Wilhelm Reich, that many functional and organic illnesses are attributable to constriction of the flow of vital energies in the body by tensions that become locked into the musculature. ◊Bioenergetics and Rolfing are related approaches.

relational database ◊database in which data are viewed as a collection of linked tables. It is the most popular of the three basic database models, the others being **network** and **hierarchical**.

relative in computing (of a value), variable and calculated from a base value. For example, a **relative address** is a memory location that is found by adding a variable to a base (fixed) address, and a **relative cell reference** locates a cell in a spreadsheet by its position relative to a base cell – perhaps directly to the left of the base cell or three columns to the right of the base cell. The opposite of relative is ◊absolute.

relative atomic mass the mass of an atom relative to one-twelfth the mass of an atom of carbon-12. It depends primarily on the number of protons and neutrons in the atom, the electrons having negligible mass. If more than one ◊isotope of the element is present, the relative atomic mass is calculated by taking an average that takes account of the relative proportions of each isotope, resulting in values that are not whole numbers. The term **atomic weight**, although commonly used, is strictly speaking incorrect.

relative density the density (at 20°C/68°F) of a solid or liquid relative to (divided by) the maximum density of water (at 4°C/39.2°F). The relative density of a gas is its density divided by the density of hydrogen (or sometimes dry air) at the same temperature and pressure.

relative humidity the concentration of water vapour in the air. It is expressed as the ratio of the partial pressure of the water vapour to its saturated vapour pressure at the same temperature. The higher the temperature, the higher the saturated vapour pressure.

relative molecular mass the mass of a molecule, calculated relative to one-twelfth the mass of an atom of carbon-12. It is found by adding the relative atomic masses of the atoms that make up the molecule.

The term **molecular weight** is often used, but strictly this is incorrect.

relativity in physics, the theory of the relative rather than absolute character of motion and mass, and the interdependence of matter, time, and space, as developed by German-born US physicist Albert Einstein in two phases.

Special theory of relativity (1905) Starting with the premises that (1) the laws of nature are the same for all observers in unaccelerated motion, and (2) the speed of light is independent of the motion of its source, Einstein arrived at some rather unexpected consequences. Intuitively familiar concepts, like mass, length, and time, had to

be modified. For example, an object moving rapidly past the observer will appear to be both shorter and heavier than when it is at rest (that is, at rest relative to the observer), and a clock moving rapidly past the observer will appear to be running slower than when it is at rest. These predictions of relativity theory seem to be foreign to everyday experience merely because the changes are quite negligible at speeds less than about 1,500 km s^{-1}, and they only become appreciable at speeds approaching the speed of light.

General theory of relativity (1915) The geometrical properties of space-time were to be conceived as modified locally by the presence of a body with mass. A planet's orbit around the Sun (as observed in three-dimensional space) arises from its natural trajectory in modified space-time; there is no need to invoke, as Isaac Newton did, a force of ◊gravity coming from the Sun and acting on the planet. Einstein's general theory accounts for a peculiarity in the behaviour of the motion of the perihelion of the orbit of the planet Mercury that cannot be explained in Newton's theory. The new theory also said that light rays should bend when they pass by a massive object. The predicted bending of starlight was observed during the eclipse of the Sun in 1919. A third corroboration is found in the shift towards the red in the spectra of the Sun and, in particular, of stars of great density – white dwarfs such as the companion of Sirius.

Einstein showed that, for consistency with the above premises (1) and (2), the principles of dynamics as established by Newton needed modification; the most celebrated new result was the equation $E = mc^2$, which expresses an equivalence between mass (m) and ◊energy (E), c being the speed of light in a vacuum. In 'relativistic mechanics', conservation of mass is replaced by the new concept of conservation of 'mass-energy'.

Although since modified in detail, general relativity remains central to modern ◊astrophysics and ◊cosmology; it predicts, for example, the possibility of ◊black holes. General relativity theory was inspired by the simple idea that it is impossible in a small region to distinguish between acceleration and gravitation effects (as in a lift one feels heavier when the lift accelerates upwards), but the mathematical development of the idea is formidable. Such is not the case for the special theory, which a non-expert can follow up to $E = mc^2$ and beyond.

relaxation therapy development of regular and conscious control of physiological processes and their related emotional and mental states, and of muscular tensions in the body, as a way of relieving stress and its results. Meditation, ◊hypnotherapy, ◊autogenics, and ◊biofeedback are techniques commonly employed.

relaxin hormone produced naturally by women during pregnancy that assists childbirth. It widens the pelvic opening by relaxing the ligaments, inhibits uterine contractility, so preventing premature labour, and causes dilation of the cervix. A synthetic form was pioneered by the Howard Florey Institute in Australia, and this drug has possible importance in helping the birth process and avoiding surgery or forceps delivery.

relay in electrical engineering, an electromagnetic switch. A small current passing through a coil of wire wound around an iron core attracts an ◊armature whose movement closes a pair of sprung contacts to complete a secondary circuit, which may carry a large current or activate other devices. The solid-state equivalent is a thyristor switching device.

rem acronym of *roentgen equivalent man* unit of radiation dose equivalent.

remission in medicine, temporary disappearance of symptoms during the course of a disease.

remote sensing gathering and recording information from a distance. Space probes have sent back photographs and data about planets as distant as Neptune. In archaeology, surface survey techniques provide information without disturbing subsurface deposits.

Satellites such as *Landsat* have surveyed all the Earth's surface from orbit. Computer processing of data obtained by their scanning instruments, and the application of so-called false colours (generated by the computer), have made it possible to reveal surface features invisible in ordinary light. This has proved valuable in agriculture, forestry, and urban planning, and has led to the discovery of new deposits of minerals.

remote terminal in computing, a terminal that communicates with a computer via a modem (or acoustic coupler) and a telephone line.

REM sleep (acronym for *rapid-eye-movement* sleep) Phase of sleep that recurs several times nightly in humans and is associated with dreaming. The eyes flicker quickly beneath closed lids.

renal pertaining to the ◊kidneys.

renewable energy power from any source that replenishes itself. Most renewable systems rely on ◊solar energy directly or through the weather cycle as ◊wave power, ◊hydroelectric power, or wind power via ◊wind turbines, or solar energy collected by plants (alcohol fuels, for example). In addition, the gravitational force of the Moon can be harnessed through ◊tidal power stations, and the heat trapped in the centre of the Earth is used via ◊geothermal energy systems.

renewable resource natural resource that is replaced by natural processes in a reasonable amount of time. Soil, water, forests, plants, and animals are all renewable

resources as long as they are properly conserved. Solar, wind, wave, and geothermal energies are based on renewable resources.

rennet extract, traditionally obtained from a calf's stomach, that contains the enzyme rennin, used to coagulate milk in the cheesemaking process. The enzyme can now be chemically produced.

rennin or *chymase* enzyme found in the gastric juice of young mammals, used in the digestion of milk.

repellent anything whose smell, taste, or other properties discourages nearby creatures. *Insect repellent* is usually a chemical substance that keeps, for example, mosquitoes at bay; natural substances include citronella, lavender oil, and eucalyptus oils. A device that emits ultrasound waves is also claimed to repel insects and small mammals.

repetitive strain injury (RSI) inflammation of tendon sheaths, mainly in the hands and wrists, which may be disabling. It is found predominantly in factory workers involved in constant repetitive movements, and in those who work with computer keyboards. The symptoms include aching muscles, weak wrists, tingling fingers and in severe cases, pain and paralysis. Some victims have successfully sued their employers for damages.

replication in biology, production of copies of the genetic material DNA; it occurs during cell division (◊mitosis and ◊meiosis). Most mutations are caused by mistakes during replication.

During replication the paired strands of DNA separate, exposing the bases. Nucleotides floating in the cell matrix pair with the exposed bases, adenine pairing with thymine, and cytosine with guanine.

repression in psychology, a mental process that ejects and excludes from consciousness ideas, impulses, or memories that would otherwise threaten emotional stability.

In the Austrian psychiatrist Sigmund Freud's early writing, repression is controlled by the censor, a hypothetical mechanism or agency that allows ideas, memories, and so on from the unconscious to emerge into

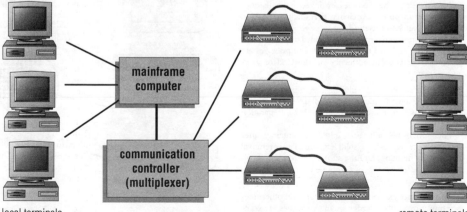

local terminals   remote terminals

remote terminal Remote computer terminals communicate with the central mainframe via modems and telephone lines. The controller allocates computer time to the terminals according to predetermined priority rules. The multiplexer allows more than one terminal to use the same communications link at the same time (multiplexing).

consciousness only if distorted or disguised, as for example in dreams.

reproduction in biology, process by which a living organism produces other organisms similar to itself. There are two kinds: ◊asexual reproduction and ◊sexual reproduction. The ability to reproduce is considered one of the fundamental attributes of living things.

Reproduction

The mating technique of African velvet worms is bizarre. The male places sperm bundles anywhere along the female's body. Beneath the sperm bundle the skin dissolves forming a hole. The sperm can thus enter the female's body cavity, where it makes its way to her ovaries.

reptile any member of a class (Reptilia) of vertebrates. Unlike amphibians, reptiles have hard-shelled, yolk-filled eggs that are laid on land and from which fully formed young are born. Some snakes and lizards retain their eggs and give birth to live young. Reptiles are cold-blooded, and their skin is usually covered with scales. The metabolism is slow, and in some cases (certain large snakes) intervals between meals may be months. Reptiles date back over 300 million years.

Many extinct forms are known, including the orders Pterosauria, Plesiosauria, Ichthyosauria, and Dinosauria. The chief living orders are the Chelonia (tortoises and turtles), Crocodilia (alligators and crocodiles), and Squamata, divided into three suborders: Lacertilia (lizards), Ophidia or Serpentes (snakes), and Amphisbaenia (worm lizards). The order Rhynchocephalia has one surviving species, the lizardlike tuatara of New Zealand.

A four-year study of rainforest in E Madagascar revealed 26 new reptile species in 1995.

research the primary activity in science, a combination of theory and experimentation directed towards finding scientific explanations of phenomena. It is commonly classified into two types: **pure research**, involving theories with little apparent relevance to human concerns; and **applied research**, concerned with finding solutions to problems of social importance – for instance in medicine and engineering. The two types are linked in that theories developed from pure research may eventually be found to be of great value to society.

Scientific research is most often funded by government and industry, and so a nation's wealth and priorities are likely to have a strong influence on the kind of work undertaken.

residue in chemistry, a substance or mixture of substances remaining in the original container after the removal of one or more components by a separation process.

The nonvolatile substance left in a container after ◊evaporation of liquid, the solid left behind after removal of liquid by filtration, and the substances left in a distillation flask after removal of components by ◊distillation, are all residues.

resin substance exuded from pines, firs, and other trees in gummy drops that harden in air. Varnishes are common products of the hard resins, and ointments come from the soft resins.

Rosin is the solid residue of distilled turpentine, a soft resin. The name 'resin' is also given to many synthetic

products manufactured by polymerization; they are used in adhesives, plastics, and varnishes.

resistance in physics, that property of a conductor that restricts the flow of electricity through it, associated with the conversion of electrical energy to heat; also the magnitude of this property. Resistance depends on many factors, such as the nature of the material, its temperature, dimensions, and thermal properties; degree of impurity; the nature and state of illumination of the surface; and the frequency and magnitude of the current. The SI unit of resistance is the ohm.

resistivity in physics, a measure of the ability of a material to resist the flow of an electric current. It is numerically equal to the ◊resistance of a sample of unit length and unit cross-sectional area, and its unit is the ohm metre (symbol Ωm). A good conductor has a low resistivity (1.7×10^{-8} Ωm for copper); an insulator has a very high resistivity (10^{15} Ωm for polyethane).

resistor in physics, any component in an electrical circuit used to introduce ◊resistance to a current. Resistors are often made from wire-wound coils or pieces of carbon. ◊Rheostats and ◊potentiometers are variable resistors.

resolution in computing, the number of dots per unit length in which an image can be reproduced on a screen or printer. A typical screen resolution for colour monitors is 75 dpi (dots per inch). A ◊laser printer will typically have a printing resolution of 300 dpi, and ◊dot matrix printers typically have resolutions from 60 dpi to 180 dpi. Photographs in books and magazines have a resolution of 1,200 dpi or 2,400 dpi.

resonance rapid amplification of a vibration when the vibrating object is subject to a force varying at its ◊natural frequency. In a trombone, for example, the length of the air column in the instrument is adjusted until it resonates with the note being sounded. Resonance effects are also produced by many electrical circuits. Tuning a radio, for

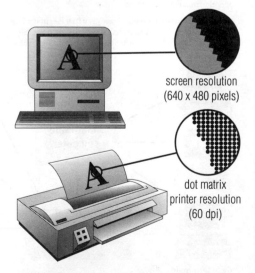

screen resolution
(640 x 480 pixels)

dot matrix
printer resolution
(60 dpi)

resolution An example of typical resolutions of screens and printers. The resolution of a screen image when printed can only be as high as the resolution supported by the printer itself.

example, is done by adjusting the natural frequency of the receiver circuit until it coincides with the frequency of the radio waves falling on the aerial.

Resonance has many physical applications. Children use it to increase the size of the movement on a swing, by giving a push at the same point during each swing. Soldiers marching across a bridge in step could cause the bridge to vibrate violently if the frequency of their steps coincided with its natural frequency.

Resonance was the cause of the collapse of the Tacoma Narrows Bridge, USA, in 1940, when the frequency of the wind gusts coincided with the natural frequency of the bridge.

resources materials that can be used to satisfy human needs. Because human needs are diverse and extend from basic physical requirements, such as food and shelter, to ill-defined aesthetic needs, resources encompass a vast range of items. The intellectual resources of a society – its ideas and technologies – determine which aspects of the environment meet that society's needs, and therefore become resources. For example, in the 19th century, uranium was used only in the manufacture of coloured glass. Today, with the advent of nuclear technology, it is a military and energy resource. Resources are often categorized into **human resources**, such as labour, supplies, and skills, and **natural resources**, such as climate, fossil fuels, and water. Natural resources are divided into ◊nonrenewable resources and ◊renewable resources.

Nonrenewable resources include minerals such as coal, copper ores, and diamonds, which exist in strictly limited quantities. Once consumed they will not be replenished within the time span of human history. In contrast, water supplies, timber, food crops, and similar resources can, if managed properly, provide a steady yield virtually forever; they are therefore replenishable or renewable resources. Inappropriate use of renewable resources can lead to their destruction, as for example the cutting down of rainforests, with secondary effects, such as the decrease

in oxygen and the increase in carbon dioxide and the ensuing ◊greenhouse effect. Some renewable resources, such as wind or solar energy, are continuous; supply is largely independent of people's actions.

Demands for resources made by rich nations are causing concern that the present and future demands of industrial societies cannot be sustained for more than a century or two, and that this will be at the expense of the Third World and the global environment. Other authorities believe that new technologies will emerge, enabling resources that are now of little importance to replace those being exhausted.

The earth only has so much bounty to offer and inventing ever larger and more notional prices for that bounty does not change its real value.

On **resources** Ben Elton *Stark*, 'Dinner in Los Angeles'

respiration biochemical process whereby food molecules are progressively broken down (oxidized) to release energy in the form of ◊ATP. In most organisms this requires oxygen, but in some bacteria the oxidant is the nitrate or sulphate ion instead. In all higher organisms, respiration occurs in the ◊mitochondria. Respiration is also used to mean breathing, in which oxygen is exchanged for carbon dioxide in the lung alveoli, though this is more accurately described as a form of ◊gas exchange.

respiratory distress syndrome (RDS) formerly **hyaline membrane disease** condition in which a newborn baby's lungs are insufficiently expanded to permit adequate oxygenation. Premature babies are most at risk. Such babies survive with the aid of intravenous fluids and oxygen, sometimes with mechanically assisted ventilation.

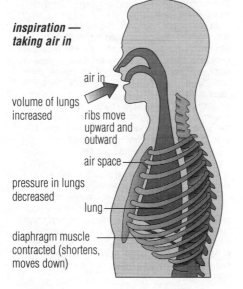

*inspiration —
taking air in*

air in

volume of lungs
increased

ribs move
upward and
outward

air space

pressure in lungs
decreased

lung

diaphragm muscle
contracted (shortens,
moves down)

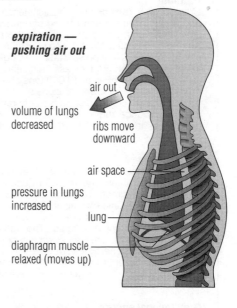

*expiration —
pushing air out*

air out

volume of lungs
decreased

ribs move
downward

air space

pressure in lungs
increased

lung

diaphragm muscle
relaxed (moves up)

Respiration

Energy resources

Humans are using up the world's energy resources in a way no other animal has ever done. We use them to provide light and heating in our homes, to plough the land, to cook our food, to travel, to run our factories, and in countless other ways. Whether we are rural workers in a developing country or urban workers in a wealthy industrial country, we all need energy, although the sources of the energy and the amounts used vary greatly from one society to another. Energy resources fall into two broad groups: renewable and nonrenewable. Renewable resources are those which replenish themselves naturally and will either always be available – hydroelectric power, solar energy, wind and wave power, tidal energy, and geothermal energy – or will continue to be available provided supplies are given sufficient time to replenish themselves – peat and firewood. Nonrenewable resources are those of which there are limited supplies and which once used are gone forever. These include coal, oil, natural gas, and uranium.

Fossil fuels
Coal, oil, and natural gas are called fossil fuels because they are the fossilized remains of plants and animals that lived hundreds of millions of years ago. Burning fossil fuels releases chemicals that cause acid rain, and is gradually increasing the carbon dioxide in the atmosphere, causing global warming.

Uranium
Uranium is a radioactive metallic element and a very concentrated source of energy; large reserves are found in Australia, North America, and South Africa. Used to produce electricity in a nuclear power station, a single ton of uranium can produce as much energy as 15,000 tons of coal, or 10,000 tons of oil. Although nuclear power stations do not produce carbon dioxide or cause acid rain, they do produce radioactive waste that is dangerous and difficult to process or store safely.

Solar energy
Many renewable resources take advantage of the energy in sunlight. The Sun's energy can be tapped directly by photovoltaic cells that convert light into electricity. Other solar energy plants use mirrors to direct sunlight onto pipes containing a liquid. The liquid boils and is used to drive an electricity generator. The Sun's energy also drives the wind and waves, so energy produced by wind farms and wave-driven generators is also derived from the Sun.

Gravitational energy
Hydroelectricity and tidal power stations make use of gravitational forces. The Earth's gravity pulls water downward through the turbines in a hydroelectric power station. In a tidal power station, the Moon's gravity lifts water as the tides rise, giving the water potential energy (energy due to position) which is released as the water flows through a turbine. Geothermal energy (the heat energy of hot rocks deep beneath the Earth's surface) is due to gravity compressing and heating the rocks when the Earth formed.

The worldwide energy pie
Globally, the largest contributions to current energy resources come from oil (31%), coal (26%), and natural gas (19%). Renewable energy currently supplies about 20% of the world's energy needs, with hydroelectricity supplying 6% of the world's needs and traditional biofuels (firewood, crop wastes, peat, and dung) supplying 12%. A small contribution is made by new renewables, such as the conversion of crops such as sugar into alcohol fuel and the burning of waste material.

Future demands and solutions
It is clear that, in the future, demand for energy will be higher than at present, due to population growth and increased industrialization. Furthermore, the energy available must be at a reasonable cost or economic growth will be held back. This is especially important for developing countries, where the inability to meet high-energy costs hinders development.

In principle, known resources of nonrenewable energy should be sufficient for several hundred years or more. However, in practice, the outlook is uncertain. Increasing concern about pollution might make dirty coal-fired power stations unacceptable in the future.

One alternative is to make greater use of nuclear power, moving to fast reactors and then developing nuclear fusion plants that would mimic the power production process found in the Sun. However, anxiety about safety and waste disposal is already limiting the use of nuclear energy, so it is unlikely to provide an answer in the future.

There is considerable room for development in the use of renewable resources, but with most of the world's energy production based around nonrenewable fuel supplies, the widespread introduction of efficient renewable energy will require a complete restructuring of the ways we produce and use energy.

Peter Lafferty

Normal inflation of the lungs requires the presence of a substance called a surfactant to reduce the surface tension of the alveoli (air sacs) in the lungs. In premature babies, the surfactant is deficient and the lungs become hard and glassy. As a result, the breathing is rapid, laboured, and shallow, and there is the likelihood of ◊asphyxia. Artificial surfactant is administered to babies at risk.

respiratory surface area used by an organism for the exchange of gases, for example the lungs, gills or, in plants, the leaf interior. The gases oxygen and carbon dioxide are both usually involved in respiration and photosynthesis. Although organisms have evolved different types of respiratory surface according to need, there are certain features in common. These include thinness and moistness, so that the gas can dissolve in a membrane and then diffuse into the body of the organism. In many animals the gas is then transported away from the surface and towards interior cells by the blood system.

response time in computing, the delay between entering a command and seeing its effect.

rest mass in physics, the mass of a body when its velocity is zero. For subatomic particles, it is their mass at rest or at velocities considerably below that of light. According to the theory of ◊relativity, at very high velocities, there is a relativistic effect that increases the mass of the particle.

restriction enzyme bacterial ◊enzyme that breaks a chain of ◊DNA into two pieces at a specific point; used in ◊genetic engineering. The point along the DNA chain at which the enzyme can work is restricted to places where a specific sequence of base pairs occurs. Different restriction enzymes will break a DNA chain at different points. The overlap between the fragments is used in determining the sequence of base pairs in the DNA chain.

resuscitation steps taken to revive anyone on the brink of death. The most successful technique for life-threatening emergencies, such as electrocution, near-drowning, or heart attack, is mouth-to-mouth resuscitation. Medical and paramedical staff are trained in cardiopulmonary resuscitation (CPR): the use of specialized equipment and techniques to attempt to restart the breathing and/or heartbeat and stabilize the patient long enough for more definitive treatment. CPR has a success rate of less than 30%.

retina light-sensitive area at the back of the ◊eye connected to the brain by the optic nerve. It has several layers and in humans contains over a million rods and cones, sensory cells capable of converting light into nervous messages that pass down the optic nerve to the brain.

The **rod cells**, about 120 million in each eye, are distributed throughout the retina. They are sensitive to low levels of light, but do not provide detailed or sharp images, nor can they detect colour. The **cone cells**, about 6 million in number, are mostly concentrated in a central region of the retina called the **fovea**, and provide both detailed and colour vision. The cones of the human eye contain three visual pigments, each of which responds to a different primary colour (red, green, or blue). The brain can interpret the varying signal levels from the three types of cone as any of the different colours of the visible spectrum.

The image actually falling on the retina is highly distorted; research into the eye and the optic centres within the brain has shown that this poor quality image is processed to improve its quality. The retina can become separated from the wall of the eyeball as a result of a trauma, such as a boxing injury. It can be reattached by being 'welded' into place by a laser.

retinol or **vitamin A** fat-soluble chemical derived from β-carotene and found in milk, butter, cheese, egg yolk, and liver. Lack of retinol in the diet leads to the eye disease **xerophthalmia**.

retinopathy in medicine, a general term to describe diseases of the retina. Such diseases are often complications of other diseases, such as hypertensive retinopathy, which occurs in people with high blood pressure and diabetic retinopathy, which occurs in patients with diabetes mellitus. The underlying condition needs to be treated to prevent the progress of the retinopathy.

retrograde in astronomy, the orbit or rotation of a ◊planet or ◊satellite if the sense of rotation is opposite to the general sense of rotation of the Solar System. On the ◊celestial sphere, it refers to motion from east to west against the background of stars.

retrovirus any of a family of ◊viruses (Retroviridae) containing the genetic material ◊RNA rather than the more usual ◊DNA.

For the virus to express itself and multiply within an infected cell, its RNA must be converted to DNA. It does this by using a built-in enzyme known as reverse transcriptase (since the transfer of genetic information from DNA to RNA is known as ◊transcription, and retroviruses do the reverse of this). Retroviruses include those causing ◊AIDS and some forms of leukaemia. See ◊immunity.

Retroviruses are used as vectors in ◊genetic engineering, but they cannot be used to target specific sites on the chromosome. Instead they incorporate their genes at random sites.

reuse multiple use of a product (often a form of packaging), by returning it to the manufacturer or processor each time. Many such returnable items are sold with a deposit which is reimbursed if the item is returned. Reuse is usually more energy- and resource-efficient than ◊recycling unless there are large transport or cleaning costs.

reverberation in acoustics, the multiple reflections, or echoes, of sounds inside a building that merge and persist a short while (up to a few seconds) before fading away. At each reflection some of the sound energy is absorbed, causing the amplitude of the sound wave and the intensity of the sound to reduce a little.

Too much reverberation causes sounds to become confused and indistinct, and this is particularly noticeable in empty rooms and halls, and such buildings as churches and cathedrals where the hard, unfurnished surfaces do not absorb sound energy well. Where walls and surfaces absorb sound energy very efficiently, too little reverberation may cause a room or hall to sound dull or 'dead'. Reverberation is a key factor in the design of theatres and concert halls, and can be controlled by lining ceilings and walls with materials possessing specific sound-absorbing properties.

reverse osmosis the movement of solvent (liquid) through a semipermeable membrane from a more concentrated solution to a more dilute solution. The solvent's direction of movement is opposite to that which it would experience during ◊osmosis, and is achieved by applying an external pressure to the solution on the more concentrated side of the membrane. The technique is used in desalination plants, when water (the solvent) passes from brine (a salt solution) into fresh water via a semipermeable filtering device.

reversible reaction chemical reaction that proceeds in both directions at the same time, as the product decomposes back into reactants as it is being produced. Such reactions do not run to completion, provided that no substance leaves the system. Examples include the manufacture of ammonia from hydrogen and nitrogen, and the oxidation of sulphur dioxide to sulphur trioxide.

Reye's syndrome rare disorder of the metabolism causing fatty infiltration of the liver and ◊encephalitis. It occurs mainly in children and has been linked with aspirin therapy, although its cause is still uncertain. The mortality rate is 50%.

Reynolds number number used in ◊fluid mechanics to determine whether fluid flow in a particular situation (through a pipe or around an aircraft body or a fixed structure in the sea) will be turbulent or smooth. The Reynolds number is calculated using the flow velocity, density and viscosity of the fluid, and the dimensions of the flow channel. It is named after British engineer Osborne Reynolds.

RGB (abbreviation of *red–green–blue*) method of connecting a colour screen to a computer, involving three separate signals: red, green, and blue. All the colours displayed by the screen can be made up from these three component colours.

rhe unit of fluidity equal to the reciprocal of the ◊poise.

rhenium heavy, silver-white, metallic element, symbol Re, atomic number 75, relative atomic mass 186.2. It has chemical properties similar to those of manganese and a very high melting point (3,180°C/5,756°F), which makes it valuable as an ingredient in alloys.

It was identified and named in 1925 by German chemists W Noddack (1893–1960), I Tacke, and O Berg from the Latin name for the river Rhine.

rheostat in physics, a variable ◊resistor, usually consisting of a high-resistance wire-wound coil with a sliding contact. It is used to vary electrical resistance without interrupting the current (for example, when dimming lights). The circular type in electronics (which can be used, for example, as the volume control of an amplifier) is also known as a ◊potentiometer.

rhesus factor group of ◊antigens on the surface of red blood cells of humans which characterize the rhesus blood group system. Most individuals possess the main rhesus factor (Rh+), but those without this factor (Rh−) produce ◊antibodies if they come into contact with it. The name comes from rhesus monkeys, in whose blood rhesus factors were first found.

If an Rh− mother carries an Rh+ fetus, she may produce antibodies if fetal blood crosses the ◊placenta. This is not normally a problem with the first infant because antibodies are only produced slowly. However, the antibodies continue to build up after birth, and a second Rh+ child may be attacked by antibodies passing from mother to fetus, causing the child to contract anaemia, heart failure, or brain damage. In such cases, the blood of the infant has

to be changed for Rh− blood; a badly affected fetus may be treated in the womb (see ◊fetal therapy). The problem can be circumvented by giving the mother anti-Rh globulin just after the first pregnancy, preventing the formation of antibodies.

rheumatic fever or *acute rheumatism* acute or chronic illness characterized by fever and painful swelling of joints. Some victims also experience involuntary movements of the limbs and head, a form of ◊chorea. It is now rare in the developed world.

Rheumatic fever, which strikes mainly children and young adults, is always preceded by a streptococcal infection such as ◊scarlet fever or a severe sore throat, usually occurring a couple of weeks beforehand. It is treated with bed rest, antibiotics, and painkillers. The most important complication of rheumatic fever is damage to the heart and its valves, producing rheumatic heart disease many years later, which may lead to disability and death.

rheumatism nontechnical term for a variety of ailments associated with inflammation and stiffness of the joints and muscles.

rheumatoid arthritis inflammation of the joints; a chronic progressive disease, it begins with pain and stiffness in the small joints of the hands and feet and spreads to involve other joints, often with severe disability and disfigurement. There may also be damage to the eyes, nervous system, and other organs. The disease is treated with a range of drugs and with surgery, possibly including replacement of major joints.

Rheumatoid arthritis most often develops between the ages of 30 and 40, and is three times more common in women than men. In the West it affects 2–3% of women and nearly 1% of men; an estimated 165 million people worldwide suffered from rheumatoid arthritis in 1995. In children rheumatoid arthritis is known as Still's disease.

rhinoplasty in medicine, operation to repair the nose or modify its shape. It is usually performed for cosmetic reasons. The operation may involve alteration of the bony skeleton of the nose or of the septum.

rhizoid hairlike outgrowth found on the ◊gametophyte generation of ferns, mosses and liverworts. Rhizoids anchor the plant to the substrate and can absorb water and nutrients.

They may be composed of many cells, as in mosses, where they are usually brownish, or may be unicellular, as in liverworts, where they are usually colourless. Rhizoids fulfil the same functions as the ◊roots of higher plants but are simpler in construction.

rhizome or *rootstock* horizontal underground plant stem. It is a ◊perennating organ in some species, where it is generally thick and fleshy, while in other species it is mainly a means of ◊vegetative reproduction, and is therefore long and slender, with buds all along it that send up new plants. The potato is a rhizome that has two distinct parts, the tuber being the swollen end of a long, cordlike rhizome.

rhm (abbreviation of *roentgen–hour–metre*) the unit of effective strength of a radioactive source that produces gamma rays. It is used for substances for which it is difficult to establish radioactive disintegration rates.

rhodium hard, silver-white, metallic element, symbol Rh, atomic number 45, relative atomic mass 102.905. It is one of the so-called platinum group of metals and is resistant to tarnish, corrosion, and acid. It occurs as a free metal in the natural alloy osmiridium and is used in jewellery, electroplating, and thermocouples.

rhombus in geometry, an equilateral (all sides equal) ◊parallelogram. Its diagonals bisect each other at right angles, and its area is half the product of the lengths of the two diagonals. A rhombus whose internal angles are 90° is called a ◊square.

rhyolite ◊igneous rock, the fine-grained volcanic (extrusive) equivalent of granite.

rhythm method method of natural contraception that relies on refraining from intercourse during ◊ovulation.

The time of ovulation can be worked out by the calendar (counting days from the last period), by temperature changes, or by inspection of the cervical mucus. All these methods are unreliable because it is possible for ovulation to occur at any stage of the menstrual cycle.

ria long narrow sea inlet, usually branching and surrounded by hills. A ria is deeper and wider towards its mouth, unlike a ◊fjord. It is formed by the flooding of a river valley due to either a rise in sea level or a lowering of a landmass.

rib long, usually curved bone that extends laterally from the ◊spine in vertebrates. Most fishes and many reptiles have ribs along most of the spine, but in mammals they are found only in the chest area. In humans, there are 12 pairs of ribs. The ribs protect the lungs and heart, and allow the chest to expand and contract easily.

At the rear, each pair is joined to one of the vertebrae of the spine. The upper seven ('true' or vertebro-sternal ribs) are joined by ◊cartilage directly to the breast bone (sternum). The next three ('false' or vertebro-costal ribs) are joined by cartilage to the end of the rib above. The last two ('floating ribs') are not attached at the front. The diaphragm and muscles between adjacent ribs are responsible for the respiratory movements which fill the lungs with air.

ribbon lake long, narrow lake found on the floor of a ◊glacial trough. A ribbon lake will often form in an elongated hollow carved out by a glacier, perhaps where it came across a weaker band of rock. Ribbon lakes can also form when water ponds up behind a terminal moraine or a landslide. The English Lake District is named after its many ribbon lakes, such as Lake Windermere and Coniston Water.

riboflavin or *vitamin B₂* ◊vitamin of the B complex important in cell respiration. It is obtained from eggs, liver, and milk. A deficiency in the diet causes stunted growth.

ribonucleic acid full name of ◊RNA.

ribosome in biology, the protein-making machinery of the cell. Ribosomes are located on the endoplasmic reticulum (ER) of eukaryotic cells, and are made of proteins and a special type of ◊RNA, ribosomal RNA. They receive messenger RNA (copied from the ◊DNA) and ◊amino acids, and 'translate' the messenger RNA by using its chemically coded instructions to link amino acids in a specific order, to make a strand of a particular protein.

rice principal ◊cereal of the wet regions of the tropics, derived from wild grasses probably native to India and SE Asia. Rice is unique among cereal crops in that it is grown standing in water. The yield is very large, and rice is said to be the staple food of one-third of the world's population. (*Oryza sativa.*)

Rice takes 150–200 days to mature in warm, wet conditions. During its growing period, it needs to be flooded either by the heavy monsoon rains or by irrigation. This restricts the cultivation of swamp rice, the usual kind, to level land and terraces. A poorer variety, known as hill rice, is grown on hillsides. Outside Asia, there is some rice production in the Po Valley in Italy and in Louisiana, the Carolinas, and California in the USA.

In Oct 1994 the International Rice Research Institute announced a new rice variety that can potentially increase rice yields by 25%. It produces more seed heads than the standard crop and each seed head contains 200 rice grains, compared with the present 100. The plant is also more compact, enabling it to be planted more densely.

Rice contains 8–9% protein. Brown, or unhusked, rice has valuable B-vitamins that are lost in husking or polishing. Most of the rice eaten in the world is, however, sold in the polished white form.

Rice has been cultivated since prehistoric days in the East. New varieties with greatly increased protein content have been developed by gamma radiation for commercial cultivation, and yields are now higher than ever before (see ◊green revolution).

Rice husks when burned provide a ◊silica ash that, mixed with lime, produces an excellent cement.

Richter scale scale based on measurement of seismic waves, used to determine the magnitude of an ◊earthquake at its epicentre. The magnitude of an earthquake differs from its intensity, measured by the ◊Mercalli scale, which is subjective and varies from place to place for the same earthquake. The scale is named after US seismologist Charles Richter.

An earthquake's magnitude is a function of the total amount of energy released, and each point on the Richter scale represents a thirtyfold increase in energy over the previous point. The greatest earthquake ever recorded, in 1920 in Gansu, China, measured 8.6 on the Richter scale.

ricin extremely poisonous extract from the seeds of the castor-oil plant. When incorporated into ◊monoclonal antibodies, ricin can attack cancer cells, particularly in the treatment of lymphoma and leukaemia.

rickets defective growth of bone in children due to an insufficiency of calcium deposits. The bones, which do not harden adequately, are bent out of shape. It is usually caused by a lack of vitamin D and insufficient exposure to sunlight. Renal rickets, also a condition of malformed bone, is associated with kidney disease.

ridge of high pressure elongated area of high atmospheric pressure extending from an anticyclone. On a synoptic weather chart it is shown as a pattern of lengthened isobars. The weather under a ridge of high pressure is the same as that under an anticyclone.

rift valley valley formed by the subsidence of a block of the Earth's ◊crust between two or more parallel ◊faults. Rift valleys are steep-sided and form where the crust is being pulled apart, as at ◊ocean ridges, or in the Great Rift Valley of E Africa.

Rigel or *Beta Orionis* brightest star in the constellation Orion. It is a blue-white supergiant, with an estimated diameter 50 times that of the Sun. It is 900 light years

Richter Scale

Magnitude	Relative amount of energy released	Examples	Year
1	1		
2	31		
3	960		
4	30,000	Carlisle, England (4.7)	1979
5	920,000	Wrexham, Wales (5.1)	1990
		northern Afghanistan (5.5–6.1)	1998
6	29,000,000	San Fernando (CA) (6.5)	1971
		northern Armenia (6.8)	1988
7	890,000,000	Loma Prieta (CA) (7.1)	1989
		Kobe, Japan (7.2)	1995
		Rasht, Iran (7.7)	1990
		San Francisco (CA) (7.7–7.9)[1]	1906
8	28,000,000,000	Tangshan, China (8.0)	1976
		Gansu, China (8.6)	1920
		Lisbon, Portugal (8.7)	1755
9	850,000,000,000	Prince William Sound (AK) (9.2)	1964

[1] Richter's original estimate of a magnitude of 8.3 has been revised by two recent studies carried out by the California Institute of Technology and the US Geological Survey.

from Earth, and is intrinsically the brightest of the first-magnitude stars, its luminosity being about 100,000 times that of the Sun. It is the seventh-brightest star in the sky.

right-angled triangle triangle in which one of the angles is a right angle (90°). It is the basic form of triangle for defining trigonometrical ratios (for example, sine, cosine, and tangent) and for which ◊Pythagoras' theorem holds true. The longest side of a right-angled triangle is called the hypotenuse; its area is equal to half the product of the lengths of the two shorter sides.

Any triangle constructed with its hypotenuse as the diameter of a circle, with its opposite vertex (corner) on the circumference, is a right-angled triangle. This is a fundamental theorem in geometry, first credited to the Greek mathematician Thales about 580 BC.

right ascension in astronomy, the coordinate on the ◊celestial sphere that corresponds to longitude on the surface of the Earth. It is measured in hours, minutes, and seconds eastwards from the point where the Sun's path, the ecliptic, once a year intersects the celestial equator; this point is called the ***vernal equinox***.

rigor medical term for shivering or rigidity. ***Rigor mortis*** is the stiffness that ensues in a corpse soon after death, owing to chemical changes in muscle tissue.

rinderpest acute viral disease of cattle (sometimes also of sheep and goats) characterized by fever and bloody diarrhoea, due to inflammation of the intestines. It can be fatal. Almost eliminated in the 1960s, it revived in Africa in the 1980s.

Rinderpest belongs to the Paramyxoviridae virus family and has a mortality rate of 95%. Its highest incidence is in the Horn of Africa where it kills 50–80% of cattle.

ring circuit household electrical circuit in which appliances are connected in series to form a ring with each end of the ring connected to the power supply. It superseded the radial circuit.

ringworm any of various contagious skin infections due to related kinds of fungus, usually resulting in circular, itchy, discoloured patches covered with scales or blisters. The scalp and feet (athlete's foot) are generally involved. Treatment is with antifungal preparations.

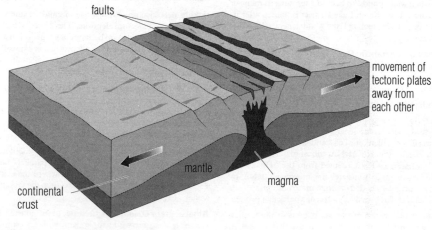

rift valley

ripple tank in physics, shallow water-filled tray used to demonstrate various properties of waves, such as reflection, refraction, diffraction, and interference.

RISC (acronym for *reduced instruction-set computer*) in computing, a microprocessor (processor on a single chip) that carries out fewer instructions than other (◊CISC) microprocessors in common use in the 1990s. Because of the low number of ◊machine code instructions, the processor carries out those instructions very quickly.

ritualization in ethology, a stereotype that occurs in certain behaviour patterns when these are incorporated into displays. For example, the exaggerated and stylized head toss of the goldeneye drake during courtship is a ritualization of the bathing movement used to wet the feathers; its duration and form have become fixed.

Ritualization may make displays clearly recognizable, so ensuring that individuals mate only with members of their own species.

river long water course that flows down a slope along a channel. It originates at a point called its *source*, and enters a sea or lake at its *mouth*. Along its length it may be joined by smaller rivers called *tributaries*. A river and its tributaries are contained within a drainage basin.

One way of classifying rivers is using their stage of development. A youthful stream is typified by a narrow V-shaped valley with numerous waterfalls, lakes, and rapids. When maturity is reached the river is said to be graded; erosion and deposition are delicately balanced as the river meanders across the extensive floodplain. At this stage the floodplain is characterized by extensive ◊meanders, ◊ox-bow lakes, and ◊levées.

middle course
The river flows through a broad valley, floored with sediments and changes its course quite frequently. It cuts into the bank on the outsides of the curves where the current flows fast and deep. Along the inside of the curves, sand and gravel deposits build up. When the river washes against a valley spur it cuts it back into a steep bank, or bluff.

upper course
The river begins its descent through a narrow V-shaped valley. Falling steeply over a short distance, it follows a zig-zag course and produces interlocking spurs.

Loops and oxbow lakes form where the changing course of a river cuts off a meander.

lower course
The river meanders from side to side across a flat plain on which deep sediments lie. Often the water level is higher than that of the plain. This is caused by the deposition of sediment forming high banks and levées, particularly at times of flood.

Sand and mud deposited at the river mouth form sand banks and may produce a delta.

river

river blindness another name for ◊onchocerciasis, a disease prevalent in some Third World countries.

riveting method of joining metal plates. A hot metal pin called a rivet, which has a head at one end, is inserted into matching holes in two overlapping plates, then the other end is struck and formed into another head, holding the plates tight. Riveting is used in building construction, boilermaking, and shipbuilding.

RNA (abbreviation for **ribonucleic acid**) nucleic acid involved in the process of translating the genetic material ◊DNA into proteins. It is usually single-stranded, unlike the double-stranded DNA, and consists of a large number of nucleotides strung together, each of which comprises the sugar ribose, a phosphate group, and one of four bases (uracil, cytosine, adenine, or guanine). RNA is copied from DNA by the formation of ◊base pairs, with uracil taking the place of thymine.

RNA occurs in three major forms, each with a different function in the synthesis of protein molecules. **Messenger RNA** (mRNA) acts as the template for protein synthesis. Each ◊codon (a set of three bases) on the RNA molecule is matched up with the corresponding amino acid, in accordance with the ◊genetic code. This process (translation) takes place in the ribosomes, which are made up of proteins and **ribosomal RNA** (rRNA). **Transfer RNA** (tRNA) is responsible for combining with specific amino acids, and then matching up a special 'anticodon' sequence of its own with a codon on the mRNA. This is how the genetic code is translated.

Although RNA is normally associated only with the process of protein synthesis, it makes up the hereditary material itself in some viruses, such as ◊retroviruses.

road specially constructed route for wheeled vehicles to travel on. Reinforced tracks became necessary with the invention of wheeled vehicles in about 3000 BC and most ancient civilizations had some form of road network. The Romans developed engineering techniques that were not equalled for another 1,400 years.

Until the late 18th century most European roads were haphazardly maintained, making winter travel difficult. In the UK the turnpike system of collecting tolls created some improvement. The Scottish engineers Thomas Telford and John McAdam introduced sophisticated construction methods in the early 19th century. Recent developments have included durable surface compounds and machinery for rapid ground preparation.

robot any computer-controlled machine that can be programmed to move or carry out work. Robots are often used in industry to transport materials or to perform repetitive tasks. For instance, robotic arms, fixed to a floor or workbench, may be used to paint machine parts or assemble electronic circuits. Other robots are designed to work in situations that would be dangerous to humans – for example, in defusing bombs or in space and deep-sea exploration.

Some robots are equipped with sensors, such as touch sensors and video cameras, and can be programmed to make simple decisions based on the sensory data received.

Roche limit in astronomy, the distance from a planet within which a large moon would be torn apart by the planet's gravitational force, creating a system of rings. The Roche limit lies at approximately 2.5 times the planet's radius (the distance from its centre to its surface).

rock constituent of the Earth's crust, composed of minerals and/or materials of organic origin consolidated into a hard mass as ◊igneous, ◊sedimentary, or ◊metamorphic rocks.

rocket projectile driven by the reaction of gases produced by a fast-burning fuel. Unlike jet engines, which are also reaction engines, modern rockets carry their own oxygen supply to burn their fuel and do not require any surrounding atmosphere. For warfare, rocket heads carry an explosive device.

Rockets have been valued as fireworks over the last seven centuries, but their intensive development as a means of propulsion to high altitudes, carrying payloads, started only in the interwar years with the state-supported work in Germany (primarily by Wernher von Braun) and of Robert Hutchings Goddard (1882–1945) in the USA. Being the only form of propulsion available that can function in a vacuum, rockets are essential to exploration in outer space. ◊Multistage rockets have to be used, consisting of a number of rockets joined together.

Two main kinds of rocket are used: one burns liquid propellants, the other solid propellants. The fireworks rocket uses gunpowder as a solid propellant. The ◊space shuttle's solid rocket boosters use a mixture of powdered aluminium in a synthetic rubber binder. Most rockets, however, have liquid propellants, which are more powerful and easier to control. Liquid hydrogen and kerosene are common fuels, while liquid oxygen is the most common oxygen provider, or oxidizer. One of the biggest rockets ever built, the Saturn V Moon rocket, was a three-stage design, standing 111 m/365 ft high. It weighed more than 2,700 tonnes/3,000 tons on the launch pad, developed a takeoff thrust of some 3.4 million kg/ 7.5 million lb, and could place almost 140 tonnes/150 tons into low Earth orbit. In the early 1990s, the most powerful rocket system was the Soviet Energiya, capable of placing 100 tonnes/110 tons into low Earth orbit. The US space shuttle can put only 24 tonnes/26 tons into orbit. See ◊missile.

rodent any mammal of the worldwide order Rodentia, making up nearly half of all mammal species. Besides ordinary 'cheek teeth', they have a single front pair of incisor teeth in both upper and lower jaw, which continue to grow as they are worn down.

roentgen or **röntgen** unit (symbol R) of radiation exposure, used for X-rays and gamma rays. It is defined in terms of the number of ions produced in one cubic centimetre of air by the radiation. Exposure to 1,000 roentgens gives rise to an absorbed dose of about 870 rads (8.7 grays), which is a dose equivalent of 870 rems (8.7 sieverts).

rolling common method of shaping metal. Rolling is carried out by giant mangles, consisting of several sets, or stands, of heavy rollers positioned one above the other. Red-hot metal slabs are rolled into sheet and also (using shaped rollers) girders and rails. Metal sheets are often cold-rolled finally to impart a harder surface.

ROM (acronym for **read-only memory**) in computing, a memory device in the form of a collection of integrated circuits (chips), frequently used in microcomputers. ROM chips are loaded with data and programs during manufacture and, unlike ◊RAM (random-access memory) chips, can subsequently only be read, not written to,

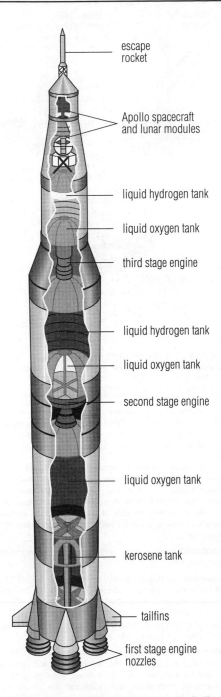

escape rocket

Apollo spacecraft and lunar modules

liquid hydrogen tank

liquid oxygen tank

third stage engine

liquid hydrogen tank

liquid oxygen tank

second stage engine

liquid oxygen tank

kerosene tank

tailfins

first stage engine nozzles

rocket The three-stage Saturn V rocket used in the Apollo moonshots of the 1960s and 1970s. It stood 111 m/365 ft high, as tall as a 30-storey skyscraper, weighed 2,700 tonnes/3,000 tons when loaded with fuel, and developed a power equivalent to 50 Boeing 747 jumbo jets.

by computer. However, the contents of the chips are not lost when the power is switched off, as happens in RAM.

ROM is used to form a computer's permanent store of vital information, or of programs that must be readily available but protected from accidental or deliberate change by a user. For example, a microcomputer ◊operating system is often held in ROM memory.

Roman numerals ancient European number system using symbols different from Arabic numerals (the ordinary numbers 1, 2, 3, 4, 5, and so on). The seven key symbols in Roman numerals, as represented today, are I (1), V (5), X (10), L (50), C (100), D (500), and M (1,000). There is no zero, and therefore no place-value as is fundamental to the Arabic system. The first ten Roman numerals are I, II, III, IV (or IIII), V, VI, VII, VIII, IX, and X. When a Roman symbol is preceded by a symbol of equal or greater value, the values of the symbols are added (XVI = 16).

When a symbol is preceded by a symbol of less value, the values are subtracted (XL = 40). A horizontal bar over a symbol indicates a multiple of 1,000 ($\bar{X} = 10,000$). Although addition and subtraction are fairly straightforward using Roman numerals, the absence of a zero makes other arithmetic calculations (such as multiplication) clumsy and difficult.

röntgen alternative spelling for ◊roentgen, unit of X- and gamma-ray exposure.

root the part of a plant that is usually underground, and whose primary functions are anchorage and the absorption of water and dissolved mineral salts. Roots usually grow downwards and towards water (that is, they are positively geotropic and hydrotropic; see ◊tropism). Plants such as epiphytic orchids, which grow above ground, produce aerial roots that absorb moisture from the atmosphere. Others, such as ivy, have climbing roots arising from the stems, which serve to attach the plant to trees and walls.

Root

In a study of a four-month-old rye plant, the root system was estimated to have 14 billion root hairs, with a total surface area of over 600 sq m/6,450 sq ft. This was 130 times more than the surface area of the plant above ground.

The absorptive area of roots is greatly increased by the numerous slender root hairs formed near the tips. A calyptra, or root cap, protects the tip of the root from abrasion as it grows through the soil.

Symbiotic associations occur between the roots of certain plants, such as clover, and various bacteria that fix nitrogen from the air (see ◊nitrogen fixation). Other modifications of roots include ◊contractile roots, ◊pneumatophores, ◊taproots, and ◊prop roots. *See illustration overleaf.*

root of an equation, a value that satisfies the equality. For example, $x = 0$ and $x = 5$ are roots of the equation $x^2 - 5x = 0$.

root crop plant cultivated for its swollen edible root (which may or may not be a true root). Potatoes are the major temperate root crop; the major tropical root crops are cassava, yams, and sweet potatoes. Root crops are second in importance only to cereals as human food. Roots have a high carbohydrate content, but their protein

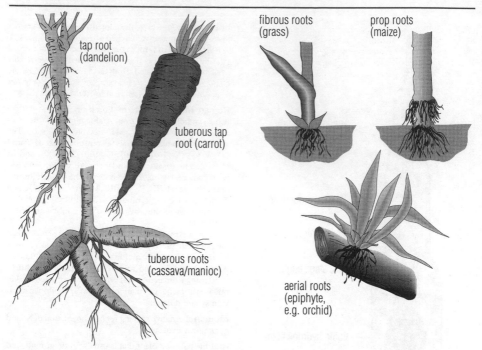

fibrous roots (grass)

prop roots (maize)

tap root (dandelion)

tuberous tap root (carrot)

tuberous roots (cassava/manioc)

aerial roots (epiphyte, e.g. orchid)

root Types of root. Many flowers (dandelion) and vegetables (carrot) have swollen tap roots with smaller lateral roots. The tuberous roots of the cassava are swollen parts of an underground stem modified to store food. The fibrous roots of the grasses are all of equal size. Prop roots grow out from the stem and then grow down into the ground to support a heavy plant. Aerial roots grow from stems but do not grow into the ground; many absorb moisture from the air.

content rarely exceeds 2%. Consequently, communities relying almost exclusively upon roots may suffer from protein deficiency. Food production for a given area from roots is greater than from cereals.

Root crops are also used as animal feed, and may be processed to produce starch, glue, and alcohol. Sugar beet has largely replaced sugar cane as a source of sugar in Europe.

root directory in computing, the top directory in a tree-and-branch filing system. It contains all the other directories.

root hair tiny hairlike outgrowth on the surface cells of plant roots that greatly increases the area available for the absorption of water and other materials. It is a delicate structure, which survives for a few days only and does not develop into a root.

New root hairs are continually being formed near the root tip, one of the places where plants show the most active growth to replace the ones that are lost. The majority of land plants have root hairs. The layer of the root's epidermis that produces root hairs is known as the *piliferous layer*.

root mean square (rms) in mathematics, value obtained by taking the ◊square root of the mean of the squares of a set of values; for example the rms value of four quantities a, b, c, and d is $\sqrt{[(a^2 + b^2 + c^2 + d^2)/4]}$.

root nodule clearly visible swelling that develops in the roots of members of the bean family, the Leguminosae. The cells inside this tumorous growth have been invaded by the bacteria Rhizobium, a soil microbe capable of converting gaseous nitrogen into nitrate. The nodule is therefore an association between a plant and a bacterium, with both partners benefiting. The plant obtains nitrogen compounds while the bacterium obtains nutrition and shelter.

Nitrogen fixation by bacteria is one of the main ways by which the nitrogen in the atmosphere is cycled back into living things. The economic value of the process is so great that research has been carried out into the possibility of stimulating the formation of root nodules in crops such as wheat, which do not normally form an association with rhizobium.

rootstock another name for ◊rhizome, an underground plant organ.

rope stout cordage with circumference over 2.5 cm/1 in. Rope is made similarly to thread or twine, by twisting yarns together to form strands, which are then in turn twisted around one another in the direction opposite to that of the yarns. Although ◊hemp is still used to make rope, nylon is increasingly used.

Rorschach test in psychology, a method of diagnosis involving the use of inkblot patterns that subjects are asked to interpret, to help indicate personality type, degree of intelligence, and emotional stability. It was invented by the Swiss psychiatrist Hermann Rorschach (1884–1922).

rosacea or *acne rosacea* in medicine, condition characterized by flushing of the face and the formation of red papules (hard pimples). Initially, the flushing is exacerbated by sunlight, large meals, or excessive consumption of alcohol. Progress of the disease results in enlargement of the sebaceous glands and the redness becoming permanent. The nose may become very enlarged in severe cases. Rosacea is treated with long courses of tetracycline antibiotics or by the application of metronidazole gel to the face.

ROSAT joint US/German/UK satellite launched in 1990 to study cosmic sources of X-rays and extremely short ultraviolet wavelengths, named after Wilhelm Röntgen, the discoverer of X-rays.

Rosetta in astronomy, a project of the ◊European Space Agency, due for launch in 2003, to send a spacecraft to Comet Wirtanen. *Rosetta* is expected to go into orbit around the comet in 2011 and land two probes on the nucleus a year later. The spacecraft will stay with the comet as it makes its closest approach to the Sun in October 2013.

rotavirus in medicine, one of a group of viruses that infect the small intestine and are a common cause of gastroenteritis in young children.

rotifer any of the tiny invertebrates, also called 'wheel animalcules', of the phylum Rotifera. Mainly freshwater, some marine, rotifers have a ring of ◊cilia that carries food to the mouth and also provides propulsion. They are the smallest of multicellular animals – few reach 0.05 cm/0.02 in.

roughage alternative term for dietary ◊fibre, material of plant origin that cannot be digested by enzymes normally present in the human ◊gut.

roup contagious respiratory disease of poultry and game birds. It is characterized by swelling of the head and purulent catarrh.

Royal Botanic Gardens, Kew botanic gardens in Richmond, Surrey, England, popularly known as ◊Kew Gardens.

Royal Greenwich Observatory the national astronomical observatory of the UK, founded in 1675 at Greenwich, SE London, England, to provide navigational information for sailors. After World War II it was moved to Herstmonceux Castle, Sussex; in 1990 it was transferred to Cambridge. It also operates telescopes on La Palma in the Canary Islands, including the 4.2 m/165 in William Herschel Telescope, commissioned 1987.

The observatory was founded by King Charles II. The eminence of its work resulted in Greenwich Time and the Greenwich Meridian being adopted as international standards of reference in 1884.

Royal Horticultural Society British society established in 1804 for the improvement of horticulture. The annual Chelsea Flower Show, held in the grounds of the Royal Hospital, London, is also a social event, and another flower show is held at Vincent Square, London. There are gardens, orchards, and trial grounds at Wisley, Surrey, and the Lindley Library has one of the world's finest horticultural collections.

Royal Institution of Great Britain organization for the promotion, diffusion, and extension of science and knowledge, founded in London in 1799 by the Anglo-American physicist Count Rumford (1753–1814).

Royal Society oldest and premier scientific society in Britain, originating in 1645 and chartered in 1662; Robert Boyle, Christopher Wren, and Isaac Newton were prominent early members. Its Scottish equivalent is the *Royal Society of Edinburgh* (1783).

RSI abbreviation for ◊*repetitive strain injury*, a condition affecting workers, such as typists, who repeatedly perform certain movements with their hands and wrists.

RU486 another name for ◊mifepristone, an abortion pill.

rubber slang term for a ◊condom.

rubber coagulated ◊latex of a variety of plants, mainly from the New World. Most important is Para rubber, which comes from the tree *Hevea brasiliensis*, belonging to the spurge family. It was introduced from Brazil to SE Asia, where most of the world supply is now produced, the chief exporters being Peninsular Malaysia, Indonesia, Sri Lanka, Cambodia, Thailand, Sarawak, and Brunei. At about seven years the tree, which may grow to 20 m/60 ft, is ready for tapping. Small cuts are made in the trunk and the latex drips into collecting cups. In pure form, rubber is white and has the formula $(C_5H_8)_n$.

Other sources of rubber are the Russian dandelion *Taraxacum koksagyz*, which grows in temperate climates and can yield about 45 kg/100 lb of rubber per tonne of roots, and guayule *Parthenium argentatum*, a small shrub which grows in the southwestern USA and Mexico.

Early uses of rubber were limited by its tendency to soften on hot days and harden on colder ones, a tendency that was eliminated by Charles Goodyear's invention of ◊vulcanization in 1839.

In the 20th century, world production of rubber increased a hundredfold, and World War II stimulated the production of synthetic rubber to replace the supplies from Malaysian sources overrun by the Japanese. There are an infinite variety of synthetic rubbers adapted to special purposes, but economically foremost is SBR (styrene-butadiene rubber). Cheaper than natural rubber, it is preferable for some purposes, for example in car tyres, where its higher abrasion resistance is useful, and it is either blended with natural rubber or used alone for industrial moulding and extrusions, shoe soles, hoses, and latex foam.

rubefacient in medicine, substance that produces irritation of the skin (counterirritation) and subsequently alleviates pain, whether deep seated or superficial. Creams and gels containing rubefacients, such as salicylates, are useful in relieving muscle, tendon, and joint pains and also some types of rheumatic pain.

rubella technical term for ◊German measles.

rubidium soft, silver-white, metallic element, symbol Rb, atomic number 37, relative atomic mass 85.47. It is one of the ◊alkali metals, ignites spontaneously in air, and reacts violently with water. It is used in photocells and vacuum-tube filaments.

Rubidium was discovered spectroscopically by German physicists Robert Bunsen and Gustav Kirchhoff in 1861, and named after the red lines in its spectrum.

ruby the red transparent gem variety of the mineral ◊corundum Al_2O_3, aluminium oxide. Small amounts of chromium oxide, Cr_2O_3, substituting for aluminium oxide, give ruby its colour.

Natural rubies are found mainly in Myanmar (Burma), but rubies can also be produced artificially and such synthetic stones are used in ◊lasers.

ruminant any even-toed hoofed mammal with a rumen, the 'first stomach' of its complex digestive system. Plant food is stored and fermented before being brought back to the mouth for chewing (chewing the cud) and is then swallowed to the next stomach. Ruminants include cattle, antelopes, goats, deer, and giraffes, all with a four-chambered stomach. Camels are also ruminants, but they have a three-chambered stomach.

runner or *stolon* in botany, aerial stem that produces new plants.

rupture in medicine, another name for ◊hernia.

Russian Academy of Sciences society founded in 1725 by Catherine the Great in St Petersburg. The academy has been responsible for such achievements as the ◊Sputnik satellite, and has daughter academies in the

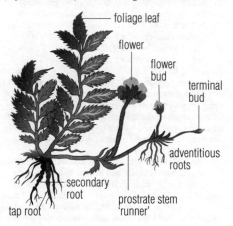

runner A runner, or stolon, grows horizontally near the base of some plants, such as the strawberry. It produces roots along its length and new plants grow at these points.

Ukraine (welding, cybernetics), Armenia (astrophysics), and Georgia (mechanical engineering).

rust reddish-brown oxide of iron formed by the action of moisture and oxygen on the metal. It consists mainly of hydrated iron(III) oxide ($Fe_2O_3.H_2O$) and iron(III) hydroxide ($Fe(OH)_3$).

Paints that penetrate beneath any moisture, and plastic compounds that combine with existing rust to form a protective coating, are used to avoid rusting.

ruthenium hard, brittle, silver-white, metallic element, symbol Ru, atomic number 44, relative atomic mass 101.07. It is one of the so-called platinum group of metals; it occurs in platinum ores as a free metal and in the natural alloy osmiridium. It is used as a hardener in alloys and as a catalyst; its compounds are used as colouring agents in glass and ceramics.

rutherfordium synthesized, radioactive, metallic element, symbol Rf. It is the first of the ◊transactinide series, atomic number 104, relative atomic mass 262. It is produced by bombarding californium with carbon nuclei and has ten isotopes, the longest-lived of which, Rf-262, has a half-life of 70 seconds.

Two institutions claim to be the first to have synthesized it: the Joint Institute for Nuclear Research in Dubna, Russia, in 1964; and the University of California at Berkeley, USA, in 1969.

rutile titanium oxide mineral, TiO_2, a naturally occurring ore of titanium. It is usually reddish brown to black, with a very bright (adamantine) surface lustre. It crystallizes in the tetragonal system. Rutile is common in a wide range of igneous and metamorphic rocks and also occurs concentrated in sands; the coastal sands of E and W Australia are a major source. It is also used as a pigment that gives a brilliant white to paint, paper, and plastics.

Rydberg constant in physics, a constant that relates atomic spectra to the ◊spectrum of hydrogen. Its value is 1.0977×10^7 per metre.

S abbreviation for *south*.

saccharide another name for a ◊sugar molecule.

saccharin or *ortho-sulpho benzimide* $C_7H_5NO_3S$ sweet, white, crystalline solid derived from coal tar and substituted for sugar. Since 1977 it has been regarded as potentially carcinogenic. Its use is not universally permitted and it has been largely replaced by other sweetening agents.

sadism tendency to derive pleasure (usually sexual) from inflicting physical or mental pain on others. The term is derived from the Marquis de Sade.

sadomasochism sexual behaviour that combines ◊sadism and ◊masochism. The term was coined in 1907 by sexologist Richard von Krafft-Ebing.

safe period in medicine, that part of the menstrual cycle during which fertilization of the ovum is unlikely to occur. Ovulation occurs about 15 days before the onset of the menstrual period and a woman is usually fertile for five days before ovulation, on the day of ovulation and for five days afterwards. Therefore, the safe period is the first week and the last ten days of the menstrual cycle. The safe period can be used as a form of contraception, the ◊rhythm method.

safety glass glass that does not splinter into sharp pieces when smashed. **Toughened glass** is made by heating a glass sheet and then rapidly cooling it with a blast of cold air; it shatters into rounded pieces when smashed. **Laminated glass** is a 'sandwich' of a clear plastic film between two glass sheets; when this is struck, it simply cracks, the plastic holding the glass in place.

safety lamp portable lamp designed for use in places where flammable gases such as methane may be encountered; for example, in coal mines. The electric head lamp used as a miner's working light has the bulb and contacts in protected enclosures. The flame safety lamp, now used primarily for gas detection, has the wick enclosed within a strong glass cylinder surmounted by wire gauzes. English chemist Humphrey Davy and English engineer George Stephenson each invented flame safety lamps.

Sagittarius bright zodiac constellation in the southern hemisphere, represented as a centaur aiming a bow and arrow at neighbouring Scorpius. The Sun passes through Sagittarius from mid-Dec to mid-Jan, including the winter solstice, when it is farthest south of the Equator. The constellation contains many nebulae and ◊globular clusters, and open ◊star clusters. Kaus Australis and Nunki are its brightest stars. The centre of our Galaxy, the ◊Milky Way, is marked by the radio source Sagittarius A. In astrology, the dates for Sagittarius are about 22 Nov–21 Dec (see ◊precession).

sago starchy material obtained from the pith of the sago palm *Metroxylon sagu*. It forms a nutritious food and is used for manufacturing glucose and sizing textiles.

Sahel marginal area to the south of the Sahara, from Senegal to Somalia, which experiences desert-like conditions during periods of low rainfall. The desertification is partly due to climatic fluctuations but has also been caused by the pressures of a rapidly expanding population, which have led to overgrazing and the destruction of trees and scrub for fuelwood. In recent years many famines have taken place in the area.

St Elmo's fire bluish, flamelike electrical discharge that sometimes occurs above ships' masts and other pointed objects or about aircraft in stormy weather. Although high voltage, it is low current and therefore harmless. St Elmo (or St Erasmus) is the patron saint of sailors.

St Vitus' dance archaic name for Sydenham's ◊chorea, a transient disorder of childhood and adolescence, often associated with ◊rheumatic fever. St Vitus, martyred under the Roman emperor Diocletian, was the patron saint of dancers.

salicylic acid HOC_6H_4COOH the active chemical constituent of aspirin, an analgesic drug. The acid and its salts (salicylates) occur naturally in many plants; concentrated sources include willow bark and oil of wintergreen.

When purified, salicylic acid is a white solid that crystallizes into prismatic needles at 318°F/159°C. It is used as an antiseptic, in food preparation and dyestuffs, and in the preparation of aspirin.

saliva in vertebrates, an alkaline secretion from the salivary glands that aids the swallowing and digestion of food in the mouth. In mammals, it contains the enzyme amylase, which converts starch to sugar. The salivary glands of mosquitoes and other blood-sucking insects produce ◊anticoagulants.

salivary gland or *parotid gland* in mammals, one of two glands situated near the mouth responsible for the manufacture of saliva and its secretion into the mouth. The salivary glands are stimulated to produce saliva during a meal. Saliva contains an enzyme, ptyalin, and mucous which are essential for the mastication and initial digestion of food.

salmonella any of a very varied group of bacteria, genus *Salmonella* that colonize the intestines of humans and some animals. Some strains cause typhoid and paratyphoid fevers, while others cause salmonella ◊food poisoning, which is characterized by stomach pains, vomiting, diarrhoea, and headache. It can be fatal in elderly people, but others usually recover in a few days without antibiotics. Most cases are caused by contaminated animal products, especially poultry meat.

Human carriers of the disease may be well themselves but pass the bacteria on to others through unhygienic preparation of food. Domestic pets can also carry the bacteria while appearing healthy.

salt in chemistry, any compound formed from an acid and a base through the replacement of all or part of the

Salt: Common Ions That Form Salts

Positive ions	Negative ions
silver Ag^+	bromide Br^-
aluminium Al^{3+}	chloride Cl^-
barium Ba^{2+}	carbonate $CO_3{}^{2-}$
calcium Ca^{2+}	fluoride F^-
copper Cu^{2+}	hydrogencarbonate HCO_3^-
iron(II) Fe^{2+}	hydrogensulphate HSO_4^-
iron(III) Fe^{3+}	iodide I^-
hydrogen H^+	nitrate NO_3^-
potassium K^+	oxide O^{2-}
lithium Li^+	hydroxide OH^-
magnesium Mg^{2+}	sulphide S^{2-}
sodium Na^+	sulphite $SO_3{}^{2-}$
ammonium NH_4^+	sulphate $SO_4{}^{2-}$
lead Pb^{2+}	
zinc Zn^{2+}	

hydrogen in the acid by a metal or electropositive radical. **Common salt** is sodium chloride (see ◊salt, common).

A salt may be produced by chemical reaction between an acid and a base, or by the displacement of hydrogen from an acid by a metal (see ◊displacement activity). As a solid, the ions normally adopt a regular arrangement to form crystals. Some salts only form stable crystals as hydrates (when combined with water). Most inorganic salts readily dissolve in water to give an electrolyte (a solution that conducts electricity).

salt, common or **sodium chloride** NaCl white crystalline solid, found dissolved in sea water and as rock salt (the mineral halite) in large deposits and salt domes. Common salt is used extensively in the food industry as a preservative and for flavouring, and in the chemical industry in the making of chlorine and sodium.

salt marsh wetland with halophytic vegetation (tolerant to sea water). Salt marshes develop around estuaries and on the sheltered side of sand and shingle spits. Salt marshes usually have a network of creeks and drainage channels by which tidal waters enter and leave the marsh.

saltpetre former name for potassium nitrate (KNO_3), the compound used in making gunpowder (from about 1500). It occurs naturally, being deposited during dry periods in places with warm climates, such as India.

Salyut series of seven space stations launched by the USSR 1971–82. Salyut was cylindrical in shape, 15 m/50 ft long, and weighed 19 tonnes/21 tons. It housed two or three cosmonauts at a time, for missions lasting up to eight months.

Salyut 1 was launched on 19 April 1971. It was occupied for 23 days in June 1971 by a crew of three, who died during their return to Earth when their ◊Soyuz ferry craft depressurized. In 1973 *Salyut 2* broke up in orbit before occupation. The first fully successful Salyut mission was a 14-day visit made to *Salyut 3* in July 1974. In 1984–85 a team of three cosmonauts endured a record 237-day flight in *Salyut 7*. In 1986 the Salyut series was superseded by ◊*Mir*, an improved design capable of being enlarged by additional modules sent up from Earth.

samara in botany, a winged fruit, a type of ◊achene.

samarium hard, brittle, grey-white, metallic element of the ◊lanthanide series, symbol Sm, atomic number 62, relative atomic mass 150.4. It is widely distributed in nature and is obtained commercially from the minerals

monzanite and bastnaesite. It is used only occasionally in industry, mainly as a catalyst in organic reactions. Samarium was discovered by spectroscopic analysis of the mineral samarskite and named in 1879 by French chemist Paul Lecoq de Boisbaudran (1838–1912) after its source.

San Andreas fault geological fault stretching for 1,125 km/700 mi NW–SE through the state of California, USA. It marks a conservative plate margin, where two plates slide past each other (see ◊plate tectonics).

Friction is created as the coastal Pacific plate moves northwest, rubbing against the American continental plate, which is moving slowly southeast. The relative movement is only about 5 cm/2 in a year, which means that Los Angeles will reach San Francisco's latitude in 10 million years. The friction caused by the tectonic movement gives rise to frequent, destructive ◊earthquakes. For example, in 1906 an earthquake originating from the fault almost destroyed San Francisco and killed about 700 people.

sand loose grains of rock, sized 0.0625–2.00 mm/ 0.0025–0.08 in in diameter, consisting most commonly of ◊quartz, but owing their varying colour to mixtures of other minerals. Sand is used in cement-making, as an abrasive, in glass-making, and for other purposes.

Sands are classified into marine, freshwater, glacial, and terrestrial. Some 'light' soils contain up to 50% sand. Sands may eventually consolidate into ◊sandstone.

sandbar ridge of sand built up by the currents across the mouth of a river or bay. A sandbar may be entirely underwater or it may form an elongated island that breaks the surface. A sandbar stretching out from a headland is a **sand spit**. Coastal bars can extend across estuaries to form **bay bars**.

sandstone ◊sedimentary rocks formed from the consolidation of sand, with sand-sized grains (0.0625–2 mm/ 0.0025–0.08 in) in a matrix or cement. Their principal component is quartz. Sandstones are commonly permeable and porous, and may form freshwater ◊aquifers. They are mainly used as building materials.

Sandstones are classified according to the matrix or cement material (whether derived from clay or silt; for example, as calcareous sandstone, ferruginous sandstone, siliceous sandstone).

Santa Ana periodic warm Californian ◊wind.

sap the fluids that circulate through ◊vascular plants, especially woody ones. Sap carries water and food to plant tissues. Sap contains alkaloids, protein, and starch; it can be milky (as in rubber trees), resinous (as in pines), or syrupy (as in maples).

saponification in chemistry, the ◊hydrolysis (splitting) of an ◊ester by treatment with a strong alkali, resulting in the liberation of the alcohol from which the ester had been derived and a salt of the constituent fatty acid. The process is used in the manufacture of soap.

sapphire deep-blue, transparent gem variety of the mineral ◊corundum Al_2O_3, aluminium oxide. Small amounts of iron and titanium give it its colour. A corundum gem of any colour except red (which is a ruby) can be called a sapphire; for example, yellow sapphire.

saprotroph (formerly **saprophyte**) organism that feeds on the excrement or the dead bodies or tissues of others. They include most fungi (the rest being parasites); many bacteria and protozoa; animals such as dung beetles and vultures; and a few unusual plants, including several orchids. Saprotrophs cannot make food for themselves, so they are a type of ◊heterotroph. They are useful scavengers, and in sewage farms and refuse dumps break down organic matter into nutrients easily assimilable by green plants.

sarcoidosis chronic disease of unknown cause involving enlargement of the lymph nodes and the formation of small fleshy nodules in the lungs. It may also affect the eyes, and skin, and (rarely) other tissue. Many cases resolve spontaneously or may be successfully treated using ◊corticosteroids.

sarcoma malignant ◊tumour arising from the fat, muscles, bones, cartilage, or blood and lymph vessels and connective tissues. Sarcomas are much less common than ◊carcinomas.

sard yellow or red-brown variety of ◊chalcedony.

sarin poison gas 20 times more lethal to humans than potassium cyanide. It cripples the central nervous system, blocking the action of an enzyme that removes acetylcholine, the chemical that transmits signals. Sarin was developed in Germany during World War II.

Sarin was used in 1995 in a terrorist attack on the Tokyo underground by a Japanese sect. There are no known safe disposal methods.

satellite any small body that orbits a larger one, either natural or artificial. Natural satellites that orbit planets are called moons. The first **artificial satellite**, *Sputnik 1*, was launched into orbit around the Earth by the USSR in 1957. Artificial satellites are used for scientific purposes, communications, weather forecasting, and military applications. The brightest artificial satellites can be seen by the naked eye.

At any time, there are several thousand artificial satellites orbiting the Earth, including active satellites, satellites that have ended their working lives, and discarded sections of rockets. Artificial satellites eventually re-enter the Earth's atmosphere. Usually they burn up by friction, but sometimes debris falls to the Earth's surface, as with ◊*Skylab* and *Salyut 7*. In 1997 there were 300 active artificial satellites in orbit around Earth, the majority used in communications.

Satellite

The Moon is the sixth largest satellite in the Solar System. It is 3,476 km/2,160 mi in diameter. The larger satellites are: Jupiter's moons, Io, Ganymede, and Callisto; Saturn's Titan; and Neptune's Triton.

---

satellite applications the uses to which artificial satellites are put. These include:

Scientific experiments and observation Many astronomical observations are best taken above the disturbing effect of the atmosphere. Satellite observations have been carried out by *IRAS* (*Infrared Astronomical Satellite*, launched in 1983) which made a complete infrared survey of the skies, and *Solar Max* launched in 1980, which observed solar flares. The *Hipparcos* satellite, launched in 1989, measured the positions of many stars. The *ROSAT* (Roentgen Satellite), launched in June 1990, examined UV and X-ray radiation. In 1992, the COBE (Cosmic Background Explorer) satellite detected

details of the Big Bang that mark the first stage in the formation of galaxies. Medical experiments have been carried out aboard crewed satellites, such as the Soviet *Mir* and the US *Skylab*.

Reconnaissance, land resource, and mapping applications Apart from military use and routine mapmaking, the US *Landsat*, the French *SPOT*, and equivalent USSR satellites have provided much useful information about water sources and drainage, vegetation, land use, geological structures, oil and mineral locations, and snow and ice.

Weather monitoring The US NOAA series of satellites, and others launched by the European space agency, Japan, and India, provide continuous worldwide observation of the atmosphere.

Navigation The US Global Positioning System, features 24 Navstar satellites that enable users (including walkers and motorists) to find their position to within 100 m/328 ft.

Communications A complete worldwide communications network is now provided by satellites such as the US-run Intelsat system.

satellite television transmission of broadcast signals through artificial communications satellites. Mainly positioned in ◊geostationary orbit, satellites have been used since the 1960s to relay television pictures around the world.

Higher-power satellites have more recently been developed to broadcast signals to cable systems or directly to people's homes.

saturated compound organic compound, such as propane, that contains only single covalent bonds. Saturated organic compounds can only undergo further reaction by ◊substitution reactions, as in the production of chloropropane from propane.

saturated fatty acid ◊fatty acid in which there are no double bonds in the hydrocarbon chain.

saturated solution in physics and chemistry, a solution obtained when a solvent (liquid) can dissolve no more of a solute (usually a solid) at a particular temperature. Normally, a slight fall in temperature causes some of the solute to crystallize out of solution. If this does not happen the phenomenon is called supercooling, and the solution is said to be **supersaturated**.

Saturn in astronomy, the second-largest planet in the Solar System, sixth from the Sun, and encircled by bright and easily visible equatorial rings. Viewed through a telescope it is ochre. Its polar diameter is 12,000 km/7,450 mi smaller than its equatorial diameter, a result of its fast rotation and low density, the lowest of any planet. Its mass is 95 times that of Earth, and its magnetic field 1,000 times stronger.

Mean distance from the Sun 1.427 billion km/0.886 billion mi.

Equatorial diameter 120,000 km/75,000 mi.

Rotational period 10 hr 14 min at equator, 10 hr 40 min at higher latitudes.

Year 29.46 Earth years.

Atmosphere visible surface consists of swirling clouds, probably made of frozen ammonia at a temperature of $-170°C/-274°F$, although the markings in the clouds are not as prominent as Jupiter's. The space probes *Voyager 1* and *2* found winds reaching 1,800 kph/1,100 mph.

Surface Saturn is believed to have a small core of rock and iron, encased in ice and topped by a deep layer of liquid hydrogen.

Satellites at least 20 known moons, more than for any other planet. The largest moon, ◊Titan, has a dense atmosphere. Other satellites include Epimetheus, Janus, Pandor, and Prometheus. The rings visible from Earth begin about 14,000 km/9,000 mi from the planet's cloud-tops and extend out to about 76,000 km/47,000 mi. Made of small chunks of ice and rock (averaging 1 m/3 ft across), they are 275,000 km/170,000 mi rim to rim, but only 100 m/300 ft thick. The Voyager probes showed that the rings actually consist of thousands of closely spaced ringlets, looking like the grooves in a gramophone record.

Saturn rocket family of large US rockets, developed by Wernher von Braun (1912–1977) for the ◊Apollo project. The two-stage *Saturn IB* was used for launching Apollo spacecraft into orbit around the Earth. The three-stage *Saturn V* sent Apollo spacecraft to the Moon, and launched the ◊*Skylab* space station. The liftoff thrust of a *Saturn V* was 3,500 tonnes. After Apollo and *Skylab*, the Saturn rockets were retired in favour of the ◊space shuttle.

savanna or **savannah** extensive open tropical grass-lands, with scattered trees and shrubs. Savannas cover large areas of Africa, North and South America, and N Australia. The soil is acidic and sandy and generally considered suitable only as pasture for low-density grazing.

A new strain of rice suitable for savanna conditions was developed in 1992. It not only grew successfully under test conditions in Colombia but also improved pasture quality so that grazing numbers could be increased twentyfold.

scabies contagious infection of the skin caused by the parasitic itch mite *Sarcoptes scabiei*, which burrows under the skin to deposit eggs. Treatment is by antiparasitic creams and lotions.

scalar quantity in mathematics and science, a quantity that has magnitude but no direction, as distinct from a ◊vector quantity, which has a direction as well as a magnitude. Temperature, mass, and volume are scalar quantities.

scale in chemistry, ◊calcium carbonate deposits that form on the inside of a kettle or boiler as a result of boiling ◊hard water.

scandium silver-white, metallic element of the ◊lanthanide series, symbol Sc, atomic number 21, relative atomic mass 44.956. Its compounds are found widely distributed in nature, but only in minute amounts. The metal has little industrial importance.

Scandium is relatively more abundant in the Sun and other stars than on Earth. Scandium oxide (scandia) is used as a catalyst, in making crucibles and other ceramic parts, and scandium sulphate (in very dilute aqueous solution) is used in agriculture to improve seed germination.

scanner in computing, a device that can produce a digital image of a document for input and storage in a computer. It uses technology similar to that of a photo-copier. Small scanners can be passed over the document surface by hand; larger versions have a flat bed, like that of a photocopier, on which the input document is placed and scanned.

Scanners are widely used to input graphics for use in ◊desktop publishing. If text is input with a scanner, the image captured is seen by the computer as a single digital picture rather than as separate characters. Consequently, the text cannot be processed by, for example, a word processor unless suitable optical character-recognition software is available to convert the image to its constituent characters. Scanners vary in their resolution, typical hand-held scanners ranging from 75 to 300 dpi. Types include flat-bed, drum, and overhead.

scanning in medicine, the non-invasive examination of body organs to detect abnormalities of structure or function. Detectable waves – for example, ◊ultrasound, gamma, or ◊X-rays – are passed through the part to be scanned. Their absorption pattern is recorded, analysed by computer, and displayed pictorially on a screen.

scanning electron microscope (SEM) electron microscope that produces three-dimensional images, magnified 10–200,000 times. A fine beam of electrons, focused by electromagnets, is moved, or scanned, across the specimen. Electrons reflected from the specimen are collected by a detector, giving rise to an electrical signal, which is then used to generate a point of brightness on a television-like screen. As the point moves rapidly over the screen, in phase with the scanning electron beam, an image of the specimen is built up.

The resolving power of an SEM depends on the size of the electron beam – the finer the beam, the greater the resolution. Present-day instruments typically have a resolution of 7–10 nm.

The first scanning electron picture was produced in 1935 by Max Knoll of the German company Telefunken, though the first commercial SEM (produced by the Cambridge Instrument Company in the UK) did not go on sale until 1965.

scanning transmission electron microscope (STEM) electron microscope that combines features of the ◊scanning electron microscope (SEM) and the ◊transmission electron microscope (TEM). First built in the USA in 1966, the microscope has both the SEM's contrast characteristics and lack of aberrations and the high resolution of the TEM. Magnifications of over 90 million times can be achieved, enough to image single atoms.

A fine beam of electrons, 0.3 nm in diameter, moves across the specimen, as in an SEM. However, because the specimen used is a thin slice, the beam also passes through the specimen (as in a TEM). The reflected electrons and those that penetrated the specimen are collected to form an electric signal, which is interpreted by computer to form an image on screen.

scanning tunnelling microscope (STM) microscope that produces a magnified image by moving a tiny tungsten probe across the surface of the specimen. The tip of the probe is so fine that it may consist of a single atom, and it moves so close to the specimen surface that electrons jump (or tunnel) across the gap between the tip and the surface.

The magnitude of the electron flow (current) depends on the distance from the tip to the surface, and so by measuring the current, the contours of the surface can be determined. These can be used to form an image on a computer screen of the surface, with individual atoms resolved. Magnifications up to 100 million times are possible.

The STM was invented in 1981 by Gerd Binning from Germany and Heinrich Rohrer from Switzerland at the IBM Zurich Research Laboratory. With Ernst Ruska, who invented the transmission electron microscope in 1933, they were awarded the Nobel Prize for Physics in 1986.

scapolite group of white or greyish minerals, silicates of sodium, aluminium, and calcium, common in metamorphosed limestones and forming at high temperatures and pressures.

scapula or **shoulder blade** large, flat, triangular bone which lies over the second to seventh ribs on the back, forming part of the pectoral girdle, and assisting in the articulation of the arm with the chest region. Its flattened shape allows a large region for the attachment of muscles.

scarlet fever or **scarlatina** acute infectious disease, especially of children, caused by the bacteria in the *Streptococcus pyogenes* group. It is marked by fever, vomiting, sore throat, and a bright red rash spreading from the upper to the lower part of the body. The rash is followed by the skin peeling in flakes. It is treated with antibiotics.

scarp and dip in geology, the two slopes formed when a sedimentary bed outcrops as a landscape feature. The scarp is the slope that cuts across the bedding plane; the dip is the opposite slope which follows the bedding plane. The scarp is usually steep, while the dip is a gentle slope.

scatter diagram or **scattergram** diagram whose purpose is to establish whether or not a relationship or ◊correlation exists between two variables; for example, between life expectancy and gross national product. Each observation is marked with a dot in a position that shows the value of both variables. The pattern of dots is then examined to see whether they show any underlying trend by means of a **line of best fit** (a straight line drawn so that its distance from the various points is as short as possible).

scattering in physics, the random deviation or reflection of a stream of particles or of a beam of radiation such as light.

Alpha particles scattered by a thin gold foil provided the first convincing evidence that atoms had very small, very dense, positive nuclei. From 1906 to 1908 Ernest Rutherford carried out a series of experiments from which he estimated that the closest approach of an alpha particle to a gold nucleus in a head-on collision was about 10^{-14} m. He concluded that the gold nucleus must be no larger than this. Most of the alpha particles fired at the gold foil passed straight through undeviated; however, a few were scattered in all directions and a very small fraction bounced back towards the source. This result so surprised Rutherford that he is reported to have commented: 'It was almost as if you fired a 15-inch shell at a piece of tissue paper and it came back and hit you'.

Light is scattered from a rough surface, such as that of a sheet of paper, by random reflection from the varying angles of each small part of the surface. This is responsible for the dull, flat appearance of such surfaces and their inability to form images (unlike mirrors). Light is also scattered by particles suspended in a gas or liquid. The red and yellow colours associated with sunrises and sunsets are due to the fact that red light is scattered to a lesser extent than is blue light by dust particles in the

atmosphere. When the Sun is low in the sky, its light passes through a thicker, more dusty layer of the atmosphere, and the blue light radiated by it is scattered away, leaving the red sunlight to pass through to the eye of the observer.

scent gland gland that opens onto the outer surface of animals, producing odorous compounds that are used for communicating between members of the same species (◊pheromones), or for discouraging predators.

schist ◊metamorphic rock containing ◊mica or another platy or elongate mineral, whose crystals are aligned to give a foliation (planar texture) known as schistosity. Schist may contain additional minerals such as ◊garnet.

schistosomiasis another name for ◊bilharzia.

schizocarp dry ◊fruit that develops from two or more carpels and splits, when mature, to form separate one-seeded units known as mericarps.

The mericarps may be dehiscent, splitting open to release the seed when ripe, as in *Geranium*, or indehiscent, remaining closed once mature, as in mallow *Malva* and plants of the Umbelliferae family, such as the carrot *Daucus carota* and parsnip *Pastinaca sativa*.

schizophrenia mental disorder, a psychosis of unknown origin, which can lead to profound changes in personality, behaviour, and perception, including delusions and hallucinations. It is more common in males and the early-onset form is more severe than when the illness develops in later life. Modern treatment approaches include drugs, family therapy, stress reduction, and rehabilitation.

Schizophrenia implies a severe divorce from reality in the patient's thinking. Although the causes are poorly understood, it is now recognized as an organic disease, associated with structural anomalies in the brain. In 1995, Canadian researchers identified a protein in the brain, PSA-NCAM, that plays a part in filtering sensory information. The protein is significantly reduced in the brains of schizophrenics, supporting the idea that schizophrenia occurs when the brain is overwhelmed by sensory information.

There is some evidence that early trauma, either in the womb or during delivery, may play a part in causation. There is also a genetic contribution.

Schmidt Telescope reflecting telescope used for taking wide-angle photographs of the sky. Invented in 1930 by Estonian astronomer Bernhard Schmidt (1879–1935), it has an added corrector lens to help focus the incoming light. Examples are the 1. 2 m/48 in Schmidt telescope on ◊Mount Palomar and the UK Schmidt telescope, of the same size, at ◊Siding Spring.

Schwarzschild radius in ◊astrophysics, the radius of the event horizon surrounding a ◊black hole within which light cannot escape its gravitational pull.

For a black hole of mass M, the Schwarzschild radius Rs is given by $Rs = 2gm/c^2$, where g is the gravitational constant and c is the speed of light. The Schwarzschild radius for a black hole of solar mass is about 3 km/1.9 mi. It is named after Karl Schwarzschild, the German mathematician who deduced the possibility of black holes from Einstein's general theory of relativity in 1916.

sciatica persistent pain in the back and down the outside of one leg, along the sciatic nerve and its branches. Causes of sciatica include inflammation of the nerve or pressure of a displaced disc on a nerve root leading out of the lower spine.

science any systematic field of study or body of knowledge that aims, through experiment, observation, and deduction, to produce reliable explanations of phenomena, with reference to the material and physical world.

The important thing in science is not so much to obtain new facts as to discover new ways of thinking about them.

On **science** Lawrence Bragg

Activities such as healing, star-watching, and engineering have been practised in many societies since ancient times. Pure science, especially physics (formerly called natural philosophy), had traditionally been the main area of study for philosophers. The European scientific revolution between about 1650 and 1800 replaced speculative philosophy with a new combination of observation, experimentation, and rationality.

Progress in science depends on new techniques, new discoveries, and new ideas, probably in that order.

On **science** Sydney Brenner *Nature* 1980

Today, scientific research involves an interaction between tradition, experiment and observation, and deduction. The subject area called philosophy of science investigates the nature of this complex interaction, and the extent of its ability to gain access to the truth about the material world. It has long been recognized that induction from observation cannot give explanations based on logic. In the 20th century Karl Popper has described scientific method as a rigorous experimental testing of a scientist's ideas or hypotheses (see ◊hypothesis). The origin and role of these ideas, and their interdependence with observation, have been examined, for example, by the US thinker Thomas S Kuhn, who places them in a historical and sociological setting.

Science without religion is lame. Religion without science is blind.

On **scientific method** Albert Einstein, quoted in A Pais *'Subtle is the Lord...'*: *The Science and the Life of Albert Einstein* 1982

The sociology of science investigates how scientific theories and laws are produced, and questions the possibility of objectivity in any scientific endeavour. One controversial point of view is the replacement of scientific realism with scientific relativism, as proposed by Paul K Feyerabend. Questions concerning the proper use of science and the role of science education are also restructuring this field of study.

scientific law in science, principles that are taken to be universally applicable.

Laws (for instance, ◊Boyle's law and ◊Newton's laws of motion) form the basic theoretical structure of the physical sciences, so that the rejection of a law by the scientific community is an almost inconceivable event. On occasion though, a law may be modified, as was the case when Einstein showed that Newton's laws of motion

do not apply to objects travelling at speeds close to that of light.

scintillation counter instrument for measuring very low levels of radiation. The radiation strikes a scintillator (a device that emits a unit of light when a charged elementary particle collides with it), whose light output is 'amplified' by a ◊photomultiplier; the current pulses of its output are in turn counted or added by a scaler to give a numerical reading.

sclerenchyma plant tissue whose function is to strengthen and support, composed of thick-walled cells that are heavily lignified (toughened). On maturity the cell inside dies, and only the cell walls remain.

Sclerenchyma may be made up of one or two types of cells: *sclereids*, occurring singly or in small clusters, are often found in the hard shells of fruits and in seed coats, bark, and the stem cortex; *fibres*, frequently grouped in bundles, are elongated cells, often with pointed ends, associated with the vascular tissue (◊xylem and ◊phloem) of the plant.

Some fibres provide useful materials, such as flax from *Linum usitatissimum* and hemp from *Cannabis sativa*.

scleroderma in medicine, condition characterized by thickening of the skin. This results in the skin becoming hard, the joints stiffening, and a gradual wasting of the muscles. It is thought to be caused by malfunctioning of the immune system.

sclerosis any abnormal hardening of body tissues, especially the nervous system or walls of the arteries. See ◊multiple sclerosis and ◊atherosclerosis.

scoliosis lateral (sideways) deviation of the spine. It may be congenital or acquired (through bad posture, illness, or other deformity); or it may be idiopathic (of unknown cause). Treatments include mechanical or surgical correction, depending on the cause.

scopolamine alternative name for ◊hyoscine, a sedative drug.

Scorpius bright zodiacal constellation in the southern hemisphere between ◊Libra and ◊Sagittarius, represented as a scorpion. The Sun passes briefly through Scorpius in the last week of Nov. The heart of the scorpion is marked by the bright red supergiant star ◊Antares. Scorpius contains rich ◊Milky Way star fields, plus the strongest ◊X-ray source in the sky, Scorpius X-1. The whole area is rich in clusters and nebulae. In astrology, the dates for Scorpius are about 24 Oct–21 Nov (see ◊precession).

Scottish Natural Heritage Scottish nature conservation body formed in 1991 after the break-up of the national ◊Nature Conservancy Council. It is government-funded.

scrapie fatal disease of sheep and goats that attacks the central nervous system, causing deterioration of the brain cells, and leading to the characteristic staggering gait and other behavioural abnormalities, before death. It is caused by the presence of an abnormal version of the brain protein PrP and is related to ◊bovine spongiform encephalopathy, the disease of cattle known as 'mad cow disease', and Creutzfeldt–Jakob disease in humans. It is a transmissible spongiform encephalopathy.

In 1996 Dutch researchers announced a test for detecting abnormal PrP in the tonsils of affected sheep before the symptoms of scrapie become apparent.

scree pile of rubble and sediment that collects at the foot of a mountain range or cliff. The rock fragments that form scree are usually broken off by the action of frost (◊freeze-thaw weathering).

With time, the rock waste builds up into a heap or sheet of rubble that may eventually bury even the upper cliffs, and the growth of the scree then stops. Usually, however, erosional forces remove the rock waste so that the scree stays restricted to lower slopes.

screen in computing, another name for monitor.

screen dump in computing, the process of making a printed copy of the current VDU screen display. The screen dump is sometimes stored as a data file instead of being printed immediately.

screening or **health screening** the systematic search for evidence of a disease, or of conditions that may precede it, in people who are at risk but not suffering from any symptoms. The aim of screening is to try to limit ill health from preventable diseases that might otherwise go undetected in the early stages. Examples are hypothyroidism and phenylketonuria, for which all newborn babies in Western countries are screened; breast cancer (◊mammography) and cervical cancer; and stroke, for which high blood pressure is a known risk factor.

The criteria for a successful screening programme are that the disease should be important and treatable, the population at risk identifiable, the screening test acceptable, accurate, and cheap, and that the results of screening should justify the costs involved.

scuba (acronym for **self-contained underwater breathing apparatus**) another name for ◊aqualung.

scurvy disease caused by deficiency of vitamin C (ascorbic acid), which is contained in fresh vegetables and fruit. The signs are weakness and aching joints and muscles, progressing to bleeding of the gums and other spontaneous haemorrhage, and drying-up of the skin and hair. It is reversed by giving the vitamin.

scythe harvesting tool with long wooden handle and sharp, curving blade.

It is similar to a ◊sickle. The scythe was in common use in the Middle East and Europe from the dawn of agriculture until the early 20th century, by which time it had generally been replaced by machinery.

Until the beginning of the 19th century, the scythe was used in the hayfield for cutting grass, but thereafter was applied to cereal crops as well, because it was capable of a faster work rate than the sickle. One person could mow 0.4 hectares/1 acre of wheat in a day with a scythe. Next came a team of workers to gather and bind the crop into sheaves and stand them in groups, or stooks, across the field.

sea anemone invertebrate marine animal of the phylum Cnidaria with a tubelike body attached by the base to a rock or shell. The other end has an open 'mouth' surrounded by stinging tentacles, which capture crustaceans and other small organisms. Many sea anemones are beautifully coloured, especially those in tropical waters.

seaborgium synthesized radioactive element of the ◊transactinide series, symbol Sg, atomic number 106, relative atomic mass 263. It was first synthesized in 1974 in the USA and given the temporary name unnilhexium. The discovery was not confirmed until 1993. It was officially named in 1997 after US nuclear chemist Glenn Seaborg.

seafloor spreading growth of the ocean ◊crust outwards (sideways) from ocean ridges. The concept of seafloor spreading has been combined with that of continental drift and incorporated into ◊plate tectonics.

Seafloor spreading was proposed in 1960 by US geologist Harry Hess (1906–1969), based on his observations of ocean ridges and the relative youth of all ocean beds. In 1963, British geophysicists Fred Vine and Drummond Matthews observed that the floor of the Atlantic Ocean was made up of rocks that could be arranged in strips, each strip being magnetized either normally or reversely (due to changes in the Earth's polarity when the North Pole becomes the South Pole and vice versa, termed ◊polar reversal). These strips were parallel and formed identical patterns on both sides of the ocean ridge. The implication was that each strip was formed at some stage in geological time when the magnetic field was polarized in a certain way. The seafloor magnetic-reversal patterns could be matched to dated magnetic reversals found in terrestrial rock. It could then be shown that new rock forms continuously and spreads away from the ocean ridges, with the oldest rock located farthest away from the midline. The observation was made independently in 1963 by Canadian geologist Lawrence Morley, studying an ocean ridge in the Pacific near Vancouver Island.

seaplane aeroplane capable of taking off from, and landing on, water. There are two major types, floatplanes and flying boats. The floatplane is similar to an ordinary aeroplane but has floats in place of wheels; the flying boat has a broad hull shaped like a boat and may also have floats attached to the wing tips.

Seaplanes depend on smooth water for a good landing, and since World War II few have been built, although they were widely used in both world wars and the first successful international airlines, such as Pan Am, relied on a fleet of flying boats in the 1920s and 1930s.

searching in computing, extracting a specific item from a large body of data, such as a file or table. The method used depends on how the data are organized. For example, a binary search, which requires the data to be in sequence, involves first deciding which half of the data contains the required item, then which quarter, then which eighth, and so on until the item is found.

season period of the year having a characteristic climate. The change in seasons is mainly due to the change in attitude of the Earth's axis in relation to the Sun, and hence the position of the Sun in the sky at a particular place. In temperate latitudes four seasons are recognized: spring, summer, autumn (fall), and winter. Tropical regions have two seasons – the wet and the dry. Monsoon areas around the Indian Ocean have three seasons: the cold, the hot, and the rainy.

The northern temperate latitudes have summer when the southern temperate latitudes have winter, and vice versa. During winter, the Sun is low in the sky and has less heating effect because of the oblique angle of incidence and because the sunlight has further to travel through the atmosphere. The differences between the seasons are more marked inland than near the coast, where the sea has a moderating effect on temperatures. In polar regions the change between summer and winter is abrupt; spring and autumn are hardly perceivable. In tropical regions, the

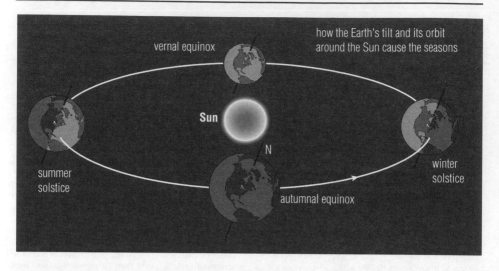

how the Earth's tilt and its orbit
around the Sun cause the seasons

vernal equinox

Sun

N

summer
solstice

winter
solstice

autumnal equinox

season The cause of the seasons. As the Earth orbits the Sun, its axis of rotation always points in the same direction. This means that, during the northern hemisphere summer solstice (21 June), the Sun is overhead in the northern hemisphere. At the northern hemisphere winter solstice (22 December), the Sun is overhead in the southern hemisphere.

belt of rain associated with the trade winds moves north and south with the Sun, as do the dry conditions associated with the belts of high pressure near the tropics. The monsoon's three seasons result from the influence of the Indian Ocean on the surrounding land mass of Asia in that area.

seasonal affective disorder (SAD) form of depression which occurs in winter and is relieved by the coming of spring. Its incidence decreases closer to the Equator. One type of SAD is associated with increased sleeping and appetite.

It has been suggested that SAD may be caused by changes in the secretion of melatonin, a hormone produced by the ◊pineal body in the brain. Melatonin secretion is inhibited by bright daylight.

sea water the water of the seas and oceans, covering about 70% of the Earth's surface and comprising about 97% of the world's water (only about 3% is fresh water). Sea water contains a large amount of dissolved solids, the most abundant of which is sodium chloride (almost 3% by mass); other salts include potassium chloride, bromide, and iodide, magnesium chloride, and magnesium sulphate. It also contains a large amount of dissolved carbon dioxide, and thus acts as a carbon 'sink' that may help to reduce the greenhouse effect.

seborrhoeic eczema common skin disease affecting any sebum-(natural oil) producing area of the skin. It is thought to be caused by the yeast *Pityrosporum*, and is characterized by yellowish-red, scaly areas on the skin, and dandruff. Antidandruff shampoos are often helpful.

sebum oily secretion from the sebaceous glands that acts as a skin lubricant. ◊Acne is caused by inflammation of the sebaceous glands and over-secretion of sebum.

sec or **s** abbreviation for **second**, a unit of time.

secant in trigonometry, the function of a given angle in a right-angled triangle, obtained by dividing the length of the hypotenuse (the longest side) by the length of the side

adjacent to the angle. It is the ◊reciprocal of the ◊cosine (sec = 1/cos).

second basic ◊SI unit (symbol sec or s) of time, one-sixtieth of a minute. It is defined as the duration of 9,192,631,770 cycles of regulation (periods of the radiation corresponding to the transition between two hyperfine levels of the ground state) of the cesium-133 isotope. In mathematics, the second is a unit (symbol ') of angular measurement, equalling one-sixtieth of a minute, which in turn is one-sixtieth of a degree.

secondary data information that has been collected by another agency. Examples of secondary data include government reports and statistics, company reports and accounts, and weather reports in newspapers.

secondary emission in physics, an emission of electrons from the surface of certain substances when they are

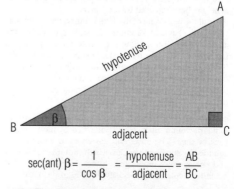

$$\sec(\text{ant})\ \beta = \frac{1}{\cos \beta} = \frac{\text{hypotenuse}}{\text{adjacent}} = \frac{AB}{BC}$$

secant The secant of an angle is a function used in the mathematical study of the triangle. If the secant of angle B is known, then the hypotenuse can be found given the length of the adjacent side, or the adjacent side can be found from the hypotenuse.

struck by high-speed electrons or other particles from an external source. It can be detected with a ◊photomultiplier.

secondary growth or ***secondary thickening*** increase in diameter of the roots and stems of certain plants (notably shrubs and trees) that results from the production of new cells by the ◊cambium. It provides the plant with additional mechanical support and new conducting cells, the secondary ◊xylem and ◊phloem. Secondary growth is generally confined to ◊gymnosperms and, among the ◊angiosperms, to the dicotyledons. With just a few exceptions, the monocotyledons (grasses, lilies) exhibit only primary growth, resulting from cell division at the apical ◊meristems.

secondary sexual characteristic in biology, an external feature of an organism that is indicative of its gender (male or female), but not the reproductive organs themselves. They include facial hair in men and breasts in women, combs in cockerels, brightly coloured plumage in many male birds, and manes in male lions. In many cases, they are involved in displays and contests for mates and have evolved by ◊sexual selection. Their development is stimulated by sex hormones.

secretin ◊hormone produced by the small intestine of vertebrates that stimulates the production of digestive secretions by the pancreas and liver.

secretion in biology, any substance (normally a fluid) produced by a cell or specialized gland, for example, sweat, saliva, enzymes, and hormones. The process whereby the substance is discharged from the cell is also known as secretion.

sector in geometry, part of a circle enclosed by two radii and the arc that joins them.

sedative any drug that has a calming effect, reducing anxiety and tension.

Sedatives will induce sleep in larger doses. Examples are ◊barbiturates, ◊narcotics, and ◊benzodiazepines.

sediment any loose material that has 'settled' – deposited from suspension in water, ice, or air, generally as the water current or wind speed decreases. Typical sediments are, in order of increasing coarseness, clay, mud, silt, sand, gravel, pebbles, cobbles, and boulders.

Sediments differ from sedimentary rocks in which deposits are fused together in a solid mass of rock by a process called ◊lithification. Pebbles are cemented into ◊conglomerates; sands become sandstones; muds become mudstones or shales; peat is transformed into coal.

sedimentary rock rock formed by the accumulation and cementation of deposits that have been laid down by water, wind, ice, or gravity. Sedimentary rocks cover more than two-thirds of the Earth's surface and comprise three major categories: clastic, chemically precipitated, and organic (or biogenic). Clastic sediments are the largest group and are composed of fragments of pre-existing rocks; they include clays, sands, and gravels.

Chemical precipitates include some limestones and evaporated deposits such as gypsum and halite (rock salt). Coal, oil shale, and limestone made of fossil material are examples of organic sedimentary rocks.

Most sedimentary rocks show distinct layering (stratification), caused by alterations in composition or by changes in rock type. These strata may become folded or fractured by the movement of the Earth's crust, a process known as **deformation**.

Seebeck effect in physics, the generation of a voltage in a circuit containing two different metals, or semiconductors, by keeping the junctions between them at different temperatures. Discovered by the German physicist Thomas Seebeck (1770–1831), it is also called the thermoelectric effect, and is the basis of the ◊thermocouple. It is the opposite of the ◊Peltier effect (in which current flow causes a temperature difference between the junctions of different metals).

seed the reproductive structure of higher plants (◊angiosperms and ◊gymnosperms). It develops from a fertilized ovule and consists of an embryo and a food store, surrounded and protected by an outer seed coat, called the testa. The food store is contained either in a specialized nutritive tissue, the ◊endosperm, or in the ◊cotyledons of the embryo itself. In angiosperms the seed is enclosed within a ◊fruit, whereas in gymnosperms it is usually naked and unprotected, once shed from the female cone. Following ◊germination the seed develops into a new plant.

Seeds may be dispersed from the parent plant in a number of different ways. Agents of dispersal include animals, as with ◊burs and fleshy edible fruits, and wind, where the seed or fruit may be winged or plumed. Water can disperse seeds or fruits that float, and various mechanical devices may eject seeds from the fruit, as in the pods of some leguminous plants (see ◊legume).

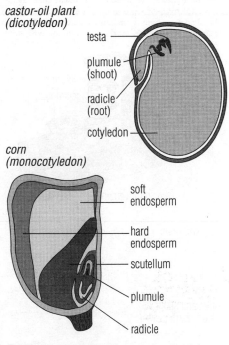

castor-oil plant (dicotyledon)

testa
plumule (shoot)
radicle (root)
cotyledon

corn (monocotyledon)

soft endosperm
hard endosperm
scutellum
plumule
radicle

seed The structure of seeds. The castor is a dicotyledon, a plant in which the developing plant has two leaves, developed from the cotyledon. In corn, a monocotyledon, there is a single leaf developed from the scutellum.

There may be a delay in the germination of some seeds to ensure that growth occurs under favourable conditions (see ◊after-ripening, ◊dormancy). Most seeds remain viable for at least 15 years if dried to about 5% water and kept at −20°C/−4°F, although 20% of them will not survive this process.

Seed

The largest seeds in the world are those of the coco de mer coconut palm *Lodoicea maldivica*. Each weighs up to 18 kg/40 lb. The smallest seeds are those of tree-living orchids, of which 992 million seeds are needed to make up a gram (35 million to make up an ounce).

seed drill machine for sowing cereals and other seeds, developed by Jethro Tull in England in 1701, although simple seeding devices were known in Babylon circa 2000 BC.

The seed is stored in a hopper and delivered by tubes into furrows in the ground. The furrows are made by a set of blades, or coulters, attached to the front of the drill. A ◊harrow is drawn behind the drill to cover up the seeds.

seed plant any seed-bearing plant; also known as a **spermatophyte**.

The seed plants are subdivided into two classes: the ◊angiosperms, or flowering plants, and the ◊gymnosperms, principally the cycads and conifers. Together, they comprise the major types of vegetation found on land.

Angiosperms are the largest, most advanced, and most successful group of plants at the present time, occupying a highly diverse range of habitats. There are estimated to be about 250,000 different species. Gymnosperms differ from angiosperms in their ovules which are borne unprotected (not within an ◊ovary) on the scales of their cones.

The arrangement of the reproductive organs, and their more simplified internal tissue structure, also distinguishes them from the flowering plants. In contrast to the gymnosperms, the ovules of angiosperms are enclosed within an ovary and many species have developed highly specialized reproductive structures associated with ◊pollination by insects, birds, or bats.

segment in geometry, part of a circle cut off by a straight line or ◊chord, running from one point on the circumference to another. All angles in the same segment are equal.

seiche a pendulous movement seen in large areas of water resembling a ◊tide. It was originally observed on Lake Geneva and is created either by the wind, earth tremors or other atmospheric phenomena.

seismograph instrument used to record the activity of an ◊earthquake. A heavy inert weight is suspended by a spring and attached to this is a pen that is in contact with paper on a rotating drum. During an earthquake the instrument frame and drum move, causing the pen to record a zigzag line on the paper; the pen does not move.

seismology study of earthquakes and how their shock waves travel through the Earth. By examining the global pattern of waves produced by an earthquake, seismologists can deduce the nature of the materials through which they have passed. This leads to an understanding of the Earth's internal structure.

On a smaller scale artificial earthquake waves, generated by explosions or mechanical vibrators, can be used to search for subsurface features in, for example, oil or mineral exploration. Earthquake waves from underground nuclear explosions can be distinguished from

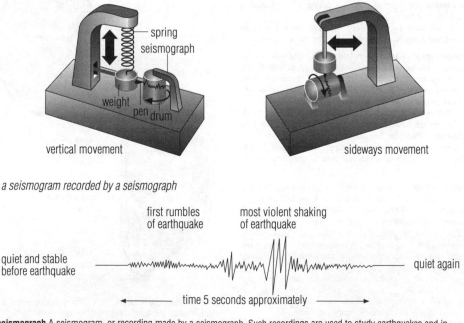

seismograph A seismogram, or recording made by a seismograph. Such recordings are used to study earthquakes and in prospecting.

natural waves by their shorter wavelength and higher frequency.

selective serotonin-reuptake inhibitor (SSRI) in medicine, one of a group of drugs that prevent the uptake of serotonin in neurones, resulting in an increase in its concentration in the brain. They are used in the treatment of depression. Fluoxetine (Prozac) is an example of a commonly prescribed SSRI.

selenium grey, nonmetallic element, symbol Se, atomic number 34, relative atomic mass 78.96. It belongs to the sulphur group and occurs in several allotropic forms that differ in their physical and chemical properties. It is an essential trace element in human nutrition.

Obtained from many sulphide ores and selenides, it is used as a red colouring for glass and enamel.

Because its electrical conductivity varies with the intensity of light, selenium is used extensively in photoelectric devices. It was discovered in 1817 by Swedish chemist Jöns Berzelius and named after the Moon because its properties follow those of tellurium, whose name derives from Latin *Tellus* 'Earth'.

self-inductance or *self-induction* in physics, the creation of an electromotive force opposing the current. See ◊inductance.

self-poisoning in medicine, deliberate consumption of an overdose in order to commit suicide. Many attempts at self-poisoning involve the use of drugs such as tranquillizers and antidepressants. Carbon monoxide, ◊paracetamol, paraquat, and organophosphorus insecticides are also used in self-poisoning. The highest incidence is in people in their twenties and it is more common in women.

Self-posioning with ◊paracetamol is a particular problem because of the ease of obtaining it and the serious consequences of small overdoses.

Sellafield site of a nuclear power station on the coast of Cumbria, NW England. It was known as **Windscale** until 1971, when the management of the site was transferred from the UK Atomic Energy Authority to British Nuclear Fuels Ltd. It reprocesses more than 1,000 tonnes of spent fuel from nuclear reactors annually. The plant is the world's greatest discharger of radioactive waste: between 1968 and 1979, 180 kg/400 lb of plutonium was discharged into the Irish Sea.

In 1990 a scientific study revealed an increased risk of leukaemia in children whose fathers worked at Sellafield 1950–85.

In 1996, British Nuclear Fuels was fined £25,000 after admitting 'serious and significant' failures in safety that left a Sellafield plant worker contaminated with radioactivity.

For accidents, see ◊nuclear safety.

selva equatorial rainforest, such as that in the Amazon basin in South America.

semelparity in biology, the occurrence of a single act of reproduction during an organism's lifetime. Most semelparous species produce very large numbers of offspring when they do reproduce, and normally die soon afterwards. Examples include the Pacific salmon and the pine looper moth. Many plants are semelparous, or ◊monocarpic. Repeated reproduction is called ◊iteroparity.

semen fluid containing ◊sperm from the testes and secretions from various sex glands (such as the prostate gland) that is ejaculated by male animals during copulation. The secretions serve to nourish and activate the sperm cells, and prevent them clumping together.

semiconductor material with electrical conductivity intermediate between metals and insulators and used in a wide range of electronic devices. Certain crystalline materials, most notably silicon and germanium, have a small number of free electrons that have escaped from the bonds between the atoms. The atoms from which they have escaped possess vacancies, called holes, which are similarly able to move from atom to atom and can be regarded as positive charges. Current can be carried by both electrons (negative carriers) and holes (positive carriers). Such materials are known as *intrinsic semiconductors*.

Conductivity can be enhanced by doping the material with small numbers of impurity atoms which either release free electrons (making an *n-type semiconductor* with more electrons than holes) or capture them (a *p-type semiconductor* with more holes than electrons). When p-type and n-type materials are brought together to form a p-n junction, an electrical barrier is formed which conducts current more readily in one direction than the other. This is the basis of the ◊semiconductor diode, used for rectification, and numerous other devices including ◊transistors, rectifiers, and ◊integrated circuits (silicon chips).

semiconductor diode or *p-n junction diode* in electronics, a two-terminal semiconductor device that allows electric current to flow in only one direction, the *forward-bias* direction. A very high resistance prevents current flow in the opposite, or *reverse-bias*, direction. It is used as a ◊rectifier, converting alternating current (AC) to direct current (DC).

senile dementia ◊Dementia associated with old age, often caused by ◊Alzheimer's disease.

sense organ any organ that an animal uses to gain information about its surroundings. All sense organs have specialized receptors (such as light receptors in the eye) and some means of translating their response into a nerve impulse that travels to the brain. The main human sense organs are the eye, which detects light and colour (different wavelengths of light); the ear, which detects sound (vibrations of the air) and gravity; the nose, which detects some of the chemical molecules in the air; and the tongue, which detects some of the chemicals in food, giving a sense of taste. There are also many small sense organs in the skin, including pain, temperature, and pressure sensors, contributing to our sense of touch.

Research suggests that our noses may also be sensitive to magnetic forces, giving us an innate sense of direction. This sense is well developed in other animals, as are a variety of senses that we do not share. Some animals can detect small electrical discharges, underwater vibrations, minute vibrations of the ground, or sounds that are below (infrasound) or above (ultrasound) our range of hearing. Sensitivity to light varies greatly. Most mammals cannot distinguish different colours, whereas some birds can detect the polarization of light. Many insects can see light in the ultraviolet range, which is beyond our spectrum, while snakes can form images of infrared radiation (radiant heat). In many animals, light is also detected by another organ, the ◊pineal body, which 'sees' light filtering through the skull, and measures the length of the day to keep track of the seasons.

sensor in computing, a device designed to detect a physical state or measure a physical quantity, and produce an input signal for a computer. For example, a sensor may detect the fact that a printer has run out of paper or may measure the temperature in a kiln.

The signal from a sensor is usually in the form of an analogue voltage, and must therefore be converted to a digital signal, by means of an ◊analogue-to-digital converter, before it can be input.

sepal part of a flower, usually green, that surrounds and protects the flower in bud. The sepals are derived from modified leaves, and are collectively known as the ◊calyx.

In some plants, such as the marsh marigold *Caltha palustris*, where true ◊petals are absent, the sepals are brightly coloured and petal-like, taking over the role of attracting insect pollinators to the flower.

sepsis general term for infectious change in the body caused by bacteria or their toxins.

septicaemia general term for any form of ◊blood poisoning.

septic shock life-threatening fall in blood pressure caused by blood poisoning (septicaemia). Toxins produced by bacteria infecting the blood induce a widespread dilation of the blood vessels throughout the body, and it is this that causes the patient's collapse (see ◊shock). Septic shock can occur following bowel surgery, after a penetrating wound to the abdomen, or as a consequence of infection of the urinary tract. It is usually treated in an intensive care unit and has a high mortality rate.

sequence-control register or *program counter* in computing, a special memory location used to hold the address of the next instruction to be fetched from the immediate access memory for execution by the computer (see ◊fetch-execute cycle). It is located in the control unit of the ◊central processing unit.

sequencing in biochemistry, determining the sequence of chemical subunits within a large molecule. Techniques for sequencing amino acids in proteins were established in the 1950s, insulin being the first for which the sequence was completed. The ◊Human Genome Project is attempting to determine the sequence of the 3 billion base pairs within human ◊DNA.

sere plant ◊succession developing in a particular habitat. A *lithosere* is a succession starting on the surface of bare rock. A *hydrosere* is a succession in shallow freshwater, beginning with planktonic vegetation and the growth of pondweeds and other aquatic plants, and ending with the development of swamp. A *plagiosere* is the sequence of communities that follows the clearing of the existing vegetation.

serial device in computing, a device that communicates binary data by sending the bits that represent each character one by one along a single data line, unlike a ◊parallel device.

series circuit electrical circuit in which the components are connected end to end, so that the current flows through them all one after the other.

serotype in medicine, serotyping is the classification of

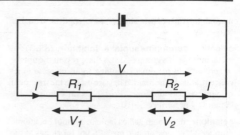

series circuit In a series circuit, the components of the circuit are connected end to end, so that the current passes through each component one after the other, without division or branching into parallel circuits.

substances according to the antigens they contain or the antibodies they provoke. Bacterial serotyping is used to classify bacteria of the same species that produce different immunological responses.

Serpens constellation on the celestial equator (see ◊celestial sphere), represented as a serpent coiled around the body of ◊Ophiuchus. It is the only constellation divided into two halves: *Serpens Caput*, the head (on one side of Ophiuchus), and *Serpens Cauda*, the tail (on the other side). Its main feature is the Eagle nebula.

serpentine group of minerals, hydrous magnesium silicate, $Mg_3Si_2O_5(OH)_4$, occurring in soft ◊metamorphic rocks and usually dark green. The fibrous form *chrysotile* is a source of ◊asbestos; other forms are *antigorite* and *lizardite*. Serpentine minerals are formed by hydration of ultramafic rocks during metamorphism. Rare snake-patterned forms are used in ornamental carving.

serum clear fluid that separates out from clotted blood. It is blood plasma with the anticoagulant proteins removed, and contains ◊antibodies and other proteins, as well as the fats and sugars of the blood. It can be produced synthetically, and is used to protect against disease.

servomechanism automatic control system used in aircraft, motor cars, and other complex machines. A specific input, such as moving a lever or joystick, causes a specific output, such as feeding current to an electric motor that moves, for example, the rudder of the aircraft. At the same time, the position of the rudder is detected and fed back to the central control, so that small adjustments can continually be made to maintain the desired course.

sessile in botany, a leaf, flower, or fruit that lacks a stalk and sits directly on the stem, as with the sessile acorns of certain oaks. In zoology, it is an animal that normally stays in the same place, such as a barnacle or mussel. The term is also applied to the eyes of ◊crustaceans when these lack stalks and sit directly on the head.

set or *class* in mathematics, any collection of defined things (elements), provided the elements are distinct and that there is a rule to decide whether an element is a member of a set. It is usually denoted by a capital letter and indicated by curly brackets {}.

For example, L may represent the set that consists of all the letters of the alphabet. The symbol ε stands for 'is a member of'; thus p ε L means that p belongs to the set consisting of all letters, and 4 ε/L means that 4 does not belong to the set consisting of all letters.

There are various types of sets. A *finite set* has a

Venn diagram of two intersecting sets

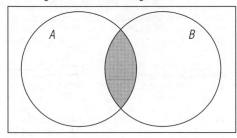

Venn diagram showing the whole numbers from 1 to 20 and the subsets of the prime and odd

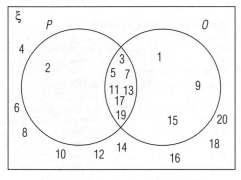

ξ = set of whole numbers from 1 to 20
O = set odd numbers
P = set of prime numbers
the intersection of *P* and *O* (*P* ∩ *O*) contains all the prime numbers that are also odd

set Venn diagrams showing the various relationships between sets A and B and sets P and O.

limited number of members, such as the letters of the alphabet; an **infinite set** has an unlimited number of members, such as all whole numbers; an **empty** or **null set** has no members, such as the number of people who have swum across the Atlantic Ocean, written as {} or ø; a **single-element set** has only one member, such as days of the week beginning with M, written as {Monday}. **Equal sets** have the same members; for example, if *W* = {days of the week} and *S* = {Sunday, Monday, Tuesday, Wednesday, Thursday, Friday, Saturday}, it can be said that *W* = *S*. Sets with the same number of members are **equivalent sets**. Sets with some members in common are **intersecting sets**; for example, if *R* = {red playing cards} and *F* = {face cards}, then *R* and *F* share the members that are red face cards. Sets with no members in common are **disjoint sets**. Sets contained within others are **subsets**; for example, *V* = {vowels} is a subset of *L* = {letters of the alphabet}.

Sets and their interrelationships are often illustrated by a ◊Venn diagram.

set-aside scheme policy introduced by the European Community (now the European Union) in the late 1980s, as part of the Common Agricultural Policy, to reduce overproduction of certain produce. Farmers are paid not to use land but to keep it ◊fallow. The policy may bring environmental benefits by limiting the amount of fertilizers and pesticides used.

SETI (abbreviation for **search for extra-terrestrial intelligence**) In astronomy, a programme originally launched by ◊NASA in 1992, using powerful ◊radio telescopes to search the skies for extra-terrestrial signals. NASA cancelled the SETI project in 1993, but other privately funded SETI projects continue.

severe combined immune deficiency (SCID) rare condition caused by a gene malfunction in which a baby is born unable to produce the enzyme ADA. Without ADA the T cells involved in fighting infection are poisoned; untreated infants usually die before the age of two. The child must be kept within a germ-free 'bubble', a transparent plastic tent, until a matched donor can provide a bone-marrow transplant (bone marrow is the source of disease-fighting cells in the body).

There have been promising results from experimental gene therapy for this condition, pioneered in the United States in 1990; in 1993 doctors inserted the ADA gene into stem cells from the umbilical cords of three babies born with SCID and reintroduced them. The gene was still present in blood cells in 1995, though only in 1% of T cells. The percentage will need to increase to a minimum of 10% for additional drug treatment to become possibly unnecessary.

sewage disposal the disposal of human excreta and other waterborne waste products from houses, streets, and factories. Conveyed through sewers to sewage works, sewage has to undergo a series of treatments to be acceptable for discharge into rivers or the sea, according to various local laws and ordinances. Raw sewage, or sewage that has not been treated adequately, is one serious source of water pollution and a cause of ◊eutrophication.

In the industrialized countries of the West, most industries are responsible for disposing of their own wastes. Government agencies establish industrial waste-disposal standards. In most countries, sewage works for residential areas are the responsibility of local authorities. The solid waste (sludge) may be spread over fields as a fertilizer or, in a few countries, dumped at sea. A significant proportion of bathing beaches in densely populated regions have unacceptably high bacterial content, largely as a result of untreated sewage being discharged into rivers and the sea. This can, for example, cause stomach upsets in swimmers.

The use of raw sewage as a fertilizer (long practised in China) has the drawback that disease-causing microorganisms can survive in the soil and be transferred to people or animals by consumption of subsequent crops. Sewage sludge is safer, but may contain dangerous levels of heavy metals and other industrial contaminants.

sex determination process by which the sex of an organism is determined. In many species, the sex of an individual is dictated by the two sex chromosomes (X and Y) it receives from its parents. In mammals, some plants, and a few insects, males are XY, and females XX; in birds, reptiles, some amphibians, and butterflies the reverse is the case. In bees and wasps, males are produced from unfertilized eggs, females from fertilized eggs.

Environmental factors can affect some fish and reptiles, such as turtles, where sex is influenced by the temperature at which the eggs develop. In 1991 it was shown that maleness is caused by a single gene, 14 base pairs long, on the Y chromosome.

sex hormone steroid hormone produced and secreted by the gonads (testes and ovaries). Sex hormones control development and reproductive functions and influence sexual and other behaviour.

sex linkage in genetics, the tendency for certain characteristics to occur exclusively, or predominantly, in one sex only. Human examples include red-green colour blindness and haemophilia, both found predominantly in males. In both cases, these characteristics are ◊recessive and are determined by genes on the ◊X chromosome.

Since females possess two X chromosomes, any such recessive ◊allele on one of them is likely to be masked by the corresponding allele on the other. In males (who have only one X chromosome paired with a largely inert ◊Y chromosome) any gene on the X chromosome will automatically be expressed. Colour blindness and haemophilia can appear in females, but only if they are ◊homozygous for these traits, due to inbreeding, for example.

sextant navigational instrument for determining latitude by measuring the angle between some heavenly body and the horizon. It was invented in 1730 by John Hadley (1682–1744) and can be used only in clear weather.

When the horizon is viewed through the right-hand side **horizon glass**, which is partly clear and partly mirrored, the light from a star can be seen at the same time in the mirrored left-hand side by adjusting an **index mirror**. The angle of the star to the horizon can then be read on a calibrated scale.

sexually transmitted disease (STD) any disease transmitted by sexual contact, involving transfer of body fluids. STDs include not only traditional ◊venereal disease, but also a growing list of conditions, such as ◊AIDS and scabies, which are known to be spread primarily by sexual contact. Other diseases that are transmitted sexually include viral ◊hepatitis. The WHO estimate that there are 356,000 new cases of STDs daily worldwide (1995).

sexual reproduction reproductive process in organisms that requires the union, or ◊fertilization, of gametes (such as eggs and sperm). These are usually produced by two different individuals, although self-fertilization occurs in a few ◊hermaphrodites such as tapeworms. Most organisms other than bacteria and cyanobacteria (◊blue-green algae) show some sort of sexual process. Except in some lower organisms, the gametes are of two distinct types called eggs and sperm. The organisms producing the eggs are called females, and those producing the sperm, males. The fusion of a male and female gamete produces a **zygote**, from which a new individual develops.

female reproductive system

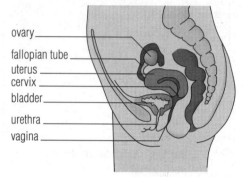

male reproductive system

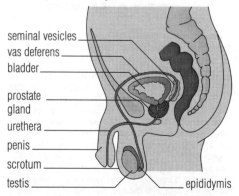

sexual reproduction The human reproductive organs. In the female, gametes called ova are released regularly in the ovaries after puberty. The Fallopian tubes carry the ova to the uterus or womb, in which the baby will develop. In the male, sperm is produced inside the testes after puberty; about 10 million sperm cells are produced each day, enough to populate the world in six months. The sperm duct or vas deferens, a continuation of the epididymis, carries sperm to the urethra during ejaculation.

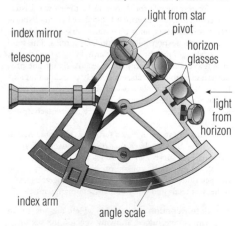

sextant The geometry of the sextant. When the light from a star can be seen at the same time as light from the horizon, the angle A can be read from the position of the index arm on the angle scale.

reproductive organs in flowering plants

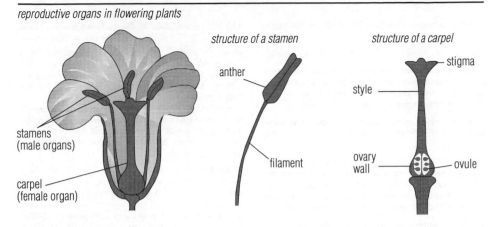

structure of a stamen

anther

filament

structure of a carpel

stigma

style

ovary wall

ovule

stamens (male organs)

carpel (female organ)

sexual reproduction in plants

sexual selection process similar to ◊natural selection but relating exclusively to success in finding a mate for the purpose of sexual reproduction and producing offspring. Sexual selection occurs when one sex (usually but not always the female) invests more effort in producing young than the other. Members of the other sex compete for access to this limited resource (usually males competing for the chance to mate with females).

Sexual selection often favours features that increase a male's attractiveness to females (such as the pheasant's tail) or enable males to fight with one another (such as a deer's antlers). More subtly, it can produce hormonal effects by which the male makes the female unreceptive to other males, causes the abortion of fetuses already conceived, or removes the sperm of males who have already mated with a female.

Seyfert galaxy galaxy whose small, bright centre is caused by hot gas moving at high speed around a massive central object, possibly a ◊black hole. Almost all Seyferts are spiral galaxies. They seem to be closely related to ◊quasars, but are about 100 times fainter. They are named after their discoverer Carl Seyfert (1911–1960).

shackle obsolete unit of length, used at sea for measuring cable or chain. One shackle is 15 fathoms (90 ft/27 m).

shadoof or **shaduf** machine for lifting water, consisting typically of a long, pivoted wooden pole acting as a lever, with a weight at one end. The other end is positioned over a well, for example. The shadoof was in use in ancient Egypt and is still used in Arab countries today.

shadow area of darkness behind an opaque object that cannot be reached by some or all of the light coming from a light source in front. Its presence may be explained in terms of light rays travelling in straight lines and being unable to bend round obstacles. A point source of light produces an ◊umbra, a completely black shadow with sharp edges. An extended source of light produces both a central umbra and a ◊penumbra, a region of semidarkness with blurred edges where darkness gives way to light.

◊Eclipses are caused by the Earth passing into the Moon's shadow or the Moon passing into the Earth's shadow.

shale fine-grained and finely layered ◊sedimentary rock composed of silt and clay. It is a weak rock, splitting easily along bedding planes to form thin, even slabs (by contrast, mudstone splits into irregular flakes). Oil shale contains kerogen, a solid bituminous material that yields ◊petroleum when heated.

sharecropping farming someone else's land, where the farmer gives the landowner a proportion of the crop instead of money. This system of rent payment was common in the USA, especially the South, until after World War II. It is still common in parts of the developing world; for example, in India. Often the farmer is left with such a small share of the crop that he or she is doomed to poverty.

sheath another name for a ◊condom.

sheep scab highly contagious disease of sheep, caused by mites that penetrate the animal's skin. Painful irritation, infection, loss of fleece, and death may result. The disease is notifiable in the UK.

shelf sea relatively shallow sea, usually no deeper than 200 m/650 ft, overlying the continental shelf around the coastlines. Most fishing and marine mineral exploitations are carried out in shelf seas.

shell the hard outer covering of a wide variety of invertebrates. The covering is usually mineralized, normally

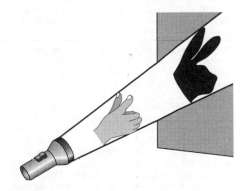

shadow

with large amounts of calcium. The shell of birds' eggs is also largely made of calcium.

shellac resin derived from secretions of the lac insect.

shell shock or *combat neurosis* or *battle fatigue* any of the various forms of mental disorder that affect soldiers exposed to heavy explosions or extreme ◊stress. Shell shock was first diagnosed during World War I.

SHF in physics, abbreviation for *superhigh* ◊*frequency*. SHF radio waves have frequencies in the range 3–30 GHz.

shiatsu in alternative medicine, Japanese method of massage derived from ◊acupuncture and sometimes referred to as 'acupressure', which treats organic or physiological dysfunctions by applying finger or palm pressure to parts of the body remote from the affected part.

shield in geology, alternative name for ◊craton, the ancient core of a continent.

shield in technology, any material used to reduce the amount of radiation (electrostatic, electromagnetic, heat, nuclear) reaching from one region of space to another, or any material used as a protection against falling debris, as in tunnelling.

Electrical conductors are used for electrostatic shields, soft iron for electromagnetic shields, and poor conductors of heat for heat shields. Heavy materials such as lead and concrete are used for protection against X-rays and nuclear radiation. See also ◊biological shield and ◊heat shield.

shifting cultivation farming system where farmers move on from one place to another when the land becomes exhausted. The most common form is *slash-and-burn* agriculture: land is cleared by burning, so that crops can be grown. After a few years, soil fertility is reduced and the land is abandoned. A new area is cleared while the old land recovers its fertility.

Slash-and-burn is practised in many tropical forest areas, such as the Amazon region, where yams, cassava, and sweet potatoes can be grown. This system works well while population levels are low, but where there is overpopulation, the old land will be reused before soil fertility has been restored. A variation of this system, found in parts of Africa, is rotational bush fallowing that involves a more permanent settlement and crop rotation.

shingles common name for ◊herpes zoster, a disease characterized by infection of sensory nerves, with pain and eruption of blisters along the course of the affected nerves.

Shinkansen fast railway network operated by Japanese Railways, on which the 'bullet' trains run. The network, opened in 1964, uses specially built straight and level track, on which average speeds of 160 kph/100 mph are attained.

In March 1997, the new Japanese bullet train, *Nozomi*-503, hit an average speed of 261.8kph/163mph between Hiroshima and Kokura, breaking the current rail speed record. Equipped with a long-nose lead carriage and new sound-proofing for a faster, quieter ride, the *Nozomi*-503 carried 1,300 passengers in 16 carriages from Osaka to Fukuoka, hitting speeds of up to 300kph/187mph. The official record for average speed between two stations of 252.6kph/157mph had been held by the French TGV.

The *Nozomi* also matched the top speed reached by the TGV.

shock in medicine, circulatory failure marked by a sudden fall of blood pressure and resulting in pallor, sweating, fast (but weak) pulse, and sometimes complete collapse. Causes include disease, injury, and psychological trauma.

In shock, the blood pressure falls below that necessary to supply the tissues of the body, especially the brain. Treatment depends on the cause. Rest is needed, and, in the case of severe blood loss, restoration of the normal circulating volume.

shock absorber in technology, any device for absorbing the shock of sudden jarring actions or movements. Shock absorbers are used in conjunction with coil springs in most motor-vehicle suspension systems and are usually of the telescopic type, consisting of a piston in an oil-filled cylinder. The resistance to movement of the piston through the oil creates the absorbing effect.

shock wave narrow band or region of high pressure and temperature in which the flow of a fluid changes from subsonic to supersonic.

Shock waves are produced when an object moves through a fluid at a supersonic speed. See ◊sonic boom.

shoot in botany, the parts of a ◊vascular plant growing above ground, comprising a stem bearing leaves, buds, and flowers. The shoot develops from the ◊plumule of the embryo.

shooting star another name for a ◊meteor.

short circuit unintended direct connection between two points in an electrical circuit.

Its relatively low resistance means that a large current flows through it, bypassing the rest of the circuit, and this may cause the circuit to overheat dangerously.

short-sightedness nontechnical term for ◊myopia.

shrub perennial woody plant that typically produces several separate stems, at or near ground level, rather than the single trunk of most trees. A shrub is usually smaller than a tree, but there is no clear distinction between large shrubs and small trees.

shunt in electrical engineering, a conductor of very low resistance that is connected in parallel to an ◊ammeter in order to enable it to measure larger electric currents. Its low resistance enables it to act like a bypass, diverting most of the current through itself and away from the ammeter.

SI abbreviation for *Système International d'Unités* (French 'International System of Metric Units'); see ◊SI units.

sial in geochemistry and geophysics, the substance of the Earth's continental ◊crust, as distinct from the ◊sima of the ocean crust. The name, now used rarely, is derived from silica and alumina, its two main chemical constituents. Sial is often rich in granite.

sick building syndrome malaise diagnosed in the early 1980s among office workers and thought to be caused by such pollutants as formaldehyde (from furniture and insulating materials), benzene (from paint), and the solvent trichloroethene, concentrated in air-conditioned buildings. Symptoms include headache, sore throat, tiredness, colds, and flu. Studies have found that it can cause a 40% drop in productivity and a 30% rise in absenteeism.

Work on improving living conditions of astronauts showed that the causes were easily and inexpensively removed by potplants in which interaction is thought to take place between the plant and microorganisms in its roots. Among the most useful are chrysanthemums (counteracting benzene), English ivy and the peace lily (trichloroethene), and the spider plant (formaldehyde).

sickle harvesting tool of ancient origin characterized by a curving blade with serrated cutting edge and short wooden handle. It was widely used in the Middle East and Europe for cutting wheat, barley, and oats from about 10,000 BC to the 19th century.

sickle-cell disease hereditary chronic blood disorder common among people of black African descent; also found in the E Mediterranean, parts of the Persian Gulf, and in NE India. It is characterized by distortion and fragility of the red blood cells, which are lost too rapidly from the circulation. This often results in ◊anaemia.

People with this disease have abnormal red blood cells (sickle cells), containing a defective ◊haemoglobin. The presence of sickle cells in the blood is called *sicklemia*.

The disease is caused by a recessive allele. Those with two copies of the allele suffer debilitating anaemia; those with a single copy paired with the normal allele, suffer with only mild anaemia and have a degree of protection against ◊malaria because fewer normal red blood cells are available to the parasites for infection.

Bone marrow transplantation can provide a cure, but the risks (a fatality rate of 10% and a complications rate of 20% are so great that it is only an option for the severely ill. US researchers announced in April 1995 that patients treated with a drug called hydroxyurea showed a reduction in the number of sickle cells. The drug works by reducing the amount of defective haemoglobin produced, and reviving the production of fetal haemoglobin. Fetal haemoglobin is not affected by sickling.

sidereal period the orbital period of a planet around the Sun, or a moon around a planet, with reference to a background star. The sidereal period of a planet is in effect a 'year'. A ◊synodic period is a full circle as seen from Earth.

sidereal time in astronomy, time measured by the rotation of the Earth with respect to the stars. A sidereal day is the time taken by the Earth to turn once with respect to the stars, namely 23 h 56 min 4 s. It is divided into sidereal hours, minutes, and seconds, each of which is proportionally shorter than the corresponding SI unit.

Siding Spring Mountain peak 400 km/250 mi NW of Sydney, site of the UK Schmidt Telescope, opened in 1973, and the 3.9 m/154 in *Anglo-Australian Telescope*, opened in 1975, which was the first big telescope to be fully computer-controlled. It is one of the most powerful telescopes in the southern hemisphere.

SIDS acronym for sudden infant death syndrome, the technical name for ◊cot death.

siemens SI unit (symbol S) of electrical conductance, the reciprocal of the ◊resistance of an electrical circuit. One siemens equals one ampere per volt. It was formerly called the mho or reciprocal ohm.

sievert SI unit (symbol Sv) of radiation dose equivalent. It replaces the rem (1 Sv equals 100 rem). Some types of radiation do more damage than others for the same absorbed dose – for example, an absorbed dose of alpha radiation causes 20 times as much biological damage as the same dose of beta radiation. The equivalent dose in sieverts is equal to the absorbed dose of radiation in rays multiplied by the relative biological effectiveness. Humans can absorb up to 0.25 Sv without immediate ill effects; 1 Sv may produce radiation sickness; and more than 8 Sv causes death.

Sigma Octantis the star closest to the south celestial pole (see ◊celestial sphere), in effect the southern equivalent of ◊Polaris, although far less conspicuous. Situated just less than 1° from the south celestial pole in the constellation Octans, Sigma Octantis is 120 light years away.

signal any sign, gesture, sound, or action that conveys information.

Examples include the use of flags (semaphore), light (traffic and railway signals), radio telephony, radio telegraphy (◊Morse code), and electricity (telecommunications and computer networks).

The International Code of Signals used by shipping was drawn up by an international committee and published in 1931. The codes and abbreviations used by aircraft are dealt with by the International Civil Aviation Organization, were established in 1944.

signal-to-noise ratio ratio of the power of an electrical signal to that of the unwanted noise accompanying the signal. It is expressed in ◊decibels.

In general, the higher the signal-to-noise ratio, the better. For a telephone, an acceptable ratio is 40 decibels; for television, the acceptable ratio is 50 decibels.

significant figures the figures in a number that, by virtue of their place value, express the magnitude of that number to a specified degree of accuracy. The final significant figure is rounded up if the following digit is greater than 5. For example, 5,463,254 to three significant figures is 5,460,000; 3.462891 to four significant figures is 3.463; 0.00347 to two significant figures is 0.0035.

silage fodder preserved through controlled fermentation in a ◊silo, an airtight structure that presses green crops. It is used as a winter feed for livestock. The term also refers to stacked crops that may be preserved indefinitely.

silencer (North American *muffler*) device in the exhaust system of cars and motorbikes. Gases leave the engine at supersonic speeds, and the exhaust system and silencer are designed to slow them down, thereby silencing them.

Some silencers use baffle plates (plates with holes, which disrupt the airflow), others use perforated tubes and an expansion box (a large chamber that slows down airflow).

silica silicon dioxide, SiO_2, the composition of the most common mineral group, of which the most familiar form is quartz. Other silica forms are ◊chalcedony, chert, opal, tridymite, and cristobalite.

Common sand consists largely of silica in the form of quartz.

silicate one of a group of minerals containing silicon and oxygen in tetrahedral units of SiO_4, bound together in various ways to form specific structural types. Silicates are the chief rock-forming minerals. Most rocks are composed, wholly or in part, of silicates (the main exception being limestones). Glass is a manufactured complex polysilicate material in which other elements (boron in borosilicate glass) have been incorporated.

Generally, additional cations are present in the structure, especially Al^{3+}, Fe^{2+}, Mg^{2+}, Ca^{2+}, Na^+, K^+, but quartz and other polymorphs of SiO_2 are also considered to be silicates; stishovite (a high pressure form of SiO_2) is a rare exception to the usual tetrahedral coordination of silica and oxygen.

In *orthosilicates*, the oxygens are all ionically bonded to cations such as Mg^{2+} or Fe^{2+} (as olivines), and are not shared between tetrahedra. All other silicate structures involve some degree of oxygen sharing between adjacent tetrahedra. For example, beryl is a *ring silicate* based on tetrahedra linked by sharing oxygens to form a circle. Pyroxenes are single *chain silicates*, with chains of linked tetrahedra extending in one direction through the structure; amphiboles are similar but have double chains of tetrahedra. In micas, which are *sheet silicates*, the tetrahedra are joined to form continuous sheets that are stacked upon one another. *Framework silicates*, such as feldspars and quartz, are based on three-dimensional frameworks of tetrahedra in which all oxygens are shared.

silicon brittle, nonmetallic element, symbol Si, atomic number 14, relative atomic mass 28.086. It is the second-most abundant element (after oxygen) in the Earth's crust and occurs in amorphous and crystalline forms. In nature it is found only in combination with other elements, chiefly with oxygen in silica (silicon dioxide, SiO_2) and the silicates. These form the mineral ◊quartz, which makes up most sands, gravels, and beaches.

Pottery glazes and glassmaking are based on the use of silica sands and date from prehistory. Today the crystalline form of silicon is used as a deoxidizing and hardening agent in steel, and has become the basis of the electronics industry because of its ◊semiconductor properties, being used to make 'silicon chips' for microprocessors.

The element was isolated by Swedish chemist Jöns Berzelius in 1823, having been named in 1817 by Scottish chemist Thomas Thomson by analogy with boron and carbon because of its chemical resemblance to these elements.

silicon chip ◊integrated circuit with microscopically small electrical components on a piece of silicon crystal only a few millimetres square.

One may contain more than a million components. A chip is mounted in a rectangular plastic package and linked via gold wires to metal pins, so that it can be connected to a printed circuit board for use in electronic devices, such as computers, calculators, television sets, car dashboards, and domestic appliances.

Silicon Valley nickname given to a region of S California, approximately 32 km/20 mi long, between Palo Alto and San Jose. It is the site of many high-technology electronic firms, whose prosperity is based on the silicon chip.

silicosis chronic disease of miners and stone cutters who inhale ◊silica dust, which makes the lung tissues fibrous and less capable of aerating the blood. It is a form of ◊pneumoconiosis.

sill sheet of igneous rock created by the intrusion of magma (molten rock) between layers of pre-existing rock. (A ◊dyke, by contrast, is formed when magma cuts *across* layers of rock.) An example of a sill in the UK is the Great Whin Sill, which forms the ridge along which Hadrian's Wall was built.

A sill is usually formed of *dolerite*, a rock that is extremely resistant to erosion and weathering, and often forms ridges in the landscape or cuts across rivers to create ◊waterfalls.

sillimanite aluminium silicate, Al_2SiO_5, a mineral that occurs either as white to brownish prismatic crystals or as minute white fibres. It is an indicator of high temperature conditions in metamorphic rocks formed from clay sediments. Andalusite, kyanite, and sillimanite are all polymorphs of Al_2SiO_5.

silo in farming, an airtight tower in which ◊silage is made by the fermentation of freshly cut grass and other forage crops. In military technology, a silo is an underground chamber for housing and launching a ballistic missile.

silt sediment intermediate in coarseness between clay and sand; its grains have a diameter of 0.002–0.02 mm/0.00008–0.0008 in. Silt is usually deposited in rivers, and so the term is often used generically to mean a river deposit, as in the silting-up of a channel.

Silurian period of geological time 439–409 million years ago, the third period of the Palaeozoic era. Silurian sediments are mostly marine and consist of shales and limestone. Luxuriant reefs were built by coral-like organisms. The first land plants began to evolve during this period, and there were many ostracoderms (armoured jawless fishes). The first jawed fishes (called acanthodians) also appeared.

silver white, lustrous, extremely malleable and ductile, metallic element, symbol Ag (from Latin *argentum*), atomic number 47, relative atomic mass 107.868. It occurs in nature in ores and as a free metal; the chief ores are sulphides, from which the metal is extracted by smelting with lead. It is one of the best metallic conductors of both heat and electricity; its most useful compounds are the chloride and bromide, which darken on exposure to light and are the basis of photographic emulsions.

Silver is used ornamentally, for jewellery and tableware, for coinage, in electroplating, electrical contacts, and dentistry, and as a solder. It has been mined since prehistory; its name is an ancient non-Indo-European one, *silubr*, borrowed by the Germanic branch as *silber*.

silver plate silverware made by depositing a layer of silver on another metal, usually copper, by the process of ◊electroplating.

sima in geochemistry and geophysics, the substance of the Earth's oceanic ◊crust, as distinct from the ◊sial of the continental crust. The name, now used rarely, is derived from silica and magnesia, its two main chemical constituents.

simple harmonic motion (SHM) oscillatory or vibrational motion in which an object (or point) moves so that its acceleration towards a central point is proportional to its distance from it. A simple example is a pendulum, which also demonstrates another feature of SHM, that the maximum deflection is the same on each side of the central point.

A graph of the varying distance with respect to time is a sine curve, a characteristic of the oscillating current or voltage of an alternating current (AC), which is another example of SHM.

simultaneous equations in mathematics, one of two or more algebraic equations that contain two or more unknown quantities that may have a unique solution. For example, in the case of two linear equations with two unknown variables, such as

(i) $x + 3y = 6$

and

(ii) $3y - 2x = 4$,

the solution will be those unique values of x and y that are valid for both equations. Linear simultaneous equations can be solved by using algebraic manipulation to eliminate one of the variables, ◊coordinate geometry, or matrices (see ◊matrix).

For example, by using algebra, both sides of equation (i) could be multiplied by 2, which gives $2x + 6y = 12$. This can be added to equation (ii) to get $9y = 16$, which is easily solved: $y = 16/9$. The variable x can now be found by inserting the known y value into either original equation and solving for x. Another method is by plotting the equations on a graph, because the two equations represent straight lines in coordinate geometry and the coordinates of their point of intersection are the values of x and y that are true for both of them. A third method of solving linear simultaneous equations involves manipulating matrices. If the equations represent either two parallel lines or the same line, then there will be no solutions or an infinity of solutions respectively.

sine in trigonometry, a function of an angle in a right-angled triangle which is defined as the ratio of the length of the side opposite the angle to the length of the hypotenuse (the longest side).

Various properties in physics vary sinusoidally; that is, they can be represented diagrammatically by a sine wave (a graph obtained by plotting values of angles against the values of their sines). Examples include ◊simple harmonic motion, such as the way alternating current (AC) electricity varies with time.

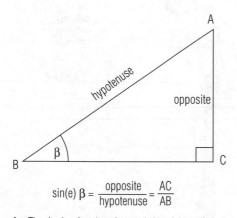

$$\sin(e)\ \beta = \frac{\text{opposite}}{\text{hypotenuse}} = \frac{AC}{AB}$$

sine The sine is a function of an angle in a right-angled triangle found by dividing the length of the side opposite the angle by the length of the hypotenuse (the longest side). Sine (usually abbreviated sin) is one of the fundamental trigonometric ratios.

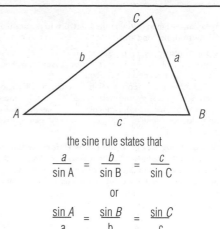

the sine rule states that

$$\frac{a}{\sin A} = \frac{b}{\sin B} = \frac{c}{\sin C}$$

or

$$\frac{\sin A}{a} = \frac{\sin B}{b} = \frac{\sin C}{c}$$

sine rule The sine rule relates the sides and angles of a triangle, stating that the ratio of the length of each side and the sine of the angle opposite is constant.

sine rule in trigonometry, a rule that relates the sides and angles of a triangle, stating that the ratio of the length of each side and the sine of the angle opposite is constant (twice the radius of the circumscribing circle). If the sides of a triangle are a, b, and c, and the angles opposite are A, B, and C, respectively, then the sine rule may be expressed as

$$a/\sin A = b/\sin B = c/\sin C.$$

singularity in astrophysics, the point in ◊space–time at which the known laws of physics break down. Singularity is predicted to exist at the centre of a black hole, where infinite gravitational forces compress the infalling mass of a collapsing star to infinite density. It is also thought, according to the Big Bang model of the origin of the universe, to be the point from which the expansion of the universe began.

sinus in biology, a sinus is a narrow cavity in the body. The paranasal sinuses in the skull communicate with the nose. Bacterial infection of these sinuses results in pain and inflammation and requires treatment with an antibiotic. Other examples of sinuses include the cavities in the membrane covering the brain through which the blood circulates.

sinusitis painful inflammation of one of the sinuses, or air spaces, that surround the nasal passages. Most cases clear with antibiotics and nasal decongestants, but some require surgical drainage.

Sinusitis most frequently involves the maxillary sinuses, within the cheek bones, producing pain around the eyes, toothache, and a nasal discharge.

siphon tube in the form of an inverted U with unequal arms. When it is filled with liquid and the shorter arm is placed in a tank or reservoir, liquid flows out of the longer arm provided that its exit is below the level of the surface of the liquid in the tank.

The liquid flows through the siphon because low pressure develops at the apex as liquid falls freely down the long arm. The difference between the pressure at the tank surface (atmospheric pressure) and the pressure at the apex causes liquid to rise in the short arm to replace that falling from the long arm.

siphonogamy plant reproduction in which a pollen tube grows to enable male gametes to pass to the ovary without leaving the protection of the plant.

Siphonogamous reproduction is found in all angiosperms and most gymnosperms, and has enabled these plants to reduce their dependency on wet conditions for reproduction, unlike the zoidogamous plants (see ◊zoidogamy).

Sirius or *Dog Star* or *Alpha Canis Majoris* the brightest star in the night sky, 8.6 light years from Earth in the constellation ◊Canis Major. Sirius is a white star with a mass 2.3 times that of the Sun, a diameter 1.8 times that of the Sun, and a luminosity of 23 Suns. It is orbited every 50 years by a ◊white dwarf, Sirius B, also known as the Pup.

Sirius is a double star with an orbital period of 50 years. Its eighth-magnitude companion is sometimes known as 'the Dark Companion' as it was first detected by Friedrich Bessel from its gravitational effect on the proper motion of Sirius. It was seen for the first time in 1862 but it was only in the 1920s that it was recognized as the first known example of a white dwarf.

sirocco hot, normally dry and dust-laden wind that blows from the deserts of N Africa across the Mediterranean into S Europe. It occurs mainly in the spring. The name 'sirocco' is also applied to any hot oppressive wind.

site of special scientific interest (SSSI) in the UK,

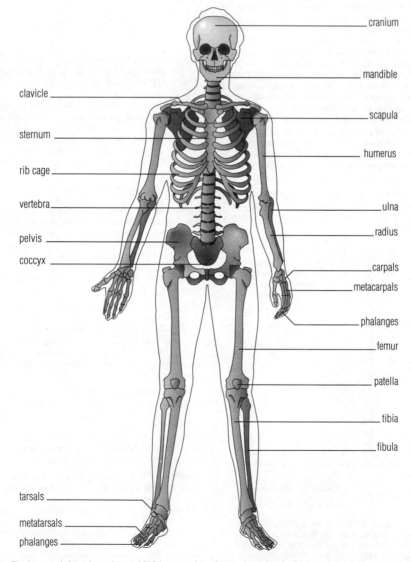

skeleton The human skeleton is made up of 206 bones and provides a strong but flexible supportive framework for the body.

land that has been identified as having animals, plants, or geological features that need to be protected and conserved. From 1991 these sites were designated and administered by English Nature, Scottish Natural Heritage, and the Countryside Council for Wales.

SI units (French *Système International d'Unités*) standard system of scientific units used by scientists worldwide.

Originally proposed in 1960, it replaces the ◊m.k.s., ◊c.g.s., and ◊f.p.s. systems. It is based on seven basic units: the metre (m) for length, kilogram (kg) for mass, second (s) for time, ampere (A) for electrical current, kelvin (K) for temperature, mole (mol) for amount of substance, and candela (cd) for luminosity.

skeleton the rigid or semi-rigid framework that supports and gives form to an animal's body, protects its internal organs, and provides anchorage points for its muscles. The skeleton may be composed of bone and cartilage (vertebrates), chitin (arthropods), calcium carbonate (molluscs and other invertebrates), or silica (many protists). The human skeleton is composed of 206 bones, with the ◊vertebral column (spine) forming the central supporting structure.

A skeleton may be internal, forming an ◊**endoskeleton**, or external, forming an ◊**exoskeleton**, as in the shells of insects or crabs. Another type of skeleton, found in invertebrates such as earthworms, is the **hydrostatic skeleton**. This gains partial rigidity from fluid enclosed within a body cavity. Because the fluid cannot be compressed, contraction of one part of the body results in extension of another part, giving peristaltic motion.

skin the covering of the body of a vertebrate. In mammals, the outer layer (epidermis) is dead and its cells are constantly being rubbed away and replaced from below; it helps to protect the body from infection and to prevent dehydration. The lower layer (dermis) contains blood vessels, nerves, hair roots, and sweat and sebaceous glands, and is supported by a network of fibrous and elastic cells. The medical speciality concerned with skin diseases is called dermatology.

Skin grafting is the repair of injured skin by placing pieces of skin, taken from elsewhere on the body, over the injured area.

skin graft in medicine, an operation in which large areas of damaged skin are replaced by skin from other areas of the body, usually from the arm or the thigh.

skull in vertebrates, the collection of flat and irregularly shaped bones (or cartilage) that enclose the brain and the organs of sight, hearing, and smell, and provide support for the jaws. In most mammals, the skull consists of 22 bones joined by fibrous immobile joints called sutures. The floor of the skull is pierced by a large hole (*foramen magnum*) for the spinal cord and a number of smaller apertures through which other nerves and blood vessels pass.

The skull comprises the cranium (brain case) and the bones of the face, which include the upper jaw, enclosing the sinuses, and form the framework for the nose, eyes,

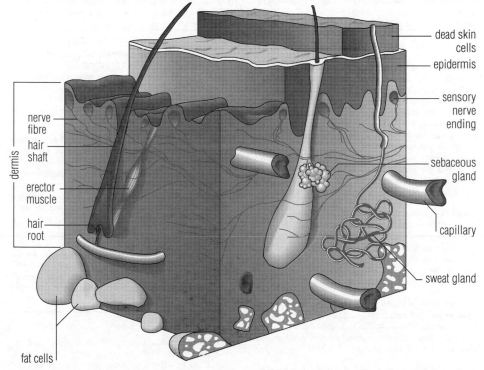

skin The skin of an adult man covers about 1.9 sq m/20 sq ft; a woman's skin covers about 1.6 sq m/17 sq ft. During our lifetime, we shed about 18 kg/40 lb of skin.

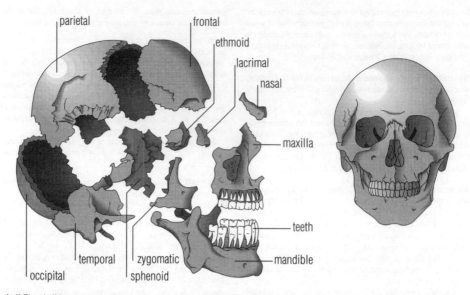

parietal — frontal — ethmoid — lacrimal — nasal — maxilla — teeth — mandible — zygomatic — sphenoid — temporal — occipital

skull The skull is a protective box for the brain, eyes, and hearing organs. It is also a framework for the teeth and flesh of the face. The cranium has eight bones: occipital, two temporal, two parietal, frontal, sphenoid, ethmoid. The face has 14 bones, the main ones being two maxillae, two nasal, two zygoma, two lacrimal, and the mandible.

and the roof of the mouth cavity. The lower jaw is hinged to the middle of the skull at its lower edge. The opening to the middle ear is located near the jaw hinge. The plate at the back of the head is jointed at its lower edge with the upper section of the spine. Inside, the skull has various shallow cavities into which fit different parts of the brain.

Skylab US space station launched on 14 May 1973, made from the adapted upper stage of a Saturn V rocket. At 75 tonnes/82.5 tons, it was the heaviest object ever put into space, and was 25.6 m/84 ft long. *Skylab* contained a workshop for carrying out experiments in weightlessness, an observatory for monitoring the Sun, and cameras for photographing the Earth's surface. Damaged during launch, it had to be repaired by the first crew of astronauts. Three crews, each of three astronauts, occupied *Skylab* for periods of up to 84 days, at that time a record duration for human spaceflight. *Skylab* finally fell to Earth on 11 July 1979, dropping debris on Western Australia.

slag in chemistry, the molten mass of impurities that is produced in the smelting or refining of metals.

The slag produced in the manufacture of iron in a ◊blast furnace floats on the surface above the molten iron. It contains mostly silicates, phosphates, and sulphates of calcium. When cooled, the solid is broken up and used as a core material in the foundations of roads and buildings.

slaked lime $Ca(OH)_2$ (technical name *calcium hydroxide*) substance produced by adding water to quicklime (calcium oxide, CaO). Much heat is given out and the solid crumbles as it absorbs water. A solution of slaked lime is called ◊limewater.

slash and burn simple agricultural method whereby natural vegetation is cut and burned, and the clearing then farmed for a few years until the soil loses its fertility,

whereupon farmers move on and leave the area to regrow. Although this is possible with a small, widely dispersed population, it becomes unsustainable with more people and is now a form of ◊deforestation.

slate fine-grained, usually grey metamorphic rock that splits readily into thin slabs along its ◊cleavage planes. It is the metamorphic equivalent of ◊shale.

Slate is highly resistant to atmospheric conditions and can be used for writing on with chalk (actually gypsum). Quarrying slate takes such skill and time that it is now seldom used for roof and sill material except in restoring historic buildings.

sleep state of natural unconsciousness and activity that occurs at regular intervals in most mammals and birds, though there is considerable variation in the amount of time spent sleeping. Sleep differs from hibernation in that it occurs daily rather than seasonally, and involves less drastic reductions in metabolism. The function of sleep is unclear. People deprived of sleep become irritable, uncoordinated, forgetful, hallucinatory, and even psychotic.

In humans, sleep is linked with hormone levels and specific brain electrical activity, including delta waves, quite different from the brain's waking activity. REM (rapid eye movement) phases, associated with dreams, occur at regular intervals during sleep, when the eyes move rapidly beneath closed lids.

Sleep

The koala sleeps on average 22 hours a day. It is the only animal more slothful than the sloth, which sleeps only 20 hours a day. Some say that the koala's diet of eucalyptus leaves means that it is in an almost permanently drugged sleep.

sleep apnoea in medicine, temporary cessation of breathing during sleep. It is usually due to a transient obstruction of the airway between the soft palate and the larynx when the airway dilator muscles relax too much. Muscle relaxation is accentuated by the use of sedative drugs and alcohol. Sleep apnoea is more common in obese people because the chances of airway obstruction are greater. The obstruction is overcome during each episode of apnoea by vigorous restoration of respiration accompanied by ◊snoring and the person is aroused from sleep.

sleeping pill any ◊sedative that induces sleep; in small doses, such drugs may relieve anxiety.

sleeping sickness or *trypanosomiasis* infectious disease of tropical Africa, a form of ◊trypanosomiasis. Early symptoms include fever, headache, and chills, followed by ◊anaemia and joint pains. Later, the disease attacks the central nervous system, causing drowsiness, lethargy, and, if left untreated, death. Sleeping sickness is caused by either of two trypanosomes, *Trypanosoma gambiense* or *T. rhodesiense*. Control is by eradication of the tsetse fly, which transmits the disease to humans.

Sleeping sickness in cattle is called nagana.

slide rule mathematical instrument with pairs of logarithmic sliding scales, used for rapid calculations, including multiplication, division, and the extraction of square roots. It has been largely superseded by the electronic calculator.

It was invented in 1622 by the English mathematician William Oughtred. A later version was devised by the French army officer Amédée Mannheim (1831–1906).

slime mould or *myxomycete* extraordinary organism that shows some features of ◊fungus and some of ◊protozoa. Slime moulds are not closely related to any other group, although they are often classed, for convenience, with the fungi. There are two kinds, cellular slime moulds and plasmodial slime moulds, differing in their complex life cycles.

Cellular slime moulds go through a phase of living as single cells, looking like amoebae, and feed by engulfing the bacteria found in rotting wood, dung, or damp soil. When a food supply is exhausted, up to 100,000 of these amoebae form into a colony resembling a single sluglike animal and migrate to a fresh source of bacteria. The colony then takes on the aspect of a fungus, and forms long-stalked fruiting bodies which release spores. These germinate to release amoebae, which repeat the life cycle.

Plasmodial slime moulds have a more complex life cycle involving sexual reproduction. They form a slimy mass of protoplasm with no internal cell walls, which slowly spreads over the bark or branches of trees.

slow virus in medicine, one of a group of viruses that cause disorders of the central nervous system and take many years to develop. Gradual and widespread damage of nerve tissue is followed eventually by a loss of brain function and death. Slow viruses are implicated in Creutzfeldt-Jakob disease and a type of ◊meningitis.

SLR abbreviation for *single-lens reflex*, a type of ◊camera in which the image can be seen through the lens before a picture is taken.

A small mirror directs light entering the lens to the viewfinder.

When a picture is taken the mirror moves rapidly aside to allow the light to reach the film. The SLR allows different lenses, such as close-up or zoom lenses, to be used because the photographer can see exactly what is being focused on.

slurry form of manure composed mainly of liquids. Slurry is collected and stored on many farms, especially when large numbers of animals are kept in factory units (see ◊factory farming). When slurry tanks are accidentally or deliberately breached, large amounts can spill into rivers, killing fish and causing ◊eutrophication. Some slurry is spread on fields as a fertilizer.

smack slang term for ◊heroin, an addictive depressant drug.

smallpox acute, highly contagious viral disease, marked by aches, fever, vomiting, and skin eruptions leaving pitted scars. Widespread vaccination programmes have eradicated this often fatal disease.

It was endemic in Europe until the development of vaccination by Edward Jenner in about 1800, and remained so in Asia, where a virulent form of the disease (variola major) entailed a fatality rate of 30% until the World Health Organization (WHO) campaign from 1967, which resulted in its virtual eradication by 1980. The campaign was estimated to have cost $300 million/£200 million, and was the organization's biggest health success to date.

smart card plastic card with an embedded microprocessor and memory. It can store, for example, personal data, identification, and bank-account details, to enable it to be used as a credit or debit card. The card can be loaded with credits, which are then spent electronically, and reloaded as needed. Possible other uses range from hotel door 'keys' to passports.

The smart card was invented by French journalist Juan Moreno in 1974. By the year 2000 it will be possible to make cards with as much computing power as the leading personal computers of 1990.

smart drug any drug or combination of nutrients (vitamins, amino acids, minerals, and sometimes herbs) said to enhance the functioning of the brain, increase mental energy, lengthen the span of attention, and improve the memory. As yet there is no scientific evidence to suggest that these drugs have any significant effect on healthy people.

Some smart drugs consist of food additives which are precursors of the neurotransmitter ◊acetylcholine. Most, however, are experimental drugs devised by pharmaceutical companies to treat aspects of dementia, in particular, memory loss. The description is also applied to existing drugs claimed to improve mental performance but which are legally prescribed for other purposes. These include beta-blockers (prescribed for some heart disease), phenytoin (epilepsy), and L-dopa (Parkinson's disease).

smart fluid or *electrorheological fluid* liquid suspension that solidifies to form a jellylike solid when a high-voltage electric field is applied across it and that returns to the liquid state when the field is removed. Most smart fluids are ◊zeolites or metals coated with polymers or oxides.

smell sense that responds to chemical molecules in the air. It works by having receptors for particular chemical groups, into which the airborne chemicals must fit to trigger a message to the brain.

Smell

Human mothers can identify their babies by smell alone.

A sense of smell is used to detect food and to communicate with other animals (see ◊pheromone and ◊scent gland). Humans can distinguish between about 10,000 different smells. Aquatic animals can sense chemicals in water, but whether this sense should be described as 'smell' or 'taste' is debatable. See also ◊nose.

smelling salts or *sal volatile* a mixture of ammonium carbonate, bicarbonate, and carbamate together with other strong-smelling substances, formerly used as a restorative for dizziness or fainting.

smelting processing a metallic ore in a furnace to produce the metal. Oxide ores such as iron ore are smelted with coke (carbon), which reduces the ore into metal and also provides fuel for the process.

A substance such as limestone is often added during smelting to facilitate the melting process and to form a slag, which dissolves many of the impurities present.

smog natural fog containing impurities, mainly nitrogen oxides (NO_x) and volatile organic compounds (VOCs) from domestic fires, industrial furnaces, certain power stations, and internal-combustion engines (petrol or diesel). It can cause substantial illness and loss of life, particularly among chronic bronchitics, and damage to wildlife.

photochemical smog is mainly prevalent in the summer as it is caused by chemical reaction between strong sunlight and vehicle exhaust fumes. Such smogs create a build-up of ozone and nitrogen oxides which cause adverse symptoms, including coughing and eye irritation, and in extreme cases can kill.

The London smog of 1952 lasted for five days and killed more than 4,000 people from heart and lung diseases. The use of smokeless fuels, the treatment of effluent, and penalties for excessive smoke from poorly maintained and operated vehicles can be effective in reducing smog but it still occurs in many cities throughout the world.

smokeless fuel fuel that does not give off any smoke when burned, because all the carbon is fully oxidized to carbon dioxide (CO_2). Natural gas, oil, and coke are smokeless fuels.

smoker or *hydrothermal vent* crack in the ocean floor, associated with an ◊ocean ridge, through which hot, mineral-rich ground water erupts into the sea, forming thick clouds of suspended material. The clouds may be dark or light, depending on the mineral content, thus producing 'white smokers' or 'black smokers'.

Sea water percolating through the sediments and crust is heated in the active area beneath and dissolves minerals from the hot rocks.

As the charged water is returned to the ocean, the sudden cooling causes these minerals to precipitate from solution, so forming the suspension. The chemical-rich water around a smoker gives rise to colonies of bacteria, and these form the basis of food chains that can be sustained without sunlight and photosynthesis. Strange animals that live in such regions include huge tube worms 2 m/6 ft long, giant clams, and species of crab, anemone, and shrimp found nowhere else.

smoking method of preserving fresh oily meats (such as pork and goose) or fish (such as herring and salmon). Before being smoked, the food is first salted or soaked in brine, then hung to dry. Meat is hot-smoked over a fast-burning wood fire, which is covered with sawdust, producing thick smoke and partly cooking the meat. Fish may be hot-smoked or cold-smoked over a slow-burning wood fire, which does not cook it. Modern refrigeration techniques mean that food does not need to be smoked to help it keep, so factory-smoked foods tend to be smoked just enough to give them a smoky flavour, with colours added to give them the appearance of traditionally smoked food.

smoking inhaling the fumes from burning substances, generally tobacco in the form of cigarettes. The practice is habit-forming and is dangerous to health, since carbon monoxide and other toxic materials result from the combustion process. A direct link between lung cancer and tobacco smoking was established 1950; the habit is also linked to respiratory and coronary heart diseases. In the West, smoking is now forbidden in many public places because even *passive smoking* – breathing in fumes from other people's cigarettes – can be harmful.

smooth muscle involuntary muscle capable of slow contraction over a period of time. It is present in hollow organs, such as the intestines, stomach, bladder, and blood vessels. Its presence in the wall of the alimentary canal allows slow rhythmic movements known as ◊peristalsis, which cause food to be mixed and forced along the gut. Smooth muscle has a microscopic structure distinct from other forms.

snellen unit expressing the visual power of the eye.

Snell's law of refraction in optics, the rule that when a ray of light passes from one medium to another, the sine of the angle of incidence divided by the sine of the angle of refraction is equal to the ratio of the indices of refraction in the two media. For a ray passing from medium 1 to medium 2:

$$N_2/n_1 = \sin i/\sin r$$

Where n_1 and n_2 are the refractive indices of the two media. The law was devised by the Dutch physicist, Willebrord Snell.

snoring loud noise during sleep made by vibration of the soft palate (the rear part of the roof of the mouth), caused by streams of air entering the nose and mouth at the same time. It is most common when the nose is partially blocked.

Sleep apnoea causes loud snoring that wakes the sufferer repeatedly throughout the night, causing chronic tiredness.

SNR (abbreviation for *supernova remnant*) in astronomy, the glowing remains of a star that has been destroyed in a ◊supernova explosion. The brightest and most famous example is the ◊Crab Nebula.

soap mixture of the sodium salts of various ◊fatty acids: palmitic, stearic, and oleic acid. It is made by the action of sodium hydroxide (caustic soda) or potassium hydroxide (caustic potash) on fats of animal or vegetable origin. Soap makes grease and dirt disperse in water in a similar manner to a ◊detergent.

soapstone compact, massive form of impure ◊talc.

social behaviour in zoology, behaviour concerned with altering the behaviour of other individuals of the same species. Social behaviour allows animals to live harmoniously in groups by establishing hierarchies of dominance to discourage disabling fighting. It may be aggressive or submissive (for example, cowering and other signals of appeasement), or designed to establish bonds (such as social grooming or preening).

The social behaviour of mammals and birds is generally more complex than that of lower organisms, and involves relationships with individually recognized animals. Thus, courtship displays allow individuals to choose appropriate mates and form the bonds necessary for successful reproduction. In the social systems of bees, wasps, ants, and termites, an individual's status and relationships with others are largely determined by its biological form, as a member of a caste of workers, soldiers, or reproductives; see ◊eusociality.

sociobiology study of the biological basis of all social behaviour, including the application of population genetics to the evolution of behaviour. It builds on the concept of ◊inclusive fitness, contained in the notion of the 'selfish gene'. Contrary to some popular interpretations, it does not assume that all behaviour is genetically determined.

soda ash former name for ◊sodium carbonate.

soda lime powdery mixture of calcium hydroxide and sodium hydroxide or potassium hydroxide, used in medicine and as a drying agent.

sodium soft, waxlike, silver-white, metallic element, symbol Na, atomic number 11, relative atomic mass 22.989. It is one of the ◊alkali metals and has a very low density, being light enough to float on water. It is the sixth-most abundant element (the fourth-most abundant

metal) in the Earth's crust. Sodium is highly reactive, oxidizing rapidly when exposed to air and reacting violently with water. Its most familiar compound is sodium chloride (common salt), which occurs naturally in the oceans and in salt deposits left by dried-up ancient seas.

sodium carbonate or **soda ash** Na_2CO_3 anhydrous white solid. The hydrated, crystalline form $(Na_2CO_3.10H_2O)$ is also known as washing soda.

It is made by the ◊Solvay process and used as a mild alkali, as it is hydrolysed in water.

$$CO_3{}^{2-}{}_{(aq)} + H_2O_{(l)} \rightarrow HCO_3{}^-{}_{(aq)} + OH^-{}_{(aq)}$$

It is used to neutralize acids, in glass manufacture, and in water softening.

sodium chloride or **common salt** or **table salt** NaCl white, crystalline compound found widely in nature. It is a a typical ionic solid with a high melting point (801°C/1,474°F); it is soluble in water, insoluble in organic solvents, and is a strong electrolyte when molten or in aqueous solution. Found in concentrated deposits, is widely used in the food industry as a flavouring and preservative, and in the chemical industry in the manufacture of sodium, chlorine, and sodium carbonate.

sodium hydrogencarbonate chemical name for ◊bicarbonate of soda.

sodium hydroxide or **caustic soda** NaOH the commonest alkali. The solid and the solution are corrosive. It is used to neutralize acids, in the manufacture of soap, and in oven cleaners. It is prepared industrially from sodium chloride by the ◊electrolysis of concentrated brine.

software in computing, a collection of programs and procedures for making a computer perform a specific task, as opposed to ◊hardware, the physical components

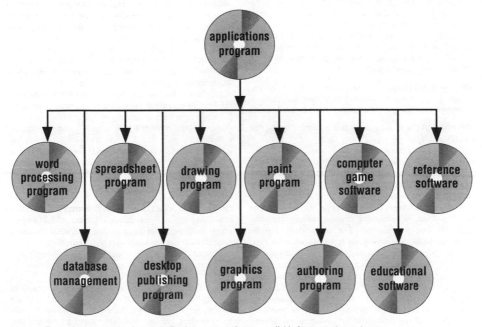

software The various types of software application program that are available for computer systems.

of a computer system. Software is created by programmers and is either distributed on a suitable medium, such as the ◊floppy disc, or built into the computer in the form of ◊firmware. Examples of software include ◊operating systems, ◊compilers, and applications programs, such as payrolls. No computer can function without some form of software.

soft water water that contains very few dissolved metal ions such as calcium (Ca^{2+}) or magnesium (Mg^{2+}). It lathers easily with soap, and no ◊scale is formed inside kettles or boilers. It has been found that the incidence of heart disease is higher in soft-water areas.

SOHO (abbreviation for *Solar and Heliospheric Observatory*) space probe launched in December 1995 by the ◊European Space Agency to study the ◊solar wind of atomic particles streaming towards the Earth from the Sun. It also observes the Sun in ultraviolet and visible light, and measures slight oscillations on the Sun's surface that can reveal details of the structure of the Sun's interior. It is positioned 1.5 million km/938,000 mi from Earth towards the Sun.

soil loose covering of broken rocky material and decaying organic matter overlying the bedrock of the Earth's surface. Various types of soil develop under different conditions: deep soils form in warm wet climates and in valleys; shallow soils form in cool dry areas and on slopes. *Pedology*, the study of soil, is significant because of the relative importance of different soil types to agriculture.

The organic content of soil is widely variable, ranging from zero in some desert soils to almost 100% in peats.

Soil Association pioneer British ecological organization founded in 1946, which campaigns against pesticides and promotes organic farming.

soil creep gradual movement of soil down a slope in response to gravity. This eventually results in a mass downward movement of soil on the slope.

Evidence of soil creep includes the formation of terracettes (steplike ridges along the hillside), leaning walls and telegraph poles, and trees that grow in a curve to counteract progressive leaning.

soil depletion decrease in soil quality over time. Causes include loss of nutrients caused by overfarming, erosion by wind, and chemical imbalances caused by acid rain.

soil erosion the wearing away and redistribution of the Earth's soil layer.

It is caused by the action of water, wind, and ice, and also by improper methods of ◊agriculture. If unchecked, soil erosion results in the formation of deserts (◊desertification). It has been estimated that 20% of the world's cultivated topsoil was lost between 1950 and 1990.

If the rate of erosion exceeds the rate of soil formation (from rock and decomposing organic matter), then the land will become infertile. The removal of forests (◊deforestation) or other vegetation often leads to serious soil erosion, because plant roots bind soil, and without them the soil is free to wash or blow away, as in the American ◊dust bowl. The effect is worse on hillsides, and there has been devastating loss of soil where forests have been cleared from mountainsides, as in Madagascar.

Improved agricultural practices such as contour ploughing are needed to combat soil erosion. Windbreaks, such as hedges or strips planted with coarse grass,

are valuable, and organic farming can reduce soil erosion by as much as 75%.

Soil degradation and erosion are becoming as serious as the loss of the rainforest. It is estimated that more than 10% of the world's soil lost a large amount of its natural fertility during the latter half of the 20th century. Some of the worst losses are in Europe, where 17% of the soil is damaged by human activity such as mechanized farming and fallout from acid rain. Mexico and Central America have 24% of soil highly degraded, mostly as a result of deforestation.

soil mechanics branch of engineering that studies the nature and properties of the soil. Soil is investigated during construction work to ensure that it has the mechanical properties necessary to support the foundations of dams, bridges, and roads.

sol ◊colloid of very small solid particles dispersed in a liquid that retains the physical properties of a liquid.

solar cycle in astronomy, the variation of activity on the ◊Sun over an 11-year period indicated primarily by the number of ◊sunspots visible on its surface. The next period of maximum activity is expected around the year 2001.

solar energy energy derived from the Sun's radiation. The amount of energy falling on just 1 sq km/0.3861 sq mi is about 4,000 megawatts, enough to heat and light a small town. In one second the Sun gives off 13 million times more energy than all the electricity used in the USA in one year. *Solar heaters* have industrial or domestic uses. They usually consist of a black (heat-absorbing) panel containing pipes through which air or water, heated by the Sun, is circulated, either by thermal ◊convection or by a pump.

Solar energy may also be harnessed indirectly using *solar cells* (photovoltaic cells) made of panels of ◊semiconductor material (usually silicon), which generate electricity when illuminated by sunlight. Although it is difficult to generate a high output from solar energy compared to sources such as nuclear or fossil fuels, it is a major nonpolluting and renewable energy source used as far north as Scandinavia as well as in the SW USA and in Mediterranean countries.

A solar furnace, such as that built in 1970 at Odeillo in the French Pyrenees, has thousands of mirrors to focus the Sun's rays; it produces uncontaminated intensive heat (up to 3,000°C/5,4000°F) for industrial and scientific or experimental purposes. The world's first solar power station connected to a national grid opened in 1991 at Adrano in Sicily. Scores of giant mirrors move to follow the Sun throughout the day, focusing the rays into a boiler. Steam from the boiler drives a conventional turbine. The plant generates up to 1 megawatt. A similar system, called Solar 1, has been built in the Mojave Desert near Daggett, California, USA. It consists of 1,818 computer-controlled mirrors arranged in circles around a central boiler tower 91 m/300 ft high. Advanced schemes have been proposed that would use giant solar reflectors in space to harness solar energy and beam it down to Earth in the form of ◊microwaves.

In March 1996 the first solar power plant capable of storing heat was switched on in California's Mojave Desert. Solar 2, part of a three-year government sponsored project, consists of 2,000 motorized mirrors that will focus the Sun's rays on to a 91 m/300 ft metal tower containing molten nitrate salt. When the salt reaches 565°C/1049°F it boils water to drive a 10-megawatt steam turbine. The molten salt retains its heat for up to 12 hours.

Despite their low running costs, their high installation cost and low power output have meant that solar cells have found few applications outside space probes and artificial satellites. Solar heating is, however, widely used for domestic purposes in many parts of the world, and is an important renewable source of energy.

solar pond natural or artificial 'pond', such as the Dead Sea, in which salt becomes more soluble in the Sun's heat. Water at the bottom becomes saltier and hotter, and is insulated by the less salty water layer at the top. Temperatures at the bottom reach about 100°C/212°F and can be used to generate electricity.

solar radiation radiation given off by the Sun, consisting mainly of visible light, ◊ultraviolet radiation, and ◊infra-

red radiation, although the whole spectrum of ◊electromagnetic waves is present, from radio waves to X-rays. High-energy charged particles such as electrons are also emitted, especially from solar ◊flares. When these reach the Earth, they cause magnetic storms (disruptions of the Earth's magnetic field), which interfere with radio communications.

Solar System the ◊Sun (a star) and all the bodies orbiting it: the nine ◊planets (Mercury, Venus, Earth, Mars, Jupiter, Saturn, Uranus, Neptune, and Pluto), their moons, the asteroids, and the comets. The Sun contains 99.86% of the mass of the Solar System.

The Solar System gives every indication of being a strongly unified system having a common origin and development. It is isolated in space; all the planets go round the Sun in orbits that are nearly circular and co-planar, and in the same direction as the Sun itself rotates;

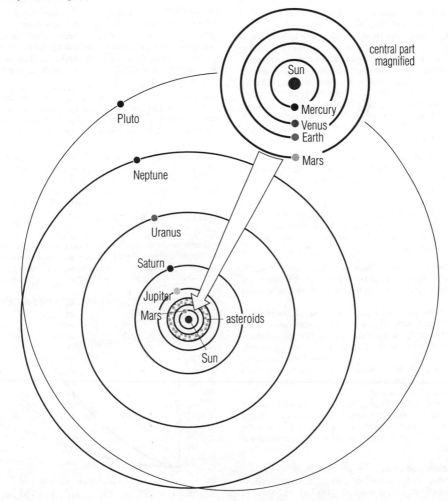

Solar System Most of the objects in the Solar System lie close to the plane of the ecliptic. The planets are tiny compared to the Sun. If the Sun were the size of a basketball, the planet closest to the Sun, Mercury, would be the size of a mustard seed 15 m/48 ft from the Sun. The most distant planet, Pluto, would be a pinhead 1.6 km/1 mi away from the Sun. The Earth, which is the third planet out from the Sun, would be the size of a pea 32 m/100 ft from the Sun.

moreover this same pattern is continued in the regular system of satellites that accompany Jupiter, Saturn, and Uranus. It is thought to have formed by condensation from a cloud of gas and dust in space about 4.6 billion years ago.

> *... in my studies of astronomy and philosophy I hold this opinion about the universe, that the Sun remains fixed in the centre of the circle of heavenly bodies, without changing its place: and the Earth, turning upon itself moves round the Sun.*
>
> On the **Solar System**
> Galileo, in letter to Cristina di Lorena 1615

solar time in astronomy, the time of day as determined by the position of the ◊Sun in the sky.

Apparent solar time, the time given by a sundial, is not uniform because of the varying speed of the Earth in its elliptical orbit. **Mean solar time** is a uniform time that coincides with apparent solar time at four instants through the year. The difference between them is known as the equation of time, and is greatest in early November when the Sun is more than 16 minutes fast on mean solar time. Mean solar time on the Green meridian is known as ◊Greenwich Mean Time and is the basis of civil timekeeping.

solar wind stream of atomic particles, mostly protons and electrons, from the Sun's corona, flowing outwards at speeds of between 300 kps/200 mps and 1,000 kps/600 mps.

The fastest streams come from 'holes' in the Sun's corona that lie over areas where no surface activity occurs. The solar wind pushes the gas of comets' tails away from the Sun, and 'gusts' in the solar wind cause geomagnetic disturbances and aurorae on Earth.

solder any of various alloys used when melted for joining metals such as copper, its common alloys (brass and bronze), and tin-plated steel, as used for making food cans.

Soft solders (usually alloys of tin and lead, sometimes with added antimony) melt at low temperatures (about 200°C/392°F), and are widely used in the electrical industry for joining copper wires. Hard (or brazing) solders, such as silver solder (an alloy of copper, silver, and zinc), melt at much higher temperatures and form a much stronger joint. ◊Printed circuit boards for computers are assembled by soldering.

A necessary preliminary to making any solder joint is thorough cleaning of the surfaces of the metal to be joined (to remove oxide) and the use of a flux (to prevent the heat applied to melt the solder from reoxidizing the metal).

solenoid coil of wire, usually cylindrical, in which a magnetic field is created by passing an electric current through it (see ◊electromagnet). This field can be used to move an iron rod placed on its axis.

Mechanical valves attached to the rod can be operated by switching the current on or off, so converting electrical energy into mechanical energy. Solenoids are used to relay energy from the battery of a car to the starter motor by means of the ignition switch.

solid in physics, a state of matter that holds its own shape (as opposed to a liquid, which takes up the shape of its container, or a gas, which totally fills its container). According to ◊kinetic theory, the atoms or molecules in a solid are not free to move but merely vibrate about fixed positions, such as those in crystal lattices.

solidification change of state from liquid (or vapour) to solid that occurs at the freezing point of a substance.

solid-state circuit electronic circuit where all the components (resistors, capacitors, transistors, and diodes) and interconnections are made at the same time, and by the same processes, in or on one piece of single-crystal silicon. The small size of this construction accounts for its use in electronics for space vehicles and aircraft.

solifluction the downhill movement of topsoil that has become saturated with water. Solifluction is common in periglacial environments (those bordering glacial areas) during the summer months, when the frozen topsoil melts to form an unstable soggy mass. This may then flow slowly downhill under gravity to form a **solifluction lobe** (a tonguelike feature).

solstice either of the days on which the Sun is farthest north or south of the celestial equator each year. The **summer solstice**, when the Sun is farthest north, occurs around the 21 June; the **winter solstice** around the 22 Dec.

solubility measure of the amount of solute (usually a solid or gas) that will dissolve in a given amount of solvent (usually a liquid) at a particular temperature. Solubility may be expressed as grams of solute per 100 grams of solvent or, for a gas, in parts per million (ppm) of solvent.

solute substance that is dissolved in another substance (see ◊solution).

solution two or more substances mixed to form a single, homogenous phase. One of the substances is the **solvent** and the others (**solutes**) are said to be dissolved in it.

comparative solubility curves for copper (II) sulphate and potassium nitrate

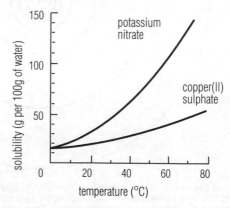

solubility

The constituents of a solution may be solid, liquid, or gaseous. The solvent is normally the substance that is present in greatest quantity; however, if one of the constituents is a liquid this is considered to be the solvent even if it is not the major substance.

solution in earth science, the dissolving in water of minerals within a rock. It may result in weathering (for example, when weakly acidic rainfall causes carbonation) and erosion (when flowing water passes over rocks).

Solution commonly affects limestone and chalk, and may be responsible for forming features such as sink holes.

Solvay process industrial process for the manufacture of sodium carbonate.

It is a multistage process in which carbon dioxide is generated from limestone and passed through ◊brine saturated with ammonia. Sodium hydrogen carbonate is isolated and heated to yield sodium carbonate. All intermediate by-products are recycled so that the only ultimate by-product is calcium chloride.

solvent substance, usually a liquid, that will dissolve another substance (see ◊solution). Although the commonest solvent is water, in popular use the term refers to low-boiling-point organic liquids, which are harmful if used in a confined space. They can give rise to respiratory problems, liver damage, and neurological complaints.

Typical organic solvents are petroleum distillates (in glues), xylol (in paints), alcohols (for synthetic and natural resins such as shellac), esters (in lacquers, including nail varnish), ketones (in cellulose lacquers and resins), and chlorinated hydrocarbons (as paint stripper and dry-cleaning fluids). The fumes of some solvents, when inhaled (◊glue-sniffing), affect mood and perception. In addition to damaging the brain and lungs, repeated inhalation of solvent from a plastic bag can cause death by asphyxia.

sonar (acronym for *sound navigation and ranging*) method of locating underwater objects by the reflection of ultrasonic waves. The time taken for an acoustic beam to travel to the object and back to the source enables the distance to be found since the velocity of sound in water is known. Sonar devices, or *echo sounders*, were developed in 1920, and are the commonest means of underwater navigation.

sone unit of subjective loudness. A tone of 40 decibels above the threshold of hearing with a frequency of 1,000 hertz is defined as one sone; any sound that seems twice as loud as this has a value of two sones, and so on. A loudness of one sone corresponds to 40 ◊phons.

sonic boom noise like a thunderclap that occurs when an aircraft passes through the ◊sound barrier, or begins to travel faster than the speed of sound. It happens when the cone-shaped shock wave caused by the plane touches the ground.

sorbic acid $CH_3CH=CHCH=CHCOOH$ tasteless acid found in the fruit of the mountain ash (genus *Sorbus*) and prepared synthetically. It is widely used in the preservation of food – for example, cider, wine, soft drinks, animal feeds, bread, and cheese.

sorting in computing, arranging data in sequence. When sorting a collection, or file, of data made up of several different ◊fields, one must be chosen as the *key field* used to establish the correct sequence. For example, the data in a company's mailing list might include fields for each customer's first names, surname, address, and telephone number. For most purposes the company would wish the records to be sorted alphabetically by surname; therefore, the surname field would be chosen as the key field.

The choice of sorting method involves a compromise between running time, memory usage, and complexity. Those used include *selection sorting*, in which the smallest item is found and exchanged with the first item, the second smallest exchanged with the second item, and so on; *bubble sorting*, in which adjacent items are continually exchanged until the data are in sequence; and *insertion sorting*, in which each item is placed in the correct position and subsequent items moved down to make a place for it.

sorus in ferns, a group of sporangia, the reproductive structures that produce ◊spores. They occur on the lower surface of fern fronds.

sound physiological sensation received by the ear, originating in a vibration that communicates itself as a pressure variation in the air and travels in every direction, spreading out as an expanding sphere. All sound waves in air travel with a speed dependent on the temperature; under ordinary conditions, this is about 330 m/1,070 ft per second. The pitch of the sound depends on the number of vibrations imposed on the air per second, but the speed is unaffected. The loudness of a sound is dependent primarily on the amplitude of the vibration of the air.

The lowest note audible to a human being has a frequency of about 20 ◊hertz (vibrations per second), and the highest one of about 20,000 Hz; the lower limit of this range varies little with the person's age, but the upper range falls steadily from adolescence onwards.

sound barrier concept that the speed of sound, or sonic speed (about 1,220 kph/760 mph at sea level), constitutes a speed limit to flight through the atmosphere, since a badly designed aircraft suffers severe buffeting at near sonic speed owing to the formation of shock waves. US test pilot Chuck Yeager first flew through the 'barrier' in 1947 in a Bell X-1 rocket plane. Now, by careful design, such aircraft as Concorde can fly at supersonic speed with ease, though they create in their wake a ◊sonic boom.

sound synthesis the generation of sound (usually music) by electronic synthesizer.

sound therapy in alternative medicine, treatment based on the finding that human blood cells respond to sound frequencies by changing colour and shape, and the hypothesis that therefore sick or rogue cells can be healed or harmonized by sound. In 1991 the therapy was being developed and researched by French musician and acupuncturist Fabien Maman and US physicist Joel Sternheimer. It is claimed that sound frequencies applied to acupuncture points are as effective as needles.

soundtrack band at one side of a cine film on which the accompanying sound is recorded. Usually it takes the form of an optical track (a pattern of light and shade). The pattern is produced on the film when signals from the recording microphone are made to vary the intensity of a light beam. During playback, a light is shone through the track on to a photocell, which converts the pattern of light falling on it into appropriate electrical signals. These signals are then fed to loudspeakers to recreate the original sounds.

source language in computing, the language in which a program is written, as opposed to ◊machine code, which is the form in which the program's instructions are carried out by the computer. Source languages are classified as either ◊high-level languages or ◊low-level languages, according to whether each notation in the source language stands for many or only one instruction in machine code.

Programs in high-level languages are translated into machine code by either a ◊compiler or an ◊interpreter program. Low-level programs are translated into machine code by means of an ◊assembler program. The program, before translation, is called the *source program*; after translation into machine code it is called the *object program*.

souring change that occurs to wine on prolonged exposure to air. The ethanol in the wine is oxidized by the air (oxygen) to ethanoic acid. It is the presence of the ethanoic (acetic) acid that produces the sour taste.

$$CH_3CH_2OH_{(aq)} + O_{2(g)} \rightarrow CH_3COOH_{(aq)} + H_2O_{(l)}$$

Southern Cross popular name for the constellation ◊Crux.

southern lights common name for the ◊aurora australis, coloured light in southern skies.

Southern Ocean corridor linking the Pacific, Atlantic, and Indian oceans, all of which receive cold polar water from the world's largest ocean surface current, the Antarctic Circumpolar Current, which passes through the Southern Ocean.

Soyuz Russian 'union' Soviet series of spacecraft, cap-

Space Flight: Chronology

1903	Russian scientist Konstantin Tsiolkovsky published the first practical paper on astronautics.
1926	US engineer Robert Goddard launched the first liquid-fuel rocket.
1937–45	In Germany, Wernher von Braun developed the V2 rocket.
1957	The first space satellite, *Sputnik 1* (USSR, Russian 'fellow-traveller'), orbited the Earth at a height of 229–898 km/142–558 mi in 96.2 min. *Sputnik 2* was launched carrying a dog, 'Laika'; it died on board seven days later.
1958	*Explorer 1*, the first US satellite, discovered the Van Allen radiation belts.
1961	The first crewed spaceship, *Vostok 1* (USSR), with Yuri Gagarin on board, was recovered after a single orbit of 89.1 min at a height of 142–175 km/88–109 mi.
1962	John Glenn in *Friendship 7* (USA) became the first American to orbit the Earth. *Telstar* (USA), a communications satellite, sent the first live television transmission between the USA and Europe.
1963	Valentina Tereshkova in *Vostok 1* (USSR) became the first woman in space.
1967	US astronauts Virgil Grissom, Edward White, and Roger Chaffee were killed during a simulated countdown when a flash fire swept through the cabin of *Apollo 1*. Vladimir Komarov was the first person to be killed on a mission, when his ship, *Soyuz 1* (USSR), crash-landed on the Earth.
1969	Neil Armstrong of *Apollo 11* (USA) was the first person to walk on the Moon.
1970	*Luna 17* (USSR) was launched; its space probe, *Lunokhod*, took photographs and made soil analyses of the Moon's surface.
1971	*Salyut 1* (USSR), the first orbital space station, was established; it was later visited by the *Soyuz 11* crewed spacecraft.
1973	*Skylab 2*, the first US orbital space station, was established.
1975	*Apollo 18* (USA) and *Soyuz 19* (USSR) made a joint flight and linked up in space.
1979	The European Space Agency's satellite launcher, *Ariane 1*, was launched.
1981	The first reusable crewed spacecraft, the space shuttle *Columbia* (USA), was launched.
1986	Space shuttle *Challenger* (USA) exploded shortly after take-off, killing all seven crew members.
1988	US shuttle programme resumed with launch of *Discovery*. Soviet shuttle *Buran* was launched from the rocket *Energiya*. Soviet cosmonauts Musa Manarov and Vladimir Titov in space station *Mir* spent a record 365 days 59 min in space.
1990	Hubble Space Telescope (USA) was launched from Cape Canaveral.
1991	The Gamma Ray Observatory was launched from the space shuttle *Atlantis* to survey the sky at gamma-ray wavelengths. Astronaut Helen Sharman, the first Briton in space, was launched with Anatoli Artsebarsky and Sergei Krikalev to *Mir* space station, returning to Earth 26 May in *Soyuz TM-11* with Viktor Afanasyev and Musa Manarov. Manarov set a record for the longest time spent in space, 541 days, having also spent a year aboard *Mir* 1988.
1992	European satellite *Hipparcos*, launched 1989 to measure the position of 120,000 stars, failed to reach geostationary orbit and went into a highly elliptical orbit, swooping to within 500 km/308 mi of the Earth every ten hours. The mission was later retrieved. Space shuttle *Endeavour* returned to Earth after its first voyage. During its mission, it circled the Earth 141 times and travelled 4 million km/2.5 million mi. *LAGEOS II* (Laser Geodynamics Satellite) was released from the space shuttle *Columbia* into an orbit so stable that it will still be circling the Earth in billions of years. Space shuttle *Endeavour* successfully carried out mission to replace the Hubble Space Telescope's solar panels and repair its mirror.
1994	Japan's heavy-lifting *H-2* rocket was launched successfully, carrying an uncrewed shuttle craft.
1995	The US space shuttle *Atlantis* docked with *Mir*, exchanging crew members.
1996	The *Ariane 5* rocket disintegrated almost immediately after takeoff, destroying the four Cluster satellites.
1997	NASA ceased operations using the space probe *Pioneer 10* after 25 years of sending signals from the edge of the Solar System. *Mir* underwent increasing difficulties, following a collision with a cargo ship in June that depressurized one of its modules, Spektr.

able of carrying up to three cosmonauts. Soyuz spacecraft consist of three parts: a rear section containing engines; the central crew compartment; and a forward compartment that gives additional room for working and living space. They are now used for ferrying crews up to space stations, though they were originally used for independent space flight.

Space isn't remote at all. It's only an hour's drive away if your car could go straight upwards.

On **space** Fred Hoyle *Observer* Sept 1979

space or *outer space* the void that exists beyond Earth's atmosphere. Above 120 km/75 mi, very little atmosphere remains, so objects can continue to move quickly without extra energy. The space between the planets is not entirely empty, but filled with the tenuous gas of the ◊solar wind as well as dust specks.

Space is out of this world.

On **space** Helen Sharman, in a speech May 1991

Spacelab small space station built by the European Space Agency, carried in the cargo bay of the US space shuttle, in which it remains throughout each flight, returning to Earth with the shuttle. Spacelab consists of a pressurized module in which astronauts can work, and a series of pallets, open to the vacuum of space, on which equipment is mounted.

Spacelab is used for astronomy, Earth observation, and experiments utilizing the conditions of weightlessness and vacuum in orbit. The pressurized module can be flown with or without pallets, or the pallets can be flown

Space Probe: Chronology

1959	*Luna 2* (USSR) hit the Moon, the first craft to do so.
	Luna 3 photographed the far side of the Moon.
1962	*Mariner 2* (USA) flew past Venus; launch date 26 Aug 1962.
1964	*Ranger 7* (USA) hit the Moon, having sent back 4,316 pictures before impact.
1965	*Mariner 4* flew past Mars; launch date 28 Nov 1964.
1966	*Luna 9* achieved the first soft landing on the Moon, having transmitted 27 close up panoramic photographs; launch date 31 Jan 1966.
	Surveyor 1 (USA) landed softly on the Moon and returned 11,150 pictures; launch date 30 May 1965.
1971	*Mariner 9* entered orbit of Mars; launch date 30 May 1971.
1973	*Pioneer 10* (USA) flew past Jupiter; launch date 3 March 1972.
1974	*Mariner 10* flew past Mercury; launch date 3 Nov 1973.
1975	*Venera 9* (USSR) landed softly on Venus and returned its first pictures; launch date 8 June 1975.
1976	*Viking 1* (USA) first landed on Mars; launch date 20 Aug 1975.
	Viking 2 transmitted data from the surface of Mars.
1977	*Voyager 2* (USA) launched. (5 Sept) *Voyager 1* launched.
1978	*Pioneer-Venus 1* (USA) orbited Venus; launch date 20 May 1978.
1979	*Voyager 1* and *Voyager 2* encountered Jupiter, respectively.
1980	*Voyager 1* reached Saturn.
1981	*Voyager 2* flew past Saturn.
1982	*Venera 13* transmitted its first colour pictures of the surface of Venus; launch date 30 Oct 1981.
1983	*Venera 15* mapped the surface of Venus from orbit; launch date 2 June 1983.
1985	*Giotto* (European Space Agency) launched to Halley's comet.
1986	*Voyager 2* encountered Uranus.
	Giotto met Halley's comet, closest approach 596 km/370 mi, at a speed 50 times faster than that of a bullet.
1989	*Magellan* (USA) launched from space shuttle *Atlantis* on a 15-month cruise to Venus across 15 million km/9 million mi of space.
	Voyager 2 reached Neptune (4,400 million km/2,700 million mi from Earth), approaching it to within 4,850 km/3,010 mi.
	Galileo (USA) launched from space shuttle *Atlantis* for six-year journey to Jupiter.
1990	*Magellan* arrived at Venus and transmitted its first pictures 16 Aug 1990.
	Ulysses (European Space Agency) launched from space shuttle *Discovery*, to study the Sun.
1991	*Galileo* made the closest-ever approach to an asteroid, Gaspra, flying within 1,600 km/990 mi.
1992	*Ulysses* flew past Jupiter at a distance of 380,000 km/236,000 mi from the surface, just inside the orbit of Io and closer than 11 of Jupiter's 16 moons.
	Giotto (USA) flew at a speed of 14 kms/8.5 mps to within 200 km/124 mi of comet Grigg-Skellerup, 12 light years (240 million km/150 mi) away from Earth.
	Mars Observer (USA) launched from Cape Canaveral, the first US mission to Mars for 17 years.
	Pioneer-Venus 1 burned up in the atmosphere of Venus.
1993	*Mars Observer* disappeared three days before it was due to drop into orbit around Mars.
	Galileo flew past the asteroid Ida.
1995	*Galileo's* probe entered the atmosphere of Jupiter. It radioed information back to the orbiter for 57 minutes before it was destroyed by atmospheric pressure.
1996	NASA's Near Earth Asteroid Rendezvous (NEAR) was launched to study Eros.
1997	The US spacecraft *Mars Pathfinder* lands on Mars. Two days later the probe's rover *Sojourner*, a six-wheeled vehicle that is controlled by an Earth-based operator, begins to explore the area around the spacecraft.

on their own, in which case the astronauts remain in the shuttle's own crew compartment. All the sections of Spacelab can be reused many times. The first Spacelab mission, consisting of a pressurized module and pallets, lasted for ten days in Nov–Dec 1983.

space probe any instrumented object sent beyond Earth to collect data from other parts of the Solar System and from deep space. The first probe was the Soviet *Lunik 1*, which flew past the Moon in 1959. The first successful planetary probe was the US *Mariner 2*, which flew past Venus in 1962, using ◊transfer orbit. The first space probe to leave the Solar System was *Pioneer 10* in 1983. Space probes include *Galileo, Giotto, Magellan, Mars Observer, Ulysses, Cassini*, the ◊Moon probes, and the Mariner, Pioneer, Viking, and Voyager series.

space shuttle reusable crewed spacecraft. The first was launched on 12 April 1981 by the USA. It was developed by NASA to reduce the cost of using space for commercial, scientific, and military purposes. After leaving its payload in space, the space-shuttle orbiter can be flown back to Earth to land on a runway, and is then available for reuse.

Four orbiters were built: *Columbia, Challenger, Discovery*, and *Atlantis. Challenger* was destroyed in a midair explosion just over a minute after its tenth launch on 28 Jan 1986, killing all seven crew members, the result of a failure in one of the solid rocket boosters. Flights resumed with redesigned boosters in Sept 1988. A replacement orbiter, *Endeavour*, was built, which had its maiden flight in May 1992. At the end of the 1980s, an average of $375 million had been spent on each space-shuttle mission.

The USSR produced a shuttle of similar size and appearance to the US one. The first Soviet shuttle, *Buran*, was launched without a crew by the Energiya rocket on 15 Nov 1988.

space sickness or **space adaptation syndrome** feeling of nausea, sometimes accompanied by vomiting, experienced by about 40% of all astronauts during their first few days in space. It is akin to travel sickness, and is thought to be caused by confusion of the body's balancing mechanism, located in the inner ear, by weightlessness. The sensation passes after a few days as the body adapts.

space station any large structure designed for human occupation in space for extended periods of time. Space stations are used for carrying out astronomical observations and surveys of Earth, as well as for biological studies and the processing of materials in weightlessness. The first space station was ◊*Salyut 1*, and the USA has launched ◊*Skylab*.

NASA plans to build a larger space station, to be called *Alpha* in cooperation with other countries, including the European Space Agency, which is building a module called *Columbus*; Russia and Japan are also building modules.

space suit protective suit worn by astronauts and cosmonauts in space. It provides an insulated, air-conditioned cocoon in which people can live and work for hours at a time while outside the spacecraft. Inside the suit is a cooling garment that keeps the body at a comfortable temperature even during vigorous work. The suit provides air to breathe, and removes exhaled carbon dioxide and moisture. The suit's outer layers insulate the occupant from the extremes of hot and cold in space (−150°C/ −240°F in the shade to +180°C/+350°F in sunlight), and

from the impact of small meteorites. Some space suits have a jet-propelled backpack, which the wearer can use to move about.

space-time in physics, combination of space and time used in the theory of ◊relativity. When developing relativity, Albert Einstein showed that time was in many respects like an extra dimension (or direction) to space. Space and time can thus be considered as entwined into a single entity, rather than two separate things.

Space-time is considered to have four dimensions: three of space and one of time. In relativity theory, events are described as occurring at points in space-time. The **general theory of relativity** describes how space-time is distorted by the presence of material bodies, an effect that we observe as gravity.

spadix in botany, an ◊inflorescence consisting of a long, fleshy axis bearing many small, stalkless flowers. It is partially enclosed by a large bract or ◊spathe. A spadix is characteristic of plants belonging to the family Araceae, including the arum lily *Zantedeschia aethiopica*.

spark chamber electronic device for recording tracks of charged subatomic particles, decay products, and rays. In combination with a stack of photographic plates, a spark chamber enables the point where an interaction has taken place to be located, to within a cubic centimetre. At its simplest, it consists of two smooth threadlike ◊electrodes that are positioned 1–2 cm/0.5–1 in apart, the space between being filled by an inert gas such as neon. Sparks jump through the gas along the ionized path created by the radiation. See ◊particle detector.

spark plug plug that produces an electric spark in the cylinder of a petrol engine to ignite the fuel mixture. It consists essentially of two electrodes insulated from one another. High-voltage (18,000 V) electricity is fed to a central electrode via the distributor. At the base of the electrode, inside the cylinder, the electricity jumps to another electrode earthed to the engine body, creating a spark. See also ◊ignition coil.

spasm in medicine, a sudden, involuntary contraction of a muscle or of a hollow organ with a muscular wall. A **tonic spasm** describes a spasm in which the contraction persists for some time followed by sudden or gradual relaxation, whilst **clonic spasm** is characterized by a series of contractions and relaxations of the muscle. Cramps are spasms that affect the limbs and ◊colic describes spasms of the stomach and intestinal walls.

spastic term applied generally to limbs with impaired movement, stiffness, and resistance to passive movement, and to any body part (such as the colon) affected with spasm.

spathe in flowers, the single large bract surrounding the type of inflorescence known as a ◊spadix. It is sometimes brightly coloured and petal-like, as in the brilliant scarlet spathe of the flamingo plant *Anthurium andreanum* from South America; this serves to attract insects.

speciation emergence of a new species during evolutionary history. One cause of speciation is the geographical separation of populations of the parent species, followed by reproductive isolation and selection for different environments so that they no longer produce viable offspring when they interbreed. Other causes are ◊assortative mating and the establishment of a ◊polyploid population.

species in biology, a distinguishable group of organisms that resemble each other or consist of a few distinctive types (as in ◊polymorphism), and that can all interbreed to produce fertile offspring. Species are the lowest level in the system of biological classification.

Related species are grouped together in a genus. Within a species there are usually two or more separate ◊populations, which may in time become distinctive enough to be designated subspecies or varieties, and could eventually give rise to new species through ◊speciation. Around 1.4 million species have been identified so far, of which 750,000 are insects, 250,000 are plants, and 41,000 are vertebrates. In tropical regions there are roughly two species for each temperate-zone species. It is estimated that one species becomes extinct every day through habitat destruction.

specific gravity alternative term for ◊relative density.

specific heat capacity in physics, quantity of heat required to raise unit mass (1 kg) of a substance by one ◊kelvin (1°C). The unit of specific heat capacity in the SI system is the ◊joule per kilogram kelvin (J kg^{-1} K^{-1}).

specific latent heat in physics, the heat that changes the physical state of a unit mass (one kilogram) of a substance without causing any temperature change.

The **specific latent heat of fusion** of a solid substance is the heat required to change one kilogram of it from solid to liquid without any temperature change. The **specific latent heat of vaporization** of a liquid substance is the heat required to change one kilogram of it from liquid to vapour without any temperature change.

speckle interferometry technique whereby large telescopes can achieve high resolution of astronomical objects despite the adverse effects of the atmosphere through which light from the object under study must pass. It involves the taking of large numbers of images, each under high magnification and with short exposure times. The pictures are then combined to form the final picture. The technique was introduced by the French astronomer Antoine Labeyrie in 1970.

spectacles pair of lenses fitted in a frame and worn in front of the ◊eyes to correct or assist defective vision. Common defects of the eye corrected by spectacle lenses are short sight (myopia) by using concave (spherical) lenses, long sight (hypermetropia) by using convex (spherical) lenses, and astigmatism by using cylindrical lenses.

Spherical and cylindrical lenses may be combined in one lens. Bifocal spectacles correct vision both at a distance and for reading by combining two lenses of different curvatures in one piece of glass. Varifocal spectacles have the same effect without any visible line between the two types of lens.

Spectacles are said to have been invented in the 13th century by a Florentine monk. Few people found the need for spectacles until printing was invented, when the demand for them increased rapidly. Using photosensitive glass, lenses can be produced that darken in glare and return to normal in ordinary light conditions. Lightweight plastic lenses are also common. The alternative to spectacles is contact lenses.

spectator ion in a chemical reaction that takes place in solution, an ion that remains in solution without taking part in the chemical change. For example, in the precipitation of barium sulphate from barium chloride and sodium sulphate, the sodium and chloride ions are spectator ions.

$$BaCl_{2 (aq)} + Na_2SO_{4 (aq)} \rightarrow BaSO_{4 (s)} + 2NaCl_{(aq)}$$

spectral classification in astronomy, the classification of stars according to their surface temperature and ◊luminosity as determined from their spectra. Stars are assigned a spectral type (or class) denoted by the letters O, B, A, F, G, K and M, where O stars (about 40,000 K) are the hottest and M stars (about 3,000 K) are the coolest.

Each letter may be further divided into ten subtypes, B0, B1, B2, etc. Stars are also assigned a luminosity class denoted by a Roman numeral attached to the spectral type: I (◊supergiants), II (bright giants), III (giants), IV (subgiants), V (main sequence), VI (subdwarfs), or VII (◊white dwarfs). The Sun is classified as type G2V. See also ◊Hertzsprung–Russell diagram.

spectrometer in physics and astronomy, an instrument used to study the composition of light emitted by a source. The range, or ◊spectrum, of wavelengths emitted by a source depends upon its constituent elements, and may be used to determine its chemical composition.

The simpler forms of spectrometer analyse only visible light. A **collimator** receives the incoming rays and produces a parallel beam, which is then split into a spectrum by either a ◊diffraction grating or a prism mounted on a turntable. As the turntable is rotated each of the constituent colours of the beam may be seen through a **telescope**, and the angle at which each has been deviated may be measured on a circular scale. From this information the wavelengths of the colours of light can be calculated.

Spectrometers are used in astronomy to study the electromagnetic radiation emitted by stars and other celestial bodies. The spectral information gained may be used to determine their chemical composition, or to measure the ◊red shift of colours associated with the expansion of the universe and thereby calculate the speed with which distant stars are moving away from the Earth.

spectrometry in analytical chemistry, a technique involving the measurement of the spectrum of energies (not necessarily electromagnetic radiation) emitted or absorbed by a substance.

spectroscopy study of spectra (see ◊spectrum) associated with atoms or molecules in solid, liquid, or gaseous phase. Spectroscopy can be used to identify unknown compounds and is an invaluable tool in science, medicine, and industry (for example, in checking the purity of drugs).

Emission spectroscopy is the study of the characteristic series of sharp lines in the spectrum produced when an ◊element is heated. Thus an unknown mixture can be analysed for its component elements. Related is **absorption spectroscopy**, dealing with atoms and molecules as they absorb energy in a characteristic way. Again, dark lines can be used for analysis. More detailed structural information can be obtained using **infrared spectroscopy** (concerned with molecular vibrations) or **nuclear magnetic resonance (NMR) spectroscopy** (concerned with interactions between adjacent atomic nuclei). **Supersonic jet laser beam spectroscopy** enables the isolation and study of clusters in the gas phase.

The fastest of them all

The speed of light in a vacuum, c, is a fundamental constant of nature. It occurs in many scientific equations, such as Einstein's famous $E = mc^2$. Numerous attempts have been made to determine c as accurately as possible.

First estimates

The first reasonable estimate was made in 1675 by a young Danish astronomer, Ole Roemer, at the Royal Observatory in Paris. Roemer was studying the motion of Jupiter's moons, which take between 1.7 and 16.7 days to orbit Jupiter. The orbital time of a moon can be measured by timing its eclipse by Jupiter's shadow. Roemer noticed that the interval between successive eclipses gradually decreased when Jupiter and the Earth were approaching one another and increased when they were receding. He explained this by assuming that light travels through space at a finite though very great speed. Thus if Jupiter is approaching the Earth, the time the light from a moon takes to reach the Earth will gradually decrease as Jupiter and the Earth get closer, and the interval between eclipses will appear to decrease. From the differences in orbital times and the known rates of motion of Jupiter and the Earth, Roemer estimated that the speed of light must be about 300,000 km/186,400 mi per second.

In 1728 British Astronomer Royal James Bradley (1693–1762) made another estimate using a different approach. Tradition has it that Bradley devised his method while sailing on the River Thames. He noticed how the apparent wind direction depended on the boat's movement and the windspeed. Bradley reasoned that if starlight was regarded as a wind blowing earthwards, its direction should appear to change throughout the year as the Earth moved round the Sun. By observing the change he was able to calculate the speed of light and verify Roemer's result.

Laboratory measurements

The first laboratory measurement was made by French physicist Armand Fizeau in 1849. He used a bright beam of light which travelled to a distant mirror and was reflected back along its path. Using a partly silvered mirror, he could look along the light path and see the reflected beam. A large square-toothed wheel in front of the mirror cut the beam into a sequence of flashes as it turned. The speed of rotation could be carefully controlled. With the wheel spinning, Fizeau could only see a flash of light if it passed through the gap between two teeth. He increased the speed of the wheel until the first gap coincided with the returning flash. It

was then simple, using the known rotation speed and the dimensions of the teeth, to calculate how long it had taken the tooth to move out of the beam path. The light had travelled 17.2 km/10.7 mi, there were 720 equal teeth in the wheel, and the wheel speed was 12.6 revolutions per second. Fizeau's result, therefore, for the speed of light was 313,300 km/194,700 mi per second.

A year later another French physicist, Léon Foucault, improved on Fizeau's experiment. A beam of light was reflected by a rotating mirror, travelled to a distant mirror and then returned to the rotating mirror. The rotating mirror had turned a little, so the returned beam appeared to be in a slightly different position from when it set out. This change could be measured accurately with a micrometer and used to calculate the rotation speed of the mirror more accurately than in Fizeau's experiment. Foucault used a baseline of only 20 m/65 ft, but achieved a slightly more accurate result: 298,000 km/185,200 mi per second.

All done by mirrors

In 1926 US physicist Albert Michelson (1852–1931) refined Foucault's experiment. He used a much longer baseline, starting at the Mount Wilson Observatory and travelling to a mirror on nearby Mount San Antonio. The baseline was measured precisely by the US Coast and Geodetic Survey as 35,385.53 m/116,135.3 ft, accurate to within 3 mm/0.1 in. Michelson used a multi-sided octagonal mirror which could be spun at 550 revolutions per second. The exact speed was measured by stroboscopic comparison with an electrically driven tuning fork.

Light from an electric arc passed through a narrow slit and was reflected from one facet of the rotating mirror. The beam travelled to the Mount San Antonio station, was reflected from a flat mirror and returned to Mount Wilson. For the reflection to be observed, the octagonal mirror had to have made one-eighth of a rotation while the light travelled along the baseline and returned. The round trip took 0.00023 seconds when the mirror spun at 528 revolutions per second. The calculated speed of light, after correction for slowing of the beam by air along its path, was 299,863 km/186,335 mi per second. In 1929 Michelson repeated the experiment inside a long excavated tube, to eliminate the uncertainty caused by variations in air density along the path. His result was $c = 299,774$ km/186,280 mi per second: within 0.006% of the modern value.

Peter Lafferty

A laser vaporizes a small sample, which is cooled in helium, and ejected into an evacuated chamber. The jet of clusters expands supersonically, cooling the clusters to near absolute zero, and stabilizing them for study in a ◊mass spectrometer.

spectrum (plural *spectra*) in physics, an arrangement of frequencies or wavelengths when electromagnetic radiations are separated into their constituent parts. Visible light is part of the ◊electromagnetic spectrum and most sources emit waves over a range of wavelengths that can be broken up or 'dispersed'; white light can be separated into red, orange, yellow, green, blue, indigo, and violet. The visible spectrum was first studied in 1672 by Isaac Newton, who showed how white light could be broken up into different colours.

speculum (plural *specula*) medical instrument to aid examination of an opening into the body; for example, the nose or vagina. The speculum allows the opening to be widened, permitting the passage of instruments. Many specula also have built-in lights to illuminate the examined cavity.

speech disorder in medicine, condition that makes speech difficult or impossible. Speech disorders occur for a variety of physical and psychological reasons. Congenital defects, such as ◊cleft palate, require surgical correction before speech is intelligible. Deafness is the most frequent cause of dumbness, the inability to pronounce the sounds that make up words. A combination of training in lip reading, speech therapy, psychotherapy, and specialist education may be required in children with congenital deafness. Stroke, head injury, and some severe psychiatric disorders may be responsible for speech disorders in adults.

A child may have difficulties in understanding language or self-expression if the language development area of the brain develops abnormally and specialist education and speech therapy may then be required.

speech recognition or *voice input* in computing, any technique by which a computer can understand ordinary speech. Spoken words are divided into 'frames', each lasting about one-thirtieth of a second, which are converted to a wave form. These are then compared with a series of stored frames to determine the most likely word. Research into speech recognition started in 1938, but the technology did not become sufficiently developed for commercial applications until the late 1980s.

There are three types: *separate word recognition* for distinguishing up to several hundred separately spoken words; *connected speech recognition* for speech in which there is a short pause between words; and *continuous speech recognition* for normal but carefully articulated speech.

speech synthesis or *voice output* computer-based technology for generating speech. A speech synthesizer is controlled by a computer, which supplies strings of codes representing basic speech sounds (phonemes); together these make up words. Speech-synthesis applications include children's toys, car and aircraft warning systems, and talking books for the blind.

speech writing system computing system that enables data to be input by voice. It includes a microphone, and soundcard that plugs into the computer and converts the analogue signals of the voice to digital signals. Examples include DragonDictate, and IBM's Personal Dictation System released in 1994.

The user must read sample sentences to the computer on first use to familiarize it with individual pronunciation. Early speech writers were very inaccurate and slow but by the mid-1990s speeds of 80 words per minute with 95–99% accuracy were achievable.

speed common name for ◊amphetamine, a stimulant drug.

speed the rate at which an object moves. The average speed v of an object may be calculated by dividing the distance s it has travelled by the time t taken to do so, and may be expressed as:

$$v = s/t$$

The usual units of speed are metres per second or kilometres per hour.

Speed is a scalar quantity in which direction of motion is unimportant (unlike the vector quantity ◊velocity, in which both magnitude and direction must be taken into consideration).

speed of light speed at which light and other ◊electromagnetic waves travel through empty space. Its value is 299,792,458 m/186,281 mi per second. The speed of light is the highest speed possible, according to the theory of ◊relativity, and its value is independent of the motion of its source and of the observer. It is impossible to accelerate any material body to this speed because it would require an infinite amount of energy.

Speed of light

Light takes: 0.14 seconds to travel around the world; 1.25 seconds to travel from the Moon to the Earth; 8 minutes 27 seconds to travel from the Sun; 6 hours to travel from Pluto; 4.3 years to travel from the nearest star, Proxima Centauri; 75,000 years to travel from the most distant stars in our galaxy; 160,000 years to travel from the nearest small galaxy, the Larger Magellanic Cloud; 2,200,000 years to travel from the nearest large galaxy, the Andromeda galaxy; 13 billion years to travel from quasars, the most distant objects in the Universe.

---

speed of sound speed at which sound travels through a medium, such as air or water. In air at a temperature of 0°C/32°F, the speed of sound is 331 m/1,087 ft per second. At higher temperatures, the speed of sound is greater; at 18°C/64°F it is 342 m/1,123 ft per second.

It is greater in liquids and solids; for example, in water it is around 1,440 m/4,724 ft per second, depending on the temperature.

speedometer instrument attached to the transmission of a vehicle by a flexible drive shaft, which indicates the speed of the vehicle in miles or kilometres per hour on a dial easily visible to the driver.

speleology scientific study of caves, their origin, development, physical structure, flora, fauna, folklore, exploration, mapping, photography, cave-diving, and rescue work. *Pot holing*, which involves following the course of underground rivers or streams, has become a popular sport.

Speleology first developed in France in the late 19th century, where the Société de Spéléologie was founded in 1895.

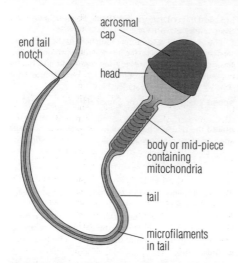

end tail notch

acrosmal cap

head

body or mid-piece containing mitochondria

tail

microfilaments in tail

sperm Only a single sperm is needed to fertilize a female egg, or ovum. Yet up to 500 million may start the journey towards the egg. Once a sperm has fertilized an egg, the egg's wall cannot be penetrated by other sperm. The unsuccessful sperm die after about three days.

sperm or **spermatozoon** in biology, the male ◊gamete of animals. Each sperm cell has a head capsule containing a nucleus, a middle portion containing ◊mitochondria (which provide energy), and a long tail (flagellum). See ◊sexual reproduction.

In most animals, the sperm are motile, and are propelled by a long flagellum, but in some (such as crabs and lobsters) they are nonmotile. Sperm cells are produced in the testes (see ◊testis). From there they pass through the sperm ducts via the seminal vesicles and the ◊prostate gland, which produce fluids called semen that give the sperm cells energy and keep them moving after they leave the body. Hundreds of millions of sperm cells are contained in only a small amount of semen. The human sperm is 0.005 mm/0.0002 in long and can survive inside the female for 2–9 days. Mammalian sperm have receptor cells identical to some of those found in the lining of the nose. These may help in navigating towards the egg.

The term is sometimes applied to the motile male gametes (◊antherozoids) of lower plants.

Sperm

A human male produces 1,000 sperm cells each second, or 86 million each day. This is enough to populate the whole world in less than two months.

spermatophore small capsule containing ◊sperm and other nutrients produced in invertebrates, newts, and cephalopods.

spermatophyte in botany, another name for a ◊seed plant.

spermatozoon in medicine, another name for ◊sperm.

spermicide any cream, jelly, pessary, or other preparation that kills the ◊sperm cells in semen. Spermicides are used for contraceptive purposes, usually in combination with a ◊condom or ◊diaphragm. Sponges impreg-

nated with spermicide have been developed but are not yet in widespread use. Spermicide used alone is only 75% effective in preventing pregnancy.

sphalerite mineral composed of zinc sulphide with a small proportion of iron, formula $(Zn,Fe)S$. It is the chief ore of zinc. Sphalerite is brown with a nonmetallic lustre unless an appreciable amount of iron is present (up to 26% by weight). Sphalerite usually occurs in ore veins in limestones, where it is often associated with galena. It crystallizes in the cubic system but does not normally form perfect cubes.

sphere in mathematics, a perfectly round object with all points on its surface the same distance from the centre. This distance is the radius of the sphere. For a sphere of radius r, the volume $V = 4/3\pi r^3$ and the surface area $A = 4\pi r^2$.

sphincter ring of muscle, such as is found at various points in the ◊alimentary canal, that contracts and relaxes to open and close the canal and control the movement of food. The **pyloric sphincter**, at the base of the stomach, controls the release of the gastric contents into the ◊duodenum. After release the sphincter contracts, closing off the stomach. The **external anal sphincter** closes the ◊anus; the **internal anal sphincter** constricts the rectum; the **sphincter vesicae** controls the urethral orifice of the bladder. In the eye the **sphincter pupillae** contracts the pupil in response to bright light.

sphygmomanometer instrument for measuring blood pressure. Consisting of an inflatable arm cuff joined by a rubber tube to a pressure-recording device (incorporating a column of mercury with a graduated scale), it is used, together with a stethoscope, to measure arterial blood pressure.

Spica or **Alpha Virginis** the brightest star in the constellation Virgo and the 16th brightest star in the sky. First-magnitude Spica has a true luminosity of over 1,500 times that of the Sun, and is 140 light years from Earth. It is also a spectroscopic binary star, the components of which orbit each other every four days.

spicules, solar in astronomy, short-lived jets of hot gas in the upper ◊chromosphere of the Sun. Spiky in appearance, they move at high velocities along lines of magnetic force to which they owe their shape, and last for a few minutes each. Spicules appear to disperse material into the ◊corona.

spikelet in botany, one of the units of a grass ◊inflorescence. It comprises a slender axis on which one or more flowers are borne.

Each individual flower or floret has a pair of scalelike bracts, the glumes, and is enclosed by a membranous lemma and a thin, narrow palea, which may be extended into a long, slender bristle, or **awn**.

spin in physics, the intrinsic angular momentum of a subatomic particle, nucleus, atom, or molecule, which continues to exist even when the particle comes to rest. A particle in a specific energy state has a particular spin, just as it has a particular electric charge and mass. According to ◊quantum theory, this is restricted to discrete and indivisible values, specified by a spin ◊quantum number. Because of its spin, a charged particle acts as a small magnet and is affected by magnetic fields.

spina bifida congenital defect in which part of the spinal cord and its membranes are exposed, due to incomplete development of the spine (vertebral column). It is a neural tube defect.

Spina bifida, usually present in the lower back, varies in severity. The most seriously affected babies may be paralysed below the waist. There is also a risk of mental retardation and death from hydrocephalus, which is often associated. Surgery is performed to close the spinal lesion shortly after birth, but this does not usually cure the disabilities caused by the condition. Spina bifida can be diagnosed prenatally.

spinal anaesthetic in medicine, ◊anaesthetic injected into the spinal cord, as in an epidural, for example.

spinal cord major component of the ◊central nervous system in vertebrates, encased in the spinal column. It consists of bundles of nerves enveloped in three layers of membrane (the meninges).

spinal tap another term for ◊lumbar puncture, a medical test.

spine backbone of vertebrates. In most mammals, it contains 26 small bones called vertebrae, which enclose and protect the spinal cord (which links the peripheral nervous system to the brain).

The spine articulates with the skull, ribs, and hip bones, and provides attachment for the back muscles.

In humans it is made up of individual vertebrae, separated by intervertebral discs. In the adult there are seven cervical **vertebrae** in the neck; twelve thoracic in the upper trunk; five lumbar in the lower back; the sacrum (consisting of five rudimentary vertebrae fused together, joined to the hipbones); and the coccyx (four vertebrae, fused into a tailbone). The human spine has four curves (front to rear), which allow for the increased size of the chest and pelvic cavities, and permit springing, to minimize jolting of the internal organs.

spinel any of a group of 'mixed oxide' minerals consisting mainly of the oxides of magnesium and aluminium, $MgAl_2O_4$ and $FeAl_2O_4$. Spinels crystallize in the cubic system, forming octahedral crystals. They are found in high-temperature igneous and metamorphic rocks. The aluminium oxide spinel contains gem varieties, such as the ruby spinels of Sri Lanka and Myanmar (Burma).

spiracle in insects, the opening of a ◊trachea, through which oxygen enters the body and carbon dioxide is expelled. In cartilaginous fishes (sharks and rays), the same name is given to a circular opening that marks the remains of the first gill slit.

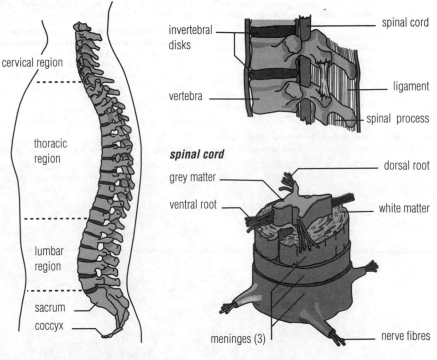

side view of backbone

cervical region

thoracic region

lumbar region

sacrum

coccyx

vertebral joints

invertebral disks

vertebra

spinal cord

ligament

spinal process

spinal cord

grey matter

ventral root

meninges (3)

dorsal root

white matter

nerve fibres

spine The human spine extends every night during sleep. During the day, the cartilage discs between the vertebra are squeezed when the body is in a vertical position, standing or sitting, but at night, with pressure released, the discs swell and the spine lengthens by about 8 mm/0.3 in.

In tetrapod vertebrates, the spiracle of early fishes has evolved into the Eustachian tube, which connects the middle ear cavity with the pharynx.

spiral a plane curve formed by a point winding round a fixed point from which it distances itself at regular intervals, for example the spiral traced by a flat coil of rope. Various kinds of spirals can be generated mathematically – for example, an equiangular or logarithmic spiral (in which a tangent at any point on the curve always makes the same angle with it) and an involute. Spirals also occur in nature as a normal consequence of accelerating growth, such as the spiral shape of the shells of snails and some other molluscs.

spiral galaxy in astronomy, one of the main classes of ◊galaxy in the Hubble classification comprising up to 30% of known galaxies. Spiral galaxies are characterized by a central bulge surrounded by a flattened disc containing (normally) two spiral arms composed of hot young stars and clouds of dust and gas. In about half of spiral galaxies (barred spirals) the arms originate at the ends of a bar across the central bulge. The bar is not a rigid object but consists of stars in motion about the centre of the galaxy.

spiritual healing or ***psychic healing*** transmission of energy from or through a healer, who may practise hand healing or absent healing through prayer or meditation.

In religions worldwide, from shamanism to latter-day charismatic Christianity, healing powers have been attributed to gifted individuals, and sometimes to particular locations (Delphi, Lourdes) or objects (religious relics), and the anecdotal evidence for the reality of spiritual healing is substantial and cross-cultural. Since both healers and beneficiaries can only adduce metaphysical explanations for the effects, medical science remains sceptical, at most allowing that in exceptional cases faith and will may bring about inexplicable cures or remissions, which, however, also occur in cases where no spiritual contribution is claimed.

spirochaete spiral-shaped bacterium. Some spirochaetes are free-living in water, others inhabit the intestines and genital areas of animals. The sexually transmitted disease syphilis is caused by a spirochaete.

spit ridge of sand or shingle projecting from the land into a body of water. It is deposited by waves carrying material from one direction to another across the mouth of an inlet (longshore drift). Deposition in the brackish water behind a spit may result in the formation of a ◊salt marsh.

spleen organ in vertebrates, part of the reticuloendothelial system, which helps to process ◊lymphocytes. It also regulates the number of red blood cells in circulation by destroying old cells, and stores iron. It is situated on the left side of the body, behind the stomach.

splenectomy surgical removal of the ◊spleen. It may be necessary following injury to the spleen, or in the course of some blood diseases.

spongiform encephalopathy in medicine, another term for transmissable spongiform encephalopathy.

spontaneous combustion burning that is not initiated by the direct application of an external source of heat. A number of materials and chemicals, such as hay and sodium chlorate, can react with their surroundings, usually by oxidation, to produce so much internal heat that combustion results.

Special precautions must be taken for the storage and handling of substances that react violently with moisture or air. For example, phosphorus ignites spontaneously in the presence of air and must therefore be stored under water; sodium and potassium are stored under kerosene in order to prevent their being exposed to moisture.

spontaneous generation or ***abiogenesis*** erroneous belief that living organisms can arise spontaneously from non-living matter. This survived until the mid-19th century, when the French chemist Louis Pasteur demonstrated that a nutrient broth would not generate microorganisms if it was adequately sterilized. The theory of ◊biogenesis holds that spontaneous generation cannot now occur; it is thought, however, to have played an essential role in the origin of ◊life on this planet 4 billion years ago.

spooling in computing, the process in which infor-

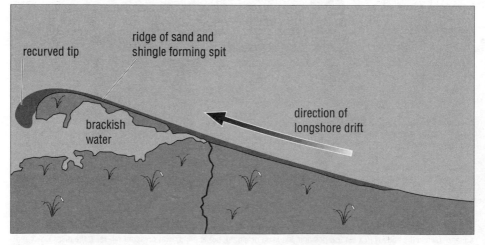

recurved tip

ridge of sand and shingle forming spit

brackish water

direction of longshore drift

spit

spreadsheet A typical spreadsheet software package. The data it contains may be output in a graphical form, enabling the production of charts and diagrams.

mation to be printed is stored temporarily in a file, the printing being carried out later. It is used to prevent a relatively slow printer from holding up the system at critical times, and to enable several computers or programs to share one printer.

sporangium structure in which ◊spores are produced.

spore small reproductive or resting body, usually consisting of just one cell. Unlike a ◊gamete, it does not need to fuse with another cell in order to develop into a new organism. Spores are produced by the lower plants, most fungi, some bacteria, and certain protozoa. They are generally light and easily dispersed by wind movements.

Plant spores are haploid and are produced by the sporophyte, following ◊meiosis; see ◊alternation of generations.

sporophyte diploid spore-producing generation in the life cycle of a plant that undergoes ◊alternation of generations.

spreadsheet in computing, a program that mimics a sheet of ruled paper, divided into columns and rows. The user enters values in the sheet, then instructs the program to perform some operation on them, such as totalling a column or finding the average of a series of numbers.

Highly complex numerical analyses may be built up from these simple steps.

Spreadsheets are widely used in business for forecasting and financial control. The first spreadsheet program, Software Arts' VisiCalc, appeared in 1979. The best known include ◊Lotus 1–2–3 and Microsoft Excel.

spring device, usually a metal coil, that returns to its original shape after being stretched or compressed. Springs are used in some machines (such as clocks) to store energy, which can be released at a controlled rate. In other machines (such as engines) they are used to close valves.

In vehicle-suspension systems, springs are used to cushion passengers from road shocks. These springs are used in conjunction with ◊shock absorbers to limit their amount of travel. In bedding and upholstered furniture springs add comfort.

spring in geology, a natural flow of water from the ground, formed at the point of intersection of the water table and the ground's surface. The source of water is rain that has percolated through the overlying rocks. During its underground passage, the water may have dissolved mineral substances that may then be precipitated at the spring (hence, a mineral spring).

A spring may be continuous or intermittent, and depends on the position of the water table and the topography (surface features).

sprite in computing, a graphics object made up of a pattern of ◊pixels (picture elements) defined by a computer programmer. Some ◊high-level languages and ◊applications programs contain routines that allow a user to define the shape, colours, and other characteristics of individual graphics objects. These objects can then be manipulated and combined to produce animated games or graphic screen displays.

spur ridge of rock jutting out into a valley or plain. In mountainous areas rivers often flow around interlocking

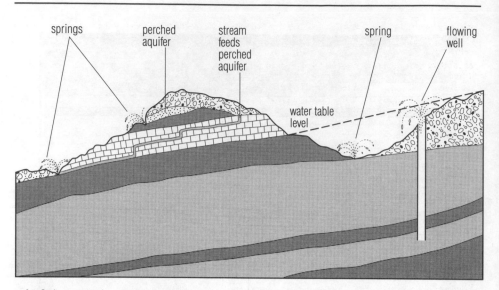

spring Springs occur where water-laden rock layers (aquifers) reach the surface. Water will flow from a well whose head is below the water table.

spurs because they are not powerful enough to erode through the spurs. Spurs may be eroded away by large and powerful glaciers to form truncated spurs.

Sputnik series of ten Soviet Earth-orbiting satellites. *Sputnik 1* was the first artificial satellite, launched on 4 Oct 1957. It weighed 84 kg/185 lb, with a 58 cm/23 in diameter, and carried only a simple radio transmitter which allowed scientists to track it as it orbited Earth. It burned up in the atmosphere 92 days later. Sputniks were superseded in the early 1960s by the Cosmos series.

Sputnik 2, launched on 3 Nov 1957, weighed about 500 kg/1,100 lb including the dog 'Laika', the first living creature in space. Unfortunately, there was no way to return the dog to Earth, and it died in space. Later Sputniks were test flights of the ◊Vostok spacecraft.

sq abbreviation for *square* (measure).

SQL (abbreviation for *structured query language*) high-level computer language designed for use with ◊relational databases. Although it can be used by programmers in the same way as other languages, it is often used as a means for programs to communicate with each other. Typically, one program (called the 'client') uses SQL to request data from a database 'server'.

Although originally developed by IBM, SQL is now widely used on many types of computer.

square in geometry, a quadrilateral (four-sided) plane figure with all sides equal and each angle a right angle. Its diagonals bisect each other at right angles. The area A of a square is the length l of one side multiplied by itself ($A = l \times l$).

Also, any quantity multiplied by itself is termed a square, represented by an ◊exponent of power 2; for example, $4 \times 4 = 4^2 = 16$ and $6.8 \times 6.8 = 6.8^2 = 46.24$.

An algebraic term is squared by doubling its exponent and squaring its coefficient if it has one; for example, $(x^2)^2 = x^4$ and $(6y^3)^2 = 36y^6$.

A number that has a whole number as its ◊square root is

known as a *perfect square*; for example, 25, 144 and 54,756 are perfect squares (with roots of 5, 12 and 234, respectively).

square root in mathematics, a number that when squared (multiplied by itself) equals a given number. For example, the square root of 25 (written $\sqrt{25}$) is ± 5, because $5 \times 5 = 25$, and $(-5) \times (-5) = 25$. As an ◊exponent, a square root is represented by 1/2, for example, $16^{1/2} = 4$.

Negative numbers (less than 0) do not have square roots that are ◊real numbers. Their roots are represented by ◊complex numbers, in which the square root of −1 is given the symbol i (that is, ± i² = −1). Thus the square root of −4 is $\sqrt{[(-1) \times 4]} = \sqrt{-1} \times \sqrt{4} = 2i$.

squint or *strabismus* common condition in which one eye deviates in any direction. A squint may be convergent (with the bad eye turned inwards), divergent (outwards), or, in rare cases, vertical. A convergent squint is also called *cross-eye*.

There are two types of squint: *paralytic,* arising from disease or damage involving the extraocular muscles or their nerve supply; and *nonparalytic,* which may be inherited or due to some refractive error within the eye. Nonparalytic (or concomitant) squint is the typical condition seen in small children. It is treated by corrective glasses, exercises for the eye muscles, or surgery.

SRAM (acronym for *static random-access memory*) computer memory device in the form of a silicon chip used to provide ◊immediate access memory. SRAM is faster but more expensive than ◊DRAM (dynamic random-access memory).

DRAM loses its contents unless they are read and rewritten every 2 milliseconds or so. This process is called *refreshing* the memory. SRAM does not require such frequent refreshing.

SSSI abbreviation for ◊*Site of Special Scientific Interest.*

stability measure of how difficult it is to move an object from a position of balance or ◊equilibrium with respect to gravity.

An object displaced from equilibrium does not remain in its new position if its weight, acting vertically downwards through its ◊centre of mass, no longer passes through the line of action of the contact force (the force exerted by the surface on which the object is resting), acting vertically upwards through the object's new base. If the lines of action of these two opposite but equal forces do not coincide they will form a couple and create a moment (see ◊moment of a force) that will cause the object either to return to its original rest position or to topple over into another position.

An object in **stable equilibrium** returns to its rest position after being displaced slightly. This form of equilibrium is found in objects that are difficult to topple over; these usually possess a relatively wide base and a low

stable equilibrium

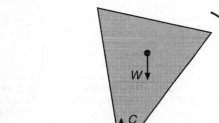

unstable equilibrium

neutral equilibrium

W stands for weight; *C* stands for contact

stability

centre of mass – for example, a cone resting on its flat base on a horizontal surface. When such an object is tilted slightly its centre of mass is raised and the line of action of its weight no longer coincides with that of the contact force exerted by its new, smaller base area. The moment created will tend to lower the centre of mass and so the cone will fall back to its original position.

An object in **unstable equilibrium** does not remain at rest if displaced, but falls into a new position; it does not return to its original rest position. Objects possessing this form of equilibrium are easily toppled and usually have a relatively small base and a high centre of mass – for example, a cone balancing on its point, or apex, on a horizontal surface. When an object such as this is given the slightest push its centre of mass is lowered and the displacement of the line of action of its weight creates a moment. The moment will tend to lower the centre of mass still further and so the object will fall on to another position.

An object in **neutral equilibrium** stays at rest if it is moved into a new position – neither moving back to its original position nor on any further. This form of equilibrium is found in objects that are able to roll, such as a cone resting on its curved side placed on a horizontal surface. When such an object is rolled its centre of mass remains in the same position, neither rising nor falling, and the line of action of its weight continues to coincide with the contact force; no moment is created and so its equilibrium is maintained.

stabilizer one of a pair of fins fitted to the sides of a ship, especially one governed automatically by a ◊gyroscope mechanism, designed to reduce side-to-side rolling of the ship in rough weather.

stack in computing, a method of storing data in which the most recent item stored will be the first to be retrieved. The technique is commonly called 'last in, first out'.

Stacks are used to solve problems involving nested structures; for example, to analyse an arithmetical expression containing subexpressions in parentheses, or to work out a route between two points when there are many different paths.

stack isolated pillar of rock that has become separated from a headland by ◊coastal erosion. It is usually formed by the collapse of an ◊arch. Further erosion will reduce it to a stump, which is exposed only at low tide.

stain in chemistry, a coloured compound that will bind to other substances. Stains are used extensively in micro-biology to colour microorganisms and in histochemistry to detect the presence and whereabouts in plant and animal tissue of substances such as fats, cellulose, and proteins.

stainless steel widely used ◊alloy of iron, chromium, and nickel that resists rusting. Its chromium content also gives it a high tensile strength. It is used for cutlery and kitchen fittings, and in surgical instruments. Stainless steel was first produced in the UK in 1913 and in Germany in 1914.

stalactite and stalagmite cave structures formed by the deposition of calcite dissolved in ground water. **Stalactites** grow downwards from the roofs or walls and can be icicle-shaped, straw-shaped, curtain-shaped, or formed as terraces. **Stalagmites** grow upwards from the cave floor and can be conical, fir-cone-shaped, or resemble a stack of saucers. Growing stalactites and stalagmites may meet to form a continuous column from floor to ceiling.

Stalactites are formed when ground water, hanging as a drip, loses a proportion of its carbon dioxide into the air of the cave. This reduces the amount of calcite that can be held in solution, and a small trace of calcite is deposited. Successive drips build up the stalactite over many years. In stalagmite formation the calcite comes out of the sol-ution because of agitation – the shock of a drop of water hitting the floor is sufficient to remove some calcite from the drop. The different shapes result from the splashing of the falling water.

stamen male reproductive organ of a flower. The stamens are collectively referred to as the ◊androecium. A typical stamen consists of a stalk, or filament, with an anther, the pollen-bearing organ, at its apex, but in some primitive plants, such as *Magnolia*, the stamen may not be markedly differentiated.

The number and position of the stamens are significant in the classification of flowering plants. Generally the more advanced plant families have fewer stamens, but

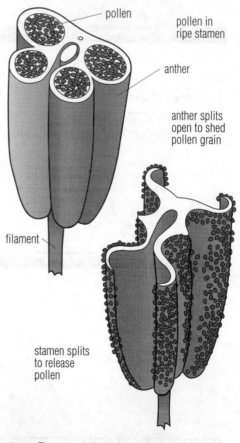

pollen

pollen in ripe stamen

anther

anther splits open to shed pollen grain

filament

stamen splits to release pollen

stamen The stamen is the male reproductive organ of a flower. It has a thin stalk called a filament with an anther at the tip. The anther contains pollen sacs, which split to release tiny grains of pollen.

they are often positioned more effectively so that the likelihood of successful pollination is not reduced.

standard atmosphere alternative term for ◊atmosphere, a unit of pressure.

standard deviation in statistics, a measure (symbol σ or s) of the spread of data. The deviation (difference) of each of the data items from the mean is found, and their values squared. The mean value of these squares is then calculated. The standard deviation is the square root of this mean.

If n is the number of items of data, x is the value of each item, and $\bar{x}$ is the mean value, the standard deviation σ may be given by the formula:

$$\sigma = \sqrt{[\Sigma(x - \bar{x})^2/n]}$$

where Σ indicates that the differences between the value of each item of data and the mean should be summed.

To simplify the calculations, the formula may be rearranged to

$$\sigma = \sqrt{[\Sigma x^2/n - \bar{x}^2]}.$$

As a result, it becomes necessary only to calculate Σx and Σx^2.

For example, if the ages of a set of children were 4, 4.5, 5, 5.5, 6, 7, 9, and 11, Σx would be 52, $\bar{x}$ would be $52/n = 52/8 = 6.5$, and Σx^2 would be 378.5 ($= 4^2 + 4.5^2 + 5^2 + 5.5^2 + 6^2 + 7^2 + 9^2 + 11^2$). Therefore, the standard deviation σ would be $\sqrt{[378.5/8 - (6.5)^2]} = \sqrt{5.0625} = 2.25$.

standard form or **scientific notation** method of writing numbers often used by scientists, particularly for very large or very small numbers. The numbers are written with one digit before the decimal point and multiplied by a power of 10. The number of digits given after the decimal point depends on the accuracy required. For example, the ◊speed of light is 2.9979×10^8 m/1.8628×10^5 mi per second.

standard gravity acceleration due to gravity, generally taken as 9.81274 m/32.38204 ft per second per second. See also ◊g scale.

standard illuminant any of three standard light intensities, A, B, and C, used for illumination when phenomena involving colour are measured. A is the light from a filament at 2,848K (2,575°C/4,667°F), B is noon sunlight, and C is normal daylight. B and C are defined with respect to A. Standardization is necessary because colours appear different when viewed in different lights.

standard model in physics, the modern theory of ◊elementary particles and their interactions. According to the standard model, elementary particles are classified as leptons (light particles, such as electrons), ◊hadrons (particles, such as neutrons and protons, that are formed from quarks), and gauge bosons. Leptons and hadrons interact by exchanging ◊gauge bosons, each of which is responsible for a different fundamental force: photons mediate the electromagnetic force, which affects all charged particles; gluons mediate the strong nuclear force, which affects quarks; gravitons mediate the force of gravity; and the weakons (intermediate vector bosons) mediate the weak nuclear force. See also ◊forces, fundamental, ◊quantum electrodynamics, and ◊quantum chromodynamics.

standard temperature and pressure (STP) in chemistry, a standard set of conditions for experimental measurements, to enable comparisons to be made between sets of results. Standard temperature is 0°C/32°F (273K) and standard pressure is equivalent to 1 atmosphere (101,325 Pa).

standard volume in physics, the volume occupied by one kilogram molecule (the molecular mass in kilograms) of any gas at standard temperature and pressure. Its value is approx 22.414 cubic metres.

standing crop in ecology, the total number of individuals of a given species alive in a particular area at any moment. It is sometimes measured as the weight (or ◊biomass) of a given species in a sample section.

standing wave in physics, a wave in which the positions of ◊nodes (positions of zero vibration) and antinodes (positions of maximum vibration) do not move. Standing waves result when two similar waves travel in opposite directions through the same space.

For example, when a sound wave is reflected back along its own path, as when a stretched string is plucked, a standing wave is formed. In this case the antinode remains fixed at the centre and the nodes are at the two ends. Water and ◊electromagnetic waves can form standing waves in the same way.

staphylococcus spherical bacterium that occurs in clusters. It is found on the skin and mucous membranes of humans and other animals. It can cause abscesses and systemic infections that may prove fatal.

Staphylococcus aureus is a very common bacterium, present in the nose in 30% of people. Normally it gives no trouble, but, largely due to over-prescribing of antibiotics, strains have arisen that are resistant to the drugs used to treat them, principally methicillin, a semi-synthetic form of penicillin. Methicillin-resistant *S. aureus* (MRSA) strains represent a serious hazard to the critically ill or immunosuppressed.

MRSA normally responds to two antibiotics which are considered too toxic for use in any but life-threatening infections (vancomycin and teicoplanin) but it still causes fatalities.

star luminous globe of gas, mainly hydrogen and helium, which produces its own heat and light by nuclear reactions. Although stars shine for a very long time – many billions of years – they are not eternal, and have been found to change in appearance at different stages in their lives.

The smallest mass possible for a star is about 8% that of the Sun (80 times that of ◊Jupiter), otherwise nuclear reactions do not occur. Objects with less than this critical mass shine only dimly, and are termed **brown dwarfs**.

Star

Neutron stars are so condensed that a fragment the size of a sugar cube would weigh as much as all the people on Earth put together.

---

starburst galaxy in astronomy, a spiral galaxy that appears unusually bright in the infrared part of the spectrum due to a recent burst of star formation, possibly triggered by the gravitational influence of a nearby companion galaxy.

starch widely distributed, high-molecular-mass ◊carbohydrate, produced by plants as a food store; main dietary sources are cereals, legumes, and tubers, including

Stars: the top twenty brightest stars

Common name	Scientific name	Apparent magnitude
Sirius	α Canis Majoris	−1.46
Canopus	α Carinae	−0.72
Rigil Kent	α Centauri	−0.27*
Arcturus	α Boötis	−0.4
Vega	α Lyrae	+0.03
Capella	α Aurigae	0.08
Rigel	β Orionis	0.12
Procyon	α Canis Minoris	0.38
Achernar	α Eridani	0.46
Betelgeuse	α Orionis	0.50**
Hadar	β Centauri	0.61
Altair	α Aquilae	0.77
Acrux	α Crucis	0.79*
Aldebaran	α Tauri	0.85**
Antares	α Scorpii	0.96**
Spica	α Virginis	0.98
Pollux	β Geminorum	1.14
Fomalhaut	α Piscis Austrini	1.16
Deneb	α Cygni	1.25
Mimosa	β Crucis	1.25

* combined magnitude of double star.
** variable.

potatoes. It consists of varying proportions of two ◊glucose polymers (◊polysaccharides): straight-chain (amylose) and branched (amylopectin) molecules.

Purified starch is a white powder used to stiffen textiles and paper and as a raw material for making various chemicals. It is used in the food industry as a thickening agent. Chemical treatment of starch gives rise to a range of 'modified starches' with varying properties. Hydrolysis (splitting) of starch by acid or enzymes generates a variety of 'glucose syrups' or 'liquid glucose' for use in the food industry. Complete hydrolysis of starch with acid generates the ◊monosaccharide glucose only. Incomplete hydrolysis or enzymic hydrolysis yields a mixture of glucose, maltose, and non-hydrolysed fractions called dextrins.

star cluster group of related stars, usually held together by gravity. Members of a star cluster are thought to form together from one large cloud of gas in space. *Open clusters* such as the ◊Pleiades contain from a dozen to many hundreds of young stars, loosely scattered over several light years. ◊Globular clusters are larger and much more densely packed, containing perhaps 100,000 old stars.

The more conspicuous clusters were originally catalogued with the nebulae, and are usually known by their Messier or NGC numbers. A few clusters like the Pleiades, Hyades, and Praesepe are also known by their traditional names.

Stardust US project to obtain a sample of dust and gas from the head of a comet. Due for launch in February 1999, the *Stardust* space probe will fly through the head Comet Wild 2 in January 2004, passing within 100 km/ 62 mi of the 4 km/2.5 mi nucleus. It will return to Earth with its samples in January 2006.

stasis in biology, the cessation of blood flow in the blood vessels or the passage of food in the gastrointestinal tract.

states of matter forms (solid, liquid, or gas) in which

material can exist. Whether a material is solid, liquid, or gas depends on its temperature and the pressure on it. The transition between states takes place at definite temperatures, called melting point and boiling point.

◊Kinetic theory describes how the state of a material depends on the movement and arrangement of its atoms or molecules. A hot ionized gas or ◊plasma is often called the fourth state of matter, but liquid crystals, ◊colloids, and glass also have a claim to this title.

state symbol symbol used in chemical equations to indicate the physical state of the substances present. The symbols are: (s) for solid, (l) for liquid, (g) for gas, and (aq) for aqueous.

static electricity ◊electric charge that is stationary, usually acquired by a body by means of electrostatic induction or friction. Rubbing different materials can produce static electricity, as seen in the sparks produced on combing one's hair or removing a nylon shirt. In some processes static electricity is useful, as in paint spraying where the parts to be sprayed are charged with electricity of opposite polarity to that on the paint droplets, and in ◊xerography.

statics branch of mechanics concerned with the behaviour of bodies at rest and forces in equilibrium, and distinguished from ◊dynamics.

statistical mechanics branch of physics in which the properties of large collections of particles are predicted by considering the motions of the constituent particles. It is closely related to ◊thermodynamics.

statistics branch of mathematics concerned with the collection and interpretation of data. For example, to determine the ◊mean age of the children in a school, a statistically acceptable answer might be obtained by calculating an average based on the ages of a representative sample, consisting, for example, of a random tenth of the pupils from each class. ◊Probability is the branch of statistics dealing with predictions of events.

status asthmaticus in medicine, repeated attacks of ◊asthma with no respite between the spasms. The condition requires urgent treatment in hospital.

staurolite silicate mineral, $(Fe,Mg)_2(Al,Fe)_9Si_4O_{20}(OH)_2$. It forms brown crystals that may be twinned in the form of a cross. It is a useful indicator of medium grade (moderate temperature and pressure) metamorphism in metamorphic rocks formed from clay sediments.

STD abbreviation for ◊sexually transmitted disease.

steady-state theory in astronomy, a rival theory to that of the ◊Big Bang, which claims that the universe has no origin but is expanding because new matter is being created continuously throughout the universe. The theory was proposed in 1948 by Hermann Bondi, Thomas Gold and Fred Hoyle, but was dealt a severe blow in 1965 by the discovery of ◊cosmic background radiation (radiation left over from the formation of the universe) and is now largely rejected.

steam in chemistry, a dry, invisible gas formed by vaporizing water.

The visible cloud that normally forms in the air when water is vaporized is due to minute suspended water particles. Steam is widely used in chemical and other industrial processes and for the generation of power.

steam engine engine that uses the power of steam to produce useful work. It was the principal power source during the British Industrial Revolution in the 18th century. The first successful steam engine was built in 1712 by English inventor Thomas Newcomen at Dudley, West Midlands; it was developed further by Scottish mining engineer James Watt from 1769 and by English mining engineer Richard Trevithick, whose high-pressure steam engine of 1802 led to the development of the steam locomotive.

In Newcomen's engine, steam was admitted to a cylinder as a piston moved up, and was then condensed by a spray of water, allowing air pressure to force the piston downwards. James Watt improved Newcomen's engine in 1769 by condensing the steam outside the cylinder (thus saving energy formerly used to reheat the cylinder) and by using steam to force the piston upwards. Watt also introduced the **double-acting engine**, in which steam is alternately sent to each side of the piston. The **compound engine** (1781) uses the exhaust from one cylinder to drive the piston of another. A later development was the steam ◊turbine, still used today to power ships and generators in power stations. In other contexts, the steam engine was superseded by the ◊internal-combustion engine.

stearic acid $CH_3(CH_2)_{16}COOH$ saturated long-chain ◊fatty acid, soluble in alcohol and ether but not in water. It is found in many fats and oils, and is used to make soap and candles and as a lubricant. The salts of stearic acid are called stearates.

stearin mixture of stearic and palmitic acids, used to make soap.

steel alloy or mixture of iron and up to 1.7% carbon, sometimes with other elements, such as manganese, phosphorus, sulphur, and silicon. The USA, Russia, Ukraine, and Japan are the main steel producers. Steel has innumerable uses, including ship and car manufacture, skyscraper frames, and machinery of all kinds.

Steels with only small amounts of other metals are called **carbon steels**. These steels are far stronger than pure iron, with properties varying with the composition. **Alloy steels** contain greater amounts of other metals. Low-alloy steels have less than 5% of the alloying material; high-alloy steels have more. Low-alloy steels containing up to 5% silicon with relatively little carbon have a high electrical resistance and are used in power transformers and motor or generator cores, for example. **Stainless steel** is a high-alloy steel containing at least 11% chromium. Steels with up to 20% tungsten are very hard and are used in high-speed cutting tools. About 50% of the world's steel is now made from scrap.

Steel is produced by removing impurities, such as carbon, from raw or pig iron, produced by a ◊blast furnace. The main industrial process is the ◊ **basic–oxygen process**, in which molten pig iron and scrap steel is placed in a container lined with heat-resistant, alkaline (basic) bricks. A pipe or lance is lowered near to the surface of the molten metal and pure oxygen blown through it at high pressure. The surface of the metal is disturbed by the blast and the impurities are oxidized (burned out). The **open-hearth process** is an older steelmaking method in which molten iron and limestone are placed in a shallow bowl or hearth (see ◊open-hearth furnace). Burning oil or gas is blown over the surface of the metal, and the impurities are

oxidized. High-quality steel is made in an **electric furnace**. A large electric current flows through electrodes in the furnace, melting a charge of scrap steel and iron. The quality of the steel produced can be controlled precisely because the temperature of the furnace can be maintained exactly and there are no combustion by-products to contaminate the steel. Electric furnaces are also used to refine steel, producing the extra-pure steels used, for example, in the petrochemical industry.

The steel produced is cast into ingots, which can be worked when hot by hammering (forging) or pressing between rollers to produce sheet steel. Alternatively, the **continuous-cast process**, in which the molten metal is fed into an open-ended mould cooled by water, produces an unbroken slab of steel.

Stefan–Boltzmann law in physics, a law that relates the energy, E, radiated away from a perfect emitter (a black body), to the temperature, T, of that body. It has the form $M = \sigma T^4$, where M is the energy radiated per unit area per second, T is the temperature, and σ is the **Stefan–Boltzmann constant**. Its value is 5.6705×10^{-8} W m^{-2} K^{-4}. The law was derived by Austrian physicists Joseph Stefan and Ludwig Boltzmann.

stellar population in astronomy, a classification of stars according to their chemical composition as determined by ◊spectroscopy.

Population I stars have a relatively high abundance of elements heavier than hydrogen and helium, and are confined to the spiral arms and disc of the Galaxy. They are believed to be young stars formed from material that has already been enriched with elements created by ◊nuclear fusion in earlier generations of stars. Examples include open clusters and ◊supergiants.

Population II stars have a low abundance of heavy elements and are found throughout the Galaxy but especially in the central bulge and outer halo. They are among the oldest objects in the Galaxy, and include ◊globular clusters. The Sun is a Population I star.

stem main supporting axis of a plant that bears the leaves, buds, and reproductive structures; it may be simple or branched. The plant stem usually grows above ground, although some grow underground, including ◊rhizomes, ◊corms, ◊rootstocks, and ◊tubers. Stems contain a continuous vascular system that conducts water and food to and from all parts of the plant.

The point on a stem from which a leaf or leaves arise is called a node, and the space between two successive nodes is the internode. In some plants, the stem is highly modified; for example, it may form a leaf-like ◊cladode or it may be twining (as in many climbing plants), or fleshy and swollen to store water (as in cacti and other succulents). In plants exhibiting ◊secondary growth, the stem may become woody, forming a main trunk, as in trees, or a number of branches from ground level, as in shrubs.

steppe the temperate grasslands of Europe and Asia. Sometimes the term refers to other temperate grasslands and semi-arid desert edges.

stepper motor electric motor that can be precisely controlled by signals from a computer. The motor turns through a precise angle each time it receives a signal pulse from the computer. By varying the rate at which signal pulses are produced, the motor can be run at different speeds or turned through an exact angle and then stopped. Switching circuits can be constructed to allow the computer to reverse the direction of the motor.

By combining two or more motors, complex movement control becomes possible. For example, if stepper motors are used to power the wheels of a small vehicle, a computer can manoeuvre the vehicle in any direction.

Stepper motors are commonly used in small-scale applications where computer-controlled movement is required. In larger applications, where greater power is necessary, pneumatic or hydraulic systems are usually preferred.

steradian SI unit (symbol sr) of measure of solid (three-dimensional) angles, the three-dimensional equivalent of the ◊radian. One steradian is the angle at the centre of a sphere when an area on the surface of the sphere equal to the square of the sphere's radius is joined to the centre.

stereophonic sound system of sound reproduction using two complementary channels leading to two loudspeakers, which gives a more natural depth to the sound. Stereo recording began with the introduction of two-track magnetic tape in the 1950s. See ◊hi-fi.

sterilization the killing or removal of living organisms such as bacteria and fungi. A sterile environment is necessary in medicine, food processing, and some scientific experiments. Methods include heat treatment (such as boiling), the use of chemicals (such as disinfectants), irradiation with gamma rays, and filtration. See also ◊asepsis.

sterilization any surgical operation to terminate the possibility of reproduction. In women, this is normally achieved by sealing or tying off the ◊Fallopian tubes (tubal ligation) so that fertilization can no longer take place. In men, the transmission of sperm is blocked by ◊vasectomy.

According to the results of a long-term US study released in 1996, the failure rate for female sterilization is 1 in 50, higher than previously believed, with some pregnancies occurring as long as 14 years after the operation.

Sterilization may be encouraged by governments to limit population growth or as part of a selective-breeding policy (see ◊eugenics).

sterling silver ◊alloy containing 925 parts of silver and 75 parts of copper. The copper hardens the silver, making it more useful.

sternum or **breastbone**, the large flat bone, 15–20 cm/5.9–7.8 in long in the adult, at the front of the chest, joined to the ribs. It gives protection to the heart and lungs. During open-heart surgery the sternum must be split to give access to the thorax.

steroid in biology, any of a group of cyclic, unsaturated alcohols (lipids without fatty acid components), which, like sterols, have a complex molecular structure consisting of four carbon rings. Steroids include the sex hormones, such as ◊testosterone, the corticosteroid hormones produced by the ◊adrenal gland, bile acids, and ◊cholesterol.

The term is commonly used to refer to ◊anabolic steroid. In medicine, synthetic steroids are used to treat a wide range of conditions.

Steroids also found in plants. The most widespread are the **brassinosteroids**, necessary for normal plant growth.

sterol any of a group of solid, cyclic, unsaturated alcohols, with a complex structure that includes four carbon rings; cholesterol is an example. Steroids are derived from sterols.

stethoscope instrument used to ascertain the condition of the heart and lungs by listening to their action. It consists of two earpieces connected by flexible tubes to a small plate that is placed against the body. It was invented in 1819 in France by René Théophile Hyacinthe Laënnec.

Stevens-Johnson syndrome in medicine, syndrome is characterized by skin lesions that develop into blisters. Severe ulceration of the eyes and mucous membranes can also occur. Certain drugs, such as the ◊sulphonamide group of antibiotics, cause this reaction in individuals who are hypersensitive to them and their use is now restricted due to the severity of this condition.

stigma in a flower, the surface at the tip of a ◊carpel that receives the ◊pollen. It often has short outgrowths, flaps, or hairs to trap pollen and may produce a sticky secretion to which the grains adhere.

stimulant any substance that acts on the brain to increase alertness and activity; for example, ◊amphetamine. When given to children, stimulants may have a paradoxical, calming effect. Stimulants cause liver damage, are habit-forming, have limited therapeutic value, and are now prescribed only to treat narcolepsy and severe obesity.

stipule outgrowth arising from the base of a leaf or leaf stalk in certain plants. Stipules usually occur in pairs or fused into a single semicircular structure.

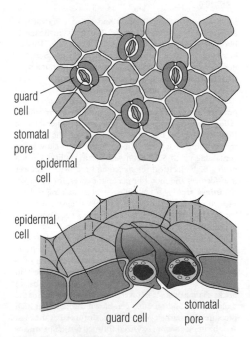

guard cell

stomatal pore

epidermal cell

epidermal cell

stomatal pore

guard cell

stoma The stomata, tiny openings in the epidermis of a plant, are surrounded by pairs of crescent-shaped cells, called guard cells. The guard cells open and close the stoma by changing shape.

Stirling engine a hot-air external combustion engine invented by Scottish priest Robert Stirling in 1816. The engine operates by adapting to the fact that the air in its cylinders heats up when it is compressed and cools when it expands. The engine will operate on any fuel, is non-polluting and relatively quiet. It was used fairly widely in the 19th century before the appearance of small, powerful, and reliable electric motors. Attempts have also been made in recent times to use Stirling's engine to power a variety of machines.

STOL (acronym for *short takeoff and landing*) aircraft fitted with special devices on the wings (such as sucking flaps) that increase aerodynamic lift at low speeds. Small passenger and freight STOL craft may become common with the demand for small airports, especially in difficult terrain.

stolon in botany, a type of ◊runner.

stoma (plural *stomata*) in botany, a pore in the epidermis of a plant. Each stoma is surrounded by a pair of guard cells that are crescent-shaped when the stoma is open but can collapse to an oval shape, thus closing off the opening between them. Stomata allow the exchange of carbon dioxide and oxygen (needed for ◊photosynthesis and ◊respiration) between the internal tissues of the plant and the outside atmosphere. They are also the main route by which water is lost from the plant, and they can be closed to conserve water, the movements being controlled by changes in turgidity of the guard cells.

stomach the first cavity in the digestive system of animals. In mammals it is a bag of muscle situated just below the diaphragm. Food enters it from the oesophagus, is digested by the acid and ◊enzymes secreted by the stomach lining, and then passes into the duodenum. Some plant-eating mammals have multichambered stomachs that harbour bacteria in one of the chambers to assist in the digestion of ◊cellulose.

The gizzard is part of the stomach in birds.

stone (plural *stone*) imperial unit (abbreviation st) of mass. One stone is 14 pounds (6.35 kg).

Stone Age the developmental stage of humans in ◊prehistory before the use of metals, when tools and weapons were made chiefly of stone, especially flint. The Stone Age is subdivided into the Old or **Palaeolithic**, when flint implements were simply chipped into shape; the Middle or **Mesolithic**; and the New or **Neolithic**, when implements were ground and polished. Paleolithic people were hunters and gatherers; by the Neolithic period people were taking the first steps in agriculture, the domestication of animals, weaving, and pottery.

Recent research has been largely directed towards the relationship of the Palaeolithic period to ◊geochronology (the measurement of geological time) and to the clarification of an absolute chronology based upon geology. The economic aspects of the Neolithic cultures have attracted as much attention as the typology of the implements and pottery, and the study of chambered tombs.

Stonehenge old English 'hanging stones' Megalithic monument on Salisbury Plain, 3 km/1.9 mi W of Amesbury, Wiltshire, England. The site developed over various periods from a simple henge (earthwork circle and ditch), dating from about 3000 BC, to a complex stone structure, from about 2100 BC, which included a circle of 30 upright stones, their tops linked by lintel stones to form a continuous circle about 30 m/100 ft across.

Within this sarsen *peristyle* was a horseshoe arrangement of five sarsen *trilithons* (two uprights plus a lintel, set as five separate entities), and the so-called 'Altar Stone' – an upright pillar – on the axis of the horseshoe at the open, NE end, which faces in the direction of the rising sun. A further horseshoe and circle within the sarsen peristyle were constructed from bluestone relocated from previous outer circles.

It has been suggested that Stonehenge was constructed as an observatory.

stool in medicine, alternative term for ◊faeces.

storm surge abnormally high tide brought about by a combination of a deep atmospheric depression (very low pressure) over a shallow sea area, high spring tides, and winds blowing from the appropriate direction. A storm surge can cause severe flooding of lowland coastal regions and river estuaries.

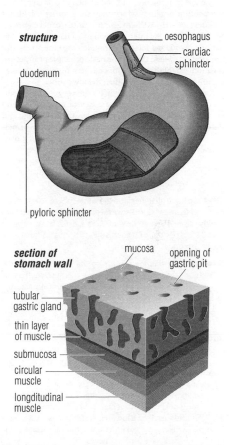

structure

oesophagus

cardiac sphincter

duodenum

pyloric sphincter

section of stomach wall

mucosa

opening of gastric pit

tubular gastric gland

thin layer of muscle

submucosa

circular muscle

longditudinal muscle

stomach The human stomach can hold about 1.5 l/2.6 pt of liquid. The digestive juices are acidic enough to dissolve metal. To avoid damage, the cells of the stomach lining are replaced quickly – 500,000 cells are replaced every minute, and the whole stomach lining every three days.

Bangladesh is particularly prone to surges, being sited on a low-lying ◊delta where the Indian Ocean funnels into the Bay of Bengal. In May 1991, 125,000 people were killed there in such a disaster. In Feb 1953 more than 2,000 died when a North Sea surge struck the Dutch and English coasts.

STP abbreviation for ◊standard temperature and pressure.

strabismus technical term for a ◊squint.

strain in the science of materials, the extent to which a body is distorted when a deforming force (stress) is applied to it. It is a ratio of the extension or compression of that body (its length, area, or volume) to its original dimensions (see ◊Hooke's law). For example, linear strain is the ratio of the change in length of a body to its original length.

strata (singular *stratum*) layers or ◊beds of sedimentary rock.

stratigraphy branch of geology that deals with the sequence of formation of ◊sedimentary rock layers and the conditions under which they were formed. Its basis was developed by William Smith, a British canal engineer. Stratigraphy in the interpretation of archaeological excavations provides a relative chronology for the levels and the artefacts within them. The basic principle of superimposition establishes that upper layers or deposits have accumulated later in time than the lower ones.

Stratigraphy involves both the investigation of sedimentary structures to determine ancient geographies and environments, and the study of fossils for identifying and dating particular beds of rock. Stratigraphy is the principal means by which the context of archaeological deposits is evaluated. Chronologies are constructed and events are sequenced. It is invaluable for interpreting the sequence of deposition of the site and thereby the relative age of artefacts, features, and other phenomena in the site.

stratosphere that part of the atmosphere 10–40 km/6–25 mi from the Earth's surface, where the temperature slowly rises from a low of −55°C/−67°F to around 0°C/32°F. The air is rarefied and at around 25 km/15 mi much ◊ozone is concentrated.

streamlining shaping a body so that it offers the least resistance when travelling through a medium such as air or water. Aircraft, for example, must be carefully streamlined to reduce air resistance, or ◊drag.

High-speed aircraft must have swept-back wings, supersonic craft a sharp nose and narrow body.

strength of acids and bases in chemistry, the ability of ◊acids and ◊bases to dissociate in solution with water, and hence to produce a low or high ◊pH respectively.

A strong acid is fully dissociated in aqueous solution, whereas a weak acid is only partly dissociated. Since the ◊dissociation of acids generates hydrogen ions, a solution of a strong acid will have a high concentration of hydrogen ions and therefore a low pH. A strong base will have a high pH, whereas a weaker base will not dissociate completely and will have a pH of nearer 7.

streptokinase enzyme produced by *Streptococcus* bacteria that is capable of digesting fibrin, the protein making up blood clots. It is used to treat pulmonary ◊embolism and heart attacks, reducing mortality.

streptomycin antibiotic drug discovered in 1944, active against a wide range of bacterial infections.

Streptomycin is derived from a soil bacterium *Streptomyces griseus* or synthesized.

stress in psychology, any event or situation that makes heightened demands on a person's mental or emotional resources. Stress can be caused by overwork, anxiety about exams, money, job security, unemployment, bereavement, poor relationships, marriage breakdown, sexual difficulties, poor living or working conditions, and constant exposure to loud noise.

Many changes that are apparently 'for the better', such as being promoted at work, going to a new school, moving to a new house, and getting married, are also a source of stress. Stress can cause, or aggravate, physical illnesses, among them psoriasis, eczema, asthma, stomach and mouth ulcers. Apart from removing the source of stress, acquiring some control over it and learning to relax when possible are the best responses.

stress and strain in the science of materials, measures of the deforming force applied to a body (stress) and of the resulting change in its shape (◊strain). For a perfectly elastic material, stress is proportional to strain (◊Hooke's law).

stress fracture in medicine, bone fracture caused by excessive amounts of exercise. It is marked by the gradual onset of pain and swelling and a lump may develop over the affected site. Stress fractures cannot be detected by X-rays in the early stages. They are treated by rest, external support, and analgesics to control the pain.

stridulatory organs in insects, organs that produce sound when rubbed together. Crickets rub their wings together, but grasshoppers rub a hind leg against a wing. Stridulation is thought to be used for attracting mates, but may also serve to mark territory.

string in computing, a group of characters manipulated as a single object by the computer. In its simplest form a string may consist of a single letter or word – for example, the single word SMITH might be established as a string for processing by a computer. A string can also consist of a combination of words, spaces, and numbers – for example, 33 HIGH STREET ANYTOWN ALLSHIRE could be established as a single string.

Most high-level languages have a variety of string-handling ◊functions. For example, functions may be provided to read a character from any given position in a string or to count automatically the number of characters in a string.

strobilus in botany, a reproductive structure found in most ◊gymnosperms and some ◊pteridophytes, notably the club mosses. In conifers the strobilus is commonly known as a ◊cone.

stroboscope instrument for studying continuous periodic motion by using light flashing at the same frequency as that of the motion; for example, rotating machinery can be optically 'stopped' by illuminating it with a stroboscope flashing at the exact rate of rotation.

stroke or *cerebrovascular accident* or *apoplexy* interruption of the blood supply to part of the brain due to a sudden bleed in the brain (cerebral haemorrhage) or ◊embolism or ◊thrombosis. Strokes vary in severity from producing almost no symptoms to proving rapidly fatal.

In between are those (often recurring) that leave a wide range of impaired function, depending on the size and location of the event.

Strokes involving the right side of the brain, for example, produce weakness of the left side of the body. Some affect speech. Around 80% of strokes are **ischaemic strokes**, caused by a blood clot blocking an artery transporting blood to the brain. Transient ischaemic attacks, or 'mini-strokes', with effects lasting only briefly (less than 24 hours), require investigation to try to forestall the possibility of a subsequent full-blown stroke.

The disease of the arteries that predisposes to stroke is ◊atherosclerosis. High blood pressure (◊hypertension) is also a precipitating factor – a worldwide study in 1995 estimated that high blood pressure before middle age gives a tenfold increase in the chance of having a stroke later in life.

Strokes can sometimes be prevented by surgery (as in the case of some ◊aneurysms), or by use of ◊anticoagulant drugs or vitamin E or daily aspirin to minimize the risk of stroke due to blood clots. According to the results of a US trial announced in Dec 1995, the clot-buster drug tPA, if administered within three hours of a stroke, can cut the number of stroke victims experiencing lasting disability by 50%.

strong nuclear force one of the four fundamental ◊forces of nature, the other three being the electromagnetic force, gravity, and the weak nuclear force. The strong nuclear force was first described by Japanese physicist Hideki Yukawa in 1935. It is the strongest of all the forces, acts only over very small distances (within the nucleus of the atom), and is responsible for binding together ◊quarks to form ◊hadrons, and for binding together protons and neutrons in the atomic nucleus. The particle that is the carrier of the strong nuclear force is the ◊gluon, of which there are eight kinds, each with zero mass and zero charge.

strontium soft, ductile, pale-yellow, metallic element, symbol Sr, atomic number 38, relative atomic mass 87.62. It is one of the ◊alkaline-earth metals, widely distributed in small quantities only as a sulphate or carbonate. Strontium salts burn with a red flame and are used in fireworks and signal flares.

The radioactive isotopes Sr-89 and Sr-90 (half-life 25 years) are some of the most dangerous products of the nuclear industry; they are fission products in nuclear explosions and in the reactors of nuclear power plants. Strontium is chemically similar to calcium and deposits in bones and other tissues, where the radioactivity is damaging. The element was named in 1808 by English chemist Humphry Davy, who isolated it by electrolysis, after Strontian, a mining location in Scotland where it was first found.

structured programming in computing, the process of writing a program in small, independent parts. This makes it easier to control a program's development and to design and test its individual component parts. Structured programs are built up from units called **modules**, which normally correspond to single ◊procedures or ◊functions. Some programming languages, such as PASCAL and Modula-2, are better suited to structured programming than others.

strychnine $C_{21}H_{22}O_2N_2$ bitter-tasting, poisonous alkaloid. It is a poison that causes violent muscular spasms, and is usually obtained by powdering the seeds of plants of the genus *Strychnos* (for example *S. nux vomica*). Curare is a related drug.

stupor in medicine, state of partial unconsciousness in which the patient can be roused by pain or may respond vaguely to questions. The deeper state of unconsciousness, in which nothing rouses the patient, is ◊coma.

stutter in medicine, a disruption to the normal flow of speech. Stuttering usually becomes apparent between the ages of two and five years. It is more common in boys and genetic and environmental factors contribute to its development. Stuttering can be improved considerably by speech therapy. Stuttering rarely develops after puberty and it usually occurs as a result of brain damage.

style in flowers, the part of the ◊carpel bearing the ◊stigma at its tip. In some flowers it is very short or completely lacking, while in others it may be long and slender, positioning the stigma in the most effective place to receive the pollen.

subatomic particle in physics, a particle that is smaller than an atom. Such particles may be indivisible ◊elementary particles, such as the ◊electron and ◊quark, or they may be composites, such as the ◊proton, ◊neutron, and ◊alpha particle. See also ◊particle physics.

The assumption of a state of matter more finely subdivided than the atom of an element is a somewhat startling one.

On **subatomic particle** J J Thomson, in Royal Institution Lecture 1897

subconscious state of being partially conscious. The mind can perceive impressions or events that may be forgotten at the time but can continue to influence the conscious mind or may enter the conscious mind a long time after the event has taken place. Psychoanalysis places much importance on the influence of painful and unpleasant experiences that, although forgotten, continue to have a detrimental influence on the mind.

subcutaneous tissue in medicine, loose cellular tissue beneath the skin. Injection into the subcutaneous tissue is a route of administration for some drugs, such as insulin.

subduction zone region where two plates of the Earth's rigid lithosphere collide, and one plate descends below the other into the semiliquid asthenosphere. Subduction occurs along ocean trenches, most of which encircle the Pacific Ocean; portions of the ocean plate slide beneath other plates carrying continents.

Ocean trenches are usually associated with volcanic ◊island arcs and deep-focus earthquakes (more than 185 mi/300 km below the surface), both the result of disturbances caused by the plate subduction. The Aleutian Trench bordering Alaska is an example of an active subduction zone, which has produced the Aleutian Island arc.

sublimation in chemistry, the conversion of a solid to vapour without passing through the liquid phase.

Sublimation depends on the fact that the boiling-point of the solid substance is lower than its melting-point at atmospheric pressure. Thus by increasing pressure, a substance which sublimes can be made to go through a liquid stage before passing into the vapour state.

Some substances that do not sublime at atmospheric pressure can be made to do so at low pressures. This is the principle of freeze-drying, during which ice sublimes at low pressure.

subliminal message any message delivered beneath the human conscious threshold of perception. It may be visual (words or images flashed between the frames of a cinema or TV film), or aural (a radio message broadcast constantly at very low volume).

submarine underwater warship. The first underwater boat was constructed in 1620 for James I of England by the Dutch scientist Cornelius van Drebbel (1572–1633). A naval submarine, or submersible torpedo boat, the *Gymnote*, was launched by France in 1888. The conventional submarine of World War I was driven by diesel engine on the surface and by battery-powered electric motors underwater.

The diesel engine also drove a generator that produced electricity to charge the batteries.

In 1954 the USA launched the first nuclear-powered submarine, the *Nautilus*. The US nuclear submarine *Ohio*, in service from 1981, is 170 m/560 ft long and carries 24 Trident missiles, each with 12 independently targetable nuclear warheads. The nuclear warheads on US submarines have a range that is being extended to 11,000 km/6,750 mi. Three Vanguard-class Trident missile-carrying submarines, which when armed will each wield more firepower than was used in the whole of World War II, are being built in the 1990s in the UK. Operating depth is usually up to 300 m/1,000 ft, and nuclear-powered speeds of 30 knots (55 kph/34 mph) are reached. As in all nuclear submarines, propulsion is by steam turbine driving a propeller. The steam is raised using the heat given off by the nuclear reactor (see ◊nuclear energy). In oceanography, salvage, and pipe-laying, smaller submarines called **submersibles** are used. They are also being developed for tourism.

submersible vessel designed to operate under water, especially a small submarine used by engineers and research scientists as a ferry craft to support diving operations. The most advanced submersibles are the so-called lock-out type, which have two compartments: one for the pilot, the other to carry divers. The diving compartment is pressurized and provides access to the sea.

subroutine in computing, a small section of a program that is executed ('called') from another part of the program. Subroutines provide a method of performing the same task at more than one point in the program, and also of separating the details of a program from its main logic. In some computer languages, subroutines are similar to ◊functions or ◊procedures.

subsistence farming farming when the produce is enough to feed only the farmer and family and there is no surplus to sell.

substitution reaction in chemistry, the replacement of one atom or ◊functional group in an organic molecule by another.

substrate in biochemistry, a compound or mixture of compounds acted on by an enzyme. The term also refers to a substance such as ◊agar that provides the nutrients for the metabolism of microorganisms. Since the enzyme systems of microorganisms regulate their metabolism, the essential meaning is the same.

succession in ecology, a series of changes that occur in the structure and composition of the vegetation in a given area from the time it is first colonized by plants (**primary succession**), or after it has been disturbed by fire, flood, or clearing (**secondary succession**).

If allowed to proceed undisturbed, succession leads naturally to a stable ◊climax community (for example, oak and hickory forest or savannah grassland) that is determined by the climate and soil characteristics of the area.

Having in the natural history of this earth, seen a succession of worlds, we may conclude that there is a system in nature. ... The result, therefore of our present enquiry is, that we find no vestige of a beginning – no prospect of an end.

On **succession** James Hutton
Transactions of the Royal Society of Edinburgh
1788

succulent plant thick, fleshy plant that stores water in its tissues; for example, cacti and stonecrops *Sedum*. Succulents live either in areas where water is very scarce, such as deserts, or in places where it is not easily obtainable

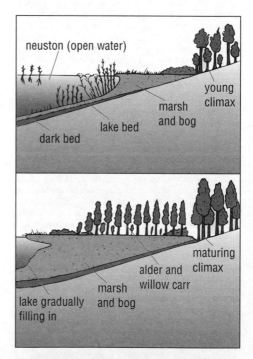

succession The succession of plant types along a lake. As the lake gradually fills in, a mature climax community of trees forms inland from the shore. Extending out from the shore, a series of plant communities can be discerned with small, rapidly growing species closest to the shore.

because of the high concentrations of salts in the soil, as in salt marshes. Many desert plants are ◊xerophytes.

suckering in plants, reproduction by new shoots (suckers) arising from an existing root system rather than from seed. Plants that produce suckers include elm, dandelion, and members of the rose family.

sucrose or **cane sugar** or **beet sugar** $C_{12}H_{22}O_{11}$ a sugar found in the pith of sugar cane and in sugar beets. It is popularly known as ◊sugar.

Sucrose is a disaccharide sugar, each of its molecules being made up of two simple sugar (monosaccharide) units: glucose and fructose.

sudden infant death syndrome (SIDS) in medicine, the technical term for ◊cot death.

sugar or **sucrose** sweet, soluble crystalline carbohydrate found in the pith of sugar cane and in sugar beet. It is a **disaccharide** sugar, each of its molecules being made up of two simple-sugar (**monosaccharide**) units: glucose and fructose. Sugar is easily digested and forms a major source of energy in humans, being used in cooking and in the food industry as a sweetener and, in high concentrations, as a preservative. A high consumption is associated with obesity and tooth decay. In the UK, sucrose may not be used in baby foods.

The main sources of sucrose sugar are tropical sugar cane *Saccharum officinarum*, which accounts for two-thirds of production, and temperate sugar beet *Beta vulgaris*. Minor quantities are produced from the sap of maple trees, and from sorghum and date palms. Raw sugar crystals obtained by heating the juice of sugar canes are processed to form brown sugars, such as Muscovado and Demerara, or refined and sifted to produce white sugars, such as granulated, caster, and icing sugar. The syrup that is drained away from the raw sugar is molasses; it may be processed to form golden syrup or treacle, or fermented to form rum. Molasses obtained from sugar beet juice is too bitter for human consumption. The fibrous residue of sugar cane, called bagasse, is used in the manufacture of paper, cattle feed, and fuel; and new types of cane are being bred for low sugar and high fuel production.

sulphate SO_4^{2-} salt or ester derived from sulphuric acid. Most sulphates are water soluble (the exceptions are lead, calcium, strontium, and barium sulphates), and require a very high temperature to decompose them.

The commonest sulphates seen in the laboratory are copper(II) sulphate ($CuSO_4$), iron(II) sulphate ($FeSO_4$), and aluminium sulphate ($Al_2(SO_4)_3$). The ion is detected in solution by using barium chloride or barium nitrate to precipitate the insoluble sulphate.

sulphide compound of sulphur and another element in which sulphur is the more electronegative element (see ◊electronegativity). Sulphides occur in a number of minerals. Some of the more volatile sulphides have extremely unpleasant odours (hydrogen sulphide smells of bad eggs).

sulphite SO_3^{2-} salt or ester derived from sulphurous acid.

sulphonamide any of a group of compounds containing the chemical group sulphonamide (SO_2NH_2) or its derivatives, which were, and still are in some cases, used to treat bacterial diseases. Sulphadiazine ($C_{10}H_{10}N_4O_2S$) is an example.

Sulphonamide was the first commercially available antibacterial drug, the forerunner of a range of similar drugs. Toxicity and increasing resistance have limited their use chiefly to the treatment of urinary-tract infection.

sulphur brittle, pale-yellow, nonmetallic element, symbol S, atomic number 16, relative atomic mass 32.064. It occurs in three allotropic forms: two crystalline (called rhombic and monoclinic, following the arrangements of the atoms within the crystals) and one amorphous. It burns in air with a blue flame and a stifling odour. Insoluble in water but soluble in carbon disulphide, it is a good electrical insulator. Sulphur is widely used in the manufacture of sulphuric acid (used to treat phosphate rock to make fertilizers) and in making paper, matches, gunpowder and fireworks, in vulcanizing rubber, and in medicines and insecticides.

It is found abundantly in nature in volcanic regions combined with both metals and nonmetals, and also in its elemental form as a crystalline solid. It is a constituent of proteins, and has been known since ancient times.

sulphur dioxide SO_2 pungent gas produced by burning sulphur in air or oxygen. It is widely used for disinfecting food vessels and equipment, and as a preservative in some food products. It occurs in industrial flue gases and is a major cause of ◊acid rain.

sulphuric acid or **oil of vitriol** H_2SO_4 a dense, viscous, colourless liquid that is extremely corrosive. It gives out heat when added to water and can cause severe burns. Sulphuric acid is used extensively in the chemical industry, in the refining of petrol, and in the manufacture of fertilizers, detergents, explosives, and dyes. It forms the acid component of car batteries.

sulphurous acid H_2SO_3 solution of sulphur dioxide (SO_2) in water. It is a weak acid.

summer time practice introduced in the UK in 1916 whereby legal time from spring to autumn is an hour in advance of Greenwich mean time.

Continental Europe 'puts the clock back' a month earlier than the UK in autumn. British summer time was permanently in force from Feb 1940–Oct 1945 and from Feb 1968–Oct 1971. Double summer time (2 hours in advance) was in force during the summers of 1941–45 and 1947.

In North America the practice is known as *daylight saving time*.

rhombic sulphur crystal   *monoclinic sulphur crystal*

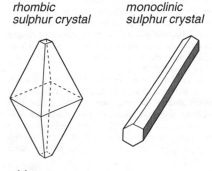

sulphur

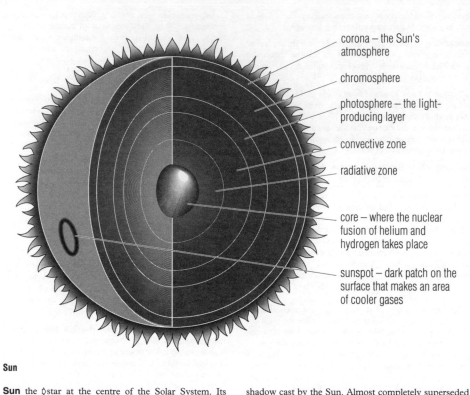

corona – the Sun's atmosphere

chromosphere

photosphere – the light-producing layer

convective zone

radiative zone

core – where the nuclear fusion of helium and hydrogen takes place

sunspot – dark patch on the surface that makes an area of cooler gases

Sun

Sun the ☉ star at the centre of the Solar System. Its diameter is 1,392,000 km/865,000 mi; its temperature at the surface is about 5,800K (5,500°C/9,900°F), and at the centre 15,000,000K (about 15,000,000°C/27,000,000°F). It is composed of about 70% hydrogen and 30% helium, with other elements making up less than 1%. The Sun's energy is generated by nuclear fusion reactions that turn hydrogen into helium at its centre. The gas core is far denser than mercury or lead on Earth. The Sun is about 4.7 billion years old, with a predicted lifetime of 10 billion years.

sunburn in medicine, skin damage caused by overexposure to sunlight. Individuals with fair skin are particularly susceptible to the effects of sunlight due to the relative lack of melanin, a pigment that is made in certain cells in the skin. Skin reddening, due to dilation of blood vessels, occurs shortly after exposure to sunlight. This is followed by tanning as melanin production is stimulated. Overexposure to ultraviolet light, especially of medium wavelength (UVB), produces itching and tingling. Swelling of the superficial layer of the skin can be followed by blister formation and peeling. Headache and fever may also occur if sunburn is severe.

Exposure to the ultraviolet light produced by the sun results in thickening of the superficial layer of the skin and premature ageing. The longer wavelengths of ultraviolet light (UVA) are responsible for photosensitivity reactions and long-term damage to the skin. Prolonged exposure to the sun, especially in individuals with fair skin, is a risk factor in the development of skin cancers, including melanomas.

sundial instrument measuring time by means of a shadow cast by the Sun. Almost completely superseded by the proliferation of clocks, it survives ornamentally in gardens. The dial is marked with the hours at graduated distances, and a style or gnomon (parallel to Earth's axis and pointing to the north) casts the shadow.

sun protection factor (SPF) in medicine, the multiples of protection provided by sunscreens against burning compared with the unprotected skin. Sunscreens are used to protect the skin from the harmful effects of excessive exposure to ultraviolet radiation. They should be applied regularly and most do not offer complete protection. Proprietary sunscreens are graded according to the degree of protection provided against medium wavelengths of ultraviolet radiation (UVB). Sunscreens do not prevent the long-term damage due to long wavelengths of ultraviolet radiation (UVA).

sunshine recorder device for recording the hours of sunlight during a day. The *Campbell-Stokes sunshine recorder* consists of a glass sphere that focuses the sun's rays on a graduated paper strip. A track is burned along the strip corresponding to the time that the Sun is shining.

sunspot dark patch on the surface of the Sun, actually an area of cooler gas, thought to be caused by strong magnetic fields that block the outward flow of heat to the Sun's surface. Sunspots consist of a dark central *umbra*, about 4,000K (3,700°C/6,700°F), and a lighter surrounding *penumbra*, about 5,500K (5,200°C/9,400°F). They last from several days to over a month, ranging in size from 2,000 km/1,250 mi to groups stretching for over 100,000 km/62,000 mi.

Sunspots are more common during active periods in

the Sun's magnetic cycle, when they are sometimes accompanied by nearby ◊flares. The number of sunspots visible at a given time varies from none to over 100, in a cycle averaging 11 years. There was a lull in sunspot activity between 1645–1715, known as the Maunder minimum, that coincided with a cold spell in Europe.

sunstroke ◊heatstroke caused by excessive exposure to the Sun.

superactinide any of a theoretical series of superheavy, radioactive elements, starting with atomic number 113, that extend beyond the ◊transactinide series in the periodic table. They do not occur in nature and none has yet been synthesized.

It is postulated that this series has a group of elements that have half-lives longer than those of the transactinide series.

This group, centred on element 114, is referred to as the 'island of stability', based on the nucleon arrangement. The longer half-lives will, it is hoped, allow enough time for their chemical and physical properties to be studied when they have been synthesized.

supercluster in astronomy, a grouping of several clusters of galaxies to form a structure about 100–300 million light years across. Our own Galaxy and its neighbours lie on the edge of the local supercluster of which the ◊Virgo cluster is the dominant member.

supercomputer the fastest, most powerful type of computer, capable of performing its basic operations in picoseconds (thousand-billionths of a second), rather than nanoseconds (billionths of a second), like most other computers.

To achieve these extraordinary speeds, supercomputers use several processors working together and techniques such as cooling processors down to nearly ◊absolute zero temperature, so that their components conduct electricity many times faster than normal. Supercomputers are used in weather forecasting, fluid dynamics and aerodynamics. Manufacturers include Cray, Fujitsu, and NEC.

superconductivity in physics, increase in electrical conductivity at low temperatures. The resistance of some metals and metallic compounds decreases uniformly with decreasing temperature until at a critical temperature (the superconducting point), within a few degrees of absolute zero (0K/−273.15°C/−459.67°F), the resistance suddenly falls to zero. The phenomenon was discovered by Dutch scientist Heike Kamerlingh Onnes in 1911.

Some metals, such as platinum and copper, do not become superconductive; as the temperature decreases, their resistance decreases to a certain point but then rises again. Superconductivity can be nullified by the application of a large magnetic field.

In the superconducting state, an electric current will continue indefinitely once started, provided that the material remains below the superconducting point. In 1986 IBM researchers achieved superconductivity with some ceramics at −243°C/−405°F), opening up the possibility of **'high-temperature' superconductivity**; Paul Chu at the University of Houston, Texas, achieved superconductivity at −179°C/−290°F, a temperature that can be sustained using liquid nitrogen. In 1992, two Japanese researchers developed a material which becomes superconducting at around −103°C/−153°F.

The Los Alamos National Laboratory, New Mexico,

produced a flexible superconducting film in 1995. The film, which can carry enough current for practical applications such as electromagnets, works at the relatively high temperature of 77K (−196°C/−320°F). It is made of yttrium barium copper oxide with a backing of nickel tape coated with zirconia.

supercooling the cooling of a liquid below its freezing point without freezing taking place; or the cooling of a ◊saturated solution without crystallization taking place, to form a supersaturated solution. In both cases supercooling is possible because of the lack of solid particles around which crystals can form. Crystallization rapidly follows the introduction of a small crystal (seed) or agitation of the supercooled solution.

superego in Freudian psychology, the element of the human mind concerned with the ideal, responsible for ethics and self-imposed standards of behaviour. It is characterized as a form of conscience, restraining the ◊ego, and responsible for feelings of guilt when the moral code is broken.

superfluid fluid that flows without viscosity or friction and has a very high thermal conductivity. Liquid helium at temperatures below 2K (−271°C/−456°F) is a superfluid: it shows unexpected behaviour; for instance, it flows uphill in apparent defiance of gravity and, if placed in a container, will flow up the sides and escape.

supergiant the largest and most luminous type of star known, with a diameter of up to 1,000 times that of the Sun and absolute magnitudes of between −5 and −9. Supergiants are likely to become ◊supernovae.

superheterodyne receiver the most widely used type of radio receiver, in which the incoming signal is mixed with a signal of fixed frequency generated within the receiver circuits. The resulting signal, called the intermediate-frequency (i.f.) signal, has a frequency between that of the incoming signal and the internal signal. The intermediate frequency is near the optimum frequency of the amplifier to which the i.f. signal is passed.

This arrangement ensures greater gain and selectivity. The superheterodyne system is also used in basic television receivers.

superior planet planet that is farther away from the Sun than the Earth is: that is, Mars, Jupiter, Saturn, Uranus, Neptune, and Pluto.

supernova the explosive death of a star, which temporarily attains a brightness of 100 million Suns or more, so that it can shine as brilliantly as a small galaxy for a few days or weeks. Very approximately, it is thought that a supernova explodes in a large galaxy about once every 100 years. Many supernovae remain undetected because of obscuring by interstellar dust – astronomers estimate some 50%.

The name 'supernova' was coined by US astronomers Fritz Zwicky and Walter Baade in 1934. Zwicky was also responsible for the division into types I and II. **Type I** supernovae are thought to occur in ◊binary star systems, in which gas from one star falls on to a ◊white dwarf, causing it to explode. **Type II** supernovae occur in stars ten or more times as massive as the Sun, which suffer runaway internal nuclear reactions at the ends of their lives, leading to explosions. These are thought to leave behind ◊neutron stars and ◊black holes. Gas ejected by such an explosion causes an expanding radio source, such as the ◊Crab nebula. Supernovae are thought to be the main source of elements heavier than hydrogen and helium.

superphosphate phosphate fertilizer made by treating apatite with sulphuric or phosphoric acid. The commercial mixture contains largely monocalcium phosphate. Single-superphosphate obtained from apatite and sulphuric acid contains 16–20% available phosphorus, as P_2O_5; triple-superphosphate, which contains 45–50% phosphorus, is made by treating apatite with phosphoric acid.

supersaturation in chemistry, the state of a solution that has a higher concentration of ◊solute than would normally be obtained in a ◊saturated solution.

Many solutes have a higher ◊solubility at high temperatures. If a hot saturated solution is cooled slowly, sometimes the excess solute does not come out of solution. This is an unstable situation and the introduction of a small solid particle will encourage the release of excess solute.

supersonic speed speed greater than that at which sound travels, measured in ◊Mach numbers. In dry air at 0°C/32°F, sound travels at about 1,170 kph/727 mph, but decreases its speed with altitude until, at 12,000 m/39,000 ft, it is only 1,060 kph/658 mph.

When an aircraft passes the ◊sound barrier, shock waves are built up that give rise to ◊sonic boom, often heard at ground level. US pilot Captain Charles Yeager was the first to achieve supersonic flight, in a Bell VS-1 rocket plane on 14 Oct 1947.

superstring theory in physics, a mathematical theory developed in the 1980s to explain the properties of ◊elementary particles and the forces between them (in particular, gravity and the nuclear forces) in a way that combines ◊relativity and ◊quantum theory.

In string theory, the fundamental objects in the universe are not point-like particles but extremely small stringlike objects. These objects exist in a universe of ten dimensions, although, for reasons not yet understood, only three space dimensions and one dimension of time are discernible.

There are many unresolved difficulties with superstring theory, but some physicists think it may be the ultimate 'theory of everything' that explains all aspects of the universe within one framework.

supersymmetry in physics, a theory that relates the two classes of elementary particle, the ◊fermions and the ◊bosons. According to supersymmetry, each fermion particle has a boson partner particle, and vice versa. It has not been possible to marry up all the known fermions with the known bosons, and so the theory postulates the existence of other, as yet undiscovered fermions, such as the photinos (partners of the photons), gluinos (partners of the gluons), and gravitinos (partners of the gravitons). Using these ideas, it has become possible to develop a theory of gravity – called **supergravity** – that extends Einstein's work and considers the gravitational, nuclear, and electromagnetic forces to be manifestations of an underlying superforce. Supersymmetry has been incorporated into the ◊superstring theory, and appears to be a crucial ingredient in the 'theory of everything' sought by scientists.

support environment in computing, a collection of programs (◊software) used to help people design and write other programs. At its simplest, this includes a ◊text editor (word-processing software) and a ◊compiler for translating programs into executable form; but it can also include interactive debuggers for helping to locate faults, data dictionaries for keeping track of the data used, and rapid prototyping tools for producing quick, experimental mock-ups of programs.

suprarenal gland alternative name for the ◊adrenal gland.

surd expression containing the root of an ◊irrational number that can never be exactly expressed – for example, $\sqrt{3} = 1.732050808...$.

surface area the area of the outer surface of a three-dimensional shape, or solid. *See illustration opposite.*

surface-area-to-volume ratio the ratio of an animal's surface area (the area covered by its skin) to its total volume. This is high for small animals, but low for large animals such as elephants.

The ratio is important for endothermic (warm-blooded) animals because the amount of heat lost by the body is proportional to its surface area, whereas the amount generated is proportional to its volume. Very small birds and mammals, such as hummingbirds and shrews, lose a lot of heat and need a high intake of food to maintain their body temperature. Elephants, on the other hand, are in danger of overheating, which is why they have no fur.

surface tension in physics, the property that causes the surface of a liquid to behave as if it were covered with a weak elastic skin; this is why a needle can float on water. It is caused by the exposed surface's tendency to contract to the smallest possible area because of cohesive forces between ◊molecules at the surface. Allied phenomena include the formation of droplets, the concave profile of a meniscus, and the ◊capillary action by which water soaks into a sponge.

surfactant (contraction of **surface-active agent**) substance added to a liquid in order to increase its wetting or spreading properties. Detergents are examples.

surge abnormally high tide; see ◊storm surge.

surgery branch of medicine concerned with the treatment of disease, abnormality, or injury by operation. Traditionally it has been performed by means of cutting instruments, but today a number of technologies are used to treat or remove lesions, including ultrasonic waves and laser surgery.

Surgery is carried out under sterile conditions using an ◊anaesthetic. There are many specialized fields, including cardiac (heart), orthopaedic (bones and joints), ophthalmic (eye), neuro (brain and nerves), thoracic (chest), and renal (kidney) surgery; other specialities include plastic and reconstructive surgery, and ◊transplant surgery.

Historically, surgery for abscesses, amputation, dental problems, trepanning, and childbirth was practised by the ancient civilizations of both the Old World and the New World.

During the Middle Ages, Arabic surgeons passed their techniques on to Europe, where, during the Renaissance, anatomy and physiology were pursued. By the 19th century, anaesthetics and Joseph Lister's discovery of antiseptics became the basis for successful surgical prac-

surface area of common three-dimensional shapes

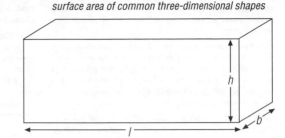

surface area of a **cube**
(faces are identical)
= 6 × area of each surface
= 6 l^2

surface area of a **cuboid**
(opposite faces are identical)
= area of two end faces + area of two sides
   + area of top and base
= 2lh + 2hb + 2lb
= 2(lh + hb + lb)

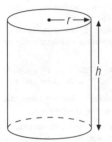

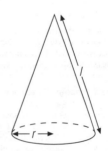

surface area of a **cylinder**
= area of a curved surface + area of top
   and base
= 2π × (radius of cross-section × height)
   + 2π (radius of cross-section)^2
= 2πrh + 2πr^2
= 2πr(h + r)

surface area of a **cone**
= area of a curved surface + area base
= π × (radius of cross-section × slant height)
   + π (radius of cross-section)^2
= πrl + πr^2
= πr(l + r)

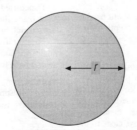

surface area of a **sphere**
= 4π × radius^2
= 4πr^2

surface area

tices. The 20th century's use of antibiotics and blood ◊transfusions has made surgery safer and more effective.

Some early operations, such as thoracoplasty (causing partial collapse of a lung) for tuberculosis, have been replaced by other treatments. Also, the need for exploratory surgery has been reduced by the introduction of non-invasive imaging techniques, such as ◊ultrasound

and ◊CAT scans. The practice of endoscopy (examination of the interior of the body by direct viewing) has enabled the development of minimally invasive keyhole surgery.

surgical spirit ◊ethanol to which has been added a small amount of methanol to render it unfit to drink. It is used to sterilize surfaces and to cleanse skin abrasions and sores.

surrogacy practice whereby a woman is sought, and usually paid, to bear a child for an infertile couple or a single parent.

surveying the accurate measuring of the Earth's crust, or of land features or buildings. It is used to establish boundaries, and to evaluate the topography for engineering work. The measurements used are both linear and angular, and geometry and trigonometry are applied in the calculations.

suspension mixture consisting of small solid particles dispersed in a liquid or gas, which will settle on standing. An example is milk of magnesia, which is a suspension of magnesium hydroxide in water.

suspensory ligament in the ◊eye, a ring of fibre supporting the lens. The ligaments attach to the ciliary muscles, the circle of muscle mainly responsible for changing the shape of the lens during ◊accommodation. If the ligaments are put under tension, the lens becomes flatter, and therefore able to focus on objects in the far distance.

sustainable capable of being continued indefinitely. For example, the sustainable yield of a forest is equivalent to the amount that grows back. Environmentalists made the term a catchword, in advocating the sustainable use of resources.

sustained-yield cropping in ecology, the removal of surplus individuals from a ◊population of organisms so that the population maintains a constant size. This usually requires selective removal of animals of all ages and both sexes to ensure a balanced population structure. Taking too many individuals can result in a population decline, as in overfishing.

Excessive cropping of young females may lead to fewer births in following years, and a fall in population size. Appropriate cropping frequencies can be determined from an analysis of a life table.

suture any thread or wire used in surgery to stitch together the edges of a wound or incision. Also, the stitch itself.

swab in medicine, a small piece of gauze or lint used to dry wounds. It also describes a sterile piece of cotton wool that is used to obtain samples from wounds or the throat for bacteriological examination.

swamp region of low-lying land that is permanently saturated with water and usually overgrown with vegetation; for example, the everglades of Florida, USA. A swamp often occurs where a lake has filled up with sediment and plant material. The flat surface so formed means that runoff is slow, and the water table is always close to the surface. The high humus content of swamp soil means that good agricultural soil can be obtained by draining.

S-wave (abbreviation of *secondary wave*) in seismology, a class of seismic wave that passes through the Earth in the form of transverse shear waves. S-waves from an earthquake travel at roughly half the speed of P-waves and arrive later at monitoring stations (hence secondary waves) though with greater amplitude. They can travel through solid rock but not through the liquid outer core of the Earth.

sweat gland ◊gland within the skin of mammals that produces surface perspiration. In primates, sweat glands are distributed over the whole body, but in most other mammals they are more localized; for example, in cats and dogs, they are restricted to the feet and around the face.

sweetener any chemical that gives sweetness to food. Caloric sweeteners are various forms of ◊sugar; non-caloric, or artificial, sweeteners are used by dieters and diabetics and provide neither energy nor bulk. Questions have been raised about the long-term health effects from several artificial sweeteners.

Sweeteners are used to make highly processed foods attractive, whether sweet or savoury. Most of the non-caloric sweeteners do not have E numbers. Some are banned for baby foods and for young children: thaumatin, aspartame, acesulfame-K, sorbitol, and mannitol. Cyclamate is banned in the UK and the USA; acesulfame-K is banned in the USA.

swim bladder thin-walled, air-filled sac found between the gut and the spine in bony fishes. Air enters the bladder from the gut or from surrounding capillaries (see ◊capillary), and changes of air pressure within the bladder maintain buoyancy whatever the water depth.

In evolutionary terms, the swim bladder of higher fishes is a derivative of the lungs present in all primitive fishes (not just lungfishes).

swine fever virus disease (hog cholera) of pigs, almost eradicated in the UK from 1963 by a slaughter policy.

swine flu virulent, highly contagious form of influenza, infecting pigs and communicable to people.

swine vesicular disease virus disease (porcine enterovirus) closely resembling foot and mouth disease, and communicable to humans. It may have originated in the infection of pigs by a virus that causes flu-like symptoms in people.

swing wing correctly *variable-geometry wing* aircraft wing that can be moved during flight to provide a suitable configuration for either low-speed or high-speed flight. The British engineer Barnes Wallis developed the idea of the swing wing, first used on the US-built Northrop X-4, and since used in several aircraft, including the US F-111, F-114, and the B-1, the European Tornado, and several Soviet-built aircraft.

These craft have their wings projecting nearly at right angles for takeoff and landing and low-speed flight, and swung back for high-speed flight.

syenite grey, crystalline, plutonic (intrusive) ◊igneous rock, consisting of feldspar and hornblende; other minerals may also be present, including small amounts of quartz.

symbiosis any close relationship between two organisms of different species, and one where both partners benefit from the association. A well-known example is the pollination relationship between insects and flowers, where the insects feed on nectar and carry pollen from one flower to another. This is sometimes known as ◊mutualism.

symbolic address in computing, a symbol used in ◊assembly language programming to represent the binary ◊address of a memory location.

symbolic processor computer purpose-built to run so-called symbol-manipulation programs rather than programs involving a great deal of numerical computation. They exist principally for the ◊artificial intelligence lan-

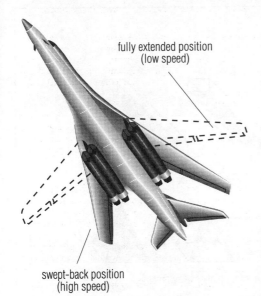

fully extended position
(low speed)

swept-back position
(high speed)

swing wing The swing-wing fighter aircraft forms a delta shape for supersonic flight. There are considerable engineering problems involved in hinging the wings, but there are several types of swing-wing craft now in use such as the American Rockwell B1 bomber illustrated here.

guage ◊LISP, although some have also been built to run ◊PROLOG.

symmetry exact likeness in shape about a given line (axis), point, or plane. A figure has symmetry if one half can be rotated and/or reflected onto the other. (Symmetry preserves length, angle, but not necessarily orientation.) In a wider sense, symmetry exits if a change in the system leaves the essential features of the system unchanged; for example, reversing the sign of electric charges does not change the electrical behaviour of an arrangement of charges. *See illustration overleaf.*

sympathetic nervous system in medicine, a division of the ◊autonomic nervous system.

symptom any change or manifestation in the body suggestive of disease as perceived by the sufferer. Symptoms are subjective phenomena. In strict usage, *symptoms* are events or changes reported by the patient; *signs* are noted by the doctor during the patient's examination.

synapse junction between two ◊nerve cells, or between a nerve cell and a muscle (a neuromuscular junction), across which a nerve impulse is transmitted. The two cells are separated by a narrow gap called the *synaptic cleft*. The gap is bridged by a chemical ◊neurotransmitter, released by the nerve impulse.

The threadlike extension, or ◊axon, of the transmitting nerve cell has a slightly swollen terminal point, the *synaptic knob*. This forms one half of the synaptic junction and houses membrane-bound vesicles, which contain a chemical neurotransmitter. When nerve impulses reach the knob, the vesicles release the transmitter and this flows across the gap and binds itself to special receptors on the receiving cell's membrane. If the receiving cell is a nerve cell, the other half of the synaptic junction will be one or more extensions called ◊dendrites; these will be stimulated by the neurotransmitter to set up an impulse, which will then be conducted along the length of the nerve cell and on to its own axons. If the receiving cell is a muscle cell, it will be stimulated by the neurotransmitter to contract.

Synapsida group of mammal-like reptiles living 315–195 million years ago, whose fossil record is largely complete, and who were for a while the dominant land animals, before being replaced by the dinosaurs. The true mammals are their descendants.

synchronous orbit another term for ◊geostationary orbit.

synchronous rotation in astronomy, another name for ◊captured rotation.

synchrotron particle ◊accelerator in which particles move, at increasing speed, around a hollow ring. The particles are guided around the ring by electromagnets, and accelerated by electric fields at points around the ring. Synchrotrons come in a wide range of sizes, the smallest being about a metre across while the largest is 27 km across. The Tevatron synchrotron at ◊Fermilab is some 6 km in circumference and accelerates protons and antiprotons to 1 TeV.

The European Synchrotron Radiation Facility (ESRF) opened in Grenoble, France, in September 1994.

syncline geological term for a fold in the rocks of the Earth's crust in which the layers or ◊beds dip inwards, thus forming a trough-like structure with a sag in the middle. The opposite structure, with the beds arching upwards, is an ◊anticline.

syncope medical term for any temporary loss of consciousness, as in ◊fainting.

syndrome in medicine, a set of signs and symptoms that always occur together, thus characterizing a particular condition or disorder.

synergy in medicine, the 'cooperative' action of two or more drugs, muscles, or organs; applied especially to drugs whose combined action is more powerful than their simple effects added together.

synodic period the time taken for a planet or moon to return to the same position in its orbit as seen from the Earth; that is, from one ◊opposition to the next. It differs from the ◊sidereal period because the Earth is moving in orbit around the Sun.

synovial fluid viscous colourless fluid that bathes movable joints between the bones of vertebrates. It nourishes and lubricates the ◊cartilage at the end of each bone.

Synovial fluid is secreted by a membrane, the synovium, that links movably jointed bones. The same kind of fluid is found in bursae, the membranous sacs that buffer some joints, such as in the shoulder and hip region.

synovitis inflammation of the membranous lining of a joint, or of a tendon sheath, caused by injury or infection.

synthesis in chemistry, the formation of a substance or compound from more elementary compounds. The synthesis of a drug can involve several stages from the initial material to the final product; the complexity of these stages is a major factor in the cost of production.

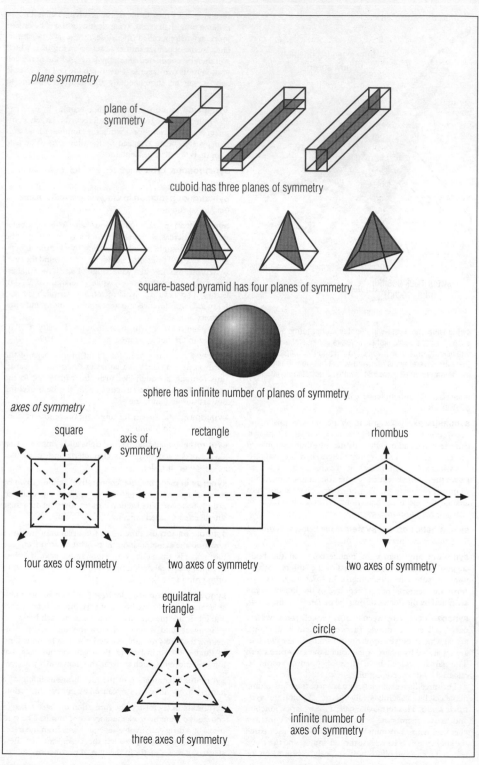

plane symmetry

plane of symmetry

cuboid has three planes of symmetry

square-based pyramid has four planes of symmetry

sphere has infinite number of planes of symmetry

axes of symmetry

square

axis of symmetry

rectangle

rhombus

four axes of symmetry

two axes of symmetry

two axes of symmetry

equilatral triangle

circle

three axes of symmetry

infinite number of axes of symmetry

symmetry

synthetic any material made from chemicals. Since the 1900s, more and more of the materials used in everyday life are synthetics, including plastics (polythene, polystyrene), ◊synthetic fibres (nylon, acrylics, polyesters), synthetic resins, and synthetic rubber. Most naturally occurring organic substances are now made synthetically, especially pharmaceuticals.

synthetic fibre fibre made by chemical processes, unknown in nature. There are two kinds. One is made from natural materials that have been chemically processed in some way; ◊rayon, for example, is made by processing the cellulose in wood pulp. The other type is the true synthetic fibre, made entirely from chemicals. ◊Nylon was the original synthetic fibre, made from chemicals obtained from petroleum (crude oil).

syphilis sexually transmitted disease caused by the spiral-shaped bacterium (spirochete) *Treponema pallidum*. Untreated, it runs its course in three stages over many years, often starting with a painless hard sore, or chancre, developing within a month on the area of infection (usually the genitals). The second stage, months later, is a rash with arthritis, hepatitis, and/or meningitis. The third stage, years later, leads eventually to paralysis, blindness, insanity, and death. The Wassermann test is a diagnostic blood test for syphilis.

With widespread availability of antibiotics, syphilis is now increasingly cured in the industrialized world, at least to the extent that the final stage of the disease is rare. The risk remains that the disease may go undiagnosed or that it may be transmitted by a pregnant woman to her fetus.

syrinx the voice-producing organ of a bird. It is situated where the trachea divides in two and consists of vibrating membranes, a reverberating capsule, and numerous controlling muscles.

Système International d'Unités official French name for ◊SI units.

systemic in medicine, relating to or affecting the body as a whole. A systemic disease is one where the effects are present throughout the body, as opposed to local disease, such as ◊conjunctivitis, which is confined to one part.

systems analysis in computing, the investigation of a business activity or clerical procedure, with a view to deciding if and how it can be computerized. The analyst discusses the existing procedures with the people involved, observes the flow of data through the business, and draws up an outline specification of the required computer system. The next step is ◊systems design.

Systems in use in the 1990s include Yourdon, SSADM (Structured Systems Analysis and Design Methodology), and Soft Systems Methodology.

systems design in computing, the detailed design of an applications package. The designer breaks the system down into component programs, and designs the required input forms, screen layouts, and printouts. Systems design forms a link between systems analysis and ◊programming.

systems program in computing, a program that performs a task related to the operation and performance of the computer system itself. For example, a systems program might control the operation of the display screen, or control and organize backing storage. In contrast, an ◊applications program is designed to carry out tasks for the benefit of the computer user.

System X in communications, a modular, computer-controlled, digital switching system used in telephone exchanges.

systole in biology, the contraction of the heart. It alternates with diastole, the resting phase of the heart beat.

systolic pressure in medicine, the measurement due to the pressure of blood against the arterial wall during systole (the contraction of the heart). It is the highest ◊blood pressure during the cardiac cycle. The average systolic blood pressure in young adults is 120 mmHg. It increases with age as the arteries get thicker and harder and it may rise to more than 200 mmHg in people with untreated hypertension (high blood pressure).

syzygy in astronomy, the alignment of three celestial bodies, usually the ◊Sun, Earth, and ◊Moon or the Sun, Earth, and another planet. A syzygy involving the Sun, Earth, and Moon usually occurs during solar and lunar ◊eclipses.

The term also refers to the Moon or another planet when it is in ◊conjunction or ◊opposition.

t symbol for ◊tonne, ◊ton.

tablet in medicine, preparation containing drugs and inert substances that is compressed to form a disc with flat or convex surfaces. Tablets enable accurate doses of drugs to be given and they are a convenient form of medication for the patient to take.

tachograph combined speedometer and clock that records a vehicle's speed (on a small card disc, magnetic disc, or tape) and the length of time the vehicle is moving or stationary. It is used to monitor a lorry driver's working hours.

taiga or *boreal forest* Russian name for the forest zone south of the ◊tundra, found across the northern hemisphere. Here, dense forests of conifers (spruces and hemlocks), birches, and poplars occupy glaciated regions punctuated with cold lakes, streams, bogs, and marshes. Winters are prolonged and very cold, but the summer is warm enough to promote dense growth.

The varied fauna and flora are in delicate balance because the conditions of life are so precarious. This ecology is threatened by mining, forestry, and pipeline construction.

talc $Mg_3Si_4O_{10}(OH)_2$, mineral, hydrous magnesium silicate. It occurs in tabular crystals, but the massive impure form, known as *steatite* or *soapstone*, is more common. It is formed by the alteration of magnesium compounds and is usually found in metamorphic rocks. Talc is very soft, ranked 1 on the Mohs' scale of hardness. It is used in powdered form in cosmetics, lubricants, and as an additive in paper manufacture.

French chalk and potstone are varieties of talc. Soapstone has a greasy feel to it, and is used for carvings such as Inuit sculptures.

Talgai skull cranium of a pre-adult male, dated from 10,000–20,000 years ago, found at Talgai station, S Queensland, Australia. It is one of the earliest human archaeological finds in Australia, having been made in 1886. Its significance was not realized, however, until the work of Edgeworth David and others after 1914. The skull is large with heavy eyebrow ridges and cheekbones.

tamponade in medicine, compression of the heart due to the accumulation of fluid within the smooth membrane that surrounds it. The condition is potentially life threatening and it is characterized by tachycardia (increase in heart rate), low blood pressure, and abnormally quiet heart sounds.

Tanegashima Space Centre Japanese rocket-launching site on a small island off S Kyushu.

Tanegashima is run by the National Space Development Agency (NASDA), responsible for the practical applications of Japan's space programme (research falls under a separate organization based at Kagoshima Space Centre). NASDA, founded in 1969, has headquarters in Tokyo; a tracking and testing station, the Tsukuba Space Centre, in E central Honshu; and an Earth observation centre near Tsukuba.

tangent in geometry, a straight line that touches a curve and gives the gradient of the curve at the point of contact. At a maximum, minimum, or point of inflection, the

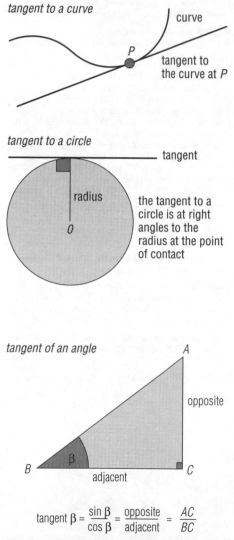

tangent to a curve

tangent to a circle

tangent of an angle

$$\text{tangent } \beta = \frac{\sin \beta}{\cos \beta} = \frac{\text{opposite}}{\text{adjacent}} = \frac{AC}{BC}$$

tangent The tangent of an angle is a mathematical function used in the study of right-angled triangles. If the tangent of an angle β is known, then the length of the opposite side can be found given the length of the adjacent side, or vice versa.

tangent to a curve has zero gradient. Also, in trigonometry, a function of an acute angle in a right-angled triangle, defined as the ratio of the length of the side opposite the angle to the length of the side adjacent to it; a way of expressing the gradient of a line.

tangram puzzle made by cutting up a square into seven pieces.

tannic acid or **tannin** $C_{14}H_{10}O_9$ yellow astringent substance, composed of several ◊phenol rings, occurring in the bark, wood, roots, fruits, and galls (growths) of certain trees, such as the oak. It precipitates gelatin to give an insoluble compound used in the manufacture of leather from hides (tanning).

tanning treating animal skins to preserve them and make them into leather. In vegetable tanning, the prepared skins are soaked in tannic acid. Chrome tanning, which is much quicker, uses solutions of chromium salts.

tantalum hard, ductile, lustrous, grey-white, metallic element, symbol Ta, atomic number 73, relative atomic mass 180.948. It occurs with niobium in tantalite and other minerals. It can be drawn into wire with a very high melting point and great tenacity, useful for lamp filaments subject to vibration. It is also used in alloys, for corrosion-resistant laboratory apparatus and chemical equipment, as a catalyst in manufacturing synthetic rubber, in tools and instruments, and in rectifiers and capacitors.

It was discovered and named in 1802 by Swedish chemist Anders Ekeberg (1767–1813) after the mythological Greek character Tantalos.

tape recording, magnetic method of recording electric signals on a layer of iron oxide, or other magnetic material, coating a thin plastic tape. The electrical signals from the microphone are fed to the electromagnetic recording head, which magnetizes the tape in accordance with the frequency and amplitude of the original signal. The impulses may be audio (for sound recording), video (for television), or data (for computer). For playback, the tape is passed over the same, or another, head to convert magnetic into electrical signals, which are then amplified for reproduction. Tapes are easily demagnetized (erased) for reuse, and come in cassette, cartridge, or reel form.

tape streamer in computing, a backing storage device consisting of a continuous loop of magnetic tape. Tape streamers are largely used to store dumps (rapid backup copies) of important data files (see ◊data security).

tapeworm any of various parasitic flatworms of the class Cestoda. They lack digestive and sense organs, can reach 15 m/50 ft in length, and attach themselves to the host's intestines by means of hooks and suckers. Tapeworms are made up of hundreds of individual segments, each of which develops into a functional hermaphroditic reproductive unit capable of producing numerous eggs. The larvae of tapeworms usually reach humans in imperfectly cooked meat or fish, causing anaemia and intestinal disorders.

taproot in botany, a single, robust, main ◊root that is derived from the embryonic root, or ◊radicle, and grows vertically downwards, often to considerable depth. Taproots are often modified for food storage and are common in biennial plants such as the carrot *Daucus carota*, where they act as ◊perennating organs.

tar dark brown or black viscous liquid obtained by the destructive distillation of coal, shale, and wood. Tars consist of a mixture of hydrocarbons, acids, and bases. ◊Creosote and ◊paraffin are produced from wood tar. See also ◊coal tar.

tartaric acid $HOOC(CHOH)_2COOH$ organic acid present in vegetable tissues and fruit juices in the form of salts of potassium, calcium, and magnesium. It is used in carbonated drinks and baking powders.

tartrazine (E102) yellow food colouring produced synthetically from petroleum. Many people are allergic to foods containing it. Typical effects are skin disorders and respiratory problems. It has been shown to have an adverse effect on hyperactive children.

taste sense that detects some of the chemical constituents of food. The human ◊tongue can distinguish only four basic tastes (sweet, sour, bitter, and salty) but it is supplemented by the sense of smell. What we refer to as taste is really a composite sense made up of both taste and smell.

tau ◊elementary particle with the same electric charge as the electron but a mass nearly double that of a proton. It has a lifetime of around 3×10^{-13} seconds and belongs to the ◊lepton family of particles – those that interact via the electromagnetic, weak nuclear, and gravitational forces, but not the strong nuclear force.

Tau Ceti one of the nearest stars visible to the naked eye, 11.9 light years from Earth in the constellation Cetus. It has a diameter slightly less than that of the Sun, and an actual luminosity of about 45% of the Sun's. Its similarity to the Sun is sufficient to suggest that Tau Ceti may possess a planetary system, although observations have yet to reveal evidence of this.

Taurus conspicuous zodiacal constellation in the northern hemisphere near ◊Orion, represented as a bull. The Sun passes through Taurus from mid-May to late June. In astrology, the dates for Taurus are between about 20 April and 20 May (see ◊precession).

The V-shaped Hyades open ◊star cluster forms the bull's head, with ◊Aldebaran as the red eye. The ◊Pleiades open cluster is in the shoulder. Taurus also contains the ◊Crab nebula, the remnants of the supernova of AD 1054, which is a strong radio and X-ray source and the location of one of the first ◊pulsars to be discovered.

tautomerism form of isomerism in which two interconvertible ◊isomers are in equilibrium. It is often specifically applied to an equilibrium between the keto (–C–C=O) and enol (–C=C–OH) forms of carbonyl compounds.

taxis (plural **taxes**) or **tactic movement** in botany, the movement of a single cell, such as a bacterium, protozoan, single-celled alga, or gamete, in response to an external stimulus. A movement directed towards the stimulus is described as positive taxis, and away from it as negative taxis. The alga *Chlamydomonas*, for example, demonstrates positive **phototaxis** by swimming towards a light source to increase the rate of photosynthesis.

Chemotaxis is a response to a chemical stimulus, as seen in many bacteria that move towards higher concentrations of nutrients.

taxonomy another name for the ◊classification of living organisms.

Tay-Sachs disease inherited disorder, due to a defective gene, causing an enzyme deficiency that leads to blindness, retardation, and death in infancy. It is most common in people of E European Jewish descent.

TB abbreviation for the infectious disease ◊ *tuberculosis*.

T cell or **T lymphocyte** immune cell (see ◊immunity and ◊lymphocyte) that plays several roles in the body's defences. T cells are so called because they mature in the ◊thymus.

There are three main types of T cells: T helper cells (Th cells), which allow other immune cells to go into action; T suppressor cells (Ts cells), which stop specific immune reactions from occurring; and T cytotoxic cells (Tc cells), which kill cells that are cancerous or infected with viruses. Like ◊B cells, to which they are related, T cells have surface receptors that make them specific for particular antigens.

tear gas any of various volatile gases that produce irritation and watering of the eyes, used by police against crowds and used in chemical warfare. The gas is delivered in pressurized, liquid-filled canisters or grenades, thrown by hand or launched from a specially adapted rifle. Gases (such as Mace) cause violent coughing and blinding tears, which pass when the victim breathes fresh air, and there are no lasting effects.

tears salty fluid exuded by lacrimal glands in the eyes. The fluid contains proteins that are antibacterial, and also absorbs oils and mucus. Apart from cleaning and disinfecting the surface of the eye, the fluid supplies nutrients to the cornea, which does not have a blood supply.

If insufficient fluid is produced, as sometimes happens in older people, the painful condition of 'dry-eye' results and the eye may be damaged.

technetium silver-grey, radioactive, metallic element, symbol Tc, atomic number 43, relative atomic mass 98.906. It occurs in nature only in extremely minute amounts, produced as a fission product from uranium in ◊pitchblende and other uranium ores. Its longest-lived isotope, Tc-99, has a half-life of 216,000 years. It is a superconductor and is used as a hardener in steel alloys and as a medical tracer.

It was synthesized in 1937 (named in 1947) by Italian physicists Carlo Perrier and Emilio Segrè, who bombarded molybdenum with deuterons, looking to fill a missing spot in the ◊periodic table of the elements (at that time it was considered not to occur in nature). It was later isolated in large amounts from the fission-product debris of uranium fuel in nuclear reactors.

technology the use of tools, power, and materials, generally for the purposes of production. Almost every human process for getting food and shelter depends on complex technological systems, which have been developed over a 3-million-year period. Significant milestones include the advent of the ◊steam engine in 1712, the introduction of ◊electricity and the ◊internal combustion engine in the mid-1870s, and recent developments in communications, ◊electronics, and the nuclear and space industries. The **advanced technology** (highly automated and specialized) on which modern industrialized society depends is frequently contrasted with the **low technology** (labour-intensive and unspecialized) that characterizes some developing countries. ◊Intermediate technology is an attempt to adapt scientifically advanced inventions to less developed areas by using local materials

and methods of manufacture. Appropriate technology refers to simple and small-scale tools and machinery of use to developing countries.

Science clears the fields on which technology can build.

On **technology** Werner Heisenberg

---

tectonics in geology, the study of the movements of rocks on the Earth's surface. On a small scale tectonics involves the formation of ◊folds and ◊faults, but on a large scale ◊plate tectonics deals with the movement of the Earth's surface as a whole.

teething the erupting of the teeth. In children, it can accompanied by irritability, loss of sleep, salivation, and failure to feed. Similar, but less severe, symptoms may occur in adults when the wisdom teeth erupt.

Teflon trade name for polytetrafluoroethylene (PTFE), a tough, waxlike, heat-resistant plastic used for coating nonstick cookware and in gaskets and bearings.

tektite small, rounded glassy stone, found in certain regions of the Earth, such as Australasia. Tektites are probably the scattered drops of molten rock thrown out by the impact of a large ◊meteorite.

telecommunications communications over a distance, generally by electronic means.

Long-distance voice communication was pioneered in 1876 by Scottish scientist Alexander Graham Bell when he invented the telephone. Today it is possible to communicate with most countries by telephone cable or by satellite or microwave link, with over 100,000 simultaneous conversations and several television channels being carried by the latest satellites.

The first mechanical telecommunications systems were the semaphore and heliograph (using flashes of sunlight), invented in the mid-19th century, but the forerunner of the present telecommunications age was the electric telegraph. The earliest practicable telegraph instrument was invented by William Cooke and Charles Wheatstone in Britain in 1837 and used by railway companies. In the USA, Samuel Morse invented a signalling code, ◊Morse code, which is still used, and a recording telegraph, first used commercially between England and France in 1851.

Following German physicist Heinrich Hertz's discoveries using electromagnetic waves, Italian inventor Guglielmo Marconi pioneered a 'wireless' telegraph, ancestor of the radio. He established wireless communication between England and France in 1899 and across the Atlantic by 1901.

The modern telegraph uses teleprinters to send coded messages along telecommunications lines. Telegraphs are keyboard-operated machines that transmit a five-unit Baudot code (see ◊baud). The receiving teleprinter automatically prints the received message.

The drawback to long-distance voice communication via microwave radio transmission is that the transmissions follow a straight line from tower to tower, so that over the sea the system becomes impracticable. A solution was put forward in 1945 by the science-fiction writer Arthur C Clarke, when he proposed a system of communications satellites in an orbit 35,900 km/22,300 mi above the Equator, where they would circle the Earth in

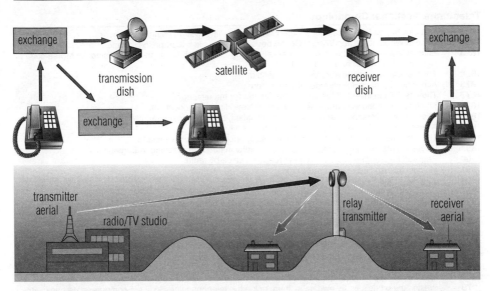

telecommunications The international telecommunications system relies on microwave and satellite links for long-distance international calls. Cable links are increasingly made of optical fibres. The capacity of these links is enormous. The TDRS-C (tracking data and relay satellite communications) satellite, the world's largest and most complex satellite, can transmit in a single second the contents of a 20-volume encyclopedia, with each volume containing 1,200 pages of 2,000 words. A bundle of optical fibres, no thicker than a finger, can carry 10,000 phone calls – more than a copper wire as thick as an arm.

exactly 24 hours, and thus appear fixed in the sky. Such a system is now in operation internationally, by Intelsat. The satellites are called geostationary satellites (syncoms). The first to be successfully launched, by Delta rocket from Cape Canaveral, was *Syncom 2* in July 1963. Many such satellites are now in use, concentrated over heavy traffic areas such as the Atlantic, Indian, and Pacific oceans. Telegraphy, telephony, and television transmissions are carried simultaneously by high-frequency radio waves. They are beamed to the satellites from large dish antennae or Earth stations, which connect with international networks.

◊Integrated-Services Digital Network (ISDN) makes videophones and high-quality fax possible; the world's first large-scale centre of ISDN began operating in Japan in 1988. ISDN is a system that transmits voice and image data on a single transmission line by changing them into digital signals. The chief method of relaying long-distance calls on land is microwave radio transmission.

Fibre-optic cables consisting of fine glass fibres present an alternative to the usual copper cables for telephone lines. The telecommunications signals are transmitted along the fibres as pulses of laser light.

Genius is one per cent inspiration and ninety-nine per cent perspiration.

On **telecommunications** Thomas Alva Edison
Life ch 24

---

telegraphy transmission of coded messages along wires by means of electrical signals. The first modern form of telecommunication, it now uses printers for the transmission and receipt of messages. Telex is an international telegraphy network.

Overland cables were developed in the 1830s, but early attempts at underwater telegraphy were largely unsuccessful until the discovery of the insulating gum gutta-percha in 1843 enabled a cable to be laid across the English Channel in 1851. **Duplex telegraph** was invented in the 1870s, enabling messages to be sent in both directions simultaneously.

telepathy 'the communication of impressions of any kind from one mind to another, independently of the recognized channels of sense', as defined by the English essayist F W H Myers (1843–1901), cofounder of the Psychical Research Society, in 1882, who coined the term. It is a form of ◊extrasensory perception.

telephone instrument for communicating by voice over long distances, developed by US inventor Alexander Graham Bell in 1876. The transmitter (mouthpiece) consists of a carbon microphone, with a diaphragm that vibrates when a person speaks into it. The diaphragm vibrations compress grains of carbon to a greater or lesser extent, altering their resistance to an electric current passing through them. This sets up variable electrical signals, which travel along the telephone lines to the receiver of the person being called. There they cause the magnetism of an electromagnet to vary, making a diaphragm above the electromagnet vibrate and give out sound waves, which mirror those that entered the mouthpiece originally.

The standard instrument has a handset, which houses the transmitter (mouthpiece), and receiver (earpiece), resting on a base, which has a dial or push-button mechanism for dialling a telephone number. Some telephones combine a push-button mechanism and mouthpiece and earpiece in one unit. A cordless telephone is of this kind, connected to a base unit not by wires but by radio. It can be used at distances up to about 100 m/330 ft from the base unit. In 1988 Japan and in 1991 Britain introduced

Telecommunications: Chronology

1794	Claude Chappe in France built a long-distance signalling system using semaphore.
1839	Charles Wheatstone and William Cooke devised an electric telegraph in England.
1843	Samuel Morse transmitted the first message along a telegraph line in the USA, using his Morse code of signals – short (dots) and long (dashes).
1858	The first transatlantic telegraph cable was laid.
1876	Alexander Graham Bell invented the telephone.
1877	Thomas Edison invented the carbon transmitter for the telephone.
1878	The first telephone exchange was opened at New Haven, Connecticut.
1884	The first long-distance telephone line was installed, between Boston and New York.
1891	A telephone cable was laid between England and France.
1892	The first automatic telephone exchange was opened, at La Porte, Indiana.
1894	Guglielmo Marconi pioneered wireless telegraphy in Italy, later moving to England.
1900	Reginald Fessenden in the USA first broadcast voice by radio.
1901	Marconi transmitted the first radio signals across the Atlantic.
1904	John Ambrose Fleming invented the thermionic valve.
1907	Charles Krumm introduced the forerunner of the teleprinter.
1920	Stations in Detroit and Pittsburgh began regular radio broadcasts.
1922	The BBC began its first radio transmissions, for the London station 2LO.
1932	The Telex was introduced in the UK.
1956	The first transatlantic telephone cable was laid.
1962	Telstar pioneered transatlantic satellite communications, transmitting live TV pictures.
1966	Charles Kao in England advanced the idea of using optical fibres for telecommunications transmissions.
1969	Live TV pictures were sent from astronauts on the Moon back to Earth.
1975	Prestel, the world's first viewdata system, using the telephone lines to link a computer data bank with the TV screen, was introduced in the UK.
1977	The first optical fibre cable was installed in California.
1984	First commercial cellphone service started in Chicago, USA.
1988	International Services Digital Network (ISDN), an international system for sending signals in digital format along optical fibres and coaxial cable, launched in Japan.
1989	The first transoceanic optical fibre cable, capable of carrying 40,000 simultaneous telephone conversations, was laid between Europe and the USA.
1991	ISDN introduced in the UK.
1992	Videophones, made possible by advances in image compression and the development of ISDN, introduced in the UK.
1993	Electronic version of the *Guardian* newspaper, for those with impaired vision, launched in the UK. The newspaper is transmitted to the user's home and printed out in braille or spoken by a speech synthesizer.
1995	The USA's main carrier of long-distance telecommunications, AT&T, processed 160 million calls daily, an increase of 50% since 1990.
1996	Researchers in Tokyo and California used lasers to communicate with an orbiting satellite, the first time lasers have been used to provide two-way communications with space.
1996	Work began on laying the world's longest fibreoptic cable, which will follow a 17,000 mi/27,300 km route around the globe from Europe, through the Suez Canal, across the Indian Ocean, and around the Pacific.
1996	Computer scientists in Japan and the USA transmitted information along optical fibre at a rate of one trillion bits. At this rate, 300 years' worth of a daily newspaper could be transmitted in one second.

an ◊Integrated Services Digital Network (see ◊telecommunications), providing fast transfer of computerized information.

> *Mr Watson, come here; I want you.*
>
> Alexander Graham Bell, first complete sentence spoken over the **telephone** March 1876

telephone tapping or *telephone bugging* listening in on a telephone conversation, without the knowledge of the participants; in the UK and the USA this is a criminal offence if done without a warrant or the consent of the person concerned.

teleprinter or *teletypewriter* transmitting and receiving device used in telecommunications to handle coded messages. Teleprinters are automatic typewriters keyed telegraphically to convert typed words into electrical signals (using a five-unit Baudot code, see ◊baud) at the transmitting end, and signals into typed words at the receiving end.

telescope optical instrument that magnifies images of faint and distant objects; any device for collecting and focusing light and other forms of electromagnetic radiation. It is a major research tool in astronomy and is used to sight over land and sea; small telescopes can be attached to cameras and rifles. A telescope with a large aperture, or opening, can distinguish finer detail and fainter objects than one with a small aperture. The *refracting telescope* uses lenses, and the *reflecting telescope* uses mirrors. A third type, the *catadioptric telescope*, is a combination of lenses and mirrors. See also ◊radio telescope.

In a refractor, light is collected by a ◊lens called the *object glass* or *objective*, which focuses light down a tube, forming an image magnified by an *eyepiece*. Invention of the refractor is attributed to a Dutch

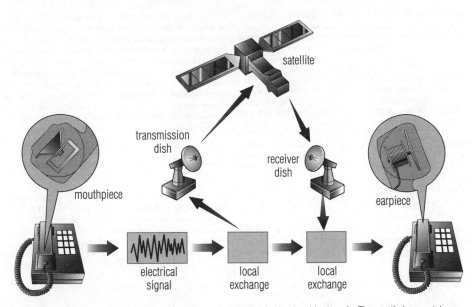

telephone In the telephone sound vibrations are converted to an electric signal and back again. The mouthpiece contains a carbon microphone that produces an electrical signal which varies in step with the spoken sounds. The signal is routed to the receiver via local or national exchanges. The earpiece contains an electromagnetic loudspeaker which reproduces the sounds by vibrating a diaphragm.

optician, Hans Lippershey in 1608. Hearing of the invention in 1609, Galileo quickly constructed one for himself and went on to produce a succession of such instruments which he used from 1610 onwards for astronomical observations. The largest refracting telescope in the world, at ◊Yerkes Observatory, Wisconsin, USA, has an aperture of 102 cm/40 in.

In a reflector, light is collected and focused by a concave mirror. The first reflector was built in about 1670 by Isaac Newton. Large mirrors are cheaper to make and easier to mount than large lenses, so all the largest telescopes are reflectors. The largest reflector with a single mirror, 6 m/236 in, is at ◊Zelenchukskaya, Russia. Telescopes with larger apertures composed of numerous smaller segments have been built, such as the ◊Keck Telescope on ◊Mauna Kea. A *multiple-mirror telescope* was installed in 1979 on Mount Hopkins, Arizona, USA. It consists of six mirrors of 1.8 m/72 in aperture, which perform like a single 4.5 m/176 in mirror. ◊*Schmidt telescopes* are used for taking wide-field photographs of the sky. They have a main mirror plus a thin lens at the front of the tube to increase the field of view.

The *liquid-mirror telescope* is a reflecting telescope constructed with a rotating mercury mirror. In 1995 NASA completed a 3 m/9.8 ft liquid mirror telescope at its Orbital Debris Observatory in New Mexico, USA.

Large telescopes can now be placed in orbit above the distorting effects of the Earth's atmosphere. Telescopes in space have been used to study infrared, ultraviolet, and X-ray radiation that does not penetrate the atmosphere but carries much information about the births, lives, and deaths of stars and galaxies. The 2.4 m/94 in ◊Hubble

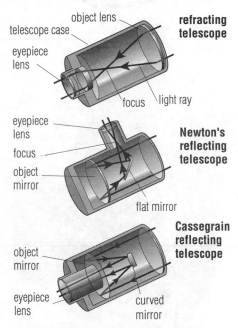

telescope Three kinds of telescope. The refracting telescope uses a large objective lens to gather light and form an image which the smaller eyepiece lens magnifies. A reflecting telescope uses a mirror to gather light. The Cassegrain telescope uses curved mirrors to direct light to a focus behind the primary lens. All modern, large astronomical telescopes are of the Cassegrain type.

Space Telescope, launched in 1990, can see the sky more clearly than can any telescope on Earth.

Telescope

Modern astronomers rarely look through their telescopes. For years most observations were recorded on photographic plates and, more recently, on electronic devices called charge-coupled devices, or as computer data to be analysed later. The computer screen provides a visual picture.

teletext broadcast system of displaying information on a television screen. The information – typically about news items, entertainment, sport, and finance – is constantly updated. Teletext is a form of ◊videotext, pioneered in Britain by the British Broadcasting Corporation (BBC) with Ceefax and by Independent Television with Teletext.

television (TV) reproduction at a distance by radio waves of visual images. For transmission, a television camera converts the pattern of light it takes in into a pattern of electrical charges. This is scanned line by line by a beam of electrons from an electron gun, resulting in variable electrical signals that represent the picture. These signals are combined with a radio carrier wave and broadcast as electromagnetic waves. The TV aerial picks up the wave and feeds it to the receiver (TV set). This separates out the vision signals, which pass to a cathode-ray tube where a beam of electrons is made to scan across the screen line by line, mirroring the action of the electron gun in the TV camera. The result is a recreation of the pattern of light that entered the camera. Twenty-five pictures are built up each second with interlaced scanning in Europe (30 in North America), with a total of 625 lines in Europe (525 lines in North America and Japan).

In addition to transmissions received by all viewers, the 1970s and 1980s saw the growth of pay-television cable networks, which are received only by subscribers, and of devices, such as those used in the Qube system (USA), which allow the viewers' opinions to be transmitted instantaneously to the studio via a response button, so that, for example, a home viewing audience can vote in a talent competition. The number of programme channels

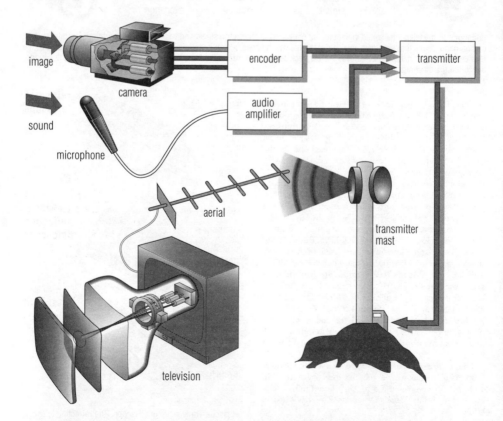

television Simplified block diagram of a complete colour television system – transmitting and receiving. The camera separates the picture into three colours – red, blue, green – by using filters and different camera tubes for each colour. The audio signal is produced separately from the video signal. Both signals are transmitted from the same aerial using a special coupling device called a diplexer. There are four sections in the receiver: the aerial, the tuners, the decoders, and the display. As in the transmitter, the audio and video signals are processed separately. The signals are amplified at various points.

continues to increase, following the introduction of satellite-beamed TV signals.

Further use of TV sets has been brought about by ◊videotext and the use of video recorders to tape programmes for playback later or to play pre-recorded video cassettes, and by their use as computer screens and for security systems. Extended-definition television gives a clear enlargement from a microscopic camera and was first used in 1989 in neurosurgery to enable medical students to watch brain operations.

In 1873 it was realized that, since the electrical properties of the nonmetallic chemical element selenium vary according to the amount of light to which it is exposed, light could be converted into electrical impulses, making it possible to transmit such impulses over a distance and then reconvert them into light. The chief difficulty was seen to be the 'splitting of the picture' so that the infinite variety of light and shade values might be transmitted and reproduced. In 1908 it was found that cathode-ray tubes would best effect transmission and reception. Mechanical devices were used at the first practical demonstration of television, given by Scottish electrical engineer John Logie Baird in London on 27 Jan 1926, and cathode-ray tubes were used experimentally in the UK from 1934. The first high-definition television service in the world began in Nov 1936 with the opening of the BBC's station at Alexandra Palace, London.

Baird was an early pioneer in this area, and one of the first techniques developed employed a system whereby the normal frame frequency was increased by a factor of 3, each successive frame containing the material for one primary colour. The receiver used revolving colour discs in front of the viewing screen, synchronized with the correct frame colours at the camera. A similar system replaced the colour discs by three superimposed projected pictures corresponding to the three primary colours. Baird demonstrated colour TV in London in 1928, but it was not until Dec 1953 that the first successful system was adopted for broadcasting, in the USA. This is called the NTSC system, since it was developed by the National Television System Committee, and variations of it have been developed in Europe; for example, SECAM (sequential and memory) in France and Eastern Europe, and PAL (phase alternation by line) in most of Western Europe. The three differ only in the way colour signals are prepared for transmission, the scanning rate, and the number of lines used. There was no agreement on a universal European system so, in 1967 the UK, West Germany, the Netherlands, and Switzerland adopted PAL while France and the USSR adopted SECAM. In 1989 the European Community (now the European Union) agreed to harmonize TV channels from 1991, allowing any station to show programmes anywhere in the EC.

The method of colour reproduction is related to that used in colour photography and printing. It uses the principle that any colours can be made by mixing the primary colours red, green, and blue in appropriate proportions. (This is different from the mixing of paints, where the primary colours are red, yellow, and blue.) In colour television the receiver reproduces only three basic colours: red, green, and blue. The effect of yellow, for example, is reproduced by combining equal amounts of red and green light, while white is formed by a mixture of all three basic colours.

Signals indicate the amounts of red, green, and blue

light to be generated at the receiver. To transmit each of these three signals in the same way as the single brightness signal in black and white television would need three times the normal band width and reduce the number of possible stations and programmes to one-third of that possible with monochrome television. The three signals are therefore coded into one complex signal, which is transmitted as a more or less normal black and white signal and produces a satisfactory – or compatible – picture on black and white receivers. A fraction of each primary red, green, and blue signal is added together to produce the normal brightness, or luminance, signal. The minimum of extra colouring information is then sent by a special subcarrier signal, which is superimposed on the brightness signal. This extra colouring information corresponds to the hue and saturation of the transmitted colour, but without any of the fine detail of the picture. The impression of sharpness is conveyed only by the brightness signal, the colouring being added as a broad colour wash. The various colour systems differ only in the way in which the colouring information is sent on the subcarrier signal. The colour receiver has to amplify the complex signal and decode it back to the basic red, green, and blue signals; these primary signals are then applied to a colour cathode-ray tube.

The colour display tube is the heart of any colour receiver. Many designs of colour picture tubes have been invented; the most successful of these is known as the 'shadow mask tube'. It operates on similar electronic principles to the black and white television picture tube, but the screen is composed of a fine mosaic of over 1 million dots arranged in an orderly fashion. One-third of the dots glow red when bombarded by electrons, one-third glow green, and one-third blue. There are three sources of electrons, respectively modulated by the red, green, and blue signals. The tube is arranged so that the shadow mask allows only the red signals to hit red dots, the green signals to hit green dots, and the blue signals to hit blue dots. The glowing dots are so small that from a normal viewing distance the colours merge into one another and a picture with a full range of colours is seen.

telex (acronym for *tel*eprinter *ex*change) international telecommunications network that handles telegraph messages in the form of coded signals. It uses ◊teleprinters for transmitting and receiving, and makes use of land lines (cables) and radio and satellite links to make connections between subscribers.

tellurium silver-white, semi-metallic (◊metalloid) element, symbol Te, atomic number 52, relative atomic mass 127.60. Chemically it is similar to sulphur and selenium, and it is considered one of the sulphur group. It occurs naturally in telluride minerals, and is used in colouring glass blue-brown, in the electrolytic refining of zinc, in electronics, and as a catalyst in refining petroleum.

It was discovered in 1782 by Austrian mineralogist Franz Müller (1740–1825), and named in 1798 by German chemist Martin Klaproth.

Telstar US communications satellite, launched on 10 July 1962, which relayed the first live television transmissions between the USA and Europe. *Telstar* orbited the Earth in 158 minutes, and so had to be tracked by ground stations, unlike the geostationary satellites of today.

Table of Equivalent Temperatures

Celsius and Fahrenheit temperatures can be interconverted as follows:

$$C = (F - 32) \times 100/180;$$
$$F = (C \times 180/100) + 32.$$

°C	°F	°C	°F	°C	°F	°C	°F
100	212.0	70	158.0	40	104.0	10	50.0
99	210.2	69	156.2	39	102.2	9	48.2
98	208.4	68	154.4	38	100.4	8	46.4
97	206.6	67	152.6	37	98.6	7	44.6
96	204.8	66	150.8	36	96.8	6	42.8
95	203.0	65	149.0	35	95.0	5	41.0
94	201.2	64	147.2	34	93.2	4	39.2
93	199.4	63	145.4	33	91.4	3	37.4
92	197.6	62	143.6	32	89.6	2	35.6
91	195.8	61	141.8	31	87.8	1	33.8
90	194.0	60	140.0	30	86.0	0	32.0
89	192.2	59	138.2	29	84.2	−1	30.2
88	190.4	58	136.4	28	82.4	−2	28.4
87	188.6	57	134.6	27	80.6	−3	26.6
86	186.8	56	132.8	26	78.8	−4	24.8
85	185.0	55	131.0	25	77.0	−5	23.0
84	183.2	54	129.2	24	75.2	−6	21.2
83	181.4	53	127.4	23	73.4	−7	19.4
82	179.6	52	125.6	22	71.6	−8	17.6
81	177.8	51	123.8	21	69.8	−9	15.8
80	176.0	50	122.0	20	68.0	−10	14.0
79	174.2	49	120.2	19	66.2	−11	12.2
78	172.4	48	118.4	18	64.4	−12	10.4
77	170.6	47	116.6	17	62.6	−13	8.6
76	168.8	46	114.8	16	60.8	−14	6.8
75	167.0	45	113.0	15	59.0	−15	5.0
74	165.2	44	111.2	14	57.2	−16	3.2
73	163.4	43	109.4	13	55.4	−17	1.4
72	161.6	42	107.6	12	53.6	−18	−0.4
71	159.8	41	105.8	11	51.8	−19	−2.2

temperature degree or intensity of heat of an object and the condition that determines whether it will transfer heat to another object or receive heat from it, according to the laws of ◊thermodynamics. The temperature of an object is a measure of the average kinetic energy possessed by the atoms or molecules of which it is composed. The SI unit of temperature is the kelvin (symbol K) used with the Kelvin scale. Other measures of temperature in common use are the Celsius scale and the Fahrenheit scale.

The normal temperature of the human body is about 36.9°C/98.4°F. Variation by more than a degree or so indicates ill health, a rise signifying excessive activity (usually due to infection), and a decrease signifying deficient heat production (usually due to lessened vitality).

temperature regulation the ability of an organism to control its internal body temperature; in warm-blooded animals this is known as ◊homeothermy.

Although some plants have evolved ways of resisting extremes of temperature, sophisticated mechanisms for maintaining the correct temperature are found in multicellular animals. Such mechanisms may be behavioural, as when a lizard moves into the shade in order to cool down, or internal, as in mammals and birds, where temperature is regulated by the ◊medulla.

tempering heat treatment for improving the properties of metals, often used for steel alloys. The metal is heated to a certain temperature and then cooled suddenly in a water or oil bath.

tendon or **sinew** in vertebrates, a cord of very strong, fibrous connective tissue that joins muscle to bone. Tendons are largely composed of bundles of fibres made of the protein collagen, and because of their inelasticity are very efficient at transforming muscle power into movement.

tendril in botany, a slender, threadlike structure that supports a climbing plant by coiling around suitable supports, such as the stems and branches of other plants. It may be a modified stem, leaf, leaflet, flower, leaf stalk, or stipule (a small appendage on either side of the leaf stalk), and may be simple or branched. The tendrils of Virginia creeper *Parthenocissus quinquefolia* are modified flower heads with suckerlike pads at the end that stick to walls, while those of the grapevine *Vitis* grow away from the light and thus enter dark crevices where they expand to anchor the plant firmly.

tension reaction force set up in a body that is subjected to stress. In a stretched string or wire it exerts a pull that is equal in magnitude but opposite in direction to the stress

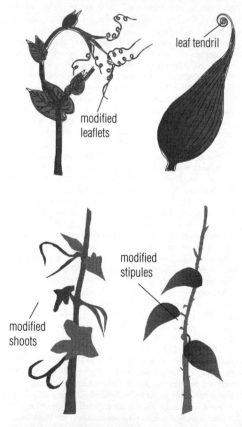

leaf tendril

modified leaflets

modified stipules

modified shoots

tendril Tendrils are specially modifed leaves, shoots, or stems. They support the plant by twining around the stems of other plants nearby as in the pea, or they may attach themselves to suitable surfaces by means of suckers as in the virginia creeper.

being applied at its ends. Tension originates in the net attractive intermolecular force created when a stress causes the mean distance separating a material's molecules to become greater than the equilibrium distance. It is measured in newtons.

teratogen any substance or agent that can induce deformities in the fetus if absorbed by the mother during pregnancy. Teratogens include some drugs (notably alcohol and ◊thalidomide), other chemicals, certain disease organisms, and radioactivity.

terbium soft, silver-grey, metallic element of the ◊lanthanide series, symbol Tb, atomic number 65, relative atomic mass 158.925. It occurs in gadolinite and other ores, with yttrium and ytterbium, and is used in lasers, semiconductors, and television tubes. It was named in 1843 by Swedish chemist Carl Mosander (1797–1858) for the town of Ytterby, Sweden, where it was first found.

terminal in computing, a device consisting of a keyboard and display screen (◊VDU) to enable the operator to communicate with the computer. The terminal may be physically attached to the computer or linked to it by a telephone line (remote terminal). A 'dumb' terminal has no processor of its own, whereas an 'intelligent' terminal has its own processor and takes some of the processing load away from the main computer.

terminal moraine linear, slightly curved ridge of rocky debris deposited at the front end, or snout, of a glacier. It represents the furthest point of advance of a glacier, being formed when deposited material (till), which was pushed ahead of the snout as it advanced, became left behind as the glacier retreated.

terminal velocity or **terminal speed** the maximum velocity that can be reached by a given object moving through a fluid (gas or liquid). As the speed of the object increases so does the total magnitude of the forces resisting its motion. Terminal velocity is reached when the resistive forces exactly balance the applied force that has caused the object to accelerate; because there is now no resultant force, there can be no further acceleration.

For example, an object falling through air will reach a terminal velocity and cease to accelerate under the influence of gravity when the air resistance equals the object's weight.

Parachutes are designed to increase air resistance so that the acceleration of a falling person or package ceases more rapidly, thereby limiting terminal velocity to a safe level.

terminal voltage the potential difference (pd) or voltage across the terminals of a power supply, such as a battery of cells. When the supply is not connected in circuit its terminal voltage is the same as its ◊electromotive force (emf); however, as soon as it begins to supply current to a circuit its terminal voltage falls because some electric potential energy is lost in driving current against the supply's own internal resistance. As the current flowing in the circuit is increased the terminal voltage of the supply falls.

terrane in geology, a tract of land with a distinct geological character. The term **exotic terrane** is commonly used to describe a rock mass that has a very different history from others near by. The exotic terranes of the Western ◊Cordillera of North America represent old island chains that have been brought to the North American continent by the movements of plate tectonics, and welded to its edge.

terrestrial planet in astronomy, any of the four small, rocky inner ◊planets of the Solar System: ◊Mercury, ◊Venus, ◊Earth, and ◊Mars. The ◊Moon is sometimes also included, although it is a satellite of the Earth and not strictly a planet.

territorial behaviour in biology, any behaviour that serves to exclude other members of the same species from a fixed area or ◊territory. It may involve aggressively driving out intruders, marking the boundary (with dung piles or secretions from special scent glands), conspicuous visual displays, characteristic songs, or loud calls.

territory in animal behaviour, a fixed area from which an animal or group of animals excludes other members of the same species. Animals may hold territories for many different reasons; for example, to provide a constant food supply, to monopolize potential mates, or to ensure access to refuges or nest sites.

The size of a territory depends in part on its function: some nesting and mating territories may be only a few square metres, whereas feeding territories may be as large as hundreds of square kilometres.

Tertiary period of geological time 65–1.64 million years ago, divided into five epochs: Palaeocene, Eocene, Oligocene, Miocene, and Pliocene. During the Tertiary, mammals took over all the ecological niches left vacant by the extinction of the dinosaurs, and became the prevalent land animals. The continents took on their present positions, and climatic and vegetation zones as we know them became established. Within the geological time column the Tertiary follows the Cretaceous period and is succeeded by the Quaternary period.

Terylene trade name for a polyester synthetic fibre produced by the chemicals company ICI. It is made by polymerizing ethylene glycol and terephthalic acid. Cloth made from Terylene keeps its shape after washing and is hard-wearing.

Terylene was the first wholly synthetic fibre invented in Britain. It was created by the chemist J R Whinfield of Accrington in 1941. In 1942 the rights were sold to ICI (Du Pont in the USA) and bulk production began in 1955. Since 1970 it has been the most widely produced synthetic fibre, often under the generic name polyester.

tesla SI unit (symbol T) of ◊magnetic flux density. One tesla represents a flux density of one ◊weber per square metre, or 10^4 ◊gauss. It is named after the Croatian engineer Nikola Tesla.

testa the outer coat of a seed, formed after fertilization of the ovule. It has a protective function and is usually hard and dry. In some cases the coat is adapted to aid dispersal, for example by being hairy. Humans have found uses for many types of testa, including the fibre of the cotton seed.

test cross in genetics, a breeding experiment used to discover the genotype of an individual organism. By crossing with a double recessive of the same species, the offspring will indicate whether the test individual is homozygous or heterozygous for the characteristic in question. In peas, a tall plant under investigation would be crossed with a double recessive short plant with known genotype tt. The results of the cross will be all tall plants if the test plant is TT. If the individual is in fact Tt then there will be some short plants (genotype tt) among the offspring.

test data data designed to test whether a new computer program is functioning correctly. The test data are carefully chosen to ensure that all possible branches of the program are tested. The expected results of running the data are written down and are then compared with the actual results obtained using the program.

testis (plural *testes*) the organ that produces ◊sperm in male (and hermaphrodite) animals. In vertebrates it is one of a pair of oval structures that are usually internal, but in mammals (other than elephants and marine mammals), the paired testes (or testicles) descend from the body cavity during development, to hang outside the abdomen in a scrotal sac. The testes also secrete the male sex hormone ◊androgen.

Testes

The testes of the tiny male Australian fairy wren account for one-tenth of his body weight. He can produce more sperm than any other bird, and ejaculates 8.3 billion sperm at a time. (A human male manages about 4 million.)

testosterone in vertebrates, hormone secreted chiefly by the testes, but also by the ovaries and the cortex of the adrenal glands. It promotes the development of secondary sexual characteristics in males. In animals with a breeding season, the onset of breeding behaviour is accompanied by a rise in the level of testosterone in the blood.

Synthetic or animal testosterone is used to treat inadequate development of male characteristics or (illegally) to aid athletes' muscular development. Like other sex hormones, testosterone is a ◊steroid.

tetanus or *lockjaw* acute disease caused by the toxin of the bacillus *Clostridium tetani*, which usually enters the body through a wound. The bacterium is chiefly found in richly manured soil. Untreated, in seven to ten days tetanus produces muscular spasm and rigidity of the jaw spreading to other parts of the body, convulsions, and death. There is a vaccine, and the disease may be treatable with tetanus antitoxin and antibiotics.

tetany in medicine, muscle spasm caused by reduction in the concentration of calcium circulating in the blood. In the absence of calcium, the muscles become hyperexcitable and will go into spasm on the slightest stimulus. It is treated with calcium salts and vitamin D.

Tethys Sea sea that once separated ◊Laurasia from ◊Gondwanaland. It has now closed up to become the Mediterranean, the Black, the Caspian, and the Aral seas.

tetrachloromethane CCl₄ or *carbon tetrachloride* chlorinated organic compound that is a very efficient solvent for fats and greases, and was at one time the main constituent of household dry-cleaning fluids and of fire extinguishers used with electrical and petrol fires. Its use became restricted after it was discovered to be carcinogenic and it has now been largely removed from educational and industrial laboratories.

tetracycline one of a group of antibiotic compounds having in common the four-ring structure of chlortetracycline, the first member of the group to be isolated. They are prepared synthetically or obtained from certain bacteria of the genus *Streptomyces*. They are broad-spectrum antibiotics, effective against a wide range of disease-causing bacteria.

tetraethyl lead Pb(C₂H₅)₄ compound added to leaded petrol as a component of ◊anti-knock to increase the efficiency of combustion in car engines. It is a colourless liquid that is insoluble in water but soluble in organic solvents such as benzene, ethanol, and petrol.

tetrahedron (plural *tetrahedra*) in geometry, a solid figure (◊polyhedron) with four triangular faces; that is, a ◊pyramid on a triangular base. A regular tetrahedron has equilateral triangles as its faces.

In chemistry and crystallography, tetrahedra describe the shapes of some molecules and crystals; for example, the carbon atoms in a crystal of diamond are arranged in space as a set of interconnected regular tetrahedra.

tetrapod type of ◊vertebrate. The group includes mammals, birds, reptiles, and amphibians. Birds are included because they evolved from four-legged ancestors, the forelimbs having become modified to form wings. Even snakes are tetrapods, because they are descended from four-legged reptiles.

text editor in computing, a program that allows the user to edit text on the screen and to store it in a file. Text editors are similar to ◊word processors, except that they lack the ability to format text into paragraphs and pages and to apply different typefaces and styles.

textured vegetable protein manufactured meat substitute; see ◊TVP.

thalassaemia or *Cooley's anaemia* any of a group of chronic hereditary blood disorders that are widespread in the Mediterranean countries, Africa, the Far East, and the Middle East. They are characterized by an abnormality of the red blood cells and bone marrow, with enlargement of the spleen. The genes responsible are carried by about 100 million people worldwide. The diseases can be diagnosed prenatally.

thalidomide ◊hypnotic drug developed in the 1950s for use as a sedative. When taken in early pregnancy, it caused malformation of the fetus (such as abnormalities

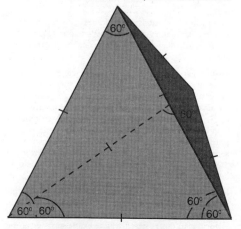

tetrahedron A regular tetrahedron is a pyramid on a triangular base with all its sides equal in length.

in the limbs) in over 5,000 recognized cases, and the drug was withdrawn.

In 1995 US researchers announced trials using thalidomide derivatives for a variety of immune disorders, including AIDS, cancer, and graft-versus-host disease. It is already used routinely in the treatment of leprosy.

thallium soft, bluish-white, malleable, metallic element, symbol Tl, atomic number 81, relative atomic mass 204.38. It is a poor conductor of electricity. Its compounds are poisonous and are used as insecticides and rodent poisons; some are used in the optical-glass and infrared-glass industries and in photocells.

Discovered spectroscopically in 1861 by its green line, thallium was isolated and named by William Crookes.

thallus any plant body that is not divided into true leaves, stems, and roots. It is often thin and flattened, as in the body of a seaweed, lichen, or liverwort, and the gametophyte generation (◊prothallus) of a fern.

Some flowering plants (◊angiosperms) that are adapted to an aquatic way of life may have a very simple plant body which is described as a thallus (for example, duckweed *Lemna*).

thaumatin naturally occurring, non-carbohydrate sweetener derived from the bacterium *Thaumatococcus danielli*. Its sweetness is not sensed as quickly as that of other sweeteners, and it is not as widely used in the food industry.

theodolite instrument for the measurement of horizontal and vertical angles, used in surveying. It consists of a small telescope mounted so as to move on two graduated circles, one horizontal and the other vertical, while its axes pass through the centre of the circles. See also ◊triangulation.

theorem mathematical proposition that can be deduced by logic from a set of axioms (basic facts that are taken to be true without proof). Advanced mathematics consists almost entirely of theorems and proofs, but even at a simple level theorems are important.

theory in science, a set of ideas, concepts, principles, or methods used to explain a wide set of observed facts. Among the major theories of science are ◊relativity, ◊quantum theory, ◊evolution, and ◊plate tectonics.

Science is built up with facts, as a house is with stones. But a collection of facts is no more a science than a heap of stones is a house.

On **theory** Jules Henri Poincaré
Science and Hypothesis 1905

---

therm unit of energy defined as 10^5 British thermal units; equivalent to 1.055×10^8 joules. It is no longer in scientific use.

thermal capacity another name for ◊heat capacity

thermal conductivity in physics, the ability of a substance to conduct heat. Good thermal conductors, like good electrical conductors, are generally materials with many free electrons (such as metals).

Thermal conductivity is expressed in units of joules per second per metre per kelvin ($J\,s^{-1}\,m^{-1}\,K^{-1}$). For a block of material of cross-sectional area a and length l, with temperatures T_1 and T_2 at its end faces, the thermal conductivity λ equals $Hl/at(T_2 - T_1)$, where H is the amount of heat transferred in time t.

thermal dissociation reversible breakdown of a compound into simpler substances by heating it (see ◊dissociation). The splitting of ammonium chloride into ammonia and hydrogen chloride is an example. On cooling, they recombine to form the salt.

thermal expansion in physics, expansion that is due to a rise in temperature. It can be expressed in terms of linear, area, or volume expansion.

The coefficient of linear expansion α is the fractional increase in length per degree temperature rise; area, or superficial, expansion β is the fractional increase in area per degree; and volume, or cubic, expansion γ is the fractional increase in volume per degree. To a close approximation, $\beta = 2\alpha$ and $\gamma = 3\alpha$.

thermal prospection an expensive remote-sensing method used in aerial reconnaissance, based on weak variations in temperature which can be found above buried structures whose thermal properties are different from those of their surroundings.

thermal reactor nuclear reactor in which the neutrons released by fission of uranium-235 nuclei are slowed down in order to increase their chances of being captured by other uranium-235 nuclei, and so induce further fission. The material (commonly graphite or heavy water) responsible for doing so is called a **moderator**. When the fast newly-emitted neutrons collide with the nuclei of the moderator's atoms, some of their kinetic energy is lost and their speed is reduced. Those that have been slowed down to a speed that matches the thermal (heat) energy of the surrounding material are called **thermal neutrons**, and it is these that are most likely to induce fission and ensure the continuation of the chain reaction. See ◊nuclear reactor and ◊nuclear energy.

thermic lance cutting tool consisting of a tube of mild steel, enclosing tightly packed small steel rods and fed with oxygen. On ignition temperatures above 3,000°C/5,400°F are produced and the thermic lance becomes its own sustaining fuel. It rapidly penetrates walls and a 23 cm/9 in steel door can be cut through in less than 30 seconds.

thermionics branch of electronics dealing with the emission of electrons from matter under the influence of heat.

thermistor semiconductor device whose electrical ◊resistance falls as temperature rises. The current passing through a thermistor increases rapidly as its temperature rises, and so they are used in electrical thermometers.

thermite process method used in incendiary devices and welding operations. It uses a powdered mixture of aluminium and (usually) iron oxide, which, when ignited, gives out enormous heat. The oxide is reduced to iron, which is molten at the high temperatures produced. This can be used to make a weld. The process was discovered in 1895 by German chemist Hans Goldschmidt (1861–1923).

thermocouple electric temperature-measuring device consisting of a circuit having two wires made of different metals welded together at their ends. A current flows in the circuit when the two junctions are maintained at different temperatures (◊Seebeck effect). The electromotive force generated – measured by a millivoltmeter – is proportional to the temperature difference.

thermodynamics branch of physics dealing with the transformation of heat into and from other forms of energy. It is the basis of the study of the efficient working of engines, such as the steam and internal-combustion engines. The three laws of thermodynamics are (1) energy can be neither created nor destroyed, heat and mechanical work being mutually convertible; (2) it is impossible for an unaided self-acting machine to convey heat from one body to another at a higher temperature; and (3) it is impossible by any procedure, no matter how idealized, to reduce any system to the ◊absolute zero of temperature (0 K/−273°C/−459°F) in a finite number of operations. Put into mathematical form, these laws have widespread applications in physics and chemistry.

thermography photographic recording of heat patterns. It is used medically as an imaging technique to identify 'hot spots' in the body – for example, tumours, where cells are more active than usual. Thermography was developed in the 1970s and 1980s by the military to assist night vision by detecting the body heat of an enemy or the hot engine of a tank. It uses a photographic method (using infrared radiation) employing infrared-sensitive films.

thermoluminescence (TL) release, in the form of a light pulse, of stored nuclear energy (electrons) in a mineral substance when heated to 500°C by ◊irradiation. The energy results from the radioactive decay of uranium and thorium, which is absorbed by crystalline inclusions within the mineral matrix, such as quartz and feldspar. The release of TL from these crystalline substances is used in archaeology to date pottery, and by geologists in studying terrestrial rocks and meteorites.

Crystalline substances find their way into the clay fabric of ancient pottery as additives designed to strengthen the material and allow it to breathe during kiln-firing at 600°C and above. Firing erases the huge level of TL energy accrued in geological times and sets a 'time-zero' for fresh energy accumulation over archaeological times, the TL intensity measured today being proportional to the pottery's age. TL can date inorganic materials, including stone tools left as burnt flint, older than about 50,000–80,000 years, although it is regarded as less precise in its accuracy.

thermometer instrument for measuring temperature. There are many types, designed to measure different temperature ranges to varying degrees of accuracy. Each makes use of a different physical effect of temperature. Expansion of a liquid is employed in common **liquid-in-glass thermometers**, such as those containing mercury or alcohol. The more accurate **gas thermometer** uses the effect of temperature on the pressure of a gas held at constant volume. A **resistance thermometer** takes advantage of the change in resistance of a conductor (such as a platinum wire) with variation in temperature. Another electrical thermometer is the ◊**thermocouple**. Mechanically, temperature change can be indicated by the change in curvature of a **bimetallic strip** (as commonly used in a ◊thermostat).

thermopile instrument for measuring radiant heat, consisting of a number of ◊thermocouples connected in series with alternate junctions exposed to the radiation. The current generated (measured by an ◊ammeter) is proportional to the radiation falling on the device.

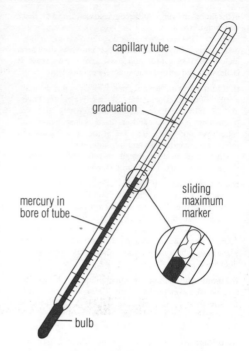

capillary tube

graduation

mercury in bore of tube

sliding maximum marker

bulb

thermometer Maximum and minimum thermometers are universally used in weather-reporting stations. The maximum thermometer, shown here, includes a magnet that fits tightly inside a capillary tube and is moved up it by the rising mercury. When the temperature falls, the magnet remains in position, thus enabling the maximum temperature to be recorded.

thermoplastic or **thermosoftening plastic** type of ◊plastic that always softens on repeated heating. Thermoplastics include polyethylene (polyethene), polystyrene, nylon, and polyester.

thermoset type of ◊plastic that remains rigid when set, and does not soften with heating. Thermosets have this property because the long-chain polymer molecules cross-link with each other to give a rigid structure. Examples include Bakelite, resins, melamine, and urea-formaldehyde resins.

thermosphere layer in the Earth's ◊atmosphere above the mesosphere and below the exosphere. Its lower level is about 80 km/50 mi above the ground, but its upper level is undefined. The ionosphere is located in the thermosphere. In the thermosphere the temperature rises with increasing height to several thousand degrees Celsius. However, because of the thinness of the air, very little heat is actually present.

thermostat temperature-controlling device that makes use of feedback. It employs a temperature sensor (often a bimetallic strip) to operate a switch or valve to control electricity or fuel supply. Thermostats are used in central heating, ovens, and car engines.

At the required preset temperature (for example of a room or gas oven), the movement of the sensor switches off the supply of electricity to the room heater or gas to the

Monomer	Polymer	Name	Uses
$CH_2=CH_2$ ethene	$[CH_2-CH_2]_n$	poly(ethene), polythene	bottles, packaging, insulation, pipes

$CH_2=CH-CH_3$ propene	$[CH_2-CH]_n$
	     $\mid$
	     CH_3
		poly(propene), polypropylene	mouldings, film, fibres

$CH_2=CH-CL$ chloroethene (vinyl chloride)	$[CH_2-CH]_n$
	     $\mid$
	     Cl
		polyvinylchloride (PVC), poly(chloroethene)	insulation, flooring, household fabric

$CH_2=CH-C_6H_5$ phenylethene (styrene)	$[CH_2-CH]_n$
	     $\mid$
	     C_6H_5
		polystyrene, poly(phenylethene)	insulation, packaging

$CF_2=CF_2$ tetrafluoroethene	$[CF_2-CF_2]_n$	poly(tetrafluoroethylene) (PTFE)	high resistance to chemical and electrical reaction, low-friction applications
	$(n = 1000+)$		

thermoplastic

oven. As the room or oven cools down, the sensor turns back on the supply of electricity or gas.

thiamine or ***vitamin B₁*** a water-soluble vitamin of the B complex. It is found in seeds and grain. Its absence from the diet causes the disease ◊beriberi.

35 mm width of photographic film, the most popular format for the camera today. The 35 mm camera falls into two categories, the ◊SLR and the ◊rangefinder.

thorax in four-limbed vertebrates, the part of the body containing the heart and lungs, and protected by the ribcage; in arthropods, the middle part of the body, between the head and abdomen.

In mammals the thorax is separated from the abdomen by the muscular diaphragm. In insects the thorax bears the legs and wings. The thorax of spiders and crustaceans, such as lobsters, is fused with the head, to form the cephalothorax.

thorium dark-grey, radioactive, metallic element of the ◊actinide series, symbol Th, atomic number 90, relative atomic mass 232.038. It occurs throughout the world in small quantities in minerals such as thorite and is widely distributed in monazite beach sands. It is one of three fissile elements (the others are uranium and plutonium), and its longest-lived isotope has a half-life of 1.39×10^{10}

years. Thorium is used to strengthen alloys. It was discovered by Jöns Berzelius in 1828 and was named by him after the Norse god Thor.

threadworm kind of ◊nematode.

Three Age System the division of prehistory into the ◊Stone Age, ◊Bronze Age, and ◊Iron Age, first proposed by the Danish archaeologist Christian Thomsen between 1816–19. Subsequently, the Stone Age was subdivided into the Old (Palaeolithic) and the New (Neolithic). The Middle (Mesolithic) Stone Age was added even later, and the Copper (Chalcolithic) Age inserted between the New Stone Age and the Bronze Age.

The system was first published as a classification of Danish antiquities in 1836. Thomsen's pupil Jens Worsaae pioneered the adoption of the system in Europe and added further subdivisions within each age. Although a valuable and valid classification system for prehistoric material, the Three Age System did not provide dates but a relative sequence of developmental stages, which were not necessarily followed in that order by different societies.

threshing agricultural process of separating cereal grains from the plant. Traditionally, the work was carried out by hand in winter months using the flail, a jointed beating

Monomer I	Monomer II	Polymer name	Uses
formaldehyde (methanal)	phenol	PF resins (Bakelites)	electrical fittings, radio cabinets
formaldehyde	urea	UF resins	electrical fittings, insulation, adhesives
formaldehyde	melamine	melamines	laminates for furniture

thermoset

stick. Today, threshing is done automatically inside the combine harvester at the time of cutting.

From the late 18th century, through the work of Andrew Meikle and others, machine threshing slowly overtook the flail and made rapid progress after 1850.

throat in human anatomy, the passage that leads from the back of the nose and mouth to the ◊trachea and ◊oesophagus. It includes the ◊pharynx and the ◊larynx, the latter being at the top of the trachea. The word 'throat' is also used to mean the front part of the neck, both in humans and other vertebrates; for example, in describing the plumage of birds. In engineering, it is any narrowing entry, such as the throat of a carburettor.

thrombocyte in medicine, another name for a ◊platelet.

thromboembolism in biology, the formation of a blood clot in one part of the cardiovascular system from which it becomes detached and then lodges in another part of the cardiovascular system. The most common form of thromboembolism involves the formation of a clot in the veins of the legs that migrates to the lungs. This condition requires urgent treatment with anticoagulants and/or surgery.

thrombolysis in medicine, the breakdown of blood clots due to the activity of enzymes. Naturally occurring enzymes limit the size of the clot. Enzymes, such as streptokinase, may be given to dissolve clots following thromboembolism.

thrombosis condition in which a blood clot forms in a vein or artery, causing loss of circulation to the area served by the vessel. If it breaks away, it often travels to the lungs, causing pulmonary embolism.

Thrombosis increases the risk of heart attack and stroke. It is treated by surgery and/or anticoagulant drugs.

thrush infection usually of the mouth (particularly in infants), but also sometimes of the vagina, caused by a yeastlike fungus (◊*Candida*). It is seen as white patches on the mucous membranes.

Thrush, also known as **candidiasis**, may be caused by antibiotics removing natural antifungal agents from the body. It is treated with a further antibiotic.

Thrust 2 jet-propelled car in which British driver Richard Noble set a new world land-speed record in the Black Rock Desert of Nevada, USA, 4 Oct 1983. The record speed was 1,019.4 kph/633.468 mph. In 1996 Noble attempted to break the sound barrier in *Thrust SCC*. *Thrust SCC* has two Rolls-Royce Spey engines (the same kind used in RAF Phantom jets) that provide 110,000 horsepower; it weighs 6,350 kg/13,970 lb, and is 16.5 m/54 ft in length. It was driven by RAF fighter pilot Andy Green to break the sound barrier in Sept 1997, setting a speed of 1,149.272 kph/714.144 mph.

thulium soft, silver-white, malleable and ductile, metallic element, of the ◊lanthanide series, symbol Tm, atomic number 69, relative atomic mass 168.94. It is the least abundant of the rare-earth metals, and was first found in gadolinite and various other minerals. It is used in arc lighting.

The X-ray-emitting isotope Tm-170 is used in portable X-ray units. Thulium was named by French chemist Paul Lecoq de Boisbaudran in 1886 after the northland, Thule.

thunderstorm severe storm of very heavy rain, thunder,

and lightning. Thunderstorms are usually caused by the intense heating of the ground surface during summer. The warm air rises rapidly to form tall cumulonimbus clouds with a characteristic anvil-shaped top. Electrical charges accumulate in the clouds and are discharged to the ground as flashes of lightning. Air in the path of lightning becomes heated and expands rapidly, creating shock waves that are heard as a crash or rumble of thunder.

The rough distance between an observer and a lightning flash can be calculated by timing the number of seconds between the flash and the thunder. A gap of 3 seconds represents about a kilometre; 5 seconds represents about a mile.

thymus organ in vertebrates, situated in the upper chest cavity in humans. The thymus processes ◊lymphocyte cells to produce T-lymphocytes (T denotes 'thymus-derived'), which are responsible for binding to specific invading organisms and killing them or rendering them harmless.

The thymus reaches full size at puberty, and shrinks thereafter; the stock of T-lymphocytes is built up early in life, so this function diminishes in adults, but the thymus continues to function as an ◊endocrine gland, producing the hormone thymosin, which stimulates the activity of the T-lymphocytes.

thyristor type of ◊rectifier, an electronic device that conducts electricity in one direction only. The thyristor is composed of layers of ◊semiconductor material sandwiched between two electrodes called the anode and cathode. The current can be switched on by using a third electrode called the gate.

Thyristors are used to control mains-driven motors and in lighting dimmer controls.

thyroid ◊endocrine gland of vertebrates, situated in the neck in front of the trachea. It secretes several hormones, principally thyroxine, an iodine-containing hormone that stimulates growth, metabolism, and other functions of the body. The thyroid gland may be thought of as the regulator gland of the body's metabolic rate. If it is overactive, as in ◊hyperthyroidism, the sufferer feels hot and sweaty, has an increased heart rate, diarrhoea, and weight loss. Conversely, an underactive thyroid leads to **myxoedema**, a condition characterized by sensitivity to the cold, constipation, and weight gain. In infants, an underactive thyroid leads to **cretinism**, a form of mental retardation.

thyrotoxicosis synonym for ◊hyperthyroidism.

thyroxine in medicine, a hormone containing iodine that is produced by the thyroid gland. It is used to treat conditions that are due to deficiencies in thyroid function, such as myxoedema.

tibia the anterior of the pair of bones in the leg between the ankle and the knee. In humans, the tibia is the shinbone. It articulates with the ◊femur above to form the knee joint, the ◊fibula externally at its upper and lower ends, and with the talus below, forming the ankle joint.

tidal energy energy derived from the tides. The tides mainly gain their potential energy from the gravitational forces acting between the Earth and the Moon. If water is trapped at a high level during high tide, perhaps by means of a barrage across an estuary, it may then be gradually released and its associated gravitational potential energy exploited to drive turbines and generate electricity. Several schemes have been proposed for the Bristol Channel, in SW England, but environmental concerns as well as construction costs have so far prevented any decision from being taken.

tidal heating in astrophysics, a process in which one body is heated internally by tidal stresses set up by the gravitational pull of another body. Tidal heating is common among the moons of the giant planets, and is the heat source for volcanic activity on ◊Io, one of the moons of ◊Jupiter.

tidal power station ◊hydroelectric power plant that uses the 'head' of water created by the rise and fall of the ocean tides to spin the water turbines. The world's only large tidal power station is located on the estuary of the river Rance in the Gulf of St Malo, Brittany, France, which has been in use since 1966. It produces 240 megawatts and can generate electricity on both the ebb and flow of the tide.

tidal wave common name for a ◊tsunami.

Tidbinbilla space tracking station and nature reserve in Australia, just S of Canberra. It provides tracking facilities and command transmissions in support of NASA manned and unmanned spacecraft including the ◊Voyager probes.

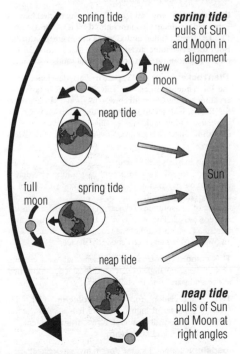

spring tide

spring tide pulls of Sun and Moon in alignment

new moon

neap tide

full moon

spring tide

Sun

neap tide

neap tide pulls of Sun and Moon at right angles

tide The gravitational pull of the Moon is the main cause of the tides. Water on the side of the Earth nearest the Moon feels the Moon's pull and accumulates directly under the Moon. When the Sun and the Moon are in line, at new and full moon, the gravitational pull of Sun and Moon are in line and produce a high spring tide. When the Sun and Moon are at right angles, lower neap tides occur.

tide the rhythmic rise and fall of sea level in Earth's oceans and their inlets and estuaries due to the gravitational attraction of the Moon and, to a lesser extent, the Sun, affecting regions of Earth unequally as it rotates. Water on the side of Earth nearest the Moon feels the Moon's pull and accumulates directly below it producing high tide.

High tide occurs at an interval of 12 hr 24 min 30 sec. The maximum high tides, or spring tides, occur at or near new and full Moon when the Moon and Sun are in line and exert the greatest combined gravitational pull. Lower high tides, or neap tides, occur when the Moon is in its first or third quarter and the Moon and Sun are at right angles to each other.

till or **boulder clay** deposit of clay, mud, gravel, and boulders left by a ◊glacier. It is unsorted, with all sizes of fragments mixed up together, and shows no stratification; that is, it does not form clear layers or ◊beds.

tilt-rotor aircraft type of vertical takeoff aircraft, also called a ◊convertiplane.

timber wood used in construction, furniture, and paper pulp. **Hardwoods** include tropical mahogany, teak, ebony, rosewood, temperate oak, elm, beech, and eucalyptus. All except eucalyptus are slow-growing, and world supplies are almost exhausted. **Softwoods** comprise the conifers (pine, fir, spruce, and larch), which are quick to grow and easy to work but inferior in quality of grain. **White woods** include ash, birch, and sycamore; all have light-coloured timber, are fast-growing, and can be used as veneers on cheaper timber.

time continuous passage of existence, recorded by division into hours, minutes, and seconds. Formerly the measurement of time was based on the Earth's rotation on its axis, but this was found to be irregular. Therefore the second, the standard ◊SI unit of time, was redefined in 1956 in terms of the Earth's annual orbit of the Sun, and in 1967 in terms of a radiation pattern of the element caesium.

time-sharing in computing, a way of enabling several users to access the same computer at the same time. The computer rapidly switches between user ◊terminals and programs, allowing each user to work as if he or she had sole use of the system.

Time-sharing was common in the 1960s and 1970s before the spread of cheaper computers, and looks set to make a comeback with the advent of **network computing**.

tin soft, silver-white, malleable and somewhat ductile, metallic element, symbol Sn, atomic number 50, relative atomic mass 118.69. Tin exhibits ◊allotropy, having three forms: the familiar lustrous metallic form above 13.2°C/55.8°F; a brittle form above 161°C/321.8°F; and a grey powder form below 13.2°C/55.8°F (commonly called tin pest or tin disease). The metal is quite soft (slightly harder than lead) and can be rolled, pressed, or hammered into extremely thin sheets; it has a low melting point. In nature it occurs rarely as a free metal. It resists corrosion and is therefore used for coating and plating other metals.

Tin and copper smelted together form the oldest desired alloy, bronze; since the Bronze Age (3500 BC) that alloy has been the basis of both useful and decorative materials. Tin is also alloyed with metals other than copper to make solder and pewter. It was recognized as an element by Antoine Lavoisier, but the name is very old and comes from the Germanic form *zinn*. The mines of Cornwall were the principal western source of tin until the 19th century, when rich deposits were found in South America, Africa, SE Asia, and Australia. Tin production is concentrated in Malaysia, Indonesia, Brazil, and Bolivia.

Tinea in medicine, species of fungus that causes ◊ringworm.

tinnitus in medicine, constant buzzing or ringing in the ears. The phenomenon may originate from prolonged exposure to noisy conditions (drilling, machinery, or loud music) or from damage to or disease of the middle or inner ear. The victim may become overwhelmed by the relentless noise in the head.

In some cases there is a hum at a frequency of about 40 Hz, which resembles that heard by people troubled by environmental ◊hum but may include whistles and other noises resembling a machine workshop. Being in a place where external noises drown the internal ones gives some relief, and devices may be worn that create pleasant, soothing sounds to override them.

Objective tinnitus is a very rare form in which other people can also hear the noises. This may be caused by muscle spasms in the inner ear or throat, or abnormal resonance of the eardrum and ossicles.

tin ore mineral from which tin is extracted, principally cassiterite, SnO_2. The world's chief producers are Malaysia, Thailand, and Bolivia.

tinplate milled steel coated with tin, the metal used for most 'tin' cans. The steel provides the strength, and the tin provides the corrosion resistance, ensuring that the food inside is not contaminated. Tinplate may be made by ◊electroplating or by dipping in a bath of molten tin.

tissue in biology, any kind of cellular fabric that occurs in an organism's body. Several kinds of tissue can usually be distinguished, each consisting of cells of a particular kind bound together by cell walls (in plants) or extracellular matrix (in animals). Thus, nerve and muscle are different kinds of tissue in animals, as are ◊parenchyma and ◊sclerenchyma in plants.

tissue culture process by which cells from a plant or animal are removed from the organism and grown under controlled conditions in a sterile medium containing all the necessary nutrients. Tissue culture can provide information on cell growth and differentiation, and is also used in plant propagation and drug production.

tissue plasminogen activator (tPA) naturally occurring substance in the body tissues that activates the enzyme plasmin that is able to dissolve blood clots. Human tPA, produced in bacteria by genetic engineering, has, like ◊streptokinase, been used to dissolve blood clots in the coronary arteries of heart-attack victims. It has been shown to be more effective than streptokinase when used in conjunction with heparin, but it is much more expensive.

Titan in astronomy, largest moon of the planet Saturn, with a diameter of 5,150 km/3,200 mi and a mean distance from Saturn of 1,222,000 km/759,000 mi. It was discovered in 1655 by Dutch mathematician and astronomer Christiaan Huygens, and is the second largest

moon in the Solar System (Ganymede, of Jupiter, is larger).

Titan is the only moon in the Solar System with a substantial atmosphere (mostly nitrogen), topped with smoggy orange clouds that obscure the surface, which may be covered with liquid ethane lakes. Its surface atmospheric pressure is greater than Earth's. Radar signals suggest that Titan has dry land as well as oceans (among the planets, only Earth has both in the Solar System).

titanium strong, lightweight, silver-grey, metallic element, symbol Ti, atomic number 22, relative atomic mass 47.90. The ninth-most abundant element in the Earth's crust, its compounds occur in practically all igneous rocks and their sedimentary deposits. It is very strong and resistant to corrosion, so it is used in building high-speed aircraft and spacecraft; it is also widely used in making alloys, as it unites with almost every metal except copper and aluminium. Titanium oxide is used in high-grade white pigments.

Titanium bonds with bone in a process called **osseo-integration**. As the body does not react to the titanium it is valuable for permanent implants such as prostheses.

The element was discovered in 1791 by English mineralogist William Gregor (1761–1817) and was named by German chemist Martin Klaproth in 1796 after the Titans, the giants of Greek mythology. It was not obtained in pure form until 1925.

titanium ore any mineral from which titanium is extracted, principally ilmenite ($FeTiO_3$) and rutile (TiO_2). Brazil, India, and Canada are major producers. Both these ore minerals are found either in rock formations or concentrated in heavy mineral sands.

Titan rocket family of US space rockets, developed from the Titan intercontinental missile. Two-stage Titan rockets launched the ◊Gemini crewed missions. More powerful Titans, with additional stages and strap-on boosters, were used to launch spy satellites and space probes, including the ◊Viking and ◊Voyager probes and ◊*Mars Observer*.

titration in analytical chemistry, a technique to find the concentration of one compound in a solution by determining how much of it will react with a known amount of another compound in solution.

One of the solutions is measured by ◊pipette into the reaction vessel. The other is added a little at a time from a burette. The end-point of the reaction is determined with an ◊indicator or an electrochemical device.

TLR camera twin-lens reflex camera that has a viewing lens of the same angle of view and focal length mounted above and parallel to the taking lens.

TNT (abbreviation for *trinitrotoluene*) $CH_3C_6H_2(NO_2)_3$, a powerful high explosive. It is a yellow solid, prepared in several isomeric forms from ◊toluene by using sulphuric and nitric acids.

toadstool common name for many umbrella-shaped fruiting bodies of fungi (see ◊fungus). The term is normally applied to those that are inedible or poisonous.

tocopherol or *vitamin E* fat-soluble chemical found in vegetable oils. Deficiency of tocopherol leads to multiple adverse effects on health. In rats, vitamin E deficiency has been shown to cause sterility.

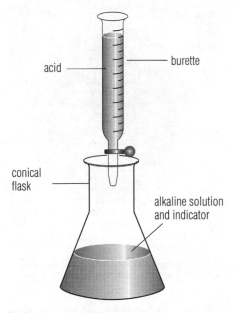

acid — burette

conical
flask —

alkaline solution
and indicator

titration

tog unit of measure of thermal insulation used in the textile trade; a light summer suit provides 1.0 tog.

The tog-value of an object is equal to ten times the temperature difference (in °C) between its two surfaces when the flow of heat is equal to one watt per square metre; one tog equals 0.645 ◊clo.

tokamak acronym for toroidal magnetic chamber, an experimental machine conceived by Soviet physicist Andrei Sakharov and developed in the Soviet Union to investigate controlled nuclear fusion. It consists of a doughnut-shaped chamber surrounded by electromagnets capable of exerting very powerful magnetic fields. The fields are generated to confine a very hot (millions of degrees) ◊plasma of ions and electrons, keeping it away from the chamber walls. See also ◊JET.

toluene or **methyl benzene** $C_6H_5CH_3$ colourless, inflammable liquid, insoluble in water, derived from petroleum. It is used as a solvent, in aircraft fuels, in preparing phenol (carbolic acid, used in making resins for adhesives, pharmaceuticals, and as a disinfectant), and the powerful high explosive ◊TNT.

tomography the technique of using X-rays or ultrasound waves to procure images of structures deep within the body for diagnostic purposes. In modern medical imaging there are several techniques, such as the ◊CAT scan (computerized axial tomography).

ton imperial unit of mass. The **long ton**, used in the UK, is 1,016 kg/2,240 lb; the **short ton**, used in the USA, is 907 kg/2,000 lb. The **metric ton** or **tonne** is 1,000 kg/2,205 lb.

ton in shipping, unit of volume equal to 2.83 cubic metres/100 cubic feet. **Gross tonnage** is the total internal volume of a ship in tons; **net register tonnage** is the volume used for carrying cargo or passengers. **Displacement tonnage** is the weight of the vessel, in terms of the number of imperial tons of seawater displaced when the ship is loaded to its load line; it is used to describe warships.

tongue in tetrapod vertebrates, a muscular organ usually attached to the floor of the mouth. It has a thick root attached to a U-shaped bone (hyoid), and is covered with a ◊mucous membrane containing nerves and taste buds. It is the main organ of taste. The tongue directs food to the teeth and into the throat for chewing and swallowing. In humans, it is crucial for speech; in other animals, for lapping up water and for grooming, among other functions. In some animals, such as frogs, it can be flipped forwards to catch insects; in others, such as anteaters, it serves to reach for food found in deep holes.

Tongue

The sensors on the feet of a red admiral butterfly are 200 times more sensitive to sugar than the human tongue.

---

tonne the metric ton of 1,000 kg/2,204.6 lb; equivalent to 0.9842 of an imperial ◊ton.

tonsillitis inflammation of the ◊tonsils.

tonsils in higher vertebrates, masses of lymphoid tissue situated at the back of the mouth and throat (palatine tonsils), and on the rear surface of the tongue (lingual tonsils). The tonsils contain many ◊lymphocytes and are part of the body's defence system against infection.

The ◊adenoids are sometimes called pharyngeal tonsils.

tooth in vertebrates, one of a set of hard, bonelike structures in the mouth, used for biting and chewing food, and in defence and aggression. In humans, the first set (20 milk teeth) appear from age six months to two and a half years. The permanent ◊dentition replaces these from the sixth year onwards, the wisdom teeth (third molars) sometimes not appearing until the age of 25 or 30. Adults have 32 teeth: two incisors, one canine (eye tooth), two premolars, and three molars on each side of each jaw. Each tooth consists of an enamel coat (hardened calcium deposits), dentine (a thick, bonelike layer), and an inner pulp cavity, housing nerves and blood vessels. Mammalian teeth have roots surrounded by cementum, which fuses them into their sockets in the jawbones.

The neck of the tooth is covered by the ◊gum, while the enamel-covered crown protrudes above the gum line.

The chief diseases of teeth are misplacements resulting from defect or disturbance of the tooth-germs before birth, eruption out of their proper places, and caries (decay).

Tooth

About one in every 2,000 babies is born with a tooth. Louis XIV of France was born with two teeth, which may explain why he had had eight wet-nurses by the time he moved on to solid foods.

---

toothache in medicine, dental pain commonly due to inflammation of the pulp or the tooth or the fibrous tissue supporting the tooth in the bone socket. It is most frequently due to ◊caries when the cavity is close to the pulp. Infection of the pulp can lead to the formation of an abscess. Chronic toothache can occur when the substance

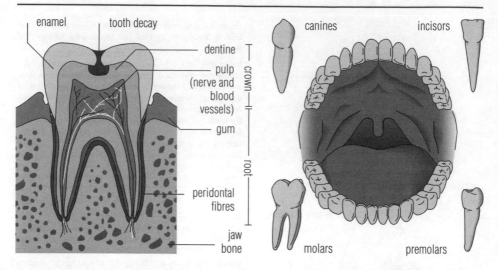

tooth Adults have 32 teeth: two incisors, one canine, two premolars, and three molars on each side of each jaw. Each tooth has three parts: crown, neck, and root. The crown consists of a dense layer of mineral, the enamel, surrounding hard dentine with a soft centre, the pulp.

of the tooth beneath the enamel (dentine) is unprotected because of loss of enamel due to decay or receding gums. The pain is often associated with temperature changes or sweet foods.

topaz mineral, aluminium fluorosilicate, $Al_2(F_2SiO_4)$. It is usually yellow, but pink if it has been heated, and is used as a gemstone when transparent. It ranks 8 on the Mohs' scale of hardness.

topography the surface shape and composition of the landscape, comprising both natural and artificial features, and its study. Topographical features include the relief and contours of the land; the distribution of mountains, valleys, and human settlements; and the patterns of rivers, roads, and railways.

topology in computing, the arrangement of devices in a ◊network. Common topologies include *star networks*, where a central computer manages network access, and *ring networks*, where users can establish direct connections with other work stations.

topology branch of geometry that deals with those properties of a figure that remain unchanged even when the figure is transformed (bent, stretched) – for example, when a square painted on a rubber sheet is deformed by distorting the sheet.

Topology has scientific applications, as in the study of turbulence in flowing fluids.

The topological theory, proposed in 1880, that only four colours are required in order to produce a map in which no two adjoining countries have the same colour, inspired extensive research, and was proved in 1972 by Kenneth Appel and Wolfgang Haken.

topsoil the upper, cultivated layer of soil, which may vary in depth from 8 to 45 cm/3 to 18 in. It contains organic matter – the decayed remains of vegetation, which plants need for active growth – along with a variety of soil organisms, including earthworms.

tor isolated mass of rock, often granite, left upstanding on

a hilltop after the surrounding rock has been broken down. Weathering takes place along the joints in the rock, reducing the outcrop into a mass of rounded blocks.

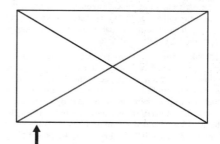

this figure is topologically equivalent to this one

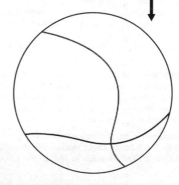

topology Topology is often called 'rubber sheet geometry'. It is the branch of mathematics which studies the properties of a geometric figure which are unchanged when the figure is distorted; for example, the two lines still intersect when the rectangle is distorted.

tornado extremely violent revolving storm with swirling, funnel-shaped clouds, caused by a rising column of warm air propelled by strong wind. A tornado can rise to a great height, but with a diameter of only a few hundred metres or less. Tornadoes move with wind speeds of 160–480 kph/100–300 mph, destroying everything in their path. They are common in central USA and Australia.

torpor in medicine, a condition of mental and physical inactivity which interferes with normal life. It is found in people suffering from fever and in elderly people with arterial diseases.

torque the turning effect of force on an object. A turbine produces a torque that turns an electricity generator in a power station. Torque is measured by multiplying the force by its perpendicular distance from the turning point.

torr unit of pressure equal to 1/760 of an ◊atmosphere, used mainly in high-vacuum technology.

One torr is equivalent to 133.322 pascals, and for practical purposes is the same as the millimetre of mercury. It is named after Evangelista Torricelli.

torsion in physics, the state of strain set up in a twisted material; for example, when a thread, wire, or rod is twisted, the torsion set up in the material tends to return the material to its original state. The **torsion balance**, a sensitive device for measuring small gravitational or magnetic forces, or electric charges, balances these against the restoring force set up by them in a torsion suspension.

torsion in biology, the process in which organs or tumours that are attached to other structures in the body by a narrow piece of tissue, become twisted. This results in a narrowing of the blood vessels supplying the organ or tumour.

torus ring with a D-shaped cross-section used to contain ◊plasma in nuclear fusion reactors such as the Joint European Torus (◊JET) reactor.

total internal reflection the complete reflection of a beam of light that occurs from the surface of an optically 'less dense' material. For example, a beam from an underwater light source can be reflected from the surface of the water, rather than escaping through the surface. Total internal reflection can only happen if a light beam hits a surface at an angle greater than the critical angle for that particular pair of materials. Total internal reflection is used as a means of reflecting light inside ◊prisms and ◊optical fibres. Light is contained inside an optical fibre not by the cladding around it, but by the ability of the internal surface of the glass-fibre core to reflect 100% of the light, thereby keeping it trapped inside the fibre.

touch sensation produced by specialized nerve endings in the skin. Some respond to light pressure, others to heavy pressure. Temperature detection may also contribute to the overall sensation of touch. Many animals, such as nocturnal ones, rely on touch more than humans do. Some have specialized organs of touch that project from the body, such as whiskers or antennae.

touch screen in computing, an input device allowing the user to communicate with the computer by touching a display screen with a finger. In this way, the user can point to a required ◊menu option or item of data. Touch screens are used less widely than other pointing devices such as the ◊mouse or ◊joystick.

Typically, the screen is able to detect the touch either because the finger presses against a sensitive membrane or because it interrupts a grid of light beams crossing the screen surface.

touch sensor in a computer-controlled ◊robot, a device used to give the robot a sense of touch, allowing it to manipulate delicate objects or move automatically about a room. Touch sensors provide the feedback necessary for the robot to adjust the force of its movements and the pressure of its grip. The main types include the strain gauge and the microswitch.

Tourette's syndrome or *Gilles de la Tourette syndrome* rare neurological condition characterized by multiple tics and vocal phenomena such as grunting, snarling, and obscene speech, named after French physician Georges Gilles de la Tourette (1859–1904). It affects one to five people per 10,000, with males outnumbering females by four to one, and the onset is usually around the age of six. There are no convincing explanations of its cause, and it is usually resistant to treatment.

tourmaline hard, brittle mineral, a complex silicate of various metals, but mainly sodium aluminium borosilicate.

Small tourmalines are found in granites and gneisses. The common varieties range from black (schorl) to pink, and the transparent gemstones may be colourless (achromatic), rose pink (rubellite), green (Brazilian emerald), blue (indicolite, verdelite, Brazilian sapphire), or brown (dravite).

tourniquet in medicine, a cuff that can be tightened around the upper end of the thigh or the arm to close the arteries to the limb and stop the flow of blood. Tourniquets are rarely used in modern medicine because gangrene may result. Direct pressure to the point of bleeding is simpler, safer, and at least as effective as a tourniquet.

toxaemia another term for ◊blood poisoning; *toxaemia of pregnancy* is another term for ◊pre-eclampsia.

toxicity tests tests carried out on new drugs, cosmetics, food additives, pesticides, and other synthetic chemicals to see whether they are safe for humans to use. They aim to identify potential toxins, carcinogens, teratogens, and mutagens.

Traditionally such tests use live animals such as rats, rabbits, and mice. Animal tests have become a target for criticism by antivivisection groups, and alternatives have been sought. These include tests on human cells cultured in a test tube and on bacteria.

toxicology branch of medicine dealing with the study of poisons. This includes the chemical nature of poisons, their origin and preparation, their physiological action, tests to recognize them, pathological changes due to their presence, their antidotes, and the recognition of them by post-mortem evidence.

toxic shock syndrome rare condition marked by rapid onset of fever, vomiting, and low blood pressure, sometimes leading to death. It is caused by a toxin of the bacterium *Staphylococcus aureus*, normally harmlessly present in the body. It is seen most often in young women using tampons during menstruation.

toxic syndrome fatal disease for which the causes are not confirmed. In an outbreak in Spain in the early 1980s, more than 20,000 people became ill and 600–700 died.

One theory held that the toxic syndrome was caused by adulterated industrial oil, illegally imported from France into Spain, re-refined, and sold for human consumption from 1981. However, other studies pointed more convincingly to the use of a 'plaguicide' product, Nemacur (produced by Bayer), which contains organophosphates, and can be dangerous if fruit and vegetables are eaten too soon after its application. Animal tests have shown symptoms similar to toxic syndrome, with damage to the nervous system, after ingesting organophosphates.

toxic waste dumped ◊hazardous waste.

toxin any poison produced by another living organism (usually a bacterium) that can damage the living body. In vertebrates, toxins are broken down by ◊enzyme action, mainly in the liver.

Toxin

The hairs on stinging nettle leaves operate like mini hypodermic syringes. When the leaf is touched, the pressure in the hair falls as the fragile tip is broken away and a dose of nerve toxin is released.

toxocariasis human infestation with the larvae of roundworms, sometimes present in dogs and cats. Symptoms include fever, rash, aching joints and muscles, vomiting and convulsions; there may be enlargement of the liver and damage to the lungs. Small children are most at risk.

toxoplasmosis disease transmitted to humans by animals, often in pigeon or cat excrement, or in undercooked meat. It causes flu-like symptoms and damages the central nervous system, eyes, and visceral organs. It is caused by a protozoan, *Toxoplasma gondii*. Congenital toxoplasmosis, transmitted from an infected mother to her unborn child, can lead to blindness and retardation.

trace in computing, a method of checking that a computer program is functioning correctly by causing the changing values of all the ◊variables involved to be displayed while the program is running. In this way it becomes possible to narrow down the search for a bug, or error, in the program to the exact instruction that causes the variables to take unexpected values.

trace element chemical element necessary in minute quantities for the health of a plant or animal. For example, magnesium, which occurs in chlorophyll, is essential to photosynthesis, and iodine is needed by the thyroid gland of mammals for making hormones that control growth and body chemistry.

tracer a small quantity of a radioactive ◊isotope (form of an element) used to follow the path of a chemical reaction or a physical or biological process. The location (and possibly concentration) of the tracer is usually detected by using a Geiger-Muller counter.

For example, the activity of the thyroid gland can be monitored by giving the patient an injection containing a small dose of a radioactive isotope of iodine, which is selectively absorbed from the bloodstream by the gland.

trachea tube that forms an airway in air-breathing animals. In land-living ◊vertebrates, including humans, it is also known as the **windpipe** and runs from the larynx to the upper part of the chest. Its diameter is about 1.5 cm/0.6 in and its length 10 cm/4 in. It is strong and flexible, and reinforced by rings of ◊cartilage. In the upper chest, the trachea branches into two tubes: the left and right

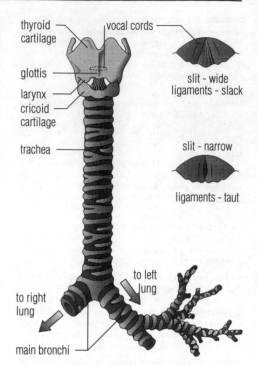

trachea The human trachea, or windpipe. The larynx, or voice box, lies at the entrance to the trachea. The two vocal cords are membranes that normally remain open and still. When they are drawn together, the passage of air makes them vibrate and produce sounds.

bronchi, which enter the lungs. Insects have a branching network of tubes called tracheae, which conduct air from holes (◊spiracles) in the body surface to all the body tissues. The finest branches of the tracheae are called tracheoles.

Some spiders also have tracheae but, unlike insects, they possess gill-like lungs (book lungs) and rely on their circulatory system to transport gases throughout the body.

tracheid cell found in the water-conducting tissue (◊xylem) of many plants, including gymnosperms (conifers) and pteridophytes (ferns). It is long and thin with pointed ends.

The cell walls are thickened by ◊lignin, except for numerous small rounded areas, or pits, through which water and dissolved minerals pass from one cell to another. Once mature, the cell itself dies and only its walls remain.

tracheotomy or **tracheostomy** surgical opening in the windpipe (trachea), usually created for the insertion of a tube to enable the patient to breathe. It is done either to bypass an airway impaired by disease or injury, or to safeguard it during surgery or a prolonged period of mechanical ventilation.

trachoma chronic eye infection, resembling severe ◊conjunctivitis. The conjunctiva becomes inflamed, with scarring and formation of pus, and there may be damage to the cornea. It is caused by a virus-like organism

(◊chlamydia), and is a disease of dry tropical regions. Although it responds well to antibiotics, numerically it remains the biggest single cause of blindness worldwide.

track in computing, part of the magnetic structure created on a disc surface during ◊disc formatting so that data can be stored on it. The disc is first divided into circular tracks and then each circular track is divided into a number of sectors.

traction in medicine, the application of a system of weights and pulleys to the distal part of a fracture to allow the fracture to heal with the bone in correct alignment.

trade wind prevailing wind that blows towards the Equator from the northeast and southeast. Trade winds are caused by hot air rising at the Equator and the consequent movement of air from north and south to take its place. The winds are deflected towards the west because of the Earth's west-to-east rotation.

The unpredictable calms known as the ◊doldrums lie at their convergence.

The trade-wind belts move north and south about 5° with the seasons. The name is derived from the obsolete expression '*blow trade*' meaning to blow regularly, which indicates the trade winds' importance to navigation in the days of cargo-carrying sailing ships.

tramway transport system for use in cities, where wheeled vehicles run along parallel rails. Trams are powered either by electric conductor rails below ground or by conductor arms connected to overhead wires. Greater manoeuvrability is achieved with the ◊trolley bus, similarly powered by conductor arms overhead but without tracks.

trance mental state in which the subject loses the ordinary perceptions of time and space, and even of his or her own body.

In this highly aroused state, often induced by rhythmic music, 'speaking in tongues' (glossolalia) may occur; this usually consists of the rhythmic repetition of apparently meaningless syllables, with a euphoric return to normal consciousness. It is also practised by Native American and Australian Aboriginal healers, Afro-Brazilian spirit mediums, and in shamanism.

tranquillizer common name for any drug for reducing anxiety or tension (◊anxiolytic), such as ◊benzodiazepines, barbiturates, antidepressants, and beta-blockers. The use of drugs to control anxiety is becoming much less popular, because most of the drugs available are capable of inducing dependence.

transactinide element any of a series of eight radioactive, metallic elements with atomic numbers that extend beyond the ◊actinide series, those from 104 (rutherfordium) to 111 (unununium). They are grouped because of their expected chemical similarities (they are all bivalent), the properties differing only slightly with atomic number. All have ◊half-lives that measure less than two minutes.

Trans-Alaskan Pipeline one of the world's greatest civil engineering projects, the construction of a pipeline to carry petroleum (crude oil) 1,285 km/800 mi from N Alaska to the ice-free port of Valdez. It was completed in 1977 after three years' work and much criticism by ecologists. In 1997 the Pipeline delivered more than 20% of US oil production.

The engineers had to elevate nearly half the pipeline on supports above ground level to avoid thawing the permafrost (permanently frozen ground), which would have caused much environmental damage. They also had to cross 600 rivers and streams and two mountain ranges, and allow for earthquakes.

transcription in living cells, the process by which the information for the synthesis of a protein is transferred from the ◊DNA strand on which it is carried to the messenger ◊RNA strand involved in the actual synthesis.

It occurs by the formation of ◊base pairs when a single strand of unwound DNA serves as a template for assembling the complementary nucleotides that make up the new RNA strand.

transducer device that converts one form of energy into another. For example, a thermistor is a transducer that converts heat into an electrical voltage, and an electric motor is a transducer that converts an electrical voltage into mechanical energy. Transducers are important components in many types of ◊sensor, converting the physical quantity to be measured into a proportional voltage signal.

transfer orbit elliptical path followed by a spacecraft moving from one orbit to another, designed to save fuel although at the expense of a longer journey time.

Space probes travel to the planets on transfer orbits. A probe aimed at Venus has to be 'slowed down' relative to the Earth, so that it enters an elliptical transfer orbit with its perigee (point of closest approach to the Sun) at the same distance as the orbit of Venus; towards Mars, the vehicle has to be 'speeded up' relative to the Earth, so that it reaches its apogee (furthest point from the Sun) at the same distance as the orbit of Mars. *Geostationary transfer orbit* is the highly elliptical path followed by satellites to be placed in ◊geostationary orbit around the Earth (an orbit coincident with Earth's rotation). A small rocket is fired at the transfer orbit's apogee to place the satellite in geostationary orbit.

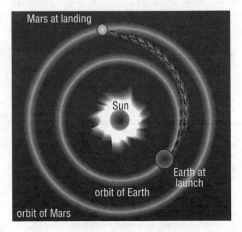

transfer orbit The transfer orbit used by a spacecraft when travelling from Earth to Mars. The orbit is chosen to minimize the fuel needed by the spacecraft; the craft is in free fall for most of the journey.

transformation in mathematics, a mapping or ◊function, especially one which causes a change of shape or position in a geometric figure. Reflection, rotation, enlargement, and translation are the main geometrical transformations.

transformer device in which, by electromagnetic induction, an alternating current (AC) of one voltage is transformed to another voltage, without change of ◊frequency. Transformers are widely used in electrical apparatus of all kinds, and in particular in power transmission where high voltages and low currents are utilized.

A transformer has two coils, a primary for the input and a secondary for the output, wound on a common iron core. The ratio of the primary to the secondary voltages (and currents) is directly (and inversely) proportional to the number of turns in the primary and secondary coils.

transfusion intravenous delivery of blood or blood products (plasma, red cells) into a patient's circulation to make up for deficiencies due to disease, injury, or surgical intervention. Cross-matching is carried out to ensure the patient receives the right blood group. Because of worries about blood-borne disease, there is a growing interest in autologous transfusion with units of the patient's own blood 'donated' over the weeks before an operation.

Blood is rarely transfused whole. Blood cells and platelets are separated and resuspended in solution. Plasma can be frozen and is used to treat clotting deficiencies.

Blood transfusion, first successfully pioneered in humans in 1818, remained highly risky until the discovery of blood groups, by Austrian-born immunologist Karl Landsteiner in 1900, indicated the need for compatibility of donated blood.

transhumance seasonal movement by pastoral farmers of their livestock between areas of different climate. There are three main forms: in **Alpine** regions, such as Switzerland, cattle are moved to high-level pastures in summer and returned to milder valley pastures in winter; in **Mediterranean** lands, summer heat and drought make it necessary to move cattle to cooler mountain slopes; in **W Africa**, the nomadic herders of the Fulani peoples move cattle south in search of grass and water in the dry season and north in the wet season to avoid the tsetse fly.

transistor solid-state electronic component, made of ◊semiconductor material, with three or more ◊electrodes, that can regulate a current passing through it. A transistor can act as an amplifier, ◊oscillator, ◊photocell, or switch, and (unlike earlier thermionic valves) usually operates on a very small amount of power. Transistors commonly consist of a tiny sandwich of ◊germanium or ◊silicon, alternate layers having different electrical properties because they are impregnated with minute amounts of different impurities. A crystal of pure germanium or silicon would act as an insulator (nonconductor). By introducing impurities in the form of atoms of other materials (for example, boron, arsenic, or indium) in minute amounts, the layers may be made either **n-type**, having an excess of electrons, or **p-type**, having a deficiency of electrons. This enables electrons to flow from one layer to another in one direction only.

Transistors have had a great impact on the electronics industry, and thousands of millions are now made each year. They perform many of the functions of the thermionic valve, but have the advantages of greater reliability, long life, compactness, and instantaneous action, no warming-up period being necessary. They are widely used in most electronic equipment, including portable radios and televisions, computers, and satellites, and are the basis of the ◊integrated circuit (silicon chip). They were invented at Bell Telephone Laboratories in the USA in 1948 by John Bardeen and Walter Brattain, developing the work of William Shockley.

transistor–transistor logic (TTL) in computing, the type of integrated circuit most commonly used in building electronic products. In TTL chips the bipolar transistors are directly connected (usually collector to base). In mass-produced items, large numbers of TTL chips are commonly replaced by a small number of uncommitted logic arrays (ULAs), or logic gate arrays.

transit in astronomy, the passage of a smaller object across the visible disc of a larger one. Transits of the inferior planets occur when they pass directly between the Earth and the Sun, and are seen as tiny dark spots against the Sun's disc.

Other forms of transit include the passage of a satellite or its shadow across the disc of Jupiter and the passage of planetary surface features across the central ◊meridian of that planet as seen from Earth. The passage of an object in the sky across the observer's meridian is also known as a transit.

transition metal any of a group of metallic elements that have incomplete inner electron shells and exhibit variable valency – for example, cobalt, copper, iron, and molybdenum. They are excellent conductors of electricity, and generally form highly coloured compounds.

translation in living cells, the process by which proteins are synthesized. During translation, the information coded as a sequence of nucleotides in messenger ◊RNA is transformed into a sequence of amino acids in a peptide chain. The process involves the 'translation' of the ◊genetic code. See also ◊transcription.

translation program in computing, a program that translates another program written in a high-level language or assembly language into the machine-code instructions that a computer can obey. See ◊assembler, ◊compiler, and ◊interpreter.

translocation in genetics, the exchange of genetic material between chromosomes. It is responsible for congenital abnormalities, such as Down's syndrome.

transmission electron microscope (TEM) the most powerful type of ◊electron microscope, with a resolving power ten times better than that of a ◊scanning electron microscope and a thousand times better than that of an optical microscope. A fine electron beam passes through the specimen, which must therefore be sliced extremely thinly – typically to about one-thousandth of the thickness of a sheet of paper (100 nanometres). The TEM can resolve objects 0.001 micrometres (0.04 millionth of an inch) apart, a gap that is 100,000 times smaller than the unaided eye can see.

A TEM consists of a tall evacuated column at the top of which is a heated filament that emits electrons. The electrons are accelerated down the column by a high voltage (around 100,000 volts) and pass through the slice of specimen at a point roughly half-way down. Because the density of the specimen varies, the 'shadow' of the beam falls on a fluorescent screen near the bottom of the column and forms an image. A camera is mounted beneath the screen to record the image.

The electron beam is controlled by magnetic fields produced by electric coils, called electron lenses. One electron lens, called the condenser, controls the beam size and brightness before it strikes the specimen. Another electron lens, called the objective, focuses the beam on the specimen and magnifies the image about 50 times. Other electron lenses below the specimen then further magnify the image.

The **high voltage transmission electron microscope** (HVEM) uses voltages of up to 3 million volts to accelerate the electron beam. The largest of these instruments is as tall as a three-storey building.

The first experimental TEM was built in 1931 by German scientists Max Knoll and Ernest Ruska of the Technische Hochschule, Berlin, Germany. They produced a picture of a platinum grid magnified 117 times. The first commercial electron microscope was built in England in 1936.

transparency in photography, a picture on slide film. This captures the original in a positive image (direct reversal) and can be used for projection or printing on positive-to-positive print material, for example by the Cibachrome or Kodak R-type process.

Slide film is usually colour but can be obtained in black and white.

transpiration the loss of water from a plant by evaporation. Most water is lost from the leaves through pores known as ◊stomata, whose primary function is to allow ◊gas exchange between the plant's internal tissues and the atmosphere. Transpiration from the leaf surfaces causes a continuous upward flow of water from the roots via the ◊xylem, which is known as the transpiration stream.

transplant in medicine, the transfer of a tissue or organ from one human being to another or from one part of the body to another (skin grafting). In most organ transplants, the operation is for life-saving purposes, though the immune system tends to reject foreign tissue. Careful matching and immunosuppressive drugs must be used, but these are not always successful.

Corneal grafting, which may restore sight to a diseased or damaged eye, was pioneered in 1905, and is the oldest successful human transplant procedure. Of the internal organs, kidneys were first transplanted successfully in the early 1950s and remain most in demand. Modern transplantation also encompasses the heart, lungs, liver, pancreatic tissue, bone, and bone-marrow.

Most transplant material is taken from cadaver donors, usually those suffering death of the ◊brainstem, or from frozen tissue banks. In rare cases, kidneys, corneas, and part of the liver may be obtained from living donors. Besides the shortage of donated material, the main problem facing transplant surgeons is rejection of the donated organ by the recipient's body. The 1990 Nobel Prize for Medicine or Physiology was awarded to two US surgeons, Donnall Thomas and Joseph Murray, for their pioneering work on organ and tissue transplantation.

The first experiments to use genetically altered animal organs in humans were given US government approval in July 1995 – genetically altered pig livers would be attached to the circulatory systems of patients who were near death or whose livers had failed. Need for the tests

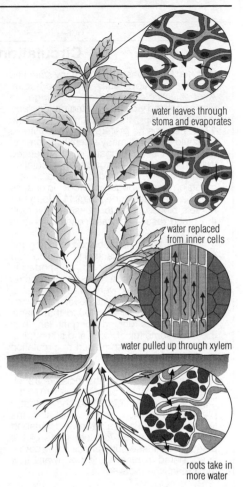

water leaves through
stoma and evaporates

water replaced
from inner cells

water pulled up through xylem

roots take in
more water

transpiration The loss of water from a plant by evaporation is known as transpiration. Most of the water is lost through the surface openings, or stomata, on the leaves. The evaporation produces what is known as the transpiration stream, a tension that draws water up from the roots through the xylem, water-carrying vessels in the stem.

had arisen due to a shortage of human organs available for transplant. *See chronology overleaf.*

transputer in computing, a member of a family of microprocessors designed for parallel processing, developed in the UK by Inmos. In the circuits of a standard computer the processing of data takes place in sequence; in a transputer's circuits processing takes place in parallel, greatly reducing computing time for those programs that have been specifically written for it.

The transputer implements a special programming language called OCCAM, which Inmos based on CSP (communicating sequential processes), developed by C A R Hoare of Oxford University Computing Laboratory.

transsexual person who identifies himself or herself completely with the opposite sex, believing that the wrong

Circulation in plants?

Stephen Hales was a man of many scientific interests and, apparently, of settled habits. After scientific education at Cambridge, he was appointed vicar of Teddington, England in 1709, holding the post until he died in 1761. Perhaps he preferred a settled life so he could follow his many scientific interests. Amongst his investigations, he refined William Harvey's ideas on blood circulation in animals. This involved gruesome experiments on horses, dogs, and frogs. In one, he attached a vertical tube 3.3 m/11 ft long to a horse's blood vessel to measure how high the blood rose – he was measuring blood pressure.

In about 1724, Hales began studying the circulation of sap in plants. He showed how water is drawn in at the roots, transported to the leaves, and then transpired. Several pieces of this jigsaw were already in place. After the microscope was developed in the mid 17th century, it was found that plants contained tubes running along the length of the stem, some full of water and some full of air. This suggested some sort of circulation system, comparable to that in animals. Common sense suggested that plants take in water through the roots, the moisture travelling to all parts of the plant. Around 1670 the Italian Marcello Malpighi had shown that a plant's food substances were built up in the leaves. There must be a downward transport of these substances to other parts of the plant. Since these substances were sometimes stored in tubers, the downward flow must reach the roots.

Hales' experiments
The questions Hales addressed were: was there a closed circulation, as in animals, and what powered it? He undertook three crucial experiments. In the first, he established that the water was drawn upwards by the leaves. His notebook records:

'July 27 (1716). I fixed an apple branch ... to a tube. I filled the tube with water, and then immersed the whole branch ... into a vessel full of water. The water subsided 6 inches in the first two hours (being the filling of the sap vessels) and 6 inches the following night.'

Hales concluded that this showed the great drawing power of 'perspiration'. Evaporation from the leaves pulled the sap upwards, rather than pressure pushing from the roots. The next step was simple: by putting a leafy plant in a closed vessel, and collecting the transpired liquid, he confirmed that the material that transpired into the air was water. Water was indeed flowing up the stem and being expired from the leaves.

Hales' apple branch
The third experiment was more complicated.

Hales needed to discover whether the upward movement of the water was balanced by an equal downward flow; water might be absorbed from the air by the leaves, making the flow a continuous circulation. His notebook records a beautifully simple and conclusive experiment:

'August 20 (1716). At 1 p.m. I took an apple branch nine foot long, 13/4 inch diameter, with proportional lateral branches. I cemented it fast to one end of a vertical U tube; but first I cut away the bark, and last year's ringlet of wood, near the bottom of the branch. I then filled the tube with water, which was 22 feet long and 1/2 inch diameter. The water was very freely imbibed, at a rate of 31/2 inch in a minute. In half an hour's time I could plainly perceive the lower part of the cut to be moister than before; when at the same time the upper part of the wound looked white and dry.'

'The water must necessarily ascend from the tube, through the innermost wood, because the last year's wood was cut away, for 3 inches all round the stem; and consequently, if the sap in its natural course descended by the last year's ringlet of wood, between that and the bark (as many have thought) the water should have descended by last year's wood, or the bark, and so have moistened the upper part of the gap; but, on the contrary, the lower part was moistened, and not the upper part.'

There was no sign of descending water, and hence no circulation of water within the plant. Hales had already shown that the sunflower transpires water 17 times faster than a man, bulk for bulk; if there was a circulation, it would have to be enormously fast.

Further discoveries about the workings of plants
Hales continued his work on plant physiology. He showed that plants take in part of the air (carbon dioxide) which is used in their nutrition. He measured growth rates. His interests were wide-ranging: from preservation of foods, water purification, and ventilation of ships and buildings, to the best way of supporting pie crusts.

In 1779 Dutch doctor Jan Ingen-Housz established two distinct processes in plants: respiration, in which oxygen is absorbed and carbon dioxide is exhaled, as in animal respiration; and photosynthesis, in which carbon dioxide is taken in to make food, and oxygen is given out. It is the daily ebb and flow of these processes that transports water and food around a plant.

Peter Lafferty

Transplant: Chronology

1771	Scottish surgeon John Hunter describes his experiments in the transplantation of tissues, including a human tooth into a cock's comb in *Treatise on the Natural History of Human Teeth*.
1905	Corneal grafting, which may restore sight to a diseased or damaged eye, was pioneered.
1950s	The kidneys were first transplanted successfully; kidney transplants were pioneered by British surgeon Roy Calne. Peter Medawar conducted vital research into the body's tolerance of transplanted organs and skin grafts.
1964	Chimpanzee kidneys were transplanted into humans in the USA, but with little success; in the UK a pig's heart valve was transplanted successfully and the operation became routine.
1967	South African surgeon Christiaan Barnard performed the first human heart transplant. The 54-year-old patient lived for 18 days.
1969	The world's first heart and lung transplant is performed at the Stanford Medical Centre, California, USA.
1970	The first successful nerve transplant is achieved in West Germany.
1978	Cyclosporin, an immunosuppressive drug derived from a fungus revolutionized transplant surgery by reducing the incidence and severity of rejection of donor organs.
1982	Jarvik 7, an artificial heart made of plastic and aluminium was transplanted; the recipient lived another 112 days.
1986	British surgeons John Wallwork and Roy Calne perform the first triple transplant – heart, lung, and liver – at Papworth Hospital, Cambridge, England.
1987	The world's longest-surviving heart-transplant patient died in France, 18 years after his operation. A three-year-old girl in the USA receives a new liver, pancreas, small intestine, and parts of the stomach and colon; the first successful five-organ transplant.
1989	Grafts of fetal brain tissue were first used to treat Parkinson's disease.
1990	Nobel Prize for Medicine or Physiology was awarded to two US surgeons, Donnall Thomas and Joseph Murray, for their pioneering work on organ and tissue transplantation.
1995	The first experiments to use genetically altered animal organs in humans were given US government approval July – genetically altered pig livers were attached to the circulatory systems of patients whose livers had failed. An AIDS patient received a bone marrow transplant from a baboon but the graft failed to take.

sex was assigned at birth. Unlike *transvestites*, who desire to dress in clothes traditionally worn by the opposite sex; transsexuals think and feel emotionally in a way typically considered appropriate to members of the opposite sex, and may undergo surgery to modify external sexual characteristics.

In 1995 Dutch researchers identified a structural difference between the brains of transsexual men and other men. Within the brain a small cluster of cells, the bed nucleus of the stria terminalis (BST), is smaller in women than men. Transsexual males in the study were found to have a female-sized BST. Research is continuing into the significance of this finding.

transuranic element or *transuranium element* chemical element with an atomic number of 93 or more – that is, with a greater number of protons in the nucleus than has uranium. All transuranic elements are radioac-

tive. Neptunium and plutonium are found in nature; the others are synthesized in nuclear reactions.

transverse wave ◊wave in which the displacement of the medium's particles is at right-angles to the direction of travel of the wave motion.

trapezium (North American *trapezoid*) in geometry, a four-sided plane figure (quadrilateral) with two of its sides parallel. If the parallel sides have lengths a and b and the perpendicular distance between them is h (the height of the trapezium), its area $A = \frac{1}{2}h(a + b)$.

An isosceles trapezium has its sloping sides equal, and is symmetrical about a line drawn through the midpoints of its parallel sides.

trauma in psychiatry, a painful emotional experience or shock with lasting psychic consequences; in medicine, any physical damage or injury.

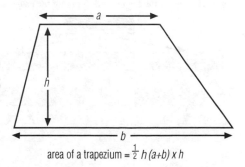

area of a trapezium $= \frac{1}{2} h (a+b) \times h$

trapezium A trapezium is a four-sided plane figure with two of its sides parallel.

direction of
travel of wave

direction of
displacement of
particles

transverse wave The diagram illustrates the motion of a transverse wave. Light waves are examples of transverse waves: they undulate at right angles to the direction of travel and are characterized by alternating crests and troughs. Simple water waves, such as the ripples produced when a stone is dropped into a pond, are also examples of transverse waves.

In psychiatric terms a trauma may have long-lasting effects, during which an insignificant event triggers the original distress. A person then may have difficulties in normal life, such as in establishing relationships or sleeping. In psychological terms this is known as **post-traumatic stress disorder**. It can be treated by ◊psychotherapy.

travel medicine see ◊immunization for travel.

travel sickness nausea and vomiting caused by the motion of cars, boats, or other forms of transport. Constant vibration and movement may stimulate changes in the fluid of the semicircular canals (responsible for balance) of the inner ear, to which the individual fails to adapt, and to which are added visual and psychological factors. Some proprietary remedies contain ◊antihistamine drugs. **Space sickness** is a special case: in weightless conditions normal body movements result in unexpected and unfamiliar signals to the brain. Astronauts achieve some control of symptoms by wedging themselves in their bunks.

treadmill wheel turned by foot power (often by a domesticated animal) and used, for instance, to raise water from a well or grind grain.

tree perennial plant with a woody stem, usually a single stem (trunk), made up of ◊wood and protected by an outer layer of ◊bark. It absorbs water through a ◊root system. There is no clear dividing line between shrubs and trees, but sometimes a minimum achievable height of 6 m/20 ft is used to define a tree.

Tree

The tallest tree ever measured was an Australian eucalyptus (*Eucalyptus regnans*), reported in 1872. It was 132 m/435 ft tall.

Angiosperms A tree-like form has evolved independently many times in different groups of plants. Among the ◊angiosperms, or flowering plants, most trees are ◊dicotyledons. This group includes trees such as oak, beech, ash, chestnut, lime, and maple, and they are often referred to as broad-leaved trees because their leaves are broader than those of conifers, such as pine and spruce. In temperate regions angiosperm trees are mostly ◊deciduous (that is, they lose their leaves in winter), but in the tropics most angiosperm trees are evergreen. There are fewer trees among the ◊monocotyledons, but the palms

and bamboos (some of which are tree-like) belong to this group.

Gymnosperms The ◊gymnosperms include many trees and they are classified into four orders: Cycadales (including cycads and sago palms), Coniferales (the conifers), Ginkgoales (including only one living species, the ginkgo, or maidenhair tree), and Taxales (including yews). Apart from the ginkgo and the larches (conifers), most gymnosperm trees are evergreen.

tree-and-branch filing system in computing, a filing system where all files are stored within directories, like folders in a filing cabinet. These directories may in turn be stored within further directories. The root directory contains all the other directories and may be thought of as equivalent to the filing cabinet. Another way of picturing the system is as a tree with branches from which grow smaller branches, ending in leaves (individual files).

tree diagram in probability theory, a branching diagram consisting only of arcs and nodes (but not loops curving back on themselves), which is used to establish probabilities.

tree rings rings visible in the wood of a cut tree; see ◊annual rings.

tremor minor ◊earthquake.

triangle in geometry, a three-sided plane figure, the sum of whose interior angles is 180°. Triangles can be classified by the relative lengths of their sides. A **scalene triangle** has three sides of unequal length; an **isosceles triangle** has at least two equal sides; an **equilateral triangle** has three equal sides (and three equal angles of 60°).

triangle of forces method of calculating the force (the resultant) produced by two other forces. It is based on the fact that if three forces acting at a point can be represented by the sides of a triangle, the forces are in equilibrium. See ◊parallelogram of forces.

triangulation technique used in surveying and navigation to determine distances, using the properties of the triangle. To begin, surveyors measure a certain length exactly to provide a base line. From each end of this line they then measure the angle to a distant point, using a ◊theodolite. They now have a triangle in which they know the length of one side and the two adjacent angles. By simple trigonometry they can work out the lengths of the other two sides. To make a complete survey of the region, they repeat the process, building on the first triangle.

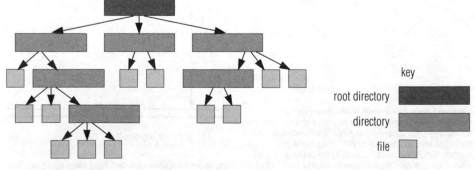

key
root directory
directory
file

tree-and-branch filing system

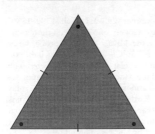

Equilateral triangle: all the sides are the same length; all the angles are equal to 60°

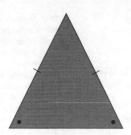

Isosceles triangle: two sides and two angles are the same

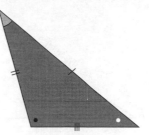

Scalene triangle: all the sides and angles are different

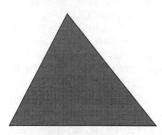

Acute-angle triangle: each angle is acute (less than 90°)

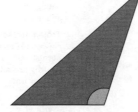

Obtuse-angle triangle: one angle is obtuse (more than 90°)

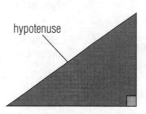

hypotenuse

Right-angle triangle: one angle is 90°, the *hypotenuse* is the side opposite the right angle

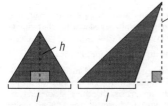

Area of triangle = $^1/_2 lh$

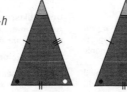

Triangles are *congruent* if corresponding sides and corresponding angles are equal

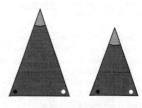

Similar triangles have corresponding angles that are equal; they therefore have the same shape

triangle Types of triangles.

Triassic period of geological time 245–208 million years ago, the first period of the Mesozoic era. The continents were fused together to form the world continent ◊Pangaea. Triassic sediments contain remains of early dinosaurs and other reptiles now extinct. By late Triassic times, the first mammals had evolved.

tributyl tin (TBT) chemical used in antifouling paints on ships' hulls and other submarine structures to deter the growth of barnacles. The tin dissolves in sea water and enters the food chain. It can cause reproductive abnormalities – exposed female whelks develop penises; the use of TBT has therefore been banned in many countries, including the UK.

trichloromethane technical name for ◊chloroform.

tricuspid valve flap of tissue situated on the right side of the ◊heart between the atrium and the ventricle. It prevents blood flowing backwards when the ventricle contracts.

As in all valves, its movements are caused by pressure changes during the beat rather than by any intrinsic muscular activity. As the valve snaps shut, a vibration passes through the chest cavity and is detectable as the first sound of the heartbeat.

triglyceride chemical name for ◊fat comprising three fatty acids reacted with a glycerol.

trigonometry branch of mathematics that solves problems relating to plane and spherical triangles. Its principles are based on the fixed proportions of sides for a particular angle in a right-angled triangle, the simplest of

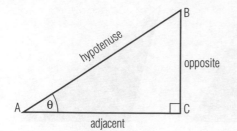

for any right-angled triangle with angle θ as shown the trigonometrical ratios are

$$\sin\theta = \frac{BC}{AB} = \frac{opposite}{hypotenuse}$$

$$\cos\theta = \frac{AC}{AB} = \frac{adjacent}{hypotenuse}$$

$$\tan\theta = \frac{BC}{AC} = \frac{opposite}{adjacent}$$

trigonometry At its simplest level, trigonometry deals with the relationships between the sides and angles of triangles. Unknown angles or lengths are calculated by using trigonometrical ratios such as sine, cosine, and tangent. The earliest applications of trigonometry were in the fields of navigation, surveying, and astronomy, and usually involved working out an inaccessible distance such as the distance of the Earth from the Moon.

which are known as the ◊sine, ◊cosine, and ◊tangent (so-called trigonometrical ratios). It is of practical importance in navigation, surveying, and simple harmonic motion in physics.

Invented by Hipparchus, trigonometry was developed by Ptolemy of Alexandria and was known to early Hindu and Arab mathematicians.

trimester in medicine, a stage of pregnancy lasting three months. A normal human pregnancy consists of three trimesters.

triple bond three covalent bonds between adjacent atoms, as in the ◊alkynes (–C≡C–).

triticale cereal crop of recent origin that is a cross between wheat *Triticum* and rye *Secale*. It can produce heavy yields of high-protein grain, principally for use as animal feed.

tritium radioactive isotope of hydrogen, three times as heavy as ordinary hydrogen, consisting of one proton and two neutrons. It has a half-life of 12.5 years.

Triton in astronomy, the largest of Neptune's moons. It has a diameter of 2,700 km/1,680 mi, and orbits Neptune every 5.88 days in a retrograde (east to west) direction. It takes the same time to rotate about its own axis as it does to make one revolution of Neptune.

It is slightly larger than the planet Pluto, which it is thought to resemble in composition and appearance. Probably Triton was formerly a separate body like Pluto but was captured by Neptune. Triton was discovered in 1846 by British astronomer William Lassell (1799–1880) only weeks after the discovery of Neptune. Triton's surface, as revealed by the *Voyager 2* space probe, has a temperature of 38K (−235°C/−391°F), making it the coldest known place in the solar system. It is covered with frozen nitrogen and methane, some of which evaporates to form a tenuous atmosphere with a pressure only 0.00001 that of the Earth at sea level. Triton has a pink south polar cap, probably coloured by the effects of solar radiation on methane ice. Dark streaks on Triton are thought to be formed by geysers of liquid nitrogen. The surface has few impact craters (the largest is the Mazomba, with a diameter of 27 km/17 mi), indicating that many of the earlier craters have been erased by the erupting and freezing of water (cryovulcanism).

troglodyte ancient Greek term for a cave dweller, designating certain pastoral peoples of the Caucasus, Ethiopia, and the S Red Sea coast of Egypt.

Trojan horse in computing, a ◊virus program that appears to function normally but, while undetected by the normal user, causes damage to other files or circumvents security procedures.

The earliest appeared in the UK in about 1988.

trolley bus bus driven by electric power collected from overhead wires. It has greater manoeuvrability than a tram (see ◊tramway).

trophic level in ecology, the position occupied by a species (or group of species) in a ◊food chain. The main levels are *primary producers* (photosynthetic plants), *primary consumers* (herbivores), *secondary consumers* (carnivores), and *decomposers* (bacteria and fungi).

trophoblast in medicine, the outer layer of the ovum that supplies nutrition to the embryo and attaches the ovum to the wall of the uterus.

tropical cyclone another term for ◊hurricane.

tropical disease any illness found mainly in hot climates. The most important tropical diseases worldwide are ◊malaria, ◊leishmaniasis, ◊sleeping sickness, lymphatic filiarasis, and ◊schistosomiasis. Other major scourges are ◊Chagas's disease, ◊leprosy, and ◊river blindness. Malaria kills about 1.5 million people each year, and produces chronic anaemia and tiredness in 100 times as many, while schistosomiasis is responsible for 1 million deaths a year. All the main tropical diseases are potentially curable, but the facilities for diagnosis and treatment are rarely adequate in the countries where they occur.

tropics the area between the tropics of Cancer and Capricorn, defined by the parallels of latitude approximately 23°30′ N and S of the Equator. They are the limits of the area of Earth's surface in which the Sun can be directly overhead. The mean monthly temperature is over 20°C/68°F.

Climates within the tropics lie in parallel bands. Along the Equator is the ◊intertropical convergence zone, characterized by high temperatures and year-round heavy rainfall. Tropical rainforests are found here. Along the tropics themselves lie the tropical high-pressure zones, characterized by descending dry air and desert conditions. Between these, the conditions vary seasonally between wet and dry, producing the tropical grasslands.

tropism or *tropic movement* the directional growth of a plant, or part of a plant, in response to an external stimulus such as gravity or light. If the movement is directed towards the stimulus it is described as positive; if away from it, it is negative. *Geotropism* for example, the response of plants to gravity, causes the root (positively geotropic) to grow downwards, and the stem (negatively geotropic) to grow upwards.

Phototropism occurs in response to light, *hydrotropism* to water, *chemotropism* to a chemical stimulus, and *thigmotropism*, or *haptotropism*, to physical contact, as in the tendrils of climbing plants when they touch a support and then grow around it.

Tropic movements are the result of greater rate of growth on one side of the plant organ than the other. Tropism differs from a ◊nastic movement in being influenced by the direction of the stimulus.

troposphere lower part of the Earth's ◊atmosphere extending about 10.5 km/6.5 mi from the Earth's surface, in which temperature decreases with height to about −60°C/−76°F except in local layers of temperature inversion. The *tropopause* is the upper boundary of the troposphere, above which the temperature increases slowly with height within the atmosphere. All of the Earth's weather takes place within the troposphere.

troy system system of units used for precious metals and gems. The pound troy (0.37 kg) consists of 12 ounces (each of 120 carats) or 5,760 grains (each equal to 65 mg).

truncation error in computing, an ◊error that occurs when a decimal result is cut off (truncated) after the maximum number of places allowed by the computer's level of accuracy.

trypanosomiasis any of several debilitating long-term diseases caused by a trypanosome (protozoan of the genus *Trypanosoma*). They include sleeping sickness (nagana) in Africa, transmitted by the bites of tsetse flies, and ◊Chagas's disease in Central and South America, spread by assassin bugs.

Trypanosomes can live in the bloodstream of humans and other vertebrates. Millions of people are affected in warmer regions of the world; the diseases also affect cattle, horses, and wild animals, which form a reservoir of infection.

trypsin an enzyme in the vertebrate gut responsible for the digestion of protein molecules. It is secreted by the pancreas but in an inactive form known as trypsinogen. Activation into working trypsin occurs only in the small intestine, owing to the action of another enzyme enterokinase, secreted by the wall of the duodenum. Unlike the digestive enzyme pepsin, found in the stomach, trypsin does not require an acid environment.

tsunami ocean wave generated by vertical movements of the sea floor resulting from ◊earthquakes or volcanic activity. Unlike waves generated by surface winds, the entire depth of water is involved in the wave motion. In the open ocean the tsunami takes the form of several successive waves, rarely in excess of 1 m/3 ft in height but travelling at speeds of 650-800 kph/400-500 mph. In the coastal shallows tsunamis slow down and build up producing huge swells over 15 m/45 ft high in some cases and over 30 m/90 ft in rare instances. The waves sweep inland causing great loss of life and property. On 26 May 1983,

an earthquake in the Pacific Ocean caused tsunamis up to 14 m/42 ft high, which killed 104 people along the western coast of Japan near Minehama, Honshu.

Before each wave there may be a sudden withdrawal of water from the beach. Used synonymously with tsunami, the popular term 'tidal wave' is misleading: tsunamis are not caused by the gravitational forces that affect ◊tides.

tuber swollen region of an underground stem or root, usually modified for storing food. The potato is a *stem tuber*, as shown by the presence of terminal and lateral buds, the 'eyes' of the potato. *Root tubers*, for example dahlias, developed from adventitious roots (growing from the stem, not from other roots) lack these. Both types of tuber can give rise to new individuals and so provide a means of ◊vegetative reproduction.

Unlike a bulb, a tuber persists for one season only; new tubers developing on a plant in the following year are formed in different places. See also ◊rhizome.

tuberculosis (TB) formerly known as *consumption* or *phthisis* infectious disease caused by the bacillus *Mycobacterium tuberculosis*. It takes several forms, of which pulmonary tuberculosis is by far the most common. A vaccine, ◊BCG, was developed around 1920 and the first antituberculosis drug, streptomycin, became available in 1944. The bacterium is mostly kept in check by the body's immune system; about 5% of those infected develop the disease. Treatment of patients with a combination of anti-TB medicines for 6–8 months produces a cure rate of 80%.

In pulmonary TB, a patch of inflammation develops in the lung, with formation of an abscess. Often, this heals spontaneously, leaving only scar tissue. The dangers are of rapid spread through both lungs (what used to be called 'galloping consumption') or the development of miliary

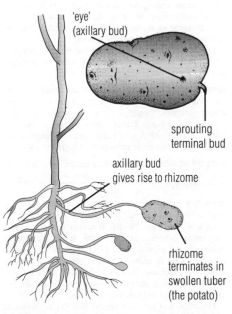

'eye'
(axillary bud)

sprouting
terminal bud

axillary bud
gives rise to rhizome

rhizome
terminates in
swollen tuber
(the potato)

tuber Tubers are produced underground from stems, as in the potato, or from roots, as in the dahlia. Tubers can grow into new plants.

tuberculosis (spreading in the bloodstream to other sites) or tuberculous ◊meningitis.

Over the last 15 years there has been a sharp resurgence in countries where the disease was in decline. The increase has been most marked in deprived inner city areas, particularly in the USA, and here there is a clear link between TB and HIV, the virus which causes AIDS. TB is the main cause of death in HIV positive individuals.

The last decade has seen the spread of drug-resistant strains of the TB bacterium. Many strains are now resistant to the two frontline drugs, isoniazid and rifampicin, and some are multi-drug resistant (MDR). Rare until its recent appearance in the USA, MDR TB is now spreading through a number of developing countries. It is untreatable and many of its victims have died. According to a 1996 WHO estimate there may be as many as 50 million people worldwide with the drug-resistant form of TB (Britain had its first case in 1995).

Tucana constellation of the southern hemisphere, represented as a toucan. It contains the second most prominent ◊globular cluster in the sky, 47 Tucanae, and the Small ◊Magellanic Cloud.

Tucana is one of the 11 constellations named by Johann Bayer early in the 17th century to complement the 65 constellations delineated in ancient times.

tufa or **travertine** soft, porous, ◊limestone rock, white in colour, deposited from solution from carbonate-saturated ground water around hot springs and in caves.

tumour overproduction of cells in a specific area of the body, often leading to a swelling or lump. Tumours are classified as **benign** or **malignant** (see ◊cancer). Benign tumours grow more slowly, do not invade surrounding tissues, do not spread to other parts of the body, and do not usually recur after removal. However, benign tumours can be dangerous in areas such as the brain. The most familiar types of benign tumour are warts on the skin. In some cases, there is no sharp dividing line between benign and malignant tumours.

In 1995, US doctors managed to halt tumour growth in mice by firing gold beads coated with DNA from a gene gun.

tundra region of high latitude almost devoid of trees, resulting from the presence of ◊permafrost. The vegetation consists mostly of grasses, sedges, heather, mosses, and lichens. Tundra stretches in a continuous belt across N North America and Eurasia. Tundra is also used to describe similar conditions at high altitudes.

tung oil oil used in paints and varnishes, obtained from tung trees native to China. (Genus *Aleurites*, family Euphorbiaceae.)

tungsten hard, heavy, grey-white, metallic element, symbol W (from German *Wolfram*), atomic number 74, relative atomic mass 183.85. It occurs in the minerals wolframite, scheelite, and hubertite. It has the highest melting point of any metal (3,410°C/6,170°F) and is added to steel to make it harder, stronger, and more elastic; its other uses include high-speed cutting tools, electrical elements, and thermionic couplings. Its salts are used in the paint and tanning industries.

Tungsten was first recognized in 1781 by Swedish chemist Karl Scheele in the ore scheelite. It was isolated in 1783 by Spanish chemists Fausto D'Elhuyar (1755–1833) and his brother Juan José (1754–1796).

tungsten ore either of the two main minerals, wolframite ($FeMn)WO_4$ and scheelite, $CaWO_4$, from which tungsten is extracted. Most of the world's tungsten reserves are in China, but the main suppliers are Bolivia, Australia, Canada, and the USA.

Tunguska Event explosion at Tunguska, central Siberia, Russia, in June 1908, which devastated around 6,500 sq km/2,500 sq mi of forest. It is thought to have been caused by either a cometary nucleus or a fragment of ◊Encke's comet about 200 m/220 yards across, or possibly an asteroid. The magnitude of the explosion was equivalent to an atom bomb (10–20 megatons) and produced a colossal shock wave; a bright falling object was seen 600 km/375 mi away and was heard up to 1,000 km/625 mi away.

tunnel passageway through a mountain, under a body of water, or underground. Tunnelling is a significant branch of civil engineering in both mining and transport. The difficulties naturally increase with the size, length, and depth of tunnel, but with the mechanical appliances now available no serious limitations are imposed. Granite or other hard rock presents little difficulty to modern power drills. In recent years there have been notable developments in linings (for example, concrete segments and steel liner plates), and in the use of rotary diggers and cutters and explosives.

turbine engine in which steam, water, gas, or air (see ◊windmill) is made to spin a rotating shaft by pushing on angled blades, like a fan. Turbines are among the most powerful machines. Steam turbines are used to drive generators in power stations and ships' propellers; water turbines spin the generators in hydroelectric power plants; and gas turbines (as jet engines; see ◊jet propulsion) power most aircraft and drive machines in industry.

The high-temperature, high-pressure steam for **steam turbines** is raised in boilers heated by furnaces burning coal, oil, or gas, or by nuclear energy. A steam turbine consists of a shaft, or rotor, which rotates inside a fixed casing (stator). The rotor carries 'wheels' consisting of blades, or vanes. The stator has vanes set between the vanes of the rotor, which direct the steam through the rotor vanes at the optimum angle. When steam expands through the turbine, it spins the rotor by ◊reaction. The steam engine of Hero of Alexandria (130 BC), called the *aeolipile*, was the prototype of this type of turbine, called a **reaction turbine**. Modern development of the reaction turbine is largely due to English engineer Charles Parsons. Less widely used is the **impulse turbine**, patented in 1882 by Carl Gustaf Patrick de Laval (1845–1913). It works by directing a jet of steam at blades on a rotor. Similarly there are reaction and impulse water turbines: impulse turbines work on the same principle as the water wheel and consist of sets of buckets arranged around the edge of a wheel; reaction turbines look much like propellers and are fully immersed in the water.

In a **gas turbine** a compressed mixture of air and gas, or vaporized fuel, is ignited, and the hot gases produced expand through the turbine blades, spinning the rotor. In the industrial gas turbine, the rotor shaft drives machines. In the jet engine, the turbine drives the compressor, which supplies the compressed air to the engine, but most of the power developed comes from the jet exhaust in the form of propulsive thrust.

turbocharger turbine-driven device fitted to engines to

force more air into the cylinders, producing extra power. The turbocharger consists of a 'blower', or ◊compressor, driven by a turbine, which in most units is driven by the exhaust gases leaving the engine.

turbofan jet engine of the type used by most airliners, so called because of its huge front fan. The fan sends air not only into the engine for combustion but also around the engine for additional thrust. This results in a faster and more fuel-efficient propulsive jet (see ◊jet propulsion).

turbojet jet engine that derives its thrust from a jet of hot exhaust gases. Pure turbojets can be very powerful but use a lot of fuel.

A single-shaft turbojet consists of a shaft (rotor) rotating in a casing. At the front is a multiblade ◊compressor, which takes in and compresses air and delivers it to one or more combustion chambers. Fuel (kerosene) is then sprayed in and ignited. The hot gases expand through a nozzle at the rear of the engine after spinning a ◊turbine. The turbine drives the compressor. Reaction to the backward stream of gases produces a forward propulsive thrust.

turboprop jet engine that derives its thrust partly from a jet of exhaust gases, but mainly from a propeller powered by a turbine in the jet exhaust. Turboprops are more economical than turbojets but can be used only at relatively low speeds.

A turboprop typically has a twin-shaft rotor. One shaft carries the compressor and is spun by one turbine, while the other shaft carries a propeller and is spun by a second turbine.

turbulence irregular fluid (gas or liquid) flow, in which vortices and unpredictable fluctuations and motions occur. ◊Streamlining reduces the turbulence of flow around an object, such as an aircraft, and reduces drag. Turbulent flow of a fluid occurs when the ◊Reynolds number is high.

turgor the rigid condition of a plant caused by the fluid contents of a plant cell exerting a mechanical pressure against the cell wall. Turgor supports plants that do not have woody stems.

turnip biennial plant widely cultivated in temperate regions for its edible white- or yellow-fleshed root and young leaves, which are used as a green vegetable. Closely allied to it is the ◊swede (*B. napus*). (*Brassica rapa*, family Cruciferae.)

turpentine solution of resins distilled from the sap of conifers, used in varnish and as a paint solvent but now largely replaced by ◊white spirit.

turquoise mineral, hydrous basic copper aluminium phosphate, $CuAl_6(PO_4)_4(OH)_8.5H_2O$. Blue-green, blue, or green, it is a gemstone. Turquoise is found in Australia, Egypt, Ethiopia, France, Germany, Iran, Turkestan, Mexico, and southwestern USA. It was originally introduced into Europe through Turkey, from which its name is derived.

turtle small computer-controlled wheeled robot. The turtle's movements are determined by programs written by a computer user, typically using the high-level programming language ◊LOGO.

TVP (abbreviation for *texturized vegetable protein*) meat substitute usually made from soya beans. In manufacture, the soya-bean solids (what remains after oil has been removed) are ground finely and mixed with a binder to form a sticky mixture. This is forced through a spinneret and extruded into fibres, which are treated with salts and flavourings, wound into hanks, and then chopped up to resemble meat chunks.

twin one of two young produced from a single pregnancy. Human twins may be genetically identical (monozygotic), having been formed from a single fertilized egg that splits into two cells, both of which became implanted. Nonidentical (fraternal or dizygotic) twins are formed when two eggs are fertilized at the same time.

two's complement number system number system, based on the ◊binary number system, that allows both positive and negative numbers to be conveniently represented for manipulation by computer.

In the two's complement system the most significant column heading (the furthest to the left) is always taken to represent a negative number.

For example, the four-column two's complement number 1101 stands for -3, since: $-8 + 4 + 1 = -3$.

two-stroke cycle operating cycle for internal combustion piston engines. The engine cycle is completed after just two strokes (up or down) of the piston, which distinguishes it from the more common ◊four-stroke cycle. Power mowers and lightweight motorcycles use two-stroke petrol engines, which are cheaper and simpler than four-strokes.

Most marine diesel engines are also two-stroke. In a typical two-stroke motorcycle engine, fuel mixture is drawn into the crankcase as the piston moves up on its first stroke to compress the mixture above it. Then the compressed mixture is ignited, and hot gases are produced, which drive the piston down on its second stroke. As it moves down, it uncovers an opening (port) that allows the fresh fuel mixture in the crankcase to flow into the combustion space above the piston. At the same time, the exhaust gases leave through another port.

tympanum in mammals, the middle ear. It contains three small bones, the malleus, anvil and stapes, which connect the ear drum to the internal ear. They convert air waves that strike on the drum into mechanical movements that are essential for hearing.

typesetting means by which text, or copy, is prepared for ◊printing, now usually carried out by computer.

Text is keyed on a typesetting machine in a similar way to typing. Laser or light impulses are projected on to light-sensitive film that, when developed, can be used to make plates for printing.

typewriter keyboard machine that produces characters on paper. The earliest known typewriter design was patented by Henry Mills in England in 1714. However, the first practical typewriter was built in 1867 in Milwaukee, Wisconsin, USA, by Christopher Sholes, Carlos Glidden, and Samuel Soulé. By 1873 Remington and Sons, US gunmakers, produced under contract the first typing machines for sale and in 1878 patented the first with lower-case as well as upper-case (capital) letters.

The first typewriter patented by Sholes included an alphabetical layout of keys, but Remington's first commercial typewriter had the QWERTY keyboard, designed by Sholes to slow down typists who were too fast for their mechanical keyboards.

Other layouts include the Dvorak keyboard developed by John Dvorak in 1932, in which the most commonly used letters are evenly distributed between left and right, and are positioned under the strongest fingers. Later developments included tabulators from about 1898, portable machines from about 1907, the gradual introduction of electrical operation (allowing increased speed, since the keys are touched, not depressed), proportional spacing in 1940, and the rotating typehead with stationary plates in 1962. More recent typewriters work electronically, are equipped with a memory, and can be given an interface that enables them to be connected to a computer. The ◊wordprocessor has largely replaced the typewriter for typing letters and text.

typhoid fever acute infectious disease of the digestive tract, caused by the bacterium *Salmonella typhi*, and usually contracted through a contaminated water supply. It is characterized by bowel haemorrhage and damage to the spleen. Treatment is with antibiotics.

The symptoms begin 10–14 days after ingestion and include fever, headache, cough, constipation, and rash. The combined TAB vaccine protects both against typhoid and the milder, related condition known as *paratyphoid fever*.

typhoon violent revolving storm, a ◊hurricane in the W Pacific Ocean.

typhus any one of a group of infectious diseases caused by bacteria transmitted by lice, fleas, mites, and ticks. Symptoms include fever, headache, and rash. The most serious form is epidemic typhus, which also affects the brain heart, lungs, and kidneys and is associated with insanitary overcrowded conditions. Treatment is by antibiotics.

The small bacteria responsible are of the genus *Rickettsia*, especially *R. pronazekii*. A preventive vaccine exists.

Typhus

Typhus claimed fewer victims in World War II than fighting. This was the first war in which this was the case since the first documented typhus epidemic in 1489.

---

tyre (North American *tire*) inflatable rubber hoop fitted round the rims of bicycle, car, and other road-vehicle wheels. The first pneumatic rubber tyre was patented in 1845 by the Scottish engineer Robert William Thomson (1822–73), but it was Scottish inventor John Boyd Dunlop of Belfast who independently reinvented pneumatic tyres for use with bicycles between 1888–89. The rubber for car tyres is hardened by ◊vulcanization.

tyrosine in medicine, an ◊amino acid. It is important in the production of ◊catecholamines, such as adrenaline, noradrenaline, and dopamine, which are involved in the transmission of nerve impulses. It is also important in the production of thyroid hormones, such as thyroxine.

Tyuratam site of the ◊Baikonur Cosmodrome in Kazakhstan.

UHF (abbreviation for *ultra high frequency*) referring to radio waves of very short wavelength, used, for example, for television broadcasting.

UHT abbreviation for *ultra-heat treated* or ◊*ultra-heat treatment*.

ulcer any persistent breach in a body surface (skin or mucous membrane). It may be caused by infection, irritation, or tumour and is often inflamed. Common ulcers include aphthous (mouth), gastric (stomach), duodenal, decubitus ulcers (pressure sores), and those complicating varicose veins.

Treatment of ulcers depends on the site. Drugs are the first line of attack against peptic ulcers (those in the digestive tract), though surgery may become necessary. Bleeding stomach ulcers can be repaired without an operation by the use of endoscopy: a flexible fibre-optic tube is passed into the stomach and under direct vision fine instruments are used to repair the tissues.

Ulcers

Headless bedbugs were applied to ulcers in the 16th century, and crushed bedbugs were still believed to be effective in treating ulcers in 18th-century China. It was not until this century that positive evidence was found to support the antibacterial properties of insect haemolymph.

---

ulna one of the two bones found in the lower limb of the tetrapod (four-limbed) vertebrate. It articulates with the shorter radius and ◊humerus (upper arm bone) at one end and with the radius and wrist bones at the other.

ultrabasic in geology, an igneous rock with a lower silica content than basic rocks (less than 45% silica). Part of a system of classification based on the erroneous concept of silica acidity and basicity. Once used widely it has now been largely replaced by the term **ultramafic**.

ultrafiltration process by which substances in solution are separated on the basis of their molecular size. A solution is forced through a membrane with pores large enough to permit the passage of small solute molecules but not large ones.

Ultrafiltration is a vital mechanism in the vertebrate kidney: the cell membranes lining the Bowman's capsule act as semipermeable membranes allowing water and low-molecular-weight substances such as urea and salts to pass through into the urinary tubules but preventing the larger proteins from being lost from the blood.

ultra-heat treatment (UHT) preservation of milk by raising its temperature to 132°C/269°F or more. It uses higher temperatures than pasteurization, and kills all bacteria present, giving the milk a long shelf life but altering the flavour.

ultrasonics branch of physics dealing with the theory and application of ultrasound: sound waves occurring at frequencies too high to be heard by the human ear (about 20 kHz).

The earliest practical application of ultrasonics was the detection of submarines during World War I by reflecting pulses of sound from them (see ◊sonar). Similar principles are now used in industry for nondestructive testing of materials and in medicine to produce images of internal organs and developing fetuses (◊ultrasound scanning). High-power ultrasound can be used for cleaning, welding plastics, and destroying kidney stones without surgery.

ultrasound scanning or **ultrasonography** in medicine, the use of ultrasonic pressure waves to create a diagnostic image. It is a safe, noninvasive technique that often eliminates the need for exploratory surgery.

The sound waves transmitted through the body are absorbed and reflected to different degrees by different body tissues.

ultraviolet astronomy study of cosmic ultraviolet emissions using artificial satellites. The USA has launched a series of satellites for this purpose, receiving the first useful data in 1968. Only a tiny percentage of solar ultraviolet radiation penetrates the atmosphere, this being the less dangerous longer-wavelength ultraviolet. The dangerous shorter-wavelength radiation is absorbed by gases in the ozone layer high in the Earth's upper atmosphere.

The US Orbiting Astronomical Observatory (OAO) satellites provided scientists with a great deal of information regarding cosmic ultraviolet emissions. *OAO-1*, launched in 1966, failed after only three days, although *OAO-2*, put into orbit in 1968, operated for four years instead of the intended one year, and carried out the first ultraviolet observations of a supernova and also of Uranus. *OAO-3* (*Copernicus*), launched in 1972, continued transmissions into the 1980s and discovered many new ultraviolet sources. The *International Ultraviolet Explorer* (*IUE*), launched in Jan 1978 and ceased operation in Sept 1996, observed all the main objects in the Solar System (including Halley's comet), stars, galaxies, and the interstellar medium.

ultraviolet radiation electromagnetic radiation invisible to the human eye, of wavelengths from about 400 to 4 nm (where the ◊X-ray range begins). Physiologically, ultraviolet radiation is extremely powerful, producing sunburn and causing the formation of vitamin D in the skin.

Levels of ultraviolet radiation have risen an average of 6.8% a decade in the northern hemisphere and 9.9% in the southern hemisphere, according to data gathered by the Total Ozone Mapping Spectrometer on the *Nimbus 7* satellite.

Ultraviolet rays are strongly germicidal and may be produced artificially by mercury vapour and arc lamps for therapeutic use. The radiation may be detected with ordinary photographic plates or films. It can also be studied by its fluorescent effect on certain materials. The desert iguana *Disposaurus dorsalis* uses it to locate the boundaries of its territory and to find food.

Ulysses space probe to study the Sun's poles, launched in 1990 by a US space shuttle. It is a joint project by NASA and the European Space Agency. In Feb 1992, the gravity of Jupiter swung *Ulysses* on to a path that looped it first under the Sun's south pole in 1994 and then over the north pole in 1995 to study the Sun and solar wind at latitudes not observable from the Earth.

umbilical cord connection between the ◊embryo and the ◊placenta of placental mammals. It has one vein and two arteries, transporting oxygen and nutrients to the developing young, and removing waste products. At birth, the connection between the young and the placenta is no longer necessary. The umbilical cord drops off or is severed, leaving a scar called the navel.

umbra central region of a ◊shadow that is totally dark because no light reaches it, and from which no part of the light source can be seen (compare ◊penumbra). In astronomy, it is a region of the Earth from which a complete ◊eclipse of the Sun or Moon can be seen.

uncertainty principle or *indeterminacy principle* in quantum mechanics, the principle that it is meaningless to speak of a particle's position, momentum, or other parameters, except as results of measurements; measuring, however, involves an interaction (such as a ◊photon of light bouncing off the particle under scrutiny), which must disturb the particle, though the disturbance is noticeable only at an atomic scale. The principle implies that one cannot, even in theory, predict the moment-to-moment behaviour of such a system.

It was established by German physicist Werner Heisenberg, and gave a theoretical limit to the precision with which a particle's momentum and position can be measured simultaneously: the more accurately the one is determined, the more uncertainty there is in the other.

In effect, we have redefined the task of science to be the discovery of laws that will enable us to predict events up to the limits set by the uncertainty principle.

On **uncertainty principle** Stephen Hawking

unconformity surface of erosion or nondeposition eventually overlain by younger ◊sedimentary rock strata and preserved in the geologic record. A surface where the ◊beds above and below lie at different angles is called an *angular unconformity*. The boundary between older igneous or metamorphic rocks that are truncated by erosion and later covered by younger sedimentary rocks is called a *nonconformity*.

unconscious in psychoanalysis, a part of the personality of which the individual is unaware, and which contains impulses or urges that are held back, or repressed, from conscious awareness.

underground (North American *subway*) rail service that runs underground. The first underground line in the world was in London, opened in 1863; it was essentially a roofed-in trench. The London Underground is still the longest, with over 400 km/250 mi of routes. Many large cities throughout the world have similar systems, and Moscow's underground, the Metro, handles up to 6.5 million passengers a day.

undernourishment condition that results from consuming too little food over a period of time. Like *malnutrition* – the result of a diet that is lacking in certain nutrients (such as protein or vitamins) – undernourishment is common in poor countries. Both lead to a reduction in mental and physical efficiency, a lowering of resistance to disease in general, and often to deficiency diseases such as beriberi or anaemia. In the Third World, lack of adequate food is a common cause of death.

unicellular organism animal or plant consisting of a single cell. Most are invisible without a microscope but a few, such as the giant ◊amoeba, may be visible to the naked eye. The main groups of unicellular organisms are bacteria, protozoa, unicellular algae, and unicellular fungi or yeasts. Some become disease-causing agents, ◊pathogens.

unidentified flying object or *UFO* any light or object seen in the sky whose immediate identity is not apparent. Despite unsubstantiated claims, there is no evidence that UFOs are alien spacecraft. On investigation, the vast majority of sightings turn out to have been of natural or identifiable objects, notably bright stars and planets, meteors, aircraft, and satellites, or to have been perpetrated by pranksters. The term *flying saucer* was coined in 1947.

unified field theory in physics, the theory that attempts to explain the four fundamental forces (strong nuclear, weak nuclear, electromagnetic, and gravity) in terms of a single unified force (see ◊particle physics).

Research was begun by Albert Einstein and, by 1971, a theory developed by US physicists Steven Weinberg and Sheldon Glashow, Pakistani physicist Abdus Salam, and others, had demonstrated the link between the weak and electromagnetic forces. The next stage is to develop a theory (called the ◊grand unified theory) that combines the strong nuclear force with the electroweak force. The final stage will be to incorporate gravity into the scheme. Work on the ◊superstring theory indicates that this may be the ultimate 'theory of everything'.

uniformitarianism in geology, the principle that processes that can be seen to occur on the Earth's surface today are the same as those that have occurred throughout geological time. For example, desert sandstones containing sand-dune structures must have been formed under conditions similar to those present in deserts today. The principle was formulated by James Hutton and expounded by Charles Lyell.

unit standard quantity in relation to which other quantities are measured. There have been many systems of units. Some ancient units, such as the day, the foot, and the pound, are still in use. ◊SI units, the latest version of the metric system, are widely used in science.

universal indicator in chemistry, a mixture of ◊pH indicators, used to gauge the acidity or alkalinity of a solution. Each component changes colour at a different pH value, and so the indicator is capable of displaying a range of colours, according to the pH of the test solution, from red (at pH 1) to purple (at pH 13).

universal joint flexible coupling used to join rotating shafts; for example, the drive shaft in a car. In a typical universal joint the ends of the shafts to be joined end in U-shaped yokes. They dovetail into each other and pivot flexibly about an X-shaped spider. This construction allows side-to-side and up-and-down movement, while still transmitting rotary motion.

universe all of space and its contents, the study of which is called ◊cosmology. The universe is thought to be between 10 billion and 20 billion years old, and is mostly empty space, dotted with ◊galaxies for as far as telescopes can see. The most distant detected galaxies and ◊quasars lie 10 billion light years or more from Earth, and are moving farther apart as the universe expands. Several theories attempt to explain how the universe came into being and evolved; for example, the ◊Big Bang theory of an expanding universe originating in a single explosive event, and the contradictory ◊steady-state theory.

Apart from those galaxies within the ◊Local Group, all the galaxies we see display ◊red shifts in their spectra, indicating that they are moving away from us. The farther we look into space, the greater are the observed red shifts, which implies that the more distant galaxies are receding at ever greater speeds.

This observation led to the theory of an expanding universe, first proposed by Edwin Hubble in 1929, and to Hubble's law, which states that the speed with which one galaxy moves away from another is proportional to its distance from it. Current data suggest that the galaxies are moving apart at a rate of 50–100 kps/30–60 mps for every million ◊parsecs of distance.

UNIX multiuser ◊operating system designed for minicomputers but becoming increasingly popular on large microcomputers, workstations, mainframes, and supercomputers. It was developed by AT&T's Bell Laboratories in the USA during the late 1960s, using the programming language ◊C. It could therefore run on any machine with a C compiler, so ensuring its wide portability. Its wide range of functions and flexibility, in addition to the fact that it was available free between 1976–83, have made it widely used by universities and in commercial software.

unleaded petrol petrol manufactured without the addition of ◊antiknock. It has a slightly lower octane rating than leaded petrol, but has the advantage of not polluting the atmosphere with lead compounds. Many cars can be converted to running on unleaded petrol by altering the timing of the engine, and most new cars are designed to do so.

Cars fitted with a ◊catalytic converter must use unleaded fuel.

Aromatic hydrocarbons and alkenes are added to unleaded petrol instead of lead compounds to increase the octane rating. After combustion the hydrocarbons produce volatile organic compounds. These have been linked to cancer, and are involved in the formation of phytochemical smog. A low-lead fuel is less toxic than unleaded petrol for use in cars that are not fitted with a catalytic converter.

The use of unleaded petrol has been standard in the USA for some years.

Unnilennium

The rarest element of all is the synthesized element unnilennium. A single atom of this element was produced by German scientists in 1982. Unfortunately, the atom vanished by radioactive decay almost instantly.

---

unnilennium temporary name assigned to ◊meitnerium, atomic number 109.

unnilhexium temporary identification assigned to the element ◊seaborgium between 1974–97.

unniloctium temporary name for ◊hassium atomic number 108.

unnilpentium temporary name assigned to ◊dubnium.

unnilquadium temporary identification assigned to the element ◊rutherfordium between 1964–97.

unnilseptium temporary identification assigned to the element ◊bohrium between 1964–97.

unsaturated compound chemical compound in which two adjacent atoms are bonded by a double or triple covalent bond.

Examples are ◊alkenes and ◊alkynes, where the two adjacent atoms are both carbon, and ◊ketones, where the unsaturation exists between atoms of different elements (carbon and oxygen). The laboratory test for unsaturated compounds is the addition of bromine water; if the test substance is unsaturated, the bromine water will be decolorized.

unsaturated solution solution that is capable of dissolving more solute than it already contains at the same temperature.

unnunnilium synthesized radioactive element of the ◊transactinide series, symbol Uun, atomic number 110, relative atomic mass 269. It was discovered in Oct 1994, detected for a millisecond, at the GSI heavy-ion cyclotron, Darmstadt, Germany, while lead atoms were bombarded with nickel atoms.

unununium synthesized radioactive element of the ◊transactinide series, symbol Uuu, atomic number 111, relative atomic mass 272. It was detected at GSI heavy-ion cyclotron, Darmstadt, Germany, in Dec 1994, when they bombarded bismuth-209 with nickel.

upthrust upward force experienced by all objects that are totally or partially immersed in a fluid (liquid or gas). It acts against the weight of the object, and, according to Archimedes' principle, is always equal to the weight of the fluid displaced by that object. An object will float when the upthrust from the fluid is equal to its weight.

uraninite uranium oxide, UO_2, an ore mineral of uranium, also known as **pitchblende** when occurring in massive form. It is black or brownish black, very dense, and radioactive. It occurs in veins and as massive crusts, usually associated with granite rocks.

uranium hard, lustrous, silver-white, malleable and ductile, radioactive, metallic element of the ◊actinide series, symbol U, atomic number 92, relative atomic mass 238.029. It is the most abundant radioactive element in the Earth's crust, its decay giving rise to essentially all radioactive elements in nature; its final decay product is the stable element lead. Uranium combines readily with most elements to form compounds that are extremely poisonous. The chief ore is ◊pitchblende, in which the element was discovered by German chemist Martin Klaproth in 1789; he named it after the planet Uranus, which had been discovered in 1781.

Small amounts of certain compounds containing uranium have been used in the ceramics industry to make orange-yellow glazes and as mordants in dyeing; however, this practice was discontinued when the dangerous effects of radiation became known.

Uranium is one of three fissile elements (the others are thorium and plutonium). It was long considered to be the element with the highest atomic number to occur in nature. The isotopes U-238 and U-235 have been used to help determine the age of the Earth.

Uranium-238, which comprises about 99% of all naturally occurring uranium, has a half-life of 4.51×10^9 years. Because of its abundance, it is the isotope from which fissile plutonium is produced in breeder ◊nuclear reactors. The fissile isotope U-235 has a half-life of 7.13×10^8 years and comprises about 0.7% of naturally occurring uranium; it is used directly as a fuel for nuclear reactors and in the manufacture of nuclear weapons.

uranium ore material from which uranium is extracted, often a complex mixture of minerals. The main ore is uraninite (or pitchblende) UO_2, which is commonly found with sulphide minerals. The USA, Canada, and South Africa are the main producers in the West.

Uranus the seventh planet from the Sun, discovered by William Herschel in 1781. It is twice as far out as the sixth planet, Saturn. Uranus has a mass 14.5 times that of Earth. The spin axis of Uranus is tilted at 98°, so that one pole points towards the Sun, giving extreme seasons.

Mean distance from the Sun 2.9 billion km/1.8 billion mi.

Equatorial diameter 50,800 km/31,600 mi.

Rotation period 17.2 hr.

Year 84 Earth years.

Atmosphere deep atmosphere composed mainly of hydrogen and helium.

Surface composed primarily of hydrogen and helium but may also contain heavier elements, which might account for Uranus's mean density being higher than Saturn's

Satellites 15 moons; 11 thin rings around the planet's equator were discovered in 1977.

Uranus has a peculiar magnetic field, whose axis is tilted at 60° to its axis of spin, and is displaced about one-third of the way from the planet's centre to its surface. Uranus spins from east to west, the opposite of the other planets, with the exception of Venus and possibly Pluto. The rotation rate of the atmosphere varies with latitude, from about 16 hours in mid-southern latitudes to longer than 17 hours at the equator.

Uranus's equatorial ring system comprises 11 rings. The ring furthest from the planet centre (51,000 km/31,800 mi), Epsilon, is 100 km/62 mi at its widest point. In 1995, US astronomers determined the ring particles contained long-chain hydrocarbons. Looking at the brightest region of Epsilon, they were also able to calculate the ◊precession of Uranus as 264 days, the fastest known precession in the Solar System.

urea $CO(NH_2)_2$ waste product formed in the mammalian liver when nitrogen compounds are broken down. It is filtered from the blood by the kidneys, and stored in the bladder as urine prior to release. When purified, it is a white, crystalline solid. In industry it is used to make urea-formaldehyde plastics (or resins), pharmaceuticals, and fertilizers.

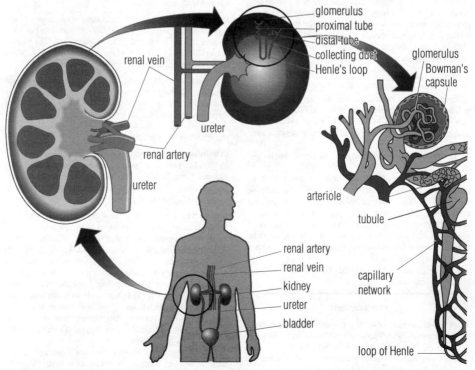

urinary system The human urinary system.

ureter tube connecting the kidney to the bladder. Its wall contains fibres of smooth muscle whose contractions aid the movement of urine out of the kidney.

urethra in mammals, a tube connecting the bladder to the exterior. It carries urine and, in males, semen.

uric acid $C_5H_4N_4O_3$ nitrogen-containing waste substance, formed from the breakdown of food and body protein. It is only slightly soluble in water. Uric acid is the normal means by which most land animals that develop in a shell (birds, reptiles, insects, and land gastropods) deposit their waste products. The young are unable to get rid of their excretory products while in the shell and therefore store them in this insoluble form.

Humans and other primates produce some uric acid as well as urea, the normal nitrogenous waste product of mammals, adult amphibians, and many marine fishes. If formed in excess and not excreted, uric acid may be deposited in sharp crystals in the joints and other tissues, causing gout; or it may form stones (calculi) in the kidneys or bladder.

Uric acid

Pigs never suffer from gout. In humans, this painful disease is caused by a build-up of uric acid. Pigs have an enzyme that breaks uric acid into soluble components. Humans do not have this enzyme, so they suffer from gout and pigs do not.

urinary system system of organs that removes nitrogenous waste products and excess water from the bodies of animals. In vertebrates, it consists of a pair of kidneys, which produce urine; ureters, which drain the kidneys; and (in bony fish, amphibians, some reptiles, and mammals) a bladder that stores the urine before its discharge. In mammals, the urine is expelled through the urethra; in other vertebrates, the urine drains into a common excretory chamber called a ◊cloaca, and the urine is not discharged separately.

urine amber-coloured fluid filtered out by the kidneys from the blood. It contains excess water, salts, proteins, waste products in the form of urea, a pigment, and some acid. The kidneys pass it through two fine tubes (ureters) to the bladder, which may act as a reservoir for up to 0.7l/1.5 pt at a time. In mammals, it then passes into the urethra, which opens to the outside by a sphincter (constricting muscle) under voluntary control. In reptiles and birds, nitrogenous wastes are discharged as an almost solid substance made mostly of ◊uric acid, rather than urea.

urology medical speciality concerned with diseases of the urinary tract.

Ursa Major the third largest constellation in the sky, in the north polar region. Its seven brightest stars make up the familiar shape or asterism of the **Big Dipper** or **Plough**. The second star of the handle of the dipper, called Mizar, has a companion star, Alcor.

urine Urine consists of excess water and waste products that have been filtered from the blood by the kidneys; it is stored in the bladder until it can be expelled from the body via the urethra. Analysing the composition of an individual's urine can reveal a number of medical conditions, such as poorly functioning kidneys, kidney stones, and diabetes.

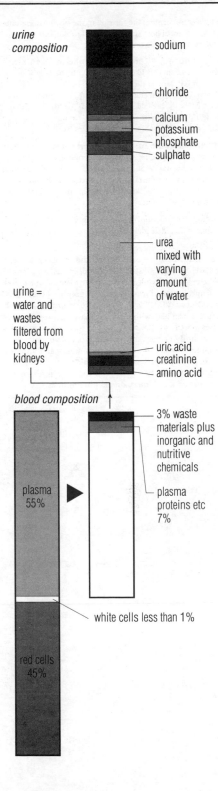

urine composition

- sodium
- chloride
- calcium
- potassium
- phosphate
- sulphate
- urea mixed with varying amount of water
- uric acid
- creatinine
- amino acid

urine = water and wastes filtered from blood by kidneys

blood composition

- 3% waste materials plus inorganic and nutritive chemicals
- plasma proteins etc 7%

plasma 55%

white cells less than 1%

red cells 45%

Two stars forming the far side of the bowl act as pointers to the north pole star, ◊Polaris. Dubhe, one of them, is the constellation's brightest star.

Ursa Minor small constellation of the northern hemisphere. It is shaped like a dipper, with the bright north pole star ◊Polaris at the end of the handle.

Two other bright stars in this group, Beta and Gamma Ursae Minoris, are called 'the Guards' or 'the Guardians of the Pole'. The constellation also contains the orange subgiant Kochab, about 95 light years from Earth.

urticaria or ***nettle rash*** or ***hives*** irritant skin condition characterized by itching, burning, stinging, and the spontaneous appearance of raised patches of skin. Treatment is usually by ◊antihistamines or steroids taken orally or applied as lotions. Its causes are varied and include allergy and stress.

user ID (contraction of ***user id***entification) name or nickname that identifies the user of a computer system or network.

user interface in computing, the procedures and methods through which the user operates a program. These might include ◊menus, input forms, error messages, and keyboard procedures. A ◊graphical user interface (GUI or WIMP) is one that makes use of icons (small pictures) and allows the user to make menu selections with a mouse.

The study of the ways in which people interact with computers is a sub-branch of ergonomics. It aims to make it easier for people to use computers effectively and comfortably, and has become a focus of research for many national and international programmes.

US Naval Observatory US government observatory in Washington DC, which provides the nation's time service and publishes almanacs for navigators, surveyors, and astronomers. It contains a 66 cm/26 in refracting telescope opened 1873. A 1.55 m/61 in reflector for measuring positions of celestial objects was opened in 1964 at Flagstaff, Arizona.

UTC abbreviation for ***coordinated universal time***, the standard measurement of ◊time.

uterus hollow muscular organ of female mammals, located between the bladder and rectum, and connected to the Fallopian tubes above and the vagina below. The embryo develops within the uterus, and in placental mammals is attached to it after implantation via the ◊placenta and umbilical cord. The lining of the uterus changes during the ◊menstrual cycle. In humans and other higher primates, it is a single structure, but in other mammals it is paired.

The outer wall of the uterus is composed of smooth muscle, capable of powerful contractions (induced by hormones) during childbirth.

utility program in computing, a systems program designed to perform a specific task related to the operation of the computer when requested to do so by the computer user. For example, a utility program might be used to complete a screen dump, format a disc, or convert the format of a data file so that it can be accessed by a different applications program.

UV in physics, abbreviation for ***ultraviolet***.

v in physics, symbol for *velocity*.

V roman numeral for *five*; in physics, symbol for *volt*.

vaccine any preparation of modified pathogens (viruses or bacteria) that is introduced into the body, usually either orally or by a hypodermic syringe, to induce the specific ◊antibody reaction that produces ◊immunity against a particular disease.

In 1796, Edward Jenner was the first to inoculate a child successfully with cowpox virus to produce immunity to smallpox. His method, the application of an infective agent to an abraded skin surface, is still used in smallpox inoculation. *See feature overleaf.*

vacuole in biology, a fluid-filled, membrane-bound cavity inside a cell. It may be a reservoir for fluids that the cell will secrete to the outside, or may be filled with excretory products or essential nutrients that the cell needs to store.

vacuum in general, a region completely empty of matter; in physics, any enclosure in which the gas pressure is considerably less than atmospheric pressure (101,325 pascals).

vacuum cleaner cleaning device invented in 1901 by the Scot Hubert Cecil Booth (1871–1955). Having seen an ineffective dust-blowing machine, he reversed the process so that his machine (originally on wheels, and operated from the street by means of tubes running into the house) operated by suction.

vacuum flask or *Dewar flask* or *Thermos flask* container for keeping things either hot or cold. It has two silvered glass walls with a vacuum between them, in a metal or plastic outer case. This design reduces the three forms of heat transfer: radiation (prevented by the silvering), conduction, and convection (both prevented by the vacuum). A vacuum flask is therefore equally efficient at keeping cold liquids cold or hot liquids hot.

vagina the lower part of the reproductive tract in female mammals, linking the uterus to the exterior. It admits the

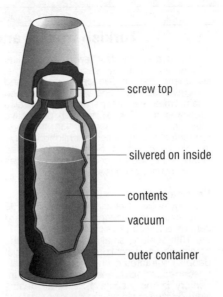

vacuum flask The vacuum flask allows no heat to escape from or enter its contents. It has double walls with a vacuum between to prevent heat loss by conduction. Radiation is prevented by silvering the walls. The vacuum flask was invented by Scottish chemist James Dewar in about 1872.

The labels on the figure read:
- screw top
- silvered on inside
- contents
- vacuum
- outer container

penis during sexual intercourse, and is the birth canal down which the baby passes during delivery.

valence electron in chemistry, an electron in the outermost shell of an ◊atom. It is the valence electrons that are involved in the formation of ionic and covalent bonds (see ◊molecule). The number of electrons in this outermost shell represents the maximum possible valence for many elements and matches the number of the group that the element occupies in the ◊periodic table of the elements.

valency in chemistry, the measure of an element's ability to combine with other elements, expressed as the number of atoms of hydrogen (or any other standard univalent element) capable of uniting with (or replacing) its atoms. The number of electrons in the outermost shell of the atom dictates the combining ability of an element.

The elements are described as uni-, di-, tri-, and tetravalent when they unite with one, two, three, and four univalent atoms respectively. Some elements have *variable valency*: for example, nitrogen and phosphorus have a valency of both three and five. The valency of oxygen is two: hence the formula for water, H_2O (hydrogen being univalent).

validation in computing, the process of checking input data to ensure that it is complete, accurate, and reason-

Valency Shell

group number	I	II	III	IV	V	VI	VII
element	Na	Mg	Al	Si	P	S	Cl
atomic number	11	12	13	14	15	16	17
electron arrangement	2.8.1	2.8.2	2.8.3	2.8.4	2.8.5	2.8.6	2.8.7
valencies	1	2	3	4(2)	5(3)	6(2)	7(1)

Turkish travels and English milkmaids

Today, during their first few months of life, infants are routinely vaccinated against diphtheria, tetanus, whooping cough, and poliomyelitis. After one year, vaccination against measles, mumps, and rubella is also recommended. Thanks to such preventive medicine and to proper nutrition, the common infectious diseases of childhood have largely disappeared from the developed world; if they do occur, the consequences are not usually serious.

Vaccinations

The regular use of vaccination began in 1796 as a result of the pioneering work of Edward Jenner, a British physician (1749–1823). In the 18th century, smallpox was one of the commonest and most deadly diseases. Most people contracted it, and a face completely unscarred by the disease was rare. In 17th-century London, some 10% of all deaths were due to smallpox.

Because the disease was so common, it was well known that an earlier, non-fatal attack of smallpox conferred immunity in any following epidemics. In Eastern countries, people were deliberately exposed to mild forms of the disease. This method was brought back to England in 1721 by Lady Mary Wortley Montagu, wife of the British Ambassador to Turkey. She had her own children 'vaccinated' (the procedure was then called *variolation*), encountering much prejudice as a result.

Kill or cure?

In England, an epidemic of smallpox swept Gloucestershire in 1788, and variolation (first described in 1713) was widely practised. It involved scratching a vein in the arm of a healthy person, and working into it a small amount of matter from a smallpox pustule taken from a person with a mild attack of the disease. This risky procedure had two major disadvantages. The inoculated subject, unless isolated, was likely to start a fresh smallpox epidemic; and if the dose was too virulent, the resulting disease was fatal.

Edward Jenner had a country medical practice, and was familiar with both human and animal diseases. He noted that milkmaids often caught the disease cowpox, and inquired further. He saw that milking was regularly done by both men and maidservants. The men, after changing the dressings on horses suffering from a disease called 'the grease', went on to milk the cows, thus infecting them at the same time. In due course, the cows became diseased, and in turn the milkmaids who milked them caught cowpox. Jenner wrote: 'Thus the disease makes its progress from the horse to the nipple of the cow, and from the cow to the human subject.' He went on: 'What makes the cowpox virus so extremely singular is that the person who has been thus affected is for ever secure from the infection of the smallpox.' Jenner then describes a great number of instances, in proof of his observations. Here are some of them.

Case I. Joseph Merret, the undergardener to the Earl of Berkeley, had been a farmer's servant in 1770, and sometimes helped with the milking. He also attended the horses. Merret caught cowpox. In April 1795, he was treated during a general inoculation that took place. Jenner found that despite repeatedly inserting variolous matter into Merret's arm, it was impossible to infect him with smallpox. During the whole time his family had smallpox, Merret stayed with them, and remained perfectly healthy. Jenner was at pains to make sure that Merret had at no time previously caught smallpox.

Case II. Sarah Portlock nursed one of her own children who, in 1792, had accidentally caught smallpox. She considered herself safe from infection, for as a farmer's servant 27 years previously, she had contracted cowpox. She remained in the same room as her child, and as in the previous case, variolation produced no disease.

Jenner then performed the experiment that was to make him famous, and give medicine vaccination, one of the most powerful weapons against disease.

The first vaccination

Case XVII. On 14 May 1796, Jenner selected a healthy eight-year-old boy and inoculated him with cowpox, taken from a sore in the hand of a dairymaid. He inserted the infected matter in two incisions, each about 25 mm/1 in long. The boy showed only mild symptoms: on the seventh day he complained of a slight headache, became a little chilly and suffered loss of appetite.

Then on 1 July 1796, Jenner inoculated the boy with variolous matter, inserting it in several slight punctures and incisions in both arms. No disease followed. The only symptoms the boy showed were those of someone who had recovered from smallpox, or had previously suffered from cowpox. Several months later, the boy was inoculated again with similar results.

Jenner published his results privately on the advice of Fellows of the Royal Society, who considered that he should not risk his reputation by presenting anything 'so much at variance with established knowledge'. However, in a few years, vaccination was a widespread practice.

Julian Rowe

able. Although it would be impossible to guarantee that only valid data are entered into a computer, a suitable combination of validation checks should ensure that most errors are detected.

Common validation checks include:

Character-type check Each input data item is checked to ensure that it does not contain invalid characters.

For example, an input name might be checked to ensure that it contains only letters of the alphabet, or an input six-figure date might be checked to ensure it contains only numbers.

Field-length check The number of characters in an input field is checked to ensure that the correct number of characters has been entered. For example, a six-figure date field might be checked to ensure that it does contain exactly six digits.

Control-total check The arithmetic total of a specific field from a group of records is calculated – for example, the hours worked by a group of employees might be added together – and then input with the data to which it refers. The program recalculates the control total and compares it with the one entered to ensure that entry errors have not been made.

Hash-total check An otherwise meaningless control total is calculated – for example, by adding together account numbers. Even though the total has no arithmetic meaning, it can still be used to check the validity of the input account numbers.

Parity check Parity bits are added to binary number codes to ensure that each number in a set of data has the same ◊parity (that each binary number has an even number of 1s, for example). The binary numbers can then be checked to ensure that their parity remains the same. This check is often applied to data after it has been transferred from one part of the computer to another; for example, from a disc drive into the immediate-access memory.

Check digit A digit is calculated from the digits of a code number and then added to that number as an extra digit. For example, in the ISBN (International Standard Book Number) 0 631 90057 8, the 8 is a check digit calculated from the book code number 063190057 and then added to it to make the full ISBN. When the full code number is input, the computer recalculates the check digit and compares it with the one entered. If the entered and calculated check digits do not match, the computer reports that an entry error of some kind has been made.

Range check An input numerical data item is checked to ensure that its value falls in a sensible range. For example, an input two-digit day of the month might be checked to ensure that it is in the range 01 to 31.

Valley of Ten Thousand Smokes valley in SW Alaska, on the Alaska Peninsula, where in 1912 Mount Katmai erupted in one of the largest volcanic explosions ever known, though without loss of human life since the area was uninhabited. The valley was filled with ash to a depth of 200 m/660 ft. It was dedicated as the Katmai National Monument in 1918. Thousands of fissures on the valley floor continue to emit steam and gases.

valve in animals, a structure for controlling the direction of the blood flow. In humans and other vertebrates, the contractions of the beating heart cause the correct blood flow into the arteries because a series of valves prevent back flow. Diseased valves, detected as 'heart murmurs', have decreased efficiency. The tendency for low-pressure venous blood to collect at the base of limbs under the influence of gravity is counteracted by a series of small valves within the veins. It was the existence of these valves that prompted the 17th-century physician William Harvey to suggest that the blood circulated around the body.

valve or **electron tube** in electronics, a glass tube containing gas at low pressure, which is used to control the flow of electricity in a circuit.

Three or more metal electrodes are inset into the tube. By varying the voltage on one of them, called the **grid electrode**, the current through the valve can be controlled, and the valve can act as an amplifier. Valves have been replaced for most applications by ◊transistors.

However, they are still used in high-power transmitters and amplifiers, and in some hi-fi systems.

valve device that controls the flow of a fluid. Inside a valve, a plug moves to widen or close the opening through which the fluid passes. The valve was invented by US radio engineer Lee de Forest (1873–1961).

Common valves include the cone or needle valve, the globe valve, and butterfly valve, all named after the shape of the plug. Specialized valves include the one-way valve, which permits fluid flow in one direction only, and the safety valve, which cuts off flow under certain conditions.

valvular heart disease damage to the heart valves, leading to either narrowing of the valve orifice when it is open (stenosis) or leaking through the valve when it is closed (regurgitation). Worldwide, rheumatic fever is the commonest cause of damage to the heart valves, but in industrialized countries it is being replaced by bacterial infection of the valves themselves (infective endocarditis) and ischaemic heart disease as the main causes. Valvular heart disease is diagnosed by hearing heart murmurs with a stethoscope, or by cardiac ◊ultrasound.

vanadium silver-white, malleable and ductile, metallic element, symbol V, atomic number 23, relative atomic mass 50.942. It occurs in certain iron, lead, and uranium ores and is widely distributed in small quantities in igneous and sedimentary rocks. It is used to make steel alloys, to which it adds tensile strength.

Spanish mineralogist Andrés del Rio (1764–1849) and Swedish chemist Nils Sefström (1787–1845) discovered vanadium independently, the former in 1801 and the latter in 1831. Del Rio named it 'erythronium', but was persuaded by other chemists that he had not in fact discovered a new element; Sefström gave it its present name, after the Norse goddess of love and beauty, Vanadis (or Freya).

Van Allen radiation belts two zones of charged particles around the Earth's magnetosphere, discovered in 1958 by US physicist James Van Allen. The atomic particles come from the Earth's upper atmosphere and the ◊solar wind, and are trapped by the Earth's magnetic field. The inner belt lies 1,000–5,000 km/620–3,100 mi above the Equator, and contains ◊protons and ◊electrons. The outer belt lies 15,000–25,000 km/9,300–15,500 mi above the Equator, but is lower around the magnetic poles. It contains mostly electrons from the solar wind. *See illustration overleaf.*

van de Graaff generator electrostatic generator capable of producing a voltage of over a million volts. It consists of a continuous vertical conveyor belt that carries electrostatic charges (resulting from friction) up to a large hollow sphere supported on an insulated stand. The lower end of

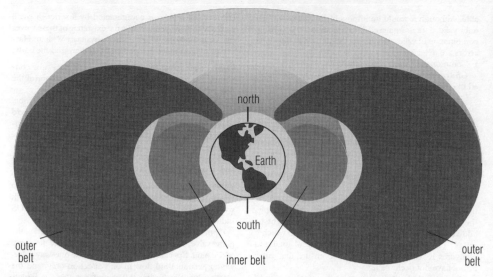

Van Allen radiation belts The Van Allen belts of trapped charged particles are a hazard to spacecraft, affecting on-board electronics and computer systems. Similar belts have been discovered around the planets Mercury, Jupiter, Saturn, Uranus, and Neptune.

the belt is earthed, so that charge accumulates on the sphere. The size of the voltage built up in air depends on the radius of the sphere, but can be increased by enclosing the generator in an inert atmosphere, such as nitrogen.

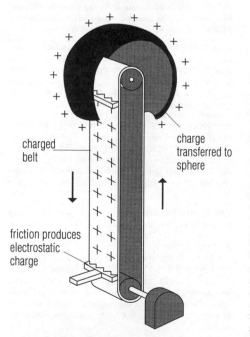

van de Graaff generator US physicist Robert Jemison van de Graaff developed this high-powered generator that can produce more than a million volts. Experiments involving charged particles make use of van de Graaff generators as particle accelerators.

van der Waals' law modified form of the ◊gas laws that includes corrections for the non-ideal behaviour of real gases (the molecules of ideal gases occupy no space and exert no forces on each other). It is named after Dutch physicist J D van der Waals (1837–1923).

The equation derived from the law states that:

$$(P + a/V^2)(V - b) = RT$$

where P, V, and T are the pressure, volume, and temperature (in kelvin) of the gas, respectively; R is the ◊gas constant; and a and b are constants for that particular gas.

Vanguard early series of US Earth-orbiting satellites and their associated rocket launcher. *Vanguard 1* was the second US satellite, launched on 17 March 1958 by the three-stage Vanguard rocket. Tracking of its orbit revealed that Earth is slightly pear-shaped. The series ended in Sept 1959 with *Vanguard 3*.

vapour one of the three states of matter (see also ◊solid and ◊liquid). The molecules in a vapour move randomly and are far apart, the distance between them, and therefore the volume of the vapour, being limited only by the walls of any vessel in which they might be contained. A vapour differs from a ◊gas only in that a vapour can be liquefied by increased pressure, whereas a gas cannot unless its temperature is lowered below its critical temperature; it then becomes a vapour and may be liquefied.

vapour density density of a gas, expressed as the ◊mass of a given volume of the gas divided by the mass of an equal volume of a reference gas (such as hydrogen or air) at the same temperature and pressure. It is equal approximately to half the relative molecular weight (mass) of the gas.

vapour pressure pressure of a vapour given off by (evaporated from) a liquid or solid, caused by vibrating atoms or molecules continuously escaping from its surface. In an enclosed space, a maximum value is reached when the number of particles leaving the surface is in equilibrium with those returning to it; this is known as the **saturated vapour pressure** or **equilibrium vapour pressure**.

variable in computing, a quantity that can take different values. Variables can be used to represent different items of data in the course of a program.

A computer programmer will choose a symbol to represent each variable used in a program. The computer will then automatically assign a memory location to store the current value of each variable, and use the chosen symbol to identify this location.

For example, the letter *P* might be chosen by a programmer to represent the price of an article. The computer would automatically reserve a memory location with the symbolic address *P* to store the price being currently processed.

Different programming languages place different restrictions on the choice of symbols used to represent variables. Some languages only allow a single letter followed, where required, by a single number. Other languages allow a much freer choice, allowing, for example, the use of the full word 'price' to represent the price of an article.

A **global variable** is one that can be accessed by any program instruction; a **local variable** is one that can only be accessed by the instructions within a particular subroutine.

variable in mathematics, a changing quantity (one that can take various values), as opposed to a ◊constant. For example, in the algebraic expression $y = 4x^3 + 2$, the variables are x and y, whereas 4 and 2 are constants.

A variable may be dependent or independent. Thus if y is a ◊function of x, written $y = f(x)$, such that $y = 4x^3 + 2$, the domain of the function includes all values of the **independent variable** x while the range (or co-domain) of the function is defined by the values of the **dependent variable** y.

variable star in astronomy, a star whose brightness changes, either regularly or irregularly, over a period ranging from a few hours to months or years. The ◊Cepheid variables regularly expand and contract in size every few days or weeks.

Stars that change in size and brightness at less precise intervals include **long-period variables**, such as the red giant ◊Mira in the constellation ◊Cetus (period about 330 days), and **irregular variables**, such as some red supergiants. **Eruptive variables** emit sudden outbursts of light. Some suffer flares on their surfaces, while others, such as a ◊nova, result from transfer of gas between a close pair of stars. A ◊supernova is the explosive death of a star. In an ◊ **eclipsing binary**, the variation is due not to any change in the star itself, but to the periodic eclipse of a star by a close companion. The different types of variability are closely related to different stages of stellar evolution.

variance in statistics, the square of the ◊standard deviation, the measure of spread of data. Population and sample variance are denoted by σ^2 or s^2, respectively. Variance provides a measure of the dispersion of a set of statistical results about the mean or average value.

variation in biology, a difference between individuals of the same species, found in any sexually reproducing population. Variations may be almost unnoticeable in some cases, obvious in others, and can concern many aspects of the organism. Typically, variation in size, behaviour, biochemistry, or colouring may be found. The cause of the variation is genetic (that is, inherited), environmental, or more usually a combination of the two. The origins of variation can be traced to the recombination of the genetic material during the formation of the gametes, and, more rarely, to mutation.

varicose veins or **varicosis** condition where the veins become swollen and twisted. The veins of the legs are most often affected; other vulnerable sites include the rectum (◊haemorrhoids) and testes.

Some people have an inherited tendency to varicose veins, and the condition often appears in pregnant women, but obstructed blood flow is the direct cause. They may cause a dull ache or may be the site for ◊thrombosis, infection, or ulcers. The affected veins can be injected with a substance that causes them to shrink, or surgery may be needed.

variegation description of plant leaves or stems that exhibit patches of different colours. The term is usually applied to plants that show white, cream, or yellow on their leaves, caused by areas of tissue that lack the green pigment ◊chlorophyll. Variegated plants are bred for their decorative value, but they are often considerably weaker than the normal, uniformly green plant. Many will not breed true and require ◊vegetative reproduction.

The term is sometimes applied to abnormal patchy colouring of petals, as in the variegated petals of certain tulips, caused by a virus infection. A mineral deficiency in the soil may also be the cause of variegation.

varve in geology, a pair of thin sedimentary beds, one coarse and one fine, representing a cycle of thaw followed by an interval of freezing, in lakes of glacial regions.

Each couplet thus constitutes the sedimentary record of a year, and by counting varves in glacial lakes a record of absolute time elapsed can be determined. Summer and winter layers often are distinguished also by colour, with lighter layers representing summer deposition, and darker layers the result of dark clay settling from water while the lake was frozen.

vascular bundle in botany, strand of primary conducting tissue (a 'vein') in vascular plants, consisting mainly of water-conducting tissues, metaxylem and protoxylem, which together make up the primary ◊xylem, and nutrient-conducting tissue, ◊phloem. It extends from the roots to the stems and leaves. Typically the phloem is situated nearest to the epidermis and the xylem towards the centre of the bundle. In plants exhibiting ◊secondary growth, the xylem and phloem are separated by a thin layer of vascular ◊cambium, which gives rise to new conducting tissues.

vascular plant plant containing vascular bundles. ◊Pteridophytes (ferns, horsetails, and club mosses), ◊gymnosperms (conifers and cycads), and ◊angiosperms (flowering plants) are all vascular plants.

vas deferens in male vertebrates, a tube conducting sperm from the testis to the urethra. The sperm is carried in a fluid secreted by various glands, and can be transported very rapidly when the smooth muscle in the wall of the vas deferens undergoes rhythmic contraction, as in sexual intercourse.

vasectomy male sterilization; an operation to cut and tie the ducts (see ◊vas deferens) that carry sperm from the testes to the penis. Vasectomy does not affect sexual performance, but the semen produced at ejaculation no longer contains sperm.

VBLA (abbreviation for *very long baseline array*) in astronomy, a group of ten 25 m/82.5 ft ◊radio telescopes spread across North America and Hawaii which operate as a single instrument using the technique of very long baseline interferometry (VBLI). The longest baseline (distance between pairs of telescopes) is about 8,000 km/ 4,970 mi.

VDU abbreviation for ◊ *visual display unit*.

vector graphics computer graphics that are stored in the computer memory by using geometric formulas. Vector graphics can be transformed (enlarged, rotated, stretched, and so on) without loss of picture resolution. It is also possible to select and transform any of the components of a vector-graphics display because each is separately defined in the computer memory. In these respects vector graphics are superior to ◊raster graphics. Vector graphics are typically used for drawing applications, allowing the user to create and modify technical diagrams such as designs for houses or cars.

vector quantity any physical quantity that has both magnitude and direction (such as the velocity or acceleration of an object) as distinct from ◊scalar quantity (such as speed, density, or mass), which has magnitude but no direction. A vector is represented either geometrically by an arrow whose length corresponds to its magnitude and points in an appropriate direction, or by two or three numbers representing the magnitude of its components.

Vectors can be added graphically by constructing a parallelogram of vectors (such as the ◊parallelogram of forces commonly employed in physics and engineering).

If two forces p and q are acting on a body at A, then the parallelogram of forces is drawn to determine the resultant force and direction r.

p, q, and r are vectors. In technical writing, a vector is denoted by **bold** type, underlined AB, or overlined AB.

Vega or **Alpha Lyrae** brightest star in the constellation ◊Lyra and the fifth brightest star in the night sky. It is a blue-white star, 25 light years from Earth, with a luminosity 50 times that of the Sun.

In 1983 the Infrared Astronomy Satellite (IRAS) discovered a ring of dust around Vega, possibly a disc from which a planetary system is forming.

vegetative reproduction type of ◊asexual reproduction in plants that relies not on spores, but on multicellular

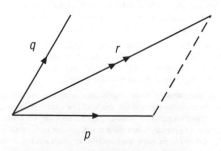

vector quantity A parallelogram of vectors. Vectors can be added graphically using the parallelogram rule. According to the rule, the sum of vectors p and q is the vector r which is the diagonal of the parallelogram with sides p and q.

structures formed by the parent plant. Some of the main types are ◊stolons and runners, ◊gemmae, ◊bulbils, sucker shoots produced from roots (such as in the creeping thistle *Cirsium arvense*), ◊tubers, ◊bulbs, ◊corms, and ◊rhizomes. Vegetative reproduction has long been exploited in horticulture and agriculture, with various methods employed to multiply stocks of plants. See also ◊plant propagation.

vein in animals with a circulatory system, any vessel that carries blood from the body to the heart. Veins contain valves that prevent the blood from running back when moving against gravity. They always carry deoxygenated blood, with the exception of the veins leading from the lungs to the heart in birds and mammals, which carry newly oxygenated blood.

The term is also used more loosely for any system of channels that strengthens living tissues and supplies them with nutrients – for example, leaf veins (see ◊vascular bundle), and the veins in insects' wings.

Vela bright constellation of the southern hemisphere near Carina, represented as the sails of a ship. It contains large wisps of gas – called the Gum nebula after its discoverer, the Australian astronomer Colin Gum (1924–1960) – believed to be the remains of one or more ◊supernovae. Vela also contains the second optical ◊pulsar (a pulsar that flashes at a visible wavelength) to be discovered.

Vela was originally regarded as part of Argo. Its four brightest stars are second-magnitude, one of them being Suhail, about 490 light years from Earth.

veldt subtropical grassland in South Africa, equivalent to the Pampas of South America.

velocity speed of an object in a given direction. Velocity is a ◊vector quantity, since its direction is important as well as its magnitude (or speed).

The velocity at any instant of a particle travelling in a curved path is in the direction of the tangent to the path at the instant considered. The velocity v of an object travelling in a fixed direction may be calculated by dividing the distance s it has travelled by the time t taken to do so, and may be expressed as:

$$v = s/t$$

vena cava either of the two great veins of the trunk, returning deoxygenated blood to the right atrium of the ◊heart. The **superior vena cava**, beginning where the arches of the two innominate veins join high in the chest, receives blood from the head, neck, chest, and arms; the **inferior vena cava**, arising from the junction of the right and left common iliac veins, receives blood from all parts of the body below the diaphragm.

venereal disease (VD) any disease mainly transmitted by sexual contact, although commonly the term is used specifically for gonorrhoea and syphilis, both occurring worldwide, and chancroid ('soft sore') and lymphogranuloma venerum, seen mostly in the tropics. The term ◊ **sexually transmitted disease** (STD) is more often used to encompass a growing list of conditions passed on primarily, but not exclusively, by sexual contact.

Venn diagram in mathematics, a diagram representing a ◊set or sets and the logical relationships between them. The sets are drawn as circles. An area of overlap between two circles (sets) contains elements that are common to both sets, and thus represents a third set. Circles that do

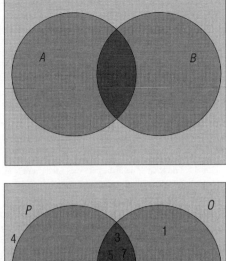

ξ = set of whole numbers from 1 to 20
O = set of odd numbers
P = set of prime numbers

Venn diagram Sets and their relationships are often represented by Venn diagrams. The sets are drawn as circles – the area of overlap between the circles shows elements that are common to each set, and thus represent a third set. Here (a) is a Venn diagram of two intersecting sets and (b) a Venn diagram showing the set of whole numbers from 1 to 20 and the subsets P and O of prime and odd numbers, respectively. The intersection of P and O contains all the prime numbers that are also odd.

not overlap represent sets with no elements in common (disjoint sets). The method is named after the British logician John Venn.

ventral surface the front of an animal. In vertebrates, the side furthest from the backbone; in invertebrates, the side closest to the ground. The positioning of the main nerve pathways on the ventral side is a characteristic of invertebrates.

ventricle in zoology, either of the two lower chambers of the heart that force blood into the circulation by contraction of their muscular walls. The term also refers to any of four cavities within the brain in which cerebrospinal fluid is produced.

Venturi tube device for measuring the rate of fluid flow through a pipe. It consists of a tube with a constriction (narrowing) in the middle of its length. The constriction causes a drop in pressure in the fluid flowing in the pipe. A

pressure gauge attached to the constriction measures the pressure drop and this is used to find the rate of fluid flow.

Venturi tubes are also used in the carburettor of a motor car to draw petrol into the engine.

Venus second planet from the Sun. It can approach Earth to within 38 million km/24 million mi, closer than any other planet. Its mass is 0.82 that of Earth. Venus rotates on its axis more slowly than any other planet, from east to west, the opposite direction to the other planets (except Uranus and possibly Pluto).
  Mean distance from the Sun 108.2 million km/ 67.2 million mi.
  Equatorial diameter 12,100 km/7,500 mi.
  Rotation period 243 Earth days.
  Year 225 Earth days.
  Atmosphere Venus is shrouded by clouds of sulphuric acid droplets that sweep across the planet from east to west every four days. The atmosphere is almost entirely carbon dioxide, which traps the Sun's heat by the ◊greenhouse effect and raises the planet's surface temperature to 480°C/900°F, with an atmospheric pressure of 90 times that at the surface of the Earth.
  Surface consists mainly of silicate rock and may have an interior structure similar to that of Earth: an iron nickel core, a ◊mantle composed of more mafic rocks (rocks made of one or more ferromagnesian, dark-coloured minerals), and a thin siliceous outer ◊crust. The surface is dotted with deep impact craters. Some of Venus's volcanoes may still be active.
  Satellites no moons.
  The first artificial object to hit another planet was the Soviet probe *Venera 3*, which crashed on Venus on 1 March 1966. Later Venera probes parachuted down through the atmosphere and landed successfully on its surface, analysing surface material and sending back information and pictures. In Dec 1978 a US ◊Pioneer Venus probe went into orbit around the planet and mapped most of its surface by radar, which penetrates clouds. In 1992 the US space probe Magellan mapped 99% of the planet's surface to a resolution of 100 m/ 330 ft.
  The largest highland area is Aphrodite Terra near the equator, half the size of Africa. The highest mountains are on the northern highland region of Ishtar Terra, where the massif of Maxwell Montes rises to 10,600 m/35,000 ft above the average surface level. The highland areas on Venus were formed by volcanoes.

Venus

It was once thought that the 'evening star' and the 'morning star' were different bodies. The evening star was called Hesperus and the morning star Phosporous. It was eventually discovered that they were one and the same object – the planet Venus.

verdigris green-blue coating of copper ethanoate that forms naturally on copper, bronze, and brass. It is an irritating, greenish, poisonous compound made by treating copper with ethanoic acid, and was formerly used in wood preservatives, antifouling compositions, and green paints.

verification in computing, the process of checking that data being input to a computer have been accurately copied from a source document.

This may be done visually, by checking the original copy of the data against the copy shown on the VDU screen. A more thorough method is to enter the data twice, using two different keyboard operators, and then to check the two sets of input copies against each other. The checking is normally carried out by the computer itself, any differences between the two copies being reported for correction by one of the the the keyboard operators.

Where large quantities of data have to be input, a separate machine called a **verifier** may be used to prepare fully verified tapes or discs for direct input to the main computer.

vermilion HgS red form of mercuric sulphide; a scarlet that occurs naturally as the crystalline mineral ◊cinnabar.

vernal equinox see spring ◊equinox.

vernalization the stimulation of flowering by exposure to cold. Certain plants will not flower unless subjected to low temperatures during their development. For example, winter wheat will flower in summer only if planted in the previous autumn. However, by placing partially germinated seeds in low temperatures for several days, the cold requirement can be supplied artificially, allowing the wheat to be sown in the spring.

vernier device for taking readings on a graduated scale to a fraction of a division. It consists of a short divided scale that carries an index or pointer and is slid along a main scale. It was invented by Pierre Vernier.

verruca growth on the skin; see ◊wart.

vertebra in vertebrates, an irregularly shaped bone that forms part of the ◊vertebral column. Children have 33 vertebrae, 5 of which fuse in adults to form the sacrum and 4 to form the coccyx. There are 7 cervical vertebrae in the neck, 12 thoracic vertebrae in the thorax with the ribs attached, and 5 lumbar vertebrae in the lower back.

vertebral column the backbone, giving support to an animal and protecting its spinal cord. It is made up of a series of bones or vertebrae running from the skull to the tail, with a central canal containing the nerve fibres of the spinal cord. In tetrapods the vertebrae show some specialization with the shape of the bones varying according to position. In the chest region the upper or thoracic vertebrae are shaped to form connections to the ribs. The backbone is only slightly flexible to give adequate rigidity to the animal structure.

vertebrate any animal with a backbone. The 41,000 species of vertebrates include mammals, birds, reptiles, amphibians, and fishes. They include most of the larger animals, but in terms of numbers of species are only a tiny proportion of the world's animals. The zoological taxonomic group Vertebrata is a subgroup of the ◊phylum Chordata.

A giant conodont (an eel-like organism from the Cambrian period) was discovered in South Africa in 1995, and is believed to be one of the first vertebrates. Conodonts evolved 520 million years ago, predating the earliest fish by about 50 million years.

vertex (plural **vertices**) in geometry, a point shared by three or more sides of a solid figure; the point farthest from a figure's base; or the point of intersection of two sides of a plane figure or the two rays of an angle.

vertical takeoff and landing craft (VTOL) aircraft that can take off and land vertically. ◊Helicopters, airships, and balloons can do this, as can a few fixed-wing aeroplanes, like the ◊convertiplane.

vertigo dizziness; a whirling sensation accompanied by a loss of any feeling of contact with the ground. It may be due to temporary disturbance of the sense of balance (as in spinning for too long on one spot), psychological reasons, or disease such as ◊labyrinthitis, or intoxication.

Very Large Array (VLA) largest and most complex single-site radio telescope in the world. It is located on the Plains of San Augustine, 80 km/50 mi west of Socorro, New Mexico. It consists of 27 dish antennae, each 25 m/82 ft in diameter, arranged along three equally spaced arms forming a Y-shaped array. Two of the arms are 21 km/13 mi long, and the third, to the north, is 19 km/11.8 mi long. The dishes are mounted on railway tracks enabling the configuration and size of the array to be altered as required.

Pairs of dishes can also be used as separate interferometers (see ◊radio telescope), each dish having its own individual receivers that are remotely controlled, enabling many different frequencies to be studied. There are four standard configurations of antennae ranging from A (the most extended) through B and C to D. In the A configuration the antennae are spread out along the full extent of the arms and the VLA can map small, intense radio sources with high resolution. The smallest configuration, D, uses arms that are just 0.6 km/0.4 mi long for mapping larger sources. Here the resolution is lower, although there is greater sensitivity to fainter, extended fields of radio emission.

vestigial organ in biology, an organ that remains in diminished form after it has ceased to have any significant function in the adult organism. In humans, the appendix is vestigial, having once had a digestive function in our ancestors.

vetch any of a group of trailing or climbing plants belonging to the pea family, usually having seed pods and purple, yellow, or white flowers, and including the fodder crop alfalfa (*Medicago sativa*). (Several genera, especially *Vicia*, family Leguminosae.)

VHF (abbreviation for *very high frequency*) referring to radio waves that have very short wavelengths (10 m–1 m). They are used for interference-free FM transmissions (see ◊frequency modulation). VHF transmitters have a relatively short range because the waves cannot be reflected over the horizon like longer radio waves.

video camera or *camcorder* portable television camera that takes moving pictures electronically on magnetic tape. It produces an electrical output signal corresponding to rapid line-by-line scanning of the field of view. The output is recorded on video cassette and is played back on a television screen via a video cassette recorder.

video cassette recorder (VCR) device for recording on and playing back video cassettes; see ◊videotape recorder.

video disc disc with pictures and sounds recorded on it, played back by laser. The video disc is a type of ◊compact disc.

The video disc (originated by Scottish inventor John Logie Baird in 1928; commercially available from 1978) is chiefly used to provide commercial films for private viewing. Most systems use a 30 cm/12 in rotating vinyl

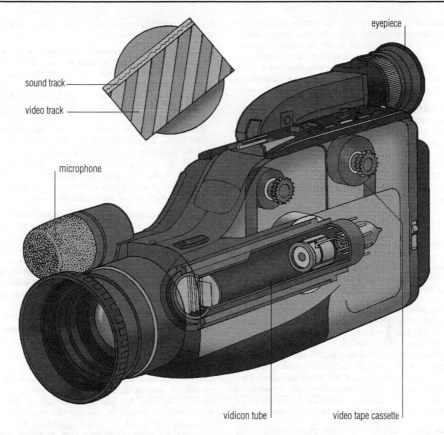

sound track

video track

microphone

eyepiece

vidicon tube

video tape cassette

video camera The heart of the video camera is the vidicon tube which converts light entering the front lens to an electrical signal. An image is formed on a light-sensitive surface at the front of the tube. The image is then scanned by an electron beam, to give an output signal corresponding to the image brightness. The signal is recorded as a magnetic track traversing the tape diagonally. The sound track, which records the sounds picked up by the microphone, runs along the edge of the tape.

disc coated with a reflective material. Laser scanning recovers picture and sound signals from the surface where they are recorded as a spiral of microscopic pits.

video game electronic game played on a visual-display screen or, by means of special additional or built-in components, on the screen of a television set. The first commercially sold was a simple bat-and-ball game developed in the USA in 1972, but complex variants are now available in colour and with special sound effects.

videotape recorder (VTR) device for recording pictures and sound on cassettes or spools of magnetic tape. The first commercial VTR was launched in 1956 for the television broadcasting industry, but from the late 1970s cheaper models developed for home use, to record broadcast programmes for future viewing and to view rented or owned video cassettes of commercial films.

Video recording works in the same way as audio ◊tape recording: the picture information is stored as a line of varying magnetism, or track, on a plastic tape covered with magnetic material. The main difficulty – the huge amount of information needed to reproduce a picture – is overcome by arranging the video track diagonally across the tape. During recording, the tape is wrapped around a

drum in a spiral fashion. The recording head rotates inside the drum.

The combination of the forward motion of the tape and the rotation of the head produces a diagonal track. The audio signal accompanying the video signal is recorded as a separate track along the edge of the tape.

Two video cassette systems were introduced to the mass market by Japanese firms in the 1970s, but the Betamax standard was first introduced in 1965. The Sony Betamax was technically superior, but Matsushita's VHS had larger marketing resources behind it and after some years became the sole system on the market. Super-VHS is an improved version of the VHS system, launched 1989, with higher picture definition and colour quality.

Videotape

The screening of videotapes in buses was banned in Turkey in June 1984 after a bus driver had tried to censor a love scene by holding his arm in front of the television while he was driving. Seventeen people died in the ensuing accident.

videotext system in which information (text and simple pictures) is displayed on a television (video) screen.

There are two basic systems, known as ◊teletext and ◊viewdata. In the teletext system information is broadcast with the ordinary television signals, whereas in the viewdata system information is relayed to the screen from a central data bank via the telephone network. Both systems require the use of a television receiver (or a connected VTR) with special decoder.

viewdata system of displaying information on a television screen in which the information is extracted from a computer data bank and transmitted via the telephone lines. It is one form of ◊videotext. The British Post Office (now British Telecom) developed the first viewdata system, Prestel, in 1975. Similar systems are now in widespread use in other countries. Users have access to a large store of information, presented on the screen in the form of 'pages'.

Prestel has hundreds of thousands of pages, presenting all kinds of information, from local weather and restaurant menus to share prices and airport timetables. Since viewdata uses telephone lines, it can become a two-way interactive information system, making possible, for example, home banking and shopping. In contrast, the only user input allowed by the ◊teletext system is to select the information to be displayed.

Viking probes two US space probes to Mars, each one consisting of an orbiter and a lander. They were launched on 20 Aug and 9 Sept 1975. They transmitted colour pictures and analysed the soil.

Viking 1 carried life detection labs and landed in the Chryse lowland area on 20 July 1976 for detailed research and photos. Designed to work for 90 days, it operated for six and a half years, going silent in Nov 1982. *Viking 2* was similar in set-up to *Viking 1*; it landed in Utopia on 3 Sept 1976 and functioned for three and a half years.

villus (plural *villi*) small fingerlike projection extending into the interior of the small intestine and increasing the absorptive area of the intestinal wall. Digested nutrients, including sugars and amino acids, pass into the villi and are carried away by the circulating blood.

vincristine ◊alkaloid extracted from the blue periwinkle plant (*Vinca rosea*). Developed as an anticancer agent, it has revolutionized the treatment of childhood acute leukaemias; it is also included in ◊chemotherapy regimens for some lymphomas (cancers arising in the lymph tissues) and lung and breast cancers. Side effects, such as nerve damage and loss of hair, are severe but usually reversible.

Virgo zodiacal constellation of the northern hemisphere, the second largest in the sky. It is represented as a maiden holding an ear of wheat, marked by first-magnitude ◊Spica, Virgo's brightest star. The Sun passes through Virgo from late Sept to the end of Oct. In astrology, the dates for Virgo are between about 23 Aug and 22 Sept (see ◊precession).

Virgo contains the nearest large cluster of galaxies to us, 50 million light years away, consisting of about 3,000 galaxies centred on the giant elliptical galaxy M87. Also in Virgo is the nearest ◊quasar, 3C 273, an estimated 3 billion light years away.

virtual in computing, without physical existence. Some computers have virtual memory, making their immediate-access memory seem larger than it is. ◊Virtual reality is a computer simulation of a whole physical environment.

virtual memory in computing, a technique whereby a portion of the computer backing storage, or external, ◊memory is used as an extension of its immediate-access, or internal, memory. The contents of an area of the immediate-access memory are stored on, say, a hard disc while they are not needed, and brought back into main memory when required.

The process, called paging or segmentation, is controlled by the computer ◊operating system and is hidden from the programmer, to whom the computer's internal memory appears larger than it really is. The technique can be successfully implemented only if very fast backing store is available, so that 'pages' of memory can be rapidly switched into and out of the immediate-access memory.

virtual reality advanced form of computer simulation, in which a participant has the illusion of being part of an artificial environment. The participant views the environment through two tiny television screens (one for each eye) built into a visor. Sensors detect movements of the participant's head or body, causing the apparent viewing position to change. Gloves (datagloves) fitted with sensors may be worn, which allow the participant seemingly to pick up and move objects in the environment.

The technology is still under development but is expected to have widespread applications; for example, in military and surgical training, architecture, and home entertainment.

virus in computing, a piece of ◊software that can replicate and transfer itself from one computer to another, without the user being aware of it. Some viruses are relatively harmless, but others can damage or destroy data. They are written by anonymous programmers, often maliciously, and are spread on ◊floppy discs, on local networks, and more recently on the ◊Internet. Antivirus software can be used to detect and destroy well-known viruses, but new viruses continually appear and these may bypass existing antivirus programs.

Computer viruses may be programmed to operate on a particular date, such as the Michelangelo Virus, which was triggered on 6 March 1992 (the anniversary of the birthday of Italian artist Michelangelo) and erased hard discs.

Virus

An average virus is only around 50 nm (50 millionths of a millimetre/2 millionths of an inch) across. Hundreds of thousands of viruses could fit on the full stop at the end of this sentence.

virus infectious particle consisting of a core of nucleic acid (DNA or RNA) enclosed in a protein shell. Viruses are acellular and able to function and reproduce only if they can invade a living cell to use the cell's system to replicate themselves. In the process they may disrupt or alter the host cell's own DNA. The healthy human body reacts by producing an antiviral protein, ◊interferon, which prevents the infection spreading to adjacent cells.

Many viruses mutate continuously so that the host's body has little chance of developing permanent resistance; others transfer between species, with the new host similarly unable to develop resistance. The viruses that cause ◊AIDS and ◊Lassa fever are both thought to have 'jumped' to humans from other mammalian hosts.

Among diseases caused by viruses are canine distemper, chickenpox, common cold, herpes, influenza, rabies, smallpox, yellow fever, AIDS, and many plant diseases. Recent evidence implicates viruses in the development of some forms of cancer (see ◊oncogenes). *Bacteriophages* are viruses that infect bacterial cells.

Retroviruses are of special interest because they have an RNA genome and can produce DNA from this RNA by a process called reverse transcription.

Viroids, discovered in 1971, are even smaller than viruses; they consist of a single strand of nucleic acid with no protein coat. They may cause stunting in plants and some rare diseases in animals, including humans.

It is debatable whether viruses and viroids are truly living organisms, since they are incapable of an independent existence. Outside the cell of another organism they remain completely inert. The origin of viruses is also unclear, but it is believed that they are degenerate forms of life, derived from cellular organisms, or pieces of nucleic acid that have broken away from the genome of some higher organism and taken up a parasitic existence.

Antiviral drugs are difficult to develop because viruses replicate by using the genetic machinery of host cells, so that drugs tend to affect the host cell as well as the virus. Acyclovir (used against the herpes group of diseases) is one of the few drugs so far developed that is successfully selective in its action. It is converted to its active form by an enzyme that is specific to the virus, and it then specifically inhibits viral replication. Some viruses have shown developing resistance to the few antiviral drugs available.

Viruses have recently been found to be very abundant in seas and lakes, with between 5 and 10 million per millilitre of water at most sites tested, but up to 250 million per millilitre in one polluted lake. These viruses infect bacteria and, possibly, single-celled algae. They may play a crucial role in controlling the survival of bacteria and algae in the plankton.

viscera general term for the organs contained in the chest and abdominal cavities.

viscose yellowish, syrupy solution made by treating cellulose with sodium hydroxide and carbon disulphide. The solution is then regenerated as continuous filament for the making of ◊rayon and as cellophane.

viscosity in physics, the resistance of a fluid to flow, caused by its internal friction, which makes it resist flowing past a solid surface or other layers of the fluid. It applies to the motion of an object moving through a fluid as well as the motion of a fluid passing by an object.

Fluids such as pitch, treacle, and heavy oils are highly viscous; for the purposes of calculation, many fluids in physics are considered to be perfect, or nonviscous.

vision defect any abnormality of the eye that causes less-than-perfect sight. Common defects are ◊shortsightedness or myopia; ◊long-sightedness or hypermetropia; lack of ◊accommodation or presbyopia; and ◊astigmatism. Other eye defects include colour blindness.

visual display unit (VDU) computer terminal consisting of a keyboard for input data and a screen for displaying output. The oldest and the most popular type of VDU screen is the ◊cathode-ray tube (CRT), which uses essentially the same technology as a television screen.

Other types use plasma display technology and ◊liquid-crystal displays.

vitalistic medicine in alternative medicine, generic term for a range of therapies that base their practice on the theory that disease is engendered by energy deficiency in the organism as a whole or a dynamic dysfunction in the affected part.

Such deficiencies or dysfunctions are regarded as antecedent to the biochemical effects in which disease becomes manifest and upon which orthodox medicine focuses. ◊Acupuncture, crystal therapy, ◊homoeopathy, ◊magnet therapy, ◊naturopathy, radionics, and ◊Reichian therapy are all basically vitalistic.

vitamin any of various chemically unrelated organic compounds that are necessary in small quantities for the normal functioning of the human body. Many act as coenzymes, small molecules that enable ◊enzymes to function effectively. Vitamins must be supplied by the diet because the body cannot make them. They are normally present in adequate amounts in a balanced diet. Deficiency of a vitamin may lead to a metabolic disorder ('deficiency disease'), which can be remedied by sufficient intake of the vitamin. They are generally classified as *water-soluble* (B and C) or *fat-soluble* (A, D, E, and K). See separate entries for individual vitamins, also ◊nicotinic acid, ◊folic acid, and ◊pantothenic acid.

Scurvy (the result of vitamin C deficiency) was observed at least 3,500 years ago, and sailors from the 1600s were given fresh sprouting cereals or citrus-fruit juice to prevent or cure it. The concept of scurvy as a deficiency disease, however, caused by the absence of a specific substance, emerged later. In the 1890s a Dutch doctor, Christiaan Eijkman, discovered that he could cure hens suffering from a condition like beriberi by feeding them on whole-grain, rather than polished, rice. In 1912 Casimir Funk, a Polish-born biochemist, had proposed the existence of what he called 'vitamines' (vital amines), but it was not fully established until about 1915 that several deficiency diseases were preventable and curable by extracts from certain foods. By then it was known that two groups of factors were involved, one being water-soluble and present, for example, in yeast, rice-polishings, and wheat germ, and the other being fat-soluble and present in egg yolk, butter, and fish-liver oils. The water-soluble substance, known to be effective against beriberi, was named vitamin B. The fat-soluble vitamin complex was at first called vitamin A. As a result of analytical techniques these have been subsequently separated into their various components, and others have been discovered.

vitamin A another name for ◊retinol.

vitamin B₁ another name for ◊thiamine.

vitamin B₁₂ another name for ◊cyanocobalamin.

vitamin B₂ another name for ◊riboflavin.

vitamin B₆ another name for ◊pyridoxine.

vitamin C another name for ◊ascorbic acid.

vitamin D another name for ◊cholecalciferol.

vitamin E another name for ◊tocopherol.

vitamin H another name for ◊biotin.

vitamin K another name for ◊phytomenadione.

Vitamins

Vitamin	Name	Main dietary sources	Established benefit	Deficiency symptoms
A	retinol	dairy products, egg yolk, liver; also formed in body from beta-carotene, a pigment present in some leafy vegetables	aids growth; prevents night blindness and xerophthalmia (a common cause of blindness among children in developing countries); helps keep the skin and mucous membranes resistant to infection	night blindness; rough skin; impaired bone growth
B_1	thiamin	germ and bran of seeds, grains; yeast	essential for carbohydrate metabolism and health of nervous system	beriberi, Korsakov's syndrome
B_2	riboflavin	eggs, liver, milk, poultry, broccoli, mushrooms	involved in energy metabolism; protects skin, mouth, eyes, eyelids, mucous membranes	inflammation of tongue and lips; sores in corners of the mouth
B_6	pyridoxine/pantothenic acid/biotin	meat, poultry, fish, fruits, nuts, whole grains, leafy vegetables, yeast extract	important in the regulation of the central nervous system and in protein metabolism; helps prevent anaemia, skin lesions, nerve damage	dermatitis, neurological problems, kidney stones
B_{12}	cyanocobalamin	liver, meat, fish, eggs, dairy products, soybeans	involved in synthesis of nucleic acids, maintenance of myelin sheath around nerve fibres; efficient use of folic acid	anaemia, neurological disturbance
	folic acid	green leafy vegetables, liver, peanuts; cooking and processing can cause serious losses in food	involved in synthesis of nucleic acids; helps protect against cervical dysplasia (precancerous changes in the cells of the uterine cervix)	megaloblastic anaemia
	nicotinic acid (or niacin)	meat, yeast extract, some cereals; also formed in the body from the amino acid tryptophan	maintains the health of the skin, tongue, and digestive system	pellagra
C	ascorbic acid	citrus fruits, green vegetables, tomatoes, potatoes; losses occur during storage and cooking	prevents scurvy, loose teeth; fights haemorrhage; important in synthesis of collagen (constituent of connective tissue); aids in resistance of some types of virus and bacterial infections	scurvy
D	calciferol, cholecalciferol	liver, fish oil, dairy products, eggs; also produced when skin is exposed to sunlight	promotes growth and mineralization of bone	rickets in children; osteomalacia in adults
E	tocopherol	vegetable oils, eggs, butter, some cereals, nuts	prevents damage to cell membranes	anaemia
K	phytomenadione, menaquinone	green vegetables, cereals, fruits, meat, dairy products	essential for blood clotting	haemorrhagic problems

vitreous humour transparent jellylike substance behind the lens of the vertebrate ◊eye. It gives rigidity to the spherical form of the eye and allows light to pass through to the retina.

vitriol any of a number of sulphate salts. Blue, green, and white vitriols are copper, ferrous, and zinc sulphate, respectively. *Oil of vitriol* is sulphuric acid.

viviparous in animals, a method of reproduction in which the embryo develops inside the body of the female from which it gains nourishment (in contrast to ◊oviparous and ◊ovoviviparous). Vivipary is best developed in placental mammals, but also occurs in some arthropods, fish, amphibians, and reptiles that have placenta-like structures. In plants, it is the formation of young plantlets or bulbils instead of flowers. The term also describes seeds that germinate prematurely, before falling from the parent plant. Premature germination is common in mangrove trees, where the seedlings develop sizable spearlike roots before dropping into the swamp below; this prevents their being washed away by the tide.

vivisection literally, cutting into a living animal. Used originally to mean experimental surgery or dissection practised on a live subject, the term is often used by antivivisection campaigners to include any experiment on animals, surgical or otherwise.

VLBI (abbreviation for *very long baseline interferometry*) In radio astronomy, a method of obtaining high-resolution images of astronomical objects by combining simultaneous observations made by two or more radio telescopes thousands of kilometres apart. The maximum resolution that can be achieved is proportional to the longest baseline in the array (the distance between any pair of telescopes), and inversely proportional to the radio wavelength being used.

There are people who do not object to eating mutton chop – people who do not even object to shooting pheasant with the considerable chance that it may be only wounded and may have to die after lingering pain, unable to obtain its proper nutriment – and yet consider it something monstrous to introduce under the skin of a guinea pig a little inoculation of some microbe to ascertain its action. These seem to me the most inconsistent views.

On **vivisection** Joseph Lister
British Medical Journal 1897

VLF in physics, abbreviation for *very low ◊frequency*. VLF radio waves have frequencies in the range 3–30 kHz.

vocal cords the paired folds, ridges, or cords of tissue within a mammal's larynx, and a bird's syrinx. Air constricted between the folds or membranes makes them vibrate, producing sounds. Muscles in the larynx change the pitch of the sounds produced, by adjusting the tension of the vocal cords.

vol abbreviation for *volume*.

volatile in chemistry, term describing a substance that readily passes from the liquid to the vapour phase. Volatile substances have a high ◊vapour pressure.

volatile memory in computing, ◊memory that loses its contents when the power supply to the computer is disconnected.

volcanic rock another name for ◊extrusive rock, igneous rock formed on the Earth's surface.

volcano crack in the Earth's crust through which hot magma (molten rock) and gases well up. The magma is termed lava when it reaches the surface. A volcanic mountain, usually cone shaped with a crater on top, is formed around the opening, or vent, by the build-up of solidified lava and ashes (rock fragments). Most volcanoes arise on plate margins (see ◊plate tectonics), where the movements of plates generate magma or allow it to rise from the mantle beneath. However, a number are

composite
volcano

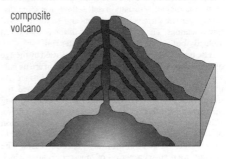

cinder
cone

shield volcano

volcano There are two main types of volcano but three distinctive cone shapes. Composite volcanoes emit a stiff, rapidly solidifying lava which forms high, steep-sided cones. Volcanoes that regularly throw out ash build up flatter domes known as cinder cones. The lava from a shield volcano is not ejected violently, flowing over the crater rim forming a broad low profile.

found far from plate-margin activity, on 'hot spots' where the Earth's crust is thin.

There are two main types of volcano:

Composite volcanoes, such as Stromboli and Vesuvius in Italy, are found at destructive plate margins (areas where plates are being pushed together), usually in association with island arcs and coastal mountain chains. The magma is mostly derived from plate material and is rich in silica. This makes a very stiff lava such as andesite, which solidifies rapidly to form a high, steep-sided volcanic mountain. The magma often clogs the volcanic vent, causing violent eruptions as the blockage is blasted free, as in the eruption of Mount St Helens, USA, in 1980. The crater may collapse to form a ◊caldera.

Shield volcanoes, such as Mauna Loa in Hawaii, are found along the rift valleys and ocean ridges of constructive plate margins (areas where plates are moving apart), and also over hot spots. The magma is derived from the Earth's mantle and is quite free-flowing. The lava formed from this magma – usually basalt – flows for some distance over the surface before it sets and so forms broad low volcanoes. The lava of a shield volcano is not ejected violently but simply flows over the crater rim.

The type of volcanic activity is also governed by the age of the volcano. The first stages of an eruption are usually vigorous as the magma forces its way to the surface. As the pressure drops and the vents become established, the main phase of activity begins, composite volcanoes giving ◊pyroclastic debris and shield volcanoes giving lava flows. When the pressure from below ceases, due to exhaustion of the magma chamber, activity wanes and is confined to the emission of gases and in time this also ceases. The volcano then enters a period of quiescence, after which activity may resume after a period of days, years, or even thousands of years. Only when the root zones of a volcano have been exposed by erosion can a volcano be said to be truly extinct.

Many volcanoes are submarine and occur along mid-ocean ridges. The chief terrestrial volcanic regions are around the Pacific rim (Cape Horn to Alaska); the central Andes of Chile (with the world's highest volcano, Guallatiri, 6,060 m/19,900 ft); North Island, New Zealand; Hawaii; Japan; and Antarctica. There are more than 1,300 potentially active volcanoes on Earth. Volcanism has helped shape other members of the Solar System, including the Moon, Mars, Venus, and Jupiter's moon Io.

An undersea volcano erupted in June 1995, adding an island of about one hectare in size and 15 m/49 ft above sea level to the kingdom of Tonga.

volt SI unit of electromotive force or electric potential, symbol V. A small battery has a potential of 1.5 volts, whilst a high-tension transmission line may carry up to 765,000 volts. The domestic electricity supply in the UK is 230 volts (lowered from 240 volts in 1995); it is 110 volts in the USA.

The absolute volt is defined as the potential difference necessary to produce a current of one ampere through an electric circuit with a resistance of one ohm. It can also be defined as the potential difference that requires one joule of work to move a positive charge of one coulomb from the lower to the higher potential. It is named after the Italian scientist Alessandro Volta.

voltage commonly used term for ◊potential difference (pd) or ◊electromotive force (emf).

Voltage

Bees exposed to the electrical fields of high-voltage power cables produce their honey in an erratic manner. As a consequence, bee-keepers have learned not to place their hives under power cables.

voltage amplifier electronic device that increases an input signal in the form of a voltage or ◊potential difference, delivering an output signal that is larger than the input by a specified ratio.

voltmeter instrument for measuring potential difference (voltage). It has a high internal resistance (so that it passes only a small current), and is connected in parallel with the component across which potential difference is to be measured. A common type is constructed from a sensitive current-detecting moving-coil ◊galvanometer placed in series with a high-value resistor (multiplier). To measure an AC (◊alternating current) voltage, the circuit must usually include a rectifier; however, a moving-iron instrument can be used to measure alternating voltages without the need for such a device.

volume in geometry, the space occupied by a three-dimensional solid object. A prism (such as a cube) or a cylinder has a volume equal to the area of the base multiplied by the height. For a pyramid or cone, the volume is equal to one-third of the area of the base multiplied by the perpendicular height. The volume of a sphere is equal to $4/3 \times \pi r^3$, where r is the radius. Volumes of irregular solids may be calculated by the technique of ◊integration.

volumetric analysis procedure used for determining the concentration of a solution. A known volume of a solution of unknown concentration is reacted with a solution of known concentration (standard). The standard solution is delivered from a burette so the volume added is known. This technique is known as ◊titration. Often an indicator is used to show when the correct proportions have reacted. This procedure is used for acid–base, ◊redox, and certain other reactions involving solutions.

Voskhod Soviet spacecraft used in the mid-1960s; it was modified from the single-seat Vostok, and was the first spacecraft capable of carrying two or three cosmonauts. During *Voskhod 2's* flight in 1965, Aleksi Leonov made the first space walk.

Vostok first Soviet spacecraft, used between 1961–63. Vostok was a metal sphere 2.3 m/7.5 ft in diameter, capable of carrying one cosmonaut. It made flights lasting up to five days. *Vostok 1* carried the first person into space, Yuri Gagarin.

Voyager probes two US space probes. *Voyager 1*, launched on 5 Sept 1977, passed Jupiter in March 1979, and reached Saturn in Nov 1980. *Voyager 2* was launched earlier, on 20 Aug 1977, on a slower trajectory that took it past Jupiter in July 1979, Saturn in Aug 1981, Uranus in Jan 1986, and Neptune in Aug 1989. Like the ◊Pioneer probes, the *Voyagers* are on their way out of the Solar System; at the start of 1995, *Voyager 1* was 8.8 billion km/5.5 billion mi from Earth, and *Voyager 2* was 6.8 billion km/4.3 billion mi from Earth. Their tasks now include helping scientists to locate the position of the heliopause, the boundary at which the influence of the Sun gives way to the forces exerted by other stars. Both *Voyagers* carry specially coded long-playing rec-

volume of common three-dimensional shapes

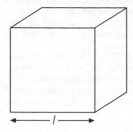

volume of a ***cube***
= length³
= l^3

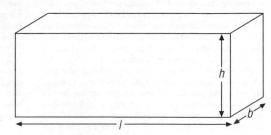

volume of a ***cuboid***
= length × breadth × height
= $l \times b \times h$

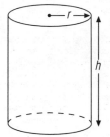

volume of a ***cylinder***
= π × (radius of cross-section)² × height
= $\pi r^2 h$

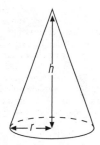

volume of a ***cone***
= ⅓ π × (radius of cross-section)² × height
= $\frac{1}{3} \pi r^2 h$

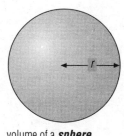

volume of a ***sphere***
= ⅘ π radius³
= $\frac{4}{3} \pi r^3$

volume

ords called *Sounds of Earth* for the enlightenment of any other civilizations that might find them.

V-shaped valley river valley with a V-shaped cross-section. Such valleys are usually found near the source of a river, where the steeper gradient means that there is a great deal of ◊corrasion (grinding away by rock particles) along the stream bed and erosion cuts downwards more than it does sideways. However, a V-shaped valley may also be formed in the lower course of a river when its powers of downward erosion become renewed by a fall in sea level, a rise in land level, or the capture of another river.

VSTOL abbreviation for ***vertical/short takeoff and landing***, aircraft capable of taking off and landing either vertically or using a very short length of runway (see ◊STOL). Vertical takeoff requires a vector-control system that permits the thrust of the aircraft engine to be changed from horizontal to vertical for takeoff and back again to horizontal to permit forward flight. An alternative VSTOL technology developed in the USA involves tilting the wings of the aircraft from vertical to horizontal and along with them the aircraft propellers, thus changing from vertical lift to horizontal thrust.

The British ◊Harrier fighter bomber was the first VSTOL aircraft. It is now manufactured under licence in the USA and provides integral air support for the US Marines. In addition to the UK's Royal Air Force and Royal Navy, the Indian, Spanish, and Italian navies are equipped with the Harrier. It was used in the 1982 Falklands conflict and the 1991 Gulf War.

vulcanization technique for hardening rubber by heating and chemically combining it with sulphur. The process also makes the rubber stronger and more elastic. If the sulphur content is increased to as much as 30%, the product is the inelastic solid known as ebonite. More expensive alternatives to sulphur, such as selenium and tellurium, are used to vulcanize rubber for specialized products such as vehicle tyres. The process was discovered accidentally by US inventor Charles Goodyear in 1839 and patented in 1844.

Accelerators can be added to speed the vulcanization process, which takes from a few minutes for small objects to an hour or more for vehicle tyres. Moulded objects are often shaped and vulcanized simultaneously in heated moulds; other objects may be vulcanized in hot water, hot air, or steam.

Vulpecula small constellation in the northern hemisphere just south of ◊Cygnus, represented as a fox. It contains a major planetary ◊nebula, the Dumbbell, and the first ◊pulsar (pulsating radio source) to be discovered.

W abbreviation for *west*; in physics, symbol for *watt*.

wadi in arid regions of the Middle East, a steep-sided valley containing an intermittent stream that flows in the wet season.

Waldsterben tree decline related to air pollution, common throughout the industrialized world. It appears to be caused by a mixture of pollutants; the precise chemical mix varies between locations, but it includes acid rain, ozone, sulphur dioxide, and nitrogen oxides.

Waldsterben was first noticed in the Black Forest of Germany during the late 1970s, and is spreading to many Third World countries, such as China.

Wallace line imaginary line running down the Lombok Strait in SE Asia, between the island of Bali and the islands of Lombok and Sulawesi. It was identified by English naturalist Alfred Russel Wallace as separating the S Asian (Oriental) and Australian biogeographical regions, each of which has its own distinctive animals.

Subsequently, others have placed the boundary between these two regions at different points in the Malay archipelago, owing to overlapping migration patterns.

wall pressure in plants, the mechanical pressure exerted by the cell contents against the cell wall. The rigidity (turgor) of a plant often depends on the level of wall pressure found in the cells of the stem. Wall pressure falls if the plant cell loses water.

Wankel engine rotary petrol engine developed by the German engineer Felix Wankel (1902–1988) in the 1950s. It operates according to the same stages as the ◊four-stroke petrol engine cycle, but these stages take place in different sectors of a figure-eight chamber in the space between the chamber walls and a triangular rotor. Power is produced once on every turn of the rotor. The Wankel engine is simpler in construction than the four-stroke piston petrol engine, and produces rotary power directly (instead of via a crankshaft). Problems with rotor seals have prevented its widespread use.

warfarin poison that induces fatal internal bleeding in rats; neutralized with sodium hydroxide, it is used in medicine as an anticoagulant in the treatment of ◊thrombosis: it prevents blood clotting by inhibiting the action of vitamin K. It can be taken orally and begins to act several days after the initial dose.

Warfarin is a crystalline powder, $C_{19}H_{16}O_4$. ◊Heparin may be given in treatment at the same time and discontinued when warfarin takes effect. It is often given as a preventive measure, to reduce the risk of ◊thrombosis or ◊embolism after major surgery.

warning coloration in biology, an alternative term for ◊aposematic coloration.

wart protuberance composed of a local overgrowth of skin. The common wart (*verruca vulgaris*) is due to a virus infection. It usually disappears spontaneously within two years, but can be treated with peeling applications, burning away (cautery), freezing (cryosurgery), or laser treatment.

washing soda $Na_2CO_3.10H_2O$ (chemical name *sodium carbonate decahydrate*) substance added to washing water to 'soften' it (see ◊hard water).

Washington Convention alternative name for ◊ *CITES*, the international agreement that regulates trade in endangered species.

waste materials that are no longer needed and are discarded. Examples are household waste, industrial waste (which often contains toxic chemicals), medical waste (which may contain organisms that cause disease), and ◊nuclear waste (which is radioactive). By ◊recycling, some materials in waste can be reclaimed for further use.

There has been a tendency to increase the amount of waste generated per person in industrialized countries, particularly through the growth in packaging and disposable products, creating a 'throwaway society'.

waste disposal depositing waste. Methods of waste disposal vary according to the materials in the waste and include incineration, burial at designated sites, and dumping at sea.

Organic waste can be treated and reused as fertilizer (see ◊sewage disposal). ◊Nuclear waste and ◊toxic waste are usually buried or dumped at sea, although this does not negate the danger.

Waste disposal is an increasing problem in the late 20th century. Environmental groups, such as Greenpeace and Friends of the Earth, are campaigning for more recycling, a change in lifestyle so that less waste (from packaging and containers to nuclear materials) is produced, and safer methods of disposal.

Although incineration cuts down on landfill and can produce heat as a useful by-product it is still a wasteful method of disposal in comparison with recycling. For example, recycling a plastic bottle saves twice as much energy as is obtained by burning it.

The USA burns very little of its rubbish as compared with other industrialized countries. Most of its waste, 80%, goes into land fills. Many of the country's landfill sites will have to close in the 1990s because they do not meet standards to protect groundwater.

watch portable timepiece. In the early 20th century increasing miniaturization, mass production, and, in World War I, the advantages of the wristband led to the watch moving from the pocket to the wrist. Watches were also subsequently made waterproof, antimagnetic, self-winding, and shock-resistant. In 1957 the electric watch was developed, and in the 1970s came the digital watch, which dispensed with all moving parts.

Traditional mechanical watches with analogue dials (hands) are based on the invention by Peter Henlein (1480–1542) of the mainspring as the energy store. By 1675 the invention of the balance spring allowed watches to be made small enough to move from waist to pocket. By the 18th century pocket-watches were accurate, and by the 20th century wristwatches were introduced. In the 1950s battery-run electromagnetic watches were developed; in the 1960s electronic watches were marketed, which use the ◊piezoelectric oscillations of a quartz crystal to mark time and an electronic circuit to drive the hands. In the 1970s quartz watches without moving parts were developed – the solid-state watch with a display of digits. Some include a tiny calculator and such functions as date, alarm, stopwatch, and reminder beeps.

water H_2O liquid without colour, taste, or odour. It is an oxide of hydrogen with a relative molecular mass of 18. Water begins to freeze at 0°C or 32°F, and to boil at 100°C or 212°F. When liquid, it is virtually incompressible; frozen, it expands by 1/11 of its volume. At 4°C/39.2°F, one cubic centimetre of water has a mass of one gram; this is its maximum density, forming the unit of specific gravity. It has the highest known specific heat, and acts as an efficient solvent, particularly when hot. Most of the world's water is in the sea; less than 0.01% is fresh water.

Water covers 70% of the Earth's surface and occurs as standing (oceans, lakes) and running (rivers, streams) water, rain, and vapour and supports all forms of life on Earth.

According to two UN reports published in Jan 1997 large areas of the globe will start running critically short of water in the next 30 years. Total worldwide water consumption has been growing at 2.5% a year, roughly twice as fast as population, and by 1997 had reached 4,200 cubic kilometres annually. Water consumption has risen sixfold during the 20th century. Growing demands for this resource could lead to future conflicts as many rivers cross national boundaries.

water-borne disease disease associated with poor water supply. In the Third World four-fifths of all illness is caused by water-borne diseases, with diarrhoea being the leading cause of childhood death. Malaria, carried by mosquitoes dependent on stagnant water for breeding, affects 400 million people every year and kills 5 million. Polluted water is also a problem in industrialized nations, where industrial dumping of chemical, hazardous, and radioactive wastes causes a range of diseases from headache to cancer.

water closet (WC) alternative name for ◊toilet.

water cycle or ***hydrological cycle*** in ecology, the natural circulation of water through the ◊biosphere. Water is lost from the Earth's surface to the atmosphere either by evaporation caused by the Sun's heat on the surface of lakes, rivers, and oceans, or through the transpiration of plants. This atmospheric water is carried by the air moving across the Earth, and condenses as the air cools to form clouds, which in turn deposit moisture on the land and sea as rain or snow. The water that collects on land flows to the ocean in streams and rivers.

Water cycle

A drop of water may travel thousands of kilometres or miles between the time it evaporates and the time it falls to Earth again as rain, sleet, or snow.

---

waterfall cascade of water in a river or stream. It occurs when a river flows over a bed of rock that resists erosion; weaker rocks downstream are worn away, creating a steep, vertical drop and a plunge pool into which the water falls. Over time, continuing erosion causes the

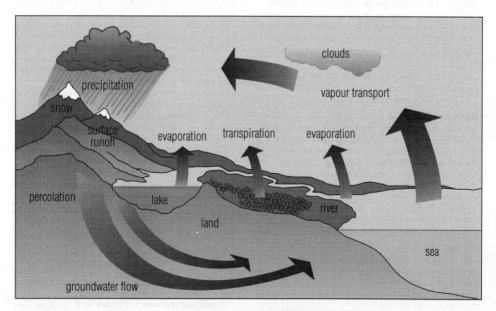

water cycle About one-third of the solar energy reaching the Earth is used in evaporating water. About 380,000 cubic km/ 95,000 cubic mi is evaporated each year. The entire contents of the oceans would take about one million years to pass through the water cycle.

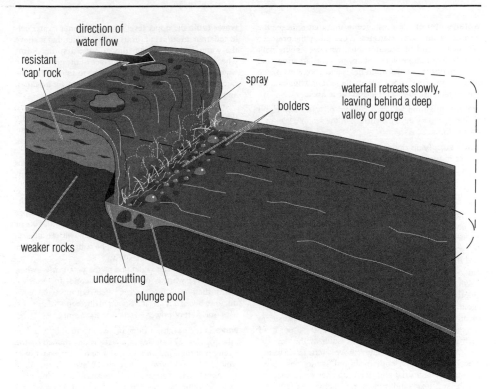

direction of
water flow

resistant
'cap' rock

spray

bolders

waterfall retreats slowly,
leaving behind a deep
valley or gorge

weaker rocks

undercutting

plunge pool

waterfall

waterfall to retreat upstream forming a deep valley, or
◊gorge.

water gas fuel gas consisting of a mixture of carbon
monoxide and hydrogen, made by passing steam over
red-hot coke. The gas was once the chief source of hydrogen for chemical syntheses such as the Haber process for
making ammonia, but has been largely superseded in this
and other reactions by hydrogen obtained from natural
gas.

water glass common name for sodium metasilicate
(Na_2SiO_3). It is a colourless, jellylike substance that
dissolves readily in water to give a solution used for preserving eggs and fireproofing porous materials such as
cloth, paper, and wood. It is also used as an adhesive for
paper and cardboard and in the manufacture of soap and
silica gel, a substance that absorbs moisture.

water meadow irrigated meadow. By flooding the land
for part of each year, increased yields of hay are obtained.
Water meadows were common in Italy, Switzerland, and
England (from 1523) but have now largely disappeared.

water mill machine that harnesses the energy in flowing
water to produce mechanical power, typically for milling
(grinding) grain. Water from a stream is directed against
the paddles of a water wheel to make it turn. Simple
gearing transfers this motion to the millstones. The
modern equivalent of the water wheel is the water turbine,
used in ◊hydroelectric power plants.

Although early step wheels were used in ancient China

and Egypt, and parts of the Middle East, the familiar
vertical water wheel came into widespread use in Roman
times. There were two types: **undershot**, in which the
wheel simply dipped into the stream, and the more
powerful **overshot**, in which the water was directed at the
top of the wheel. The Domesday Book records over 7,000
water mills in Britain. Water wheels remained a prime
source of mechanical power until the development of a
reliable steam engine in the 1700s, not only for milling,
but also for metalworking, crushing and grinding operations, and driving machines in the early factories. The
two were combined to form paddlewheel steamboats in
the 18th century.

water of crystallization water chemically bonded to a
salt in its crystalline state. For example, in copper(II)
sulphate, there are five moles of water per mole of copper
sulphate: hence its formula is $CuSO_4.5H_2O$. This water is
responsible for the colour and shape of the crystalline
form. When the crystals are heated gently, the water is
driven off as steam and a white powder is formed.

$$CuSO_4.5H_2O_{(s)} \rightarrow CuSO_{4\,(s)} + 5H_2O_{(g)}$$

water pollution any addition to fresh or sea water that
disrupts biological processes or causes a health hazard.
Common pollutants include nitrates, pesticides, and sewage (see ◊sewage disposal), though a huge range of industrial contaminants, such as chemical byproducts and
residues created in the manufacture of various goods, also
enter water – legally, accidentally, and through illegal
dumping.

water softener any substance or unit that removes the hardness from water. Hardness is caused by the presence of calcium and magnesium ions, which combine with soap to form an insoluble scum, prevent lathering, and cause deposits to build up in pipes and cookware (kettle fur). A water softener replaces these ions with sodium ions, which are fully soluble and cause no scum.

waterspout funnel-shaped column of water and cloud that is drawn by from the surface of the sea or a lake by a ◊tornado.

water supply distribution of water for domestic, municipal, or industrial consumption. Water supply in sparsely populated regions usually comes from underground water rising to the surface in natural springs, supplemented by pumps and wells. Urban sources are deep artesian wells, rivers, and reservoirs, usually formed from enlarged lakes or dammed and flooded valleys, from which water is conveyed by pipes, conduits, and aqueducts to filter beds. As water seeps through layers of shingle, gravel, and sand, harmful organisms are removed and the water is then distributed by pumping or gravitation through mains and pipes.

Often other substances are added to the water, such as chlorine and fluoride; aluminium sulphate, a clarifying agent, is the most widely used chemical in water treatment. In towns, domestic and municipal (road washing, sewage) needs account for about 135 l/30 gal per head each day. In coastal desert areas, such as the Arabian peninsula, desalination plants remove salt from sea water. The Earth's waters, both fresh and saline, have been polluted by industrial and domestic chemicals, some of which are toxic and others radioactive (see ◊water pollution).

A period of prolonged dry weather can disrupt water supply and lead to drought. The area of the world subject to serious droughts, such as the Sahara, is increasing because of destruction of forests, overgrazing, and poor agricultural practices. A World Bank report in 1995 warned that a global crisis was imminent: chronic water shortages were experienced by 40% of the world's population, notably in the Middle East, northern and sub-Saharan Africa, and central Asia. By 1997, 1.4 billion people (25% of population) still had no access to safe drinking water.

water table the upper level of ground water (water collected underground in porous rocks). Water that is above the water table will drain downwards; a spring forms where the water table cuts the surface of the ground. The water table rises and falls in response to rainfall and the rate at which water is extracted, for example, for irrigation and industry.

In many irrigated areas the water table is falling due to the extraction of water. Below N China, for example, the water table is sinking at a rate of 1 m/3 ft a year. Regions with high water tables and dense industrialization have problems with ◊pollution of the water table. In the USA, New Jersey, Florida, and Louisiana have water tables contaminated by both industrial ◊wastes and saline seepage from the ocean.

watt SI unit (symbol W) of power (the rate of expenditure or consumption of energy) defined as one joule per second. A light bulb, for example, may use 40, 60, 100, or 150 watts of power; an electric heater will use several kilowatts (thousands of watts). The watt is named after the Scottish engineer James Watt.

The absolute watt is defined as the power used when one joule of work is done in one second. In electrical terms, the flow of one ampere of current through a conductor whose ends are at a potential difference of one volt uses one watt of power (watts = volts × amperes).

wave in the oceans, a ridge or swell formed by wind or other causes. The power of a wave is determined by the strength of the wind and the distance of open water over which the wind blows (the fetch). Waves are the main agents of ◊coastal erosion and deposition: sweeping away or building up beaches, creating ◊spits and berms, and wearing down cliffs by their hydraulic action and by the corrasion of the sand and shingle that they carry. A ◊tsunami (misleadingly called a 'tidal wave') is formed after a submarine earthquake.

As a wave approaches the shore it is forced to break as a result of friction with the sea bed. When it breaks on a beach, water and sediment are carried up the beach as *swash*; the water then drains back as *backwash*.

A *constructive wave* causes a net deposition of material on the shore because its swash is stronger than its backwash. Such waves tend be low and have crests that spill over gradually as they break. The backwash of a

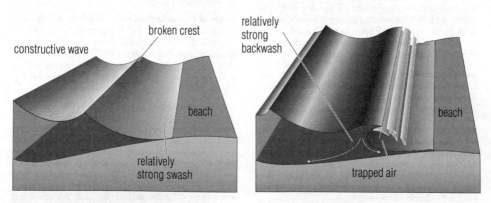

wave

destructive wave is stronger than its swash, and therefore causes a net removal of material from the shore. Destructive waves are usually tall and have peaked crests that plunge downwards as they break, trapping air as they do so.

If waves strike a beach at an angle the beach material will be gradually moved along the shore (longshore drift), causing a deposition of material in some areas and erosion in others.

Atmospheric instability caused by the ◊greenhouse effect and global warming appears to be increasing the severity of Atlantic storms and the heights of the ocean waves. Waves in the South Atlantic are shrinking – they are on average half a metre smaller than in the mid-1980s – and those in the Northeast Atlantic have doubled in size over the last 40 years. As the height of waves affects the supply of marine food, this could affect fish stocks, and there are also implications for shipping and oil and gas rigs in the North Atlantic, which will need to be strengthened if they are to avoid damage.

wave in physics, a disturbance travelling through a medium (or space). There are two types: in a ◊longitudinal wave, such as a sound wave, the disturbance is parallel to the wave's direction of travel; in a ◊transverse wave, such as an electromagnetic wave, it is perpendicular. The medium (for example the Earth, for seismic waves) is not permanently displaced by the passage of a wave. See also ◊standing wave.

wave-cut platform gently sloping rock surface found at the foot of a coastal cliff. Covered by water at high tide but exposed at low tide, it represents the last remnant of an eroded headland (see ◊coastal erosion).

waveguide hollow metallic tube, either empty or containing a ◊dielectric used to guide a high-frequency electromagnetic wave (microwave) travelling within it. The wave is reflected from the internal surfaces of the guide. Waveguides are extensively used in radar systems.

wavelength the distance between successive crests of a ◊wave. The wavelength of a light wave determines its colour; red light has a wavelength of about 700 nanometres, for example. The complete range of wavelengths of electromagnetic waves is called the electromagnetic ◊spectrum.

wave power power obtained by harnessing the energy of water waves. Various schemes have been advanced since 1973, when oil prices rose dramatically and an energy shortage threatened. In 1974 the British engineer Stephen Salter developed the duck – a floating boom whose segments nod up and down with the waves. The nodding motion can be used to drive pumps and spin generators. Another device, developed in Japan, uses an oscillating water column to harness wave power. A major breakthrough will be required if wave power is ever to contribute significantly to the world's energy needs, although several ideas have reached prototype stage.

A 75-kilowatt wave-power generator off the Scottish island of Islay is one of only three in the world connected to a power grid.

The pilot for the world's first commercial wave-powered electricity generator was launched on the River Clyde, Scotland, in Aug 1995 and connected to the grid at Dounreay. The plant, called Osprey, is 20 m/66 ft high and expected to produce 2 megawatts of electricity from wave power, and another 1.5 from a connected wind turbine.

wax solid fatty substance of animal, vegetable, or mineral origin.

Waxes are composed variously of ◊esters, ◊fatty acids, free ◊alcohols, and solid hydrocarbons.

Mineral waxes are obtained from petroleum and vary in hardness from the soft petroleum jelly (or petrolatum) used in ointments to the hard paraffin wax employed for making candles and waxed paper for drinks cartons.

Animal waxes include beeswax, the wool wax lanolin, and spermaceti from sperm-whale oil; they are used mainly in cosmetics, ointments, and polishes. Another animal wax is tallow, a form of suet obtained from cattle and sheep's fat, once widely used to make candles and soap. Sealing wax is made from lac or shellac, a resinous substance obtained from secretions of scale insects.

Vegetable waxes, which usually occur as a waterproof coating on plants that grow in hot, arid regions, include carnauba wax (from the leaves of the carnauba palm) and candelilla wax, both of which are components of hard polishes such as car waxes.

weak acid acid that only partially ionizes in aqueous solution (see ◊dissociation). Weak acids include ethanoic acid and carbonic acid. The pH of such acids lies between pH 3 and pH 6.

weak base base that only partially ionizes in aqueous solution (see ◊dissociation); for example, ammonia. The pH of such bases lies between pH 8 and pH 10.

weak nuclear force or **weak interaction** one of the four fundamental forces of nature, the other three being gravity, the electromagnetic force, and the strong force. It causes radioactive decay and other subatomic reactions. The particles that carry the weak force are called ◊weakons (or intermediate vector bosons) and comprise the positively and negatively charged W particles and the neutral Z particle.

weakon or **intermediate vector boson** in physics, a ◊gauge boson that carries the weak nuclear force, one of the fundamental forces of nature. There are three types of weakon, the positive and negative W particle and the neutral Z particle.

weapon any implement used for attack and defence, from simple clubs, spears, and bows and arrows in prehistoric times to machine guns and nuclear bombs in modern times. The first revolution in warfare came with the invention of ◊gunpowder and the development of cannons and shoulder-held guns. Many other weapons now exist, such as grenades, shells, torpedoes, rockets, and guided missiles. The ultimate in explosive weapons are the atomic (fission) and hydrogen (fusion) bombs. They release the enormous energy produced when atoms split or fuse together. There are also chemical and bacteriological weapons, which release poisons or disease. *See chronology overleaf.*

We turned the switch, saw the flashes, watched for ten minutes, then switched everything off and went home. That night I knew the world was headed for sorrow.

On **Nuclear fission** and the **atom bomb**
Leo Szilard 1939

Weapons: Chronology

1000 BC	Clubs, hammers, axes daggers, spears, swords, slings, and bows are all common weapons by this time. Armour is also in use for personal protection.
700 BC	The Assyrians develop the ram for battering down walls.
500 BC	The crossbow appears in China.
600 AD	The stirrup, introduced into Europe from the East, gives riders better control and allows the more effective use of lances and swords from the saddle.
668	First reported use of 'Greek Fire', an incendiary compound.
13th century	Gunpowder brought to the West from China (where it was long in use but only for fireworks).
1232	First reliable report of the use of rockets by the Chinese.
1242	First written record of gunpowder.
c.1300	Guns invented by the Arabs, with bamboo muzzles reinforced with iron.
1326	First mention of firearms, in England and Italy, in the form of small cannon.
1346	Battle of Crécy in which gunpowder was probably used in battle for the first time.
1376	Explosive shells used in Venice.
1378	Cannon mounted in warships by the Venetians.
1400	Hand firearms – 'hand-gonnes' – appear in Europe.
1411	First mention of the matchlock to ignite hand weapons.
1421	First record of gunpowder-filled explosive shells fired from cannon.
1450	Casting of cannon becomes common.
1451	Mortars, fired at a high angle to drop projectiles on to the enemy, are developed by the Turks.
1500	The arquebus and hackbut, the first shoulder-fired arms.
1530	The wheel lock appears, allowing guns to be carried ready for firing, leading to the development of the pistol, a single-handed weapon.
1547	Earliest flintlocks developed in Spain.
1570	Rifled firearms developed in Germany.
1579	First use of red-hot shot from cannon.
1596	Introduction of a wooden, gunpowder-filled fuse to ignite explosive shells so as to burst them in the air over the target.
17th century	Widespread use of guns and cannon in the Thirty Years' War and English Civil War.
1625	Introduction of wheeled carriages for field artillery.
1635	The perfected French flintlock mechanism, which becomes the standard ignition method in universal use by 1660.
1647	The French introduce the bayonet.
1750	Introduction of horse artillery, in which the gunners are mounted, giving greater mobility than ordinary horse-drawn artillery where the gunners march alongside their guns.
1776	British adopt the Firguson breech-loading rifle for use during the American Revolution.
1778	The Carronade, a light, short-range gun developed for naval use.
1787	Lt Shrapnel begins development of the 'spherical case shot' which eventually becomes the 'shrapnel' shell.
1792	The Indians under Tipu use rockets against the British at Seringapatam.
1800	Henry Shrapnel invented shrapnel for the British army.
1806	British use the Congreve rocket against Boulogne.
1807	Rev. Forsyth patents the percussion principle for ignition of firearms.
1812	Self-contained cartridges for small arms.
1813	British artillery first fires over the heads of assaulting infantry in order to suppress enemy fire until the infantry are close enough to charge.
1814	Development of the percussion cap.
1818	Collier and Wheeler develop a hand-rotated flintlock revolver.
1836	Samuel Colt patents his percussion revolver.
1841	The 'Needle Gun', first bolt-action breech-loading rifle with self-contained cartridge, adopted by Prussia.
1846	Invention of guncotton, the first 'modern' explosive, by Schonbein.
1849	Beginning of a move to convert muzzle-loading muskets into breech-loaders.
1855	Rifled muzzle-loading cannon adopted in France.
1857	Smith and Wesson patent the first metallic rim-fire cartridge revolver.
1859	Britain adopts Armstrong's rifled breech-loading cannon.
1858	French introduce ironclad warships.
1861	The Gatling gun, first successful mechanical machine gun.
1862	American Civil War introduces breech-loading small arms into military use.
	First battle between ironclad warships (*Monitor* and *Merrimac*) and first use in combat of a gun turret on a warship (*Monitor*).
1863	TNT discovered by German chemist J Wilbrand.
1864	The British revert to rifled muzzle-loading cannon in order to produce guns of sufficient power to attack ironclad ships.
1865	Invention of smokeless powder by Schultz.
1866	Whitehead's self-propelled naval torpedo.
	Alfred Nobel invents dynamite.
1868	General adoption of electrically-fired sea mines for harbour protection.

Weapons: Chronology – *continued*

1870	German development of an anti-aircraft gun to shoot at balloons escaping from besieged Paris.
1880	General adoption of breech-loading artillery.
	TNT (high explosive) perfected.
1884	Hiram Maxim develops the first automatic machine gun.
1885	The Brennan wire-guided torpedo adopted in Britain for coastal defence.
1886	France adopts the Lebel rifle, the first bolt-action, magazine repeater of small caliber.
1888	First use of the Maxim machine gun in war.
1890	Bolt action magazine rifles such as the Mauser, Lee-Enfield, Mannlicher, and Krag-Jorgensen models become the universal infantry weapon.
	General adoption of cast picric acid as a high explosive filling for artillery shells under various names such as 'Lyddite', 'Schneiderite'.
1893	The Borchardt, first practical automatic pistol.
1897	The French introduce the 75 mm quick-firing gun, the first field artillery piece to have on-carriage recoil control, self-contained ammunition, and a shield to protect the gunners.
1898	Development of the Madsen light machine gun, the first single operator magazine-fed machine gun.
	First practical submarine.
1904	Russo-Japanese war sees the revival of the hand grenade.
1906	First successful flight of a Zeppelin airship.
1909	Self-propelled anti-aircraft guns developed in Germany.
1914	Development of rifle-propelled grenades.
	First use of firearms from aircraft.
	Armoured cars in use in Belgium.
1915	First use of chemical gas in war: the Germans use tear-gas shells against the Russians at Bolimov.
	Introduction of synchronised machine guns capable of firing through a spinning propeller without hitting the blades.
	First use of cloud poison gas in war by the Germans against the British at Ypres.
	German introduction of the flamethrower.
	Stokes' trench mortar is developed; it is the precursor of all mortars since then.
1916	Introduction of the first sub-machine gun.
	Tanks used for the first time by the British at Cambrai.
1917	First recoilless gun, the Davis cannon, used from aircraft against submarines.
1925	The Oerlikon aircraft cannon.
1931	Germany begins development of rockets for air defence and long-range bombardment.
1932	First studies in biological warfare – the use of disease germs – in Europe, the USA, and Japan.
1936	US Army adopt the Garand automatic rifle, the first such weapon to become standard infantry issue.
	Discovery of nerve gases in Germany.
1938	British defensive radar system goes into operation.
1940	First use of the shaped-charge principle, in demolition charges against Eben Emael fort in Belgium.
	The British use a shaped-charge rifle grenade.
	The British introduce air defence rockets.
	The Germans introduce lightweight recoilless guns for field use.
1942	The Germans develop the 'assault rifle'.
	US development of the shoulder-fired 'Bazooka' anti-tank rocket.
	Germany introduces rocket-boosted artillery shells.
1943	First use of proximity fuse against aircraft by US Navy.
	First guided aircraft bombs.
1944	Germany introduces pilotless flying bombs ('V1').
	Germany introduces ballistic missile (the 'V2' rocket).
	First air-defence guided missiles developed in Germany.
	First wire-guided anti-tank missile developed in Germany.
1945	First test explosion and military use of atom bomb by the USA against Japan.
1950	First military use of helicopters, for casualty evacuation (Korea).
1954–73	Vietnam War, use of chemical warfare (defoliants and other substances) by the USA.
1955	First strategic guided missiles (Soviet SS-3) enter service.
1965	First anti-missile missile (the US 'Sprint').
1982	Anti-ship missiles are first used in the Falklands War.
1983	Star Wars or Strategic Defense Initiative research announced by the USA to develop space laser and particle-beam weapons as a possible future weapons system in space.
1991	'Smart' weapons used by the USA and allied powers in the Gulf War; equipped with computers (using techniques such as digitized terrain maps) and laser guidance, they reached their targets with precision accuracy (for example a 'smart' bomb destroyed the Ministry of Air Defence in Baghdad by flying into an air shaft).

weather day-to-day variation of climatic and atmospheric conditions at any one place, or the state of these conditions at a place at any one time. Such conditions include humidity, precipitation, temperature, cloud cover, visibility, and wind. To a meteorologist the term 'weather' is limited to the state of the sky, precipitation, and visibility as affected by fog or mist. A region's ◊climate is derived from the average weather conditions over a long period of time. See also ◊meteorology.

Weather forecasts, in which the likely weather is predicted for a particular area, based on meteorological readings, may be short-range (covering a period of one or two days), medium-range (five to seven days), or long-range (a month or so). Readings from a series of scattered recording stations are collected and compiled on a weather map. Such a procedure is called synoptic forecasting. The weather map uses conventional symbols to show the state of the sky, the wind speed and direction, the kind of precipitation, and other details at each gathering station. Points of equal atmospheric pressure are joined by lines called isobars (lines joining places of equal pressure). The trends shown on such a map can be extrapolated to predict what weather is coming.

weather area any of the divisions of the sea around the British Isles for the purpose of weather forecasting for shipping. The areas are used to indicate where strong or gale-force winds are expected.

weathering process by which exposed rocks are broken down on the spot by the action of rain, frost, wind, and other elements of the weather. It differs from ◊erosion in that no movement or transportation of the broken-down material takes place. Two types of weathering are recognized: physical (or mechanical) and chemical. They usually occur together.

Physical weathering This includes such effects as freeze–thaw (the splitting of rocks by the alternate freezing and thawing of water trapped in cracks) and exfoliation, or onion-skin weathering (flaking caused by the alternate expansion and contraction of rocks in response to extreme changes in temperature).

Weathering

Physical weathering	
temperature changes	weakening rocks by expansion and contraction
frost	wedging rocks apart by the expansion of water on freezing
unloading	the loosening of rock layers by release of pressure after the erosion and removal of those layers above

Chemical weathering	
carbonation	the breakdown of calcite by reaction with carbonic acid in rainwater
hydrolysis	the breakdown of feldspar into china clay by reaction with carbonic acid in rainwater
oxidation	the breakdown of iron-rich minerals due to rusting
hydration	the expansion of certain minerals due to the uptake of water

Chemical weathering Involving a chemical change in the rocks affected, the most common form is caused by rainwater that has absorbed carbon dioxide from the atmosphere and formed a weak carbonic acid. This then reacts with certain minerals in the rocks and breaks them down. Examples are the solution of caverns in limestone terrains, and the breakdown of feldspars in granite to form china clay or kaolin.

Although physical and chemical weathering normally occur together, in some instances it is difficult to determine which type is involved. For example, exfoliation, which produces rounded ◊inselbergs in arid regions, such as Ayers Rock in central Australia, may be caused by the daily physical expansion and contraction of the surface layers of the rock in the heat of the Sun, or by the chemical reaction of the minerals just beneath the surface during the infrequent rains of these areas.

weber SI unit (symbol Wb) of ◊magnetic flux (the magnetic field strength multiplied by the area through which the field passes). It is named after German chemist Wilhelm Weber. One weber equals 10^8 ◊maxwells.

A change of flux at a uniform rate of one weber per second in an electrical coil with one turn produces an electromotive force of one volt in the coil.

wedge block of triangular cross-section that can be used as a simple machine. An axe is a wedge: it splits wood by redirecting the energy of the downward blow sideways, where it exerts the force needed to split the wood.

weedkiller or *herbicide* chemical that kills some or all plants. Selective herbicides are effective with cereal crops because they kill all broad-leaved plants without affecting grasslike leaves. Those that kill all plants include sodium chlorate and ◊paraquat; see also ◊Agent Orange. The widespread use of weedkillers in agriculture has led to an increase in crop yield but also to pollution of soil and water supplies and killing of birds and small animals, as well as creating a health hazard for humans.

weight the force exerted on an object by ◊gravity. The weight of an object depends on its mass – the amount of material in it – and the strength of the Earth's gravitational pull, which decreases with height. Consequently, an object weighs less at the top of a mountain than at sea level. On the Moon, an object has only one-sixth of its weight on Earth, because the pull of the Moon's gravity is one-sixth that of the Earth.

weightlessness the apparent loss in weight of a body in ◊free fall. Astronauts in an orbiting spacecraft do not feel any weight because they are falling freely in the Earth's gravitational field. The same phenomenon can be experienced in a falling lift or in an aircraft deliberately imitating the path of a freely falling object.

weights and measures see under ◊c.g.s. system, ◊f.p.s. system, ◊m.k.s. system, ◊SI units.

Weil's disease or *leptospirosis* infectious disease of animals that is occasionally transmitted to human beings, usually by contact with water contaminated with rat urine. It is characterized by acute fever, and infection may spread to the brain, liver, kidneys, and heart. It has a 10% mortality rate.

The usual form occurring in humans is caused by a spiral-shaped bacterium (spirochete) that is a common parasite of rats. The condition responds poorly to antibiotics, and death may result.

weir low wall built across a river to raise the water level. The oldest surviving weir in England is at Chester, across the river Dee, dating from around 1100.

welding joining pieces of metal (or nonmetal) at faces rendered plastic or liquid by heat or pressure (or both). The principal processes today are gas and arc welding, in which the heat from a gas flame or an electric arc melts the faces to be joined. Additional 'filler metal' is usually added to the joint.

Forge (or hammer) welding, employed by blacksmiths since early times, was the only method available until the late 19th century. Resistance welding is another electric method in which the weld is formed by a combination of pressure and resistance heating from an electric current. Recent developments include electric-slag, electron-beam, high-energy laser, and the still experimental radio-wave energy-beam welding processes.

Westerlies prevailing winds from the west that occur in both hemispheres between latitudes of about 35° and 60°. Unlike the ◊trade winds, they are very variable and produce stormy weather.

The Westerlies blow mainly from the SW in the northern hemisphere and the NW in the southern hemisphere, bringing moist weather to the W coast of the landmasses in these latitudes.

wetland permanently wet land area or habitat. Wetlands include areas of ◊marsh, fen, ◊bog, flood plain, and shallow coastal areas. Wetlands are extremely fertile. They provide warm, sheltered waters for fisheries, lush vegetation for grazing livestock, and an abundance of wildlife. Estuaries and seaweed beds are more than 16 times as productive as the open ocean.

The term is often more specifically applied to a naturally flooding area that is managed for agriculture or wildlife. A water meadow, where a river is expected to flood grazing land at least once a year thereby replenishing the soil, is a traditional example.

wetted perimeter the length of that part of a river's cross-section that is in contact with the water. The wetted perimeter is used to calculate a river's hydraulic radius, a measure of its channel efficiency.

whaling the hunting of whales. Whales have been killed by humans at least since the middle ages. There were hundreds of thousands of whales at the beginning of the 20th century, but the invention of the harpoon in 1864 and improvements in ships and mechanization have led to the near-extinction of several species of whale. Commercial whaling was largely discontinued 1986, although Norway and Japan have continued commercial whaling.

Traditional whaling areas include the coasts of Greenland and Newfoundland, but the Antarctic, in the summer months, supplied the bulk of the catch.

Practically the whole of the animal can be utilized in one form or another: whales are killed for whale oil (made from the thick layer of fat under the skin called 'blubber'), used as a lubricant, or for making soap, candles, and margarine; for the large reserve of oil in the head of the sperm whale, used in the leather industry; and for **ambergris**, a waxlike substance from the intestines of the sperm whale, used in making perfumes. Whalebone was

Whaling: Chronology

11th/12th century	Basque villagers began hunting right whales in the Bay of Biscay, severely depleting local stocks by the 13th century.
1610	Greenland bowhead whale fishery began. Stocks were reduced to the brink of extinction by the 19th century.
1789	The first commercial whalers entered the Pacific.
1848	Arctic bowhead whales discovered and quickly exploited; stocks were almost totally depleted by 1914.
1864	Norwegian engineer Sven Foyn invents the gun-launched harpoon with an explosive head. It permits hunting the faster and more plentiful fin, sei, and blue whales and ushers in the era of modern whaling.
1890	Californian gray whale believed to be extinct after 40 years of hunting; however, some individuals survived and are now protected.
1904	Whaling operations began in the Southern Ocean.
1925	First factory ships used by whalers, doing away with the need to return to shore-based stations and so increasing catches.
1946	International Whaling Commission (IWC) established.
1950s	First commercial whale-watching tours began, observing migrating gray whales off southern California.
1960–61	All-time peak in whaling; 64,000 cetaceans caught worldwide.
1962	Whaling has reduced the humpback whale popoulation to an estimated 1,000.
1967	IWC prohibits the hunting of blue, right, grey, and humpback whales.
1979	IWC bans the hunting of sperm whales.
1985	IWC officially banned all commercial whaling; however, whaling continued under the guise of 'scientific' expeditions.
1988–89	Boycott of Icelandic fishing products, in protest against the country's continuation of commercial whaling, cost the Icelandic fishing industry $30 million.
1993	Greenpeace conducted a boycott campaign against Norwegian products to discourage Norway's continued whaling.
1994	Russia revealed that the former Soviet Union fleet had secretly continued large-scale commercial whaling for 40 years, without declaring catch numbers to IWC. The IWC establishes a whale sanctuary in Antarctica.
1995	Japan and Norway continued their commercial whaling, despite an international moratorium. Norway was openly whaling commercially; Japan claimed it was doing so for scientific research.

used by corset manufacturers and in the brush trade; there are now synthetic substitutes for all these products. Whales have also been killed for use in petfood manufacture in the USA and Europe, and as a food in Japan. The flesh and ground bones are used as soil fertilizers.

wheat cereal plant derived from the wild *Triticum*, a grass native to the Middle East. It is the chief cereal used in breadmaking and is widely cultivated in temperate climates suited to its growth. Wheat is killed by frost, and damp makes the grains soft, so warm, dry regions produce the most valuable grain.

The main wheat-producing areas of the world are the Ukraine, the prairie states of the USA, the Punjab in India, the prairie provinces of Canada, parts of France, Poland, S Germany, Italy, Argentina, and SE Australia. Flour is milled from the nutritious tissue surrounding the embryonic plant in the grain (the ◊endosperm); the coatings of the grain produce bran. Semolina is also prepared from wheat; it is a by-product from the manufacture of fine flour.

wheel and axle simple machine with a rope wound round an axle connected to a larger wheel with another rope attached to its rim. Pulling on the wheel rope (applying an effort) lifts a load attached to the axle rope. The velocity ratio of the machine (distance moved by load divided by distance moved by effort) is equal to the ratio of the wheel radius to the axle radius.

Early wheels probably evolved from log rollers used to aid the progress of a heavily laden sled or stone block, a slice of log attached solidly to an axle giving way to a crude disc wheel revolving on a fixed axle. Spokes were introduced about 2000 BC in Near East regions, an improvement which lightened the structure and provided levers to propel the vehicle when necessary.

whey watery by-product of the cheesemaking process, which is drained off after the milk has been heated and ◊rennet (a curdling agent) added to induce its coagulation.

In Scandinavia, especially Norway, whey is turned into cheese, *mysost* and (from goat's whey) *gjetost*. The flavour of whey cheese is sweet from added brown sugar and is an acquired taste.

whiplash injury damage to the neck vertebrae and their attachments caused by a sudden backward jerk of the head and neck. It is most often seen in vehicle occupants as a result of the rapid deceleration experienced in a crash.

whirlwind rapidly rotating column of air, often synonymous with a ◊tornado. On a smaller scale it produces the dust-devils seen in deserts.

white blood cell or *leucocyte* one of a number of different cells that play a part in the body's defences and give immunity against disease. Some (◊phagocytes and ◊macrophages) engulf invading microorganisms, others kill infected cells, while ◊lymphocytes produce more specific immune responses. White blood cells are colourless, with clear or granulated cytoplasm, and are capable of independent amoeboid movement. They occur in the blood, ◊lymph, and elsewhere in the body's tissues.

Unlike mammalian red blood cells, they possess a nucleus. Human blood contains about 11,000 leucocytes to the cubic millimetre – about one to every 500 red cells.

White blood cell numbers may be reduced (leucopenia) by starvation, pernicious anaemia, and certain infections, such as typhoid and malaria. An increase in their numbers (leucocytosis) is a reaction to normal events such as digestion, exertion, and pregnancy, and to abnormal ones such as loss of blood, cancer, and most infections.

white dwarf small, hot ◊star, the last stage in the life of a star such as the Sun. White dwarfs make up 10% of the stars in the Galaxy; most have a mass 60% of that of the Sun, but only 1% of the Sun's diameter, similar in size to the Earth. Most have surface temperatures of 8,000°C/14,400°F or more, hotter than the Sun. Yet, being so small, their overall luminosities may be less than 1% of that of the Sun. The Milky Way contains an estimated 50 billion white dwarfs.

whiteout 'fog' of grains of dry snow caused by strong winds in temperatures of between −18°C/0°F and −1°C/30°F. The uniform whiteness of the ground and air causes disorientation in humans.

white spirit colourless liquid derived from petrol; it is used as a solvent and in paints and varnishes.

WHO acronym for *World Health Organization*.

whooping cough or *pertussis* acute infectious disease, seen mainly in children, caused by colonization of the air passages by the bacterium *Bordetella pertussis*. There may be catarrh, mild fever, and loss of appetite, but the main symptom is violent coughing, associated with the sharp intake of breath that is the characteristic 'whoop', and often followed by vomiting and severe nose bleeds. The cough may persist for weeks. Although debilitating, the disease is seldom serious in older children, but infants are at risk both from the illness itself and from susceptibility to other conditions, such as ◊pneumonia.

wide-angle lens photographic lens of shorter focal length than normal, taking in a wider angle of view.

wide area network in computing, a ◊network that connects computers distributed over a wide geographical area.

Wien's displacement law in physics, a law of radiation stating that the wavelength carrying the maximum energy is inversely proportional to the absolute temperature of a black body: the hotter a body is, the shorter the wavelength. It has the form $\lambda_{max} T$ = constant, where λ_{max} is the wavelength of maximum intensity and T is the temperature. The law is named after German physicist Wilhelm Wien.

wilderness area of uninhabited land that has never been disturbed by humans, usually located some distance from towns and cities. According to estimates by US group Conservation International, 52% (90 million sq km/ 35 million sq mi) of the Earth's total land area was still undisturbed 1994.

wildlife trade international trade in live plants and animals, and in wildlife products such as skins, horns, shells, and feathers. The trade has made some species virtually extinct, and whole ecosystems (for example, coral reefs) are threatened. Wildlife trade is to some extent regulated by ◊CITES (Convention on International Trade in Endangered Species).

Species almost eradicated by trade in their products include many of the largest whales, crocodiles, marine turtles, and some wild cats. Until recently, some 2 million snake skins were exported from India every year. Populations of black rhino and African elephant have collapsed because of hunting for their horns and tusks (◊ivory), and poaching remains a problem in cases where trade is prohibited.

wild type in genetics, the naturally occurring gene for a particular character that is typical of most individuals of a given species, as distinct from new genes that arise by mutation.

will-o'-the-wisp light sometimes seen over marshy ground, believed to be burning gas containing methane from decaying organic matter. An Australian scientist put forward the hypothesis in 1995 that the phenomenon may be caused by barn owls, which sometimes develop a ghostly glow due to a light-emitting honey fungus that they pick up from rotting trees.

willy-willy Australian Aboriginal term for a cyclonic whirlwind.

Wilms's tumour or **nephroblastoma** one of the rare cancers of infancy, arising in the kidneys. Often the only symptom is abdominal swelling. Treatment is by removal of the affected kidney (nephrectomy), followed by radiotherapy and ◊cytotoxic drugs.

wilting the loss of rigidity (◊turgor) in plants, caused by a decreasing wall pressure within the cells making up the supportive tissues. Wilting is most obvious in plants that have little or no wood.

WIMP (acronym for **windows, icons, menus, pointing device**) in computing, another name for ◊graphical user interface (GUI).

Winchester drive in computing, an old-fashioned term for ◊hard disc.

wind the lateral movement of the Earth's atmosphere from high-pressure areas (anticyclones) to low-pressure areas (depression). Its speed is measured using an ◊anemometer or by studying its effects on, for example, trees by using the ◊Beaufort scale. Although modified by features such as land and water, there is a basic worldwide system of ◊trade winds, ◊westerlies, and polar easterlies.

A belt of low pressure (the ◊doldrums) lies along the Equator. The trade winds blow towards this from the horse latitudes (areas of high pressure at about 30° N and 30° S of the Equator), blowing from the NE in the northern hemisphere, and from the SE in the southern. The Westerlies (also from the horse latitudes) blow north of the Equator from the SW and south of the Equator from the NW.

Cold winds blow outwards from high-pressure areas at the poles. More local effects result from landmasses heating and cooling faster than the adjacent sea, producing onshore winds in the daytime and offshore winds at night.

The ◊monsoon is a seasonal wind of S Asia, blowing from the SW in summer and bringing the rain on which crops depend. It blows from the NE in winter.

Famous or notorious warm winds include the **chinook** of the eastern Rocky Mountains, North America; the **föhn** of Europe's Alpine valleys; the **sirocco** (Italy)/ **khamsin** (Egypt)/**sharav** (Israel), spring winds that bring warm air from the Sahara and Arabian deserts across the Mediterranean; and the **Santa Ana**, a periodic warm wind from the inland deserts that strikes the California coast.

The dry northerly **bise** (Switzerland) and the **mistral**, which strikes the Mediterranean area of France, are unpleasantly cold winds.

wind-chill factor or **wind-chill index** estimate of how much colder it feels when a wind is blowing. It is the sum of the temperature (in °F below zero) and the wind speed (in miles per hour). So for a wind of 15 mph at an air temperature of −5°F, the wind-chill factor is 20.

wind farm array of windmills or ◊wind turbines used for generating electrical power. A wind farm at Altamont Pass, California, USA, consists of 300 wind turbines, the smallest producing 60 kW and the largest 750 kW of electricity. Wind farms supply about 1.5% of California's electricity needs. To produce 1,200 megawatts of electricity (an output comparable with that of a nuclear power station), a wind farm would need to occupy around 370 sq km/140 sq mi.

Denmark built the world's first offshore wind farm, off Vindeby on Lolland Island in the North Sea.

windmill mill with sails or vanes that, by the action of wind upon them, drive machinery for grinding corn or pumping water, for example. Wind turbines, designed to use wind power on a large scale, usually have a propeller-type rotor mounted on a tall shell tower. The turbine drives a generator for producing electricity.

Windmills were used in the East in ancient times, and in Europe they were first used in Germany and the Netherlands in the 12th century. The main types of traditional windmill are the **post mill**, which is turned around a post when the direction of the wind changes, and the **tower mill**, which has a revolving turret on top. It usually has a device (fantail) that keeps the sails pointing into the wind. In the USA windmills were used by the colonists and later a light type, with steel sails supported on a long steel girder shaft, was introduced for use on farms.

window in computing, a rectangular area on the screen of a ◊graphical user interface. A window is used to display data and can be manipulated in various ways by the computer user.

windpipe in medicine, another name for the ◊trachea.

wind power the harnessing of wind energy to produce power. The wind has long been used as a source of energy: sailing ships and windmills are ancient inventions. After the energy crisis of the 1970s ◊wind turbines began to be used to produce electricity on a large scale.

By the year 2000, 10% of Denmark's energy is expected to come from wind power. Denmark now supplies 1.5% of its energy from wind energy alone. Since 1979 the Danish government has been giving grants of up to 30% of the cost of turbines.

wind tunnel test tunnel in which air is blown over, for example, a stationary model aircraft, motor vehicle, or locomotive to simulate the effects of movement. Lift, drag, and airflow patterns are observed by the use of special cameras and sensitive instruments. Wind-tunnel testing assesses aerodynamic design, preparatory to full-scale construction.

wind turbine windmill of advanced aerodynamic design connected to an electricity generator and used in wind-power installations. Wind turbines can be either large propeller-type rotors mounted on a tall tower, or flexible metal strips fixed to a vertical axle at top and bottom.

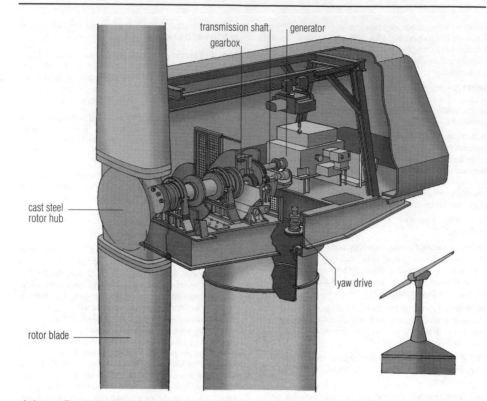

wind power The wind turbine is the modern counterpart of the windmill. The rotor blades are huge – up to 100 m/330 ft across – in order to extract as much energy as possible from the wind. Inside the turbine head, gears are used to increase the speed of the turning shaft so that the electricity generation is as efficient as possible.

The world's largest wind turbine is on Hawaii, in the Pacific Ocean. It has two blades 50 m/160 ft long on top of a tower 20 storeys high. An example of a propeller turbine is found at Tvind in Denmark and has an output of some 2 megawatts. Other machines use novel rotors, such as the 'egg-beater' design developed at Sandia Laboratories in New Mexico, USA.

wing in biology, the modified forelimb of birds and bats, or the membranous outgrowths of the ◊exoskeleton of insects, which give the power of flight. Birds and bats have two wings. Bird wings have feathers attached to the fused digits ('fingers') and forearm bones, while bat wings consist of skin stretched between the digits. Most insects have four wings, which are strengthened by wing veins.

The wings of butterflies and moths are covered with scales. The hind pair of a fly's wings are modified to form two knoblike balancing organs (halteres).

wire thread of metal, made by drawing a rod through progressively smaller-diameter dies. Fine-gauge wire is used for electrical power transmission; heavier-gauge wire is used to make load-bearing cables.

Gold, silver, and bronze wire has been found in the ruins of Troy and in ancient Egyptian tombs. From early times to the 14th century, wire was made by hammering metal into sheets, cutting thin strips, and making the strips round by hammering them. The Romans made wire by hammering heated metal rods.

Wire drawing was introduced in Germany in the 14th century. In this process, a metal rod is pulled (drawn) through a small hole in a mould (die). Until the 19th century this was done by hand; now all wire is drawn by machine. Metal rods are pulled through a series of progressively smaller tungsten carbide dies to produce large-diameter wire, and through diamond dies for very fine wire. The die is funnel-shaped, with the opening smaller than the diameter of the rod. The rod, which is pointed at one end, is coated with a lubricant to allow it to slip through the die. Pincers pull the rod through until it can be wound round a drum. The drum then rotates, drawing the wire through the die and winding it into a coil.

There are many kinds of wire for different uses: galvanized wire (coated with zinc), which does not rust; barbed wire and wire mesh for fencing; and wire cable, made by weaving thin wires into ropes. Needles, pins, nails, and rivets are made from wire.

wireless original name for a radio receiver. In early experiments with transmission by radio waves, notably by Italian inventor Guglielmo Marconi in Britain, signals were sent in Morse code, as in telegraphy. Radio, unlike the telegraph, used no wires for transmission, and the means of communication was termed 'wireless telegraphy'.

wolfram alternative name for ◊tungsten

wolframite iron manganese tungstate, $(Fe,Mn)WO_4$, an ore mineral of tungsten. It is dark grey with a submetallic surface lustre, and often occurs in hydrothermal veins in association with ores of tin.

womb common name for the ◊uterus.

wood the hard tissue beneath the bark of many perennial plants; it is composed of water-conducting cells, or secondary ◊xylem, and gains its hardness and strength from deposits of ◊lignin. *Hardwoods*, such as oak, and *softwoods*, such as pine, have commercial value as structural material and for furniture.

The central wood in a branch or stem is known as *heartwood* and is generally darker and harder than the outer wood; it consists only of dead cells. As well as providing structural support, it often contains gums, tannins, or pigments which may impart a characteristic colour and increased durability. The surrounding *sapwood* is the functional part of the xylem that conducts water.

The *secondary xylem* is laid down by the vascular ◊cambium which forms a new layer of wood annually, on the outside of the existing wood and visible as an ◊annual ring when the tree is felled; see ◊dendrochronology.

Commercial wood can be divided into two main types: hardwood, containing xylem vessels and obtained from angiosperms (for example, oak), and softwood, containing only ◊tracheids, obtained from gymnosperms (for example, pine).

Although in general softwoods are softer than hardwoods, this is not always the case: balsa, the softest wood known, is a hardwood, while pitch pine, very dense and hard, is a softwood. A superhard wood is produced in wood-plastic combinations (WPC), in which wood is impregnated with liquid plastic (monomer) and the whole is then bombarded with gamma rays to polymerize the plastic.

woodland area in which trees grow more or less thickly; generally smaller than a forest. Temperate climates, with four distinct seasons a year, tend to support a mixed woodland habitat, with some conifers but mostly broadleaved and deciduous trees, shedding their leaves in autumn and regrowing them in spring. In the Mediterranean region and parts of the southern hemisphere, the trees are mostly evergreen.

Temperate woodlands grow in the zone between the cold coniferous forest and the tropical forests of the hotter climates near the Equator. They develop in areas where the closeness of the sea keeps the climate mild and moist.

Old woodland can rival tropical rainforest in the number of species it supports, but most of the species are hidden in the soil. A study in Oregon, USA, in 1991 found that the soil in a single woodland location contained 8,000 arthropod species (such as insects, mites, centipedes, and millipedes), compared with only 143 species of reptile, bird, and mammal in the woodland above.

wood pitch by-product of charcoal manufacture, made from *wood tar*, the condensed liquid produced from burning charcoal gases. The wood tar is boiled to produce the correct consistency. It has been used since ancient times for caulking wooden ships (filling in the spaces between the hull planks to make them watertight).

wool the natural hair covering of the sheep, and also of the llama, angora goat, and some other ◊mammals. The domestic sheep *Ovis aries* provides the great bulk of the fibres used in textile production. Lanolin is a by-product.

word in computing, a group of bits (binary digits) that a computer's central processing unit treats as a single working unit. The size of a word varies from one computer to another and, in general, increasing the word length leads to a faster and more powerful computer.

In the late 1970s and early 1980s, most microcomputers were 8-bit machines. During the 1980s 16-bit microcomputers were introduced and 32-bit microcomputers are now available. Mainframe computers may be 32-bit or 64-bit machines.

word processor in computing, a program that allows the input, amendment, manipulation, storage, and retrieval of text; or a computer system that runs such software. Since word-processing programs became available to microcomputers, the method has largely replaced the typewriter for producing letters or other text. Typical facilities include insert, delete, cut and paste, reformat, search and replace, copy, print, mail merge, and spelling check.

The leading word processing programs include Lotus Word Pro, Microsoft Word, and WordPerfect.

work in physics, a measure of the result of transferring energy from one system to another to cause an object to move. Work should not be confused with ◊energy (the capacity to do work, which is also measured in joules) or with ◊power (the rate of doing work, measured in joules per second).

Work is equal to the product of the force used and the distance moved by the object in the direction of that force. If the force is F newtons and the distance moved is d metres, then the work W is given by:

$W=Fd$ For example, the work done when a force of 10 newtons moves an object 5 metres against some sort of resistance is 50 joules (50 newton-metres).

workstation high-performance desktop computer with strong graphics capabilities, traditionally used for engineering (◊CAD and ◊CAM), scientific research, and desktop publishing. Frequently based on fast RISC (reduced instruction-set computer) chips, workstations generally offer more processing power than microcomputers (although the distinction between workstations and the more powerful microcomputer models is becoming increasingly blurred). Most workstations use UNIX as their operating system, and have good networking facilities.

World Meteorological Organization agency, part of the United Nations since 1950, that promotes the international exchange of weather information through the establishment of a worldwide network of meteorological stations. It was founded as the International Meteorological Organization in 1873, and its headquarters are now in Geneva, Switzerland.

World Wide Fund for Nature (WWF, formerly the *World Wildlife Fund*) international organization established in 1961 to raise funds for conservation by public appeal. Projects include conservation of particular species, for example, the tiger and giant panda, and special areas, such as the Simen Mountains, Ethiopia.

World Wildlife Fund former and US name of the *World Wide Fund for Nature*.

worm any of various elongated limbless invertebrates belonging to several phyla. Worms include the ◊flatworms, such as ◊flukes and ◊tapeworms; the round-worms or ◊nematodes, such as the eelworm and the hook-worm; the marine ribbon worms or nemerteans; and the segmented worms or ◊annelids.

In 1979, giant sea worms about 3 m/10 ft long, living within tubes created by their own excretions, were dis-covered in hydrothermal vents 2,450 m/8,000 ft beneath the Pacific NE of the Galápagos Islands.

WORM (acronym for *write once read many times*) in computing, a storage device, similar to ◊CD-ROM. The computer can write to the disc directly, but cannot later erase or overwrite the same area. WORMs are mainly used for archiving and backup copies.

W particle in physics, an ◊elementary particle, one of the weakons responsible for transmitting the ◊weak nuclear force.

write protection device on discs and tapes that provides ◊data security by allowing data to be read but not deleted, altered, or overwritten.

wrought iron fairly pure iron containing some beads of slag, widely used for construction work before the days of cheap steel. It is strong, tough, and easy to machine. It is made in a puddling furnace, invented by Henry Colt in England in 1784. Pig iron is remelted and heated strongly in air with iron ore, burning out the carbon in the metal, leaving relatively pure iron and a slag containing impur-ities. The resulting pasty metal is then hammered to remove as much of the remaining slag as possible. It is still used in fences and grating.

wt abbreviation for *weight*.

WWF abbreviation for ◊ *World Wide Fund for Nature* (formerly World Wildlife Fund).

WYSIWYG (acronym for *what you see is what you get*) in computing, a program that attempts to display on the screen a faithful representation of the final printed output. For example, a WYSIWYG ◊word processor would show actual page layout – line widths, page breaks, and the sizes and styles of type.

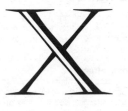

xanthophyll yellow pigment in plants that, like ◊chlorophyll, is responsible for the production of carbohydrates by photosynthesis.

X chromosome larger of the two sex chromosomes, the smaller being the ◊Y chromosome. These two chromosomes are involved in sex determination. Females have two X chromosomes, males have an X and a Y. Genes carried on the X chromosome produce the phenomenon of ◊sex linkage.

Early in the development of a female embryo, one of the X chromosomes becomes condensed so that most of its genes are inactivated. If this inactivation is incomplete, skeletal defects and mental retardation result.

xenon colourless, odourless, gaseous, non-metallic element, symbol Xe, atomic number 54, relative atomic mass 131.30. It is grouped with the ◊inert gases and was long believed not to enter into reactions, but is now known to form some compounds, mostly with fluorine. It is a heavy gas present in very small quantities in the air (about one part in 20 million).

Xenon is used in bubble chambers, light bulbs, vacuum tubes, and lasers. It was discovered in 1898 in a residue from liquid air by Scottish chemists William Ramsay and Morris Travers.

xenophobia fear (◊phobia) or strong dislike of strangers or anybody foreign or different.

xerography dry, electrostatic method of producing images, without the use of negatives or sensitized paper, invented in the USA by Chester Carlson in 1938 and applied in the Xerox ◊photocopier.

An image of the document to be copied is projected on to an electrostatically charged photo-conductive plate. The charge remains only in the areas corresponding to its image. The latent image on the plate is then developed by contact with ink powder, which adheres only to the image, and is then usually transferred to ordinary paper or some other flat surface, and quickly heated to form a permanent print.

Applications include document copying, enlarging from microfilm, preparing printing masters for offset litho printing and dyeline machines, making X-ray pictures, and printing high-speed computer output.

xerophyte plant adapted to live in dry conditions. Common adaptations to reduce the rate of ◊transpiration include a reduction of leaf size, sometimes to spines or scales; a dense covering of hairs over the leaf to trap a layer of moist air (as in edelweiss); water storage cells; sunken stomata; and permanently rolled leaves or leaves that roll up in dry weather (as in marram grass). Many desert cacti are xerophytes.

X-ray band of electromagnetic radiation in the wavelength range 10^{-11} to 10^{-9} m (between gamma rays and ultraviolet radiation; see ◊electromagnetic waves). Applications of X-rays make use of their short wavelength (as in ◊X-ray diffraction) or their penetrating power (as in medical X-rays of internal body tissues). X-rays are dangerous and can cause cancer.

X-rays with short wavelengths pass through most body tissues, although dense areas such as bone prevent their passage, showing up as white areas on X-ray photographs. The X-rays used in ◊radiotherapy have very short wavelengths that penetrate tissues deeply and destroy them.

X-rays were discovered by German experimental physicist Wilhelm Röntgen in 1895 and formerly called roentgen rays. They are produced when high-energy electrons from a heated filament cathode strike the surface of a target (usually made of tungsten) on the face of a massive heat-conducting anode, between which a high alternating voltage (about 100 kV) is applied.

X-ray astronomy detection of X-rays from intensely hot gas in the universe. Such X-rays are prevented from reaching the Earth's surface by the atmosphere, so detectors must be placed in rockets and satellites. The first celestial X-ray source, Scorpius X-1, was discovered by a rocket flight in 1962.

Since 1970, special satellites have been orbited to study X-rays from the Sun, stars, and galaxies. Many X-ray sources are believed to be gas falling on to ◊neutron stars and ◊black holes.

X-ray diffraction method of studying the atomic and molecular structure of crystalline substances by using ◊X-rays. X-rays directed at such substances spread out as they pass through the crystals owing to ◊diffraction (the slight spreading of waves around the edge of an opaque object) of the rays around the atoms. By using measurements of the position and intensity of the diffracted waves, it is possible to calculate the shape and size of the atoms in the crystal. The method has been used to study substances such as ◊DNA that are found in living material.

X-ray diffraction

Over 4,000 new chemical structures are determined each year by X-ray diffraction.

X-ray diffraction analysis the use of X-rays to study the atomic and molecular structure of crystalline substances such as ceramics, stone, sediments, and weathering products on metals. The sample, as a single crystal or ground to powder, is exposed to X-rays at various angles; the diffraction patterns produced are then compared with reference standards for identification.

X-ray fluorescence spectrometry technique used to determine the major and trace elements in the chemical composition of such materials as ceramics, obsidian, and glass. A sample is bombarded with X-rays, and the wavelengths of the released energy, or fluorescent X-rays, are detected and measured. Different elements have unique wavelengths, and their concentrations can be estimated from the intensity of the released X-rays. This analysis

GREAT EXPERIMENTS AND DISCOVERIES

The bones in his wife's hands

On 8 Nov 1895, German physicist Wilhelm Conrad Röntgen (1845–1923) discovered X-rays – by accident. This key discovery in atomic physics transformed medical diagnosis, and later provided a powerful tool in cancer therapy.

Röntgen was investigating the effects of electricity discharged through gases at low pressures to produce a beam of cathode rays (or electrons). Röntgen used a Crookes tube: an improved vacuum tube invented by British chemist and physicist William Crookes (1832–1919). He directed a narrow beam of rays from the tube, which was covered with black cardboard, onto a screen in a darkened room. He noticed a faint light on a nearby bench, caused by fluorescence from another screen.

The unknown rays

Röntgen knew that cathode rays could only travel a few centimetres through air. This other screen was about a metre away from his Crookes tube; he realized that he had discovered a new phenomenon. For six weeks, he worked feverishly to find out all about the new rays. Because their nature was unknown, he called them X-rays. On 2 Dec 1895, he made the first X-ray photograph, of his wife's hand. It shows that the bones in her hands were deformed by polyarthritis.

Röntgen announced his discovery on 28 Dec 1895. He accurately described many of the properties of X-rays. They are produced at the walls of the discharge tube by cathode rays; like light, they travel in straight lines and cast shadows; most bodies are transparent to them in some degree; they can blacken photographic plates, and cause some chemical substances to fluoresce; unlike cathode rays, they are not deflected by magnetic fields.

The uses of X-rays

All this created tremendous scientific excitement. In 1901, Röntgen received the first Nobel Prize for Physics for his discovery. Initially, X-rays were used to examine the bones of the skeleton: effectively the start of radiology, or the use of X-rays for medical purposes. Fractures could be seen clearly, as could embedded foreign bodies such as bullets or needles. Accidentally swallowed objects were accurately located. For the first time, surgeons knew in advance of operating exactly what to do.

The barium meal

Solid objects and bones could be clearly visualized on photographic film, recently developed by George Eastman (1854–1932) in the USA. However, it was hard to distinguish the soft tissues of the body. US physician Walter Bradford Cannon (1871–1945), who researched digestive problems and pioneered the use of X-rays in studying the alimentary canal, invented the 'barium meal' in 1897. Barium salts are relatively impervious to X-rays; by swallowing a solution of them before radiology, the patient's digestive system is selectively thrown into relief on the X-ray film. By the 1930s, X-ray examination was routine medical practice, particularly in examining the lungs for tuberculosis.

The damaging effects

As X-rays traverse living tissue, they also damage it. This danger to health was only understood much later, and both Röntgen and his assistant suffered X-ray poisoning. The destructive nature of radiation can be harnessed therapeutically. By focusing X-radiation on a tumour, its tissue is destroyed. When a cancer is not widely disseminated within the body, radiation therapy is a powerful weapon in the fight against cancer. Other forms of radiation, from radioactive sources for example, can be similarly employed.

Tomography

Traditional radiology now accounts for no more than half of medical photography. In tomography, an interesting extension of the medical uses for X-rays, they are used to photograph a selected plane of the human body. Tomography was first demonstrated successfully in 1928. In 1973, British engineer Godfrey Newbold Hounsfield invented computerized axial tomography (CAT), in which a computer assembles a high-resolution X-ray picture of a 'slice' through the body (or head) of a patient from information provided by detectors rotating around the patient. CAT can detect very small changes in the density of living tissue.

At first, such scanners were used to diagnose diseases of the brain: now, improved machines can scan an entire body. With Allan Macleod Cormack, the South African physicist who independently developed the mathematical basis for computer-assisted X-ray tomography, Hounsfield received the 1979 Nobel Prize for Physiology or Medicine for the development of CAT.

From X-rays to radioactivity

Röntgen's discovery of X-rays led directly to the discovery of radioactivity by French physicist Antoine Henri Becquerel (1852–1908). In 1896 Becquerel was searching for X-rays in the fluorescence seen when certain salts absorb ultraviolet radiation: from the Sun, for example. He placed his test salt on a photographic plate wrapped in black paper, reasoning that any X-radiation given off by a fluorescing salt would fog the film. By chance, he left a test salt in a drawer near a wrapped plate. When the plate was developed, it had fogged.

The salt could not have fluoresced in the dark; some radiation other than X-rays must be responsible. His test salt was potassium uranyl sulphate, a salt of uranium; Becquerel had discovered radioactivity.

Julian Rowe

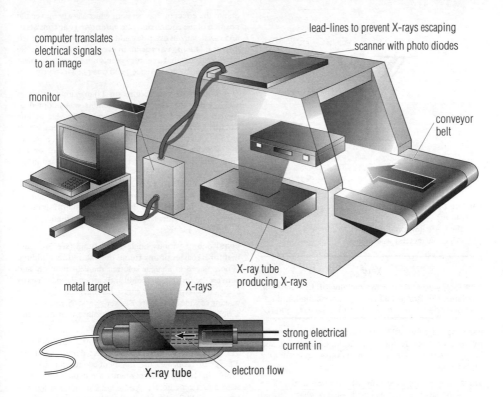

computer translates
electrical signals
to an image

lead-lines to prevent X-rays escaping

scanner with photo diodes

monitor

conveyor
belt

metal target   X-rays

X-ray tube
producing X-rays

strong electrical
current in

X-ray tube   electron flow

X-ray An X-ray image. The X-rays are generated by high-speed electrons impinging on a tungsten target. The rays pass through the specimen and on to a photographic plate or imager.

may, for example, help an archaeologist in identifying the source of the material.

X-ray telescope in astronomy, a ◊telescope designed to receive ◊electromagnetic waves in the X-ray part of the spectrum. X-rays cannot be focussed by lenses or mirrors in the same way as visible light, and a variety of alternative techniques are used to form images. Because X-rays cannot penetrate the Earth's atmosphere, X-ray telescopes are mounted on ◊satellites, ◊rockets, or high-flying ◊balloons.

xylem tissue found in ◊vascular plants, whose main function is to conduct water and dissolved mineral nutrients from the roots to other parts of the plant. Xylem is composed of a number of different types of cell, and may

include long, thin, usually dead cells known as ◊tracheids; fibres (schlerenchyma); thin-walled ◊parenchyma cells; and conducting vessels.

In most ◊angiosperms (flowering plants) water is moved through these vessels. Most ◊gymnosperms and ◊pteridophytes lack vessels and depend on tracheids for water conduction.

Non-woody plants contain only primary xylem, derived from the procambium, whereas in trees and shrubs this is replaced for the most part by secondary xylem, formed by ◊secondary growth from the actively dividing vascular ◊cambium. The cell walls of the secondary xylem are thickened by a deposit of ◊lignin, providing mechanical support to the plant; see ◊wood.

yard unit (symbol yd) of length, equivalent to three feet (0.9144 m).

In the USA, it is sometimes used to denote a cubic yard (0.7646 cubic meters) as of topsoil.

Yard

The foot, yard and inch owe their length to Henry I of England. He decreed that one yard was the length of his extended arm, measured from nose to fingertip.

yardang ridge formed by wind erosion from a dried-up riverbed or similar feature, as in Chad, China, Peru, and North America. On the planet Mars yardangs occur on a massive scale.

yaws contagious tropical disease common in the West Indies, W Africa, and some Pacific islands, characterized by red, raspberry-like eruptions on the face, toes, and other parts of the body, sometimes followed by lesions of the bones; these may progress to cause gross disfigurement. It is caused by a spirochete (*Treponema pertenue*), a bacterium related to the one that causes ◊syphilis. Treatment is by antibiotics.

Y chromosome smaller of the two sex chromosomes. In male mammals it occurs paired with the other type of sex chromosome (X), which carries far more genes. The Y chromosome is the smallest of all the mammalian chromosomes and is considered to be largely inert (that is, without direct effect on the physical body). See also ◊sex determination.

In humans, about one in 300 males inherits two Y chromosomes at conception, making him an XYY triploid. Few if any differences from normal XY males exist in these individuals, although at one time they were thought to be emotionally unstable and abnormally aggressive. In 1989 the gene determining that a human being is male was found to occur on the X as well as on the Y chromosome; however, it is not activated in the female.

yd abbreviation for ◊*yard*.

year unit of time measurement, based on the orbital period of the Earth around the Sun. The *tropical year*, from one spring ◊equinox to the next, lasts 365.2422 days. It governs the occurrence of the seasons, and is the period on which the calendar year is based. The *sidereal year* is the time taken for the Earth to complete one orbit relative to the fixed stars, and lasts 365.2564 days (about 20 minutes longer than a tropical year). The difference is due to the effect of ◊precession, which slowly moves the position of the equinoxes. The *calendar year* consists of 365 days, with an extra day added at the end of Feb each leap year. *Leap years* occur in every year that is divisible by four, except that a century year is not a leap year unless it is divisible by 400. Hence 1900 was not a leap year, but 2000 will be.

A *historical year* begins on 1 Jan, although up to 1752, when the Gregorian ◊calendar was adopted in England, the civil or legal year began on 25 March. The English *fiscal/financial year* still ends on 5 April, which is 25 March plus the 11 days added under the reform of the calendar in 1752. The *regnal year* begins on the anniversary of the sovereign's accession; it is used in the dating of acts of Parliament.

The *anomalistic year* is the time taken by any planet in making one complete revolution from perihelion to perihelion; for the Earth this period is about 5 minutes longer than the sidereal year due to the gravitational pull of the other planets.

yeast one of various single-celled fungi (see ◊fungus) that form masses of tiny round or oval cells by budding. When placed in a sugar solution the cells multiply and convert the sugar into alcohol and carbon dioxide. Yeasts are used as fermenting agents in baking, brewing, and the making of wine and spirits. Brewer's yeast (*S. cerevisiae*) is a rich source of vitamin B. (Especially genus *Saccharomyces*; also other related genera.)

yeast artificial chromosome (YAC) fragment of ◊DNA from the human genome inserted into a yeast cell. The yeast replicates the fragment along with its own DNA. In this way the fragments are copied to be preserved in a gene library. YACs are characteristically between 250,000 and 1 million base pairs in length. A cosmid works in the same way.

yellow fever or *yellow jack* acute tropical viral disease, prevalent in the Caribbean area, Brazil, and on the west coast of Africa. The yellow fever virus is an arbovirus transmitted by mosquitoes. Its symptoms include a high fever, headache, joint and muscle pains, vomiting, and yellowish skin (jaundice, possibly leading to liver failure); the heart and kidneys may also be affected. Mortality is high in serious cases.

Before the arrival of Europeans, yellow fever was not a problem because indigenous people had built up an immunity. The disease was brought under control after the discovery that it is carried by the mosquito *Aëdes aegypti*. The first effective vaccines were produced by Max Theiler (1899–1972) of South Africa, for which he was awarded the 1951 Nobel Prize for Medicine.

Yerkes Observatory astronomical centre in Wisconsin, USA, founded by George Hale in 1897. It houses the world's largest refracting optical ◊telescope, with a lens of diameter 102 cm/40 in.

yield point or *elastic limit* the stress beyond which a material deforms by a relatively large amount for a small increase in stretching force. Beyond this stress, the material no longer obeys ◊Hooke's law.

yolk store of food, mostly in the form of fats and proteins, found in the ◊eggs of many animals. It provides nourishment for the growing embryo.

yolk sac sac containing the yolk in the egg of most vertebrates. The term is also used for the membranous sac formed below the developing mammalian embryo and connected with the umbilical cord.

ytterbium soft, lustrous, silvery, malleable, and ductile element of the ◊lanthanide series, symbol Yb, atomic number 70, relative atomic mass 173.04. It occurs with (and resembles) yttrium in gadolinite and other minerals, and is used in making steel and other alloys.

In 1878 Swiss chemist Jean-Charles de Marignac gave the name ytterbium (after the Swedish town of Ytterby, near where it was found) to what he believed to be a new element. In 1907, French chemist Georges Urbain (1872–1938) discovered that this was in fact a mixture of two elements: ytterbium and lutetium.

yttrium silver-grey, metallic element, symbol Y, atomic number 39, relative atomic mass 88.905. It is associated with and resembles the rare earth elements (◊lanthanides), occurring in gadolinite, xenotime, and other minerals. It is used in colour-television tubes and to reduce steel corrosion.

The name derives from the Swedish town of Ytterby, near where it was first discovered in 1788. Chemist Carl Mosander (1797–1858) isolated the element in 1843.

tube before the ovum has undergone its first cell division. This mimics the natural processes of fertilization (which normally occurs in the Fallopian tube) and implantation, more effectively than older techniques.

zinc hard, brittle, bluish-white, metallic element, symbol Zn, atomic number 30, relative atomic mass 65.37. The principal ore is sphalerite or zinc blende (zinc sulphide, ZnS). Zinc is little affected by air or moisture at ordinary temperatures; its chief uses are in alloys such as brass and in coating metals (for example, galvanized iron). Its compounds include zinc oxide, used in ointments (as an astringent) and cosmetics, paints, glass, and printing ink.

Zinc is an essential trace element in most animals; adult humans have 2–3 g/0.07–0.1 oz zinc in their bodies. There are more than 300 known enzymes that contain zinc.

Zinc has been used as a component of brass since the Bronze Age, but it was not recognized as a separate metal until 1746, when it was described by German chemist Andreas Sigismund Marggraf (1709–1782). The name derives from the shape of the crystals on smelting.

zinc ore mineral from which zinc is extracted, principally sphalerite $(Zn,Fe)S$, but also zincite, ZnO_2, and smithsonite, $ZnCO_3$, all of which occur in mineralized veins. Ores of lead and zinc often occur together, and are common worldwide; Canada, the USA, and Australia are major producers.

zinc oxide ZnO white powder, yellow when hot, that occurs in nature as the mineral zincite. It is used in paints and as an antiseptic in zinc ointment; it is the main ingredient of calamine lotion.

zinc sulphide ZnS yellow-white solid that occurs in nature as the mineral sphalerite (also called zinc blende). It is the principal ore of zinc, and is used in the manufacture of fluorescent paints.

zip fastener fastening device used in clothing, invented in the USA by Whitcomb Judson in 1891, originally for doing up shoes. It has two sets of interlocking teeth, meshed by means of a slide that moves up and down. It did not become widely used in the clothing industry till the 1930s.

zircon zirconium silicate, $ZrSiO_4$, a mineral that occurs in small quantities in a wide range of igneous, sedimentary, and metamorphic rocks. It is very durable and is resistant to erosion and weathering. It is usually coloured brown, but can be other colours, and when transparent may be used as a gemstone.

Zircons contain abundant radioactive isotopes of uranium and so are useful for uranium–lead dating to determine the ages of rocks.

zirconium lustrous, greyish-white, strong, ductile, metallic element, symbol Zr, atomic number 40, relative atomic mass 91.22. It occurs in nature as the mineral zircon (zirconium silicate), from which it is obtained commercially. It is used in some ceramics, alloys for wire and filaments, steel manufacture, and nuclear reactors, where its low neutron absorption is advantageous.

It was isolated in 1824 by Swedish chemist Jöns Berzelius. The name was proposed by English chemist Humphry Davy in 1808.

zodiac zone of the heavens containing the paths of the Sun, Moon, and planets. When this was devised by the

---

Z in physics, the symbol for *impedance* (electricity and magnetism).

Zelenchukskaya site of the world's largest single-mirror optical telescope, with a mirror of 6 m/236 in diameter, in the Caucasus Mountains of Russia. At the same site is the RATAN 600 radio telescope, consisting of radio reflectors in a circle of 600 m/2,000 ft diameter. Both instruments are operated by the Academy of Sciences in St Petersburg.

zenith uppermost point of the celestial horizon, immediately above the observer; the ◊nadir is below, diametrically opposite. See ◊celestial sphere.

zeolite any of the hydrous aluminium silicates, also containing sodium, calcium, barium, strontium, or potassium, chiefly found in igneous rocks and characterized by a ready loss or gain of water. Zeolites are used as 'molecular sieves' to separate mixtures because they are capable of selective absorption. They have a high ion-exchange capacity and can be used to make petrol, benzene, and toluene from low-grade raw materials, such as coal and methanol.

Zhubov scale scale for measuring ice coverage, developed in the USSR. The unit is the *ball*; one ball is 10% coverage, two balls 20%, and so on.

zidovudine (formerly *AZT*) antiviral drug used in the treatment of ◊AIDS. It is not a cure for AIDS but is effective in prolonging life; it does not, however, delay the onset of AIDS in people carrying the virus.

Zidovudine was developed in the mid-1980s and approved for use by 1987. Taken every four hours, night and day, it reduces the risk of opportunistic infection and relieves many neurological complications. However, frequent blood monitoring is required to control anaemia, a potentially life-threatening side effect of zidovudine. Blood transfusions are often necessary, and the drug must be withdrawn if bone-marrow function is severely affected.

A US trial in 1994 showed that the drug does provide some protection to babies born to HIV-positive mothers. The number of babies infected was reduced by two-thirds where mothers received zidovudine during pregnancy. Long-term affects of zidovudine on the babies' health remain to be determined.

ZIFT abbreviation for *zygote inter-Fallopian transfer* modified form of ◊in vitro fertilization in which the fertilized ovum is re-introduced into the mother's ◊Fallopian

ancient Greeks, only five planets were known, making the zodiac about 16° wide. In astrology, the zodiac is divided into 12 signs, each 30° in extent: Aries, Taurus, Gemini, Cancer, Leo, Virgo, Libra, Scorpio, Sagittarius, Capricorn, Aquarius, and Pisces. These do not cover the same areas of sky as the astronomical constellations.

The 12 astronomical constellations are uneven in size and do not between them cover the whole zodiac, or even the line of the ecliptic, much of which lies in Ophiuchus.

zodiacal light cone-shaped light sometimes seen extending from the Sun along the ◊ecliptic, visible after sunset or before sunrise. It is due to thinly spread dust particles in the central plane of the Solar System. It is very faint, and requires a dark, clear sky to be seen.

zoidogamy type of plant reproduction in which male gametes (antherozoids) swim in a film of water to the female gametes. Zoidogamy is found in algae, bryophytes, pteridophytes, and some gymnosperms (others use ◊siphonogamy).

zone system in photography, a system of exposure estimation invented by Ansel Adams that groups infinite tonal gradations into ten zones, zone 0 being black and zone 10 white. An ◊f-stop change in exposure is required from zone to zone.

zone therapy alternative name for ◊reflexology.

zoo (abbreviation for **zoological gardens**) place where animals are kept in captivity. Originally created purely for visitor entertainment and education, zoos have become major centres for the breeding of endangered species of animals; a 1984 report identified 2,000 vertebrate species in need of such maintenance. The Arabian oryx has already been preserved in this way; it was captured in 1962, bred in captivity, and released again in the desert in 1972, where it has flourished.

Many groups object to zoos because they keep animals in unnatural conditions alien to their habitat.

zoology branch of biology concerned with the study of animals. It includes any aspect of the study of animal form and function – description of present-day animals, the study of evolution of animal forms, ◊anatomy, ◊physiology, ◊embryology, behaviour, and geographical distribution.

Anyone who has got any pleasure at all should try to put something back. Life is like a superlative meal and the world is like the maitre d'hotel. What I am doing is the equivalent of leaving a reasonable tip.

On **zoo** Gerald Durrell *Guardian* 1971

---

zoom lens photographic lens that, by variation of focal length, allows speedy transition from long shots to close-ups.

zoonosis any infectious disease that can be transmitted to humans by other vertebrate animals. Probably the most feared example is ◊rabies. The transmitted microorganism sometimes causes disease only in the human host, leaving the animal host unaffected.

Z particle in physics, an ◊elementary particle, one of the weakons responsible for carrying the ◊weak nuclear force.

zwitterion ion that has both a positive and a negative charge, such as an ◊amino acid in neutral solution. For example, glycine contains both a basic amino group (NH^2) and an acidic carboxyl group (-COOH); when both these are ionized in aqueous solution, the acid group loses a proton to the amino group, and the molecule is positively charged at one end and negatively charged at the other.

zygote ◊ovum (egg) after ◊fertilization but before it undergoes cleavage to begin embryonic development.

Appendices

Units in the Metric System

Length

1 centimetre	= 10 millimetres	
1 decimetre	= 10 centimetres	= 100 millimetres
1 metre	= 10 decimetres	= 1,000 millimetres
1 decametre	= 10 metres	
1 hectometre	= 10 decametres	= 100 metres
1 kilometre	= 10 hectometres	= 1,000 metres

Area

1 square centimetre	= 100 square millimetres	
1 square metre	= 10,000 square centimetres	= 1,000,000 square millimetres
1 are	= 100 square metres	
1 hectare	= 100 ares	= 10,000 square metres
1 square kilometre	= 100 hectares	= 1,000,000 square metres

Mass (Avoirdupois)

1 centigram	= 10 milligrams	
1 decigram	= 10 centigrams	= 100 milligrams
1 gram	= 10 decigrams	= 1,000 milligrams
1 decagram	= 10 grams	
1 hectogram	= 10 decagrams	= 100 grams
1 kilogram	= 10 hectograms	= 1,000 grams
1 metric ton	= 1,000 kilograms	

Volume

1 cubic centimetre	= 1,000 cubic millimetres	
1 cubic decimetre	= 1,000 cubic centimetres	= 1,000,000 cubic millimetres
1 cubic metre	= 1,000 cubic decimetres	= 1,000,000,000 cubic millimetres

Capacity

1 centilitre	= 10 millilitres	
1 decilitre	= 10 centilitres	= 100 millilitres
1 litre	= 10 decilitres	= 1,000 millilitres
1 decalitre	= 10 litres	
1 hectolitre	= 10 decalitres	= 100 litres
1 kilolitre	= 10 hectolitres	= 1,000 litres

SI Units

Quantity	SI unit	Symbol
absorbed radiation dose	gray	Gy
amount of substance	mole*	mol
electric capacitance	farad	F
electric charge	coulomb	C
electric conductance	siemens	S
electric current	ampere*	A
energy or work	joule	J
force	newton	N
frequency	hertz	Hz
illuminance	lux	lx
inductance	henry	H
length	metre*	m
luminous flux	lumen	lm
luminous intensity	candela*	cd
magnetic flux	weber	Wb
magnetic flux density	tesla	T
mass	kilogram*	kg
plane angle	radian	rad
potential difference	volt	V
power	watt	W
pressure	pascal	Pa
radiation dose equivalent	sievert	Sv
radiation exposure	roentgen	R
radioactivity	becquerel	Bq
resistance	ohm	Ω
solid angle	steradian	sr
sound intensity	decibel	dB
temperature	°Celsius	°C
temperature, thermodynamic	kelvin*	K
time	second*	s

* SI base unit.

Astronomical Constants

Constant	Value
Astronomical unit (au)	149,597,870 km
Speed of light in a vacuum (c)	299,792.458 km/sec
Solar parallax	8.794148 arc seconds
Mass of the Sun	1.9891×10^{30} kg
Mass of the Earth	5.9742×10^{24} kg
Mass of the Moon	7.3483×10^{22} kg
Light year (ly)	9.4605×10^{12} km = 0.30660 pc
Parsec (pc)	30.857×10^{12} km = 3.26161 ly
Obliquity of the ecliptic (2000)	23 deg 26 min 21.448 sec
General precession (2000)	50.290966 arc seconds/year
Constant of nutation (2000)	9.2025 arc seconds
Constant of aberration (2000)	20.49552 arc seconds

Greek Language: Alphabet

A	α	alpha
B	β	beta
Γ	γ	gamma
Δ	δ	delta
E	ε	epsilon
Z	ζ	zeta
H	η	eta
Θ	θ	theta
I	ι	iota
K	κ	kappa
Λ	λ	lambda
M	μ	mu
N	ν	nu
Ξ	ξ	xi
O	o	omicron
Π	π	pi
P	ρ	rho
Σ	σ	sigma
T	τ	tau
Y	υ	upsilon
Φ	φ	phi
X	χ	chi
Ψ	ψ	psi
Ω	ω	omega

Physical Constants

Constant	Symbol	Value in SI units
acceleration of free fall	g	9.80665 m s^{-2}
Avogadro's constant	N_A	6.0221367×10^{23} mol^{-1}
Boltzmann's constant	k	1.380658×10^{-23} J K^{-1}
elementary charge	e	$1.60217733 \times 10^{-19}$ C
electronic rest mass	m_e	$9.1093897 \times 10^{-31}$ kg
Faraday's constant	F	9.6485309×10^{4} C mol^{-1}
gas constant	R	8.314510 J K^{-1} mol^{-1}
gravitational constant	G	6.672×10^{-11} N m^2 kg^{-2}
Loschmidt's number	N_L	2.686763×10^{25} m^{-3}
neutron rest mass	m_n	$1.6749286 \times 10^{-27}$ kg
Planck's constant	h	$6.6260755 \times 10^{-34}$ J s
proton rest mass	m_p	$1.6726231 \times 10^{-27}$ kg
speed of light in a vacuum	c	2.99792458×10^{8} m s^{-1}
standard atmosphere	atm	1.01325×10^{5} Pa
Stefan–Boltzmann constant	σ	5.67051×10^{-8} W m^{-2} K^{-4}

Physical constants, or fundamental constants, are standardized values whose parameters do not change.

Imperial System: Units

Length

1 foot	= 12 inches
1 yard	= 3 feet
1 rod	= 5½ yards
1 chain	= 4 rods (= 22 yards)
1 furlong	= 10 chains (= 220 yards)
1 mile	= 5,280 feet
1 mile	= 1,760 yards
1 mile	= 8 furlongs

Nautical

1 fathom	= 6 feet
1 cable length	= 100 fathoms
1 nautical mile	= 6,080 feet

Area

1 square foot	= 144 square inches
1 square yard	= 9 square feet
1 square rod	= 304¼ square yards
1 rood	= 40 square rods
1 acre	= 4 roods
1 acre	= 4,840 square yards
1 square mile	= 640 acres

Volume

1 cubic foot	= 1,728 cubic inches
1 cubic yard	= 27 cubic feet
1 bulk barrel	= 5.8 cubic feet

Shipping

1 register ton	= 100 cubic feet

Capacity

1 fluid ounce	= 8 fluid drahms
1 gill	= 5 fluid ounces
1 pint	= 4 gills
1 quart	= 2 pints
1 gallon	= 4 quarts
1 peck	= 2 gallons
1 bushel	= 4 pecks
1 quarter	= 8 bushels
1 bulk barrel	= 36 gallons

Mass (avoirdupois)

1 ounce	= 437½ grains
1 ounce	= 16 drams
1 pound	= 16 ounces
1 stone	= 14 pounds
1 quarter	= 28 pounds
1 hundredweight	= 4 quarters
1 ton	= 20 hundredweight

Bones of the human body

Bone	Number	Bone	Number
Skull		**Carpus (Wrist)**	
occipital	1	*scaphoid*	1
parietal: one pair	2	*lunate*	1
sphenoid	1	*triquetral*	1
ethmoid	1	*pisiform*	1
inferior nasal conchae: two	2	*trapezium*	1
frontal: one pair, fused	1	*trapezoid*	1
nasal: one pair	2	*capitate*	1
lacrimal: one pair	2	*hamate*	1
temporal: one pair	2	*metacarpals*	5
maxilla: one pair	2		
zygomatic: one pair	2	**Phalanges (Fingers)**	
vomer	1	*first digit*	2
palatine: one pair	2	*second digit*	3
mandible: one pair, fused (jawbone)	1	*third digit*	3
Total	22	*fourth digit*	3
		fifth digit	3
Ears		**Total**	30
malleus (hammer)	2		
incus (anvil)	2	**Pelvic Girdle**	
stapes (stirrups)	2	*ilium, ischium, and pubis (combined):*	
Total	6	*one pair of hip bones, innominate*	2
Vertebral Column (Spine)		**Lower Extremity (Each Leg)**	
cervical	7	*femur (thighbone)*	1
thoracic	12	*tibia*	1
lumbar	5	*fibula*	1
sacral: five, fused to form the sacrum	1	*patella (kneebone)*	1
coccyx: between three and five, fused	1		
Total	26	**Tarsus (Ankle)**	
		talus	1
Ribs		*calcaneus*	1
ribs, 'true': seven pairs	14	*navicular*	1
ribs, 'false': five pairs of which two		*cuneiform, medial*	1
pairs are floating	10	*cuneiform, intermediate*	1
Total	24	*cuneiform, lateral*	1
		cuboid	1
Sternum (Breastbone)		*metatarsals*	5
manubrium	1		
sternebrae	1	**Phalanges (Toes)**	
xiphisternum	1	*first digit*	2
Total	3	*second digit*	3
		third digit	3
Throat		*fourth digit*	3
hyoid	1	*fifth digit*	3
		Total	30
Pectoral Girdle			
clavicle: one pair (collar bone)	2	**Summary**	
scapula (including coracoid): one pair		*skull*	22
(shoulder blade)	2	*ears*	6
Total	4	*vertebrae*	26
		vertebral ribs	24
Upper Extremity (Each Arm)		*sternum*	3
Forearm		*throat*	1
humerus	1	*pectoral girdle*	4
radius	1	*upper extremity (arms): 2 × 30*	60
ulna	1	*hip bones*	2
		lower extremity (legs): 2 × 30	60
		Total	208

Nobel Prize for Chemistry

Year	Winner(s)	Awarded for
1901	Jacobus van't Hoff (Netherlands)	laws of chemical dynamics and osmotic pressure
1902	Emil Fischer (Germany)	sugar and purine syntheses
1903	Svante Arrhenius (Sweden)	electrolytic theory of dissociation
1904	William Ramsay (UK)	inert gases in air and their locations in the periodic table
1905	Adolf von Baeyer (Germany)	organic dyes and hydroaromatic compounds
1906	Henri Moissan (France)	isolation of fluorine and adoption of electric furnace
1907	Eduard Buchner (Germany)	discovery of cell-free fermentation
1908	Ernest Rutherford (New Zealand)	atomic disintegration and radioactive substances
1909	Wilhelm Ostwald (Germany)	catalysis, principles of equilibria, rates of reaction
1910	Otto Wallach (Germany)	alicyclic compounds
1911	Marie Curie (Poland)	discovery of radium and polonium
1912	Victor Grignard (France)	discovery of Grignard reagent
	Paul Sabatier (France)	catalytic hydrogenation of organic compounds
1913	Alfred Werner (Switzerland)	bonding of atoms within molecules
1914	Theodore Richards (USA)	determination of the atomic masses of elements
1915	Richard Willstäter (Germany)	research into plant pigments, especially chlorophyll
1916–17	no awards	
1918	Fritz Haber (Germany)	synthesis of ammonia from its elements
1919	no award	
1920	Walther Nernst (Germany)	work on thermochemistry
1921	Frederick Soddy (UK)	work on radioactive substances, especially isotopes
1922	Francis Aston (UK)	mass spectrometry of isotopes of radioactive elements, and enunciation of the whole-number rule
1923	Fritz Pregl (Austria)	microanalysis of organic substances
1924	no award	
1925	Richard Zsigmondy (Austria)	heterogeneity of colloids
1926	Theodor Svedberg (Sweden)	investigation of dispersed systems
1927	Heinrich Wieland (Germany)	constitution of bile acids and related substances
1928	Adolf Windaus (Germany)	constitution of sterols and related vitamins
1929	Arthur Harden (UK) and Hans von Euler-Chelpin (Germany)	fermentation of sugar, and fermentative enzymes
1930	Hans Fischer (Germany)	analysis of haem and chlorophyll
1931	Carl Bosch (Germany) and Friedrich Bergius (Germany)	invention and development of chemical high-pressure methods
1932	Irving Langmuir (USA)	surface chemistry
1933	no award	
1934	Harold Urey (USA)	discovery of deuterium (heavy hydrogen)
1935	Irène and Frédéric Joliot-Curie (France)	synthesis of new radioactive elements
1936	Peter Debye (Netherlands)	work on molecular structures by investigation of dipole moments and the diffraction of X-rays and electrons in gases
1937	Norman Haworth (UK)	work on carbohydrates and ascorbic acid (vitamin C)
	Paul Karrer (Switzerland)	work on carotenoids, flavins, retinol (vitamin A) and riboflavin (vitamin B_2)
1938	Richard Kuhn (Austria)	carotenoids and vitamins
1939	Adolf Butenandt (Germany)	work on sex hormones
	Leopold Ružička (Switzerland)	polymethylenes and higher terpenes
1940–42	no awards	
1943	Georg von Hevesy (Sweden)	use of isotopes as tracers in chemical processes
1944	Otto Hahn (Germany)	discovery of nuclear fission
1945	Artturi Virtanen (Finland)	agriculture and nutrition, especially fodder preservation
1946	James Sumner (USA)	crystallization of enzymes
	John Northrop (USA) and Wendell Stanley (USA)	preparation of pure enzymes and virus proteins
1947	Robert Robinson (UK)	biologically important plant products, especially alkaloids
1948	Arne Tiselius (Sweden)	electrophoresis and adsorption analysis, and discoveries concerning serum proteins
1949	William Giauque (USA)	chemical thermodynamics, especially at very low temperatures
1950	Otto Diels and Kurt Alder (Germany)	discovery and development of diene synthesis
1951	Edwin McMillan and Glenn Seaborg (USA)	chemistry of transuranic elements
1952	Archer Martin and Richard Synge (UK)	invention of partition chromatography
1953	Hermann Staudinger (West Germany)	discoveries in macromolecular chemistry
1954	Linus Pauling (USA)	nature of chemical bonds
1955	Vincent Du Vigneaud (USA)	investigations into sulphur compounds, and the first synthesis of a polypeptide hormone

Nobel Prize for Chemistry – *continued*

Year	Winner(s)	Awarded for
1956	Cyril Hinshelwood (UK) and Nikoly Semenov (USSR)	mechanism of chemical reactions
1957	Alexander Todd (UK)	nucleotides and nucleotide coenzymes
1958	Frederick Sanger (UK)	structure of proteins, especially insulin
1959	Jaroslav Heyrovský (Czechoslovakia)	polarographic methods of chemical analysis
1960	Willard Libby (USA)	radiocarbon dating
1961	Melvin Calvin (USA)	assimilation of carbon dioxide by plants
1962	Max Perutz (UK) and John Kendrew (UK)	structures of globular proteins
1963	Karl Ziegler (West Germany) and Giulio Natta (Italy)	chemistry and technology of high polymers
1964	Dorothy Crowfoot Hodgkin (UK)	crystallographic determination of the structures of biochemical compounds, notably penicillin and cyanocobalamin (vitamin B_{12})
1965	Robert Woodward (USA)	organic synthesis
1966	Robert Mulliken (USA)	molecular orbital theory of chemical bonds and structures
1967	Manfred Eigen (West Germany), Ronald Norrish (UK), and George Porter (UK)	investigation of rapid chemical reactions by means of very short pulses of energy
1968	Lars Onsager (USA)	discovery of reciprocal relations
1969	Derek Barton (UK) and Odd Hassel (Norway)	concept and applications of conformation
1970	Luis Federico Leloir (Argentina)	discovery of sugar nucleotides and their role in carbohydrate biosynthesis
1971	Gerhard Herzberg (Canada)	electronic structure and geometry of molecules
1972	Christian Anfinsen (USA), Stanford Moore (USA), and William Stein (USA)	amino-acid structure and biological activity of the enzyme ribonuclease
1973	Ernst Fischer (West Germany) and Geoffrey Wilkinson (UK)	chemistry of organometallic sandwich compounds
1974	Paul Flory (USA)	physical chemistry of macromolecules
1975	John Cornforth (Australia)	stereochemistry of enzyme-catalysed reactions
	Vladimir Prelog (Yugoslavia)	stereochemistry of organic molecules and their reactions
1976	William N Lipscomb (USA)	structure and chemical bonding of boranes
1977	Ilya Prigogine (USSR)	thermodynamics of irreversible and dissipative processes
1978	Peter Mitchell (UK)	biological energy transfer and chemiosmotic theory
1979	Herbert Brown (USA) and Georg Wittig (West Germany)	use of boron and phosphorus compounds, respectively, in organic syntheses
1980	Paul Berg (USA)	biochemistry of nucleic acids, especially DNA
	Walter Gilbert (USA) and Frederick Sanger (UK)	base sequences in nucleic acids
1981	Kenichi Fukui (Japan) and Roald Hoffmann (USA)	theories concerning chemical reactions
1982	Aaron Klug (UK)	crystallographic electron microscopy
1983	Henry Taube (USA)	electron-transfer reactions in inorganic chemistry
1984	Bruce Merrifield (USA)	chemical syntheses on a solid matrix
1985	Herbert A Hauptman (USA) and Jerome Karle (USA)	methods of determining crystal structures
1986	Dudley Herschbach (USA), Yuan Lee (USA), and John Polanyi (Canada)	dynamics of chemical elementary processes
1987	Donald Cram (USA), Jean-Marie Lehn (France), and Charles Pedersen (USA)	molecules with highly selective structure-specific interactions
1988	Johann Deisenhofer, Robert Huber, and Hartmut Michel (West Germany)	three-dimensional structure of the reaction centre of photosynthesis
1989	Sidney Altman and Thomas Cech (USA)	discovery of catalytic function of RNA
1990	Elias James Corey (USA)	new methods of synthesizing chemical compounds
1991	Richard R Ernst (Switzerland)	improvements in the technology of nuclear magnetic resonance (NMR) imaging
1992	Rudolph A Marcus (USA)	theoretical discoveries relating to reduction and oxidation reactions
1993	Kary Mullis (USA)	invention of the polymerase chain reaction technique for amplifying DNA
	Michael Smith (Canada)	development of techniques for splicing foreign genetic segments into an organism's DNA in order to modify the proteins produced
1994	George A Olah (USA)	development of technique for examining hydrocarbon molecules

Nobel Prize for Chemistry – *continued*

Year	Winner(s)	Awarded for
1995	F Sherwood Roland (USA), Mario Molina (Mexico), and Paul Crutzen (Netherlands)	explaining the chemical process of the ozone layer
1996	Robert F Curl, Jr (USA), Sir Harold W Kroto (UK), and Richard E Smalley (USA)	discovery of fullerenes
1997	Paul D. Boyer (USA) and John E. Walker (UK)	progressing our understanding of the enzymatic mechanism underlying the synthesis of adenosine triphosphate (ATP)
	Jens C. Skou (Denmark)	first discovery of an ion-transporting enzyme, Na$^+$, K$^+$-ATPase

Nobel Prize for Physics

Year	Winner(s)	Awarded for
1901	Wilhelm Röntgen (Germany)	discovery of X-rays
1902	Hendrik Lorentz (Netherlands) and Pieter Zeeman (Netherlands)	influence of magnetism on radiation phenomena
1903	Antoine Becquerel (France)	discovery of spontaneous radioactivity
	Pierre Curie (France) and Marie Curie (Poland)	research on radiation phenomena
1904	John Strutt (Lord Rayleigh, UK)	densities of gases and discovery of argon
1905	Philipp von Lenard (Germany)	work on cathode rays
1906	Joseph J Thomson (UK)	theoretical and experimental work on the conduction of electricity by gases
1907	Albert Michelson (USA)	measurement of the speed of light through the design and application of precise optical instruments such as the interferometer
1908	Gabriel Lippmann (France)	photographic reproduction of colours by interference
1909	Guglielmo Marconi (Italy) and Karl Braun (Germany)	development of wireless telegraphy
1910	Johannes van der Waals (Netherlands)	equation describing the physical behaviour of gases and liquids
1911	Wilhelm Wien (Germany)	laws governing radiation of heat
1912	Nils Dalen (Sweden)	invention of light-controlled valves, which allow lighthouses and buoys to operate automatically
1913	Heike Kamerlingh Onnes (Netherlands)	studies of properties of matter at low temperatures
1914	Max von Laue (Germany)	discovery of diffraction of X-rays by crystals
1915	William Bragg (UK) and Lawrence Bragg (UK)	X-ray analysis of crystal structures
1916	no award	
1917	Charles Barkla (UK)	discovery of characteristic X-ray emission of the elements
1918	Max Planck (Germany)	formulation of quantum theory
1919	Johannes Stark (Germany)	discovery of Doppler effect in rays of positive ions, and splitting of spectral lines in electric fields
1920	Charles Guillaume (Switzerland)	precision measurements through anomalies in nickel–steel alloys
1921	Albert Einstein (Switzerland)	theoretical physics, especially law of photoelectric effect
1922	Niels Bohr (Denmark)	structure of atoms and radiation emanating from them
1923	Robert Millikan (USA)	discovery of the electric charge of an electron, and study of the photoelectric effect
1924	Karl Siegbahn (Sweden)	X-ray spectroscopy
1925	James Franck (USA) and Gustav Hertz (Germany)	laws governing the impact of an electron upon an atom
1926	Jean Perrin (France)	confirmation of the discontinuous structure of matter
1927	Arthur Compton (USA)	transfer of energy from electromagnetic radiation to a particle
	Charles Wilson (UK)	invention of the Wilson cloud chamber, by which the movement of electrically charged particles may be tracked
1928	Owen Richardson (UK)	thermionic phenomena and associated law
1929	Louis Victor de Broglie (France)	discovery of wavelike nature of electrons
1930	ChandrasekaraV Raman (India)	discovery of the scattering of single-wavelength light when it is passed through a transparent substance
1931	no award	
1932	Werner Heisenberg (Germany)	creation of quantum mechanics
1933	Erwin Schrödinger (Austria) and Paul Dirac (UK)	development of quantum mechanics
1934	no award	
1935	James Chadwick (UK)	discovery of the neutron
1936	Victor Hess (Austria)	discovery of cosmic radiation
	Carl Anderson (USA)	discovery of the positron
1937	Clinton Davisson (USA) and George Thomson (UK)	diffraction of electrons by crystals
1938	Enrico Fermi (USA)	use of neutron irradiation to produce new elements, and discovery of nuclear reactions induced by slow neutrons
1939	Ernest O Lawrence (USA)	invention and development of cyclotron, and production of artificial radioactive elements
1940–42	no awards	
1943	Otto Stern (Germany)	molecular-ray method of investigating elementary particles, and discovery of magnetic moment of proton

Nobel Prize for Physics – *continued*

Year	Winner(s)	Awarded for
1944	Isidor Isaac Rabi (USA)	resonance method of recording the magnetic properties of atomic nuclei
1945	Wolfgang Pauli (Austria)	discovery of the exclusion principle
1946	Percy Bridgman (USA)	development of high-pressure physics
1947	Edward Appleton (UK)	physics of the upper atmosphere
1948	Patrick Blackett (UK)	application of the Wilson cloud chamber to nuclear physics and cosmic radiation
1949	Hideki Yukawa (Japan)	theoretical work predicting existence of mesons
1950	Cecil Powell (UK)	use of photographic emulsion to study nuclear processes, and discovery of pions (pi mesons)
1951	John Cockcroft (UK) and Ernest Walton (Ireland)	transmutation of atomic nuclei by means of accelerated subatomic particles
1952	Felix Bloch (USA) and Edward Purcell (USA)	precise nuclear-magnetic measurements
1953	Frits Zernike (Netherlands)	invention of phase-contrast microscope
1954	Max Born (Germany)	statistical interpretation of wave function in quantum mechanics
	Walther Bothe (Germany)	coincidence method of detecting the emission of electrons
1955	Willis Lamb (USA)	structure of hydrogen spectrum
	Polykarp Kusch (USA)	determination of magnetic moment of the electron
1956	William Shockley (USA), John Bardeen (USA), and Walter Houser Brattain (USA)	study of semiconductors, and discovery of transistor effect
1957	Yang Chen Ning (USA) and Lee Tsung-Dao (China)	investigations of weak interactions between elementary particles
1958	Pavel Cherenkov (USSR), Ilya Frank (USSR), and Igor Tamm (USA)	discovery and interpretation of Cherenkov radiation
1959	Emilio Segrè (Italy) and Owen Chamberlain (USA)	discovery of the antiproton
1960	Donald Glaser (USA)	invention of the bubble chamber
1961	Robert Hofstadter (USA)	scattering of electrons in atomic nuclei, and structure of protons and neutrons
	Rudolf Mössbauer (Germany)	resonance absorption of gamma radiation
1962	Lev Landau (USSR)	theories of condensed matter, especially liquid helium
1963	Eugene Wigner (USA)	discovery and application of symmetry principles in atomic physics
	Maria Goeppert-Mayer (USA) and Hans Jensen (Germany)	discovery of the shell-like structure of atomic nuclei
1964	Charles Townes (USA), Nikolai Basov (USSR), and Aleksandr Prokhorov (USSR)	quantum electronics leading to construction of oscillators and amplifiers based on maser–laser principle
1965	Sin-Itiro Tomonaga (Japan), Julian Schwinger (USA), and Richard Feynman (USA)	quantum electrodynamics
1966	Alfred Kastler (France)	development of optical pumping, whereby atoms are raised to higher energy levels by illumination
1967	Hans Bethe (USA)	theory of nuclear reactions, and discoveries concerning production of energy in stars
1968	Luis Alvarez (USA)	elementary-particle physics, and discovery of resonance states, using hydrogen bubble chamber and data analysis
1969	Murray Gell-Mann (USA)	classification of elementary particles, and study of their interactions
1970	Hannes Alfvén (Sweden)	magnetohydrodynamics and its applications in plasma physics
	Louis Néel (France)	antiferromagnetism and ferromagnetism in solid-state physics
1971	Dennis Gabor (UK)	invention and development of holography
1972	John Bardeen (USA), Leon Cooper (USA), and John Robert Schrieffer (USA)	theory of superconductivity
1973	Leo Esaki (Japan) and Ivar Giaver (USA)	tunnelling phenomena in semiconductors and superconductors
	Brian Josephson (UK)	theoretical predictions of the properties of a supercurrent through a tunnel barrier
1974	Martin Ryle (UK) and Antony Hewish (UK)	development of radioastronomy, particularly aperture-synthesis technique, and the discovery of pulsars

Nobel Prize for Physics – *continued*

Year	Winner(s)	Awarded for
1975	Aage Bohr (Denmark), Ben Mottelson (Denmark), and James Rainwater (USA)	discovery of connection between collective motion and particle motion in atomic nuclei, and development of theory of nuclear structure
1976	Burton Richter (USA) and Samuel Ting (USA)	discovery of the psi meson
1977	Philip Anderson (USA), Nevill Mott (UK), and John Van Vleck (USA)	electronic structure of magnetic and disordered systems
1978	Pyotr Kapitza (USSR)	low-temperature physics
	Arno Penzias (Germany) and Robert Wilson (USA)	discovery of cosmic background radiation
1979	Sheldon Glashow (USA), Abdus Salam (Pakistan), and Steven Weinberg (USA)	unified theory of weak and electromagnetic fundamental forces, and prediction of the existence of the weak neutral current
1980	James W Cronin (USA) and Val Fitch (USA)	violations of fundamental symmetry principles in the decay of neutral kaon mesons
1981	Nicolaas Bloemergen (USA) and Arthur Schawlow (USA)	development of laser spectroscopy
	Kai Siegbahn (Sweden)	high-resolution electron spectroscopy
1982	Kenneth Wilson (USA)	theory for critical phenomena in connection with phase transitions
1983	Subrahmanyan Chandrasekhar (USA)	theoretical studies of physical processes in connection with structure and evolution of stars
	William Fowler (USA)	nuclear reactions involved in the formation of chemical elements in the universe
1984	Carlo Rubbia (Italy) and Simon van der Meer (Netherlands)	contributions to the discovery of the W and Z particles (weakons)
1985	Klaus von Klitzing (Germany)	discovery of the quantized Hall effect
1986	Erns Ruska (Germany)	electron optics, and design of the first electron microscope
	Gerd Binnig (Germany) and Heinrich Rohrer (Switzerland)	design of scanning tunnelling microscope
1987	Georg Bednorz (Germany) and Alex Müller (Switzerland)	superconductivity in ceramic materials
1988	Leon M Lederman (USA), Melvin Schwartz (USA), and Jack Steinberger (Germany)	neutrino-beam method, and demonstration of the doublet structure of leptons through discovery of muon neutrino
1989	Norman Ramsey (USA)	measurement techniques leading to discovery of caesium atomic clock
	Hans Dehmelt (USA) and Wolfgang Paul (Germany)	ion-trap method for isolating single atoms
1990	Jerome Friedman (USA), Henry Kendall (USA), and Richard Taylor (Canada)	experiments demonstrating that protons and neutrons are made up of quarks
1991	Pierre-Gilles de Gennes (France)	work on disordered systems including polymers and liquid crystals; development of mathematical methods for studying the behaviour of molecules in a liquid on the verge of solidifying
1992	Georges Charpak (Poland)	invention and development of detectors used in high-energy physics
1993	Joseph Taylor (USA) and Russell Hulse (USA)	discovery of first binary pulsar (confirming the existence of gravitational waves)
1994	Clifford G Shull (USA) and Bertram N Brockhouse (Canada)	development of technique known as 'neutron scattering' which led to advances in semiconductor technology
1995	Frederick Reines (USA)	discovery of the neutrino
	Martin L Perl (USA)	discovery of the tau lepton
1996	David M Lee (USA), Douglas D Osheroff (USA), and Robert C Richardson (USA)	discovery of superfluidity in helium-3
1997	Steven Chu (USA), Claude Cohen-Tannoudji (France), and William D Phillips (USA)	development of methods to cool and trap atoms with laser light

Nobel Prize for Physiology or Medicine

Year	Winner(s)	Awarded for
1901	Emil von Behring (Germany)	discovery that the body produces antitoxins, and development of serum therapy for diseases such as diphtheria
1902	Ronald Ross (UK)	role of the *Anopheles* mosquito in transmitting malaria
1903	Niels Finsen (Denmark)	use of ultraviolet light to treat skin diseases
1904	Ivan Pavlov (Russia)	physiology of digestion
1905	Robert Koch (Germany)	investigations and discoveries in relation to tuberculosis
1906	Camillo Golgi (Italy) and Santiago Ramón y Cajal (Spain)	fine structure of the nervous system
1907	Charles Laveran (France)	discovery that certain protozoa can cause disease
1908	Ilya Mechnikov (Russia) and Paul Ehrlich (Germany)	work on immunity
1909	Emil Kocher (Switzerland)	physiology, pathology, and surgery of the thyroid gland
1910	Albrecht Kossel (Germany)	study of cell proteins and nucleic acids
1911	Allvar Gullstrand (Sweden)	refraction of light through the different components of the eye
1912	Alexis Carrel (USA)	techniques for connecting severed blood vessels and transplanting organs
1913	Charles Richet (France)	allergic responses
1914	Robert Bárány (Austria)	physiology and pathology of the equilibrium organs of the inner ear
1915–18	no awards	
1919	Jules Bordet (Belgium)	work on immunity
1920	August Krogh (Denmark)	discovery of mechanism regulating the dilation and constriction of blood capillaries
1921	no award	
1922	Archibald Hill (UK)	production of heat in contracting muscle
	Otto Meyerhof (Germany)	relationship between oxygen consumption and metabolism of lactic acid in muscle
1923	Frederick Banting (Canada) and John Macleod (UK)	discovery and isolation of the hormone insulin
1924	Willem Einthoven (Netherlands)	invention of the electrocardiograph
1925	no award	
1926	Johannes Fibiger (Denmark)	discovery of a parasite *Spiroptera carcinoma* that causes cancer
1927	Julius Wagner-Jauregg (Austria)	use of induced malarial fever to treat paralysis caused by mental deterioration
1928	Charles Nicolle (France)	role of the body louse in transmitting typhus
1929	Christiaan Eijkman (Netherlands)	discovery of a cure for beriberi, a vitamin-deficiency disease
	Frederick Hopkins (UK)	: discovery of trace substances, now known as vitamins, that stimulate growth
1930	Karl Landsteiner (USA)	discovery of human blood groups
1931	Otto Warburg (Germany)	discovery of respiratory enzymes that enable cells to process oxygen
1932	Charles Sherrington (UK) and Edgar Adrian (UK)	function of neurons (nerve cells)
1933	Thomas Morgan (USA)	role of chromosomes in heredity
1934	George Whipple (USA), George Minot (USA), and William Murphy (USA)	treatment of pernicious anaemia by increasing the amount of liver in the diet
1935	Hans Spemann (Germany)	organizer effect in embryonic development
1936	Henry Dale (UK) and Otto Loewi (Germany)	chemical transmission of nerve impulses
1937	Albert Szent-Györgyi (Hungary)	investigation of biological oxidation processes and of the action of ascorbic acid (vitamin C)
1938	Corneille Heymans (Belgium)	mechanisms regulating respiration
1939	Gerhard Domagk (Germany)	discovery of the first antibacterial sulphonamide drug
1940–42	no awards	
1943	Carl Dam (Denmark)	discovery of vitamin K
	Edward Doisy (USA)	chemical nature of vitamin K
1944	Joseph Erlanger (USA) and Herbert Gasser (USA)	transmission of impulses by nerve fibres
1945	Alexander Fleming (UK)	discovery of the bactericidal effect of penicillin
	Ernst Chain (UK) and Howard Florey (Australia)	isolation of penicillin and its development as an antibiotic drug
1946	Hermann Muller (USA)	discovery that X-ray irradiation can cause mutation
1947	Carl Cori (USA) and Gerty Cori (USA)	production and breakdown of glycogen (animal starch)

Nobel Prize for Physiology or Medicine – *continued*

Year	Winner(s)	Awarded for
	Bernardo Houssay (Argentina)	function of the pituitary gland in sugar metabolism
1948	Paul Müller (Switzerland)	discovery of the first synthetic contact insecticide DDT
1949	Walter Hess (Switzerland)	mapping areas of the midbrain that control the activities of certain body organs
	Antonio Egas Moniz (Portugal)	therapeutic value of prefrontal leucotomy in certain psychoses
1950	Edward Kendall (USA), Tadeus Reichstein (Poland), and Philip Hench (USA)	structure and biological effects of hormones of the adrenal cortex
1951	Max Theiler (South Africa)	discovery of a vaccine against yellow fever
1952	Selman Waksman (USA)	discovery of streptomycin, the first antibiotic effective against tuberculosis
1953	Hans Krebs (UK)	discovery of the Krebs cycle
	Fritz Lipmann (USA)	discovery of coenzyme A, a nonprotein compound that acts in conjunction with enzymes to catalyse metabolic reactions leading up to the Krebs cycle
1954	John Enders (USA), Thomas Weller (USA), and Frederick Robbins (USA)	cultivation of the polio virus in the laboratory
1955	Hugo Theorell (Sweden)	nature and action of oxidation enzymes
1956	André Cournand (USA), Werner Forssmann (Germany), and Dickinson Richards Jr (USA)	technique for passing a catheter into the heart for diagnostic purposes
1957	Daniel Bovet (Switzerland)	discovery of synthetic drugs used as muscle relaxants in anaesthesia
1958	George Beadle (USA) and Edward Tatum (USA)	discovery that genes regulate precise chemical effects
	Joshua Lederberg (USA)	genetic recombination and the organization of bacterial genetic material
1959	Severo Ochoa (USA) and Arthur Kornberg (USA)	discovery of enzymes that catalyse the formation of RNA (ribonucleic acid) and DNA (deoxyribonucleic acid)
1960	Macfarlane Burnet (Australia) and Peter Medawar (UK)	acquired immunological tolerance of transplanted tissues
1961	Georg von Békésy (USA)	investigations into the mechanism of hearing within the cochlea of the inner ear
1962	Francis Crick (UK), James Watson (USA), and Maurice Wilkins (UK)	discovery of the double-helical structure of DNA and of the significance of this structure in the replication and transfer of genetic information
1963	John Eccles (Australia), Alan Hodgkin (UK), and Andrew Huxley (UK)	ionic mechanisms involved in the communication or inhibition of impulses across neuron (nerve cell) membranes
1964	Konrad Bloch (USA) and Feodor Lynen (West Germany)	cholesterol and fatty-acid metabolism
1965	François Jacob (France), André Lwoff (France), and Jacques Monod (France)	genetic control of enzyme and virus synthesis
1966	Peyton Rous (USA)	discovery of tumour-inducing viruses
	Charles Huggins (USA)	hormonal treatment of prostatic cancer
1967	Ragnar Granit (Sweden), Haldan Hartline (USA), and George Wald (USA)	physiology and chemistry of vision
1968	Robert Holley (USA), Har Gobind Khorana (USA), and Marshall Nirenberg (USA)	interpretation of genetic code and its function in protein synthesis
1969	Max Delbruck (USA), Alfred Hershey (USA), and Salvador Luria (USA)	replication mechanism and genetic structure of viruses
1970	Bernard Katz (UK), Ulf von Euler (Austria), and Julius Axelrod (USA)	storage, release, and inactivation of neurotransmitters
1971	Earl Sutherland (USA)	discovery of cyclic AMP, a chemical messenger that plays a role in the action of many hormones
1972	Gerald Edelman (USA) and Rodney Porter (UK)	chemical structure of antibodies
1973	Karl von Frisch (Austria), Konrad Lorenz (Austria), and Nikolaas Tinbergen (Netherlands)	animal behaviour patterns
1974	Albert Claude (USA), Christian de Duve (Belgium), and George Palade (USA)	structural and functional organization of the cell
1975	David Baltimore (USA), Renato Dulbecco (USA), and Howard Temin (USA)	interactions between tumour-inducing viruses and the genetic material of the cell
1976	Baruch Blumberg (USA) and Carleton Gajdusek (USA)	new mechanisms for the origin and transmission of infectious diseases

Nobel Prize for Physiology or Medicine – *continued*

Year	Winner(s)	Awarded for
1977	Roger Guillemin (USA) and Andrew Schally (USA)	discovery of hormones produced by the hypothalamus region of the brain
	Rosalyn Yalow (USA)	radioimmunoassay techniques by which minute quantities of hormone may be detected
1978	Werner Arber (Switzerland), Daniel Nathans (USA), and Hamilton Smith (USA)	discovery of restriction enzymes and their application to molecular genetics
1979	Allan Cormack (USA) and Godfrey Hounsfield (UK)	development of the CAT scan
1980	Baruj Benacerraf (USA), Jean Dausset (France), and George Snell (USA)	genetically determined structures on the cell surface that regulate immunological reactions
1981	Roger Sperry (USA)	functional specialization of the brain's cerebral hemispheres
	David Hubel (USA) and Torsten Wiesel (Sweden)	visual perception
1982	Sune Bergström (Sweden), Bengt Samuelson (Sweden), and John Vane (UK)	discovery of prostaglandins and related biologically active substances
1983	Barbara McClintock (USA)	discovery of mobile genetic elements
1984	Niels Jerne (Denmark), Georges Köhler (West Germany), and César Milstein (UK)	work on immunity and discovery of a technique for producing highly specific, monoclonal antibodies
1985	Michael Brown (USA) and Joseph L Goldstein (USA)	regulation of cholesterol metabolism
1986	Stanley Cohen (USA) and Rita Levi-Montalcini (Italy)	discovery of factors that promote the growth of nerve and epidermal cells
1987	Susumu Tonegawa (Japan)	process by which genes alter to produce a range of different antibodies
1988	James Black (UK), Gertrude Elion (USA), and George Hitchings (USA)	principles governing the design of new drug treatment
1989	Michael Bishop (USA) and Harold Varmus (USA)	discovery of oncogenes, genes carried by viruses that can trigger cancerous growth in normal cells
1990	Joseph Murray (USA) and Donnall Thomas (USA)	pioneering work in organ and cell transplants
1991	Erwin Neher (Germany) and Bert Sakmann (Germany)	discovery of how gatelike structures (ion channels) regulate the flow of ions into and out of cells
1992	Edmund Fisher (USA) and Erwin Krebs (USA)	isolating and describing the action of the enzyme responsible for reversible protein phosphorylation, a major biological control mechanism
1993	Phillip Sharp (USA) and Richard Roberts (UK)	discovery of split genes (genes interrupted by nonsense segments of DNA)
1994	Alfred Gilman (USA) and Martin Rodbell (USA)	discovery of a family of proteins (G proteins) that translate messages – in the form of hormones or other chemical signals – into action inside cells
1995	Edward B Lewis and Eric F Wieschaus (USA), and Christiane Nüsslein-Volhard (Germany)	discovery of genes which control the early stages of the body's development
1996	Peter C Doherty (Australia) and Rolf M Zinkernagel (Switzerland)	discovery of how the immune system recognizes virus-infected cells
1997	Stanley B Prusiner (USA)	discovery of the infectious agents prions that cause transmissible spongiform encephalopathies

Scientific Discoveries (A–Z)

Discovery	Date	Discoverer	Nationality
Absolute zero, concept	1851	William Thomson, 1st Baron Kelvin	Irish
Adrenalin, isolation	1901	Jokichi Takamine	Japanese
Alizarin, synthesized	1869	William Perkin	English
Allotropy (in carbon)	1841	Jöns Jakob Berzelius	Swedish
Alpha rays	1899	Ernest Rutherford	New Zealand-born British
Alternation of generations (ferns and mosses)	1851	Wilhelm Hofmeister	German
Aluminium, extraction by electrolysis of aluminium oxide	1886	Charles Hall, Paul Héroult	US, French
Aluminium, improved isolation	1827	Friedrich Wöhler	German
Anaesthetic, first use (ether)	1842	Crawford Long	US
Anthrax vaccine	1881	Louis Pasteur	French
Antibacterial agent, first specific (Salvarsan for treatment of syphilis)	1910	Paul Ehrlich	German
Antiseptic surgery (using phenol)	1865	Joseph Lister	English
Argon	1892	William Ramsay	Scottish
Asteroid, first (Ceres)	1801	Giuseppe Piazzi	Italian
Atomic theory	1803	John Dalton	English
Australopithecus	1925	Raymond Dart	Australian-born South African
Avogadro's hypothesis	1811	Amedeo Avogadro	Italian
Bacteria, first observation	1683	Anton van Leeuwenhoek	Dutch
Bacteriophages	1916	Felix D'Herelle	Canadian
Bee dance	1919	Karl von Frisch	Austrian
Benzene, isolation	1825	Michael Faraday	English
Benzene, ring structure	1865	Friedrich Kekulé	German
Beta rays	1899	Ernest Rutherford	New Zealand-born British
Big-Bang theory	1948	Ralph Alpher, George Gamow	US
Binary arithmetic	1679	Gottfried Leibniz	German
Binary stars	1802	William Herschel	German-born English
Binomial theorem	1665	Isaac Newton	English
Blood, circulation	1619	William Harvey	English
Blood groups, ABO system	1900	Karl Landsteiner	Austrian-born US
Bode's law	1772	Johann Bode, Johann Titius	German
Bohr atomic model	1913	Niels Bohr	Danish
Boolean algebra	1854	George Boole	English
Boyle's law	1662	Robert Boyle	Irish
Brewster's law	1812	David Brewster	Scottish
Brownian motion	1827	Robert Brown	Scottish
Cadmium	1817	Friedrich Strohmeyer	German
Caesium	1861	Robert Bunsen	German
Carbon dioxide	1755	Joseph Black	Scottish
Charles's law	1787	Jacques Charles	French
Chlorine	1774	Karl Scheele	Swedish
Complex numbers, theory	1746	Jean d'Alembert	French
Conditioning	1902	Ivan Pavlov	Russian
Continental drift	1912	Alfred Wegener	German
Coriolis effect	1834	Gustave-Gaspard Coriolis	French
Cosmic radiation	1911	Victor Hess	Austrian
Decimal fractions	1576	François Viète	French
Dinosaur fossil, first recognized	1822	Mary Ann Mantell	English
Diphtheria bacillus, isolation	1883	Edwin Krebs	US
DNA	1869	Johann Frederick Miescher	Swiss
DNA and RNA	1909	Phoebus Levene	Russian-born US
DNA, double-helix structure	1953	Francis Crick, James Watson	English, US
Doppler effect	1842	Christian Doppler	Austrian
Earth's magnetic pole	1546	Gerardus Mercator	Flemish
Earth's molten core	1916	Albert Michelson	German-born US
Earth's molten core, proof	1906	Richard Oldham	Welsh
Earth's rotation, demonstration	1851	Léon Foucault	French
Eclipse, prediction	585 BC	Thales of Miletus	Greek
Electrolysis, laws	1833	Michael Faraday	English
Electromagnetic induction	1831	Michael Faraday	English
Electromagnetism	1819	Hans Christian Oersted	Danish
Electron	1897	J J Thomson	English

Scientific Discoveries (A–Z) – *continued*

Discovery	Date	Discoverer	Nationality
Electroweak unification theory	1967	Sheldon Lee Glashow, Abdus Salam, Steven Weinberg	US, Pakistani, US
Endorphins	1975	John Hughes	US
Enzyme, first animal (pepsin)	1836	Theodor Schwann	German
Enzyme, first (diastase from barley)	1833	Anselme Payen	French
Enzymes, 'lock and key' hypothesis	1899	Emil Fischer	German
Ether, first anaesthetic use	1842	Crawford Long	US
Eustachian tube	1552	Bartolomeo Eustachio	Italian
Evolution by natural selection	1858	Charles Darwin	English
Exclusion principle	1925	Wolfgang Pauli	Austrian-born Swiss
Fallopian tubes	1561	Gabriello Fallopius	Italian
Fluorine, preparation	1886	Henri Moissan	French
Fullerines	1985	Harold Kroto, David Walton	English
Gay-Lussac's law	1808	Joseph-Louis Gay-Lussac	French
Geometry, Euclidean	300 BC	Euclid	Greek
Germanium	1886	Clemens Winkler	German
Germ theory	1861	Louis Pasteur	French
Global temperature and link with atmospheric carbon dioxide	1896	Svante Arrhenius	Swedish
Gravity, laws	1687	Isaac Newton	English
Groups, theory	1829	Evariste Galois	French
Gutenberg discontinuity	1914	Beno Gutenberg	German-born US
Helium, production	1896	William Ramsay	Scottish
Homo erectus	1894	Marie Dubois	Dutch
Homo habilis	1961	Louis Leakey, Mary Leakey	Kenyan, English
Hormones	1902	William Bayliss, Ernest Starling	English
Hubble's law	1929	Edwin Hubble	US
Hydraulics, principles	1642	Blaise Pascal	French
Hydrogen	1766	Henry Cavendish	English
Iapetus	1671	Giovanni Cassini	Italian-born French
Infrared solar rays	1801	William Herschel	German-born English
Insulin, isolation	1921	Frederick Banting, Charles Best	Canadian
Insulin, structure	1969	Dorothy Hodgkin	English
Interference of light	1801	Thomas Young	English
Irrational numbers	450 BC	Hipparcos	Greek
Jupiter's satellites	1610	Galileo	Italian
Kinetic theory of gases	1850	Rudolf Clausius	German
Krypton	1898	William Ramsay, Morris Travers	Scottish, English
Lanthanum	1839	Carl Mosander	Swedish
Lenses, how they work	1039	Ibn al-Haytham Alhazen	Arabic
Light, finite velocity	1675	Ole Römer	Danish
Light, polarization	1678	Christiaan Huygens	Dutch
Linnaean classification system	1735	Linnaeus	Swedish
'Lucy', hominid	1974	Donald Johanson	US
Magnetic dip	1576	Robert Norman	English
Malarial parasite in *Anopheles* mosquito	1897	Ronald Ross	British
Malarial parasite observed	1880	Alphonse Laveran	French
Mars, moons	1877	Asaph Hall	US
Mendelian laws of inheritance	1866	Gregor Mendel	Austrian
Messenger RNA	1960	Sydney Brenner, François Jacob	South African, French
Microorganisms as cause of fermentation	1856	Louis Pasteur	French
Monoclonal antibodies	1975	César Milstein, George Köhler	Argentine-born British, German
Motion, laws	1687	Isaac Newton	English
Natural selection	1859	Charles Darwin	English
Neon	1898	William Ramsay, Morris Travers	Scottish, English
Neptune	1846	Johann Galle	German
Neptunium	1940	Edwin McMillan, Philip Abelson	US
Nerve impulses, electric nature	1771	Luigi Galvani	Italian
Neutron	1932	James Chadwick	English
Nitrogen	1772	Daniel Rutherford	Scottish
Normal distribution curve	1733	Abraham De Moivre	French
Nuclear atom, concept	1911	Ernest Rutherford	New Zealand-born British

Scientific Discoveries (A–Z) – *continued*

Discovery	Date	Discoverer	Nationality
Nuclear fission	1938	Otto Hahn, Fritz Strassman	German
Nucleus, plant cell	1831	Robert Brown	Scottish
Ohm's law	1827	Georg Ohm	German
Organic substance, first synthesis (urea)	1828	Friedrich Wöhler	German
Oxygen	1774	Joseph Priestley	English
Oxygen, liquefication	1894	James Dewar	Scottish
Ozone layer	1913	Charles Fabry	French
Palladium	1803	William Hyde Wollaston	English
Pallas (asteroid)	1802	Heinrich Olbers	German
Pendulum, principle	1581	Galileo	Italian
Penicillin	1928	Alexander Fleming	Scottish
Penicillin, widespread preparation	1940	Ernst Chain, Howard Florey	German, Australian
Pepsin	1836	Theodor Schwann	German
Periodic law for elements	1869	Dmitri Mendeleyev	Russian
Period–luminosity law	1912	Henrietta Swan	US
Phosphorus	1669	Hennig Brand	German
Piezoelectric effect	1880	Pierre Curie	French
Pi meson (particle)	1947	Cecil Powell, Giuseppe Occhialini	English, Italian
Pistils, function	1676	Nehemiah Grew	English
Planetary nebulae	1790	William Herschel	German-born English
Planets, orbiting Sun	1543	Copernicus	Polish
Pluto	1930	Clyde Tombaugh	US
Polarization of light by reflection	1808	Etienne Malus	French
Polio vaccine	1952	Jonas Salk	US
Polonium	1898	Marie and Pierre Curie	French
Positron	1932	Carl Anderson	US
Potassium	1806	Humphry Davy	English
Probability, theory	1654	Blaise Pascal, Pierre de Fermat	French
Probability theory, expansion	1812	Pierre Laplace	French
Proton	1914	Ernest Rutherford	New Zealand-born British
Protoplasm	1846	Hugo von Mohl	German
Pulsar	1967	Jocelyn Bell Burnell	English
Pythagoras' theorem	550 BC	Pythagoras	Greek
Quantum chromodynamics	1972	Murray Gell-Mann	US
Quantum electrodynamics	1948	Richard Feynman, Seymour Schwinger, Shin'chiro Tomonaga	US, US, Japanese
Quark, first suggested existence	1963	Murray Gell-mann, George Zweig	US
Quasar	1963	Maarten Schmidt	Dutch-born US
Rabies vaccine	1885	Louis Pasteur	French
Radioactivity	1896	Henri Becquerel	French
Radio emissions, from Milky Way	1931	Karl Jansky	US
Radio waves, production	1887	Heinrich Hertz	German
Radium	1898	Marie and Pierre Curie	French
Radon	1900	Friedrich Dorn	German
Refraction, laws	1621	Willibrord Snell	Dutch
Relativity, general theory	1915	Albert Einstein	German-born US
Relativity, special theory	1905	Albert Einstein	German-born US
Rhesus factor	1940	Karl Landsteiner, Alexander Wiener	Austrian, US
Rubidium	1861	Robert Bunsen	German
Sap circulation	1846	Giovanni Battista	Italian
Sap flow in plants	1733	Stephen Hales	English
Saturn, 18th moon	1990	Mark Showalter	US
Saturn's satellites	1656	Christiaan Huygens	Dutch
Smallpox inoculation	1796	Edward Jenner	English
Sodium	1806	Humphry Davy	English
Stamens, function	1676	Nehemiah Grew	English
Stars, luminosity sequence	1905	Ejnar Hertzsprung	Danish
Stereochemistry, foundation	1848	Louis Pasteur	French
Stratosphere	1902	Léon Teisserenc	French
Sunspots	1611	Galileo, Christoph Scheiner	Italian, German
Superconductivity	1911	Heike Kamerlingh-Onnes	Dutch
Superconductivity, theory	1957	John Bardeen, Leon Cooper, John Schrieffer	US
Thermodynamics, second law	1834	Benoit-Pierre Clapeyron	French
Thermodynamics, third law	1906	Hermann Nernst	German

Scientific Discoveries (A–Z) – *continued*

Discovery	Date	Discoverer	Nationality
Thermoelectricity	1821	Thomas Seebeck	German
Thorium-X	1902	Ernest Rutherford, Frederick Soddy	New Zealand-born British, English
Titius–Bode law	1772	Johan Bode, Johann Titius	German
Tranquillizer, first (reserpine)	1956	Robert Woodward	US
Transformer	1831	Michael Faraday	English
Troposphere	1902	Léon Teisserenc	French
Tuberculosis bacillus, isolation	1883	Robert Koch	German
Tuberculosis vaccine	1923	Albert Calmette, Camille Guérin	French
Uranus	1781	William Herschel	German-born English
Urea cycle	1932	Hans Krebs	German
Urease, isolation	1926	James Sumner	US
Urea, synthesis	1828	Friedrich Wöhler	German
Valves, in veins	1603	Geronimo Fabricius	Italian
Van Allen radiation belts	1958	James Van Allen	US
Virus, first identified (tobacco mosaic disease, in tobacco plants)	1898	Martinus Beijerinck	Dutch
Vitamin A, isolation	1913	Elmer McCollum	US
Vitamin A, structure	1931	Paul Karrer	Russian-born Swiss
Vitamin B, composition	1955	Dorothy Hodgkin	English
Vitamin B, isolation	1925	Joseph Goldberger	Austrian-born US
Vitamin C	1928	Charles Glen King, Albert Szent-Györgi	US, Hungarian-born US
Vitamin C, isolation	1932	Charles Glen King	US
Vitamin C, synthesis	1933	Tadeus Reichstein	Polish-born Swiss
Wave mechanics	1926	Erwin Schrödinger	Austrian
Xenon	1898	William Ramsay, Morris Travers	Scottish, English
X-ray crystallography	1912	Max von Laue	German
X-rays	1895	Wilhelm Röntgen	German

Inventions (A–Z)

Invention	Date	Inventor	Nationality
Achromatic lens	1733	Chester Moor Hall	English
Adding machine	1642	Blaise Pascal	French
Aeroplane, powered	1903	Orville and Wilbur Wright	US
Air conditioning	1902	Willis Carrier	US
Air pump	1654	Otto Guericke	German
Airship, first successful	1852	Henri Giffard	French
Airship, rigid	1900	Ferdinand von Zeppelin	German
Amniocentesis test	1952	Douglas Bevis	English
Aqualung	1943	Jacques Cousteau	French
Arc welder	1919	Elihu Thomson	US
Armillary ring	125	Zhang Heng	Chinese
Aspirin	1899	Felix Hoffman	German
Assembly line	1908	Henry Ford	US
Autogiro	1923	Juan de la Cierva	Spanish
Automatic pilot	1912	Elmer Sperry	US
Babbitt metal	1839	Isaac Babbitt	US
Bakelite, first synthetic plastic	1909	Leo Baekeland	US
Ballpoint pen	1938	Lazlo Biró	Hungarian
Barbed wire	1874	Joseph Glidden	US
Bar code system	1970	Monarch Marking, Plessey Telecommunications	US, English
Barometer	1642	Evangelista Torricelli	Italian
Bathysphere	1934	Charles Beebe	US
Bessemer process	1856	Henry Bessemer	British
Bicycle	1839	Kirkpatrick Macmillan	Scottish
Bifocal spectacles	1784	Benjamin Franklin	US
Binary calculator	1938	Konrad Zuse	German
Bottling machine	1895	Michael Owens	US
Braille	1837	Louis Braille	French
Bunsen burner	1850	Robert Bunsen	German
Calculator, pocket	1971	Texas Instruments	US
Camera film (roll)	1888	George Eastman	US
Camera obscura	1560	Battista Porta	Italian
Carbon fibre	1963	Leslie Phillips	English
Carbon–zinc battery	1841	Robert Bunsen	German
Carburettor	1893	Wilhelm Maybach	German
Car, four-wheeled	1887	Gottlieb Daimler	German
Car, petrol-driven	1885	Karl Benz	German
Carpet sweeper	1876	Melville Bissell	US
Cash register	1879	James Ritty	US
Cassette tape	1963	Philips	Dutch
Catapult	c. 400 BC	Dionysius of Syracuse	Greek
Cathode ray oscilloscope	1897	Karl Braun	German
CD-ROM	1984	Sony, Fujitsu, Philips	Japanese, Japanese, Dutch
Cellophane	1908	Jacques Brandenberger	Swiss
Celluloid	1869	John Wesley Hyatt	US
Cement, Portland	1824	Joseph Aspidin	English
Centigrade scale	1742	Anders Celsius	Swedish
Chemical symbols	1811	Jöns Jakob Berzelius	Swedish
Chronometer, accurate	1762	John Harrison	English
Cinematograph	1895	Auguste and Louis Lumière	French
Clock, pendulum	1656	Christiaan Huygens	Dutch
Colt revolver	1835	Samuel Colt	US
Compact disc	1972	RCA	US
Compact disc player	1984	Sony, Philips	Japanese, Dutch
Compass, simple	1088	Shen Kua	Chinese
Computer, bubble memory	1967	A H Bobeck and Bell Telephone Laboratories team	US
Computer, first commercially available (UNIVAC 1)	1951	John Mauchly, John Eckert	US
Computerized axial tomography (CAT) scanning	1972	Godfrey Hounsfield	English
Contraceptive pill	1954	Gregory Pincus	US
Cotton gin	1793	Eli Whitney	US
Cream separator	1878	Carl de Laval	Swedish

Inventions (A–Z) – *continued*

Invention	Date	Inventor	Nationality
Crookes tube	1878	William Crookes	English
Cyclotron	1931	Ernest O Lawrence	US
DDT	1940	Paul Müller	Swiss
Diesel engine	1892	Rudolf Diesel	German
Difference engine (early computer)	1822	Charles Babbage	English
Diode valve	1904	Ambrose Fleming	English
Dynamite	1866	Alfred Nobel	Swedish
Dynamo	1831	Michael Faraday	English
Electric cell	1800	Alessandro Volta	Italian
Electric fan	1882	Schuyler Wheeler	US
Electric generator, first commercial	1867	Zénobe Théophile Gramme	French
Electric light bulb	1879	Thomas Edison	US
Electric motor	1821	Michael Faraday	English
Electric motor, alternating current	1888	Nikola Tesla	Croatian-born US
Electrocardiography	1903	Willem Einthoven	Dutch
Electroencephalography	1929	Hans Berger	German
Electromagnet	1824	William Sturgeon	English
Electron microscope	1933	Ernst Ruska	German
Electrophoresis	1930	Arne Tiselius	Swedish
Fahrenheit scale	1714	Gabriel Fahrenheit	Polish-born Dutch
Felt-tip pen	1955	Esterbrook	English
Floppy disc	1970	IBM	US
Flying shuttle	1733	John Kay	English
FORTRAN	1956	John Backus, IBM	US
Fractal images	1962	Benoit Mandelbrot	Polish-born French
Frozen food	1929	Clarence Birdseye	US
Fuel cell	1839	William Grove	Welsh
Galvanometer	1820	Johann Schwiegger	German
Gas mantle	1885	Carl Welsbach	Austrian
Geiger counter	1908	Hans Geiger, Ernest Rutherford	German, New Zealand-born British
Genetic fingerprinting	1985	Alec Jeffreys	British
Glider	1877	Otto Lilienthal	German
Gramophone	1877	Thomas Edison	US
Gramophone (flat discs)	1887	Emile Berliner	German
Gyrocompass	1911	Elmer Sperry	US
Gyroscope	1852	Jean Foucault	French
Heart, artificial	1982	Robert Jarvik	US
Heart–lung machine	1953	John Gibbon	US
Helicopter	1939	Igor Sikorsky	US
Holography	1947	Dennis Gabor	Hungarian-born British
Hovercraft	1955	Christopher Cockerell	English
Hydrogen bomb	1952	US government scientists	US
Hydrometer	1675	Robert Boyle	Irish
Iconoscope	1923	Vladimir Zworykin	Russian-born US
Integrated circuit	1958	Jack Kilby, Texas Instruments	US
Internal-combustion engine, four-stroke	1877	Nikolaus Otto	German
Internal-combustion engine, gas-fuelled	1860	Etienne Lenoir	Belgian
In vitro fertilization	1969	Robert Edwards	Welsh
Jet engine	1930	Frank Whittle	English
Jumbo jet	1969	Joe Sutherland and Boeing team	US
Laser, prototype	1960	Theodore Maiman	US
Lightning rod	1752	Benjamin Franklin	US
Linoleum	1860	Frederick Walton	English
Liquid crystal display (LCD)	1971	Hoffmann-LaRoche Laboratories	Swiss
Lock (canal)	980	Ciao Wei-yo	Chinese
Lock, Yale	1851	Linus Yale	US
Logarithms	1614	John Napier	Scottish
Loom, power	1785	Edmund Cartwright	English
Machine gun	1862	Richard Gatling	US
Magnifying glass	1250	Roger Bacon	English
Map	*c.* 510 BC	Hecataeus	Greek
Map, star	*c.* 350 BC	Eudoxus	Greek

Inventions (A–Z) – *continued*

Invention	Date	Inventor	Nationality
Maser	1953	Charles Townes and Arthur Schawlow	US
Mass-spectrograph	1918	Francis Aston	English
Microscope	1590	Zacharias Janssen	Dutch
Miners' safety lamp	1813	Humphry Davy	English
Mohs' scale for mineral hardness	1822	Frederick Mohs	German
Morse code	1838	Samuel Morse	US
Motorcycle	1885	Gottlieb Daimler	German
Neutron bomb	1977	US military	US
Nylon	1934	Wallace Carothers	US
Paper chromatography	1944	Archer Martin, Richard Synge	English
Paper, first	105	Ts'ai Lun	Chinese
Particle accelerator	1932	John Cockcroft, Ernest Walton	English, Irish
Pasteurization (wine)	1864	Louis Pasteur	French
Pen, fountain	1884	Lewis Waterman	US
Photoelectric cell	1904	Johann Elster	German
Photograph, first colour	1881	Frederic Ives	US
Photograph, first (on a metal plate)	1827	Joseph Niepce	French
Piano	1704	Bartelommeo Cristofori	Italian
Planar transistor	1959	Robert Noyce	US
Plastic, first (Parkesine)	1862	Alexander Parkes	English
Plough, cast iron	1785	Robert Ransome	English
Punched-card system for carpet-making loom	1805	Joseph-Marie Jacquard	French
Radar, first practical equipment	1935	Robert Watson-Watt	Scottish
Radio	1901	Guglielmo Marconi	Italian
Radio interferometer	1955	Martin Ryle	English
Radio, transistor	1952	Sony	Japanese
Razor, disposable safety	1895	King Gillette	US
Recombinant DNA, technique	1973	Stanley Cohen, Herbert Boyer	US
Refrigerator, domestic	1918	Nathaniel Wales, E J Copeland	US
Richter scale	1935	Charles Richter	US
Road locomotive, steam	1801	Richard Trevithick	English
Road vehicle, first self-propelled (steam)	1769	Nicolas-Joseph Cugnot	French
Rocket, powered by petrol and liquid oxygen	1926	Robert Goddard	US
Rubber, synthetic	1909	Karl Hoffman	German
Scanning tunnelling microscope	1980	Heinrich Rohrer, Gerd Binning	Swiss, German
Seed drill	1701	Jethro Tull	English
Seismograph	1880	John Milne	English
Shrapnel shell	1784	Henry Shrapnel	English
Silicon transistor	1954	Gordon Teal	US
Silk, method of producing artificial	1887	Hilaire, Comte de Chardonnet	French
Spinning frame	1769	Richard Arkwright	English
Spinning jenny	1764	James Hargreaves	English
Spinning mule	1779	Samuel Crompton	English
Stainless steel	1913	Harry Brearley	English
Steam engine	50 BC	Heron of Alexandria	Greek
Steam engine, first successful	1712	Thomas Newcomen	English
Steam engine, improved	1765	James Watt	Scottish
Steam locomotive, first effective	1814	George Stephenson	English
Steam turbine, first practical	1884	Charles Parsons	English
Steel, open-hearth production	1864	William Siemens, Pierre Emile Martin	German, French
Submarine	1620	Cornelius Drebbel	Dutch
Superheterodyne radio receiver	1918	Edwin Armstrong	US
Tank	1914	Ernest Swinton	English
Telephone	1876	Alexander Graham Bell	Scottish-born US
Telescope, binocular	1608	Johann Lippershey	Dutch
Telescope, reflecting	1668	Isaac Newton	English
Television	1926	John Logie Baird	Scottish
Terylene (synthetic fibre)	1941	John Whinfield, J T Dickson	English
Thermometer	1607	Galileo	Italian
Thermometer, alcohol	1730	Réné Antoine Ferchault de Réaumur	French
Thermometer, mercury	1714	Gabriel Fahrenheit	Polish-born Dutch
TNT	1863	J Willbrand	German
Toaster, pop-up	1926	Charles Strite	US

Inventions (A–Z) – *continued*

Invention	Date	Inventor	Nationality
Toilet, flushing	1778	Joseph Bramah	English
Transistor	1948	John Bardeen, Walter Brattain, William Shockley	US
Triode valve	1906	Lee De Forest	US
Tunnel diode	1957	Leo Esaki, Sony	Japanese
Tupperware	1944	Earl Tupper	US
Type, movable earthenware	1045	Pi Shêng	Chinese
Type, movable metal	1440	Johannes Gutenberg	German
Ultrasound, first use in obstetrics	1958	Ian Donald	Scottish
Velcro	1948	Georges de Mestral	Swiss
Video, home	1975	Matsushita, JVC, Sony	Japanese
Viscose	1892	Charles Cross	English
Vulcanization of rubber	1839	Charles Goodyear	US
Wind tunnel	1932	Ford Motor Company	US
Wireless telegraphy	1895	Guglielmo Marconi	Italian
Word processor	1965	IBM	US
Zinc–carbon battery	1868	George Leclanché	French
Zip	1891	Whitcombe Judson	US